D1476380

VARIATIONAL AND EXTREMUM PRINCIPLES IN MACROSCOPIC SYSTEMS

Elsevier Internet Homepage - http://www.elsevier.com

Consult the Elsevier homepage for full catalogue information on all books, major reference works, journals, electronic products and services.

Elsevier Titles of Related Interest

James Hartnett et al.
Advances in Heat Transfer, 38
2004, 0-12-020038-4

Tuncer Cebeci
Analysis of Turbulent Flows
2004, 0-08-044350-8

Ira Cohen
Fluid Mechanics
2004, 0-12-178253-0

Related Journals:

Elsevier publishes a wide-ranging portfolio of high quality research journals, including papers detailing the research and development of variational and extremum principles in macroscopic systems. A sample journal issue is available online by visiting the Elsevier web site (details at the top of this page). Leading titles include:

Applied Mathematics and Computation
International Journal of Solids and Structures
Journal of Sound and Vibration
Acta Materialia
Journal of Fluids and Structures
International Journal of Heat and Mass Transfer
European Journal of Mechanics - A/Solids
European Journal of Mechanics - B/Fluids
Experimental Thermal and Fluid Science
International Journal of Thermal Science
Computers and Fluids
International Journal of Heat and Fluid Flow
Applied Thermal Engineering
International Journal of Multiphase Flow
European Journal of Operational Research

All journals are available online via ScienceDirect: www.sciencedirect.com

To contact the Publisher

Elsevier welcomes enquiries concerning publishing proposals: books, journal special issues, conference proceedings, etc. All formats and media can be considered. Should you have a publishing proposal you wish to discuss, please contact, without obligation, the publisher responsible for Elsevier's Mechanics programme:

Arno Schouwenburg
Senior Publishing Editor
Elsevier Ltd
The Boulevard, Langford Lane
Kidlington, Oxford
OX5 1GB, UK

Phone: +44 1865 84 3879
Fax: +44 1865 84 3987
E.mail: a.schouwenburg@elsevier.com

General enquiries, including placing orders, should be directed to Elsevier's Regional Sales Offices – please access the Elsevier homepage for full contact details (homepage details at the top of this page).

VARIATIONAL AND EXTREMUM PRINCIPLES IN MACROSCOPIC SYSTEMS

Edited by

Stanislaw Sieniutycz

Faculty of Chemical & Process Engineering,
Warsaw University of Technology,
Warsaw, Poland

Henrik Farkas

Institute of Physics,
Budapest University of Technology and Economics,
Budapest, Hungary

2005

ELSEVIER

AMSTERDAM – BOSTON – HEIDELBERG – LONDON – NEW YORK – OXFORD
PARIS – SAN DIEGO – SAN FRANCISCO – SINGAPORE – SYDNEY – TOKYO

ELSEVIER B.V.
Radarweg 29
P.O. Box 211, 1000 AE Amsterdam
The Netherlands

ELSEVIER Inc.
525 B Street, Suite 1900
San Diego, CA 92101-4495
USA

ELSEVIER Ltd
The Boulevard, Langford Lane
Kidlington, Oxford OX5 1GB
UK

ELSEVIER Ltd
84 Theobalds Road
London WC1X 8RR
UK

First edition 2005

Library of Congress Cataloging in Publication Data
A catalog record is available from the Library of Congress.

British Library Cataloguing in Publication Data
A catalogue record is available from the British Library.

ISBN: 0-08-044488-1

♾ The paper used in this publication meets the requirements of ANSI/NISO Z39.48-1992 (Permanence of Paper). Printed in The Netherlands.

FOREWORD

The concept of a variational principle as a fundamental characteristic of all phenomena is remarkable, if one steps back and considers what conceptual basis it implies. At the first encounter, it may seem obvious and almost trivial: of course, every process is an extremum of *something*, because we think we can find a suitable "something", a set of constraints, that makes the answer come out right. That is, we are inclined to think that we could find constraints—and a suitable variational function or functional—for any given problem that would assure that the process satisfied a variational principle. The not-so-subtle difficulty, in many situations easy to recognize, but usually not at all easy to resolve, is determining what the relevant constraints are.

This line of reasoning leads to a question that underlies many of the chapters in this volume. Is it true that every process can be described by some pathway that makes the entropy change an extremum? That is, of course, not quite how the question has to be expressed. Rather, we should ask this: *Is it true that all processes can be described by a pathway that makes the entropy change an extremum, subject to a specific set of constraints*? Is there a variational principle for all possible processes, if we can determine the constraints suitably? And if so, what quantities can be the basic ones for such a principle? We have principles of least action, of least time and of least (or most) entropy production. Are there relations among these that are stronger than mere analogies? Are there other quantities for which there are variational principles?

But let us turn to our original statement: the concept, valid in a remarkable range of situations, that we can find and determine *some* quantitative property that is an extremum for many, possibly all processes we can describe, is indeed amazing. We are so familiar with this idea that we take it very much as a natural, given property of the physical world, even extending it as is done here to ecosystems and economic behavior. Using it as we do, we may appreciate the beauty of any variational principle, yet, even if we appreciate its beauty, that very familiarity hides from us the amazement we should have at our capacity to discover it at all. We should realize how remarkable it is that the human mind could recognize and then generalize the concept. Perhaps the easiest to discover was the principle of least time; this minimizes something in our direct experience, even though it is not so obvious that we might conceive of time trajectories different from the ones we observe. But variational principles for action and for entropy take us beyond direct experience. We have no sensors in our bodies for action, no organ that responds to entropy. These are abstract creations of human imagination that we can quantify. Then to go beyond that stage and discover that they are the bases for variational principles is a thing for us to see with wonder.

One further observation: most of the chapters in this volume deal with *descriptive* variational methods, some with methodology and some with applications. A few of the chapters follow the other kind of use of variational methods, what we may call the *prescriptive*. This direction is perhaps epitomized by optimal control theory. Here we use the power of variational calculus to tell us what *we* should do in order to achieve some goal. Here, we can sometimes open new directions by recognizing that some variable previously considered beyond our control could, in fact, become a control variable and enable us to construct entirely new kinds of devices. Perhaps, in using this volume, some readers will see ways innovating by pursuing directions such as that, beginning with what they learn from the rich material presented here.

R. Stephen Berry
The University of Chicago

VOLUME EDITORS' PREFACE

This volume comprises the theory and applications of the variational calculus and theory of extrema in macroscopic physics, chemistry and engineering with a few relevant excursions made to the self-organization theory, ecology and econophysics. The vast range of topics comprises the problems found in nonequilibrium thermodynamics, described by classical or extended entropies, transport theory in continua (reversible or irreversible), elasticity theory, and theory of propagation of chemical fronts. Some attention is also given to macroscopic relativistic systems. The authors search for conditions under which macroscopic equations admit a variational formulation and discuss details of Lagrangian or Hamiltonian formalisms for the kinetic and balance equations of processes with transport of heat, mass, and momentum in chemically active or inactive systems. The book outlines a broad spectrum of applications (including simple and complex fluids, energy and chemical systems, liquid crystals, chemical waves, polymeric liquids, composites, and radiation fluids) useful for practical systems with materials possessing unusual flow properties. The book can be used as a source or supplementary text in graduate courses on theory and applications of the variational calculus in applied mathematics, theoretical and applied physics, fluid mechanics, nonequilibrium thermodynamics, transport phenomena, chemical, mechanical and environmental engineering, theory of synergetic systems, theoretical ecology, econophysics, and optimal control. As a reference text for further investigations it will attract researchers working in various branches of applied mathematics and macroscopic physics/chemistry, especially those in mechanics of continua, thermodynamics (classical and extended), wave phenomena, energy and mass transfer, solar energy systems, etc. Applied mathematicians will welcome the use of the field (Lagrangian and Hamiltonian) formalisms in complex continua and the exposition of novel approaches dealing with formulation of variational principles in spaces adjoint with respect to the original space of the system.

Stanislaw Sieniutycz
Henrik Farkas

PREFACE

Variational approaches are central in formulating and solving problems of optimal behavior in the natural world. By the Noether theorem and symmetry requirements they lead to conservation laws, of which those for conservation of energy, momentum, mass, and electric charge belong to the most fundamental laws of Nature. A variational formulation increases our insight and understanding of the process, its brevity and rigor mean that the achievement of the formulation is often identified with the main result attributed to the investigated problem. Variational approaches can also serve to develop efficient numerical procedures characterized by high-accuracy solutions (via the so-called direct variational methods).

An increasing trend was recently observed to derive models of macroscopic phenomena encountered in the fields of physics, chemistry, engineering, ecology, self-organization theory, and econophysics from various variational or extremum principles. Through the link between the integral extremum of a functional and the local extremum of a function (explicit, for example, in Pontryagin's principle and in the Hamilton–Jacobi–Bellman equation) variational and extremum formulations are mutually related, so it makes sense to consider them within a common context. The main task of the present two-part book is to display in a way as coherent as possible various recent (variational and extremum) formulations describing both theoretical aspects and applications for systems of the macroscopic world. The first part of this book is focused on the theory, whereas the second focuses on applications. The demarcation line between these parts (the editors' choice and pain) is necessarily fuzzy. The assignment of a chapter to the first part was decided by taking as the criterion the novelty and power of the mathematical methodology (in the field of functional or function extrema) rather than the depth of physical penetration; so the reader can find a number of chapters with sophisticated physical theory in the second part as well. Moreover, a number of examples in Part II contribute significantly to the critical, indepth understanding of many mathematical aspects of part one.

Throughout the book the unifying power of variational calculus is used to derive the balance or conservation equations, phenomenological equations linking fluxes and forces and equations of change for processes with coupled transfer of energy and substances. Also presented is a reconstruction and revalidation of a few basic, pioneering variational principles in thermo-hydrodynamics, in particular Gibbs, Natanson's, and Eckart's principles. For the limit of ideal continuum, variational formulations in Euclidean four-dimensional space-time for mechanics of solid-type continua (Green and Naghdi model of thermoelasticity) lead to the canonical balance laws of energy and material momentum. In approaches of this sort applied to fluids with heat flow and mass diffusion canonical balance laws also follow in the limit of a fluid thermal superconductor. Also discussed are subtle theoretical issues on how Hamilton's Principle can be fitted into the theoretical context of irreversible thermodynamics and how to accomplish quantization of thermal fields. Linearly damped hydrodynamic systems, with frictional forces proportional to the flow velocity, are incorporated into the framework of Lagrangian field theory, to describe the seepage of fluids in porous media (Darcy's law). Through Hamilton's Principle and the resulting noncanonical Poisson bracket origin of rotational motions in media of complex rheology (liquid crystals) is investigated, to illustrate that the variational approaches are effective in even very complicated fluidic materials. Special attention is given to chemical and biological reactions and propagation of chemical fronts, described by variational formulations involving conservation laws, path integrals, speed of wave fronts and Fermat-type principles for chemical waves. Analyses are presented of macroscopic relativistic systems, aimed to treat elastic continua by gauge-type theories and clarify the relation between gravitational entropy and the geometric properties of space-time.

One of the primary mathematical features of the book is the exposition of the so-called potential representations of physical fields. These representations, that have their methodological roots in electromagnetic field theory and perfect fluid hydrodynamics, are helpful in constructing variational principles with final Lagrangians expressed in terms of certain potentials and their derivatives rather than original physical variables. With this idea in mind, several chapters analyze, for unsteady state processes, the effect of constraints and accepted kinetic potential in the inverse problem of variational calculus. An attractive issue are the mathematical results showing that these analyses can be useful to obtain variational formulations for some transport equations and equations of change describing irreversible thermal fields. A new topic refers to recent results in statistical mechanics where constrained extremization of Fisher's information measure yields the thermostatistics, usually derived microscopically from Shannon's information measure. The new procedure has the advantage of dealing with both equilibrium and nonequilibrium processes on an equal footing, as illustrated by examples. It is also shown that generalized (nonextensive) entropies and their applications can be linked with variational methods. Engineering applications of variational approaches are exemplified by a number of analyses. In several examples, frictional incompressible and compressible flows are associated with the extrema of entropy production, flow resistance or viscous drag. The problem of impinging streams is treated, and a variational principle for the two impinging unequal streams in potential flow is derived. In this context, a general class of free boundary-value problems associated with variational principles is reviewed. The results of the drag minimization are highlighted by a theorem of Hellmann–Feynman-type for the drag of arbitrarily shaped bodies. A variational formulation based on the Ritz method is used in a stability analysis of composite structures to set an eigenvalue problem, and, by using different buckling deformation functions, to obtain the solutions for buckling polymers. Field variational principles are constructed for irreversible transfer of energy and matter, involving suitably constructed potentials rather than original physical variables. The hyperbolic transport of energy is also analyzed in the framework of the path-integral approach. With the minimum entropy generation principle, a new formulation of the boundary-value problems in steady-state heat conduction is proposed. With the help of restricted variational approaches (of Rosen-type) standard equations describing heat, mass, and momentum transfer in moving, chemically reactive continua are recovered including the Navier–Stokes equation. Also reviewed are some important recent developments in the nonequilibrium thermodynamics of radiation and matter–radiation mixtures; these include variational principles for nonequilibrium steady states of photons, neutrinos, and matter in local thermodynamic equilibrium. Finally, problems are displayed showing the role of extremum principles in control thermodynamics, ecology, self-organization theory, and econophysics, treated, as a rule, from the viewpoint of the optimal control or systems theory.

The topics discussed in Part I of the book comprise theoretical problems of existence, objectivity, structure, symmetries, and construction of variational principles and their interpretation in nonequilibrium thermodynamics, theory of transport phenomena, (thermo)elasticity, and theory of chemical systems. Chapter I.1 is the guiding introductory text that discusses in detail the content of this multiauthor volume and highlights the most important results. Chapter I.2 characterizes fundamental physical features of the Lagrange formalism in thermodynamics of irreversible processes governed by the Second Law of Thermodynamics. Chapters I.3–I.5 discuss various aspects of mathematical issues associated with the inverse problem of variational calculus and macroscopic relativistic gauge-type theories, where the role of gauge transformations is played by diffeomorphisms of the material space. Chapter I.6 supplements the importance of the entropy view in Chapter I.1 by considering the relation between gravitational entropy and the geometric properties of space-time. Various aspects of the variational theory and conservation laws in nonreacting reversible and irreversible continua are treated in Chapters I.3–I.14. Whereas the variational mathematics of propagation of chemical fronts in reaction–diffusion problems is covered in Chapters I.15–I.17.

The considered physical systems are governed by either classical entropies (Chapters I.1–I.17) or extended entropies and other information measures (Chapters II.1, II.2, II.7, and II.8). In Chapter II.1 equilibrium and nonequilibrium features are derived from a constrained Fisher extremization process where equilibrium corresponds to the ground-state solution and disequilibria—to superpositions of the ground state with excited states. In Chapter II.2 the Hamiltonian approach is applied to derive the generalized thermostatistics of Tsallis when (following the usual routine used in classical statistical mechanics) one starts with a finite Hamiltonian system composed by N weakly interacting elementary subsystems and writes its micro-canonical and canonical distributions in the phase space. In the thermodynamics limit of an infinite N, one recovers the traditional Boltzmann–Gibbs statistics.

In Chapters I.1–I.3 and I.10–I.13, and in the more application-oriented Chapters II.7–II.10, the authors ask under what conditions irreversible macroscopic equations admit a variational formulation and discuss details of Hamiltonian or Lagrangian descriptions for the kinetic and balance equations of processes with transport of heat, mass, and chemical reactions. For the Eulerian (field) description of transport phenomena the relevant method is associated with variational principles involving suitably constructed potentials rather than original physical variables, Chapters I.3, I.12, I.13, II.7 and II.8. Approaches using restricted variations can be found in Chapters II.9 and II.10. Simple hyperbolic heat transfer as well as coupled parabolic transfer of heat, mass and electric charge are described in terms of exact variational principles, Chaps. I.12, I.13, II.7 and II.8. Therein, various gradient or nongradient representations of original physical fields are applied with potentials that are quantities similar to those that Clebsch used in his representation of hydrodynamic velocity. It is stressed (Chapter II.7) that the application of Lagrange multipliers in action-type functionals with constraints adjoined to a kinetic potential automatically generates a canonical system in the extended space spanned by these multipliers and state coordinates, thus making possible a variational formulation in this extended space, although it may not exist in the system's own space (spanned on the original state variables). Action-type criteria of Hamilton's type, or criteria of thermodynamic evolution of Onsager's type, and suitable extensions of both, can thus be constructed, and corresponding Lagrangian and Hamiltonian formalisms developed.

In Chapter I.5 the dynamics of the macroscopic theory is formulated within an exact (gauged, relativistic) framework in terms of independent, hyperbolic, second-order partial differential equations imposed on independent gauge potentials. Components of the energy-momentum tensor are evaluated for suitable Lagrangians (Chapters I.2, I.3, I.5, I.8, I.9, II.7, and references therein). With the total variation one calculates the canonically conjugated quantities, the Hamiltonian, and the Poisson structure both for parabolic and hyperbolic differential equations (Chapter I.12). In this context, Onsager's symmetry is a consequence of a dynamical symmetry. This also opens the road towards canonical quantization (Chapter I.13) that means that operators can be introduced for physical quantities, i.e. energy and quasiparticle number operators for thermal processes. Geometrical and dynamical symmetries provide a deeper insight into the theory. The results imply that the Hamiltonian structure, possessed especially by micromotions, is a richer feature than reversibility. They also suggest that (at least for some processes; Chapter II.7) the thermodynamic irreversibility does not change either the kinetic potential or the action functional, it only complicates potential representations of physical fields in comparison with those describing the reversible evolution. Under the condition that only positive-definite transformations are allowed for the time co-ordinate, extended version of Nöther's theorem shows that a strictly increasing quantity has to exist giving evidence for entropy and the Second Law of Thermodynamics (Chaps. I.12 and I.13).

Mesoscopic descriptions also shed light on the structure of the macroscopic equations. At the mesoscopic level master equations and path-integral formulations of variational principles are relevant. The corresponding theory is developed for diffusion–reaction equations that arise naturally in problems of population dynamics, chemical reactions, flame propagation and others (Chapter I.15). The significance of relative entropy is meaningful. With the path-integral formulation of the Hamilton – Jacobi approximation of the master equation it is possible to calculate rate constants for the transition from one well to another of the information potential and give estimates of mean exit times.

The results presented prove that the potential of Lagrangian and Hamiltonian approaches is not yet fully exploited in contemporary scientific and technical disciplines and that new applications are expected in the fields of nonequilibrium thermodynamics, complex material engineering, heat- and mass-transfer operations, and in processes with chemical reactions. Recent results for speeds of traveling fronts (Chapter I.16) show that for one-dimensional reaction–diffusion equations the speed of the front is characterized by an integral variational principle (generic upper and lower bounds). The effect of density-dependent diffusion and convective terms on the speed of the fronts can be determined as well. Chemical waves are derived from reaction–diffusion systems that have traveling-wave solutions. In this case the Fermat principle of minimum propagation time is valid (Chapter I.17). The essential features of the evolution of chemical fronts are derived from the geometric theory of waves. An example of the obstacle problem in a plane is instructive, where the fronts take the form of involutes of the obstacle curve whenever the propagation velocity is constant. Applications in Chapter I.7 refer to systems composed of rigid particles, such as liquid crystals, suspensions, slurries, polymers, and composites. These materials often possess unusual flow properties, which make them challenging to process in their respective industries. With Hamilton's principle, rotational motion is incorporated into fluid models with the same degree of rigor as the translational motion, so that the rotational component can rigorously be treated as a part of the dynamical response of complex fluids (Chapter I.7).

Observations can be made regarding the extremum properties of entropy production, and therefore, the available energy's destruction, in practical and industrial processes, both transient and steady state (Chapter II.3). They show that a process would typically arrange itself in such a way that the entropy production, or rate of entropy production, is either minimized or maximized. For certain processes, whether there is minimization or maximization depends upon the boundary conditions of the problem. The cases considered include various versions of frictional incompressible flow and of compressible flow (of Fanno and Rayleigh) as well as other cases such as normal shock, combustion and detonation (Chapter II.3). Hydrodynamics of fluids governed by Darcy's law, which are under the influence of frictional force densities proportional to the flow velocity, is treated by variational principles constructed on the basis of an analogy between the field equations of classical hydrodynamics and the hydrodynamic (Madelung) picture of the quantum theory (Chapter I.11). A variational approach serves to calculate the drag on a submerged body in various linear approximations of the Navier–Stokes equation (Chapter II.4). A variational principle holds for the impinging streams, where the Euler equations and natural boundary conditions result from making stationary a functional associated with the potential energy of the whole flow field (Chapter II.5). Variational principles are also applied to buckling analysis of composite laminated structures (Chapter II.6). Principles based on potentials rather than original physical variables are used to describe hyperbolic heat transfer and coupled parabolic transfer of heat, mass, and electric charge (Chapter II.7).

Restricted variational approaches (with some variables frozen) are found for extended thermodynamic systems; they are illustrated for the cases of the soil–water system, an ionized gas and heat transport in solids (Chapter II.8). The restricted principles based on a variational equation describing heat, mass, and momentum transfer in moving chemically reactive continuous media can be constructed and then applied to various purposes, in particular to describe the formation of a solid structure in the course of the eutectic solidification (Chapter II.9). Also, restricted variational formulations can be applied to the Navier–Stokes equation (Chapters II.9 and II.10).

Under the assumption of the minimum entropy generation the structure of heat-source terms is tested in Chapter II.11. Entropy and entropy production criteria govern the analysis of radiation systems and matter–radiation mixtures (Chapter II.12). Optimal control applications refer to the energy generators driven by heat flux and/or mass fluxes (Chapter II.13). In this spirit, also, very complex processes, e.g. aerothermoservoelectroelasticity, that exhibit behavior influenced by numerous couplings, are treated (Chapter II.14).

Thermodynamic aspects of ecological systems are associated with the self-organization paradigm that influences ecosystem dynamics on different time scales. The hypothesis of 'maximization of exergy' in ecosystems is promoted, and it is discussed how these ideas can be used in environmental management (Chapter II.15). In the realm of self-organizing systems, derivation of self-organized criticality from the extremum principle is studied using the example of avalanche formation (Chapter II.16). Extremum criteria are formulated for nonequilibrium patterns and self-organization in economical systems (Chapter II.17). An analogy between irreversible processes in thermodynamic and microeconomic systems is shown (Chapter II.18). The notion of generalized exergy is introduced as the limiting work that can be extracted from the system subject to various constraints, including that on the process duration. Profitability, defined as the maximal capital extractable from the system, provides an analog of the generalized exergy in microeconomics where the resources and capital are distributed between subsystems in such a way that capital dissipation is minimum (Chapter II.18).

We hope that the book will be valuable for mathematicians, physicists, chemists, and engineers, in particular those involved in the application of the mathematical and thermodynamic knowledge to systems with energy generation and transport, solar radiation, chemical waves, liquid crystals, thermoelastic media, composites, multiphase flows, porous media, membrane transfer, microeconomics, etc. Biological systems open new avenues for applications that probably will be summarized soon.

Stanislaw Sieniutycz
Faculty of Chemical & Process Engineering,
Warsaw University of Technology, Warsaw, Poland

Acknowledgements

Acknowledgments constitute the last and most pleasant part of this preface. The editors who participated since 1993 in the so-called Carnet (Carnot Network), express their gratitude to the European Commission under the auspices of which the earliest part of the related research was made possible in the framework of the second Copernicus program (Contract no ICOP-DISS-2168-96). The research towards variational principles was later continued by one of the editors (SS) under the support of the US NSF Grant conducted by R.S. Berry in the Chemistry Department of The University of Chicago in 2001. Currently, this research is conducted by both editors, in the framework of two national grants, the KBN grant 3 T09C 02426 from the Polish Committee of National Research and the Hungarian OTKA grant T—42708. In preparing this volume the editors received help and guidance from R. Stephen Berry (The University of Chicago), Leon Gradoń (Warsaw TU), Elisabeth Sieniutycz (University of Warsaw), Lynne Honnigmann, Arno Schouwenburg and Joyce Happee (Elsevier), and Fiona Woodman and her production team at Alden Prepress Services. The editors, furthermore, owe a debt of gratitude to all contributors. Working with these stimulating colleagues has been a privilege and a very satisfying experience. A critical part of writing any book is the review process, thus the authors and editors are very much obliged to the researchers who patiently helped them read through subsequent chapters and who made valuable suggestions.

S. Sieniutycz
Warsaw, June 2004

LIST OF CONTRIBUTORS

Murilo Pereira de Almeida
Departamento de Física, Universidade Federal do Ceará, C.P. 6030, 60455-900 Fortaleza, Ceará, Brazil
E-mail: murilo@fisica.ufc.br

Karl-Heinz Anthony
Universität Paderborn, Theoretische Physik, D-33095 Paderborn, Germany
anthony.schwaney@t-online.de

Janusz Badur
Institute of Fluid Flow Machinery, Polish Academy of Sciences, ul. Fiszera 14, 81-852 Gdańsk, Poland
jb@imp.gda.pl

Jordan Badur
Institute of Oceanography, University of Gdańsk, ul. Piłsudskiego 46, 81-378 Gdynia, Poland
jorb@kdm.task.gda.pl

Rafael D. Benguria
Department of Physics, P. Universidad Católica de Chile, Casilla 306, Santiago 22, Chile
rbenguri@fis.puc.cl

Sidney A. Bludman
Department of Physics, University of Pennsylvania, Philadelphia, PA 19104, USA and Deutsches Elektron Synchrotron Notkestrasse 85, D-22603 Hamburg, Germany
bludman@mail.desy.de

B.I.M. ten Bosch
Shell Global Solutions International, PO Box 38000, 1030 BN Amsterdam, The Netherlands
dick.tenbosch@shell.com

M. Casas
Departament de Fisica and IMEDEA, Universitat de les Illes Balears 07122 Palma de Mallorca, Spain
montse.casas@uib.es

G. Caviglia
Department of Mathematics, Via Dodecaneso 35, I-16146 Genova, Italy
caviglia@dima.unige.it

J.P. Curtis
QinetiQ, Fort Halstead, Building Q13, Sevenoaks, Kent TN14 7BP, UK
jpcurtis@qinetiq.com

M. Cristina Depassier
Department of Physics, P. Universidad Católica de Chile, Casilla 306, Santiago 22, Chile
mcdepass@fis.puc.cl

Janusz Donizak
AGH University of Science and Technology, Department of Theoretical Metallurgy, Al. Mickiewicza 30, 30059 Kraków, Poland
szczur@uci.agh.edu.pl

Brian J. Edwards
Department of Chemical Engineering,
The University of Tennessee,
Knoxville, TN 37996, USA
bjedwards@chem.engr.utk.edu

Christopher Essex
Department of Applied Mathematics,
University of Western Ontario,
London, Ontario,
Canada N6A 5B7
essex@uwo.ca

Henrik Farkas
Department of Chemical Physics,
Institute of Physics,
Budapest University of Technology
and Economics, Budafoki út. 8,
1521 Budapest, Hungary
farkashe@goliat.eik.bme.hu

L. Fatibene
Department of Mathematics,
University of Torino,
Via C. Alberto 10,
10123 Torino, Italy
fatibene@dm.unito.it

Mauro Francaviglia
Department of Mathematics,
University of Torino,
Via C. Alberto 10,
10123 Torino, Italy
mauro.francaviglia@unito.it

Richard A. Gaggioli
Department of Mechanical and
Industrial Engineering,
Marquette University,
1515 West Wisconsin Avenue,
Milwaukee, WI 52333, USA
richard.gaggioli@marquette.edu

Katalin Gambár
Department of Atomic Physics, Roland
Eötvös University, Pázmány Péter sétány
1/A, H-1117 Budapest, Hungary
gambar@ludens.elte.hu

Bernard Gaveau
UFR de Mathématique 925,
Université Pierre & Marie Curie,
75252 Paris Cedex 05, France
gaveau@ccr.jussieu.fr

Ernest S. Geskin
New Jersey Institute of Technology,
Newark, NJ 07102, USA
geskin@njit.edu

Mircea Gligor
Department of Physics,
National College "Roman Voda",
Roman-5550 Neamt, Romania
mgligor_13@yahoo.com

M. López de Haro
Centro de Investigación en Energía,
Universidad Nacional Autónoma
de México, A.P. 34, 62580 Temixco,
Mor. Mexico
malopez@servidor.unam.mx

Ji-Huan He
College of Science, Donghua University,
1882 Yan'an Xilu Road,
P.O. Box 471, Shanghai 200051,
People's Republic of China
jhhe@dhu.edu.cn

Adam Hołda
AGH University of Science and
Technology, Department of Theoretical
Metallurgy, Al. Mickiewicza 30,
30059 Kraków, Poland
adam@uci.agh.edu.pl

Jerzy Hubert
Institute of Nuclear Physics,
ul. Radzikowskiego 152,
31342 Kraków, Poland
jerzy.hubert@ifj.edu.pl

Sven Erik Jørgensen
The Danish University of Pharmaceutical Sciences, Environmental Chemistry,
University Park 2, 2100 Copenhagen Ø,
Denmark
sej@dfh.dk

Vassilios K. Kalpakides
Division of Applied Mathematics and Mechanics, University of Ioannina,
GR-45110, Ioannina, Greece
vkalpak@cc.uoi.gr

Kristóf Kály-Kullai
Department of Chemical Physics,
Institute of Physics,
Budapest University of Technology and Economics, Budafoki út. 8,
1521 Budapest, Hungary
kk222@ural2.hszk.bme.hu

Vladimir Kazakov
School of Finance & Economics,
University of Technology, Sydney,
P.O. Box 123, Broadway,
NSW 2007, Australia
vladimir.kazakov@uts.edu.au

Dallas C. Kennedy
The MathWorks, Inc.,
3 Apple Hill Drive, Natick,
MA 01760, USA
dkennedy@mathworks.com

Dmitrii O. Kharchenko
Sumy State University,
Rimskii-Korsakov St. 2,
40007, Sumy, Ukraine
dikh@sumdu.edu.ua

Jerzy Kijowski
Centre for Theoretical Physics, Polish Academy of Sciences,
Al. Lotników 32/46;
02-668 Warsaw, Poland
kijowski@cft.edu.pl

Zygmunt Kolenda
AGH University of Science and Technology, Department of Theoretical Metallurgy, Al. Mickiewicza 30,
30059 Kraków, Poland
kolenda@uci.agh.edu.pl

Michael Lauster
University of the German Armed Forces Munich, Faculty for Economics,
Werner-Heisenberg-Weg 39,
85577 Neubiberg, Germany
michael.a.lauster@t-online.de

Giulio Magli
Dipartimento di Matematica,
del Politecnico di Milano, Piazza Leonardo da Vinci 32, 20133 Milano, Italy
magli@mate.polimi.it

Ferenc Márkus
Institute of Physics, Budapest University of Technology and Economics, Budafoki út 8.,
H-1521 Budapest, Hungary
markus@phy.bme.hu

Tamás Matolcsi
Department of Applied Mathematics,
Eötvös Loránd University,
Pázmány P. s. 1/C, 1117 Budapest, Hungary
matolcsi@ludens.elte.hu

Gérard A. Maugin
Laboratoire de Modélisation en Mécanique (UMR CNRS 7607), Université Pierre et Marie Curie, Case 162, 4 place Jussieu,
75252 Paris Cedex 05, France
gam@ccr.jussieu.fr

Michel Moreau
Laboratoire des Physique Theoretique des Liquides, Université Pierre & Marie Curie, 75252 Paris Cedex 05, France
moreau@lptl.jussieu.fr

A. Morro
University of Genova, DIBE, Via Opera Pia 11a, I-16145 Genova, Italy
morro@dibe.unige.it

Alexander I. Olemskoi
Sumy State University,
Rimskii-Korsakov St. 2,
40007, Sumy, Ukraine
alex@ufn.ru

Hayrani Öz
Department of Aerospace Engineering, The Ohio State University, Columbus, OH 43210, USA
oz.1@osu.edu

David M. Paulus
Institut für Energietechnik,
Technische Universität Berlin,
Marchstraße 18 10587,
Berlin, Germany
d.paulus@iet.tu-berlin.de

A. Plastino
Universidad Nacional de La Plata,
C.C. 727, 1900 La Plata, Argentina,
Argentine National Research Center (CONICET), La Plata,
Argentina
plastino@sinectis.com.ar

A.R. Plastino
Universidad Nacional de La Plata,
C.C. 727, 1900 La Plata, Argentina,
Argentine National Research Center (CONICET), La Plata Argentina
and Department of Physics,
University of Pretoria,
Pretoria 0002, South Africa
arplastino@maple.up.ac.za

Pizhong Qiao
Department of Civil Engineering,
The University of Akron, Akron,
OH 44325-3905, USA
qiao@uakron.edu

M. Raiteri
Department of Mathematics,
University of Torino,
Via C. Alberto 10,
10123 Torino, Italy
raiteri@dm.unito.it

Harald Ries
Philipps-University Marburg,
and OEC AG, Munich,
D-35032 Marburg, Germany
harald.ries@physik.uni-marburg.de

J.A. del Río
Centro de Investigación en Energía,
Universidad Nacional Autónoma
de México, A.P. 34,
62580 Temixco,
Mor. Mexico
antonio@servidor.unam.mx

Enrico Sciubba
Department of Mechanical and
Aeronautical Engineering,
University of Roma 1-'La Sapienza',
via eudossiana 18-00184 Roma, Italy
enrico.sciubba@uniroma1.it

Luyang Shan
Department of Civil Engineering,
The University of Akron, Akron,
OH 44325-3905, USA
ls30@uakron.edu

Stanislaw Sieniutycz
Warsaw University of Technology,
Faculty of Chemical and Process
Engineering, Warsaw, Poland
sieniutycz@ichip.pw.edu.pl

Wolfgang Spirkl
Infineon Technologies A.G.,
Balanstr. 73, 81541 Munich, Germany
Wolfgang.Spirkl@infineon.com

János Tóth
Department of Analysis,
Institute of Mathematics,
Budapest University of Technology
and Economics,
Egry J. út.1, Budapest H-1521,
Hungary
jtoth@math.bme.hu

A.M. Tsirlin
Program Systems Institute,
Russian Academy of Sciences,
et.'Botic', Perejaslavl-Zalesky
152140, Russia
tsirlin@sarc.botik.ru

P. Ván
Department of Chemical Physics,
Budapest University of Technology
and Economics, Budafoki út 8,
1521 Budapest, Hungary
vpet@phyndi.fke.bme.hu

F. Vázquez
Facultad de Ciencias,
Universidad Autónoma del Estado
de Morelos, A. P. 396-3,
62250 Cuernavaca, Mor. Mexico
vazquez@servm.fc.uaem.mx

J. Verhás
Department of Chemical Physics,
Budapest University of Technology
and Economics, Budafoki út. 8,
1521 Budapest, Hungary
verhas@phyndi.fke.bme.hu

Heinz-Jürgen Wagner
Theoretische Physik, Universität
Paderborn, Pohlweg 55,
33098 Paderborn, Germany
wagner@phys.uni-paderborn.de

A.J. Weisenborn
Shell Exploration and Production UK Ltd,
1 Altens Farm Road, Nigg,
Aberdeen, AB12 3FY, UK
toon.a.j.weisenborn@shell.com

Juan Zhang
College of Science, Donghua University,
1882 Yan'an Xilu Road,
P.O. Box 471, Shanghai 200051,
People's Republic of China
zhangjuan@dhu.edu.cn

CONTENTS

Part I

THEORY

Chapter 1

PROGRESS IN VARIATIONAL FORMULATIONS FOR MACROSCOPIC PROCESSES

Stanislaw Sieniutycz

Warsaw University of Technology, Faculty of Chemical and Process Engineering, Warsaw, Poland

Henrik Farkas

Department of Chemical Physics, Institute of Physics, Budapest University of Technology and Economics, Budafoki út. 8, 1521 Budapest, Hungary

Abstract

The Preface and this introductory chapter constitute two guiding texts that discuss the content of the multiauthored volume on Variational and Extremum Principles in Macroscopic Systems. The Preface does its job synthetically, presenting the results from the perspective of the whole book while outlining the circumstances of its accomplishment. On the other hand, this chapter, based in large part on the authors' abstracts, treats the matter analytically; it discusses the content of subsequent chapters in some detail, after a suitable introduction. The volume comprises the theory and applications of the variational calculus and theory of extrema in macroscopic physics, chemistry and engineering with a few relevant excursions made to the self-organization theory, ecology and econophysics. The vast range of topics comprises the problems found in nonequilibrium thermodynamics (described by classical or extended entropies), transport theory in continua (reversible or irreversible), elasticity theory, and the theory of propagation of chemical fronts. Some attention is also given to macroscopic relativistic systems. The authors search for conditions under which macroscopic equations admit a variational formulation and discuss details of Hamiltonian or Lagrangian formalisms for the kinetic and balance equations of processes with transport of heat, mass and momentum in chemically active or inactive systems. The book outlines a rich spectrum of applications (including simple and complex fluids, energy and chemical systems, liquid crystals, suspensions, polymeric liquids, and composites) useful for practical systems with materials possessing unusual flow properties.

Keywords: Hamilton's principle; variational calculus; Lagrangians; theory of extrema; nonequilibrium thermodynamics

1. Theoretical aspects

Lagrange formalism (LF) based on Hamilton's Principle of stationary action is the most concise form of a theory for dynamical processes. It represents a unifying and universal method, i.e. it applies methodically in the same way to every physical system (Chapter I.2).

E-mail address: sieniutycz@ichip.pw.edu.pl (S. Sieniutycz); farkashe@goliat.eik.bme.hu (H. Farkas).

The complete information on the dynamics of a system is included in one function only, namely in its Lagrangian, i.e. once the Lagrangian of a physical system is known the whole dynamics of the system is defined and conservation laws follow from Noether's theorem [1]. Numerous applications of LF in dynamics of adiabatic 'perfect' media (with constant specific entropies of particles) are not discussed here. For adiabatic fluids we adduce here only the first correct variational formulation of Hamilton's Principle in Eulerian frames [2] and the associated way to conservation laws [3]. For nonadiabatic media diffusional fluxes of energy, momentum and matter complicate variational formulations greatly, especially in irreversible processes (those not invariant with respect to the time reversal, $t \rightarrow -t$). In a number of research papers published in the last decade it was explicitly shown that the familiar difficulties to set variational principles for irreversible processes can be omitted if the process is transferred from its own space (in which Vainberg's test [4] of Frechet symmetry [5] is violated) into another space, sometimes called adjoint space or space of process potentials.

For the Eulerian (field) description of transport phenomena the relevant method is associated with description of thermal fields by variational principles involving suitably constructed 'potentials' rather than original physical variables. For example, hyperbolic and parabolic heat transfer as well as coupled transfer of heat, mass and electric charge can be described in terms of variational principles based on potentials that are quantities similar to those that Clebsch used in his representation of hydrodynamic velocity. Various gradient or nongradient representations of original physical fields in terms of potentials may be useful. In fact, a group of sister methods was proposed each leading to the same primary purpose of getting exact variational formulation. Representative while quite diverse methods in this group (Chapter I.3) include complex fields [6], integrating operators [7,8] and Chapter I.11, additional state variables and/or coordinates (enlarged spaces) [9–14] and variable transformations [15,16]. Use of functionals with higher-order derivatives can also be treated as working in this group insofar that suitable substitutions lead to enlarged spaces. For processes with reversible energy fluxes both kinetic equations and physical conservation laws can be obtained from Hamilton's principle [9,11,12–14], for irreversible processes [15–17] the simultaneous satisfaction of both kinds of equations still remains an open question (Chapter II.7). Synthesizing reviews appeared that aimed to summarise the main methodological issues [18–21]. Simultaneously a group of variational approaches appeared for reaction–diffusion systems, directed towards determining speeds of chemical wave fronts [22,23] and the geometrical theory of chemical waves [24–26]. Constraints in these problems were usually treated by the method of Lagrange multipliers [18,26]. Yet, the so-called semi-inverse method appeared [27] successful in omitting Lagrange multipliers in composite principles.

The contemporary trend towards the variational settings can thus be regarded as broadening of original spaces to minimally increased dimensions; this may be observed for final functionals resulting from composite variational principles. The presence of Lagrange multipliers in composite functionals (with adjoined constraints) automatically generates a canonical system in the extended space spanned by these multipliers and state coordinates, thus making possible a variational formulation in this extended space, although it may not exist in the system's own space spanned on the original state variables [18,28]. Action-type criteria, of Hamilton's type, or criteria of thermodynamic evolution, of Onsager's type, and suitable extensions of both, can be exploited, and corresponding Lagrangian and Hamiltonian formalisms developed. Within an exact (gauged, relativistic) framework the dynamics of macroscopic theories may be formulated in terms of independent, hyperbolic, partial differential equations imposed on independent gauge potentials [29]. Symmetry principles can also be considered, and components of energy-momentum tensors evaluated for suitable Lagrangians [18,29]. With the total variation one may calculate the canonically conjugated quantities, the Hamiltonian, and the Poisson structure [14] both for parabolic and hyperbolic differential equations [19]. This also opens the road towards canonical quantization (Chapter I.13) that means that the operators can be introduced for physical quantities, e.g. energy and quasiparticle number operators for thermal processes. Geometrical and dynamical symmetries can be examined to provide a deeper insight into the theory; in these formalisms, Onsager's symmetry is a consequence of a dynamical symmetry [19]. These results show that the Hamiltonian structure, possessed especially by micromotion, is a richer feature than reversibility. In many variational settings of this sort a limiting reversible process appears as a suitable reference frame to a more general family of irreversible processes [30]. Evaluation of physical conservation laws (canonical structures) is the best in reversible frames (Ref. [31] and Chapter I.9).

Mesoscopic and microscopic approaches help to understand the structure of the macroscopic equations. At the mesoscopic level master equations and path-integral formulations of variational principles are appropriate [32]. With the path-integral formulation of the Hamilton–Jacobi approximation of the master equation one can give estimates of mean exit times and calculate rate constants. In this volume a mesoscopic theory is developed for diffusion–reaction equations arising in problems of population dynamics, chemical reactions, flame propagation, etc. (Chapter I.15).

2. Applications

In the realm of statistical descriptions the Hamiltonian approach may be applied to derive generalized thermostatistics for finite Hamiltonian systems composed of N weakly interacting elementary subsystems, resulting in microcanonical and canonical distributions in the phase space. In the thermodynamics limit of an infinite N the classical Boltzmann–Gibbs statistics is recovered (Chapter II.1). Equilibrium and nonequilibrium properties can also be derived from a constrained Fisher extremization, where equilibrium corresponds to the ground-state solution and disequilibria—to superpositions of the ground state with excited states. For generalized thermostatistics, see Chapter II.2.

Recent results, including those presented in this volume, suggest that the potential of Lagrangian and Hamiltonian approaches is not fully exploited in contemporary science and engineering, and that the growth of applications is expected soon, especially in the fields of nonequilibrium thermodynamics, complex material engineering, heat and mass transfer separation science, and in processes with chemical reactions in situations when one searches for speeds of traveling fronts. In fact, for 1D reaction–diffusion equations the speed of the front is described by an integral variational principle (generic upper and lower bounds; Chapter I.16). The effect of density-dependent diffusion and convective terms on the speed of the fronts can also be determined. Chemical waves are described in detail by reaction–diffusion systems that have traveling-wave solutions. The basic properties of the evolution of chemical wave fronts follow from the geometric theory of waves.

Applications are possible to systems composed of rigid particles, such as liquid crystals, suspensions, slurries, polymeric liquids, and composites. The size scales of the inherent particles in these media range from Angstroms to centimeters. Unusual flow properties of these materials make them interesting in their respective industries. With Hamilton's principle, rotational motion can be incorporated into fluid models with the same exactness as the translational motion, allowing the knowledge of the complete dynamical response to be gained (Chapter I.7).

Other applications may also be specified. The nature of the entropy-production extrema in practical and industrial processes is generally complex. A process would typically arrange itself in such a way that the rate of entropy production is either a minimum or a maximum depending upon the boundary conditions of the system (Chapters II.3 and II.11). Yet, the entropy considerations can be omitted for fluids governed by Darcy's law that are under the influence of frictional-force densities proportional to the flow velocity (Chapter I.11). Quasiclassical variational principles can then be constructed on the basis of an analogy between the field equations of classical hydrodynamics and the hydrodynamic picture of the quantum theory. Variational principles can be used to calculate the drag on a submerged body in various linear approximations of the Navier–Stokes equation (Chapter II.4) and for the problem of impinging streams (Chapter II.5). Variational approaches can also be formulated in buckling analysis of composite laminated structures (Chapter II.6). Coupled processes of heat, mass and electric-charge transfer can be treated by variational methods based on potentials (Chapters II.7 and II.8). Many applications arise in radiation systems and matter–radiation mixtures and in the theory of energy generators driven by the heat flux or coupled fluxes of heat and mass (Chapters II.12 and II.13).

Still other applications are found in ecological systems, associated with the self-organization paradigm that influences the ecosystem dynamics. An ascendancy concept and extremum behavior in complex ecosystems can be postulated. The hypothesis of 'maximization of exergy' in ecosystems is often promoted to be used in environmental management (Chapter II.15). For self-organization problems derivation of self-organized criticality from an extremum principle can be studied using the example of steady-state analysis of avalanche formation (Chapter II.16). Extremum criteria may be formulated for nonequilibrium patterns and self-organization in economical systems (Chapter II.17). Analogy between irreversible processes in thermodynamic and microeconomic systems can be considered to show that the distribution of resources and capital between subsystems is associated with minimum dissipation (Chapter II.18).

Finally, a group of applications not considered in this volume can be predicted. These might involve multiphase flows, non-Newtonian media, reactors with decaying catalysts, electrochemical cells and biological systems. Their implementation would require interaction between physicists, applied mathematicians, chemists, and engineers, in particular those familiar with systems of energy generation and transport, multiphase flows, systems with matter transport in biological membranes, etc.

3. Discussion

The main results contained in the chapters of this volume are reviewed below. The first part focuses on theory and the second on applications. The borderline between these parts is necessarily fuzzy. The assignment of a chapter to the first part is based on the generality of mathematical methodology in the field of functional or function extrema rather than on the depth of physical penetration, so in the second part the reader can find a number of chapters with subtle physical issues as well. In fact, a number of examples in the second part contribute significantly to the critical understanding of many theoretical aspects contained in the first part.

Chapter I.1 (the present one) is the guiding text that discusses the content of the volume *Variational and Extremum Principles in Macroscopic Systems*, focusing on theory and applications of variational calculus and theory of extrema in mathematics, macroscopic physics and chemistry. Special attention is given to problems found in nonequilibrium thermodynamics (described by classical or extended entropies), transport theory in continua (reversible or irreversible), elasticity theory, and the theory of propagation of chemical fronts. Representative problems in the self-organization theory, ecology and econophysics are also briefly described.

In *Chapter I.2*, K.-H. Anthony reviews LF for irreversible thermodynamics with particular attention paid to the Second Law of Thermodynamics and the Principle of Least Entropy Production. In traditional thermodynamics of irreversible processes the latter principle is restricted to stationary processes in quite special (linear, or quasilinear) systems. Anthony shows that, associated with the entropy concept, a general Principle of Least Entropy Production holds within the LF for macroscopic systems. The entropy is implicitly defined by an inhomogeneous balance equation expressing the second Law of Thermodynamics. The latter essentially results from a universal structure of LF and the Lagrangian invariance with respect to a group of common gauge transformations applied to the fundamental field variables of the system. Thermodynamics of irreversible processes is included in the framework of LF that presents a unified method for reversible and irreversible processes.

Within LF Hamilton's Action Principle is the very fundamental basis of the system's dynamical theory. In Anthony's approach this primary variational principle is called *Action Principle of the first Kind* (AP1). Virtual variations are applied to real processes. Together with a particular Lie group of invariance requirements, the relevant set of *observables* of the system is defined by means of Noether's Theorem. The observables are implicitly derived from the Lagrangian as density functions, flux-density functions, and production-rate densities, all entering the balance equations (Noether balances) associated with their particular observables. The important example is the observable of energy and its conservation, both being associated with time-translational invariance of all physics, i.e. with the requirement that all physics has to be reproducible at any time. The first Law of Thermodynamics is thus involved in the LF. An analogous statement holds for the second Law of Thermodynamics. However, the entropy balance is related to the properties of the second variation rather than the first; it also refers to the system's stability and the Jacobi sufficient condition of extremum. This case deals with the variations of perturbations of a given (real) reference process. The Lagrangian and thus the fundamental field equations and the boundary conditions are kept fixed. Associated Euler equations are the perturbation equations of the system. Entropy, S, while still essential, appears as a quantity somewhat to the side of Hamilton's action principle. This role of entropy S differs from that known for perfect fluids, where it is a basic field variable satisfying a sourceless balance. Yet, the different treatment of the entropy constraint in action principles of irreversible processes seems to be implied by other approaches (Chapters II.7 and II.8).

In *Chapter I.3*, T. Matolcsi, P. Ván and J. Verhá's review contemporary mathematical problems of variational principles focusing on their objectivity, symmetries and construction. In particular, they investigate the relation between governing equations, space–time symmetries and fundamental balances in nonrelativistic field theories. Equivalent Lagrangians, LF in field theory, symmetries, and the inverse problem of variational calculus (construction of variational principle for a given equation) are considered in detail. First, the authors present an outline of the nonrelativistic space–time model and the formulation of covariance. Next, as an example of handling equations in a space–time model, they discuss the LF and generalized Noether theorem for point masses. Furthermore, they consider the LF and the corresponding Noether theorem in nonrelativistic field theories. An inherently covariant (observer-independent) LF and the corresponding symmetry principles are given.

The inverse problem of variational calculus is the main topic of this chapter. A differential operator Θ is used to characterize a given equation, and a Hamilton principle is sought with Θ as its Euler–Lagrange equation. A mathematical theorem states that if Θ is symmetric then a Hamilton principle exists along with a suitable action (with $\Theta = 0$ for the Euler–Lagrange equation). In the case of nonsymmetric Θ there are several ways to circumvent this theorem if the condition $\Theta = 0$ does not need to be reproduced exactly, and only an equivalent equation with the same solution is sufficient. The appropriate possibilities are classified and discussed by the authors. In particular, they discuss ways to construct Lagrangians for dissipative systems and to develop a uniform theory that generates the balance equations and takes into account the second law. Several representative methods are analyzed including complex field variables, integrating operators, additional coordinates, and variable transformation. They also exemplify the role of symmetries by constructing a general variational principle for Maxwell's equations. The same is accomplished in the case of Gyarmati's principle, as an example of non-Hamiltonian variational principles of nonequilibrium thermodynamics. The advantages and disadvantages of the so-called quasivariational formulations are surveyed.

In *Chapter I.4*, Ji-H. He and J. Zhang present a semi-inverse method for establishment of variational principles with applications for incremental thermoelasticity with voids. An introduction to the inverse method of calculus of variations is given that complements the considerations on inverse problems presented in the previous chapter. The method proposed by the first of the authors has been proven to be effective and convenient to search for generalized variational principles from field equations and boundary conditions. An advantage of the method is that it can provide a powerful mathematical tool to establishing a family of variational principles for a rather wide class of physical problems without using the Lagrange multipliers. In the chapter, the semi-inverse method applied to the search for variational formulae directly from the field equations and boundary conditions is systematically illustrated. As an example, a family of variational principles for thermoelasticity with voids is deduced systematically without using the Lagrange multiplier methods.

In *Chapter I.5*, J. Kijowski and G. Magli consider variational formulations for relativistic elasticity and thermoelasticity. Existing, equivalent variational formulations of relativistic elasticity theory are reviewed. Emphasis is put on the formulation based on the parametrization of material configurations in terms of unconstrained degrees of freedom. In this description elasticity can be treated as a gauge-type theory, where the role of gauge transformations is played by diffeomorphisms of the material space. Within this framework it is shown that the dynamics of the theory can be formulated in terms of three independent, hyperbolic, second-order partial differential equations imposed on three independent gauge potentials. Canonical (unconstrained) momenta conjugate to the three configuration variables and the Hamiltonian of the system is easily found. The extension of the theory to nondissipative thermoelasticity and the applications of the theory in astrophysics and quantum gravity are briefly discussed.

In *Chapter I.6*, L. Fatibene, M. Francaviglia and M. Raiteri set a geometric variational framework for entropy in general relativity. They first review the basic frameworks to define the entropy of gravitational systems and discuss the relation among different perspectives. Particular attention is paid to macroscopic prescriptions in order to clarify the relation between gravitational entropy and the geometric properties of space–times. The main aim of this review is to discuss the fundamental contributions that geometry and calculus of variations offer in characterizing the entropy of black-hole solutions as well as more general singular solutions in general relativity. The authors conclude that in the space–time entropy turns out to be associated to the observer's unseen degrees of freedom, namely to the regions that the observers have no physical access to. Extending Beckenstein's original idea to a broader context, the authors state that entropy measures the information content beyond the hidden regions. They stress that the hidden regions they are referring to comprise the standard black-hole horizons but encompass also a much wider class of singularities.

In *Chapter I.7*, B. J. Edwards examines the fundamental origin of rotational motion in rigid particle dynamics and demonstrates that rotational motion can be incorporated into variational models of fluid motion with the same degree of rigor as the translational motion. To date, many different theories have been developed that describe the dynamical properties of complex fluids, but the vast majority of these have neglected an important part of the dynamical response of such fluids: the rotational motion. Edwards' analysis treats translational and rotational motion of a uniaxial liquid crystal as derived using Hamilton's Principle of least action. The conservative dynamics of the Leslie–Ericksen (LE) model of a uniaxial liquid crystal are derived using Hamilton's Principle. This derivation is performed for two purposes: first, to illustrate that variational principles apply to even very complicated fluidic materials, and, second, to derive the noncanonical Poisson bracket that generates the conservative dynamics of the LE model. Although dissipative mechanisms in the LE model can and have been treated through variational formulations (see references in the chapter), dissipation is neglected therein to focus on the origin of rotational motion in the model. This demonstrates that the convective evolution of the microstructure in the model originates solely from the translational and rotational motion of the constituent particles, and not, as is commonly believed, by affine or nonaffine convection and deformation with respect to the local velocity field. The derivation leads naturally to a canonical Poisson bracket, incorporating both translational and rotational particle motions. The canonical Poisson bracket may then be reduced to a noncanonical structure for deriving continuum-scale models for material behavior that incorporate rotational motion in a much more fundamental way than through the use of nonaffine motion. Applications of this approach are remarkable as many fluids are composed of rigid particles, such as liquid crystals, suspensions, slurries, polymeric liquids, and composites. (The size scales of the inherent particles in these fluids range from Angstroms to centimeters.) These materials possess very unusual flow properties, which make them challenging to process in their respective industries.

In *Chapter I.8*, J. Badur and J. Badur Jr. give an introduction to variational derivation of the pseudomomentum conservation in thermohydrodynamics. They provide a reconstruction and revaluation of a few basic, pioneering variational principles in thermohydrodynamics. In particular, most of the authors' attention is aimed at Gibbs', Natanson's and Eckart's principles. All of them either contain preliminary forms of pseudomomentum and pseudomomentum flux notions or are natural ground for them. They also present a method to obtain pseudomomentum balance equations from properly constructed variational principles. The study is completed with analysis of other variational principles and variational grounds for a 'Generalized Lagrangian Mean' (GLM) approach recently applied in atmospheric and ocean turbulence descriptions. The authors conclude that the variational approach is most useful in the case of new, unrecognized physical phenomena. The role of the variational approach in the proper examination of pseudomomentum flux is quite the same as in the case of the variational approach to the proper definition of the momentum flux in so-called nonsimple continua.

In *Chapter I.9*, G. A. Maugin and V. K. Kalkpakides treat variational mechanics of continua of the solid type. Inhomogeneous hyperelasticity provides the standard of the formulation. The main elements of variational mechanics are given in the theory where principal attention is focused on the balance laws derived by use of Noether's theorem. A possible road to including relevant dissipative mechanisms (heat conduction, thermodynamically irreversible anelasticity) is outlined. Their approach exploits the notion of thermacy (Lagrange multiplier of the entropy balance) and shows how the limiting thermoelasticity 'without dissipation' of Green and Naghdi follows therefrom. The variational formulation is shown to yield the classical thermoelasticity of anelastic conductors in the appropriate reduction. A Euclidean 4D space–time formulation of the canonical balance laws of energy and material momentum is then given. Finally, an analogy between the obtained canonical balance laws and the Hamiltonian equations that give the kinetic-wave theory of nonlinear dispersive waves in inhomogeneous materials is established. Further, generalizations to the material behavior involving coupled fields, nonlocality and evolving microstructure are outlined.

In *Chapter I.10*, M. Lauster presents an extension of the classical formulation of the Principle of Least Action (PLA) to meet the needs of a new theory of nonequilibrium processes called the alternative theory of nonequilibrium processes (AT). Based on the idea that natural processes run with a minimum amount of action a mathematical formulation for the minimum of a suitable action integral follows. In the context of the AT the extremum procedure delivers local conservation laws for energy, linear and angular momentum as well as the electrical charge and shows—via Noether's theorem—how the respective coordinates for space and time have to be constructed. Those are closely connected to the so-called Callen principle that is used to choose suitable extensive quantities for the system under consideration. The chapter deals with the application of the PLA to a new and promising theory of nonequilibrium processes. It is shown how Hamilton's Principle, can be fitted into the theoretical context of irreversible thermodynamics. Also, a formulation for the natural trend to equilibrium is derived from the PLA. The variational approach results in an extended version of Noether's theorem allowing expression of the conservation laws for certain extensive quantities in connection with symmetry transformations of the space and time coordinates. Under the condition that only positive-definite transformations are allowed for the time coordinate it can be shown with the PLA that a strictly increasing quantity has to exist giving evidence for entropy and the Second Law of Thermodynamics.

In *Chapter I.11*, H.-J. Wagner deals with variational principles for the linearly damped flow of barotropic and Madelung-type fluids. His analysis contributes to the hydrodynamic theory of fluids that are under the influence of frictional force densities proportional to the flow velocity. These models approximately describe the seepage of fluids in porous media (Darcy's law). One particular aim is the incorporation of such linearly damped hydrodynamic systems into the framework of Lagrangian field theory, i.e. the construction and discussion of variational principles for these systems. In fact, the search for variational principles greatly benefits from the fact that there exists a far-reaching analogy between the field equations of classical hydrodynamics and of the so-called hydrodynamic (Madelung) picture of quantum theory. This analogy often allows for an immediate transfer of strategies and concepts.

Introduction of linear frictional force densities into the Madelung picture leads to certain well-known phenomenological extensions of the ordinary Schrödinger equation such as the Caldirola–Kanai equation and Kostin's nonlinear Schrödinger–Langevin equation. They have often been employed for the description of various frictional effects in quantum systems including, e.g. inelastic nucleon–nucleon scattering, diffusion of interstitial impurity atoms, radiation damping, and dissipative tunneling. Despite several failed attempts in the literature, the construction of variational principles for these equations turns out to be relatively straightforward. The main tools used include the method of Fréchet differentials as well as the so-called Schrödinger quantization procedure. Utilizing the above-mentioned analogy between hydrodynamics and the Madelung picture, the results for the dissipative quantum systems can immediately be carried over to the analogous frictional system in classical hydrodynamics. However, some additional considerations are still needed due to the possibility of rotational (vorticity) degrees of freedom that can be present in ordinary hydrodynamics but are usually foreign to the Madelung picture.

In *Chapter I.12*, K. Gambar writes on a least action principle for dissipative processes. Her premise is that the least action principle can be extended for dissipative processes and that the Lagrange–Hamilton formalism can be completely worked out. For a mathematical model of nonequilibrium thermodynamics the examinations of symmetries lead to Onsager's reciprocity relations. On the basis of the Poisson structure, the stochastic behavior of processes can be described in the phase space. The mathematical difficulties associated with the thermodynamic irreversibility (odd partial derivatives with respect to time) are solved by introducing special potential functions (scalar fields), and by the help of these a complete Hamilton–Lagrange theory is constructed. The potential of total variation allows calculation of the canonically conjugated quantities, the Hamiltonian, and the Poisson structure both for parabolic and hyperbolic differential equations, i.e. both for diffusive processes (e.g. heat conduction as a completely dissipative process) and damped-wave processes (e.g. absorption of electromagnetic waves in a medium). Geometrical and dynamical symmetries are examined to provide a deeper insight into the theory structure. Onsager's symmetry relations are proved as a consequence of a dynamical symmetry. Physical quantities (e.g. entropy, entropy current and entropy production) are introduced; this shows the compatibility of the variational theory with nonequilibrium thermodynamics. The concept of a new phase field arises, based on the space of canonically conjugated quantities. It enables one to examine the stochastic behavior of processes. The constructed Poisson structure leads to Onsager's regression hypothesis.

In *Chapter I.13*, F. Markus applies a Hamiltonian formulation to accomplish the quantization of thermal fields. His chapter introduces and then applies the concepts of modern field theories to such a theory where dissipative processes prevail. With the example of heat conduction, the chapter shows how the quantization procedure can be developed. The author discusses the energy and number operators, the fluctuations, the description of Bose systems and q-boson approximation. The generalized Hamilton–Jacobi equation and special potentials—classical and quantum–thermodynamical—are also calculated. The author points out an interesting connection of dissipation with the Fisher and the extreme physical information. To accomplish the thermal quantization the variational description of the Fourier heat conduction is examined. The Lagrangian, describing the process in the whole space, is constructed by a potential function that is expressed by the coefficients of Fourier series. These coefficients are the generalized coordinates by which the canonical momenta, Hamiltonian and the Poisson brackets are calculated. This mathematical construction opens the way towards the canonical quantization that means that the operators can be introduced for physical quantities, e.g. energy and quasiparticle number operators for thermal processes. The elaborated quantization procedure is applied for weakly interacting boson systems where the nonextensive thermodynamical behavior and the q-algebra can be taken into account and can be successfully built into the theory. The Hamilton–Jacobi equation, the action and the kernel can be calculated in the space of generalized coordinates. This quantization method—which is called Feynman quantization—shows that repulsive potentials are working in the heat process, similar to a classical and a quantum–thermodynamical potential.

In *Chapter I.14*, A. Morro and G. Caviglia treat the problem of conservation laws and variational conditions for wave propagation in planarly stratified media. Time-harmonic fields, satisfying the Helmholtz (or Schrodinger) equation, are investigated through a conservation law in the form of a first integral. The fields describe oblique incidence on a planarly stratified medium. By way of examples, elastic, electromagnetic, and acoustic waves are considered. Possible interfaces are allowed and the boundary or jump conditions are derived through the variational approach and balance equations. Attention is mainly addressed to the reflection–transmission process generated by a multilayer between homogeneous half-spaces. A number of consequences follow. The occurrence of turning points does not preclude wave propagation; while a jump of the ray between two turning points is provided by the ray theory, no jump is allowed for the exact solution. If the amplitude of the incident wave is nonzero then the field is nonzero everywhere. Yet, depending on the value of the parameters in the half-spaces, a total-reflection condition occurs. Irrespective of the form of the multilayer, a relation between the amplitudes of the reflected–transmitted waves follows that is a generalization of that for two half-spaces in contact.

In *Chapter I.15*, J. Tóth, M. Moreau and B. Gaveau investigate master equations and path-integral formulation of variational principles for diffusion–reaction problems. The mesoscopic nonequlibrium thermodynamics of a reaction–diffusion system is described by the master equation. The information potential is defined as the logarithm of the stationary distribution. The Fokker–Planck and the Wentzel–Kramers–Brillouin methods give very different results. The information potential is shown to obey a Hamilton–Jacobi equation, and from this fact general properties of this potential are derived. The Hamilton–Jacobi equation is shown to have a unique regular solution. With the path-integral formulation of the Hamilton – Jacobi approximation of the master equation it is possible to calculate rate constants for the transition from one well to another of the information potential and give estimates of mean exit times. In progress variables, the Hamilton–Jacobi equation always has a simple solution that is a state function if and only if there exists a thermodynamic equilibrium for the system. An inequality between energy and information dissipation is studied, and the notion of relative entropy is investigated. A specific two-variable system and systems with a single chemical species are investigated in detail, where all the defined, relevant quantities can be calculated explicitly.

In *Chapter I.16*, R. D. Benguria and M. C. Depassier discuss variational principles for the speed of traveling fronts of reaction–diffusion equations. Reaction–diffusion equations arise naturally in problems of population dynamics, chemical reactions, flame propagation and others. A characteristic feature of these systems is the appearance of propagating fronts when the system is suddenly quenched into an unstable state. A small localized perturbation grows to eventually cover the whole space. An important problem is the determination of the speed at which the front moves into the undisturbed region. In this chapter the authors review recent results that show that for 1D reaction–diffusion equations the speed of the front is characterized by an integral variational principle. Generic upper and lower bounds as well as accurate estimations in specific cases are obtained in a unified way from this principle. The effect of density-dependent diffusion and convective terms on the speed of the fronts is determined as well. The methods can be applied to a class of hyperbolic equations that describe diffusion in systems with memory.

In *Chapter I.17*, H. Farkas, K. Kály-Kullai and S. Sieniutycz write on the Fermat principle for chemical waves. The chapter outlines the historical background as well as the different formulations of the Fermat principle. The most common formulation of the Fermat principle in optics ensures that the light propagates in such a way that the propagation time is a minimum. Yet, the Fermat principle is valid for any traveling wave, not only for light. In optics, the underlying equation is the wave equation, which describes all details of propagation. Geometrical optics is a limiting case when wavelength tends to zero. The opposite limiting case—when wavelength tends to infinity—is the subject of geometrical wave theory. This theory is based on the Fermat principle as well as on the dual concepts of rays and fronts.

Chemical waves are described in detail by reaction–diffusion equations. These equations may have traveling wave solutions. The essential features of the evolution of chemical wave fronts can also be derived from the geometric theory of waves. Some recent formulations of the Fermat principle require only stationarity of extremals instead of the stricter requirement of minimum propagation time. A discussion reveals that for chemical waves only the requirement of minimum is relevant; maximum and 'inflexion-type' local stationarity does not play a role. For chemical waves and their prairie-fire picture only the quickest rays have effects. Nevertheless, there are singularities, when paths of rays with the same propagation time exist; the related fields of singularities and caustics are then relevant.

Special attention is paid to aplanatic surfaces, where all the reflected or refracted paths require the same time, that is stationarity holds globally. Nonaplanatic refraction is discussed, too: a special example of a refracting sphere is treated. Finally, the shape of a chemical lens is derived.

In *Chapter II.1*, A. Plastino, A. R. Plastino and M. Casas write on the Fisher variational principle and thermodynamics. (In the book, this is the first of two chapters of Part II that show applications of variational approaches in statistical physics and thermodynamics.) The authors prove that standard thermostatistics, usually derived microscopically from Shannon's information measure via Jaynes' Maximum Entropy procedure, can equally be obtained from a constrained extremization of Fisher's information measure that results in a Schrödinger-like wave equation. The new procedure has the advantage of dealing on an equal footing with both equilibrium and offequilibrium processes. Equilibrium corresponds to the ground-state solution, nonequilibrium to superpositions of the gas with excited states. As an example, the authors illustrate these properties with reference to material currents in diluted gases, provided that the ground state corresponds to the usual case of a Maxwell–Boltzmann distribution.

In *Chapter II.2*, M. P. de Almeida displays the connection between the generalized entropy and the Hamiltonian structure of statistical mechanics. More explicitly: he adopts a Hamiltonian approach to derive the generalized thermostatistics of Tsallis. Following the usual routine used in classical statistical mechanics, he begins with a finite Hamiltonian system composed by N weakly interacting elementary subsystems and writes its microcanonical and canonical distributions in terms of its structure function in phase space. The exponential and the power-law canonical distributions emerge naturally from a unique differential relation of the structure function. More precisely, if the heat capacity of the heat bath is infinite, the canonical distribution is exponential, while if it is constant and finite, the canonical density follows a power-law behavior. He then derives microscopic analogs of the main thermodynamic quantities (temperature and entropy) and shows the validity of the thermodynamics relations. The finite-dimensional physical entropy turns out to be a version of the so-called generalized entropy of Tsallis. In the thermodynamic limit ($N \rightarrow \infty$) he recovers the traditional Boltzmann–Gibbs statistics. Furthermore, he shows that the classical thermodynamics relation, $S = U/T - k \ln \rho_0$, is independent of the explicit form of the canonical distribution. Finally, he discusses the issue of additivity of the physical entropy and presents theoretical and numerical examples of Hamiltonian systems obeying the generalized statistics.

In *Chapter II.3*, D. M. Paulus, Jr. and R. A. Gaggioli analyze the nature of entropy extrema in physical processes, mainly in the field of hydrodynamics and continuum mechanics. Their chapter applies the Second Law of Thermodynamics to investigate why processes behave in a certain manner. The extremization of entropy's production (and therefore available energy's destruction) is studied in kinetic processes, including both incompressible and compressible flow, and in static situations, such as chemical equilibrium. Their 'case studies' are directed to gain insight to discover a general extremum principle in thermodynamics (for example, a variational principle similar to Hamilton's principle). The majority of these 'case studies' are processes for which there are well-known mathematical models. In several of them, entropy extremization has been recognized before. The authors find that a process would typically arrange itself in such a way that the entropy produced, or rate of entropy production, is either minimized or maximized. For certain processes, whether there is minimization or maximization depends upon the boundary conditions of the problem. The cases considered include: frictional incompressible flow (Case 1: Boundary condition with specified flowrate; Case 2: Boundary condition with specified pressure drop; Case 3: Two parallel ducts) and compressible flow (Case 4: Fanno flow; Case 5: Rayleigh flow; Other cases, such as normal shock, combustion and detonation).

In *Chapter II.4*, B. I. M. ten Bosch and A. J. Weisenborn work out a variational principle for the drag in linear hydrodynamics (drag on an object in Oseen flow). Their important result is a succinct formula for the variation of the drag. Familiar variational principles in fluid flow generally either include inertia but no viscous dissipation, or, conversely, include viscous dissipation while neglecting inertia. The variational principle for the drag in Oseen flow as first put forward in the previous papers by the authors and further developed in this chapter is a hybrid structure in that both viscous and inertial terms play their own role. After a brief review of the variational formulation, the authors investigate the heat and momentum transfer as a function of the Reynolds number for the case of Oseen flow around a disk. The highlight of the their results is a theorem of Hellmann–Feynman type for the drag of arbitrarily shaped bodies. The original Hellmann–Feynman theorem in quantum mechanics provides a rule to calculate the derivative of the ground-state energy with respect to a parameter in the system Hamiltonian. In the hydrodynamic context the theorem provides a short and simple proof of Brenner's lowest-order Oseen correction formula for the drag. As an application, the authors provide a derivation of the first-order Oseen correction for the drag on an arbitrarily shaped body. For small Reynolds numbers, a correspondence is established with an Oseen resistance put forward by Brenner.

The method of induced forces is shown to be a suitable tool to calculate the drag on a submerged body in various linear approximations of the Navier–Stokes equation. The authors derive it from a variational principle. The stationary value of the appropriate functional is the drag. The derivation of the set of equations for the induced force moments can be applied for a number of hydrodynamically relevant problems. These problems include a sphere moving slowly along the axis of a rotating viscous fluid and a sphere, and a disk in Oseen flow. This variational scheme may be applied to study the influence of momentum convection on the drag on a sphere for small Reynolds and Taylor numbers. Already with a simple approximation for the induced force density, the results are in good agreement with experimental data. The possibility of a variational scheme for lift calculations can also be investigated. Finally, an extension of the principle to yield the full Navier–Stokes equation can be discussed, under the condition of stationarity and the Navier–Stokes drag as the stationary value.

In *Chapter II.5*, J.P. Curtis presents a variational principle for the impinging streams problem. The problem of impinging streams has interested researchers since early last century and is still under investigation today. The historical and current literature on this problem is reviewed. A general class of free boundary-value problems associated with variational principles is then reviewed, following an explanatory 1D analog formulation.

In a synthesis of these two previously unrelated research areas a variational principle for the problem of impinging unequal streams in potential flow is derived under conditions of planar flow. A coupled pair of boundary-value problems defined in two flow regions is obtained comprising the Euler equations and the original and natural boundary conditions associated with this variational principle. The Euler equations and natural boundary conditions result from making stationary a functional associated with the pressure integral over the whole flow field. In each region Laplace's equation emerges as the Euler equation, and the natural boundary conditions correspond to the appropriate physical boundary conditions on the free surfaces and at the interface between the regions. The conservation equations and the stationarity conditions provide the equations necessary to close the problem.

In *Chapter II.6*, P. Qiao and L. Shan discuss the role of variational principles in stability analysis of composite structures. Variational principles, as an important part of the theory of elasticity, have been extensively used in stability analysis of structures made of fiber-reinforced polymer (FRP) composites. In the chapter, variational principles in buckling analysis of FRP composite structures are presented. A survey of variational principles in stability analysis of composite structures is first given, followed by a brief introduction of the theoretical background of variational principles in elasticity. A variational formulation of the Ritz method is used to establish an eigenvalue problem, and, by using different buckling deformation functions, the solutions of buckling of FRP structures are obtained. As application examples, the local and global buckling of FRP thin-walled composite structural shapes is analyzed using the variational principles of total potential. For the local buckling of FRP composite shapes, the flange or web of the beams is considered as a discrete anisotropic laminated plate subjected to rotational restraints at the flange–web connections, and by enforcing the equilibrium condition to the first variation of the total potential energy, the explicit solutions for local buckling of the plates with various unloaded edge boundary conditions are developed. For the global buckling of FRP composite beams, the second variation of the total potential energy (based on nonlinear plate theory) is applied, and the formulation includes shear effect and beam bending–twisting coupling. In summary, the application of variational principles is a vital tool in buckling analysis of composites, and the present explicit formulations using the variational principles of the Ritz method can be applied to determine buckling capacities of composite structures and facilitate the buckling analysis, design and optimization of FRP structural profiles.

In *Chapter II.7*, S. Sieniutycz treats field-variational principles for irreversible energy and mass transfer in formulations similar to those in the optimal control theory, where given constraints are adjoined to a kinetic potential. The considered processes are: hyperbolic heat transfer and coupled parabolic transfer of heat, mass and electric charge, both in convection-free systems. In this chapter (the first in the part on *Transport Phenomena and Energy Conversion*) the author demonstrates the violation of the standard, canonical structure of conservation laws in formulations similar but not equivalent to Hamilton's Principle that typically uses the sourceless entropy constraint. As Hamilton's form of the kinetic potential L is applied, the difference between the two formulations is due to the applied constraints that comprise here the entropy balance and the energy-representation counterpart of the Cattaneo equation called Kaliski's equation. Despite the generally noncanonical conservation laws (obtained via Noether's theorem) the method that adjoints a given constraining set to a kinetic potential L works efficiently, i.e. it leads to an exact variational formulation for the constraints in the potential space of Lagrange multipliers, implying that the appropriateness of the set should be verified by physical rather than mathematical criteria. For the field (Eulerian) description of heat conduction, equations of the thermal field follow from a variational principle involving potentials rather than original physical variables. With various gradient or nongradient representations of physical fields in terms of potentials (quantities similar to those used by Clebsch in his representation of hydrodynamic velocity) actions and extremum conditions are found in the entropy representation (variables $\mathbf{q}$ and ρ_e) or in the energy representation (variables $\mathbf{j}_s$ and ρ_s). Symmetry principles are considered and components of the formal energy–momentum tensor are evaluated. Attention is given to the source terms in balances of internal energy and entropy. When sources are present, a variational formulation does not exist in the original 'physical' space, and, if one insists on exploiting this space plus possibly a necessary part of the potential space, an action is obtained in an enlarged space spanned on the multiplier of a nonconserved balance constraint as an extra variable. Inhomogeneous waves are found for variational adjoints in both entropy and energy representations in which physical densities are sources of the thermal field. These may be contrasted with homogeneous equations for the state variables (telegraphers equations). Similarly as in gravitational and electromagnetic systems where the specification of sources (electric four-current or matter tensor, respectively) defines the behavior of the field potentials, thermal dynamics can be broken down to potentials.

In *Chapter II.8*, F. Vázquez, J. A. del Rio and M. L. de Haro write more on variational principles for irreversible hyperbolic transport. Restricted variational principles as applied to extended irreversible thermodynamics are illustrated for the cases of the soil–water system and heat transport in solids. This kind of restricted variational principles leads to the time-evolution equations for the nonconserved variables as extremum conditions. In particular, as has been noted in the case of heat transport, this perspective provides interesting generalizations of the well-known Maxwell–Cattaneo–Vernotte forms. In order to show how a Poissonian structure may be obtained, a formulation in terms of the variational potentials is developed and used to derive the time evolution of the fluctuations in hyperbolic transport. These fluctuations are shown to obey the Chapman–Kolmogorov equation. A process of relativistic heat transport is discussed as an example of such formulation. The hyperbolic transport is also analyzed in the framework of the path-integral approach. This latter methodology allows for the consideration of nonlinear hyperbolic transport, in contrast with what occurs in the case of the variational potentials scheme.

In *Chapter II.9*, E. S. Geskin presents a restricted variational principle for transport processes in continuous systems. In a personal treatment, he develops a routine procedure for derivation of variational equations representing a wide range of continuous system. The procedure involves the use of the generalized variables and fluxes for system description. The application of this procedure is demonstrated by constructing the variational equations for both dissipative and reversible processes. A variational equation describing heat, mass and momentum transfer in a moving, chemically reactive continuous media is constructed using the proposed routine. The Euler–Lagrange equations following from the constructed variational equation are identical to the balance equations for entropy, momentum and mass. A Lagrangian density, relating the rate of the energy change in the system with energy dissipation, work and entropy production, is constructed. The use of this Lagrangian is demonstrated by its application to the formation of a solid structure in the course of a eutectic solidification. It is assumed that the liquid and solid phases are divided by the plane boundary and the solidification rate is constant. Under these conditions solidification results in the development of the lamellar structure in the solid and the process is defined by the lamellar spacing and the rate of motion of the liquid/solid interface. The variational equation is suitably readjusted to describe the process in question. A solution of this equation determines the relationship between the solidification rate and the lamellar spacing. The obtained results comply with available experimental data.

In *Chapter II.10*, E. Sciubba poses a challenging question: Do the Navier–Stokes equations admit a variational formulation? As in the previous chapter, the answer is affirmative only in the realm of approaches using restricted variations. Mature analysis of the past literature material is the virtue of the chapter. The problem of finding a variational formulation for the Navier–Stokes equations has been debated for a long time, since the fundamental statements of Hermann von Helmholtz and John William Strutt, Lord Rayleigh. There is a remarkable lack of agreement among different authors even on the theoretical possibility of the existence of such a statement, leave alone its practical derivation. On the other hand, there is a similarly remarkable sequence of consistent attempts to solve the problem, all based on what appears to be a common intuition: that the driving mechanism is indeed some sort of entropy-based functional. This chapter is divided into two parts: in the first, the author tries to put into proper perspective both this longstanding debate and its possible formal and practical implications; in the second, he discusses a novel procedure for deriving the incompressible Navier–Stokes equations from a Lagrangian density based on the exergy 'accounting' of a control volume. The exergy-balance equation, which includes its kinetic, pressure–work, diffusive, and dissipative parts (the latter due to viscous irreversibility) is written for a steady, quasiequilibrium and isothermal flow of an incompressible fluid. It is shown that, under the given assumptions, and without recourse to the concept of 'local potential', the Euler–Lagrange equations of a formal minimisation of the exergy variation (=destruction) result, in fact, in the Navier–Stokes equations of motion.

In *Chapter II.11*, Z. Kolenda, J. Donizak and J. Hubert investigate the entropy-generation minimization in the steady-state heat conduction. On the basis of this process they propose a new formulation of the boundary-value problems in the transfer processes based on the second law. (Traditionally, boundary-value problems are formulated on the basis of the first law of thermodynamics.) To obtain consistent results when the method of irreversible thermodynamics is applied an additional assumption of relatively small temperature gradients over the whole domain is introduced. Such an assumption also means that $|T(x_i) - T_{\mathrm{avg}}|/T_{\mathrm{avg}} \ll 1$ for all x_i $(i = 1, 2, 3)$, where $T(x_i)$ is temperature at x_i and T_{avg} is the average temperature of the solid. The new formulation of the boundary-value problem uses the principle of the minimum entropy generation. Applying the variational formalism based on the Euler–Lagrange equation, a new mathematical form of heat-conduction equation with additional heat-source terms is derived. It follows that the entropy-generation rate of the process can significantly be reduced, associated with the decrease of the irreversibility ratio according to the Gouy–Stodola theorem. By introducing additional heat sources it is shown that the minimization of entropy generation in heat-conduction processes is always possible. The most important result derived from these theoretical considerations is directly connected with the solution of the boundary-value problems for solids with temperature-dependent heat-conduction coefficients. In these cases, additional internal heat sources can be arbitrarily chosen as positive or negative, which enables one to extend the applications presented in the literature. Heat conduction in anisotropic solids is also discussed as a suitable extension.

In *Chapter II.12*, Ch. Essex, D. C. Kennedy and S. Bludman investigate the role of variational formulations in the nonequilibrium thermodynamics of radiation interaction. They review some important recent developments in the nonequilibrium thermodynamics of radiation and matter–radiation mixtures. These include variational principles for nonequilibrium steady states of photons, neutrinos, and matter in local thermodynamic equilibrium. These variational principles can be extended to include mass and chemical potential. The general nature of radiation entropy, entropy production, equilibrium, and nonequilibrium is also discussed. The distinctions between fermionic and bosonic, massive and massless, conserved and nonconserved quanta are the basis for the broadly different thermodynamics of matter and radiation. Much that appears nonlocal in position space is local if one keeps in mind the full mode space, with both momentum and position space coordinates, as well as spin and charge labels. The alternative descriptions of classical field and quantum counting are possible for bosons because of the existence of bosonic coherent states. A broadened framework allows for the correct reformulation of classical nonequilibrium thermodynamics in terms of elementary quanta, including fundamental and possibly massless and free-streaming bosons.

In *Chapter II.13*, W. Spirkl and H. Ries treat optimal finite-time endoreversible processes (processes where irreversibilities are lumped in external subsystems). Using the variational calculus, they treat the general problem of transferring a system from a given initial state to a given final state in a given finite time such that the produced entropy or the loss of availability is minimized. This problem leads to a second-order differential equation similar to the Euler–Lagrange equation. However, while mechanical systems naturally follow the trajectory that minimizes the action, a thermodynamic system does not tend to minimize dissipation, rather an external control is required, for which they give the equations. In fact, they give exact equations for the optimal process for the general case of a nonlinear system with several state variables, and show solved examples for the case of two state variables. Not only the speed but also the path depends on the available time. For linear processes, e.g. in the limit of slow processes or if the Onsager coefficients do not depend on the fluxes, they find a constant entropy-production rate or a constant loss rate of availability and an optimal path independent of the available time. For nonlinear cases, the entropy-production rate or loss rate of availability is generally not constant in an optimal process.

In *Chapter II.14*, H. Öz writes on the evolutionary energy method (EEM) and presents its aerothermoservoelectroelastic application. The chapter presents a novel theoretical foundation introduced by the author as the EEM finding its root in the natural law of energy conservation, specifically the first law of thermodynamics. To this end, the law of evolutionary energy (LEE) is introduced as the encompassing foundational evolutionary equation, where the evolutionary operator D– is a directional change operation via parameter alterations on the energy quantities satisfying the energy-conservation law along the actual dynamic path, and acts on the total evolving energy, which is defined as the time integral of the total actual energy interactions in a dynamic system. The EEM is an algebraic—direct—energy method; i.e. it uses and needs no knowledge of differential equations of the system for response and/or control studies of dynamic systems. Introduction of the concept of assumed time modes (ATM) for the generalized response variables and generalized control inputs of a dynamic system in conjunction with the law of evolutionary energy culminates in elimination of time from the system dynamics completely, yielding the algebraic evolutionary energy description of the system dynamics for response and control studies. As an application of the EEM, an aerothermoservoelectroelastic system is described completely algebraically to study the optimal distributed control of both structural displacements and skin temperature by using only temperature feedback via electroelastic and thermoelectric actuation in high-speed flight. Skin-temperature control of a flat panel by distributed actuation in Mach 10 flight is simulated to demonstrate the capabilities of the method.

In *Chapter II.15*, S.-E. Jørgensen introduces the reader to ecological extrema by considering the maximization of ecoexergy in ecosystems. Ecoexergy, as usually applied exergy (maximum work from a system and the environment) expresses the work capacity; but ecoexergy uses the same system at the same temperature and pressure at thermodynamic equilibrium as the reference state. The chapter shows how ecoexergy can be found for organisms and discusses why living systems have a particularly high exergy, associated with their high information content. A maximization hypothesis is proposed and supported by observations. If a system receives an exergy input, it may utilize this exergy to move further from thermodynamic equilibrium. If more than one pathway to depart from equilibrium is offered the one yielding the largest gradients, and exergy storage under the prevailing conditions, to give the most ordered structure furthest from equilibrium, will tend to be selected. This formulation may be considered a translation of Darwin's ideas to thermodynamics: the organisms that have the properties that are fitted to the prevailing conditions in the ecosystem, will be able to contribute most to biomass and information and thereby give the ecosystem most exergy—they are the best survivors. An argument is given that the principle of exergy maximization in ecosystems is consistent with the description of ecosystem development. Also, it is discussed to that extent the exergy-maximization principle is in accordance with other proposed principles as minimum entropy production, maximum power (throughflow of energy), maximum ecoexergy destruction and maximum energy residence time.

In *Chapter II.16*, A. I. Olemskoi and D. O. Kharchenko consider the extrema in the theory of self-organization: they treat self-organized criticality within the framework of a variational principle. The theory of a steady-state flux related to avalanche formation is presented for the simplest model of a sand pile within the framework of a Lorenz-type system. The system with a self-organized criticality regime is studied within the realm of Euclidean field theory. The authors consider a self-similar behavior introduced as a fractional feedback in a three-parameter model of Lorenz-type. The main modes to govern the system dynamics are: an avalanche size, an energy of moving grains and a complexity (entropy) of the avalanche ensemble. They account for the additive noise of the energy to investigate the process of a nondriven avalanche-formation process in the presence of the energy noise that plays a crucial role. The kinetics of the system is studied in detail on the basis of the variational principle. This distribution is shown to be a solution of both fractional and linear Fokker–Planck equations. Relations between the exponent of the size distribution, fractal dimension of phase space, characteristic exponent of multiplicative noise, and number of governing equations are obtained. Making use of the supposition that the avalanche-formation process has the properties of anomalous diffusion the authors set the main relations between exponents of anomalous diffusion, the exponent of the theory and number of governing stochastic equations to represent the system displaying self-organized criticality. They show that the type of fluctuations (white or colored) does not change the universality class of the system in the supposition of diffusion process of a avalanche formation.

In *Chapter II.17*, M. Gligor develops the econophysics and extremum criteria for nonequilibrium states of dissipative macroeconomic systems. The first part of his work is centered upon the concept of entropy in the dynamical description of some socioeconomic systems. Examination of the logarithm of price distribution from several catalogs indicates that this distribution is very close to the Gaussian distribution and, as such, it can be derived from the maximizing of the entropy functional associated with this variable. The exponential distribution of incomes, reported in the literature to be valid for the great majority of the population, also maximizes the entropy functional when the variable is positive-definite and the mean value is fixed. A kinetic approach developed in the next sections brings substantial support to clarify the exponential distribution of wealth and income, and also enlarges the framework of the analysis including nonequilibrium steady states. Another extremum criterion, namely the minimum production of entropy is discussed in the context of the economic systems in near-to-equilibrium steady states. The final sections discuss the multiplicity of equilibria, the economic cycles and the effect of the random fluctuations. These phenomena, which arise in the description of the economic systems in far-from-equilibrium states, are investigated by methods of nonlinear dynamics and stochastic theory. Nonperiodicity of economic depressions may be caused by the randomness of the noise-induced transitions between nonequilibrium steady states. For far-from-equilibrium states, dissipative structures, instabilities and crashes are observed in socioeconomic systems. According to the irreversible thermodynamics, such systems may develop nonequilibrium patterns and self-organization phenomena. In financial markets, the dissipative structures appear as speculative bubbles (of market prices and financial indices), which can break, the system undergoing transitions towards new steady states. The minimum of an 'effective potential' appears to separate the normal random-walk-like regime of parameters from the crash regime.

In *Chapter II.18*, A. M. Tsirlin and V. Kazakov compare extremum principles and limiting possibilities in open thermodynamic and economic systems. In their chapter Prigogine's minimum-entropy principle is generalized to thermodynamic and micro-economic systems that include active subsystems (heat engine or economic intermediary). The notion of capital dissipation for microeconomic systems, similar to the notion of entropy production in thermodynamics, is introduced. The economic counterpart states that in an open microeconomic system that consists of subsystems in internal equilibrium with flows that depend linearly on the difference of resources' estimates the resources are distributed in such a way that capital dissipation attains a minimum with respect to free variables. The authors derive new bounds on the limiting possibilities of an open system with an active subsystem, including the bound on the productivity of the heat-driven separation. They also derive economic analogies of Onsager's reciprocity conditions.

Economic systems differ from thermodynamic ones in many respects including the voluntary, discretional nature of exchange, production in addition to exchange, competition in various forms, etc. Yet, thermodynamic and economic systems are both macrosystems, which explains the analogies between them, including the irreversibility of respective processes. In particular, this refers to steady states of open thermodynamic and microeconomic systems that include internally equilibrium subsystems with fixed intensive variables (reservoirs) and subsystems whose intensive variables are free (determined by exchange flows). For linear dependence of flows on driving forces an analogy is observed between the conditional minimum of the entropy production (in thermodynamics) and the conditional minimum of capital dissipation (in microeconomics).

Acknowledgements

This research was supported by two national grants, from the Polish Committee of National Research, the KBN grant 3 T09C 02426 (SS), and the Hungarian OTKA grant T(42708 (HF).

References

[1] Nöther, E. (1918), Invariante Variationsprobleme, Nachr. Akad. Wiss. Gottingen. Math-Phys. Kl. II, 235–257.

[2] Herivel, J.W. (1955), The derivation of the equations of motion of an ideal fluid by Hamilton's principle, Proc. Camb. Philos. Soc. 51, 344–349.

[3] Stephens, J.J. (1967), Alternate forms of the Herrivel–Lin variational principle, Phys. Fluids 10, 76–77.

[4] Vainberg, M.M. (1964), Variational Methods for the Study of Nonlinear Operators, Holden-Day, San Francisco.

[5] Finlayson, B.A. (1972), The Method of Weighted Residuals and Variational Principles, Academic Press, New York.
[6] Anthony, K.H. (1981), A new approach describing irreversible processes. In Continuum Models of Discrete Systems, (Brulin, O., Hsieh, R.K.T., eds.), Vol. 4, North-Holland, Amsterdam.
[7] Sieniutycz, S. (1984), Variational approach to extended irreversible thermodynamics of heat and mass transfer, J. Non-equilib. Thermodyn. 9, 61–70.
[8] Sieniutycz, S. (1987), From a least action principle to mass action law and extended affinity, Chem. Eng. Sci. 11, 2697–2711.
[9] Sieniutycz, S. (1986), Heat processes and Hamilton's Principle (in Polish), Rep. Poznan Technol. Univer., Chem. Chem. Eng. 19, 311–324.
[10] Anthony, K.H. (1988), Entropy and dynamical stability—a method due to Lagrange formalism as applied to TIP. Trends in Applications of Mathematics to Mechanics (Besseling, J.F., Eckhaus, W., eds.), Springer, Berlin.
[11] Sieniutycz, S. (1988), Hamiltonian energy–momentum tensor in extended thermodynamics of one-component fluid (in Polish), Inż. Chemiczna i Procesowa 4, 839–861.
[12] Sieniutycz, S. and Berry, R.S. (1989), Conservation laws from Hamilton's principle for nonlocal thermodynamic equilibrium fluids with heat flow, Phys. Rev. A 40, 348–361.
[13] Kupershmidt, B.A. (1990), Hamiltonian formalism for reversible non-equilibrium fluids with heat flow, J. Phys. A: Math. Gen. 23, L529–L532.
[14] Jezierski, J. and Kijowski, J. (1990), Thermo-hydrodynamics as a field theory. In Nonequilibrium Theory and Extremum Principles, Advances in Thermodynamics, (Sieniutycz, S., Salamon, P., eds.), Vol. 3, Taylor and Francis, New York, pp. 282–317.
[15] Markus, F. and Gambar, K. (1991), A Variational principle in thermodynamics, J. Non-equilib. Thermodyn. 16, 27–31.
[16] Nyiri, B. (1991), On the construction of potentials and variational principles in thermodynamics and physics, J. Non-equilib. Thermodyn. 16, 39–55.
[17] Sieniutycz, S. and Berry, R.S. (1993), Canonical formalism, fundamental equation and generalized thermomechanics for irreversible fluids with heat transfer, Phys. Rev. E. 47, 1765–1783.
[18] Sieniutycz, S. (1994), Conservation Laws in Variational Thermo-Hydrodynamics, Kluwer Academic Publishers, Dordrecht.
[19] Gambár, K. and Márkus, F. (1994), Hamilton–Lagrange formalism of nonequilibrium thermodynamics, Phys. Rev. E 50, 1227–1231.
[20] Ván, P. and Muschik, W. (1995), The structure of variational principles in nonequilibrium thermodynamics, Phys. Rev. E 52, 3584–3590.
[21] Vazquez, F., del Rio, J.A., Gambar, K. and Markus, F. (1996), Comments on the existence of Hamiltonian principles for non-selfadjoint operators, J. Non-equilib. Thermodyn. 21, 357–360.
[22] Benguria, R.D., Cisternas, J. and Depassier, M.C. (1995), Variational calculations for thermal combustion waves, Phys. Rev. E 52, 4410–4413.
[23] Benguria, R.D. and Depassier, M.C. (1995), Variational principle for the asymptotic speed of fronts of the density-dependent diffusion-reaction equation, Phys. Rev. E 52, 3285–3287.
[24] Lazar, A., Noszticzius, Z. and Farkas, H. (1995), Involutes: the geometry of chemical waves rotating in annular membranes, Chaos 5, 443–447.
[25] Simon, P.L. and Farkas, H. (1996), Geometric theory of trigger waves—a dynamical system approach, J. Math. Chem. 19, 301–315.
[26] Sieniutycz, S. and Farkas, H. (1997), Chemical waves in confined regions by Hamilton–Jacobi–Bellman theory, Chem. Eng. Sci. 52, 2927–2945.
[27] He, J.H. (1997), Semi-inverse method of establishing generalized variational principles for fluid mechanics with emphasis on turbomachinery aerodynamics, Int. J. Turbo Jet-Engines 14(1), 23–28.
[28] Van, P. and Nyiri, B. (1999), Hamilton formalism and variational principle construction, Ann. Phys (Leipzig) 8, 331–354.
[29] Kijowski, J. and Magli, G. (1999), Unconstrained Hamiltonian formulation of general relativity with thermo-elastic sources, Class. Quantum Gravity 15, 3891–3916.

[30] Sieniutycz, S. and Berry, R.S. (2002), Variational theory for thermodynamics of thermal waves, Phys. Rev. E 65(46132), 1–11.

[31] Kalpakides, V.K. and Maugin, G.A. (2004), Canonical formulation and conservation laws of thermoelasticity without dissipation, Rep. Math. Phys., 51.

[32] Gaveau, B., Moreau, M. and Tóth, J. (1999), Variational nonequilibrium thermodynamics of reaction-diffusion systems. I. The information potential, J. Chem. Phys. 111, 7736–7747. II. Path integrals, large fluctuations and rate constants, J. Chem. Phys. 111 (1999) 7748–7757. III. Progress variables and dissipation of energy and information, J. Chem. Phys., 115 (2001) 680–690.

Chapter 2

LAGRANGE FORMALISM AND IRREVERSIBLE THERMODYNAMICS: THE SECOND LAW OF THERMODYNAMICS AND THE PRINCIPLE OF LEAST ENTROPY PRODUCTION

Karl-Heinz Anthony

Universität Paderborn, Theoretische Physik, D-33095 Paderborn, Germany

Abstract

In traditional thermodynamics of irreversible processes (TIP) a Principle of Least Entropy Production is discussed. This principle is verified for stationary processes of very special systems only. A general formulation of the principle is still missing.

In Lagrange formalism as applied to TIP, a general concept of entropy is involved in quite a natural and straightforward way. It is implicitly defined by an inhomogeneous balance equation—the Second Law of Thermodynamics—which essentially results from a universal structure of Lagrange formalism and from the invariance of the Lagrangian with respect to a group of common gauge transformations applied to the complex-valued fundamental field variables of the thermodynamical system.

Within Lagrange formalism a general Principle of Least Entropy Production is associated with the entropy concept. On the background of the traditional, very confined concept the general concept in Lagrange formalism is presented in some detail.

Keywords: thermodynamics of irreversible processes; Lagrange formalism; Hamilton's action principle in state space (= AP1); action principle in perturbation space (= AP2); first and second kind observables; balance equations; the Second Law of Thermodynamics; the Principle of Least Entropy Production.

1. The traditional structure of thermodynamics of irreversible processes versus universal structures of Lagrange formalism (LF)—a challenge

The Principle of Least Entropy Production is due to that design of thermodynamics of irreversible processes (TIP) that is related with the names Onsager, Prigogine, De Groot, Gyarmati [1–6]: the structure of thermo*statics* has been conceptually taken over to thermo*dynamics*!

1.1. The Principle of Local Equilibrium

The traditional TIP is associated with the *Principle of Local Equilibrium* (PLE) [5,6]. Following this principle thermo*dynamical* processes are assumed to run locally through

E-mail address: anthony.schwaney@t-online.de (K.-H. Anthony).

equilibrium states. Formally it is firstly based on the ad hoc generalization of *Gibbs' fundamental form* of thermostatics for homogeneous systems at rest[1]

$$\mathrm{d}U = T\,\mathrm{d}S - p\,\mathrm{d}V + \sum_{j=1}^{N} \mu_j\,\mathrm{d}N_j, \tag{1}$$

and secondly on the use of the state functions of thermostatics that are related by

$$U = TS - pV + \sum_{j=1}^{N} \mu_j N_j. \tag{2}$$

Switching over to local densities these equations have to be substituted by

$$\mathrm{d}u = T\,\mathrm{d}s + \sum_{j=1}^{N} \mu_j\,\mathrm{d}n_j, \tag{3}$$

and

$$u = Ts - p + \sum_{j=1}^{N} \mu_j n_j. \tag{4}$$

U, S, V, N_j are the internal energy, the entropy, the volume and the partial masses of chemical constituents S_j of a homogeneous system (given in moles). u, s, n_j are the respective densities ($u = U/V$, etc.) and $T, p, \mu_j, j = 1, \ldots, N$, the absolute temperature, the hydroelastic pressure, and the chemical potentials of the constituents S_j. On the basis of Eqs. (3) and (4) the formal essence of the PLE is the following procedure: switching over from thermostatics to thermodynamics one has to replace the differential form (3) by the differential equation

$$\partial_t u = T\,\partial_t s + \sum_{j=1}^{N} \mu_j\,\partial_t n_j \qquad \Rightarrow \text{first assumption} \tag{5}$$

preserving locally relation (4) for dynamical processes $\Rightarrow$ second assumption (6)

i.e. the quantities u, s, n_j, T, p, μ_j are now field quantities depending on space and time coordinates x and t in the volume V of the thermodynamical system. They are *locally taken over from thermostatics*. ∂_t indicates partial, (i.e. local) derivation with respect to time.

1.2. The entropy and the Second Law of Thermodynamics, First and Second Part—traditional view

Especially the entropy being primarily a state quantity of thermostatics is conceptually taken over into thermodynamics! Starting form the PLE, Eq. (5), and using the energy-balance equation (First Law of Thermodynamics) and the mass-balance equations for the diffusing and chemically reacting constituents S_j (diffusion-reaction equations) one gets an

[1] For simplicity the discussion is confined to gaseous systems.

inhomogeneous balance equation [5,6]

$$\partial_t s + \nabla \cdot \vec{J}_{(s)} = \sigma_{(s)}, \tag{7}$$

which for thermodynamical processes is the First Part of the Second Law of Thermodynamics. The concept of entropy is implicitly defined by this equation. The *entropy density* s is taken over from thermostatics of the respective system (4) and (6), whereas the entropy flux density $\vec{J}_{(s)}$ and the *entropy-production-rate density* $\sigma_{(s)}$ straightforwardly result from the procedure outlined:

$$\vec{J}_{(s)} = \frac{1}{T}\left(\vec{J}_{(u)} - \sum_{j=1}^{N} \mu_j \vec{J}_j\right) = \frac{\vec{J}_{(Q)}}{T}, \tag{8}$$

$$\sigma_{(s)} = \vec{J}_{(u)} \cdot \nabla\left(\frac{1}{T}\right) + \sum_{j=1}^{N} \vec{J}_j \cdot \nabla\left(-\frac{\mu_j}{T}\right) + \sum_{j=1}^{N} \sigma_j\left(-\frac{\mu_j}{T}\right), \tag{9}$$

$$\sigma_{(s)} = \vec{J}_{(Q)} \cdot \nabla\left(\frac{1}{T}\right) + \sum_{j=1}^{N} \vec{J}_j \cdot \frac{\nabla(-\mu_j)}{T} + \sum_{j=1}^{N} \sigma_j\left(-\frac{\mu_j}{T}\right). \tag{10}$$

$\vec{J}_{(u)}$, $\vec{J}_j$, $\vec{J}_{(Q)}$ are the flux densities of internal energy, of diffusion of constituents S_j and of heat, respectively. σ_j is the mass production-rate density of constituent S_j during chemical reaction. Subsuming the quantities σ_j under the set of *thermodynamical fluxes*,

$$\{J_\alpha,\ \alpha = 1,\ldots,f\} = \{\vec{J}_{(Q)}, \vec{J}_j, \sigma_j,\ j = 1,\ldots,N\} \tag{11}$$

the entropy production takes a remarkable bilinear form

$$\sigma_{(s)} = \sum_{\alpha} J_\alpha X_\alpha, \tag{12}$$

where X_α is the *thermodynamicalforce* conjugated to the *thermodynamicalflux* J_α (e.g. $\vec{X}_{(Q)} = \nabla(1/T)$, etc.).

Concerning the *Second Part of the Second Law of Thermodynamics* another ad hoc assumption is coming into the play, which is again suggested by thermostatics: The *entropy-production-rate density has to be positive-definite*,

$$\sigma_{(s)} \geq 0 \qquad \Rightarrow \text{third assumption} \tag{13}$$

the inequality being associated with *irreversible* and the equality with *reversible* processes.

The thermodynamical flux densities (11) are driven by the thermodynamical forces, which formally is described by means of a set of *constitutive equations*. Except for the validity of the restriction (13) applied to Eq. (10) there is a great deal of freedom in establishing constitutive equations. *Linear systems* are associated with *linear constitutive equations*,

$$J_\alpha = \sum_{\beta=1}^{f} L_{\alpha\beta} X_\beta \qquad \Rightarrow \text{fourth assumption} \tag{14}$$

the coefficients of which have to be subjected to *Onsager's reciprocal relations* [1,2,5,6]:

$$L_{\alpha\beta} = L_{\beta\alpha}. \tag{15}$$

Let us sum up: obviously from the conceptual as well as from the formal point of view and in spite of its great merits the TIP outlined above is but an *extrapolation of thermostatics* into the neighboring dynamical region. Its merits are confined to a sufficiently small neighborhood of the dynamical processes to equilibrium states; i.e. the driving forces have to be sufficiently small, which means that gradients of the temperature field $T(x,t)$, of mass densities $n_j(x,t)$, etc. have to be sufficiently small. However, *the theory fails to be a self-contained theory for thermodynamical systems*. The question arises if this situation can fundamentally be improved.

1.3. The Principle of Least Entropy Production—traditional view

Nevertheless, there has arisen the *Principle of Least Entropy Production* [4–6]. The entropy production being associated with dissipation of nonthermal forms of energy, it is quite natural to ask for conditions of minimal dissipation and thus of minimal entropy production. The traditional form of the *Principle of Least Entropy Production* reads:

$$\Sigma_{(s)}(t) = \int_V \sigma_{(s)}(X_\alpha(x,t), \alpha = 1,\ldots,f)\mathrm{d}^3x$$

$$= \text{extremum by free variation of all driving forces } X_\alpha(x,t). \tag{16}$$

$\Sigma_{(s)}(t)$ is the (instantaneous) time-dependent total entropy-production-rate of the whole system. The traditional answer: there are real processes with least instantaneous entropy production, namely *stationary processes*, provided the system is a *linear one* (see Eq. (14)) and the processes are of *lower rank than 2*. The latter feature means that the relevant thermodynamical fluxes are scalars or vectors, such as those joined together in Eq. (11); i.e. the principle holds, e.g. for stationary heat transport, stationary diffusion, and stationary chemical reactions—of course, for the case of linear constitutive equations only. Outside the scope of these restrictions there exists no Principle of Least Entropy Production in the traditional theory, or—preferably—there exists no proof for a Principle of Least Entropy Production. This statement especially holds for viscous flow. An additional remark: the stationary processes turn out to be dynamically stable, i.e. nonstationary processes tend asymptotically towards stationary processes.

A few additional remarks might be helpful for the methodical ranking of the traditional Principle of Least Entropy Production: of course one may ask in any case for the minimal entropy production in the sense of the variational principle (16). Insofar that a 'Principle of Least Entropy Production' does always trivially 'exist' and the associated Euler–Lagrange that equations can always be determined. However, in general it is not ensured that the solutions of these equations are contained in the set of real processes of the system, i.e. the variational principle (16) might be meaningless for a particular system. So, in traditional thermodynamics the variational principle (16) has to be understood as a variational principle with constraints, the latter being the fundamental equations of the systems processes. Instead of asking: "Does a PLEP exist for a given system?" one should better ask: "Do, for a given system, particular processes exist with minimal entropy production?" In traditional thermodynamics the PLEP is no universal, fundamental law! It is a special feature of particular systems. Again one may look for a structure, where the entropy production is associated with a universal variational principle that is not confined to special systems and special processes.

1.4. Lagrange formalism—a universal and unified procedure for physical dynamics. Introductory remarks

Lagrange formalism (LF) based on *Hamilton's Action Principle* [7–15] is the most concise form of a field theory for dynamical processes. It presents a *universal* and *unified method*, i.e. it applies methodically in the same way to each physical system. The *whole information* on the dynamics of a system is included in one function only, namely in its *Lagrangian*; i.e. once the Lagrangian of a physical system is defined the whole dynamics of the system is defined.

Within Lagrange formalism *Hamilton's Action Principle* is the very fundamental basis. This *Primary Variational Principle* will be called *Action Principle of the First Kind (AP1)*: virtual variations are applied to the (real) processes,[2] which are solutions of this variational problem. One should carefully note the term 'action', which is meant in its original physical sense: 'action = energy × time'.[3] On this level, different physical systems may be coupled within the unified method of LF. The variational procedure of the AP1, i.e. virtual variation of (real) processes, results in the *fundamental field equations for the (real) processes* of the system as the Euler–Lagrange equations (ELeqs) of the variational principle and in its *relevant boundary conditions*.

Of course, the AP1 implicitly contains all informations on the (real) perturbations of each (real) process, too: explicitly the *perturbations* can be associated with an *Action Principle of the Second Kind (AP2)* that is derived from that of the first kind, and the ELeqs of which are the *perturbation equations* of the system. The latter are known in the literature as *Jacobi equations* [15]. This *secondary variational principle* refers to *virtual variations of the perturbations of a given (real) reference process*, the latter playing in AP2 the role of a control variable during variation.

The knowledge on dynamical stability or instability of processes is an essential feature of the dynamics of a system [15,16]. The AP2 and the associated perturbation equations are fundamental for these questions. As a remarkable fact a stability theory in the sense of Ljapunov's direct method can be derived from the AP2, which, on the one hand, has its own merits and, on the other hand, is closely related with the *Second Part of the Second Law of Thermodynamics*, namely with the *positive-definiteness of entropy production*.[4]

---

[2] One should carefully distinguish between 'real processes' and their 'real perturbations', on the one hand, and 'virtual variations' of real processes or real perturbations, on the other hand. In this paragraph this distinction is occasionally supported by the adjective 'real' in brackets, which, of course, is principally superfluous.

[3] In the literature the term 'Hamilton's Principle' occurs frequently, however, in most cases in the sense of a variational principle only.

[4] This rather complicated theory goes beyond the scope of this chapter. The reader is referred to Refs. [15,18].

Hamilton's Action Principle, i.e. the AP1, and thus the Lagrangian itself have to be subjected to a particular *Lie group of invariance requirements*, which altogether are methodically motivated [15]. Of course, the invariance requirements apply to the AP2 too. On the basis of the AP1 as well as on the AP2 and by means of *Noether's Theorem* these invariance requirements give rise to the definition of a variety of *observables* of the system [8,15,17]. These are implicitly defined from the Lagrangian in the form of *constitutive equations* for *densities, flux density*, and *production-rate densities*, which all together enter into *balance equations (Noether balances)* associated with their particular observables.

The most prominent example is the *observable energy and its universal conservation*, both being associated with *time-translational invariance* of all physics, i.e. with the methodically motivated requirement that all physics has to be reproducible at any time. Thus, the *universal energy conservation* is a straightforward outcome from the AP1 on the grounds of the universal time-translational invariance of isolated physical systems.

Furthermore, the algebraic form of the integral kernel of the AP2 gives rise to an outstanding *universal, inhomogeneous balance equations*, which holds for each perturbation of each process. It will be called the *Central Balance Equation (in Perturbation Space)*. It plays an important rule in stability theory quite generally as well as for the First Part of the Second Law of Thermodynamics [15].

Summing up: in Lagrange formalism the complete dynamics of a system can be derived from its Lagrangian and its relevant, universal invariance requirements by straightforward procedures. 'Complete dynamics' means the fundamental field equations, the relevant balances and its associated constitutive equations as well as the more sophisticated structures of perturbation and stability theory. The methods involved in the self-contained theory hold for each system in the same way. In LF no restrictions on the range of the dynamical processes are implied.

1.5. Thermodynamics of irreversible processes in the framework of Lagrange formalism

TIP can be included in the framework of Lagrange formalism as a self-contained dynamical theory [15,18–23]. Especially thermostatics is involved as a special class of real processes, all the thermodynamical fluxes of which vanish. Thermostatics plays no outstanding role. As a remarkable fact the *First and the Second Law of Thermodynamics* are *derived* in Lagrange formalism by means of straightforward procedures on the basis of particular invariance requirements:

(a) The *First Law of Thermodynamics* is the above-mentioned Noether's universal energy balance associated with time-translational invariance of thermodynamical systems.
(b) The *Second Law of Thermodynamics* is contained in LF on the basis of the following structures:
  - Taking into account *complex-valued variables* as *fundamental field variables for irreversible processes* a *common gauge transformation* of all complex field variables comes into play as a methodical invariance requirement: Phase differences between the various complex field variables are physically relevant rather than the phases themselves (interferences!)

- By the common gauge transformations a particular and outstanding perturbation of the process is induced, which by definition fulfils the central balance equation. In principle, the central balance equation works in perturbation space. However, as a peculiarity of the gauge transformations in this case the central balance equation is completely drawn back into the state space of the real processes. This result is universal whenever complex field variables and a common gauge invariance comes into play. In the case of thermodynamics it is interpreted as the *First Part of the Second Law*. Evaluating the density, the flux density and the production-rate density of the central balance equation by means of the gauge-induced perturbation one gets constitutive equations for the *density*, the *flux density* and the *production-rate density of the entropy*.
- As far as the *Second Part of the Second Law* is concerned, namely the *positive-definiteness of the entropy-production-rate*, we have to look at the *dynamical stability* of the processes. It is supposed that the positive-definite entropy-production-rate is associated with the asymptotic stability of the irreversible processes in the sense of Ljapunov's direct method [15,16]. For particular cases this statement holds; the general proof, however, is still missing.

(c) Finally, combining the AP2 with the concept of entropy LF results in quite a natural way in a *Principle of Least Entropy Production*, which in contrast to the well-known classical principle is generally valid, i.e. which—in contrast to the traditional principle—is no more confined to stationary irreversible processes and to linear constitutive equations for the fluxes of a few particular systems.

2. Preparatories: a few formal structures of Lagrange formalism

2.1. Hamilton's Action Principle—the Action Principle of the First Kind (AP1), the fundamental field equations, boundary conditions, state space S [15]

Let a physical system operate in the region V of the three-dimensional space, referred to the coordinates $x = \{x^1, x^2, x^3\}$, and let its processes be defined by the set $\psi = \{\psi_j, j = 1, \ldots, f\}$ of *fundamental field variables*. For simplicity let us assume the volume V to be constant in time. Then a real physical process $|\psi(t)\rangle = \{\psi(x,t)\} = \{\psi_j(x,t), j = 1, \ldots, f\}$[5] running in the time interval $[t_1, t_2]$ is distinguished as a solution of *Hamilton's Action Principle*, which in the present context will be called the

Action Principle of the First Kind:

$$A_{(1)} = \int_{t_1}^{t_2} \int_V \ell(\psi(x,t), \partial_t \psi(x,t), \nabla \psi(x,t); x, t) \mathrm{d}V \, \mathrm{d}t = \text{extremum} \tag{17}$$

[5] The notation $|\psi(t)\rangle$ indicates the system's 'state vector' in analogy with the customs in quantum mechanics.

with respect to free variations $\delta\psi(x,t)$ of the real process $\psi(x,t)$ keeping fixed its values at the beginning and the end of the process in volume V and on its boundary $\partial V : \delta\psi(x,t_{1,2}) = 0$.

The integral $A_{(1)}$ is the well-known *action integral*; in the subsequent context the index (1) indicates 'action integral of the First kind'. Its kernel $\ell(\cdot)$ is the *Lagrange density function* or the *Lagrangian* of the system; it has to be a *real-valued function*. The particular analytical form of ℓ—including fixed values of characteristic system parameters (e.g. material moduli)—is associated with the system once and for all: once ℓ is defined the complete dynamics of the system is defined.

For the Lagrangian ℓ depending on the fundamental field variables ψ and on all its first derivatives we are dealing with Lagrange formalism of the *first order*. Furthermore, the density function ℓ at site x and at time t depending on the fields $\psi(x,t)$ at the same site x and time t the theory is a *local and instantaneous one*.

Looking at the arguments of the Lagrangian in Eq. (17) the variable ψ and its derivatives are associated with the *internal eigendynamics* of the system, whereas the *explicit variables* x and t are due to the *intervention of the outer world* into the system. The dynamics of the outer world does not take part in the variational procedure of Hamilton's Principle! Its variables are strictly controlled from outside, giving rise to the explicit dependence of the Lagrangian on x and t. This chapter is confined to *isolated systems* that by no means interact with their outer world neither in its volume V, which implies the special structure

$$\ell = \ell(\psi, \partial_t\psi, \nabla\psi) \tag{18}$$

of the Lagrangian, nor across its boundary ∂V, which is associated with

vanishing physical fluxes of any kind across the boundary. (19)

The ELeqs[6]

$$\partial_\alpha \frac{\partial \ell}{\partial(\partial_\alpha \psi_k)} - \frac{\partial \ell}{\partial \psi_k} = 0, \qquad k = 1,\ldots,f, \tag{20}$$

are necessary conditions for an extremum of the AP1 (17) due to free variation of the fields $\psi_k(x,t)$ within volume V of the system. They are the *basic field equations* of the system. These equations have to be supplemented by *boundary conditions*, which fit with the special situation of the system. In the simplest case of an *isolated system* they can be

[6] Throughout the chapter the following notations are used.

$$\{x^\alpha\} = \{x^0 = t, x^1, x^2, x^3\}, \qquad \partial_t = \frac{\partial}{\partial t} = \partial_0 = \frac{\partial}{\partial x^0}, \qquad \partial_\alpha = \frac{\partial}{\partial x^\alpha}, \qquad \nabla = \{\partial_1, \partial_2, \partial_3\}.$$

Summation convention is used only for summation over the indices α of space-time coordinates. All other summations are denoted explicitly.

obtained from the variational principle, too:

$$\vec{n}\cdot\frac{\partial \ell}{\partial(\nabla\psi_k)} = 0, \qquad k = 1,\ldots,f. \tag{21}$$

They are due to *free variations of the fields* $\psi_k(x,t)$ *at the boundary* ∂V of the system.[7] $\vec{n}$ is the outer unit normal vector of the boundary. Eqs. (20) and (21) are *fulfilled by the real processes* of an isolated system, which may be represented by trajectories $|\psi(t)\rangle = \{\psi(x,t)\} = \{\psi_j(x,t), j = 1,\ldots,f\}$ in state space S.[8] The latter is a function space spanned by all functions ψ_j.[9]

2.2. *The perturbation or Jacobi equations, boundary conditions, perturbation space P, the Action Principle of the Second Kind (AP2)*

Let us look at a one-parameter class $|\psi(t;\pi)\rangle = \{\psi_k(x,t;\pi), k = 1,\ldots,f\}$ of *solutions of the fundamental field equations* (20) in a particular neighborhood of $\pi = 0$ of the real parameter π. The functions $\psi_k(x,t;\pi)$ are assumed to be analytical with respect to π. Referred to the *reference process* $|\psi(t)\rangle = |\psi(t;\pi = 0)\rangle$ the processes $|\psi(t;\pi)\rangle$ are regarded as a *class of perturbed processes* with π as *perturbation parameter*. The exact perturbation

$$\Delta|\psi(t;\pi)\rangle = |\psi(t;\pi)\rangle - |\psi(t)\rangle, \tag{22}$$

is approximated by its *main linear part*,

$$\Delta|\psi(t;\pi)\rangle \underset{\text{linear}}{\approx} \left.\frac{\partial|\psi(t;\pi)\rangle}{\partial\pi}\right|_{\pi=0} \times \pi, \tag{23}$$

giving rise to the definition of the set of functions

$$|\eta(t)\rangle = \left.\frac{\partial|\psi(t;\pi)\rangle}{\partial\pi}\right|_{\pi=0} = \left\{\eta_k(x,t) = \left.\frac{\partial\psi_k(x,t;\pi)}{\partial\pi}\right|_{\pi=0}, k = 1,\ldots,f\right\}, \tag{24}$$

which is an element of the *accompanying tangent or perturbation space* $P(t)$ running along the trajectory of the reference process $|\psi(t)\rangle$ in state space (see Fig. 1). Let us call $|\eta(t)\rangle$ the

---

[7] Closing the boundary of a system is physically characterized by vanishing fluxes $\vec{J}$ across the boundary:

$$\vec{n}\cdot\vec{J} = 0. \tag{$*$}$$

From Noether's Theorem it becomes apparent that Eq. (21) results in Eq. ($*$) for all relevant fluxes. Thus *closing the boundary of a system is associated with free variation of field variables at the boundary* of the system.

[8] The symbol $|\mathcal{T}(t)\rangle$ may be regarded as a moving state vector in state space S. See footnote 5.

[9] See footnote 5. Look at Fig. 1.

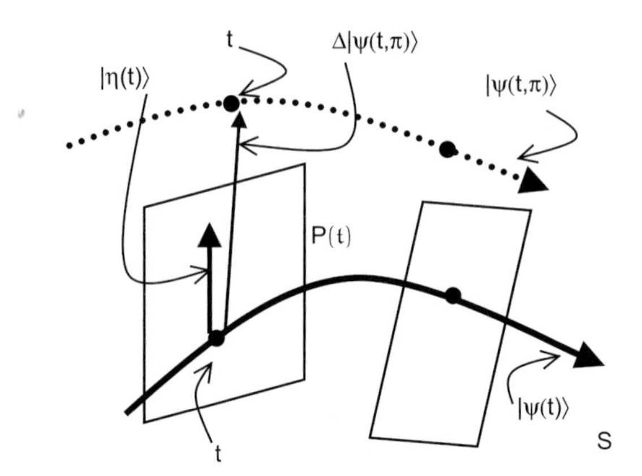

Fig. 1. Accompanying tangent or perturbation space $P(t)$ along the trajectory $|\psi(t)\rangle$ of the process in state space S. Perturbation: $\Delta|\psi(t;\pi)\rangle$, representative: $|\eta(t)\rangle$, being approximately parallel to $\Delta|\psi(t;\pi)\rangle$.

representative of the class $|\psi(t;\pi)\rangle$. It is a representative of the real perturbation (21) of the reference process $|\psi(t)\rangle$. Analogously to the state space S, the perturbation space P is a function space spanned by all functions η_j.

It should be noted that we are not dealing with perturbations of the *system* but of the *processes*! The Lagrangian, and thus the fundamental field equations (20) and the boundary conditions (21) are kept fixed! Only the process $|\psi(t)\rangle$ suffers from a perturbation. For example, the perturbation parameter π might be associated with a perturbation of the initial values of the process. However, there are other, more sophisticated possibilities with regard to invariance properties of the Lagrangian (see below).

For each parameter π the real processes $|\psi(t;\pi)\rangle$ are solutions of the field equations (20) and boundary conditions (21). Thus, based on definition (24) we get the *perturbation* or *Jacobi's equations*[10,11] in volume V,

$$\partial_\alpha \frac{\partial \Omega(\cdot)}{\partial(\partial_\alpha \eta_k)} - \frac{\partial \Omega(\cdot)}{\partial \eta_k} = 0, \qquad k = 1,\dots,f, \tag{25}$$

and for the closed system the associated boundary conditions

$$\vec{n}\cdot\frac{\partial \Omega(\cdot)}{\partial(\nabla \eta_k)} = 0, \qquad k = 1,\dots,f. \tag{26}$$

[10] Insert $|\psi(t;\pi)\rangle$ in Eqs. (20) and (21); apply the operation

$$\frac{\partial}{\partial \pi}\cdots\Big|_{\pi=0}.$$

[11] These equations are often called 'variational equations'. However, one should carefully distinguish between (virtual) variations, applied to a variational problem, and (real) perturbations within a class of real processes. So, the term 'perturbation equations' is preferred.

The kernel Ω is derived uniquely from the Lagrangian ℓ:[12]

$$2\Omega(\psi(x,t),\partial\psi(x,t)|\eta(x,t),\partial\eta(x,t))$$
$$=\sum_{i,j=1}^{f}\left[\frac{\partial^2\ell(\psi,\partial\psi)}{\partial\psi_i\,\partial\psi_j}\eta_i\eta_j+2\frac{\partial^2\ell(\psi,\partial\psi)}{\partial\psi_i\partial(\partial_\alpha\psi_j)}\eta_i\,\partial_\alpha\eta_j+\frac{\partial^2\ell(\psi,\partial\psi)}{\partial(\partial_\alpha\psi_i)\partial(\partial_\beta\psi_j)}\partial_\alpha\eta_i\,\partial_\beta\eta_j\right]. \tag{27}$$

Two kinds of field variables are involved. In Eq. (27) the arguments of Ω are accordingly separated by a vertical bar: ψ denotes the *given* reference process and η the *associated* representative of the perturbation class. Ω—and thus the Jacobi equations (25) and the boundary conditions (26)—contain the complete information on all perturbations of each given reference process of the isolated system. Both sets (25) and (26) of equations are exact. For a given reference process $|\psi(t)\rangle$, which enters into the equations, too; they are by definition fulfilled by the representative $|\eta(t)\rangle$. In particular, the Jacobi equations are linear, second-order partial differential equations for the representative $|\eta(t)\rangle$.

Obviously the perturbation equations (25) and the boundary conditions (26) can formally be associated with a variational procedure of a *closed* system.[13,14] Thus, we introduce the

Action Principle of the Second Kind (AP2):[15]

$$A_{(2)}=\int_{t_1}^{t_2}\int_V 2\Omega(\psi,\partial_t\psi,\nabla\psi|\eta,\partial_t\eta,\nabla\eta)\mathrm{d}V\,\mathrm{d}t=\text{extremum} \tag{28}$$

for a fixed reference process $|\psi(t)\rangle$ by free (virtual) variations $|\delta\eta(t)\rangle$ of the representative $|\eta(t)\rangle$ within volume V and on its boundary ∂V, keeping fixed its values at the beginning and the end of the process: $|\delta\eta(t_{1,2})\rangle=0$.

Eqs. (25) and (26) are the associated ELeqs and boundary conditions. One should carefully note that the variational procedure does not affect the given reference process $|\psi(t)\rangle$. The latter formally plays the role of control variables. In the AP2 three entities are in play: the real reference process $|\psi(t)\rangle$, its real perturbation represented by the representative $|\eta(t)\rangle$ and the virtual variations $|\delta\eta(t)\rangle$ of the latter.

A physical interpretation of the AP2: from the point of view of the perturbation space P the AP2 is Hamilton's Action Principle for representatives $|\eta(t)\rangle$, which suffer from an intervention by the given reference process $|\psi(t)\rangle$, i.e. we are formally dealing with an open system in perturbation space P the 'Lagrangian'[16] which depends explicitly on x and t.[17]

[12] '∂' is a shorthand for all derivatives with respect to time *and* space coordinates.

[13] Compare with Eqs. (17), (20) and (21).

[14] Note: in the present context only closure of the boundary is implied rather than the total isolation of the whole system. Via the reference process $|\psi(t)\rangle$ there is an explicit dependence on x and t involved in the kernel Ω.

[15] The factor '2' in the integral is conventional.

[16] Read 'Ω'.

[17] Via $|\psi(t)\rangle$.

On handling the kernel Ω one should take care of a sophisticated and very important structure: Ω as defined by formula (27) is not *invariant with respect to a change* $\psi \Rightarrow \psi'$ *of the field variables.* The crucial points are the second derivatives of the Lagrangian ℓ appearing in definition (26).[18] In detail: let us regard two different sets

$$\psi = \{\psi_k, k = 1,\dots,f\} \quad \text{and} \quad \psi' = \{\psi'_k, k = 1,\dots,f\} \tag{29}$$

of field variables, which are in a one-to-one correspondence

$$\psi'_j = \psi'_j(\psi_k, k = 1,\dots,f), \qquad j = 1,\dots,f. \tag{30}$$

Let us further regard one and the same class of perturbations with respect to both sets of field variables. Then the associated representatives, $\eta = \{\eta_k, k = 1,\dots,f\}$ and $\eta' = \{\eta'_k, k = 1,\dots,f\}$, defined by Eq. (24), respectively, are in a linear relationship

$$\eta'_j = \sum_{k=1}^{f} \frac{\partial \psi'_j(\psi)}{\partial \psi_k} \eta_k, \tag{31}$$

due to Eq. (30). Of course, the Lagrangian of the system is invariant when changing the set of field variables:

$$\ell(\psi, \partial\psi) = \ell'(\psi', \partial\psi')|_{\psi' = \psi'(\psi),\ \text{Eq. (30)}}. \tag{32}$$

However, the same is not true for the kernel Ω of the AP2:

$$\Omega(\psi, \partial\psi|\eta, \partial\eta) \neq \Omega'(\psi', \partial\psi'|\eta', \partial\eta')|_{\substack{\psi = \psi'(\psi),\ \text{Eq. (30)} \\ \eta' = \eta'(\psi,\eta),\ \text{Eq. (31)}}}. \tag{33}$$

Ω and Ω' being defined in the respective manner by Eq. (27) one finds after a lengthy calculation the relation

$$\Omega(\psi, \partial\psi|\eta, \partial\eta) = \left\{ \Omega'(\psi', \partial\psi'|\eta', \partial\eta') + \partial_\alpha \left(\sum_{k,i,j=1}^{f} \frac{\partial \ell'(\psi', \partial\psi')}{\partial(\partial_\alpha \psi'_k)} \frac{\partial^2 \psi'_k(\psi)}{\partial\psi_i \, \partial\psi_j} \eta_i \eta_j \right) \right\} \Bigg|_{\substack{\psi' = \psi'(\psi),\ \text{Eq. (30)} \\ \eta' = \eta'(\psi,\eta),\ \text{Eq. (31)}}} \tag{34}$$

[18] The situation in state space S is analogous to that of differential geometry in ordinary Euclidean space R^3: being interested in invariant quantities (or objects) in R^3 one has to introduce covariant derivatives instead of ordinary partial derivatives, if the space is referred to curvilinear coordinates. (Ricci-calculus is the relevant keyword.) Only with respect to rectilinear coordinates is the use of ordinary derivatives justified. This situation holds in function space S, too: in order to define an invariant object Ω based on second derivatives of the Lagrangian one should introduce *in state space a* 'covariant Ricci-calculus', which of course causes a lot of mathematical difficulty. However, preferring once and for all one particular set ψ of field variables above all other sets ψ' we arrive at the same goal. This preferred set of variables will be in a way the 'rectilinear coordinates' in the function space. For the concept of entropy these things become essential (see below).

As defined by Eq. (27) *the kernel Ω of the AP2 is ambigious*! However, as a remarkable fact the second term on the right-hand side of Eq. (34) is a *complete 4-divergence*. Thus the ambiguity of Ω in the AP2 does not affect the Jacobi equations![19] From the practical point of view this term is quite convenient for calculating entropy quantities (see below).

2.3. The kernel of the AP2 and the central balance equation in perturbation space P

The kernel Ω (Eq. (27)) of the AP2, being a homogeneous form of second degree with respect to the variables η and $\partial\eta$, may be subjected to Euler's formula:

$$2\Omega(\psi_k, \partial_\alpha \psi_k | \eta_k, \partial_\alpha \eta_k) = \sum_{j=1}^{f} \left(\frac{\partial \Omega(\cdot)}{\partial \eta_j} \partial \eta_j + \frac{\partial \Omega(\cdot)}{\partial(\partial_\alpha \eta_j)} \partial_\alpha \eta_j \right). \tag{35}$$

Substituting the first term on the right-hand side by means of Eq. (25) the relation will be confined to solutions of the Jacobi equations. The right-hand side can be condensed into a 4-divergence resulting in the *central balance equation in perturbation space P*,

$$\partial_t \varpi + \nabla \cdot \vec{J}_{(\varpi)} = \sigma_{(\varpi)}, \tag{36}$$

the density, flux density, and production-rate density of which are defined by the *constitutive equations*

$$\varpi = \varpi(\psi, \partial\psi | \eta, \partial\eta) = \sum_{j=1}^{f} \frac{\partial \Omega(\cdot)}{\partial(\partial_t \eta_j)} \eta_j, \tag{37}$$

$$\vec{J}_{(\varpi)} = \vec{J}_{(\varpi)}(\psi, \partial\psi | \eta, \partial\eta) = \sum_{j=1}^{f} \frac{\partial \Omega(\cdot)}{\partial(\nabla \eta_j)} \eta_j, \tag{38}$$

$$\sigma_{(\varpi)} = \sigma_{(\varpi)}(\psi, \partial\psi | \eta, \partial\eta) = 2\Omega(\psi, \partial\psi | \eta, \partial\eta). \tag{39}$$

The central balance (36) is *universal*, i.e. it holds within LF for each system! It is due to the algebraic structure of the kernel Ω and to the Jacobi equations only, both being universal structures of LF. It is *fulfilled by the representative $|\eta(t)\rangle$ of any perturbation of each real process $|\psi(t)\rangle$*!

One should keep in mind that the formulae (37)–(39) are affected by the ambiguity of the kernel Ω, i.e. they depend on the choice of the fundamental field variables ψ. Turning over to a different set ψ' of field variables the formula (34) has to be taken into account. This problem will become important in the context of entropy (see below).

[19] A 4-divergence in the kernel of a four-dimensional variational problem—such as the AP1 or the AP2—does not affect the Euler–Lagrange equations.

2.4. Lie-invariance group of the Lagrangian—particular perturbations induced by the invariance group—first and second kind balance equations—first and second kind observables

A G-parametric Lie group $\mathscr{L}_{\varepsilon_1,\varepsilon_2,\ldots,\varepsilon_G}$,[20,21]

$$\begin{aligned} t &\Rightarrow \bar{t} = f^0(x,t,\varepsilon), \\ x^\alpha &\Rightarrow \bar{x}^\alpha = f^\alpha(x,t,\varepsilon), \qquad \alpha = 1,\ldots,3, \\ \psi_k &\Rightarrow \bar{\psi}_k = F_k(\psi,x,t,\varepsilon), \quad k = 1,\ldots,f, \end{aligned} \tag{40}$$

is called the *invariance or symmetry group of the Lagrangian* ℓ if it transforms each *real* process $|\psi(t)\rangle$ into other *real* processes $|\bar{\psi}(t;\varepsilon)\rangle$. In Eq. (40) the abbreviation $\varepsilon = \{\varepsilon_K, K = 1,\ldots,G\}$ is used for the *set of group parameters*. The 1-element of the group is identified with $\varepsilon = 0$.

For the invariance of the Lagrangian the sufficient *invariance criterion*,

$$\ell(\psi_k,\partial_\alpha\psi_k) \underset{x,\psi,\partial\psi,\varepsilon}{\equiv} \mathrm{Det}|\partial_\alpha f^\beta(x,\varepsilon)|\ell(F_k(\psi,x,\varepsilon),F_{\alpha k}(\psi,\partial\psi,x,\varepsilon)) + \partial_\alpha \Xi^\alpha(\psi,x,\varepsilon), \tag{41}$$

has to be fulfilled identically with a well-defined 4-vector $\Xi = \{\Xi^\alpha(\psi,x,\varepsilon)\}$. The identity refers to the arguments noted below the identity symbol. The term $F_{\alpha k}$ involved on the right-hand side is defined by

$$F_{\alpha k}(\psi,\partial\psi,x,\varepsilon) = \left.\frac{\partial \overset{-1}{f^\beta}(\bar{x},\varepsilon)}{\partial \bar{x}^\alpha}\right|_{\bar{x}\Rightarrow x} \partial_\beta F_k(\psi,x,\varepsilon). \tag{42}$$

The criterion ensures that a real process $|\psi(t)\rangle$ is transformed into another real process $|\bar{\psi}(t,\varepsilon)\rangle$ again:[22]

$$\begin{aligned} |\psi(t)\rangle &= \{\psi_k(x,t), k = 1,\ldots,f\} \Rightarrow |\bar{\psi}(t,\varepsilon) \\ &= \left\{ \bar{\psi}_k(x,t,\varepsilon) = F_k(\psi(x,t),x,t\,\varepsilon)\Big|_{(x,t) = \overset{-1}{f}(\bar{x},\bar{t},\varepsilon)}\Big|_{(\bar{x},\bar{t})\Rightarrow(x,t)}, k = 1,\ldots,f \right\}. \end{aligned} \tag{43}$$

By means of Noether's Theorem [15,17] each group parameter ε_K is associated with a *homogeneous balance equation*, which is fulfilled by the processes of the system and that

[20] Here, and in the following equations, the joint abbreviation x for the set of the four space and time coordinates is frequently used in the functions arguments.

[21] The transformation is separated into two parts, the first concerning the space-time region of the system only, and the second concerning the functional degrees of freedom depends on space and time coordinates too. This is usually the actual physical situation.

[22] Both processes are solutions of Hamilton's Principle simultaneously. Comparing the variational principles for both processes and taking account of the ambiguity of the Lagrangian concerning an additive 4-divergence the criterion can be established. One should further keep in mind that we are concerned with transformations in space and time and not with coordinate transformations: in a fixed reference frame the process is 'deformed'. This is also the reason for the final substitution ($\bar{x} \Rightarrow x$) at the end of formula (43).

in an isolated system gives rise to a conservation law. These balance equations operate in state space S. Referring via Eq. (41) to the Lagrangian ℓ immediately and to the AP1 these are called *balance equations of the first kind.* In the case of the first-order Lagrange formalism the balance equations of the first kind involve the fundamental field variables up to their second derivatives only.

The most important invariance requirements are due to *universal methodical* reasons, e.g. to

time-translational invariance,
space-translational invariance,
rotational invariance, (44)
Galilei-invariance (in the case of classical physics),
gauge invariance in the case of complex fundamental field variables.

These universal invariance requirements give rise to implicit and universal definitions of *observables of the first kind* via their respective Noether balances. Among these the *time-translational invariance* is the most prominent one; it is associated with the first kind observable 'energy' and thus—in the case of thermodynamics—with the *First Law of Thermodynamics* as a straightforward outcome of the time-translational invariant Lagrange formalism. Time-translational invariance is the mathematical correlate for the universally required reproducibility of each physical process.

An insertion for completeness: the invariance group (40) of the Lagrangian ℓ induces a Lie-invariance group for the kernel Ω of the AP2: the transformation equations (40) have to be supplemented by the transformations

$$\eta_k \Rightarrow \tilde{\eta}_k = G_k(\psi_l, \eta_l, x^\beta, \varepsilon_K) = \sum_{j=1}^{f} \frac{\partial F_k(\cdot)}{\partial \psi_j} \eta_j, \qquad k = 1, \ldots, f \tag{45}$$

for the representatives of perturbation classes.[23] Now a procedure analogous to Eqs. (38)–(41) holds for the kernel Ω of the AP2 giving rise to an *inhomogeneaous* Noether balance equation for each group parameter ε_K. These balances operate on the perturbation space P; they are fulfilled by the representative $|\eta(t)\rangle$ of each perturbation class $|\psi(t;\pi)\rangle$ of a given process $|\psi(t)\rangle$.[24]

All balance equations working primarily in perturbation space P, can in various ways be pulled back into the state space S by means of the Lagrangian's invariance group (40), thus giving rise to a variety of *balance equations of the second kind in state space S. This statement holds especially for the central-balance equation (36)*! In detail, by the invariance group $\mathscr{L}_{\varepsilon_1,\varepsilon_2,\ldots,\varepsilon_G}$ *real* processes $|\bar{\psi}(t,\varepsilon)\rangle$ have been constructed from the *real* process $|\psi(t)\rangle = |\bar{\psi}(t,\varepsilon=0)\rangle$ (see Eqs. (40) and (43)). Thus, confining ourselves to a one-parametric subgroup $\mathcal{L}_{\varepsilon_p}$,

$$\varepsilon = \{\varepsilon_1, \varepsilon_2, \ldots, \varepsilon_P, \ldots, \varepsilon_G\} = \{0, 0, \ldots, \pi, 0, \ldots, 0\}, \tag{46}$$

[23] Insert $|\psi(t;\pi)\rangle$ in Eq. (38) and apply definition (24).

[24] Within LF these balance equations are essential for a stability theory in the sense of Ljapunov's direct method.

we get a one-parametric class of real processes

$$|\psi(t,\pi)\rangle_P = |\bar{\psi}(t,\varepsilon = \{0,0,\ldots,\pi,0,\ldots,0\})\rangle, \tag{47}$$

which obviously can be regarded as a *one-parametric class of perturbations* of the reference process $|\psi(t)\rangle$ in the sense of Section 2.2. It is *induced by an invariance group of the Lagrangian.* Confining ourselves to the cases (44) of universal invariance groups the respective *representatives* are now *induced by the group generators* of the respective one-parametric invariance subgroups $\mathcal{L}_{\varepsilon_P}$:

$$|\eta(t)\rangle_P = \left.\frac{\partial|\psi(t,\pi)\rangle_P}{\partial\pi}\right|_{\pi=0} = \left.\frac{\partial|\bar{\psi}(t,\varepsilon=\{0,0,\ldots,\pi,0,\ldots,0\})\rangle}{\partial\pi}\right|_{\pi=0} = \underline{\eta_P}|\psi(t)\rangle. \tag{48}$$

$\underline{\eta_P}$ is a linear operator—the *generator of the subgroup* $\mathcal{L}_{\varepsilon_P}$—which *applies to the reference process* $|\psi(t)\rangle$! Thus the representative of the induced perturbation class is pulled back to the level of state space S. It fulfils, by definition, Jacobis' equations. Inserting Eq. (48) into the inhomogeneous balance equations of perturbation space P, these are also completely pulled back to the level of the state space S giving rise to inhomogeneous *balance equations of the second kind.* By definition each balance equation of the second kind is fulfilled by each real process $|\psi(t)\rangle$ of the system. These balance equations implicitly define *observables of the second kind*, which, due to the production terms, are no longer conserved quantities, even for an isolated system.

Two simple examples are as follows.

(a) Let us specify the structures of this paragraph for the *combination of the central-balance equation and the time-translational invariance* of a given Lagrangian, which due to the invariance criterion (41) does not explicitly depend on time,[25] i.e. the associated system of which is energetically isolated in its volume from the external world. The one-parameter Lie group $\mathcal{L}_\varepsilon^{(\text{time translation})}$,

$$\begin{aligned} \bar{t} = t + t_0 \Rightarrow \quad \bar{x}^0 &= x^0 + \varepsilon =: f^0(x,\varepsilon) \\ \bar{x}^\alpha &= x^\alpha =: f^\alpha(x,\varepsilon), \qquad \alpha = 1,2,3, \\ \bar{\psi}_k &= \psi_k =: F_k(\psi,x,\varepsilon), \quad k = 1,\ldots,f, \end{aligned} \tag{49}$$

is assumed to apply to the time coordinate only; the functional degrees of freedom are assumed to be invariant. Then, starting from the real process $|\psi(t)\rangle$ we get by means of Eq. (43) the real processes $|\bar{\psi}(t,\varepsilon)\rangle = |\psi(t-\varepsilon)\rangle$, which are shifted in time, and that via Eq. (47) finally result in the invariance-induced perturbation class of real processes[26]

$$|\psi(t;\pi)\rangle = |\psi(t-\pi)\rangle, \tag{50}$$

[25] See Eq. (18).

[26] The minus sign in Eq. (50) corresponds with the plus sign in Eq. $(49)_1$. It is due to the substitution $(x,t) = f^{-1}(\bar{x},\bar{t},\varepsilon)$ in Eq. (43)—here, $t = \bar{t} - \varepsilon$. t and $\bar{t}$ are measured with the *same* clocks; the process is shifted forward in time by ε.

referred to real process $|\psi(t)\rangle$ as reference process. By means of Eq. (48) the associated representative reads

$$|\eta(t)\rangle = \left.\frac{\partial|\psi(t-\pi)\rangle}{\partial\pi}\right|_{\pi=0} = -\frac{\partial}{\partial t}|\psi(t)\rangle = \underline{\eta}|\psi(t)\rangle, \tag{51}$$

which is pulled back to the time derivative of the given reference process $|\psi(t)\rangle$. (The operator $\underline{\eta} = -\partial/\partial t$ is the well-known *generator of time translations.*) Inserting Eq. (51) into the central-balance equations (36)–(39) we get the *inhomogeneous balance equation of the second kind*,

$$\partial_t \varpi^{\mathrm{I}} + \nabla\cdot\vec{J}^{\mathrm{I}}_{(\varpi)} = \sigma^{\mathrm{I}}_{(\varpi)}, \tag{52}$$

the density, flux density, and production-rate density of which is given by the *constitutive equations*

$$\varpi^{\mathrm{I}} = \varpi(\psi,\partial\psi|\eta,\partial\eta)|_{|\eta\rangle=-\partial_t|\psi(t)\rangle} = \left.\left\{\sum_{j=1}^{f}\frac{\partial\Omega(\psi,\partial\psi|\eta,\partial\eta)}{\partial(\partial_t\eta_j)}\eta_j\right\}\right|_{|\eta\rangle=-\partial_t|\psi(t)\rangle}, \tag{53}$$

$$\vec{J}^{\mathrm{I}}_{(\varpi)} = \vec{J}_{(\varpi)}(\psi,\partial\psi|\eta,\partial\eta)\Big|_{|\eta\rangle=-\partial_t|\psi(t)\rangle} = \left.\left\{\sum_{j=1}^{f}\frac{\partial\Omega(\psi,\partial\psi|\eta,\partial\eta)}{\partial(\nabla\eta_j)}\eta_j\right\}\right|_{|\eta\rangle=-\partial_t|\psi(t)\rangle}, \tag{54}$$

$$\sigma^{\mathrm{I}}_{(\varpi)} = \sigma_{(\varpi)}(\psi,\partial\psi|\eta,\partial\eta)\Big|_{|\eta\rangle=-\partial_t|\psi(t)\rangle} = 2\Omega(\psi,\partial\psi|\eta,\partial\eta)|_{|\eta\rangle=-\partial_t|\psi(t)\rangle}. \tag{55}$$

This no-name balance equation of the second kind completely operates in state space S; it is fulfilled by each real process $|\psi(t)\rangle$ of an isolated system.

(b) Next let us combine the central-balance equation with the gauge invariance of a given system, the fundamental field variables of which are complex-valued. Because of the reality of the Lagrangian the set of variables must contain with each variable ψ its complex-conjugated counterpart ψ^*, too. Let us put this set of $f = 2f'$ field variables into the following order:

$$\psi = \{\psi_1,\ldots,\psi_{f'},\psi_{f'+1},\ldots,\psi_{2f'}\} \qquad \text{with } \psi_k = \psi^*_{k+f'} \text{ for } k = 1,\ldots,f'. \tag{56}$$

The Lagrangian is now assumed to be invariant with respect to the one-parameter Lie group $\mathcal{L}^{(\mathrm{gauge})}_{\varepsilon}$ of *common gauge transformations*,

$$\begin{aligned}
\bar{x}^0 &= x^0 =: f^0(x,\varepsilon), \\
\bar{x}^\alpha &= x^\alpha =: f^\alpha(x,\varepsilon), & \alpha &= 1,2,3, \\
\bar{\psi}_k &= \psi_k\,\mathrm{e}^{\mathrm{i}\varepsilon} =: F_k(\psi,x,\varepsilon), & k &= 1,\ldots,f', \\
\bar{\psi}_k &= \psi_k\,\mathrm{e}^{-\mathrm{i}\varepsilon} =: F_k(\psi,x,\varepsilon), & k &= f'+1,\ldots,2f',
\end{aligned} \tag{57}$$

which apply to the field variables only. In a similar way as in the preceding example this group gives rise to an invariance-induced perturbation class,

$$|\psi(t;\pi)\rangle = \{\psi_1(x,t)e^{i\pi},\dots,\psi_{f'}(x,t)e^{i\pi},\psi_{f'+1}(x,t)e^{-i\pi},\dots,\psi_{2f'}(x,t)e^{-i\pi}\} \tag{58}$$

associated with the real process $|\psi(t)\rangle$ as the reference process. Eq. (48) now results in the representative

$$\begin{aligned}
|\eta(t)\rangle &= \left.\frac{\partial|\psi(t;\pi)\rangle}{\partial\pi}\right|_{\pi=0} = \{\eta_1(x,t),\dots,\eta_{f'}(x,t),\eta_{f'+1}(x,t),\dots,\eta_{2f'}(x,t)\} \\
&= \{i\psi_1(x,t),\dots,i\psi_{f'}(x,t),-i\psi_{f'+1}(x,t),\dots,-i\psi_{2f'}(x,t)\} \\
&= \begin{pmatrix} i & 0 & 0 & \cdot & \cdot & \cdot \\ 0 & i & 0 & \cdot & \cdot & \cdot \\ \cdot & \cdot & \cdot & \cdot & \cdot & \cdot \\ \cdot & \cdot & \cdot & -i & 0 & \cdot \\ \cdot & \cdot & \cdot & 0 & -i & \cdot \\ \cdot & \cdot & \cdot & \cdot & \cdot & \cdot \end{pmatrix} \begin{pmatrix} \psi_1(x,t) \\ \cdot \\ \cdot \\ \psi_{f'+1}(x,t) \\ \cdot \\ \cdot \end{pmatrix} = \underline{\eta} \begin{pmatrix} \psi_1(x,t) \\ \cdot \\ \cdot \\ \psi_{f'+1}(x,t) \\ \cdot \\ \cdot \end{pmatrix},
\end{aligned} \tag{59}$$

which is again pulled back to the given reference process $|\psi(t)\rangle = \{\psi_1(x,t),\dots,\psi_{2f'}(x,t)\}$. (Now the matrix-operator

$$\underline{\eta} = \begin{pmatrix} i & \cdot & \cdot \\ \cdot & \cdot & \cdot \\ \cdot & \cdot & \cdot \end{pmatrix}$$

is the *generator of gauge group*.) Inserting Eq. (58) into the central-balance equation (36)–(39) we get another *inhomogeneous balance equation of the second kind for the reference process* $|\psi(t)\rangle$,

$$\partial_t \varpi^{\text{II}} + \nabla\cdot\vec{J}^{\text{II}}_{(\varpi)} = \sigma^{\text{II}}_{(\varpi)}, \tag{60}$$

the density, flux density, and production-rate density of which are given by the *constitutive equations*

$$\begin{aligned}
\varpi^{\text{II}} &= \varpi(\psi,\partial\psi|\eta,\partial\eta)\big|_{|\eta\rangle=\{i\psi_1(x,t),\dots,-i\psi_{f'+1}(x,t),\dots\}} \\
&= \left.\left\{\sum_{j=1}^{f}\frac{\partial\Omega(\psi,\partial\psi|\eta,\partial\eta)}{\partial(\partial_t\eta_j)}\eta_j\right\}\right|_{|\eta\rangle=\{i\psi_1(x,t),\dots,-i\psi_{f'+1}(x,t),\dots\}},
\end{aligned} \tag{61}$$

$$\vec{J}^{\mathrm{II}}_{(\varpi)} = \vec{J}_{\varpi}(\psi, \partial\psi|\eta, \partial\eta)\Big|_{|\eta\rangle=\{\mathrm{i}\psi_1(x,t),\ldots,-\mathrm{i}\psi_{f'+1}(x,t),\ldots\}}$$

$$= \left\{\sum_{j=1}^{f} \frac{\partial\Omega(\psi, \partial\psi|\eta, \partial\eta)}{\partial(\nabla\eta_j)}\eta_j\right\}\Bigg|_{|\eta\rangle=\{\mathrm{i}\psi_1(x,t),\ldots,-\mathrm{i}\psi_{f'+1}(x,t),\ldots\}} \tag{62}$$

$$\sigma^{\mathrm{II}}_{(\varpi)} = \sigma_{(\varpi)}(\psi, \partial\psi|\eta, \partial\eta)\Big|_{|\eta\rangle=\{\mathrm{i}\psi_1(x,t),\ldots,-\mathrm{i}\psi_{f'+1}(x,t),\ldots\}}$$

$$= 2\Omega(\psi, \partial\psi|\eta, \partial\eta)\big|_{|\eta\rangle=\{\mathrm{i}\psi_1(x,t),\ldots,-\mathrm{i}\psi_{f'+1}(x,t),\ldots\}} \tag{63}$$

Again this balance equation of the second kind completely operates in state space S; it is fulfilled by each real process $|\psi(t)\rangle$ of a gauge-invariant system.

Comparing both balance equations (52)–(55) and (60)–(63) one should keep in mind the following facts:

- On the basis of the central-balance equation both operate in state space S as balance equations of the second kind.
- However, among all these balance equations of the second kind[27] that of (60)–(63) is the only one that operates on the same level as all balance equations of the first kind do: there are the fundamental field variables involved and their first and second derivatives only. Contrasting with this feature the fundamental field variables are involved up to their third derivatives in Eqs. (52)–(55).[28]

3. Lagrange formalism as applied to thermodynamics of irreversible processes

The structure of Lagrange formalism presented in the previous sections will completely be taken over; it will be interpreted for the sake of TIP. For demonstration it will be exemplified by the relatively simple problem of heat transport in a rigid body that is fixed in space.[29]

3.1. Specific basics of the theory

First statement Lagrange formalism based on Hamilton's *Action* Principle applies to TIP. This implies the existence of a Lagrangian for a given system.

Of course, there is no proof of the statement. However, there are good methodical reasons for it: in all nondissipative physics[30] LF as a universal and methodically unified dynamical theory. In particular the universal definition of observables such as energy,

[27] Of course, according to other invariance requirements there are a few more balance equations of the second kind based on the central balance.

[28] See the substitutions involved in Eqs. (61)–(63) and in Eqs. (53)–(55), respectively.

[29] For other—much more complicated—examples the reader is referred to former publications [13–15,21–23].

[30] For example, mechanics, quantum mechanics, electrodynamics, relativity, particle physics.

linear, and angular momentum, mass, charge are due to universal invariance requirements. For example, the universal time translational invariance is the methodical reason why different manifestations of energy in mechanics (kinetic, potential energies due to elasticity or gravitation) and in electrodynamics (field energy due to the electromagnetic fields), belong to one and the same observable 'energy'. Furthermore, it is the reason for the energy transfer between qualitatively quite different physical subsystems—or more generally speaking—Lagrange formalism allows for coupling of different subsystems on the same methodical level. As far as the traditional view of thermodynamics is concerned it is in fact outside the scope of these universal methods. However, there is no principle reason to exclude thermodynamics from methodical universality. At least—as a pragmatic argument—one can say: define a Lagrangian and look for the results, i.e. practice 'trial and error' as usual.

In fact, the construction of a thermodynamical Lagrangian is no trivial problem. This problem is closely related with the *Inverse Problem of Variational Calculus*. As a warning one should keep in mind that the usual set of balance equations, which are the ad hoc starting point of the traditional TIP, are not necessarily the fundamental field equations, i.e. the ELeqs of the problem.[31]

Looking at the ELeqs (20) as the fundamental field equations of a thermodynamical system, the reader might ask for the appearance of *dissipation potentials*, which are used in traditional TIP. However, one should keep in mind that the concept of dissipation potentials is primarily due to traditional TIP and its particular structure.[32] It is no a priori concept in LF! Nevertheless, it can be introduced in LF too: the difference of the Lagrangians of the dissipative system and of the respective nondissipative system may be regarded as a sort of dissipation potential $W : \ell_{\text{dissipative}} = \ell_{\text{nondissipative}} + W$.[33]

Second statement Irreversible processes have to be described by means of complex-valued fundamental field variables.

In Lagrange formalism the usual real-valued variables of traditional TIP, such as temperature and mass densites, are secondary quantities.

Third statement The Lagrangian has to be invariant with respect to a *common* gauge transformation applied to all complex field variables.

Statements 2 and 3 are motivated as follows: on the one hand, irreversible processes are associated with entropy production. On the other hand, a universal subformalism is involved in LF that in the case of complex-valued field variables successfully leads to the entropy concept (see below).

Fourth statement The Lagrangian is assumed to be explicitly independent of time t.

[31] These things are outside the scope of this chapter. The reader is referred to Ref. [10,11,13,15] and to forthcoming papers.

[32] See, e.g. Gyarmati [6], Chapters V and VI.

[33] See, e.g. the example 'heat conduction' in Section 3.4 (Eq. (84)): $\ell_{\text{dissipative}} = \ell$, $\ell_{\text{nondissipative}} =$ first row, $W =$ second row.

Via time-translational invariance this implies the *First Law of Thermodynamics* in the form of a homogeneous Noether's energy-balance equation[34]

$$\partial_t u + \nabla \cdot \vec{J}_{(u)} = 0, \tag{64}$$

with well-defined constitutive equations for the energy density u and the energy flux density $\vec{J}_{(u)}$ [15].

Fifth statement According to the irreversibility of real thermodynamical processes the Lagrangian *may not be invariant with respect to time reversal*:[35]

$$\begin{aligned} &t \Rightarrow \bar{t} = -t, \\ &x^\alpha \Rightarrow \bar{x}^\alpha = x^\alpha, \qquad \alpha = 1, \ldots, 3, \\ &\psi_k \Rightarrow \bar{\psi}_k = F_k(\psi), \quad k = 1, \ldots, f, \end{aligned} \tag{65}$$

i.e. there exist *no* functions F_k, $k = 1, \ldots, f$, such that Eq. (65) is an invariance transformation in the sense of criterion (41).

If this statement did not hold, i.e. if the Lagrangian were invariant with respect to time reversal, a real process could be transformed into another real process that runs (in a specified sense) in the reverse direction, i.e.—symbolically speaking—a movie, once running forward and once running backward, would represent real processes in *both* cases. However, such a situation is associated with reversibility that by definition is excluded from TIP as a dynamical theory of *real* processes.[36,37]

The kernel 2Ω of the AP2 gives rise to the definition of the functional

$$\mathrm{L}[\psi(\cdot, t), \eta(\cdot, t)] = \int_V 2\Omega(\psi(x,t), \partial_t \psi(x,t), \nabla \psi(x,t) | \eta(x,t), \partial_t \eta(x,t), \nabla \eta(x,t) \mathrm{d}V, \tag{66}$$

which is already involved in the AP2,[38] and that will be taken as a basis for a stability theory in the sense of Ljapunov's direct method. It is supposed that this functional results in a Ljapunov functional[39] via the following statement.

[34] For details see Ref. [15].

[35] The time reversal is an element of a discrete group $\{1, -1\}$ of order 2. The functions F_k must be chosen appropriately. The argumentation concerning invariance of the Lagrangian is the same as in the case of continuous Lie groups. The Noether Theorem, however, is missing.

[36] The described fictive situation belongs to a reversible thermodynamics, which never exists in reality. The so-called 'reversible process' of thermostatics is no real dynamical process; it is but a (timeless) path in the equilibrium manifold of the system.

[37] A counter example: the theory of Schrödinger's matter wave (one-particle quantum mechanics) is based on the complex-valued matter wave $\psi(x,t)$. The theory is a reversible one, i.e. the associated Lagrangian [9] is invariant with respect to time-reversal in the sense of Eq. (65). The third line in Eq. (65) reads: $\bar{\psi}(x,t) = \psi(x,-t)^*$.

[38] See Eqs. (27) and (28).

[39] Investigations towards this goal are in progress. It might be that statements 5 and 6 are not quite independent. The reader is referred to Ref. [15] and to forthcoming papers. A qualitative argument: the kernel Ω contains all information on the perturbations of the system. Asymptotic stability of a dynamical system is necessarily associated with irreversibility of the system.

Sixth statement The functional (66) is assumed to be positive-definite with respect to the set of functions $\{\eta_j(x,t), j=1,\dots,f\}$ for each reference process $\{\psi_j(x,t), j=1,\dots,f\}$:

$$L[\psi(\cdot,t), \eta(\cdot,t)] \geq 0. \tag{67}$$

Résumé and perspective for TIP in the framework of LF: once a Lagrangian is defined following the statements mentioned before the complete thermodynamics of the system is defined, it can be evaluated along the universal formal line of LF.[40]

3.2. The Second Law of Thermodynamics

According to statements 2 and 3 the Lagrangian for a thermodynamical system depends on $f = 2f'$ complex-valued fundamental field variables and is invariant with respect to the one-parameter Lie group of common gauge transformations (see Eq. (57)). Thus, the balance equations (60)–(63) of the second kind, which results from a combination of the central balance (36)–(39) and the common gauge group (57), holds for each process. It is identified with the *First Part of the Second Law of Thermodynamics*:

$$\partial_t \tilde{s} + \nabla\cdot\vec{J}_{(\tilde{s})} = \sigma_{(\tilde{s})}, \tag{68}$$

which is based on the following *constitutive equations* (see Eqs. (61)–(63)):

entropy density

$$\tilde{s}(x,t) = \varpi(\psi,\partial\psi|\eta,\partial\eta)\big|_{|\eta\rangle=\{\eta_1,\dots,\eta_{f'},\eta_1^*,\dots,\eta_{f'}^*\}=\{i\psi_1(x,t),\dots,i\psi_{f'}(x,t),-i\psi_1(x,t)^*,\dots,-i\psi_{f'}(x,t)^*\}}$$

$$= \left\{\sum_{j=1}^{2f'} \frac{\partial\Omega(\psi,\partial\psi|\eta,\partial\eta)}{\partial(\partial_t\eta_j)}\eta_j\right\}\Bigg|_{|\eta\rangle=\{\eta_1,\dots,\eta_{f'},\eta_1^*,\dots,\eta_{f'}^*\}=\{i\psi_1(x,t),\dots,i\psi_{f'}(x,t),-i\psi_1(x,t)^*,\dots,-i\psi_{f'}(x,t)^*\}}, \tag{69}$$

entropy flux density

$$\vec{J}_{(\tilde{s})}(x,t) = \vec{J}_{(\varpi)}(\psi,\partial\psi|\eta,\partial\eta)\big|_{|\eta\rangle=\{i\psi_1(x,t),\dots,-i\psi_1(x,t)^*,\dots\}}$$

$$= \left\{\sum_{j=1}^{2f'} \frac{\partial\Omega(\psi,\partial\psi|\eta,\partial\eta)}{\partial(\nabla\eta_j)}\right\}\Bigg|_{|\eta\rangle=\{i\psi_1(x,t),\dots,-i\psi_1(x,t)^*,\dots\}}, \tag{70}$$

entropy-production-rate density

$$\sigma_{(\tilde{s})}(x,t) = \sigma_{(\varpi)}(\psi,\partial\psi|\eta,\partial\eta)\big|_{|\eta\rangle=\{i\psi_1(x,t),\dots,-i\psi_1(x,t)^*,\dots\}}$$

$$= 2\Omega(\psi,\partial\psi|\eta,\partial\eta)\big|_{|\eta\rangle=\{i\psi_1(x,t),\dots,-i\psi_1(x,t)^*,\dots\}}. \tag{71}$$

[40] The situation is quite analogous to that of Rational Thermodynamics [24]: once the relevant set of balance equations and its associated constitutive equations are defined in accordance with the constraint of a positive-definite entropy-production-rate a thermodynamics is completely defined. The advantage of LF, however, is its most concise form: the Lagrangian exclusively contains all information.

One should keep in mind that the balance equations (7) and (68) are due to quite different roots: Eq. (7) results from the PLE. As a consequence the entropy concept based on Eqs. (7)–(10) is confined to a neighborhood of thermodynamical equilibrium. On the other hand the entropy concept based on Eqs. (68)–(71) belongs to an unrestricted dynamics within LF, i.e. it is not restricted to a neighborhood of thermodynamical equilibrium. It is due to a Lagrangian following the second and third statements of Section 3.1. An identification of both concepts (i.e. $s = \tilde{s}$, $\vec{J}_{(s)} = \vec{J}_{(\tilde{s})}$, $\sigma_{(s)} = \sigma_{(\tilde{s})}$) holds if the Lagrangian is fitted to the PLE, as will be done in Section 3.4.

Furthermore, the reader is reminded of the ambiguity of the kernel Ω that is associated with a change of the field variables ψ (see the remarks related with Eqs. (29)–(34)). This feature appears in the entropy quantities again. However,

> "the concept of entropy has to be invariant with respect to the choice of field variables."

Summing up: either we have to introduce an absolute differential calculus in state space or we have to distinguish a particular set of *fundamental* field variables once and for all. For simplicity we choose the latter option in this chapter. From the point of view of particular systems[41] as well as from a general point of view of stability theory the following statement turned out to be successful:

> "The concept of entropy has to be based on the set of complex field variables!"

This is a strong motivation for the second statement in Section 3.1. Referring to the analogy mentioned in Section 2.2 the complex field variables define so to speak the 'Cartesian frame' in state space.[42] In thermodynamics this set is called *the set of fundamental field variables*. Of course we can use an equivalent set of field variables. Then, however, we have to pay some price concerning the calculation of the entropy quantities (see Eqs. (30) and (34)).

The *Second Part of the Second Law* as applied to the expression (71)

$$\sigma_{(\tilde{s})}(x,t) = 2\Omega(\psi, \partial\psi | \eta, \partial\eta)\big|_{|\eta\rangle = \{\mathrm{i}\psi_1(x,t),\ldots,-\mathrm{i}\psi_1(x,t)^*,\ldots\}} \geq 0 \tag{72}$$

follows from the sixth statement and follows (Eqs. (66) and (67)). However, it is still open if the positive-definiteness of the entropy-production-rate can finally be related with dynamical stability.[43]

---

[41] See Section 3.4.

[42] Keyword: 'Euclidean metric' in perturbation space in the context of Ljapunov's stability theory:

$$\|\eta\|^2 = \int_V \left(\sum_{j=1}^{f''} (\eta_j(x,t)\eta_j(x,t)^* + \partial_t\eta_j(x,t)\partial_t\eta_j(x,t)^* + \nabla\eta_j(x,t)\cdot\nabla\eta_j(x,t)^*) \right) \mathrm{d}V.$$

[43] An alternative approach: giving up the sixth statement and following the traditional line, the requirement (72) can be stated exclusively without relating it with dynamical stability. However, relating the Second Part of the Second Law with dynamical stability is a most attractive goal. Such investigations are in progress.

3.3. The Principle of Least Entropy Production

Instead of looking at the instantaneous total entropy-production-rate $\Sigma_{(s)}(t)$ as the traditional principle does (see Eq. (16)) we are looking at the total entropy production during the whole duration $t_1 \cdots t_2$ of the process:

$$P_{(\bar{s})} = \int_{t_1}^{t_2} \Sigma_{(\bar{s})}(t)\mathrm{d}t = \int_{t_1}^{t_2} \int_V \sigma_{(\bar{s})}(x,t)\mathrm{d}V\,\mathrm{d}t. \tag{73}$$

Due to the genesis of the entropy-production-rate density (71) the following extremum principle for the quantity $P_{(s)}$ is suggested:

$$P_{(\bar{s})} = \int_{t_1}^{t_2} \int_V \sigma_{(\bar{s})}(\psi,\partial\psi|\eta,\partial\eta)\Big|_{|\eta\rangle = \{\underset{\uparrow}{\mathrm{i}\psi_1(x,t)},\ \ldots,\ \underset{\uparrow}{-\mathrm{i}\psi_1(x,t)^*},\ \ldots\}} \mathrm{d}V\,\mathrm{d}t = \text{extremum} \tag{74}$$

by free variation of the given reference process $|\psi(t)\rangle$ at those sites in Eq. (74) that are marked by arrows. The free variation $\partial|\psi(t)\rangle$ applies in volume V and at its boundary ∂V keeping fixed its values at the beginning and at the end of the process: $\delta\psi_j(x,t_{1,2}) = 0$.

Within LF this is the *quite general Principle of Least Entropy Production.* At first glance it is seemingly quite a strange variational principle, because the variables ψ and $\partial\psi$ appear twice, once associated with the fixed reference process (first two arguments in $\sigma_{(s)}$) and once via substituting η (last two arguments in $\sigma_{(s)}$). The variation, however, applies at the latter site only! Looking at the genesis of the entropy-production-rate density, the variational principle (74) is nothing other than the AP2 (28) evaluated at $|\eta\rangle = \{\mathrm{i}\psi_1(x,t),\ldots,-\mathrm{i}\psi_1(x,t)^*,\ldots\}$:

$$P_{(s)} = \text{extremum}$$

$$\Leftrightarrow \left[A_{(2)} = \int_{t_1}^{t_2}\int_V \begin{array}{c} 2\Omega(\psi,\partial\psi|\eta,\partial\eta)\mathrm{d}V\,\mathrm{d}t = \text{extremum} \\ \text{by free variation of } \eta\ldots \end{array} \right]\Bigg|_{|\eta\rangle=\{\mathrm{i}\psi_1(x,t),\ldots,-\mathrm{i}\psi_1(x,t)^*,\ldots\}}. \tag{75}$$

Instead of asking the question: "which are the solutions of the variational problem?", we have now to ask: "does the *given* reference process $|\psi(t)\rangle$ solve the variational principle by means of $|\eta(t)\rangle = \{\mathrm{i}\psi_1(x,t),\ldots,-\mathrm{i}\psi_1(x,t)^*,\ldots\}$?" The answer: yes, it does always, i.e. for each real process $|\psi(t)\rangle$ of the system. The proof is quite easy: the Jacobi equations (25) and (26) are the ELeqs of Eqs. (74) and (75), and $|\eta(t)\rangle = \{\mathrm{i}\psi_1(x,t),\ldots,-\mathrm{i}\psi_1(x,t)^*,\ldots\}$ being the representative of the gauge group does always solve Jacobi's equation (see Section 2.4).

Does the principle (74) and (75) define a minimum or a maximum of the total entropy production $P_{(s)}$? We have to look at the second variation $\delta^{(2)}A_{(2)}$ of the functional $A_{(2)}$:

The Taylor expansion of the total variation of $A_{(2)}$ associated with respect to the variation $\delta\eta = \{\delta\eta_1,\ldots,\delta\eta_f\}$ reads

$$\delta A_{(2)} = \delta^{(1)}A_{(2)} + \delta^{(2)}A_{(2)} + \cdots. \tag{76}$$

The two terms on the right-hand side indicate the first- and second-order variations. The series (76) breaks after the second term because the functional $A_{(2)}$ is a quadratic form in η and $\partial\eta$. The first variation vanishes because of the Jacobi equations (25) and (26). The second variation reads explicitly:

$$\delta^{(2)}A_{(2)} = \frac{1}{2}\int_{t_1}^{t_2}\int_V \sum_{i,j=1}^{f} \left[\begin{array}{l} \dfrac{\partial^2 2\Omega(\psi(x,t), \partial\psi(x,t) | \eta(x,t), \partial\eta(x,t))}{\partial\eta_i\,\partial\eta_j}\delta\eta_i(x,t)\delta\eta_j(x,t) \\ +2\dfrac{\partial^2\, 2\Omega(\cdot)}{\partial\eta_i\,\partial(\partial_\alpha\eta_j)}\delta\eta_i(\cdot)\delta\,\partial_\alpha\eta_j(\cdot) \\ +\dfrac{\partial^2\, 2\Omega(\cdot)}{\partial(\partial_\alpha\eta_i)\partial(\partial_\beta\eta_j)}\delta\,\partial_\alpha\eta_i(\cdot)\delta\,\partial_\beta\eta_j(\cdot) \end{array} \right] \mathrm{d}V\,\mathrm{d}t \tag{77}$$

Using the definition (27) twice

$$\begin{aligned} \delta^{(2)}A_{(2)} &= \frac{1}{2}\int_{t_1}^{t_2}\int_V \sum_{i,j=1}^{f} \left[\begin{array}{l} \dfrac{\partial^2\ell(\psi(x,t), \partial\psi(x,t))}{\partial\psi_i\,\partial\psi_j}\delta\eta_i(x,t)\delta\eta_j(x,t) \\ +2\dfrac{\partial^2\ell(\cdot)}{\partial\psi_i\,\partial(\partial_\alpha\psi_j)}\delta\eta_i(\cdot)\delta\,\partial_\alpha\eta_j(\cdot) \\ +\dfrac{\partial^2\ell(\cdot)}{\partial(\partial_\alpha\psi_i)\partial(\partial_\beta\psi_j)}\delta\,\partial_\alpha\eta_i(\cdot)\delta\,\partial_\beta\eta_j(\cdot) \end{array} \right] \mathrm{d}V\,\mathrm{d}t \\ &= \frac{1}{2}\int_{t_1}^{t_2}\left\{\int_V 2\Omega(\psi(x,t), \partial\psi(x,t) | \delta\eta(x,t), \delta\eta(x,t))\mathrm{d}V\right\}\mathrm{d}t, \end{aligned} \tag{78}$$

we get, together with the sixth statement (66) and (67) the final result

$$\delta^{(2)}A_{(2)} = \frac{1}{2}\int_{t_1}^{t} \mathrm{L}[\psi(\cdot,t), \delta\eta(\cdot,t)]\mathrm{d}t \geq 0, \tag{79}$$

i.e. the extremum in Eqs. (74) and (75) is a minimum and we are really dealing with a Principle of *Least* Entropy Production. When the sixth statement does not hold for a particular system, the situation is much more involved; the minimum cannot be asserted. However, the extremum is still preserved.

Stationary processes are involved as a special case: the entropy-production-rate density $\sigma_{(\tilde{s})}$ will become constant in time and the time integral in Eq. (74) becomes irrelevant:[44]

$$\begin{aligned} P_{(\tilde{s})} &= (t_2 - t_1)\int_V \sigma_{(\tilde{s})}(\psi, \partial\psi | \eta, \partial\eta)\Big|_{|\eta\rangle = \left\{ \substack{\mathrm{i}\psi_1(x,t) \\ \uparrow} \; \substack{,\ldots, \\ ,\ldots,} \; \substack{-\mathrm{i}\psi_1(x,t)^* \\ \uparrow} \; \substack{,\ldots \\ ,\ldots} \right\}} \mathrm{d}V \\ &= (t_2 - t_1)\Sigma_{(\tilde{s})} = \text{extremum}. \end{aligned} \tag{80}$$

[44] See Eq. (63).

We arrive for stationary processes *seemingly* at the same result as the traditional Principle of Least Entropy Production (see Eq. (16)):

$$\Sigma_{(\tilde{s})} = \text{extremum}. \tag{81}$$

However, no additional restrictions on the type of processes and on constitutive equations are involved here. Furthermore, we should carefully note the fact that in spite of the same global quantity $\Sigma_{(s)}$ the traditional principle and the present principle are completely different variational principles: the first one refers to variations of the driving forces, the latter to variations of the fundamental, complex-valued field variables at those sites marked by arrows in Eq. (80).

Summing up: Lagrange formalism being based on the statements of Section 3.1 results in the Second Law of Thermodynamics, i.e. in an entropy concept for dynamical systems, and in a Principle of Least Entropy Production in a quite natural and straightforward way and for quite general thermodynamical systems.

3.4. An example: heat transport in a rigid body

The theory is presented in brief outline in order to visualize the general considerations of this chapter only. For refined details the reader is referred to former papers; he is invited to reproduce the relevant calculations.

Heat transport is described by means of the complex-valued *field of thermal excitation* or *thermion field*[45]

$$\psi(x,t) = \sqrt{T(x,t)}\mathrm{e}^{\mathrm{i}\varphi(x,t)}. \tag{82}$$

From ψ and its complex conjugate ψ^* the positive-definite *absolute temperature*,

$$T(x,t) = \psi^*(x,t)\psi(x,t) \geq 0, \tag{83}$$

is defined as a *secondary quantity*. As an additional information ψ contains the *thermal phase* φ that is *associated with the deviation of the process from local equilibrium.*[46] The complex-valued set $|\psi\rangle = \{\psi_1, \psi_2\} = \{\psi, \psi^*\}$ will be used as a *set of fundamental field variables*, on which the entropy concept is based. The equivalent set of real-valued field variables $|\psi'\rangle = \{\psi'_1, \psi'_2\} = \{T, \varphi\}$ will be used too. It has some advantage with respect to calculations.

The Lagrangian will be used in the two equivalent forms

$$\ell = \ell(\psi,\psi^*,\partial\psi,\partial\psi^*) = -c\psi\psi^* - \frac{c}{\omega}\left[\frac{1}{2\mathrm{i}}(\psi^*\partial_t\psi - \psi\partial_t\psi^*)\right] - \frac{1}{\omega}\partial_t g(\psi\psi^*)$$
$$+\frac{\lambda}{\omega}\left[\frac{1}{2\mathrm{i}}\{(\psi^*\nabla\psi)\otimes(\psi^*\nabla\psi) - (\psi\nabla\psi^*)\otimes(\psi\nabla\psi^*)\} + \frac{T_0}{2(\psi\psi^*)^2}\nabla(\psi\psi^*)\otimes\nabla(\psi\psi^*)\right], \tag{84}$$

[45] This term is chosen with regard to a 'field quantization' [25,26].

[46] This statement becomes apparent not before the end of the whole Lagrange procedure.

and

$$\ell = \ell'(T, \partial T, \partial \varphi) = -\frac{1}{\omega}\{cT\partial_t[\varphi - \varphi^0(t,T)] + (-\underline{\lambda}\cdot\nabla T)\cdot\nabla[\varphi - \varphi^0(\cdot)] + \partial_t g(T)\}. \tag{85}$$

Involved abbreviations and symbols:

$$\varphi^0(t,T) = -\omega t + \frac{T_0}{2T}, \tag{86}$$

$$g(T) = \frac{T_0}{2T}\int_{T_0/T}^{1} \frac{p(\vartheta T)}{\vartheta^2} d\vartheta. \tag{87}$$

$p(T)$ is the hydroelastic pressure, which in the rigid body has to be regarded as a rigid-body reaction. It will finally be fitted to the Principle of Local Equilibrium (5) and (6), which in the case of pure heat conduction reduces to

$$\partial_t u = T\partial_t s, \quad p = Ts - u. \tag{88}$$

ω is a frequency that is necessary for dimensional reasons; it drops out of all relevant equations. c and λ are the specific heat and the tensor of heat conductivity. For simplicity both are assumed to be constants. T_0 is a reference temperature. The term $\partial_t g(T)$ being a total time derivative is irrelevant for the variational principles, for the AP1 as well as for the AP2. However, in general it is relevant for the implicit definition of observables by means of balance equations, especially for the entropy concept.

There is no time reversal involved in the Lagrangian (84) and (85); i.e. the dynamics is irreversible. The Lagrangian being not explicitly dependent on time t allows for the First Law of Thermodynamics. Being, furthermore, gauge invariant[47] it allows for the Second Law of Thermodynamics. A Principle of Least Entropy Production is automatically involved.

The bracket $[\cdot]$ in the second row of Eq. (84) is a symmetric tensor of rank 2. This implies symmetry of the heat conductivity tensor $\underline{\lambda}$ in front of the bracket. This feature coincides with *Onsager's reciprocal relation* [5].

The ELeqs associated with the AP1 and based on Lagrangian (85) are due to variations of the variables φ and T:[48]

$$\delta\varphi : c\,\partial_t T - \underline{\lambda}\cdot\cdot\nabla\otimes\nabla T = 0, \tag{89}$$

$$\partial T : \frac{\partial \varphi^0(t,T)}{\partial T}[c\,\partial_t T - \underline{\lambda}\cdot\cdot\nabla\otimes\nabla T] + [c\,\partial_t(\varphi - \varphi^0(\cdot)) + \underline{\lambda}\cdot\cdot\nabla\otimes\nabla(\varphi - \varphi^0(\cdot))] = 0. \tag{90}$$

[47] See Eq. (57). In the Lagrangian in the form (85) the phase variable φ is contained by its derivatives only.

[48] The ELeqs due to variations of ψ^* and ψ are rather big. They are left to the reader.

Using Eq. (89) the latter simplifies to

$$c\,\partial_t(\varphi - \varphi^0(t,T)) + \underline{\lambda}\cdot\cdot\nabla\otimes\nabla(\varphi - \varphi^0(\cdot)) = 0 \tag{91}$$

Eq. (89) is *Fourier's Law of heat conduction*. Within LF it is supplemented by Eq. (91) that is solved by

$$\varphi = \varphi^0(t,T) = -\omega t + \frac{T_0}{2T}. \tag{92}$$

With regard to the various observables (see below) *this particular solution is associated with the Principle of Local Equilibrium*. Having this in mind it becomes apparent that the Lagrangian (84) and (85) reproduces the traditional TIP within the framework of LF.

A side remark: relating deviations from the solution (92), i.e. $\varphi - \varphi^0(t,T) \neq 0$, with deviations of a process from local equilibrium, the Lagrangian in the form (85) can be interpreted as the first term of a Taylor series of a more general Lagrangian in the neighborhood of local equilibrium. This Lagrangian can be generalized in order to take into account processes running outside of local equilibrium [23].

The *boundary conditions* for the isolated system follow from the variational procedure, too:

$$\delta\varphi: \qquad \vec{n}\cdot(-\underline{\lambda}\cdot\nabla T) = 0 \tag{93}$$

$$\delta T: \qquad \vec{n}\cdot\left[-\lambda\nabla(\varphi - \varphi^0(\cdot)) + (-\underline{\lambda}\cdot\nabla T)\frac{T_0}{2T^2}\right] = 0. \tag{94}$$

Eq. (93) in physical terms: there is *no heat flux*[49] *across the boundary*. Eq. (94) is fulfilled due to Eqs. (93) and (92).

Summing up: a heat conduction process following local equilibrium is associated with a *thermion wave*

$$\psi(x,t) = \sqrt{T(x,t)}\exp\left[i\left(-\omega t + \frac{T_0}{2T(x,t)}\right)\right], \tag{95}$$

which is the physical carrier of heat transport, where $T(x,t)$ solves Fouriers equation.

Without detailed comments:[50] time-translational invariance of Lagrangian (84) and (85) results with the solution (92) in the First Law (64) for heat transport: the associated constitutive equations for the energy density and the energy flux density turn out to be

$$u = cT, \tag{96}$$

$$\vec{J}_{(u)} = -\underline{\lambda}\cdot\nabla T, \tag{97}$$

which coincide with those of traditional linear TIP.

[49] See Eq. (97).

[50] See Section 3.4 in Ref. [15].

The entropy concept is essentially based on the complex-valued fundamental field variables ψ and ψ^* as the reference system. So, starting from the Lagrangian in the form (84) the first step is the calculation of the kernel Ω of the AP2. The calculations can be done either directly with a lot of effort or—more elegantly—on the detours (34) using intermediately the reference system $|\psi'\rangle = \{\psi'_1, \psi'_2\} = \{T, \varphi\}$. The latter way in some detail.

Let $|\eta\rangle = \{\eta_1, \eta_2\} = \{\eta, \eta^*\}$ and $|\eta'\rangle = \{\eta'_1, \eta'_2\} = \{\eta_T, \eta_\varphi\}$ be the representatives of the same process perturbation associated with the reference systems $|\psi\rangle$ and $|\psi'\rangle$, respectively. Due to Eqs. (82) and (24) both are related by

$$\eta_T = \psi^*\eta + \psi\eta^*, \qquad \eta_\varphi = \frac{1}{2\mathrm{i}}\frac{\psi^*\eta - \psi\eta^*}{T}. \tag{98}$$

Let further, $\Omega(\psi, \psi^*, \partial\psi, \partial\psi^* | \eta, \eta^*, \partial\eta, \partial\eta^*)$ and $\Omega'(T, \partial T, \partial\varphi | \eta_T, \partial\eta_T, \partial\eta_\varphi)$ be the kernels of the AP2 associated with the Lagrangian in the forms (84) and (85), respectively. Then both are related by Eq. (34), which in the present case reads

$$\begin{aligned}
&\Omega(\psi, \psi^*, \partial\psi, \partial\psi^* | \eta, \eta^*, \partial\eta, \partial\eta^*)\\
&\quad = \Omega'(T, \partial T, \partial\varphi | \eta_T, \partial\eta_T, \partial\eta_\varphi)\\
&\quad + \partial_\alpha\left(\begin{aligned}&\frac{\partial\ell'(T,\partial T,\partial\varphi)}{\partial(\partial_\alpha T)}\left(\frac{\partial^2 T(\psi,\psi^*)}{\partial\psi\,\partial\psi}\eta\eta + 2\frac{\partial^2 T(\psi,\psi^*)}{\partial\psi\,\partial\psi^*}\eta\eta^* + \frac{\partial^2 T(\psi,\psi^*)}{\partial\psi^*\,\partial\psi^*}\eta^*\eta^*\right)\\
&+\frac{\partial\ell'(T,\partial T,\partial\varphi)}{\partial(\partial_\alpha\varphi)}\left(\frac{\partial^2\varphi(\psi,\psi^*)}{\partial\psi\,\partial\psi}\eta\eta + 2\frac{\partial^2\varphi(\psi,\psi^*)}{\partial\psi\,\partial\psi^*}\eta\eta^* + \frac{\partial^2\varphi(\psi,\psi^*)}{\partial\psi^*\,\partial\psi^*}\eta^*\eta^*\right)\end{aligned}\right).
\end{aligned} \tag{99}$$

Substitute on the right-hand side: T, φ by ψ, ψ^* and η_T, η_φ by η, η^* by means of Eqs. (81) and (97).

Applying definition (27) to the Lagrangian in the form (85) the term Ω'can be calculated relatively easily:[51]

$$\begin{aligned}
&2\Omega'(T, \partial T, \partial\varphi | \eta_T, \partial\eta_T, \partial\eta_\varphi)\\
&\quad = -\frac{1}{\omega}\left\{\begin{aligned}&\left[\left(\frac{cT_0}{T^3} + \frac{\mathrm{d}^3 g(T)}{\mathrm{d}T^3}\right)\partial_t T - 3\frac{T_0}{T^4}\underline{\lambda}\cdot\cdot\nabla T\otimes\nabla T\right]\eta_T\eta_T\\
&-\frac{T_0}{T^2}\underline{\lambda}\cdot\cdot\nabla\eta_T\otimes\nabla\eta_T - 2\underline{\lambda}\cdot\cdot\nabla\eta_T\otimes\nabla\eta_\varphi\end{aligned}\right\}.
\end{aligned} \tag{100}$$

From Eq. (99) we get the Jacobi or perturbation equations (25) for heat transport; they are rather lengthy.

---

[51] Because of the gauge invariance of the Lagrangian the variables φ, η_φ are involved via their derivatives only.

For the sake of entropy the quantities (99) and (100) must be evaluated for the representative of the gauge transformation (57), which in the present problem reads

$$\bar{\psi} = \psi\,\mathrm{e}^{\mathrm{i}\varepsilon}, \quad \bar{\psi}^* = \psi^*\,\mathrm{e}^{-\mathrm{i}\varepsilon} \qquad \text{or} \qquad \bar{T} = T, \quad \bar{\varphi} = \varphi + \varepsilon. \tag{101}$$

The associated representative is

$$|\eta\rangle = \{\eta = \mathrm{i}\psi, \eta^* = -\mathrm{i}\psi^*\} \qquad \text{or} \qquad |\eta'\rangle = \{\eta_T = 0, \eta_\varphi = 1\}. \tag{102}$$

Thus, the quantity (100) vanishes and Eq. (99) already presents the *First Part of the Second Law* (Eq. (68)). From the 4-divergence being left over on the right-hand side of Eq. (99) the *constitutive equations* (69) and (70) for the *entropy density* and the *flux density* can directly be read off. Taking into account Eqs. (87), (92), and (102) and using Eq. (97) we get

$$\tilde{s} = \frac{1}{\omega}\left(cT_0 + 2T\frac{\mathrm{d}g(T)}{\mathrm{d}T}\right) = \frac{T_0}{\omega}\frac{cT + p(T)}{T}, \tag{103}$$

$$\vec{J}_{(\tilde{s})} = \frac{T_0}{\omega}\frac{-\underline{\lambda}\cdot\nabla T}{T}. \tag{104}$$

The factor (T_0/ω) appearing in all the terms of the entropy balance may be deleted. Using the quantities (96) and (97) of the internal energy we arrive at

$$\tilde{s} = \frac{u + p}{T}, \tag{105}$$

$$\vec{J}_{(\tilde{s})} = \frac{-\underline{\lambda}\cdot\nabla T}{T} = \frac{\vec{J}_{(u)}}{T}. \tag{106}$$

Inserting Eqs. (105) and (106) into the entropy balance we finally get the *entropy-production-rate density*

$$\sigma_{(\tilde{s})} = \frac{\nabla T\cdot\underline{\lambda}\cdot\nabla T}{T^2} = \vec{J}_{(u)}\cdot\nabla\left(\frac{1}{T}\right). \tag{107}$$

The quantities (105) and (107) are well known in traditional, linear theory of heat conduction. Obviously Eq. (105) coincides with the Principle of Local Equilibrium $(88)_2$. Eq. (88) together with Eq. (96) result, for the entropy density, in

$$\tilde{s} = c\ln\frac{T}{T_0}, \tag{108}$$

and for the hydroelastic pressure in

$$p(T) = cT\left(\ln\left(\frac{T}{T_0}\right) - 1\right). \tag{109}$$

Together with definition (87) the yet unspecified term $\partial_t g(T)$ in the Lagrangian (84) and (85) is now determined:

$$\partial_t g(T) = \frac{c}{2}\frac{T_0}{T}\left(\ln\left(\frac{T}{T_0}\right) - 1\right)\partial_t T. \tag{110}$$

Thus, finally the Lagrangian is completely fitted to the traditional, linear theory of heat conduction. Of course, substituting in advance the term (110) in the Lagrangian we shall get the expression (108) immediately.

The positive-definiteness of the functional (65) and (66) (sixth statement) for the kernel (99) and (100) is still open. It will be important with regard to stability theory. In the present context, however, we may be satisfied with the weaker assumption of a positive-definite entropy production (107), i.e.—as usual—with a positive definite tensor $\underline{\lambda}$ of heat conductivity giving rise to the *Second Part of the Second Law*.

Concerning the Principle of Least Entropy Production specialized for the kernel (99) and (100) we must carefully look at the form (74) and (75) of the variational procedure! It applies for the variables η and η^* and its derivatives exclusively! The procedure has to be evaluated for the choice $(102)_1$ of the variables η and η^*. The associated ELeqs are the Jacobi equations (25) evaluated for that choice. They are automatically fulfilled.

References

[1] Onsager, L. (1931), Phys. Rev. 37, 405.

[2] Onsager, L. (1931), Phys. Rev. 38, 2265.

[3] Prigogine, I. (1947), Études Thermodynamique des Phénomènes Irréversibles, Desoer, Liège.

[4] Glansdorff, P. and Prigogine, I. (1954), Physica 20, 773.

[5] de Groot, S.R. and Mazur, P. (1969), Non-Equilibrium Thermodynamics, North Holland, Amsterdam.

[6] Gyarmati, I. (1970), Non-equilibrium Thermodynamics, Field Theory and Variational Principles, Springer, Berlin.

[7] Hamel, G. (1949), Theoretische Mechanik, Springer, Berlin.

[8] Corson, E.M. (1953), Introduction to Tensors, Spinors and Relativistic Wave-Equations, Blackie, London.

[9] Fick, E. (1968), Einführung in die Grundlagen der Quantenmechanik, Akademische Verlagsgesellschaft Geest & Portig, Leipzig.

[10] Santilli, R.M. (1978), Foundations of Theoretical Mechanics I—The Inverse Problem in Newtonian Mechanics, Springer, Berlin.

[11] Santilli, R.M. (1977), Necessary and sufficient conditions for the existence of a lagrangian in field theory, I: Ann. Phys. 103, 354–408; II: Ann. Phys (1977) 103, 409–468; III: Ann. Phys (1977) 105, 227–258.

[12] Anthony, K.-H. (1989), Unification of continuum mechanics and thermodynamics by means of lagrange formalism—present status of the theory and presumable applications, Arch. Mech. (Warsaw) 41, 511–534.

[13] Anthony, K.-H. (1997), Lagrangian field theory of plasticity and dislocation dynamics—attempts towards a unification with thermodynamics of irreversible processes, Arch. Mech. (Warsaw) 50, 345–365.

[14] Scholle, M (1999), Das Hamilton'sche Prinzip in der Kontinuumstheorie nichtdissipativer und dissipativer Systeme. Dissertation Universität Paderborn.

[15] Anthony, K.-H. (2001), Hamilton's Action Principle and thermodynamics of irreversible processes—a unifying procedure for reversible and irreversible processes, J. Non-Newton. Fluid Mech.

[16] Zubov, V.I. (1964), Methods of A.M. Lyapunov and their Application, Verlag Noordhoff, Groningen.

[17] Schmutzer, E. (1972), Symmmetrien und Erhaltungssätze der Physik, Akademie-Verlag, Berlin.

[18] Anthony, K.-H. (1988), Entropy and dynamical stability—a method due to lagrange formalism as applied to thermodynamics of irreversible processes. In Trends in Applications of Mathematics to Mechanics (Besseling, J.F., Eckhaus, W., eds.), Springer, Berlin, pp. 297–320.

[19] Anthony, K.-H. (1981), A new approach describing irreversible processes. In Continuum Models of Discrete Systems 4 (Brulin, O., Hsieh, R.T.K., eds.), North Holland, Amsterdam, pp. 481–494.

[20] Anthony, K.-H. (1986), A new approach to thermodynamics of irreversible processes by means of Lagrange formalism. In Disequilibrium and Self-Organization (Kilmister, C.W., eds.), Reidel, Dordrecht, pp. 75–92.

[21] Anthony, K.-H. (1990), Phenomenological thermodynamics of irreversible processes within lagrange formalism, Acta Phys. Hungarica 67, 321–340.

[22] Anthony, K.-H. (1991), Defect dynamics and Lagrangian thermodynamics of irreversible processes. In Continuum Models and Discrete Systems 6, (Maugin, G.A., eds.), Vol. 2, Longman Scientific & Technical, pp. 231–242.

[23] Anthony, K.-H. and Knoppe, H. (1987), Phenomenological thermodynamics of irreversible processes and lagrange formalism—hyperbolic equations for heat transport. In Kinetic Theory and Extended Thermodynamics (Müller, I., Ruggeri, T., eds.), Pitagora Editrice, Bologna, pp. 15–30.

[24] Truesdell, C. and Noll, W. (1965), The non-linear field theories of mechanics. In Handbuch der Physik (Flügge), Bd. III/3, Springer, Berlin.

[25] Azirhi, A. (1993), Thermodynamik und Quantenfeldtheorie—Ein quantenfeldtheoretisches Modell der Wärmeleitung. Diploma-Thesis, Universität Paderborn.

[26] Schargott, M. (1997), Trägheitseffekte in der Wärmeleitung—Beiträge zur Wärme-leitungstheorie im Thermionenformalismus. Diploma-Thesis, Universität Paderborn.

Chapter 3

FUNDAMENTAL PROBLEMS OF VARIATIONAL PRINCIPLES: OBJECTIVITY, SYMMETRIES AND CONSTRUCTION

Tamás Matolcsi

Department of Applied Mathematics, Eötvös Loránd University, Pázmány P. s. 1/C, 1117 Budapest, Hungary

P. Ván and J. Verhás

Department of Chemical Physics, Budapest University of Technology and Economics, Budafoki út. 8, 1521 Budapest, Hungary

Abstract

Some aspects of the relation of governing equations, spacetime symmetries and fundamental balances in nonrelativistic field theories are investigated. Special emphasis is given to variational principle construction methods for a given differential equation.

Keywords: Variational principles; symmetry; objectivity; construction of variational principles

1. Introduction

A physical theory is more than the collection of relevant physical quantities and the differential equations that determine their processes in spacetime. Mechanics, the best-developed discipline of physics gives a clear example of the above fact. The governing differential equations are Euler–Lagrange equations of a Hamiltonian variational principle and spacetime symmetries play an important role: the fundamental physical quantities (energy, momentum, etc.) are generators of representations of the spacetime symmetries and their balances are derived from the invariance requirement of the variational principle in 'free motion'. Mechanics is a brilliant example for other disciplines of physics to construct a uniform theory with deep interrelations. The program to find the fundamental structures, similar to those of mechanics in other field theories is one of the strongest of the deep currents in physical research.

Mechanics is the science of movements. The most important general principle that one can learn from mechanics is to formulate the principles and equations to characterize the system itself, independently of other motions especially the motion of an observer.

E-mail address: matolcsi@ludens.elte.hu (T. Matolcsi); rpet@phyndi.fke.bme.hu (P. Ván); verhas@phyndi.fke.bme.hu (J. Verhás).

The above-mentioned structure, the existence of proper Lagrange functions, is connected to ideal systems without dissipation. In the case of dissipation the governing equations cannot be derived from Hamiltonian variational principles (without any further ado) and the whole ideal structure of mechanics seems to fail. The basic principles are the balances of the fundamental physical quantities and we are to find the governing equations specifying the material functions according to another fundamental physical law, the Second Law of thermodynamics.

Thermodynamics is the science of materials. The most important general principle that one can learn from thermodynamics is to formulate the equations in accordance with the Second Law, in such a way that the stability of some special solutions expressing neutral, insulated conditions be ensured.

Models of dissipative thermodynamic systems should be objective, independent of observers; models of mechanical systems should be stable, i.e. in the case of dissipation asymptotically stable.

There are several ways to harmonize these views. The most accepted strategy is to build hierarchical models and think that on a deeper (molecular, atomic) level there is no dissipation, everything is governed by ideal, moreover mechanical equations. On the other hand, we should construct theories based on the same principles, without contradictions at any level of the hierarchy.

We may exemplify the structure of mechanics and find variational principles for dissipative systems. However, in this case one should take care not only on the governing equations, but also on the whole structure of the theory. For a given equation one can create several basically different variational principles non-Hamiltonian or Hamiltonian (see, e.g. in Ref. [1,2]). However, to create a principle is not enough, different variational principles for the same or equivalent equations can result in very different outlooks regarding spacetime symmetries [3,4]. The relation of spacetime symmetries, variational principles and conservative quantities of the dynamic equations is not simple [5].

One can favor a definite way of creating a Lagrangian for dissipative systems and to develop a uniform theory that generates the balances and takes the Second Law into account, too. This is done in a complete form by Anthony and coworkers (see Ref. [6] and the references within). They favor a special technique of constructing the Lagrangian with additional variables introducing complex fields in analogy to quantum mechanics. Another remarkable example where symmetry principles and conservation laws play a primary role is given by Sieniutycz and Berry [7,8]. A third example is based on the construction method of variable transformations developed by Gambár and Márkus. They also investigate their theory from the point of view of spacetime symmetries and the Second Law [9–11].

We mention here only those most developed research lines that investigate (more or less) the role of the symmetries of the corresponding theory beyond the equations in some specific dissipative systems. There are several interesting variational principles and construction methods that reproduce given equations for special purposes and do not take account of other parts of a theory. Unfortunately, the interrelation between these methods is rather poor. The representatives of the different ideas try to validate their proposition through applications, mostly reproducing classical equations. Only a few of them pay attention general aspects or to other construction methods. We think that the evaluation, the development and also the search of new physics with variational methods would be easier with a true attention to other ideas and to fundamental aspects of a physical theory.

In this chapter we call attention to some ways in which these fundamental aspects can be evaluated best. We show examples how different variational principles can be constructed while paying attention to objectivity in nonrelativistic spacetime. This is an aspect that the above-mentioned construction schemes do not consider. Moreover, we investigate how spacetime symmetries can be exploited in different construction schemes.

The structure of the paper is the following. In the next section we give a short outline of the nonrelativistic spacetime model and the formulation of covariance. As an example of handling equations in a spacetime model, a Lagrange formalism and generalized Noether theorem for point masses are given in the third section. In the fourth one a Lagrange formalism and the corresponding Noether theorem are given in nonrelativistic field theories. In the fifth section we briefly discuss variational principle construction methods in general and investigate the role of symmetries creating a general variational principle for Maxwell's equations. In the sixth section the same will be accomplished in the case of the Governing Principle of Dissipative Processes (GPDP), as an example of non-Hamiltonian variational principles. Finally we briefly discuss the results.

2. Objectivity in nonrelativistic mechanics

Lagrange formalism is a frequently used and efficient method both in point mechanics and in field theory. It is conceived, in general, that Lagrange formalism is essentially the same in those two branches of physics; nevertheless, they differ in some important aspects. One of them is the relation between symmetries and conservation laws. Another is the question of covariance: Lagrange formalism in relativistic field theory is required to be Lorentz covariant but in nonrelativistic field theory Galilean covariance does not arise (which resulted in the problem of material objectivity [12]); on the other hand, Galilean covariance appears in the Lagrange formalism of nonrelativistic particle mechanics [13] while Lorentz covariance fails manifestly in the Lagrange formalism of relativistic particle mechanics [14].

Now we compose unified covariant Lagrange formalisms both in particle mechanics and in field theory; then we treat symmetries and conservation laws in the same framework. Let us remark that our treatment considers the consequences that not a single Lagrangian, but an equivalence class of Lagrangians ((s)equivalence of Lopuszanski [4]) is relevant from a physical point of view. Here we restrict ourselves in nonrelativistic spacetime. The treatment of some relativistic aspects is given in Ref. [15]. However, we should recall how the notion of covariance is formulated in special relativity to give a proper formulation for the case.

Lorentz covariance is based on the use of four-vectors (x^i) and covectors (p_i), the Einstein summation $(p_i x^i)$ and the Lorentz transformation rule of vectors and covectors according to the relative velocity $\mathbf{v}$:

$$
\begin{aligned}
x^0 &\mapsto \frac{1}{\sqrt{1-v^2}}\left(x^0 - v_\alpha x^\alpha\right), \quad x^\alpha \mapsto \frac{1}{\sqrt{1-v^2}}\left(x^\alpha - v^\alpha x^0\right), \\
p_0 &\mapsto \frac{1}{\sqrt{1-v^2}}(p_0 + v^\alpha p_\alpha), \quad p_\alpha \mapsto \frac{1}{\sqrt{1-v^2}}(p_\alpha + v_\alpha p_0).
\end{aligned} \tag{1}
$$

The last and important feature of Lorentz covariance is the possibility of moving the indices up and down (equivalence of vectors and covectors: $x_i = g_{ik}x^k$, $p^i = g^{ik}p_k$).

As above, in the following too, indices will be denoted by roman letters $i, k, \ldots$ having values 0,1,2,3 and by Greek letters $\alpha, \beta, \ldots$ having values 1,2,3.

Galilean covariance, too, is based on the use of four-vectors (x^i) and covectors (p_i), the Einstein summation $(p_i x^i)$, and the Galilean transformation rule of vectors and covectors according to the relative velocity $\mathbf{v}$:

$$
\begin{aligned}
x^0 &\mapsto x^0, \quad x^\alpha \mapsto x^\alpha - v^\alpha x^0, \\
p_0 &\mapsto p_0 + v^\alpha p_\alpha, \quad p_\alpha \mapsto p_\alpha.
\end{aligned} \tag{2}
$$

Galilean covariance differs essentially from Lorentz covariance in that the equivalence of vectors and covectors is missing (it is impossible to move the indices up and down). Velocity and acceleration are vectors; the derivative of a scalar function f is a covector: $\partial_i f$; the electromagnetic scalar potential and vector potential together form a covector.

The above transformation laws show clearly that the time-like component of vectors and the space-like components of covectors are *absolute*, i.e. independent of observers. Moreover, vectors whose time-like component is zero are absolute space-like and covectors whose space-like components are zero are absolute time-like.

Covariance can be best formulated if coordinates are not used at all; namely, it is evident that spacetime and events in spacetime are absolute (independent of observers), coordinates are not inherent objects of spacetimes. Having this trivial remark as a guideline, we can construct absolute spacetime models.

2.1. Nonrelativistic spacetime model

The basic concepts of nonrelativistic spacetime are formulated, e.g. in Ref. [16–18], and are completely expounded in Ref. [19], which is recapitulated below briefly.

The nonrelativistic spacetime model is $(M, I, \tau, \mathbf{D}, \mathbf{b})$, where

- M is *spacetime*, a 4D affine space over the vector space $\mathbf{M}$ (i.e. the difference of two points in M is an element of $\mathbf{M}$),
- I is *time*, a 1D affine space over $\mathbf{I}$; the latter is *the measure line of time intervals*,
- $\tau : M \rightarrow I$ is the *time evaluation*, an affine surjection over the linear map $\tau : \mathbf{M} \rightarrow \mathbf{I}$,
- $\mathbf{D}$ is *the measure line of distances*, a 1D vector space,
- $\mathbf{b} : \mathbf{E} \times \mathbf{E} \rightarrow \mathbf{D} \otimes \mathbf{D}$ is the *Euclidean structure*, a positive-definite symmetric bilinear mapping where $\mathbf{E} := \mathrm{Ker}\boldsymbol{\tau}$ is the *subspace of the space-like vectors*.

Vectors are elements of $\mathbf{M}$, covectors are elements of $\mathbf{M}^*$, the dual of $\mathbf{M}$ (in other words, covectors are linear functionals on $\mathbf{M}$). Though $\mathbf{M}$ and $\mathbf{M}^*$ are isomorphic (both are 4D vector spaces), there is no distinguished isomorphism between them, so vectors and covectors are essentially different objects. The absolute time-like component of a vector $\mathbf{x}$ is $\boldsymbol{\tau}\mathbf{x}$ and the absolute space-like component of a covector $\mathbf{p}$ is denoted by $\mathbf{i}^*\mathbf{p}$. A vector is future-like if its absolute time-like component is positive.

The history of a point mass is described by a *world line function*, a continuously differentiable function $r : I \to M$ defined on an interval, with $\tau(r(t)) = t$ for any $t \in \mathrm{Dom}r$. Using a construction similar to the one of tensor products, one can define the 4D tensor quotient space $\mathbf{M/I}$. Then $\dot{r}(t) \in \mathbf{M/I}$ and $\boldsymbol{\tau}(\dot{r}(t)) = 1$. Hence

$$V(1) := \left\{\mathbf{u} \in \frac{\mathbf{M}}{\mathbf{I}} : \boldsymbol{\tau}(\mathbf{u}) = 1\right\},$$

is accepted as the set of *absolute velocity values*. $V(1)$ is an affine space over $\mathbf{E/I}$.

Observers and coordinate systems are defined in this framework and the Galilean transformation rules for coordinates of vectors and covectors are obtained.

The isomorphisms of spacetime are constituted by Galilean transformations and Noether transformations. A Galilean transformation $\mathbf{L}$ is a linear bijection $\mathbf{M} \to \mathbf{M}$ such that $\boldsymbol{\tau}\cdot\mathbf{L} = \boldsymbol{\tau}$ and the restriction of $\mathbf{L}$ onto $\mathbf{E}$ is a rotation. A Noether transformation L is an affine bijection $M \to M$ such that the underlying linear map is a Galilean transformation.

The Galilean transformations form a 6D Lie group having the Lie algebra

$$\{\mathbf{H} : \mathbf{M} \to \mathbf{M} | \mathbf{H} \text{ is linear, } \boldsymbol{\tau}\cdot\mathbf{H} = 0,\ \mathbf{H}|_{\mathbf{E}}^{+} = -\mathbf{H}|_{\mathbf{E}}\}, \tag{3}$$

and the Noether transformations form a 10D Lie group whose Lie algebra consists of affine maps $H : M \to \mathbf{M}$ such that the underlying $\mathbf{H}$ is in the Lie algebra of the Galilean group.

2.2. *What is covariance?*

It is worth underlying the following: In usual treatments observers and coordinate systems are primitive (undefined) objects, Galilean (or Lorentz) transformation rules for coordinates are postulated. Covariance is the requirement that describing phenomena one must use correctly transforming quantities.

In the above recapitulated treatment spacetime models are postulated, observers and coordinate systems are defined (there are no undefined objects at all), Galilean (or Lorentz) transformation rules for coordinates are obtained as a result; covariance is substituted with the requirement that to describe phenomena one must use *absolute objects*, i.e. objects in spacetime corresponding only to the phenomena in question and not referring to any observers or coordinates.

Lastly, we call attention to the fact that we have to distinguish clearly between Galilean (or Lorentz) transformations (3) (of spacetime vectors) and Galilean transformation rules (2) (for coordinates); the former are absolute objects, the latter are not.

3. Lagrange formalism

The basic object of the description of a system of point masses in mechanics is the *evolution space*: the set of possible spacetime positions and velocity values. For the sake of simplicity, here we shall consider a single point mass (the general case can be treated similarly); the evolution space of a point mass is $M \times V(1)$ in nonrelativistic spacetime (or in relativistic spacetime).

3.1. Point masses

The Lagrange formalism of a nonrelativistic point mass consists in the following. There is given a continuous Lagrangian function $\mathfrak{L}$ defined on $M \times V(1)$ that, for given spacetime points x_0 and x_1, defines the action

$$S(r) := \int_{t_0}^{t_1} \mathfrak{L}(r(t), \dot{r}(t))\mathrm{d}t \tag{4}$$

for world-line functions r, where $t_0 := \tau(x_0)$ and $t_1 := \tau(x_1)$; the point mass is supposed propagating from x_0 to x_1 along a world line, for which the 'variation' of the action is zero.

The precise definition of 'variation' requires the definition of differentiability of the action, i.e. of the mapping $r \mapsto S(r)$. The simplest possibility is the following. Taking a norm $|\ |_{\mathbf{M}}$ on $\mathbf{M}$ and a norm $|\ |_{\mathbf{M/I}}$ on $\mathbf{M/I}$ (any two norms on a finite-dimensional vector space are equivalent), we introduce the vector space

$$\mathbf{V} := \{\mathbf{r} : [t_0, t_1] \to \mathbf{E} | \mathbf{r} \text{ is continuously differentiable, } \mathbf{r}(t_0) = \mathbf{r}(t_1) = 0\}$$

endowed with the norm

$$\|r\| := \max_{t \in [t_0, t_1]} \left(|\mathbf{r}(t)|_{\mathbf{M}} + |\dot{\mathbf{r}}(t)|_{\frac{\mathbf{M}}{\mathbf{I}}} \right). \tag{5}$$

Then

$$V := \{r : [t_0, t_1] \to M | \mathbf{r} \text{ is a world-line function } r(t_0) = x_0, r(t_1) = x_1\}$$

is an affine space over $\mathbf{V}$. Hence differentiability of $S : V \to \mathbb{R}$ is well defined [20].

If S is differentiable, the world-line function realized by the point mass is selected by $DS(r) = 0$, i.e. by the zero value of the derivative of S. It is known that if the Lagrangian $\mathfrak{L}$ is twice continuously differentiable, then S is differentiable, and in this case, $DS(r) = 0$ is equivalent to twice continuous differentiability of r satisfying the Euler–Lagrange equation

$$\mathbf{i}^* \partial_r \mathfrak{L}(r(t), \dot{r}(t)) - \frac{\mathrm{d}}{\mathrm{d}t} \partial_{\dot{r}} \mathfrak{L}(r(t), \dot{r}(t)) = 0. \tag{6}$$

Here, ∂_r and $\partial_{\dot{r}}$ stand for the partial derivative according to the first variable (the spacetime variable in M) and to the second variable (the velocity variable in $V(1)$), respectively.

In order to compare our formulae with those of usual treatments, we write our expressions in the usual ambiguous (but convenient) notations and in coordinates, too.

$V(1)$ is a 3D affine subspace over space-like vectors, hence the coordinates of a derivative by elements of $V(1)$ only contains indices 1,2,3. Accordingly, we have the Euler–Lagrange equation in the form

$$\frac{\partial \mathfrak{L}(r,\dot{r})}{\partial r^{\alpha}} - \frac{\mathrm{d}}{\mathrm{d}t}\left(\frac{\partial \mathfrak{L}(r,\dot{r})}{\partial \dot{r}^{\alpha}}\right) = 0.$$

3.2. *Equivalent Lagrangians*

A Lagrange function is destined to describe the histories of a point mass under the action of an external force. Evidently, different Lagrange functions can give rise to the same differential equation: for instance, $\mathfrak{L}$ and $\lambda\mathfrak{L}$ for arbitrary nonzero real number λ.

Furthermore, a function $\mathfrak{f}$ defined on $M \times V(1)$ is called a *full time-derivative* if there exists a continuously differentiable function ϕ defined on M, such that $\mathfrak{f}(x,u) = \mathrm{D}\phi(x)u$ for all $x \in M$ and $u \in V(1)$. Then it is a simple fact that $\mathfrak{L}$ and $\mathfrak{L} + \mathfrak{f}$ result in the same Euler–Lagrange equation. Let us note here that it would be possible to extend the Lagrangian to the future-like vectors in **M/I** and that extension could be important in the treatment of the relativistic case [15,21].

We say that $\mathfrak{L}$ and $\mathfrak{L}'$ are *equivalent*, if $\mathfrak{L}' - \mathfrak{L}$ is a full time-derivative. Note that $\mathfrak{L}$ and $\lambda\mathfrak{L}$ are not equivalent.

Then we conceive that a point mass in an external force is completely characterized by an equivalence class of Lagrangians. This is a far-reaching assumption whose meaning is the following: point masses, though, described by the same differential equation (Newtonian equation), are physically different if their Lagrangians are not equivalent.

Let us consider a point mass having mass m, in an external force having the (four-)potential $K : M \rightarrow \mathbf{M}^*$. Its nonrelativistic Lagrangian is given by an arbitrary $\mathbf{c} \in V(1)$ with the form

$$\mathfrak{L}(x,\mathbf{u}) = \frac{m|\mathbf{u} - \mathbf{c}|^2}{2} + K(x)\cdot\mathbf{u}$$

for $x \in M$ and $\mathbf{u} \in V(1)$.

Here $\mathbf{c}$ is a parameter alien to the point mass and the potential; thus, this Lagrangian is not absolute in this sense. Of course, $\mathbf{c}$ drops out from the Euler–Lagrange equation. Moreover, two such Lagrangians with $\mathbf{c}$ and $\mathbf{c}'$ are equivalent; thus, the equivalence class of such Lagrangians is absolute. The fact that this parameter cannot be avoided in absolute Lagrangians means that the kinetic energy is expressed by relative velocities and has far-reaching consequences beyond classical mechanics [22].

The systems corresponding to masses m and m' and potentials K and K' are equivalent (physically equal) if and only if $m' = m$ and $K' = K + \mathrm{D}\phi$ for an arbitrary continuously differentiable function ϕ. Thus free point masses (the potential is zero) are equivalent (physically equal) if and only if their mass values coincide, though their Newtonian equations are the same $\ddot{x} = 0$.

3.3. Symmetries

Let $F : M \rightarrow M$ be a continuously differentiable map so that $\mathrm{D}F(x)[V(1)] \subset V(1)$ for all $x \in M$. We say that F is a *symmetry* of the physical system described by the Lagrangian $\mathfrak{L}$ if $\mathfrak{L}$ and its transform by F are equivalent; more closely, if there exists a function ϕ_F such that

$$\mathfrak{L}(F(x), \mathrm{D}F(x)\mathbf{u}) = \mathfrak{L}(x, \mathbf{u}) + \mathrm{D}\phi_F(x)\mathbf{u}$$

for all $x \in M$ and $\mathbf{u} \in V(1)$. In particular, we shall consider spacetime isomorphisms: a Noether transformation L is a symmetry of $\mathfrak{L}$ if and only if

$$\mathfrak{L}(Lx, \mathbf{L}\mathbf{u}) = \mathfrak{L}(x, \mathbf{u}) + \mathrm{D}\phi_L(x)\mathbf{u}.$$

Let us suppose that a one-parameter subgroup $s \mapsto e^{sH}$ is a symmetry (H is in the Lie algebra of the spacetime isomorphism group). Then taking e^{sH} in the role of L, differentiating with respect to s at zero, we get

$$\mathfrak{L}(x, \mathbf{u}) + \frac{\partial \mathfrak{L}(x, \mathbf{u})}{\partial x} \cdot H(x) + \frac{\partial \mathfrak{L}(x, \mathbf{u})}{\partial \mathbf{u}} \cdot \mathbf{H}\mathbf{u} = \frac{\partial g(H, x)}{\partial x} \cdot \mathbf{u}$$

where $g(H, x) := \frac{\mathrm{d}}{\mathrm{d}s} \phi_{\exp(sH)}(x)|_{s=0}$.

If we coordinatize spacetime and the linear map $\mathbf{H}$ has coordinates H^i_j, then $H(x)$ has coordinates $H^i_j x^j + h^i$ where $h := H(o)$, o being the spacetime origin of the coordinatization. Thus the above equality in coordinates reads

$$\mathfrak{L} + \frac{\partial \mathfrak{L}}{\partial x^i}\left(H^i_j x^j + h^i\right) + \frac{\partial \mathfrak{L}}{\partial u^i} H^i_j u^j = \frac{\partial g}{\partial x^i} u^i.$$

Exploiting the Euler–Lagrange equations we obtain Noether's theorem in this framework: if the one-parameter subgroup generated by the Lie-algebra element H of (either the nonrelativistic or the relativistic) spacetime group is a symmetry, then

$$(x, \mathbf{u}) \mapsto \frac{\partial \mathfrak{L}(x, \mathbf{u})}{\partial \mathbf{u}} \cdot H(x) - g(H, x)$$

is a conserved quantity (constant of motion) where the function g is defined as above.

4. Lagrange formalism in field theory

Similarly to Lagrangian mechanics, we can expound nonrelativistic Lagrangian field theory.

4.1. The Euler–Lagrange equation

Field quantities are functions defined on spacetime having values in a finite-dimensional real or complex vector space $\mathbf{V}$. If $R : M \rightarrow \mathbf{V}$ is a field quantity, then its derivative is a function $\mathrm{D}R : M \rightarrow \mathbf{V} \otimes \mathbf{M}^*$.

A Lagrangian $\mathfrak{L}$ is a scalar-valued continuous function defined on $\mathbf{V} \times (\mathbf{V} \otimes \mathbf{M}^*)$.

Given a bounded open subset Ω in M, having a smooth boundary, and a continuous function $\eta : \partial\Omega \to \mathbf{V}$, we consider the affine space B of continuously differentiable functions $R : \bar{\Omega} \to \mathbf{V}$ satisfying $R|_{\partial\Omega} = \eta$ over the vector space of continuously differentiable functions $\mathbf{R} : \bar{\Omega} \to \mathbf{V}$ satisfying $\mathbf{R}|_{\partial\Omega} = 0$. The norm of such a function is defined to be

$$\max_{x\in\bar{\Omega}} |\mathbf{R}(x)|_{\mathbf{V}} + \max_{x\in\bar{\Omega}} |\mathrm{D}\mathbf{R}(x)|_{\mathbf{V}\otimes\mathbf{M}^*},$$

where $|\cdot|_{\mathbf{V}}$ and $|\cdot|_{\mathbf{V}\otimes\mathbf{M}^*}$ are arbitrary norms on $\mathbf{V}$ and $\mathbf{V}\otimes\mathbf{M}^*$, respectively. The linear space of the functions $\mathbf{R}$ endowed with the above maximum norm will be denoted by $\mathbf{B}_R$.

The action for the field quantities is defined to be

$$S(R) := \int_{\Omega} \mathfrak{L}(R(x), \mathrm{D}R(x))\mathrm{d}x$$

and the field that comes out in Ω with boundary condition η is the one for which the derivative of the action is zero.

If the Lagrangian is twice-continuously differentiable, then $\mathrm{D}S(R) = 0$ is equivalent to the twice-continuous differentiability of R satisfying the Euler–Lagrange equation

$$\frac{\partial \mathfrak{L}(R, \mathrm{D}R)}{\partial R} - \mathrm{D}\cdot\frac{\partial \mathfrak{L}(R, \mathrm{D}R)}{\partial(\mathrm{D}R)} = 0, \tag{7}$$

where the usual ambiguous (but convenient) notation is used. In coordinates,

$$\frac{\partial \mathfrak{L}(R, \mathrm{D}R)}{\partial R^{\sigma}} - \partial_i\frac{\partial \mathfrak{L}(R, \mathrm{D}R)}{\partial(\partial_i R^{\sigma})} = 0,$$

where i labels spacetime coordinates and σ labels coordinates in $\mathbf{V}$.

4.2. *Equivalent Lagrangians*

A function $\mathfrak{f}$ defined on $\mathbf{V}\times(\mathbf{V}\otimes\mathbf{M}^*)$ is called a *full divergence* if a continuously differentiable function $\Phi : \mathbf{V} \to \mathbf{M}$ exists such that $\mathfrak{f}(X, \mathbf{W}) = \mathrm{D}\Phi(\mathrm{X}) : \mathbf{W}$ for all $X \in \mathbf{V}$ and $\mathbf{W} \in \mathbf{V}\otimes\mathbf{M}^*$; in coordinates, $f(X, \mathbf{W}) = \frac{\partial \Phi^i(X)}{\partial X^{\sigma}} W_i^{\sigma}$.

Then it is a simple fact that $\mathfrak{L}$ and $\mathfrak{L} + \mathfrak{f}$ result in the same Euler–Lagrange equation.

We say that $\mathfrak{L}$ and $\mathfrak{L}'$ are *equivalent*, if $\mathfrak{L}' - \mathfrak{L}$ is a full divergence.

Then we conceive that a field is completely characterized by an equivalence class of Lagrangians. This is a far-reaching assumption whose meaning is the following: fields, though, described by the same differential equation, are physically different if their Lagrangian are not equivalent.

4.3. *Symmetries*

Up to now Lagrangian formalism in particle mechanics and that in field theory are strictly analogous. The meaning of symmetries and conservation laws are different in the two theories.

Namely, symmetries in mechanics are connected with the variables of the Lagrangians only (with $M \times V(1)$), whereas symmetries in field theory are, besides the variables of the Lagrangian (besides $\mathbf{V} \times (\mathbf{V} \otimes \mathbf{M}^*)$), connected with the variable of the field (with M), too.

We say that a pair (Z, L) is a *symmetry* of the physical system described by the Lagrangian $\mathfrak{L}$ if $Z : \mathbf{V} \to \mathbf{V}$ is a linear bijection and L is a spacetime isomorphism such that there exists a continuously differentiable function $\Phi_{Z,L} : \mathbf{V} \to \mathbf{M}$ and

$$\mathfrak{L}(ZR(L^{-1}(x)), Z\mathrm{D}R(l^{-1}(x))\mathbf{L}^{-1}) = \mathfrak{L}(R(x), \mathrm{D}R(x)) + \mathrm{D}\Phi_{Z,L}(R(x)) : \mathrm{D}R(x) \tag{8}$$

for all $x \in M$ and for all fields R satisfying the corresponding Euler–Lagrange equation. Here Z stands for the so-called 'dynamical' symmetries.

Let $A : \mathbf{V} \to \mathbf{V}$ be a linear map and H an element of the Lie algebra of the spacetime isomorphism group. Let us suppose that the couple of the one-parameter subgroups $s \mapsto (e^{sA}, e^{sH})$ is a symmetry. Then taking e^{sA} and e^{sH} in the role of Z and L, respectively, differentiating with respect to s at zero we easily get that

$$\begin{aligned} &\frac{\partial \mathfrak{L}}{\partial R}(AR + \mathrm{D}RH) + \frac{\partial \mathfrak{L}}{\partial(\mathrm{D}R)}(A\mathrm{D}R + \mathrm{D}R\mathbf{H} + \mathrm{D}^2RH) \\ &\quad = \frac{d}{ds}(\mathrm{D}\Phi_{\exp(sA),\exp(sH)}(R) : DR)|_{s=0}. \end{aligned} \tag{9}$$

Taking into account the Euler–Lagrange equation, we get that the divergence of

$$\left(\mathfrak{L}(R, \mathrm{D}R) - \frac{\partial \mathfrak{L}(R, \mathrm{D}R)}{\partial(\mathrm{D}R)}\mathrm{D}R\right)H + \frac{\partial \mathfrak{L}(R, \mathrm{D}R)}{\partial(\mathrm{D}R)}AR - g(A, H, R) \tag{10}$$

is zero where $g(A, H, R(x)) := \frac{\mathrm{d}}{\mathrm{d}s}\Phi_{\exp(sA),\exp(sH)}(R(x))|_{s=0}$.

In coordinates,

$$\begin{aligned} \partial_i\Bigg(&\left(\mathfrak{L}(R, \mathrm{D}R)\delta^i_k - \frac{\partial \mathfrak{L}(R, \mathrm{D}R)}{\partial(\partial_i R^\sigma)}\partial_k R^\sigma\right)\left(H^k_l x^l + h^k\right) \\ &+ \frac{\partial \mathfrak{L}(R, \mathrm{D}R)}{\partial(\partial_i R^\sigma)}A^\sigma_\rho R^\rho - g^i(A, H, R)\Bigg) = 0. \end{aligned} \tag{11}$$

Thus, the function given in Eq. (10) is a locally conserved quantity. Let us recall here a sometimes convenient but even more ambiguous notation, usual in traditional physics books (e.g. Ref. [23]): $\delta R^\rho = \partial_k R^\rho(H^k_l x^l + h^k)$ stands for the 'variation' of R under spacetime 'transformations' and $\tilde{\delta} R^\sigma = A^\sigma_\rho R^\rho(x^k)$ stands for the 'variation' of R under 'dynamic' symmetries. Sometimes the 'variation' of the coordinates under spacetime symmetries is denoted as $\delta x = H^k_l x^l + h^k$.

5. Construction of variational principles

After the concise formulation of an inherently covariant (observer independent) Lagrange formalism and the corresponding symmetry principles we switch our attention to the inverse problem: construction of a variational principle for a given equation in a nonrelativistic spacetime. The function space of field quantities $R : M \to \mathbf{V}$ and the norm is given as in the previous section. Let us introduce a suitable function space B_R of the functions R. A differential operator $\Theta : B_R \rightarrow \mathbf{B}_R^{\prime *}$ (where $*$ denotes the dual), will be used to define a given governing equation as

$$\Theta(R) = 0. \tag{12}$$

The consequent considerations are valid without any further ado if $\mathbf{B}_R$ and $\mathbf{B}_R^{\prime}$ are Banach spaces. For example, in the case of pure mechanics, where Θ is a second-order differential equation $\mathbf{B}_R$ is the Banach space of two-times continuously differentiable functions with a suitable maximum norm.

We are looking for a Hamilton principle with Eq. (12) as the Euler–Lagrange equation. A mathematical theorem states that if Θ is symmetric, then there is a Hamiltonian principle, there exists a suitable action $S(R)$ with Eq. (12) as the Euler–Lagrange equation. In the case of nonsymmetric Θ there are several ways to circumvent this theorem if we do not want to reproduce Eq. (12) exactly, only an equivalent equation with the same solutions. The possibilities are classified in Fig. 1.

With Hamiltonian constructing methods, we are looking for Hamiltonian variational principles. Using Hamiltonian variational principles means we get the Euler–Lagrange equations with the help of the derivation of the action functional with a suitable maximum norm, as in the previous sections); with modification methods, we are constructing non-Hamiltonian, 'quasivariational' principles. All Hamiltonian methods have a common origin: they are doubling the fields. In the methods of integrating operators and variable transformations the additional fields are eliminated. The method of the elimination is based on a kind of 'operator Hamilton' formalism, where the given differential equation (12) is considered as one of the two Hamiltonian equations. The whole classification, a general construction scheme, suitable algorithmic construction methods for linear operators and examples are developed in Ref. [2]. However, the question of symmetries is not investigated there.

Among the Hamiltonian methods the method of variable transformations is the generalization of the idea behind the traditional variational principle of the Maxwell equations. The electromagnetic field quantities are transformed, derived from a (scalar and vector) 'potential' in such a way that the governing equations for the potential functions are Euler–Lagrange equations of a variational principle in the new variables. This idea

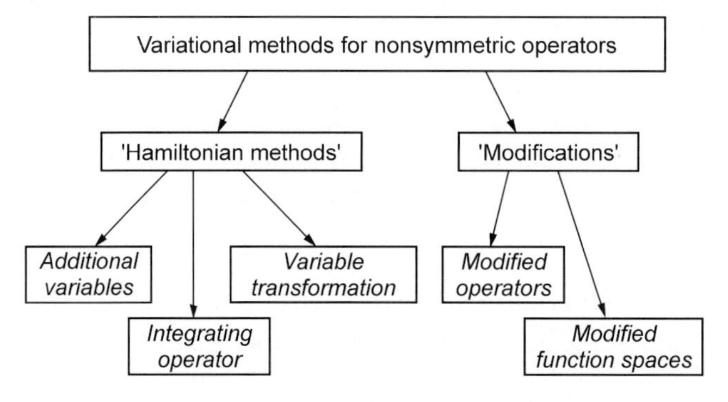

Fig. 1. Possible routes to construct a variational principle.

was suggested for linear transport equations by Márkus and Gambár [24] and generalized by Nyíri for linear and recursively quasilinear equations [25]. The general construction method gives the Lagrangian only up to an undetermined function. However, in the case of nonlinear base equations (12) the construction is not straightforward and requires further investigations.

We will see in the following that the undetermined function gives us the freedom to consider additional physical assumptions in deriving balances from spacetime symmetries and some physical expectations lead to traditional quadratic Lagrangians. We will not investigate the question in full generality, we restrict ourselves to construct a variational principle for the Maxwell equations. Let us note that one can easily construct variational principles for several forms of the Maxwell equations (with or without considering specific material laws for polarization and conduction, for the whole set of the equations and for some of them, etc.). Some possibilities are given to demonstrate the applicability of the method and the special assumptions behind the traditional choice in Ref. [2]. However, there is no *general* construction method that would go beyond the possibilities of recursively quasilinear $\boldsymbol{\Theta}$ operators. Here we do not use a frame-independent, objective formalism (for the sake of simplicity), moreover, we apply the usual terminology.

5.1. Two Maxwell equations without polarization

We will construct the variational principle for the first two Maxwell equations (13) without considering specific materials.

$$\nabla\cdot\mathbf{D} = \rho, \quad \nabla\times\mathbf{H} = \mathbf{j} + \frac{\partial \mathrm{D}}{\partial t}, \tag{13}$$

$$\nabla\cdot\mathbf{B} = 0, \quad \nabla\times\mathbf{E} = -\frac{\partial \mathbf{B}}{\partial t}. \tag{14}$$

Now we introduce a scalar-valued function ϕ for the first equation (13) and a vector-valued function $\mathbf{A}$ for the second. Let us emphasize that we did not introduce the functions as solutions of the last two equations! Our starting point is different, $(\varphi, \mathbf{A})$ are candidates of a variational potential. Now we are looking for a Lagrangian in the form

$$F(\phi, \mathbf{A}) = f\left(\nabla\phi, \phi, \nabla\times\mathbf{A}, \mathbf{A}, \frac{\partial A}{\partial t}\right), \tag{15}$$

where f is considered as an algebraic function on the appropriate domain. Let us remember that here the different variables were introduced for the different terms in Eqs. (13) and (14). The following adjoint differential operators are used: $(-\nabla\cdot)^* = \nabla, \left(-\frac{\partial}{\partial t}\right)^* = \frac{\partial}{\partial t}$ and $(\nabla\times)^* = \nabla\times$. Now the partial variation of Eq. (15) results in the following differential equations

$$-\nabla\cdot\partial_1 f + \partial_2 f = 0,$$

$$\nabla\times\partial_3 f + \partial_4 f - \frac{\partial(\partial_5 f)}{\partial t} = 0,$$

where ∂_i is the partial derivative of f according to its i-th variable. The identification of the different terms with the help of the Maxwell equations (13) yields:

$$\mathbf{D} := \partial_1 f = \partial_5 f, \quad \rho := \partial_2 f, \quad \mathbf{H} := \partial_3 f, \quad \mathbf{j} := -\partial_4 f.$$

Therefore a general Lagrangian of our equations (containing an arbitrary four-variable function) follows:

$$F(\phi, \mathbf{A}) = \tilde{f}\left(\nabla\phi + \frac{\partial \mathbf{A}}{\partial t}, \phi, \nabla \times \mathbf{A}, \mathbf{A}\right). \tag{16}$$

This Lagrangian together with the identifications of the original physical quantities gives a Hamiltonian variational principle with Eq. (13) as Euler–Lagrange equations. We can recover with the identification $\mathbf{B} = \nabla \times \mathbf{A}$ and $\mathbf{E} = \partial_t \mathbf{A} + \nabla\varphi$ a solution of the last two Maxwell equations (14).

$\tilde{f}$ is degenerate (does not contain the time derivative of φ), because the original problem is degenerate, too. We can consider the space symmetries on this general form. The invariance under spacetime translations results in the next energy-momentum tensor

$$\mathbf{T}^{\mathbf{M}} = \begin{pmatrix} \rho_H^M & \mathbf{j}_H^M \\ \rho_p^M & \mathbf{j}_p^M \end{pmatrix} = \begin{pmatrix} \tilde{f} - \partial_1 \tilde{f} \partial_0 \mathbf{A} & -\partial_1 \tilde{f} \nabla\varphi - \partial_3 \tilde{f} \times \partial_t \mathbf{A} \\ -\partial_1 \tilde{f} \times (\nabla \times \mathbf{A}) & \tilde{f}\mathbf{I} - \partial_1 \tilde{f} \nabla\varphi - \partial_3 \tilde{f} \otimes \nabla \times \mathbf{A} \end{pmatrix}. \tag{17}$$

Here, $\rho_H, \rho_p, \mathbf{j}_H$ and $\mathbf{j}_p$ are the energy and moment densities and current densities, respectively. $\mathbf{I}$ is the second-order space-like unit tensor. Let us note, that the Hamiltonian ρ_H has nothing to do with the Hamiltonians of the general operator Hamiltonian formalism.

One can prove that in the case of linear polarization ($\mathbf{D} = \epsilon\mathbf{E}$ and $\mathbf{B} = \mu\mathbf{H}$), with the usual (quadratic) choice of $\tilde{f}$ and considering a Coulomb gauge and dropping the longitudinal terms, one can get back the traditional energy-momentum from Eq. (17). However, the general case is independent of the particular material equations and shows well the role of a particular polarization and the role of the special quadratic starting Lagrangian in the traditional treatments. Moreover, weakening the simplifications one can get different generalizations. For example, if a particular polarization was not applied one recovered the energy momentum tensor of Minkowski [26]. We cannot go into a more detailed treatment here but it is worth nothing that this kind of inverse, constructive approach can contribute to solving the old problem of energy momentum in polarizable electromagnetic media. On the other hand, we have demonstrated here that an inverse scheme did not forbid the treatment of proper symmetry principles and Noether theorems, moreover, it can be a tool to identify the most convenient (right?) Lagrangian.

6. Spacetime symmetries of dissipative systems

Our other example is a more involved one. We will show that one can apply the symmetry principles in the case of non-Hamiltonian variational principles, too. These kind of approaches are not Hamiltonian ones either from a mathematical point of view or from that of symmetry principles. Application of nontraditional function spaces (strange variational prescriptions) do not forbid the use and derivation of Noether theorems. Moreover, the results can be evidently clear from a physical point of view (but disappointing). Our example will be the Governing Principle of Dissipative Processes (GPDP) [27,28]. GPDP is constructed to recover the quasilinear constitutive equations and the transport equations of nonequilibrium thermodynamics. Due to its Gaussian structure, it is useful to treat complex constrained problems [29] and stable and powerful direct solution methods [30,31]. GPDP is based on a mixed construction scheme, because it

- introduces additional variables,
- applies the integrating operator method by using the Gauss form,
- modifies the functions spaces with the variational prescription 'time derivatives are not varied'.

Our starting points are the mentioned balance and constitutive equations of irreversible thermodynamics. The balance of the extensives a and the quasilinear constitutive equations are

$$\varrho\dot{a}(\Gamma) + \nabla\cdot\mathbf{J} = \sigma(\Gamma), \tag{18}$$

$$\mathbf{J} = L(\Gamma)\cdot\nabla\Gamma. \tag{19}$$

Here (Γ are the intensive variables, $\mathbf{J}$ denotes the current densities of the extensives a, L is the Onsagerian conductivity matrix and σ denotes the source term. Substituting Eq. (19) into Eq. (18) one can get the transport equations

$$\varrho\dot{a}(\Gamma) + \nabla\cdot(L(\Gamma)\nabla\Gamma) = \sigma(\Gamma). \tag{20}$$

The variational potential of the principle to derive the transport equation (20) and constitutive equations Eq. (19) is:

$$S(\Gamma,\mathbf{J}) = \int_U \mathfrak{L}^G(\Gamma,\mathbf{J})\mathrm{d}V = \int_U (\sigma_s(\Gamma,\mathbf{J}) - \Phi(\Gamma,\mathbf{J}) - \Psi(\Gamma,\mathbf{J}))\mathrm{d}V$$

$$= -\int_U \frac{1}{2}(\mathbf{J} - \mathbf{L}(\Gamma)\cdot\nabla\Gamma)\cdot(\mathbf{L}^{-1}(\Gamma)\cdot\mathbf{J} - \nabla\Gamma)\mathrm{d}V. \tag{21}$$

Here, σ^s, Φ and Ψ denote the entropy production and the dissipation potentials, respectively. U is a suitable subset of 3D Euclidean space $\mathbf{E}$ (relative to a given observer). The independent field quantities here are the intensives Γ and the current densities $\mathbf{J}$. There the variational prescription is that the 'time derivatives are not varied', time is a parameter in the above functions and the corresponding norm is considered in this sense, too. The time parametrization is an observer-dependent, relative concept. A time parametric form of a function $f : \mathbf{I}\times\mathbf{E}\to\mathbf{V}$, $(t,\mathbf{x})\mapsto f(t,\mathbf{x})$ is a function $\tilde{f} : \mathbf{I}\to \mathrm{Fun}(\mathbf{E},\mathbf{V})$, $t\mapsto\tilde{f}(t)$, where $\mathbf{I}$ is the time and Fun($\mathbf{E},\mathbf{V}$) denotes a space of functions $\mathbf{E}\to\mathbf{V}$. Therefore it is evident that $\frac{\mathrm{d}\tilde{f}}{\mathrm{d}t}(t) \neq \frac{\partial f}{\partial t}(t)$ (see Ref. [32] for a detailed treatment). This is why the integration is regarded on the space variables only in the original formulation of the principle. However, one may integrate according to the parameter, keeping in mind that space and time variables are treated separately in the functions and considering that the Lagrangian is independent of the time derivatives.

The Euler–Lagrange equations due to the variation of the principle are

$$\delta_{\mathbf{J}} S : 0 = L^{-1} \cdot \mathbf{J} - \nabla \Gamma, \tag{22a}$$

$$\delta_{\Gamma} S : 0 = \nabla \cdot (\mathbf{J} - L \cdot \nabla \Gamma) + \frac{1}{2} (\mathbf{J} \cdot L^{-1} + \nabla \Gamma) \cdot L' \cdot (L^{-1} \cdot \mathbf{J} - \nabla \Gamma). \tag{22b}$$

As can be seen Eq. (22a) is equivalent to the constitutive equations (19), but the derivation of the transport equation is more involved. Let us recognize that Eqs. (22a) and (22b) are not independent of each other. However, according to the usual procedure we should consider Eq. (18) as a *subsidiary condition*. Therefore we use it to eliminate the divergence of the current from Eq. (22b). After that we put Eq. (22a) in Eq. (22b) to eliminate the second term (this is called the 'supplementary theorem' of Gyarmati [27]). Because time is a parameter and the time derivative an independent variable, Eq. (18) is independent of Eqs. (22a) and (22b). This is why the balances are not constraints but subsidiary conditions.

Now we may consider the Noether theorems for GPDP. First let us note that the canonical formalism of Verhás and Vojta [33,34] does not look like a good starting point. There the gradients are playing the role of time derivatives and one introduces generalized moments with the help of them (that is the special case of the operator Hamilton formalism, where canonical momenta are considered for any linear operators [2]). On the other hand, in our treatment the symmetries were considered by using only the Lagrangians, they had nothing to do with an invariance of any action! Our formulation of the Noether theorems shows that they are independent of the method of 'variation'! To get the balances we should calculate Eq. (9) and apply the corresponding symmetries and Euler–Lagrange equations. The Euler–Lagrange equations are not necessarily Eq. (7) calculated from a Hamiltonian principle.

Considering the previous remarks and substituting the particular Lagrangian of GPDP into Eq. (9) in the case of spacetime translations one can get

$$\mathbf{T}^{\mathbf{G}} = \begin{pmatrix} \rho_H^G & \mathbf{j}_H^G \\ \rho_p^G & \mathbf{j}_p^G \end{pmatrix} = \begin{pmatrix} \mathfrak{L}^G & 2(\mathbf{J} - L \cdot \nabla \Gamma) \dfrac{\partial \tilde{\Gamma}}{\partial t} \\ 0 & \mathfrak{L}^G \mathbf{I} - 2(\mathbf{J} - L \cdot \nabla \Gamma) \otimes \nabla \Gamma \end{pmatrix}. \tag{23}$$

It is easy to recognize that along the particular motions all components of the above matrix are zero, due to the particular, Gaussian form of the Lagrangian. This is not very surprising if we recall that the GPDP is a variational principle for the constitutive functions, to deal with pure dissipation and the fundamental balances – expressing not material but spacetime relations – are subsidiary conditions in the principle.

7. Discussion

We have given an absolute formulation of the traditional Lagrange formalism in nonrelativistic spacetime both in particle mechanics and in field theory. Coordinatized expressions appeared only for comparing our formulae with those of usual treatments.

The treatment in relativistic spacetime is completely analogous to the one given here (see Ref. [15] for more details).

We emphasize that symmetries have nothing to do with coordinates. One frequently says that the spacetime symmetries mean that the origin of the coordinate system and the direction of the coordinate axis can be chosen arbitrarily. This has to be replaced with the assertion that a spacetime symmetry means: if the system is translated and rotated in spacetime (i.e. a spacetime isomorphism is applied to it) then we get the same (or an equivalent) system.

The Lagrange formalism in particle mechanics is based on the 7D evolution space, an absolute object, while the Hamiltonian formalism is based on the 6D phase space, a relative object, obtained from the evolution space in such a way that time is ruled out according to an observer.

The Lagrangian formalism in field theory uses the (absolute) derivative of field quantities, while the Hamiltonian formalism is based on the use of time derivatives (according to an observer).

The formulation of objectivity by an absolute formalism is fundamentally important because it shows clearly that Hamiltonian formalism, arising from the Lagrange formalism, is inevitably relative (frame dependent). As a consequence several important methods and concepts in physics are relative, e.g. canonical quantization, phase space, etc. This is a fact that requires further investigations.

Furthermore, we have investigated the possibility to consider symmetry principles with different variational principle construction methods. The method of variable transformations was treated using the example of Maxwell equations. We have concluded that the requirements of physically meaningful Noether balances can fix the freedom of the construction (the arbitrary function f), and this freedom can help us to develop the existing balances to treat more general problems, e.g. to find variational principles for the Maxwell equations in materials with nonlinear polarization.

As a different example we have investigated the possible role of spacetime symmetries in the case of the GPDP, a non-Hamiltonian variational principle with complex structure. Our Lagrangian-based formulation of symmetry principles showed that they can be applied in the case of modified function spaces, too. We found that the objectivity of the principle (a possibility to formulate in a frame-independent manner) is questionable. The resulted energy-momentum is identically zero because of the quadratic Gaussian structure of the principle. These properties seem to be physically reasonable for dissipative systems.

Finally, let us recall that the original problem was to find methods to build a whole theory given by the symmetry requirements beyond the governing equations. In our investigations the starting point was a variational principle and the fundamental balances were consequences of that.

$$\text{Hamilton principle} + \text{symmetry requirements} \Rightarrow \text{fundamental balances.}$$

However, one may ask: why should we use a Lagrangian dynamics? What kind of systems are excluded if we restrict ourselves to governing equations that are deduced from a variational principle?

One may think on the contrary: let us start from the fundamental balances and find such constitutive functions that lead to governing equations that can be derived from a variational principle. Here the Second Law should play a decisive role: in the case of special systems (internal variable dynamics [35], generalized, weakly nonlocal fluids [36]) and zero dissipation (entropy production) the governing equations can be derived from a Hamiltonian variational principle. Moreover, the nondissipative part can be *calculated* considering that condition. That is, one can get the Lagrangian from the structure of the governing balances. One can investigate in general

$$\text{Fundamental balances} + \text{Second Law} \Rightarrow \text{Hamilton principle?}$$

Acknowledgements

This research was supported by OTKA T034715 and T034603.

References

[1] Ván, P. and Muschik, W. (1995), The structure of variational principles in nonequilibrium thermodynamics, Phys. Rev. E 52(4), 3584–3590.
[2] Ván, P. and Nyíri, B. (1999), Hamilton formalism and variational principle construction, Ann. Phys (Leipzig) 8, 331–354.
[3] Anthony, K.H. (1997), Remarks on dissipation in Lagrange formalism and attempts to overcome the difficulties. In Proceedings of Second Workshop on Dissipation in Physical Systems, Borków, pp. 55–77.
[4] Lopuszanski, J. (1999), The Inverse Variational Problem in Classical Mechanics, World Scientific, Singapore.
[5] Havas, P. (1973), The connection between conservation laws and invariance groups: folklore, fiction, fact, Acta Phys. Aust. 38, 145–167.
[6] Anthony, K.-H. (2001), Hamilton's action principle and thermodynamics of irreversible processes—a unifying procedure for reversible and irreversible processes, J. Non-Newtonian Fluid Mech. 96, 291–339.
[7] Sieniutycz, S. and Berry, R.S. (1989), Conservation laws from Hamilton's principle for nonlocal thermodynamic equilibrium fluids with heat flow, Phys. Rev. E 40(1), 348–361.
[8] Sieniutycz, S. (1994), Conservation Laws in Variational Thermo-Hydrodynamics (Mathematics and its Applications, V279), Kluwer Academic Publishers, Berlin.
[9] Gambár, K. and Márkus, F. (1993), On the global symmetry of thermodynamics and Onsager's reciprocity relations, J. Non-equilib. Thermodyn. 18, 51–57.
[10] Gambár, K. and Márkus, F. (1994), Hamilton-Lagrange formalism of nonequilibrium thermodynamics, Phys. Rev. E 50(2), 1227–1231.
[11] Gambár, K., Martinás, K. and Márkus, F. (1997), Examination of phenomenological coefficient matrices within the canonical model of field theory of thermodynamics, Phys. Rev. E 55(5), 5581–5586.
[12] Matolcsi, T. (1986), On material frame-indifference, Arch. Rational Mech. Anal. 91(2), 99–118.
[13] Landau, L.D. and Lifshitz, E.M. (1976), Mechanics. In Course of Theoretical Physics, 3rd edn, Vol. 1, Pergamon Press, Oxford.
[14] Landau, L.D. and Lifshitz, E.M. (1959), The Classical Theory of Fields. In Course of Theoretical Physics, 2nd edn, Vol. 2, Pergamon Press, Oxford.
[15] M. Balázs, P. Ván, Lagrange formalism of point-masses (mat-ph/0205040), 2002.
[16] Weyl, H. (1922), Space-Time-Matter, Dover, New York.
[17] Arnold, V.I. (1978), Mathematical Methods of Classical Mechanics, Springer, Berlin.
[18] Friedman, M. (1983), Foundations of Space-Time Theories. In Relativistic Physics and Philosophy of Science, Princeton University Press, Princeton, New Jersey.

[19] Matolcsi, T. (1993), Spacetime Without Reference Frames, Akadémiai Kiadó Publishing House of the Hungarian Academy of Sciences, Budapest.
[20] Choquet-Bruhat, Y., DeWitt-Morette, C. and Dillard-Bleick, M. (1982), Analysis, Manifolds and Physics, 2nd, revised edn, North-Holland Publishing Company, Amsterdam.
[21] Rund, H. (1966), The Hamilton-Jacobi Theory in the Calculus of Variations, D. Van Nostrand Company Ltd, London.
[22] T. Fülöp, S.D. Katz. (1998), A frame and gauge free formulation of quantum mechanics. quant-ph/9806067.
[23] Sudarshan, E.C.G. and Mukunda, N. (1974), Classical Dynamics: A Modern Perspective, Wiley, New York, etc.
[24] Márkus, F. and Gambár, K. (1991), A variational principle in thermodynamics, J. Non-equilib. Thermodyn. 16(1), 27–31.
[25] Nyíri, B. (1991), On the construction of potentials and variational principles in thermodynamics and physics, J. Non-equilib. Thermodyn. 16, 39–55.
[26] Minkowski, H. (1908), Nachr. Ges. Wiss. Göttingen, 53.
[27] Gyarmati, I. (1969), On the Governing Principle of Dissipative Processes and its extension to nonlinear problems, Ann. Phys. 23, 353–378.
[28] Gyarmati, I. (1970), Non-equilibrium Thermodynamics. Field Theory and Variational Principles, Springer, Berlin.
[29] Verhás, J. (1997), Thermodynamics and Rheology, Akadémiai Kiadó and Kluwer Academic Publishers, Budapest.
[30] Stark, A. (1974), Approximation methods for the solution of heat conduction problems using Gyarmati's principle, Ann. Phys. 31(1), 53–75.
[31] Mészáros, Cs., Farkas, Á. and Bálint, I. (2001), A new application of percolation theory for coupled transport phenomena through porous media, Math. Comput. Simulation 56(4-5), 395–404.
[32] Ván, P. (1996), On the structure of the 'Governing Principle of Dissipative Processes', J. Non-equilibr. Thermodyn. 21(1), 17–29.
[33] Verhás, J. (1967), On the Gyarmati principle and the canonical equations of thermodynamics, Ann. Phys. 20(1/2), 90–94.
[34] Vojta, G. (1967), Hamiltonian formalism in the thermodynamic theory of irreversible processes in continuous systems, Acta Chim. Hungarica 54(1), 55–64.
[35] Ván, P. (2002), Weakly nonlocal continuum physics—the Ginzburg-Landau equation (cond-mat/0210402).
[36] Ván, P. and Fulop, T. (2003), Weakly nonlocal fluid mechanics — the Schrödinger equation (quant-ph/0304062).

Chapter 4

SEMI-INVERSE METHOD FOR ESTABLISHMENT OF VARIATIONAL THEORY FOR INCREMENTAL THERMOELASTICITY WITH VOIDS

Ji-Huan He and Juan Zhang

College of Science, Donghua University, 1882 Yan'an Xilu Road, P.O. Box 471, Shanghai 200051, People's Republic of China

Abstract

The semi-inverse method applied to the search for variational formulae directly from the field equations and boundary conditions is introduced, a family of variational principles for thermoelasticity with voids is deduced systematically without using the Lagrange multiplier method.

Keywords: Variational principle; semi-inverse method; incremental thermoelasticity

1. A brief introduction to inverse method of calculus of variations

As pointed out by Noor [1], variational methods such as Rayleigh–Ritz and Bubnov–Galerkin techniques have been, and continue to be, popular tools for nonlinear analysis. When contrasted with finite-element methods, direct variational methods combine the following two advantages [1]: (1) they provide physical insight into the nature of the solution of the problem; and (2) they generally have a higher rate of (asymptotic) convergence and result in a much smaller system of (nonsparse) equations. At present and despite decades of concentrated effort, we do not have a rational method that could deduce a variational formula for a real-life problem directly from field equations and boundary conditions/initial conditions. In this chapter, we will introduce the semi-inverse method applied to the search for various variational formulae for the discussed problem.

To begin our study, we give a brief introduction to the inverse method of calculus of variations. We consider Newton's law of motion

$$m\ddot{\mathbf{r}} - \mathbf{F} = 0. \tag{1}$$

The first step in the derivation consists in taking the scalar (dot) product of the variation $\delta\mathbf{r}$ into Eq. (1), to obtain the relation ([2], p. 148):

$$\int_{t_1}^{t_2} (m\ddot{\mathbf{r}} - \mathbf{F})\cdot\delta\mathbf{r} = 0. \tag{2}$$

E-mail address: jhhe@dhu.edu.cn (J.-H. He); zhangjuan@dhu.edu.cn (J. Zhang).

Integrating by parts, and noting that $\delta\mathbf{r}|_{t=t_1} = \delta\mathbf{r}|_{t=t_2} = 0$, we obtain Hamilton's principle in its most general form

$$\int_{t_1}^{t_2} (\delta T + \mathbf{F}\cdot\delta\mathbf{r}) = 0, \qquad T = \frac{1}{2}m\dot{\mathbf{r}}^2. \tag{3}$$

Let us consider a general ordinary differential equation

$$\Phi(t, x, \dot{x}, \ddot{x}) = 0. \tag{4}$$

Due to the dissipation term ($\dot{x}$), it is not easy to obtain Eq. (4) from a functional:

$$J(x) = \int_{t_0}^{t_1} F(t, x, \dot{x})\mathrm{d}t. \tag{5}$$

However, Eq. (4) admits a Lagrangian if the following relation holds [3]

$$(G\Phi)_{\dot{y}} - \frac{\mathrm{d}}{\mathrm{d}t}(G\Phi)_{\ddot{y}} = 0, \tag{6}$$

where G is a nonzero factor, the subscript denotes the differential with respect to the variable.

The proof is straightforward, the detailed proof can be found in Ref. [3].

Consider an example

$$\ddot{x} + p(t)\dot{x} + q(t)x = 0. \tag{7}$$

According to the above theorem, an energy integration requires

$$(G\Phi)_{\dot{x}} - \frac{\mathrm{d}}{\mathrm{d}t}(G\Phi)_{\ddot{x}} = Gp - \dot{G} = 0, \tag{8}$$

from which the factor G can be identified as

$$G = \exp\left(\int p\mathrm{d}t\right). \tag{9}$$

Its energy functional reads as

$$J = \frac{1}{2}\int(-G\dot{x}^2 + Gqx^2)\mathrm{d}t. \tag{10}$$

Now consider the Laplace equation ([2], p. 138)

$$\nabla^2 u = 0. \tag{11}$$

Suppose that u satisfies Eq. (11) everywhere in a region R, and takes on prescribed values on the boundary of that region. We may then multiply both sides of Eq. (11) by any continuously differentiable variation δu that vanishes on the boundary of R, and integrate the results over R to obtain

$$\iiint_R \delta u\cdot\nabla^2 u\mathrm{d}x\mathrm{d}y\mathrm{d}z = 0. \tag{12}$$

The equation is transformed, by integration by parts, into the following equation

$$\delta \iiint_R \frac{1}{2}(u_x^2 + u_y^2 + u_z^2)\mathrm{d}x\mathrm{d}y\mathrm{d}z = 0. \tag{13}$$

We, therefore, obtain the following functional

$$J = \iiint_R \frac{1}{2}(u_x^2 + u_y^2 + u_z^2)\mathrm{d}x\mathrm{d}y\mathrm{d}z. \tag{14}$$

2. The semi-inverse method

The semi-inverse method was first proposed in 1997 [4].The method has been proven to be effective and convenient to search for needed generalized variational principles from field equations and boundary conditions. An advantage of the suggested method is that it can provide a powerful mathematical tool to establishing a family of variational principles for a rather wide class of physical problems without using the Lagrange multipliers that might fail due to the variational crises (see Eqs. (61) and (69) or Ref. [3,5,6]).

To best illustrate the basic idea of the semi-inverse method, we consider the 2D incompressible inviscid potential flow. The governing equations are

$$\frac{\partial u}{\partial x} + \frac{\partial v}{\partial y} = 0, \tag{15}$$

$$\frac{\partial v}{\partial x} - \frac{\partial u}{\partial y} = 0. \tag{16}$$

In view of the semi-inverse method, we can suppose that there exists an unknown functional [7]

$$J = \iint F\mathrm{d}x\mathrm{d}y, \tag{17}$$

under the constraint of Eq. (15). By the Lagrange multiplier method, we have

$$J(u, v, \Phi) = \iint \left\{ F + \Phi\left(\frac{\partial u}{\partial x} + \frac{\partial v}{\partial y} \right) \right\} \mathrm{d}x\mathrm{d}y, \tag{18}$$

where Φ is a Lagrange multiplier.

The stationary conditions of the above functional are Eq. (15) and the following

$$\frac{\delta F}{\delta u} - \frac{\partial \Phi}{\partial x} = 0, \tag{19}$$

$$\frac{\delta F}{\delta v} - \frac{\partial \Phi}{\partial y} = 0. \tag{20}$$

Here $\delta F/\delta u$ is called the variational derivative of F with respect to u, defined as

$$\frac{\delta F}{\delta u} = \frac{\partial F}{\partial u} - \frac{\partial}{\partial x}\left(\frac{\partial F}{\partial u_x}\right) - \frac{\partial}{\partial y}\left(\frac{\partial F}{\partial u_y}\right).$$

From Eqs. (19) and (20), we have

$$\frac{\partial}{\partial y}\left(\frac{\delta F}{\delta u}\right) - \frac{\partial}{\partial x}\left(\frac{\delta F}{\delta v}\right) = 0, \tag{21}$$

which should be the field equation, Eq. (16), so we set

$$\frac{\delta F}{\delta u} = u \quad \text{and} \quad \frac{\delta F}{\delta v} = v, \tag{22}$$

from which we identify the unknown F as follows

$$F = \frac{1}{2}(u^2 + v^2). \tag{23}$$

Hence, we obtain the following variational principle:

$$J(\Phi) = \iint \frac{1}{2}(u^2 + v^2)\mathrm{d}x\mathrm{d}y, \tag{24}$$

which is under the constraint of Eq. (16), or equivalently the following equations:

$$\frac{\partial \Phi}{\partial x} = u \tag{25a}$$

and

$$\frac{\partial \Phi}{\partial y} = v. \tag{25b}$$

From physical understanding Φ is the potential function.

The generalized variational principle reads

$$J(u, v, \Phi) = \iint \left\{\frac{1}{2}(u^2 + v^2) + \Phi\left(\frac{\partial u}{\partial x} + \frac{\partial v}{\partial y}\right)\right\}\mathrm{d}x\mathrm{d}y. \tag{26}$$

Integrating by parts, and neglecting the boundary terms, we obtain

$$J(u, v, \Phi) = \iint \left\{\frac{1}{2}(u^2 + v^2) - u\frac{\partial \Phi}{\partial x} - v\frac{\partial \Phi}{\partial y}\right\}\mathrm{d}x\mathrm{d}y. \tag{27}$$

To search for a generalized variational principle, we always begin with an energy-like trial functional with an unknown function F. For example, we can construct the following trial functional

$$J(u, v, \Phi) = \iint \left\{u\frac{\partial \Phi}{\partial x} + v\frac{\partial \Phi}{\partial y} + F\right\}\mathrm{d}x\mathrm{d}y, \tag{28}$$

where Φ is the potential function, F is an unknown function of u, v, and/or their derivatives.

The advantage of the above trial functional, Eq. (28), is that the stationary condition with respect to Φ results in one of the field equations, Eq. (15). There exist many alternative approaches to construction of trial functionals, illustrative examples can be found in Refs. [8–17]. Calculating the variation of functional (28) with respect to u and v, we have

$$\frac{\partial \Phi}{\partial x} + \frac{\delta F}{\delta u} = 0, \tag{29}$$

$$\frac{\partial \Phi}{\partial y} + \frac{\delta F}{\delta u} = 0. \tag{30}$$

We search for such an F so that the above two equations turn out to be Eqs. (25a) and (25b), respectively, to this end, we write $F = -(u^2 + v^2)/2$. We, therefore, obtain the following functional:

$$J(u, v, \Phi) = \int\int\left\{u\frac{\partial \Phi}{\partial x} + v\frac{\partial \Phi}{\partial y} - \frac{1}{2}(u^2 + v^2)\right\}\mathrm{d}x\mathrm{d}y. \tag{31}$$

Now consider the constrained functional (25a) and (25b), by the Lagrange multipliers, we have

$$J(\Phi, u, v, \lambda_1, \lambda_2) = \int\int\left\{\frac{1}{2}(u^2 + v^2) + \lambda_1\left(u - \frac{\partial \Phi}{\partial x}\right) + \lambda_2\left(v - \frac{\partial \Phi}{\partial y}\right)\right\}\mathrm{d}x\mathrm{d}y. \tag{32}$$

Generally speaking the multipliers, after identification, can be expressed in the functions of the original fields, u, v and Φ (in this problem $\lambda_1 = -u$, $\lambda_2 = -v$). The semi-inverse method introduces an unknown F to replace the terms including the multipliers, i.e.

$$F = \lambda_1\left(u - \frac{\partial \Phi}{\partial x}\right) + \lambda_2\left(v - \frac{\partial \Phi}{\partial y}\right), \tag{33}$$

where F is an unknown function of u, v, Φ and/or there derivatives, i.e.

$$F = F(\Phi, u, v, \Phi_x, \Phi_y, u_x, u_y, v_x, v_y, \ldots). \tag{34}$$

So we can construct the following trial functional:

$$J(\Phi, u, v) = \int\int\left\{\frac{1}{2}(u^2 + v^2) + F\right\}\mathrm{d}x\mathrm{d}y. \tag{35}$$

Now making the above trial functional stationary with respect to u, v and Φ, we have

$$\delta_u J = \int\int(u + \delta_u F)\delta u\mathrm{d}x\mathrm{d}y = 0, \tag{36}$$

$$\delta_v J = \int\int(v + \delta_v F)\delta v\mathrm{d}x\mathrm{d}y = 0, \tag{37}$$

$$\delta_\Phi J = \int\int \delta_\Phi F\delta\Phi\mathrm{d}x\mathrm{d}y = 0, \tag{38}$$

leading to the following Euler–Lagrange equations

$$u + F_u = 0, \tag{39}$$

$$v + F_v = 0, \tag{40}$$

$$F_\Phi = 0. \tag{41}$$

where F_Φ denotes the variational derivative with respect to Φ, i.e.

$$F_\Phi = \delta_\Phi F = \frac{\delta F}{\delta \Phi} = \frac{\partial F}{\partial \Phi} - \frac{\partial}{\partial x}\left(\frac{\partial F}{\partial \Phi_x}\right) - \frac{\partial}{\partial y}\left(\frac{\partial F}{\partial \Phi_y}\right).$$

We search for such an F so that Eqs. (39)–(41) satisfy the field equations (15), (25a) and (25b), respectively. If we assume that

$$F_u = -\Phi_x \quad \text{and} \quad F_v = -\Phi_y, \tag{42}$$

then Eqs. (39) and (40) turn out to be the field equations (25a) and (25b). From the above relation, we can identify F in the form

$$F = -u\Phi_x - v\Phi_y + F_1(\Phi), \tag{43}$$

where F_1 is a newly introduced function of Φ and/or its derivative. So the trial functional (35) is updated as

$$J(\Phi, u, v) = \iint \frac{1}{2}(u^2 + v^2 - u\Phi_x - v\Phi_y + F_1)\mathrm{d}x\mathrm{d}y. \tag{44}$$

Now the stationary condition with respect to Φ reads as

$$u_x + v_y + F_{1\Phi} = 0, \tag{45}$$

which should be the field equation (15), so we set $F_{1\Phi} = 0$, i.e. $F_1 = 0$. We, therefore, obtain the following generalized functional

$$J(\Phi, u, v) = \iint \frac{1}{2}(u^2 + v^2 - u\Phi_x - v\Phi_y)\mathrm{d}x\mathrm{d}y. \tag{46}$$

We can also begin with the following trial functional

$$J(\Phi, u, v) = \iint \left\{\frac{1}{2}(\Phi_x^2 + \Phi_y^2) + F\right\}\mathrm{d}x\mathrm{d}y. \tag{47}$$

Its Euler–Lagrange equations read

$$F_u = 0, \tag{48}$$

$$F_v = 0, \tag{49}$$

$$-\Phi_{xx} - \Phi_{yy} + F_\Phi = 0. \tag{50}$$

If we set

$$F_u = C_1(\Phi_x - u), \tag{51}$$

$$F_v = C_2(\Phi_y - v), \tag{52}$$

where C_1 and C_2 are constants, then Eqs. (48) and (49) become two field equations. From Eqs. (51) and (52), we can identify F in the form

$$F = C_1\left(u\Phi_x - \frac{1}{2}u^2\right) + C_2\left(v\Phi_y - \frac{1}{2}v^2\right). \tag{53}$$

Substituting Eq. (149) into Eq. (50), we have

$$-\Phi_{xx} - \Phi_{yy} - C_1 u_x - C_2 v_y = 0. \tag{54}$$

The matching condition of Eq. (15) requires that $C_1 = C_2 = C \neq -1$, so the following generalized variational principle is obtained

$$J(\Phi, u, v) = \iint \frac{1}{2}(\Phi_x^2 + \Phi_y^2)\mathrm{d}x\mathrm{d}y + C\iint\left\{\left(u\Phi_x - \frac{1}{2}u^2\right) + \left(v\Phi_y - \frac{1}{2}v^2\right)\right\}\mathrm{d}x\mathrm{d}y, \tag{55}$$

where C is a nonzero constant but $C \neq -1$.

Though the Lagrange multiplier method is widely applied to eliminating constraints, it might be invalid for some special cases. Now consider a constrained functional

$$J(\Phi) = \iint\left\{\frac{1}{2}(u\Phi_x + v\Phi_y)\right\}\mathrm{d}x\mathrm{d}y, \tag{56}$$

which is under the constraints of Eqs. (25a) and (25b). By the Lagrange multipliers, we have

$$J(\Phi, u, v, \lambda, \eta) = \iint \frac{1}{2}(u\Phi_x + v\Phi_y)\mathrm{d}x\mathrm{d}y + \iint[\lambda(\Phi_x - u) + \eta(\Phi_y - v)]\mathrm{d}x\mathrm{d}y. \tag{57}$$

Calculating the first variation with respect to u and v results in

$$\delta_u J = \iint\left\{\frac{1}{2}\Phi_x - \lambda\right\}\delta u\mathrm{d}x\mathrm{d}y = 0, \tag{58}$$

$$\delta_v J = \iint\left\{\frac{1}{2}\Phi_y - \eta\right\}\delta v\mathrm{d}x\mathrm{d}y = 0, \tag{59}$$

so we identify the multipliers in the forms

$$\lambda = \frac{1}{2}\Phi_x \qquad \text{and} \qquad \eta = \frac{1}{2}\Phi_y. \tag{60}$$

The substitution of the identified multipliers into Eq. (57) results in

$$\begin{aligned} J(\Phi, u, v) &= \iint\left\{\frac{1}{2}(u\Phi_x + v\Phi_y) + \frac{1}{2}[\Phi_x(\Phi_x - u) + \Phi_y(\Phi_y - v)]\right\}\mathrm{d}x\mathrm{d}y \\ &= \iint \frac{1}{2}\left\{\Phi_x^2 + \Phi_y^2\right\}\mathrm{d}x\mathrm{d}y. \end{aligned} \tag{61}$$

We find the constraints have not been eliminated by the Lagrange multiplier method, leading to the second kind of variational crisis [6,18].

The variational crisis can be easily eliminated by the semi-inverse method. Constructing the following trial functional

$$J(\Phi, u, v) = \int\int\left\{\frac{1}{2}(u\Phi_x + v\Phi_y) + F\right\}\mathrm{d}x\mathrm{d}y, \tag{62}$$

we obtain the following Euler–Lagrange equations

$$\frac{1}{2}\Phi_x + F_u = 0, \tag{63}$$

$$\frac{1}{2}\Phi_y + F_v = 0, \tag{64}$$

$$-\frac{1}{2}(u_x + v_y) + F_\Phi = 0. \tag{65}$$

If we set $F_u = -u/2$, $F_v = -v/2$ and $F_\Phi = 0$, then the above equations turn out to be the field equations. We, therefore, immediately identify the unknown F in the form

$$F = -\frac{1}{4}(u^2 + v^2), \tag{66}$$

leading to the required generalized variational principle

$$J(\Phi, u, v) = \int\int\left\{\frac{1}{2}(u\Phi_x + v\Phi_y) - \frac{1}{4}(u^2 + v^2)\right\}\mathrm{d}x\mathrm{d}y. \tag{67}$$

Now consider the following functional

$$J(\Phi, u) = \int\int\left\{-\frac{1}{2}u^2 + u\Phi_x + \frac{1}{2}\Phi_y^2\right\}\mathrm{d}x\mathrm{d}y, \tag{68}$$

which is obtained from the generalized variational principle, Eq. (46), by constraining it with the field equation (25b). Elimination of the constraint by the Lagrange multiplier results in

$$J(\Phi, u, v, \lambda) = \int\int\left\{-\frac{1}{2}u^2 + u\Phi_x + \frac{1}{2}\Phi_y^2 + \lambda(\Phi_y - v)\right\}\mathrm{d}x\mathrm{d}y. \tag{69}$$

Calculating the variation with respect to v results in $\lambda = 0$. To eliminate the variational crisis ($\lambda = 0$), we construct the following trial functional

$$J(\Phi, u, v) = \int\int\left\{-\frac{1}{2}u^2 + u\Phi_x + \frac{1}{2}\Phi_y^2 + F\right\}\mathrm{d}x\mathrm{d}y, \tag{70}$$

its Euler–Lagrange equations read

$$F_v = 0, \tag{71}$$

$$-u + \Phi_x + F_u = 0, \tag{72}$$

$$-u_x - \Phi_{yy} + F_\Phi = 0. \tag{73}$$

We set

$$F_v = C_2(\Phi_y - v), \quad (C_2 \neq 0), \tag{74}$$

$$F_u = C_1(\Phi_x - u), \quad (C_1 \neq -1). \tag{75}$$

From Eqs. (74) and (75), we can identify F in the form

$$F = C_1\left(u\Phi_x - \frac{1}{2}u^2\right) + C_2\left(v\Phi_y - \frac{1}{2}v^2\right) + F_1, \tag{76}$$

where F_1 is a function of Φ and/or its derivatives.

Substituting Eq. (76) into Eq. (73) yields

$$-u_x - \Phi_{yy} - C_1 u_x - C_2 v_y + \delta_\Phi F_1 = 0, \tag{77}$$

which should satisfy the left field equation, Eq. (15), this requires that $C_1 = C_2 = C \neq -1$ and $F_1 = 0$, accordingly the following generalized variational principle is obtained

$$J(\Phi, u, v) = \int\int\left\{-\frac{1}{2}u^2 + u\Phi_x + \frac{1}{2}\Phi_y^2\right\}\mathrm{d}x\mathrm{d}y + C\int\int\left\{u\Phi_x + v\Phi_y - \frac{1}{2}(u^2 + v^2)\right\}\mathrm{d}x\mathrm{d}y, \tag{78}$$

where C is a nonzero constant, but $C \neq -1$.

The semi-inverse method is also used to introduce some special functions, we consider 3D incompressible potential flow:

$$\begin{cases} \dfrac{\partial u}{\partial x} + \dfrac{\partial v}{\partial y} + \dfrac{\partial w}{\partial z} = 0, \\ \dfrac{\partial v}{\partial x} - \dfrac{\partial u}{\partial y} = 0, \\ \dfrac{\partial u}{\partial z} - \dfrac{\partial w}{\partial x} = 0. \end{cases} \tag{79}$$

In view of the semi-inverse method, we write a trial functional in the form

$$J(u, v, w, \mu_1, \mu_2) = \int\int\int\left\{F + \mu_1\left(\frac{\partial v}{\partial x} - \frac{\partial u}{\partial y}\right) + \mu_2\left(\frac{\partial u}{\partial z} - \frac{\partial w}{\partial x}\right)\right\}\mathrm{d}V. \tag{80}$$

The stationary conditions of the functional, Eq. (80), read

$$\begin{cases} \dfrac{\delta F}{\delta u} + \dfrac{\partial \mu_1}{\partial y} - \dfrac{\partial \mu_2}{\partial z} = 0, \\ \dfrac{\delta F}{\delta v} - \dfrac{\partial \mu_1}{\partial x} = 0, \\ \dfrac{\delta F}{\delta w} + \dfrac{\partial \mu_1}{\partial x} = 0. \end{cases} \tag{81}$$

Note that

$$\frac{\partial}{\partial x}\left(\frac{\delta F}{\delta u}\right)+\frac{\partial}{\partial y}\left(\frac{\delta F}{\delta v}\right)+\frac{\partial}{\partial z}\left(\frac{\delta F}{\delta w}\right)=0, \tag{82}$$

which should be the first equation of Eq. (79), we, therefore, set

$$\frac{\delta F}{\delta u}=u,\quad \frac{\delta F}{\delta v}=v \text{ and } \frac{\delta F}{\delta w}=w, \tag{83}$$

from which we identify the unknown F as follows

$$F=\frac{1}{2}(u^2+v^2+w^2). \tag{84}$$

So we obtain the following variational principle

$$J=\iiint \frac{1}{2}(u^2+v^2+w^2)\mathrm{d}V, \tag{85}$$

and the following generalized variational principle

$$J(u,v,w,\mu_1,\mu_2)=\iiint\left\{\frac{1}{2}(u^2+v^2+w^2)+\mu_1\left(\frac{\partial v}{\partial x}-\frac{\partial u}{\partial y}\right)+\mu_2\left(\frac{\partial u}{\partial z}-\frac{\partial w}{\partial x}\right)\right\}\mathrm{d}V. \tag{86}$$

Hereby μ_1 and μ_2 can be considered as generalized stream functions for 3D incompressible potential flow, defined as

$$u=\frac{\partial \mu_2}{\partial z}-\frac{\partial \mu_1}{\partial y},\quad v=\frac{\partial \mu_1}{\partial x},\quad w=-\frac{\partial \mu_2}{\partial z}. \tag{87}$$

3. Mathematical formulation of thermoelasticity with voids

The development of the concept of volume fractions dates back to 1794 [19]. Modern mixture theories for porous bodies emanates from the work by Goodman and Cowin [20], Nunziato and Cowin [21], and Cowin and Nunziato [22].

In recent years potential applications have brought about a renewed interest in the material with voids. Principally these applications arise in situations where the medium has porosity, e.g. rock and soil, bone, manufactured porous materials. This theory enables us to analyze the behavior of granular and porous solids.

The original idea has been further generalized and studied by many authors. Iesan [23] introduced a linear theory of thermoelastic materials with voids and established uniqueness, reciprocal and variational theorems; Pi and Jin [24] obtained a generalized variational principle for static linear elastic materials with voids; Ciarletta and Scalia [25] studied a linear thermoelastic theory of material with voids, and established uniqueness and reciprocal theorems; Dhaliwal and Wang [26] proposed a domain of influence theorem for the linear theory of elastic materials with voids; Martinez and Quintanilla [27] applied the theory of semigroups of linear operators to obtain some results on existence and uniqueness of incremental thermoelasticity with voids; Martinez and Quintanilla [28] also studied uniqueness, existence and asymptotic behavior for viscoelasticity with voids. A comprehensive account of the present state of the subject, with numerous related references has been given by Ciarletta and Iesan [29] and Schrefler [30].

The rapid development of computer science and the finite-element applications reveals the importance of searching for a classical variational principle for the discussed problem, which is the theoretical basis of the finite-element methods [31]. Furthermore, such variational formulations have served as a basis for development of a variety of approximate methods of analysis. The recent research reveals that variational theory is also a powerful tool for the meshless method or the element-free method [32].

However, the thermoelasticity with voids has not been frequently viewed from a variational point of view. In this chapter, we will apply the semi-inverse method to establishing a variational model for the entitled problem.

We consider a body that at time $t = 0$ occupies a regular region Ω_0 of 3D Euclidean space. Let $\partial\Omega_0$ be the boundary of Ω_0. We refer the motion to the reference configuration Ω_0 and a fixed system of rectangular Cartesian axes $Ox_i(i = 1\text{–}3)$.

Now we consider two other states: the primary state Ω, and the secondary state Ω^*. We designate as incremental quantities associated with the difference of motions between the current states. If the point $\mathbf{X}$ in the reference configuration Ω_0 moves to $\mathbf{x}$ in the primary state Ω and to $\mathbf{y}$ in the secondary state Ω^*, then $\mathbf{u} = \mathbf{y} - \mathbf{x}$ is the incremental displacement. An incremental quantity, Ξ^{I}, is defined as $\Xi^{\mathrm{I}} = \Xi^* - \Xi$, the quantity at secondary state is denoted by an asterisk, and Ξ is the quantity at primary state (we ignore the superscript I in the equations below).

The governing equations for linear thermoelasticity with voids, using the Einstein summation and differentiation conventions, take the following forms [28]:

(1) Equations of motion (balance of linear momentum)

$$\sigma_{ij,j} + \rho_0 f_i = \rho_0 \ddot{u}_i, \tag{88}$$

where σ_{ij} is the incremental first Piola–Kirchhoff stress tensor, $\sigma_{ij,j} = \partial\sigma_{ij}/\partial x_j$, u_i is the incremental displacement vector, $\ddot{u}_i = \partial^2 u_i/\partial t^2$, f_i represents the incremental vector of mechanical body force, ρ_0 is the density in the reference configuration.

(2) Balance of equilibrated forces (balance of momentum associated with the porosity)

$$M_{i,i} + \chi + \rho_0 L = \rho_0 k\ddot{\varphi}, \tag{89}$$

where M_i is the incremental intrinsic equilibrated stress vector, χ is the incremental intrinsic equilibrated body force, L is the incremental extrinsic equilibrated body force, φ is the incremental volume fraction, for the physical significance of this kinematic variable, see Ref. [22], k, equilibrated inertia, accounts for the inertia associated with the presence of voids.

(3) Strain–displacement relations

$$e_{ij} = \frac{1}{2}(u_{i,j} + u_{j,i}), \tag{90}$$

where e_{ij} is the infinitesimal strain tensor.

(4) For simplicity, we introduce a potential for the incremental volume fraction:

$$\Phi_i = \varphi_{,i}. \tag{91}$$

(5) Constitutive equations

$$\sigma_{ij} = A_{ijkl}e_{kl} + D_{ijk}\Phi_k + B_{ij}\varphi - \beta_{ij}\theta, \tag{92}$$

$$M_k = D_{ijk}e_{ij} + A_{jk}\Phi_j + B_k\varphi - A_k\theta, \tag{93}$$

$$\gamma = \beta_{ij}e_{ij} + A_i\Phi_i + B\varphi + A\theta, \tag{94}$$

$$\chi = -B_{ij}e_{ij} - B_i\Phi_i - \xi\varphi + B\theta, \tag{95}$$

where γ is the incremental entropy measured per unit volume, the thermoelastic coefficients are functions of the deformation, temperature and volume fraction of the body in the primary state.

(6) Heat equation

$$\rho_0(T\dot{\gamma}^* - T^*\dot{\gamma}) = \tilde{q}_{i,i} + \tilde{Q}. \tag{96a}$$

In view of Eq. (94), the above equation becomes

$$A\dot{\theta} + \beta_{ij}\dot{e}_{ij} + A_i\dot{\Phi}_i + B\dot{\varphi} = \frac{1}{\rho_0 T^*}(T^* - \theta)\dot{\gamma}^* + (q_{i,i} + Q), \tag{96b}$$

where $\theta = T^* - T$, T and T^* are, respectively, the absolute temperature at primary and secondary states, γ^* is the entropy at the secondary state, $\tilde{Q}$ is the incremental strength of the extrinsic heat source, $\tilde{q}$ is the incremental heat flux vector. $q = -\tilde{q}/(\rho_0 T^*)$, $Q = -\tilde{Q}/(\rho_0 T^*)$.

(7) Modified Fourier's law: In order to simplify the deduction procedure, we write Fourier's law in the form [12]

$$\theta_{,i} = -K_{ij}q_j, \tag{97}$$

where K_{ij} is the heat-conduction coefficient.

3.1. Boundary conditions

On A_1 surface displacement is prescribed

$$u_i = \bar{u}_i, \text{ (on } A_1), \tag{98}$$

and on the complementary part A_2 the traction is given

$$\sigma_{ij}n_j = \bar{p}_i, \text{ (on } A_2), \tag{99}$$

where $A_1 \cup A_2 = A = \partial\Omega$ covers the total boundary surface, $\partial\Omega$.

On A_3 and A_4, we prescribe the following boundary conditions:

$$M_i n_i = \bar{M}, \text{ (on } A_3), \tag{100}$$

$$\varphi = \bar{\varphi}, \text{ (on } A_4), \tag{101}$$

where $A_3 \cup A_4 = A = \partial\Omega$ covers the total boundary surface.

On A_5, A_6 and A_7 surfaces ($A_5 \cup A_6 \cup A_7 = A = \partial\Omega$), we give the temperature conditions

$$\theta = \bar{\theta}, \text{ (on } A_5), \tag{102}$$

$$q_i n_i = \bar{q}_n, \text{ (on } A_6), \tag{103}$$

$$q_i n_i + Bi\theta = Bi\Theta(t), \text{ (on } A_7), \tag{104}$$

where Bi is the Biot number and Θ is a known function for the direct problem.

3.2. *Initial conditions*

When $t = 0$, we have the following initial conditions

$$u_i = u_i^0(x_1, x_2, x_3), \text{ (on } \Omega_0), \tag{105}$$

$$\dot{u}_i = \dot{u}_i^0(x_1, x_2, x_3), \text{ (on } \Omega_0), \tag{106}$$

$$\varphi_i = \varphi_i^0(x_1, x_2, x_3), \text{ (on } \Omega_0), \tag{107}$$

$$\dot{\varphi}_i = \dot{\varphi}_i^0(x_1, x_2, x_3), \text{ (on } \Omega_0), \tag{108}$$

$$q_i = q_i^0(x_1, x_2, x_3), \text{ (on } \Omega_0), \tag{109}$$

$$\dot{q}_i = \dot{q}_i^0(x_1, x_2, x_3), \text{ (on } \Omega_0), \tag{110}$$

where Ω_0 is the initial configuration.

Our aim is to search for a generalized variational principle, the stationary conditions of the functional satisfying all governing equations and boundary/initial conditions.

4. Variational formulas for incremental thermoelasticity with voids

The essence of the proposed method is to construct an energy-like functional with a certain unknown function, which can be identified step by step. An energy-like trial functional with eight kinds of independent variations (σ_{ij}, e_{ij}, u_i, M_i, φ, Φ, θ, and q_i) can be constructed as follows

$$J(\sigma_{ij}, e_{ij}, u_i, M_i, \varphi, \Phi, \theta, q_i) = \int_0^t \int_V \Pi \mathrm{d}V \mathrm{d}t + IB, \tag{111a}$$

where

$$IB = \sum_{k=1}^{7} \int_0^t \int_{A_k} G_k \mathrm{d}A \mathrm{d}t + \int_{\Omega_0} G_8 \mathrm{d}V, \tag{111b}$$

in which G_i ($i = 1$–8) are unknown functions, Π is a trial Lagrangian. There exist many approaches to the construction of a trial Lagrangian, illustrating examples can be found in the author's previous publications.

We begin with the following trial Lagrangian:

$$\Pi = \frac{1}{2}\rho_0 \dot{u}_i^2 + \frac{1}{2}\rho_0 k \dot{\varphi}^2 + (\sigma_{ij,j} + \rho_0 f_i)u_i + \Big(M_{i,i} - B_{ij}e_{ij} - B_i\Phi_i - \frac{1}{2}\xi\varphi + B\theta + \rho_0 L\Big)\varphi + F, \tag{112}$$

where F is an unknown function to be further determined, and it is free from the variables u_i and φ.

The stationary conditions with respect to u_i and φ, respectively, are

$$\delta u_i: \quad -\rho_0 \ddot{u}_i + \sigma_{ij,j} + \rho_0 f_i = 0, \tag{113}$$

$$\delta\varphi: \quad -\rho_0 k\ddot{\varphi} + M_{i,i} - B_{ij}e_{ij} - B_i\Phi_i - \xi\varphi + B\theta + \rho_0 L = 0. \tag{114}$$

It is obvious that Eq. (113) is the field equation (88), and in view of Eq. (95), Eq. (114) reduces to the field equation (89).

Now calculating the variation of the functional with respect to σ_{ij} and M_i, we obtain the following Euler–Lagrange equations

$$\delta\sigma_{ij}: \quad -\frac{1}{2}(u_{i,j} + u_{j,i}) + \frac{\delta F}{\delta\sigma_{ij}} = 0, \tag{115}$$

$$\delta M_i: \quad -\varphi_{,i} + \frac{\delta F}{\delta M_i} = 0. \tag{116}$$

We search for such an F that the above Euler–Lagrange equations (115) and (116) satisfy, respectively, two of field equations, i.e. Eqs. (90) and (91). Accordingly we can identify the unknown F in the form

$$F = \sigma_{ij}e_{ij} + M_k\Phi_k + F_1. \tag{117}$$

The Lagrangian, Eq. (112), therefore, can be renewed as follows

$$\Pi = \frac{1}{2}\rho_0 \dot{u}_i^2 + \frac{1}{2}\rho_0 k \dot{\varphi}^2 + (\sigma_{ij,j} + \rho_0 f_i)u_i + \Big(M_{i,i} - B_{ij}e_{ij} - B_i\Phi_i - \frac{1}{2}\xi\varphi + B\theta + \rho_0 L\Big)\varphi + \sigma_{ij}e_{ij} + M_k\Phi_k + F_1, \tag{118}$$

where F_1 is a newly introduced unknown function with less variables. The absence of the variables u_i, φ, σ_{ij} and M_i in F_1 makes the identification of F_1 simple.

By the same operation, the Euler equations for δe_{ij} and $\delta\Phi_k$ read as

$$\delta e_{ij}: \quad -B_{ij}\varphi + \sigma_{ij} + \frac{\delta F_1}{\delta e_{ij}} = 0, \tag{119}$$

$$\delta\Phi_k: \quad -B_k\varphi + M_k + \frac{\delta F_1}{\delta\Phi_k} = 0. \tag{120}$$

If we set

$$\frac{\delta F_1}{\delta e_{ij}} = B_{ij}\varphi - \sigma_{ij} = -A_{ijkl}e_{kl} - D_{ijk}\Phi_k + \beta_{ij}\theta, \tag{121}$$

and

$$\frac{\delta F_1}{\delta \Phi_k} = B_k\varphi - M_k = -D_{ijk}e_{ij} - A_{jk}\Phi_j + A_k\theta, \tag{122}$$

then the above two Euler–Lagrange equations (119) and (120) turn out to be the field equations (92) and (93), respectively.

From the above two relations, Eqs. (121) and (122), we can easily identify F_1 in the form

$$F_1 = -\frac{1}{2}A_{ijkl}e_{kl}e_{ij} - \frac{1}{2}A_{jk}\Phi_j\Phi_k - D_{ijk}\Phi_k e_{ij} + \beta_{ij}\theta e_{ij} + A_k\theta\Phi_k + F_2, \tag{123}$$

where F_2 is an unknown function of θ and q_i.

The trial Lagrangian can be further updated as

$$\begin{aligned} \Pi = {} & \frac{1}{2}\rho_0\dot{u}_i^2 + \frac{1}{2}\rho_0 k\dot{\varphi}^2 + (\sigma_{ij,j} + \rho_0 f_i)u_i \\ & + \left(M_{i,i} - B_{ij}e_{ij} - B_i\Phi_i - \frac{1}{2}\xi\varphi + B\theta + \rho_0 L\right)\varphi + \sigma_{ij}e_{ij} + M_k\Phi_k \\ & - \frac{1}{2}A_{ijkl}e_{kl}e_{ij} - \frac{1}{2}A_{jk}\Phi_j\Phi_k - D_{ijk}\Phi_k e_{ij} + \beta_{ij}\theta e_{ij} + A_k\theta\Phi_k + F_2. \end{aligned} \tag{124}$$

Now the Euler–Lagrange equation for $\delta\theta$ reads as

$$\delta\theta: \ B\varphi + \beta_{ij}e_{ij} + A_k\Phi_k + \frac{\delta F_2}{\delta\theta} = 0. \tag{125}$$

Differentiating the equation with respect to time, and applying Eq. (96b), we obtain the result

$$\frac{\mathrm{d}}{\mathrm{d}t}\left(\frac{\delta F_2}{\delta\theta}\right) = -(\beta_{ij}\dot{e}_{ij} + A_i\dot{\Phi}_i + B\dot{\varphi}) = A\dot{\theta} - \frac{1}{\rho_0 T^*}(T^* - \theta)\dot{\gamma}^* - (q_{i,i} + Q). \tag{126}$$

Integrating Eq. (126) with respect to time, we have

$$\frac{\delta F_2}{\delta\theta} = A\theta - \int_0^t \frac{1}{\rho_0 T^*}\dot{\gamma}^*(T^* - \theta)\mathrm{d}t - \int_0^t (q_{i,i} + Q)\mathrm{d}t. \tag{127}$$

In implementing the numerical simulation, time is in a discretization form $(t_0, t_1, t_2, \ldots, t_n)$. So we have approximately the following Newton–Cotes formulae

$$\int_0^{t_1} f\mathrm{d}t = \frac{1}{2}t_1[f(t_0) + f(t_1)],$$

$$\int_0^{t_2} f\mathrm{d}t = \frac{1}{6}t_2[f(t_0) + 4f(t_1) + f(t_2)],$$

$$\int_0^{t_3} f\mathrm{d}t = \frac{1}{8}t_2[f(t_0) + 3f(t_1) + 3f(t_2) + f(t_3)],$$

$$\int_0^{t_4} f\mathrm{d}t = \frac{1}{90}t_4[7f(t_0) + 32f(t_1) + 12f(t_2) + 32f(t_3) + 7f(t_4)].$$

We write $t = t_n$, and all variables are calculated before t_n, so we can write the Newton–Cotes formulation in the following general form

$$\int_0^{t} f\mathrm{d}t = \lambda t[f_0 + f(t)], \tag{128}$$

where λ is a constant depending upon the time steps, and $f_0 = f_0(t_0, t_1, t_2, \ldots, t_{n-1})$. Accordingly, Eq. (127) reduces to the form

$$\frac{\delta F_2}{\delta \theta} = \left(A + \frac{1}{\rho_0 T^*}\dot{\gamma}^*\lambda t\right)\theta - \lambda t(q_{i,i} + Q) + F_0, \tag{129}$$

where F_0 is a known function depending on the Newton–Cotes formulae, which is defined as

$$F_0 = -\frac{1}{\rho_0 T_0^*}\dot{\gamma}_0^*\lambda t(T_0^* - \theta_0 + T^*) + \frac{1}{\rho_0 T^*}\lambda t(q_{0i,i} + Q_0),$$

in which the quantities with subscript '0' are known functions derived from the Newton–Cotes formulae.

From Eq. (129), we have

$$F_2 = \frac{1}{2}\left(A + \frac{1}{\rho_0 T^*}\dot{\gamma}^*\lambda t\right)\theta^2 - \lambda t\theta(q_{i,i} + Q) + F_0\theta + F_3. \tag{130}$$

The Lagrangian, Eq. (124), can be written in a more complete form

$$\begin{aligned}\Pi = {} & \frac{1}{2}\rho_0\dot{u}_i^2 + \frac{1}{2}\rho_0 k\dot{\varphi}^2 + (\sigma_{ij,j} + \rho_0 f_i)u_i \\ & + \left(M_{i,i} - B_{ij}e_{ij} - B_i\Phi_i - \frac{1}{2}\xi\varphi + B\theta + \rho_0 L\right)\varphi + \sigma_{ij}e_{ij} + M_k\Phi_k \\ & - \frac{1}{2}A_{ijkl}e_{kl}e_{ij} - \frac{1}{2}A_{jk}\Phi_j\Phi_k - D_{ijk}\Phi_k e_{ij} + \beta_{ij}\theta e_{ij} + A_k\theta\Phi_k \\ & + \frac{1}{2}\left(A + \frac{1}{\rho_0 T^*}\dot{\gamma}^*\lambda t\right)\theta^2 - \lambda t\theta(q_{i,i} + Q) + F_0\theta + F_3,\end{aligned}$$

where F_3 is only a function of q_i.

Now the stationary condition with respect to q_i reads as

$$\delta q_i : \ \lambda t\theta_{,i} + \frac{\delta F_3}{\delta q_i} = 0. \tag{131}$$

In view of the field equation, Eq. (97), we set

$$\frac{\delta F_3}{\delta q_i} = K_{ij}\lambda t q_j, \tag{132}$$

from which the unknown F_3 can be identified as

$$F_3 = \frac{1}{2} K_{ij} \lambda t q_i q_j. \tag{133}$$

Now we finally obtain the following required Lagrangian:

$$\begin{aligned}
\Pi = & \frac{1}{2}\rho_0 \dot{u}_i^2 + \frac{1}{2}\rho_0 k \dot{\varphi}^2 + (\sigma_{ij,j} + \rho_0 f_i) u_i \\
& + \left(M_{i,i} - B_{ij} e_{ij} - B_i \Phi_i - \frac{1}{2}\xi\varphi + B\theta + \rho_0 L \right)\varphi + \sigma_{ij} e_{ij} + M_k \Phi_k \\
& - \frac{1}{2} A_{ijkl} e_{kl} e_{ij} - \frac{1}{2} A_{jk} \Phi_j \Phi_k - D_{ijk} \Phi_k e_{ij} + \beta_{ij} \theta e_{ij} + A_k \theta \Phi_k \\
& + \frac{1}{2}\left(A + \frac{1}{\rho_0 T^*} \dot{\gamma}^* \lambda t \right)\theta^2 - \lambda t \theta (q_{i,i} + Q) + F_0 \theta + \frac{1}{2} K_{ij} \lambda t q_i q_j.
\end{aligned} \tag{134}$$

The unknown functions G_i ($i = 1-8$) can be determined in the same way as illustrated in my previous publications [11,12]:

$$G_1 = -\sigma_{ij} n_j \bar{u}_i, \tag{135}$$

$$G_2 = -u_i(\sigma_{ij} n_j - \bar{p}_i), \tag{136}$$

$$G_3 = -\varphi(M_i n_i - \bar{M}), \tag{137}$$

$$G_4 = -\bar{\varphi} M_i n_i, \tag{138}$$

$$G_5 = \lambda t \bar{\theta} q_i n_i, \tag{139}$$

$$G_6 = \lambda t \theta (q_i n_i - \bar{q}_i), \tag{140}$$

$$G_7 = \lambda t \theta q_i n_i + \lambda t B i \theta \left(\frac{1}{2}\theta - \Theta \right), \tag{141}$$

$$G_8 = -\frac{1}{2}\rho_0 \dot{u}_i^0 u_i - \frac{1}{2}\rho_0 \dot{u}_i (u_i - u_i^0) - \frac{1}{2}\rho_0 k \dot{\varphi}^0 \varphi - \frac{1}{2}\rho_0 k \dot{\varphi}(\varphi - \varphi^0). \tag{142}$$

Note: $\dot{u}_i$ and $\dot{\varphi}$ in G_8 are independent variables during variation. On Ω_0 (reference configuration when $t = 0$), according to Eq. (142), we have the following Euler equations:

$$\delta u_i : \ \rho_0 \dot{u}_i - \frac{1}{2}\rho_0 \dot{u}_i^0 - \frac{1}{2}\rho_0 \dot{u}_i = 0.$$

$$\delta \dot{u}_i : \ -\frac{1}{2}\rho_0 (u_i - u_i^0) = 0$$

$$\delta \varphi : \ \rho_0 k \dot{\varphi} - \frac{1}{2}\rho_0 k \dot{\varphi}^0 - \frac{1}{2}\rho_0 k \dot{\varphi} = 0.$$

$$\delta \dot{\varphi} : \ -\frac{1}{2}\rho_0 k (\varphi - \varphi^0) = 0.$$

The initial conditions, Eqs. (105)–(108), become the natural initial conditions of the obtained variational functional, providing some convenience when implementing finite element simulations.

We introduce a generalized quadratic energy density function, π, defined as

$$\pi = \frac{1}{2}A_{ijkl}e_{kl}e_{ij} + \frac{1}{2}A_{jk}\Phi_j\Phi_k + \frac{1}{2}\xi\varphi^2 - \frac{1}{2}A\theta^2 + D_{ijk}\Phi_k e_{ij} - \beta_{ij}\theta e_{ij}$$
$$- A_k\theta\Phi_k + B_{ij}e_{ij}\varphi + B_i\Phi_i\varphi - B\theta\varphi. \tag{143}$$

From Eq. (143), we have the following relations

$$\frac{\partial \pi}{\partial e_{ij}} = A_{ijkl}e_{kl} + D_{ijk}\Phi_k - \beta_{ij}\theta + B_{ij}\varphi = \sigma_{ij}, \tag{144}$$

$$\frac{\partial \pi}{\partial \Phi_k} = A_{jk}\Phi_j + D_{ijk}e_{ij} - A_k\theta + B_k\varphi = M_k, \tag{145}$$

$$\frac{\partial \pi}{\partial \varphi} = \xi\varphi + B_{ij}e_{ij} + B_i\Phi_i - B\theta = -\chi, \tag{146}$$

$$\frac{\partial \pi}{\partial \theta} = -A\theta - \beta_{ij}e_{ij} - A_k\Phi_k - B\varphi = -\gamma. \tag{147}$$

We further introduce a complementary energy density function, Γ, defined as

$$\Gamma = \sigma_{ij}e_{ij} + M_k\Phi_k - \pi. \tag{148}$$

The obtained Lagrangian can be written in a more concise form:

$$\Pi = \frac{1}{2}\rho_0\dot{u}_i^2 + \frac{1}{2}\rho_0 k\dot{\varphi}^2 + (\sigma_{ij,j} + \rho_0 f_i)u_i + (M_{i,i} + \rho_0 L)\varphi + \sigma_{ij}e_{ij} + M_k\Phi_k$$
$$- \Pi + \frac{1}{2\rho_0 T^*}\dot{\gamma}^*\lambda t\theta^2 - \lambda t\theta(q_{i,i} + Q) + F_0\theta + \frac{1}{2}K_{ij}\lambda t q_i q_j \tag{149}$$

or

$$\Pi = \frac{1}{2}\rho_0\dot{u}_i^2 + \frac{1}{2}\rho_0 k\dot{\varphi}^2 + (\sigma_{ij,j} + \rho_0 f_i)u_i + (M_{i,i} + \rho_0 L)\varphi$$
$$+ \Gamma + \frac{1}{2\rho_0 T^*}\dot{\gamma}^*\lambda t\theta^2 - \lambda t\theta(q_{i,i} + Q) + F_0\theta + \frac{1}{2}K_{ij}\lambda t q_i q_j. \tag{150}$$

Further eliminating the constraints of Eqs. (94), (95), (109) and (110), we have the following generalized functional:

$$J(\sigma_{ij}, e_{ij}, u_i, M_i, \varphi, \Phi, \theta, q_i, \gamma, \chi)$$
$$= \int_0^t \int_\Omega \tilde{\Pi} \mathrm{d}V\mathrm{d}t + \sum_{k=1}^{7} \int_0^t \int_{A_k} G_k \mathrm{d}A\mathrm{d}t + \int_{\Omega_0} G_8 \mathrm{d}V + \int_0^t \int_{A_7} [c(\dot{q}_i - \dot{q}_i^0)^2$$
$$+ \mathrm{d}(q_i - q_i^0)^2]\mathrm{d}A\mathrm{d}t, \tag{151}$$

where G_i ($i = 1$–8) are defined, respectively, as Eqs. (135)–(142), and $\tilde{\Pi}$ is defined in the form

$$\tilde{\Pi} = \Pi + a(\beta_{ij}e_{ij} + A_i\Phi_i + B\varphi + A\theta - \gamma)^2 + b(\chi + B_{ij}e_{ij} + B_i\Phi_i + \xi\varphi - B\theta)^2, \quad (152)$$

where a, b, c and d are nonzero constants.

Constraining the generalized Lagrangian (149) or (150) by selected field equations yields various variational formulations under constraints [33,34]. We only write the most important Hamilton principle

$$J(u_i, \varphi) = \int_0^t \int_\Omega (T + \rho_0 f_i u_i + \rho_0 L\varphi - \pi)\mathrm{d}V\mathrm{d}t + \sum_{k=1}^{7} \int_0^t \int_{A_k} G_k \mathrm{d}A\mathrm{d}t$$
$$+ \int_{\Omega_0} G_8 \mathrm{d}V + \int_0^t \int_{A_7} c(\dot{q}_i - \dot{q}_i^0)^2 + \mathrm{d}(q_i - q_i^0)^2 \mathrm{d}A\mathrm{d}t, \quad (153)$$

where T is defined as

$$T = \frac{1}{2}\rho_0 \dot{u}_i^2 + \frac{1}{2}\rho_0 k\dot{\varphi}^2. \quad (154)$$

We can also derive a parameterized variational principle with the same operation as illustrated in Refs. [3,12]:

$$J(\sigma_{ij}, e_{ij}, u_i, M_i, \varphi, \Phi, \theta, q_i, \gamma, \chi) = \int_0^t \int_\Omega \hat{\Pi}\mathrm{d}V\mathrm{d}t + \sum_{k=1}^{7} \int_0^t \int_{A_k} G_k \mathrm{d}A\mathrm{d}t + \int_{\Omega_0} G_8 \mathrm{d}V$$
$$+ \int_0^t \int_{A_7} c(\dot{q}_i - \dot{q}_i^0)^2 + \mathrm{d}(q_i - q_i^0)^2 \mathrm{d}A\mathrm{d}t, \quad (155)$$

where

$$\hat{\Pi} = \frac{1}{2}\rho_0 \dot{u}_i^2 + \frac{1}{2}\rho_0 k\dot{\varphi}^2 + (\alpha\sigma_{ij,j} + \rho_0 f_i)u_i - \frac{1}{2}(1-\alpha)\sigma_{ij}(u_{i,j} + u_{j,i})$$
$$+ (\beta M_{i,i} + \rho_0 L)\varphi - (1-\beta)M_i\varphi_{,i} + \sigma_{ij}e_{ij} + M_k\Phi_k - \Pi - \lambda t\theta[F_0 + (\xi q_{i,i} + Q)]$$
$$+ (1-\xi)\lambda t\theta q_i\theta_{,i} + a(\beta_{ij}e_{ij} + A_i\Phi_i + B\varphi + A\theta - \gamma)^2$$
$$+ b(\chi + B_{ij}e_{ij} + B_i\Phi_i + \xi\varphi - B\theta)^2, \quad (156)$$

in which α, β, ξ, a and b are constants, the free parameters are very useful and advantageous in finite element simulation. It is a powerful tool for high-performance FEM derivations [33,34]. $\hat{G}_i$ $(i = 1\text{–}8)$ are, respectively, expressed in the forms:

$$\hat{G}_1 = -\sigma_{ij}n_j\bar{u}_i + (1-\alpha)\sigma_{ij}u_i n_j, \quad (157)$$

$$G_2 = -u_i(\sigma_{ij}n_j - \bar{p}_i) + (1-\alpha)\sigma_{ij}u_i n_j, \quad (158)$$

$$\hat{G}_3 = -\varphi(M_i n_i - \bar{M}) + (1-\beta)\varphi M_i n_i, \quad (159)$$

$$\hat{G}_4 = -\bar{\varphi}M_i n_i + (1-\beta)\varphi M_i n_i, \quad (160)$$

$$\hat{G}_i = G_i \, (i = 5\text{–}8). \quad (161)$$

We only prove the stationary conditions on A_1:

$$\delta\sigma_{ij}:\ \alpha u_i n_j - n_j\bar{u}_i + (1-\alpha)u_i n_j = 0, \tag{162}$$

$$\delta u_i:\ -(1-\alpha)\sigma_{ij}n_j + (1-\alpha)\sigma_{ij}n_j = 0. \tag{163}$$

The Euler equation, Eq. (162), is the boundary condition, Eq. (98), while Eq. (163) is an identity.

The pertinent functionals represent the generalizations to the case of porous thermoelastic bodies of the corresponding well-known functionals of the classical theory of linear elasticity.

5. Conclusion

Several new variational formulations for the discussed problem have been proposed. These formulations are the bases for numerical approximation techniques. Herein the generalized variational principles without or with fewer constraints are emphasized, from which various variational formulations under constraints can easily be obtained by constraining the functional by selective field equations or boundary/initial conditions.

A quadratic energy density function π and a complementary energy density function Γ are introduced, producing the obtained functionals with concise forms and physical understanding.

References

[1] Noor, A.K. (1984), Recent advances in the application of variational methods to nonlinear problems. Unification of Finite Element Methods (Kardestuncer, H., ed.), Elsevier, Netherlands, pp. 275–302.
[2] Hildebrand, F.B. (1965), Methods of Applied Mathematics, 2nd edn, Prentice-Hall, Englewood Cliffs, NJ.
[3] He, J.H. (2003), Generalized Variational Principles in Fluids (Liuti Lixue Guangyi Bianfen Yuanli), Science and Culture Publishing House of China, Hongkong, (in Chinese).
[4] He, J.H. (1997), Semi-inverse method of establishing generalized variational principles for fluid mechanics with emphasis on turbomachinery aerodynamics, Int. J. Turbo Jet Engines 14(1), 23–28.
[5] Chien, W.Z. (1983), Method of higher-order Lagrange multiplier and generalized variational principles of elasticity with more general forms, Appl. Math. Mech. 4(2), 143–157.
[6] He, J.H. (1997), Modified Lagrange multiplier method and generalized variational principles in fluid mechanics, J. Shanghai Univ. 1(2), 117–122. (English edition).
[7] He, J.H. (2001), Derivation and transformation of variational principles with emphasis on inverse and hybrid problems in fluid mechanics: a systematic approach, Acta Mech. 149(1–4), 247–249.
[8] He, J.H. (2000), Generalized Hellinger–Reissner Principle, ASME J. Appl. Mech. 67(2), 326–331.
[9] He, J.H. (2000), A variational approach to electroelastic analysis of piezoelectric ceramics with surface electrodes, Mech. Res. Commun. 27, 445–450.
[10] He, J.H. (2000), Coupled variational principles of piezoelectricity, Int. J. Eng. Sci. 39, 323–341.
[11] He, J.H. (2001), Hamilton principle and generalized variational principles of linear thermopiezoelectricity, ASME J. Appl. Mech. 68, 666–667.
[12] He, J.H. (2003), A family of variational principles for linear micromorphic elasticity, Comp. Struct. 81(21), 2079–2085.
[13] He, J.H. (2001), Variational theory for linear magneto-electro-elasticity, Int. J. Non-Linear Sci. Numer. Simul. 2, 309–316.

[14] He, J.H. (2002), Generalized variational principles for thermopiezoelectricity, Arch. Appl. Mech. 72, 248–256.
[15] Hao, T.H. (2003), Application of Lagrange multiplier method and semi-inverse method to the search for generalized variational principle in quantum mechanics, Int. J. Nonlinear Sci. Numer. Simul. 4(3), 311–312.
[16] Liu, H.M. (2004), Variational approach to nonlinear electrochemical system, Int. J. Nonlinear Sci. Numer. Simul. 5(1), 95–96.
[17] Mosconi, M. (2002), Mixed variational formulations for continua with microstructure, Int. J. Solids Struct. 39, 4181–4195.
[18] He, J.H. (1999), A new method to elimination of variational crisis, J. Shanghai Univ. 5(5), 453–455. (in Chinese).
[19] Woltman, R. (1794), Beitraege zur hydraulischen Architektur, Dritter Band, Johann Christian Dietrich.
[20] Goodman, M.A. and Cowin, S.C. (1972), A continuum theory for granular materials, Arch. Rat. Mech. Anal. 44, 249–266.
[21] Nunziato, J.W. and Cowin, S.C. (1979), A nonlinear theory of elastic materials with voids, Arch. Rat. Mech. Anal. 72, 175–201.
[22] Cowin, S.C. and Nunziato, J.W. (1982), Linear elastic materials with voids, J. Elasticity 13, 125–147.
[23] Iesan, D. (1986), A theory of thermoelastic materials with voids, Acta Mech. 60, 67–89.
[24] Pi, D.-H. and Jin, L.-H. (1991), Structural function theory of generalized variational principles for linear elastic materials with voids, Appl. Math. Mech. 12(6), 565–575.
[25] Ciarletta, M. and Scalia, A. (1992), On uniqueness and reciprocity in linear thermoelasticity of materials with voids, J. Elasticity 32, 1–17.
[26] Dhaliwal, R.S. and Wang, J. (1994), Domain of influence theorem in the theory of elastic materials with voids, Int. J. Eng. Sci. 32(11), 1823–1828.
[27] Martinez, F. and Quintanilla, R. (1995), Existence, uniqueness solutions to the equations of incremental thermoelasticity with voids. Trends in Applications of Mathematics to Mechanics. In Pitman Monographs and Surveys in Pure and Applied Mathematics, (Marques, M.M., Rodrigues, J.F., eds.), pp. 45–56.
[28] Martinez, F. and Quintanilla, R. (1998), Existence, uniqueness and asymptotic behaviour of solutions to the equations of viscoelasticity with voids, Int. J. Solids Struct. 35, 3347–3361.
[29] Ciarletta, M. and Iesan, D (1992), Non-classical Elastic Solids. Pitman Research Notes in Mathematical Series 293.
[30] Schrefler, B.A. (2002), Mechanics and thermodynamics of saturated/unsaturated porous materials and quantitative solutions, Appl. Mech. Rev. 55, 351–388.
[31] He, J.H. (1999), Treatment of shocks in transonic aerodynamics in meshless method, Int. J. Turbo Jet Engines 16(1), 19–26.
[32] Zienkiewicz, O.C. and Taylor, R.L. (1989), The Finite Element Method, 4th edn, Vol. 1, McGraw-Hill, London.
[33] Felippa, C.A. (1989), Parametrized multifield variational principles in elasticity: I. Mixed functionals, Commun. Appl. Numer. Methods 5, 89–98.
[34] Felippa, C.A. (1992), Parametrized variational principles encompassing compressible and incompressible elasticity, Int. J. Solids Struct. 29, 57–68.

Chapter 5

VARIATIONAL FORMULATIONS OF RELATIVISTIC ELASTICITY AND THERMOELASTICITY

Jerzy Kijowski

Centre for Theoretical Physics, Polish Academy of Sciences, Al. Lotników 32/46; 02-668 Warsaw, Poland

Giulio Magli

Dipartimento di Matematica, del Politecnico di Milano, Piazza Leonardo da Vinci 32, 20133 Milano, Italy

Abstract

Existing, equivalent variational formulations of relativistic elasticity theory are reviewed. Emphasis is put on the formulation based on the parameterization of material configurations in terms of unconstrained degrees offreedom. In this description, elasticity can be treated as a gauge-type theory, where the role of gauge transformations is played by diffeomorphisms of the material space. Within this framework it is shown that the dynamics of the theory can be formulated in terms of three independent, hyperbolic, second-order partial differential equations imposed on three independent gauge potentials. Canonical (unconstrained) momenta conjugate to the three configuration variables and the Hamiltonian of the system are easily found. The extension of the theory to nondissipative thermoelasticity and the applications of the theory in astrophysics and quantum gravity are briefly discussed.

Keywords: relatavistic elasticity; canonical relativity; non-linear elasticity

1. Introduction

Interaction between the gravitational field and anelastic solid body became quite an important subject in astrophysical applications, since the discovery that, due to a process of crystallization of dense neutron matter, neutron stars and white dwarfs are solid (or partly solid) objects (see, e.g. Refs. [1,2]).

There are many different formulations of relativistic mechanics of continua in the literature, namely those given by DeWitt [3], Souriau [4], Hernandez [5], Maugin [6–9], Glass and Winicour [10], Carter and Quintana [11], Cattaneo [12], Carter [13], and Bressan [14]. Recently, we reconsidered this theory ([15–17]) formulating it as a gauge-type one. The main advantage of this formulation consists in removing all the constraints and working with three independent degrees of freedom of the material (four, if thermal

E-mail address: kijowski@cft.edu.pl (J. Kijowski); magli@mate.polimi.it (G. Magli).

adiabatic processes are allowed—see Section 4). In this formulation, relativistic elasticity is derived from an *unconstrained* variational principle, and the dynamics can be formulated in terms of independent, second-order hyperbolic partial differential equations imposed on the gauge potentials. All physical quantities (like, e.g. the stress and the strain tensors, the matter current and so on) are defined in terms of first-order derivatives of the field potentials, and all the compatibility conditions of the theory are automatically satisfied. In this way, *canonical* Poisson brackets between the potentials and their conjugate momenta are obtained when passing to the Hamiltonian description of the dynamics. This description in terms of unconstrained degrees of freedom is well suited for the applications to problems of astrophysical relevance, like, e.g. the study of interior solutions of the Einstein field equations in elastic media [18–20].

In the present chapter, we give an overview of the unconstrained variational formulation of relativistic elasticity theory together with a discussion of (virtually) all other variational formulations of the theory present in the literature so far. We also review the extension of the unconstrained formulation to the case of nondissipative thermoelastic materials (see also [21] for the extension to discontinuous materials).

2. Unconstrained variational formulation of relativistic elasticity

We begin our discussion on equivalent variational formulations of relativistic elasticity starting from the 'gauge-type' one. The nonrelativistic counterpart of this approach to continuum mechanics is known as the Piola (or inverse-motion) description (see, e.g. Ref. [22]). For a complete formulation of finite elasticity in a language to which that of the present chapter is close see Maugin's book [23].

In this section, the pseudo-Riemannian geometry $g_{\mu\nu}$ ($\mu, \nu = 0, 1, 2, 3$, signature $(-, +, +, +)$) of the general-relativistic space–time $\mathcal{M}$ is considered as given a priori. To formulate the dynamical theory of a continuous material moving within $\mathcal{M}$, denote by $\mathcal{B}$ the collection of all the idealized points ('molecules') of the material, organized in an abstract 3D manifold, the *material space*. The space–time configuration of the material is completely described by a mapping $\mathcal{G} : \mathcal{M} \to \mathcal{B}$, assigning to each space–time point x the material point ξ (a specific 'molecule') that passes through this point. Each molecule $\xi \in \mathcal{B}$ follows, therefore, the space–time trajectory defined as the inverse image $\mathcal{G}^{-1}(\xi) \subset \mathcal{M}$. Given a coordinate system (ξ^a) $(a = 1, 2, 3)$ in $\mathcal{B}$ and a coordinate system (x^μ) in $\mathcal{M}$, the configuration may, thus, be described by three fields $\xi^a = \xi^a(x^\mu)$ depending on four variables x^μ. We will show how to formulate the physical laws governing the mechanical properties of the material in terms of a system of second-order, hyperbolic partial differential equations imposed on these fields. In this way, the mechanics of continua becomes a field theory and we may use its standard tools as variational principles, Noether theorem, Hamiltonian formulation with the underlying canonical (symplectic) structure of the phase space of Cauchy data, etc.

As a first step, we show that the kinematic quantities characterizing the space–time configuration of the material, like the four-velocity u^μ, the matter current J^μ and the state of strain, can be encoded in the first derivatives of the fields. Consider the tangent mapping $\mathcal{G}_* : T_x\mathcal{M} \mapsto T_{\xi(x)}\mathcal{B}$, described by the (3×4)-matrix $(\xi^a_\mu) := (\partial_\mu \xi^a)$. We assume this matrix to have maximal rank. Moreover, we assume that the 1D kernel of (ξ^a_μ) is time-like (in fact, the dynamical equations of the theory prevent the fields from violating these conditions in the future, once they are fulfilled by the Cauchy data). Vectors belonging to the kernel of (ξ^a_μ) are tangent to the world lines of the material, because the value of ξ^a remains constant on these lines. It follows that the velocity field u^μ can be defined as the unique future-oriented, normalized (i.e. $u^\mu u_\mu = -1$) vector field satisfying the orthogonality condition: $u^\mu \xi^a_\mu = 0$. These four conditions allow us to calculate u^μ uniquely in terms of the fields' derivatives and the metric (an explicit formula will be given in the following).

Given a space–time configuration of the material, consider the push-forward of the contravariant physical metric $g^{\mu\nu}$ from the space–time $\mathcal{M}$ to the material space $\mathcal{B}$ [9]

$$G^{ab} := g^{\mu\nu}\xi^a_\mu \xi^b_\nu. \tag{1}$$

Our assumption about the time-like character of the kernel of ξ^a_μ implies that this symmetric tensor must be positive-definite. It defines, therefore, a (time-dependent) Riemannian metric in $\mathcal{B}$, carrying the information about the actual distances of adjacent particles of the material, measured in the local rest frame. Comparing this metric with an appropriate, pre-existing, geometric structure of $\mathcal{B}$, describing the mechanical structure of the material (like, e.g. volume rigidity or shape rigidity) we can 'decode' information about the local state of strain of the material at each instant of time: the more the structure inherited from space–time $\mathcal{M}$ (via the tensor G) differs from the pre-existing structure of $\mathcal{B}$, the higher is the state of strain of the material under consideration.

Below we give three different examples of such internal structures of $\mathcal{B}$, corresponding to fluids, isotropic elastic media and anisotropic (crystalline) materials, respectively. These structures *are not* dynamical objects of the theory: they are given a priori for any specific material. We stress, however, that the dynamical theory we are going to formulate in the following, is universal and applies to any material whose physical properties may be described in terms of an appropriate geometric structure of $\mathcal{B}$.

2.1. Examples

1. *Volume structure*. A three-form (a scalar density)

$$\omega = r(\xi^a)\mathrm{d}\xi^1 \wedge \mathrm{d}\xi^2 \wedge \mathrm{d}\xi^3, \tag{2}$$

enables us to measure the quantity of matter (number of particles or *moles*) contained in a volume $\mathcal{D} \subset \mathcal{B}$ by integration over D. This 'volume structure' is sufficient to describe the mechanical properties of a perfect fluid. Indeed, the ratio between the material's own volume form ω and the one inherited from space–time via G^{ab}, i.e. the number

$$\rho := r\sqrt{\det G^{ab}}, \tag{3}$$

describes the actual density of the material (moles per cm^3), measured in the rest frame. Its inverse $v := 1/\rho$ is equal to the local, rest-frame specific volume of the fluid (cm^3 per mole). The Pascal law, which is derived later, implies that ρ contains the complete information about the state of strain of the fluid.

2. *Metric structure*. Elastic materials, displaying not only volume rigidity but also shape rigidity, are equipped with a Riemannian metric γ_{ab}, the *material metric*. It describes

the 'would be' rest-frame space distances between neighboring 'molecules', measured in the locally relaxed state of the material. (To obtain such a locally relaxed state, we have to extract an 'infinitesimal' portion from the bulk of the material. In this way, the influence of the rest of the material—possibly prestressed—is eliminated. Such an influence could otherwise make the relaxation impossible.)

The state of strain of the material is described by the 'ratio' between the material metric γ and the physical metric inherited from the space–time $\mathcal{M}$ via its actual configuration. This 'ratio' may be measured by the tensor

$$S_a^b := \gamma_{ac} G^{cb}. \tag{4}$$

The material is locally relaxed at the point x if and only if both structures coincide at x, i.e. if the actual, physical distances between material points in the vicinity of x agree with their material distances. This happens if and only if the strain tensor is equal to the identity tensor δ_a^b.

The simplest example of a material metric is obviously the flat, Euclidean metric, corresponding to nonprestressed materials. A material carrying such a metric displays no 'internal' or 'frozen' stresses and can be embedded into flat Minkowski space *without generating any strain*. Such embedding is impossible if the material metric has a nonvanishing curvature. Materials corresponding to curved metrics are, therefore, *prestressed* (in what follows, no specific assumption about γ_{ab} will be necessary).

Denote by u_{I} the amount of internal mechanical energy (per mole of the material), accumulated in an infinitesimal portion of the material during its deformation from the locally relaxed state to the actual state of strain. If the material is isotropic, this function may depend on the deformation only via invariants of its strain tensor. Since the metric γ automatically carries a volume structure $r := \sqrt{\det \gamma}$, we can take as one of these invariants the rest-frame matter density ρ (or its inverse ν), defined exactly as for fluids

$$\rho = \sqrt{\det \gamma}\,\sqrt{\det G^{ab}} = \sqrt{\det S_a^b}. \tag{5}$$

As the remaining two invariants of S we can take, e.g. its trace and the trace of its square

$$s = S_a^a, \qquad q = S_a^b S_b^a. \tag{6}$$

The physical meaning of these invariants is easily recognized if one considers a weak-strain limit (Hookean approximation). In this case, the function u_{I} coincides with the standard formula of linear elasticity

$$u_{\text{I}} = \lambda(\nu)s^2 + 2\mu(\nu)q,$$

where λ and μ are the Lamé coefficients, and s and q are the linear and the quadratic invariants of the strain, respectively.

3. *Privileged deformation axis*. For an anisotropic material (like a crystal) the energy of a deformation may depend upon its orientation with respect to a specific axis, reflecting the microscopic composition of the material. The information about the existence of such an axis may be encoded in a vector field E^a 'frozen' in $\mathcal{B}$. We may, therefore, admit

an additional dependence of the energy u_I upon the orientation of G with respect to one or several vectors E^a, i.e. upon the quantities $(G^{-1})_{ab}E^aE^b$.

To give an explicit formula for the velocity u^μ in terms of the fields ξ^a, consider the pull-back of the material volume form from the material space to the space–time. This pull-back is a differential three-form in the 4D manifold $\mathcal{M}$, i.e. a vector density J that we call the *material current*. We have

$$J := \mathcal{G}^*\omega = r(\xi)\mathrm{d}\xi^1(x) \wedge \mathrm{d}\xi^2(x) \wedge \mathrm{d}\xi^3(x) = r(\xi)\xi^1_\nu\xi^2_\mu\xi^3_\sigma\,\mathrm{d}x^\nu \wedge \mathrm{d}x^\mu \wedge \mathrm{d}x^\sigma.$$

On the other hand, every three-form in $\mathcal{M}$ may be written in the 'vector-density' representation as

$$J = J^\mu(\partial_\mu \lrcorner \mathrm{d}x^0 \wedge \mathrm{d}x^1 \wedge \mathrm{d}x^2 \wedge \mathrm{d}x^3). \tag{7}$$

This gives us the following formula for the components J^μ in terms of the fields ξ and their derivatives:

$$J^\mu = r(\xi)\varepsilon^{\mu\nu\rho\sigma}\xi^1_\nu\xi^2_\rho\xi^3_\sigma, \tag{8}$$

(here we denote by $\varepsilon^{\mu\nu\rho\sigma}$ the standard Levi–Civita tensor density). The vector density J is a priori conserved due to its geometric construction. Indeed, the exterior derivative of J is equal to the pull-back of the exterior derivative of ω, and the latter vanishes identically being a four-form in the 3D space $\mathcal{B}$:

$$(\partial_\mu J^\mu)\mathrm{d}x^0 \wedge \mathrm{d}x^1 \wedge \mathrm{d}x^2 \wedge \mathrm{d}x^3 = \mathrm{d}J = \mathrm{d}(\mathcal{G}^*\omega) = \mathcal{G}^*(\mathrm{d}\omega) = 0, \tag{9}$$

or, equivalently

$$\partial_\mu J^\mu \equiv 0.$$

Observe now that $J^\mu\xi^a_\mu \equiv 0$, since $\varepsilon^{\mu\nu\rho\sigma}\xi^1_\nu\xi^2_\rho\xi^3_\sigma\xi^a_\mu$ is the determinant of a matrix with two identical columns. This means that J^μ is proportional to the velocity field and, therefore, may be written in the standard form

$$J^\mu = \sqrt{-g}\,\rho u^\mu, \tag{10}$$

where the scalar $\rho = \sqrt{J^\mu J_\mu/g}$, with J^μ given by Eq. (8), is a nonlinear function of ξ^a_μ. Dividing J^μ given in Eq. (8) by '$\sqrt{-g}\rho$' defined above, we obtain an explicit formula for u^μ in terms of ξ^a_μ.

Being a scalar, the quantity ρ can be calculated, e.g. in the material rest frame, where $u^\mu = (1/\sqrt{-g_{00}}, 0, 0, 0)$. In this frame we have $\xi^a_0 = 0$ and, therefore,

$$\rho = \frac{J^0}{u^0\sqrt{-g}} = \frac{r\det(\xi^a_k)\sqrt{-\det g^{\mu\nu}}}{\sqrt{-g_{00}}} = r\det(\xi^a_k)\sqrt{\det g^{kl}} = r\sqrt{\det G^{ab}}.$$

This proves that the quantity ρ defined in this way coincides, indeed, with the rest-frame matter density defined by Eq. (3).

In the present approach, the dynamical equations governing the evolution of the material under consideration will be derived from the Lagrangian density $\Lambda := -\sqrt{-g}\varepsilon = -\sqrt{-g}\rho e$, where $\varepsilon = \rho e$ denotes the rest-frame energy per unit volume of the material and e denotes the molar rest-frame energy. The mechanical properties of each specific material are completely encoded in the constitutive function $e = e(G^{ab})$, which describes the dependence of its energy upon its state of strain. This function plays, therefore, the role of *equation of state* of the material. According to the general principles of relativity theory, it must contain also the molar rest mass m, i.e. we have $e = m + u_{\rm I}$. In generic situations, the equation of state depends also upon the point ξ via the volume structure, metric structure, specific deformation axis or any other structure, which one may find necessary to describe the specific physical properties of the material. By abuse of notation we will, however, write $e = e(G^{ab})$ (instead of $e = e(\xi^a, G^{ab})$) whenever it does not lead to any misunderstanding. The density Λ is, therefore, a first-order Lagrangian, depending upon the unknown fields ξ^a, their first derivatives ξ^a_μ (which enter through G^{ab}) and—possibly—the independent variables x^μ (which enter via the components $g_{\mu\nu}$ of the space–time metric). The dynamical equations of the theory are, therefore, those generated by the variational principle $\delta I = 0$ where the action I is defined as

$$I := \int \Lambda \mathrm{d}^4 x \tag{11}$$

and the variation is performed with respect to three unconstrained degrees of freedom ξ^a. In this way, the physical laws describing the evolution of the elastic material will assume the form of the Euler–Lagrange equations

$$\partial_\mu p^\mu_a = \frac{\partial \Lambda}{\partial \xi^a}, \tag{12}$$

where we have introduced the momentum canonically conjugate to ξ^a :

$$p^\mu_a := \frac{\partial \Lambda}{\partial \xi^a_\mu}, \tag{13}$$

(for historical reasons we call it the Piola–Kirchhoff momentum density).

The following identities (cf. [24,25]) may be immediately checked in the framework of the above theory (see also [15,16]):

Proposition 1. ***(Belinfante–Rosenfeld identity)*** *The canonical energy-momentum tensor-density*

$$-\mathcal{T}^\mu{}_\nu := p^\mu_a \xi^a_\nu - \delta^\mu_\nu \Lambda, \tag{14}$$

coincides with the symmetric energy-momentum tensor-density

$$T_{\mu\nu} = -2 \frac{\partial \Lambda}{\partial g^{\mu\nu}}. \tag{15}$$

(This identity is a straightforward consequence of the relativistic invariance of Λ. It may be checked explicitly by inspection if we take into account that both ξ^a_μ and $g_{\mu\nu}$ enter into Λ through their combination (1) only). □

Proposition 2. ***(Noether identity)***

$$-\nabla_\mu \mathcal{T}^\mu_\nu \equiv \left(\partial_\mu \frac{\partial \Lambda}{\partial \xi^a_\mu} - \frac{\partial \Lambda}{\partial \xi^a} \right) \xi^a_\nu. \tag{16}$$

Proof: Differentiating Eq. (14) we obtain

$$-\partial_\mu \mathcal{T}^\mu_\nu = (\partial_\mu p^\mu_a)\xi^a_\nu + p^\mu_a \xi^a_{\nu\mu} - \partial_\nu \Lambda.$$

But

$$\partial_\nu \Lambda = \frac{\partial \Lambda}{\partial \xi^a} \xi^a_\nu + \frac{\partial \Lambda}{\partial \xi^a_\mu} \xi^a_{\mu\nu} + \frac{\partial \Lambda}{\partial g^{\sigma\kappa}} \frac{\partial g^{\sigma\kappa}}{\partial x^\nu}.$$

Taking into account the definition (13) of momenta, the symmetry of the second derivatives ($\xi^a_{\nu\mu} \equiv \xi^a_{\mu\nu}$) and the Belinfante–Rosenfeld identity, we obtain

$$-\partial_\mu \mathcal{T}^\mu_\nu = \left(\partial_\mu \frac{\partial \Lambda}{\partial \xi^a_\mu} - \frac{\partial \Lambda}{\partial \xi^a} \right) \xi^a_\nu + \frac{1}{2} \mathcal{T}_{\sigma\kappa} \frac{\partial g^{\sigma\kappa}}{\partial x^\nu}.$$

Expressing the derivatives of the metric in terms of the connection coefficients we see that the last term gives exactly the contribution that is necessary to convert the partial derivative on the left-hand side into the covariant derivative. This ends the proof. □

We stress that the above identities are purely kinematical. They hold also for configurations that *do not* fulfill the dynamical equations. In particular, the Noether identity proves that the latter are actually *equivalent* to the energy-momentum conservation $\nabla_\mu \mathcal{T}^\mu_\nu = 0$. Indeed, due to identity $\xi^a_\mu \equiv 0$, the right-hand side of Eq. (16) is automatically orthogonal to the matter velocity u^ν. This observation reduces the number of *independent* conservation laws from four to three—exactly the number of the Euler–Lagrange equations.

To see, therefore, that the above theory describes correctly the laws of continuum mechanics, it is sufficient to calculate the energy-momentum tensor-density $\mathcal{T}$ and to identify it with the energy-momentum carried by the material under consideration. For this purpose we define, at each point of $\mathcal{B}$ separately, the *response tensor* of the material

$$Z_{ab} := 2 \frac{\partial e}{\partial G^{ab}},$$

or, equivalently

$$\mathrm{d}e(G^{ab}) = \tfrac{1}{2} Z_{ab} \mathrm{d}G^{ab}. \tag{17}$$

As an example consider an isotropic elastic material, whose energy depends only upon the invariants (v, h, q) of the strain. Consequently, the response tensor may be fully characterized by the following response parameters

$$p = -\frac{\partial e}{\partial v}, \qquad B = \frac{2}{v} \frac{\partial e}{\partial s}, \qquad C = \frac{2}{v} \frac{\partial e}{\partial q}, \tag{18}$$

according to the formula

$$Z_{ab} = v(p(G^{-1})_{ab} + B\gamma_{ab} + CG_{ab}).$$

The generating formula (17) reduces, in this case, to

$$de(v, s, q) = -p\mathrm{d}v + \tfrac{1}{2}vB\mathrm{d}s + \tfrac{1}{2}vC\mathrm{d}q. \tag{19}$$

The response parameters defined above describe the reaction of the material to the strain. In particular, p describes the isotropic stress while B and C describe the anisotropic response as in the ordinary, nonrelativistic elasticity. The particular case of a perfect fluid materials, corresponding to a constitutive function e that depends only on the specific volume v, may be characterized by the vanishing of both the anisotropic responses, i.e. by equations $B = C \equiv 0$. Consequently, the response tensor Z_{ab} for fluids fulfils the Pascal law: it is proportional to the physical metric $(G^{-1})_{ab}$. The generating formula (17) for fluids reduces, therefore, to $\mathrm{d}e(v) = -p\mathrm{d}v$.

For a general (not necessarily isotropic) material we have the following:

Proposition 3. *The energy-momentum tensor-density of the above field theory is equal to*

$$\mathcal{T}_{\mu\nu} = \sqrt{-g}\rho(e\, u_\mu u_\nu + z_{\mu\nu}), \tag{20}$$

where $z_{\mu\nu}$ is the pull-back of the response tensor Z_{ab} from $\mathcal{B}$ to $\mathcal{M}$:

$$z_{\mu\nu} := Z_{ab}\xi^a_\mu \xi^b_\nu. \tag{21}$$

Proof: We have

$$\mathcal{T}_{\mu\nu} = 2\frac{\partial}{\partial g^{\mu\nu}}(\sqrt{-g}\rho e). \tag{22}$$

But

$$\frac{\partial\sqrt{-g}}{\partial g^{\mu\nu}} = -\frac{1}{2}\sqrt{-g}g_{\mu\nu}, \tag{23}$$

$$\frac{\partial\rho}{\partial g^{\mu\nu}} = \frac{1}{2\rho}\frac{\partial}{\partial g^{\mu\nu}}\left(\frac{J^\rho J^\sigma g_{\rho\sigma}}{g}\right) = \frac{1}{2}\rho(g_{\mu\nu} + u_\mu u_\nu),$$

$$\frac{\partial e}{\partial g^{\mu\nu}} = \frac{\partial e}{\partial G^{ab}}\frac{\partial G^{ab}}{\partial g^{\mu\nu}} = \frac{1}{2}Z_{ab}\frac{\partial(g^{\rho\sigma}\xi^a_\rho\xi^b_\sigma)}{\partial g^{\mu\nu}} = \frac{1}{2}z_{\mu\nu}.$$

Inserting the above results in Eq. (22) we obtain Eq. (20). □

We recognize in formula (20) the standard energy momentum-tensor of continuum mechanics, composed of two components: the energy component '$\varepsilon\, u_\mu u_\nu$', proportional to the velocity, and the *stress tensor* $\rho z_{\mu\nu}$, which is automatically orthogonal to the velocity, due to Eq. (21). Eqs. (2) and (21), which give the stress in terms of the strain (*stress–strain relations*) are uniquely implied by the constitutive equation $e = e(G^{ab})$ of the material.

The above formulation of continuum mechanics is, of course, invariant with respect to reparameterizations of the material space. Such reparameterizations may be interpreted as gauge transformations of the theory. They form the group of all the diffeomorphisms of $\mathcal{B}$. Physically, such transformations consist in merely changing the 'labels' ξ^a assigned to the molecules of the material. Correspondingly, the fields ξ^a may be regarded as gauge potentials for the 'elastic field strength' G^{ab} that is already gauge invariant. The three gauge potentials describe the three degrees of freedom of the material.

To describe the interacting 'matter + gravity' system, we treat both mechanical configuration ξ^a and the space–time metric $g_{\mu\nu}$ as dynamical field variables. The gauge group of the resulting theory is the product of the group of space–time diffeomorphisms (which is the gauge group of general relativity) by the group of diffeomorphisms of the material space. Einstein equations describing the dynamics of gravitational field may be derived from the total Lagrangian density $\Lambda_{\text{tot}} = \Lambda_{\text{grav}} + \Lambda$, where Λ_{grav} is the gravitational Hilbert Lagrangian, whereas Λ is the mechanical Lagrangian described above.

3. Constrained variational principles

The stress tensor S of the material, defined by formula (4), measures the 'ratio' between the material distances frozen in the material space (and measured by γ) and the physical space–time distances, measured by g. In the theory presented above, the confrontation between the two objects has been performed in the material space, and for this reason the physical metric was transported from $\mathcal{M}$ to $\mathcal{B}$, according to Eq. (1). There is, however, another possibility: to confront both metrics in the physical space, after having transported ('pulled back') γ from $\mathcal{B}$ to $\mathcal{M} : h := \mathcal{G}^* g$ or, in terms of coordinates,

$$h_{\mu\nu} = \gamma_{ab}\xi^a_\mu \xi^b_\nu. \tag{24}$$

This quantity is a priori orthogonal to the velocity field ($h_{\mu\nu}u^\mu \equiv 0$) and, therefore, contains the entire information about u as its only normalized, future-oriented eigenvector with vanishing eigenvalue.

The degenerate metric h describes the material distance between two adjacent particles. Because we describe only materials without hysteresis (no memory about previous strain configurations!), these distances remain constant during the evolution. More precisely, this condition means the following: extracting an infinitesimal portion of the material and letting it relax leads to the same distance between the particles, independently of the moment at which the portion has been extracted. Mathematically, this means that the metric h is 'frozen' in the material, i.e. its Lie derivative with respect to u vanishes

$$\mathcal{L}_u h_{\mu\nu} = 0, \tag{25}$$

where

$$\mathcal{L}_u h_{\mu\nu} := u^\lambda \partial_\lambda h_{\mu\nu} + h_{\mu\lambda}\partial_\nu u^\lambda + h_{\nu\lambda}\partial_\mu u^\lambda. \tag{26}$$

As u is also a function of h, Eq. (25) has to be considered as a constraint imposed on h alone; only those h describe physically admissible configurations of the material that fulfill this condition. But the Lie derivative of h is automatically orthogonal to the velocity, due to identity

$$u^{\mu}\mathcal{L}_u h_{\mu\nu} = \mathcal{L}_u(u^{\mu}h_{\mu\nu}) - h_{\mu\nu}\mathcal{L}_u u^{\mu} \equiv 0.$$

Hence, there are only six independent conditions among 10 equations (Eq. (25)). These six conditions are imposed on the nine independent components of h. We conclude that the quantity $h_{\mu\nu}$ carries three independent degrees of freedom of the material, which are defined implicitly by the constraints (Eq. (25)).

The dynamically admissible configurations are those that correspond to the stationary value of the action (11) with respect to all the kinematically admissible configurations. Here arises the possibility of different formulations of the theory, each of them corresponding to a specific parameterization of the kinematically admissible configurations via different mathematical objects. For instance, one could parameterize the configurations by u^{μ} and $h_{\mu\nu}$, mutually orthogonal and subject to constraints (25). In this formulation, one has to calculate the constrained variation of the Lagrangian, each of the constraints giving rise to a Lagrange multiplier.

Another possibility consists in solving the algebraic constraints and treating u^{μ} as a function of $h_{\mu\nu}$. Now, the variation with respect to $h_{\mu\nu}$ alone has to be performed. In this formulation, we have the differential constraints (25) together with a single (residual) algebraic constraint $\det h = 0$, which assures the existence of the zero-eigenvalue of h.

The method used by Maugin [6] is still less constrained. Here, the only kinematic constraint left unsolved is the conservation of the matter current. Also the variational principle proposed by Salié [26] and using velocity potentials belongs to this family of approaches.

To illustrate the role of the above 'constrained' variational principles consider different variational formulations of Maxwell electrodynamics. Here, the field configurations are described by the electromagnetic, skew-symmetric tensor $F_{\mu\nu}$. This object plays a role analogous to that of the (symmetric) tensor $h_{\mu\nu}$ in elastodynamics. It is subject to the kinematic constraint condition, given by the first pair of Maxwell equations

$$\varepsilon^{\lambda\mu\rho\sigma}\,\partial_{\mu}F_{\rho\sigma} = 0, \tag{27}$$

fully analogous to constraint (26) in elastodynamics. The Lagrangian equals:

$$\Lambda = -\frac{1}{4}\sqrt{-g}F_{\rho\sigma}F^{\rho\sigma}. \tag{28}$$

Dynamically admissible configurations are those that correspond to the stationary value of the action with respect to all the kinematically admissible configurations. The 'standard' method of deriving the dynamical equations consists in solving Eq. (27) completely by introducing the (unconstrained!) electromagnetic potentials A_{μ} and expressing the Maxwell field in terms of them

$$F_{\mu\nu} = \partial_{\mu}A_{\nu} - \partial_{\nu}A_{\mu}.$$

In this way, constraint equations are *automatically* satisfied. Next, we vary the Lagrangian (28) with respect to the potentials and derive the second pair of Maxwell equations. This method is analogous to our approach, presented in the previous section: we solve constraints (25) completely by introducing unconstrained 'potentials' ξ^{a} and expressing the 'elastomechanical field' $h_{\mu\nu}$ in terms of them using formula (24). In this way, constraint equations are *automatically* satisfied and we obtain the field dynamics from the unconstrained variation of the Lagrangian with respect to the potentials.

It is, however, obvious that the introduction of potentials in electrodynamics is not necessary, since the same result will be obtained when varying Eq. (28) directly with respect to all the tensors $F_{\rho\sigma}$ fulfilling constraint (27). Indeed, such a variation gives us

$$-\frac{1}{2}\sqrt{-g}F^{\rho\sigma} = \varepsilon^{\lambda\mu\rho\sigma}\partial_{\mu}C_{\lambda}, \tag{29}$$

where C_{λ} is a Lagrange multiplier corresponding to the constraint (27). The above equation obviously implies the Maxwell equation

$$\nabla_{\rho}F^{\rho\sigma} = 0. \tag{30}$$

A similar procedure is also possible in elastomechanics, but the constrained variational principles obtained in this way are poorly adapted for the Hamiltonian formulation of the theory (similarly as in electrodynamics, whose canonical formulation, otherwise very complicated, extremely simplifies due to the use of potentials).

Finally, we mention further possibilities to derive dynamical equations of relativistic elasticity form a variational principle. The method used by Carter and Quintana [11] consists in considering a priori the elastic body and the gravitational field interacting with it. The total Lagrangian density of such a system is

$$\Lambda_{\text{total}} = -\sqrt{-g}\left(\frac{1}{16\pi}R + \varepsilon\right), \tag{31}$$

where ε is again the rest-frame energy density of the material and R is the scalar curvature of the space–time. Keeping the material configuration fixed and varying Λ_{total} with respect to the gravitational field we obtain the Einstein equations

$$G_{\mu\nu} = 8\pi T_{\mu\nu}, \tag{32}$$

where $T_{\mu\nu}$ is the symmetric energy-momentum tensor (15). In this approach, the dynamical equations for h arise as the compatibility conditions

$$\nabla_{\mu}T^{\mu\nu} = 0 \tag{33}$$

of Eq. (32) with the Bianchi identities, satisfied by the Einstein tensor G. Among Eq. (33), only three are independent. Within this formalism the proofs are rather involved and require the introduction of the concept of *convected derivative*.

To illustrate the Carter–Quintana approach one can again resort to Maxwell electrodynamics. For this purpose, consider the Lagrangian of the system composed of both the electromagnetic and the gravitational field:

$$\Lambda_{\text{total}} = -\sqrt{-g}\left(\frac{1}{16\pi}R + \frac{1}{4}F_{\rho\sigma}F^{\rho\sigma}\right). \tag{34}$$

Keeping the electromagnetic field F fixed and varying Λ_{total} with respect to the gravitational field g we obtain Einstein equations

$$G^{\mu}_{\nu} = 8\pi(F^{\mu\rho}\,F_{\rho\nu} + \tfrac{1}{4}\,\delta^{\mu}_{\nu}F^{\sigma\rho}\,F_{\sigma\rho}).$$

Now, Maxwell equations are obtained as the compatibility conditions for the above system. Indeed, Bianchi identities imply

$$F_{\rho\nu}\nabla_{\mu}F^{\mu\rho} + F^{\mu\rho}\nabla_{\mu}F_{\rho\nu} + \tfrac{1}{2}F^{\rho\sigma}\nabla_{\nu}F_{\rho\sigma} = 0.$$

Using the skew-symmetry of F and the kinematic equations (27), it is easy to see that the last two terms cancel. Hence, for a generic, nonsingular F, we obtain the second pair of Maxwell equations (30).

The approach described so far may be called 'Eulerian' in the following sense: the configuration of the material is described by assigning to each space–time point the particle that passes through this point. Similarly, as in the standard nonrelativistic continuum mechanics, the inverse picture is also possible. It consists in assigning to each particle its space–time position. This approach to relativistic elasticity was first developed by DeWitt [3]. In this approach, a 'material space–time' is considered. It is parameterized by four coordinates (t, ξ^a) (in fact, DeWitt considers only the case of a flat material metric γ_{ab}, but the generalization to an arbitrarily curved metric—i.e. to an arbitrarily prestrained material—is straightforward). The configuration of the material is described by four functions $x^{\mu} = x^{\mu}(t, \xi^a)$. Numerically, the Lagrangian is always the same, equal to Eq. (11), but all the quantities have now to be expressed in terms of the first derivatives of the functions x^{μ}. Hence, the mechanical part of the action $\mathcal{A}$ is written as

$$\mathcal{A}_{\text{mech.}} = \int \mathcal{L}(x^{\mu}, \dot{x}^{\mu}, x^{\mu}_{a})\mathrm{d}t\mathrm{d}^3\xi$$

and may be immediately translated to the 'Eulerian' form

$$\mathcal{A}_{\text{mech.}} = \int \mathcal{L}(x^{\mu}, \dot{x}^{\mu}, x^{\mu}_{a})\frac{\partial(t, \xi^a)}{\partial(x^0, x^k)}\mathrm{d}^4x.$$

Expressing everything in terms of the inverse functions $t = t(x^{\mu})$ and $\xi^a = \xi^a(x^{\mu})$ and denoting

$$\Lambda := \mathcal{L}\frac{\partial(t, \xi^a)}{\partial(x^0, x^k)},$$

one obtains an approach that is *almost* equivalent to the others described so far. The only difference is that a priori we have to perform here the variation with respect to four unknown functions t and ξ^a. It is, however, worthwhile to note that, due to the structure of the DeWitt's Lagrangian L, the additional variable t becomes cyclic: neither t nor its derivatives appear in Λ. We conclude that (as it must be) among the four Euler–Lagrange equations obtained in the DeWitt picture, only three are independent. This is why we prefer the 'Eulerian' picture where we may simply eliminate unnecessary cyclic variables and consider only three independent dynamical equations imposed on three independent degrees of freedom ξ^a.

A variational principle in which the world lines of the particles are used as independent variables has also been proposed by Carter [13] who extended it to the case of electromagnetically coupled systems [27]. Carter's method may be described as follows. Once the variation of the Lagrangian density (31), with respect to the total variation of space–time metric has been calculated, one can rewrite it in terms of the *Eulerian* (fixed point) variation of the metric, which vanishes if the gravitational field is held fixed, plus the variation due to a displacement of the world lines in $\mathcal{M}$, the difference between the total and the Eulerian variation being the Lie derivative of the metric itself. Using the displacement of the world lines as the independent variable, one obtains a variational principle in which the equations of motions for elastic bodies (33) arise as 'Noether identities' associated with the invariance of the action with respect to infinitesimal displacements.

4. Thermodynamics of isentropic flows

It is relatively easy to extend the above theory to the thermodynamics of isentropic flows (no heat conductivity!) For this purpose, we begin with the generating formula

$$\mathrm{d}e(G^{ab}, S) = \tfrac{1}{2} Z_{ab}\, \mathrm{d}G^{ab} + T\mathrm{d}S, \tag{35}$$

which generalizes Eq. (2) to the case of 'thermodynamically sensitive' materials. Performing the Legendre transformation $T\mathrm{d}S = \mathrm{d}(TS) - S\mathrm{d}T$, we obtain an equivalent formula with the Helmholtz free energy $f := e - TS$ playing the role of the generating function

$$\mathrm{d}f(G^{ab}, T) = \tfrac{1}{2} Z_{ab}\, \mathrm{d}G^{ab} - S\mathrm{d}T. \tag{36}$$

In the particular case of perfect fluids the above formulae reduce to $\mathrm{d}e(v, S) = -p\mathrm{d}v + T\mathrm{d}S$ and to $\mathrm{d}f(v, T) = -p\mathrm{d}v - S\mathrm{d}T$, respectively.

Eq. (36) suggests that the temperature T may be interpreted as a strain and the entropy as the corresponding stress. This goes far beyond a formal analogy since it is possible to express the temperature in terms of derivatives of a new potential (ξ^0, say), corresponding to a new, time-like dimension—the 'material time'—in the material space. The configurations of the material turn out, therefore, to be described by *four* fields $\xi^\alpha = \xi^\alpha(x^\mu)(\alpha = 0, 1, 2, 3)$, and the 'thermal strain T' can be described in a way similar to that given by Eq. (1) for the elastic strain. The required formula for the temperature is the following [28–31]:

$$T = \beta u^\lambda \xi^0_\lambda, \tag{37}$$

where β is a dimension-fixing constant.

The ansatz (37) can be viewed at a purely phenomenological level. Indeed, in the case of fluids the field theory derived from the Lagrangian $\Lambda = -\sqrt{-g}\rho f$ (where $f = f(v, T)$ stands for the molar free energy) describes correctly the relativistic hydrodynamics of isentropic flows (Eq. (23)), and we shall show below that the same ansatz works for a generic elastic material as well. However, the potential ξ^0 also has a natural microscopic interpretation as the retardation of the proper time of the molecules with respect to the physical time calculated over averaged space–time trajectories of the idealized continuum material. Indeed, consider the mean kinetic energy of the motion of the molecules of the material, calculated with respect to its rest frame. For temperatures not too high and average velocities $\mathbf{v}$ much smaller than one (i.e. than the velocity of light) this energy equals $(1/2)m\mathbf{v}^2 = (3/2)\kappa T$ where κ is the Boltzmann constant. Consequently, the proper time τ of the particles is retarded with respect to the 'physical' time x^0 (the affine parameter along the tangent to u^μ) according to

$$\tau = \int \sqrt{1 - v^2} \mathrm{d}x^0 \approx \left(1 - \frac{3}{2}\frac{\kappa T}{m}\right) x^0.$$

It follows that, defining the retardation $\xi^0 = x^0 - \tau$, we have

$$T = \beta \frac{\partial \xi^0}{\partial x^0},$$

where $\beta = 2m/3\kappa$. Passing from the material rest frame to a general frame we get Eq. (37).

To show that the Lagrangian

$$\Lambda = -\sqrt{-g}\rho f(G^{ab}, T)$$

describes correctly the thermomechanical behavior of the material, observe that this is again a first-order Lagrangian that depends upon the first derivatives ξ^α_μ of the potentials via the mechanical strains and the temperature. Now, we have four independent Euler–Lagrange equations corresponding to the variation with respect to four potentials

$$\partial_\mu P^\mu_\alpha = \frac{\partial \Lambda}{\partial \xi^\alpha}, \tag{38}$$

where the momenta canonically conjugate to ξ^α (generalized Piola–Kirchhoff momenta) are defined as usual

$$P^\mu_\alpha := \frac{\partial \Lambda}{\partial \xi^\alpha_\mu}.$$

Again, we have the following

Proposition 4. *Both the Belinfante–Rosenfeld identity*

$$\mathcal{T}^\mu_\nu := -(P^\mu_\alpha \xi^\alpha_\nu - \delta^\mu_\nu \Lambda) \equiv -g^{\mu\sigma}\left(2\frac{\partial \Lambda}{\partial g^{\sigma\nu}}\right), \tag{39}$$

and the Noether identity

$$-\nabla_\mu \mathcal{T}^\mu_\nu \equiv \left(\partial_\mu \frac{\partial \Lambda}{\partial \xi^\alpha_\mu} - \frac{\partial \Lambda}{\partial \xi^\alpha}\right)\xi^\alpha_\nu, \tag{40}$$

are valid for any field configuration of the above theory, not necessarily fulfilling the field equations.

(The proof is a simple generalization of the previous proofs.)

The Noether identity implies equivalence between the dynamical equations (38) of the theory and the energy-momentum conservation because the deformation gradient ξ^{α}_{ν} is a (4×4)-nondegenerate matrix. The variation with respect to ξ^0 produces the entropy conservation law. Indeed, due to Eqs. (36) and (37) we have

$$P^{\mu}_{0} = -\sqrt{-g}\rho \frac{\partial f}{\partial T}\frac{\partial T}{\partial \xi^{0}_{\mu}} = \sqrt{-g}\beta S \rho u^{\mu} = \beta S J^{\mu}. \tag{41}$$

Because ξ^0 does not enter into the Lagrangian, the corresponding Euler–Lagrange equation reads

$$0 = \partial_{\mu} P^{\mu}_{0} = \partial_{\mu}(\beta S J^{\mu}) = \beta J^{\mu} \partial_{\mu} S,$$

which means that the amount of entropy contained in each portion of the material remains constant during the evolution.

The physical interpretation of the remaining three Euler–Lagrange equations is given by the following:

Proposition 5. *The energy-momentum tensor of the above theory is given by the same formula* (20)*, where the function e is defined by the Legendre transformation from Eq.* (36) *back to Eq.* (35)*, i.e. by the formula* $e := f + TS$.

Proof: Due to the Rosenfeld–Belinfante identity we have

$$\mathcal{T}^{\mu}_{\nu} = 2\frac{\partial}{\partial g^{\mu\nu}}(\sqrt{-g}\rho f).$$

Calculating the above derivative we obtain the same terms as in the proof of Proposition 3 and, moreover, a term arising from the dependence of the Lagrangian upon T. Therefore, we have

$$\mathcal{T}^{\mu}_{\nu} = \sqrt{-g}\rho(f u_{\mu} u_{\nu} + z_{\mu\nu}) + 2\sqrt{-g}\rho \frac{\partial f}{\partial T}\frac{\partial T}{\partial g^{\mu\nu}}. \tag{42}$$

Now

$$\frac{\partial T}{\partial g^{\mu\nu}} = \frac{\partial}{\partial g^{\mu\nu}}\left(\frac{\beta J^{\lambda}\xi^{0}_{\lambda}}{\sqrt{-g}\rho}\right) = \beta J^{\lambda}\xi^{0}_{\lambda}\frac{\partial}{\partial g^{\mu\nu}}\frac{1}{\sqrt{-g}\rho} = -\frac{1}{2}T u_{\mu} u_{\nu},$$

where the first two formulae of Eq. (23) have been used. Inserting the above result in Eq. (42) and recalling that $\partial f/\partial T = -S$ we obtain formula (20) with $f + ST$ playing the role of e. □

5. Hamiltonian version of the unconstrained formulation

The above-described formulation of relativistic thermomechanics of continua as a Lagrangian field theory leads in a natural way to its Hamiltonian counterpart (see also [32,33]). In the Hamiltonian formalism the infinite-dimensional phase space of Cauchy data for the fields on a given Cauchy surface $\{t = \text{constant}\}$ is described by the four configurations variables ξ^α and their canonical conjugate momenta π_α, defined as derivatives of the Lagrangian with respect to the 'velocities' $\dot{\xi}^\alpha = \xi^\alpha_0$ (see [34])

$$\pi_\alpha := \frac{\partial \Lambda}{\partial \xi^\alpha_0} = P^0_\alpha. \tag{43}$$

The Poisson bracket between configurations and momenta assumes its canonical delta form

$$\{\pi_\beta(x), \xi^\alpha(y)\} = \delta^\alpha_\beta \delta(x, y).$$

Assuming Eq. (43) to be invertible with respect to the $\dot{\xi}^\alpha$, the Hamiltonian $H_{\text{el.}}$ of the theory can be obtained by performing the Legendre transformation

$$H_{\text{el.}} := \pi_\alpha \dot{\xi}^\alpha - \Lambda = -\mathcal{T}^0_0 = -\sqrt{-g}T^0_0,$$

and, therefore, it is numerically equal to the energy density. Once expressed in terms of the canonical variables (ξ^α, π_β) and their spatial derivatives, $H_{\text{el.}}$ generates the Hamiltonian version of the field equations

$$\dot{f} = \{H_{\text{el.}}, f\}.$$

Coupling the theory to gravity, it is possible to construct the canonical Hamiltonian formulation for the complete (gravity + matter) system [17,35,36]. It turns out that it is possible to impose a gauge condition that allows us to encode the six degrees of freedom (two gravitational and four thermomechanical ones), together with their conjugate momenta, in the Riemannian metric q_{ij} and its conjugate ADM momentum P^{ij}. These variables are not subject to constraints. The Hamiltonian of this system is equal to the total matter entropy, and generates uniquely the dynamics once expressed as a function of the canonical variables. The Hamiltonian fulfils a system of three, first-order, partial differential equations of the Hamilton-Jacobi-type in the variables (q_{ij}, P^{ij}). These equations are universal and do not depend upon the properties of the material, the equation of state entering only as a boundary condition (we refer the reader to Ref. [17] for details).

6. Concluding remarks

We have presented a consistent formulation of the relativistic elasticity theory, admitting also adiabatic thermal processes. This formulation fits perfectly into the framework of relativistic gauge field theories. The field configuration of the system is described in terms of three field potentials ξ^a, geometrically organized into a differential mapping $\mathcal{G} : \mathcal{M} \to \mathcal{B}$ from the space–time $\mathcal{M}$ into an abstract material space $\mathcal{B}$. The role of gauge transformations is played by diffeomorphisms (reparameterizations) of $\mathcal{B}$. They correspond to different labeling of the particles of the material. Physical properties of the material are fully described by the pre-existing geometrical structure of the material space. Dynamical quantities characterizing the actual physical state of the material (like strain tensor, velocity field, matter current, density) are all encoded in the first-order derivatives of the potentials. Dynamical equations of the theory are derived from an (unconstraint) variational principle, with the (minus) rest-frame energy playing the role of the Lagrangian function. They are hyperbolic and the initial-value problem is well posed. Canonical momenta of the theory are equal to the components of the mechanical momentum in an appropriate representation. The Hamiltonian description of the dynamics is easily obtained via a Legendre transformation.

To incorporate also the thermal phenomena (but only the reversible ones!), an additional time-like potential ξ^0 is necessary. It measures the retardation of the material time (i.e. age of the particles) with respect to the external time parameter along particle trajectories. The nonvanishing retardation is a purely relativistic phenomenon ('Gemini Paradox'), due to the thermal motion of the particles. Its ratio with respect to the external time uniquely encodes the temperature of the material.

In the rest frame of the material, its energy-momentum tensor reduces to the standard, nonrelativistic expression. This observation constitutes the *correspondence principle*. Conservation equations (four in the thermoelasticity case and only three independent equations in the pure elasticity case) are equivalent to the dynamical equations (respectively, four or three).

The role of different (constrained) formulations of the theory has been illustrated by analogous formulations of classical Maxwell electrodynamics.

References

[1] Haensel, P. (1995), Solid interiors of neutron stars and gravitational radiation. Astrophysical Sources of Gravitational Radiation (Les Houches) (Marck, J.A., Lasota, J.P., eds.).

[2] Shapiro, S.L. and Teukolsky, S.A. (1983), In Black Holes, White Dwarfs and Neutron Stars, Wiley, New York.

[3] De Witt, B. (1962), The quantization of geometry. Gravitation: An Introduction to Current Research (Witten, L., eds.), Wiley, New York, pp. 266–381.

[4] Souriau, J.M. (1964), Géométrie et Relativité, Hermann, Paris.

[5] Hernandez, W.C. (1970), Elasticity in general relativity, Phys. Rev. D 1, 1013–1017.

[6] Maugin, G.A. (1971), Magnetized deformable media in general relativity, Ann. Inst. H. Poincaré' A15, 275–302.

[7] Maugin, G.A. (1977), Infinitesimal discontinuities in initially stressed relativistic elastic solids, Commun. Math. Phys. 53, 233–256.

[8] Maugin, G.A. (1978), Exact relativistic theory of wave propagation in prestressed nonlinear elastic solids, Ann. Inst. H. Poincaré' A28, 155–185.

[9] Maugin, G.A. (1978), On the covariant equations of the relativistic electrodynamics of continua iii. elastic solids, J. Math. Phys. 19, 1212–1219.

[10] Glass, E.N. and Winicour, J. (1972), General relativistic elastic systems, J. Math. Phys. 13, 1934–1940.

[11] Carter, B. and Quintana, H. (1972), Foundations of general relativistic high pressure elasticity theory, Proc. R. Soc. Lond. A 331, 57–83.

[12] Cattaneo, C. (1973), Elasticité Relativiste, Symp. Math. 12, 337–352.

[13] Carter, B. (1973), Elastic perturbation theory in general relativity and a variation principle for a rotating star, Commun. Math. Phys. 30, 261–286.

[14] Bressan, A. (1978), Relativistic Theories of Materials, Springer, Berlin.

[15] Kijowski, J. and Magli, G. (1992), Relativistic elastomechanics as a Lagrangian field theory, Geom. Phys. 9, 207–223.
[16] Kijowski, J. and Magli, G. (1997), Unconstrained variational principle and canonical structure for relativistic theory, Rep. Math. Phys. 39, 99–112.
[17] Kijowski, J. and Magli, G. (1998), Unconstrained Hamiltonian formulation of General Relativity with thermoelastic sources, Class, Quantum Gravity 15(12), 3891–3916.
[18] Giambo', R., Giannoni, F., Magli, G. and Piccione, P. (2003), New solutions of Einstein equations in spherical symmetry: the Cosmic Censor to court, Commun. Math. Phys. 235, 545–563.
[19] Magli, G. (1997), Gravitational collapse with non-vanishing tangential stresses: a generalization of the Tolman-Bondi model, Class. Quantum Gravity 14, 1937–1953.
[20] Magli, G. (1998), Gravitational collapse with non-vanishing tangential stresses ii: a laboratory for cosmic censorship experiments, Class. Quantum Gravity 15, 3215–3228.
[21] Hajicek, P. and Kijowski, J. (1998), Lagrangian and Hamiltonian formalism for discontinuous fluids and gravitational field, Phys. Rev. D 57, 914–935.
[22] Truesdell, C. and Toupin, R. (1960), The classical field theories. In Handbuch der Physik, Springer, Berlin, ch. Bd. III/1, pp. 226–793.
[23] Maugin, G.A. (1993), Material Inhomogeneities in Elasticity, Chapman & Hall, London, p. 1993.
[24] Belinfante, F.J. (1940), On the current and the density of the electric charge, the energy, the linear momentum and the angular momentum of arbitrary fields, Physica 7, 449.
[25] Rosenfeld, L. (1940), Sur le tenseur d'impulsion-energie, Acad. Roy. Belg. 18, 1.
[26] Salié, N. (1974), Allgemeinrelativistiches Variationsprinzip für Flüssigkeiten und Festkörper in Eulerschen Koordinaten, Czech. J. Phys. B 24, 831.
[27] Carter, B. (1980), Rheometric structure theory, convective differentiation and continuum electrodynamics, Proc. R. Soc. Lond. A 372, 169.
[28] Kijowski, J., Pawlik, B. and Tulczyjew, W.M. (1979), A variational formulation of nongravitating and gravitating hydrodynamics, Bull. Acad. Polon. Sci. 27, 163–170.
[29] Kijowski, J. and Tulczyjew, W.M. (1982), Relativistic hydrodynamics of isentropic flows, Mem. Acad. Sci. Torino V(6), 3–17.
[30] Kijowski, J., Smólski, A. and Górnicka, A. (1990), Hamiltonian theory of self-gravitating perfect fluids and a method of effective deparameterization of Einstein theory of gravitation, Phys. Rev. D 41, 1875–1884.
[31] Jezierski, J. and Kijowski, J. (1991), Thermo-hydrodynamics as a field theory. Hamiltonian Thermodynamics (Sieniutycz, S., Salamon, P., eds.), Taylor and Francis, New York, Bristol, PA, Washington, DC, London.
[32] Holm, D. (1989), Hamilton techniques for relativistic fluid dynamics and stability theory. Relativistic Fluid Dynamics. In Lecture Notes in Math, (Anile, A.M., Choquet Bruhat, Y., eds.), Vol. 1385, Springer, Berlin, Heidelberg, New York, Tokyo, pp. 65–151.
[33] Künzle, H.P. and Nester, J.M. (1984), Hamiltonian formulation of gravitating perfect fluids and the Newtonian limit, J. Math. Phys. 25, 1009–1018.
[34] Kijowski, J. and Tulczyjew, W.M. (1979), A symplectic framework for field theories. In Lecture Notes in Physics, Vol. 107, Springer, Berlin.
[35] Kijowski, J. (1997), A simple derivation of canonical structure and quasi-local Hamiltonian in General Relativity, Gen. Rel. Grav. 29, 307–343.
[36] Kuchar, K.V. (1993), Canonical quantum gravity. In General Relativity and gravitation-Cordoba 1992, Inst. Phys. Publ., Bristol, UK, pp. 119–150.

Chapter 6

THE GEOMETRIC VARIATIONAL FRAMEWORK FOR ENTROPY IN GENERAL RELATIVITY

L. Fatibene, Mauro Francaviglia and M. Raiteri

Department of Mathematics, University of Torino, Via C. Alberto 10, 10123 Torino, Italy

Abstract

We first review the basic frameworks to define entropy of gravitational systems and we discuss the relation among different perspectives. Particular attention is paid to macroscopic prescriptions in order to clarify the relation between gravitational entropy and the geometric properties of spacetimes. The main aim of this review is to discuss the fundamental contributions that Geometry and Calculus of Variations offer in characterizing the entropy of black-hole solutions as well as more general singular solutions in General Relativity.

Keywords: general relativity; black holes entropy; conserved qualities

1. Introduction

In the 1970s well-known papers by Bekenstein and Hawking [1–3] based on previous work by Christodoulou and Ruffini [4,5] gradually established that an entropy must be associated with black-hole solutions of field equations of General Relativity.

Christodoulou and Ruffini were the first to study the possibility of extracting energy from a black hole by describing in detail the physics of particles around a Kerr black hole. Bekenstein, assuming these as paradigmatic 'Gedanken' experiments, slightly later noticed a striking correspondence between the dynamical laws of (nonextreme) black holes and the classical laws of thermodynamics. The entropy (modulo a universal factor) was identified with the area A of the (outer) horizon (proportional, in turn, to the irreducible mass previously introduced by Christodoulou and Ruffini when studying energy extraction) while the thermodynamical potential associated to it (i.e. the temperature) was identified with the surface gravity at the horizon. Hawking finally discovered that a black hole can emit particles because of quantum effects. Black holes hence emit a radiation with a black-body spectrum at a temperature T (far below the 3 K for stellar black holes, but rapidly increasing as the mass decreases) which substantially

E-mail address: fatibene@dm.unito.it (L. Fatibene); mauro.francaviglia@unito.it (M. Francaviglia); raiteri@dm.unito.it (M. Raiteri).

agrees with the original prediction by Bekenstein. The identification of the Hawking temperature with a physical temperature is considered as the cornerstone for the identification of black-hole entropy with a real, physical entropy.

Both the horizon area and the surface gravity are macroscopic geometrical quantities associated with the black hole. On the other hand, entropy is known to be fundamentally related to information theory and microscopic state counting. The relation between irreducible mass and information was already stated by Christodoulou. The fundamental origin of entropy was recognized by Bekenstein to be a measure of our ignorance about the internal status of a black hole, considered as a black box perfectly hiding the microscopic status of matter falling into it. This double identity of entropy both as a macroscopic and as a microscopic notion is not peculiar to black-hole entropy: gas entropy was originally introduced as a macroscopic quantity and only later was it more fundamentally recognized as a microscopic quantity.

In information theory the notion of entropy for a stream of characters taken from a fixed alphabet was introduced by Shannon and Weaver [6] as a quantification of the (lack of) information transmitted by the stream. For example, the entropy associated with a constant stream (i.e. when a single character is repeated on and on) reaches its minimum (being independent of the length of the message) while the entropy of a random stream is maximal since one is never able to predict the next character regardless of how long the past stream is known. If p_i denotes the probability (or the frequency) of the occurrence of the i-th character in the stream then Shannon's entropy is defined as

$$S = -\sum_i p_i \ln p_i. \tag{1}$$

It is positive-definite, vanishing when the stream is constant and maximal for completely random sequences.

Despite the fact that Shannon theory does not seem to have much to do with physical entropies, these have been recognized to be more or less directly a byproduct of Shannon entropy. For example, gas macrostates can be realized by many different microstates and entropy measures our ignorance about the status of the system. The basic axioms a system should have to support Shannon entropy are

(i) microstates are defined by setting their volume so that they result to be *equiprobable*;
(ii) *ergodic*: the system should fill a dense open set of phase space and the probability of being in a region is proportional to the region volume;
(iii) the phase space accessible to the system should be *finite*, so that volumes can be normalized to estimate probabilities.

Under these simple assumptions microstate counting (normalized to the accessible volume of phase space) within a macrostate is a direct measure of its probability. As a consequence Shannon entropy is a measure of our capacity (or better, our incapacity) to predict the microstate of the system knowing its past history. Probably it is trivial to mention that being able to predict the state of the system would enable us to extract energy from the system (which is the link with the classical thermodynamical definition of entropy). The more information we know the more energy we will able to extract (imagine if we were able to predict when a gas in a box will happen to be wholly contained in the right half of the box!).

In General Relativity it is still difficult to implement this direct and fundamental approach. A microscopic model of General Relativity is necessarily related to quantum gravity (or at least to its classical preliminary step, i.e. a Hamiltonian formulation). Unfortunately our knowledge about quantum gravity is still preliminary: just partial knowledge about a few simple examples is available and even in these simple situations the agreement on details is not universal. The same happened for the Hamiltonian formulation that is not universally established. Despite some simple scenarios producing very interesting predictions in limited situations, the ability to deal with entropy in generic situations by microstate counting within an accepted framework is not likely to be available in the near future.

From a macroscopic viewpoint many interesting results have been obtained. The relation between the area of the black-hole horizon and entropy is well established in many general situations (however, there is no general consensus on how to deal on a macroscopic scale with extreme black holes that are most commonly and *simply* treated from a microscopic viewpoint). Relying on these results it was proposed to universally extend the identification $S = \frac{1}{4}A$ (in Planck units). Once again Hawking, together with Page and Hunter [7], substantially contributed to the topics by studying Taub-bolt solutions. Taub-bolt is a singular solution [8–10] that is not enclosed within a Killing horizon (due to the so-called *Misner string* running along a z-axis); despite these peculiar features and despite the physical interpretation of the solution not being universally accepted (in the first place the solution was considered in the purely Euclidean sector) it was clearly shown that an entropy should be associated with the solution. Furthermore, this case exhibits a deviation from the simple one-quarter of the area law. Hawking and coworkers proposed in turn an interpretation that relates the entropy to the obstruction to define a global foliation of space-time.

An independent research program based on a Hamiltonian formulation for spatially confined gravitational system (confinement being related with axiom (iii) above) has been developed by Brown and York [11,12]. This approach is deeply related to a statistical approach based on a microcanonical ensemble formulation. In this way macroscopical quantities were defined such as quasilocal energy and entropy. The main advantage of these approaches is a definition of conserved quantities for finite spatially extended systems that improves substantially the standard ADM results [13,14] to compute conserved quantities for asymptotically flat solutions.

2. The geometry of entropy in General Relativity

Motivated by microstate-counting methods, the Taub-bolt example and statistical approaches, it was perceived that a geometrical definition of entropy in General Relativity might have lost its initial central role: the identification between entropy and horizon area was still considered important for (physically reasonable) stationary black holes but in more general situations the geometric character of entropy was more difficult to be recognized. Iyer and Wald [15] have recently proposed a prescription intrinsically and deeply geometrical (though based on strong and unnecessary hypotheses that partially hide the geometrical content of the prescription); but nonstationary black holes seemed to definitely behave in a nongeometrical way. In the last few years we have generalized Wald's prescription, revealing its full geometrical nature, relaxing the unnecessary hypotheses and showing that all cases examined above can be easily handled in this way [16–19]. This shows that geometry is present also when it is not easily recognized and it provides a unifying viewpoint.

The geometrization of entropy in General Relativity can be understood only after the problem has been split into two parts, one being deeply geometrical provided one gives up control on the other part. This is not a peculiarity of General Relativity. In any macroscopical approach to entropy, e.g. when dealing with classical thermodynamics of gases, one has to predict thermodynamical potentials (e.g. the temperature). It can be easily argued that no classical way of computing the temperature in a macroscopic thermodynamics exists unless all other thermodynamical potentials are known. Similarly, the best way to predict temperature in General Relativity is Hawking radiation, which is in turn based on a semiclassical approximation of quantum mechanics on a curved background or some equivalent classical result about geodesics geometry. The further identifications between the temperature and other geometrical entities (surface gravity of the horizon, the period of time compactification in the Euclidean sector to avoid conical singularities and so on) are necessarily less fundamental than Hawking radiation and they basically represent a coincidence. It is worthwhile to mention that such a coincidence is deeply related to the no-hair theorem: *if the status of a black hole is described by very few parameters that in turn determine its geometrical features*, then it is trivial that any status function is necessarily a function of these few parameters (or, equivalently, it is a function of the geometrical parameters of the black hole). If one considers a one-parameter family of black-hole solutions (e.g. Schwarzschild solutions) such that the horizon area can be considered as a parameter for that family, than *any* physical quantity associated with those black holes can be trivially expressed as a function of the area of the horizon.

When the thermodynamical potentials are provided by some other physical equations, then one can write down the first law of thermodynamics that relates all the relevant quantities; the usual form of this fundamental principle is

$$\delta m = T\delta S + \Omega\delta J + b\delta q + \ldots, \tag{2}$$

where m is a measure of the energetic content of the gravitational system; T, Ω, b are the temperature, the angular velocity of the horizon and an electromagnetic thermodynamical potential, respectively. The other quantities S, J and q are the entropy, the angular momentum and the electric charge. Other terms can of course appear depending on how liberal we decide to be in allowing other physical quantities (e.g. when more general gauge fields are present). The deformation operator δ denotes any infinitesimal variation of parameters in the space of solutions, e.g. along a (restricted) family of arbitrarily fixed solutions (as it happens, e.g. along Kerr–Newman solutions).

The main content of the Wald prescription was to use the first law of thermodynamics to define S whenever all the other quantities are already known. We note that δm, δJ, $\delta q, \ldots$ can be defined via the Nöther theorem (or other variational techniques) as boundary-asymptotic quantities. Hence the stronger the tools we have to compute conservation laws the more general situations we are able to deal with. In this view variational calculus has a prominent role being the natural setting for studying conservation laws in a global and covariant way and to guide us towards their physical interpretation.

3. A variational setting for conserved currents and entropy in relativistic field theories

A relativistic field theory is a generally covariant Lagrangian theory in which fields have enough information to determine the geometry (or precisely a metric structure) of space-time. The details of the formalism can be found in Ref. [20]. How exactly space-time geometry is encoded by dynamical fundamental fields actually depends on the context and on the framework adopted (e.g. purely metric, Palatini metric affine, purely affine, purely tetrad, tetrad affine, Ashtekar's variables, Chern–Simons formulations, just to mention the most commonly used).

Fields are sections of a natural (or gauge natural) bundle $\pi : C \rightarrow M$ over the space-time base manifold M, called the *configuration bundle*; see Ref. [20]. If the configuration bundle is natural then space-time diffeomorphisms act on sections and the Lagrangian is required to be generally covariant. If, more generally, the configuration bundle is gauge natural then it is associated to a principal structure bundle $p : \mathcal{P} \rightarrow M$ that selects a class of transformations, called *gauge transformations* and denoted by $\mathrm{Aut}(\mathcal{P})$, acting on sections. In these cases (e.g. when tetrads, Ashtekar's variables or Chern–Simons formulations are used) the Lagrangian is required to be gauge covariant; see Refs. [16,20].

We shall also use hereafter jet bundle prolongations, which provide the geometrical framework for Calculus of Variations on bundles. For any bundle C an infinite chain of bundles can be defined that are denoted by $J^k C$. If we denote by (x^μ, y^i) fibered coordinates on C, fibered coordinates on $J^k C$ are $(x^\mu, y^i, y^i_\mu, \ldots, y^i_{\mu_1\ldots\mu_k})$ where the extra coordinates are meant to represent partial derivatives of fields up to order k and hence to be symmetric in the lower indices. Any section, fibered morphism or vector field on C uniquely and functorially induces a section, a fibered morphism or a vector field on $J^k C$.

A *Lagrangian* is a bundle morphism $L : J^k C \rightarrow A_m(M)$, m being $\dim(M)$ and $A_m(M)$ the bundle of m-forms over M. Locally it is given by

$$L = \mathcal{L}(x^\mu, y^i, y^i_\mu, \ldots, y^i_{\mu_1\ldots\mu_k})\mathrm{d}s, \tag{3}$$

where $\mathrm{d}s = \mathrm{d}x^1 \wedge \mathrm{d}x^2 \wedge \ldots \wedge \mathrm{d}x^m$ is the canonical pointwise basis of m-forms induced by coordinates x^μ chosen (locally) on M.

Given a vertical vector field X on C, locally represented as $X = \delta y^i \partial_i$, both the sections and the Lagrangian can be dragged along its flow. One can hence define a (global) bundle morphism $\delta L : J^k C \rightarrow V^*(J^k C) \otimes A_m(M)$. The *Lagrangian deformation* usually defined in physics is related to the bundle morphism δL as follows

$$\delta_X L = \langle \delta L | j^k X \rangle, \tag{4}$$

where $\langle \cdot | \cdot \rangle$ denotes the natural duality between $V(J^k C)$ and $V^*(J^k C)$.

Inspired by Spencer cohomology [21,22], we define a *variational morphism* to be any global bundle morphism $\mathbb{A} : J^k C \rightarrow J^h E^* \otimes A_{m-n}(M)$ for some vector bundle E. Due to the canonical isomorphisms $V(J^k C) \simeq J^k V(C)$, the morphism δL is, in fact, a variational morphism with $E = V(C)$, $h = k$ and $n = 0$. It can be proven [20] that each variational morphism $\mathbb{A}$ can be globally and almost canonically split to define two global morphisms

$$\begin{cases} \mathbb{V} : J^{k+h}C \to J^{h'}E^* \otimes A_{m-n}(M) \\ \mathbb{S} : J^{k+h-1}C \to J^{h-1}E^* \otimes A_{m-n-1}(M), \end{cases} \tag{5}$$

where h' is a function of h and n ($h' = 0$ if $n = 0$ while $h' = h - 1$ when $n > 0$). These morphisms are such that the following splitting property holds for any section X of E

$$\langle \mathbb{A} | j^h X \rangle = \langle \mathbb{V} | j^{h'} X \rangle + \mathrm{Div} \langle \mathbb{S} | j^{h-1} X \rangle. \tag{6}$$

In Eq. (6) Div denotes the *formal divergence operator on forms*, defined by

$$\mathrm{Div}(f) \circ j^{k+1} \sigma = \mathrm{d}(f \circ j^k \sigma), \qquad f : J^k Y \to A(M), \quad \sigma : M \to (C),$$

$d(\cdot)$ being the exterior differential operator on forms, σ being a section and $A(M) \equiv \oplus_i A_i(M)$ denoting the bundle of all forms over M. The splitting Eq. (6) is simply the globalization of the usual algorithm for computing field equations out of the variation of the Lagrangian.

The variational morphism $\mathbb{V}$ is uniquely determined, while $\mathbb{S}$ belongs, in general, to a family determined by an extra choice (e.g. on a connection used to globalize the integration by parts involved in the splitting algorithm); in all cases analyzed below $\mathbb{S}$ will be, however, uniquely defined, due to the low order of the Lagrangian ($k = 1$ or $k = 2$); see Ref. [23]. Variational morphisms are an algebraization of the classical variational calculus; they capture the geometrical essence of it and provide the tools sufficient to deal with global properties of field equations, dependence on boundary conditions and conservation laws, see Refs. [16,20,24–26] for more details.

When $\mathbb{A} = \delta L$ then the two morphisms obtained by the splitting are denoted, respectively, by $\mathbb{V} = \mathbb{E}(L)$ and $\mathbb{S} = \mathbb{F}(L)$ and they are called the *Euler–Lagrange morphism* and the *Poincaré–Cartan morphism*, respectively:

$$\langle \delta L | j^k X \rangle = \langle \mathbb{E}(L) | X \rangle + \mathrm{Div} \langle \mathbb{F}(L) | j^{k-1} X \rangle. \tag{7}$$

We stress that despite the Poincaré–Cartan morphism being irrelevant for field equations, conservation laws obtained via the Nöther theorem are strongly dependent on it.

In both cases of natural and gauge natural theories there exists a vector bundle the sections of which are in one-to-one correspondence with infinitesimal symmetries. A section Ξ of this *symmetry bundle* (the tangent bundle of the space-time M in the natural case, or the bundle of right invariant vector fields on the structure bundle in the gauge natural case) acts on fields and hence defines Lie derivatives $\mathcal{L}_\Xi y$ of fields. The condition for Ξ being an infinitesimal symmetry is the following

$$\langle \delta L | j^k \mathcal{L}_\Xi y \rangle = \mathrm{Div}(\alpha(\Xi)) \tag{8}$$

for some suitable variational morphism $\alpha(\Xi)$ linear with respect to Ξ (e.g. in natural theories with Ξ equal to a space-time vector field ξ we have simply $\alpha(\xi) = i_\xi L$). This definition, which is equivalent to the (gauge or general) covariance of the Lagrangian, contains all computational details of the Nöther theorem. In fact, it is easy to use the standard variational techniques introduced above, together with the fact that

Lie derivatives $\mathcal{L}_{\Xi}y$ take their values in the space of vertical vector fields, to recast the covariance condition into the form of a conservation law

$$\mathrm{Div}\ \mathcal{E}(L,\Xi) = \mathcal{W}(L,\Xi) \begin{cases} \mathcal{E}(L,\Xi) = \langle \mathcal{F}(L) | j^{k-1}\mathcal{L}_{\Xi}y \rangle - \alpha(\Xi) \\ \mathcal{W}(L,\Xi) = -\langle \mathbb{E}(L) | \mathcal{L}_{\Xi}y \rangle. \end{cases} \tag{9}$$

Of course $\mathcal{W}$ vanishes when computed along a solution of field equations $y = \sigma(x)$; hence $\mathcal{E}$ is conserved on-shell.

Being infinitesimal symmetries Ξ generic sections of a suitable defined vector bundle, in both natural and gauge natural cases both the Nöther current $\mathcal{E}$ and the *work form* $\mathcal{W}$ can be regarded as variational morphisms themselves and split to define the so-called *Bianchi identities* $\mathcal{B}$, *reduced current* $\tilde{\mathcal{E}}$ and the *superpotential* $\mathcal{U}$:

$$\begin{cases} \mathcal{E}(L,\Xi) = \tilde{\mathcal{E}}(L,\Xi) + \mathrm{Div}\ \mathcal{U}(L,\Xi) \\ \mathcal{W}(L,\Xi) = \mathcal{B}(L,\Xi) + \mathrm{Div}\ \tilde{\mathcal{E}}(L,\Xi), \end{cases} \tag{10}$$

where $\mathcal{B} = 0$ off-shell (this follows from Eq. (9); see Refs. [16,20,22,27]), $\tilde{\mathcal{E}} \sim 0$ (i.e. $\tilde{\mathcal{E}} = 0$ on-shell), while $\mathcal{U}$ will enter into the definition of conserved quantities (later appearing in the definition of the entropy) in a Gauss-like fashion. We stress that the content of the splitting used to define superpotentials and hence conserved quantities is variational in nature and, relying on Spencer cohomology (6), it is also algorithmic. Moreover, one of the most relevant results that can be inferred from the jet-bundle approach to conserved quantities relies on the fact that the Nöther current $\mathcal{E}$ is exact on-shell regardless of the space-time topology (*weak conservation law*) while the difference $\mathcal{E} - \tilde{\mathcal{E}}$ is exact even off-shell (*strong conservation law*); see Eq. (10).

Before defining conserved quantities, i.e. physical observables, it is still necessary to proceed one step further with the geometric Variational Calculus machinery. Indeed, it is well known that definitions of conserved quantities that rely on direct integration of the superpotential $\mathcal{U}$ lead, in gravitational theories, to the wrong conserved quantities (e.g. the so-called anomalous factor in asymptotically flat space-times [28]) and, even worse, they usually give divergent results when solutions are not asymptotically flat [16] (see also Ref. [29]).

We then proceed as follows. Let X be a vertical vector tangent to the space of solutions. By considering the variation along X of the Nöther current $\mathcal{E}$, defined according to Eq. (10) and to Eq. (9), one can define [16,25] the *variation* $\delta_X Q_B(\Xi)$ of the conserved quantity $Q_B(\Xi)$ to be the following quantity:

$$\delta_X Q_B(\Xi) = \int_B \delta_X \mathcal{U}(L,\Xi) - i_\xi \langle \mathbb{F}(L) | j^{k-1} X \rangle, \tag{11}$$

where i_ξ denotes contraction along the space-time projection ξ of Ξ, B is the (outer) boundary of a (space-like) $(m-1)$-dimensional surface Σ and the evaluation along a solution σ is understood. If Eq. (11) is integrable, then $\delta_X Q_B(\Xi) = \delta_X[Q_B(\Xi)]$ is the *true* variation of a true conserved quantity $Q_B(\Xi)$. Otherwise Eq. (11) is still defined and called, by an abuse, the variation of a conserved quantity; see e.g. Refs. [30,31].

We have at least two leading and convincing arguments to support such a definition.

(1) First, the integrand in the right-hand side of Eq. (11) does not depend on the variational cohomology class $[L]$ to which the Lagrangian L belongs, where two

elements of $[L]$ differ only by the addition of divergence terms. This is a rather welcome property and we shall analyze its far-reaching consequences later on.

(2) The second reason relies instead on the fact (see Refs. [30–32]) that the variation of the conserved quantity (11) satisfies the symplectic relation:

$$\delta_X Q_{\partial\Sigma}(\Xi) = \int_{\Sigma} \omega(X, \Xi) \mathrm{d}^{m-1}x, \tag{12}$$

where

$$\omega(X, \xi) = \delta_X \langle \mathbb{F}(L) | j^{k-1} \mathcal{L}_{\Xi} \sigma \rangle - \mathcal{L}_{\xi} \langle \mathbb{F}(L) | j^{k-1} X \rangle \tag{13}$$

is the (pre)symplectic current (see Ref. [33]), while $\partial\Sigma$ denotes the boundary of a $(m-1)$-dimensional region Σ. The relation (12) shows that the conserved quantity $Q(\Xi)$, if it exists, can be truly considered as the Hamiltonian that generates the evolution of fields along the vector field Ξ.

In natural theories the information on the energy-momentum content inside a region bounded by a surface B is obtained by integrating the right-hand side of Eq. (11) on the closed surface B with respect to generators of diffeomorphisms, i.e. space-time vector fields ξ. In particular, when ξ is time-like we refer to $\delta_X Q_B(\xi)$ as the *variation of the energy* inside B and relative to the vector field ξ.

In gauge natural theories a connection on the structure bundle can be built directly out of fields so that gauge transformations can be formally split into vertical and horizontal gauge transformations (*horizontal* and *vertical* with respect to the dynamical connection that will be known when field equations have been solved). In this case horizontal symmetries define energy-momentum while vertical symmetries define *gauge charges*.

It is important to stress that we could have obtained expression (11) leaving a direct Lagrangian formulation of the theory out of consideration. This is a desirable property both from a mathematical and a physical point of view. Let us indeed point out that what is really relevant in the description of a field theory are its field equations (and the Lagrangian is just a—powerful—mathematical prescription to obtain them; but all the Lagrangians belonging to the same equivalence class $[L]$—see item (1) above—give rise to the same field equations). Field equations select the admissible configurations and dictate their evolution. However, the solutions of field equations represent just a small part of the information that can be extracted out of a field theory. Other relevant information is encoded in the field symmetries; the knowledge on how fields are dragged along infinitesimal generators of symmetries is hence a further fundamental detail. The two main ingredients that have to enter into a satisfactory definition of conserved quantities, namely the field equations content and the symmetry information, can be joined together into a unique structure. One can in fact define *the variational Lagrangian* [26]:

$$L'(\Xi) = -\langle \mathbb{E} | \mathcal{L}_{\Xi} y \rangle \tag{14}$$

through the contraction of field equations with the Lie derivatives of the fields (it corresponds to the variational morphism $\mathcal{W}$ of Eq. (9) and, obviously, it is vanishing on-shell). This is a true Lagrangian and for the information it encodes, it is then in a good position to represent the fundamental object out of which we derive conserved quantities.

As L' is a Lagrangian it can in fact be handled by means of the tools of the Calculus of Variations in jet bundles we have introduced so far. In particular we can make use of Eq. (7) to obtain the splitting:

$$\langle \delta L'(\Xi) | j^a X \rangle = \langle \mathbb{E}(L'(\Xi)) | X \rangle + \mathrm{Div} \langle \mathbb{F}(L'(\Xi)) | j^{a-1} X \rangle = \mathrm{Div} \langle \mathbb{F}(L'(\Xi)) | j^{a-1} X \rangle, \quad (15)$$

where a is a finite integer depending on the theory (for simplicity, a can be assumed to be the order of field equations). The last equality in Eq. (15) is due to the fact that L' is a pure divergence and therefore variationally trivial: $\mathbb{E}(L') = 0$; see Eq. (10) and Ref. [26]. Since L' is vanishing on-shell we obtain the (weak) conservation law:

$$\mathrm{Div} \langle \mathbb{F}(L'(\Xi)) | j^{a-1} X \rangle = 0 \ \text{ on-shell.}$$

Moreover, the morphism $\langle \mathbb{F}(L'(\Xi)) | j^{a-1} X \rangle$ depends linearly on the vector Ξ together with its derivatives up to a finite order. As Ξ is a generic section of a well-defined vector bundle we can again implement Spencer cohomology (6) with respect to the vector Ξ (roughly speaking, we integrate by parts with respect to the vector field Ξ) and we end up with the canonical splitting:

$$\langle \mathbb{F}(L'(\Xi)) | j^{a-1} X \rangle = \delta_X \tilde{\mathcal{E}}(\Xi) + \mathrm{Div}\ \mathcal{V}(X, \Xi),$$

where $\tilde{\mathcal{E}}$ is exactly the reduced current (9), while $\mathcal{V}(X, \Xi)$ agrees with the right-hand side of Eq. (11) (we refer the interested reader to Ref. [26] for the technical details). In this way, we could then say that the definition (11) captures the fundamental mathematical essence of the theory we are dealing with and it endows it with a direct physical interpretation. Namely, *theories that are equivalent as far as their field equation content is considered produce conserved quantities that are equivalent when computed along equivalent solutions*. Indeed definition (11), being built directly from the field equations via the variational Lagrangian (14), turns out to be related only to the solutions and not to the specific mathematical rule (i.e. the Lagrangian) we choose to describe the theory. For example, the energy enclosed in a region surrounding a black-hole solution will eventually depend, as is physically expected, only on the geometry of space-time and not on the specific configuration variables (e.g. metric, tetrad, connection) or the Lagrangians (e.g. Hilbert, tetrad-affine, Chern–Simons Lagrangians) we initially choose to describe it.

In particular it has been shown that in gravitational theories that admit a solution with horizon, a symmetry generator Ξ can be found so that its associated conserved quantity $\delta_X Q_B(\Xi)$ is $\delta_X m - \Omega \delta_X J - b \delta_X q$. According to Eq. (2), $\delta_X Q_B(\Xi)$ defines directly $T \delta_X S$ that can be integrated for S once the temperature T is known. In this way the expected entropy has been recovered not only for asymptotically flat black holes (Kerr–Newman [16]) but also for asymptotically anti-de Sitter black holes [17,18,29,34], asymptotically de Sitter black holes (Schwarzschild–de Sitter solution [30]), asymptotically locally flat and asymptotically locally anti-de Sitter solutions (Taub-bolt [19,35]) and, recently, also isolated [36] and causal [30] horizons.

These techniques for conservation laws have been shown to capture the covariant and global aspects of the physical conserved quantities that enter into the first law of thermodynamics (2). The Kerr–Newman, BTZ and Taub-bolt solutions, as well as Yang–Mills theories, spinor physics, Chern–Simons and Lovelock theories have been directly and successfully treated covering many different examples with different boundary conditions.

Moreover, from Eqs. (3.10) and (3.11) it directly follows that

$$\delta_X Q_{\partial\Sigma}(\Xi) = 0, \tag{16}$$

whenever Ξ is a Killing vector field (a condition that is obviously encountered in stationary space-times). We stress that this property reverberates on the fact that entropy is defined as a homological quantity that can, but need not, be defined on horizons (any surface B homolog to an horizon cross section H, i.e. $B - H = \partial\Sigma$, may play the same role). This fact drastically reduces the requirement on existence and type of horizons. We also mention that the general case of isolated horizons, where a global Killing vector does not exist, can be handled inside the same framework developed so far.

4. Boundary conditions and reference backgrounds in General Relativity

Of course, through definition (11) we are, in general, able to produce just the *variation of a conserved quantity* and not the *conserved quantity* on its own. To do that, a position on boundary conditions has, in principle, to be taken. This problem has been solved in many equivalent but different ways (we stress by the way that variational calculus is essential to prove equivalence of different prescriptions). One way is to break covariance to define the so-called energy-momentum pseudotensors [37]. A prescription that is more in the spirit of General Relativity (principle of general covariance) is obtained by introducing a reference background in the variational principle (of course some sort of flat background is used also in the definition of asymptotically flat solutions, which can be seen as deformations of a Minkowski flat space-time). It is, in fact, known [17,20,28,38,39] that one can define a covariant Lagrangian depending on two fields g and $\bar{g}$ by setting:

$$L_{g\bar{g}} = (R\sqrt{g} - d_\mu(\sqrt{g}g^{\alpha\beta}w^\mu_{\alpha\beta}) - \bar{R}\sqrt{\bar{g}})\mathrm{d}s, \tag{17}$$

where we set

$$\begin{array}{ccc} g_{\mu\nu} & \bar{g}_{\mu\nu} & \text{metrics} \\ \Gamma^\alpha_{\beta\nu} & \bar{\Gamma}^\alpha_{\beta\nu} & \text{connections} \\ R^\alpha_{\beta\mu\nu} & \bar{R}^\alpha_{\beta\mu\nu} & \text{Riemann tensors} \\ R_{\mu\nu} & \bar{R}_{\mu\nu} & \text{Ricci tensors} \\ R = g^{\mu\nu}R_{\mu\nu} & \bar{R} = \bar{g}^{\mu\nu}\bar{R}_{\mu\nu} & \text{scalar curvatures} \\ u^\mu_{\alpha\beta} = \Gamma^\mu_{\alpha\beta} - \Gamma^\epsilon_{\epsilon(\alpha}\delta^\mu_{\beta)} & \bar{u}^\mu_{\alpha\beta} = \bar{\Gamma}^\mu_{\alpha\beta} - \bar{\Gamma}^\epsilon_{\epsilon(\alpha}\delta^\mu_{\beta)} & \end{array} \tag{18}$$

and

$$w^\mu_{\alpha\beta} = u^\mu_{\alpha\beta} - \bar{u}^\mu_{\alpha\beta}. \tag{19}$$

This new Lagrangian has a corrected superpotential with respect to the superpotential ensuing from the usual Hilbert Lagrangian (see Refs. [16,20,28,38,40]); it is known that this corrected superpotential provides a conserved quantity that depends on both g and $\bar{g}$ and satisfactorily represents the relative energy-momentum between the two field configurations. In this way $\bar{g}$ can be fixed to be any solution of field equations (e.g. Minkowski, pure de Sitter or anti-de Sitter, Taub NUT) in order to study solutions (e.g. Kerr–Newman, Schwarzschild–de Sitter, BTZ or Taub-bolt) with different asymptotic behaviors (asymptotically flat, asymptotically de Sitter, asymptotically anti-de Sitter or asymptotically locally flat, respectively).

A 'philosophical' comment is in order at this point. We are well aware that the use of backgrounds does not meet universal consensus in current literature, especially in loop quantum gravity [41]. For this reason we would like to briefly comment on this issue before going on to discuss other equivalent ways of dealing with different boundary conditions. First, we completely agree that backgrounds should not directly enter the description of physics at a fundamental level; we agree with all considerations about the need for a background-free quantization scheme.

However, in our opinion, more attention should be given to *what* exactly a background is. In particular, we claim that the reference background entering Eq. (17) is completely different in nature from the background assumed in standard string theory or from the role of Minkowski background in particle theory. The differences are both in the physical interpretation and in the technical variational behavior of the background fields and we believe such a difference deserves more attention.

In particle physics the background is endowed with a direct physical interpretation (for example, free particles are supposed to move along its geodesics) and some of the fundamental physical structures explicitly depend on it (e.g. the inner product of the Hilbert space of quantum states). From a technical point of view the background is considered fixed by both variations and symmetries in order to kill some terms in the conservation laws by the condition $\mathcal{L}_\xi \bar{g} = 0$ (i.e. restricting to Killing symmetries). These extra terms need to be killed because they do not vanish on-shell, which in turn is directly due to the fact that backgrounds are not varied and one has less field equations than one usually would have in an unconstrained variational principle. This position about the variation of the background is forced because the field equation one would obtain by varying the background is in direct conflict with the decision of keeping the background fixed.

In Eq. (17) the situation is completely different (as well as, in our opinion, completely satisfactory from a physical point of view). First, the background $\bar{g}$ is not endowed with a direct physical interpretation; free particles move along geodesics of g and both g and $\bar{g}$ are unknown until field equations are solved (so that, for example, the Hilbert structure on quantum states cannot depend on them). Let us note that the choice of $\bar{g}$ does not influence the dynamics of g due to the particular form of the Lagrangian (17) (g and $\bar{g}$ do not interact). The role of $\bar{g}$ is physically just to account for the observer freedom in choosing a vacuum state with respect to which one can compute some physical quantity (e.g. the energy). It is, in fact, known that just variations of the energy are endowed with a direct fundamental role while the absolute values of conserved quantities are observer dependent and, as such, noncovariant.

We also stress that in many 'simple' field theories the configuration bundle (e.g. when it is a vector bundle) has a canonical section (e.g. the zero section) which can be selected as a reference background. Hence, although one could certainly look for a reference-background-depending variational principle, in all these cases one can escape it. On the contrary, when the configuration bundle has no canonical sections (e.g. when it is $C = \text{Lor}(M)$, the bundle of Lorentzian metrics over M) such ambiguities have to be dealt with explicitly.

Also from a technical point of view the metric $\bar{g}$ in Eq. (17) is considered in a more elegant way: both fields g and $\bar{g}$ are treated on exactly an equal footing and as absolutely normal variational stance. Variations are general (they do not fix $\bar{g}$) and symmetries are not required to keep the background fixed. On the contrary, symmetries drag both g and $\bar{g}$ equally. Field equations are doubled (we obtain field equations for both g and $\bar{g}$) meaning that $\bar{g}$ can be, in principle, chosen obeying Einstein field equation. Of course field equations show that g and $\bar{g}$ do not interact so that they can be chosen independently. It is the particular form of the Lagrangian that ensures that no extra terms survive in the conservation laws once both field equations for g and $\bar{g}$ are taken into account. Hence we have no need of requiring the symmetry to be a Killing symmetry of the dynamical field g nor of the reference background.

Let us finally mention that the superpotential $\mathcal{U}(L_{g\bar{g}}, \xi)$ obtained from the Lagrangian (17) reproduces the well-known Regge–Teitelboim prescription (see Ref. [14]) once it is specialized to an asymptotically flat solution and a Minkowski background, so that the variational principle (17) can be regarded as a generalization of this fairly accepted prescription. Indeed, the variation $\delta_X \int_B \mathcal{U}(L_{g\bar{g}}, \xi)$ reproduces the variational formula (11) when Dirichlet boundary conditions are imposed on the metric, namely when the metric g approaches the background $\bar{g}$ on the domain of integration B (which may or may not coincide with spatial infinity); see Ref. [40]. We obtain in this way a formula to define the quasilocal energy, i.e. the energetic content inside a bounded region of space-time, which is general enough to be adapted to many (every?) situations, provided the metric and its background are suitably chosen and matched.

Nevertheless, as already explained in the introduction, we have nowadays convincing arguments to state that gravitational systems are truly thermodynamic systems. As such, we should be able to attribute to them different kinds of energies, such as, e.g. internal energy or free energy. To meet this requirement we have to again take into consideration formula (11). We stress again that in specific applications it is more convenient for calculations to consider just variations of physical observables without performing any a priori integration. In this way complicated calculations related to background choices and background matching conditions are avoided in a first approach. The background will instead come into play just at the end, in a more manageable fashion, as a 'constant of integration'. Note indeed that, once we fix B in the expression (11), there exist many 'energies', depending both on the symmetry vector field Ξ and on the variational vector field X. The former vector (or better, its projection ξ onto the space-time manifold M) determines, via its flow parameter, how the observers located on B evolve as the 'time' flows. The vector field X fixes instead the boundary conditions on the dynamical fields (e.g. Dirichlet or Neumann boundary conditions). Each choice of the pair (ξ, X) gives rise to a particular realization of a physical system endowed thence with its own 'energy content'. It was indeed argued in Refs. [11,12,42] that boundary conditions in General Relativity exactly correspond to boundary conditions of a thermodynamical ensemble and that dynamical fields that are conjugated to each other in a symplectic sense, can be also considered as being thermodynamically conjugated. Accordingly, different ensembles can be realized by exchanging the fields that are kept fixed as boundary data with their respective boundary conjugated fields. From this point of view the variational vector field X turns out to be a variation into the ensemble realized with specified boundary conditions (e.g. microcanonical, canonical or grand canonical ensemble), namely X fixes among the dynamical variables, which ones are the intensive and which ones are the extensive variables. The internal energy obtained from the Lagrangian (17), as already noted, corresponds to Dirichlet boundary conditions, i.e. boundary metric kept fixed. It was recently suggested [30] that Neumann boundary conditions (consisting in keeping fixed the momenta that are conjugated to the boundary metric) lead to another kind of energy that we refer to as the 'gravitational heat', namely the TS contribution to the internal energy of the thermodynamic system ($T =$ temperature, $S =$ entropy). We stress that the definition of entropy as it arises from this prescription still depends on the vector ξ, i.e. it is *observer dependent*. This in turn implies that entropy contributions arise whenever an obstruction exists to globally foliating space-time into surfaces of constant time (and this property perfectly agrees with the perspective of Hawking et al. [7]). In other words, entropy turns out to be closely related to the coordinate singularities of the solution describing space-time. Let us better clarify this point on a physical ground. Given a time-like vector ξ we can define a set of (local) observers evolving with velocity ξ and we can choose an observer-adapted system of coordinates in which the observers themselves are at rest. In this reference frame the metric coordinate singularities correspond to homological boundaries and, according to the homological properties our definition of entropy is endowed with (and that are basically inherited from the property (16)) each boundary gives rise to its own contribution to the total entropy. However, from a physical point of view, the metric coordinate singularities correspond to regions of inaccessibility for the observers (one-way membranes). Entropy then turns out to be associated with the observer's unseen degrees of freedom, namely to the regions that the observers have no physical access to; see Ref. [43]. Extending Beckenstein's original idea to a much broader context, we are then induced to state that *entropy measures the information content beyond the hidden regions*. We indeed stress that the hidden regions we are referring to comprise the standard black-hole horizons but encompass also a much wider class of singularities such as causal horizons (e.g. cosmological or Rindler horizons) as well as Taub-bolt and Taub-NUT singularities.

5. Conclusion and perspectives

We note that, due to cohomological properties, the entropy as defined via the first law (2) can be geometrically calculated, e.g. as an integral at space infinity; hence the geometrical properties of the horizons are certainly irrelevant. In particular there is no need for the horizon to be Killing (even less to be a bifurcate Killing horizon!). This traditional assumption is related to the wish of reducing the entropy to be an integral at the horizon with the consequent need to control the limit behavior at the horizon itself, although these features turn out to be unnecessary and produce too stringent requirements that are not justified by phenomenology. For example, in Taub-bolt solutions the Misner string is not dressed by a 'good' horizon, although the entropy can be properly defined as a quantity at infinity. The traditional physical quantities at singularities can be recovered by (artificially) splitting the domain of integration into the union of a number of regions surrounding the singularities themselves.

The present situation is fairly satisfactory and covers a wide number of examples, extremely different in the space-time dimension, in asymptotic behavior or structure of singularities. We have successfully dealt with Kerr–Newmann, BTZ, Taub-bolt solutions, black holes in higher dimension; interesting results have also been found for nonstationary solutions (even if specific examples suffer from the lack of exact solutions that are at the same time geometrically well defined). Moreover, entropy is a geometrical quantity the cohomological properties of which reverberate in the first laws of thermodynamics. We already compared our approach to the Brown–York prescription and to Hamiltonian-based formulations, showing that they can be understood within our geometrical framework. An interesting future investigation will establish a correspondence between our geometrical framework and the prescriptions based on counterterms inspired by the AdS/CFT correspondence.

Another interesting example to be coped with will be the case of naked singularities. They have been proved to form during (not completely odd) dust collapses; Witten and Vaz [44] proved, moreover, that they emit particles (of course with an infinite power flux), in agreement with the usual interpretations of singularities as quantum-unstable situations. In principle, studying the entropy of such solutions is extremely interesting since it could give us new hints about the role of horizons for entropy in General Relativity. In the 'golden age' papers in the area, back in the 1970s, horizons were qualitatively essential in the interpretation of the entropy. They provided a black box hiding the internal degrees of freedom, making it impossible for us to investigate the internal status of black holes. As we said, entropy was introduced and first motivated as a measure of our ignorance due to this fundamental effect. Of course this does not mean quantitatively that all gravitational contributions to the entropy should be of this kind.

Certainly, black holes (as well as other kinds of coordinate singularities) are a great source of entropy since horizons in general *absolutely* prevent us from knowing *anything* about what is going on within it, although some contribution can come from a less absolute difficulty in getting new information. For example, it seems reasonable that looking at what happens very close to a naked singularity (just in the middle of an infinite power flux of particles emitted because of Hawking-like radiation) might be fairly difficult as well. Of course we expect naked singularities to be *less* entropic with respect to black holes, at least because of the fact that a naked singularity should tend to dress with an horizon as time passes by and its energy decreases because of Hawking radiation. For these reasons and because of the second principle of thermodynamics (although beyond the scope for which it was originally proved in black-holes dynamics) it seems reasonable to expect that the entropy of naked singularities should be less than that of the black hole they could eventually produce. This is particularly interesting, since despite the obvious difficulties to associate a (finite) temperature with the infinite-power emission spectrum of naked singularities, it seems to indicate that in any case one should expect a finite entropy as a final result.

References

[1] Bekenstein, J.D. (1973), Phys. Rev. D 7(6), 2333.
[2] Bekenstein, J.D. (1974), Phys. Rev. D 9(12), 3292.
[3] Hawking, S.W. (1975), Commun. Math. Phys. 43, 199.
[4] Christodoulou, D. (1970), Phys. Rev. Lett. 25.
[5] Christodoulou, D. and Ruffini, R. (1971), Phys. Rev. D 4.
[6] Shannon, C.E. and Weaver, W (1949), The Mathematical Theory of Communication, University of Illinois Press, Urbana.
[7] Hawking, S.W., Hunter, C.J. and Page, D.N. (1999), Phys. Rev. D 59, 044033 hep-th/9809035.
[8] Taub, A.H. (1951), Ann. Math. 53, 472.
[9] Newman, E.T., Tamburino, L. and Unti, T. (1963), J. Math. Phys. 4.
[10] Misner, C.W. (1963), J. Math. Phys. 4, 924.
[11] Brown, J.D. and York, J.W. (1993), Phys. Rev. D 47(4), 1407.
[12] Brown, J.D. and York, J.W. (1993), Phys. Rev. D 47(4), 1420.
[13] Arnowitt, R., Deser, S. and Misner, C.W. (1962), Gravitation: an Introduction to Current Research (Witten, L., eds.), Wiley, New York, p. 227.
[14] Regge, T. and Teitelboim, C. (1974), Ann. Phys. 88, 286.
[15] Iyer, V. and Wald, R. (1994), Phys. Rev. D 50 gr-qc/9403028.
[16] Fatibene, L., Ferraris, M., Francaviglia, M. and Raiteri, M. (1999), Ann. Phys. 275, 27.
[17] Fatibene, L., Ferraris, M., Francaviglia, M. and Raiteri, M. (1999), Phys. Rev. D 60, 124012.
[18] Fatibene, L., Ferraris, M., Francaviglia, M. and Raiteri, M. (1999), Phys. Rev. D 60, 124013.
[19] Fatibene, L., Ferraris, M., Francaviglia, M. and Raiteri, M. (2000), Ann. Phys. 284, 197.
[20] Fatibene, L. and Francaviglia, M. (2003), Natural and Gauge Natural Framework for Classical Field Theories, Kluwer, Dordrecht.
[21] Goldschmidt, H. and Spencer, D. (1978), J. Diff. Geom. 13, 455.
[22] Ferraris, M., Francaviglia, M. and Robutti, O. (1987), Géométrie et Physique. In Proceedings of the Journées Relativistes 1985 (Marseille, 1985), (Choquet-Bruhat, Y., Coll, B., Kerner, R., Lichnerowicz, A., eds.), Hermann, Paris, pp. 112–125.
[23] Ferraris, M (1983), in: Geometrical Methods in Physics, Proceedings of the Conference on Differential Geometry and its Applications, Nové Město na Moravě, Czechoslovakia, September 5–9, 1983.
[24] Trautman, A. (1962), Gravitation: an Introduction to Current Research (Witten, L., eds.), Wiley, New York, p. 168; Trautman, A. (1967), Commun. Math. Phys. 6, 248.
[25] Ferraris, M. and Francaviglia, M. (1989), Atti Sem. Mat. Univ. Modena 37, 61; Ferraris, M., Francaviglia, M. and Sinicco, I. (1992), Nuovo Cimento 107B(11), 1303; Ferraris, M. and Francaviglia, M (1986), 7th Italian Conference on General Relativity and Gravitational Physics, Rapallo (Genoa), September 3–6, 1986.
[26] Ferraris, M., Francaviglia, M. and Raiteri, M. (2003), Class. Quantum Grav. 20, 4043–4066. gr-qc/0305047.
[27] Ferraris, M. and Francaviglia, M. (1991), Mechanics, Analysis and Geometry: 200 Years after Lagrange (Francaviglia, M., eds.), Elsevier, Amsterdam, p. 451.
[28] Katz, J. (1985), Class. Quantum Grav. 2, 423; Katz, J., Bicak, J. and Lynden-Bell, D. (1997), Phys. Rev. D 55(10), 5957.
[29] Allemandi, G., Francaviglia, M. and Raiteri, M. (2003), Class. Quantum Grav. 20, 483–506. gr-qc/0211098.
[30] Francaviglia, M. and Raiteri, M. gr-qc/0402080.
[31] Iyer, V. and Wald, R.M. (1995), Phys. Rev. D 52, 4430–4439. gr-qc/9503052.
[32] Francaviglia, M. and Raiteri, M. (2002), Class. Quantum Grav. 19.
[33] Burnett, G. and Wald, R.M. (1990), Proc. Roy. Soc. Lond. A 430, 56; Lee, J. and Wald, R.M. (1990), J. Math. Phys. 31, 725.
[34] Allemandi, G., Francaviglia, M. and Raiteri, M. (2003), Class. Quantum Grav. 20, 5103–5120. gr-qc/0308019.
[35] Clarkson, R., Fatibene, L. and Mann, R.B. (2003), Nucl. Phys. B 652, 348–382. hep-th/0210280.
[36] Allemandi, G., Francaviglia, M. and Raiteri, M. (2002), Class. Quantum Grav. 19, 2633–2655. gr-qc/0110104.

[37] Misner, C.W., Thorne, K.S. and Wheeler, J.A. (1973), Gravitation, Freeman, San Francisco.
[38] Ferraris, M. and Francaviglia, M. (1990), Gen. Rel. Grav. 22(9), 965; Ferraris, M. and Francaviglia, M. (1988), 8th Italian Conference on General Relativity and Gravitational Physics, World Scientific, Singapore, Cavalese (Trento), August 30–September 3, p. 183.
[39] Chen, C.M. and Nester, J.M. (2000), Gravit. Cosmol. 6, 257 gr-qc/0001088.
[40] Fatibene, L., Ferraris, M., Francaviglia, M. and Raiteri, M. (2001), J. Math. Phys. 42(3), 1173 gr-qc/0003019.
[41] Rovelli, C (1998), Loop quantum gravity, Living Rev. Relativity 1; http://relativity.livingreviews.org/Articles/lrr-1998-1.
[42] Brown, J.D., Comer, G.L., Martinez, E.A., Melmed, J., Whiting, B.F. and York, J.W. (1990), Class. Quantum Grav. 7, 1433–1444.
[43] Padmanabhan, T., gr-qc/0311036; Padmanabhan, T., gr-qc/0309053; Padmanabhan, T., gr-qc/0308070.
[44] Vaz, C. and Witten, L. (1998), Phys. Lett. B 442, 90–96. gr-qc/9804001.

Chapter 7

TRANSLATIONAL AND ROTATIONAL MOTION OF A UNIAXIAL LIQUID CRYSTAL AS DERIVED USING HAMILTON'S PRINCIPLE OF LEAST ACTION

Brian J. Edwards

Department of Chemical Engineering, The University of Tennessee, Knoxville, TN 37996, USA

If you fancy a fluid with rotation
The problem goes beyond notation
Pondering stress
Is sure to depress
And lead to excessive potation

Abstract

The conservative dynamics of the Leslie–Ericksen (LE) Model of a uniaxial liquid crystal are rederived using Hamilton's Principle of Least Action. This derivation is performed for two purposes: first, to illustrate that variational principles apply to even very complicated fluidic materials, and, secondly, to derive the noncanonical Poisson bracket that generates the conservative dynamics of the LE Model. Although the dissipative mechanisms in the LE Model can be and have been derived through variational formulations (see references below), dissipation is neglected herein to focus on an examination of the origin of rotational motion in this model. This demonstrates that the convective evolution of the microstructure in this model originates solely from the translational and rotational motion of the constituent particles, and not, as is commonly believed, by affine or nonaffine convection and deformation with respect to the local velocity field.

Keywords: Liquid crystals; Poisson bracket; Hamiltonian; rotational motion; Leslie–Ericksen theory; reduction methodology

1. Introduction

Variational principles can be used to derive the *conservative*, time-reversible dynamics of any complex material, regardless of the complexity of its microstructure. This is accomplished through the application of Hamilton's Principle of Least Action to the canonical coordinates in terms of particle positions and momenta, and subsequent reduction of those coordinates to appropriate variables representing the inherent microstructure of the material. One thus proceeds in a rigorous fashion from the heart of classical mechanics to a useful result, without at any time possessing knowledge of what the final result (e.g. a set of continuum-level evolution equations) should be. To accomplish this, a systematic reduction procedure must be implemented.

E-mail address: bjedwards@chem.engr.utk.edu (B.J. Edwards).

Fortunately, such a procedure exists, and has been used for this purpose in the recent past [1–4]. In older literature on this subject, variational principles were applied to derive continuum-level evolution equations using inferior methodology, wherein knowledge of the evolution equations was required a priori: the method of constraints involving Lagrange multipliers was used 'to derive' the conservation equations, which had to be guessed from the start as system constraints—see Refs. [5–7] for a few examples.

In this chapter, the rigorous methodology, mentioned above, is illustrated for a complex fluid that is particularly difficult to model: a liquid crystal. This type of fluid possesses a substantial degree of the long-range order of a crystalline solid, but still maintains the ability to deform continuously under the application of a finite shear strain. The difficulty arises due to the influence of multiple physical effects on the dynamics of these fluids. One must cope with external electric and magnetic fields, imposed velocity fields, excluded-volume effects, elastic distortion fields, etc., all simultaneously and consistently. To maintain the focus on the application of the variational principle, and not on the particularly complicated example chosen, attention is paid herein to one of the simplest models for liquid-crystalline dynamics: the Leslie–Ericksen (LE) Model.

Only the conservative part of the LE Model will be discussed here. This is because the dissipative dynamics cannot, as yet, be *rigorously* derived from first principles using variational methodology (which is the subject of this book). Rather than introduce a lack of rigor, this chapter focuses on that which can be done rigorously from first principles. As discussed below, the restriction to conservative dynamics does not mean that all of the interesting physics of liquid crystals has been eliminated from the problem formulation. For the first full treatment of dissipative as well as conservative dynamics in liquid crystals from a variational perspective, the reader can refer to Refs. [13,14], wherein the microstructure of the material was quantified using a second-rank tensor field.

The LE Model [8–11] of liquid-crystalline dynamics has enjoyed over three decades of success in describing the flow behavior, pattern formation, and rheology of primarily low molecular weight liquid crystals. Although it neglects excluded-volume effects and can only describe uniaxial liquid crystals, it still displays many of the rich and diverse dynamical phenomena exhibited experimentally. The microstructure of the material is represented by a single unit vector, $\mathbf{n}(\mathbf{x}, t)$, often called the *director*; this limits the applicability of the LE Model to uniaxial nematic liquid crystals. Hence the material is viewed as being composed of thin, rod-like particles; each particle within a given macroscopic fluid element at position $\mathbf{x}$ and time t is assumed to have the same orientation, quantified by the director. The conservative dynamics of the liquid crystal are then quantified by two field equations, one for conservation of linear momentum and one for conservation of angular momentum. These are, respectively

$$\rho \dot{\mathbf{v}}_\alpha = F_\alpha - \nabla_\alpha p - \nabla_\beta \left(\rho \frac{\partial W}{\partial (\nabla_\beta n_\gamma)} \nabla_\alpha n_\gamma \right), \tag{1}$$

$$\begin{aligned} \hat{I} \ddot{n}_\alpha = & - \hat{I} \dot{n}_\beta \dot{n}_\beta n_\alpha + G_\alpha - G_\beta n_\beta n_\alpha - \frac{\partial W}{\partial n_\alpha} + \frac{\partial W}{\partial n_\beta} n_\beta n_\alpha \\ & + \nabla_\beta \left(\frac{\partial W}{\partial (\nabla_\beta n_\alpha)} \right) - \nabla_\gamma \left(\frac{\partial W}{\partial (\nabla_\gamma n_\beta)} \right) n_\beta n_\alpha, \end{aligned} \tag{2}$$

in which $\dot{\mathbf{a}}$ denotes the substantial time derivative of the vector field **a** with respect to the imposed velocity field, $v(\mathbf{x}, t)$:

$$\dot{a}_\alpha \equiv \frac{\partial a_\alpha}{\partial t} + v_\beta \nabla_\beta a_\alpha. \tag{3}$$

Note that the Einstein convention of summing over repeated indices has been assumed in the above equations, as well as in all that follows.

The first equation (Eq. (1)) is the Cauchy momentum equation, wherein p is the isotropic pressure, ρ is the mass density, $W(\mathbf{n}, \nabla\mathbf{n})$ is the Oseen–Frank distortion energy, and **F** is the body force resulting from an external magnetic field, **H**,

$$F_\alpha = \rho(\chi_\perp H_\beta + (\chi_\Downarrow - \chi_\perp)H_\gamma n_\gamma n_\beta)\nabla_\alpha H_\beta, \tag{4}$$

in which $\chi_\perp$ and $\chi_\Downarrow$ denote the magnetic susceptibilities perpendicular and parallel to the director, respectively. Note that the distortion energy, arising from spatial variations in the director field, gives rise to additional elastic stress within the fluid, which is known as the *Ericksen stress*—see Eq. (1).

The second equation (Eq. (2)) is the angular-momentum balance. In this expression, $\hat{I}$ is the inertial constant of the material, $\ddot{\mathbf{n}}$ denotes the substantial time derivative of $\dot{\mathbf{n}}$, and **G** represents the body couple

$$G_\alpha = (\chi_\Downarrow - \chi_\perp)n_\beta H_\beta H_\alpha. \tag{5}$$

Again, the dissipative contributions to the above equations have been neglected, since they cannot be derived through a variational principle in a rigorous way, but must be introduced therein phenomenologically. The feasibility of such a practice has been demonstrated sufficiently in the past [4,12]. For liquid crystals specifically, this practice has yielded one of the most far-reaching continuum theories published to date [13,14]. For the present purpose of illustrating the role of rotational motion in complex fluid theory, it is perfectly apt to neglect the dissipative contributions to the above equation set. Note, again, that this statement does not imply that no interesting physics can be described with the conservative equations: as shown in Refs. [13,14], for example, the nature of the director dynamics in shear flow that leads to sign transitions in the first normal stress difference can be traced to inertial effects arising in the angular momentum balance (Eq. (2)).

The purpose of this chapter thus boils down to deriving the above-stated set of equations using Hamilton's Principle of Least Action in terms of canonical coordinates as the starting point. To do this rigorously, it is convenient to derive a noncanonical Poisson bracket, which describes the dynamics of the LE Model in the spatial description of continuum mechanics. This Poisson bracket was introduced in Refs. [13,14], but was not derived there. A derivation can be found as a special case of the theory in Ref. [15], but it is not explicitly written down in that paper. After completing the derivation, certain aspects of the above equation set will become readily apparent, such as the rotational origin of the director field motion under an imposed velocity field. To get the most out of this derivation, we must begin with the simplest system possible, i.e. discrete, rigid particles, and work up from this humble starting point to the more complicated continuum system of Eqs. (1)–(5).

2. Lagrangian and Hamiltonian dynamics of discrete rigid particles

In this section, we study the evolution in time of the dynamical state of a system composed of ν rigid particles, which are allowed to translate with respect to a coordinate system fixed in space, and also to rotate about a point within the body of each particle. This section largely follows the presentation in Ref. [15]. No generality is lost by considering the center of mass as the fixed point of rotation for each rigid particle. Thus the entire problem reduces to a consideration of the translation of the center of mass for each particle and the rotation about its center of mass; this concept is known as *Chasle's Theorem*. The tasks of this section are thus to derive the Lagrangian equations of motion for this system using the Principle of Least Action, and then to derive Hamilton's equations of motion and the corresponding Poisson bracket.

2.1. The Lagrangian equations of motion

For simplicity, we consider only Cartesian coordinate systems in this work. Roman or Greek characters denote quantities of tensorial (first-rank or greater) character, and italicized characters denote scalars, which include the components of tensorial quantities. Roman and Greek subscripts or superscripts denote the particle number and spatial component, respectively, of the appropriate quantity. As an example, $\mathbf{x}_i$ denotes the vector coordinate of the center of mass of particle i, and x_i^α denotes the α component of $\mathbf{x}_i$. Remember that the Einstein summation convention applies over repeated Greek indices; however, this summation is not implied over repeated Roman indices, but will be stated explicitly, when applicable, using conventional mathematical notation.

Let a space-fixed coordinate system be denoted by $\mathbf{x} = (x^1, x^2, x^3)^{\mathrm{T}}$. The position of the center of mass of particle i, whose total mass is m_i, is then denoted by the function $\mathbf{x}_i(t) = \left(x_i^1(t), x_i^2(t), x_i^3(t)\right)^{\mathrm{T}}$, and the translational velocity of the center of mass is denoted by $\dot{\mathbf{x}}_i(t)$. The rotational motion of each particle is described using two orthogonal coordinate systems, $\mathbf{r}_i$ and ξ_i. The origin of $\mathbf{r}_i$ translates with the center of mass of particle i, but the axes are required to be parallel always to the corresponding axes of $\mathbf{x}$. The other coordinate system, ξ_i, also has its origin fixed at the center of mass of particle i, but this coordinate system is fixed in the body of the particle so that it rotates along with the particle. At any instant of time, there is an orthogonal transformation that relates the two-coordinate systems, as expressed in terms of the Euler angles defined in Fig. 1. This transformation is

$$\xi_i^\alpha = A_i^{\alpha\gamma} r_i^\gamma, \quad i = 1, 2, \ldots, \nu, \tag{6}$$

in which the 3×3 matrix $\mathbf{A}_i$ is

$$\mathbf{A}_i = \begin{pmatrix} c\psi_i c\phi_i - c\theta_i s\phi_i s\psi_i & c\psi_i s\phi_i + c\theta_i c\phi_i s\psi_i & s\psi_i s\theta_i \\ -s\psi_i c\phi_i - c\theta_i s\phi_i c\psi_i & -s\psi_i s\phi_i + c\theta_i c\phi_i c\psi_i & c\psi_i s\theta_i \\ s\theta_i s\phi_i & s\theta_i c\psi_i & c\theta_i \end{pmatrix}. \tag{7}$$

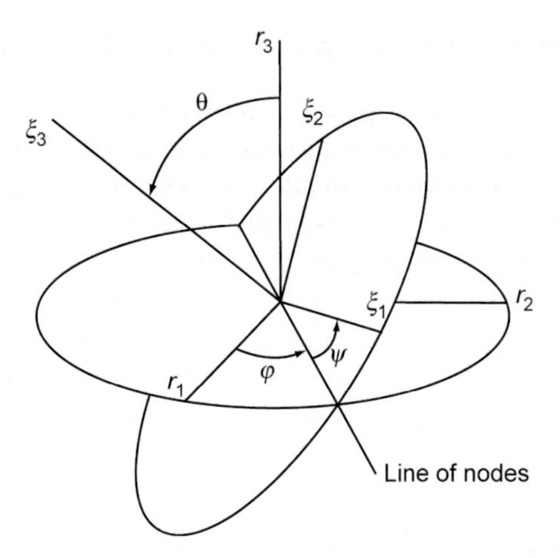

Fig. 1. The definition of the Euler angles for an arbitrary particle.

Note that the symbols and $ca \equiv \cos a$ and $sa \equiv \sin a$. Each $\mathbf{A}_i$ is the linear product of three successive rotation matrices, each with respect to one of the axes that defines the rotations of the Euler angles. Therefore, $A_i^{\alpha\beta} = B_i^{\alpha\gamma} C_i^{\gamma\varepsilon} D_i^{\varepsilon\beta}$, where the three new rotation matrices are defined as

$$\mathbf{B}_i = \begin{pmatrix} c\psi_i & s\psi_i & 0 \\ -s\psi_i & c\psi_i & 0 \\ 0 & 0 & 1 \end{pmatrix}, \tag{8}$$

$$\mathbf{C}_i = \begin{pmatrix} 1 & 0 & 0 \\ 0 & c\theta_i & s\theta_i \\ 0 & -s\theta_i & c\theta_i \end{pmatrix}, \tag{9}$$

$$\mathbf{D}_i = \begin{pmatrix} c\phi_i & s\phi_i & 0 \\ -s\phi_i & c\phi_i & 0 \\ 0 & 0 & 1 \end{pmatrix}. \tag{10}$$

Note that $\mathbf{A}_i$ and each of the matrices defined above satisfy the orthogonality relationship $\mathbf{A}_i^{-1} = \mathbf{A}_i^{\mathrm{T}}$.

It is well known that a proper set of generalized coordinates for describing the motion of a particle in space must be an independent set. For the translation of a rigid body in space, the spatial coordinates of the center of mass relative to **x** serve well. For the rotation of a rigid body about a fixed point, in this case the center of mass, a set of generalized coordinates is given by the Euler angles, expressed concisely in a pseudovectorial form as $\Theta_i = (\theta_i, \phi_i, \psi_i)^{\mathrm{T}}$, and their time derivatives, $\dot{\Theta}_i = (\dot{\theta}_i, \dot{\phi}_i, \dot{\psi}_i)^{\mathrm{T}}$, which may be viewed as angular velocities. The reader is cautioned, however, that although Θ_i might retain some of the characteristics of a vector, it is not correct to label it as a true vector per se, but merely as a notational convenience. In particular, due to the way in which the Euler angles are defined, Θ_i does not act as a vector under a linear transformation of coordinates of the form of Eq. (6).

As the starting point for the derivation of the equations of motion for this system of particles, we wish to apply the Principle of Least Action to the Lagrangian of the system. The Lagrangian is written as a function of all of the generalized coordinates of the system, $L(\mathbf{x}_1, \ldots, \mathbf{x}_\nu, \dot{\mathbf{x}}_1, \ldots, \dot{\mathbf{x}}_\nu, \Theta_1, \ldots, \Theta_\nu, \dot{\Theta}_1, \ldots, \dot{\Theta}_\nu)$, and it may be written as the difference between the kinetic and potential energies of the system:

$$L = \sum_{i=1}^{\nu} [T_i^t(\mathbf{x}_i, \dot{\mathbf{x}}_i) + T_i^r(\Theta_i, \dot{\Theta}_i)] - V(\mathbf{x}_1, \ldots, \mathbf{x}_\nu, \Theta_1, \ldots, \Theta_\nu). \tag{11}$$

(For the derivation of the LE Model, it is not necessary to consider an explicit time dependence in the corresponding Lagrangian or Hamiltonian; consequently, such a complication is not considered in this chapter.) In this expression, T_i^r is the kinetic energy of particle i due to its rotation, T_i^t is its translational kinetic energy, and V is the potential energy of the entire system of particles. It is intuitive, given Chasle's Theorem, that the total kinetic energy may be decoupled in the above-stated fashion; however, the potential energy is not split analogously since it seems apparent that the orientation of asymmetrical particles might affect the positional (center-of-mass) potential energy, and vice versa.

The variational problem for this particular Lagrangian can now be examined. Let t_1 and t_2 be two fixed times, with $t_1 < t_2$, such that the state of the system is known completely at these times. Let an admissible motion of the system be specified as the set of functions $\{\mathbf{x}_i^*(t), \Theta_i^*(t)\}$, $i = 1, \ldots, \nu$, which fulfill the required system configurations at times t_1 and t_2, and that have continuous second derivatives in the interval $[t_1, t_2]$. The Principle of Least Action asserts that, among admissible motions, the actual system motion, $\{\mathbf{x}_i(t), \Theta_i(t)\}$, is such that the integral

$$\Im = \int_{t_1}^{t_2} L \, \mathrm{d}t \tag{12}$$

is stationary with respect to neighboring admissible motions. Since the set $\{\mathbf{x}_i(t), \Theta_i(t)\}$ is the actual system motion, the integral in Eq. (12) is stationary when $\{\mathbf{x}_i^*(t, \varepsilon), \Theta_i^*(t, \varepsilon)\} = \{\mathbf{x}_i(t), \Theta_i(t)\}$, where ε is a small constant parameter. Performing the necessary calculations provides the Lagrangian equations of motion for the system in terms of the set $\{\mathbf{x}_i(t), \Theta_i(t)\}$:

$$\frac{\mathrm{d}}{\mathrm{d}t}\left(\frac{\partial T_i^t}{\partial \dot{\mathbf{x}}_i}\right) - \frac{\partial T_i^t}{\partial \mathbf{x}_i} = -\frac{\partial V}{\partial \mathbf{x}_i}, \quad i = 1, \ldots, \nu \text{ on } [t_1, t_2], \tag{13}$$

$$\frac{\mathrm{d}}{\mathrm{d}t}\left(\frac{\partial T_i^r}{\partial \dot{\Theta}_i}\right) - \frac{\partial T_i^r}{\partial \Theta_i} = -\frac{\partial V}{\partial \Theta_i}, \quad i = 1, \ldots, \nu \text{ on } [t_1, t_2]. \tag{14}$$

The translational kinetic energy of particle i is well known to be a function of $\dot{\mathbf{x}}_i$ only:

$$T_i^t = \frac{1}{2} m_i \dot{\mathbf{x}}_i \cdot \dot{\mathbf{x}}_i, \quad i = 1, \ldots, \nu. \tag{15}$$

Hence the Lagrangian equations of Eq. (13) are equivalent to Newton's equations

$$m_i \frac{\mathrm{d}\dot{\mathbf{x}}_i}{\mathrm{d}t} = \mathbf{F}_i, \quad i = 1, \ldots, \nu, \tag{16}$$

wherein the force has the usual definition, $\mathbf{F}_i = -\partial V / \partial \mathbf{x}_i$.

The rotational kinetic energy of a given particle is difficult to express in terms of the set of variables $\{\Theta_i, \dot{\Theta}_i\}$, chiefly because the Θ_i are not real vectors. In other words, the Θ_i do not represent mutually consistent variables because the Euler angles are defined with respect to coordinate rotations of different intermediate axes—see Eqs. (8)–(10). The net result is some indeterminacy in the T_i^r as expressed relative to the rotations of the Euler angles. Furthermore, the inconsistency of the Θ_i results in a system of equations (14) that is not expressed relative to a specific coordinate system, such as $\mathbf{r}_i$ or ξ_i, and may not be transformed to such via a transformation relation of the type of Eq. (6). Therefore, at this stage one is forced to introduce the concept of *quasicoordinates*, which are treated in some of the older literature [16], in order to express the rotational equations of Eq. (14) relative to a specific coordinate system. First, however, it is necessary to examine which set of specific coordinate axes is the most convenient for the expression of the equations of rotational motion (Eq. (14)).

The convenience noted above is often dictated by the functional form that the kinetic energy has with respect to a specific set of coordinates. It is, therefore, necessary to examine the form of the rotational kinetic energy in terms of the angular velocities relative to a given set of coordinate axes and the inertia tensor of the rotating body.

An inertia tensor may be written for each particle, relative to the $\mathbf{r}_i$, as

$$\mathbf{I}_i = \int_{V_i} \rho_i(\mathbf{r}_i)[\mathrm{tr}(\mathbf{r}_i\mathbf{r}_i)\delta - \mathbf{r}_i\mathbf{r}_i]\mathrm{d}^3 r_i, \tag{17}$$

wherein ρ_i is the mass–density function of the rigid particle, V_i is its volume, δ is the unit tensor, and $\mathrm{d}^3 r_i \equiv \mathrm{d}r_i^1\, \mathrm{d}r_i^2\, \mathrm{d}r_i^3$. An analogous expression applies for calculating the $\mathbf{I}_i$ relative to any other set of coordinates, including the body coordinates, ξ_i.

It is now assumed that the body coordinates, ξ_i, $i = 1, \ldots, \nu$, are those that diagonalize the inertia tensors; i.e. the transformation of Eq. (6) is the principal-axis transformation. This is not in any way a limiting assumption, since such a transformation is always possible. Therefore, the only nonzero components of $\mathbf{I}_i$ in the body coordinates are the *constant* diagonal elements, or principal moments, I_i^1, I_i^2, and I_i^3, each associated with the corresponding axis of ξ_i.

Since the components of the ξ_i are now identified with the principal axes, the rotational kinetic energy takes the simple form

$$T_i^r = \tfrac{1}{2} I_i^\alpha \omega_i^\alpha \omega_i^\alpha, \quad i = 1, \ldots, \nu. \tag{18}$$

Therefore, it is most convenient to work in terms of the body coordinates, ξ_i, and to express the angular velocities relative to this reference frame. These quantities will be denoted as ω_i, and in terms of the Euler angles, they may be written as [17]

$$\omega_i^1 = \dot{\phi}_i s\theta_i s\psi_i + \dot{\theta}_i c\psi_i, \qquad \omega_i^2 = \dot{\phi}_i s\theta_i c\psi_i + \dot{\theta}_i s\psi_i, \qquad \omega_i^3 = \dot{\phi}_i c\theta_i + \dot{\psi}_i. \tag{19}$$

In the above expressions, note that ω_i is not related to $\dot{\Theta}_i$ via a transformation relation of the form of Eq. (6), although at first glance it might appear to be so. The reason it does not is because the different components of ω_i are obtained with different transformation matrices: $\dot{\theta}_i$ is transformed with the matrix $\mathbf{B}_i$, $\dot{\phi}_i$ is transformed with the matrix $\mathbf{A}_i$, and $\dot{\psi}_i$ requires no transformation. To be sure, a composite transformation matrix may be constructed, however, as shown below, the result is not a proper transformation matrix; i.e. the resulting composite matrix is not, among other things, orthogonal.

The concept of quasicoordinates allows one to represent the inconsistent coordinates, Θ_i (called the *true coordinates*), with respect to a more convenient set of reference axes, such as those represented by the principal axes of the inertia tensor, as described above. Consider the rotational equation for particle i only (the others are analogous) following Whittaker [16]. Rewriting Eq. (14) as

$$\frac{\mathrm{d}}{\mathrm{d}t}\left(\frac{\partial T_i^r}{\partial \dot{\Theta}_i}\right) - \frac{\partial T_i^r}{\partial \Theta_i} = N_i^\alpha, \quad \alpha = 1, 2, 3, \tag{20}$$

in which $\mathbf{N}_i \equiv -(\partial V/\partial \Theta_i)$ is the torque on particle i, analogous to the force appearing in Newton's equations of translational motion (Eq. (16)). Consequently, the work performed on the system during a small arbitrary displacement $(\Delta\theta_i, \Delta\phi_i, \Delta\psi_i)$ is $N_i^1\Delta\theta_i + N_i^2\Delta\phi_i + N_i^3\Delta\psi_i$. Let the ω_i of Eq. (19) be independent linear combinations of the velocities of the Euler angles, defined by

$$\omega_i^\alpha \equiv \alpha_i^{1\alpha}\dot{\theta}_i + \alpha_i^{2\alpha}\dot{\phi}_i + \alpha_i^{3\alpha}\dot{\psi}_i, \quad \alpha = 1, 2, 3, \tag{21}$$

wherein the $\alpha_i^{\alpha\beta}$ are functions of the Euler angles. From Eq. (19), α_i is defined through

$$\alpha_i \equiv \begin{pmatrix} \alpha_i^{11} & \alpha_i^{21} & \alpha_i^{31} \\ \alpha_i^{12} & \alpha_i^{22} & \alpha_i^{32} \\ \alpha_i^{13} & \alpha_i^{23} & \alpha_i^{33} \end{pmatrix} = \begin{pmatrix} c\psi_i & s\theta_i s\psi_i & 0 \\ -s\psi_i & s\theta_i c\psi_i & 0 \\ 0 & c\theta_i & 1 \end{pmatrix}, \tag{22}$$

and it is necessary to define $\mathrm{d}\pi_i^1$, $\mathrm{d}\pi_i^2$, and $\mathrm{d}\pi_i^3$ as linear combinations of the differentials $\mathrm{d}\theta_i$, $\mathrm{d}\phi_i$, and $\mathrm{d}\psi_i$:

$$\mathrm{d}\pi_i^\alpha \equiv \alpha_i^{1\alpha}\,\mathrm{d}\theta_i + \alpha_i^{2\alpha}\,\mathrm{d}\phi_i + \alpha_i^{3\alpha}\,\mathrm{d}\psi_i, \quad \alpha = 1, 2, 3, \tag{23}$$

The $\mathrm{d}\pi_i^\alpha$ are called *differentials of quasicoordinates*. Now the components of α_i can be viewed as

$$\alpha_i^{\alpha\beta} = \frac{\partial \pi_i^\beta}{\partial \Theta_i^\alpha}, \quad \alpha, \beta = 1, 2, 3. \tag{24}$$

From this equation, it is apparent that Eq. (23) would be immediately integrable if

$$\frac{\partial \alpha_i^{\alpha\beta}}{\partial \Theta_i^\gamma} = \frac{\partial \alpha_i^{\gamma\beta}}{\partial \Theta_i^\alpha}, \quad \text{for all } \alpha, \beta, \gamma, \tag{25}$$

and, in this case, there would exist true coordinates, π_i^α, $\alpha = 1, 2, 3$. It is also apparent from Eq. (22) that true coordinates do not exist for the present case. Therefore, this procedure amounts to assuming that the small arbitrary displacement $(\Delta\theta_i, \Delta\phi_i, \Delta\psi_i)$ of the particle is equivalent to the resultant of small rotations $(\Delta\pi_i^1, \Delta\pi_i^2, \Delta\pi_i^3)$ about the axes of ξ_i [16]. In essence, the ω_i were defined to be components of the angular velocity about the body axes in such a way that the $\mathrm{d}\pi_i^\alpha$, $\alpha = 1, 2, 3$, are the differentials of quasicoordinates that correspond to the components of ω_i. (This statement might be interpreted by the reader as a circular argument; if it is unclear, please examine Ref. [16] for clarification.) This alleviates the problems associated with the inconsistency of the transformation relations by relieving the transformation matrix of the physical attribute of corresponding to a set of true coordinates.

By taking the inverse of α_i,

$$\alpha_i^{-1} \equiv \beta_i = \begin{pmatrix} \beta_i^{11} & \beta_i^{12} & \beta_i^{13} \\ \beta_i^{21} & \beta_i^{22} & \beta_i^{23} \\ \beta_i^{31} & \beta_i^{32} & \beta_i^{33} \end{pmatrix} = \begin{pmatrix} c\psi_i & -s\psi_i & 0 \\ \dfrac{s\psi_i}{s\theta_i} & \dfrac{c\psi_i}{s\theta_i} & 0 \\ -\dfrac{s\psi_i c\theta_i}{s\theta_i} & -\dfrac{c\psi_i c\theta_i}{s\theta_i} & 1 \end{pmatrix}, \tag{26}$$

expressions for the components of $\dot{\theta}_i$ can be written as linear functions of the components of ω_i :

$$\dot{\Theta}_i^\alpha = \beta_i^{\alpha 1}\omega_i^1 + \beta_i^{\alpha 2}\omega_i^2 + \beta_i^{\alpha 3}\omega_i^3, \quad \alpha = 1, 2, 3. \tag{27}$$

Note that $\beta_i = \alpha_i^{-1} \neq \alpha^{\mathrm{T}}$, which implies that the α_i are not orthogonal, as alluded to earlier.

By expressing the rotational kinetic energy in terms of ω_i as

$$\frac{\partial T_i^r}{\partial \dot{\Theta}_i^\alpha} = \frac{\partial \overline{T}_i^r}{\partial \omega_i^\beta}\alpha_i^{\alpha\beta}, \tag{28}$$

the rotational Lagrangian equations of motion of Eq. (20) for each particle become

$$\frac{\mathrm{d}}{\mathrm{d}t}\left(\frac{\partial \overline{T}_i^r}{\partial \omega_i^\beta}\right) + \frac{\partial \overline{T}_i^r}{\partial \omega_i^r}\omega_i^\eta\left[\beta_i^{\alpha\beta}\beta_i^{\varepsilon\eta}\left(\frac{\partial \alpha_i^{\alpha\gamma}}{\partial \Theta_i^\varepsilon} - \frac{\partial \alpha_i^{\varepsilon\gamma}}{\partial \Theta_i^\alpha}\right)\right] - \beta_i^{\alpha\beta}\frac{\partial \overline{T}_i^r}{\partial \Theta_i^\alpha} = \Pi_i^\beta, \tag{29}$$

in which $\Pi_i^\beta = \beta_i^{\alpha\beta}N_i^\alpha$ is the torque acting on particle i projected into the body-fixed coordinate system. Note that $\beta_i^{\alpha\beta}(\partial\overline{T}_i^r/\partial\Theta_i^\alpha) = (\partial\overline{T}_i^r/\partial\Theta_i^\alpha)(\partial\Theta_i^\alpha/\partial\pi_i^\beta)$, which would just be $\partial\overline{T}_i^r/\partial\pi_i^\beta$ if the π_i were true coordinates. Furthermore, the term inside the square brackets in Eq. (29) depends only on the relationship between the true coordinates and the differentials of the quasicoordinates. When the π_i are true coordinates, this term vanishes and Eq. (29) reduces to the Lagrangian equations of Eq. (20) in terms of π_i and $\dot{\pi}_i$.

Substitution of the kinetic energy (18) into Eq. (29), realizing that Eq. (18) provides $\overline{T}_i^r$, eventually yields

$$I_i^1\frac{\mathrm{d}\omega_i^1}{\mathrm{d}t} - \omega_i^2\omega_i^3(I_i^2 - I_i^3) = \Pi_i^1, \qquad I_i^2\frac{\mathrm{d}\omega_i^2}{\mathrm{d}t} - \omega_i^1\omega_i^3(I_i^3 - I_i^1) = \Pi_i^2,$$

$$I_i^3\frac{\mathrm{d}\omega_i^3}{\mathrm{d}t} - \omega_i^2\omega_i^1(I_i^1 - I_i^2) = \Pi_i^3 \tag{30}$$

for all i. These equations are known as *Euler's equations* for rotation about a fixed point, in this case, the center of mass of each particle. Together with Eq. (16), there is now a system of six ordinary differential equations to solve for the system motion. Consequently, the first task of this section has been achieved; namely, to derive the equations of motion for the rigid-particle system using the Principle of Least Action. However, the equations of motion are still in Lagrangian form, and thus need to be expressed in Hamiltonian form in terms of a Poisson bracket before the section is complete.

2.2. Interlude

In the derivation above, the body coordinates were chosen as the frame of reference because of the simplicity of the rotational kinetic energy expression when written in terms of these coordinates (Eq. (18)). But what if the coordinates are the fixed axes, $\mathbf{r}_i$? In this case, it is apparent that the motion of the system is still described by the general equations (16) and (29); however, in the latter expression, the $\overline{T}_i^r$, ω_i, β_i, α_i, and Π_i are no longer those of the preceding case. The new expressions for the ω_i can be determined, from which α_i and β_i (and subsequently Π_i) can be calculated easily. The expression for $\overline{T}_i^r$ is now more complicated than before [17]

$$\overline{T}_i^r = \frac{1}{2}\omega_i^\alpha I_i^{\alpha\beta}\omega_i^\beta, \tag{31}$$

wherein the $\mathbf{I}_i$ are no longer constants, but vary with the orientation of the body with respect to the $\mathbf{r}_i$. These inertia tensors (i.e. their components) may be considered as functions of the Euler angles, but, as mentioned above, are assumed to have no explicit time dependence.

At this point, the equations of rotational motion for this system can be written down in accordance with Eq. (29); however, the new equations turn out to be much more complicated than the Euler equations of Eq. (30) due to the dependence of the $\mathbf{I}_i$ on Θ_i. The point of this section is to show that one can, in practice, write down the rotational motion for the particles with respect to any given coordinate system; however, it is prudent to consider which of the possibilities possesses the greatest degree of symmetry and simplicity for the problem under study. For instance, the Euler equations might look simple, but the dynamical variables, ω_i, and the torques, Π_i, are projected onto a time-dependent, body-fixed coordinate system, which implies that their integration only quantifies the motion as seen by an observer on the body. The solution for an inertial observer would be much more complicated. Furthermore, the moments, Π_i, must be evaluated consistently. These facts restrict the practical utility of the Euler equations to torque-free motions, or those in which the principal axes are partially restrained, thereby reducing the number of degrees of freedom.

Although the new set of equations is more complicated than the Euler equations due to the Θ_i, dependence of the $\mathbf{I}_i$, the equations are expressed relative to laboratory-fixed coordinates, and, therefore, the resulting Π_i have a simpler physical interpretation. It is apparent that a compromise between the two cases, where the inertia tensors are constant and diagonal yet the dynamical equations are projected onto an inertial reference frame, would prove extremely useful. This line of thought will be taken up in Section 3.

2.3. Hamilton's equations and the Poisson bracket

The next pertinent question is whether or not Eqs. (13), (14), and (29) will suffer expression in terms of Poisson brackets. In order to answer this question, it is first necessary to express these equations in Hamiltonian form by applying the variational principle to the Legendre transform of the system Lagrangian. Consider first Eqs. (13) and (14). In this case, the Lagrangian is expressed in functional form as $L(\mathbf{x}_1, \dots, \mathbf{x}_\nu, \dot{\mathbf{x}}_1, \dots, \dot{\mathbf{x}}_\nu, \Theta_1, \dots, \Theta_\nu, \dot{\Theta}_1, \dots, \Theta_\nu)$. A Legendre transformation can be performed on the velocity and angular velocity variables, for each particle i, keeping the position and orientation variables passive. This defines the particle momentum and moment variables as

$$\mathbf{p}_i \equiv \frac{\partial L}{\partial \dot{\mathbf{x}}_i}, \qquad \Xi_i \equiv \frac{\partial L}{\partial \dot{\Theta}_i}, \qquad i = 1, \dots, \nu, \tag{32}$$

and the system Hamiltonian, $H(\mathbf{x}_1, \dots, \mathbf{x}_\nu, \mathbf{p}_1, \dots, \mathbf{p}_\nu, \Theta_1, \dots, \Theta_\nu, \Xi_1, \dots, \Xi_\nu)$, is given by

$$H = \sum_{i=1}^{\nu} [\mathbf{p}_i \cdot \dot{\mathbf{x}}_i + \Xi_i \cdot \dot{\Theta}_i] - L. \tag{33}$$

The variational equation then yields four Hamiltonian equations for each particle [15]

$$\dot{\mathbf{x}}_i = \frac{\partial H}{\partial \mathbf{p}_i}, \qquad \dot{\mathbf{p}}_i = -\frac{\partial H}{\partial \mathbf{x}_i}, \qquad \dot{\Theta}_i = \frac{\partial H}{\partial \Xi_i}, \qquad \dot{\Xi}_i = -\frac{\partial H}{\partial \Theta_i}. \tag{34}$$

Note that these equations possess canonical form, and reduce to the set of Eqs. (13) and (14) under the inverse Legendre transformation; i.e. expressing Eq. (34) by the substitution of Eq. (33).

The above expressions define a bracket structure

$$\frac{\mathrm{d}F}{\mathrm{d}t} = \{F, H\} \equiv \sum_{i=1}^{\nu} \left[\frac{\partial F}{\partial \mathbf{x}_i} \cdot \frac{\partial H}{\partial \mathbf{p}_i} - \frac{\partial H}{\partial \mathbf{x}_i} \cdot \frac{\partial F}{\partial \mathbf{p}_i} + \frac{\partial F}{\partial \Theta_i} \cdot \frac{\partial H}{\partial \Xi_i} - \frac{\partial H}{\partial \Theta_i} \cdot \frac{\partial F}{\partial \Xi_i} \right]. \tag{35}$$

This bracket possesses all of the properties attributed to a Poisson bracket: it is bilinear, antisymmetric (which implies the conservation of energy, $\mathrm{d}H/\mathrm{d}t = 0$), and satisfies the Jacobi identity. Hence the equations of motion for the system (Eqs. (13) and (14)) can also be expressed in Hamiltonian form and written in terms of a Poisson bracket.

The last question to answer in this section is whether or not the equations of rotational motion in terms of the quasicoordinates, Eq. (29) along with the translational motion equations (Eq. (13)) may also be expressed in Hamiltonian form with an associated bracket structure. The same procedure may be followed as above by applying a Legendre transformation to the quasi-Lagrangian, $\overline{L}(\mathbf{x}_1, \dots, \mathbf{x}_\nu, \dot{\mathbf{x}}_1, \dots, \dot{\mathbf{x}}_\nu, \Theta_1, \dots, \Theta_\nu, \omega_1, \dots, \omega_\nu)$:

$$\overline{H} = \sum_{i=1}^{\nu} [\mathbf{p}_i \cdot \dot{\mathbf{x}}_i + \Omega_i \cdot \omega_i] - \overline{L}. \tag{36}$$

In this expression, $\mathbf{p}_i$ is defined by Eq. (36) with $\overline{L}$ replacing L, and

$$\Omega_i \equiv \frac{\partial \overline{L}}{\partial \omega_i}, \qquad i = 1, \dots, \nu. \tag{37}$$

Assuming that $\overline{F}$ is a function with the same functionality as $\overline{H}$, the Poisson bracket in terms of quasicoordinates can be calculated as

$$\{\overline{F},\overline{H}\} \equiv \sum_{i=1}^{\nu}\left[\frac{\partial \overline{F}}{\partial x_i^\alpha}\frac{\partial \overline{H}}{\partial p_i^\alpha} - \frac{\partial \overline{H}}{\partial x_i^\alpha}\frac{\partial \overline{F}}{\partial p_i^\alpha} + \frac{\partial \overline{F}}{\partial \Theta_i^\alpha}\beta_i^{\alpha\gamma}\frac{\partial \overline{H}}{\partial \Omega_i^\gamma} - \frac{\partial \overline{H}}{\partial \Theta_i^\alpha}\beta_i^{\alpha\gamma}\frac{\partial \overline{F}}{\partial \Omega_i^\gamma} + \frac{\partial \overline{F}}{\partial \Omega_i^\varepsilon}\Omega_i^\alpha\frac{\partial \overline{H}}{\partial \Omega_i^\eta}M_i^{\alpha\eta\varepsilon}\right], \tag{38}$$

in which

$$M_i^{\alpha\eta\varepsilon} \equiv \frac{\partial \alpha_i^{\gamma\alpha}}{\partial \Theta_i^\beta}\left[\beta_i^{\gamma\eta}\beta_i^{\beta\varepsilon} - \beta_i^{\beta\eta}\beta_i^{\gamma\varepsilon}\right]. \tag{39}$$

The tensor $\mathbf{M}_i$ is equivalent to the negative of the quantity within the square brackets in Eq. (29). It bears the component identity relationships $M_i^{\alpha\eta\varepsilon} = -M_i^{\alpha\varepsilon\eta}$ and $M_i^{\alpha\eta\eta} = 0$, which lead one to speculate that it might be proportional to the permutation tensor. For the α_i and β_i of Eqs. (22) and (26), it turns out that this is the case. Consequently, the last term in the quasibracket may be written as $-\Omega_i\cdot(\partial\overline{F}/\partial\Omega_i \times \partial\overline{H}/\partial\Omega_i)$, which establishes an immediate connection with the commutation relationships for spin in quantum mechanics, $[\Omega_i^\alpha, \Omega_i^\beta] = i\hbar\varepsilon_{\alpha\beta\gamma}\Omega_i^\gamma$. For the α_i and β_i corresponding to the example in Section 2.2, $M_i^{\alpha\eta\varepsilon} = -\varepsilon_{\alpha\eta\varepsilon}$, so that the direct proportionality of $\mathbf{M}_i$ with ε cannot be established in general; apparently, its exact form depends on the coordinate system chosen for the system description. The apparent antisymmetry of this bracket immediately requires that $\mathrm{d}\overline{H}/\mathrm{d}t = 0$.

The brackets (35) and (38) are the culmination of this section. It is apparent now how the equations of motion for the particle system, both translational and rotational, transform between various coordinate frames, and what the bracket structures look like under general conditions. Furthermore, it has been demonstrated how these equations of motion can be derived from the Principle of Least Action, in terms of either true or quasicoordinates.

3. Rotational motion in terms of true vectors

As alluded to in Section 2.2, the rotational equations of motion can be quite difficult to solve in terms of the Euler angles for the particles. The main reason for this is that the Euler angles do not represent the components of a 'true vector;' e.g. the composite column vectors, Θ_i, do not satisfy linear transformation relationships such as Eq. (6). Consequently, the ω_i and Π_i are difficult to resolve in an inertial reference frame. In this section, it is necessary to transform the Euler angles in the principal-axis coordinate frame into the inertial coordinates $\mathbf{r}_i$ by defining two orthogonal true vectors to describe the particle orientation. This exercise is not only gratifying from the perspective of discrete particles, but is also paramount for a transferal of information from a material to a spatial description in continuum mechanics, which is the subject of the remainder of this chapter. Moreover, the use of two vectors with a total of six coordinates in order to express only three independent degrees of freedom by necessity requires the use of constraints. The introduction of constraints makes the theoretical development particularly challenging.

3.1. The Lagrangian formulation

For simplicity in this section, only the rotational motion is considered; the translational motion can be described as before without any alterations. Only one particle is considered, since all particles follow analogous equations. Hence $\nu = 1$, and the Roman subscripts are dropped and the Greek indices are lowered. The orientation of the particle is specified with two perpendicular unit vectors, **n** and **k**, which are defined in terms of the Euler angles of the principal-axis coordinate system relative to the inertial reference frame, **r**—see Fig. 1. Thus **n** and **k** are defined through the unit direction vectors $\mathbf{e}_1$, $\mathbf{e}_2$, and $\mathbf{e}_3$ pointing in the r_1, r_2, and r_3-directions, respectively, as

$$\begin{aligned}\mathbf{n} &= s\theta s\phi \mathbf{e}_1 - s\theta c\phi \mathbf{e}_2 + c\theta \mathbf{e}_3, \\ \mathbf{k} &= (c\psi c\phi - c\theta s\psi c\phi)\mathbf{e}_1 + (c\psi s\phi + c\theta s\psi c\phi)\mathbf{e}_2 + s\psi s\theta \mathbf{e}_3.\end{aligned} \tag{40}$$

It is evident that

$$\begin{aligned}&\mathbf{n}\cdot\mathbf{n} = 1, \quad \mathbf{k}\cdot\mathbf{k} = 1, \quad \mathbf{n}\cdot\mathbf{k} = 0, \quad \dot{\mathbf{n}}\cdot\mathbf{n} = 0, \quad \dot{\mathbf{k}}\cdot\mathbf{k} = 0, \\ &\dot{\mathbf{n}}\cdot\mathbf{k} + \dot{\mathbf{k}}\cdot\mathbf{n} = 0, \quad \ddot{\mathbf{n}}\cdot\mathbf{n} + \dot{\mathbf{n}}\cdot\dot{\mathbf{n}} = 0, \quad \ddot{\mathbf{k}}\cdot\mathbf{k} + \dot{\mathbf{k}}\cdot\dot{\mathbf{k}} = 0, \\ &\ddot{\mathbf{n}}\cdot\mathbf{k} + 2\dot{\mathbf{n}}\cdot\dot{\mathbf{k}} + \ddot{\mathbf{k}}\cdot\mathbf{n} = 0.\end{aligned} \tag{41}$$

The relationships (40) and (41) are enough to specify completely the vectors **n** and **k** in terms of the Euler angles of the principal-axis coordinate system; there are three degrees of freedom in either case.

In terms of the true vectors as expressed through the principal-axis coordinate system, the rotational Lagrangian component maintains the simple form

$$L = T^r(\dot{\mathbf{n}}, \dot{\mathbf{k}}, \mathbf{n}, \mathbf{k}) - V(\mathbf{n}, \mathbf{k}), \tag{42}$$

in which

$$T^r = \frac{1}{2} I_\alpha \omega_\alpha \omega_\alpha. \tag{43}$$

In the above equation, ω is again given by Eq. (19), but this time written in terms of **n** and **k** instead of the Euler angles:

$$T^r = \frac{1}{2}\left[I_1(\dot{\mathbf{n}} - (\dot{\mathbf{n}} - \mathbf{k})\mathbf{k})^2 + \frac{1}{2}I_2((\dot{\mathbf{n}}\cdot\mathbf{k})^2 + (\dot{\mathbf{k}}\cdot\mathbf{n})^2) + I_3(\dot{\mathbf{k}} - (\dot{\mathbf{k}} - \mathbf{n})\mathbf{n})^2\right]. \tag{44}$$

From this equation, it is apparent that ω_1 is the angular velocity of **n** less its component in the direction of **k**, ω_3 is the angular velocity of **k** less its component in the direction of **n**, and ω_2 is the sum of the neglected contributions. Hence the vectors **n** and **k** have a nice physical interpretation. The drawback of using **n** and **k** is that T^r now has an explicit orientation dependence, as expressed in Eq. (42). The advantage is that now one is dealing with true vectors, which satisfy all of the characteristic properties of such quantities. At the same time, a further advantage is that the inertial constants appearing in Eq. (43) are constants. As will be seen shortly, for the liquid-crystalline fluid examined herein, the disadvantage vanishes; i.e. $T^r(\dot{\mathbf{n}}, \dot{\mathbf{k}}, \mathbf{n}, \mathbf{k}) \rightarrow T^r(\dot{\mathbf{n}}, \dot{\mathbf{k}})$.

The equations of motion for the constrained vectors **n** and **k** can now be obtained from the Principle of Least Action by computing variations of

$$\delta\Im = \int_{t_1}^{t_2} [L - \lambda_1(\mathbf{k}\cdot\mathbf{k} - 1) - \lambda_2(\mathbf{n}\cdot\mathbf{n} - 1) - \lambda_3\mathbf{n}\cdot\mathbf{k}]\mathrm{d}t. \tag{45}$$

Taking independent variations $\delta\mathbf{n}$ and $\delta\mathbf{k}$, and subsequently solving for the Lagrange multipliers λ_1, λ_2, λ_3 [15], yields the rotational equations of motion for **n** and **k**:

$$\begin{aligned} I_1\ddot{n}_\alpha + \frac{I_2 - 2I_1}{2}\ddot{n}_\beta k_\beta k_\alpha &= \frac{I_1 - I_2 - I_3}{2}\dot{n}_\beta \dot{k}_\beta k_\alpha + (I_1 - I_2 + I_3)(\dot{n}_\beta k_\beta)^2 n_\alpha \\ &- I_1\dot{n}_\beta\dot{n}_\beta n_\alpha + (I_1 - I_2 + I_3)\dot{n}_\beta k_\beta \dot{k}_\alpha - \frac{\partial V}{\partial n_\alpha} + \frac{\partial V}{\partial n_\beta}n_\beta n_\alpha \\ &+ \frac{1}{2}\frac{\partial V}{\partial k_\beta}n_\beta k_\alpha + \frac{1}{2}\frac{\partial V}{\partial n_\beta}k_\beta k_\alpha, \end{aligned} \tag{46}$$

$$\begin{aligned} I_3\ddot{k}_\alpha + \frac{I_2 - 2I_3}{2}\ddot{k}_\beta n_\beta n_\alpha &= \frac{I_3 - I_2 - I_1}{2}\dot{n}_\beta \dot{k}_\beta n_\alpha + (I_1 - I_2 + I_3)(n_\beta \dot{k}_\beta)^2 k_\alpha \\ &- I_3\dot{k}_\beta\dot{k}_\beta k_\alpha + (I_1 - I_2 + I_3)n_\beta \dot{k}_\beta \dot{n}_\alpha - \frac{\partial V}{\partial k_\alpha} + \frac{\partial V}{\partial k_\beta}k_\beta k_\alpha \\ &+ \frac{1}{2}\frac{\partial V}{\partial n_\beta}k_\beta n_\alpha + \frac{1}{2}\frac{\partial V}{\partial k_\beta}n_\beta n_\alpha. \end{aligned} \tag{47}$$

Eqs. (46) and (47) do indeed yield the Euler equations (30) when $\dot{\mathbf{n}}$ and $\dot{\mathbf{k}}$ are expressed in terms of ω:

$$\dot{\mathbf{n}} = -\omega_1\mathbf{m} + \omega_2\mathbf{k}, \qquad \dot{\mathbf{k}} = \omega_3\mathbf{m} - \omega_2\mathbf{n}, \qquad \dot{\mathbf{m}} = \omega_1\mathbf{n} - \omega_3\mathbf{k}. \tag{48}$$

In these expressions, **m** is the unit vector that is perpendicular to both **n** and **k**; it points in the direction of $\mathbf{e}_2$ when **k** is pointing in the direction of $\mathbf{e}_1$. The vector **m** satisfies the constraint relationships

$$\begin{aligned} &\mathbf{m}\cdot\mathbf{n} = 0, \qquad \mathbf{m}\cdot\mathbf{m} = 1, \qquad \mathbf{m}\cdot\mathbf{k} = 0, \qquad \dot{\mathbf{m}}\cdot\mathbf{m} = 0, \\ &\dot{\mathbf{m}}\cdot\mathbf{k} = -\dot{\mathbf{k}}\cdot\mathbf{m} = -\omega_3, \qquad \dot{\mathbf{m}}\cdot\mathbf{n} = -\dot{\mathbf{n}}\cdot\mathbf{m} = \omega_1. \end{aligned} \tag{49}$$

Using Eqs. (48) and (49) in the equations of motion (46) and (47) results ultimately in the Euler equations (30).

3.2. Application to a long, thin rod

Now consider a long, rod-like particle with negligible thickness. In this case, the body coordinates, assumed to coincide with the principal-axis coordinates, are such that one axis points axially along the rod, and the other two axes are mutually perpendicular to the rod axis. Thus, evaluating the analog of Eq. (17) in terms of the body coordinates for this particle where the ξ_3-axis is taken as the rod axis, one finds that $I_3 \to 0$ and that $I_1 = I_2 = \rho_r L^3/12$ for a rod of length L and constant mass density ρ_r. But how do the true vectors **n** and **k** correspond to the body-fixed coordinate system? To determine this most conveniently, consider a particular time when the body-fixed coordinates, ξ_i and the inertial coordinates, $\mathbf{r}_i$, just happen to coincide. At this time, all of the Euler angles have vanishing values. Consequently, from Eq. (40), it is apparent that $\mathbf{n} = \mathbf{e}_3$ and $\mathbf{k} = \mathbf{e}_1$. Hence when $I_3 \to 0$ and $I_1 = I_2$, $\mathbf{n}$ is pointed along the axial direction of the rod, and $\mathbf{k}$ is perpendicular to it.

For the case of a long thin rod, where $I_3 \to 0$ and $I_1 = I_2$, the rotational kinetic energy of Eq. (44) becomes, in view of the constraint relationships of Eq. (41), $T^r = (I_1/2)\dot{\mathbf{n}}\cdot\dot{\mathbf{n}}$. Note that this rotational energy expression no longer depends on $\dot{\mathbf{k}}$, and has lost its orientation dependence; i.e. it depends on neither $\mathbf{n}$ nor $\mathbf{k}$, as alluded to above. The kinetic energy does not depend on $\dot{\mathbf{k}}$ because this vector is always perpendicular to the rod axis. Since the rod is vanishingly thin, and thus has no moment for rotation in that direction, there is no impetus (inertia) for the orientation perpendicular to the rod to change in time; i.e. all orientations perpendicular to the rod axis are equivalent.

In light of the above remarks, one would expect that $\mathbf{k}$ and $\dot{\mathbf{k}}$ would drop out of the system of equations completely. This is indeed the case. The ultimate result is a single equation for the dynamics of the director:

$$I\ddot{n}_\alpha = -I\dot{n}_\beta\dot{n}_\beta n_\alpha - \frac{\partial V}{\partial n_\alpha} + \frac{\partial V}{\partial n_\beta}n_\beta n_\alpha. \tag{50}$$

Thus, the rotational dynamics of a thin, rod-like particle are completely described in terms of a set of three (one for each component of $\mathbf{n}$) second-order ordinary differential equations, which involve a single inertial constant, I.

Other examples can be worked out as well. For instance, for a long, rectangular board of negligible thickness

$$I_1 = \frac{\rho_r}{12}W^4 m(m^2+1), \qquad I_2 = I_3 = \frac{\rho_r}{12}W^4 m, \tag{51}$$

where the long axis of the board is oriented in the ξ_3-direction. In these expressions, W is the width of the board and mW its length, with m being a positive integer.

3.3. *The Hamiltonian formulation and the Poisson bracket*

Now it is necessary to find the Hamiltonian equations corresponding to Eqs. (46) and (47), and subsequently to derive a bracket structure as well. To accomplish this, it is a prerequisite to calculate the Legendre transformation of the Lagrangian, which gives

$$H = N_\alpha\dot{n}_\alpha + K_\alpha\dot{k}_\alpha - \overline{T}^r + V = \overline{T}^r + V, \tag{52}$$

and

$$N_\alpha \equiv \frac{\partial L}{\partial \dot{n}_\alpha}, \qquad K_\alpha \equiv \frac{\partial L}{\partial \dot{k}_\alpha}. \tag{53}$$

In these expressions, $\overline{T}^r$ is the rotational kinetic energy of Eq. (44) written as a function of **N**, **K**, **n** and **k**:

$$\overline{T}^r = \frac{1}{2I_1}\left[N_\alpha - (N_\beta k_\beta)k_\alpha\right]^2 + \frac{1}{2I_3}\left[K_\alpha - (K_\beta n_\beta)n_\alpha\right]^2 + \frac{1}{I_2}\left[(N_\alpha k_\alpha)^2 + (K_\alpha n_\alpha)^2\right]. \quad (54)$$

The variation of the action can then be written analogously to Eq. (45), from which the Hamiltonian equations of motion may be derived as

$$\begin{aligned}
\dot{n}_\alpha &= \frac{\partial H}{\partial N_\alpha}, \qquad \dot{k}_\alpha = \frac{\partial H}{\partial K_\alpha},\\
\dot{N}_\alpha &= -\frac{\partial H}{\partial n_\alpha} + \frac{\partial H}{\partial n_\beta}n_\beta n_\alpha - \frac{\partial H}{\partial N_\beta}N_\beta n_\alpha\\
&\quad + \frac{1}{2}k_\alpha\left(\frac{\partial H}{\partial n_\beta}k_\beta + \frac{\partial H}{\partial k_\beta}n_\beta - \frac{\partial H}{\partial N_\beta}K_\beta - \frac{\partial H}{\partial K_\beta}N_\beta\right),\\
\dot{K}_\alpha &= -\frac{\partial H}{\partial k_\alpha} + \frac{\partial H}{\partial k_\beta}k_\beta k_\alpha - \frac{\partial H}{\partial K_\beta}K_\beta k_\alpha\\
&\quad + \frac{1}{2}n_\alpha\left(\frac{\partial H}{\partial n_\beta}k_\beta + \frac{\partial H}{\partial k_\beta}n_\beta - \frac{\partial H}{\partial N_\beta}K_\beta - \frac{\partial H}{\partial K_\beta}N_\beta\right).
\end{aligned} \quad (55)$$

Note that these equations reduce to the Lagrangian equations of motion (46) and (47), when **N** and **K** are expressed in terms of $\dot{\mathbf{n}}$ and $\dot{\mathbf{k}}$, respectively.

A bracket structure for this Hamiltonian system may be written down by noting that the evolution equation for F no longer has a simple, unconstrained form. Since **n** and **k** are perpendicular unit vectors, not all of the functionality of $F(\mathbf{N}, \mathbf{K}, \mathbf{n}, \mathbf{k})$ is independent. Thus, the Volterra derivatives necessary to construct a Poisson bracket correspond to projections into a constrained space, and may be evaluated as

$$\begin{aligned}
\frac{\delta F}{\delta N_\alpha} &= \frac{\partial F}{\partial N_\alpha}, \qquad \frac{\delta F}{\delta K_\alpha} = \frac{\partial F}{\partial K_\alpha},\\
\frac{\delta F}{\delta n_\alpha} &= \frac{\partial F}{\partial n_\alpha} - \frac{\partial F}{\partial n_\beta}n_\beta n_\alpha - \frac{1}{2}\left(\frac{\partial F}{\partial n_\beta}k_\beta + \frac{\partial F}{\partial k_\beta}n_\beta\right)k_\alpha,\\
\frac{\delta F}{\delta k_\alpha} &= \frac{\partial F}{\partial k_\alpha} - \frac{\partial F}{\partial k_\beta}k_\beta k_\alpha - \frac{1}{2}\left(\frac{\partial F}{\partial n_\beta}k_\beta + \frac{\partial F}{\partial k_\beta}n_\beta\right)n_\alpha.
\end{aligned} \quad (56)$$

Using these projections, the bracket structure corresponding to the Hamiltonian equations of rotational motion (Eq. (55)) can be derived as

$$
\begin{aligned}
\{F,H\} = {} & \frac{\delta F}{\delta n_\alpha}\frac{\delta H}{\delta N_\alpha} - \frac{\delta H}{\delta n_\alpha}\frac{\delta F}{\delta N_\alpha} + \frac{\delta F}{\delta k_\alpha}\frac{\delta H}{\delta K_\alpha} - \frac{\delta H}{\delta k_\alpha}\frac{\delta F}{\delta K_\alpha} \\
& - \frac{1}{2}\left(\frac{\delta F}{\delta N_\alpha}k_\alpha + \frac{\delta F}{\delta K_\alpha}n_\alpha\right)\left(\frac{\delta H}{\delta N_\beta}K_\beta + \frac{\delta H}{\delta K_\beta}N_\beta\right) \\
& + \frac{1}{2}\left(\frac{\delta H}{\delta N_\alpha}k_\alpha + \frac{\delta H}{\delta K_\alpha}n_\alpha\right)\left(\frac{\delta F}{\delta N_\beta}K_\beta + \frac{\delta F}{\delta K_\beta}N_\beta\right) - \frac{\delta F}{\delta N_\alpha}n_\alpha\frac{\delta H}{\delta N_\beta}N_\beta \\
& + \frac{\delta H}{\delta N_\alpha}n_\alpha\frac{\delta F}{\delta N_\beta}N_\beta - \frac{\delta F}{\delta K_\alpha}k_\alpha\frac{\delta H}{\delta K_\beta}K_\beta + \frac{\delta H}{\delta K_\alpha}k_\alpha\frac{\delta F}{\delta K_\beta}K_\beta.
\end{aligned}
\tag{57}
$$

The bracket structure of Eq. (57) is the culmination of this section. It has thus been shown how the equations of rotational motion may be expressed in terms of true vectors, and derived in both Lagrangian and Hamiltonian forms. This exercise, though largely pedagogical, will prove to be very important in subsequent sections when the continuum equations of the LE Model are derived.

3.4. The bracket structure of the long, thin rod

To simplify the analysis in subsequent sections, it is now beneficial to reduce the bracket structure of Eq. (57) to the special case of a long, thin rod. Since the LE Model is written in terms of this special case, this process will allow the neglect of a large portion of the full functionality inherent in the general rigid-body bracket structure of Eq. (57).

Remember that for the long, thin rod, $I_3 \to 0$ and $I_1 = I_2 \equiv I$. With this in mind, according to Eq. (53)

$$
K_\alpha = \frac{1}{2} I n_\beta \dot{k}_\beta n_\alpha. \tag{58}
$$

From the Hamiltonian equations of motion (55) $\dot{\mathbf{k}} = \mathbf{0}$, so that Eq. (58) mandates that

$$
\dot{K}_\alpha = 0 = \frac{1}{2} n_\alpha k_\beta \frac{\partial H}{\partial n_\beta}. \tag{59}
$$

Since $H = H(\mathbf{n}) \neq H(\mathbf{n},\mathbf{k})$, the Hamiltonian equations of motion reduce to

$$
\dot{n}_\alpha = \frac{\partial H}{\partial N_\alpha}, \qquad \dot{N}_\alpha = -\frac{\partial H}{\partial n_\alpha} + \frac{\partial H}{\partial n_\beta} n_\beta n_\alpha - \frac{\partial H}{\partial N_\beta} N_\beta n_\alpha. \tag{60}
$$

Now according to identity (59), the constrained derivatives of Eq. (56) become

$$
\frac{\delta F}{\delta N_\alpha} = \frac{\partial F}{\partial N_\alpha}, \quad \frac{\delta F}{\delta K_\alpha} = 0, \quad \frac{\delta F}{\delta n_\alpha} = \frac{\partial F}{\partial n_\alpha} - \frac{\partial F}{\partial n_\beta} n_\beta n_\alpha, \quad \frac{\delta F}{\delta k_\alpha} = 0. \tag{61}
$$

These derivatives are obtained by realizing that $\delta F/\delta \mathbf{n}$ must belong to the same operating space as $\delta H/\delta \mathbf{n}$; therefore, it is necessary that $(\delta F/\delta \mathbf{n})\cdot\mathbf{k} = 0$. Finally, the bracket structure of Eq. (57) reduces to

$$\{F,H\} = \frac{\delta F}{\delta n_\alpha}\frac{\delta H}{\delta N_\alpha} - \frac{\delta H}{\delta n_\alpha}\frac{\delta F}{\delta N_\alpha} - \frac{\delta F}{\delta N_\alpha}n_\alpha\frac{\delta H}{\delta N_\beta}N_\beta + \frac{\delta H}{\delta N_\alpha}n_\alpha\frac{\delta F}{\delta N_\beta}N_\beta. \tag{62}$$

The continuum version of this bracket structure will be the starting point of the derivation in Section 4.

4. The material description of continuous fluids of rod-like particles

In this section, the formalism refined in Section 3.4 for the special case of a long, thin rod is transferred over to a continuum description composed of fluid particles. Each fluid particle is small on a macroscopic length scale, but large on a microscopic length scale. Thus, each one is small enough to be considered as infinitesimal in size from a continuum perspective, yet large enough to contain many microscopic long, thin rods, each pointing in the same direction. From a thermodynamic point of view, each fluid particle is a tiny thermodynamic subsystem in a quasiequilibrium state [4]. Thus the potential energy of each particle is made up of an equilibrium thermodynamic potential function, as well as a contribution from external fields. The inertial constant, $\hat{I}$, is no longer that of a single rod, as in Section 3.2, but is still a constant due to the fact that the orientation of the rods within a given fluid particle is still described by the true vector, $\mathbf{n}$, called the director, which is taken relative to the principal-axis coordinate system. Hence the physical model of the system used herein is essentially one of a continuum wherein the fluid particles are allowed to translate and expand autonomously from their rotation, which is described with the director. This is nothing more than a statement of Chasle's Theorem applied to a continuum. Of course, this does not mean, as it did not mean before, that the rotational and translational motions are not coupled to each other through the potential-energy term appearing in the Hamiltonian of the system.

To derive the equations of motion for the continuous LE fluid, it is prerequisite to assert which variables are necessary for the system description. This is much more difficult for a continuum than for discrete particles. A given elemental fluid particle of the continuous medium is labeled by a position vector at time $t = 0$, $\mathbf{r}$, with a coordinate function $\mathbf{Y} = \mathbf{Y}(\mathbf{r}, t) \in \Re^3$ that specifies the trajectory of this particle in space and time. At time $t = 0$, the fluid occupies a certain volume, Ω, with boundary $\partial\Omega$, and the initial condition on the function $\mathbf{Y}$ is $\mathbf{Y}(\mathbf{r}, 0) = \mathbf{r}$. At a later time, t', the entire body of the fluid may have translated, rotated, and deformed, so that the fluid occupies a new volume, Ω', with boundary $\partial\Omega'$. Therefore, the function $\mathbf{Y}$ is viewed as the one-to-one mapping that gives the location at time t of the particle that was at position $\mathbf{r}$ at $t = 0$.

Associated with this fluid particle is a volume element at $t = 0$, d^3r. Although the mass of the fluid particle is taken as constant, its volume element is allowed to vary in space and time due to the compressible nature of the fluid. The volume element at any time can be related to the volume element at $t = 0$ by $\mathrm{d}^3Y = J\,\mathrm{d}^3r$, in which the *Jacobian* is defined as

$$J \equiv \det \mathbf{F}, \qquad F_{\alpha\beta} \equiv \frac{\partial Y_\alpha}{\partial r_\beta}. \tag{63}$$

The tensor $\mathbf{F}$ is known as the *deformation gradient*.

The distribution of the system mass at $t = 0$ can be described by a density function, $\rho_0(\mathbf{r})$, and, since the mass is conserved, it is immediately evident that $\rho\,\mathrm{d}^3Y = \rho_0\,\mathrm{d}^3r$, where ρ is the mass density at any $\mathbf{Y}(\mathbf{r}, t)$. It is evident from the preceding arguments that $\rho = \rho_0/J$, which indicates that ρ depends on the deformation gradient; hence, the mass density will be denoted as $\rho(\mathbf{F})$. The remaining variables necessary for the description of the LE fluid are $\dot{\mathbf{Y}}(\mathbf{r}, t)$, $\mathbf{n}(\mathbf{r}, t)$, and $\mathbf{N}(\mathbf{r}, t)$. These variables are the continuum analogs of the variables of Section 3.4. Note that $\mathbf{n}\cdot\mathbf{n} = 1$.

The model as defined above is based in the *material description*; i.e. in a reference frame in which the coordinates are convected along with the flow of a given fluid particle. This is not a laboratory-based reference system; such a reference system is known as the *spatial description*. The LE Model is based in the spatial description; however, in order to make a direct transformation of the theory from Section 3 (a material description) into a continuous formalism, it is necessary to transfer from the material description into the spatial description. In doing so, every quantity from Section 3, variable to bracket structure, has an analog in the material description of the continuous model. After obtaining the continuum analogs in the material description, the theory can then be reduced to the spatial description; this is the subject of Section 5.

The Lagrangian for the LE fluid in terms of the proper variables is

$$
\begin{aligned}
L &\equiv \int_{\Omega'} \rho(\mathbf{F})\lfloor \hat{T}^t(\mathbf{Y}, \dot{\mathbf{Y}}) + \hat{T}^r(\mathbf{n}, \dot{\mathbf{n}}) - \hat{e}_p(\mathbf{Y}, \mathbf{n}) - \hat{U}(\rho(\mathbf{F}), \hat{S}(\mathbf{r}))\rfloor \mathrm{d}^3Y \\
&= \int_{\Omega} \rho_0(\mathbf{r})\lfloor \hat{T}^t(\mathbf{Y}, \dot{\mathbf{Y}}) + \hat{T}^r(\mathbf{n}, \dot{\mathbf{n}}) - \hat{e}_p(\mathbf{Y}, \mathbf{n}) - \hat{U}(\rho(\mathbf{F}), \hat{S}(\mathbf{r}))\rfloor \mathrm{d}^3r.
\end{aligned}
\tag{64}
$$

As before, the Lagrangian represents the difference between the kinetic and potential energies. $\hat{T}^t$ is the translational kinetic energy per unit mass of the fluid particle:

$$\hat{T}^t = \frac{1}{2}\dot{\mathbf{Y}}\cdot\dot{\mathbf{Y}}. \tag{65}$$

$\hat{T}^r$ is the rotational kinetic energy per unit mass of the fluid particle. For the LE fluid, it is given by

$$\hat{T}^r = \frac{1}{2}\hat{I}\dot{\mathbf{n}}\cdot\dot{\mathbf{n}}, \tag{66}$$

in which $\hat{I} = I/\rho$. Note that for the LE fluid, $\hat{T}^t$ and $\hat{T}^r$ depend only on their respective time derivatives, and not on particle orientation or position.

The potential energy of Section 3, V, has been split into two terms in the Lagrangian of Eq. (64). The first term, $\hat{e}_p$, is the potential energy per mass of the particle due to external fields. It depends on the particle's location and orientation in space. In specific cases, it may decouple into a spatially dependent term and an orientationally dependent term, but, in general, this is not required for the LE Model. $\hat{U}$ is the internal energy per unit mass, and, as such, it represents a thermodynamic potential function. In this work, it is assumed that the rotation, just as translation, has no effect upon the quasiequilibrium internal energy of the fluid particle. Note that for adiabatic flow, $\hat{S}$, the entropy per mass, is constant. Although the continuous function $\hat{U}$ varies from one fluid particle to another as each has a different value of $\hat{S}$ and ρ, for a specific fluid particle, $\hat{U}$ does not depend on $\hat{S}$ since $\hat{S}$ does not change for that fluid particle. From equilibrium thermodynamics, the pressure is defined as

$$p \equiv -\left.\frac{\partial \hat{U}}{\partial \hat{V}}\right|_{\hat{S}} = -\left.\frac{\partial \hat{U}}{\partial \frac{1}{\rho}}\right|_{\hat{S}} = \rho^2 \left.\frac{\partial \hat{U}}{\partial \rho}\right|_{\hat{S}}, \tag{67}$$

in which $\hat{V}$ is the specific volume of equilibrium thermodynamics.

Hamiltonian equations and their associated bracket structure can now be derived, similarly as in Sections 3.3 and 3.4. The bracket structures are in terms of Volterra derivatives, however, but nothing else changes noticeably. Of course, the Hamiltonian, as well as the Lagrangian of Eq. (64), is now a functional of the relevant variables, as is the arbitrary functional

$$F[\mathbf{Y}, \dot{\mathbf{Y}}, \mathbf{n}, \dot{\mathbf{n}}] \equiv \int_{\Omega} f(\mathbf{Y}, \nabla\mathbf{Y}, \dot{\mathbf{Y}}, \mathbf{n}, \dot{\mathbf{n}}) \mathrm{d}^3 r. \tag{68}$$

For an unspecified variable a, the Volterra functional derivative is given by [4]

$$\frac{\delta F}{\delta a} \equiv \frac{\partial f}{\partial a} - \nabla \cdot \left(\frac{\partial f}{\partial (\nabla a)} \right) \in P. \tag{69}$$

Note that for the LE fluid, it is necessary to allow the functionals to depend on the gradient of a, as well as just a, since this theory incorporates $\rho(\mathbf{F})$ in the material description and the Frank–Oseen distortion energy, $W(\mathbf{n}, \nabla\mathbf{n})$, in the spatial description.

In view of the Lagrangian of Eq. (64), the Hamiltonian variables are $\mathbf{Y}$, $\mathbf{n}$,

$$\Pi_\alpha \equiv \frac{\delta L}{\delta \dot{Y}_\alpha} = \rho_0 \dot{Y}_\alpha, \qquad N_\alpha \equiv \frac{\delta L}{\delta \dot{n}_\alpha} = \rho_0 \hat{I} \dot{n}_\alpha. \tag{70}$$

For the constrained variables under consideration, the proper Volterra derivatives of the arbitrary functional follow analogously to Eq. (61) from Section 3.4

$$\begin{aligned} \frac{\delta F}{\delta Y_\alpha} &= \frac{\partial f}{\partial Y_\alpha} - \nabla_\beta \left(\frac{\partial f}{\partial F_{\alpha\beta}} \right), \qquad & \frac{\delta F}{\delta \Pi_\alpha} &= \frac{\partial f}{\partial \Pi_\alpha}, \\ \frac{\delta F}{\delta n_\alpha} &= \frac{\partial f}{\partial n_\alpha} - \frac{\partial f}{\partial n_\beta} n_\beta n_\alpha, \qquad & \frac{\delta F}{\delta N_\alpha} &= \frac{\partial f}{\partial N_\alpha}. \end{aligned} \tag{71}$$

Because of Eqs. (64) and (70), the Hamiltonian for the LE fluid in the material description is

$$H = \int_{\Omega} \rho_0(\mathbf{r}) \left[\frac{1}{2\rho_0^2} \Pi_\alpha \Pi_\alpha + \frac{1}{2\hat{I}\rho_0^2} N_\alpha N_\alpha + \hat{e}_p(\mathbf{Y}, \mathbf{n}) + \hat{U}(\rho(\mathbf{F}), \hat{S}(\mathbf{r})) \right] \mathrm{d}^3 r, \tag{72}$$

and thus the Hamiltonian equations of motion are

$$\frac{\partial Y_\alpha}{\partial t} = \frac{\delta H}{\delta \Pi_\alpha} = \frac{1}{\rho_0}\Pi_\alpha, \qquad \frac{\partial \Pi_\alpha}{\partial t} = -\frac{\delta H}{\delta Y_\alpha} = -\rho_0 \frac{\partial \hat{e}_p}{\partial Y_\alpha} - \frac{\rho_0}{\rho}\frac{\partial p}{\partial Y_\alpha},$$

$$\frac{\partial n_\alpha}{\partial t} = \frac{\delta H}{\delta N_\alpha} = \frac{1}{\hat{I}\rho_0}N_\alpha, \tag{73}$$

$$\frac{\partial N_\alpha}{\partial t} = -\frac{\delta H}{\delta n_\alpha} - \frac{\delta H}{\delta N_\alpha}N_\beta n_\alpha = -\rho_0\frac{\partial \hat{e}_p}{\partial n_\alpha} + \rho_0\frac{\partial \hat{e}_p}{\partial n_\beta}n_\beta n_\alpha - \frac{1}{\hat{I}\rho_0}N_\beta N_\beta n_\alpha.$$

Finally, the Poisson bracket structure corresponding to the LE fluid in the material description is

$$\begin{aligned}\{F,H\} = \int_\Omega \bigg(&\frac{\delta F}{\delta Y_\alpha}\frac{\delta H}{\delta \Pi_\alpha} - \frac{\delta H}{\delta Y_\alpha}\frac{\delta F}{\delta \Pi_\alpha} + \frac{\delta F}{\delta n_\alpha}\frac{\delta H}{\delta N_\alpha} - \frac{\delta H}{\delta n_\alpha}\frac{\delta F}{\delta N_\alpha} \\ &- \frac{\delta F}{\delta N_\alpha}n_\alpha\frac{\delta H}{\delta N_\beta}N_\beta + \frac{\delta H}{\delta N_\alpha}n_\alpha\frac{\delta F}{\delta N_\beta}N_\beta\bigg)\mathrm{d}^3 r.\end{aligned} \tag{74}$$

Note that the last term in this bracket vanishes since $\mathbf{n}\cdot(\delta H/\delta \mathbf{N}) = 0$: it has been added to make the bracket structure antisymmetric. This bracket structure will be reduced to the spatial description in Section 5.

5. The spatial description of the Leslie–Ericksen fluid

The equations of motion for the LE fluid derived in Section 4 are not in the most useful form for potential applications, and are not immediately recognizable as those presented in Section 1. This is because the equations of Section 1 were expressed relative to a coordinate system fixed in the laboratory, not one that convects along with the fluid particle.

In this section, it is necessary to transform the material equations of Section 4 into their spatial counterparts; i.e. relative to a set of coordinates that is fixed in the laboratory frame of reference. This transformation will be accomplished by a reduction of the continuum bracket structure of Eq. (74) in terms of the material coordinates of the prior section into one in terms of spatial variables. This type of reduction has been performed in various forms for various systems [1–4,15].

It is first necessary to establish the fixed coordinate system, and to specify the fundamental relationships that interrelate it with the coordinates of Section 4. This fixed coordinate system will be denoted by the field symbol, $\mathbf{x}$, also Cartesian, so that at each specific point there is a constant volume element denoted by d^3x. At a specific time, t, it is required that a certain fixed point $\mathbf{x}$ coincides with the position function $\mathbf{Y}(\mathbf{r}, t)$, and that $\mathrm{d}^3Y = \mathrm{d}^3x$ at that time. Hence, it is possible at this instant to perform the required transformations. A new function can then be written for all later times, $\tilde{t}$, $\mathbf{R}(\mathbf{x}, \tilde{t})$, which maps $\mathbf{Y}$ back to the fixed position $\mathbf{x}$: $\mathbf{x} = \mathbf{Y}(\mathbf{R}(\mathbf{x}, \tilde{t}), \tilde{t})$. Consequently, $\mathbf{R}(\mathbf{x}, \tilde{t})$ serves as a material label for the fluid particle that was at position $\mathbf{x}$ at time t. It is evident that $\mathbf{R} = \mathbf{r}$ at time t.

Since the position coordinate $\mathbf{x}$ in the spatial description has been chosen as the irreducible quantity (along with time, t), the position is no longer a problem variable, as in the material description. All other variables that describe the state of the system are now written as functions of $\mathbf{x}$ and t, instead of just t. The logical variables for this style of state description are the mass density, $\rho(\mathbf{x}, t)$, the entropy density, $s(\mathbf{x}, t) = \rho(\mathbf{x}, t)\hat{S}(\mathbf{x}, t)$, and measures of the translational particle momentum, particle orientation, and particle moment of momentum. These five (actually more, considering separate components of vectorial quantities) variable fields will completely describe the state of the system. For the transformations of the material variables to those just mentioned to be performed, the spatial variables should be density variables [4], such as ρ and s. Therefore, the three additional vector fields are defined as the translational momentum density field, $\mathbf{M}(\mathbf{x}, t) = \rho(\mathbf{x}, t)\mathbf{v}(\mathbf{x}, t)$, the orientation density field, $\tilde{\mathbf{n}}(\mathbf{x}, t) = \rho(\mathbf{x}, t)\mathbf{n}(\mathbf{x}, t)$, and the moment of momentum-density field, $\tilde{\mathbf{N}}(\mathbf{x}, t) = \rho(\mathbf{x}, t)\mathbf{N}(\mathbf{x}, t)$. The only conditions that are necessary for the following calculations are the no-penetration condition and the unit-vector constraint.

It is now possible to express the five spatial fields in terms of their material counterparts at the instant t. This is accomplished using a delta function of the form

$$\int_{\Omega'} g(\mathbf{b})\delta^3[\mathbf{b} - \mathbf{a}]\mathrm{d}^3 b \equiv \begin{cases} g(\mathbf{a}), & \mathbf{b} = \mathbf{a} \\ 0, & \mathbf{b} \neq \mathbf{a}, \end{cases} \tag{75}$$

in which g is an arbitrary function and $\mathbf{a}$, $\mathbf{b}$ are arbitrary coordinates, such as $\mathbf{x}$. This definition allows one to apply Volterra differentiation to the appropriate quantities. The mass density at time t may now be written as

$$\begin{aligned} \rho(\mathbf{x}, t) = \rho(\mathbf{F}(\mathbf{r}), t) &\Rightarrow \int_{\Omega'} \rho(\mathbf{F}(\mathbf{r}), t)\delta^3[\mathbf{Y}(\mathbf{r}, t) - \mathbf{x}]\mathrm{d}^3 Y \\ &= \int_{\Omega} \rho_0(\mathbf{r})\delta^3[\mathbf{Y}(\mathbf{r}, t) - \mathbf{x}]\mathrm{d}^3 r. \end{aligned} \tag{76}$$

Similarly, one can also write

$$\begin{aligned} s(\mathbf{x}, t) &= \int_{\Omega} s_0(\mathbf{r})\delta^3[\mathbf{Y}(\mathbf{r}, t) - \mathbf{x}]\mathrm{d}^3 r, \\ \mathbf{M}(\mathbf{x}, t) &= \int_{\Omega} \Pi(\mathbf{r}, t)\delta^3[\mathbf{Y}(\mathbf{r}, t) - \mathbf{x}]\mathrm{d}^3 r, \\ \tilde{\mathbf{n}}(\mathbf{x}, t) &= \int_{\Omega} \rho_0(\mathbf{r})\mathbf{n}(\mathbf{r}, t)\delta^3[\mathbf{Y}(\mathbf{r}, t) - \mathbf{x}]\mathrm{d}^3 r, \\ \tilde{\mathbf{N}}(\mathbf{x}, t) &= \int_{\Omega} \mathbf{N}(\mathbf{r}, t)\delta^3[\mathbf{Y}(\mathbf{r}, t) - \mathbf{x}]\mathrm{d}^3 r. \end{aligned} \tag{77}$$

The Volterra derivatives of the spatial variables with respect to the material variables may now be calculated, and the bracket structure in the spatial description may be derived according to the procedure of Ref. [15] as

$$
\begin{aligned}
\{F,H\} = &-\int_{\Omega'}\left[\frac{\delta F}{\delta\rho}\nabla_\beta\left(\frac{\delta H}{\delta M_\beta}\rho\right)-\frac{\delta H}{\delta\rho}\nabla_\beta\left(\frac{\delta F}{\delta M_\beta}\rho\right)\right]\mathrm{d}^3x \\
&-\int_{\Omega'}\left[\frac{\delta F}{\delta s}\nabla_\beta\left(\frac{\delta H}{\delta M_\beta}s\right)-\frac{\delta H}{\delta s}\nabla_\beta\left(\frac{\delta F}{\delta M_\beta}s\right)\right]\mathrm{d}^3x \\
&-\int_{\Omega'}\left[\frac{\delta F}{\delta M_\gamma}\nabla_\beta\left(\frac{\delta H}{\delta M_\beta}M_\gamma\right)-\frac{\delta H}{\delta M_\gamma}\nabla_\beta\left(\frac{\delta F}{\delta M_\beta}M_\gamma\right)\right]\mathrm{d}^3x \\
&-\int_{\Omega'}\left[\frac{\delta F}{\delta \tilde{n}_\gamma}\nabla_\beta\left(\frac{\delta H}{\delta M_\beta}\tilde{n}_\gamma\right)-\frac{\delta H}{\delta \tilde{n}_\gamma}\nabla_\beta\left(\frac{\delta F}{\delta M_\beta}\tilde{n}_\gamma\right)\right]\mathrm{d}^3x \\
&-\int_{\Omega'}\left[\frac{\delta F}{\delta \tilde{N}_\gamma}\nabla_\beta\left(\frac{\delta H}{\delta M_\beta}\tilde{N}_\gamma\right)-\frac{\delta H}{\delta \tilde{N}_\gamma}\nabla_\beta\left(\frac{\delta F}{\delta M_\beta}\tilde{N}_\gamma\right)\right]\mathrm{d}^3x \\
&-\int_{\Omega'}\left[\rho\delta_{\alpha\beta}-\frac{1}{\rho}\tilde{n}_\alpha\tilde{n}_\beta\right]\left[\frac{\delta F}{\delta\tilde{N}_\alpha}\frac{\delta H}{\delta\tilde{n}_\beta}-\frac{\delta H}{\delta\tilde{N}_\alpha}\frac{\delta F}{\delta\tilde{n}_\beta}\right]\mathrm{d}^3x \\
&-\int_{\Omega'}\frac{1}{\rho}\tilde{n}_\alpha\tilde{N}_\beta\left[\frac{\delta F}{\delta\tilde{N}_\alpha}\frac{\delta H}{\delta\tilde{N}_\beta}-\frac{\delta H}{\delta\tilde{N}_\alpha}\frac{\delta F}{\delta\tilde{N}_\beta}\right]\mathrm{d}^3x. \qquad (78)
\end{aligned}
$$

This Poisson bracket is now in a form that is suitable to derive the LE Model. It is bilinear, antisymmetric, and satisfies the Jacobi identity. Note again that some of the terms in the bracket structure are identically zero for an LE fluid, but are included in Eq. (78) for completeness. Minus the extra terms, it is essentially the same bracket postulated in Refs. [13,14].

Now that the material bracket structure has been transformed into its spatial counterpart at the instant t, the same can be done with the Hamiltonian of Eq. (72):

$$
H=\int_{\Omega'}\left[\frac{1}{2\rho}M_\alpha M_\alpha+\frac{1}{2\hat{I}\rho}\tilde{N}_\alpha\tilde{N}_\alpha+\rho\hat{e}_p(\mathbf{x};\rho,\tilde{\mathbf{n}})+\rho W(\rho,\tilde{\mathbf{n}}\nabla\tilde{\mathbf{n}})+u(\rho,s)\right]\mathrm{d}^3r. \qquad (79)
$$

Note that in this Hamiltonian, an extra term appears over the Hamiltonian of Eq. (72): the Frank–Oseen distortion energy, W. In the material description, it is very difficult to define such an expression because gradients of the director must be defined relative to the deformation gradient. In the spatial description where such gradients are easily defined relative to the fixed laboratory coordinates, this term quantifies the energetic effects of spatial distortions in the director field. Consequently, this energetic term is now inserted into the Hamiltonian, whereas it was previously neglected. Note that its presence has absolutely no effect on the derivation of Hamilton's equations of motion, or the Poisson bracket. Also, the internal energy density has been defined as $u \equiv \rho\hat{U}$. Furthermore, in the spatial description, the external field potential, $\hat{e}_p$, depends on the density as well as the orientation, and also might include an explicit spatial dependence on $\mathbf{x}$ as, for instance, due to an inhomogeneous external magnetic field. For the LE fluid, this potential energy term is given by

$$\hat{e}_p = -\frac{1}{2\rho^2}\Big[(\chi_{\Downarrow} - \chi_{\perp})(\tilde{\mathbf{n}}\cdot\mathbf{H}(\mathbf{x}))^2 + \chi_{\perp}\mathbf{H}(\mathbf{x})\cdot\mathbf{H}(\mathbf{x})\Big]. \tag{80}$$

The necessary constraints on the variables are built into the Volterra derivatives through the derivation of the bracket structure. The Volterra derivatives with respect to the Hamiltonian are thus defined as

$$\begin{aligned}
&\frac{\delta H}{\delta s} = \frac{\delta u}{\delta s}, \qquad \frac{\delta H}{\delta \tilde{N}_\alpha} = \frac{1}{\rho\hat{I}}\tilde{N}_\alpha, \qquad \frac{\delta H}{\delta M_\alpha} = \frac{1}{\rho}M_\alpha,\\
&\frac{\delta H}{\delta \rho} = -\frac{1}{2\rho^2}M_\alpha M_\alpha - \frac{1}{2\hat{I}\rho^2}\tilde{N}_\alpha\tilde{N}_\alpha + \hat{e}_p + \rho\frac{\partial \hat{e}_p}{\partial\rho} + \frac{\partial u}{\partial\rho} + W + \rho\frac{\partial W}{\partial\rho},\\
&\frac{\delta H}{\delta \tilde{n}_\alpha} = \rho\frac{\partial \hat{e}_p}{\partial \tilde{n}_\alpha} + \rho\frac{\partial W}{\partial \tilde{n}_\alpha} - \nabla_\gamma\left(\rho\frac{\partial W}{\partial(\nabla_\gamma \tilde{n}_\alpha)}\right).
\end{aligned} \tag{81}$$

Using the bracket structure of Eq. (79), the evolution equations for the variable set $\{\rho, s, \mathbf{M}, \tilde{\mathbf{n}}, \tilde{\mathbf{N}}\}$ can be calculated. These are not the most useful forms of these equations, however, since they are in terms of density variables. It is a simple matter to rewrite them in terms of the variable set $\{\rho, \hat{S}, \mathbf{v}, \hat{\mathbf{n}}, \hat{\mathbf{N}}\}$, in which the last four new variables are obtained from the former set by division with the mass density. The evolution equations in terms of this variable set reduce to

$$\frac{\partial\rho}{\partial t} = -\nabla_\beta(\rho v_\beta), \tag{82}$$

$$\frac{\partial\hat{S}}{\partial t} = -v_\beta\nabla_\beta\hat{S}, \tag{83}$$

$$\frac{\partial\hat{n}_\alpha}{\partial t} = -v_\beta\nabla_\beta\hat{n}_\alpha + \frac{1}{\hat{I}}\hat{N}_\alpha, \tag{84}$$

$$\begin{aligned}
\frac{\partial\hat{N}_\alpha}{\partial t} = &- v_\beta\nabla_\beta(\hat{N}_\alpha) - \frac{1}{\hat{I}}\hat{N}_\beta\hat{N}_\beta\hat{n}_\alpha - \frac{\partial\hat{e}_p}{\partial\hat{n}_\alpha} + \hat{n}_\alpha\hat{n}_\beta\frac{\partial\hat{e}_p}{\partial\hat{n}_\beta} - \frac{\partial W}{\partial\hat{n}_\alpha} + \frac{\partial W}{\partial\hat{n}_\beta}\hat{n}_\beta\hat{n}_\alpha\\
&+ \nabla_\gamma\left(\frac{\partial W}{\partial(\nabla_\gamma\hat{n}_\alpha)}\right) - \hat{n}_\alpha\hat{n}_\beta\nabla_\gamma\left(\frac{\partial W}{\partial(\nabla_\gamma\hat{n}_\beta)}\right),
\end{aligned} \tag{85}$$

$$\rho\frac{\partial v_\alpha}{\partial t} = -\rho v_\beta\nabla_\beta v_\alpha - \nabla_\alpha p - \rho\nabla_\alpha\hat{e}_p(\mathbf{x};\hat{\mathbf{n}})|_{\hat{\mathbf{n}}} - \nabla_\beta\left(\frac{\partial W(\hat{\mathbf{n}}, \nabla\hat{\mathbf{n}})}{\partial(\nabla_\beta\hat{n}_\gamma)}\nabla_\alpha(\rho\hat{n}_\gamma)\right). \tag{86}$$

In these expressions, the thermodynamic pressure is defined again by Eq. (67), but with an additional term due to the distortion energy:

$$p \equiv \rho^2 \frac{\partial \hat{U}}{\partial \rho}\bigg|_{\hat{s}} - \nabla_\gamma \left(\frac{\partial W(\hat{\mathbf{n}}, \nabla\hat{\mathbf{n}})}{\partial(\nabla_\gamma \hat{n}_\beta)} \rho \hat{n}_\beta \right). \tag{87}$$

Furthermore, the identities

$$\frac{\partial \hat{e}_p(\mathbf{x};\rho,\tilde{\mathbf{n}})}{\partial \rho} = -\frac{1}{\rho}\hat{n}_\beta \frac{\partial \hat{e}_p(\mathbf{x};\hat{\mathbf{n}})}{\partial \hat{n}_\beta}, \qquad \frac{\partial \hat{e}_p(\mathbf{x};\rho,\tilde{\mathbf{n}})}{\partial \tilde{n}_\alpha} = \frac{1}{\rho}\frac{\partial \hat{e}_p(\mathbf{x};\hat{\mathbf{n}})}{\partial \hat{n}_\alpha},$$
$$\frac{\partial W(\rho,\tilde{\mathbf{n}},\nabla\tilde{\mathbf{n}})}{\partial \rho} = -\frac{1}{\rho}\hat{n}_\beta \frac{\partial W(\hat{\mathbf{n}},\nabla\hat{\mathbf{n}})}{\partial \hat{n}_\beta} - \frac{1}{\rho^2}\nabla_\gamma(\rho \hat{n}_\beta)\frac{\partial W(\hat{\mathbf{n}},\nabla\hat{\mathbf{n}})}{\partial(\nabla_\gamma \hat{n}_\beta)}, \tag{88}$$
$$\frac{\partial W(\rho,\tilde{\mathbf{n}},\nabla\tilde{\mathbf{n}})}{\partial \tilde{n}_\alpha} = \frac{1}{\rho}\frac{\partial W(\hat{\mathbf{n}},\nabla\hat{\mathbf{n}})}{\partial \hat{n}_\alpha}, \qquad \frac{\partial W(\rho,\tilde{\mathbf{n}},\nabla\tilde{\mathbf{n}})}{\partial(\nabla_\beta \tilde{n}_\alpha)} = \frac{1}{\rho}\frac{\partial W(\hat{\mathbf{n}},\nabla\hat{\mathbf{n}})}{\partial(\nabla_\beta \hat{n}_\alpha)}$$

were used; i.e. assuming the special case when $\hat{e}_p(\mathbf{x};\rho,\tilde{\mathbf{n}}) = \hat{e}_p(\mathbf{x};\hat{\mathbf{n}})$ and $W(\rho,\hat{\mathbf{n}},\nabla\hat{\mathbf{n}}) = W(\hat{\mathbf{n}},\nabla\hat{\mathbf{n}})$, which is, in actuality, the case for the LE fluid. Since the time t at which the transformation was performed is arbitrary, the above equations must hold for all times. The symbol $\nabla_\alpha \hat{e}_p(\mathbf{x},\hat{\mathbf{n}})|_{\hat{\mathbf{n}}}$ in Eq. (86) denotes that the gradient is taken at constant $\hat{\mathbf{n}}$.

The LE Model is for an incompressible, isothermal material. In this case, the evolution equation for the mass density reduces to the divergence free condition, $\nabla\cdot\mathbf{v} = 0$, and the thermodynamic pressure becomes an arbitrary scalar that guarantees the satisfaction of this additional constraint. The evolution equation for the specific entropy becomes redundant. The variable set of interest for the LE Model is $\{\mathbf{v}, \mathbf{n}, \dot{\mathbf{n}}\}$, where $\mathbf{n} = \hat{\mathbf{n}}$ and $\dot{\mathbf{n}} = (\hat{\mathbf{N}}/\hat{I})$ in Eqs. (84) and (85). In terms of this variable set, the evolution equations for the LE fluid are equivalent to Eqs. (1)–(5) of the LE Theory, which were presented in Section 1. Note that the evolution equation for $\mathbf{n}$ (Eq. (85)) reduces to the definition of $\dot{\mathbf{n}}$, as given by Eq. (3). This completes the rigorous derivation of the conservative part of the LE Model starting with the Principle of Least Action applied to both rotational and translational motion.

6. Conclusion

The purpose of this chapter was to illustrate how rotational motion contributes to the conservative Hamiltonian dynamics of complex, anisotropic fluids, such as liquid crystals. This illustration was based on one of the most revered principles of classical mechanics, Hamilton's Principle of Least Action. The derivation presented followed logical steps, and was completely rigorous. At no time was any ambiguity used in the derivation. Every component of the derivation had a sound basis in standard mathematics and well-established physical principles. The new feature of the derivation was to take these sound mathematics and principles, and apply them in a way in which they had not been applied heretofore. In so doing, it became obvious how rotational motion influences the dynamics of anisotropic fluids, and why it must be accounted for explicitly when deriving models for materials with such intricate microstructure.

References

[1] Marsden, J.E., Ratiu, T. and Weinstein, A. (1984), Semidirect products and reduction in mechanics, Trans. Am. Math. Soc. 281, 147–177.

[2] Abarbanel, H.D.I., Brown, R. and Yang, Y.M. (1988), Hamiltonian formulation of inviscid flows with free boundaries, Phys. Fluids 31, 2802–2809.

[3] Edwards, B.J. and Beris, A.N. (1991), Noncanonical Poisson bracket for nonlinear elasticity with extensions to viscoelasticity, J. Phys. A: Math. Gen. 24, 2461–2480.

[4] Beris, A.N. and Edwards, B.J. (1994), Thermodynamics of Flowing Systems, Oxford University Press, Oxford.

[5] Herivel, J.W. (1955), The derivation of the equations of motion on an ideal fluid by Hamilton's principle, Proc. Camb. Phil. Soc. 51, 344–349.

[6] Lin, C.C. (1963), Hydrodynamics of helium II. Liquid Helium (Careri, G., ed.), Academic Press, New York.

[7] Seliger, R.L. and Whitham, G.B. (1968), Variational principles in continuum mechanics, Proc. R. Soc. 305A, 1–25.

[8] Leslie, F.M. (1966), Some constitutive equations for anisotropic fluids, Q. J. Mech. Appl. Math. 19, 357–370.

[9] Leslie, F.M. (1968), Some constitutive equations for liquid crystals, Arch. Rat. Mech. Anal. 28, 265–283.

[10] Leslie, F.M. (1979), Theory of flow phenomena in liquid crystals. In Advances in Liquid Crystals, (Brown, G.H., eds.), Vol. 4, Academic Press, New York.

[11] Ericksen, J.L. (1961), Conservation laws for liquid crystals, Trans. Soc. Rheol. 5, 23–34.

[12] Sieniutycz, S. (1994), Conservation Laws in Variational Thermo-hydrodynamics, Kluwer, Dordrecht, The Netherlands.

[13] Edwards, B.J (1991), The Dynamical Continuum Theory of Liquid Crystals, PhD Dissertation, University of Delaware.

[14] Edwards, B.J., Beris, A.N. and Grmela, M. (1991), The dynamical behavior of liquid crystals: a continuum description through generalized brackets, Mol. Cryst. Liq. Cryst. 201, 51–86.

[15] Edwards, B.J. and Beris, A.N. (1998), Rotational motion and Poisson bracket structures in rigid particle systems and anisotropic fluid theory, Open Sys. Inform. Dyn. 5, 333–368.

[16] Whittaker, E.T. (1937), A Treatise on the Analytical Dynamics of Particles and Rigid Bodies, 4th edn, Cambridge University Press, Cambridge.

[17] Goldstein, H. (1980), Classical Mechanics, 2nd edn, Addison-Wesley, Reading, MA.

Chapter 8

AN INTRODUCTION TO VARIATIONAL DERIVATION OF THE PSEUDOMOMENTUM CONSERVATION IN THERMOHYDRODYNAMICS

Janusz Badur

Institute of Fluid Flow Machinery, Polish Academy of Sciences, ul. Fiszera 14, 81-852 Gdańsk, Poland

Jordan Badur

Institute of Oceanography, University of Gdańsk, ul. Piłsudskiego 46, 81-378 Gdynia, Poland

Abstract

In this study we will present a reconstruction and revaluation of a few basic, pioneering variational principles in thermohydrodynamics. In particular, most of our attention will be aimed at Gibbs', Natanson's and Eckart's principles. All of them either contain preliminary forms of pseudomomentum and pseudomomentum flux notions or are natural ground for them. They also present a method to obtain pseudomomentum balance equations from properly constructed variational principles. The study is completed with analysis of other variational principles and variational grounds for a 'Generalized Lagrangian Mean' (GLM) approach recently applied in atmospheric and ocean turbulence description.

Keywords: pseudomomentum; pseudomomentum flux; the chemical-potential tensor; Lagrangian variational principles; The Gibbs; Natanson; Eckart principles; material vorticity; configurational forces; acoustic pseudomamentum; wave momentum; Generalized Lagrangian Mean

1. Introduction

Variational calculus methods are widely applied throughout various branches of physics. We should stress that their applications fairly exceed their use as a tool for solving particular problems (no matter how important the problems may be). 'Variational principles' reflect in fact the most general physical laws of various fields of physics ranging from classical mechanics to elementary particles theory.

The variational description has numerous advantages. First, we work with a functional—a notion that is simpler and closer to physical experience than a derivative. Functional properties result in its indifference of the coordinate system and enable the use of approximate methods to find a solution. Variational physical interpretations are especially useful when analytical solutions of differential equations for mass, momentum and energy balance are nonexistent or present significant difficulties.

E-mail address: jb@imp.gda.pl (Janusz Badur); jorb@kdm.task.gda.pl (Jordan Badur).

The variational approach changes our traditional opinion that a fluid continuum is naturally described in the Eulerian (or spatial) coordinates and a solid continuum is described in the Lagrangian (material) ones. Often, a physical system consists of a set of fields some of which are naturally accounted for the Eulerian description while the others are more easily expressed in the Lagrangian description. In the Eulerian description we fix the observer's reference frame in a point of space and observe characteristic parameters of a fluid medium flowing through the point. In the Lagrangian description, in turn, we fix an observer's reference frame within a material particle and we follow it looking for an actual position as a function of beginning coordinates and time.

Traditionally, the set of governing equations for thermohydrodynamics (mass, momentum and energy balances) has been formulated in the Eulerian description as a basic postulate of the Newtonian mechanics and the Joulean thermodynamics. However, starting from a variational principles' point of view, a set of conservative balances simultaneously with the Euler–Lagrange 'equations of motion' can be obtained. This grows from a generalized form of Noether's invariance theorem [1,2]. In particular, when the total Lagrangian is invariant with respect to time and space transformation, besides of the variational 'equation of motion', we can obtain the conservation of a canonical energy–momentum tensor in a form of four following equations: $T_{\alpha\beta},_{\beta} = 0,\ \alpha, \beta = t, z, y, z$ [3–7].

This quite general conservation theorem, arising in the different branches of the classical field theory, can be applied to the variational formulation of fluid thermodynamics. This will lead to conservation of the canonical energy–momentum flux that has particular names of the conservation of pseudomomentum and conservation of pseudoenergy. This set of additional equations is obtainable in a relatively easy way if one starts from a variational principle formulated in the Lagrangian description. However, such an approach results in some difficulties in physical interpretations, especially if someone seeks simple connections and equivalence between the Euler–Lagrange 'equation of motion' in the Eulerian coordinates and the Euler–Lagrange 'equation of motion' in the Lagrangian coordinates [8–12].

Variational methods also include analysis utilizing minimum-energy criterion (with constant entropy) and minimum-entropy production. Unfortunately, variational techniques in thermodynamics of discontinuities (of shock wave, phase transitions, linear defects' (eddies) type) are scarcely used in mathematical model investigations, despite the fact that they originate from Gibbs' days. Nonetheless, Gibbs' approach is still valid, as the literature shows [13].

In our study we will present a reconstruction and revaluation of a few basic, pioneering variational principles in thermohydrodynamics. In particular, most of our attention will be aimed at Gibbs' [14], Natanson's [15] and Eckart's [16] principles. All of them either contain preliminary forms of pseudomomentum and pseudomomentum flux notions or are natural ground for them. They also introduce a method to obtain pseudomomentum balance equations from properly constructed variational principles both in the Eulerian and the Lagrangian representations. The study is completed with analysis of other variational principles and variational grounds for the 'Generalized Lagrangian Mean' (GLM) approach recently applied in atmospheric and ocean turbulence description [17–19].

2. Notion of pseudomomentum vector and pseudomomentum flux tensor

The definition of the pseudomomentum and its conservation is strictly connected with the material Lagrangian L^m usually defined as a sum of kinetic T, internal ε and potential Ω energy, which is completed with the work of several constraints like, for instance, an adiabatic process:

$$L^m = \rho_0(T - \varepsilon - \Omega) - \theta(\eta,_\tau) = L^m(x_i,_\tau, x_i,_A, \pi; X_A, \tau), \tag{1}$$

which depends on a set of unknowns fields x_i, η, its derivatives $x_i,_A, x_i,_\tau$ and independent variables X_A, τ. Throughout this study we use Cartesian tensor notation; adhering to the current literature we denote Eulerian coordinates of $\vec{x}, \vec{v}, \nabla$ by lower-case indices $(i, j, k = x, y, a)$ and Lagrangian components of $\vec{X}$ by capital indices $(A, B = X, Y, Z)$. A comma followed by an index denotes partial differentiation with respect to a coordinate or time. Then, taking into account that L^m is a function only first derivatives $x_i,_A$, $x_i,_\tau$ of unknown functions, from the Noether theorem applied for the variation of Lagrangian coordinates δX_A, a general pseudomomentum conservation equation is obtained:

$$\frac{\partial}{\partial \tau} P_A + \frac{\partial}{\partial X_B} P_{AB} = F_A^{\text{inh}} \qquad \text{or} \qquad \partial_\tau \vec{P} + \text{DIV}(\vec{P}) = \vec{F}^{\text{inh}}. \tag{2}$$

The pseudomomentum vector $P_A = \rho_0 u_A$ (in m/s per unit reference volume), has been defined using specific pseudomomentum u_A (m/s per unit mass) [16,20,21]. The above quantity is strictly related with a structure of Lagrangian (1) and is defined to be ([16], Eq. (23)):

$$P_A = x_i,_A \frac{\partial L^m}{\partial x_i,_A} = x_i,_A \frac{\partial(\rho_0 T)}{\partial x_i,_\tau}, \qquad u_A = x_i,_A \frac{\partial(T)}{\partial x_i,_\tau} = x_i,_A\, x_i,_\tau. \tag{3}$$

The last form of the specific pseudomomentum u_A follows from the classical form of kinetic energy (per unit mass) $T = 1/2(\vec{v}\cdot\vec{v}) = 1/2(x_i,_\tau x_i,_\tau)$. Components of this vector have been denoted by Eckart as: $u_\alpha, u_\beta, u_\gamma$, respectively.

Next, the flux of pseudomomentum is defined to be:

$$P_{AB} = -L^m \delta_{AB} + x_i,_A \frac{\partial L^m}{\partial x_i,_B}. \tag{4}$$

In analogy with the Cauchy flux of momentum, this tensor is sometimes called the Eshelby flux tensor [22–24]. Further specification of pseudomomentum flux follows from the detailed form of Lagrangian (1), and in the case of a thermoelastic solid is [14,25]:

$$P_{AB} = \rho_0(\varepsilon + \Omega - T - \theta\eta)\delta_{AB} - F_{iA}T_{iB}, \tag{5}$$

or, in the case of elastic fluid [15,26]

$$P_{AB} = \rho_0\left(\varepsilon + \Omega + \frac{p}{\rho} - T - \theta\eta\right)\delta_{AB} = \rho_0 \mu \delta_{AB}. \tag{6}$$

In the above formulae several known objects have appeared: ρ_0 is the reference density of medium, ε the specific internal energy, T the specific kinetic energy, $\Omega(x_i)$ the specific gravitational potential, θ the thermodynamical temperature, η the specific entropy (in Gibbs' notation), μ the chemical-potential scalar, $F_{iA} = x_{i},_{A}$ the deformation gradient, $T_{iA} = \rho_0 \partial \varepsilon / \partial x_{i},_{A}$ the first Piola–Kirchhoff stress tensor.

The pseudomomentum flux tensor $\overleftrightarrow{P} = P_{AB} \vec{E}_A \otimes \vec{E}_B$ is usually addressed with a few different names; in the literature concerning nonlinear elastic solids it is called the Eshelby tensor, or material momentum flux. In thermodynamics of multiphase mixtures, the tensor $\overleftrightarrow{P}$ and its Eulerian representation $\overleftrightarrow{p} = p_{Ai} \vec{E}_A \otimes \vec{e}_i$ are called the chemical-potential tensor [27–29]. In the Maxwell electrodynamics, an analog of pseudomomentum flux is known as an ethereal stress tensor [30–35]. In the acoustics of moving media P_{AB} is called the radiation stress or radiation pressure tensor [36–41]. In the fluid wave mechanics it is called simply the pseudomomentum flux, the material momentum flux or the wave momentum flux [21,39,42–47].

A simple, consistent explanation of the role of pseudomomentum conservation is probably impossible. In order to understand the origin of pseudomomentum let us observe that our elastic fluid may by subjected to two quite distinct invariance operations, either of which might be called a 'translation'. The first one is a translation of the whole medium (together with any inhomogenities, disturbances, surface discontinuities, vortices, cavity, wave fronts) in a given space direction. If this operation leaves the action Lagrangian invariant—provided the space is homogeneous—the conserved associate quantity is physical (Newtonian) momentum.

The second translation leaves the medium itself fixed, but translates only the inhomogenity (line, surface, volume) through the material in a given space direction. This operation is an invariant symmetry only if the physical space and fluid continuum are homogeneous. This statement leads us to the practical remark that any variation of disturbance of homogeneous material leads to the appearance of the reaction in a form of the configurational force:

$$F_A^{\text{inh}} = -\left(\frac{\partial L^m}{\partial X_A} \right)_{\text{exp}}. \tag{7}$$

In the case when 'material inhomogenity' moves through the fluid the F_A^{inh} is interpreted as a material force or configurational force acting on it [48]. This notion is a generalization of special ones as: force on a elastic singularity [25], driving force acting on an interface [22] or bubble drift [49].

3. The Thomson–Tait variational principle

Let us consider the simplest inhomogeneity of the fluid medium such as a rigid body Σ immersed in the ideal fluid. The system with the rigid-body motion energy and kinetic energy of fluid is invariant under the following transformations: the whole fluid displacement and the displacement of the immersed rigid body by the same amount. The latter transformation—leading to the pseudomomentum conservation—consists also of the displacement of all physical properties related to the one point of the fluid to an adjacent one, without moving the medium. Therefore, in the case of the energy transfer from one body to another through the inhomogeneity surface, the variational principle should take into account the virtual displacement δN of the body surface $\partial \Sigma$.

The problem has been stated by Thomson and Tait [50] as:

$$\int_{t_1}^{t_2} \mathrm{d}t \int_{\partial\Sigma} p\,\delta N\,\mathrm{d}A = \int_{t_1}^{t_2} \mathrm{d}t\,\delta L^m \tag{8}$$

where p is the pressure on the body surface $\partial\Sigma$, $\delta N = \delta\vec{u}\cdot\vec{N}$ the virtual displacement that acts in the normal direction to the surface. The integral over $\partial\Sigma$ represents virtual work of pressure forces on the displacement δN. The Lagrangian $L^m = \rho_0(T - \Omega)$ represents the difference of the kinetic and potential energies for fluid and solid. After numerous developing works, for instance Kirchhoff [51], Peierlis has verified [40] that the sum of the pseudomomentum of the fluid and the momentum of the rigid body is conserved. This follows from the remark that in the case of the rigid-body motion in unbounded fluid the integral of the right-hand side term (8) excludes the volume of the body, so that boundary terms vanish, which amount to minus, the net force Q_m of the body [52,53]:

$$\sum_{m=1}^{6} Q_m q_m = \oiint p\,\delta N\,\mathrm{d}A. \tag{9}$$

4. Gibbs' principle extended formulation (1877)

Phase equilibrium conditions originally given by Gibbs [14] as valid for heterogeneous (with interphase surfaces) as well as for homogeneous (no interphase surfaces) phase transitions, are nowadays applied only to homogeneous (equilibrium) phase transitions. The homogenous parameters distribution assumption results in space independence of temperature (θ), pressure and chemical potential (μ). However, as the phases retain their properties and influence each other on the interphase surface (in a way specified in constitutive equations), Gibbs introduced the interphase surface (Fig. 1), where phase transitions conditions apply:

- thermal equilibrium: $\theta' = \theta''$,
- mechanical equilibrium: $p' = p''$,
- chemical equlibrium: $\mu' = \mu''$.

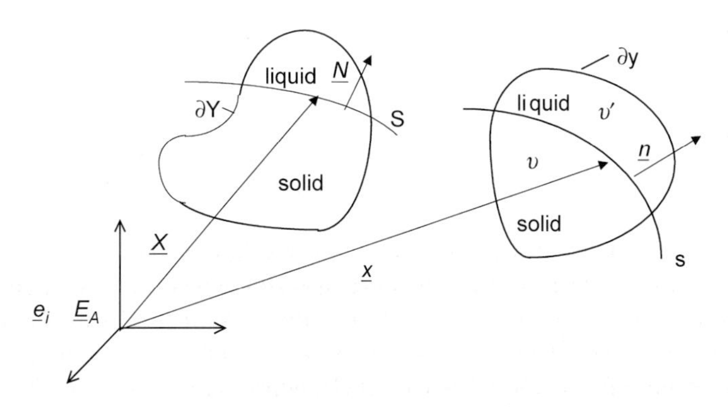

Fig. 1. Geometry of the interphase surfaces.

It turns out that the above conditions on phase equilibrium between fluid and its vapor can follow from a properly constructed variational principle. Gibbs' principle defines not only conditions for an equilibrium phase transition in fluids but also in solids, providing their deformability. The location of a point on the interphase surface S within the rigid body in the nondeformed state can be defined as:

$$\vec{X} = X\vec{E}_X + Y\vec{E}_Y + Z\vec{E}_Z = X_A\vec{E}_A, \tag{10}$$

whereas for the deformed state:

$$\vec{x} = x\vec{e}_x + y\vec{e}_y + z\vec{e}_z = x_i\vec{e}_i. \tag{11}$$

Conditions for fluid and body equilibrium can be obtained if variation of the whole system (with constant entropy) is equalized to zero:

$$\begin{aligned}&\iiint_{V_0}\left[\rho_0\theta\,\delta\eta + T_{iA}\delta F_{iA} + \rho_0 b_i\delta x_i\right]\mathrm{d}V_0 + \iint_S \rho_0\varepsilon\,\delta N'\,\mathrm{d}S\\ &+\iiint_V\left[\theta\,\delta(D\eta) - p'\,\delta(Dv_0) + \sum_k \mu_k\delta(Dm_k)\right]\mathrm{d}V = 0,\end{aligned} \tag{12}$$

where the deformation gradient expressing the rate of deformation is defined as ([14], Eq. (354)):

$$\overleftrightarrow{F} = \mathrm{GRAD}\underline{\vec{x}} = x_{i,A}\,\vec{e}_i\otimes\vec{E}_A = F_{iA}\,\vec{e}_i\otimes\vec{E}_A, \qquad J = \det\overleftrightarrow{F} = \rho_0/\rho, \tag{13}$$

whereas the two-point, first Piola–Kirchhoff's tensor T_{iA} describing purely reversible properties of the internal energy is given in the form of ([14], Eq. (355)):

$$T_{iA} = \frac{\partial\varepsilon_0}{\partial F_{iA}} = \rho_0\frac{\partial\varepsilon}{\partial F_{iA}} \qquad \text{or} \qquad \overleftrightarrow{T} = \rho_0\frac{\partial\varepsilon}{\partial\overleftrightarrow{F}}, \tag{14}$$

$\varepsilon, \rho, \theta, \eta$ in Eq. (12) denote internal energy, density, specific temperature and specific entropy, respectively. A 'prime' sign in Eq. (12) denotes that the normal of the surface $\delta N' = \delta\vec{X}_s\cdot\vec{N}$ is directed inside the fluid.

Due to constant temperature and entropy in the whole domain and the Gauss theorem, the total energy variation can be written as ([14], Eq. (369)):

$$\begin{aligned}&-\iiint_{V_0}\frac{\partial}{\partial X_A}(T_{iA})\delta x_i\mathrm{d}V_0 + \iiint_{V_0}\rho_0 b_i\delta x_i\mathrm{d}V_0 + \iint_S N_A T_{iA}\delta x_i\mathrm{d}S\\ &+\iint_S p'\vec{n}\cdot\delta\vec{x}_s\mathrm{d}S + \iint_S \rho_0(\varepsilon-\theta\eta)\delta N'\,\mathrm{d}S + \iint_S\left[p'J^{-1} + \sum_k \mu'_k\rho'_{0k}\right]\delta N'\,\mathrm{d}S = 0.\end{aligned} \tag{15}$$

Consequently, we try to extend the principle (15). We assume that the interphase surface of a two-phase flow is being formed and changed in time. In consequence, not only basic, fixed fields are variated, but also the location of the numerical region boundary, so that $X_A \to Y_A$, $\delta X_A = Y_A - X_A$. This approach is known in variational principles theory as the Noether variational approach, which leads to the comparison of the conservation laws with the Euler–Lagrange equations for the functional extreme.

Thus, let $Y_A = Y_A(X_B, \tau)$ describe the movement of boundary and the control volume itself. Each surface in X_A is subdued to small movements described by parameter τ. This can be expressed analogically in Eulerian frame as: $y_i = y_i(x_j, \tau)$. This can be interpreted as a minimal shift of volume Y into the minimally shifted region y. In consequence we can write:

$$\delta x_i = \frac{\mathrm{d}}{\mathrm{d}\tau}(y_i(Y_A, \tau), \tau)\bigg|_{\tau=0},$$

with boundary condition:

$$\begin{cases} x_i = y_i & (\tau = 0) \\ X_A = Y_A & (\tau = 0). \end{cases}$$

Variation of deformation can be expressed as a difference between variations of the base function and the boundary

$$\delta\left(\frac{\partial y_i}{\partial Y_A}\right) = \delta\left(\frac{\partial y_i}{\partial X_A}\right) - \frac{\partial x_i}{\partial X_B}\delta\left(\frac{\partial Y_B}{\partial X_A}\right).$$

Considering the fact that the internal energy of the solid body is a function of deformation gradient and entropy, whereas the internal energy of a fluid is a function of the deformation gradient determinant and entropy,

$$\varepsilon_s = \varepsilon_s(F_{iA}, \eta) \qquad \text{and} \qquad \varepsilon_l = \varepsilon_l(J, \eta),$$

on a fixed boundary we have:

$$\delta \iiint_{\bar{v}_0} \rho_0 \varepsilon\left(\frac{\partial y_i}{\partial Y_A}, \eta\right) \det\left(\frac{\partial Y}{\partial X}\right) \mathrm{d}\bar{v}_0$$

$$= \iiint_{v_0} \left\{ \rho_0 \frac{\partial \varepsilon}{\partial \left(\frac{\partial y_i}{\partial Y_B}\right)\Big|_{\tau=0}} \delta\left(\frac{\partial y_i}{\partial Y_B}\right) + \rho_0 \frac{\partial \varepsilon}{\partial \eta}\delta\eta + \rho_0 \varepsilon \frac{\partial}{\partial \eta}(\delta Y_A) \right\} \mathrm{d}v_0$$

$$= \iiint_{v_0} \left\{ \Pi_{iA}\left[\delta\left(\frac{\partial y_i}{\partial X_A}\right) - \frac{\partial x_i}{\partial X_B}\delta\left(\frac{\partial Y_B}{\partial X_A}\right)\right] + \rho_0 \frac{\partial \varepsilon}{\partial \eta}\delta\eta + \rho_0 \varepsilon \frac{\partial}{\partial X_A}(\delta Y_A) \right\} \mathrm{d}v_0$$

$$= \iiint_{v_0} \left\{ \left(-\frac{\partial}{\partial X_A} T_{iA}\right)\delta y_i - \left[\frac{\partial}{\partial X_A}\left(\rho_0 \varepsilon \delta_{AB} - \frac{\partial x_i}{\partial X_B} T_{iA}\right)\delta Y_B + \rho_0 \frac{\partial \varepsilon}{\partial \eta}\delta\eta\right] \right\} \mathrm{d}v_0$$

$$+ \iint_S \left\{ \delta y_i T_{iA} + \delta Y_B(\rho_0 \varepsilon \delta_{AB} - F_{iB} T_{iA}) \right\} N_A \mathrm{d}S + \text{com. on } \partial V_0, \tag{16}$$

where δ_{AB} is the Kroenecker delta.

Using Lagrange's multiplier Λ, the functional can be completed with constant entropy constraint:

$$\delta\left[\Lambda\iiint_{V_0}\rho_0\eta\,\mathrm{d}V_0\right]=\delta\left[\Lambda\iiint_{\bar{V}_0}\rho_0\eta(y_i,Y_A)\det\left(\frac{Y}{X}\right)\mathrm{d}\bar{V}_0\right]$$
$$=\Lambda\iiint_{V_0}\left(\rho_0\eta\frac{\partial}{\partial X_A}(\delta Y_A)+\rho_0\frac{\partial\eta}{\partial\tau}\bigg|_{\tau=0}\right)\mathrm{d}V_0$$
$$=\Lambda\iiint_{V_0}\left(-\frac{\partial}{\partial X_A}(\rho_0\eta)\delta Y_A+\rho_0\delta y\right)\mathrm{d}V_0+\Lambda\iint_S\delta Y_A(\rho_0\eta)\mathrm{d}S$$
$$+\text{expressions on boundary }\partial v_0. \tag{17}$$

Thus, following the Gibbs principle defining the minimum energy condition in the case of constant entropy, we can write:

$$\delta\left[\iiint_{V_0}\rho_0\varepsilon\,\mathrm{d}V_0+\Lambda\iiint_{V_0}\rho_0\eta\,\mathrm{d}V_0\right]=0. \tag{18}$$

Considering the definition of the pseudomomentum tensor (chemical potential) in form of:

$$P_{AB}=\rho_0(\varepsilon+\eta\Lambda)\delta_{AB}-F_{iB}T_{iA} \tag{19}$$

we obtain:

- on the side of the solid body:

$$\iiint_{V_0}\left\{\rho_0\delta\eta\left(\frac{\partial\varepsilon}{\partial\eta}+\Lambda\right)-\delta y_i[T_{iA,A}]+\frac{\partial}{\partial X_A}P_{AB}\delta Y_C\right\}\mathrm{d}V_0$$
$$+\iint_S\left[\delta y_iT_{iA}N_A+\delta Y_BP_{BA}N_A\right]\mathrm{d}S+\int_{\partial v_0}(\delta y_iT_{iA}+\delta Y_B)P_{BA}N_A\mathrm{d}(\partial V_0). \tag{20}$$

- on the side of the fluid:

$$-\iiint_V\left\{\rho_0\frac{\partial\varepsilon}{\partial J}\frac{\partial J}{\dfrac{\partial y_i}{\partial Y_B}}\delta\left(\frac{\partial y_i}{\partial Y_B}\right)+\rho_0\frac{\partial\varepsilon}{\partial\eta}\delta\eta+\rho_0\varepsilon\frac{\partial}{\partial X_A}(\delta Y_A)\right\}\mathrm{d}V_0. \tag{21}$$

For the whole region (liquid + solid) the following balance equations are obtained:

– momentum balance:

$$\delta y:\quad \mathrm{DIV}\,\vec{T}+\rho_0\vec{b}=0\qquad\text{on } V_S\cup V_1 \tag{22}$$

– pseudomomentum balance:

$$\delta Y:\qquad \mathrm{DIV}\,\vec{P}=0\qquad\text{on } V_S\cup V_1 \tag{23}$$

together with two jump conditions on the interphase surface S:

– jump condition for the momentum flux

$$\delta y^i : \qquad T_{iA}N_A|_{\text{solid}} + T_{iA}N_A|_{\text{liquid}} = 0$$

or:

$$[\![\vec{T}]\!]_l^s \vec{N} = 0, \tag{24}$$

– jump condition for the chemical potential tensor

$$\delta T_B : \qquad P_{AB}N_A|_{\text{solid}} + P_{AB}N_A|_{\text{liquid}} = 0$$

or considering, that $N_A|_{\text{solid}} = -N_A|_{\text{liquid}}$

$$[\![\vec{P}]\!]_l^s \vec{N} = 0. \tag{25}$$

Following Gibbs [14], equilibrium of the chemical potential can be expressed as the chemical-potential tensor equilibrium in the normal direction $\vec{N}$ to the rigid body.

$$[\rho_0(\varepsilon - \theta\eta) - N_A F_{iA} T_{iB} N_B]_{\text{solid}} - \left[\rho_0\left(\varepsilon - \theta\eta + \frac{p}{\rho}\right)\right]_{\text{liquid}} = 0. \tag{26}$$

In the case of coherent phase transitions, the two remaining tangent components are not significant. The F_{iA} T_{iB} term in Eq. (26) is the one that was not included in the original Gibbs' equation ([14], Eq. (369)). This element allows for better understanding of the role of the chemical potential.

5. Thermokinetic variational principle of Natanson (1899)

Natanson [15] has extended Gibbs' variational principle to cover the dynamic case by kinetic energy and mechanical forces inclusion. In this way, an extension of the classical definition of the chemical potential with the energy T and mass forces potential Ω was included.

$$\mu = \varepsilon - \theta\eta + pv - T + \Omega. \tag{27}$$

Moreover, unlike the Lagrangian Gibbs' approach, Natanson's approach is a purely Eulerian one, differing in the definition of interphase surface virtual motion.

Let us now consider the two volumes of the same fluid, divided by an interphase surface s, assuming that the fluid on both sides is in different phases (Fig. 2).

The thermokinetic Natanson principle can be written as:

$$\delta T + \delta F + \delta W + \delta' Q = 0. \tag{28}$$

Considering that the total kinetic energy is the sum of kinetic energy of all phases (neglecting kinetic energy of the interphase surface, as in this approach the interphase surface is a 'simple' dividing surface) and assuming that there is no slip between the phases (velocity of the ideal fluid transforming into the other phase is sufficiently similar to

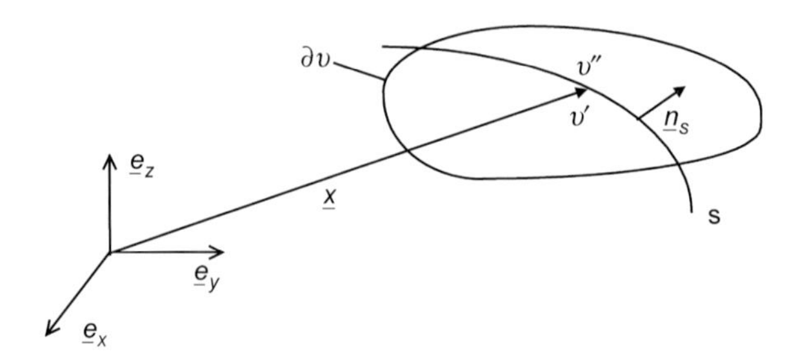

Fig. 2. Interphase geometry in the Eulerian representation.

potential flow velocity), we obtain that the variation of kinetic energy arising from 'natural inflows' into the volume υ' bounded within the surface $\partial\upsilon'$ and containing the phase-dividing surface oriented outwards is equal to ([15], Eq. (5)):

$$\delta T' = \frac{\mathrm{d}'}{\mathrm{d}t}\left\{ -\iiint_{\upsilon'}\left[\varphi'\,\delta\rho' + \varphi'\nabla\rho'\cdot\delta\vec{x}'\right]\mathrm{d}\upsilon' + \iint_{\partial\upsilon'}\rho'\varphi'\vec{n}'\cdot\delta\vec{x}'\,\mathrm{d}s + \iint_{s}\rho'\varphi'\vec{n}'_s\cdot\delta\vec{x}'\,\mathrm{d}s\right\}$$
$$-\left\{\iiint_{\upsilon'}\left[\vartheta'\,\delta\rho' + \vartheta'\nabla\rho'\cdot\delta\vec{x}'\right]\mathrm{d}\upsilon' + \iint_{\partial\upsilon'}\rho'\vartheta'\vec{n}'\cdot\delta\vec{x}'\,\mathrm{d}s + \iint_{s}\rho'\vartheta'\vec{n}'_s\cdot\delta\vec{x}'\,\mathrm{d}s\right\}, \tag{29}$$

where ϑ is the Appel acceleration potential and φ is the velocity potential. An analogous expression is obtained for the variation of kinetic energy in the volume $\partial\upsilon''$ with reversed orientation of the normal vector, that is: $\vec{n}''_s = -\vec{n}'_s$.

Because the kinetic energy balanced within the volume cannot change, displacement through the interphase surface will transport the energy from the first system to the particles of the second one. Alternatively, we can say that the system υ' will give back the following amount of energy as the result of infinitesimal change $D\vec{x}'$:

$$\frac{\mathrm{d}'}{\mathrm{d}t}\left\{\iint_{\partial\upsilon'}\rho'\varphi'(\vec{n}'\cdot D\vec{x}')\mathrm{d}s + \iint_{s}\rho'\varphi'(\vec{n}'_s\cdot D\vec{x}')\mathrm{d}s\right\} + \iint_{\partial\upsilon'}\rho'\vartheta'(\vec{n}'\cdot D\vec{x}')\mathrm{d}s$$
$$+\iint_{s}\rho'\vartheta'(\vec{n}'\cdot D\vec{x}')\mathrm{d}s. \tag{30}$$

A similar expression is valid for the system υ''.

Assuming that the phase transition of interest is isothermal, the variation of the free energy in the system υ' can be described as:

$$\delta F' = \iiint_{\upsilon'}\rho'\frac{\partial\psi'}{\partial\rho'}\delta\rho'\,\mathrm{d}\upsilon' + \iint_{\partial\upsilon'}\rho'\psi'(\vec{n}'\cdot D\vec{x}')\mathrm{d}s + \iint_{s}\rho'\psi'(\vec{n}'_s\cdot D\vec{x}')\mathrm{d}s. \tag{31}$$

An analogous expression is obtained for the system υ''.

Due to the fact that the investigated system is forced by potential forces:

$$\vec{b}' = -\mathrm{grad}\,\Omega' \qquad \text{and} \qquad \vec{b}'' = -\mathrm{grad}\Omega''$$

and the boundary loads

$$\vec{f}' = -p'_v\vec{n}' \qquad \text{and} \qquad \vec{f}'' = -p''_v\vec{n}'',$$

the variation of the work done by these forces on virtual displacements $\delta\vec{x}'$ and $D\vec{x}'$ in system υ' as well as $\delta\vec{x}''$ and $D\vec{x}''$ in system υ'' can be written as:

$$\delta W' = \iint_{\partial\upsilon'}\left[\vec{f}'(\delta\vec{x}' + D\vec{x}') + \rho'\Omega'\vec{n}'(\delta\vec{x}' + D\vec{x}')\right]\mathrm{d}s + \iint_s \rho'\Omega'\vec{n}'_s(\delta\vec{x}' + D\vec{x}')$$

$$\mathrm{d}s + \iiint_{\upsilon'} \Omega'(\delta\rho' + \operatorname{grad}\rho'\cdot\delta\vec{x}')\mathrm{d}v' \tag{32}$$

and analogously for the volume υ''.

The second principle of thermodynamics results in a non-negative increment of the uncompensated heat $\delta'Q$. As we consider only two fluids undergoing a reversible phase transition (without slip), we can take:

$$\delta'Q = 0. \tag{33}$$

The above leads to the variational formulation of the phase transition equilibrium. Taking into account:

- no-slip condition on the interphase surface,

$$\iint_s \vec{D}(D\vec{x}' - D\vec{x}'')\mathrm{d}s = 0, \tag{34}$$

- neighborhood-preserving condition for interphase surface particles

$$\iint_s \vec{A}(\delta\vec{x}' - \delta\vec{x}'')\mathrm{d}s = 0, \tag{35}$$

- mass-preserving condition:

$$\iiint_{\upsilon'}\left[B'\delta\rho' - \operatorname{grad}(\delta\vec{B}'\cdot\delta\vec{x}')\right]\mathrm{d}\upsilon' - \iint_{\partial\upsilon'}\rho'B'\vec{n}'\cdot\delta\vec{x}'\,\mathrm{d}s$$

$$-\iint_s \rho'B'\vec{n}'_s\cdot\delta\vec{x}'\mathrm{d}s = 0, \tag{36}$$

$$\iint_{\partial\upsilon'}\rho'C\vec{n}'\cdot D\vec{x}'\,\mathrm{d}s' + \iint_{\partial\upsilon''}\rho''C\vec{n}''\cdot D\vec{x}''\mathrm{d}s'' + \iint_s \rho'C\vec{n}'_s\cdot D\vec{x}'\,\mathrm{d}s$$

$$+\iint_s \rho''C\vec{n}'_s\cdot D\vec{x}''\mathrm{d}s = 0, \tag{37}$$

where $\vec{A}$, B' $(t,\vec{x})$, $\vec{D}$ are Lagrange's multipliers and $C = C(t)$ is any function of time, finally the Natanson principle is obtained as:

$$\int_{t_0}^{t_1}[I_{\upsilon'} + I_{\upsilon''} + \partial I_{\upsilon'} + \partial I_{\upsilon''} + S]\mathrm{d}t = 0, \tag{38}$$

where

$$I_{v'}: \quad -\iiint_{v'}\left[\vartheta'\delta\varphi' - \vartheta'\nabla\rho'\cdot\delta\vec{x}' + \rho'\frac{\partial\psi'}{\partial\rho'}\delta\rho' + \Omega'\delta\rho'\right]\mathrm{d}v'$$
$$+\iiint_{v'}\left[-\Omega'\nabla\rho'\cdot\delta\vec{x}' - B'\delta\rho' + \nabla(\rho'B')\cdot\delta x'\right]\mathrm{d}v'$$

$$I_{v''}: \quad -\iiint_{v''}\left[\vartheta''\delta\varphi'' - \vartheta''\nabla\rho''\cdot\delta\vec{x}'' + \rho''\frac{\partial\psi''}{\partial\rho''}\delta\rho'' + \Omega''\delta\rho''\right]\mathrm{d}v''$$
$$+\iiint_{v''}\left[\Omega''\nabla\rho''\cdot\delta\vec{x}'' - B''\delta\rho'' + \nabla(\rho''B'')\cdot\delta\vec{x}''\right]\mathrm{d}v''$$

$$\partial I_{v'}: \quad \iint_{\partial v'}\left[\rho'\vartheta'\vec{n}'\cdot(\delta\vec{x}' + D\vec{x}') + \rho'\psi'\vec{n}'\cdot D\vec{x}' + \vec{f}'\cdot(\delta\vec{x}' + D\vec{x}')\right]\mathrm{d}s'$$
$$+\iint_{\partial v'}\left[\rho'\psi'\vec{n}'\cdot(\delta\vec{x}' + D\vec{x}') - \rho'B'\vec{n}\cdot\delta\vec{x}' + \rho'C\vec{n}'\cdot D\vec{x}'\right]\mathrm{d}s'$$

$$\partial I_{v''}: \quad \iint_{\partial v''}\left[\rho''\vartheta''\vec{n}''\cdot(\delta\vec{x}'' + D\vec{x}'') + \rho''\psi''\vec{n}''\cdot D\vec{x}'' + \vec{f}''\cdot(\delta\vec{x}'' + D\vec{x}'')\right]\mathrm{d}s''$$
$$+\iint_{\partial v''}\left[\rho''\psi''\vec{n}''\cdot(\delta\vec{x}'' + D\vec{x}'') - \rho''B''\vec{n}''\cdot\delta\vec{x}'' + \rho''C\vec{n}''\cdot D\vec{x}''\right]\mathrm{d}s''$$

$$S: \quad \iint_{s}\left[\rho'\vartheta'\vec{n}'_s\cdot(\delta\vec{x}' + D\vec{x}') + \rho''\vartheta''\vec{n}''_s\cdot(\delta\vec{x}'' + D\vec{x}'') + \rho''\psi''\vec{n}''_s\cdot D\vec{x}''\right]\mathrm{d}s$$
$$+\iint_{s}\left[\rho'\psi'\vec{n}'_s\cdot D\vec{x}' + \rho'\Omega'(\vec{n}'_s\cdot\delta\vec{x}' + \vec{n}'_s\cdot D\vec{x}') + \rho''\Omega''(\vec{n}''_s\cdot\delta\vec{x}'' + \vec{n}''_s\cdot D\vec{x}'')\right]\mathrm{d}s$$
$$+\iint_{S}\left[\vec{A}\cdot(\delta\vec{x}' + \delta\vec{x}'') - \rho'B'\vec{n}'_s\cdot\delta\vec{x}' - \rho''B''\vec{n}''_s\cdot\delta\vec{x}''\right]\mathrm{d}s$$
$$+\iint_{s}\left[\vec{D}\cdot(D\vec{x}' + D\vec{x}'') + \rho'C\vec{n}'_s\cdot D\vec{x}' + \rho''C\vec{n}''_s\cdot D\vec{x}''\right]\mathrm{d}s,$$

This will lead us to the following equations on the surface s:

$$\delta\vec{x}': \quad \rho'(\vartheta' + \Omega' - B')\vec{n}'_s + \vec{A} = 0, \tag{39}$$

$$\delta\vec{x}'': \quad \rho''(\vartheta'' + \Omega'' - B'')\vec{n}''_s - \vec{A} = 0, \tag{40}$$

$$D\vec{x}': \quad \rho'(\vartheta' + \psi' + \Omega' - C)\vec{n}'_s + \vec{D} = 0, \tag{41}$$

$$D\vec{x}'': \quad \rho''(\vartheta'' + \psi'' + \Omega'' - C)\vec{n}''_s + \vec{D} = 0. \tag{42}$$

Eqs. (39) and (40) lead us, as expected, to the second Gibbs' condition:

$$\delta\vec{x}: \quad \vec{n}'_s(p' - p'') = 0 \quad \text{on the surface } s, \tag{43}$$

that is $p' = p''$.

Because the extended third Gibbs' condition is in the form of:

$$D\underline{\vec{x}}\cdot\vec{n}_s: \qquad \zeta' + \vartheta' + \Omega' = \zeta'' + \vartheta'' + \Omega'', \tag{44}$$

where $\zeta' = \psi' + p'v'$ and $\zeta'' = \psi'' + p''v''$ are free enthalpy, Eqs. (41) and (42) can be written as the jump condition:

$$[\![\psi + pv]\!]_{''}^{'} + [\![\vartheta]\!]_{''}^{'} + [\![\Omega]\!]_{''}^{'} = 0. \tag{45}$$

The presence of jump $[\![\vartheta]\!]$ allows for description of the phase transition in the flow, whereas $[\![\Omega]\!]$ takes into account the presence of mass forces.

The third Gibbs' condition has not so far received a simple interpretation, even in the case of homogenous phase transition. The literature has been dominated by the interpretation based upon Natanson's reasoning, which reads the third Gibbs' condition as a zero-entropy production requirement (that is the condition for phenomena reversibility) simplified after the heat equilibrium condition was incorporated into the expression for entropy production.

6. The Eckart variational principle

6.1. Variational statement of pseudomomentum conservation

The Lagrangian description of the flow coupled with Lagrangian variation of particle displacements provides the most natural formalism. It is a natural extension of the most powerful methods introduced in analytical mechanics by Lagrange, Hamilton and Jacobi. Most of the older work on this direction (R. Gwyther [52], A. Clebsch, L. Lichtenstein, H. Batemen, W. Weber) touch on the problem of pseudomomentum conservation, but the most excellent account of this approach is to be found in the paper by Eckart [16].

The paper under reconstruction starts from a quite general Lagrangian variational problem of finding $x_i (i = 1, 2, ..., m)$ functions of the independent variables $X_A (A = 1, 2, ..., N)$. The fundamental object is the Lagrangian density function L^m (J/m^3), i.e. by definition, the kinetic energy minus the potential energy density:

$$L^m = L^m(X_A; x_i, x_{i,A}), \qquad A = \tau, X, Y, Z, \tag{46}$$

where a convenient notation $x_{i,A} = \partial x_i / \partial X_A$ is used. The upper index m (as: material) informs no that density is referred to the undeformed (initial) volume. Since all the forms of energy depend only on first derivatives (viscosity and thermal conductivity are omitted) $x_{i,A}$ we have to deal with the so-called 'first-grade continuum', i.e. the simple continuum. If we consider the action integral I ([16], Eq. (2.3))

$$I = \int\int \cdots \int L^m \mathrm{d}X_1 \mathrm{d}X_2 \cdots \mathrm{d}X_N \tag{47}$$

over the N-dimensional volume (in time–space $N = 1$, $X_1 = \tau$, $N = 2, 3, 4$, $X_2, X_3, X_4 = X, Y, Z$). The variational Euler–Lagrange equations resulting from $\delta I = 0$

in case of fixed boundary problems, are [1]:

$$\frac{\partial}{\partial X_A}\left(\frac{\partial L^m}{\partial x_{i,A}}\right) - \frac{\partial L^m}{\partial x_i} = 0, \qquad i = 1, 2, 3, \ldots, m. \tag{48}$$

Eqs. (48) are called field equations or 'equations of motion' for fields x_i associated with the action integral (47). Lagrangian (46) is usually connected with a few conservation laws that, according to the Noether theorem, are based on certain groups of transformations of independent (X_A) and dependent variables (x_i). Eckart proposed a simple method to obtain the conservation laws. Differentiating L^m with respect to X_A he obtained the pseudoenergy-momentum conservation in the form ([16], Eq. (2.5)):

$$\frac{\partial}{\partial X_B} L_{AB} = -\left(\frac{\partial L^m}{\partial X_A}\right)_{\exp}, \tag{49}$$

where the pseudoenergy–momentum complex L_{AB} (in the case $N = 4$; X, Y, Z, it has a 4×4 array representation) is defined to be:

$$L_{AB} = x_{i,A}\frac{\partial L^m}{\partial x_{i,B}} - L^m \delta_{AB}. \tag{50}$$

Note that we have to distinguish the explicit partial derivative L^m with respect to X_A, when its other arguments x_i, $x_{i,A}$ and remaining $X_B, A \neq B$, are kept constant. Nowadays it is denoted by [48,54]:

$$\left(\frac{\partial L^m}{\partial X_A}\right)_{\exp} = \left.\frac{\partial L^m(x_i, x_{i,A}; X_A, X_B)}{\partial X_A}\right|_{x_i, x_{i,A}, X_B A \neq B}. \tag{51}$$

So far Eqs. (49)–(51) are purely mathematical relations. Eq. (49) would become conservation laws if its right-hand side were to vanish, i.e. if the Lagrangian did not depend explicitly on independent variables. This happens if the fluid medium is in the state of pure homogeneity with respect to space (X, Y, Z) and time (τ). For defects like vortices, shock waves, surface of discontinuity, surface of phase transitions, etc., which can be treated as inhomogenity in the fluid medium, the term $(\partial L^m/\partial X_A)_{\exp}$ has an interpretation of material of configurational force (if $A = X, Y, Z$) or material energy source (if $A = \tau$).

6.2. The incompressible fluid

In this case the Lagrangian consists solely of the kinetic specific energy T, and there is no internal energy. Trajectories of fluid particles are described by specifying three functions of position $\vec{x} = x_i\vec{e}_i, i = x, y, z$:

$$x_i = x_i(X, Y, Z, \tau) = x_i(X_A, \tau). \tag{52}$$

The derivatives in Eq. (46) now have their own names, respectively: the velocity $\vec{v} = v_i\vec{e}_i$:

$$v_i = \frac{\partial}{\partial \tau} x_i = x_{i,\tau}(X_A(x_i, t), \tau) \tag{53}$$

and the gradient of deformation $\overleftrightarrow{F} = F_{iA}\vec{e}_i \otimes \vec{E}_A = \text{GRAD}\vec{x}$:

$$F_{iA} = x_{i,A} = \frac{\partial x_i}{\partial X_A}. \tag{54}$$

In this case the coordinates X, Y, Z also play the role of Lagrangian position vector $\vec{X} = X_A \vec{E}_A$. Inverting the observer, one obtains the inverted deformation gradient $\overleftrightarrow{f} = f_{Ai}\vec{E}_A \otimes \vec{e}_i = \text{grad}\,\vec{X}$ written as:

$$f_{Ai} = X_{A,i} = \frac{\partial X_A}{\partial x_i}, \qquad F_{iA} f_{Aj} = \delta_{ij}, \qquad f_{Ai} F_{iB} = \delta_{AB}. \tag{55}$$

Let us denote by J and j the determinant of $\overleftrightarrow{F}$ and $\overleftrightarrow{f}$, respectively:

$$J = \det(x_{i,A}), \qquad j = \det(X_{A,i}), \qquad J = j^{-1}, \tag{56}$$

then, the following Euler–Piola–Jacobi identity takes place ([16], Eq. (2.2)):

$$(j x_{i,A})_{,i} = \text{div}(j\overleftrightarrow{F}) = 0, \qquad (J X_{A,i})_{,A} = \text{DIV}(J\overleftrightarrow{f}) = 0, \qquad \frac{\partial J}{\partial x_{i,A}} = J X_{A,i},$$

$$\text{d}v = J\text{d}V. \tag{57}$$

The connections between the Lagrangian and Eulerian changes will be explained as follows:

$$\left.\frac{\partial X_A}{\partial \tau}\right|_{\vec{X}\text{ fixed}} = \left.\frac{\partial X_A}{\partial t}\right|_{\vec{x}\text{ fixed}} + \left.\frac{\partial x_i}{\partial \tau}\right|_{\vec{X}\text{ fixed}} \frac{\partial X_A}{\partial x_i} = V_A + v_i X_{A,i} \equiv 0 \tag{58}$$

$$\left.\frac{\partial x_i}{\partial t}\right|_{\vec{x}\text{ fixed}} = \left.\frac{\partial x_i}{\partial \tau}\right|_{\vec{X}\text{ fixed}} + \left.\frac{\partial X_A}{\partial t}\right|_{\vec{x}\text{ fixed}} \frac{\partial x_i}{\partial X_A} = v_i + V_A x_{i,A} \equiv 0. \tag{59}$$

Here, $\vec{v} = v_i\vec{e}_i$ and $\vec{V} = V_A\vec{E}_A$ denote velocity and the rate of flow of material, or briefly, the material velocity.[1] The Lagrangian L^m contains, apart from kinetic energy T and gravity potential $\Omega(x_i)$, also an additional constraint related with incompressibility of fluid:

$$L^m = \rho_0[T(x_{i,\tau}) - \Omega(x_i)] + P[J(x_{i,A}) - \rho_0/\rho], \tag{60}$$

where the former Lagrange multiplier $P(X_A)$ turns out to be the pressure.

The Euler–Lagrange Eqs. (60) according to Eq. (48) are [16, Eq. (3.3)]:

$$\rho_0 x_{i,\tau\tau} + J P_{,A} X_{A,i} + \rho_0 f_i = 0, \qquad J\rho = \rho_0, \tag{61}$$

where the physical force density f_i is defined as $f_i = \partial\Omega/\partial x_i|_{\text{exp}}$.

The pseudoenergy-momentum tensor Eq. (50) now is simply determined as ([16], Eq. (3.8))

$$\text{Pseudoenergy} \qquad L_{\tau\tau} = E = \rho_0(T + Q), \tag{62}$$

$$\text{Pseudoenergy flux} \qquad L_{\tau A} = E_A = P X_{A,i} v_i = -P V_A, \tag{63}$$

$$\text{Pseudomomentum} \qquad L_{A\tau} = P_A = \rho_0 v_i x_{i,A} = \rho_0 u_A, \tag{64}$$

$$\text{Pseudomomentum flux} \qquad L_{AB} = P_{AB} = \rho_0(P/\rho - T)\delta_{AB}. \tag{65}$$

[1] For the definition see: L. Natanson (1986) On laws of irreversibility phenomena, Phil. Mag. 385–406, 41.

In Eq. (64), Eckart denotes the specific pseudomomentum vector (m/s per unit mass) by u_A. This definition is preferred in the literature of atmosphere and ocean physics [8,43–46, 55,56]. From the definition of Eq. (61), since the Lagrangian is explicitly independent of X_A and τ, it follows that the configurational force (51) both in space and time components are zero. Therefore, Eqs. (49) and (61)–(65) result in the conservation laws of pseudoenergy and pseudomomentum, respectively ([16], Eqs. (3.10) and (3.11))

$$\frac{\partial}{\partial \tau}E + \frac{\partial}{\partial X_A}(E_A) = \frac{\partial}{\partial \tau}\rho_0(T + \Omega) + \frac{\partial}{\partial X_A}(-PV_A) = 0, \tag{66}$$

$$\frac{\partial}{\partial \tau}P_A + \frac{\partial}{\partial X_B}(P_{AB}) = \frac{\partial}{\partial \tau}\rho_0 u_A + \frac{\partial}{\partial X_A}\left(\frac{P}{\rho} + \Omega - T\right) = 0. \tag{67}$$

Using Eqs. (57) and (59), the balance of pseudoenergy Eq. (66) can be transformed to ([16], Eq. (3.14))

$$\frac{\partial}{\partial \tau}\rho_0(T + P/\rho) = \in^{\alpha\beta\gamma\delta} P_{,\alpha} x_{1,\beta} x_{2,\gamma} x_{3,\delta} \qquad \alpha, \beta, \gamma, \delta = \tau, X, Y, Z. \tag{68}$$

This is a general form of the Bernoulli theorem. In the case of steady motion P is independent of time and the right side Eq. (68) vanishes, and Eq. (68) integrates at once to the usual form well known from textbooks. After the differentiation in Eq. (67) has been performed, the pseudomomentum conservation reduces to:

$$\rho_0 x_{i,\tau\tau} x_{i,A} + P_{,A} = 0, \tag{69}$$

which is the form proposed by Lamb,[2] who derived Eq. (69) using the Euler–Lagrange equations of motion multiplicated by $\overleftrightarrow{f}$. Such an approach is correct only if the Lagrangian contains one dependent function, i.e. $x_i(X_A, \tau)$.

The balance of pseudomomentum can be implemented for evaluation of the vortex conservation and vortex coordinates. Let us recall that for a closed curve Γ_0 in the initial position $\vec{X}$ the circulation is defined to be:

$$C(\tau) = \oint_{\Gamma_0} \vec{u}\cdot \mathrm{d}\vec{X} = \oint_{\Gamma_0} u_A \mathrm{d}X_A \tag{70}$$

while from Eq. (67) and the constancy of curve Γ following is obtained ([16], Eq. (4.2))

$$\frac{\partial}{\partial \tau}C(\tau) = \oint_{\Gamma_0} \mathrm{d}\left(\frac{P}{\rho} + \Omega - T\right) = 0. \tag{71}$$

It is known as the Kelvin–Helmholtz circulation theorem. If $\Gamma(\tau)$ is the instantaneous position and shape of a moving closed curve that is always composed of the same particles (i.e. is 'fixed in the fluid') then the circulation is described by:

$$C(\tau) = \oint_{\Gamma(\tau)} \vec{v}\cdot \mathrm{d}\vec{x} = \oint_{\Gamma(\tau)} v_i \mathrm{d}x_i. \tag{72}$$

[2] H. Lamb (1945) Hydrodynamics, Dover, New York, p. 13.

From Eq. (71) and from Kelvin's vorticity theorem:

$$\oint_{\Gamma} \vec{u} \cdot d\vec{X} = \iint_S \mathrm{ROT}\vec{u} \cdot d\vec{S} = \iint_S \vec{W} \cdot d\vec{S} \tag{73}$$

it follows that the material vorticity vector

$$\vec{W} = \mathrm{ROT}\vec{u} = (u_{Z,Y} - u_{Y,Z})\vec{E}_X + (u_{X,Z} - u_{Z,X})\vec{E}_Y + (u_{Y,X} - u_{X,Y})\vec{E}_Z \tag{74}$$

is conserved in time

$$\frac{\partial}{\partial \tau} \vec{W} = 0, \tag{75}$$

which means that $\vec{W}$ is a function of X_A only, independent of τ. Then, according to Clebsch's assertions it is always possible to find three scalars $A, B\phi$ such that

$$\vec{u} = A\ \mathrm{GRAD}\ B - \mathrm{GRAD}\ \phi, \tag{76}$$

which leads to an expression of the material vorticity

$$\vec{W} = (\mathrm{GRAD}\ A) \times (\mathrm{GRAD}\ B). \tag{77}$$

When $A = X, B = Y$ such a coordinate system is called the vortex coordinates.

6.3. The compressible perfect fluid

In the case of isenthalpic motion of a compressible fluid, an internal specific energy $\varepsilon = \varepsilon(v, \eta)$ should be added to the Lagrangian. According with Gibbs' notation [14]—ν is specific volume, $\nu = \rho^{-1}$ and η specific entropy. For this hyperelastic medium the constitutive equations for the thermodynamic pressure p and the thermodynamic temperature θ are given as:

$$p = -\left.\frac{\partial \varepsilon}{\partial \nu}\right|_{\eta} = \rho^2 \frac{\partial \varepsilon}{\partial \rho}, \qquad \theta = \left.\frac{\partial \varepsilon}{\partial \eta}\right|_{\nu}. \tag{78}$$

The dependence of density ρ on the deformation gradient is only via the Jacobian $J = \det(x_{i,A})$. The requirement that the motion be isenthalpic is formulated by a new subordinate condition ([16], Eq. (3.3)):

$$\frac{\partial}{\partial \tau} \eta = 0, \tag{79}$$

which means that η is a function only of X_A, $A = X, Y, Z$. It may be introduced into L^m at the beginning, so that entropy η is not a dependent variable but a given function of X_A independent of τ.

The Euler–Lagrange equations for a problem given by the Lagrangian function

$$L^m = \rho_0(T - \varepsilon - \Omega - \theta\eta) \tag{80}$$

are given as ([16], Eq. (3.5))

$$\rho_0 x_{i,\tau\tau} + Jp_{,A} X_{A,i} + \rho_0 f_i = 0, \qquad \theta = \frac{\partial \varepsilon}{\partial \eta}, \tag{81}$$

where p is defined via Eq. (78). According to definition (50) the pseudoenergy and pseudomomentum conservation are described by ([16], Eq. (5.5)–(5.8))

$$\text{Pseudoenergy} \qquad L_{\tau\tau} = E = \rho_0(T + \Omega + \varepsilon + \theta\eta), \tag{82}$$

$$\text{Pseudoenergy flux,} \qquad L_{\tau A} = E_A = PX_{A,i}v_i = -pV_A, \tag{83}$$

$$\text{Pseudomomentum} \qquad L_{A\tau} = P_A = \rho_0 v_i x_{i,A} = \rho_0 u_A, \tag{84}$$

$$\text{Pseudomomentum flux} \qquad L_{AB} = P_{AB} = \rho_0(\varepsilon + P/\rho - \theta\eta - T)\delta_{AB}. \tag{85}$$

This leads to the following pseudoenergy and pseudomomentum conservation:

$$\frac{\partial}{\partial \tau}\rho_0(T + \Omega + \varepsilon + \theta\eta) + \frac{\partial}{\partial X_A}(-pV_A) = 0, \tag{86}$$

$$\frac{\partial}{\partial \tau}\rho_0 u_A + \frac{\partial}{\partial X_A}\rho_0\left(\varepsilon + \frac{p}{\rho} + \Omega - \theta\eta - T\right) = 0. \tag{87}$$

Using the material flow u_A, and the material vorticity vector W_A, Eckart was able to prove the Helmholtz–Bjerknes theorem that asserts that the motion will be irrotational if the pressure can be expressed as a function of specific volume only—i.e. whenever the medium possesses the barotropic constitutive equation.

Eckart has also introduced a somewhat remarkable definition of a thermal potential κ ([16], Eq. (6.15)):

$$\frac{\partial}{\partial \tau}\kappa = \theta, \tag{88}$$

which leads to a novel definition of the thermodynamic circulation around the curve Γ as:

$$C_{T\kappa} = -\oint_{\Gamma} \kappa \mathrm{d}\eta \tag{89}$$

and the kinematical circulation C_{ki} as in Eq. (70). Then, interpreting pseudomomentum conservation (87) Eckart asserts that the total circulation is conserved:

$$\frac{\partial}{\partial \tau}(C_{T\kappa} + C_{ki}) = 0. \tag{90}$$

The total circulation, defined in this way, remains constant for every closed curve Γ fixed in the particles of fluid.

7. Eulerian representation of pseudomomentum

It is well known that the balance of momentum and energy are usually postulated within the framework of Newtonian mechanics and Joulean thermodynamics, respectively.

Most frequently the responsible flux of momentum:

$$\pi_{ij} = \pi_i v_j + t_{ij} - \tau_{ij} + \cdots \tag{91}$$

and the flux of energy

$$e_i = e v_i + (t_{ij} - \tau_{ij} + \cdots) v_j - q_i^{\eta} \theta + \cdots \tag{92}$$

are postulated, especially in the case when it is phenomenogicaly difficult to describe an amount of momentum and energy carried by additional fields like: radiation, progress of phase transition, mass fraction, etc. Quite similar uncertainty occurs when the medium is described via higher-order derivatives [4,5]. Therefore, looking for more precisely defined fluxes of momentum and energy, it is more comfortable to start from the variational principle based on the spatial (Eulerian) Lagrangian, say L^s [11]. Then, four equations of balance of momentum and energy can be obtained independently from the Euler–Lagrange 'equation of motion'. Both momentum and energy balances have their representations in the Lagrangian description. For instance, the flux of momentum for elastic fluid is called the Cauchy flux and its Lagrangian representation—the 1st Piola–Kirchhoff flux. Quite a similar situation concerns the pseudomomentum flux (sometimes called the Eshelby flux)—its Eulerian representation should be defined and useful in describing a few subjects stated in the Eulerian description (for instance; radiation flux pressure [39], radiation flux stress [8]).

To understand a relation between momentum and pseudomomentum in both Lagrangian and Eulerian representations, let us consider a very simple case, purely mechanical when the positions $\vec{X} = X_A \vec{E}_A$ and $\vec{x} = x_i \vec{e}_i$ are solely dependent and independent functions and exchange their role in passing from the Lagrange to the Euler description and vice versa.

Taking:

$$L^m = L^m(\tau, X_A; x_{i,A}, x_{i,\tau}, x_i) \tag{93}$$

$$L^s = L^s(t, x_i; X_{A,i}, X_{A,t}, X_A) \tag{94}$$

and using both the Euler–Lagrange equation and pseudoenergy-momentum conservation we obtain the following set of equations [48]:

Eulerian description:

Momentum conservation: $$\partial_t \pi_i + \pi_{ij,j} = \left(\frac{\partial L^s}{\partial x_i}\right)\bigg|_{\exp}, \tag{95}$$

Pseudomomentum (E–L eq.) $$\partial_t p_A + p_{Aj,j} = -\left(\frac{\partial L^s}{\partial X_A}\right)\bigg|_{\exp}, \tag{96}$$

Lagrangian description:

Momentum (E–L eq.): $$\partial_\tau \Pi_i + \Pi_{iA,A} = \left(\frac{\partial L^m}{\partial x_i}\right)\bigg|_{\exp}, \tag{97}$$

Pseudomomentum conservation $$\partial_\tau P_A + P_{AB,B} = -\left(\frac{\partial L^m}{\partial X_A}\right)\bigg|_{\exp}. \tag{98}$$

Above, a simple system of notation has been applied:

- letters π, Π (small, capital) always denote momentum (vector, tensor),
- letters p, P (small capital) always denote the pseudomomentum (vector, tensor),
- small letters (π, p) indicate on the Eulerian description,
- capital letters (Π, P) indicate on the Lagrangian description.

In the recent literature many different denotations of the above objects exist. The most frequently used ones are collected in Table 1.

Let us describe the above objects in terms of derivatives L^s and L^m, respectively. The physical momentum is defined as:

$$\pi_i = \partial L^s / \partial v_i = \rho v_i \tag{99}$$

$$\pi_{ij} = \pi_i v_j + t_{ij}, \qquad t_{ij} = x_{i,A} (\partial L^s / \partial x_{j,A}) - L^s \delta_{ij}, \tag{100}$$

$$\Pi_i = \partial L^m / \partial v_i = \rho_0 v_i = J \pi_i, \tag{101}$$

$$\Pi_{iA} = \partial L^m / \partial x_{i,A} = J t_{ij} X_{A,j} \, . \tag{102}$$

The pseudomomentum is analogously defined as:

$$p_A = \partial L^s / \partial V_A = j P_A, \tag{103}$$

$$p_{Ai} = \partial L^s / \partial X_{A,i} = j P_{AB} X_{B,i} \, , \tag{104}$$

$$P_A = \partial L^m / \partial v_i x_{i,A} \, , \tag{105}$$

$$P_{AB} = x_{i,A} (\partial L^m / \partial x_{i,B}) - L^m \delta_{AB}. \tag{106}$$

Using the above definitions, identities (57) and $L^s = jL^m$ it is simple to show (with straightforward calculations) direct relations between Lagrangian and Eulerian

Table 1
Most commonly used denotations of pseudomomentum

Author	Pseudomomentuum flux tensor		Pseudomomentuum vector	
	Lagrangian	Eulerian	Lagrangian	Eulerian
Eckart [16]	$\alpha_{\alpha\alpha}$	–	$u_\alpha, u_\beta, u_\gamma$ [a]	–
Eshelby [25]	P_{ij}	Σ_{ij}	g_i	–
Rogula [48]	$\bar{p}_{AB}$	p_{Ai}	$\bar{p}_A$	p_A
Andrews and McIntyre [17]	B_{ij}	–	A_i [a]	–
Gołębiewska–Hermann [54]	b_{ij}	B_{ik}	b_i	B_i
Grinfeld [26]	μ_{IJ}	χ_{ij} [b]	–	–
Peierlis [53]	–	–	κ_i [c]	–
Duan [57]	$B_{\mu\nu}$	$b_{i\nu}$	B_ν	b_ν
Maguin and Trimarco [23]	b_{kl}	B_{ki}	P_k	jP_k
Gurtin [24]	C_{AB}	–	–	–
Present chapter	P_{AB}	p_{Ai}	P_A	p_A

[a] Per unit mass.
[b] Symmetric—an analog of the second Piola–Kirchoff flux. Other applications of the symmetric chemical potential tensor are to be found in papers by McLellan [58] and Stuke [27].
[c] In the whole volume.

representations in the following form [54]:

$$\frac{\partial}{\partial t}\pi_i + \pi_{ij},_j = -j\left(\frac{\partial}{\partial \tau}\Pi_i + \Pi_{iA},_A\right), \tag{107}$$

$$\frac{\partial}{\partial \tau}P_A + P_{AB},_B = -J\left(\frac{\partial}{\partial t}p_A + p_{Ai},_i\right). \tag{108}$$

Quite similarly, the Eulerian representation of the material (configurational) force can be now defined using Eq. (96). In this description the Eulerian configurational force density per unit volume is defined as

$$f_A^{\text{inh}} \equiv \partial_t p_A + p_{Ai},_i = -(\partial L^s/\partial X_A)\big|_{\text{exp}}, \tag{109}$$

whereas the Lagrangian configurational force density (per unit of undeformed volume) is defined to be:

$$F_A^{\text{inh}} \equiv \partial_\tau P_A + P_{AB},_B = -(\partial L^m/\partial X_A)\big|_{\text{exp}}, \tag{110}$$

where, $f_A^{\text{inh}} = jF_A^{\text{inh}}$ owing to $L^s = jL^m$. This means that the above definitions take an exact form of the Newtonian (physical) force definitions, except that we put pseudomomentum vector and pseudomomentum flux in the places of momentum vector and momentum flux tensor, respectively [59]. The physical dimensions of momentum and pseudomomentum are the same. The fundamental difference is that physical (Newtonian) forces cause things to move and accelerate in physical space, while the configurational (material) forces will do the same with a material inhomogenity in the material space [48].

8. Capillarity extended Gibbs principle

The next chapter of Gibbs' memoir [14] is devoted to some supplementation of the model presented in Section 4 concerning the treatment of heterogeneous masses in contact. Let us recall, in Section 4 masses are in contact through a 'naked' surface of discontinuity, such that each mass is unaffected by the vicinity of the others. However, introducing a geometrical surface called the dividing surface, Gibbs had to take into account an anisotropy that occurs in the vicinity of the surface of discontinuity. Therefore, this surface possesses some superficial densities of energy ε^s specific entropy η^s and surface specific 'volume' S^s, surface specific curvatures c^s, etc. ([14], pp. 232–232).

The superficial energy density represents some excess of energy that can be added to the Lagrangian (12):

$$L^{\text{surf}} = \gamma\varepsilon^s(\eta^s, s^s, c^s), \tag{111}$$

where γ represents pseudomass density measured on a unit area of the surface. Owing to the excess of energy (111) the jump of pseudomomentum now takes the well-known form of the Laplace equation ([14], Eq. (500))

$$\delta N: \qquad \sigma(c_1 + c_2) = p' - p'', \tag{112}$$

where $\sigma = (\partial\varepsilon/\partial s^s)\big|_{\eta^s, c^s}$ is the surface tension. Buff has extended the calculation of the dividing surface variation δN on the curvature tensor and has obtained the pseudomomentum jump (112) not only with a surface tension $\overleftrightarrow{\sigma} = \sigma(\overleftrightarrow{I} - \vec{N}\otimes\vec{N})$ but also with the surface moment $\overleftrightarrow{C} = C(\overleftrightarrow{I} - \vec{N}\otimes\vec{N})$ ([60], Eq. (22)); see also Gibbs ([14], pp. 232, 232):

$$\sigma(c_1 + c_2) - (c_1^2 + c_2^2)\frac{C}{\gamma} = p' - p''. \tag{113}$$

Recently, the papers [61,62] have shown the Laplace equation to be a normal component of surface pseudomomentum conservation:

$$\delta\mathrm{iv}[\overleftrightarrow{\sigma} - \overleftrightarrow{b}\overleftrightarrow{C} + \vec{N}\otimes(\overleftrightarrow{I} - \vec{N}\otimes\vec{N})\delta\mathrm{iv}(\overleftrightarrow{C} - \overleftrightarrow{b}\overleftrightarrow{K})] = (\overleftrightarrow{P}' - \overleftrightarrow{P}'')\vec{N}, \tag{114}$$

where P'_{AB}, P''_{AB} are the bulk pseudomomentum tensor, $\overleftrightarrow{\sigma}, \overleftrightarrow{C}, \overleftrightarrow{K} = K(\overleftrightarrow{I} - \vec{N}\otimes\vec{N})$ are surface pseudomomentum fluxes, tension, moment and bimoment, respectively. The curvature tensor is defined as $\overleftrightarrow{b} = \gamma\mathrm{rad}\,\vec{N} = \vec{N}\otimes\nabla_2$. Two-dimensional divergence is denoted $\delta\mathrm{iv}(\cdot) = (\cdot)\cdot\nabla_2$. When $\overleftrightarrow{C}$ and $\overleftrightarrow{K}$ vanish the above equation coincides with one developed by Gurtin ([24], Eq. (4.2)). Our notation is based on the fundamental notions concerning physical surfaces stated by Skiba [63] and Weatherburn [64], and developed intensively in the papers by Tolman [65] and Buff [60]. The next recent generalizations of Eq. (112) are to be found in the papers by Scriven [66], Gurtin [24] and Povstenko [67]. Some important applications can be found in papers by Konorski [68] and Puzyrewski et al. [69], as well as Geurst [70] and Truskinovsky [22]. In order to examine further implications of the extended Gibbs principle it should be noted that keeping carefully to Gibbs' line of reasoning one can obtain the Laplace equation as a special simplification of the pseudomomentum jump across the dividing surface. Let us note that there are many papers (for instance see [71]) where the Laplace equation is treated as a jump of physical momentum.

Additionally, let us note that the configurational force, in comparison with Eq. (26), is differently stated for both cases ('naked' and 'superficial' surface). If we denote by F^{inh} the normal component of the configurational force that is the main 'driving force' (not the Newtonian one!) as desertion from equilibrium given by the third Gibbs' condition:

$$F^{\mathrm{inh}} = \begin{cases} \mu' - \mu'' \\ N_A(P'_{AB} - P''_{AB})N_B + \vec{N}\cdot(\overleftrightarrow{\sigma}\cdot\nabla_2) \end{cases} \tag{115}$$

for the case of 'naked' and 'superficial' surface, respectively. In the last formula we have omitted $\overleftrightarrow{C}$ and $\overleftrightarrow{K}$ tensors and a general form of $\overleftrightarrow{\sigma}$ can contain also traversal components as Gurtin has postulated ([24], Eq. (33)).

9. The acoustic pseudomomentum

Starting from research by Rayleigh [36] and Schrödinger [37] it is a well-known fact that using Lagrangian and Eulerian descriptions for an acoustic-wave statement leads to numerous discrepancies in results, such as in the case of, for instance, Rayleigh and Langevin radiation pressure [38–40]. Considerable confusion has arisen as to the precise meaning of the term 'radiation pressure' and its relevance to measurements. Even in the case when the basic fluid that is transmitting a wave is motionless, there are two different—Lagrangian and Eulerian—definitions of acoustic displacements:

$$\vec{\xi} = \xi_0 \sin(\Omega\tau - \vec{K}\cdot\vec{X}), \qquad \vec{\xi} = \xi_0 \sin(\omega t - \vec{k}\cdot\vec{x}), \tag{116}$$

which leads to different connections between wave (pseudo) momentum and wave energy. Let us recall that the most basic relations of this kind has been worked out by Rayleigh ([36], Eq. (23)):

$$\text{wave momentum} = \frac{3}{4}\frac{\text{total wave energy}}{c} \tag{117}$$

where c is the speed of wave propagation.

The question of how to formulate more general equations of acoustics for an arbitrary moving medium has been started in relation with the acoustic streaming phenomena [72–74]. The most popular method for deriving acoustic equations bas been the perturbation method that was originated in the pioneering papers by Helmholtz, Blokhintsev [75] and Eckart [72]. It is based on perturbation evaluation of every basic fields, like, for instance, density:

$$\rho = \rho_0 + \rho_1 \in + \rho_2 \in^2 + \cdots. \tag{118}$$

In 1963, a new, more correct, approach to formulation of the basic acoustic equations was formulated simultaneously by Biot [76] and Eckart [55]. In this new approach the acoustic motion is viewed as a small superimposed motion on a given fluid flow under the initial stress and initial temperature. Let us consider two moving fluids or two modes (excited and mean) of the motion of the same fluid. Let both of them differ slightly only by a small displacement $\vec{\xi}$. Such a situation can be called 'small superimposed motion'. Various concepts of superpositioning of two motions are shown in Fig. 3. The comparison of these two motions may be made in several ways—two of the simplest ways are discussed below.

If both motions ϕ and ϕ' are specified in the Lagrangian material coordinates (Fig. 3b) then a one-to-one correspondence can be set up between their particles by associating particles with the same values of $\vec{X}$. In this way the two flows may be compared using the $\vec{x}$ configuration and the description of superimposed motion by the following displacement vector $\vec{\xi} = \vec{\xi}(\vec{x}, t)$. However, if ϕ and ϕ' are specified in the Eulerian spatial coordinates then a fluid particle that actually flows through position $\vec{x}$ (Fig. 3a) can run from two, slightly different positions $\vec{X}$ and $\vec{X}^{\xi}$, which differ by a displacement vector $\vec{\xi} = \vec{\xi}(\vec{X}, \tau)$.

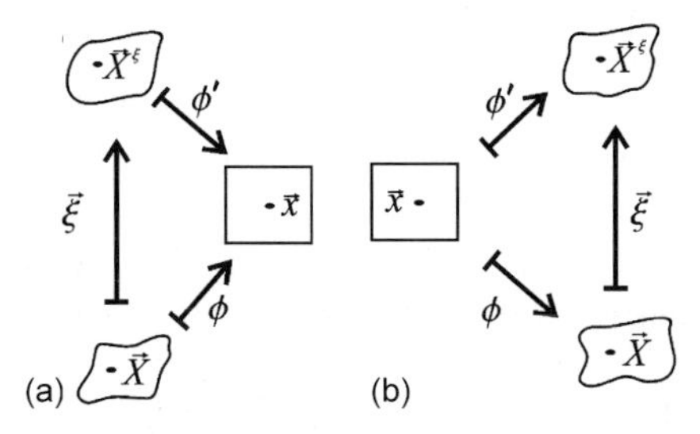

Fig. 3. Two possible approaches for comparison of the excited motion superimposed on a normal motion of fluid element.

Note that the first picture (Fig. 3a) is consistent with the Eulerian description and the second one (Fig. 3b) is consistent with the Lagrangian description that, as a result, leads to the Lagrangian mean approach. This last treatment has also been used in the description of water waves in oceans [45,47,56], turbulence [77] and electrodynamics [78,79]. If fruitfully happens that both cited papers of Biot [76] and Eckart [55] are complementary. Biot's paper concentrates on the question of what happens with internal energy ε in superimposed motion. However, Eckart concentrates on what happens with the kinetic energy in superimposed motion.

The complete reconstruction of the Biot–Eckart approach needs more attention and will be presented in a separate paper [80].

10. Generalized Lagrangian Mean approach

There are three general approaches to the formulation of turbulent motion equation that would include a mode of turbulent momentum flux. The older approach, which has its roots in the kinetic theory of gases, was formulated and developed by Reynolds [31]. Reynolds' approach is also known as the 'splitting' treatment that postulates a splitting of an arbitrary state of fluid into 'mean' and 'excited' ones. Then, any particle of fluid is composed of two subfluids that flow with different velocities, temperatures, pressures, entropies and so on. Reynolds postulated an internal transport of mass, momentum and energy between both subfluids via additional turbulent fluxes of mass, momentum and energy. These additional turbulent fluxes must be postulated as independent equations. In the original Reynolds approach there is no averaging in space or time—there is only the splitting. For instance, the turbulent flux of momentum, proposed and developed by Reynolds ([31], Eq. (222)) has, in the contemporary notation, a form [81]:

$$R_{ij} = \left(1 + \frac{\sigma}{\lambda}\right)\left\{\frac{\rho}{2}k\delta_{ij} + \frac{1}{2}\frac{\lambda k}{\sqrt{\pi}}[(\rho J_i^t - \rho v_i)_{,j} + (\rho J_j^t - \rho v_j)_{,i}]\right\} + \frac{\rho}{2}[(J_i^t + v_i)k_{,j} + (J_j^t + v_j)k_{,i}], \tag{119}$$

where λ is the mixing length, σ—the area of turbulent contact, $\vec{J}^t$—turbulent flux of mass. This flux depends only on one additional parameter k that has a meaning of kinetic turbulence energy. The model can be called the 'gradient model of turbulence'.

The second approach is based on the procedure of the Eulerian averaging in time. It is traditionally connected with the name of W.C. Reynolds. In practice, this approach is known as 'RANS' Reynolds-Averaged Navier–Stokes' [82,83]. The basic mathematical procedure on which the RANS is based, is an averaging of the momentum equation.

The third approach, introduced by Andrews and McIntyre [17] is based on the Lagrangian averaging procedure performed not on the momentum balance but on the pseudomomentum balance. This approach is known as the Generalized Lagrangian Mean (GLM), or, as proposed recently by Holm [19]—'LANS'—Lagrangian Averaged Navier–Stokes.

Atmospheric and ocean motions are strongly affected by waves that generate rectified effects and the 'mean' motion [84]. This is the case with the Langmuire circulations occurring in the oceans. When the wind blows over a water surface and generates waves, numerous streaks parallel to the wind direction may be observed. Langmuire related these streaks to convergence lines between counter-rotating vortices below the surface. Different mechanisms for generating Langmuire circulations, proposed in the literature, are based on the postulate that these vortices are induced by interactions between the wave field and mean wind-induced shear flow. A motivation for the Andrews and McIntyre model introduction was Leibovich's [85] opinion that interactions between wind-generated waves and currents are essentially better described within the pseudomomentum framework.

It should be noted that the Generalized Lagrangian Mean model describes mean motions and is therefore conceptually equivalent to Euler averaging or W.C. Reynolds averaging. However, practically, GLM describes Lagrangian aspects (pseudomomentum) in the Eulerian setting. The starting point of GLM is Fig. 3b where wave displacement $\vec{\xi} = \vec{\xi}(\vec{x}, t)$ is measured from the observer's current position. The Lagrangian mean averaging in time $[\cdot]''$ invokes any averaging operation that commutes with time or space differentiation and requires:

$$[\vec{\xi}(\vec{x}, t)]'' = 0. \tag{120}$$

This defines a particular reference trajectory, one for which the average wave displacement from the reference path of any quantity evaluated at the true Lagrangian position of a fluid particle is zero. This would seem to be a natural definition of 'mean Lagrangian' motion and the mean Lagrangian velocity $\vec{v}^{\mathrm{L}}$. This leads to the pseudomaterial derivative [17]

$$\frac{D^{\mathrm{L}}}{Dt}(\bullet) \equiv \frac{\partial}{\partial t}(\bullet) + [(\bullet)\otimes\nabla]\vec{v}^{\mathrm{L}}. \tag{121}$$

The resulting equation of pseudomomentum balance is as follows [17,86]:

$$\frac{D^{\mathrm{L}}}{Dt}(\vec{v}^{\mathrm{L}} - \vec{u}^{\mathrm{L}}) + (\vec{v}^{\mathrm{L}} - \vec{u}^{\mathrm{L}})(\vec{v}^{\mathrm{L}}\otimes\nabla) = -\nabla\pi + \vec{F}^{\mathrm{L}}, \tag{122}$$

where the pseudomomentum (per unit mass), similar to Eckart's definition (64) is [17]:

$$\vec{u}^{\mathrm{L}} = -[\xi_{j},_{i}\, \xi_{j},_{t}]''\vec{e}_{j} \tag{123}$$

and

$$\pi = p^{\mathrm{L}}/\rho^{\xi} + 1/2[\xi_{i},_{t}\, \xi_{i},_{t}]'' \tag{124}$$

is the spherical pseudomomentum flux, whereas $\vec{F}^{\mathrm{L}}$ is the viscous contribution to pseudomomentum.

The Eqs. (122) together with the definitions for $\vec{F}^{\mathrm{L}}$ govern the Lagrangian mean fields. Let as note that the role of the Reynolds turbulence flux now is being played by a single vector of pseudomomentum $\vec{u}^{\mathrm{L}}$ for which an arbitrary closure must be postulated. This suggests that the closure requires only three terms, versus six components of the Reynolds flux of momentum. Thus, there is an analogy with the Broszko model of rational turbulence [81].

11. Conclusions

Most of the potential advantages of the discussed pseudomomentum balance are yet to be explored, so we confine ourselves to just a few remarks. To illustrate the potentials of the pseudomomentum notion we have re-evaluated the original approaches proposed by Thomson and Tait, Gibbs, Natanson and Eckart. Fortunately, it follows from our investigations that the variational approach is more useful in the cases of new, unrecognized physical phenomena. Let us note that the role of variational approachs in the proper examination of pseudomomentum flux is quite the same as in the case of the variational approach to the proper definition of the momentum flux in so-called nonsimple continua [4,34,54,87,88]. This remark is also in agreement with the case of a proper definitions of momemtum and pseudomomentum within the framework of the extended irreversible thermodynamics. In that case, a line of reasoning, proposed by Sieniutycz [89], could be adopted and developed.

In our opinion, the set of the pseudomomentum balances in fluids, studied in this chapter does not exhaust the potential of the configurational force approach. Along with new applications in acoustics, turbulence and wave thermomechanics, the extension of the discussed notions with viscous and conductivity effects, is an interesting and, we believe, realistic prospect.

References

[1] Morse, P.M. and Feshbach, H. (1953), In Methods of Theoretical Physics, Vol. 1, McGraw-Hill, New York.

[2] Sieniutycz, S. (1994), Conservation Laws in Variational Thermo-Hydrodynamics, Kluwer, Dordrecht.

[3] Eckart, C. (1938), The electrodynamics of material media, Phys. Rev. 54, 920–923.

[4] Berdychevsky, V.L. (1966), Construction of models of continous media by means of the variational principle, PPM 30, 510–530.

[5] Cz. Woźniak, (1967), Thermoelasticity of bodies with micro-structures, Arch. Mech. Stos. 19, 335–365.

[6] Turski, Ł.A. (1968), Variational principles in theory of continuum with dislocations. In Metody Geometryczne w Fizyce i Technice, PWTN, Warszawa, pp. 281–294.

[7] Jezierski, J. and Kijowski, J. (1990), Thermo-hydrodynamics as a field theory. In Adv. Thermodynam., (Sieniutycz, S., Salomon, P., eds.), Vol. 3, pp. 282–317.

[8] Jones, L.W. (1973), Asymmetric wave-stress tensors and wave spin, J. Fluid Mech. 58, 737–747.

[9] Edelen, D.G.B. (1981), Aspects of variational arguments in the theory of elasticity; fact and folklore, Int. J. Solids Struct. 17, 729–740.

[10] Kobussen, J.A. and Paszkiewicz, T. (1981), Quasi-momentum, quasi-angular momentum and spin for condensed media, Helveti. Phys. Acta 54, 382–394.

[11] Capriz, C. (1984), Spatial variational principles in continuum mechanics, Arch. Rat. Mech. Anal. 85, 99–109.

[12] Grmela, M. (1985), Stress tensor in generalized hydrodynamics, Phys. Lett. 114A, 41–44.

[13] Bilicki, Z. and Mikielewicz, J. (1984), Minimum energii i minimum produkcji entropii w zastosowaniach do obliczeń stopnia zapełnienia w pęcherzykowym przepływie dwufazowym, Arch. Thermodyn. 5, 101–118.

[14] Gibbs, J. (1875/1877), On the equilibrium of heterogenous substances, Trans. Conn. Acad. Arts Sci. 3, 108–248. see also page 343–534.

[15] Natanson, W. (1899), O wpływie ruchu na zmiany stanu skupienia, RWMP 35, 220–246; Sur les chagements d'état dans un système en mouvement, BIAC (1898) 16, 103–123, 201–213; Über die Zustandsänderungen in einem in Bewegung begriffenen System, Zf. F. Phys. Chem (1898) 26, 285–305.

[16] Eckart, C. (1960), Variation principles of hydrodynamics, Phys. Fluids 3, 421–427.

[17] Andrews, D.G. and McIntyre, M.E. (1978), On wave-action and its relatives, J. Fluid Mech. 89, 647–664.

[18] McIntyre, M.E. (1981), On the wave momentum myth, J. Fluid Mech. 106, 331–360.
[19] Holm, D.D. (2002), Averaged Lagrangians and the mean effects of fluctuations in ideal fluid dynamics, Physica D 170, 253–286.
[20] Longuet-Higgins, M.S. (1964), On group velocity and energy flux in planetary wave motions, Deep-Sea Res. 11, 35–42.
[21] Hayes, W.D. (1970), Conservation of action and modal wave action, Proc. R. Soc. A320, 187–208.
[22] Truskinovsky, L. (1991), Kinks versus shocks. In Shock Induced Transitions and Phase Structure in General Media (Dunn, J.E., et al., eds.), Springer, New York, pp. 185–229.
[23] Maugin, G.A. and Trimarco, C. (1992), Pseudomomentum and material forces in nonlinear elasticity: variational formulations and application to brittle fracture, Acta Mechanica 94, 1–28.
[24] Gurtin, M.E. (1995), The nature of configurational forces, Arch. Rat. Mech. Anal. 131, 67–100.
[25] Eshelby, J.D. (1970), Energy relations and energy-momentum tensor in continuum mechanics. In Inelastic Behavior of Solids (Kanninen, M.F., Adler, W.F., Rosenfield, A.K. and Jaffee, R.I., eds.), McGraw-Hill, New York, pp. 77–115.
[26] Grinfeld, M.A. (1981), On heterogeneous equlibrium of nonlinear elastic phases and chemical potential tensors, Lett. Appl. Engng Sci. 19, 1031–1039.
[27] Stuke, B. (1975), Tensorielle chemische potentiale, Z. Naturforsch. 30a, 1433–1440.
[28] Stumpf, H. and Le, Ch.K. (1990), Variational principles of nonlinear fracture mechanics, Acta Mech. 83, 25–37.
[29] Samohyl, I. and Pabst, W. (1997), The Eshelby relation in mixtures, Int. J. Non-Linear Mech. 32, 227–233.
[30] Maxwell, J. (1861/1862), On physical lines of force, I. The theory of molecular vortices applied to magnetic phenomena, II. The theory of molecular vortices applied to electric currents, III. The theory of molecular vortices applied to statical electricity, Philos. Mag. 21–21.
[31] Reynolds, O (1903), The Sub-Mechanics of Universe, Reynolds Papers, Vol. III, pp. 1–251, Cambridge University Press, Cambridge.
[32] Burton, C.V. (1909), On the Faraday–Maxwell mechanical stress; and on aetherial stress and momentum in general, Phil. Mag. 17, 641–654.
[33] Ginzburg, V.L. and Ugarov, V.A. (1976), Remarks on forces and the energy-momentum tensor in macroscopic electrodynamics, Sov. Phys. Usp. 19, 94–101.
[34] Kijowski, J. and Rudolph, G. (1985), Hydrodynamical description of theories of gauge fields interacting with matter fields, Rep. Math. Phys. 21, 309–329.
[35] Kostro, L. (1988), Einstein's relativistic ether, its history, physical meaning and updated applications, Organon 24, 219–325.
[36] Rayleigh, L. (1914), Further calculations concerning the momentum of progressive waves, Phil. Mag. 27, 436–440.
[37] Schrödinger, E. (1917), Zur Akustik der Atmosphere, Physik Zeitschr. 19, 445–453.
[38] Brenig, W. (1955), Besitzen Schallwellen einen Impuls?, Zeit. F. Phys. 143, 168–172.
[39] Longuet-Higgins, M.S. and Steward, R.W. (1964), Radiation stresses in water waves: a physical discussion with applications, Deep-Sea Res. 11, 529–562.
[40] Peierlis, R. (1986), Momentum and pseudomomentum of light and sound, Rend. Scuola Inter. Fisica Enrico Fermi 139, 237–255.
[41] Stone, M. (2000), Acoustic energy and momentum in a moving medium, Phys. Rev. E 62, 1341–1350.
[42] Sturrock, P.A. (1961), Energy-momentum tensor for plane waves, Phys. Rev. 121, 18–19.
[43] Jones, L.W. (1971), Energy-momentum tensor for linearized waves in material media, Rev. Geophy. Space Phys. 9, 914–952.
[44] Sieniutycz, S. (1978), Zasady wariacyjne dla falowych równań przenoszenia ciepła i masy w ośrodkach ruchomych, Inżynieria Chemiczna 8, 669–690.
[45] Ripa, P. (1992), Wave-energy-momentum and pseudoenergy-momentum conservation for the layered quasi-geostrophic instability problem, J. Fluid. Mech. 235, 379–398.
[46] Ripa, P. (1995), A low-frequency approximation of a one-layer ocean model with thermodynamics, Phys. Oceanogr. 41, 39–48.
[47] Kuchner, P.J. and Shepherd, T.G. (1995), Wave-activity conservation laws and stability theorems for semi-geostrophic dynamics. Part 1. Pseudomomentum-based theory, J. Fluid Mech. 290, 67–104.

[48] Rogula, D. (1977), Forces in material space, Arch. Mech. Stos. 29, 705–717.

[49] Ben Amar, M. and Rice, J.R. (2002), Exact results with the J-integral applied to free-boundary flows, J. Fluid Mech. 461, 321–343.

[50] Thomson, W. and Tait, P. (1867), In Treatise on Natural Philosophy, Vol. 1, Cambridge Unversity Press, Cambridge.

[51] Kirchhoff, G.R. (1869), Ueber die Bewegung eines Rotaionskörperes in einer Flüssigkeit, J. Reine Angew. Math. (Crelle) 71, 237–262.

[52] Gwyther, R.F. (1880), On an adaptation of the Lagrangian form of the equations of fluid motion, Proc. Manchester Lit. Phil. Soc. 19, 203–212.

[53] Peierlis, R. (1991), More Suprises in Theoretical Physics, Princeton University Press, New Jersey.

[54] Gołebiewska-Hermann, A. (1981), On conservative laws of continuum mechanics, Int. J. Sol. Struct. 17, 1–9.

[55] Eckart, C. (1963), Some transformations of the hydrodynamic equations, Phys. Fluids 6, 1037–1041.

[56] Kuchner, P.J. and Shepherd, T.G. (1995), Wave-activity conservation laws and stability theorems for semi-geostrophic dynamics. Part 1. Pseudoenergy-based theory, J. Fluid Mech. 290, 105–130.

[57] Duan, Z.P. (1987), On the duality principle of conservation laws in finite elastodynamics, Int. J. Engng Sci. 25, 963–985.

[58] McLellan, A.G. (1968), The chemical potential in thermodynamic systems under non-hydrostatic stresses, Proc. R. Soc. London A307, 1–13.

[59] Kobussen, J.A. (1976), Local and substantial fluxes for energy, linear momentum and quasi-momentum in systems with a non-local internal interaction, Helveti. Phys. Acta 49, 599–612.

[60] Buff, F.P. (1956), Curved fluid interfaces. The generalized Gibbs–Kelvin equation, J. Chem. Phys. 23, 146–153.

[61] Badur, J. (1997), Wprowadzenie do zasad wariacyjnych równowagowych przejść fazowych. In Współczesne Kierunki w Termodynamice, (Bilicki, Z., eds.), Wyd. IMP, Gdańsk, pp. 129–146.

[62] Badur, J. and Rafalska, E. (1997), Wariacyjna interpretacja tensora potencjału chemicznego. Współczesne Kierunki w Termodynamice (Bilicki, Z., eds.), Wyd. IMP, Gdañsk, pp. 205–220.

[63] Skiba, E.W. (1869), Teoryja Zjawisk Włoskowatości, Druk, UJ Kraków.

[64] Weatherburn, C.E. (1927), On differential invariants in geometry of surfaces with some applications to mathematical physics, Q. J. Pure Appl. Math. 50, 230–269.

[65] Tolman, R.C. (1948), Considerations of the Gibbs theory of surface tension, J. Chem. Phys. 16, 758–774.

[66] Scriven, L.E. (1959), On the dynamics of phase growth, Chem. Engng Sci. 10, 1–13.

[67] Povstenko, Y.Z. (1991), Generalizations of Laplace and Young equations involving couples, J. Coll. Interface Sci. 144, 497–506.

[68] Konorski, A. (1971), Shock waves in wet steam flow, Prace Inst. Maszyn Przepływowych 57, 101–110.

[69] Puzyrewski, R., Gardzilewicz, A. and Bagiñska, M. (1973), Shock waves in condensing steram flowing through a Laval nozzle, Arch. Mech. Stos. 25, 393–409.

[70] Geurst, J.A. (1986), Variational principles and two-fluid hydrodynamics of bubbly liquid/gas mixtures, Physica 135A, 445–486.

[71] Rutkowski, J. (1981), Foundations of balancing at mobile phase interphase, Arch. Thermodyn. 2, 75–108.

[72] Eckart, C. (1948), Vortices and streams caused by sound waves, Phys. Rev. 73, 68–76.

[73] Lighthill, M.J. (1952), On sound generated aerodynamically, I. General theory, Proc. R. Soc. A211, 564–587.

[74] Lighthill, M.J. (1954), On sound generated aerodynamically. II. Turbulence as a source of sound, Proc. R. Soc. A222, 1–22.

[75] Blokhintsev, I.D. (1945), Acoustics of a Non-homogeneous Moving Medium, Gostekhizdat, Moskva, in Russian.

[76] Biot, M.A. (1963), Variational principles for acoustic-gravity waves, Phys. Fluids 6, 772–778.

[77] Zak, M. (1986), On the Lagrangian turbulence in continua, Phys. Lett. A115, 373–376.

[78] Gurevich, V.L. and Thelung, A. (1992), On the pseudomomentum of light and matter and its conservation, Physica A188, 654–674.

[79] Nelson, D.F. (1992), Wave momentum, Int. J. Engng Sci. 30, 1407–1416.

[80] Badur, J. and Badur, J (2004), Further remarks on the pseudomomentum balance in hydro- and gasodynamics, XVI Konf. Mech. Płynów, Waplewo, 20–23.

[81] Karcz, M. and Badur, J. (2003), O numerycznej implementacji modelu racjonalnej turbulencji, Zeszyty Naukowe IMP PAN 531, 1–34.

[82] Broszko, M. (1951), Sur les equations generales des mouvements turbulents des liquids, Bull. l'Acad. Polon. Sci. Lett. Cl. Cci. Math. Nat. A 1, 223–252.

[83] Batchelor, G.K. (1952), The effect of homogeneous material lines and surfaces, Proc. R. Soc. London A213, 349–366.

[84] Whitham, G.B. (1962), Mass, momentum and energy flux in water waves, J. Fluid Mech. 12, 135–147.

[85] Leibovich, S. (1992), Structural genesis in wall-bounded turbulent flows. In Studies in Turbulence (Gatski, T.B., Sarkar, S., Speziale, Ch.G., eds.), Springer, Berlin, pp. 387–411.

[86] Nordsveen, M. and Bertelsen, A.F. (1997), Wave induces secondary motions in stratified duct flow, Int. J. Multiphase Flow 23, 503–522.

[87] Kogarko, B.S. (1961), On a model of a cavitating liquid, Sov. Phys.-Dokl. 137, 1331–1333.

[88] Pommariet, J.-F. (1983), La structure de la mecanique des milieux continues, C. R. Acad. Sci. Paris 296, 517–520.

[89] Sieniutycz, S. (1988), Hamilitonowski tensor energii-pędu w rozszerzonej termodynamice płynu jednoskładnikowego, Inż. Chemiczna i Procesowa 4, 839–861.

Chapter 9

TOWARDS A VARIATIONAL MECHANICS OF DISSIPATIVE CONTINUA?

Gérard A. Maugin

Laboratoire de Modélisation en Mécanique (UMR CNRS 7607), Université Pierre et Marie Curie, Case 162, 4 place Jussieu, 75252 Paris Cedex 05, France

Vassilios K. Kalpakides

Division of Applied Mathematics and Mechanics, University of Ioannina, GR-45110, Ioannina, Greece

Abstract

Here the main elements of variational mechanics are given in the mechanics of continua of the solid type. Inhomogeneous hyperelasticity provides the standard of the formulation. Attention is focused on the balance laws derived by use of Noether's theorem. Then a possible means to include relevant dissipative mechanisms (heat conduction, thermodynamically irreversible anelasticity) is outlined. This exploits the notion of thermacy and it shows how the somewhat strange thermoelasticity 'without dissipation' of Green and Naghdi follows therefrom. The variational formulation is shown to yield the classical thermoelasticity of anelastic conductors in the appropriate reduction. A Euclidean four-dimensional space-time formulation of the canonical balance laws of energy and material momentum is then given. Finally, an analogy between the obtained canonical balance laws and the Hamiltonian equations that give the kinetic-wave theory of nonlinear dispersive waves in inhomogeneous materials is established. Furthermore, generalizations to the material behavior involving coupled fields, nonlocality and evolving microstructure are mentioned.

Keywords: Hamilton–Lagrange principle; canonical balance laws; pseudomomentum; Eshelby stress; Noether's identity; hyperelasticity; anelasticity; thermoelasticity; inhomogeneity; material forces; space–time formulation; nonlinear waves; dispersion; Hamilton–Jacobi equation; wave–particle duality

1. Introduction: the drive to a variational formulation

A recurrent dream of many mathematical physicists is to construct a variational formulation for all field equations of continuum physics *including in the presence of dissipative effects*. But we all know that this is not possible unless one uses special tricks such as introducing complex-valued functions and adjoint fields (e.g. for heat conduction).

E-mail address: gam@ccr.jussieu.fr (G.A. Maugin); vkalpak@cc.uoi.gr (V.K. Kalpakides).

Works by Gyarmati, Von Haase, Anthony and others go in that direction. Here, on the basis of recent works, we present a possible variational and canonical formulation for the nonlinear continuum theory of *thermoelastic conductors*, after presenting the pure elastic, hence thermodynamically reversible, case, and an inelastic case. This is made possible through the introduction of a rather old notion, clearly insufficiently exploited, that of *thermacy* introduced by Van Dantzig (see Ref. [1]), a field of which the time derivative is the thermodynamical temperature. It happens that one of us used such a notion in relativistic studies in the late 1960s–early 1970s, (GAM's pre-general exam Seminar at Princeton University, Spring 1969; [2,3]), a time at which it was found that thermacy is nothing but the *Lagrange multiplier* introduced to account for *isentropy* in a Lagrangian variational formulation. But, completely independently and much later, Green and Naghdi [4] formulated a strange 'thermoelasticity without dissipation'. Dascalu and Maugin [5] identified thermacy as the unknowingly used notion by Green and Naghi (unaware of works in relativistic variational formulations), and a formulation of the corresponding *canonical balance laws of momentum and energy*—e.g. of interest in the design of fracture criteria—which, contrary to the expressions of the classical theory, indeed present *no* source of dissipation and canonical momentum, e.g. no thermal source of quasi-inhomogeneities (see Ref. [6]). In recent works [7,8], the consistency between the expressions of intrinsic dissipation and source of canonical momentum in dissipative continua was exhibited. This was developed within the framework of so-called *material* or *configurational forces*, 'Eshelbian mechanics', that world of forces that, for instance, drive structural rearrangements and material defects of different types on the material manifold (for these notions see, Refs. [9–12]). Finally, an analogy of some of the analytical mechanics equations obtained with those governing some nonlinear wave processes is mentioned by way of conclusion. The contribution is completed by mentioning generalizations to the electrodynamics of continua, porous media, and media exhibiting diffusion.

2. Special attention paid to canonical balance laws: purely elastic case

We use classical elements of field theory as enunciated in several books (e.g. see Refs. [9,13,14]) The reader is referred to these works for the abstract equations that apply to all field theories. Consider first the case of pure, albeit nonlinear, anisotropic and smoothly inhomogeneous, elasticity, and thus Hamiltonian–Lagrangian densities (per unit volume of the undeformed configuration of nonlinear continuum mechanics K_R) given by the following general expression

$$L = \bar{L}(\mathbf{v}, \mathbf{F}; \mathbf{X}) = K - W, \tag{1}$$

where kinetic (K) and potential (W) energies are given by

$$K = \tfrac{1}{2}\rho_0(\mathbf{X})\mathbf{v}^2, \qquad W = \overline{W}(\mathbf{F}; \mathbf{X}), \tag{2}$$

where the former exhibits only *translational degrees of freedom*. The Lagrangian function $\bar{L}$ is a function of time and space derivatives of the basic field (a sufficiently regular function)

$$\mathbf{x} = \chi(\mathbf{X}, t), \tag{3}$$

where t and $\mathbf{X}$ are, respectively, the Newtonian time and a set of material coordinates. Here, it is supposed that the material body is an open, simply connected subset B of the *material manifold* M^3 of elements (so-called 'particles' of continuum physics) $\mathbf{X}$ equipped with a scalar mass measure $\rho_0(\mathbf{X})$, which we assume to be sufficiently smooth (here no crack, no discontinuity, etc.) in the reference configuration K_R of the body. Eq. (3) provides the time sequence of points occupied by $\mathbf{X}$ in the *actual configuration* K_t (at time t) in physical space E^3. The *physical velocity* $\mathbf{v}$ and *direct-motion gradient* $\mathbf{F}$ (usually called deformation gradient) are defined by

$$\mathbf{v} := \left.\frac{\partial \chi}{\partial t}\right|_X, \qquad \mathbf{F} := \left.\frac{\partial \chi}{\partial \mathbf{X}}\right|_t = \nabla_R \chi, \tag{4}$$

so that (T = transpose)

$$(\partial \mathbf{F}/\partial t)^{\mathrm{T}} = \nabla_R \mathbf{v}, \tag{5}$$

where ∇_R (respectively, ∇) is the gradient operator at $\mathbf{X}$ on M^3 (respectively, at $\mathbf{x}$ in E^3). From $\mathbf{F}$, we can form the following most common measures of finite strain:

$$\mathbf{C} = \mathbf{F}^{\mathrm{T}} \cdot \mathbf{F}, \qquad \mathbf{E} = \tfrac{1}{2}(\mathbf{C} - 1_R). \tag{6}$$

Reciprocally, let $J_F = \det \mathbf{F} > 0$ always (at all times t). Then the *inverse motion*

$$\mathbf{X} = \chi^{-1}(\mathbf{x}, t), \tag{7}$$

is a well-behaved function, and we can define the *material velocity* $\mathbf{V}$ and the *inverse-motion* gradient $\mathbf{F}^{-1}$ such that

$$\mathbf{V} = \left.\frac{\partial \chi^{-1}}{\partial t}\right|_x, \qquad \mathbf{F}^{-1} = \left.\frac{\partial \chi^{-1}}{\partial \mathbf{x}}\right|_t. \tag{8}$$

By using the chain rule of differentiation, we can thus check that

$$\mathbf{v} + \mathbf{F} \cdot \mathbf{V} = 0, \qquad \mathbf{V} + \mathbf{F}^{-1} \cdot \mathbf{v} = 0, \tag{9}$$

and

$$\mathbf{F}^{-1} \cdot \mathbf{F} = 1_R, \qquad \mathbf{F} \cdot \mathbf{F}^{-1} = 1, \tag{10}$$

where $\mathbf{1}_R$ and $\mathbf{1}$ are unit dyadics. For the bodies considered, the matter density ρ_0 at the reference configuration is at most a (here smooth) function of the material point. Accordingly, the local expression of the conservation of mass reads

$$\left.\frac{\partial \rho_0}{\partial t}\right|_X = 0. \tag{11}$$

This means that ρ_0 depends at most on the material point $\mathbf{X}$, i.e. that the body may present a *smooth material inhomogeneity* from the point of view of *inertia*. Similarly, for the elastic properties accounted for in the function W.

A standard variation of the direct motion generates at all regular material points the local balance of linear momentum in the Piola–Kirchhoff form (here we account for no physical forces such as gravity)

$$\left.\frac{\partial \mathbf{p}}{\partial t}\right|_X - div_R \mathbf{T} = 0, \tag{12}$$

where the linear momentum $\mathbf{p}$ (a vector in E^3) and the first Piola–Kirchhoff stress $\mathbf{T}$ (a two-point tensor field on $M^3 \times E^3$) are given by

$$\mathbf{p} = \rho_0 \mathbf{v}, \qquad \mathbf{T} = \partial \overline{W} / \partial \mathbf{F}. \tag{13}$$

Note that the direct motion itself does not intervene in L by virtue of the homogeneity of physical space, while the explicit dependence on $\mathbf{X}$ is a print left by *material inhomogeneity*. In the absence of external body force and energy supply, following general principles of field theory (Noether's theorem; see Refs. [9,14]), variations of the presently used space–time parametrization ($\mathbf{X}$, t) yield the canonical equations of energy and momentum. These are called *canonical* for they are in a frame independent of the actual placement $\mathbf{x}$ and they refer to the whole physical system under consideration and not to each degree of freedom separately as the components of (12) do. Thus the local equation of conservation of energy and the local balance of *material* or *canonical* momentum are obtained in the now classical form

$$\left.\frac{\partial H}{\partial t}\right|_X - \nabla_R \cdot \mathbf{Q}^{el} = 0, \tag{14}$$

and

$$\left.\frac{\partial \mathbf{P}}{\partial t}\right|_X - div_R \mathbf{b} = \mathbf{f}^{\mathrm{inh}}, \tag{15}$$

wherein

$$H = K + W, \ \mathbf{Q}^{el} = \mathbf{G} = \mathbf{T} \cdot \mathbf{v}, \tag{16}$$

$$\mathbf{P} = -\mathbf{p} \cdot \mathbf{F}, \ \mathbf{b} = -(\bar{L} 1_R + \mathbf{T} \cdot \mathbf{F}), \ \mathbf{f}^{\mathrm{inh}} = \left.\frac{\partial \bar{L}}{\partial \mathbf{X}}\right|_{\mathrm{expl}}, \tag{17}$$

where the last notation means the *explicit material gradient* of $\bar{L}$, i.e. the material gradient taken at fixed fields $\mathbf{v}$ and $\mathbf{F}$. The quantities defined in Eqs. (16) and (17) are, respectively, the *Hamiltonian* (energy) density H per unit reference volume, the material 'mechanical' *Poynting–Umov vector* (energy flux) $\mathbf{G}$, the *canonical momentum* or *pseudomomentum* $\mathbf{P}$ (covariant components on M^3), the *Eshelby material stress* $\mathbf{b}$ (a mixed fully material tensor), and the *material force of smooth inhomogeneities* $\mathbf{f}^{\mathrm{inh}}$

(with covariant components on M^3). The latter reads more explicitly

$$\mathbf{f}^{\mathrm{inh}} = \left(\frac{1}{2}\mathbf{v}^2\right)(\nabla_R \rho_0) - \left.\frac{\partial \overline{W}(\mathbf{F};\mathbf{X})}{\partial \mathbf{X}}\right|_{\mathbf{F}\ \text{fixed}}. \tag{18}$$

Note that by multiplying Eq. (12) on the right by $\mathbf{F}$ and performing some manipulations, we recover the so-called *Noether's identity*

$$\left(\frac{\partial \mathbf{p}}{\partial t} - div_R \mathbf{T}\right)\cdot\mathbf{F} - \left(\frac{\partial \mathbf{P}}{\partial t} - (div_R \mathbf{b} + \mathbf{f}^{\mathrm{inh}})\right) = 0. \tag{19}$$

Remark 2.1 Had we known Eq. (12) by any means, we could have deduced Eq. (15) on account of the knowledge of the functional dependency of $\overline{W}$. This points the way to be followed in a dissipative case. In particular, in quasistatics (neglect of inertial terms) and for a *materially homogeneous* elastic material, Eq. (19) reduces to *Ericksen's identity* [15] of elastostatics

$$(div_R \mathbf{T})\cdot\mathbf{F} + div_R \mathbf{b} = 0. \tag{20}$$

The general notion of the 'Ericksen–Noether identity' is dealt with in Ref. [16] for mechanics and other problems (heat and electricity conductions).

Equation (14) is the equation of conservation of energy, a strict *conservation law* because $\bar{L}$ does not depend explicitly on time. Eq. (15) is the balance equation of canonical momentum or *pseudomomentum* (or *material* (linear) *momentum*). While there exists only one representation of the local energy balance, Eqs. (12), (14) and (15) are, at all regular material points, but two statements, on different manifolds, of the *same* balance equation. The first one is often referred to as the *Lagrangian* representation, but this denomination is misleading because its components are in effect in the actual configuration (only its space–time parametrization is Lagrangian). The second one, which, according to Eq. (19) appears here as a simple pull-back of the former onto M^3, is entirely material, being projected on M^3 and using a material space–time parametrization. It is *canonical*, because of this and not only because it is derived from Noether's theorem. The Eshelby stress tensor $\mathbf{b}$ is fully material, and if there are no body couples acting on the material, satisfies the following symmetry condition (see Ref. [17]; this would be a natural consequence of the symmetry of the classical Cauchy stress); it is symmetric with respect to $\mathbf{C}$

$$\mathbf{C}\cdot\mathbf{b} = \mathbf{b}^T\cdot\mathbf{C}. \tag{21}$$

If one introduces the *second Piola–Kirchhoff stress* $\mathbf{S}$ (this is a true contravariant material tensor) by

$$\mathbf{S} = \mathbf{T}\cdot\mathbf{F}^{-T}, \tag{22}$$

it is immediately checked that $\mathbf{b}$ is given by

$$\mathbf{b} = -(\bar{L}1_R + \mathbf{M}), \tag{23}$$

where

$$\mathbf{M} = \mathbf{S}\cdot\mathbf{C} = \mathbf{T}\cdot\mathbf{F}, \tag{24}$$

is the so-called *Mandel stress* (known in finite-strain elastoplasticity). Then **b** and **M** are the same up to a sign and an 'isotropic' contribution (pressure-like term on the material manifold).

Remark 2.2 It is easily verified that the energy equation (14) can also be written in terms of Eshelby's stress as

$$\frac{\partial H}{\partial t} - \nabla_R\cdot((\mathbf{b} + L1_R)\cdot\mathbf{V}) = \frac{\partial H}{\partial t} + \nabla_R\cdot(\mathbf{M}\cdot\mathbf{V}) = 0 \tag{25}$$

Remark 2.3 The variational formulation can also be formulated on the basis of the inverse motion (7), in which case the starting point is a Lagrangian density $\hat{L}$ per unit volume of the actual configuration K_t and a density of elastic potential energy $w(\mathbf{X}, \mathbf{F}^{-1})$ per unit volume of K_t. Then $\mathbf{X}$ becomes the primary field and the space–time parametrization is given by the actual placement $\mathbf{x}$ and Newtonian time t. As a consequence, Eq. (15) or an equivalent form is obtained as *the* field equation describing the motion and Eqs. (12) and (14) are the results of the application of Noether's theorem (see Ref. [9], Section 5). This seeming equivalence between descriptions using the direct and inverse motions is the result of the special context of a purely mechanical case where the field and the space parametrization are of the same nature. It is delusive because the two formulations are *not* equivalent for problem solving since bulk and boundary data, if any, can only be prescribed in the actual configuration. However, the inverse motion description provides directly a *Hamiltonian formalism* since it can be shown that the following system of equations holds true (see Refs. [18,19])

$$\mathbf{V} \equiv \left.\frac{\partial \mathbf{X}}{\partial t}\right|_x = \frac{\delta \hat{H}}{\delta \hat{\mathbf{P}}}, \quad \left.\frac{\partial \hat{\mathbf{P}}}{\partial t}\right|_x = -\frac{\delta \hat{H}}{\delta \mathbf{X}} \text{ or } \left.\frac{\partial \hat{\mathbf{P}}}{\partial t}\right|_x + \delta_R \hat{H} = -\left.\frac{\partial \hat{H}}{\partial \mathbf{X}}\right|_{\text{expl}}, \tag{26}$$

where we have set

$$\hat{L} = J_F^{-1} L, \quad \hat{\mathbf{P}} = J_F^{-1}\mathbf{P} = \rho \mathbf{C}\cdot\mathbf{V}, \quad \hat{K} = J_F^{-1} K, \quad w = \hat{W} = J_F^{-1} W, \tag{27}$$

and

$$\hat{H} = \hat{\mathbf{P}}\cdot\mathbf{V} - \hat{L} = \frac{1}{2\rho}\hat{\mathbf{P}}\cdot\mathbf{C}^{-1}\cdot\hat{\mathbf{P}} + \hat{W}, \quad \hat{\mathbf{P}} = \frac{\partial \hat{L}}{\partial \mathbf{V}} \tag{28}$$

$$\delta_R = -\nabla\cdot\frac{\partial}{\partial \mathbf{F}^{-1}} - \left.\frac{\partial}{\partial t}\frac{\partial}{\partial \mathbf{V}}\right|_x. \tag{29}$$

We remind the reader that the continuity equation $\rho = \rho_0 J_F^{-1}$ is equivalent to Eq. (11) at all regular material points. Furthermore, studying the invariance under the group of space–time scalings for a homogeneous material with quadratic strain energy function, the authors have established the following 'balance law' for the *action* per unit current volume, $\hat{S} = \hat{\mathbf{P}}\cdot\mathbf{X} - \hat{H}t$

$$\frac{\partial}{\partial t}\hat{S} - \nabla\cdot(\mathbf{b}\cdot\mathbf{X} - \mathbf{Q}^{el}t) = 2\hat{L}, \tag{30}$$

while on account of all canonical equations they have separately shown that the following *Hamilton–Jacobi equation* holds good

$$\frac{\partial}{\partial t}\hat{S} + \hat{H} = 0, \quad \hat{H} = \tilde{H}\left(\mathbf{X}, \hat{\mathbf{P}} = \frac{\partial \hat{S}}{\partial \mathbf{X}}, \mathbf{F}^{-1}\right). \tag{31}$$

This completes the (canonical) analytical mechanics of hyperelasticity (we remind the reader that hyperelasticity is the energy-derivable pure elasticity of bodies in finite strains—usual denomination in nonlinear continuum mechanics, see Ref. [20]).

3. The case of thermoelastic conductors

The thermoelasticity of conductors involves thermal dissipation and, therefore, cannot in principle be deduced from a classical Lagrangian–Hamiltonian variational principle. This is where Remark 2.1 following Eq. (19) proves to be useful. We can thus deduce the missing pseudomomentum balance law from the classically known equations of finite strain thermoelasticity (see Refs. [6,21]). First a reminder of classical thermoelasticity in nonlinear anisotropic inhomogeneous materials is in order. For this, we refer the reader to Refs. [22,23]. Here, we simply list the relevant field equations and constitutive equations at any regular material point $\mathbf{X}$ in a material body. The Piola–Kirchhoff formulation is used from the start. Just as above, we consider no body forces for the sake of clarity. Then we have the following standard roster of classical field equations:

- Balance of mass:

$$\left.\frac{\partial \rho_0}{\partial t}\right|_{\mathbf{X}} = 0. \tag{32}$$

- Balance of linear (physical) momentum:

$$\left.\frac{\partial \mathbf{p}}{\partial t}\right|_{\mathbf{X}} - div_R\mathbf{T} = 0; \tag{33}$$

- *Balance of energy* (no body heat supply for the sake of simplicity):

$$\left.\frac{\partial H}{\partial t}\right|_{\mathbf{X}} - \nabla_R\cdot(\mathbf{T}\cdot\mathbf{v} - \mathbf{Q}) = 0; \tag{34}$$

- *Balance of entropy* (here S is the volume entropy):

$$\left.\frac{\partial S}{\partial t}\right|_{\mathbf{X}} + \nabla_R \cdot \mathbf{S} = \sigma_{\mathrm{B}}. \tag{35}$$

The *second law of thermodynamics* at all regular material points $\mathbf{X}$ in the body imposes that

$$\sigma_B \geq 0. \tag{36}$$

In the above set equations the total volume energy H, the volume kinetic energy K, the volume *internal energy* E, the volume entropy S, the material entropy flux $\mathbf{S}$ (not to be mistaken for the second Piola–Kirchhoff stress) and the bulk entropy source σ_{B} are given by

$$\begin{aligned} &H = K + E,\ K = \tfrac{1}{2}\rho_O \mathbf{v}^2,\ E = \rho_0 e, \\ &S = \rho_0 \eta,\ \mathbf{S} = \mathbf{Q}/\theta,\ \sigma_{\mathrm{B}} = -\theta^{-1}\mathbf{S}\cdot\nabla_{\mathrm{R}}\theta, \end{aligned} \tag{37}$$

where $\mathbf{Q}$ is the material heat flux, θ is the non-negative absolute temperature, and e and η are the internal energy and entropy per unit mass in the actual configuration (at time t). In the classical *thermoelasticity of conductors*, the constitutive equations (laws of state) are deduced from the *free-energy density* W by

$$W = W(\mathbf{F}, \theta; \mathbf{X}) = E - S\theta, \mathbf{T} = \frac{\partial W}{\partial \mathbf{F}}, S = -\frac{\partial W}{\partial \theta}. \tag{38}$$

The entropy flux should also satisfy a ‘continuity’ requirement such that

$$\mathbf{S}(\mathbf{F}, \theta, \nabla_R \theta; \mathbf{X}) \to 0 \text{ as } \nabla_R \theta \to 0. \tag{39}$$

Eqs. (32–39), together with a more precise expression for W and specification of some of its mathematical properties (e.g. convexity), are those to be used in studying sufficiently regular *nonlinear dynamical processes in thermoelastic conductors.* Note that the so-called *internal-energy theorem* and the *Clausius–Duhem inequality* read (tr = trace)

$$\left.\frac{\partial E}{\partial t}\right|_{\mathbf{X}} = \mathrm{tr}(\mathbf{T}\cdot(\nabla_{\mathrm{R}}\mathbf{v})^T) + \nabla_R \cdot \mathbf{Q}, \tag{40}$$

and

$$-\left(\frac{\partial W}{\partial t} + S\frac{\partial \theta}{\partial t}\right) + \mathrm{tr}(\mathbf{T}\cdot(\nabla_{\mathrm{R}}\mathbf{v})^T) - \mathbf{S}\cdot\nabla_{\mathrm{R}}\theta \geq 0. \tag{41}$$

4. The notion of thermal material force

Save for the explicit functional dependence of ρ_0 and W on $\mathbf{X}$, the above formalism does not reveal any material inhomogeneities, in particular the fact that thermoelastic properties may also be *particle dependent.* We need to write an equation of balance *directly on the material manifold* to see a manifestation of that inhomogeneity. Taking benefit of Remark 2.1, we just pull-back Eq. (33) to M^3 by applying $\mathbf{F}$ to the right and performing some manipulations on account of the functional dependence indicated in Eq. $(38)_1$ and the $\mathbf{X}$ dependence of mass density, obtaining thus the following local balance of pseudomomentum

$$\left.\frac{\partial \mathbf{P}}{\partial t}\right|_{\mathbf{X}} - div_R \mathbf{b} = \mathbf{f}^{\text{inh}} + \mathbf{f}^{\text{th}}, \tag{42}$$

in which we have defined the (material) *pseudomomentum density* $\mathbf{P}$, the Eshelby material stress tensor $\mathbf{b}$, the material force of true inhomogeneities $\mathbf{f}^{\text{inh}}$ and the material force of quasi-inhomogeneity due to thermal effects, $\mathbf{f}^{\text{th}}$, by

$$\mathbf{P} = -\mathbf{p}\cdot\mathbf{F} = \rho_0 \mathbf{C}\cdot\mathbf{V}, \tag{43}$$

$$\mathbf{b} = -(L^{\text{th}} 1_R + \mathbf{T}\cdot\mathbf{F}),\ \mathbf{f}^{\text{inh}} = \left.\frac{\partial L^{\text{th}}}{\partial \mathbf{X}}\right|_{\text{expl}}, \tag{44}$$

and

$$\mathbf{f}^{\text{th}} := S \nabla_R \theta, \tag{45}$$

together with an effective *Lagrangian density* for thermoelasticity L^{th} given by

$$L^{\text{th}} = K - W = \bar{L}^{\text{th}}(\mathbf{F}, \mathbf{v}, \theta; \mathbf{X}),\ K = \tfrac{1}{2}\rho_0 \mathbf{V}\cdot\mathbf{C}\cdot\mathbf{V}. \tag{46}$$

The fact that the 'material force' $\mathbf{f}^{\text{th}}$ arises simultaneously with $\mathbf{f}^{\text{inh}}$, and the latter may have the interpretation of a continuous distribution of dislocations, shows that, on the material manifold, a spatially nonuniform field of temperature may be viewed as a *quasimaterial inhomogeneity* or a *quasiplastic phenomenon* amenable through a geometrical interpretation (see Ref. [24]). Corresponding material forces are called *material forces of quasi-inhomogeneity*. This effect disappears once thermal equilibrium has been reached after complete diffusion of heat. Such an effect subsists in the absence of true material inhomogeneities, in which case Eq. (42) reduces to the simpler form,

$$\left.\frac{\partial \mathbf{P}}{\partial t}\right|_{\mathbf{X}} - div_R \mathbf{b} = \mathbf{f}^{\text{th}}, \tag{47}$$

before thermal equilibrium is reached. In the present exposition (proof at any regular material point) Eq. (47) is nothing but an *identity* that expresses the *Noether–Ericksen identity* for the case of inhomogeneous, anisotropic thermoelasticity in finite strains. To our knowledge, the thermal material force—not with such a denomination—first appeared in small strains in a work of Bui [25]. We must emphasize that heat conduction does not modify here the formal expressions of pseudomomentum, Eshelby stress, and true inhomogeneity force. This means that temperature, although a field in itself, is *not* an additional degree of freedom, at least in the present approach. We finally note that Eq. (35) may also be written as

$$\theta \left.\frac{\partial S}{\partial t}\right|_{\mathbf{X}} + \nabla_R \cdot \mathbf{Q} = 0, \tag{48}$$

which is the basic form of the *heat equation*. Although the right-hand side of this equation vanishes (here, there is no intrinsic dissipation), Eq. (48) is *not* a strict conservation law because of the thermal factor in the first term. Eqs. (46) and (48) may form the basis for a canonical formulation of dissipative continuum mechanics (see below). For further use in that direction, we note the following *double* Legendre transformation between H and L^{th}

$$H = \mathbf{P}\cdot\mathbf{V} + S\theta - L^{\mathrm{th}}, \quad \mathbf{P} = \frac{\partial L^{\mathrm{th}}}{\partial \mathbf{V}}, \quad S = \frac{\partial L^{\mathrm{th}}}{\partial \theta}. \tag{49}$$

5. The case of dissipative continua with nondissipative heat conduction

5.1. General statement

We propose here a canonical variational formulation of the basic equations of nonlinear continuum mechanics when the medium considered is a thermal conductor and its *anelasticity*, if any, is taken into account via *internal variables of state*. This last thermodynamic theory is quite sufficient for our purpose since it is appropriate for representing most of the known thermodynamically irreversible behaviors, while being the approach that deviates least, in both contents and methods, from the classical *theory of irreversible processes* (see Ref. [26]). The variational formulation proposed reads as follows in symbolic form:

$$\lim_{\beta \to 0} \delta \int_{E^3 \times T} L(\mathbf{v}, \mathbf{F}, \alpha, \theta = \dot{\gamma}, \beta = \nabla_R \gamma; \mathbf{X}) d^4 X = 0, \tag{50}$$

where L is the Hamiltonian–Lagrangian density per unit reference volume, **v, F,** θ, and **X** have already been introduced, α represents collectively the set of internal variables of state, γ is the so-called *thermacy* (see Section. 1; it is the time primitive—or time history—of θ or, if we like, a timewise potential for temperature, so that it is a rather unusual notion), and β is the material gradient of the latter; Space–time parametrization is represented as usual by **X** and *t;* Eq. (50) applies to the case of anisotropic, materially inhomogeneous materials. The *limit* symbolism used in Eq. (50) means that the limit as β goes to zero must be taken in the equations resulting from the variational formulation, this applying to both field equations and other consequences of the principle such as the results of the application of Noether's theorem. We claim that in this limit *all equations of the 'classical' theory of anelastic conductors of heat* are obtained, including the entropy equation and heat-propagation equation in this quite general case, a rather surprising result, we admit. This is shown in the following. The a priori introduction (no variational formulation; [5]) of γ and β yields the Green–Naghdi [4] 'dissipationless' theory of thermoelastic conductors in the absence of anelasticity. One of the important results obtained is the space–time symmetry between the equation of balance of canonical momentum (on the material manifold) and the equation of heat propagation.

5.2. Variational formulation

Here, we directly apply the abstract results of field theory to the case where the *limit* indicated in Eq. (50) is not yet taken. We directly use the abstract formulas as given, for instance, in Ref. [13] (Appendix A). Consider Hamiltonian–Lagrangian densities (per unit volume of the undeformed configuration of nonlinear continuum mechanics K_R) given by the following general expression:

$$L = \bar{L}(\mathbf{v}, \mathbf{F}, \alpha, \theta, \beta; \mathbf{X}) = K(\mathbf{v}; \mathbf{X}) - W(\mathbf{F}, \alpha, \theta, \beta; \mathbf{X}). \tag{51}$$

In the Lagrangian density Eq. (51), the *basic fields* are the *placement* $\mathbf{x}$ and the thermacy γ, both being assumed sufficiently smooth functions of the *space–time parametrization* $(\mathbf{X}, t)$, which is the one favored in the Piola–Kirchhoff formulation of nonlinear continuum mechanics. Notice that L is not an explicit function of $\mathbf{x}$ by virtue of Galilean invariance (translations in physical space of placements). Neither is it an explicit function of γ itself. Furthermore, α acts as a simple parameter. According to the general field theory, in the absence of external sources (these would be explicit functions of the fields themselves), the field equations, i.e. the Euler–Lagrange equations (see Eqs. A.7 in Ref. [13]) associated with χ and γ, are immediately given by

$$\left.\frac{\partial \mathbf{p}}{\partial t}\right|_{\mathbf{X}} - div_R \mathbf{T} = 0, \quad \left.\frac{\partial S}{\partial t}\right|_{\mathbf{X}} + \nabla_R \cdot \mathbf{S} = 0, \tag{52}$$

wherein

$$\begin{aligned} &\mathbf{p} := \rho_0 \mathbf{v} = \frac{\partial L}{\partial \dot{\chi}}, \ \mathbf{T} := \frac{\partial W}{\partial \mathbf{F}} = -\frac{\partial L}{\partial (\nabla_R \chi)}, \quad S := -\frac{\partial W}{\partial \theta} = \frac{\partial L}{\partial \dot{\gamma}}, \\ &\mathbf{S} := -\frac{\partial W}{\partial \beta} = \frac{\partial L}{\partial (\nabla_R \gamma)}, \end{aligned} \tag{53}$$

are all previously defined, but for the material entropy flux vector $\mathbf{S}$ that now formally derives from the free energy (obviously, out of equilibrium)! Invoking now Noether's theorem (see Eqs. A.11 in [13]) for the Lagrangian Eq. (51) with respect to the space–time parametrization, we obtain the following two, respectively covectorial and scalar, equations

$$\left.\frac{\partial P_L^{\text{th}}}{\partial t}\right|_{\mathbf{X}} - \left(div_R \mathbf{b}^{\text{th}}\right)_L = \left(\mathbf{f}^{\text{inh}}\right)_L, \tag{54}$$

and

$$\left.\frac{\partial H}{\partial t}\right|_{\mathbf{X}} - \nabla_R \cdot \mathbf{U} = 0, \tag{55}$$

where we have defined the *canonical momentum* (material-covariant) vector $\mathbf{P}^{\text{th}}$ of the present approach, the corresponding *canonical material stress* tensor $\mathbf{b}^{\text{th}}$, *the material force* of true inhomogeneities $\mathbf{f}^{\text{inh}}$, the *Hamiltonian density* H, the

Umov–Poynting energy-flux vector $\mathbf{U}$ by (compare the general definitions A.16, A.17, A.14 and A.15 in Ref. [13])

$$\mathbf{P}^{\text{th}} = -\nabla_R \chi \cdot \frac{\partial L}{\partial \mathbf{v}} - \nabla_R \gamma \frac{\partial L}{\partial \dot{\gamma}} = -\mathbf{p} \cdot \mathbf{F} - S\beta = \mathbf{P}^{\text{mech}} - S\beta, \tag{56}$$

$$\mathbf{b}^{\text{th}} := \left\{ b_{.L}^{K} := -\left(L\delta_L^K - \left(\gamma_{,L} \frac{\partial L}{\partial \gamma_{,K}} + \chi_{,L} \frac{\partial L}{\partial \chi_{,K}} \right) \right) = -\left(L\delta_L^K - S^K \beta_L + T_{.i}^K F_{.L}^i \right) \right\}, \tag{57}$$

$$\mathbf{f}^{\text{inh}} := \left. \frac{\partial L}{\partial \mathbf{X}} \right|_{\text{expl}} = \left(\frac{\mathbf{v}^2}{2} \right) (\nabla_R \rho_0) - \left. \frac{\partial W}{\partial \mathbf{X}} \right|_{\text{expl}}, \tag{58}$$

$$H = \dot{\gamma} \frac{\partial L}{\partial \dot{\gamma}} + \mathbf{v} \cdot \frac{\partial L}{\partial \mathbf{v}} - L = S\theta + 2K - L = K + E, \tag{59}$$

$$U^K = -\left(\dot{\gamma} \frac{\partial L}{\partial \gamma_{,K}} + v^i \frac{\partial L}{\partial \chi_{\cdot K}^i} \right) = T_{.i}^K v^i - S^K \theta, \tag{60}$$

where we have defined the *mechanical canonical* (material) momentum $\mathbf{P}^{\text{mech}}$ and the *internal energy* per unit reference volume by

$$\mathbf{P}^{\text{mech}} = -\mathbf{p} \cdot \mathbf{F}, \qquad E = W + S\theta. \tag{61}$$

The first of these is the same as Eq. $(17)_1$. The second is the usual Legendre transformation of thermodynamics. As a matter of fact, the definition (59) contains two Legendre transformations, one related to mechanical fields, and the other to thermal ones. If we assume, as in standard continuum thermodynamics, that entropy and heat flux are related by the usual relation $\mathbf{S} = \mathbf{Q}/\theta$, we have from Eq. (55) the classical form of the energy-conservation equation (34) but with an energy-derived material heat flux, i.e.

$$\left. \frac{\partial H}{\partial t} \right|_{\mathbf{X}} - \nabla_R \cdot (\mathbf{T} \cdot \mathbf{v} - \mathbf{Q}) = 0, \qquad \mathbf{Q} = -\theta \frac{\partial W}{\partial \beta}, \tag{62}$$

where both $\mathbf{T}$ *and* $\mathbf{Q}$ are given by constitutive equations. Summing up, we have deduced from the Hamiltonian–Lagrangian density (Eq. (51)) all field equations, balance laws and constitutive relations for the theory of materially inhomogeneous, finitely deformable anelastic thermoelastic conductors of heat. Note that the variable α produces neither canonical momentum nor explicit contribution to the Eshelby stress and to the energy equation $(62)_1$. It is, indeed, an *internal* variable of state. We can simply define the generalized force $A = -\partial W/\partial \alpha$ for further use. The above-given formulation is said to be 'dissipationless'(compare to Refs. [4,5,18,19,27–30].

5.3. Transition to the classical theory of anelastic conductors of heat

Eq. $(52)_1$ and Eq. $(62)_1$ are already formally in the classical form. What about Eq. $(52)_2$ and Eq. (54) whose 'classical' form is given in Refs. [7,8]. We need to isolate the contributions of the 'dissipative' variables. First we expand Eq. (54) by accounting for the expressions (56) and (57). We obtain the following equation (T = transposed)

$$\left.\frac{\partial \mathbf{P}^{\mathrm{mech}}}{\partial t}\right|_{\mathbf{X}} - div_R \bar{\mathbf{b}}^{\mathrm{mech}} = S\nabla_R\theta + \mathbf{S}\cdot(\nabla_R\beta)^{\mathrm{T}} + \mathbf{f}^{\mathrm{inh}}, \tag{63}$$

where $\bar{\mathbf{b}}^{\mathrm{mech}} = \mathbf{b}^{th} - \mathbf{S}\otimes\beta$. But this is not all because L in $\mathbf{b}^{th}$ still depends on β and α. We must isolate this dependency by writing

$$\begin{aligned}\frac{\partial W}{\partial \mathbf{X}} &= \frac{\partial W^{\mathrm{mech}}}{\partial \mathbf{X}} + \frac{\partial W}{\partial \alpha}\cdot(\nabla_R\alpha)^{\mathrm{T}} + \frac{\partial W}{\partial \beta}\cdot(\nabla_R\beta)^{\mathrm{T}} \\ &= \frac{\partial W^{\mathrm{mech}}}{\partial \mathbf{X}} - A\cdot(\nabla_R\alpha)^{T} - \mathbf{S}\cdot(\nabla_R\beta)^{\mathrm{T}},\end{aligned} \tag{64}$$

where, in essence, $W^{\mathrm{mech}} = W(\mathbf{F}, \theta, \alpha = \mathrm{const.}, \beta = 0; \mathbf{X})$ On substituting Eq. (64) into the material divergence of $\mathbf{b}^{\mathrm{mech}}$, we finally transform Eq. (63) to

$$\left.\frac{\partial \mathbf{P}^{\mathrm{mech}}}{\partial t}\right|_{\mathbf{X}} - div_R \mathbf{b}^{\mathrm{mech}} = \mathbf{f}^{\mathrm{inh}} + \mathbf{f}^{\mathrm{th}} + \mathbf{f}^{\mathrm{intr}}, \tag{65}$$

where

$$\begin{aligned}&\mathbf{b}^{\mathrm{mech}} = -(L1_R + \mathbf{T}\cdot\mathbf{F}), \qquad L = K - W^{\mathrm{mech}}(\mathbf{F}, \theta, \alpha = \mathrm{const.}; \mathbf{X}), \\ &\mathbf{f}^{\mathrm{th}} := S\nabla_R\theta, \qquad \mathbf{f}^{\mathrm{intr}} := A\cdot(\nabla_R\alpha)^{\mathrm{T}}.\end{aligned} \tag{66}$$

The last two introduced quantities are material *forces of quasi-inhomogeneity* due to a nonuniform temperature field (see [6], and Section 3) and to a nonuniform α field, respectively [7]. The presence of those terms on an equal footing with $\mathbf{f}^{\mathrm{inh}}$ means that, insofar as the material manifold is concerned, spatially nonuniform fields of α or θ are equivalent to distributed material inhomogeneities (also continuously distributed defects such as dislocations); they are *quasiplastic effects* (see Ref. [24]).

As to Eq. $(52)_2$, accounting for the kinetic-energy theorem (obtained by multiplying scalarly Eq. $(52)_1$ by $\mathbf{v}$, after multiplication by $\theta \neq 0$ and accounting for Eq. $(62)_1$ and finally making $\beta = \mathrm{const.}$ (this is equivalent to discarding β in the resulting equation and losing the connection of $\mathbf{S}$ and $\mathbf{Q}$ with β) we arrive at the *'heat-propagation' equation* in the form

$$\left.\frac{\partial (S\theta)}{\partial t}\right|_{\mathbf{X}} + \nabla_R\cdot\mathbf{Q} = S\dot{\theta} + A\cdot\dot{\alpha} \equiv \bar{\Phi}^{\mathrm{th}} + \Phi^{\mathrm{intr}}. \tag{67}$$

Then working in reverse, in this approximation one recovers the equations (compare to Ref. [7])

$$\theta\frac{\partial S}{\partial t} + \nabla_R\cdot\mathbf{Q} = \Phi^{\mathrm{intr}}, \qquad \frac{\partial S}{\partial t} + \nabla_R\cdot\mathbf{S} = \sigma^{\mathrm{th}} + \sigma^{\mathrm{intr}}, \tag{68}$$

where $\sigma^{\mathrm{th}} = -\mathbf{S}\cdot\nabla_R(\ln\,\theta)$ is the *thermal entropy source* and $\sigma^{\mathrm{intr}} = \theta^{-1}A\cdot\dot{\alpha}$ is the intrinsic entropy source, and $\Phi^{\mathrm{intr}} = \theta\sigma^{\mathrm{intr}}$ is the *intrinsic dissipation*. In the present classical limit, $\mathbf{S}$ and $\mathbf{Q} = \mathbf{S}/\theta$ are now given by a constitutive equation obtained by invoking the noncontradiction with the second law of thermodynamics that here locally reads

$$\sigma = \sigma^{\mathrm{th}} + \sigma^{\mathrm{intr}} \geq 0. \tag{69}$$

We have recovered all equations or constraints of the 'classical theory' by applying the scheme proposed in Eq. (51).

6. Canonical four-dimensional space–time formulation

Eqs. (65) and (67) present an obvious space–time symmetry (see the two right-hand sides) that hints at considering these two equations as space-like and time-like components of a unique four-dimensional equation in the appropriate space and the canonical momentum $\mathbf{P}^{\mathrm{mech}}$ and the quantity $\theta\,S$ (an energy that is the difference between internal and free energies) as complementary space–time quantities, i.e. they together form a four-dimensional canonical momentum

$$\mathbf{P}_{(4)} = (\mathbf{P}^{\mathrm{mech}},\ \mathbf{P}_4 = \theta S). \tag{70}$$

We let the reader check that Eqs. (65) and (67) can in fact be rewritten in the following pure four-dimensional or 4×4 formalism in an Euclidean 4-dim space (compare to the world-invariant kinematics in Ref. [20])

$$\frac{\partial}{\partial X^{\beta}} B^{\beta}_{\cdot\alpha} = f^{\alpha} \equiv \bar{A}\cdot\frac{\partial}{\partial X^{\alpha}}\mu + \left.\frac{\partial L}{\partial X^{\alpha}}\right|_{\mathrm{expl}} = \left.\frac{\partial L}{\partial X^{\alpha}}\right|_{(\mathbf{F},\mathbf{v})\mathrm{fixed}} \tag{71}$$

$$\bar{A} = (A, S),\ \ \mu = (\alpha, \theta), \tag{72}$$

$$X^{\alpha}(\alpha = 1,2,3,4) = \{X^K(K = 1,2,3), X^4 = t\}, \tag{73}$$

$$B^{\beta}_{\cdot\alpha} = \begin{Bmatrix} B^K_{\cdot L} = -b^K_{\cdot L} & B^4_{\cdot L} = P^{\mathrm{mech}}_L \\ B^K_{\cdot 4} = Q^K & B^4_{\cdot 4} = \theta S \end{Bmatrix}, \qquad \mathbf{P}_{(4)} = (B^4_{\cdot L}, B^4_{\cdot 4}), \tag{74}$$

or, introducing intrinsically four-dimensional gradients and divergence in E^4,

$$div_{E^4}\mathbf{B}^{\mathrm{mech}} = \nabla_{E^4} L|_{\mathrm{mech}}, \tag{75}$$

where the right-hand side means the gradient computed keeping the 'mechanical' fields $(\mathbf{F}, \mathbf{v})$ *fixed*. Eq. (75) represents the canonical form of the balance of canonical momentum and the heat-propagation equation for anelastic, anisotropic, finitely deformable solid heat conductors. The four-dimensional formalism introduced is somewhat different from that used by Maugin [8] or Herrmann and Kienzler [12]. However, in the absence of intrinsic dissipative processes and for isothermal processes, Eqs. (65) and (66) or Eq. (75), reduce to those of Kijowski and Magli [31] in isothermal thermoelasticity, the second equation reducing obviously to the simple equation $\partial(\theta S)/\partial t = 0$. This shows the closedness of the present approach with the general relativistic Hamiltonian scheme. For relativistic works akin to this one we refer the reader to Refs. [32,33].

7. Comparison with some nonlinear wave processes

We would like to conclude the present contribution with an unavoidable mention of the mechanics of quasiparticles, Hamilton–Jacobi's equations, and an analogy between the canonical equations of the mechanics of continua and the 'kinematic' theory of nonlinear wave processes. First, there is no need to emphasize the de Broglie proportionality relationships between momentum and wave vector on the one hand, and energy and frequency on the other hand in linear (quantum) wave mechanics, i.e. symbolically, $P \approx K$ and $H = E \approx \omega$, where the (same) proportionality constant is Planck's reduced constant $\hbar$. For a wave-like motion with linear phase $\varphi\ (\omega, K) = K{\cdot}X - \omega t$ and an action function $S = P{\cdot}X - Ht$, this yields the characteristic relation $S \approx \varphi$. Also, the intimate relationship between the conservation of canonical momentum $\mathbf{P}$ of some nonlinear dispersive elastic systems and the Newtonian-like equations of *quasiparticles* associated with strongly localized nonlinear wave solutions of such systems (solitary waves and solitons) is well documented (e.g. in Refs. [9], Chapter 9, and [34]). Such systems would be obtained in the present context by accounting for the presence of higher-order deformation gradients, such as $\nabla_R \mathbf{F}$, in the elastic potential energy W of Eq. $(2)_2$. One remarkable approach to such dynamical wave systems is the masterpiece of a small group of scientists under the name of *kinematic wave theory*; see Refs. [35,36]. In this theory, for dynamical *nonlinear* solutions depending only on a phase φ, one writes in general for *homogeneous* bodies $\varphi = \bar{\varphi}(\mathbf{X}, t)$. Then the wave vector $\mathbf{K}$ and the frequency ω are *defined* by

$$\mathbf{K} = \frac{\partial \bar{\varphi}}{\partial \mathbf{X}}, \qquad \omega = -\frac{\partial \bar{\varphi}}{\partial t}, \tag{76}$$

from which there follows

$$\nabla_R \times \mathbf{K} = 0, \qquad \frac{\partial \mathbf{K}}{\partial t} + \nabla_R \omega = 0. \tag{77}$$

In particular, Eqs. (76) are trivially satisfied for harmonic plane-wave solutions for which we have a dispersion relation $\omega = \Omega(K)$. For an *inhomogeneous* linear behavior with *dispersion* we have the dispersion relation $\omega = \Omega(\mathbf{K}; \mathbf{X})$. Accordingly, the conservation of wave vector $(77)_2$ becomes

$$\frac{\partial \mathbf{K}}{\partial t} + \mathbf{V}_g{\cdot}\nabla_{\mathrm{R}} \mathbf{K} = -\left.\frac{\partial \Omega}{\partial \mathbf{X}}\right|_{\mathrm{expl}}, \quad \mathbf{V}_g = \frac{\partial \Omega}{\partial \mathbf{K}}, \tag{78}$$

whence the Hamiltonian system

$$\frac{\mathrm{D}\mathbf{X}}{\mathrm{D}t} = \frac{\partial \Omega}{\partial \mathbf{K}}, \qquad \frac{\mathrm{D}\mathbf{K}}{\mathrm{D}t} = -\left.\frac{\partial \Omega}{\partial \mathbf{X}}\right|_{\mathrm{expl}}, \tag{79}$$

where we have set

$$\frac{\mathrm{D}}{\mathrm{D}t} \equiv \frac{\partial}{\partial t} + \mathbf{V}_g{\cdot}\nabla_R. \tag{80}$$

Simultaneously, we have the *Hamilton–Jacobi equation* (compare Eq. (31), and Ref. [37])

$$\frac{\partial \varphi}{\partial t} + \Omega\left(\mathbf{X}, \mathbf{K} = \frac{\partial \varphi}{\partial \mathbf{X}}\right) = 0. \tag{81}$$

If we now consider a wave in an inhomogeneous dispersive *nonlinear* material, the frequency will also depend on the amplitude. Let the n-vector of R^n α characterize this amplitude of a complex system (in general with several degrees of freedom). Thus, now, $\omega = \Omega(\mathbf{K}, \mathbf{X}, \alpha)$. Accordingly, the second of Hamilton's equations (79) will now read [38]

$$\frac{\mathrm{D}\mathbf{K}}{\mathrm{D}t} = -\left.\frac{\partial \Omega}{\partial \mathbf{X}}\right|_{\text{expl}} + \mathbf{A}\cdot(\nabla_{\mathrm{R}}\alpha)^T, \quad \mathbf{A} := -\frac{\partial \Omega}{\partial \alpha}. \tag{82}$$

We can now compare this Hamiltonian system to the one obtained in Section 2. It is clear that $\hat{\mathbf{P}}$ and $\mathbf{V}$ of hyperelasticity play the same role as $\mathbf{K}$ and $\mathrm{D}\mathbf{X}/\mathrm{D}t$ in the kinematic wave theory, and the quantity $\mathrm{D}\mathbf{X}/\mathrm{D}t$ is also a material velocity. The Hamilton–Jacobi equations (31) and (81) are obviously analogous to one another. The richest comment, however, comes from a comparison of Eq. $(82)_1$ and a possible generalization of Eq. $(26)_2$. First, Eq. $(82)_1$ does not contain in its left-hand side a term analogous to the term $\delta_R \hat{H}$. The reason for this is that the dispersion equation yielding Eq. (79) is not *itself* dispersive in the language of Newell [39]. But we know now that a good generalization of Whitham's method should seek such a generalization, see also Appendix A6 and Section 9.3 in [13]. Next, there is no term equivalent to the 'amplitude' contribution of Eq. $(82)_1$ in the right-hand side of Eq. $(26)_2$. However, if we introduce the possibility in the continuum system to have some *anelasticity* accounted for through an internal variable of state denoted by α (an n-vector of R^n), as done in Section 4 above, then we shall have in the pure Hamiltonian system an additional source term due to this thermodynamically irreversible behavior. This term, according to Section 4 will necessarily have the form $\mathbf{A}\cdot(\nabla_R \alpha)^{\mathrm{T}}$ in the equation of canonical momentum, hence a generalization of Eq. $(26)_2$ to

$$\left.\frac{\partial \hat{\mathbf{P}}}{\partial t}\right|_x + \delta_R \hat{H} = -\left.\frac{\partial \hat{H}}{\partial \mathbf{X}}\right|_{\text{expl}} + \mathbf{A}\cdot(\nabla_R \alpha)^{\mathrm{T}}, \quad \mathbf{A} := -\frac{\partial \hat{H}}{\partial \alpha}, \tag{83}$$

where the new quantity in Eq. $(83)_1$ is a pseudo-inhomogeneity force due to the fact that the dissipative internal state variable α has not yet reached a spatially uniform value. On comparing Eq. $(82)_1$ and Eq. $(83)_1$ we can draw the conclusion that the role of *nonlinearity* (dependency of dispersion relation on wave amplitude) in the Whitham–Newell kinematic wave theory is played by the *dissipation* of the internal state variable in the dynamical theory of anelasticity. Both mean that the system finally deviates from a pure Hamiltonian one. Whether the present results and comparison will help one progress in the theory of nonlinear wave propagation in complex media is, for the time being, an open question. However, the presented analogies certainly reinforce the common view of wave mechanics that we have the following equivalences

$$\hat{\mathbf{P}} \approx \mathbf{K}, \ \hat{H} \approx \Omega, \ \hat{S} \approx \varphi, \tag{84}$$

which hold for point particles in wave mechanics but are likely to hold in a much larger framework, thus contributing to an advance toward a possible nonlinear duality between the nonlinear dynamics of continua and the motion of quasiparticles as is already emphasized in soliton theory [34].

8. Conclusion

In this contribution we have given in a concise way the main elements of variational mechanics related to the mechanics of continua of the solid deformable type. A classical (nonrelativistic) framework has been used. A possible path to including relevant dissipative mechanisms such as heat-conduction and plasticity-like effects was outlined. Other related works and further generalizations are to be noted by way of conclusion: accounting for higher-order gradient elasticity theories (see Refs. [9,14]), accounting for coupled electromagnetomechanical effects (see Refs. [40,41]), introducing the influence of a dynamical microstructure such as in liquid crystals ([42], where variational principles on various manifolds and their relationships are exhibited) or in polar elastic crystals ([43] where the flow chart of balance laws and canonical equations is particularly enlightening concerning the generation of additional conservation laws in the spirit of Noether or by direct formal manipulations), and continuum solid–fluid mixtures such as in Ref. [44]. The full case of elastic systems coupled with diffusion of species as can be useful in some problems of biomechanics (growth of tissues) is also a subject to be favorably dealt with along the same lines since the biomechanical problems of this type are better formulated on the material manifold itself.

Acknowledgement

GAM benefits from a Max Planck Award for International Co-operation (2001–2005).

References

[1] von Laue, M. (1921), Relativitätstheorie, Bd. 1, Vieweg-Verlag, Leipzig.
[2] Maugin, G.A. (1971), Magnetized deformable media in general relativity, Ann. Inst. Henri Poincaré A15, 275–302.
[3] Maugin, G.A. (1972), An action principle in general relativistic magnetohydrodynamics, Ann. Inst. Henri Poincaré A16, 133–169.
[4] Green, A.E. and Naghdi, P.M. (1993), Thermoelasticity without energy dissipation, J. Elast. 31, 189–208.
[5] Dascalu, C. and Maugin, G.A. (1995), The thermal material momentum equation, J. Elast. 39, 201–212.
[6] Epstein, M. and Maugin, G.A. (1995), Thermal material forces: definition and geometric aspects, C.R. Acad. Sci. Paris II-320, 63–68.
[7] Maugin, G.A. (1997), Thermomechanics of inhomogeneous–heterogeneous systems: application to the irreversible progress of two- and three-dimensional defects, ARI (Springer) 50, 43–56.
[8] Maugin, G.A. (2000), On the universality of the thermomechanics of forces driving singular sets, Arch. Appl. Mech. 70, 31–45.
[9] Maugin, G.A. (1993), Material Inhomogeneities in Elasticity, Chapman and Hall, London.
[10] Maugin, G.A. (1995), Material forces: concepts and applications, Appl. Mech. Rev. 48, 213–245.
[11] Gurtin, M.E. (1999), Configurational Forces as Basic Concepts of Continuum Physics, Springer, New York.
[12] Kienzler, R. and Herrmann, G. (2000), Mechanics in Material Space, Springer, Berlin.

[13] Maugin, G.A. (1999a), Nonlinear Waves in Elastic Crystals, Oxford University Press, UK.
[14] Maugin, G.A. and Trimarco, C. (2001), Elements of field theory in inhomogeneous and defective materials. Configurational Mechanics of Materials. In CISM Udine Lectures, September 2000, (Kienzler, R., Maugin, G.A., eds.), Springer, Wien, pp. 55–128.
[15] Ericksen, J.D. (1977), Special topics in elastostatics. In Advances in Applied Mechanics, (Yih, C.-S., ed.), Vol. 17, Academic Press, New York, pp. 189–208.
[16] Maugin, G.A. (1996), On Ericksen's identity and material balance laws in thermoelasticity and akin phenomena. In Contemporary Research in the Mechanics and Mathematics of Materials. (Ericksen's J.L., 70th Anniversary Volume, Batra, R.C., Beatty, M.F., eds.), CIMNE, Barcelona, pp. 397–407.
[17] Epstein, M. and Maugin, G.A. (1990), The energy–momentum tensor and material uniformity in finite elasticity, Acta Mech. 83, 127–133.
[18] Kalpakides, K. and Maugin, G.A. (2002), A Hamiltonian formulation for thermoelasticity. In NonSmooth/ Nonconvex Mechanics. (Volume in the honor of the late Prof. Panagiotopoulos), (Baniatopoulos, C.C., ed.), Thessaloniki, Greece, pp. 315–322.
[19] Kalpakides, V.K. and Maugin, G.A. (2003), A Hamiltonian formulation for elasticity and thermoelasticity. In Proceedings of International Symposium on Advanced Problems of Mechanics (Grekova, E.F. and Indeitsev, D.A., eds.), IPME-RAS, St.Petersburg, Russia, pp. 312–321, Repino, Russia, 2002.
[20] Truesdell, C.A. and Toupin, R.A. (1960), Field theories of mechanics. In Handbuch der Physik, (Flügge, S., ed.), Bd. III/1, Springer, Berlin.
[21] Maugin, G.A. and Berezovski, A. (1999), Material formulation of finite-strain thermoelasticity, J. Therm. Stresses 22, 421–449.
[22] Suhubi, E.S. (1975), Thermoelasticity. In Continuum Physics, (Eringen, A.C., ed.), Vol. 2, Academic Press, New York.
[23] Maugin, G.A. (1988), Continuum Mechanics of Electromagnetic Solids, Elsevier, Amsterdam.
[24] Maugin, G.A. (2003b), Pseudo-plasticity and pseudo-inhomogeneity effects in materials mechanics, J. Elast. 71, 81–103. C.A. Truesdell Memorial Volume.
[25] Bui, H.D. (1977), Mécanique de la Rupture Fragile, Masson, Paris.
[26] Maugin, G.A. (1999b), Thermomechanics of nonlinear irreversible behaviors, World Scientific, Singapore.
[27] Maugin, G.A. and Kalpakides, V.K. (2002a), A Hamiltonian formulation for elasticity and thermoelasticity, J. Phys. A: Math. Gen. 35, 10775–10788.
[28] Maugin, G.A. and Kalpakides, V.K. (2002b), The slow march towards an analytical mechanics of dissipative materials, Technische Mechanik 22, 98–103. Proceedings of CMIRF, Berlin, 2001.
[29] Kalpakides, V.K. and Maugin, G.A. (2004), Canonical formulation and conservation laws of thermoelasticity without dissipation, Rep. Math. Phys. 53, 371–391.
[30] Kalpakides, V. and Dascalu, C. (2002), On the thermomechanical formulation of configurational forces in continua. In Geometry, Continua and Microstructures (GCM6) (Cleja-Tigoiu, S., Tigoiu, V., eds.), Editura Academiei Romane, Bucarest, pp. 133–145.
[31] Kijowski, J. and Magli, G. (1999), Unconstrained Hamiltonian formulation of general relativity with thermo-elastic sources, Class. Quantum Grav. 15, 3891–3916.
[32] Sieniutycz, S. (2002), Relativistic thermo-hydrodynamics and conservation laws in dissipative continua with thermal inertia, Rep. Math. Phys. 49, 361–370.
[33] Jezierski, J. and Kijowski, J. (1990), Thermo-hydrodynamics as a field theory. In Nonequilibrium Theory and Extremum Principles. Advances in Thermodynamics, (Sieniutycz, S., Salamon, P., eds.), Vol. 3, Taylor and Francis, New York.
[34] Maugin, G.A. and Christov, C.I. (2001), Nonlinear duality between elastic waves and quasi-particles. In Selected Topics in Nonlinear Wave Mechanics (Christov, C.I., Guran, A., eds.), Birkhäuser, Boston, pp. 101–145.
[35] Whitham, G.B. (1974a), Dispersive waves and variational principles. In Nonlinear Waves (Leibovich, S., Seebass, A.R., eds.), Cornell University Press, Ithaca, NY, pp. 139–169, Chapter 4.
[36] Whitham, G.B. (1974b), Linear and Nonlinear Waves, Wiley-Interscience, New York.
[37] Rund, H. (1966), The Hamilton–Jacobi Theory in the Calculus of Variations, Van-Nostrand-Reinhold, New York.

[38] Maugin, G.A. (2003a), Nonlinear wave mechanics of complex material systems, Proc. Estonian Acad. Sci. Phys. Math. 52, 5–11. Proceedings Euromech 435, Tallinn, 2002.

[39] Newell, A.C. (1985), Solitons in Mathematics and Physics, S.I.A.M., Philadelphia.

[40] Trimarco, C. and Maugin, G.A. (2001), Material mechanics of electromagnetic solids. In Configurational Mechanics of Materials. CISM Udine Lectures, September 2000, (Kienzler, R., Maugin, G.A., eds.), Springer, Wien, pp. 129–171.

[41] Kalpakides, V.K. and Agiasofitou, E.K. (2002), On material equations in second gradient electroelasticity, J. Elast. 67, 205–227.

[42] Maugin, G.A. and Trimarco, C. (1995), On material and physical forces in liquid crystals, Int. J. Eng. Sci. 33, 1663–1678.

[43] Maugin, G.A. (1998), On the structure of the theory of polar elasticity, Philos. Trans. R. Soc. A356, 1367–1395.

[44] Quiligotti, S. and Maugin, G.A. (2003), An eshelbian approach to the nonlinear mechanics of constrained solid-fluid mixtures, Acta Mech. 160(1–2), 45–60.

Chapter 10

ON THE PRINCIPLE OF LEAST ACTION AND ITS ROLE IN THE ALTERNATIVE THEORY OF NONEQUILIBRIUM PROCESSES

Dedicated to Howard Brenner, MIT, on the occasion of his 75th birthday

Leibniz plus Voltaire lead us to the true optimum—T. S. W. Salomon.

Michael Lauster

University of the German Armed Forces Munich, Faculty for Economics, Werner-Heisenberg-Weg 39, 85577 Neubiberg, Germany

Abstract

The principle of least action (PLA) is one of the most popular applications of the calculus of variations in physics. Based on the idea that natural processes always run with a minimum amount of action a mathematical formulation for the minimum of an appropriate 'action integral' is required. In the context of the alternative theory of nonequilibrium processes (AT) the extremum procedure delivers local conservation laws for energy, linear, and angular momentum as well as the electrical charge and shows—via Noether's celebrated theorem—how the respective coordinates for space and time have to be constructed. Those are closely connected to the so-called Callen principle that is used to choose suitable extensive quantities for the system under study. This chapter deals with the application of the PLA to a new and promising theory of nonequilibrium processes. It is shown how one of the most common mathematical formulations of the PLA, known as Hamilton's Principle, can be fitted into the theoretical context of irreversible thermodynamics. This results in an extended version of Noether's theorem allowing us to express the conservation laws for certain extensive quantities in connection with symmetry transformations of the space and time coordinates. Finally, a formulation for the natural trend to equilibrium is derived from the PLA.

Keywords: nonequilibrium; dissipation; principle of least action; alternative theory

1. Space, time, and generic physical quantities

Generally, in physics and the engineering sciences two options to formulate empirical phenomena are available.

E-mail address: michael.a.lauster@t-online.de (M. Lauster).

Following the evolution-generated experience of reality physical objects are placed in a space with length, breadth, and height.[1] This space is homogeneous and isotropic in each of its directions.[2] Objects move through the 3D space and their movement is characterized by a time coordinate flowing uniformly from past to future.[3]

The appropriate mathematical model of such a space is a 3D Euclidean space with orthogonal axis and equal scales in each direction. The origin as well as the angular position of the axis may be set arbitrarily.

As for the time coordinate the same holds: origin and unit length of the scale may be chosen arbitrarily. A space comprised of three spatial and one temporal coordinates will be called a parameter space.[4]

In the parameter space the path of objects taken during their movement is depicted by a continuous curved line called a trajectory. Provided a (positive) real number for the mass is attached to the trajectory, the mathematical model for a body is given.[5]

The second option came up when classical mechanics was given its final structure and has been intensely used when thermodynamics appeared as a new discipline in physics (see Ref. [5]).[6] It is far more abstract than the first one and does not directly refer to our sensual experiences.

Physical objects are examined with regard to the quantities they exchange with other objects of the same or different kind. The quantities exchanged may be either matter-like, e.g. like substances or immaterial like, e.g. linear or angular momentum. They should be chosen in such a way that they are constitutive for the phenomena in any discipline of physics. Quantities fulfilling this condition like, e.g. the total energy, are called generic physical quantities or simply generics.

Every generic physical quantity is mathematically represented by a variable belonging to the set of functions that may be differentiated at least twice. Depending on the number of variables, an abstract space is generated that is not accessible by our common empirical experience. It is called phase space or Gibbs space. Space and time forming the parameter space do not belong to the set of coordinates of the Gibbs space.

Each variable may attain values that are represented by real numbers. Setting all variables of a Gibbs space to a respective value, a single point of this space is referenced.

[1] There is strong evidence that our experience being (only) 3D directly corresponds to the fact that three is the least number of dimensions needed for two trajectories not to cross even when they are not strictly parallel. This might have been useful during our evolution, e.g. not to interfere with foes or to successfully use weapons like stones or spears.

[2] Undisputedly, the description given above is more or less something like an ideal case and many failings may happen to our senses while experiencing spatial and temporal phenomena. This will not be discussed here but may be looked up in the respective literature, see, e.g. Ref. [1].

[3] The meaning of time and the arrow of time in the natural sciences has been discussed widely and this is not the place to delve into a philosophical dispute. A brief summary of this discussion and an extensive compilation of literature on this topic can be found in Ref. [2].

[4] Space and time are the projection parameters used by life forms to control their biological processes in order to assure their survival, see Ref. [3].

[5] See, e.g. Ref. [4]. This is the kinematic core of classical mechanics whose task it is to calculate the trajectory of any body, provided appropriate initial and boundary conditions are given. In general, discontinuous motions of macroscopic bodies are not allowed by hypothesis. This assures the applicability of the calculus of analysis.

[6] These milestones in physics are connected to the names of Sir W.R. Hamilton and J.W. Gibbs.

The vector of values is called a state. The junction (in the sense of set theory) of all possible states for an object is called a system.

In mathematical terms a system is a relation Γ of all the variables $X_1, X_2, \ldots, X_n, n \in \mathbb{N}$, of the Gibbs space

$$\Gamma(X_1, X_2, \ldots, X_n) \equiv 0. \tag{1}$$

Γ is called Gibbs Fundamental Relation (GFR).

The set of variables (as well as the variables themselves) leading to a homogeneous relation of degree one is called extensive.[7]

Objects change their state by exchanging quantities with other objects. Such a change of state is called a process. As a hypothesis, no discontinuous processes occur for macroscopic objects. Therefore, the mathematical picture of a process is a continuous subset of the system forming a curved (1D) line. Each point of that process path, i.e. each state attained during the process, may be assigned a certain value of a time parameter. In other words: at any given time, the object attains a certain state; the inverse is not true. Thus, the process in the Gibbs spacc is the analog of the trajectory in parameter space.[8]

Both options for the description of empirical phenomena have to be combined to achieve a picture as complete as possible: while in Gibbs space the question is answered how processes run, in parameter space it is expressed where and when the processes take place.

2. Alternative theory of nonequilibrium processes

AT has been created by Straub [6]. Its primary purpose is to tackle the problems of irreversible phenomena in thermofluiddynamics. However, in the meanwhile it has been extended to microscopic as well as to electromagnetic systems (see Refs. [6] and [7]). Its mathematical core has been applied to other quantitative branches of sciences, e.g. national economics.

The mathematical basis of the AT is the formalism first found by Gibbs for thermostatic phenomena. Here, every piece information on the system under study may be obtained by differentiation. Falk extended this concept by the insight that linear and angular momentum have to be included as members of the set of variables. This changed Gibbs' thermostatics to a real dynamics.

Additionally, it is supplemented by the important conclusion of Straub that despite the usual procedures of thermostatics and classical thermodynamics the fundamental set of variables has to be extensive (see Ref. [5]).

Four primitive, i.e. not reducible elements—*generic physical quantity, value, state, and system*—in the meaning described above constitute Gibbs–Falkian dynamics (GFD).

[7] The choice of extensive variables is one of the most fundamental rules of Gibbs–Falkian dynamics. Once the basic set of variables is identified, the complete mathematical structure of the theory may be worked out rather automatically. Any extensive variable possesses a number of properties, the simplest of which is additivity. In this context, additivity means that the values of two identical extensive variables may be added if the two objects to which the variables belong are composed to form a single object.

[8] In the language of the North-American Hopi indians the word for 'time' is 'koyaanisquatsu'. Following T. S. W. Salomon, an adequate translation would be 'state of nonequilibrium tending towards equilibrium'.

The only hypothesis necessary to use this method is the possibility to define extensive variables mapping the generics for the respective object.

It is the first step and the major task of the user to identify those variables that are important for the intended description and to neglect those that are of minor interest. Once all necessary variables are identified, everything works according to the recipe given by Gibbs and Falk. A GFR exists from which any information is to be retrieved by differentiation.

Three additional items should be stressed:

- The extensivity of the set of variables results in a GFR that is homogeneous of degree one. For objects from physics the respective factor of homogeneity is the number of particles N of the object.
- As a rule for all relevant processes in thermofluiddynamics changes below the elementary-particle level do not occur. This results in the constancy of the number of baryons and leptons and therefore in the constancy of the mass m of the object. Hence, for certain applications this system parameter m may be used as the factor of homogeneity instead of N.
- Neither GFD nor the basics of AT need conservation properties as construction elements.

3. Callen's principle

To make GFD a workable concept, an important question has to be answered. If the compilation of the generics is the primary task of the construction of the theory then what—except from the observation of exchange mechanisms—could be used to find the extensive variables for the Gibbs Fundamental Equation?

Here Callen's symmetry principle, first formulated in 1974 [8], provides a powerful tool. In Callen's words, it reads

> "The primary theorem, relating symmetry considerations to physical consequences, is Noether's theorem. According to this theorem every continuous symmetry transformation … implies the existence of a conserved function. … The most primitive class of symmetries is the class of continuous space-time transformation. The (presumed) invariance of physical laws under time translation implies energy conservation; spatial translation symmetry implies conservation of momentum. [p. 62]".

and further:

> "The symmetry interpretation of thermodynamics immediately suggests, then, that energy, linear momentum and angular momentum should play fully analogous roles in thermodynamics. The equivalence of these roles is rarely evident in conventional treatments, which appear to grant the energy a misleadingly unique status …, it is evident, that in principle, the linear momentum does appear in the formalism in a form fully equivalent to the energy, … [p. 65]".

Obviously, conserved quantities play a distinguished role for the set of generics according to Callen's principle. Variables mapping conserved generics are the privileged members of the set of coordinates. Prominent examples from physics are the total energy or the linear and angular momentum obeying respective conservation laws.

However, in open systems even the values of conserved quantities may be altered by transfer processes over the boundaries of an object. Every conserved quantity possesses another extensive quantity as a partner providing these transfers. For example, the body forces are the source of changes for the linear momentum as well as the net electrical current is for the charge of the body. These partner coordinates also have to appear in the set of variables.

Furthermore, generics exist that show broken symmetries like the volume or the entropy. Volume and entropy are born members of the set of variables: the volume is needed to define the boundaries of the body over which the transfers take place. The entropy ensures that the amount of energy needed to form a coherent body from N-many particles is considered.

Thus, Callen's principle delivers conditions enabling the identification of significant co-ordinates for the GFR.

However, since neither GFD nor AT refer to conserved quantities as construction elements a principle will be required allowing conservation qualities to be introduced by means of mathematical procedures in order to avoid such characteristics as metaphysical add-ons of both the theories.

4. Matter and forces

One of the most important facts in GFD is—according to Callen's comment noted above—Falk's observation that besides the thermodynamic variables like entropy, volume, and number of particles, the linear and angular momentum are generic members in the variable list of the GFR. This turns the method from Gibbs' thermostatics formulated in Ref. [5] to a real dynamical theory that includes mechanics with all its kinematic aspects, thermodynamics, and electrodynamics.

Let us now assume a moving single-phase multicomponent body-field system that consists of polarized matter. Its respective GFR reads

$$\Gamma(E^*, \mathbf{P}, \mathbf{F}, \mathbf{L}, \mathbf{M}, S, V, N_k, Q, I^*) \equiv 0, \tag{2}$$

where $\mathbf{E}^*$ is the total energy, $\mathbf{P}$ and $\mathbf{L}$ are the linear and angular momentum, respectively. $\mathbf{F}$ and $\mathbf{M}$ denote the body forces and momenta, S is the entropy, V stands for the volume of the body and $\mathbf{N}_k$ is the particle number of the k-th component $\left(N = \sum_k N_k\right)$. Q and I^* are the electrodynamical variables: they denote the charge of the body and the electrical current, respectively.

The GFR is a simple model for polarized matter and may be resolved to any of its variables, e.g. for the total energy:

$$E^* = \hat{E}^*(\mathbf{P}, \mathbf{F}, \mathbf{L}, \mathbf{M}, S, V, N_k, Q, I^*). \tag{3}$$

With the mass m of the body as homogeneity factor, Eq. (3) can be changed from absolute quantities to the densities of the respective variables.

$$e^* = \hat{e}^*(\mathbf{i}, \mathbf{f}, \mathbf{l}, \mathbf{m}, s, \rho, \omega_k, q, I). \tag{4}$$

The structure of $\hat{e}^*$ is unknown even for the simplest objects. Nevertheless, the total derivative may be calculated. The occurring partial derivatives all have a well-known meaning

$$\begin{aligned} \mathrm{d}e^* = & \underbrace{\frac{\partial e^*}{\partial \mathbf{i}}}_{=:\mathbf{v}} \mathrm{d}\mathbf{i} + \underbrace{\frac{\partial e^*}{\partial \mathbf{f}}}_{=:\mathbf{r}} \mathrm{d}\mathbf{f} + \underbrace{\frac{\partial e^*}{\partial \mathbf{l}}}_{=:\Omega} \mathrm{d}\mathbf{l} + \underbrace{\frac{\partial e^*}{\partial \mathbf{m}}}_{=:\alpha} \mathrm{d}\mathbf{m} + \underbrace{\frac{\partial e^*}{\partial s}}_{=:T_*} \mathrm{d}s + \underbrace{\frac{\partial e^*}{\partial \rho}}_{=:-p_*} \mathrm{d}\rho + \underbrace{\frac{\partial e^*}{\partial \omega_k}}_{=:\mu_{k^*}} \mathrm{d}\omega_k \\ & + \underbrace{\frac{\partial e^*}{\partial q}}_{=:U} \mathrm{d}q + \underbrace{\frac{\partial e^*}{\partial I}}_{=:\Xi} \mathrm{d}I \end{aligned} \tag{5}$$

with $\mathbf{v}$ and Ω as the linear and angular velocity, $\mathbf{r}$ and α for the point vector and the angular orientation. T^*, p^*, and μ_{k^*} are the thermodynamic temperature, pressure, and the chemical potential of the k-th component, respectively. U and Ξ denote the local electrical force and the magnetic flux.[9]

By a three-fold Legendre transformation, concerning **f**, **m**, **I**, Eq. (5) takes a form containing the classical laws of Newtonian mechanics and electrodynamics

$$\begin{aligned} \mathrm{d}e &:= \mathrm{d}e^{*[\mathbf{f},\mathbf{m},I]} := \mathrm{d}(e^* - \mathbf{f}\cdot\mathbf{r} - \mathbf{m}\cdot\alpha - \Xi I) \\ &= \mathbf{v}\cdot\mathrm{d}\mathbf{i} - \mathbf{f}\cdot\mathrm{d}\mathbf{r} + \Omega\cdot\mathrm{d}\mathbf{l} - \mathbf{m}\cdot\mathrm{d}\alpha + T_*\mathrm{d}s - p_*\mathrm{d}\rho + \mu_{k^*}\mathrm{d}\omega_k + U\mathrm{d}q - I\mathrm{d}\Xi. \end{aligned} \tag{6}$$

The negative signs with **f**, **m**, $p*$, and I are mere convention resulting from the Legendre transformation. They indicate how these variables are used in balance equations: flows from inside the body crossing the boundary have negative values, flows to the inside have positive ones.

Let the increments now be substituted by total time derivatives and the following conditions be set

$$\frac{\mathrm{d}e}{\mathrm{d}t} = 0; \qquad \frac{\mathrm{d}s}{\mathrm{d}t} = 0; \qquad \frac{\mathrm{d}\rho}{\mathrm{d}t} = 0; \qquad \frac{\mathrm{d}\omega_k}{\mathrm{d}t} = 0, \tag{7}$$

then Eq. (6) becomes

$$0 = \mathbf{v}\frac{\mathrm{d}\mathbf{i}}{\mathrm{d}t} - \mathbf{f}\frac{\mathrm{d}\mathbf{r}}{\mathrm{d}t} + \Omega\frac{\mathrm{d}\mathbf{l}}{\mathrm{d}t} - \mathbf{m}\frac{\mathrm{d}\alpha}{\mathrm{d}t} + U\frac{\mathrm{d}q}{\mathrm{d}t} - I\frac{\mathrm{d}\Xi}{\mathrm{d}t}. \tag{8}$$

Using the *kinematic* definitions for the 'velocities' v and Ω,

$$\frac{\partial e}{\partial \mathbf{i}} = \mathbf{v} := \frac{\mathrm{d}\mathbf{r}}{\mathrm{d}t}, \qquad \frac{\partial e}{\partial \mathbf{l}} = \Omega := \frac{\mathrm{d}\alpha}{\mathrm{d}t} \tag{9}$$

the original version of motion is represented by the left side of Eq. (9), i.e. the dynamical relation of velocity, which is also true for motions without any reference to the position vector of a body; the best-known example is the wave.

[9] The asterisk indicates variables belonging to a system in motion. Of course, the same applies for the body-force density f but may be dropped if nonrelativistic velocities are considered. For the electrical force U and the electrical current I it can be shown that they are independent of the state of motion.

A similar agreement may be arranged for electrodynamical processes:

$$\frac{\partial e}{\partial I} = U := \frac{\mathrm{d}\Xi}{\mathrm{d}t}. \tag{10}$$

These so-called 'meshing conditions' combine both options for the mathematical description of empirical phenomena. While on the left side of Eqs. (9) and (10) the intensive variables appear that were obtained in the Gibbs space formulation by differentiating the energy e with respect to their conjugate extensive variables, the right side of these equations depicts the kinematic definitions of linear and angular velocity as well as the definition for the electrical force derived in the parameter space.

It should be noted that the use of partial derivatives turns the formalism into a *local* description.

Applying the meshing conditions,

$$0 = \mathbf{v}\left(\frac{\mathrm{d}}{\mathrm{d}t}\mathbf{i} - \mathbf{f}\right) + \Omega\left(\frac{\mathrm{d}}{\mathrm{d}t}\mathbf{l} - \mathbf{m}\right) + U\left(\frac{\mathrm{d}}{\mathrm{d}t}q - I\right) \tag{11}$$

results.

This equation comprises three parts. It can only be true for all times t if all three parts are equal to zero simultaneously because $\mathbf{i}$, $\mathbf{l}$, and q are independent variables.

This leads to the three well-known theorems

$$\frac{\mathrm{d}\mathbf{i}}{\mathrm{d}t} = \mathbf{f}, \qquad \frac{\mathrm{d}\mathbf{l}}{\mathrm{d}t} = \mathbf{m}, \qquad \frac{\mathrm{d}q}{\mathrm{d}t} = I \tag{12}$$

stating that either of the three conserved quantities linear momentum, angular momentum, and charge can only be changed by the flux of the respective quantities body force, moment, and electrical current flowing over the boundaries of the body. Theorems (12) are well known from classical mechanics, especially in its Hamiltonian formulation. The classical theory of mechanics is therefore contained in GFD as a special case.

However, it should be well noted that according to the assumptions (7) they only hold for isoenergetic and isentropic processes without chemical reactions where the density remains unchanged. Evidently, these processes may—if possible at all—only be established under very intricate conditions.

5. Hamiltonian theory and the principle of least action

Eqs. (9), (10), and (12) resemble the results of Hamiltonian theory from classical mechanics. Here, a set of 2s-many mutually independent variables x_i and p_i, $i = 1, 2, \ldots s$, called the canonical variables position and momentum for each of the s-many particles under study are used to describe the dynamics by a set of 2s-many coupled first-order partial differential equations of the form

$$\frac{\mathrm{d}x_i}{\mathrm{d}t} = \frac{\partial H}{\partial p_i}; \qquad \frac{\mathrm{d}p_i}{\mathrm{d}t} = \frac{\partial H}{\partial x_i}; \qquad i = 1, 2, \ldots, s. \tag{13}$$

Eqs. (13) are termed the canonical equations of motion.

The Hamilton function H represents the total energy of the particles and t is the time parameter. The evolution in time of any property of the particles represented by a variable F may be expressed with the aid of the Hamiltonian H:

$$F = \hat{F}(x_1, x_2, \ldots, x_s, p_1, p_2, \ldots, p_s) \Rightarrow \frac{\mathrm{d}F}{\mathrm{d}t} = \sum_{i=1}^{s} \left(\frac{\partial F}{\partial x_i} \frac{\mathrm{d}x_i}{\mathrm{d}t} + \frac{\partial F}{\partial p_i} \frac{\mathrm{d}p_i}{\mathrm{d}t} \right)$$

$$= \sum_{i=1}^{s} \left(\frac{\partial F}{\partial x_i} \frac{\partial H}{\partial p_i} + \frac{\partial F}{\partial p_i} \frac{\partial H}{\partial x_i} \right). \tag{14}$$

The expression on the right side of Eq. (14) is identified with the famous *Poisson bracket* by definition

$$\frac{\mathrm{d}F}{\mathrm{d}t} =: [F, H]. \tag{15}$$

Substituting the property F by H itself then with

$$[H, H] = \frac{\mathrm{d}H}{\mathrm{d}t} = 0 \tag{16}$$

a conservation law results stating that the total energy is a conserved quantity. Condition (16) shows a strong connection between the time t and the total energy given by H.[10]

Regarding the definitions from mechanics

$$\mathbf{v} := \frac{\mathrm{d}\mathbf{x}}{\mathrm{d}t}, \qquad \mathbf{f} := \frac{\mathrm{d}\mathbf{p}}{\mathrm{d}t} \tag{17}$$

it is evident that conditions (9), (10), and (12) are analogous to the canonical equations of motion. Going even further, one can say that they are an extension of (what I call) the Hamilton mechanism and that therefore the Gibbs–Falkian expression for the energy (6) is an extension to the Hamilton function of classical mechanics. This clarifies the boundary conditions under which the conservation law (16) holds: it is only valid for isoenergetic and isentropic processes without chemical reactions where the density remains unchanged.[11] In other words: the Hamiltonian H represents only a strongly restricted expression for the energy.[12]

Taking a close look at the Hamilton mechanism, the canonical equations of motion may be distinguished into two kinds of relations: $(13)_1$ is a mere definition for the kinematic velocity, while $(13)_2$ is a theorem stating how the momentum of a moving body may

---

[10] Equation (13) contains some more conservation laws: Provided that H is independent of one of the canonical variables, e.g. $\mathbf{x}_i$, then by Eq. $(13)_2$ it is evident that the momentum $\mathbf{p}_i$ is a conserved quantity.

[11] These boundary conditions apply not only for the conservation law (16) but also for the whole calculus of classical mechanics. Therefore, real processes with irreversible changes of entropy cannot be treated by the means of classical mechanics.

[12] A far-reaching conclusion concerning quantum mechanics appears. There, the *Hamilton* operator instead of the Hamilton function is used; its eigenvalues represent the values of the total energy of the quantum system. Substituting the Hamilton operator by a Gibbs–Falkian total-energy operator then dissipation and irreversibility are smoothly inserted in quantum theory. This might give an answer to the paradox how irreversible phenomena on the meso-and macrolevel may result from a dissipation-free microlevel where only reversible processes take place. The statistics resulting from such an ansatz has been treated by the author, see Ref. [7].

be changed. The same applies for the extension: Eqs. (9) and (10) are definitions or so-called meshing conditions, expressions (12) are theorems. It is most important to recall again that even though the definitions hold unconditionally, the theorems are only valid for reversible processes. The Hamilton mechanism is thus not applicable for real processes involving irreversible phenomena.

6. The least, the best, and other variations

One of the most important principles in classical mechanics is PLA. It is part of a long history of searching for basic principles and their appropriate mathematical formulation.[13] As already emphasized by Planck, its validity is not restricted to mechanical phenomena only but applies to thermal and electrodynamic events as well (see Ref. [9]).

The most popular mathematical formulation of PLA was found by Sir William Rowan Hamilton by the aid of the calculus of variations.

He raised the question of how an optimal trajectory of a moving body can be distinguished from the infinitely many (virtual) others and solved this problem by optimizing an integral over the Lagrangian of the body

$$\int_{t_1}^{t_2} L(\dot{\mathbf{r}}(t), \mathbf{r}(t))\mathrm{d}t \rightarrow \text{opt.!} \tag{18}$$

with the Lagrangian L defined as the difference between the kinetic energy and the potential of the body forces or—in the terms used here—as the Legendre transform of the Hamilton function H

$$L = H - \mathbf{v}\cdot\mathbf{P}. \tag{19}$$

In the language of the calculus of variations this reads

$$\delta\int_{t_1}^{t_2} L(\dot{\mathbf{r}}(t), \mathbf{r}(t))\mathrm{d}t = \int_{t_1}^{t_2} \delta L(\dot{\mathbf{r}}(t), \mathbf{r}(t))\mathrm{d}t = 0, \tag{20}$$

and is called Hamilton's principle (see Ref. [10]).

In the light of the considerations of the sections above and with regard to the remark of Planck it seems to be necessary to revisit the question of which quantity has to be optimized.

Analyzing Eq. (19) one recognizes that the Lagrangian L is a difference between H and the product of the conserved quantity P and its conjugate intensive variable $\mathbf{v}$. In other words: H is diminished by the energy form of the quantity P for which a meshing condition (i.e. Eq. $(17)_1$ exists. In classical mechanics, where only mass points and forces exist H stands for the total energy of the system of n-many particles.

[13] A more metaphysical formulation has been given by G.W. Leibniz according to which from all wicked worlds one can think of the one we live in is still the best.

Let us now consider the body-field system from Section 5 and its Legendre-transformed energy E with the respective total differential

$$\mathrm{d}E = \mathbf{v}{\cdot}\mathrm{d}\mathbf{P} - \mathbf{F}{\cdot}\mathrm{d}\mathbf{r} + \boldsymbol{\Omega}{\cdot}\mathrm{d}\mathbf{L} - \mathbf{M}{\cdot}\mathrm{d}\alpha + T_*\mathrm{d}S - p_*\mathrm{d}V + \mu_{k^*}\mathrm{d}N_k + U\mathrm{d}\boldsymbol{Q} - I^*\mathrm{d}\Xi. \tag{21}$$

Forming the analogy to the Lagrangian (19) it is obvious that the energy E has to be diminished by the energy forms of **P**, **L**, and **Q** for which meshing conditions exist, i.e. the linear and angular momentum as well as the charge. This is done by the respective three-fold Legendre transformation

$$E^{[\mathbf{P},\mathbf{L},\mathbf{Q}]} := E - \frac{\partial E}{\partial \mathbf{P}}{\cdot}\mathbf{P} - \frac{\partial E}{\partial \mathbf{L}}{\cdot}\mathbf{L} - \frac{\partial E}{\partial \mathbf{Q}}\mathbf{Q} = E - \mathbf{v}{\cdot}\mathbf{P} - \boldsymbol{\Omega}{\cdot}\mathbf{L} - U\mathbf{Q}. \tag{22}$$

Evidently, $E^{[\mathbf{P},\mathbf{L},\mathbf{Q}]}$ is an extension of the classical Lagrangian. It depicts the remaining part of the total energy, provided the energy forms of the generics **P**, **L**, and **Q** are subtracted, in other words: $E^{[\mathbf{P},\mathbf{L},\mathbf{Q}]}$ stands for all energy forms where dissipation is involved.

By the aid of this transformation, Eq. (21) changes to

$$\begin{aligned}\mathrm{d}E^{[\mathbf{P},\mathbf{L},\mathbf{Q}]} &= -\mathbf{F}{\cdot}d\mathbf{r} - \mathbf{P}{\cdot}\mathrm{d}\mathbf{v} - \mathbf{M}{\cdot}\mathrm{d}\alpha - \mathbf{L}{\cdot}\mathrm{d}\boldsymbol{\Omega} - I\mathrm{d}\Xi - \mathbf{Q}\mathrm{d}U \\ &\quad + T_*\mathrm{d}S - p_*\mathrm{d}V + \sum_k \mu_{k*}\mathrm{d}N_k,\end{aligned} \tag{23}$$

with

$$\hat{E}^{[\mathbf{P},\mathbf{L},\mathbf{Q}]} = \hat{E}^{[\mathbf{P},\mathbf{L},\mathbf{Q}]}(\mathbf{r}, \dot{\mathbf{r}}, \alpha, \dot{\alpha}, \Xi, \dot{\Xi}, S, V, N_k). \tag{24}$$

This function is far more complex than the Lagrangian L and includes the variables needed to describe irreversible phenomena of real processes. From its variable list it is clear that we have to consider not only the position vector and the velocity but the angular position and velocity as well as the magnetic flux and its time derivative, the entropy, the volume, and the particle numbers.

The use of the energy-like function (24) assures the compatibility of Gibbs space and parameter space that has been forced for the first time by using the meshing conditions (9) and (10).

Some remarks should be made concerning the nature of $E^{[\mathbf{P},\mathbf{L},\mathbf{Q}]}$: Its total differential (23) contains the last three energy forms constituting the total energy E^* of the system in question —(3)—by the generics S, V, and N_k. All other original energy forms vanished by Legendre transformations. Two of those last three energy forms concerning the generics S and N_k along with their conjugates T_* and μ_* can be represented by respective balances each resulting in a flux-density and production-density term. Both types of terms manifest pure dissipation effects. Additionally, the other product terms of Eq. (23) stand for dissipative effects that can be put together by algebraic manipulations in such a manner that they arise as parts of the system's two momentum balances for the linear and angular momentum. The remaining electromagnetic influences described by U and Ξ—the local electrical force and the magnetic flux—together with their conjugates Q and I—the body charge and the electrical current, respectively—also prove to be purely dissipative. A cumbersome analysis reveals the way to assign these different electromagnetic effects of irreversibility to the flux-density and production-density terms as well as to the momentum balances mentioned above.[14]

This confirms the statement that $E^{[\mathbf{P},\mathbf{L},\mathbf{Q}]}$ represents the dissipation involved in the process.

From the way expression (24) has been derived it is only logical to transform Hamilton's action integral into

$$J_* = \int_{t_0}^{t_1} \hat{E}^{[\mathbf{P},\mathbf{L},\mathbf{Q}]}(\mathbf{r}, \dot{\mathbf{r}}, \alpha, \dot{\alpha}, \Xi, \dot{\Xi}, S, V, N_k)\mathrm{d}t, \tag{25}$$

and to extend the PLA stating that this integral has to be minimized

$$\delta J_* = \delta \int_{t_0}^{t_1} \hat{E}^{[\mathbf{P},\mathbf{L},\mathbf{Q}]}(\mathbf{r}, \dot{\mathbf{r}}, \alpha, \dot{\alpha}, \Xi, \dot{\Xi}, S, V, N_k)\mathrm{d}t = 0. \tag{26}$$

This leads to a far-reaching new interpretation for the physical meaning of the PLA:

> "Without additional constraints any irreversible process in nature runs in such a way that the dissipation is minimized during the process."[15]

By this inference, Leibniz' famous thesis may be transcended from a purely mechanistic picture of the world to a realistic one: Because irreversibility and dissipation are unavoidable in real processes, the best way possible is to run them with least dissipation. Because this interpretation is not limited to stationary processes it generalizes Prigogine's statement that for stationary irreversible processes the entropy production is a minimum (see Ref. [11]).

Following the classical chain of thought the transformation characteristics of this action integral will be explored to find the symmetries and their properties. For this purpose the coordinates of space, time, angular position, and magnetic flux may now be subject to continuous transformations with r-many parameters of the form

$$\begin{aligned} &\bar{t} := \hat{Y}_t(\mathbf{r}, \alpha, \Xi, t, \varpi), \qquad \bar{r}^i := \hat{Y}^i_r(\mathbf{r}, \alpha, \Xi, t, \varpi), \qquad \bar{\alpha}^i := \hat{Y}^i_\alpha(\mathbf{r}, \alpha, \Xi, t, \varpi), \\ &\bar{\Xi} := \hat{Y}_\Xi(\mathbf{r}, \alpha, \Xi, t, \varpi), \end{aligned} \tag{27}$$

where the vector ϖ contains the r-many parameters of the transformation.

The integral (25) is said to be divergence invariant under the transformations (27) if the condition

$$\begin{aligned} &\int_{t_0}^{t_1} \hat{E}^{[\mathbf{P},\mathbf{L},\mathbf{Q}]}(\bar{\mathbf{r}}, \dot{\bar{\mathbf{r}}}, \bar{\alpha}, \dot{\bar{\alpha}}, \bar{\Xi}, \dot{\bar{\Xi}}, \bar{S}, \bar{V}, \bar{N}_k)\mathrm{d}\bar{t} - \int_{t_0}^{t_1} \hat{E}^{[\mathbf{P},\mathbf{L},\mathbf{Q}]}(\mathbf{r}, \dot{\mathbf{r}}, \alpha, \dot{\alpha}, \Xi, \dot{\Xi}, S, V, N_k)\mathrm{d}t \\ &\quad = \varpi^\nu \vartheta_\nu(\mathbf{r}, \alpha, \Xi, S, V, N_k) + O(\varpi) \end{aligned} \tag{28}$$

holds. The functions $\vartheta_\nu(\mathbf{r}, \alpha, \Xi, S, V, N_k)$ are called gauge functions.

---

[14] The mathematical procedure of how this could be done will not be shown here but may be looked up in Ref. [6], p. 170–174.

[15] Actually, the action of the irreversible parts of the total energy is subject to optimization.

Condition (28) may be formulated not only for the integrals but for the integrands as well. In its local form the invariance condition reads

$$\hat{\bar{E}}^{[\mathbf{P,L,Q}]}(\bar{\mathbf{r}}, \dot{\bar{\mathbf{r}}}, \bar{\alpha}, \dot{\bar{\alpha}}, \bar{\Xi}, \dot{\bar{\Xi}}, \bar{S}, \bar{V}, \bar{N}_k)\frac{\mathrm{d}\bar{t}}{\mathrm{d}t} - \hat{E}^{[\mathbf{P,L,Q}]}(\mathbf{r}, \dot{\mathbf{r}}, \alpha, \dot{\alpha}, \Xi, \dot{\Xi}, S, V, N_k)$$

$$= \varpi^{\nu}\frac{\mathrm{d}\vartheta_{\nu}}{\mathrm{d}t}(\mathbf{r}, \alpha, \Xi, S, V, N_k) + O(\varpi). \tag{29}$$

For further explorations, the transformation functions are expanded into series around $\varpi = 0$

$$\begin{aligned}
\bar{t} &= t + \varpi^{\nu}\tau_{\nu}(\mathbf{r}, \alpha, \Xi, t, \varpi) + \mathrm{HOT}(\varpi),\\
\bar{r}^{i}(\bar{t}) &= r^{i}(t) + \varpi^{\nu}\xi^{i}_{\nu}(\mathbf{r}, \alpha, \Xi, t, \varpi) + \mathrm{HOT}(\varpi),\\
\bar{\alpha}^{i}(\bar{t}) &= \alpha^{i}(t) + \varpi^{\nu}\lambda^{i}_{\nu}(\mathbf{r}, \alpha, \Xi, t, \varpi) + \mathrm{HOT}(\varpi),\\
\bar{\Xi}(\bar{t}) &= \Xi(t) + \varpi^{\nu}\eta_{\nu}(\mathbf{r}, \alpha, \Xi, t, \varpi) + \mathrm{HOT}(\varpi),
\end{aligned} \tag{30}$$

with the infinitesimal generators defined by

$$\begin{aligned}
\tau_{\nu}(\mathbf{r}, \alpha, \Xi, t, \varpi) &:= \frac{\partial Y_t}{\partial \varpi^{\nu}}(\mathbf{r}, \alpha, \Xi, t, \varpi)\Big|_{\varpi=0}\\
\xi^{i}_{\nu}(\mathbf{r}, \alpha, \Xi, t, \varpi) &:= \frac{\partial Y^{i}_{\mathbf{r}}}{\partial \varpi^{\nu}}(\mathbf{r}, \alpha, \Xi, t, \varpi)\Big|_{\varpi=0}\\
\lambda^{i}_{\nu}(\mathbf{r}, \alpha, \Xi, t, \varpi) &:= \frac{\partial Y^{i}_{\alpha}}{\partial \varpi^{\nu}}(\mathbf{r}, \alpha, \Xi, t, \varpi)\Big|_{\varpi=0}\\
\eta_{\nu}(\mathbf{r}, \alpha, \Xi, t, \varpi) &:= \frac{\partial Y_t}{\partial \varpi^{\nu}}(\mathbf{r}, \alpha, \Xi, t, \varpi)\Big|_{\varpi=0}.
\end{aligned} \tag{31}$$

The following steps correspond with the standard procedure for the derivation of Emmy Noether's famous theorem:

- The invariance condition (29) will be differentiated with respect to the transformation parameters to derive relations assuring the invariance; exploration at $\varpi = 0$ will suffice.
- Derivation of the necessary differentials, inclusion into the invariance condition, and
- Investigation of the necessary condition for a minimum of the action integral.

After some lengthy algebraic manipulations one arrives at the extended *fundamental invariance identity*

$$\left[\frac{\partial E^{[\mathbf{P,L,Q}]}}{\partial r_i} - \frac{\mathrm{d}}{\mathrm{d}t}\frac{\partial E^{[\mathbf{P,L,Q}]}}{\partial \dot{r}_i} + T_*\frac{\partial S}{\partial r_i} - p_*\frac{\partial V}{\partial r_i} + \mu_*\frac{\partial N_k}{\partial r_i}\right]\xi^{i}_{\nu}$$

$$+\left[\frac{\partial E^{[\mathbf{P,L,Q}]}}{\partial \alpha_i} - \frac{\mathrm{d}}{\mathrm{d}t}\frac{\partial E^{[\mathbf{P,L,Q}]}}{\partial \dot{\alpha}_i} + T_*\frac{\partial S}{\partial \alpha_i} - p_*\frac{\partial V}{\partial \alpha_i} + \mu_*\frac{\partial N_k}{\partial \alpha_i}\right]\lambda^{i}_{\nu}$$

$$
+\left[\frac{\partial E^{[\mathbf{P},\mathbf{L},\mathbf{Q}]}}{\partial \Xi_i} - \frac{\mathrm{d}}{\mathrm{d}t}\frac{\partial E^{[\mathbf{P},\mathbf{L},\mathbf{Q}]}}{\partial \dot{\Xi}_i} + T_*\frac{\partial S}{\partial \Xi_i} - p_*\frac{\partial V}{\partial \Xi_i} + \mu_*\frac{\partial N_k}{\partial \Xi_i}\right]\eta^i_\nu
$$

$$
-\left[\frac{\partial E^{[\mathbf{P},\mathbf{L},\mathbf{Q}]}}{\partial \mathbf{r}_i} - \frac{\mathrm{d}}{\mathrm{d}t}\frac{\partial E^{[\mathbf{P},\mathbf{L},\mathbf{Q}]}}{\partial \dot{\mathbf{r}}_i} + T_*\frac{\partial S}{\partial \mathbf{r}_i} - p_*\frac{\partial V}{\partial \mathbf{r}_i} + \mu_*\frac{\partial N_k}{\partial \mathbf{r}_i}\right]\dot{\mathbf{r}}^i\tau_\nu
$$

$$
-\left[\frac{\partial E^{[\mathbf{P},\mathbf{L},\mathbf{Q}]}}{\partial \alpha_i} - \frac{\mathrm{d}}{\mathrm{d}t}\frac{\partial E^{[\mathbf{P},\mathbf{L},\mathbf{Q}]}}{\partial \dot{\alpha}_i} + T_*\frac{\partial S}{\partial \alpha_i} - p_*\frac{\partial V}{\partial \alpha_i} + \mu_*\frac{\partial N_k}{\partial \alpha_i}\right]\dot{\alpha}^i\tau_\nu
$$

$$
-\left[\frac{\partial E^{[\mathbf{P},\mathbf{L},\mathbf{Q}]}}{\partial \Xi_i} - \frac{\mathrm{d}}{\mathrm{d}t}\frac{\partial E^{[\mathbf{P},\mathbf{L},\mathbf{Q}]}}{\partial \dot{\Xi}_i} + T_*\frac{\partial S}{\partial \Xi_i} - p_*\frac{\partial V}{\partial \Xi_i} + \mu_*\frac{\partial N_k}{\partial \Xi_i}\right]\dot{\Xi}\tau_\nu
$$

$$
+\frac{\mathrm{d}}{\mathrm{d}t}\left\{\left[E^{[\mathbf{P},\mathbf{L},\mathbf{Q}]} - \frac{\partial E^{[\mathbf{P},\mathbf{L},\mathbf{Q}]}}{\partial \dot{r}_i}\dot{r}^i - \frac{\partial E^{[\mathbf{P},\mathbf{L},\mathbf{Q}]}}{\partial \dot{\alpha}_i}\dot{\alpha}^i - \frac{\partial E^{[\mathbf{P},\mathbf{L},\mathbf{Q}]}}{\partial \dot{\Xi}}\dot{\Xi}\right]\tau_\nu\right.
$$

$$
\left.+\frac{\partial E^{[\mathbf{P},\mathbf{L},\mathbf{Q}]}}{\partial \dot{r}_i}\xi^i_\nu + \frac{\partial E^{[\mathbf{P},\mathbf{L},\mathbf{Q}]}}{\partial \dot{\alpha}_i}\lambda^i_\nu + \frac{\partial E^{[\mathbf{P},\mathbf{L},\mathbf{Q}]}}{\partial \dot{\Xi}}\eta_\nu - \vartheta_\nu\right\} = 0. \tag{32}
$$

Together with the abbreviation

$$
\frac{\partial E^{[\mathbf{P},\mathbf{L},\mathbf{Q}]}}{\partial z} - \frac{\mathrm{d}}{\mathrm{d}t}\frac{\partial E^{[\mathbf{P},\mathbf{L},\mathbf{Q}]}}{\partial \dot{z}} + T_*\frac{\partial S}{\partial z} - p_*\frac{\partial V}{\partial z} + \mu_*\frac{\partial N_k}{\partial z} =: E_{EL,z},
$$
$$
z \in \{r_i, \alpha_i, \Xi\}, \tag{33}
$$

and the inverse transformation of (22) finally Noether's invariance is obtained

$$
E_{EL,r_i}(\xi^i_\nu - \dot{r}^i\tau_\nu) + E_{EL,\alpha_i}(\lambda^i_\nu - \dot{\alpha}^i\tau_\nu)
$$
$$
+ E_{EL,\Xi}(\eta_i - \dot{\Xi}\tau_\nu) + \frac{\mathrm{d}}{\mathrm{d}t}\left(E\tau_\nu - P_i\xi^i_\nu - L_i\lambda^i_\nu - Q\eta_\nu - \vartheta_\nu\right) = 0. \tag{34}
$$

To fulfill the necessary conditions of the PLA for a minimum of the action integral, the Euler–Lagrange equations of the respective variational problem have to vanish identically. As can easily be calculated from Eq. (26), the expressions $E_{EL,z}$ from abbreviation (33) deliver the equations in question

$$
E_{EL,z} \overset{!}{=} 0; \qquad z \in \{r_i, \alpha_i, \Xi\}. \tag{35}
$$

This leads to an extended version of the celebrated theorem the German mathematician Emmy Noether gave in 1918

$$
E\tau_\nu - P_i\xi^i_\nu - L_i\lambda^i_\nu - Q\eta_\nu - \vartheta_\nu = \text{const.} \tag{36}
$$

To derive the final conservation laws it makes sense to distribute the scalar gauge functions according to

$$
\vartheta^{\mathbf{P}}_\nu + \vartheta^{\mathbf{L}}_\nu + \vartheta^{\mathbf{Q}}_\nu = \vartheta_\nu. \tag{37}
$$

Then, reordering of Eq. (36) results in the identity

$$E\tau_\nu - (-P_i\xi^i_\nu - \vartheta^{\mathbf{P}}_\nu) + (-L_i\lambda^i_\nu - \vartheta^{\mathbf{L}}_\nu) + (-Q\eta_\nu - \vartheta^{\mathbf{Q}}_\nu) - \text{const.} = 0. \tag{38}$$

To obtain the conservation laws first the role of the gauge functions is specified by means of allocating them to the corresponding field generics

$$\vartheta^{\mathbf{P}}_\nu := P^{\text{field}}_i \xi^i_\nu, \qquad \vartheta^{\mathbf{L}}_\nu := L^{\text{field}}_i \lambda^i_\nu, \qquad \vartheta^{\mathbf{Q}}_\nu := \mathbf{Q}^{\text{em-field}}\eta_\nu, \tag{39}$$

which are equivalent to the claim that corresponding 'fields' exist.

Because **P**, **L**, and **Q** are independent as are the infinitesimal generators, the four brackets may be treated independently; this is achieved by setting three of the four infinitesimal generators to zero at a time. Then, four equations are obtained

$$E\tau_\nu = \text{const.} \tag{40}$$

$$\begin{aligned} -\xi^i_\nu(P_i + P^{\text{field}}_i) &= \text{const.} \\ -\lambda^i_\nu(L_i + L^{\text{field}}_i) &= \text{const.} \qquad i = 1, 2, 3. \\ -\eta_\nu(Q + Q^{\text{em-field}}) &= \text{const.} \end{aligned} \tag{41}$$

The first conservation law is now readily at hand. For $\tau_\nu = 1$, i.e. an affine transformation of the time coordinate,

$$E = \text{const.} \tag{42}$$

immediately follows. Eq. (42) yields a striking result:

(1) Even for irreversible processes the conservation of the three-fold Legendre-transformed energy E can be shown.

(2) The property to be conserved is directly coupled to the use of an affine time coordinate, i.e. to a coordinate whose origin may be shifted arbitrarily.

Statement 1 indicates a certain restriction: energy conservation in this context is only a local property and may not be generalized unconditionally. The arbitrariness of the origin of time thus prevents the use of the energy conservation law (in this local form) in connection with nonlocal phenomena.[16]

From Eqs. (43) three additional conservation laws are obtained

$$\begin{aligned} P_i + P^{\text{field}}_i &= \text{const.} \\ L_i + L^{\text{field}}_i &= \text{const.} \quad i = 1, 2, 3, \\ Q + Q^{\text{em-field}} &= \text{const.} \end{aligned} \tag{43}$$

if the respective infinitesimal generator is a constant, i.e. the corresponding transformation is affine.

The interpretation of Eq. (43) is quite clear: linear and angular momentum are conserved generics of the system embedded in the assigned inertial field. In other words: the conservation is realized via the *inertial field*. Any change in the linear and

[16] On the other hand no formalism is at hand to show the global conservation of energy. This leads to severe problems in the foundations of cosmological theories like the 'Big Bang' theories.

angular momentum of the body has to be compensated by a respective change of the inertial field. The same applies for the conservation of the charge. Here, the electromagnetic field is the counterpart for the body. The conservation of linear momentum is coupled to affine spatial coordinates; the conservation of angular momentum requires an affine angular orientation of the coordinate system. For the conservation of charge, the zero point of the magnetic flux may be chosen arbitrarily.

Regarding these results an important comment is required: By means of Legendre transfomation a generalized Lagrange function L was constructed by subtracting from the (conserved) energy the energy forms of some generics assumed to be appropriate for the construction of a net of adequate coordinates of the parameter space. By a lengthy mathematical procedure a theorem was derived proving the conservation of exactly these generics used for the construction.

Has only something been confirmed that has been inserted as supposition?

The answer to this question is quite clear by taking a close look at the outcome of the PLA in connection with the extended version of Noether's theorem: The starting point, the function L, is defined by means of GFD and AT, respectively. For this reason in no way were conserved quantities used for the definition of L a priori.

PLA, along with Noether's theorem, is the tool to make some generics conservative following from an optimizing principle by mathematical construction. Thus, the usual introduction of conservative qualities by traditional metaphysics may be avoided. Moreover, the mathematical procedure delivers the instructions how the correct projection parameters, i.e. the coordinates for space and time and others have to be chosen. This is the main result of the PLA in connection with the Alternative Theory.

Space and time result from the choice of the assigned generics and the complete set of extensive variables. The latter are the real primitives, i.e. the irreducible elements of the theory. Callen's principle opens the possibility to define this set and to identify the conserved quantities. In a second step the correct projection coordinates may be calculated via PLA and Noether's theorem. Thus, in contrast to experience and historical development the extensive variables are the primary categories while space and time are mere conventions following from the calculus.

7. The principle of least action and the natural trend to equilibrium

In thermodynamics one of the most basic empirical facts is the observation that closed systems, i.e. isolated objects that do not exchange either matter or energy with their surroundings, tend to equalize the values of their intensive variables like temperature, pressure, and chemical potential over the whole section of space they occupy. Often, the closure condition is supplemented by other constraints, like, e.g. the inhibition of chemical reaction within the area of the body. The existence of gradients of extensive variables in the spatial region of the body may cause the existence of hampered equilibria. However, the entropy is always increasing its values during the time the closed system is tending towards its equilibrium state.

With the aid of the formalism presented above it is easy to show that the natural trend to equilibrium is contained in the PLA. The principle describes an optimal process running from a state at t_0 to a final state at t_1. For the equilibrium, the conditions for the final state are completely known in so far as the extensive variables attain arbitrary but fixed values. The variation needed for the PLA will be done around this final state.

The starting point is the three-fold Legendre-transformed total energy of the system according to Eq. (22) and the variation of its action integral (26).

$$\delta J_* = \delta \int_{t_0}^{t_1} E^{[\mathbf{P},\mathbf{L},\mathbf{Q}]} \mathrm{d}t = \int_{t_0}^{t_1} \delta E^{[\mathbf{P},\mathbf{L},\mathbf{Q}]} \mathrm{d}t$$
$$= \int_{t_0}^{t_1} [-\mathbf{F}\cdot\delta\mathbf{r} - \mathbf{P}\cdot\delta\mathbf{v} - \mathbf{M}\cdot\delta\alpha - \mathbf{L}\cdot\delta\Omega - I\delta\Xi - \mathbf{Q}\delta U$$
$$+ T_*\delta S - p_*\delta V + \sum_{k\mu} {}_{k*}\delta N_k]\mathrm{d}t. \tag{44}$$

The final state of the system is realized by the set of conditions[17]

$$\mathbf{P} = \text{const.};\ \mathbf{L} = \text{const.};\ \mathbf{Q} = \text{const.};\ S = \text{const.};\ V = \text{const.};\ N_k = \text{const.}, \tag{45}$$

stating that the values for the extensive variables may not change any more and that they are equally distributed over the complete spatial region of the body.[18]

According to the Lagrangian calculus, a function is constructed by multiplying restrictions (45) by their respective Lagrange multipliers and adding them to the function to be varied. The new functional then reads

$$\delta J_{**} = \int_{t_0}^{t_1} \{\delta E^{[\mathbf{P},\mathbf{L},\mathbf{Q}]} + \lambda_{\mathbf{P}}\delta\mathbf{P} + \lambda_{\mathbf{L}}\delta\mathbf{L} + \lambda_{\mathbf{Q}}\delta\mathbf{Q} + \lambda_S\delta S + \lambda_V\delta V + \lambda_{N_k}\delta N_k\}\mathrm{d}t$$
$$= \int_{t_0}^{t_1} \Bigg\{\Bigg[-\mathbf{F}\cdot\delta\mathbf{r} - \mathbf{P}\cdot\delta\mathbf{v} - \mathbf{M}\cdot\delta\alpha - \mathbf{L}\cdot\delta\Omega - I\delta\Xi - \mathbf{Q}\delta U + T_*\delta S - p_*\delta V$$
$$+ \sum_k \mu_{k*}\delta N_k\Bigg] + \lambda_{\mathbf{P}}\delta\mathbf{P} + \lambda_{\mathbf{L}}\delta\mathbf{L} + \lambda_{\mathbf{Q}}\delta\mathbf{Q} + \lambda_S\delta S + \lambda_V\delta V + \lambda_{N_k}\delta N_k\Bigg\}\mathrm{d}t. \tag{46}$$

Regarding restrictions (45), the conditions (7) for the application of the canonical equations are satisfied. Additionally, variation and differentiation may be commuted

$$\delta\mathbf{v} = \delta\frac{\mathrm{d}\mathbf{x}}{\mathrm{d}t} = \frac{\mathrm{d}}{\mathrm{d}t}\delta\mathbf{x}; \qquad \delta\mathbf{F} = \delta\frac{\mathrm{d}\mathbf{P}}{\mathrm{d}t} = \frac{\mathrm{d}}{\mathrm{d}t}\delta\mathbf{P}, \qquad \delta\Omega = \delta\frac{\mathrm{d}\alpha}{\mathrm{d}t} = \frac{\mathrm{d}}{\mathrm{d}t}\delta\alpha;$$
$$\delta\mathbf{M} = \delta\frac{\mathrm{d}\mathbf{L}}{\mathrm{d}t} = \frac{\mathrm{d}}{\mathrm{d}t}\delta\mathbf{L}, \qquad \delta U = \delta\frac{\mathrm{d}\Xi}{\mathrm{d}t} = \frac{\mathrm{d}}{\mathrm{d}t}\delta\Xi; \qquad \delta I = \delta\frac{\mathrm{d}\mathbf{Q}}{\mathrm{d}t} = \frac{\mathrm{d}}{\mathrm{d}t}\delta\mathbf{Q}. \tag{47}$$

[17] These conditions refer to the final state; it goes without saying that during the process from t_0 to t_1 the extensive variables, especially the entropy S, may change their values.

[18] Conditions for the position vector, the angular position, and the magnetic flux will result by applying the PLA.

Thus, functional (46) finally reads

$$\delta J_{**} = \int_{t_0}^{t_1} \left\{ -\left(\mathbf{F} - \frac{\mathrm{d}\mathbf{P}}{\mathrm{d}t}\right)\cdot\delta\mathbf{r} - \left(\frac{\mathrm{d}}{\mathrm{d}t}\lambda_{\mathbf{F}} + \lambda_{\mathbf{P}}\right)\cdot\delta\mathbf{P} + \frac{\mathrm{d}}{\mathrm{d}t}[\delta(\mathbf{P}\cdot\mathbf{r}) + (\lambda_{\mathbf{F}} - \mathbf{r})\cdot\delta\mathbf{P}] \right.$$
$$- \left(\mathbf{M} - \frac{\mathrm{d}\mathbf{L}}{\mathrm{d}t}\right)\cdot\delta\alpha - \left(\frac{\mathrm{d}}{\mathrm{d}t}\lambda_{\mathbf{M}} + \lambda_{\mathbf{L}}\right)\cdot\delta\mathbf{L} + \frac{\mathrm{d}}{\mathrm{d}t}[\delta(\mathbf{L}\cdot\alpha) + (\lambda_{\mathbf{M}} - \alpha)\cdot\delta\mathbf{L}]$$
$$- \left(I - \frac{\mathrm{d}\mathbf{Q}}{\mathrm{d}t}\right)\cdot\delta\Xi - \left(\frac{\mathrm{d}}{\mathrm{d}t}\lambda_I + \lambda_{\mathbf{Q}}\right)\cdot\delta\mathbf{Q} + \frac{\mathrm{d}}{\mathrm{d}t}[\delta(\mathbf{Q}\Xi) + (\lambda_I - \Xi)\cdot\delta\mathbf{Q}]$$
$$\left. + (T_* + \lambda_S)\delta S - (p_* + \lambda_V)\delta V + \sum_k (\mu_{k^*} + \lambda_{N_k})\delta N_k \right\} \mathrm{d}t. \tag{48}$$

When the PLA is now applied, the respective condition

$$\delta J_{**} = 0 \tag{49}$$

can only be fulfilled if all terms in brackets vanish identically, because the variations of the extensive variables are arbitrary. The first bracketed terms in rows one to three of Eq. (48) are the theorems (12) and therefore equal to zero.[19] Next we find

$$\lambda_{\mathbf{F}} = \mathbf{r}(= \text{const!}), \qquad \lambda_{\mathbf{M}} = \alpha(= \text{const!}), \qquad \lambda_I = \Xi(= \text{const.!}), \tag{50}$$

and (with Eq. (50) immediately applied)

$$\frac{\mathrm{d}}{\mathrm{d}t}\lambda_{\mathbf{F}} = -\lambda_{\mathbf{P}} = 0, \qquad \frac{\mathrm{d}}{\mathrm{d}t}\lambda_{\mathbf{M}} = -\lambda_{\mathbf{L}} = 0, \qquad \frac{\mathrm{d}}{\mathrm{d}t}\lambda_I = -\lambda_{\mathbf{Q}} = 0. \tag{51}$$

Thus, the position vector, the angular position, and the magnetic flux are constant, causing velocity, angular velocity, and electrical force to be zero.

Looking at the remaining terms of the first three rows of Eq. (48), the conditions of the hypothetical state at rest result

$$\mathbf{P} \to 0 \quad (\Rightarrow \mathbf{v} \to 0!), \qquad \mathbf{L} \to 0 \quad (\Rightarrow \Omega \to 0!), \qquad \mathbf{Q} \to 0 \quad (\Rightarrow U \to 0!). \tag{52}$$

Then it is easy to see that the last row of Eq. (48) finally delivers

$$T_* \to T = -\lambda_S = \text{const.} \quad p_* \to p = -\lambda_V = \text{const.} \quad \mu_{k^*} \to \mu_{k^*} = -\lambda_S = \text{const.} \tag{53}$$

This leads to the following conclusions:

- For closed systems, the PLA states that the values of the intensive variables temperature, pressure, and chemical potential are constant throughout the whole spatial area the body occupies.[20]

[19] It should be remembered that the theorems (12) are only valid for processes under very restricting conditions. However, they are fulfilled identically if only a single state as given by conditions (45) is regarded.

[20] This may be concluded from the fact that gradients of the intensive variables cause fluxes of the respective extensive variable in combination with translational and angular velocities thus contradicting results (52).

- Equilibrium can only be achieved for the hypothetical state at rest with vanishing linear and angular momentum. Additionally, no charge is allowed for this kind of equilibrium. In other words: motion and polarization are the unerring indicators for nonequilibrium processes.
- Hampered equilibria may occur, provided one or more extensive variables show gradients over the volume of the body.

8. Summary

To characterize the differences between the description of phenomena in the parameter space constituted by spatial and time coordinates and the method used in GFD the notion of a generic physical quantity has been introduced. It is the most primitive element of the theory and may not be reduced to any other element. For the alternative theory of nonequilibrium phenomena (AT) it is most important that every generic quantity is mapped to an extensive variable, i.e. a mathematical element from the set of at least twice-differentiable functions obeying a list of requirements from which additivity is the simplest and most significant.

Neither GFD nor the basics of AT need conservation properties. In order to avoid such characteristics as metaphysical elements of both these theories an optimum principle is required allowing conservation quantities together with assigned coordinates to be introduced by means of mathematical construction.

Callen's principle on the symmetry properties of extensive variables provides a powerful tool that enables the user to identify the appropriate conserved quantities and the extensive set of variables for the problem under study.

As an example the mathematical model of a single-phase multicomponent body-field system consisting of polarized matter was introduced. It could be shown that classical mechanics in the form of Hamilton's theory is contained in the AT as a special case of idealized physics. The restricting conditions for the applicability of Hamilton mechanics were derived.

The classical form of the PLA was presented for the model noted, and the question was discussed as to which property should be subjected to optimization. The three-fold Legendre-transformed energy of the system diminished by the energy forms of adequate generics is the logical extension to Hamilton's principle. This leads to the new interpretation of the PLA that irreversible processes in nature always run on least dissipation.

The analysis of the transformation properties of the new action integral delivers striking results: space and time are not the primary categories of the physical theory but mere conventions following from the choice of the conserved quantities and the set of extensive variables guided by Callen's principle. From Noether's theorem the construction orders for the spatial and time coordinates are deduced showing that they are subject to affine transformations.

Additionally, the role of the gauge functions was specified: they characterize the influence of the inertial and the electromagnetic fields on the body.

Finally, it could be shown that the natural trend to equilibrium is contained in the PLA. By use of the Lagrange formalism it was derived that an equilibrium can only exist at the hypothetical state at rest where linear and angular momentum tend to zero. As a consequence dissipation tends to zero as well. Then, the intensive variables for temperature, pressure, and chemical potential attain respective constant values valid for the whole spatial region the body occupies.

Acknowledgement

The author wishes to thank his colleague and friend Dieter Straub for the intensive discussions on this topic.

References

[1] Kline, M. (1985), Mathematics and the Search for Knowledge, Oxford University Press, New York.
[2] Straub, D. (1996), Zeitpfeile in der Natur? Versuch einer Antwort, Sonderdruck aus: Inhetveen, R., Kötter, R. (Eds): Betrachten-Beobachten-Beschreiben, Beschreibungen in Kultur- und Naturwissenschaften, Wilhelm Fink Verlag, Munich, pp. 105–146.
[3] De Broglie, L. (1950), Physik und Mikrophysik, Claassen, Hamburg.
[4] Falk, G. (1990), Physik* Zahl und Realität—Die begrifflichen und mathematischen Grundlagen einer universellen quantitativen Naturbeschreibung: Mathematische Physik und Thermodynamik, Birkhäuser, Basel.
[5] Gibbs, J.W. (1961), On the equilibrium of heterogeneous substances. The Scientific Papers, Thermodynamics, Vol. I, Dover, New York.
[6] Straub, D. (1996), Alternative Mathematical Theory of Non-equilibrium Phenomena, Academic Press, New York.
[7] Lauster, M. (1998), Statistische Grundlagen einer allgemeinen quantitativen Systemtheorie, Shaker, Aachen.
[8] Callen, H.B. (1974), A Symmetry Interpretation of Thermodynamics. Foundation of Continuum Thermodynamics (Delgado, J.J., Nina, M.N.R., Whitelaw, J.H., eds.), Macmillan Press, London.
[9] Planck, M. (1922), Das Prinzip der kleinsten Wirkung. In Physikalische Rundblicke: Gesammelte Reden und Aufsätze, Hirzel, Leipzig, pp. 103–119.
[10] Szabó, I. (1979), Geschichte der mechanischen Prinzipien, Birkhäuser, Basel.
[11] Prigogine, I. (1988), Vom Sein zum Werden: Zeit und Komplexität in den Naturwissen-schaften; überarbeitete Neuausgabe, Piper, München.

Chapter 11

VARIATIONAL PRINCIPLES FOR THE LINEARLY DAMPED FLOW OF BAROTROPIC AND MADELUNG-TYPE FLUIDS

Heinz-Jürgen Wagner

Theoretische Physik, Universität Paderborn, Pohlweg 55, 33098 Paderborn, Germany

Abstract

The present contribution deals with the hydrodynamics of fluids that are under the influence of frictional force densities proportional to the flow velocity. Such models approximately describe the seepage of fluids in porous media (Darcy's law). One particular aim is the incorporation of such linearly damped hydrodynamic systems into the framework of Lagrangian field theory, i.e. the construction and discussion of variational principles for these systems.

It turns out that the search for variational principles greatly benefits from the fact that there exists a far-reaching analogy between the field equations of classical hydrodynamics and of the so-called hydrodynamic (Madelung) picture of quantum theory. This analogy often allows for an immediate transfer of strategies and concepts.

Introduction of linear frictional force densities into the Madelung picture leads to certain well-known phenomenological extensions of the ordinary Schrödinger equation such as the Caldirola–Kanai equation and Kostin's nonlinear Schrödinger–Langevin equation. Despite several failed attempts in the literature, the construction of variational principles for these equations turns out to be relatively straightforward. The main tool used here is the so-called method of Fréchet differentials.

Utilizing the above-mentioned analogy between hydrodynamics and the Madelung picture, the results for the dissipative quantum systems can immediately be carried over to the analogous frictional system in classical hydrodynamics. However, some additional considerations are still needed due to the possibility of rotational (vorticity) degrees of freedom that can be present in ordinary hydrodynamics but are usually foreign to the Madelung picture.

Keywords: dissipative quantum theory; Madelung fluid; hydrodynamics; Clebsch transformation; Weber transformation; Darcy's law

1. Introduction

Concerning the hydrodynamic field equations of nondissipative ideal fluid flow, it is well known that there exists a large number of Lagrangians such that the corresponding Euler–Lagrange equations lead to a correct description of the system under study. To quote only a few reviews of this field, see Refs. [1–5].

E-mail address: wagner@phys.uni-paderborn.de (H.-J. Wagner).

On the other hand, the problem of finding appropriate Lagrangians for dissipative fluid flow remains notoriously difficult. One of the cases where the search has been in vain up to now consists of the hydrodynamics of viscous fluids where one has to consider the Navier–Stokes equation instead of the Euler equation. Some negative results in this respect can be found in Refs. [6–9].

As we shall see in the following, the search turns out to be more successful for cases where internal viscous friction forces are not taken into account but where external ones—such as those for fluids in porous media—are present instead. We shall deal particularly with the example of a friction-force density being proportional to the velocity field $\mathbf{u}$.

Such force densities linear in $\mathbf{u}$ are also well known in an apparently disparate field of physics, namely in dissipative quantum theory. There exist far-reaching analogies between classical hydrodynamics and the so-called hydrodynamic or Madelung picture of quantum theory and it often turns out that strategies originally developed for quantum systems can immediately be transferred to classical hydrodynamics. In the following, we therefore start with the search for proper Lagrangians in the realm of linearly damped quantum systems. Subsequently, the results are then transferred to the corresponding systems in classical hydrodynamics.

2. Dissipative quantum theory

2.1. *Phenomenological and microscopic approaches*

The strategies for the inclusion of dissipative effects within quantum theory can roughly be divided into two subclasses. The 'phenomenological approach' tries to approximate the interaction of a particle with its environment by condensing these contributions into phenomenological (mostly one-parameter) extensions of the ordinary one-particle Schrödinger equation. Such equations have often been employed for the description of various frictional effects in quantum systems including e.g. inelastic nucleon–nucleon scattering, diffusion of interstitial impurity atoms, radiation damping, and dissipative tunneling. Surveys of the whole field can be found e.g. in Refs. [10–12].

For a more satisfactory theory of quantum dissipation one has to consider the whole system of particle and environment and subsequently try to eliminate the environmental degrees of freedom by appropriate averaging processes. We, however, shall not deal with this so-called 'microscopic approach' here. For work in this direction the reader may consult e.g. Refs. [13–15].

2.2. *Caldirola–Kanai vs. Schrödinger–Langevin*

On the level of classical mechanics, the dynamical behavior of a particle under the influence of potential and linear friction forces is governed by the Newtonian equation of motion

$$m\ddot{\mathbf{x}} = -\lambda m\dot{\mathbf{x}} - \nabla V(\mathbf{x}, t). \tag{1}$$

A variational principle for this equation of motion is inferred by Bateman's Lagrangian [16]

$$L(\dot{\mathbf{x}}, \mathbf{x}, t) = \left(\frac{m}{2}\dot{\mathbf{x}}^2 - V(\mathbf{x}, t)\right)\mathrm{e}^{\lambda t}. \tag{2}$$

The corresponding Euler–Lagrange equation

$$(m\ddot{\mathbf{x}} + \lambda m\dot{\mathbf{x}} + \nabla V(\mathbf{x},t))\mathrm{e}^{\lambda t} = 0 \tag{3}$$

is obviously equivalent to Eq. (1). A Legendre transform of Eq. (2) then leads to the Hamiltonian

$$H(\mathbf{p},\mathbf{x},t) = \frac{\mathbf{p}^2}{2m}\mathrm{e}^{-\lambda t} + V(\mathbf{x},t)\mathrm{e}^{\lambda t}, \tag{4}$$

which opens the way to canonical quantization. This well-known procedure immediately leads to the so-called Caldirola–Kanai equation [17,18]

$$\mathrm{i}\hbar\frac{\partial}{\partial t}\psi(\mathbf{x},t) = -\frac{\hbar^2}{2m}\nabla^2\psi(\mathbf{x},t)\mathrm{e}^{-\lambda t} + V(\mathbf{x},t)\psi(\mathbf{x},t)\mathrm{e}^{\lambda t}. \tag{5}$$

In order to make the physical contents of such an equation more transparent, it is advantageous to re-express it in the so-called hydrodynamic (Madelung) picture of quantum theory [19]. This is accomplished by insertion of the polar decomposition

$$\psi = \sqrt{\frac{\varrho}{m}}\exp(\mathrm{i}S/\hbar) \tag{6}$$

into the field equation and splitting into real and imaginary parts.

For the case of the Caldirola–Kanai equation (5) this procedure leads to

$$\frac{\partial}{\partial t}\varrho + \nabla\cdot\left(\varrho\frac{\nabla S}{m}\mathrm{e}^{-\lambda t}\right) = 0, \tag{7}$$

$$\frac{\partial}{\partial t}S + \frac{1}{2m}(\nabla S)^2\mathrm{e}^{-\lambda t} + V\,\mathrm{e}^{\lambda t} - \frac{\hbar^2}{2m}\frac{\nabla^2\sqrt{\varrho}}{\sqrt{\varrho}}\mathrm{e}^{-\lambda t} = 0. \tag{8}$$

Assuming that the probability flow is induced by the irrotational velocity field

$$\mathbf{u} = \frac{1}{m}\nabla S\,\mathrm{e}^{-\lambda t}, \tag{9}$$

Eq. (7) takes on the form of the usual continuity equation in hydrodynamics.

Taking the gradient on both sides of the 'Bernoulli-type' equation (8) leads to an 'Euler equation' for $\mathbf{u}$

$$\frac{\mathrm{D}}{\mathrm{D}t}\mathbf{u} = -\frac{1}{m}\nabla V - \frac{1}{m}\nabla Q\,\mathrm{e}^{-2\lambda t} - \lambda\mathbf{u}, \tag{10}$$

where we have introduced the so-called substantial or material time derivative

$$\frac{\mathrm{D}}{\mathrm{D}t} = \frac{\partial}{\partial t} + \mathbf{u}\cdot\nabla, \tag{11}$$

and the 'quantum potential'

$$Q = -\frac{\hbar^2}{2m}\frac{\nabla^2\sqrt{\varrho}}{\sqrt{\varrho}}. \tag{12}$$

Comparing these results with the damping-free ($\lambda = 0$) case, one sees that the influence of friction is two-fold here. In addition to a dissipative force density linear in $\mathbf{u}$ an exponentially fast diminishing of the quantum potential appears.

If one wants to remove the latter—rather undesirable—feature from the theory, the required equations on the hydrodynamic level are

$$\frac{\partial}{\partial t}\varrho + \nabla\cdot(\varrho\mathbf{u}) = 0, \tag{13}$$

and

$$\frac{\mathrm{D}}{\mathrm{D}t}\mathbf{u} = -\frac{1}{m}\nabla V - \frac{1}{m}\nabla Q - \lambda\mathbf{u}, \tag{14}$$

where the quantum potential now shows up in the same way as it would do in the frictionless case.

On the level of wave functions ψ, ψ^*, the latter two equations can be gained from Kostin's nonlinear 'Schrödinger–Langevin equation' [20]

$$\mathrm{i}\hbar\frac{\partial}{\partial t}\psi = -\frac{\hbar^2}{2m}\nabla^2\psi + V\psi + \lambda\frac{\mathrm{i}\hbar}{2}\ln\left(\frac{\psi^*}{\psi}\right)\psi. \tag{15}$$

This can immediately be verified by inserting the polar decomposition Eq. (6) for ψ. One ends up with

$$\frac{\partial}{\partial t}\varrho + \nabla\cdot\left(\varrho\frac{\nabla S}{m}\right) = 0, \tag{16}$$

and

$$\frac{\partial}{\partial t}S + \frac{1}{2m}(\nabla S)^2 + V - \frac{\hbar^2}{2m}\frac{\nabla^2\sqrt{\varrho}}{\sqrt{\varrho}} + \lambda S = 0. \tag{17}$$

If in contrast to Eq. (9) one puts

$$\mathbf{u} = \frac{1}{m}\nabla S, \tag{18}$$

Eqs. (13) and (14) are an immediate consequence.

Conversely, insertion of Eq. (18) into Eq. (14) implies in a first step

$$\nabla\left(\frac{\partial}{\partial t}S + \frac{1}{2m}(\nabla S)^2 + V - \frac{\hbar^2}{2m}\frac{\nabla^2\sqrt{\varrho}}{\sqrt{\varrho}} + \lambda S\right) = 0, \tag{19}$$

i.e.

$$\frac{\partial}{\partial t}S + \frac{1}{2m}(\nabla S)^2 + V - \frac{\hbar^2}{2m}\frac{\nabla^2\sqrt{\varrho}}{\sqrt{\varrho}} + \lambda S = f(t). \tag{20}$$

This equation can then be carried over to Eq. (17) by using the gauge transform

$$S \to S - e^{-\lambda t} \int_{t_0}^{t} f(t')e^{\lambda t'} \, dt', \tag{21}$$

which does not affect the velocity field $\mathbf{u}$.

It might be remarked here that the two hydrodynamic equations (13) and (14) together with the potential flow ansatz (18) are the basic equations for the so-called 'stochastic quantization' of the system with the classical equation (1) [21].

To put the above results concisely: in contrast to the damping-free ($\lambda = 0$) case where canonical and stochastic quantization coincide, we are led to different results for $\lambda \neq 0$. Both equations have conceptual advantages as well as disadvantages. The Caldirola–Kanai equation is linear and remains in the realm of usual quantum theory, but it gives rise to a problematic time-dependent quantum potential. On the other hand, Kostin's Schrödinger–Langevin equation displays the nonproblematic usual quantum potential, but the wave equation is nonlinear and the superposition principle ceases to hold.

It should be noted here that a conceptual improvement can be gained if the external potential V is allowed to also contain stochastically fluctuating contributions due to the interaction with the environment. For details see e.g. Ref. [10]; a list of some further references can be found in Ref. [22].

2.3. *Fréchet differentials and variational principles*

A rather straightforward procedure to determine whether a given set of field equations has already the form of Euler–Lagrange equations of a certain variational principle consists of the so-called 'method of Fréchet differentials' (see e.g. Section 9.1 of Ref. [3] or Section 9.7 of Ref. [23]).

Provided that

$$F_k[\phi_1, \ldots, \phi_N](\mathbf{x}, t) = 0, \qquad k = 1, \ldots, N \tag{22}$$

are indeed Euler–Lagrange equations originating from an action functional

$$A = \int \mathcal{L} d^3x \, dt, \tag{23}$$

i.e.

$$F_k[\phi_1, \ldots, \phi_N](\mathbf{x}, t) = \frac{\delta A}{\delta \phi_k(\mathbf{x}, t)}, \tag{24}$$

then an appropriate Lagrangian $\mathcal{L}$ is given by

$$\mathcal{L} = \sum_{k=1}^{N} \phi_k \int_0^1 F_k[\tau\phi_1, \ldots, \tau\phi_N] d\tau. \tag{25}$$

This Lagrangian may then often be further simplified by subtraction of a suitable total divergence term (which does not affect the Euler–Lagrange equations).

For the Caldirola–Kanai equation and its complex conjugate this method is indeed successful and leads to the correct Lagrangian

$$\mathcal{L} = \frac{i\hbar}{2}\left(\psi^* \frac{\partial}{\partial t}\psi - \psi\frac{\partial}{\partial t}\psi^*\right) - \frac{\hbar^2}{2m}\nabla\psi^* \cdot \nabla\psi\, e^{-\lambda t} - V\psi^*\psi\, e^{\lambda t}. \tag{26}$$

Note that, in order to arrive at Eq. (26), the total divergence

$$\frac{\hbar^2}{4m}\nabla\cdot(\psi^*\nabla\psi + \psi\nabla\psi^*)\, e^{-\lambda t} \tag{27}$$

had to be removed.

For the Schrödinger–Langevin equation and its complex conjugate, the situation is radically different. After subtraction of the divergence term

$$\frac{\hbar^2}{4m}\nabla\cdot(\psi^*\nabla\psi + \psi\nabla\psi^*), \tag{28}$$

the above method leads to the expression

$$`\mathcal{L}' = \frac{i\hbar}{2}\left(\psi^* \frac{\partial}{\partial t}\psi - \psi\frac{\partial}{\partial t}\psi^*\right) - \frac{\hbar^2}{2m}\nabla\psi^* \cdot \nabla\psi - V\psi^*\psi - \frac{i\hbar}{2}\lambda\psi^*\psi \ln\frac{\psi^*}{\psi}. \tag{29}$$

But this is not a Lagrangian for the system under study, as it implies the wrong field equation

$$i\hbar\frac{\partial}{\partial t}\psi + \frac{\hbar^2}{2m}\nabla^2\psi - V\psi - \frac{i\hbar}{2}\lambda\psi\left(\ln\frac{\psi^*}{\psi} + 1\right) = 0, \tag{30}$$

which not only differs from the desired Schrödinger–Langevin equation (15) but also violates the conservation law of probability.

We are thus forced to conclude that there is no action functional that has exactly the Schrödinger–Langevin equation and its complex conjugate as its Euler–Lagrange equations. Fortunately, it turns out that an 'integrating factor' can be found.

$$e^{\lambda t}\left(i\hbar\frac{\partial}{\partial t}\psi + \frac{\hbar^2}{2m}\nabla^2\psi - V\psi - \frac{i\hbar}{2}\lambda\psi\ln\frac{\psi^*}{\psi}\right) = 0 \tag{31}$$

and its complex conjugate leads to the correct Lagrangian

$$\mathcal{L} = e^{\lambda t}\left\{\frac{i\hbar}{2}\left(\psi^* \frac{\partial}{\partial t}\psi - \psi\frac{\partial}{\partial t}\psi^*\right) - \frac{\hbar^2}{2m}\nabla\psi^* \cdot \nabla\psi - V\psi^*\psi - \frac{i\hbar}{2}\lambda\psi^*\psi\ln\frac{\psi^*}{\psi}\right\}. \tag{32}$$

It should be remarked that Lagrangians (26) and (32) can also be constructed by the method of 'Schrödinger quantization', which is different from the procedure given here. This has been done by the author in Ref. [24], which may also be consulted for further discussions.

In contrast to the method of Fréchet differentials, the Schrödinger quantization prescription is a rather vague one and results in a highly ambiguous procedure. Therefore, it is somewhat prone to errors and has to be carried through with extreme care in order to single out the correct result. It might thus be advisable to point out that the literature is haunted by an incorrect 'Lagrangian' for the Schrödinger–Langevin equation. For the first time it was given in Ref. [25] as an alleged result of Schrödinger quantization and it seems to have become accepted almost universally thereafter [11,21,26,27]:

$$`\mathcal{L}' = \frac{i\hbar}{2}\left(\psi^*\frac{\partial}{\partial t}\psi - \psi\frac{\partial}{\partial t}\psi^*\right) - \frac{\hbar^2}{2m}\nabla\psi^*\cdot\nabla\psi - V\psi^*\psi - \frac{i\hbar}{2}\lambda\psi^*\psi\left(\ln\frac{\psi^*}{\psi} - 1\right). \tag{33}$$

Despite its frequent use, this expression contains a serious flaw! No doubt, variation with respect to ψ^* leads to

$$i\hbar\frac{\partial}{\partial t}\psi + \frac{\hbar^2}{2m}\nabla^2\psi - V\psi - \frac{i\hbar}{2}\lambda\psi\ln\frac{\psi^*}{\psi} = 0, \tag{34}$$

and it is quite tempting to conceive this as the desired Schrödinger–Langevin equation. But variation with respect to ψ immediately reveals this point of view as mistaken:

$$-i\hbar\frac{\partial}{\partial t}\psi^* + \frac{\hbar^2}{2m}\nabla^2\psi^* - V\psi^* - \frac{i\hbar}{2}\lambda\psi^*\left(\ln\frac{\psi^*}{\psi} - 2\right) = 0. \tag{35}$$

In retrospect, this disillusioning result does not come as a complete surprise. It is nothing but a prime example of the just proven fact that it is impossible to find an action functional that has exactly the Schrödinger–Langevin equation and its complex conjugate as its Euler–Lagrange equations.

Notwithstanding the fact that the incorrectness of Eq. (33) has already been pointed out some years ago [24], the amount of confusion caused by this erroneous 'Lagrangian' is still considerable. Even in most recent papers the correctness of the misleading procedure in Ref. [25] has occasionally been taken for granted [28–32].

Another paper with a different but likewise erroneous Lagrangian for a linearly damped quantum system is Ref. [33]. For a detailed discussion the reader is again referred to Ref. [24].

2.4. *The Takabayasi–Schönberg extension of the Schrödinger–Langevin equation*

As long as no magnetic field enters the stage, the hydrodynamic picture of quantum theory, in contrast to ordinary classical hydrodynamics, only deals with potential flows, i.e. vanishing vorticity: $\nabla \times \mathbf{u} = 0$.

One of our aims consists in the transfer of strategies and concepts from the hydrodynamic picture of dissipative quantum theory to ordinary hydrodynamics. Therefore, it is quite natural to make at first a generalization of quantum theory by introducing hypothetical nonzero vorticity degrees of freedom into the Madelung picture. This leads to the so-called Takabayasi–Schönberg extension of quantum mechanics [34–39].

For reasons of simplicity we restrict ourselves in the following to the case of the Schrödinger–Langevin equation. If we write down the Schrödinger–Langevin

Lagrangian (32) using the field quantities of the Madelung picture, we get

$$\mathcal{L} = \mathrm{e}^{\lambda t}\left(-\varrho\frac{\partial}{\partial t}\phi - \frac{1}{2}\varrho(\nabla\phi)^2 - \frac{\hbar^2}{2m^2}(\nabla\sqrt{\varrho})^2 - \frac{1}{m}\varrho V - \lambda\varrho\phi\right), \tag{36}$$

with the 'velocity potential' $\phi = S/m$.

To introduce vorticity degrees of freedom we can closely follow the procedure given by Takabayasi and Schönberg for the nondissipative case [34–39]. In analogy to the Lagrangian description of ideal barotropic fluids in classical hydrodynamics (as e.g. in Ref. [40]), the velocity field is expressed by means of 'Clebsch potentials':

$$\mathbf{u} = \nabla\phi + \alpha\nabla\beta, \tag{37}$$

i.e.

$$\nabla\mathbf{u} = \nabla\alpha \times \nabla\beta. \tag{38}$$

In the Madelung picture the Lagrangian of the nondissipative Takabayasi–Schönberg extension of quantum theory has the form

$$\mathcal{L} = -\varrho\frac{\partial}{\partial t}\phi - \varrho\alpha\frac{\partial}{\partial t}\beta - \frac{1}{2}\varrho(\nabla\phi + \alpha\nabla\beta)^2 - \frac{\hbar^2}{2m^2}(\nabla\sqrt{\varrho})^2 - \frac{1}{m}\varrho V. \tag{39}$$

As the simplest joint generalization of Eqs. (36) and (39) the following expression for an extension of the Schrödinger–Langevin Lagrangian suggests itself:

$$\mathcal{L} = \mathrm{e}^{\lambda t}\left(-\varrho\frac{\partial}{\partial t}\phi - \varrho\alpha\frac{\partial}{\partial t}\beta - \frac{1}{2}\varrho(\nabla\phi + \alpha\nabla\beta)^2 - \frac{\hbar^2}{2m^2}(\nabla\sqrt{\varrho})^2 - \frac{1}{m}\varrho V - \lambda\varrho\phi\right). \tag{40}$$

The corresponding Euler–Lagrange equations are

$$\delta\varrho: \quad \mathrm{e}^{\lambda t}\left(\frac{\partial}{\partial t}\phi + \alpha\frac{\partial}{\partial t}\beta + \frac{1}{2}(\nabla\phi + \alpha\nabla\beta)^2 - \frac{\hbar^2}{2m^2}\frac{\nabla^2\sqrt{\varrho}}{\sqrt{\varrho}} + \frac{1}{m}V + \lambda\phi\right) = 0 \tag{41}$$

$$\delta\phi: \quad \frac{\partial}{\partial t}(\varrho\,\mathrm{e}^{\lambda t}) + \nabla\cdot(\varrho(\nabla\phi + \alpha\nabla\beta))\mathrm{e}^{\lambda t} - \lambda\varrho\,\mathrm{e}^{\lambda t} = 0 \tag{42}$$

$$\delta\alpha: \quad \left(\varrho\frac{\partial}{\partial t}\beta + \varrho(\nabla\phi + \alpha\nabla\beta)\cdot\nabla\beta\right)\mathrm{e}^{\lambda t} = 0 \tag{43}$$

$$\delta\beta: \quad \frac{\partial}{\partial t}(\varrho\alpha\,\mathrm{e}^{\lambda t}) + \nabla\cdot(\varrho(\nabla\phi + \alpha\nabla\beta)\alpha)\mathrm{e}^{\lambda t} = 0. \tag{44}$$

Straightforward manipulations and insertion of $\mathbf{u} = \nabla\phi + \alpha\nabla\beta$ immediately lead to the following set of equations:

$$\frac{\partial}{\partial t}\phi + \alpha\frac{\partial}{\partial t}\beta + \frac{\mathbf{u}^2}{2} + \frac{1}{m}\left(V - \frac{\hbar^2}{2m}\frac{\nabla^2\sqrt{\varrho}}{\sqrt{\varrho}}\right) + \lambda\phi = 0, \tag{45}$$

$$\frac{\partial}{\partial t}\varrho + \nabla\cdot(\varrho\mathbf{u}) = 0, \tag{46}$$

$$\frac{\partial}{\partial t}\beta + \mathbf{u}\cdot\nabla\beta = \frac{\mathrm{D}}{\mathrm{D}t}\beta = 0, \tag{47}$$

$$\frac{\partial}{\partial t}\alpha + \lambda\alpha + \mathbf{u}\cdot\nabla\alpha = \frac{\mathrm{D}}{\mathrm{D}t}\alpha + \lambda\alpha = 0. \tag{48}$$

Eq. (45) may also be rewritten as follows:

$$\frac{\mathrm{D}}{\mathrm{D}t}\phi - (\mathbf{u}\cdot\nabla)\phi + \alpha\frac{\mathrm{D}}{\mathrm{D}t}\beta - \alpha(\mathbf{u}\cdot\nabla)\beta + \frac{\mathbf{u}^2}{2} + \frac{1}{m}\left(V - \frac{\hbar^2}{2m}\frac{\nabla^2\sqrt{\varrho}}{\sqrt{\varrho}}\right) + \lambda\phi = 0. \tag{49}$$

With Eq. (47) this implies

$$\frac{\mathrm{D}}{\mathrm{D}t}\phi - \mathbf{u}\cdot(\nabla\phi + \alpha\nabla\beta) + \frac{\mathbf{u}^2}{2} + \frac{1}{m}\left(V - \frac{\hbar^2}{2m}\frac{\nabla^2\sqrt{\varrho}}{\sqrt{\varrho}}\right) + \lambda\phi = 0, \tag{50}$$

and thus

$$\frac{\mathrm{D}}{\mathrm{D}t}\phi - \frac{\mathbf{u}^2}{2} + \frac{1}{m}\left(V - \frac{\hbar^2}{2m}\frac{\nabla^2\sqrt{\varrho}}{\sqrt{\varrho}}\right) + \lambda\phi = 0. \tag{51}$$

The above set of equations comprises a meaningful extension of Kostin's Schrödinger–Langevin theory by vorticity degrees of freedom if it still implies the 'Euler equation'

$$\frac{\partial}{\partial t}\mathbf{u} + (\mathbf{u}\cdot\nabla)\mathbf{u} = \frac{\mathrm{D}}{\mathrm{D}t}\mathbf{u} = -\frac{1}{m}\nabla V - \frac{1}{m}\nabla Q - \lambda\mathbf{u} \tag{52}$$

of the hydrodynamic picture.

This is indeed the case. Using the obvious commutation relation

$$\nabla\frac{\mathrm{D}}{\mathrm{D}t} - \frac{\mathrm{D}}{\mathrm{D}t}\nabla = \sum_j (\nabla u_j)\frac{\partial}{\partial x_j} \tag{53}$$

we get

$$\begin{aligned}\frac{\mathrm{D}}{\mathrm{D}t}\mathbf{u} &= \frac{\mathrm{D}}{\mathrm{D}t}(\nabla\phi + \alpha\nabla\beta) = \frac{\mathrm{D}}{\mathrm{D}t}\nabla\phi + \left(\frac{\mathrm{D}}{\mathrm{D}t}\alpha\right)\nabla\beta + \alpha\frac{\mathrm{D}}{\mathrm{D}t}\nabla\beta \\ &= \nabla\frac{\mathrm{D}}{\mathrm{D}t}\phi - \sum_j (\nabla u_j)\frac{\partial}{\partial x_j}\phi + \left(\frac{\mathrm{D}}{\mathrm{D}t}\alpha\right)\nabla\beta + \alpha\nabla\frac{\mathrm{D}}{\mathrm{D}t}\beta - \alpha\sum_j (\nabla u_j)\frac{\partial}{\partial x_j}\beta. \end{aligned} \tag{54}$$

With Eqs. (47) and (48) this implies

$$\begin{aligned}\frac{\mathrm{D}}{\mathrm{D}t}\mathbf{u} &= \nabla\frac{\mathrm{D}}{\mathrm{D}t}\phi - \sum_j (\nabla u_j)\left(\frac{\partial}{\partial x_j}\phi + \alpha\frac{\partial}{\partial x_j}\beta\right) - \lambda\alpha\nabla\beta \\ &= \nabla\frac{\mathrm{D}}{\mathrm{D}t}\phi - \sum_j (\nabla u_j)u_j - \lambda\alpha\nabla\beta = \nabla\frac{\mathrm{D}}{\mathrm{D}t}\phi - \nabla\left(\frac{\mathbf{u}^2}{2}\right) - \lambda\mathbf{u} + \lambda\nabla\phi. \end{aligned} \tag{55}$$

Insertion of Eq. (51) then leads to the desired Euler equation

$$\frac{\mathrm{D}}{\mathrm{D}t}\mathbf{u} = -\frac{1}{m}\nabla\left(V - \frac{\hbar^2}{2m}\frac{\nabla^2\sqrt{\varrho}}{\sqrt{\varrho}}\right) - \lambda\mathbf{u}. \tag{56}$$

The Schrödinger wave functions ψ, ψ^* may be reintroduced by inserting

$$\varrho = m\psi^*\psi; \qquad \phi = \frac{S}{m} = \frac{\mathrm{i}\hbar}{2m}\ln\frac{\psi^*}{\psi}. \tag{57}$$

The Lagrangian (40) then has the following form:

$$\mathcal{L} = \mathrm{e}^{\lambda t}\left\{\frac{\mathrm{i}\hbar}{2}\left(\psi^*\frac{\partial}{\partial t}\psi - \psi\frac{\partial}{\partial t}\psi^*\right) - \frac{\hbar^2}{2m}\left(\nabla - \frac{\mathrm{i}m}{\hbar}\alpha\nabla\beta\right)\psi^*\cdot\left(\nabla + \frac{\mathrm{i}m}{\hbar}\alpha\nabla\beta\right)\psi \right.$$
$$\left. -\left(V + m\alpha\frac{\partial}{\partial t}\beta\right)\psi^*\psi - \lambda\frac{\mathrm{i}\hbar}{2}\psi^*\psi\ln\frac{\psi^*}{\psi}\right\}. \tag{58}$$

The corresponding Euler–Lagrange equations with respect to variations of ψ^* and ψ lead to the Takabayasi–Schönberg extension of the Schrödinger–Langevin equation

$$\mathrm{i}\hbar\frac{\partial}{\partial t}\psi = -\frac{\hbar^2}{2m}\left(\nabla + \frac{\mathrm{i}m}{\hbar}\alpha\nabla\beta\right)^2\psi + \left(V + m\alpha\frac{\partial}{\partial t}\beta\right)\psi + \lambda\frac{\mathrm{i}\hbar}{2}\ln\left(\frac{\psi^*}{\psi}\right)\psi \tag{59}$$

and its complex conjugate. Moreover, variations with respect to α and β lead to the additional time-evolution equations

$$\left(\frac{\partial}{\partial t} + \mathbf{u}\cdot\nabla\right)\alpha = -\lambda\alpha, \qquad \left(\frac{\partial}{\partial t} + \mathbf{u}\cdot\nabla\right)\beta = 0, \tag{60}$$

with

$$\mathbf{u} = \frac{\mathrm{i}\hbar}{2m}\nabla\ln\frac{\psi^*}{\psi} + \alpha\nabla\beta. \tag{61}$$

3. Linear damping in classical hydrodynamics

3.1. A Lagrangian for an extension of Darcy's law

A well-known Lagrangian governing the undamped flow of an ideal barotropic fluid is due to Bateman [40] and reads

$$\mathcal{L} = -\varrho\frac{\partial}{\partial t}\phi - \varrho\alpha\frac{\partial}{\partial t}\beta - \frac{1}{2}\varrho(\nabla\phi + \alpha\nabla\beta)^2 - \varrho P(\varrho) + p(\varrho) - \varrho U. \tag{62}$$

Again the velocity field is supposed to be represented by Clebsch potentials here:

$$\mathbf{u} = \nabla\phi + \alpha\nabla\beta. \tag{63}$$

Furthermore, $p(\varrho)$ denotes the unique relationship between pressure and mass density, and $P(\varrho)$ is defined as follows:

$$P(\varrho) = \int \frac{\mathrm{d}p(\varrho)}{\varrho}. \tag{64}$$

One immediately recognizes the striking similarity with the damping-free Takabayasi–Schönberg Lagrangian

$$\mathcal{L} = -\varrho\frac{\partial}{\partial t}\phi - \varrho\alpha\frac{\partial}{\partial t}\beta - \frac{1}{2}\varrho(\nabla\phi + \alpha\nabla\beta)^2 - \frac{\hbar^2}{2m^2}(\nabla\sqrt{\varrho})^2 - \frac{1}{m}\varrho V. \tag{65}$$

Thus, the transition from the Takabayasi–Schönberg extension of quantum theory to the analogous system in classical hydrodynamics is rather straightforward here. One simply has to replace the internal energy of the 'Madelung fluid' with that of a classical ideal barotropic fluid:

$$\frac{\hbar^2}{2m^2}(\nabla\sqrt{\varrho})^2 \to \varrho P(\varrho) - p(\varrho). \tag{66}$$

Furthermore, the external potential term V/m is rewritten as U.

As we are looking for a Lagrangian describing linearly damped ideal fluid flow, it is therefore tempting to make the analogous replacements in the Lagrangian (40) for the Takabayasi–Schönberg extension of the Schrödinger–Langevin equation. This leads to the following expression:

$$\mathcal{L} = \mathrm{e}^{\lambda t}\left(-\varrho\frac{\partial}{\partial t}\phi - \varrho\alpha\frac{\partial}{\partial t}\beta - \frac{1}{2}\varrho(\nabla\phi + \alpha\nabla\beta)^2 - \varrho P(\varrho) + p(\varrho) - \varrho U - \lambda\varrho\phi\right). \tag{67}$$

The corresponding Euler–Lagrange equations read

$$\delta\varrho: \quad \mathrm{e}^{\lambda t}\left(\frac{\partial}{\partial t}\phi + \alpha\frac{\partial}{\partial t}\beta + \frac{1}{2}(\nabla\phi + \alpha\nabla\beta)^2 + P + U + \lambda\phi\right) = 0 \tag{68}$$

$$\delta\phi: \quad \frac{\partial}{\partial t}(\varrho\mathrm{e}^{\lambda t}) + \nabla\cdot(\varrho(\nabla\phi + \alpha\nabla\beta))\mathrm{e}^{\lambda t} - \lambda\varrho\,\mathrm{e}^{\lambda t} = 0 \tag{69}$$

$$\delta\alpha: \quad \left(\varrho\frac{\partial}{\partial t}\beta + \varrho(\nabla\phi + \alpha\nabla\beta)\cdot\nabla\beta\right)\mathrm{e}^{\lambda t} = 0 \tag{70}$$

$$\delta\beta: \quad \frac{\partial}{\partial t}(\varrho\alpha\,\mathrm{e}^{\lambda t}) + \nabla\cdot(\varrho(\nabla\phi + \alpha\nabla\beta)\alpha)\mathrm{e}^{\lambda t} = 0. \tag{71}$$

With $\mathbf{u} = \nabla\phi + \alpha\nabla\beta$ one gets the following equations that are completely analogous to Eqs. (45)–(48):

$$\frac{\partial}{\partial t}\phi + \alpha\frac{\partial}{\partial t}\beta + \frac{\mathbf{u}^2}{2} + P + U + \lambda\phi = 0, \tag{72}$$

$$\frac{\partial}{\partial t}\varrho + \nabla\cdot(\varrho\mathbf{u}) = 0, \tag{73}$$

$$\frac{\partial}{\partial t}\beta + \mathbf{u}\cdot\nabla\beta = \frac{\mathrm{D}}{\mathrm{D}t}\beta = 0, \tag{74}$$

$$\frac{\partial}{\partial t}\alpha + \lambda\alpha + \mathbf{u}\cdot\nabla\alpha = \frac{\mathrm{D}}{\mathrm{D}t}\alpha + \lambda\alpha = 0. \tag{75}$$

Carrying out the same manipulations as in Eq. (49) ff. we are able to show that Eq. (72) can be transformed into

$$\frac{\mathrm{D}}{\mathrm{D}t}\phi - \frac{\mathbf{u}^2}{2} + P + U + \lambda\phi = 0, \tag{76}$$

which corresponds to Eq. (51).

Furthermore, the procedure analogous to Eq. (54) ff. leads to the Euler equation

$$\frac{\mathrm{D}}{\mathrm{D}t}\mathbf{u} = -\nabla(P + U) - \lambda\mathbf{u}. \tag{77}$$

Due to $\nabla P = \nabla p/\varrho$ this can also be written as

$$\varrho\frac{\mathrm{D}}{\mathrm{D}t}\mathbf{u} = -\nabla p - \varrho\nabla U - \lambda\varrho\mathbf{u}. \tag{78}$$

For very slow motion ('creeping flow') one may assume

$$\frac{\mathrm{D}}{\mathrm{D}t}\mathbf{u} \approx 0, \tag{79}$$

which leads to

$$\varrho\mathbf{u} = -\frac{1}{\lambda}(\nabla p + \varrho\nabla U). \tag{80}$$

In this latter equation one immediately recognizes Darcy's law

$$\varrho\mathbf{u} = -\frac{\kappa}{\nu}(\nabla p + \varrho\nabla U) \tag{81}$$

for the seepage of fluids through porous media (see e.g. Ref. [41]). Here, ν denotes the kinematic viscosity of the fluid and κ is a parameter that only depends on the properties of the porous medium. The Euler equation (78) therefore comprises an extension of Darcy's law, which for $\lambda = \text{const.}$ is implied by the Euler–Lagrange equations of the Lagrangian (67).

3.2. The equivalence problem

The physically relevant set of hydrodynamic field equations for the problem studied here consists of the continuity equation

$$\frac{\partial}{\partial t}\varrho + \nabla\cdot(\varrho\mathbf{u}) = 0 \tag{82}$$

together with the Euler equation

$$\frac{\mathrm{D}}{\mathrm{D}t}\mathbf{u} = -\frac{\nabla p}{\varrho} - \nabla U - \lambda\mathbf{u} = -\nabla(P + U) - \lambda\mathbf{u}. \tag{83}$$

In Section 3.1 we have seen that the continuity equation can be found within the set of Euler–Lagrange equations (68)–(71) while they do not explicitly contain the Euler equation. However, it was shown that solutions of the Euler–Lagrange equations always give rise to corresponding solutions of the Euler equation.

But to establish full equivalence between the Euler–Lagrange equations and the physically relevant field equations (82) and (83) it is also necessary to examine how far the Euler–Lagrange equations are capable of describing *all* possible solutions of the original Eqs. (82) and (83).

Complete equivalence proofs are highly nontrivial even for the nondissipative flow of an ideal barotropic fluid. In particular, they require the study of the so-called Weber and Clebsch transformations of the hydrodynamic equations (see, e.g. Ref. [1]).

For the presently studied problem a generalization of these transformations is needed. This has been carried through by the author in Ref. [42] where generalized Weber and Clebsch transformations have been introduced for all those cases where one is capable of finding field quantities $\mathbf{Y}$, χ such that the Euler equation for the system under study can be cast into the form

$$\frac{\mathrm{D}}{\mathrm{D}t}\mathbf{Y} + \sum_j Y_j \nabla u_j = \nabla \chi. \tag{84}$$

The class of hydrodynamic systems for which this is possible turns out to be rather large and comprises e.g. charged ideal fluids in external electromagnetic fields and ideal magnetohydrodynamic fluids with infinite conductivity. In particular, the case studied here is also contained.

$$\frac{\mathrm{D}}{\mathrm{D}t}\mathbf{u} = -\nabla(P+U) - \lambda\mathbf{u} \Leftrightarrow \tag{85}$$

$$\left(\frac{\mathrm{D}}{\mathrm{D}t}\mathbf{u} + \lambda\mathbf{u}\right)\mathrm{e}^{\lambda t} = -\nabla(P+U)\mathrm{e}^{\lambda t} \Leftrightarrow \tag{86}$$

$$\frac{\mathrm{D}}{\mathrm{D}t}(\mathbf{u}\mathrm{e}^{\lambda t}) = -\nabla(P+U)\mathrm{e}^{\lambda t} \Leftrightarrow \tag{87}$$

$$\frac{\mathrm{D}}{\mathrm{D}t}(\mathbf{u}\,\mathrm{e}^{\lambda t}) + \sum_i (u_i\,\mathrm{e}^{\lambda t})\nabla u_i = \nabla\left(\frac{\mathbf{u}^2}{2} - P - U\right)\mathrm{e}^{\lambda t}. \tag{88}$$

This last equation has the desired form (Eq. (84)) with

$$\mathbf{Y} = \mathbf{u}\,\mathrm{e}^{\lambda t}, \qquad \chi = \left(\frac{\mathbf{u}^2}{2} - P - U\right)\mathrm{e}^{\lambda t}. \tag{89}$$

We cannot repeat the details of Ref. [42] but only state the relevant proposition here:

Whenever $\mathbf{Y}|_{t=0}$ can be globally represented by Clebsch potentials, i.e.

$$\mathbf{Y}|_{t=0} = \nabla\tilde{\phi}_0 + \tilde{\alpha}_0\nabla\beta_0,$$

then the validity of the generalized Euler equation

$$\frac{D}{Dt}\mathbf{Y} + \sum_j Y_j \nabla u_j = \nabla \chi$$

is equivalent with the existence of the representation

$$\mathbf{Y} = \nabla\tilde{\phi} + \tilde{\alpha}\nabla\beta,$$

where $\tilde{\alpha}$, β, and $\tilde{\phi}$ are solutions of the following time-evolution equations:

$$\frac{D}{Dt}\tilde{\alpha} = 0, \qquad \frac{D}{Dt}\beta = 0, \qquad \frac{D}{Dt}\tilde{\phi} = \chi.$$

As $\mathbf{Y}|_{t=0} = \mathbf{u}|_{t=0}$ here, we therefore can conclude: whenever the initial value of the velocity field $\mathbf{u}|_{t=0}$ can be globally represented by Clebsch potentials then Eq. (88) is equivalent with the existence of the representation

$$\mathbf{u}\,e^{\lambda t} = \mathbf{Y} = \nabla\tilde{\phi} + \tilde{\alpha}\nabla\beta, \tag{90}$$

with

$$\frac{D}{Dt}\tilde{\alpha} = 0, \qquad \frac{D}{Dt}\beta = 0, \tag{91}$$

and

$$\frac{D}{Dt}\tilde{\phi} = \chi = \left(\frac{\mathbf{u}^2}{2} - P - U\right)e^{\lambda t}. \tag{92}$$

If $\tilde{\phi}$, $\tilde{\alpha}$ are replaced with

$$\phi := \tilde{\phi}\,e^{-\lambda t}, \qquad \alpha := \tilde{\alpha}\,e^{-\lambda t}, \tag{93}$$

we end up with the representation

$$\mathbf{u} = \nabla\phi + \alpha\nabla\beta, \tag{94}$$

where

$$\frac{D}{Dt}\alpha = \frac{D}{Dt}(\tilde{\alpha}\,e^{-\lambda t}) = \underbrace{\left(\frac{D}{Dt}\tilde{\alpha}\right)}_{=0} e^{-\lambda t} - \lambda\tilde{\alpha}\,e^{-\lambda t} = -\lambda\alpha, \tag{95}$$

$$\frac{D}{Dt}\beta = 0, \tag{96}$$

$$\frac{D}{Dt}\phi = \frac{D}{Dt}(\tilde{\phi}\,e^{-\lambda t}) = \underbrace{\left(\frac{D}{Dt}\tilde{\phi}\right)}_{=\chi} e^{-\lambda t} - \lambda\tilde{\phi}e^{-\lambda t} = \frac{\mathbf{u}^2}{2} - P - U - \lambda\phi. \tag{97}$$

Under the proviso that $\mathbf{u}|_{t=0}$ can be globally represented by Clebsch potentials we have therefore shown that Eqs. (95)–(97) are an equivalent substitute for the Euler equation (83). If we add the continuity equation, the conformity with the set of Eqs. (72)–(75) (and therefore with the Euler–Lagrange equations (68)–(71)) is immediately recognized.

In concluding this subsection we remark that Clebsch representations for vector fields always exist locally [43], but that there are examples of fields lacking global representability [44,45].

3.3. Other Lagrangians

The Bateman Lagrangian (62) is by no means the only Lagrangian that has been employed in the literature for the description of undamped flow of ideal barotropic fluids. For example, Bateman in his paper [40] also discussed the Lagrangian

$$\mathcal{L} = \frac{1}{2}\varrho\mathbf{u}^2 - \varrho P(\varrho) + p(\varrho) - \varrho U - \varrho\frac{\mathrm{D}}{\mathrm{D}t}\phi - \varrho\alpha\frac{\mathrm{D}}{\mathrm{D}t}\beta. \tag{98}$$

In contrast to the Lagrangian descriptions given in Section 3.1, the velocity field $\mathbf{u}$ is an independent field here. Its representation by Clebsch potentials is not presupposed but follows from the Euler–Lagrange equation with respect to variation of $\mathbf{u}$ instead.

The dissipative generalization of this Lagrangian can easily be determined and reads

$$\mathcal{L} = \mathrm{e}^{\lambda t}\left(\frac{1}{2}\varrho\mathbf{u}^2 - \varrho P(\varrho) + p(\varrho) - \varrho U - \varrho\frac{\mathrm{D}}{\mathrm{D}t}\phi - \varrho\lambda\phi - \varrho\alpha\frac{\mathrm{D}}{\mathrm{D}t}\beta\right). \tag{99}$$

The Euler–Lagrange equation with respect to variation of $\mathbf{u}$ still leads to the Clebsch representation

$$\mathbf{u} = \nabla\phi + \alpha\nabla\beta. \tag{100}$$

The remaining Euler–Lagrange equations are essentially given by Eqs. (72)–(75) again.

If the total four-divergence

$$\frac{\partial}{\partial t}(\varrho\phi\,\mathrm{e}^{\lambda t}) + \nabla\cdot(\varrho\mathbf{u}\phi\,\mathrm{e}^{\lambda t}) \tag{101}$$

is added to Eq. (99), the Euler–Lagrange equations remain unchanged and the following expression shows up:

$$\mathcal{L} = \mathrm{e}^{\lambda t}\left(\frac{1}{2}\varrho\mathbf{u}^2 - \varrho P(\varrho) + p(\varrho) - \varrho U + \phi\left(\frac{\partial}{\partial t}\varrho + \nabla\cdot(\varrho\mathbf{u})\right) - \varrho\alpha\frac{\mathrm{D}}{\mathrm{D}t}\beta\right). \tag{102}$$

This Lagrangian turns out to be a dissipative generalization of the so-called 'Davydov Lagrangian' [46]. It has a somewhat more transparent structure as the validity of $\mathrm{D}\beta/\mathrm{D}t = 0$ as well as that of the continuity equation are explicitly enforced by Lagrange multipliers here.

This last expression also gives us a hint how to generalize the above Lagrangians in order to include also the cases where $\mathbf{u}|_{t=0}$ does not have a global Clebsch representation. $\mathrm{D}\beta/\mathrm{D}t = 0$ means that β has to be a function of the material (Lagrangian) coordinates $\mathbf{x}_0(\mathbf{x}, t)$ alone

$$\beta(\mathbf{x}, t) = f(\mathbf{x}_0(\mathbf{x}, t)). \tag{103}$$

Introducing the validity of $D\beta/Dt = 0$ via Lagrange multiplier in $\mathcal{L}$ therefore simply takes into consideration the substantial time independence of one single curvilinear Lagrangian coordinate. For the damping-free case, Lin [47] has pointed out that one should instead consider the substantial time independence of all the three Lagrangian coordinates together

$$\frac{D}{Dt}\mathbf{x}_0(\mathbf{x}, t) = 0. \tag{104}$$

In our case, this procedure leads to the following 'Lin-type' Lagrangian:

$$\mathcal{L} = e^{\lambda t}\left(\frac{1}{2}\varrho\mathbf{u}^2 - \varrho P(\varrho) + p(\varrho) - \varrho U + \phi\left(\frac{\partial}{\partial t}\varrho + \nabla\cdot(\varrho\mathbf{u})\right) - \varrho\boldsymbol{\alpha}\cdot\frac{D}{Dt}\mathbf{x}_0\right). \tag{105}$$

After some trivial manipulations the corresponding set of Euler–Lagrange equations leads to

$$\mathbf{u} = \nabla\phi + \sum_i \alpha_i \nabla x_{0_i}, \tag{106}$$

$$\frac{\partial}{\partial t}\varrho + \nabla\cdot(\varrho\mathbf{u}) = 0, \tag{107}$$

$$\frac{D}{Dt}\mathbf{x}_0 = 0, \tag{108}$$

$$\frac{D}{Dt}\boldsymbol{\alpha} + \lambda\boldsymbol{\alpha} = 0, \tag{109}$$

$$\frac{D}{Dt}\phi - \frac{\mathbf{u}^2}{2} + P + U + \lambda\phi = 0. \tag{110}$$

Our problem concerning the velocity fields without global Clebsch representation is indeed solved now. Proceeding along the lines given in Section 3.2, this can immediately be inferred from the following proposition that again has been proven by the author in Ref. [42]:

The validity of the generalized Euler equation

$$\frac{D}{Dt}\mathbf{Y} + \sum_j Y_j \nabla u_j = \nabla\chi$$

is equivalent with the existence of the representation

$$\mathbf{Y} = \nabla\tilde{\phi} + \sum_i \tilde{\alpha}_i \nabla x_{0_i},$$

where $\tilde{\boldsymbol{\alpha}}$, $\mathbf{x}_0$, and $\tilde{\phi}$ are solutions of the following time-evolution equations:

$$\frac{\mathrm{D}}{\mathrm{D}t}\tilde{\boldsymbol{\alpha}} = 0, \qquad \frac{\mathrm{D}}{\mathrm{D}t}\mathbf{x}_0 = 0, \qquad \frac{\mathrm{D}}{\mathrm{D}t}\tilde{\phi} = \chi.$$

4. Summary

The present contribution has dealt with the construction and discussion of variational principles for quantum-mechanical and hydrodynamical systems under the influence of linear damping. Due to far-reaching analogies between hydrodynamics and the hydrodynamic picture of quantum theory, problems in these apparently disparate fields of physics turned out to be tractable on more or less the same footing. In particular, a variational principle that had originally been constructed for a certain linearly damped quantum system could immediately be transferred to an analogous system in classical hydrodynamics. An important intermediate step, however, consisted in an extension of ordinary dissipative quantum theory by introducing hypothetical vorticity degrees of freedom.

The Euler equation of the corresponding system in classical hydrodynamics emerged to be a nonstationary extension of Darcy's law. An interesting feature of this system consisted in its amenability to a certain class of generalized Weber and Clebsch transformations. In particular, this latter property facilitated a thorough examination of the circumstances under which one is able to establish full equivalence between the Euler–Lagrange equations of the variational principle and the physically relevant field equations of the system under study.

References

[1] Serrin, J. (1959), Handbuch der Physik VIII/1, 125–263.
[2] Seliger, R.L. and Whitham, G.B. (1968), Proc. R. Soc. Lond. A305, 1–25.
[3] Finlayson, B.A. (1972), The Method of Weighted Residuals and Variational Principles, Academic Press, New York.
[4] Yourgrau, W. and Mandelstam, S. (1979), Variational Principles in Dynamics and Quantum Theory, Dover Publications, New York.
[5] Salmon, R. (1988), Annu. Rev. Fluid Mech. 20, 225–256.
[6] Millikan, C.B. (1929), Philos. Mag. 7, 641–662.
[7] Finlayson, B.A. (1972), Phys. Fluids 15, 963–967.
[8] Deshpande, S.M. (1974), Q. Appl. Math. 31, 43–52.
[9] Mobbs, S.D. (1982), Proc. R. Soc. Lond. A381, 457–468.
[10] Messer, J. (1979), Acta Phys. Austriaca 50, 75–91.
[11] Dekker, H. (1981), Phys. Rep. 80, 1–112.
[12] Dodonov, V.V. (1993), J. Korean Phys. Soc. 26, 111–116.
[13] Leggett, A.J., Chakravarty, S., Dorsey, A.T., Fisher, M.P.A., Garg, A. and Zwerger, W. (1987), Rev. Mod. Phys. 59, 1–85.
[14] Grabert, H., Schramm, P. and Ingold, G.L. (1988), Phys. Rep. 168, 115–207.
[15] Weiss, U. (1993), Quantum Dissipative Systems, World Scientific, Singapore.
[16] Bateman, H. (1931), Phys. Rev. 38, 815–819.

[17] Caldirola, P. (1941), Nuovo Cimento 18, 393–400.
[18] Kanai, E. (1948), Prog. Theor. Phys. 3, 440–442.
[19] Madelung, E. (1926), Z. Phys. 40, 322–326.
[20] Kostin, M.D. (1972), J. Chem. Phys. 57, 3589–3591.
[21] Yasue, K. (1978), Ann. Phys. (New York) 114, 479–496.
[22] Nassar, A.B. (1986), J. Math. Phys. 27, 755–758.
[23] Soper, D.E. (1976), Classical Field Theory, Wiley, New York.
[24] Wagner, H.J. (1994), Z. Phys. B95, 261–273.
[25] Razavy, M. (1977), Z. Phys. B26, 201–206.
[26] Razavy, M. (1978), Can. J. Phys. 56, 311–320.
[27] Herrera, L., Núñez, L., Patiño, A. and Rago, H. (1986), Am. J. Phys. 54, 273–277.
[28] Bolivar, A.O. (1998), Phys. Rev. A58, 4330–4335.
[29] Wysocki, R.J. (2000), Phys. Rev. A61, 022104.
[30] Razavy, M. (2001), Iranian J. Sci. Technol. A25, 329–353.
[31] Bolivar, A.O. (2001), Physica A301, 219–240.
[32] Um, C.I., Yeon, K.H. and George, T.F. (2002), Phys. Rep. 362, 63–192.
[33] Pal, S. (1989), Phys. Rev. A39, 3825–3832.
[34] Takabayasi, T. (1952), Prog. Theor. Phys. 8, 143–182.
[35] Takabayasi, T. (1953), Prog. Theor. Phys. 9, 187–222.
[36] Schönberg, M. (1954), Nuovo Cimento 11, 674–682.
[37] Schönberg, M. (1954), Nuovo Cimento 12, 300–303.
[38] Schönberg, M. (1954), Nuovo Cimento 12, 649–667.
[39] Schönberg, M. (1955), Nuovo Cimento 1, 543–580.
[40] Bateman, H. (1929), Proc. R. Soc. Lond. A125, 598–618.
[41] Scheidegger, A.E. (1963), Handbuch der Physik VIII/2, 625–662.
[42] Wagner, H.J. (1998), Arch. Mech. 50, 645–655.
[43] Stern, D.P. (1970), Am. J. Phys. 38, 494–501.
[44] Moffatt, H.K. (1969), J. Fluid Mech. 35, 117–129.
[45] Bretherton, F.P. (1970), J. Fluid Mech. 44, 19–31.
[46] Davydov, B. (1949), Dokl. Akad. Nauk SSSR 69, 165–168.
[47] Lin, C.C. (1963), Proc. Int. School Phys. "Enrico Fermi" 21, 93–146.

Chapter 12

LEAST ACTION PRINCIPLE FOR DISSIPATIVE PROCESSES

Katalin Gambár

Department of Atomic Physics, Roland Eötvös University, Pázmány Péter sétány 1/A, H-1117 Budapest, Hungary

Abstract

We point out that the least action principle could be applied for dissipative processes as well. The Lagrange–Hamilton formalism can be completely worked out. In this mathematical model of nonequilibrium thermodynamics, the examinations of symmetries lead to the Onsager's reciprocity relations. On the basis of Poisson structure, the stochastic behavior of processes can be described in the phase space.

Keywords: least action principle; entropy balance; Onsager's symmetry; dynamical symmetries; Poisson structure; invariance properties; Lie algebra; stochastic behavior; phase space; Markovian behavior; Onsager's regression; Lagrangian for hyperbolic processes

1. Introduction

In the present chapter we show that the Hamilton's principle—the least action principle—can be applied for dissipative processes that shows that the Lagrange formalism is not restricted for reversible cases at all. The mathematical difficulties could be resolved by introducing special potential funtions (scalar fields), and by the help of these the complete Hamilton–Lagrange theory could be built up. The possibility of total variation allows calculation of the canonically conjugated quantities, the Hamiltonian, the Poisson structure both for parabolic and hyperbolic differential equations, i.e. both for diffusive processes (e.g. heat conduction as a completely dissipative process) and damped-wave processes (e.g. absorption of an electromagnetic wave in medium). Geometrical and dynamical symmetries can be examined to obtain a deeper insight into the structure of theory. A famous relation, the Onsager's symmetry relation, can be proved as a consequence of a dynamical symmetry. Physical quantities (e.g. entropy, entropy current, and production) can be introduced, and this shows that the theory is fitted to the theory of nonequilibrium thermodynamics. The concept of a new phase field is based on the space of the calculated canonically conjugated quantities and this enables us to examine the stochastic behavior of processes. The constructed Poisson structure gives us an opportunity to calculate Onsager's regression hypothesis.

E-mail address: gambar@ludens.elte.hu (K. Gambár).

2. Basic concepts

The classical irreversible thermodynamics deals with such large and continuous media of which states can be given by the field of equilibrium state variables beside the equilibrium, i.e. the hypothesis of local equilibrium is required. (Of course, this hypothesis also includes the fact that the relations of thermostatics are valid at a given point r at a given time t.) The field quantity $\rho(r,t)$ is the mass density of the continuum, by which the mass in the volume V at a time t is

$$m(t) = \int_V \rho(r,t)\mathrm{d}V. \tag{1}$$

Let G be some arbitrary extensive quantity, then

$$G = \int_V \rho(r,t)g(r,t)\mathrm{d}V, \tag{2}$$

where the specific field quantity $g(r,t)$ is the quantity G referred to unit mass. The definition of current density J_G describing the transport of quantity G is

$$\Delta_1 G = J_G \,\mathrm{d}A\,\mathrm{d}t, \tag{3}$$

where $\Delta_1 G$ is the quantity flowing through the surface $\mathrm{d}A$ during the time $\mathrm{d}t$. In connection with the quantity G we have to define the source density σ_G

$$\Delta_2 G = \sigma_G \,\mathrm{d}V\,\mathrm{d}t, \tag{4}$$

where $\Delta_2 G$ is the production of the quantity G inside the volume $\mathrm{d}V$.

Let the specific entropy be a function of all specific extensive quantities, but this does not depend on space and time explicitly

$$s = s(g_1, g_2, g_3, \ldots). \tag{5}$$

The substantial time derivate of it is[1]

$$\frac{\mathrm{d}s}{\mathrm{d}t} = \frac{\partial s}{\partial g_i}\frac{\mathrm{d}g_i}{\mathrm{d}t}. \tag{6}$$

We introduce the following notation

$$\Gamma_i = \frac{\partial s}{\partial g_i},$$

then we may write instead of Eq. (6)

$$\frac{\mathrm{d}s}{\mathrm{d}t} = \Gamma_i\frac{\mathrm{d}g_i}{\mathrm{d}t}. \tag{7}$$

The quantity $T\Gamma_i$ pertains to the intensive quantity g_i, where T is the temperature, so

$$\frac{\mathrm{d}s}{\mathrm{d}t} = \Gamma_i\frac{\mathrm{d}g_i}{\mathrm{d}t} = \frac{1}{T}\frac{\mathrm{d}u}{\mathrm{d}t} + \frac{p}{T}\frac{\mathrm{d}v}{\mathrm{d}t} - \frac{\mu_k}{T}\frac{\mathrm{d}c_k}{\mathrm{d}t}.$$

[1] We use the Einstein convention, i.e. the double indices mean summation.

On the other hand, because of the Gibbs–Duhem relation the equation

$$0 = \frac{\mathrm{d}\Gamma_i}{\mathrm{d}t} g_i \tag{8}$$

is fulfilled, the specific entropy can be given by the form

$$s = \Gamma_i g_i. \tag{9}$$

Since the specific entropy is a function of independent field quantities $g_i(r,t)$, we can write $s(g_i(r,t)) = s(r,t)$, by which the $\Gamma(g_i(r,t)) = \Gamma(r,t)$ is also obvious. The change of Γ_i in time can be given by the temporal change of g_ks

$$\frac{\partial \Gamma_i}{\partial t} = \frac{\partial \Gamma_i}{\partial g_k}\frac{\partial g_k}{\partial t} = S_{ik}\frac{\partial g_k}{\partial t}, \tag{10}$$

where S_{ik} is a symmetric coefficient matrix, which has an inverse, S_{ik}^{-1}

$$S_{ik} = \frac{\partial \Gamma_i}{\partial g_k} = \frac{\partial^2 s}{\partial g_i\,\partial g_k} = \frac{\partial^2 s}{\partial g_k\,\partial g_i} = \frac{\partial \Gamma_k}{\partial g_i} = S_{ki}, \tag{11}$$

$$S_{ik}^{-1} = \frac{\partial g_i}{\partial \Gamma_k} = \frac{\partial g_k}{\partial \Gamma_i} = S_{ki}^{-1}. \tag{12}$$

The matrix S_{ik} is a negative-definite. (The elements of S_{ik} are sometimes called generalized specific capacities.)

The general local, convection-free balance equation can be written for extensive quantities

$$\frac{\partial(\rho g)}{\partial t} + \nabla J_G = \sigma_G, \tag{13}$$

which can be simplified for

$$\rho\frac{\partial g}{\partial t} + \nabla J_G = \sigma_G \tag{14}$$

in the case of constant mass density. The currents of extensive quantities are caused by thermodynamic forces. An unambiguous relation can be given between the currents and forces

$$J_i = L_{ik}X_k, \tag{15}$$

i.e. the current of the i-th extensive quantity is determined by the linear combination of thermodynamic forces X_k, $k = (1, \ldots, n)$. These are the Onsager's linear laws (constitutive equations), L_{ik} is the Onsager coefficient matrix (conductivity matrix), for which we know from experience that

$$L_{ik} = L_{ki} \tag{16}$$

symmetry is valid close to equilibrium. This a physical consequence of the fact that the cross-effect of the transport of the different extensive quantities is symmetric, moreover, it may have a relation with stability of the processes. The connection between the forces and the intensive quantities can be given by

$$X_k = \nabla \Gamma_k. \tag{17}$$

After all, from Eqs. (10), (14), (15) and (17) we obtain the transport equations (field equations) for the quantities

$$\rho S_{ik}^{-1} \frac{\partial \Gamma_k}{\partial t} + L_{ik} \partial \Gamma_k = \sigma_i, \tag{18}$$

which are parabolic differential equations, where S_{ik}^{-1} and L_{ik} are constant coefficient matrices. We can give the balance equation for the entropy

$$\frac{\partial(\rho s)}{\partial t} + \nabla J_s = \sigma_s, \tag{19}$$

where J_s is the entropy current density, σ_s is the entropy-production density for which the relation $\sigma_s \geq 0$ is fulfilled. This expresses the second law of thermodynamics. Moreover, the entropy-production density is the sum of products of related currents and forces of the extensive quantities

$$\sigma_s = X_i J_i = L_{ik} \nabla \Gamma_i \nabla \Gamma_k. \tag{20}$$

The entropy current density equals to [1–12]

$$J_s = \Gamma_i J_i. \tag{21}$$

3. Hamilton–Lagrange formalism for linear parabolic dissipative processes

We show a canonical mathematical model for convection-free transport processes that are slow, i.e. these may be described by the spaces of equilibrium state variables. These processes are written by parabolic differential equations, which contain a first-order time derivative, i.e. nonself-adjoint operator [3,6]. So, the construction of the Lagrange density function is not an easy task. In general, the Lagrange density function is a function of space, time, the field quantities, and the space and time derivatives of these. We have to take into account this requirement if we wish to apply the Hamilton's principle [13–18].

The model is based on Hamilton's principle, the least action principle. According to the principle, the system is characterized by a given Lagrange function L_V, and the time evolution of the system fulfills the following condition

$$\delta S = \delta \int_{t_1}^{t_2} L_V \, \mathrm{d}t = 0, \tag{22}$$

where S is the action. In the case of systems with infinite degrees of freedom (continua) the principle states

$$\delta S = \delta \int_{t_1}^{t_2} \int_V L \, \mathrm{d}V \, \mathrm{d}t = 0, \tag{23}$$

where L is the Lagrange density function.

Let the Lagrange density function be given by

$$L(\varphi_i, \varphi_{i;t}, \Delta\varphi_i) = \tfrac{1}{2}(\varrho S_{ji}^{-1}\varphi_{j;t} - L_{ji}\Delta\varphi_j)^2 + \sigma_i\varphi_i \tag{24}$$

for diffusion-like coupled dissipative processes, where σ_i is a given function of space and time, S_{ji}^{-1} are L_{ji} constant coefficients. The function φ_i is a four-times differentiable scalar field, from which we require that the measurable physical quantity (intensive state variable) can be calculated in the following manner

$$\Gamma_i = \varrho S_{ji}^{-1}\frac{\partial\varphi_j}{\partial t} - L_{ji}\Delta\varphi_j, \tag{25}$$

i.e. we define φ_j through Γ_i. The introduction of this new quantity gives the possibility to resolve the mathematical difficulties.

Moreover, we define the specific entropy [19–22]

$$s = \tfrac{1}{2}\rho^{-1}p_i S_{ij} p_j, \tag{26}$$

where

$$p_i = \frac{\partial L}{\partial\varphi_{i;t}}. \tag{27}$$

The measurable extensive quantities are defined on the basis of Eq. (7).

After the variation, taking into account the Lagrange density function (24) we obtain the Euler–Lagrange equations

$$\frac{\partial L}{\partial\varphi_i} - \frac{\partial}{\partial t}\frac{\partial L}{\partial\varphi_{i;t}} + \Delta\frac{\partial L}{\partial\Delta\varphi_i} = 0. \tag{28}$$

Using the relation between Γ_i and φ_i (25) we get a parabolic field equation for Γ_i

$$\varrho S_{ik}^{-1}\Gamma_{k;t} + L_{ik}\Delta\Gamma_k = \sigma_i. \tag{29}$$

The two equations have different forms, but the content of these is not different.

Generally, the action can be given by the scalar field φ_i for dissipative processes described by second-order differential equations[2]

$$S = \int_T L(\varphi_i, \varphi_{i;\mu}, \varphi_{i;\mu\nu})\mathrm{d}^4x, \tag{30}$$

where μ, $\nu = 1, 2, 3, 4$; $x_1 = x$, $x_2 = y$, $x_3 = z$, $x_4 = t$. The total variation of action can be calculated, and the canonical tensor and canonically conjugated quantities can be

[2] These include both parabolic and hyperbolic equations. The Lagrange density function for the hyperbolic equations and related consequences will be given in Section 6.

read from its final form

$$\delta_t S = \int_{T'} L(\varphi_i + \delta\varphi_i, \varphi_{i;\mu} + \delta\varphi_{i;\mu}, \varphi_{i\mu\nu} + \delta\varphi_{i;\mu\nu}) - \int_T L(\varphi_i, \varphi_{i;\mu}, \varphi_{i;\mu\nu})$$
$$= \int_T \frac{\partial}{\partial x_\mu}(\Theta_{\mu\xi}\,\delta x_\xi + \Pi_{i\mu}\delta_t\varphi_i + \lambda_{i\mu\nu}\delta_t\varphi_{i;\nu})\mathrm{d}^4x. \tag{31}$$

(Here, for the sake of simplicity the source density σ_i is taken as zero.) The total variation of functions φ_i and $\varphi_{i;\nu}$ will be

$$\delta_t\varphi_i = \delta\varphi_i + \varphi_{i;\xi}\,\delta x_\xi, \qquad \delta_t\varphi_{i;\nu} = \delta\varphi_{i;\nu} + \varphi_{i;\xi\nu}\,\delta x_\xi, \tag{32}$$

and the domain T' arises from the domain T by an infinitesimal deformation

$$x'_\xi = x_\xi + \delta x_\xi. \tag{33}$$

The canonical tensor is

$$\Theta_{\mu\xi} = L\delta_{\mu\xi} - \varphi_{i;\xi}\frac{\partial L}{\partial\varphi_{i;\mu}} + \varphi_{i;\xi}\frac{\partial}{\partial x_\nu}\frac{\partial L}{\partial\varphi_{i;\nu\mu}} - \varphi_{i;\xi\nu}\frac{\partial L}{\partial\varphi_{i;\mu\nu}}, \tag{34}$$

and the canonical coefficients are

$$\Pi_{i\mu} = \frac{\partial L}{\partial\varphi_{i;\mu}} - \frac{\partial}{\partial x_\nu}\frac{\partial L}{\partial\varphi_{i;\nu\mu}}, \tag{35}$$

$$\lambda_{i\mu\nu} = \frac{\partial L}{\partial\varphi_{i;\mu\nu}}. \tag{36}$$

The canonical momentum density of the field is

$$p_i = \Pi_{i\mu}N_\mu = \frac{\partial L}{\partial\varphi_{i;t}} = \rho S_{ij}^{-1}\Gamma_j, \tag{37}$$

where $N_\mu = (0, 0, 0, 1)$. The p_i is the canonically conjugated quantitiy to φ_i.

We deduce the Hamilton density function from the element Θ_{44} of the canonical tensor

$$-\Theta_{44} = \varphi_{i;t}\frac{\partial L}{\partial\varphi_{i;t}} - L = H(p_i, \Delta\varphi_i) = \frac{1}{2}(\rho^{-1}S_{ij}p_j)^2 + \rho^{-1}S_{ij}p_jL_{ki}\Delta\varphi_k, \tag{38}$$

where we apply the form of p_i. The canonical equations can be calculated when we know the Hamilton density function. For this, we write dH in two different ways:

$$\mathrm{d}H = \frac{\partial H}{\partial\varphi_i}\mathrm{d}\varphi_i + \frac{\partial H}{\partial\varphi_{i;\mu}}\mathrm{d}\varphi_{i;\mu} + \frac{\partial H}{\partial\varphi_{i;\mu\nu}}\mathrm{d}\varphi_{i;\mu\nu} + \frac{\partial H}{\partial p_i}\mathrm{d}p_i, \tag{39}$$

$$\mathrm{d}H = -\frac{\partial L}{\partial\varphi_i}\mathrm{d}\varphi_i - \frac{\partial L}{\partial\varphi_{i;\mu}}\mathrm{d}\varphi_{i;\mu} - \frac{\partial L}{\partial\varphi_{i;\mu\nu}}\mathrm{d}\varphi_{i;\mu\nu} + \varphi_{i;t}\,\mathrm{d}p_i \tag{40}$$

(μ, ν = 1, 2, 3). Comparing the coefficients we obtain

$$\frac{\partial H}{\partial\varphi_i} = -\frac{\partial L}{\partial\varphi_i}, \tag{41}$$

$$\frac{\partial H}{\partial \varphi_{i;\mu}} = -\frac{\partial L}{\partial \varphi_{i;\mu}}, \tag{42}$$

$$\frac{\partial H}{\partial \varphi_{i;\mu\nu}} = -\frac{\partial L}{\partial \varphi_{i;\mu\nu}}, \tag{43}$$

$$\frac{\partial H}{\partial p_i} = \varphi_{i;t}. \tag{44}$$

The last equation is one of the canonical equations, the other one can be calculated with the help of the other three equations. Let us write the time derivatives of the canonical momentum density

$$p_{i;t} = \frac{\partial}{\partial t}\frac{\partial L}{\partial \varphi_{i;t}} = \frac{\partial L}{\partial \varphi_i} - \frac{\partial}{\partial x_\mu}\frac{\partial L}{\partial \varphi_{i;\mu}} + \frac{\partial^2}{\partial x_\mu\,\partial x_\nu}\frac{\partial L}{\partial \varphi_{i;\mu\nu}}, \tag{45}$$

$$p_{i;t} = -\frac{\partial H}{\partial \varphi_i} + \frac{\partial}{\partial x_\mu}\frac{\partial H}{\partial \varphi_{i;\mu}} - \frac{\partial^2}{\partial x_\mu\,\partial x_\nu}\frac{\partial H}{\partial \varphi_{i;\mu\nu}}. \tag{46}$$

Since H is the function of just p_i and $\Delta\varphi_i$, the second canonical equation is

$$p_{i;t} = -\Delta\frac{\partial H}{\partial \Delta\varphi_i}. \tag{47}$$

These equations can be written in their Poisson bracket forms. The definition of a Poisson bracket is

$$[p_i, H] = \frac{\delta p_i}{\delta \varphi_j}\frac{\delta H}{\delta p_j} - \frac{\delta H}{\delta \varphi_j}\frac{\delta p_i}{\delta p_j}, \tag{48}$$

where, e.g.

$$\frac{\delta H}{\delta \varphi_i} = \frac{\partial H}{\partial \varphi_i} - \frac{\partial}{\partial x_\mu}\frac{\partial H}{\partial \varphi_{i;\mu}} + \frac{\partial^2}{\partial x_\mu\,\partial x_\nu}\frac{\partial H}{\partial \varphi_{i;\mu\nu}} \tag{49}$$

is the functional derivative for H with respect to φ_i. Now, the canonical equations can be expressed by Poisson brackets [23–26]

$$\varphi_{i;t} = [\varphi_i, H], \tag{50}$$

$$p_{i;t} = [p_i, H]. \tag{51}$$

The Hamilton density function does not depend on time explicitly, so its Poisson bracket with a given physical quantity shows the change of the quantity in time. Because the entropy is an important quantity in the theory, we would like to give the entropy-balance equation within the framework of the model

$$\frac{\partial s}{\partial t} = [s, H]. \tag{52}$$

We obtain, with the help of Poisson brackets,

$$[s,H] = \left(\frac{\delta s(p_i)}{\delta \varphi_j} \frac{\delta H}{\delta p_j} - \frac{\delta H}{\delta \varphi_j} \frac{\delta s(p_i)}{\delta p_j} \right) = -\rho^{-1} S_{ij} p_j L_{ig} \Delta(\rho^{-1} S_{gk} p_k)$$

$$= -\nabla(\rho^{-1} S_{ij} p_j L_{ig} \nabla(\rho^{-1} S_{gk} p_k)) + \nabla(\rho^{-1} S_{ij} p_j) L_{ig} \nabla(\rho^{-1} S_{gk} p_k)$$

$$= -\rho^{-2} S_{ij} L_{ig} S_{gk} \nabla(p_j \nabla p_k) + \rho^{-2} S_{ij} L_{ig} S_{gk} \nabla p_j \nabla p_k. \tag{53}$$

In the calculation we used the symmetry $S_{ik} = S_{ki}$. We see in Section 4 that this assumption can be applied. Finally, the entropy-balance equation is

$$\frac{\partial s}{\partial t} + \rho^{-2} S_{ij} L_{ig} S_{gk} \nabla(p_j \nabla p_k) = \rho^{-2} S_{ij} L_{ig} S_{gk} \nabla p_j \nabla p_k, \tag{54}$$

where we find the entropy current

$$J_s = \rho^{-2} S_{ij} L_{ig} S_{gk} p_j \nabla p_k = L_{ig} \Gamma_i \nabla \Gamma_g, \tag{55}$$

and the entropy-production density is

$$\sigma = \rho^{-2} S_{ij} L_{ig} S_{gk} \nabla p_j \nabla p_k = \rho^{-1} S_{ij} \nabla p_j L_{ig} \rho^{-1} S_{gk} \nabla p_k = L_{ig} \nabla \Gamma_i \nabla \Gamma_g. \tag{56}$$

4. Invariance properties, symmetries

When $x \to x'$ is a given self-mapping transformation, then the funtion $f(x, y, \ldots)$ or a relation $O(x, y, \ldots) = A$ is invariant with respect to the transformation if the equality $f(x', y', \ldots) = f(x, y, \ldots)$ is completed for any x, y or the relation $O(x', y', \ldots) = A$ follows from $O(x, y, \ldots) = A$. We are talking about symmetry if the equations of motion are invariant with respect to certain transformation of (r, t) as well as $\varphi(r, t)$. Those transformations, for which the differential equations of motion are insensible in their forms, are called symmetry transformations.

The transformation group F_ρ is a finite continuous group, if any transformation $f \in F_\rho$ is such a special case of a most general transformation, which depends on ρ essential parameters $\varepsilon_1, \ldots, \varepsilon_\rho$. The continuous transformations arise from the set of infinitesimal transformations applying them one after the other.

Wigner dealt with the symmetries, invariances, and transformation groups in detail. He distinguished and named two kinds of symmetries. One of the groups was the geometrical invariances, the other was the so-called dynamical invariances. The geometrical symmetries are the consequences of the special properties of space-time, while the dynamical invariances follow from the special inner properties of different physical systems. The geometrical invariances are independent of interactions, i.e. these must have a general validity for all laws of nature. On the other hand, the dynamical invariances are based on the existence of special interactions, i.e. these are not generally valid for all of the basic equations [8,12,13,15,27–29].

The invariance of the action is the sufficient condition of invariance of the equations of motions. Since, the Lagrange denstity functions

$$L_1(\varphi_i, \varphi_{i;\mu}, \varphi_{i;\mu\nu}) \qquad \text{and} \qquad L_2 = L_1 + \frac{\partial l_\mu(r,t)}{\partial x_\mu}$$

(l_μ is an arbitrary function, μ, $\nu = 1, 2, 3, 4$) serve the same equation of motion, therefore, the invariance of equations of motion is expressed by the invariance of the action

$$\int_T \left(L(\varphi_i(r,t), \varphi_{i;\mu}(r,t), \varphi_{i;\mu\nu}(r,t)) + \frac{\partial l_\mu(r,t)}{\partial x_\mu} \right) \mathrm{d}^4x$$
$$= \int_{T'} L(\varphi_i'(r',t'), \varphi_{i;\mu}'(r',t'), \varphi_{i;\mu\nu}'(r',t'))\mathrm{d}^4x'. \tag{57}$$

In the sense of the first Noether's theorem, if an integral of a variational problem is invariant with respect to a finite, continuous transformation group F_ρ disregarding a divergence term, then ρ first integrals of the Euler–Lagrange equations exist, i.e. we obtain ρ conservation theorems.

In the case of the geometrical invariances of the equations of motion, we should examine the invariance of the action S. If the total variation of the action ($\delta_t S$) disappears as a result of a geometrical transformation, or equals a total divergence, then S is invariant with respect to the transformation and further conclusions for the transformation can be drawn in the sense of Noether's theorem. Since, in our examination the processes are described by classical field equations, therefore, we study the invariance with respect to the Galilei transformation group $F_\rho = G_{10}$. The Galilei group consists of four subgroups, the time and space coordinates are independent of each other, the field is Eucledian. Let us treat the invariance of the action

$$S = \int_T L(\varphi_i, \varphi_{i;\mu}, \varphi_{i;\mu\nu})\mathrm{d}^4x = \int_T \frac{1}{2}(\rho S_{ji}^{-1}\varphi_{j;t} - L_{ji}\Delta\varphi_j)^2\, \mathrm{d}^4x, \tag{58}$$

specifically, the time displacement. We take the following infinitesimal transformation

$$\delta x_1 = \delta x_2 = \delta x_3 = 0, \qquad \delta x_4 = \delta t = \varepsilon = \text{inf const}, \qquad \delta\varphi_i = 0,$$
$$\delta\varphi_{i;\mu} = 0. \tag{59}$$

In the sense of Noether's theorem, we expect that

$$\frac{\partial}{\partial x_\mu}(\Theta_{\mu\xi}\,\delta x_\xi + \pi_{i\mu}\delta_t\varphi_i + \lambda_{i\mu\nu}\,\delta_t\varphi_{i;\nu}) = \frac{\partial}{\partial x_\mu}K_\mu(\varphi_i, \varphi_{i;t}, \Delta\varphi_i, \delta x_\nu, \delta_t\varphi_i, \delta_t\varphi_{i;\nu}), \tag{60}$$

if the transformation is a symmetry transformation of the equation of motion. In the case of this transformation, the total variations of φ_i and $\varphi_{i;\mu}$ are

$$\delta_t\varphi_i = \varphi_{i;t}\,\delta t, \qquad \delta_t\varphi_{i;\nu} = \varphi_{i;t\nu}\,\delta t. \tag{61}$$

Let K_t be

$$K_t(\varphi_i, \varphi_{i;t}, \Delta\varphi_i, \delta x_\nu, \delta_t\varphi_i, \delta_t\varphi_{i;\nu}) = L(\varphi_i, \varphi_{i;t}, \Delta\varphi_i)\delta x_t, \tag{62}$$

where L is the Lagrange density function (24), then action (30) is invariant with respect to transformation (59), taking into consideration a divergence expression. Moreover,

$$\frac{\partial}{\partial x_\mu}[(\Theta_{\mu\xi}\,\delta x_\xi + \Pi_{i\mu}\delta_t\varphi_i + \lambda_{i\mu\nu}\delta_t\varphi_{i;\nu}) - K_\mu] = 0, \tag{63}$$

from which it follows

$$n(\Theta_{\mu\xi}\,\delta x_\xi + \Pi_{i\mu}\delta_t\varphi_i + \lambda_{i\mu\nu}\delta_t\varphi_{i;\nu} - K_\mu) = \text{const.} \tag{64}$$

This means a conserved quantity. In the case of transformation (59) we get

$$\begin{aligned}\Theta_{14}\varepsilon + \Theta_{24}\varepsilon + \Theta_{34}\varepsilon + \Theta_{44}\varepsilon + \Pi_{ix}\varphi_{i;t}\varepsilon + \Pi_{iy}\varphi_{i;t}\varepsilon + \Pi_{iz}\varphi_{i;t}\varepsilon + \Pi_{it}\varphi_{i;t}\varepsilon \\ + \lambda_{ixx}\varphi_{i;tx}\varepsilon + \lambda_{iyy}\varphi_{i;ty}\varepsilon + \lambda_{izz}\varphi_{i;tz}\varepsilon - L\varepsilon = \text{const.},\end{aligned} \tag{65}$$

where after the substitution of $\Theta_{\mu4}$, $\Pi_{i\mu}$, and $\lambda_{i\nu}$ we can see, that the left-hand side of the equation is zero. We can conclude that we did not obtain newer conserved quantities; i.e. since we chose the source density σ_i zero, we know that these equations must necessarily express conservation. This fact is in line that the Euler–Lagrange equations (29) are continuity equations. The K_μ can be given for translation and rotation in space, from which we obtain the same conclusion. Summarizing the results we say that the Euler–Lagrange equations by the Lagrange density function (24) are invariant with respect to the geometrical transformations, and we do not get further conserved quantity [20,21,30].

There are two coefficient matrices S_{ik} and L_{ik} defined by the conditions with Eqs. (24)–(26). We assumed that these are constant coefficients and the inverse of S_{ik} exists. We know from the Euler–Lagrange equations and the entropy-balance equation that these are exactly the phenomenological coefficient matrices. The question is, how can we conclude these symmetries within the framework of formalism, i.e. what kind of deeper physical meaning, inner symmetry may be behind the symmetry of coefficient matrices?

Let $\varphi_1(\mathbf{r},t)$, $\varphi_2(\mathbf{r},t), \ldots \varphi_K(\mathbf{r},t)$ be linearly independent four-times continuously differentiable scalar-vector functions. We consider the linear vector space ϕ_K over the field of real numbers, the basis vectors are φ_1, $\varphi_2, \ldots \varphi_K$. If we introduce multiplication among the elements of ϕ_K, which means the usual multiplication of scalar–vector functions, then we get a linear associative and commutative algebra (A_K). If we have an arbitrary associative algebra, a Lie algebra can be constructed from it by introducing a different multiplication

$$a \odot b = (ab - ba). \tag{66}$$

Let us consider the following infinitesimal and linear transformation, which transforms the basis vectors φ_i into the basis vectors φ_i'

$$\varphi_i \mapsto \varphi_i' = \delta_{il}\varphi_l - \Theta T_{il}\hat{P}\varphi_l, \tag{67}$$

the inverse of transformation $\hat{T} : \phi_K \to \phi_K'$ is

$$\hat{T}^{-1} : \phi_K' \to \phi_K, \tag{68}$$

$$\varphi_j = \delta_{jl}\varphi_l' + \Theta T_{jl}\hat{P}\varphi_l'. \tag{69}$$

The δ_{il} is the Kronecker symbol, Θ is an infinitesimal constant, T_{il} is the mixing matrix in which the elements $T_{ii} = 0$, $\hat{P}$ is the ordering operator. The ordering operator has the following properties:

(1) $\hat{P} = +1$, if in the case of $\varphi_{i_1}\varphi_{i_2}$ the ordering $i_1\ i_2$ is even (i.e. the number of inversions is 0, $i_1 > i_2$), and it leaves the ordering unchanged;

(2) $\hat{P} = 0$, if there is no ordering ($i_1 = i_2$);

(3) $\hat{P} = -1$, if the ordering $i_1 i_2$ is odd (i.e. the number of inversions is 1, $i_2 > i_1$), and it makes the ordering ($i_1 i_2 \rightarrow i_2 i_1$) inversion free.

Moreover, the operator $\hat{P}$ commutes with all other operators.

For the sake of simplicity, we consider the case $K = 3$, and we calculate the following products

$$\begin{aligned}\varphi_2'\varphi_3' &= (\varphi_2 - \Theta T_{21}\hat{P}\varphi_1 - \Theta T_{23}\hat{P}\varphi_3)(\varphi_3 - \Theta T_{31}\hat{P}\varphi_1 - \Theta T_{32}\hat{P}\varphi_2)\\ &= \varphi_2\varphi_3 - \Theta T_{21}\hat{P}\varphi_1\varphi_3 - \Theta T_{23}\hat{P}\varphi_3\varphi_3 - \varphi_2\Theta T_{31}\hat{P}\varphi_1 - \varphi_2\Theta T_{32}\hat{P}\varphi_2\\ &= \varphi_2\varphi_3 - \Theta T_{21}\varphi_1\varphi_3 + \Theta T_{31}\varphi_1\varphi_2,\end{aligned} \tag{70}$$

$$\varphi_3'\varphi_2' = \varphi_3\varphi_2 - \Theta T_{31}\varphi_1\varphi_2 + \Theta T_{21}\varphi_1\varphi_3. \tag{71}$$

We take the difference of these products

$$\varphi_2'\varphi_3' - \varphi_3'\varphi_2' = 2\Theta(T_{31}\varphi_1\varphi_2 - T_{21}\varphi_1\varphi_3), \tag{72}$$

which induces the introduction of a different algebraic product among the elements of ϕ_K'

$$\varphi_j'\odot\varphi_k' = \varphi_j'\varphi_k' - \varphi_k'\varphi_j' = 2\Theta(T_{kl}\varphi_l\varphi_j - T_{jm}\varphi_m\varphi_k), \tag{73}$$

where $l, m \neq j, k$. It is easy to prove that

$$\varphi_j'\odot\varphi_k' = -\varphi_k'\odot\varphi_j' \tag{74}$$

is completed. We can conclude that $\hat{T}(\phi_K) = \phi_K'$ linear vector space with the above-defined multiplication $\odot$ is a Lie algebra [21,31,32].

If the action is invariant with respect to a dynamical transformation as a consequence of an inner symmetry, it means mathematically that

$$L(\varphi_i(r,t), \varphi_{i;\mu}(r,t), \varphi_{i;\mu\nu}(r,t)) = L(\varphi_i'(r,t), \varphi_{i;\mu}'(r,t), \varphi_{i;\mu\nu}'(r,t)). \tag{75}$$

So, if we consider dynamical transformations only

$$\varphi(r,t) \rightarrow \varphi'(r,t), \tag{76}$$

it is sufficient to examine the invarinace of the Lagrange density function. Transforming the Lagrange density function it can be easily seen that it is invariant with respect to $\hat{T}$, i.e. $\hat{T}$ is the dynamical transformation of the system.

Taking into account that in the theory the specific entropy density and the entropy-production density have the same central role as the Lagrange density function, therefore, their invariances with respect to the transformation $\hat{T}$ are quite obvious. Let us examine this for the specific entropy density in detail:

$$
\begin{aligned}
s &= \tfrac{1}{2}\rho^{-1}[\rho S_{ik}^{-1}(\rho S_{fk}\varphi_{f;t} - L_{fk}\Delta\varphi_f)S_{ij}\rho S_{jl}^{-1}(\rho S_{gl}\varphi_{g;t} - L_{gl}\Delta\varphi_g)] \\
&= \underline{\tfrac{1}{2}\rho^3 S_{ik}^{-1}S_{fk}\varphi_{f;t}S_{ij}S_{jl}^{-1}S_{gl}\varphi_{g;t}} - \tfrac{1}{2}\rho^2 S_{ik}^{-1}L_{fk}\Delta\varphi_f S_{ij}S_{jl}^{-1}S_{gl}\varphi_{g;t} \\
&\quad - \tfrac{1}{2}\rho^2 S_{ik}^{-1}S_{fk}\varphi_{f;t}S_{ij}S_{jl}^{-1}L_{gl}\Delta\varphi_g + \tfrac{1}{2}\rho S_{ik}^{-1}L_{fk}\Delta\varphi_f S_{ij}S_{jl}^{-1}L_{gl}\Delta\varphi_g.
\end{aligned}
\tag{77}
$$

The calculation must be done for every term, but the calculation is similar, so we show the transformation for the underlined term $[s]_1$. The transformed term is

$$
\begin{aligned}
[s']_1 &= \tfrac{1}{2}\rho^3 S_{ik}^{-1}S_{fk}\varphi'_{f;t}S_{ij}S_{jl}^{-1}S_{gl}\varphi'_{g;t} \\
&= \tfrac{1}{2}\rho^3 S_{ik}^{-1}S_{fk}S_{ij}S_{jl}^{-1}S_{gl}(\delta_{fm}\varphi_{m;t} - \Theta T_{fm}\hat{P}\varphi_{m;t})(\delta_{gp}\varphi_{p;t} - \Theta T_{gp}\hat{P}\varphi_{p;t}) \\
&= \tfrac{1}{2}\rho^3 S_{ik}^{-1}S_{fk}S_{ij}S_{jl}^{-1}S_{gl}(\varphi_{f;t}\varphi_{g;t} - \Theta T_{fm}\hat{P}\varphi_{m;t}\varphi_{g;t} - \varphi_{f;t}\Theta T_{gp}\hat{P}\varphi_{p;t}) \\
&= \tfrac{1}{2}\rho^3 S_{ik}^{-1}S_{fk}S_{ij}S_{jl}^{-1}S_{gl}\varphi_{f;t}\varphi_{g;t} - \tfrac{1}{2}\rho^3 S_{ik}^{-1}S_{fk}S_{ij}S_{jl}^{-1}S_{gl}\Theta T_{fm}\hat{P}\varphi_{m;t}\varphi_{g;t} \\
&\quad - \tfrac{1}{2}\rho^3 S_{ik}^{-1}S_{fk}S_{ij}S_{jl}^{-1}S_{gl}\varphi_{f;t}\Theta T_{gp}\hat{P}\varphi_{p;t}.
\end{aligned}
\tag{78}
$$

We consider the meaning of the operator $\hat{P}$, and neglecting the second-order terms of Θ, we change the indices $m \to p$, $f \to g$, $g \to f$, $i \to j$, $j \to i$, $l \to k$ and $k \to l$ in the second term, then we obtain

$$
\begin{aligned}
[s']_1 &= \tfrac{1}{2}\rho^3 S_{ik}^{-1}S_{fk}S_{ij}S_{jl}^{-1}S_{gl}\varphi_{f;t}\varphi_{g;t} - \tfrac{1}{2}\rho^3 S_{jl}^{-1}S_{gl}S_{ji}S_{ik}^{-1}S_{fk}\Theta T_{gp}\varphi_{p;t}\varphi_{f;t} \\
&\quad + \tfrac{1}{2}\rho^3 S_{ik}^{-1}S_{fk}S_{ij}S_{jl}^{-1}S_{gl}\Theta T_{gp}\varphi_{p;t}\varphi_{f;t}.
\end{aligned}
\tag{79}
$$

We can conclude that the second and third terms eliminate each other if and only if the

$$
S_{ij} = S_{ji} \tag{80}
$$

symmetry is completed. This examination can be done for the entropy-production density function. For the calculations we use its form expressed by φ_i

$$
\begin{aligned}
\sigma(\varphi) &= \nabla(\rho S_{ji}^{-1}\varphi_{j;t} - L_{ji}\Delta\varphi_j)L_{ik}\nabla(\rho S_{mk}^{-1}\varphi_{m;t} - L_{mk}\Delta\varphi_m) \\
&= \rho S_{ji}^{-1}L_{ik}S_{mk}^{-1}\nabla\varphi_{j;t}\nabla\varphi_{m;t} - L_{ji}L_{ik}\rho S_{mk}^{-1}\nabla\Delta\varphi_j\nabla\varphi_{m;t} \\
&\quad - \rho S_{ji}^{-1}L_{ik}L_{mk}\nabla\varphi_{j;t}\nabla\Delta\varphi_m + L_{ji}L_{ik}L_{mk}\nabla\Delta\varphi_j\nabla\Delta\varphi_m.
\end{aligned}
\tag{81}
$$

The entropy-production density is the measure of irreversibility of the processes, and this is why we demand its invariance with respect to $\hat{T}$

$$
\sigma(\varphi) = \sigma(\varphi'). \tag{82}
$$

After the calculations we can see that the invariance is completed if

$$
L_{ik} = L_{ki}. \tag{83}
$$

We have shown that the description of dissipative processes in the vector space φ_i and in the Lie algebra generated by φ_i' are equivalent to each other if the phenomenological coefficient matrices are symmetric [20,21,30].

Since the symmetry of the coefficient matrix S_{ik} is the connection with the stability of the local equilibrium, the symmetry of L_{ik} is the connection with the stability of the steady state [33], we think that the transformation $\hat{T}$ may be the connection with the stability of these simple dissipative processes on the levels of the potentials φ_i.

5. Stochastic properties of parabolic processes

On the basis of previous results we examine the stochastic behavior [34] of processes described by parabolic differential equations. For the sake of simplicity we consider only one process. The Lagrange density function is

$$L = \frac{1}{2}\left(\rho S^{-1}\frac{\partial \varphi}{\partial t} - L\Delta\varphi\right)^2, \tag{84}$$

and the intensive quantity will be

$$\Gamma = \rho S^{-1}\frac{\partial \varphi}{\partial t} - L\Delta\varphi. \tag{85}$$

The Euler–Lagrange equation for Γ can be written

$$\rho S^{-1}\frac{\partial \Gamma}{\partial t} + L\Delta\Gamma = 0. \tag{86}$$

The canonically conjugated quantity for φ is

$$p = \frac{\partial L}{\partial \varphi_t} = \rho S^{-1}\Gamma = (\rho S^{-1})^2\varphi_{;t} - \rho S^{-1}L(\Delta\varphi), \tag{87}$$

where $\varphi_{;t} = \partial\varphi/\partial t$. We calculate the Hamilton density function

$$H = p\varphi_{;t} - L, \tag{88}$$

which can be expressed in another form

$$H(p, \Delta\varphi) = \tfrac{1}{2}(\rho^{-1}S)^2 p^2 + \rho^{-1}SLp\Delta\varphi \tag{89}$$

after the substitution of the time derivative $\varphi_{;t}$. The canonical equations are

$$\varphi_{;t} = \frac{\partial H}{\partial p}, \tag{90}$$

$$p_{;t} = -\Delta\frac{\partial H}{\partial \Delta\varphi}, \tag{91}$$

where $p_{;t} = \partial p/\partial t$. These may be given by Poisson-bracket expressions

$$\varphi_{;t} = [\varphi, H] = (\rho^{-1}S)^2 p + \rho^{-1}SL\Delta\varphi, \tag{92}$$

$$p_{;t} = [p, H] = -\rho^{-1} SL\Delta p. \tag{93}$$

In order to examine Eqs. (92) and (93) from the stochastic viewpoint, let us consider the quantites φ and p as fluctuating variables of the system. This assumption follows from the fact that the system is out of equilibrium, uncontrolled inner processes may exist, which may cause accidentally microscopic changes. Our aim is to construct the probability field by the paths of the system in the conjugated space.

We consider the copies of original system defined by the conjugated variables with the same initial conditions. We would like to study this ensemble in the 'phase field' (φ, p). The definition of the probability field of paths must be in line with the fact that the action S is minimal on the path given by Eqs. (92) and (93), since this represents the most probabilistic behavior of the system [35–37]. The Poisson structure allows a description of the time evolution of the process in the space of canonically conjugated quantities (φ, p) to be given. First, we examine Liouville's theorem in this phase field [38–41]. From Eqs. (92) and (93) we can see

$$\frac{\delta}{\delta\varphi}\frac{\partial\varphi}{\partial t} + \frac{\delta}{\delta p}\frac{\partial p}{\partial t} = 0. \tag{94}$$

It follows from this that the phase volume does not change in the space of conjugated variables. After all, it can be seen that the probability density $W(\varphi, p, t)$ fulfills the Liouville equation

$$\frac{\partial W}{\partial t} = [W, H]. \tag{95}$$

$W(\varphi, p, t)$ represents the set of probability densities $W_1(\varphi_1, p_1, t_1)$, $W_2(\varphi_2, p_2, t_2), \ldots$, but it is enough to know only one solution and the transition probabilities between the states

$$\begin{aligned} &W_N(\varphi_1, p_1, t_1, \ldots, \varphi_N, p_N, t_N) \\ &\quad = P(\varphi_{N-1}, p_{N-1}, t_{N-1}/\varphi_N, p_N, t_N) \times W_{N-1}(\varphi_1, p_1, t_1, \ldots, \varphi_{N-1}, p_{N-1}, t_{N-1}). \end{aligned} \tag{96}$$

So, with the help of W we can calculate the average of arbitrary quantity

$$\langle F(\varphi, p)\rangle = \frac{\int F(\varphi, p) W(\varphi, p)\mathrm{d}\Omega}{\int W(\varphi, p)\mathrm{d}\Omega}, \tag{97}$$

where $\mathrm{d}\Omega = \mathrm{d}\varphi\,\mathrm{d}p$ is the volume element. For this we need the transition probabilities [42–44]. Let the system be in the state (φ, p) tending towards the state (φ', p'). We calculate the most probabilistic change of the probability density between two states. The transition probability can be defined as a conditional probability of the fact that the system is in the state (φ', p') at time $t = \tau$ if it was in the state (φ, p) at $t = 0$, i.e. $(\varphi, p)_{t=0} \rightarrow (\varphi', p')_{t=\tau}$

$$P_\tau(\varphi'/\varphi) = \frac{\exp\left[-\frac{1}{k}S_\tau(\varphi'/\varphi)\right]\frac{\partial p}{\partial\varphi}}{\int \mathrm{d}p \exp\left[-\frac{1}{k}S_\tau(\varphi'/p)\right]}. \tag{98}$$

The action for the time interval $(0, \tau)$ is

$$S_\tau(\varphi'/\varphi) = \int_0^\tau (p\varphi_{;t} - H)\mathrm{d}t = \int_0^\tau \left(p\varphi_{;t} - \frac{1}{2}(\rho S^{-1})^2 p^2 - \rho^{-1} SLp\Delta\varphi\right)\mathrm{d}t. \tag{99}$$

To calculate S_τ we suppose that τ is finite small. The value of τ is smaller than the usual values on the hydrodynamic scale, but greater than the kinetic scale. In this way, we can obtain

$$\varphi(\tau) = \varphi + \tau\varphi_{;t} + \cdots \Rightarrow \varphi_{;t} = \frac{\varphi' - \varphi}{\tau}, \tag{100}$$

$$p = (\rho S^{-1})^2 \frac{\varphi' - \varphi}{\tau} - \rho S^{-1} L\Delta\varphi. \tag{101}$$

Then,

$$\frac{\partial p}{\partial \varphi'} = \frac{(\rho S^{-1})^2}{\tau} \tag{102}$$

is completed. With the help of this S_τ and P_τ can be calculated

$$S_\tau(\varphi'/\varphi) = \frac{1}{2}(\rho S^{-1})^2 \frac{(\varphi' - \varphi)^2}{\tau} - \rho S^{-1} L(\varphi' - \varphi)\Delta\varphi + \frac{1}{2}\tau L^2(\Delta\varphi)^2, \tag{103}$$

$$P_\tau(\varphi'/\varphi) = \sqrt{\frac{(\rho S^{-1})^2}{2k\pi\tau}} \times \exp\left(-\frac{1}{2}\frac{(\rho S^{-1})^2}{k\tau}(\varphi' - \varphi)^2 + \frac{\rho S^{-1} L}{k}(\varphi' - \varphi)\overline{\Delta\varphi} - \frac{1}{2}\frac{\tau L^2}{k}(\overline{\Delta\varphi})^2\right). \tag{104}$$

Here, we assumed that τ is infinitesimal, and now we know that all of τ', τ'' are of the same order. On the other hand, we introduce a notation, the expression $\overline{\Delta\varphi}$ is the value of $\Delta\varphi$ at the point $1/2(\varphi' + \varphi)$. Let us calculate the following integral

$$\begin{aligned}\int \mathrm{d}\varphi' P_{\tau''}(\varphi''/\varphi')P_{\tau'}(\varphi'/\varphi) &= \int \mathrm{d}\varphi' \sqrt{\frac{(\rho S^{-1})^2}{2k\pi\tau''}} \\ &\times \exp\left(-\frac{1}{2}\frac{(\rho S^{-1})^2}{k\tau''}(\varphi'' - \varphi')^2 + \frac{\rho S^{-1} L}{k}(\varphi'' - \varphi')\overline{\Delta\varphi} - \frac{1}{2}\frac{\tau'' L^2}{k}(\overline{\Delta\varphi})^2\right) \\ &\times \sqrt{\frac{(\rho S^{-1})^2}{2k\pi\tau'}}\exp\left(-\frac{1}{2}\frac{(\rho S^{-1})^2}{k\tau'}(\varphi' - \varphi)^2 + \frac{\rho S^{-1} L}{k}(\varphi' - \varphi)\overline{\Delta\varphi} - \frac{1}{2}\frac{\tau' L^2}{k}(\overline{\Delta\varphi})^2\right).\end{aligned} \tag{105}$$

In the integration, where we applied

$$\int_{-\infty}^{\infty} \exp(-ax^2 + bx)\mathrm{d}x = \sqrt{\frac{\pi}{a}}\exp\frac{b^2}{4a}, \tag{106}$$

by which we obtain

$$\sqrt{\frac{(\rho S^{-1})^2}{2k\pi(\tau' + \tau'')}}\exp\left(-\frac{1}{2}\frac{(\rho S^{-1})^2}{k\tau''}\varphi''^2 - \frac{1}{2}\frac{(\rho S^{-1})^2}{k\tau'}\varphi^2 \right.$$
$$\left. + \frac{1}{2}\frac{(\rho S^{-1})^2}{k}\left(\frac{\varphi''}{\tau''} + \frac{\varphi}{\tau'}\right)^2 \frac{\tau'\tau''}{\tau' + \tau''} + \frac{\rho S^{-1}L}{k}(\varphi'' - \varphi)\overline{\Delta\varphi} - \frac{1}{2}\frac{(\tau' + \tau'')L^2}{k}(\overline{\Delta\varphi})^2\right)$$
$$= \sqrt{\frac{(\rho S^{-1})^2}{2k\pi\tau}}\exp\left(-\frac{1}{2}\frac{(\rho S^{-1})^2}{k\tau}(\varphi'' - \varphi)^2 + \frac{\rho S^{-1}L}{k}(\varphi'' - \varphi)\overline{\Delta\varphi} - \frac{1}{2}\frac{\tau L^2}{k}(\overline{\Delta\varphi})^2\right)$$
$$= P_\tau(\varphi''/\varphi). \tag{107}$$

This is the Chapman–Kolmogorov equation. Thus, it can be seen that the parabolic processes are stochastic processes, and these are Markov processes on the new phase field.

It comes from the previous results that the Kramers–Moyal equation [45] can be written for the probabilty density

$$\tau\frac{\partial W(\varphi, t)}{\partial t} = \sum_{n=1}^{\infty}\left(-\frac{\partial}{\partial\varphi}\right)^n\left(\frac{M_n}{n!}W\right), \tag{108}$$

where M_n denotes the n-th momentum. The first two momenta are

$$M_1 = \int d\varphi'(\varphi' - \varphi)P_\tau(\varphi'/\varphi), \tag{109}$$

$$M_2 = \int d\varphi'(\varphi' - \varphi)^2 P_\tau(\varphi'/\varphi). \tag{110}$$

Taking into account the form of probability (104) we calculate these momenta. Omitting the detailed calculation we get

$$M_1 = \frac{L\tau}{\rho S^{-1}}\overline{\Delta\varphi}, \tag{111}$$

$$M_2 = \frac{k\tau}{(\rho S^{-1})^2} + \frac{L^2\tau^2}{(\rho S^{-1})^2}(\overline{\Delta\varphi})^2. \tag{112}$$

We neglect the second-order term of τ in the second momentum, moreover, those momenta, which are greater than second order, can be neglected because of the order of τ. We substitute both of the momenta into Eq. (108), and simplifying by τ we obtain

$$\frac{\partial W}{\partial t} = -\frac{L}{\rho S^{-1}}\overline{\Delta\varphi}\frac{\partial W}{\partial\varphi} + \frac{1}{2}\frac{k}{(\rho S^{-1})^2}\frac{\partial^2 W}{\partial\varphi^2}, \tag{113}$$

which is generally the Fokker–Planck equation pertaining to parabolic equations. This equation makes a connection between the potential φ and probability density, which may explain a kind of meaning of the potential at the same time. Finally, the solution of the

Fokker–Planck equation is

$$W(\varphi, t) = \left(\frac{2\pi k t}{(\rho S^{-1})^2} \right)^{-1/2} \exp\left(- \frac{\left(\varphi - \frac{L\bar{\Delta}\varphi}{\rho S^{-1}} t \right)^2}{\frac{2k}{(\rho S^{-1})^2} t} \right). \tag{114}$$

In the following, let us examine the fluctuation of the measurable quantity $\Gamma(x,t) = \Gamma(\varphi, p)$. The fluctuation is the deviation from the steady state Γ_s [2,7,33],

$$\delta\Gamma = \Gamma - \Gamma_s. \tag{115}$$

The time evolution of the fluctuation can be expressed by the Poisson bracket

$$\frac{\partial\, \delta\Gamma}{\partial t} = [\delta\Gamma, H]. \tag{116}$$

In the knowledge of $H(p, \delta\varphi)$ and $\Gamma(\varphi, p) = \rho^{-1} Sp$, moreover, Eqs. (92) and (93), we obtain the differential equation for $\delta\Gamma$

$$\frac{\partial\, \delta\Gamma}{\partial t} = -\rho^{-1} SL\Delta(\delta\Gamma), \tag{117}$$

which has the same form as that of Eq. (86). After all, we can say something about the average regression of fluctuation of measurable quantity. We define the conditional average: if the fluctuation $\delta\Gamma'$ evolves the fluctuation $\delta\Gamma$ after time τ then

$$\langle \delta\Gamma / \delta\Gamma' \rangle_\tau = \int \delta\Gamma P_\tau(\delta\Gamma_1 / \delta\Gamma') \mathrm{d}(\delta\Gamma_1), \tag{118}$$

where $\delta\Gamma_1$ is an intermediate value of the fluctuation and P_τ is the transition probability. We express the time derivative of the right-hand side

$$\int \frac{\partial\, \delta\Gamma}{\partial t} P_\tau(\delta\Gamma_1 / \delta\Gamma') \mathrm{d}(\delta\Gamma_1), \tag{119}$$

where P_τ, the transition probability does not change during the time interval τ, therefore, its derivative is zero. We express the function $P_\tau(\delta\Gamma_1/\delta\Gamma')$ to continue the calculation. Taking into account the connection between Γ and φ by Eq. (85), the related Eq. (100) for $\varphi_{;t}$ and the form of $P_\tau(\varphi'/\varphi)$ by Eq. (104) we obtain

$$P_\tau(\Gamma'/\Gamma) = \sqrt{\frac{(\rho S^{-1})^2}{2k\pi\tau}} \exp\left(-\frac{\tau}{2k} (\Gamma' - \Gamma)^2 \right). \tag{120}$$

Expressing the fluctuations

$$\delta\Gamma' = \Gamma' - \Gamma_s,$$

$$\delta\Gamma = \Gamma - \Gamma_s$$

we can immediately write $P_\tau(\delta\Gamma'/\delta\Gamma)$

$$P_\tau(\delta\Gamma'/\delta\Gamma) = \sqrt{\frac{(\rho S^{-1})^2}{2k\pi\tau}} \exp\left(-\frac{\tau}{2k} (\delta\Gamma' - \delta\Gamma)^2 \right). \tag{121}$$

Using Eqs. (117) and (121) one can get

$$\frac{\partial \langle \delta \Gamma / \delta \Gamma' \rangle}{\partial t} = \rho^{-1} S L \Delta \langle \delta \Gamma / \delta \Gamma' \rangle. \tag{122}$$

The comparison of this equation with Eq. (86) makes the Onsager's regression hypothesis evident within the present model.

6. Hyperbolic dissipative processes

To extend the theory we deal with the Lagrange formalism of such dissipative processes that are described by hyperbolic partial differential equations, the so-called telegrapher equations. In the case of these kinds of processes the different physical quantities propagate as damped waves. However, the theory of hyperbolic transports could not be based on the hypothesis of local equilibrium [9,46–52]. These kinds of differential equations also contain first-order time derivatives, which shows the fact of dissipation. Here, we construct the Lagrange function for these equations, and we apply the Hamilton's principle and the canonical formalism.

Let the Lagrangian be

$$L = \frac{1}{2}\left(\frac{\partial^2 \phi}{\partial t^2}\right)^2 + \frac{1}{2}\alpha^2\left(\frac{\partial \phi}{\partial t}\right)^2 + \frac{1}{2}\beta^2(\Delta\phi)^2 - \beta\frac{\partial^2 \phi}{\partial t^2}\Delta\phi, \tag{123}$$

where ϕ is a scalar field, α and β are constant coefficients. The connection between the field ϕ and the measurable physical quantity M is defined by

$$M = \frac{\partial^2 \phi}{\partial t^2} - \alpha\frac{\partial \phi}{\partial t} - \beta(\Delta\phi). \tag{124}$$

Applying the Hamilton's principle we obtain the Euler–Lagrange equation for ϕ

$$0 = \frac{\partial^4 \phi}{\partial t^4} - \alpha^2\frac{\partial^2 \phi}{\partial t^2} + \beta^2\Delta\Delta\phi - 2\beta\Delta\frac{\partial^2 \phi}{\partial t^2}. \tag{125}$$

Using the relation of ϕ and M we get a field equation for M

$$\frac{\partial^2 M}{\partial t^2} + \alpha\frac{\partial M}{\partial t} - \beta\Delta M = 0, \tag{126}$$

which is a hyperbolic differential equation. Following the usual method of the total variation by Eqs. (30), (31), (34)–(36) for hyperbolic case, the important quantity Θ_{44} will be

$$\Theta_{44} = L - \frac{\partial \phi}{\partial t}\frac{\partial L}{\partial \frac{\partial \phi}{\partial t}} + \frac{\partial \phi}{\partial t}\frac{\partial}{\partial t}\frac{\partial L}{\partial \frac{\partial^2 \phi}{\partial t^2}} - \frac{\partial^2 \phi}{\partial t^2}\frac{\partial L}{\partial \frac{\partial^2 \phi}{\partial t^2}}. \tag{127}$$

The canonically conjugated quantities have the form

$$P_\phi = \pi_t = \frac{\partial L}{\partial \frac{\partial \phi}{\partial t}} - \frac{\partial}{\partial t} \frac{\partial L}{\partial \frac{\partial^2 \phi}{\partial t^2}}, \tag{128}$$

$$P_\psi = \lambda_{tt} = \frac{\partial L}{\partial \frac{\partial^2 \phi}{\partial t^2}}. \tag{129}$$

Now, we define a new variable

$$\psi \doteq \frac{\partial \phi}{\partial t}. \tag{130}$$

Thus, it can be seen that we have two pairs of conjugated variables

$$\phi \rightarrow P_\phi, \tag{131}$$

$$\psi \rightarrow P_\psi. \tag{132}$$

ϕ and ψ are the generalized coordinates, P_ϕ and P_ψ are the generalized momenta. The Hamilton function relates to Θ_{44}

$$H = -\Theta_{44}, \tag{133}$$

so the Hamilton by the relevant Lagrangian (123) is

$$H = \frac{1}{2}\left(\frac{\partial^2 \phi}{\partial t^2}\right)^2 + \frac{1}{2}\alpha^2\left(\frac{\partial \phi}{\partial t}\right)^2 - \frac{1}{2}\beta^2(\Delta\phi)^2 - \frac{\partial \phi}{\partial t}\frac{\partial^3 \phi}{\partial t^3} + \beta\frac{\partial \phi}{\partial t}\Delta\frac{\partial \phi}{\partial t}. \tag{134}$$

The momenta are

$$P_\phi = \alpha^2\frac{\partial \phi}{\partial t} - \frac{\partial^3 \phi}{\partial t^3} + \beta\Delta\frac{\partial \phi}{\partial t}, \tag{135}$$

$$P_\psi = \frac{\partial^2 \phi}{\partial t^2} - \beta\Delta\phi. \tag{136}$$

The Hamilton function can be simply obtained by the conjugated variables

$$H(\phi, \Delta\phi, \psi, P_\phi, P_\psi) = \tfrac{1}{2}P_\psi^2 + \beta P_\psi \Delta\phi + P_\phi \psi - \tfrac{1}{2}\alpha^2\psi^2. \tag{137}$$

In the following we calculate the canonical equations and Poisson bracket expressions. If we have two functions $f(\phi, \nabla\phi, \Delta\phi, p, \nabla p, \Delta p)$ and $g(\phi, \nabla\phi, \Delta\phi, p, \nabla p, \Delta p)$, then the Poisson bracket of these equations is (similar to Eq. (48))

$$[f, g] = \frac{\delta f}{\delta \phi}\frac{\delta g}{\delta p} - \frac{\delta g}{\delta \phi}\frac{\delta f}{\delta p}. \tag{138}$$

Here,

$$\frac{\delta f}{\delta \phi} = \frac{\partial f}{\partial \phi} - \nabla\cdot\frac{\partial f}{\partial \nabla\phi} + \Delta\frac{\partial f}{\partial \Delta\phi} \quad \text{and} \quad \frac{\delta f}{\delta p} = \frac{\partial f}{\partial p} - \nabla\cdot\frac{\partial f}{\partial \nabla p} + \Delta\frac{\partial f}{\partial \Delta p}$$

are the functional derivatives. Since the Hamilton function does not depend explicitly on time we can calculate the time evolution of generalized coordinates and momenta in time

$$\frac{\partial \phi}{\partial t} = [\phi, H], \tag{139}$$

$$[\phi, H] = \frac{\delta \phi}{\delta \phi}\frac{\delta H}{\delta P_\phi} - \frac{\delta H}{\delta \phi}\frac{\delta \phi}{\delta P_\phi} = \frac{\delta H}{\delta P_\phi}, \tag{140}$$

$$\frac{\partial \phi}{\partial t} = \frac{\delta H}{\delta P_\phi} = \psi. \tag{141}$$

The last equation is the first canonical equation for the quantity ϕ, which is exactly the definition of ψ. The first canonical equation for the second quantity ψ can be deduced similarly

$$\frac{\partial \psi}{\partial t} = [\psi, H], \tag{142}$$

$$[\psi, H] = \frac{\delta \psi}{\delta \psi}\frac{\delta H}{\delta P_\psi} - \frac{\delta H}{\delta \psi}\frac{\delta \psi}{\delta P_\psi} = \frac{\delta H}{\delta P_\psi}, \tag{143}$$

$$\frac{\delta H}{\delta P_\psi} = P_\psi + \beta\,\Delta\phi = \frac{\partial^2 \phi}{\partial t^2} = \frac{\partial \psi}{\partial t}. \tag{144}$$

Moreover, the second canonical equation for the first momentum P_ϕ will be

$$\frac{\partial P_\phi}{\partial t} = [P_\phi, H], \tag{145}$$

$$[P_\phi, H] = \frac{\delta P_\phi}{\delta \phi}\frac{\delta H}{\delta P_\phi} - \frac{\delta H}{\delta \phi}\frac{\delta P_\phi}{\delta P_\phi} = -\frac{\delta H}{\delta \phi} = -\beta\,\delta P_\psi. \tag{146}$$

The second canonical equation for the second momentum P_ψ is

$$\frac{\partial P_\psi}{\partial t} = [P_\psi, H], \tag{147}$$

$$[P_\psi, H] = \frac{\delta P_\psi}{\delta \psi}\frac{\delta H}{\delta P_\psi} - \frac{\delta H}{\delta \psi}\frac{\delta P_\psi}{\delta P_\psi} = -\frac{\delta H}{\delta \psi} = -P_\phi + \alpha^2\psi. \tag{148}$$

It can be seen that the canonical equations conserve their usual forms

$$\frac{\partial \phi}{\partial t} = \frac{\delta H}{\delta P_\phi}, \tag{149}$$

$$\frac{\partial \psi}{\partial t} = \frac{\delta H}{\delta P_\psi}, \tag{150}$$

$$\frac{\partial P_\phi}{\partial t} = -\frac{\delta H}{\delta \phi}, \tag{151}$$

$$\frac{\partial P_{\psi}}{\partial t} = -\frac{\delta H}{\delta \psi}. \tag{152}$$

Finally, summarizing the equations we obtain

$$\frac{\partial \phi}{\partial t} = \psi, \tag{153}$$

$$\frac{\partial \psi}{\partial t} = P_{\psi} + \beta \Delta \phi, \tag{154}$$

$$\frac{\partial P_{\phi}}{\partial t} = -\beta \Delta P_{\psi}, \tag{155}$$

$$\frac{\partial P_{\psi}}{\partial t} = -P_{\phi} + \alpha^2 \psi. \tag{156}$$

We can conclude that the theory is completed and contradiction free, and we defined a new phase field $(\phi, \psi, P_{\phi}, P_{\psi})$ [53].

Acknowledgement

This work was supported by the Bolyai János Research Scholarship of the Hungarian Academy of Sciences.

References

[1] Onsager, L. (1931), Phys. Rev. 37, 405; Phys. Rev. (1931) 38, 2265.
[2] Onsager, L. and Machlup, S. (1953), Phys. Rev. 91, 1505; Phys. Rev. (1953) 91, 1512.
[3] de Groot, S.R. and Mazur, P. (1962), Non-Equilibrium Thermodynamics, North-Holland, Amsterdam.
[4] Prigogine, I. (1969), Introduction to Thermodynamics of Irreversible Processes, Interscience, New York.
[5] Gyarmati, I. (1970), Non-Equilibrium Thermodynamics, Springer, Berlin.
[6] Kreuzer, H.J. (1981), Nonequilibrium Thermodynamics and its Statistical Foundations, Clarendon Press, Oxford.
[7] Keizer, J. (1987), Statistical Thermodynamics of Nonequilibrium Processes, Springer, Berlin.
[8] Sieniutycz, S. (1994), Conservation Laws in Variational Thermo-Hydrodynamics, Kluwer, Dordrecht.
[9] Jou, D., Casas-Vázquez, J. and Lebon, G. (1996), Extended Irreversible Thermodynamics, Springer, Berlin.
[10] Miller, D.G. (1960), Chem. Rev. 60, 15.
[11] Miller, D.G. (1974), The Onsager relation: experimental evidence. Foundations of Continuum Thermodynamics (Delgado Domingos, J.J., Nina, M.N.R., Whitelaw, J.H., eds.), MacMillan, London.
[12] Morse, Ph.M. and Feschbach, H. (1953), Methods of Theoretical Physics I, II, McGraw-Hill, New York.
[13] Akhiezer, A. and Bereztetski, V.B. (1963), Quantum Electrodynamics, Wiley, New York.
[14] Bjorken, J.D. and Drell, S.D. (1965), Relativistic Quantum Fields, McGraw-Hill, New York.
[15] Courant, R. and Hilbert, D. (1966), Methods of Mathematical Physics I, II, Interscience, New York.
[16] Dirac, P.A.M. (1958), The Principles of Quantum Mechanics, Clarendon Press, Oxford.
[17] Schiff, L.I. (1955), Quantum Mechanics, McGraw-Hill, New York.
[18] Sieniutycz, S. and Berry, R.S. (1989), Phys. Rev. A 40, 348; Phys. Rev. A (1992) 46, 6359; Phys. Rev. E (1993) 47, 1765.
[19] Márkus, F. and Gambár, K. (1991), J. Non-Equilib. Thermodyn. 16, 27.
[20] Gambár, K. and Márkus, F. (1994), Phys. Rev. E 50, 1227.

[21] Gambár, K., Martinás, K. and Márkus, F. (1997), Phys. Rev. E 55, 5581.
[22] Márkus, F. and Gambár, K. (1989), J. Non-Equilib. Thermodyn. 14, 355.
[23] Márkus, F. and Gambár, K. (1993), J. Non-Equilib. Thermodyn. 18, 288.
[24] Gambár, K. and Márkus, F. (2001), Open. Sys. Inf. Dyn. 8, 369.
[25] Márkus, F. and Gambár, K. (1995), Phys. Rev. E 52, 623; Phys. Rev. E (1996) 54, 4607.
[26] Márkus, F. (1999), Physica A 274, 563.
[27] Frampton, P.H. (1987), Gauge Field Theories, The Benjamin/Cummings, Menlo Park, CA.
[28] Wigner, E.P. (1954), J. Chem. Phys. 22, 1912.
[29] Wigner, E.P. (1954), Prog. Theor. Phys. 11, 437.
[30] Gambár, K. and Márkus, F. (1993), J. Non-Equilib. Thermodyn. 18, 51.
[31] Gilmore, R. (1974), Lie Groups, Lie Algebras, and Some of Their Applications, Wiley, New York.
[32] Sattinger, P.H. and Weaver, O.L. (1986), Lie Groups and Algebras with Applications to Physics, Geometry, and Mechanics, Springer, Berlin.
[33] Glansdorff, P. and Prigogine, I. (1971), Thermodynamic Theory of Structure Stability and Fluctuations, Wiley, New York.
[34] Hänggi, P. and Thomas, H. (1982), Phys. Rep. 88, 207.
[35] Feynman, R.P. and Hibbs, A.R. (1965), Quantum Mechanics and Path Integrals, McGraw-Hill, New York.
[36] Feynman, R.P. (1994), Statistical Mechanics, Addison-Wesley, Reading, MA.
[37] McKane, A.J., Luckock, H.C. and Bray, A.J. (1990), Phys. Rev. A 41, 644.
[38] Kirkwood, J.G. (1946), J. Chem. Phys. 14, 180.
[39] Green, M.S. (1952), J. Chem. Phys. 20, 1281.
[40] Penrose, O. (1979), Rep. Prog. Phys. 42, 1937.
[41] van Kampen, N.G. (1981), Stochastic Processes in Physics and Chemistry, North-Holland, Amsterdam.
[42] Grabert, H. and Green, M.S. (1979), Phys. Rev. A 19, 1747.
[43] Grabert, H., Graham, R. and Green, M.S. (1980), Phys. Rev. A 21, 2136.
[44] Graham, R. (1977), Z. Physik B 26, 281.
[45] Risken, H. (1989), The Fokker–Planck Equation, Springer, Berlin.
[46] Eckart, C. (1940), Phys. Rev. 58, 267; Phys. Rev. (1940) 58, 269.
[47] Eu, B.C. (1992), Kinetic Theory and Irreversible Thermodynamics, Wiley, New York.
[48] Vázquez, F., del Río, J.A., Gambár, K. and Márkus, F. (1996), J. Non-Equilib. Thermodyn. 21, 357.
[49] Vázquez, F., del Río, J.A. and de Haro, L.M. (1997), Phys. Rev. E 55, 5033.
[50] Vázquez, F. and del Río, J.A. (1998), Physica A 253, 290.
[51] Gambár, K., Márkus, F. and Nyíri, B. (1991), J. Non-Equilib. Thermodyn. 16, 217.
[52] Sieniutycz, S. (2001), Open Sys. Inf. Dyn. 8, 29.
[53] Márkus, F., Gambár, K., Vázquez, F. and del Río, J.A. (1999), Physica A 268, 482.

Chapter 13

HAMILTONIAN FORMULATION AS A BASIS OF QUANTIZED THERMAL PROCESSES

Ferenc Márkus

Institute of Physics, Budapest University of Technology and Economics, Budafoki út 8., H-1521 Budapest, Hungary

Abstract

A great challange is to introduce and apply the concepts of modern field theories to a theory, where dissipative processes are going on. In this chapter, through the example of heat conduction, we show how the quantization procedure can be treated. We discuss the energy and number operators, the fluctuations, the description of Bose systems and the q-boson approximation. The generalized Hamilton–Jacobi equation, special potentials—classical and quantum-thermodynamical—are also calculated. We point out an interesting connection of dissipation with the Fisher information and the extreme physical information.

Keywords: dissipative processes; canonical quantization; commutation rule; energy and number operators; energy fluctuation; Bose system; heat capacity; critical temperature; q-boson approximation; thermal uncertainty relation; Bohmian quantum dynamics; generalized Hamiltonian–Jacobi equation; classical and quantum-thermodynamical potentials; Fisher information; extreme physical information

1. Introduction

Hamilton's variational principle has a central role in the classical and modern field theories, and this principle is the basis of the developed field theory of nonequilibrium thermodynamics. The canonical formalism can be applied for physical processes, regardless of whether these are dissipative or nondissipative. As a special example, the Fourier heat conduction is examined. The Lagrangian, describing the process in the whole space, is constructed by such a potential function that is expressed by the coefficients of Fourier series. These coefficients are the generalized coordinates by which the canonical momenta, Hamiltonian, and the Poisson brackets can be calculated. This mathematical construction opens the way towards the canonical quantization that means that the operators can be introduced for physical quantities, e.g. energy and quasiparticle number operators for thermal processes. The elaborated quantization procedure can be applied for weakly interacting boson systems where the nonextensive thermodynamical behavior and the q-algebra can be taken into account and can be successfully built into

E-mail address: markus@phy.bme.hu (F. Márkus).

the theory. The Hamilton–Jacobi equation, the action and the kernel can be calculated in the space of generalized coordinates. This quantization method—which is called Feynman quantization—shows that repulsive potentials are working during the heat process, similar to a classical and a quantum-thermodynamical potential. It can be shown that this theory can be elegantly joined to the information theory proposed by Frieden.

2. Hamiltonian of heat conduction

We would like to apply the elaborated mathematical procedure of Hamilton's variational principle [1,2] to those physical processes that were out of the focus, namely to dissipative processes [3–5]. We restrict our attention to the parabolic processes, or to be more precise we deal with the linear heat conduction. The differential equation that describes the heat conduction contains the first-order time derivative, and this itself causes some mathematical difficulties in the development of the theory. We have pointed out how to introduce scalar fields by which the Lagrangian could be written [6], Hamilton's principle could be applied [7,8] and the quantization procedure could be introduced into the heat conduction [9,10]. The Lagrangian construction enabled us to use the mathematical tools of variational calculus, methods of statistical physics, and quantum-field theories.

As has been shown [6–8] the field equation of Fourier heat conduction (a linear parabolic differential equation for temperature T with constant coefficients)

$$c_v \dot{T} - \lambda \Delta T = 0 \tag{1}$$

can be deduced from the Lagrange density function

$$L = \frac{1}{2}\dot{\varphi}^2 + \frac{1}{2}\frac{\lambda^2}{c_v^2}(\Delta\varphi)^2, \tag{2}$$

where the scalar field φ is a potential function, the dot denotes the time derivative, Δ is the Laplace operator, λ is the heat conductivity and c_v is the specific heat capacity. Moreover, in the present work, we use the so-called natural boundary conditions. This potential generates the measurable local equilibrium temperature field

$$T = -\dot{\varphi} - \frac{\lambda}{c_v}\Delta\varphi. \tag{3}$$

If we consider the potential function φ as the generalized coordinate, then the generalized momentum P can be obtained by $P = \partial L/\partial\dot{\varphi} = \dot{\varphi}$, i.e. the quantities φ and P are canonically conjugated to each other. When they are considered as operators, the following rule can be required

$$\int_V [P, \varphi]\mathrm{d}V = h^*. \tag{4}$$

Here, h^* is the unit of action of dissipativity, $S = \int L\,\mathrm{d}V\,\mathrm{d}t$ ($h^* = 2\hbar/k_\mathrm{B}$, where $\hbar$ is the Planck constant divided by 2π, k_B is the Boltzmann constant). Now, we follow the methods of the theory of quantization [11,12]. We express φ in Fourier series [9]

$$\varphi = \sum_k \sqrt{2/V}(C_k \cos kx + S_k \sin kx), \tag{5}$$

where the coefficients C_k and S_k are just the functions of time, V is the volume, and k is the wave number. The momentum can be written in the form

$$P = \dot{\varphi} = \sum_k \sqrt{2/V}(\dot{C}_k \cos kx + \dot{S}_k \sin kx). \tag{6}$$

We obtain the Lagrangian of the system

$$L_V = \int L\,\mathrm{d}V = \frac{1}{2}\sum_k\left[(\dot{C}_k^2 + \dot{S}_k^2) + \frac{\lambda^2}{c_v^2}k^4(C_k^2 + S_k^2)\right], \tag{7}$$

and the canonically conjugated quantities are

$$P_k^{(C)} = \dot{C}_k \tag{8}$$

$$P_k^{(S)} = \dot{S}_k, \tag{9}$$

i.e. $P = \sum_k \sqrt{2/V}(P_k^{(C)} \cos kx + P_k^{(S)} \sin kx)$. The Hamiltonian of the system for the whole space can be obtained as

$$H = \sum_k\left(\frac{1}{2}P_k^{(C)^2} - \frac{1}{2}\frac{\lambda^2}{c_v^2}k^4 C_k^2\right) + \sum_k\left(\frac{1}{2}P_k^{(S)^2} - \frac{1}{2}\frac{\lambda^2}{c_v^2}k^4 S_k^2\right). \tag{10}$$

We see that the Hamiltonian is the sum of the differences of terms of momentum squares and coordinate squares. This means there is an essential mathematical difference between the structure of the Hamiltonian of heat conduction and the oscillator, so we need some discussion about the Hamiltonian. If we take this expression simply mathematically we can say that it may have no minimum value because of the differences, and it may be any negative value. However, we can limit the value of the Hamiltonian from below if we take into account that it describes the heat conduction, so we can find the connection between the Hamiltonian and the internal energy. The internal energy requires positive or zero values of the Hamiltonian, and this requirement comes from the physical viewpoint. This brings the assumption of 'zero point' energy into the theory.

Let the C_k, $P_k^{(C)}$, S_k and $P_k^{(S)}$ be operators ($\mathbf{C}_k$, $\mathbf{P}_k^{(C)}$, $\mathbf{S}_k$ and $\mathbf{P}_k^{(S)}$) obeying the commutation rules $[\mathbf{P}_k^{(C)}, \mathbf{C}_l] = h^*\delta_{kl}$ and $[\mathbf{P}_k^{(S)}, \mathbf{S}_l] = h^*\delta_{kl}$. We introduce new quantities [9,10] in rather unusual forms but quite similar to the quantization of fields [11,12]

$$\mathbf{C}_k^{\mp} = \frac{\mathbf{P}_k^{(C)}}{\sqrt{2}} \mp \frac{\lambda k^2}{\sqrt{2}c_v}\mathbf{C}_k, \tag{11}$$

$$\mathbf{S}_k^{\mp} = \frac{\mathbf{P}_k^{(S)}}{\sqrt{2}} \mp \frac{\lambda k^2}{\sqrt{2}c_v}\mathbf{S}_k. \tag{12}$$

The basic difference from the description of oscillator is the missing complex unit i in the second terms. We can obtain the commutation rules of $\mathbf{C}_k^+$, $\mathbf{C}_k^-$ and $\mathbf{S}_k^+$, $\mathbf{S}_k^-$ with the help of Eqs. (11) and (12)

$$\mathbf{C}_k^-\mathbf{C}_l^+ - \mathbf{C}_l^+\mathbf{C}_k^- = \frac{\lambda}{c_v}k^2 h^*\delta_{kl}, \tag{13}$$

$$\mathbf{S}_k^- \mathbf{S}_l^+ - \mathbf{S}_l^+ \mathbf{S}_k^- = \frac{\lambda}{c_v} k^2 h^* \delta_{kl}. \tag{14}$$

The operators $\mathbf{C}_k^+$ and $\mathbf{C}_k^-$ are independent of $\mathbf{S}_k^+$ and $\mathbf{S}_k^-$, these commutate with each other. We can examine the uncertainty of momentum and coordinate integrated over the volume V

$$\int_V [P, \varphi] \mathrm{d}V = \frac{1}{2} \frac{c_v}{\lambda k^2} (\mathbf{C}_k^- \mathbf{C}_k^+ - \mathbf{C}_k^+ \mathbf{C}_k^- + \mathbf{S}_k^- \mathbf{S}_k^+ - \mathbf{S}_k^+ \mathbf{S}_k^-) = h^*, \tag{15}$$

i.e. this requirement is completed. The Hamiltonian $\mathbf{H}$ of the system based on Eq. (10) can be written in the form

$$\mathbf{H} = \frac{1}{2} \sum_k (\mathbf{C}_k^+ \mathbf{C}_k^- + \mathbf{C}_k^- \mathbf{C}_k^+ + \mathbf{S}_k^+ \mathbf{S}_k^- + \mathbf{S}_k^- \mathbf{S}_k^+), \tag{16}$$

which is an operator from this point. This form of Hamilton operator [9] is suitable to connect to the energy and quasiparticle number operator.

3. Energy and number operator of heat conduction

In the above-described theory the Hamiltonian is not necessarily an energy-like quantity. It seems obvious that the energy operator should be calculated as a well-understandable physical quantity, mainly in the case of heat conduction. In this section we introduce the energy operator and we show its eigenvalues and their connection to the eigenvalues of the temperature operator [9,10]. Let the following operators be

$$\mathbf{N}_k^{(C)+} = \frac{c_v}{\lambda k^2 h^*} \mathbf{C}_k^+ \mathbf{C}_k^-, \tag{17}$$

$$\mathbf{N}_k^{(S)+} = \frac{c_v}{\lambda k^2 h^*} \mathbf{S}_k^+ \mathbf{S}_k^-, \tag{18}$$

$$\mathbf{N}_k^{(C)-} = \frac{c_v}{\lambda k^2 h^*} \mathbf{C}_k^- \mathbf{C}_k^+, \tag{19}$$

$$\mathbf{N}_k^{(S)-} = \frac{c_v}{\lambda k^2 h^*} \mathbf{S}_k^- \mathbf{S}_k^+. \tag{20}$$

Moreover, we create certain sums of these operators to define the operators $\mathbf{b}_k^+$ and $\mathbf{b}_k^-$

$$\mathbf{b}_k^+ \mathbf{b}_k^- = \tfrac{1}{2} (\mathbf{N}_k^{(C)+} + \mathbf{N}_k^{(S)+}), \tag{21}$$

$$\mathbf{b}_k^- \mathbf{b}_k^+ = \tfrac{1}{2} (\mathbf{N}_k^{(C)-} + \mathbf{N}_k^{(S)-}). \tag{22}$$

These operators obey the following commutation rule

$$\mathbf{b}_k^- \mathbf{b}_k^+ - \mathbf{b}_k^+ \mathbf{b}_k^- = 1, \tag{23}$$

i.e. we can see that $\mathbf{b}_k^+$ and $\mathbf{b}_k^-$ are creation and annihilation operators. The Hamilton operator can also be written in terms of the operators $\mathbf{b}_k^+$ and $\mathbf{b}_k^-$ [11]

$$\mathbf{H} = \sum_k \frac{\lambda k^2 h^*}{c_v} (\mathbf{b}_k^+ \mathbf{b}_k^- + \mathbf{b}_k^- \mathbf{b}_k^+). \tag{24}$$

It follows from these results that the

$$\mathbf{N}_k = \tfrac{1}{2} (\mathbf{b}_k^+ \mathbf{b}_k^- + \mathbf{b}_k^- \mathbf{b}_k^+) \tag{25}$$

can be considered as the particle-number operator pertaining to the relevant wave number k. This new operator is similar to the particle-number operators of quantum field theories. We can determine the eigenvalue of operator $\mathbf{N}_k$ that if Ψ is an eigenstate with the eigenvalue n_k then $\mathbf{b}_k^+ \Psi$ and $\mathbf{b}_k^- \Psi$ are also eigenstates

$$\mathbf{N}_k \Psi = \tfrac{1}{2} (\mathbf{b}_k^+ \mathbf{b}_k^- + \mathbf{b}_k^- \mathbf{b}_k^+) \Psi = n_k \Psi, \tag{26}$$

$$\mathbf{N}_k \mathbf{b}_k^+ \Psi = (n_k + 1) \Psi, \tag{27}$$

$$\mathbf{N}_k \mathbf{b}_k^- \Psi = (n_k - 1) \Psi. \tag{28}$$

These really show that $\mathbf{b}_k^+$ is a creation and $\mathbf{b}_k^-$ is an annihilation operator. Then, let the energy operator of the field be

$$\mathbf{E} = \sum_k \mathbf{E}_k, \tag{29}$$

where the different operators $\mathbf{E}_k$ belong to the different wave numbers k, and these are

$$\mathbf{E}_k = \lambda k^2 h^* \mathbf{N}_k. \tag{30}$$

The energy of the system must be non-negative. Now, we can obtain the energy eigenvalues

$$\mathbf{E}_k \Psi = \lambda k^2 h^* n_k \Psi. \tag{31}$$

Let us denote the least non-negative state by Ψ_0 (ground state) for which the condition

$$\mathbf{b}_k^- \Psi_0 = 0 \tag{32}$$

must hold. In this way, we can conclude that the eigenvalues of $\mathbf{N}_k$ are $n_k = 0, 1, 2, 3, \ldots$. The forms of the Hamiltonian in Eq. (10) and (16) are different, but equivalent. We can determine from Eq. (16) on the basis of the physical picture, that it has a minimum. Thus, the 'zero point' level of energy and the Hamiltonian is ensured. On the other hand, this cutoff on the negative values of energy solves the problem of the initial value of the heat conduction, i.e. the process is going only forwards in time.

We introduce the temperature operator to support the correctness of energy operator $\mathbf{E}$. The temperature of the system can be written using Eqs. (1), (3), (8), and (9)

$$T = -\sum_k \sqrt{\frac{2}{V}} \left[\left(P_k^{(C)} - \frac{\lambda}{c_v} k^2 C_k \right) \cos kx + \left(P_k^{(S)} - \frac{\lambda}{c_v} k^2 S_k \right) \sin kx \right], \tag{33}$$

by which the temperature operator $\mathbf{T}$ of the system [9] can be obtained with the help of operators $\mathbf{C}_k^{\mp}$ and $\mathbf{S}_k^{\mp}$

$$\mathbf{T} = -\sum_k 2\sqrt{\frac{1}{V}}(\mathbf{C}_k^- \cos kx + \mathbf{S}_k^- \sin kx). \tag{34}$$

The quanta, pertaining to this operator, are

$$\frac{\lambda}{c_v} k^2 h^*. \tag{35}$$

Applying the connection $\varepsilon = c_v T$, we can obtain the value of an energy quantum

$$\varepsilon(k) = \lambda k^2 h^*. \tag{36}$$

The eigenvalues of the operator in Eq. (29) are exactly the quanta $\lambda k^2 h^*$. These show that this operator in Eq. (29) is suitable as an energy operator, and it is correlated with the Hamiltonian operator and the temperature operator correctly.

4. Description of energy fluctuation

The solution of a differential equation describes an exact path of the process in space and time. However, if we measure a physical variable, we have information about the average of this quantity because of the random fluctuations [13,14]. We can express the deviation of the field quantity $F(x,t)$

$$\Delta F = \sqrt{\overline{F^2} - \bar{F}^2}, \tag{37}$$

where $\bar{F}$ is the average of F, and $\overline{F^2}$ is the average of F^2. It is usual to calculate this kind of deviation using the concepts of statistical mechanics supposing different ensembles and the condition of equilibrium [15–17]. The system can be characterized by different distribution functions of physical variables. Here, we examine the energy-density fluctuation in Fourier heat conduction.

For the calculation of the number of quanta as a function of k in a finite volume, we express the number k— obeying the periodic boundary conditions

$$k = \frac{2\pi n}{l}, \tag{38}$$

where n may be an integer and l is the edge of a cube in which the process is going on. In local equilibrium the specific heat capacity c_v is defined with the internal energy density $u(x,t)$ and the temperature $T(x,t)$

$$\frac{\partial u}{\partial T} = c_v. \tag{39}$$

Since Eq. (1) is valid, then $u = c_v T$ linear connection holds, the c_v is independent of the temperature T. This is why we can speak about temperature or energy fluctuation at the same time in the following. The energy is an extensive quantity, thus it is more usual to sum over energy states. The possible energy states are $\varepsilon(k) = \lambda h^* k^2$, which can be written with the n_1, n_2 and n_3

$$\varepsilon_{n_1 n_2 n_3} = \lambda h^* \frac{4\pi^2}{l^2}(n_1^2 + n_2^2 + n_3^2). \tag{40}$$

The three different series of n belong to the three different space directions. We imagine a one-eighth sphere with a radius k. The numbers n are within this part of space

$$k \geq \frac{2\pi}{l}\sqrt{n_1^2 + n_2^2 + n_3^2}, \tag{41}$$

and we can give the number of the possible states $N(k)$ calculating the volume of it

$$N(k) = \frac{1}{8}\frac{4\pi}{3}\left(\frac{lk}{2\pi}\right)^3 = \frac{1}{48}\frac{l^3 k^3}{\pi^2}. \tag{42}$$

If we derive this with respect to k

$$\mathrm{d}N(k) = \frac{l^3}{16\pi^2}k^2\,\mathrm{d}k, \tag{43}$$

we will know how many possible states exist within the dk interval. Dividing it by the volume of coordinate space l^3, we obtain the density of these states

$$\mathrm{d}n(k) = \frac{1}{16\pi^2}k^2\,\mathrm{d}k. \tag{44}$$

We obtain the average energy per one quantum $\bar{\varepsilon}_1$ integrating over all of the energy states weighted by the number of states and divided by the number of these

$$\bar{\varepsilon}_1 = \frac{\int_0^\infty \varepsilon(k)\mathrm{d}n(k)}{\int_0^\infty \mathrm{d}n(k)} = \frac{\int_0^\infty \lambda h^* k^2 \frac{1}{16\pi^2}k^2\,\mathrm{d}k}{\int_0^\infty \frac{1}{16\pi^2}k^2\,\mathrm{d}k} \to \infty. \tag{45}$$

The reason for this result is that there is no limit on k. This integration supposes an infinite radius k of the sphere and this means that an infinite number of quanta exist. We can assume that the lower-energy states are more possible and we may introduce a function

$$\sim \mathrm{e}^{-\alpha k^2}, \tag{46}$$

which can decrease the number of states exponentially. We multiply the integrand of Eq. (44) by this factor and the average energy per quantum can be obtained as an infinite integral over all of the k

$$\bar{\varepsilon}_1 = \frac{\int_0^\infty \frac{\lambda h^* k^4}{16\pi^2}\mathrm{e}^{-\alpha k^2}\,\mathrm{d}k}{\int_0^\infty \frac{k^2}{16\pi^2}\mathrm{e}^{-\alpha k^2}\,\mathrm{d}k} = \frac{\frac{\lambda h^*}{2\pi^2}\frac{\Gamma(\frac{5}{2})}{2\alpha^{5/2}}}{\frac{1}{2\pi^2}\frac{\Gamma(\frac{3}{2})}{2\alpha^{3/2}}} = \frac{3}{2}\lambda h^* \frac{1}{\alpha}. \tag{47}$$

Here, we used the well-known infinite integral

$$\int_0^\infty x^n \, e^{-\alpha x^2} = \frac{\Gamma\left(\frac{n+1}{2}\right)}{2\alpha^{(n+1)/2}}. \tag{48}$$

If the whole space contains N quanta, then the energy is

$$\bar{\varepsilon} = N\bar{\varepsilon}_1 = \frac{3}{2}\lambda h^* N \frac{1}{\alpha}. \tag{49}$$

Thus, the heat capacity C_V of the whole system is

$$C_V = \frac{\partial \bar{\varepsilon}}{\partial T} = \frac{3}{2}\lambda h^* N \frac{\partial \frac{1}{\alpha}}{\partial T}. \tag{50}$$

If we suppose that C_V is a constant quantity, i.e. when the whole system is close to the equilibrium, we obtain for α

$$\alpha = \frac{3\lambda h^* N}{2C_V} \frac{1}{T}. \tag{51}$$

In this case C_V is independent of temperature T, i.e. $\bar{\varepsilon} = C_V T$. Now, we can compare this result with the result of the statistical mechanics, namely, $C_V = \frac{1}{2} f k_B N$, where f is the degrees of freedom. We get for the $\bar{\varepsilon}_1$ in Eq. (47)

$$\bar{\varepsilon}_1 = \tfrac{1}{2} f k_B T. \tag{52}$$

This result shows that our description is in line with the results of statistical physics. Or rather, these results could be obtained as a consequence of a special case of our method. To obtain the energy fluctuation we have to express the average of ε_1 square

$$\overline{\varepsilon_1^2} = \frac{\int_0^\infty \frac{\lambda^2 h^{*2}}{16\pi^2} k^6 \, e^{-\alpha k^2} \, dk}{\int_0^\infty \frac{1}{16\pi^2} k^2 \, e^{-\alpha k^2} \, dk} = \frac{15}{4}\lambda^2 h^{*2} \frac{1}{\alpha^2}. \tag{53}$$

Thus, we can express the magnitude of energy fluctuation in the sense of Eq. (37). The square of the magnitude of the deviation of $\Delta\varepsilon_1$ can be read if we use that equation

$$(\Delta\varepsilon_1)^2 = \overline{\varepsilon_1^2} - \bar{\varepsilon}_1^2. \tag{54}$$

In the case of N quanta the energy fluctuation of the system is $(\Delta\varepsilon)^2 = N(\Delta\varepsilon_1)^2$, i.e.

$$(\Delta\varepsilon)^2 = \frac{3}{2}\lambda^2 h^{*2} N \frac{1}{\alpha^2}. \tag{55}$$

Now, we can compare our result with the result of statistical mechanics again if we use the form of α in Eq. (51), and we obtain

$$(\Delta\varepsilon) = \sqrt{\tfrac{1}{6} f k_B}\sqrt{N}T. \tag{56}$$

After all, we are dealing with the examination of Eqs. (49) and (55). These equations contain the parameter α, which is not known in general and it is in the exponential of expression (46). We can calculate the relative fluctuation, which is

$$\frac{\Delta\varepsilon}{\bar{\varepsilon}} = \sqrt{\frac{2}{3}}\frac{1}{\sqrt{N}}. \tag{57}$$

It can be seen that the result is independent of α. Moreover, we are not restricted to choosing the number N as a number that is too great. The interesting case is when $\Delta\varepsilon \sim \varepsilon$, i.e. the fluctuation is comparable with the energy. This is valid when the number of quanta N is not so great, practically a few hundred or less. This means that we cannot measure the temperature exactly, because the amplitude of fluctuations will be comparable with the energy independently of the temperature. When the temperature tends to the absolute zero, the number of quanta have to decrease, consequently the relative fluctuation increases, i.e. on cooling the body with heat conduction we cannot measure temperature exactly. We can measure the average of energy and temperature but we are not sure what the exact energy and temperature are when we have a small number of quanta.

The fluctuation theory of physical processes is based on the concepts of statistical mechanics that theory takes into account the physical system as the certain ensemble of the physical variables. Here, we started from the continuum description of fields, we quantized this to obtain a discrete characteristic of it in this way. We showed how to use the quantized field of heat conduction to discuss the energy fluctuation. This method shows more possibilities to describe the fluctuations and to find a deeper connection with the statistical mechanics.

5. A Bose system

We point out how to apply the quantum field theory of thermodynamics to treat boson systems. We describe the physical behavior of Bose systems and we discuss the Bose condensation close to the equilibrium in a simple case. We make an attempt to deform infinitesimally the Bose distribution—which may mean a weak interaction inside the system—and we examine the consequences of it, especially the change of critical temperature and the nonextensivity of entropy.

We assume the general Bose distribution in the form [18]

$$F_{\mathrm{BE}}(k) = \frac{1}{f^{-1}\,\mathrm{e}^{\alpha k^2} - 1}, \tag{58}$$

where the introduced functions $f(T)$ and $\alpha(T)$ may depend on the temperature. The function $f(T)$ is the fugacity, the $\alpha(T)$ as the power of the exponent plays a special role. The number of quanta is

$$N = \int_0^\infty \frac{l^3}{16\pi^2}\frac{k^2}{f^{-1}\,\mathrm{e}^{\alpha k^2} - 1}\mathrm{d}k. \tag{59}$$

In the calculations we use

$$g_n(f)\Gamma(n) = \int_0^\infty \frac{x^{n-1}}{f^{-1}\,\mathrm{e}^x - 1}\mathrm{d}x,$$

where $g_n(f) = \sum_{k=1}^{\infty} \frac{f^k}{k^n}$, which is convergent if $0 \le f \le 1$ or $1 \le \frac{1}{f} \le \infty$. If $f = 1$ $g_n(1) = \zeta(n)$ is the Riemann zeta function. The derivative of $g_n(f)$ is $g_n'(f) = \frac{1}{f} g_{n-1}(f)$ and the Gamma function is $\Gamma(n) = \int_0^\infty \mathrm{e}^{-x}x^{n-1}\,\mathrm{d}x$.

If $T < T_\mathrm{c}$ then $f = 1$ by which we obtain the number of noncondensed quanta

$$N = \frac{l^3}{32\pi^2}\alpha^{-3/2}(T)\zeta(\tfrac{3}{2})\Gamma(\tfrac{3}{2}). \tag{60}$$

The N has a maximum value N_0 at a temperature $T = T_\mathrm{c}$

$$N_0 = \frac{l^3}{32\pi^2}\alpha^{-3/2}(T_\mathrm{c})\zeta(\tfrac{3}{2})\Gamma(\tfrac{3}{2}). \tag{61}$$

The temperature T_c is the critical value of temperature where the phase transition starts. When $T \le T_\mathrm{c}$ then $N \le N_0$ there will be a noncondensed ratio that is

$$\eta_\mathrm{nc} = \frac{N}{N_0} = \alpha^{3/2}(T_\mathrm{c})\alpha^{-3/2}(T), \tag{62}$$

and the number of noncondensed particles

$$N_\mathrm{nc}(T) = N_0\alpha^{3/2}(T_\mathrm{c})\alpha^{-3/2}(T). \tag{63}$$

The average energy of one quantum is

$$\bar{\varepsilon}_1 = \frac{\int_0^\infty \varepsilon(k)F_\mathrm{BE}(k)\mathrm{d}n(k)}{\int_0^\infty F_\mathrm{BE}(k)\mathrm{d}n(k)}, \tag{64}$$

and after substituting Eqs. (36), (44) and (58) we get

$$\bar{\varepsilon}_1 = \frac{g_{5/2}(f(T))\Gamma(\frac{5}{2})}{g_{3/2}(f(T))\Gamma(\frac{3}{2})}\lambda h^* \frac{1}{\alpha(T)}. \tag{65}$$

We obtain the average energy if we multiply the average energy of one quantum by the number of quanta $\bar{\varepsilon} = N\bar{\varepsilon}_1$.

We distinguish two cases. If $T \le T_\mathrm{c}$, i.e. the system is below the critical temperature

$$\bar{\varepsilon} = \frac{\zeta(\frac{5}{2})\Gamma(\frac{5}{2})}{\zeta(\frac{3}{2})\Gamma(\frac{3}{2})}N_\mathrm{nc}(T)\lambda h^* \frac{1}{\alpha(T)}. \tag{66}$$

The second case is if $T_\mathrm{c} \le T$, i.e. the system is above the critical temperature

$$\bar{\varepsilon} = \frac{g_{5/2}(f(T))\Gamma(\frac{5}{2})}{g_{3/2}(f(T))\Gamma(\frac{3}{2})}N_0\lambda h^* \frac{1}{\alpha(T)}. \tag{67}$$

If $T \leq T_c$ we obtain the heat capacity ($C_V = \partial \bar{\varepsilon}/\partial T$)

$$\underline{C_V(T \leq T_c)} = \frac{3}{2}\lambda h^* N_0 \alpha^{3/2}(T_c)\frac{\zeta(\frac{5}{2})}{\zeta(\frac{3}{2})}\frac{\partial}{\partial T}\alpha^{-5/2}(T). \tag{68}$$

If $T_c \leq T$ we can write

$$\underline{C_V(T_c \leq T)} = \frac{3}{2}\lambda h^* N_0 \alpha^{-1}(T)\frac{1}{f(T)}\left(1 - \frac{g_{5/2}(f(T))g_{1/2}(f(T))}{(g_{3/2}(f(T)))^2}\right)\frac{\partial f(T)}{\partial T}$$
$$+ \frac{3}{2}\lambda h^* N_0 \frac{g_{5/2}(f(T))}{g_{3/2}(f(T))}\frac{\partial}{\partial T}\alpha^{-1}(T). \tag{69}$$

These general equations contain the description of the ideal Bose gas and this helps us to determine the form of $\alpha(T)$ for this special case. We know that in this case the heat capacity should be [19–21], e.g. for $T_c \leq T$

$$C_V(T_c \leq T) = \frac{3}{2}T\frac{1}{f(T)}\left(1 - \frac{g_{5/2}(f(T))g_{1/2}(f(T))}{(g_{3/2}(f(T)))^2}\right)\frac{\partial f(T)}{\partial T} + \frac{3}{2}\frac{g_{5/2}(f(T))}{g_{3/2}(f(T))}N_0 k, \tag{70}$$

or for $T = T_c$ when $f = 1$

$$C_V(T_c) = \frac{15}{4}\frac{\zeta(\frac{5}{2})}{\zeta(\frac{3}{2})}N_0 k_B. \tag{71}$$

These equations can easily be obtained from Eq. (69) substituting for the $\alpha(T)$

$$\alpha(T) = \frac{\lambda h^*}{k_B}\frac{1}{T}, \tag{72}$$

which is the interaction-free case. Then the distribution of quanta as a function of wave number k can be expressed comparing the results with Eq. (58)

$$\frac{1}{f^{-1}\,e^{\lambda h^* k^2/k_B T} - 1}, \tag{73}$$

which is in complete agreement with the well-known form

$$\frac{1}{f^{-1}\,e^{\varepsilon(k)/k_B T} - 1}. \tag{74}$$

We may think that a minor interaction may be allowed among the particles and this effect may appear in the parameters. In Section 6 we examine the case of an infinitesimal change of the parameter α.

6. Infinitesimally deformed Bose distribution

We continue our examination of Bose systems with the introduction of an infinitesimal deformation of the Bose distribution [18,22]. The deformation parameter $\delta \ll 1$ will appear in α (see Eq. (72)). This may describe the weak interaction of particles of the system.

We write α_δ

$$\alpha_\delta = \frac{\lambda h^*}{k_B T}(1 - \delta) \tag{75}$$

for this case. Similarly to Eq. (60) we can calculate the number N as a function of temperature and δ

$$N = \frac{l^3}{32\pi^2}\alpha_\delta^{-3/2}(T)\zeta(\tfrac{3}{2})\Gamma(\tfrac{3}{2}). \tag{76}$$

The maximum value of N will be N_0 when we reach the critical temperature T_c

$$N_0 = \frac{l^3}{32\pi^2}\alpha_\delta^{-3/2}(T_c)\zeta(\tfrac{3}{2})\Gamma(\tfrac{3}{2}) \approx \frac{l^3}{32\pi^2}\zeta(\tfrac{3}{2})\Gamma(\tfrac{3}{2})\left(\frac{k_B T_c}{\lambda h^*}\right)^{3/2}\left(1 + \tfrac{3}{2}\delta\right), \tag{77}$$

where we used the approximation $(1+x)^n \approx 1 + nx$ $(x \ll 1)$ in the calculations. On the other hand, we can express the dependence of critical temperature on the parameter δ

$$T_c(\delta) = \left(\frac{32\pi^2}{l^3\zeta(\tfrac{3}{2})\Gamma(\tfrac{3}{2})}N_0\right)^{2/3}\frac{\lambda h^*}{k_B}(1 - \delta). \tag{78}$$

After all, we examine the entropy and its extensivity. We take two systems A and B below the critical temperature with the maximum number of particles N_{0A} and N_{0B} at the critical temperature calculated by Eq. (61), which means the only difference between them, i.e. the systems have the same temperature and the form of $\delta(N)$ function is the same. We can calculate the heat capacities in both cases using Eq. (68)

$$C_{VA}(T \le T_c) = \frac{15}{4}\frac{\zeta(\tfrac{5}{2})}{\zeta(\tfrac{3}{2})}N_{0A}k_B\left(\frac{T}{T_c}\right)^{3/2}(1 - \delta(N_{0A}))^{-5/2}, \tag{79}$$

$$C_{VB}(T \le T_c) = \frac{15}{4}\frac{\zeta(\tfrac{5}{2})}{\zeta(\tfrac{3}{2})}N_{0B}k_B\left(\frac{T}{T_c}\right)^{3/2}(1 - \delta(N_{0B}))^{-5/2}. \tag{80}$$

The entropy of these systems can be calculated by the equation

$$S = \int_0^T \frac{C_V}{T}\,dT, \tag{81}$$

i.e. below the critical temperature the entropies of the systems A and B are

$$S_A(T \le T_c) = \frac{5}{2}\frac{\zeta(\tfrac{5}{2})}{\zeta(\tfrac{3}{2})}N_{0A}k_B\left(\frac{T}{T_c}\right)^{3/2}(1 - \delta(N_{0A}))^{-5/2}, \tag{82}$$

$$S_B(T \le T_c) = \frac{5}{2}\frac{\zeta(\tfrac{5}{2})}{\zeta(\tfrac{3}{2})}N_{0B}k_B\left(\frac{T}{T_c}\right)^{3/2}(1 - \delta(N_{0B}))^{-5/2}. \tag{83}$$

We add the two systems resulting in the system $A + B$, and we obtain the entropy of it

$$S_{A+B} = \frac{5}{2}\frac{\zeta(\tfrac{5}{2})}{\zeta(\tfrac{3}{2})}(N_{0A} + N_{0B})k_B\left(\frac{T}{T_c}\right)^{3/2}(1 - \delta(N_{0A} + N_{0B}))^{-5/2}. \tag{84}$$

It is clear that the additivity of entropy is not completed $S_A + S_B \neq S_{A+B}$, because after substitution we can write

$$N_{0A}(1 - \delta(N_{0A}))^{-5/2} + N_{0B}(1 - \delta(N_{0B}))^{-5/2} \neq (N_{0A} + N_{0B}) \times (1 - \delta(N_{0A} + N_{0B}))^{-5/2}. \tag{85}$$

We can simplify this and finally we get

$$N_{0A}\,\delta(N_{0A}) + N_{0B}\,\delta(N_{0B}) \neq (N_{0A} + N_{0B})\delta(N_{0A} + N_{0B}). \tag{86}$$

Following Tsallis' idea [23–25] we express the equality of entropy between the initial (A and B) and final ($A + B$) states

$$S_A + S_B = S_{A+B} + (1 - q)S_A S_B, \tag{87}$$

where q is the function of N_{0A} and N_{0B}, $q(N_{0A}, N_{0B})$. When $q = 1$ is completed that means the extensivity of entropy of an ideal noninteractive system. The coefficient q, which characterizes the nonextensive entropy, can be calculated as

$$q = 1 - \frac{S_A + S_B - S_{A+B}}{S_A S_B}. \tag{88}$$

We can substitute Eqs. (82)–(84) into this expression and we obtain

$$q = 1 + \frac{1}{k_B}\frac{\zeta(\frac{3}{2})}{\zeta(\frac{5}{2})} \times \left(\frac{T_c}{T}\right)^{3/2} \frac{N_{0A}\,\delta(N_{0A}) + N_{0B}\,\delta(N_{0B}) - (N_{0A} + N_{0B})\delta(N_{0A} + N_{0B})}{N_{0A}N_{0B}(1 + \frac{5}{2}\,\delta(N_{0A}))(1 + \frac{5}{2}\,\delta(N_{0B}))}. \tag{89}$$

Let us write $\delta(N) = \beta/N$, where N should be large and the absolute value of β should be small enough because the little deformation demands this assumption. We substitute this into Eq. (89) and we obtain for q

$$q(N_{0A}, N_{0B}) = 1 - \frac{\beta}{k_B}\frac{\zeta(\frac{3}{2})}{\zeta(\frac{5}{2})}\left(\frac{T_c}{T}\right)^{3/2}\frac{1}{N_{0A}N_{0B}}, \tag{90}$$

where we used the $N_{0A}N_{0B}(1 + \frac{5}{2}\,\delta(N_{0A}))(1 + \frac{5}{2}\,\delta(N_{0B})) \approx N_{0A}N_{0B}$ approximation for the denominator. The quantity $q(N_{0A}, N_{0B})$ depends on the parameters of both systems A and B. We think that the interaction within the systems is similar to the interaction between them, thus the $q(N_{0A}, N_{0B})$ holds this property. We can take into account that the $\delta(N)$ is very small, and we use the trick

$$q_A = (q(N_{0A}, N_{0B}))^{N_{0B}} = \left(1 - \frac{\beta}{k_B}\frac{\zeta(\frac{3}{2})}{\zeta(\frac{5}{2})}\left(\frac{T_c}{T}\right)^{3/2}\frac{1}{N_{0A}N_{0B}}\right)^{N_{0B}}, \tag{91}$$

where we applied the approximation $(1 + x)^n \approx (1 + nx)$ for $x \ll 1$. We obtain a quantity q_A

$$q_A = 1 - \frac{\beta}{k_B}\frac{\zeta(\frac{3}{2})}{\zeta(\frac{5}{2})}\left(\frac{T_c}{T}\right)^{3/2}\frac{1}{N_{0A}}, \tag{92}$$

which depends only on the N_{0A}. This parameter characterizes the system A with its 'additional' interaction, which may belong to a kind of surface or/and inner interaction. A similar parameter can be expressed for the system B. We simplify the further calculations if we introduce a parameter η, which is the second term on the right-hand side

$$\eta = \frac{\beta}{k_B} \frac{\zeta(\frac{3}{2})}{\zeta(\frac{5}{2})} \left(\frac{T_c}{T}\right)^{3/2} \frac{1}{N_{0A}} = \frac{\delta(N_{0A})}{k_B} \frac{\zeta(\frac{3}{2})}{\zeta(\frac{5}{2})} \left(\frac{T_c}{T}\right)^{3/2}. \tag{93}$$

So, the parameter q_A can be written

$$q_A = 1 - \eta. \tag{94}$$

7. q-Boson approximation

We may assume that the quantity q_A pertains to the so-called q-algebra [26–28], where the creation, annihilation and number operators (a_k^+, a_k^- and N_k) satisfy the algebra

$$a_k^- a_k^+ - q_A^{-1} a_k^+ a_k^- = q_A^{N_k}, \tag{95}$$

$$a_k^+ a_k^- = (N_k)_{q_A}, \tag{96}$$

$$a_k^- a_k^+ = (N_k + 1)_{q_A}. \tag{97}$$

The eigenstates of the number operator are $N_k|n_k\rangle = [n_k]|n_k\rangle$, the $[n_k]$ are the eigenvalues. The $[n]_{qA}$ bracket symbol means

$$[n_k]_{q_A} = \frac{q_A^{n_k} - q_A^{-n_k}}{q_A - q_A^{-1}}. \tag{98}$$

Supposing that the form of energy does not change, we write the energy operator by the number operators of the q-boson approximation

$$\mathbf{E} = \sum_k \frac{1}{2} \lambda k^2 h^* ((N_k)_{q_A} + (N_k + 1)_{q_A}). \tag{99}$$

We can calculate the energy of the system

$$\begin{aligned} E_{q_A} &= \sum_k \frac{1}{2} \lambda k^2 h^* ([n_k]_{q_A} + [n_k + 1]_{q_A}) \\ &= \sum_k \left(\lambda k^2 h^* \left(n_k + \frac{1}{2} \right) + \frac{1}{12} \eta^2 \lambda k^2 h^* n_k (2n_k - 1)(n_k - 1) \right), \end{aligned} \tag{100}$$

where we used the approximated values

$$[n_k]_{q_A} = n_k + \tfrac{1}{6} \eta^2 n_k (n_k - 1)(n_k - 2),$$

$$[n_k + 1]_{q_A} = n_k + 1 + \tfrac{1}{6} \eta^2 (n_k + 1) n_k (n_k - 1).$$

It can be immediately seen that the interaction means always a positive contribution to the energy, and this term vanishes when $\eta = 0$, i.e. $q_A = 1$. We substitute the form of η from Eq. (93), thus we obtain for the energy

$$E_{q_A} = \sum_k \lambda k^2 h^* \left(n_k + \frac{1}{2} \right) + \sum_k \frac{1}{12} \frac{(\delta(N_{0A}))^2}{k_B^2} \left(\frac{\zeta(\frac{3}{2})}{\zeta(\frac{5}{2})} \right)^2 \times \left(\frac{T_c}{T} \right)^3 \lambda k^2 h^* n_k (2n_k - 1)(n_k - 1). \tag{101}$$

Now, we turn to the examination of commutation of momentum P and coordinate Q, which is expressed by $[P, Q]$ in general. We can calculate the commutation for the q-boson system if we consider the q-boson algebra by (Eqs. (95)–(97))

$$[P_{q_A}, Q_{q_A}] \rightarrow h^*([n_k + 1]_{q_A} - [n_k]_{q_A}) = h^* + \tfrac{1}{2}\eta^2 h^* n_k (n_k - 1). \tag{102}$$

The commutation of momentum and coordinate can be written using the parameter q_A

$$[P_{q_A}, Q_{q_A}] \rightarrow h^* + \tfrac{1}{2}(1 - q_A)^2 h^* n_k (n_k - 1). \tag{103}$$

Eqs. (102) and (103) show that there is an additional positive term on the right-hand side, which means a small difference from the 'usual' uncertainty relation. We can express this term by the physical parameters

$$D[P_{q_A}, Q_{q_A}] \rightarrow \frac{1}{2} \frac{(\delta(N_{0A}))^2}{k_B^2} \left(\frac{\zeta(\frac{3}{2})}{\zeta(\frac{5}{2})} \right)^2 \left(\frac{T_c}{T} \right)^3 h^* n_k (n_k - 1), \tag{104}$$

from which we can conclude that the smaller temperature and particle number mean a greater additional uncertainty in the system at a given critical temperature and quantum state. This result is in line with Biedenharn's result [27] that states that in the case of q-algebras the uncertainty relation is modified.

8. Bohmian quantum dynamics of particles

The motion of the classical ensemble can be derived from the Lagrangian L_C

$$L_C = \int p \left[\frac{\partial S}{\partial t} + \frac{1}{2m} (\nabla S)^2 + V_C \right] d^3 x, \tag{105}$$

where S is the classical action, p is the probability density($\int p \, d^3 x = 1$), V_C is the classical (mechanical) potential, m is the mass of particle, and ∇ is the gradient operator. After the variation—with respect to the variables S and p—we obtain two equations as Euler–Lagrange equations

$$\frac{\partial S}{\partial t} + \frac{1}{2m} (\nabla S)^2 + V_C = 0, \tag{106}$$

which is the classical Hamilton–Jacobi equation, and

$$\frac{\partial p}{\partial t} + \nabla\left(p\frac{1}{m}\nabla S\right) = 0, \tag{107}$$

which is a continuity equation for the probability density. Following Hall and Reginatto's work [29] a modified Lagrangian L_{QM} can be obtained as a consequence of the momentum fluctuations

$$L_{\mathrm{QM}} = \int p\left[\frac{\partial S}{\partial t} + \frac{1}{2m}(\nabla S)^2 + \frac{\hbar^2}{8m}\frac{(\nabla p)^2}{p^2} + V_{\mathrm{C}}\right]\mathrm{d}^3x. \tag{108}$$

After the calculation of the Euler–Lagrange equations one gets, on the one hand, the quantum Hamilton–Jacobi equation

$$\frac{\partial S}{\partial t} + \frac{1}{2m}(\nabla S)^2 + \frac{\hbar^2}{8m}\left[\frac{(\nabla p)^2}{p^2} - \frac{2\Delta p}{p}\right] + V_{\mathrm{C}} = 0, \tag{109}$$

and on the other hand, the continuity equation for the probability density p, which is the same as Eq. (107). Bohm deduced these [30–33] from the Schrödinger equation

$$\mathrm{i}\hbar\frac{\partial \Psi}{\partial t} = -\frac{\hbar^2}{2m}\Delta\Psi + V_{\mathrm{C}}\Psi, \tag{110}$$

where 'i' is the complex unit, Ψ is a complex function, which can be expressed as

$$\Psi = R\exp\left(\mathrm{i}\frac{S}{\hbar}\right). \tag{111}$$

Now R and S are real, and

$$p = R^2. \tag{112}$$

It is readily verified that the quantum Hamilton–Jacobi equation (109) and the continuity equation (107) can be obtained. After these, we turn to the study of a dissipative process, the linear heat conduction for which we apply Bohm's method by the needed modifications and interpretations.

9. Hamilton–Jacobi equation, the action, the kernel, and a wave function

The classical Hamilton–Jacobi equation can be written in general

$$\frac{\partial S}{\partial t} + H\left(q_1, \ldots, q_f, \frac{\partial S}{\partial q_1}, \ldots, \frac{\partial S}{\partial q_f}\right) = 0, \tag{113}$$

where the q_is are the generalized coordinates, the $\partial S/\partial q_i$s are the momenta P_is. In our special case, using Eq. (10) we can express the Hamilton–Jacobi equation of heat conduction by the Fourier coefficients as generalized coordinates

$$\frac{\partial S}{\partial t}+\sum_k\left(\frac{1}{2}\left(\frac{\partial S}{\partial C_k}\right)^2+\frac{1}{2}\left(\frac{\partial S}{\partial S_k}\right)^2\right)-\sum_k\frac{1}{2}\frac{\lambda^2}{c_v^2}k^4(C_k^2+S_k^2)=0, \tag{114}$$

where

$$V_{\mathrm{CT}}=V(C_k,S_k)=-\sum_k\frac{1}{2}\frac{\lambda^2}{c_v^2}k^4(C_k^2+S_k^2) \tag{115}$$

is the classical quadratic potential. This equation can be obtained from the variation of the integral over time and the space of generalized coordinates

$$0=\delta\int P\left[\frac{\partial S}{\partial t}+\sum_k\left(\frac{1}{2}\left(\frac{\partial S}{\partial C_k}\right)^2+\frac{1}{2}\left(\frac{\partial S}{\partial S_k}\right)^2\right)+V_{\mathrm{CT}}\right]\mathrm{d}t\,\mathrm{d}\Omega, \tag{116}$$

where P is the probability density to find the system in the state described by the set of generalized coordinates C_k and S_k : i.e. $P(C_k,S_k)$; $\mathrm{d}\Omega=\cdots\mathrm{d}C_i\cdots\mathrm{d}S_j\cdots$ is the volume element in the space of generalized coordinates. Here, the classical Lagrangian L_{CL} can be written in the following form

$$L_{\mathrm{CL}}=\int P\left[\frac{\partial S}{\partial t}+\sum_k\left(\frac{1}{2}\left(\frac{\partial S}{\partial C_k}\right)^2+\frac{1}{2}\left(\frac{\partial S}{\partial S_k}\right)^2\right)+V_{\mathrm{CT}}\right]\mathrm{d}\Omega. \tag{117}$$

After the variation a second equation appears

$$\frac{\partial P}{\partial t}+\nabla(P\nabla S)=0, \tag{118}$$

which is a continuity equation for the probability density, and here ∇ denotes the vector operator

$$\left(\ldots,\frac{\partial}{\partial C_i},\ldots,\frac{\partial}{\partial S_j},\ldots\right).$$

The system is developing from the state a at time t_a to the state b at time t_b, and we suppose $t_b>t_a$. Solving this partial differential equation, taking into account the initial and final states, the calculated action for this process is

$$S[b,a]=\sum_{k>0}\frac{\frac{\lambda}{c_v}k^2}{2\sinh\left(\frac{\lambda}{c_v}k^2(t_b-t_a)\right)}\times\left[(C_{ka}^2+C_{kb}^2+S_{ka}^2+S_{kb}^2)\cosh\left(\frac{\lambda}{c_v}k^2(t_b-t_a)\right)-2C_{ka}C_{kb}-2S_{ka}S_{kb}\right], \tag{119}$$

which is the solution of the Hamilton–Jacobi equation. In general, the kernel of a quadratic action can be written [34] (or can be calculated by the path-integral method [35–37])

$$K_F(b,a) = \prod_{k>0} \frac{\frac{\lambda}{c_v}k^2}{2\pi i h^* \sinh\left(\frac{\lambda}{c_v}k^2 t\right)} \exp\left(\frac{\frac{\lambda}{c_v}k^2 i}{2h^* \sinh\left(\frac{\lambda}{c_v}k^2 t\right)} \times \left[(C_{ka}^2 + C_{kb}^2 + S_{ka}^2 + S_{kb}^2)\cosh\left(\frac{\lambda}{c_v}k^2 t\right) - 2C_{ka}C_{kb} - 2S_{ka}S_{kb} \right] \right), \tag{120}$$

where we denote $t = t_b - t_a$. This propagator might be called the WKB propagator that is exact for Lagrangians expressed with quadratic terms, in general. It can simply be proved that this propagator is the solution of the following generalized Schrödinger-like equation

$$-\frac{h^*}{i}\frac{\partial K_F}{\partial t} = -\sum_k \frac{h^{*2}}{2}\frac{\partial^2 K_F}{\partial C_k^2} - \sum_k \frac{h^{*2}}{2}\frac{\partial^2 K_F}{\partial S_k^2} - \sum_k \frac{\lambda^2 k^4}{2c_v^2}(C_k^2 + S_k^2)K_F, \tag{121}$$

which shows the correctness of the propagator from another side. The applicability of the Feynman path-integral method shows and carries the construction of the wave function Ψ that is related to the propagator K_F. These facts may indicate the possibility of particle-wave duality in the present case. The Schrödinger-like equation for the wave function can be read as

$$-\frac{h^*}{i}\frac{\partial \Psi}{\partial t} = -\sum_k \frac{h^{*2}}{2}\frac{\partial^2 \Psi}{\partial C_k^2} - \sum_k \frac{h^{*2}}{2}\frac{\partial^2 \Psi}{\partial S_k^2} - \sum_k \frac{\lambda^2 k^4}{2c_v^2}(C_k^2 + S_k^2)\Psi. \tag{122}$$

It is obvious that the classical Hamilton–Jacobi equation (114) cannot give this equation, we should find another equation that includes the quantum behavior. We follow Bohm's idea [30] to obtain the quantum Hamilton–Jacobi equation of the problem. The wave function can be expressed

$$\Psi = R\exp\left(i\frac{S}{h^*}\right), \tag{123}$$

where $R(C_k, S_k)$ is real and

$$\Psi^* \Psi = R^2 = P. \tag{124}$$

We substitute Eq. (123) into Eq. (122), then we can separate a real and an imaginary part of the resulting equation. So, we obtain a continuity equation

$$\frac{\partial P}{\partial t} + \nabla(P\nabla S) = 0, \tag{125}$$

and the quantum-thermodynamical Hamilton–Jacobi equation

$$\frac{\partial S}{\partial t}+\sum_k\left(\frac{1}{2}\left(\frac{\partial S}{\partial C_k}\right)^2+\frac{1}{2}\left(\frac{\partial S}{\partial S_k}\right)^2\right)-\sum_k\frac{1}{2}\frac{\lambda^2}{c_v^2}k^4(C_k^2+S_k^2)$$
$$-\frac{h^{*2}}{2R}\sum_k\left(\frac{\partial^2 R}{\partial C_k^2}+\frac{\partial^2 R}{\partial S_k^2}\right)=0. \tag{126}$$

The third term is the classical potential, similar to the term in Eq. (115), the last term

$$U(C_k,S_k)=-\frac{h^{*2}}{2R}\sum_k\left(\frac{\partial^2 R}{\partial C_k^2}+\frac{\partial^2 R}{\partial S_k^2}\right) \tag{127}$$

is also a potential. This vanishes when $h^*=0$, so this pertains to a quantum-physical behavior of the process. This may be called the quantum-thermodynamical potential. We express this potential $U(C_k,S_k)$ by the probability density $P(C_k,S_k)$ introduced by Eq. (124)

$$U(C_k,S_k)=-\frac{h^{*2}}{4}\sum_k\left[\frac{1}{P}\frac{\partial^2 P}{\partial C_k^2}-\frac{1}{2P^2}\left(\frac{\partial P}{\partial C_k}\right)^2+\frac{1}{P}\frac{\partial^2 P}{\partial S_k^2}-\frac{1}{2P^2}\left(\frac{\partial P}{\partial S_k}\right)^2\right]. \tag{128}$$

10. On the thermodynamical potentials: classical and quantum-thermodynamical

We discuss the two potentials that appeared in the above sections, but we restrict our examination to the stationary case. Then, the temperature $T(x)$ is

$$T(x)=-\frac{\lambda}{c_v}\Delta\varphi. \tag{129}$$

Using the Fourier series of φ (see Eq. (5)) we obtain

$$T(x)=\frac{\lambda}{c_v}\sum_{k>0}\sqrt{2/V}k^2(C_k\cos kx+S_k\sin kx). \tag{130}$$

If we calculate the square of both sides and we integrate over the coordinate space x we get

$$\int T^2(x)\mathrm{d}^3x=\frac{\lambda^2}{c_v^2}\sum_{k>0}k^4(C_k^2+S_k^2). \tag{131}$$

It is easy to verify that in the stationary case the classical (thermodynamical) potential can be written

$$V_{\mathrm{CT}}=-\frac{1}{2}\int T^2(x)\mathrm{d}^3x, \tag{132}$$

i.e. a repulsive interaction proportional to the temperature square is working in the heat conduction.

Now, we examine the quantum potential. First, we calculate the stationary action from Eq. (119)

$$S = \sum_k \frac{1}{2} \frac{\lambda}{c_v} k^2 (C_k^2 + S_k^2 + C_{ka}^2 + S_{ka}^2), \tag{133}$$

and we take into account the stationary version of the continuity equation (125)

$$\nabla(P\nabla S) = 0. \tag{134}$$

Finding the solution, one can verify that the probability density is

$$P = \left(\prod_k C_k S_k\right)^{-1}. \tag{135}$$

We can express the quantum-thermodynamical potential (see Eq. (128)) by the generalized coordinates using the above form of the probability density

$$U(C_k, S_k) = -\frac{3h^{*2}}{8} \sum_k \left(\frac{1}{C_k^2} + \frac{1}{S_k^2}\right). \tag{136}$$

In order to write this potential using measurable parameters, we give the C_k and S_k using the form of the steady-state (stationary) temperature (Eq. (129)) and the form of φ (Eq. (5)). After the Fourier transformation we obtain

$$C_k = \frac{c_v}{\lambda k^2} \int T(x) \cos kx \, \mathrm{d}^3 x, \tag{137}$$

$$S_k = \frac{c_v}{\lambda k^2} \int T(x) \sin kx \, \mathrm{d}^3 x. \tag{138}$$

We substitute these coefficients into Eq. (136) and get the quantum potential expressed by the temperature $T(x)$

$$U = -\frac{3h^{*2}\lambda^2}{8c_v^2} \sum_k k^4 \left(\frac{1}{(\int T(x) \cos kx \, \mathrm{d}^3 x)^2} + \frac{1}{(\int T(x) \sin kx \, \mathrm{d}^3 x)^2}\right). \tag{139}$$

We mention again that this formula is valid for steady-state thermal processes. As can be seen the smaller the temperature the higher the repulsive interaction. This means that this potential should play a significant role when the temperature is close to absolute zero. When the temperature is high this potential can be neglected compared with the classical potential (Eq. (132)).

Finally, we substitute the expression of the quantum-thermodynamical potential into the expression of the quantum Hamilton–Jacobi equation (126). Now, we look for the Lagrangian that can give the Hamilton–Jacobi equation and the continuity equation for P.

It can be seen that the Lagrangian L_{QT} is

$$L_{QT} = \int P\left[\left(\frac{\partial S}{\partial t} + \frac{1}{2}(\nabla S)^2 + V_{CT}\right) + \frac{h^{*2}}{2}\frac{(\nabla P)^2}{P^2}\right] d\Omega$$

$$= L_{CL} + \frac{h^{*2}}{2}\int \frac{(\nabla P)^2}{P} d\Omega. \tag{140}$$

The last term of the integral is proportional to the so-called Fisher information

$$I_F = \int \frac{(\nabla P)^2}{P} d\Omega, \tag{141}$$

which has an interesting role in the description of quantized thermodynamical systems.

11. Fisher, bound, and extreme physical information

Information has been found to play an increasingly important role in physics, mainly since Jaynes' pioneering work [38], which is a discussion of the connection of information theory and statistical mechanics. Knowledge of probability p—representing the observer's state of knowledge about the system, rather than the state of the system itself—enables us to express the so-called Fisher information [39,40]. This is a quality of an efficient measurement procedure, and it is also a measure of the degree of system disorder, in other words, it is a form of entropy

$$I = \int \frac{(\nabla p)^2}{p} d^3x. \tag{142}$$

Here, p denotes the probability density function for the noise value x. Fisher found that it is often more convenient to calculate with a real amplitude function $q(x, t)$ [40] where

$$p = \tfrac{1}{8}q^2, \tag{143}$$

by which we can write the Fisher information

$$I = \frac{1}{2}\int (\nabla q)^2 d^3x. \tag{144}$$

Any measurement of physical parameters initiates a transformation of Fisher information. An information transition $J \rightarrow I$ takes place, where J represents the physical effect. J is the information that is intrinsic to the phenomenon. In general, information J is identified by an invariance that characterizes the measured phenomenon. As a basic case, we consider

$$J = \frac{1}{2}\int \dot{q}^2 d^3x. \tag{145}$$

The possibility of some loss of information during the information transition suggests

$$I \leq J. \tag{146}$$

The principle of Extreme Physical Information (EPI) represents a kind of game between the observer and nature. The observer wants to maximize I while nature wants to minimize it. The physical information K is

$$K = I - J = \frac{1}{2}\int (\nabla q)^2 \, \mathrm{d}x - \frac{1}{2}\int \dot{q}^2 \, \mathrm{d}^3x = \int \left(\frac{1}{2}(\nabla q)^2 - \frac{1}{2}\dot{q}^2\right)\mathrm{d}^3x, \tag{147}$$

which is a loss of information. Considering the time evolution of the process it has an extremum, which formulates a variational principle for finding the q. From this equation we can read the so-called Lagrange density function [1]

$$L = \tfrac{1}{2}(\nabla q)^2 - \tfrac{1}{2}\dot{q}^2, \tag{148}$$

which is the integrand of the above equation. From Hamilton's principle we can obtain the equation of motion as the Euler–Lagrange equation

$$\ddot{q} - \Delta q = 0, \tag{149}$$

which is a wave equation. We restrict our attention and examination to the above-described basic case. Detailed examination of the principle of EPI and several examples can be found in Refs. [39,40].

12. Another choice of probability

We can find the probability p in a quadratic form as in Eq. (143), but we give it by a different inner function. Now, let the probability be proportional to the square of gradient of a function $\varphi(x, t)$ [41]

$$p = \tfrac{1}{8}(\nabla\varphi)^2, \tag{150}$$

where φ is a generalized potential function. Using Eqs. (142) and (150) the Fisher information can be given by

$$I = \frac{1}{2}\int (\Delta\varphi)^2 \, \mathrm{d}^3x. \tag{151}$$

Similarly to the previous example, we write the bound information J in the form

$$J = \frac{1}{2}\lambda \int \dot{\varphi}^2 \, \mathrm{d}^3x, \tag{152}$$

where λ is a constant parameter. The EPI K can be formulated

$$K = I - J = \int \left(\frac{1}{2}(\Delta\varphi)^2 - \frac{1}{2}\lambda\dot{\varphi}^2\right)\mathrm{d}^3x, \tag{153}$$

from which the Lagrange density function is obtained

$$L = \tfrac{1}{2}(\Delta\varphi)^2 - \tfrac{1}{2}\lambda\dot{\varphi}^2. \tag{154}$$

The equation of motion can be calculated as the Euler–Lagrange equation

$$\lambda\ddot{\varphi} + \Delta\Delta\varphi = 0. \tag{155}$$

What may λ be? If $\lambda = 1$ (the choice of other positive numbers gives physically the same result)

$$J = \frac{1}{2}\int \dot{\varphi}^2 \, \mathrm{d}^3x. \tag{156}$$

In this case the Lagrange density function is

$$L = \tfrac{1}{2}(\Delta\varphi)^2 - \tfrac{1}{2}\dot{\varphi}^2, \tag{157}$$

by which the field equation can be calculated

$$\ddot{\varphi} + \Delta\Delta\varphi = 0. \tag{158}$$

This equation is valid for free oscillations of a thin plate or rod [42]. These waves are fundamentally different from those in a medium in all directions. Considering a monochromatic elastic wave ($\varphi \sim \exp[\mathrm{i}(kx - \omega t)]$), we obtain the dispersion relation in the form $\omega = k^2$. The propagating wave is not dissipative, in spite of the special behavior of the dispersion relation.

If $\lambda = -1$ then the bound information is always negative

$$J = -\tfrac{1}{2}\dot{\varphi}^2. \tag{159}$$

Here, we can see that the condition $I \leq J$ (see Eq (146)) is immediately violated. It is not clear what it means at all, but we can examine the mathematical results that follow from this assumption. The Lagrange density function can be written as

$$L = \tfrac{1}{2}(\Delta\varphi)^2 + \tfrac{1}{2}\dot{\varphi}^2, \tag{160}$$

which is the basic function of diffusive processes (e.g. linear heat conduction) in the field theory of nonequilibrium thermodynamics [7,8]. This proves that there exists such a physical system where this choice of J is relevant. We obtain a biparabolic differential equation as Euler–Lagrange equation

$$\ddot{\varphi} - \Delta\Delta\varphi = 0. \tag{161}$$

We have shown (in Refs. [7,8]) that a new quantity $T(x,t)$ can be introduced

$$T = -\dot{\varphi} - \Delta\varphi. \tag{162}$$

Eqs. (161) and (162) are equivalent to

$$\dot{T} - \Delta T = 0, \tag{163}$$

which is the Fourier equation. The diffusive processes are dissipative, the observed system tends to the static state. Is this the meaning of the negative bound information or is it a fortunate accident? We do not know this, yet. Here, one can ask whether it were possible to use Eqs. (156)–(158) (avoiding the assumption of negative J) to obtain the equation of heat conduction by a different substitution. It is easy to see that none of the combinations, i.e. $T = -\dot{\varphi} + \Delta\varphi$ or $T = \dot{\varphi} - \Delta\varphi$, can give Eq. (163). To understand the meaning of the different possibilities of the choice of sign of J, we turn back to the basic problem, but we write

$$J = -\frac{1}{2}\int \dot{q}^2 \, \mathrm{d}^3x. \tag{164}$$

The Lagrange density function can be obtained

$$L = \frac{1}{2}(\nabla q)^2 + \tfrac{1}{2}\dot{q}^2, \tag{165}$$

by which we calculate the equation of motion (an elliptic differential equation)

$$\ddot{q} + \Delta q = 0. \tag{166}$$

The solution of this equation can be calculated

$$q = A\,\mathrm{e}^{-\beta x}\,\mathrm{e}^{\mathrm{i}\omega t} + B\,\mathrm{e}^{-\kappa t}\,\mathrm{e}^{\mathrm{i}kx}, \tag{167}$$

which includes the dissipation in the second term. (A, B, β, ω, κ, and k are constant parameters.) It seems to us, similarly to the heat conduction, the negative bound information J may have a connection with the dissipation.

References

[1] Morse, Ph.M. and Feschbach, H. (1953), Methods of Theoretical Physics, McGraw-Hill, New York.
[2] Courant, R. and Hilbert, D. (1966), Methods of Mathematical Physics, Interscience, New York.
[3] Sieniutycz, S. and Berry, R.S. (1989), Phys. Rev. A 40, 348; Phys. Rev. A (1992) 46, 6359; Phys. Rev. E (1993) 47, 1765.
[4] Sieniutycz, S. and Shiner, J. (1992), Open Sys. Inf. Dyn. 1, 149; Open Sys. Inf. Dyn (1993) 1, 327.
[5] Sieniutycz, S. (1994), Conservation Laws in Variational Thermo-Hydrodynamics, Kluwer, Dordrecht.
[6] Márkus, F. and Gambár, K. (1991), J. Non-Equilib. Thermodyn. 16, 27.
[7] Gambár, K. and Márkus, F. (1994), Phys. Rev. E 50, 1227.
[8] Gambár, K., Martinás, K. and Márkus, F. (1997), Phys. Rev. E 55, 5581.
[9] Gambár, K. and Márkus, F. (2001), Open Sys. Inf. Dyn. 8, 369.
[10] Márkus, F. and Gambár, K. (1995), Phys. Rev. E 52, 623; Phys. Rev. E (1996) 54, 4607.
[11] Schiff, L.I. (1955), Quantum Mechanics, McGraw-Hill, New York.
[12] Dirac, P.A.M. (1958), The Principles of Quantum Mechanics, Clarendon Press, Oxford.
[13] Onsager, L. (1931), Phys. Rev. 37, 405; Phys. Rev (1931) 38, 2265.
[14] de Groot, S.R. and Mazur, P. (1962), Non-Equlibrium Thermodynamics, North-Holland, Amsterdam.
[15] Hill, T.L. (1956), Statistical Mechanics, McGraw-Hill, New York.
[16] Keizer, J. (1978), Statistical Thermodynamics of Nonequilibrium Processes, Springer, New York.
[17] Kreuzer, H.J. (1981), Nonequilibrium Thermodynamics and its Statistical Foundations, Clarendon Press, Oxford.
[18] Márkus, F. (1999), Physica A 274, 563.
[19] London, F. (1954), Superfluids II, Wiley, New York.
[20] Griffin, A. (1993), Excitations in a Bose-Condensed Liquid, Cambridge University Press, Cambridge.
[21] Greiner, W., Neise, L. and Stöcker, H. (1987), Thermodynamik und Statistische Mechanik, Harri Deutsch, Thun.
[22] Márkus, F. and Gambár, K. (2001), Physica A 293, 533.
[23] Tsallis, C. (1988), J. Stat. Phys. 52, 479.
[24] Plastino, A. and Tsallis, C. (1993), J. Phys. A: Math. Gen. 26, L893.

[25] Curado, E.M.F. and Tsallis, C. (1991), J. Phys. A: Math. Gen. 24, L69; J. Phys. A: Math. Gen (1991) 24, 3187(E); J. Phys. A: Math. Gen (1992) 25, 1019(E).
[26] Macfarlane, A.J. (1989), J. Phys. A: Math. Gen. 22, 4581.
[27] Biedenharn, L.C. (1989), J. Phys. A: Math. Gen. 22, L873.
[28] Abe, S. (1998), Phys. Lett. A 244, 229.
[29] Hall, M.J.W. and Reginatto, M. (2002), J. Phys. A: Math. Gen. 35, 3289.
[30] Bohm, D. (1952), Phys. Rev. 85, 166.
[31] Bohm, D. and Vigier, J.P. (1954), Phys. Rev. 96, 208.
[32] Takabayasi, T. (1952), Prog. Theor. Phys. 8, 143; Prog. Theor. Phys (1953) 9, 187.
[33] Holland, P.R. (1993), Phys. Rep. 224, 95.
[34] Dittrich, W. and Reuter, M. (1996), Classical and Quantum Dynamics, Springer, Berlin.
[35] Feynman, R.P. and Hibbs, A.R. (1965), Quantum Mechanics and Path Integrals, McGraw-Hill, New York.
[36] Feynman, R.P. (1994), Statistical Mechanics, Addison-Wesley, Reading, MA.
[37] Khandekar, D.C. and Lawande, S.V. (1986), Phys. Rep. 137, 115.
[38] Jaynes, E.T. (1957), Phys. Rev. 106, 620; Phys. Rev (1957) 108, 171.
[39] Frieden, B.R. (1998), Physics from Fisher Information, Cambridge University Press, Cambridge; Physica A (1993) 198, 262; Phys. Rev. A (1990) 41, 4265.
[40] Frieden, B.R. and Soffer, B.H. (1995), Phys. Rev. E 52, 2274.
[41] Márkus, F. and Gambár, K. (2003), Phys. Rev. E 68, 016121.
[42] Landau, L.D. and Lifshitz, E.M. (1975), Theory of Elasticity, Pergamon Press, Oxford.

Chapter 14

CONSERVATION LAWS AND VARIATIONAL CONDITIONS FOR WAVE PROPAGATION IN PLANARLY STRATIFIED MEDIA

A. Morro

University of Genova, DIBE, Via Opera Pia 11a, I-16145 Genova, Italy

G. Caviglia

Department of Mathematics, Via Dodecaneso 35, I-16146 Genova, Italy

Abstract

Time-harmonic fields, satisfying the Helmholtz (or Schrödinger) equation, are investigated through a conservation law in the form of a first integral. The fields describe oblique incidence on a planarly stratified medium. By way of examples, elastic, electromagnetic, and acoustic waves are considered. Possible interfaces are allowed and the boundary or jump conditions are derived through the variational approach and balance equations. Attention is mainly addressed to the reflection–transmission process generated by a multilayer between homogeneous half-spaces. A number of consequences follow. The occurrence of turning points does not preclude wave propagation; while a jump of the ray between two turning points is provided by the ray theory, no jump is allowed for the exact solution. If the amplitude of the incident wave is nonzero then the field is nonzero everywhere. Yet, depending on the value of the parameters in the half-spaces, a total-reflection condition occurs. Irrespective of the form of the multilayer, a relation between the amplitudes of the reflected–transmitted waves follows that is a generalization of that for two half-spaces in contact.

Keywords: Helmholtz equation; first integral; wave propagation; reflection–transmission; inhomogeneous media; ray; turning point; variational condition; WKB

1. Introduction

Quite often in continuum physics, the propagation of time-harmonic waves in inhomogeneous media, characterized by a single wave speed, is governed by the Helmholtz (or Schrödinger) equation

$$(\Delta + k^2 n^2)u = 0, \tag{1}$$

where Δ is the Laplacian, k is a real parameter (wave number) and n is a known function of the position $\mathbf{x}$. The unknown function u is complex-valued and is parameterized by the

E-mail address: morro@dibe.unige.it (A. Morro); caviglia@dima.unige.it (G. Caviglia).

angular frequency ω of the wave. By way of example, Eq. (1) holds for $u = p/\sqrt{\rho}$ in linear acoustics, p being the pressure and ρ the mass density [1]. Also Eq. (1) is said to hold for the electric field (components), in isotropic dielectrics, in the approximation of slowly varying density (see Ref. [2]). Moreover, in isotropic elastic solids, with the shear modulus $\mu = \mu(z)$, horizontally polarized waves, say $u_y = u_y(x, z)\exp(i\omega t)$, satisfy the equation of motion [3]

$$\nabla_\perp \cdot (\mu \nabla_\perp u_y) + \rho\omega^2 u_y = 0, \tag{2}$$

where $\nabla_\perp = (\partial_x, \partial_z)$. As shown later, an appropriate transformation makes Eq. (1) into the 1D Helmholtz equation.

Henceforth, for simplicity, we let the medium be planarly stratified (see, e.g. Ref. [4]) so that the material properties, and hence n, vary only with a coordinate, say z. However, we let u depend on x and y so that the effect of obliquely incident waves is incorporated.

Here we consider time-harmonic solutions, generated by obliquely incident waves, so that Eqs. (1) and (2) provide

$$f'' + h(z)f = 0, \tag{3}$$

for the unknown complex-valued function $f(z)$, where $h(z)$ need not be positive and a prime denotes differentiation with respect to z. Eq. (3) has the first integral

$$\Im\{f^* f'\} = \text{constant}. \tag{4}$$

The purpose of this chapter is to investigate the solution to Eqs. (1) and (3) and to determine results for the reflected and transmitted waves, in homogeneous half-spaces, generated by a planarly stratified multilayer. The motivation of the investigation is fourfold. First, Ref. [5] gives an existence and uniqueness theorem for the reflected and transmitted waves, governed by Eqs. (1) and (3), under the assumption that $h \geq \delta > 0$. The assumption is said to preclude the existence of zones where wave propagation is not possible especially if $h < 0$ somewhere. Secondly, the presence of turning points ($h = 0$) is likely to affect the wave solution both in its exact form and in approximate forms as with the WKB approximation. It is expected that the application of the first integral Eq. (4) may provide new results for the reflection–transmission process, across an inhomogeneous layer, and allow us to evaluate the effects of turning points. Thirdly, it is a standard statement in ray theory that the ray does not penetrate the region beyond the turning point but bends back toward the direction of incidence [6]. It is of interest to ascertain if the bending of the ray necessarily occurs. Lastly, we aim to obtain relations for the amplitudes of the reflected–transmitted waves, produced by a multilayer, possibly generalizing the known results for the interface between homogeneous half-spaces.

The present approach is based on a systematic application of the first integral (4) and of the boundary conditions at the possible interfaces. The boundary conditions are established by having recourse to variational properties and to global balance laws. The physical meaning of Eq. (4) is established for waves in elastic solids, in linear dielectrics, and in linear acoustics. It follows that the first integral is continuous across interfaces and hence is constant throughout. This allows us to obtain the sought relations for the amplitudes of reflected and transmitted waves, for the process originated by an inhomogeneous layer, irrespective of the occurrence of turning points. In particular, it follows that there cannot be zones where wave propagation is precluded. Moreover, in a WKB solution the upward and downward parts prove to have a constant difference of the squared amplitudes.

2. Wave propagation in planarly stratified media

The underlying medium consists of a planarly stratified layer, $z \in (0, L)$, between the homogeneous half-spaces $z < 0$, $z > L$. The function $n(z)$ is possibly discontinuous at a number of interfaces as $z \in [0, L]$, is continuous between adjacent interfaces and is constant as $z < 0$, $z > L$. We first examine how Eq. (3) can be derived.

Eq. (1) is assumed to have a separable solution

$$u(x, z) = g(x)f(z).$$

Hence Eq. (1) requires that $\mathrm{d}^2 g/\mathrm{d}x^2 = -\kappa_x^2 g$, κ_x being the separation constant. For our purposes we let κ_x be real. We then let $g(x) = \exp(i\kappa_x x)$ whence

$$u(x, z) = \exp(i\kappa_x x)f(z). \tag{5}$$

The constant κ_x is also viewed as the product $\kappa \, sin \, \alpha$ where κ is the wave number and α is the angle of the wave vector with the z-direction or $\pi/2 - \alpha$ is the grazing angle. Upon substitution in Eq. (1) we find Eq. (3) where

$$h(z) = k^2 n^2 - \kappa_x^2. \tag{6}$$

The constancy of κ_x may be viewed as the requirement of Snell's law. We now review Eqs. (5) and (6) for models of physical interest.

2.1. *Isotropic elasticity*

The assumption of a separable solution for to Eq. (2), $u_y(x, z) = g(x)U(z)$, yields again $g(x) = \exp(i\kappa_x x)$, whence

$$u_y = \exp(i\kappa_x x)U(z), \tag{7}$$

where U satisfies

$$\mu U'' + \mu' U' + NU = 0, \tag{8}$$

where

$$N = \rho\omega^2 - \mu\kappa_x^2.$$

Following Ref. [7], let

$$f(z) = U(z)\sqrt{\mu(z)}. \tag{9}$$

Substitution in Eq. (8) and some rearrangement yields Eq. (3), where

$$h(z) = \frac{\rho\omega^2}{\mu} - \kappa_x^2 - \frac{1}{2}\left(\frac{\mu'}{\mu}\right)' - \frac{1}{4}\left(\frac{\mu'}{\mu}\right)^2. \tag{10}$$

2.2. *Isotropic dielectrics*

Letting ϵ be the space-dependent permittivity and μ the constant permeability, Maxwell's equations (in differential form) read

$$\nabla\times\mathbf{E} = -\mu\partial_t\mathbf{H}, \qquad \nabla\times\mathbf{H} = \epsilon\partial_t\mathbf{E}, \qquad \nabla\cdot(\epsilon\mathbf{E}) = 0, \qquad \nabla\cdot\mathbf{H} = 0.$$

Application of the curl operator to the first equation, of time differentiation to the second one and comparison yield

$$\nabla(\nabla\cdot\mathbf{E}) - \Delta\mathbf{E} = -\mu\epsilon\partial_t^2\mathbf{E}.$$

Because

$$\epsilon\nabla\cdot\mathbf{E} + \mathbf{E}\cdot\nabla\epsilon = 0,$$

for time-harmonic waves we find that

$$\Delta\mathbf{E} + \omega^2\mu\epsilon\mathbf{E} + \nabla(\mathbf{E}\cdot\epsilon^{-1}\nabla\epsilon) = 0.$$

If $\nabla(\mathbf{E}\cdot\epsilon^{-1}\nabla\epsilon)$ is disregarded then Eq. (3) follows for each component [2]. However, Eq. (1) holds exactly if $\mathbf{E}$ and $\nabla\epsilon$ are perpendicular. Such is the case if the dielectric is planarly stratified, that is ϵ varies only with z, and $\mathbf{E}$ is horizontally polarized, $\boldsymbol{E} = E_y\boldsymbol{e}_2$. This is the model we have in mind whenever Eq. (1) is applied to electromagnetic waves. Otherwise Eq. (1) holds for any polarization in a planarly stratified dielectric provided the incidence is normal.

2.3. *Linear acoustics*

The model equation of linear acoustics is usually derived within a strong approximation. Here we follow a different scheme and eventually recover the standard equation. Let p_0, ρ_0 be the equilibrium pressure and mass density and p, ρ the present values. Also, let $p = p(\rho)$. If the body force is just the gravity force $\rho\mathbf{g}$ then the equilibrium condition is

$$\nabla p_0 + \rho_0\mathbf{g} = 0.$$

Let $\wp = p - p_0$ and $\varrho = \rho - \rho_0$. In the linear approximation, $\varrho = \wp/p_\rho$, where $p_\rho = \mathrm{d}p/\mathrm{d}\rho(\rho_0)$. The linear approximation to the continuity equation and to the equation of motion is then given by

$$\rho_0\partial_t\mathbf{v} + \nabla\wp - \varrho\mathbf{g} = 0,$$

$$\frac{1}{p_\rho}\partial_t\wp + \rho_0\nabla\cdot\mathbf{v} = 0.$$

The divergence of the equation of motion, partial time differentiation of the continuity equation, and comparison yield

$$\frac{1}{p_\rho}\partial_t^2\wp - \Delta\wp + \frac{1}{\rho_0}\nabla\rho_0\cdot\nabla\wp + \frac{1}{p_\rho}\mathbf{g}\cdot\nabla\wp - \nabla\rho_0\cdot\mathbf{g}\left(\frac{1}{\rho_0 p_\rho}\wp + \frac{1}{p_\rho^2}p_{\rho\rho}\wp\right) = 0.$$

For time-harmonic waves we have

$$\Delta\wp + \alpha\wp + \boldsymbol{\beta}\cdot\nabla\wp = 0,$$

where

$$\alpha = \frac{\omega^2}{p_\rho} - \nabla\rho_0\cdot\mathbf{g}\left(\frac{1}{\rho_0 p_\rho} + \frac{1}{p_\rho^2 p_{\rho\rho}}\right), \qquad \boldsymbol{\beta} = \frac{1}{\rho_0}\nabla\rho_0 + \frac{1}{p_\rho}\mathbf{g}.$$

Letting z be directed upward so that $\mathbf{g} = -g\nabla z,\ g = |\mathbf{g}|$, and

$$u = \wp\gamma,$$

where

$$\gamma(\mathbf{x}) = \frac{1}{\sqrt{\rho_0(z)}}\exp\left(-g\int_0^z \frac{1}{p_\rho}(\zeta)\mathrm{d}\zeta\right),$$

we find that u satisfies the Helmholtz equation (1) where

$$k^2n^2 = \alpha - \gamma''/\gamma.$$

The standard equation of linear acoustics amounts to regarding the body force as independent of the mass density. If this is done then

$$\alpha = \frac{\omega^2}{p_\rho}, \qquad \gamma = \frac{1}{\sqrt{\rho_0}}.$$

In homogeneous fluids ρ_0 is constant and

$$k^2n^2 = \frac{\omega^2}{p_\rho},$$

thus showing the well-known feature that p_ρ is the square of the speed of propagation.

2.4. Reflection–transmission problem

In the cases just examined the unknown function ultimately is required to satisfy Eq. (1). Subject to the condition Eq. (5), the pertinent unknown satisfies the 1D Helmholtz equation (3). The forms (10) and (6) of h show that given κ_x, values of z can occur where h vanishes. A point at which the coefficient h of f in Eq. (3) vanishes is called a turning point. The result (10) says that the z-coordinate of the turning point depends on both the inhomogeneity of the medium (e.g. in elastic solids, through ρ, μ and the derivatives μ', μ'') and the initial slope (through κ_x).

Whenever h is constant and positive, the solution to Eq. (3) takes the form

$$f(z) = f^{+}\exp(i\sigma z) + f^{-}\exp(-i\sigma z),$$

where $\sigma = \sqrt{h}$ and $f^{+}, f^{-} \in \mathbb{C}$; for instance,

$$\sigma = \sqrt{k^2 n^2 - \kappa_x^2},$$

if Eq. (6) holds. The corresponding function in the space–time domain is

$$f(z,t) = f^{+}\exp[i(\sigma z + \omega t)] + f^{-}\exp[i(-\sigma z + \omega t)]$$

and is the superposition of a backward-propagating wave, $f^{+}\exp[i(\sigma z + \omega t)]$, and a forward-propagating wave, $f^{-}\exp[i(-\sigma z + \omega t)]$.

In a reflection–transmission problem, an incident wave is coming from one of the half-spaces, say $z < 0$. The field f then takes the form

$$f(z) = \begin{cases} f^{i}\exp(-i\sigma_{-}z) + f^{r}\exp(i\sigma_{-}z), & z < 0, \\ \tilde{f}(z), & z \in (0, L) \\ f^{t}\exp(-i\sigma_{+}(z - L)), & z > L, \end{cases} \tag{11}$$

where

$$\sigma_{-} = \sqrt{h(0_{-})}, \qquad \sigma_{+} = \sqrt{h(L_{+})},$$

and $f^{i}, f^{r}, f^{t} \in \mathbb{C}$. The problem consists in the determination of the ratios f^{r}/f^{i} and f^{t}/f^{i}, namely the coefficients of reflection and transmission. The function $\tilde{f}$ is the solution of Eq. (3) in the layer $z \in (0, L)$. As is well known (see, e.g. Ref. [8]), such a solution exists and is unique, once we fix the initial value, say $\tilde{f}(L), \tilde{f}'(L)$, in the intervals where h is continuous.

3. Rays and turning points

Associated with the Helmholtz equation (1) is the ray equation

$$\frac{\mathrm{d}}{\mathrm{d}s}\left(n\frac{\mathrm{d}\mathbf{x}}{\mathrm{d}s}\right) = \nabla n, \qquad \left|\frac{\mathrm{d}\mathbf{x}}{\mathrm{d}s}\right| = 1, \tag{12}$$

whose solution $\mathbf{x} = \mathbf{x}(s)$ is the ray trajectory. Let n depend on $\mathbf{x}$ only through z. We can choose the x, y-axes so that the trajectory is in the x, z-plane. It then follows that

$$\frac{\mathrm{d}z}{\mathrm{d}s} = \pm\frac{1}{n}\sqrt{n^2 - a^2}, \qquad \frac{\mathrm{d}x}{\mathrm{d}z} = \pm\frac{a}{\sqrt{n^2(z) - a^2}},$$

the parameter $a > 0$ being determined by the initial (oblique) direction.

Let $z_1 \in (0, L)$ be a turning point, namely $n(z_1) = a$, whereas $n(z) > a$, $z \in (0, z_1)$. Hence we have

$$\frac{\mathrm{d}x}{\mathrm{d}z} \to \pm\infty \quad \text{as} \quad z \to z_1^{-},$$

or

$$\frac{dz}{ds} = \pm \frac{1}{n}\sqrt{n^2(z) - a^2} \to 0 \qquad \text{as} \qquad z \to z_1^-.$$

By looking at the + and − case separately, it is then inferred that the ray does not penetrate the region where $n^2 < a^2$ but bends back, at $z = z_1$, toward the direction of incidence (see Ref. [6], Section 5.2).

It is convenient to consider the special case

$$\frac{dx}{dz} = \frac{a}{(z_1 - z)^\beta}.$$

If (z_0, x_0) is a point on the ray then

$$x(z_1) - x(z_0) = a\int_{z_0}^{z_1} (z_1 - z)^{-\beta} dz.$$

If $\beta \geq 1$ then the integral approaches infinity. Hence $x(z) \to \infty$ as $z \to z_1$ namely the trajectory approaches asymptotically the line $z = z_1$. It is then an immediate consequence that the ray does not bend toward the region of incidence. Conversely, if $\beta < 1$ then the integral is finite, x approaches a finite value as $z \to z_1$ and the trajectory reaches $z = z_1$ with a finite value x_1 of x. Beyond $x = x_1$ the trajectory bends according to the branch $dx/dz = -a/\sqrt{n^2(z) - a^2}$.

In terms of $n(z)$, this argument shows that if there is $c > 0$ such that

$$0 \leq n(z) - n(z_1) < c(z_1 - z)^{2\beta}, \qquad \beta \geq 1, \tag{13}$$

in the neighborhood of z_1, then the trajectory approaches asymptotically $z = z_1$. If, instead, there is $d > 0$ such that

$$0 \leq n(z) - n(z_1) > d(z_1 - z)^{2\beta}, \qquad \beta < 1, \tag{14}$$

then the trajectory reaches $z = z_1$ with a finite value of x. It is then untrue that the ray bends back anyway. The condition Eq. (13) is sufficient for preventing the ray from bending back, whereas Eq. (14) is sufficient for the bending of the ray. In both cases the ray does not penetrate the region where $n^2 < a^2$.

If more turning points occur then these conclusions hold at any turning point. Indeed assume that z_1, z_2 are two turning points in that $n_2 > a_2$ as $z < z_1$ or $z > z_2$ and $n^2 < a^2$ as $z \in (z_1, z_2)$. According to [2], Section 24.4, the incident ray bends back at $z = z_1$, whereas another ray arises at $z = z_2$ and propagates away as in the tunneling effect of quantum mechanics. The bent ray is seen as the reflected wave, the other ray as the transmitted wave.

The ray theory is an approximation and, as such, may provide wrong results in limiting cases. By means of the first integral (4) we see at once that the behavior derived for two turning points, by the ray equation, is not true for the solution to the Helmholtz equation.

Quite obviously, turning points for rays occur at z such that $n(z) = a$. Since the constant a depends on the selected ray, it follows that turning points for rays are not uniquely related to the material properties (here n) but depend also on the ray under consideration.

4. First integrals

Let u be a solution of Eq. (1). Multiplication of Eq. (1) by the complex conjugate u^* of u allows us to write

$$\nabla \cdot (u^* \nabla u) - \nabla u^* \cdot \nabla u + k^2 n^2 u^* u = 0.$$

The second and third terms are real. Hence taking the imaginary part yields the new conservation law

$$\nabla \cdot \Im (u^* \nabla u) = 0, \tag{15}$$

where $\Im$ means imaginary part. The physical meaning of Eq. (15) depends on that of u. Anyway, we observe, in general, that by Eq. (15)

$$\mathbf{J} = \Im (u^* \nabla u)$$

is a divergence-free vector flux and, by the divergence theorem,

$$\int_S \mathbf{J} \cdot \mathbf{n} \, \mathrm{d}a = 0 \tag{16}$$

for any closed surface S.

It may happen that $\mathbf{J}$ is really 3D but the net flux through the lateral part, of any closed surface, between two planes $z =$ constant, vanishes. Such is the case for any cylindrical surface S, say $(x - x_0)^2 + (y - y_0)^2 = r^2$, $z \in [z_1, z_2]$, when $u(x, y, z) = \exp[i(\kappa_x x + \kappa_y y)] f(z)$, where κ_x and κ_y are real. Hence

$$\mathbf{J} = (\kappa_x f^* f, \kappa_y f^* f, \Im (f^* f')).$$

On the lateral surface S_L

$$\int_{S_\mathrm{L}} \mathbf{J} \cdot \mathbf{n} \, \mathrm{d}a = r \int_0^L \left[\int_0^{2\pi} (\kappa_x \cos \theta + \kappa_y \sin \theta) \mathrm{d}\theta \right] f^*(z) f(z) \mathrm{d}z.$$

Since the integral on θ vanishes it follows that

$$\int_{S_\mathrm{L}} \mathbf{J} \cdot \mathbf{n} \mathrm{d}a = 0.$$

Accordingly, Eq. (16) yields

$$\int_{z=z_1} \mathbf{J} \cdot \mathbf{e}_3 \mathrm{d}a = \int_{z=z_2} \mathbf{J} \cdot \mathbf{e}_3 \mathrm{d}a.$$

The arbitrariness of x_0, y_0, r implies that the component $\mathbf{J} \cdot \mathbf{e}_3$, in the z-direction, is constant.

If, instead, the 1D Helmholtz equation (3) holds then we find that

$$0 = f^* f'' + h f^* f = (f^* f')' - (f')^* f' + h f^* f.$$

Because $(f')^* f'$ and $hf^* f$ are real it follows that

$$\frac{\mathrm{d}}{\mathrm{d}z}\Im(f^* f') = 0,$$

whence Eq. (4). This means that the scalar field

$$\mathcal{F} := \Im(f^* f') = \mathbf{J}\cdot\mathbf{e}_3 \tag{17}$$

is constant as z varies. Moreover if, as in Eq. (5),

$$u(x, z) = \exp(i\kappa_x x) f(z), \qquad \kappa_x \in \mathbb{R},$$

then also

$$\mathcal{F} = \Im(u^* u') = \text{constant}. \tag{18}$$

We next show how the constancy of $\mathcal{F}$ provides interesting consequences on wave propagation. Of course the physical meaning of $\mathcal{F}$ depends on that of f.

4.1. *Shear waves in elastic solids*

From Eq. (9) we have

$$\mathcal{F} = \Im(U^* \mu U'), \tag{19}$$

where U is the displacement. We now show that the first integral $\mathcal{F}$ is a constant factor times the energy flux $\mathcal{P}$. A horizontally polarized displacement, $\mathbf{u} = u_y(x, z)\exp(i\omega t)\mathbf{e}_y$, produces a shear traction

$$\mathbf{t} = \mu(z)\frac{\partial u_y}{\partial z}(x, z)\exp(i\omega t)\mathbf{e}_y.$$

Consequently, from Eq. (7) we have

$$\mathbf{t}\cdot\dot{\mathbf{u}}^* = -i\omega u_y^* \mu \frac{\partial u_y}{\partial z} = -i\omega U^* \mu U' \tag{20}$$

and hence

$$\mathcal{F} = \frac{1}{\omega}\Re(\mathbf{t}\cdot\dot{\mathbf{u}}^*).$$

Since $-\Re(\mathbf{t}\cdot\dot{\mathbf{u}}^*)/2$ is the time-averaged energy flux density $\mathcal{P}$ (see Ref. [3], Section 3.4) then

$$\mathcal{F} = -\frac{2}{\omega}\mathcal{P}.$$

If, as is a common assumption, $\mathbf{t}$ and $\dot{\mathbf{u}}$ are taken to be continuous at any interface, then $\mathcal{P}$ is constant everywhere and so is $\mathcal{F}$.

4.2. Electromagnetic waves in dielectrics

The first integral (18) holds for E_y,

$$\mathcal{F} = \Im(E_y^* E_y') = \text{constant}$$

for horizontally polarized waves, $\mathbf{E} = E_y \mathbf{e}_2$. If, as in Eq. (7), $E_y = \exp(i\kappa_x x)f(z)$, $\kappa_x \in \mathbb{R}$, then

$$\nabla \times \mathbf{E} = -E_y' \mathbf{e}_1 + i\kappa_x E_y \mathbf{e}_3.$$

Hence

$$E_y' = -(\nabla \times \mathbf{E})\cdot \mathbf{e}_1 = i\mu\omega H_x,$$

and

$$\mathbf{E}^* \times \mathbf{H}\cdot \mathbf{e}_3 = \frac{i}{\mu\omega} E_y^* E_y'.$$

Consequently, because $\mu,\ \omega \in \mathbb{R}$,

$$\mathcal{F} = -\mu\omega \Re(\mathbf{E}^* \times \mathbf{H}\cdot \mathbf{e}_3).$$

This means that to within the factor $-\mu\omega$, $\mathcal{F}$ is the z-component of the real part of the complex Poynting vector.

4.3. Linear acoustic waves

Since $f = \wp/\sqrt{\rho_0}$ then

$$\mathcal{F} = \Im\left[\frac{\wp^*}{\sqrt{\rho_0}}\left(\frac{\wp}{\sqrt{\rho_0}}\right)'\right].$$

Both the perturbation pressure $\wp$ and the equilibrium density ρ_0 enter the expression of the first integral $\mathcal{F}$.

5. Wave propagation in the presence of turning points

We now look at Eq. (3) and examine whether intervals $(z_1, z_2) \subset [0, L]$, where $h > 0$, may be zones where wave propagation is not possible [5]. We let $h(z)$ be constant in the half-spaces $z < 0$ and $z > L$. Let $h(z) = h_- > 0$, $z < 0$, and $h(z) = h_+$ as $z > L$. Moreover, h may suffer jump discontinuities at a finite number of places $0 = z_1 < z_2 < \cdots < z_n = L$. The boundary conditions on f, f' at discontinuity surfaces (e.g. U, $\mu U'$ continuous) relate $f(z_-)$ and $f'(z_-)$ to $f(z_+)$ and $f'(z_+)$. As is suggested by applications, there is no significant loss of generality in assuming that, for any z where h is discontinuous,

$$f(z_-) = 0, f'(z_-) = 0 \Leftrightarrow f(z_+) = 0,\ \ f'(z_+) = 0;\quad \text{sgn}\ \mathcal{F}(z_-) = \text{sgn}\ \mathcal{F}(z_+), \qquad (21)$$

where sgn denotes the sign. Along with Eq. (4), this implies that sgn $\mathcal{F}$ is constant throughout. For example, for shear waves in the form Eq. (9) $\mathcal{F}$ is indeed constant across discontinuities.

Consider the half-space $z > L$ and let $h_+ > 0$. It is assumed that the solution to Eq. (3) satisfies

$$f(z) = f^t \exp(-i\sigma_+(z - L)), \qquad z > L, \tag{22}$$

where $\sigma_+ = \sqrt{h_+}$. This amounts to assuming that, as $z > L$, only a forward-propagating, and hence outgoing, wave occurs. Correspondingly,

$$f(L_+) = f^t, \qquad f'(L_+) = -i\sigma_+ f^t,$$

and

$$\mathcal{F} = -\sigma_+ |f^t|^2. \tag{23}$$

To save writing, it is understood that f is the solution to Eq. (3), subject to the conditions (21) and (22). Also, with a view to the reflection–transmission problem we let $h(z) = h_- > 0$ as $z > 0$. Hence

$$f(z) = f^i \exp(-i\sigma_- z) + f^r \exp(i\sigma_- z), \qquad z < 0. \tag{24}$$

Proposition 1: *If $h_+ > 0$ then f vanishes on $\mathbb{R}$ if and only if $f^t = 0$.*

Proof: If $f = 0$ on $\mathbb{R}$ then, by Eq. (22), $f^t = 0$. Conversely, if $f^t = 0$ then, from Eq. (21), $f(L_-) = 0$ and $f'(L_-) = 0$. By uniqueness, $f = 0$ as $z \in (z_{n-1}, z_n)$. Hence $f((z_{n-1})_+) = 0$ and $f'((z_{n-1})_+) = 0$. Application of Eq. (21) and iteration show that $f = 0$ on $\mathbb{R}$. □

Proposition 2: *By Eq. (24), $f = 0$ if and only if $f^i = 0, f^r = 0$.*

Proof: If $f = 0$ as $z \le 0$ then

$$0 = f^i \exp(-i\sigma_- z) + f^r \exp(i\sigma_- z), \qquad 0 = i\sigma_-[-f^i \exp(-i\sigma_- z) + f^r \exp(i\sigma_- z)].$$

Letting $z = 0$ gives $f^i + f^r = 0$, $f^r - f^i = 0$ whence $f^i = 0$, $f^r = 0$. The converse is obviously true. □

As a comment, the superposition of forward- and backward-propagating waves vanishes if and only if both waves vanish.

In applications we know f^i and ask for the possible vanishing of f^t. The answer is given as follows.

Corollary 1: *If $h_+ > 0$ then $f^i \neq 0$ implies that $f^t \neq 0$.*

Proof: If $f^t = 0$ then $f = 0$ on $\mathbb{R}$ and, by Proposition 2, $f^i = 0$. Hence the conclusion follows. □

Now let $h_+ < 0$. Hence, as $z > L$,

$$f(z) = f^t \exp\left(-\sqrt{|h_+|}(z - L)\right),$$

and $\mathcal{F} = 0$.

Proposition 3: *If $h_+ \leq 0$ then $|f^r| = |f^i|$.*

Proof: By the constancy of $\mathcal{F}$, $\mathcal{F} = 0$ everywhere. In particular, $\mathcal{F} = 0$ as $z < 0$. From Eq. (24) we find that

$$f^* f' = i\sigma_-(|f^r|^2 - |f^i|^2) + 2\Re[(f^i)^* f^r i\sigma_- \exp(2i\sigma_- z)], \tag{25}$$

whence

$$\mathcal{F} = \sigma_-(|f^r|^2 - |f^i|^2). \tag{26}$$

The vanishing of $\mathcal{F}$ implies that $|f^r| = |f^i|$.
If $h_+ = 0$ then a bounded solution f as $z > L$ is necessarily $f =$ constant whence $\mathcal{F} = 0$; again $|f^r| = |f^i|$. □

In applications we know f^i and ask for the possible vanishing of f^t. The answer is given as follows.

Corollary 1: *Irrespective of the value of h_+, $f^i \neq 0$ implies that $f \neq 0$ as $z > L$.*

Proof Let $h_+ > 0$. If $f^t = 0$ then $f = 0$ as $z > L$. By Proposition 1 $f = 0$ as $z < 0$. and hence, by Proposition 2, $f^i = 0$. Hence $f^i \neq 0$ implies that $f^t \neq 0$. Let $h_+ < 0$. If $f^t = 0$ then $f = 0$ as $z > L$. Again by Propositions 1 and 2 we find that $f^i = 0$. Finally, if $h_+ = 0$ then f is constant, say $f = b$ as $z > L$. If $b = 0$ then $f = 0$ everywhere and $f^i = 0$. The result is then proved. □

Remark: *It is natural to ask why $h_+ < 0$ implies that $\mathcal{F} = 0$, while $h < 0$ within the layer does not imply $\mathcal{F} = 0$. In the uniform half-space $z > L$ the solution $\exp(\sqrt{|h_+|}(z - L))$ is not allowed because it is unbounded. Instead, in a uniform subinterval $[z_1, z_2]$*

$$f(z) = f_+ \exp(\sqrt{|h|}(z - z_1)) + f_- \exp(-\sqrt{|h|}(z - z_1)),$$

whence

$$\mathcal{F} = \sqrt{|h|}(|f_-|^2 - |f_+|^2).$$

Consequently, in a finite subinterval $[z_1, z_2]$, h may become negative and, nevertheless, wave propagation is not ruled out.

The results of this section show also that the occurrence of an interval (z_1, z_2) where $h < 0$ does not preclude wave propagation. Nor is wave propagation precluded if $h(z) = h_+ < 0$ as $z > L$. In such a case, though, the condition $|f^r| = |f^i|$ allows for the vanishing of $\mathcal{F}$.

Remark: *The property $f^i \neq 0 \Rightarrow f \neq 0$ on $\mathbb{R}$ answers the question of the behavior between two turning points. While a jump of the ray is provided by the ray theory ([2], p. 207), no jump is allowed for the (exact) solution to Eqs. (1) and (3). In addition, irrespective of the form of h, there cannot be any interval $[z_1, z_2]$, in a layer between homogeneous half-spaces, where wave propagation is precluded [5].*

6. Variational conditions and jump relations

The Helmholtz equation (1) and the 1D version (3) are the Euler–Lagrange equations of the functionals

$$G(u) = \frac{1}{2}\int_{\Omega} [k^2 n^2 u^2 - \nabla u \cdot \nabla u] \mathrm{d}v,$$

$$\mathcal{G}(f) = \frac{1}{2}\int_a^b [hf^2 - (f')^2] \mathrm{d}z,$$

where Ω is the appropriate region and $[a, b]$ the appropriate interval.

Consider $\mathcal{G}$ and denote by

$$\mathcal{L} = \frac{1}{2}[hf^2 - (f')^2]$$

the Lagrangian density. Let $c_k \in (a, b), k = 1, \ldots, m$, be points where $\mathcal{L}$ is allowed to suffer a jump discontinuity. This means that f is continuous throughout, whereas h and f' may suffer jumps at $c_1, \ldots, c_m$ but are continuous everywhere else. Denote by

$$[\![f]\!](c_k) := f(c_{k+}) - f(c_{k-})$$

the jump of f at $z = c_k$. Hence we have

$$[\![f]\!](c_k) = 0, \qquad k = 1, \ldots, m,$$

whereas the limits $f'(c_{k-}), f'(c_{k+})$ are unrestricted. The functional $\mathcal{G}$ is extremum or stationary at f if Eq. (3) holds in the open intervals $(a, c_1), (c_1, c_2), \ldots, (c_m, b)$ and, moreover, the boundary conditions [9]

$$[\![\partial\mathcal{L}/\partial f']\!](c_k) = 0, \qquad k = 1, \ldots, m$$

hold. This in turn means that

$$[\![f']\!](c_k) = 0, \qquad k = 1, \ldots, m,$$

namely f' is required to be continuous at the interfaces $z = c_1, \ldots, c_m$.

In elasticity we can consider the 1D version $(\mu u_y')' + \rho\omega^2 u_y = 0$, of Eq. (7), and say that it is the Euler–Lagrange equation associated with

$$\mathcal{L} = \frac{1}{2}[NU^2 - \mu(U')^2].$$

The boundary conditions then become

$$[\![\mu U']\!](c_k) = 0, \qquad k = 1, \ldots, m.$$

Accordingly, if the displacement u_y is continuous then the traction $\mu u_y'$ is continuous. Hence the standard conditions of continuity for displacement and traction are consistent with the variational conditions.

In electromagnetism, if $[\![E_y]\!] = 0$ at the interfaces then the boundary conditions become

$$[\![E_y']\!](c_k) = 0, \qquad k = 1, \ldots, m.$$

We now determine the jump conditions through the balance equations in global form. In this regard, let Ω be any region that contains a surface Y such that $\Omega = \Omega_- \cup Y \cup \Omega_+$ and $\Omega_- \cap \Omega_+ = \emptyset$. Also, let $\mathbf{n}_Y$ the unit normal to Y, from Ω_- to Ω_+. If Y is a discontinuity surface for ϕ then

$$\frac{\mathrm{d}}{\mathrm{d}t}\int_{\Omega\setminus Y} \phi \,\mathrm{d}v = \int_{\Omega\setminus Y} \partial_t \phi \,\mathrm{d}v + \int_{\partial\Omega} \phi \mathbf{v}\cdot\mathbf{n}\,\mathrm{d}a - \int_Y [\![\phi]\!]\boldsymbol{\nu}\cdot\mathbf{n}\,\mathrm{d}a,$$

where $\mathbf{v}$ is the velocity of the points of Y, $\mathbf{v}$ is the velocity of the points of Ω and $[\![\phi]\!]$ is the discontinuity of ϕ across Y. For any vector field $\mathbf{w}$ we have

$$\int_{\partial\Omega} \mathbf{w}\cdot\mathbf{n}\,\mathrm{d}a = \int_{\Omega\setminus Y} \nabla\cdot\mathbf{w}\,\mathrm{d}v + \int_Y [\![\mathbf{w}]\!]\mathrm{d}a.$$

Upon the identification $\mathbf{w} = \phi\,\mathbf{v}$ we find that

$$\frac{\mathrm{d}}{\mathrm{d}t}\int_{\Omega\setminus Y} \phi\,\mathrm{d}v = \int_{\Omega\setminus Y} [\partial_t\phi + \nabla\cdot(\phi\mathbf{v})]\mathrm{d}v + \int_Y [\![\phi(\mathbf{v} - \boldsymbol{\nu})]\!]\cdot\mathbf{n}\,\mathrm{d}a.$$

Let ϕ enter a balance law for the region Ω in the form

$$\frac{\mathrm{d}}{\mathrm{d}t}\int_{\Omega\setminus Y} \phi\,\mathrm{d}v = \int_{\Omega\setminus Y} \psi\,\mathrm{d}v + \int_{\partial\Omega} \mathbf{b}\cdot\mathbf{n}\,\mathrm{d}a.$$

If Ω is a Gaussian pillbox across any interface $I_k = \{\mathbf{x} : z = c_k\}$ and $Y = \Omega \cap I_k$ then the boundedness of $\partial_t\phi + \nabla\cdot(\phi\mathbf{v})$ and ψ on $\Omega\setminus Y$ along with the limit of vanishing thickness of the pillbox provide

$$\int_Y [\![\mathbf{b} + \phi(\mathbf{v} - \boldsymbol{\nu})]\!]\cdot\mathbf{e}_3\,\mathrm{d}a = 0.$$

The arbitrariness of Y and the continuity of the integrand yield the jump relation

$$[\![\mathbf{b} + \phi(\mathbf{v} - \boldsymbol{\nu})]\!]\cdot\mathbf{e}_3 = 0$$

at any point of I_k.

Consider the mechanical context and let $\nu = 0$, which means that the interface is fixed. The mass conservation corresponds to

$$\phi = \rho, \qquad \psi = 0, \qquad \mathbf{b} = 0.$$

Hence

$$[\![\rho v_z]\!] = 0.$$

The balance of linear momentum amounts to

$$\phi = \rho\mathbf{v}, \qquad \psi = \rho\mathbf{f}, \qquad \mathbf{b} = \mathbf{T},$$

$\mathbf{T}$ being the Cauchy stress tensor. Hence

$$[\![\mathbf{t} + \rho\mathbf{v}\, v_z]\!] = 0.$$

The balance of energy amounts to

$$\phi = \rho\left(\frac{1}{2}\mathbf{v}^2 + \epsilon\right), \qquad \psi = \rho\mathbf{f}\cdot\mathbf{v} + \rho r, \qquad \mathbf{b} = \mathbf{T}\mathbf{v} - \mathbf{q},$$

where ϵ is the internal energy density, r is the heat supply, and $\mathbf{q}$ is the heat flux. Hence we have

$$\left[\!\!\left[\mathbf{v}\cdot\mathbf{t} - q_z + \rho\left(\frac{1}{2}\mathbf{v}^2 + \epsilon\right)v_z\right]\!\!\right] = 0.$$

If, furthermore, $v_z = 0$ and $q_z = 0$ then

$$[\![\mathbf{t}]\!] = 0, \qquad [\![\mathbf{v}\cdot\mathbf{t}]\!] = 0. \tag{27}$$

Since $\mathbf{v} = i\omega\mathbf{u}$ and $\mathbf{t}$ is continuous we then find that

$$[\![\mathbf{u}]\!]\cdot\mathbf{t} = 0.$$

The displacement $\mathbf{u}$ is continuous across the interface in the direction of the traction.

In electromagnetism the boundary conditions follow from Maxwell's equations in integral form. In linear dielectrics we write Gauss' laws as

$$\int_S \epsilon\mathbf{E}\cdot\mathbf{n}\, da = 0, \qquad \int_S \mathbf{H}\cdot\mathbf{n}\, da = 0,$$

for any closed surface S. By choosing S as the boundary surface of a Gaussian pillbox across the interface we eventually obtain

$$[\![\epsilon E_z]\!] = 0, \qquad [\![H_z]\!] = 0.$$

If, instead, S is any open surface with boundary ∂S we have

$$\int_{\partial S} \mathbf{E}\cdot\boldsymbol{\tau}\, dl = -i\omega\mu\int_S \mathbf{H}\cdot\mathbf{n}\, da, \qquad \int_{\partial S} \mathbf{H}\cdot\boldsymbol{\tau}\, dl = i\omega\int_S \mathbf{E}\cdot\mathbf{n}\, da,$$

where $\boldsymbol{\tau}$ is the unit tangent vector to the boundary ∂S. In particular, choosing S in the form of a rectangle, with the normal to the spanning surface tangent to the interface, we find that

$$\int_L [\![\mathbf{E}]\!] \mathrm{d}l \cdot \boldsymbol{\tau} = 0, \qquad \int_L [\![\mathbf{H}]\!] \mathrm{d}l \cdot \boldsymbol{\tau} = 0,$$

where L is the intersection of the rectangle with the interface. The continuity of $[\![\mathbf{E}]\!]$ and $[\![\boldsymbol{H}]\!]$ on L and the arbitrariness of L provide

$$[\![\mathbf{E}]\!] \cdot \boldsymbol{\tau} = 0, \qquad [\![\mathbf{H}]\!] \cdot \boldsymbol{\tau} = 0$$

at the interface. Since $\mathbf{E} = E_y \mathbf{e}_2$ the significant conclusions are

$$[\![E_y]\!] = 0, \qquad [\![H_z]\!] = 0, \qquad [\![H_x]\!] = 0.$$

Now, because

$$-i\omega\mu\mathbf{H} = \nabla \times \mathbf{E} = -E_y' \mathbf{e}_1 + i\kappa_x E_y \mathbf{e}_3$$

the continuity of $\mathbf{H}$ across the interface implies that

$$[\![E_y']\!] = 0, \qquad [\![E_y]\!] = 0.$$

Hence both E_y and E_y' are continuous across any interface.

In conclusion, the first integral $\mathcal{F}$,

$$\mathcal{F} = \Im(U^* \mu U_y') \qquad \text{or} \qquad \mathcal{F} = \Im(E_y^* E_y'),$$

is continuous across any interface in elastic solids or in dielectrics. This has the remarkable consequence that, for horizontally polarized waves, $\mathcal{F}$ is constant throughout, in elastic solids and in dielectrics.

7. Application to reflection–transmission processes

We now apply the first integral (4) to the reflection–transmission process associated with an inhomogeneous layer. As before, $h(z) = h_-$, as $z < 0$,n $h(z) = h_+$ as $z > L$ and suffers a finite number of jump discontinuities as $z \in [0, L]$. For definiteness we restrict attention to the more significant case $h_-, h_+ > 0$.

Letting the incident wave come from $z < 0$ we can write $f(z)$ as in Eq. (24), where $f^i, f^r \in \mathbb{C}$. Consequently we find that $f^* f'$ is given by Eq. (25) and $\mathcal{F}$ by Eq. (26). As $z > L$ we let only a transmitted wave occur and then take f in the form (22) where $f^t \in \mathbb{C}$. Consequently, we find that $\mathcal{F}$ is given by Eq. (23). If $\mathcal{F}$ is continuous across the discontinuities of h then $\mathcal{F}$ is constant throughout and hence

$$\sigma_-(|f^i|^2 - |f^r|^2) = \sigma_+ |f^t|^2. \tag{28}$$

The result (28) relates the amplitude of the reflected wave, $|f^r|$, to that of the transmitted one, $|f^t|$. Similar relations are given in the literature, through the scattering matrix, only when the asymptotic properties are the same ($\sigma_+ = \sigma_-$) [10].

If $\mathcal{F}$ is not continuous across the discontinuities of h then $\mathcal{F}$ is merely piecewise constant. The possible jump of $\mathcal{F}$ is determined by the boundary conditions. In such a case Eq. (28) is generalized by a relation between $\sigma_-(|f^i|^2 - |f^r|^2)$ and $\sigma_+|f^t|^2$ that originates from the discontinuities and the boundary conditions.

If, instead, h is continuous then Eq. (28) holds whatever the meaning of f and $\mathcal{F}$ can be.

For definiteness, we now examine the validity and the consequences of Eq. (28) for elastic, electromagnetic, and acoustic waves.

7.1. Reflection–transmission of elastic shear waves

Let h be continuous on $\mathbb{R}$ and set $\mu_- = \mu(0)$ and $\mu_+ = \mu(L)$. It may happen that h in Eq. (10) suffers jump discontinuities because of ρ, μ, μ', μ''. However, if as is common practice, we let **t** and **u** be continuous everywhere, because of Eq. (20), it follows that $\Im(U^* \mu U')$ is continuous across discontinuities of h. Also, from Eq. (19), $\Im(U^* \mu U')$ is constant in the open intervals. Consequently, $\Im(U^* \mu U')$ is constant everywhere and Eq. (30) holds also if jump discontinuities of h occur.

From Eq. (9), the relation Eq. (28) becomes

$$\sigma_- \mu_-(|U^i|^2 - |U^r|^2) = \sigma_+ \mu_+ |U^t|^2, \tag{29}$$

where

$$\sigma_- = \sqrt{\frac{\rho_- \omega^2}{\mu_-} - \kappa_x^2},$$

and likewise for σ_+. Letting α^i, α^t be the incidence and the transmission angles we have

$$\kappa_x = \kappa_- \sin \alpha^i = \kappa_+ \sin \alpha^t, \qquad \kappa_\pm = \sqrt{\rho_\pm / \mu_\pm}\, \omega.$$

Hence

$$\sigma_- = \sqrt{\rho_- / \mu_-}\, \omega \cos \alpha^i, \qquad \sigma_+ = \sqrt{\rho_+ / \mu_+}\, \omega \cos \alpha^t.$$

Moreover, because $\kappa_- \sin \alpha^i = \kappa_+ \sin \alpha^t$, we have

$$\cos \alpha^t = \sqrt{1 - (\kappa_- / \kappa_+)^2 \sin^2 \alpha^i}.$$

Denote by $c = \sqrt{\mu/\rho}$ the speed of the pertinent wave, in homogeneous regions. Hence Eq. (29) may be written as

$$\rho_- c_-(|U^i|^2 - |U^r|^2) \cos \alpha^i = \rho_+ c_+ |U^t|^2 \cos \alpha^t. \tag{30}$$

The form (30) of the relation between the amplitudes in a reflection–transmission process has been obtained through energy considerations only for two half-spaces without any layer (see Ref. [11], Section 3.3).

7.2. Reflection–transmission of electromagnetic waves

Because E_y and E_y' are continuous across any interface, the first integral $\mathcal{F} = \Im(E_y^* E_y)$ is constant throughout. Hence the relation Eq. (28) holds and provides

$$\sigma_-(|E_y^i|^2 - |E_y^r|^2) = \sigma_+|E_y^t|^2, \tag{31}$$

where

$$\sigma = \sqrt{\omega^2 \mu\epsilon - \kappa_x^2}.$$

It is assumed that

$$\omega^2 \mu\epsilon - \kappa_x^2 > 0$$

as $z < 0$ and $z > L$.

7.3. Reflection–transmission of acoustic waves

Since $u = \wp/\sqrt{\rho_0}$ and $h = \omega^2/p_\rho - \kappa_x^2$ the relation (28) provides

$$\left.\frac{\sigma}{\rho_0}\right|_- (|\wp^i|^2 - |\wp^r|^2) = \left.\frac{\sigma}{\rho_0}\right|_+ |\wp^t|^2, \tag{32}$$

where $\sigma = \sqrt{\omega^2/p_\rho - \kappa_x^2}$.

8. Remarks on the first integral and the WKB solution

The WKB solution is frequently considered in applications as an approximation to the solution of Eq. (3). It is then of interest to examine the connection of the WKB solution with the first integral (4) and with turning points.

Let $h > 0$. The solution f is sought in the complex form

$$f(z) = A(z)\exp(iS(z)), \tag{33}$$

where A and S, on $\mathbb{R}$, are real valued. Substitution in Eq. (3) gives

$$A'' - A(S')^2 + hA = 0, \qquad 2A'S' + AS'' = 0.$$

The second equation means that

$$A^2 S' = \text{constant}. \tag{34}$$

The solution f to Eq. (3) in the form (33) gives

$$f^* f' = AA' + iA^2 S',$$

and hence

$$\mathcal{F} = A^2 S'.$$

This means that the requirement (4) on Eq. (33) becomes Eq. (34).

In contrast to the ray method, the WKB approximation allows field penetration in that f is nonzero even beyond turning points, namely where $h < 0$. This assertion is made operative by following the Langer and Budden procedure (see Ref. [12]) for an approximate solution if h changes the sign. Assume that h is positive and slowly varying as $z < z_1$. Hence as $z < z_1$ the WKB approximation works (see Ref. [13]) and we set

$$S'(z) = \pm h^{1/2}(z), \qquad A^2(z) = A_0 h^{-1/4}(z).$$

Hence S and A are real valued. As a consequence the (approximate) solution $\hat{f}$ can be written in the form (see, e.g. Refs. [12,14])

$$\hat{f}(z) = h^{-1/4}(z)\left[a \exp\left(i \int_{z_0}^{z} h^{1/2}(\xi)\mathrm{d}\xi\right) + b \exp\left(-i \int_{z_0}^{z} h^{1/2}(\xi)\mathrm{d}\xi\right)\right], \tag{35}$$

where a and b are real constants. The two terms in Eq. (35) are called Liouville–Green functions [15].

Also, let h be negative and slowly varying as $z > z_2 > z_1$. Hence f is sought in the real form

$$f(z) = B(z)\exp(T(z)). \tag{36}$$

The conditions

$$B'' + B(T')^2 + hB = 0, \qquad 2B'T' + BT'' = 0$$

are sufficient for Eq. (36) to be a solution of Eq. (3). Again we find a conservation law, namely

$$B^2 T' = \text{constant}.$$

Neglecting B'' we find that

$$U'(z) = \pm(-h)^{1/2}, \qquad B^2(z) = B_0(-h)^{-1/4},$$

whence the approximate solution $\breve{f}$, as $z > z_2$, is

$$\breve{f}(z) = c(-h)^{-1/4}(z)\exp\left(-\int_{z_0}^{z} (-h)^{1/2}(\xi)\mathrm{d}\xi\right). \tag{37}$$

In the region $z \in [z_1, z_2]$ where h changes the sign the solution depends on the function h. We merely require that Eq. (4) hold throughout.

We now evaluate $\mathcal{F}$. Concerning $\hat{f}$, from Eq. (35) we find that

$$\hat{f}^*\hat{f}' = h^{-3/2}\left[-\frac{1}{4}h'(a^2 + b^2) + ih^{3/2}(a^2 - b^2) - \frac{1}{4}abh' \right.$$
$$\left. \times \cos\left(2\int_{z_0}^{z} h^{1/2}(\xi)\mathrm{d}\xi\right) - abh^{-3/2}\sin\left(2\int_{z_0}^{z} h^{1/2}(\xi)\mathrm{d}\xi\right)\right],$$

whence

$$\mathcal{F} = a^2 - b^2.$$

Indeed, because $\mathcal{F}$ equals $-2/\omega$ times the energy flux $\mathcal{P}$, the functions

$$ah^{-1/4}(z)\exp\left[i\left(\int_{z_0}^{z} h^{1/2}(\xi)\mathrm{d}\xi + \omega t\right)\right], \quad bh^{-1/4}(z)\exp\left[i\left(-\int_{z_0}^{z} h^{1/2}(\xi)\mathrm{d}\xi + \omega t\right)\right] \quad (38)$$

prove to propagate upward ($\mathcal{P} < 0$) and downward ($\mathcal{P} > 0$), respectively, as $\omega > 0$. Accordingly, if $h > 0$, the WKB approximation provides a solution $\hat{f}$ in planarly stratified media that is a superposition of an upward and a downward wave. The WKB solution (37), beyond the turning point, is real valued and hence the corresponding value of $\mathcal{F}$ is zero. In conclusion,

$$\mathcal{F}(\hat{f}) = a^2 - b^2, \qquad \mathcal{F}(\check{f}) = 0.$$

Different conclusions arise depending on the occurrence of turning points. If no turning point occurs then $\mathcal{F} < 0$ as $z \to \infty$. Once the amplitude of the transmitted wave, and hence $\mathcal{F}$, is fixed then $a^2 - b^2$ is known. If, instead, a turning point occurs then $\mathcal{F} = 0$, beyond the turning point, and by the constancy, $\mathcal{F}$ is zero everywhere. This in turn implies that

$$a^2 - b^2 = 0,$$

which may be regarded as a condition of total reflection, $|a| = |b|$. In addition, the field penetration of f beyond the turning point occurs through $\check{f}$ with a field that does not propagate energy ($\mathcal{F} = 0$).

9. Conclusions

Time-harmonic fields satisfying the Helmholtz equation (1) are considered. Subject to the assumption (5), which models the oblique incidence on a planarly stratified medium, we have examined the 1D counterpart Eq. (3) and found that the first integral (4) holds.

As a consequence of the first integral it follows that $f^i \neq 0 \Rightarrow f \neq 0$ everywhere. Hence, while a jump of the ray between two turning points is provided by the ray theory, no jump is allowed for the (exact) solution to Eq. (1) or Eq. (3).

By means of the variational conditions and of the appropriate balance equations the first integral $\mathcal{F}$ is shown to be continuous across interfaces (for elastic and electromagnetic waves). Hence $\mathcal{F}$ is constant throughout. This implies that, irrespective of the value of h_+, if the amplitude f^i is nonzero then the transmitted wave is nonzero and, consequently, wave propagation is not precluded. Also, if $h_+ > 0$ then Eq. (28) holds, which results in Eqs. (29), (31), and (32) for elastic, electromagnetic, and acoustic waves. If $h_+ \leq 0$ then it follows that $|f^r| = |f^i|$, namely total reflection occurs. This, in turn, gives a condition on the coefficients in the WKB approximation. If $h < 0$ within the layer then $\mathcal{F}$ need not vanish and wave propagation is not precluded.

Acknowledgements

The research leading to this chapter has been supported by the Italian MIUR through the Research Project COFIN 2002 'Mathematical Models for Materials Sciences'.

References

[1] Brekhovskikh, L.M. and Goncharov, V. (1985), Mechanics of Continua and Wave Dynamics, Springer, Berlin, Section 12.1.
[2] Brekhovskikh, L.M. (1980), Waves in Layered Media, Academic Press, New York, pp. 161–162.
[3] Caviglia, G. and Morro, A. (1992), Inhomogeneous Waves in Solids and Fluids, World Scientific, Singapore, Section 6.5.
[4] Chew, W.C. (1995), Waves and Fields in Inhomogeneous Media, IEEE Press, New York, Section 1.1.5.
[5] Chapman, P.B. and Mahony, J.J. (1978), Reflection of waves in a slowly varying medium, SIAM, J. Appl. Math. 34, 303–319.
[6] Felsen, L.B. and Marcuvitz, N. (1994), Radiation and Scattering of Waves, IEEE Press, New York, Section 5.8c.
[7] Caviglia, G. and Morro, A. (2004), First integrals and turning points for wave propagation in planarly-stratified media, Acta Mech. 169, 1–11.
[8] Coddington, E.A. and Levinson, N. (1955), Theory of Ordinary Differential Equations, McGraw-Hill, New York.
[9] Gelfand, I.M. and Fomin, S.V. (1963), Calculus of Variations, Prentice Hall, London, p. 63.
[10] Deift, P. and Trubowitz, E. (1979), Inverse scattering on the line, Commun. Pure Appl. Math. 32, 121–251.
[11] Achenbach, J.D., Gautesen, A.K. and McMaken, H. (1982), Ray Methods for Waves in Elastic Solids, Pitman, London.
[12] Wait, J.R. (1970), Electromagnetic Waves in Stratified Media, Pergamon, Oxford, Chapter IV.
[13] Aki, K. and Richards, P.G. (1980), Quantitative Seismology Theory and Methods, Freeman, San Francisco, p. 416.
[14] Nayfeh, A.H. and Nemat-Nasser, S. (1972), Elastic waves in inhomogeneoys elastic media, J. Appl. Mech. 33, 696–702.
[15] Pike, E.R. (1964), On the related-equation method of asymptotic approximation, Q. J. Mech. Appl. Math. 17, 105–124.

Chapter 15

MASTER EQUATIONS AND PATH-INTEGRAL FORMULATION OF VARIATIONAL PRINCIPLES FOR REACTIONS

Bernard Gaveau

UFR de Mathématique 925, Université Pierre & Marie Curie, 75252 Paris Cedex 05, France

Michel Moreau

Laboratoire des Physique Theoretique des Liquides, Université Pierre & Marie Curie, 75252 Paris Cedex 05, France

János Tóth

Department of Analysis, Institute of Mathematics, Budapest University of Technology and Economics, Egry J. út. 1, Budapest H-1521, Hungary

Abstract

The mesoscopic nonequlibrium thermodynamics of a reaction–diffusion system is described by the master equation. The information potential is defined as the logarithm of the stationary distribution. The Fokker–Planck approximation and the Wentzel–Kramers–Brillouin (WKB) method give very different results. The information potential is shown to obey a Hamilton–Jacobi equation, and from this fact general properties of this potential are derived. The Hamilton–Jacobi equation is shown to have a unique regular solution. Using the path-integral formulation of the Hamilton–Jacobi approximation of the master equation it is possible to calculate rate constants for the transition from one well to another of the information potential and give estimates of mean exit times. In progress variables, the Hamilton–Jacobi equation always has a simple solution that is a state function if and only if there exists a thermodynamic equilibrium for the system. An inequality between energy and information dissipation is studied, and systems with a single chemical species are investigated in detail, where all the defined relevant quantities can be calculated explicitly.

Keywords: master equation; Hamilton–Jacobi equation; chemical reaction; Fokker–Planch equation; exit time

1. Reaction–diffusion systems

A uniform reactive system can be represented by a system of volume V at temperature T containing several reactive variables (or internal species) $\mathcal{X}_m (m = 1, 2, \ldots, M)$ possibly in

E-mail address: gaveau@ccr.jussieu.fr (B. Gaveau); moreau@lptl.jussieu.fr (M. Moreau); jtoth@math.bme.hu (J. Tóth).

an inert solvent $\mathcal{X}_0$. The species can react according to an arbitrary number of reversible reaction steps, labeled by r that can be represented in the form

$$\sum_{m=1}^{M} \alpha_r^m \mathcal{X}_m \Leftrightarrow \sum_{m=1}^{M} \beta_r^m \mathcal{X}_m (r = 1, 2, ..., R), \tag{1}$$

where the **stoichiometric coefficients** α_r^m and β_r^m can be non-negative integers. This formulation can be extended to a nonhomogeneous system, if one assumes [1,2, p. 169], that the system is divided into cells small enough to be approximately homogeneous, exchanging molecules by a linear rate. The molecules of a given species receive different labels in different cells, and diffusion is represented by the pseudochemical reaction $\mathcal{X}_m \Leftrightarrow \mathcal{X}_q$, $\mathcal{X}_m$ and $\mathcal{X}_q$ representing the same species in neighboring cells. Eq. (1), however, can only occur between molecules within the same cell.

From a stochastic point of view, the system can be described by the probability $\mathcal{P}(X_1, X_2, ..., X_M, t) = \mathcal{P}(\mathbf{X}, t)$ of finding X_m particles of the m-th species ($m = 1, 2, ..., M$) at time t, where X_m denotes the number of molecules of species $\mathcal{X}_m$.

We assume that the vector $\mathbf{X} = (X_1, X_2, ..., X_M)$ changes according to a Markovian jump process.

2. Master equation and Fokker–Planck equation

2.1. *Master equation*

The system described above evolves by various subprocesses, and we call $W_\gamma(\mathbf{X})$ the probability per unit time of the transition $\mathbf{X} \to \mathbf{X} + \boldsymbol{\gamma}$ with given vectors of non-negative integers $\boldsymbol{\gamma}$. These vectors may be any of the **elementary reaction vectors** $\beta_r - \alpha_r, \alpha_r - \beta_r (r = 1, 2, ..., R)$. The state of the system at time t is described by the absolute probability distribution function $\mathcal{P}(\mathbf{X}, t)$ whose time evolution is given by the **master equation**

$$\frac{\partial \mathcal{P}(\mathbf{X}, t)}{\partial t} = (\Lambda \mathcal{P})(\mathbf{X}, t), \tag{2}$$

where Λ is the **evolution operator** [3–15]

$$(\Lambda \mathcal{P})(\mathbf{X}, t) := \sum_{\gamma} [W_\gamma(\mathbf{X} - \boldsymbol{\gamma})\mathcal{P}(\mathbf{X} - \boldsymbol{\gamma}, t) - W_\gamma(\mathbf{X})\mathcal{P}(\mathbf{X}, t)]. \tag{3}$$

To calculate the large volume limit we introduce the concentration vector by $\mathbf{x} := \mathbf{X}/V$, and define the probability density function p through

$$\mathcal{P}(B, t) = \int_B p(\mathbf{x}, t) \mathrm{d}\mathbf{x} \tag{4}$$

for all the measurable subsets B of the state space. Instead of the overall transition rates W_γ we can introduce the concentration-dependent transition rates by $w_\gamma(\mathbf{x}) := W_\gamma(\mathbf{X})/V$.

Using these notations the master equation can be rewritten for the function p as

$$\frac{\partial p(\mathbf{x},t)}{\partial t} = V\sum_{\gamma}\left[w_{\gamma}\left(\mathbf{x}-\frac{\boldsymbol{\gamma}}{V}\right)p\left(\mathbf{x}-\frac{\boldsymbol{\gamma}}{V},t\right) - w_{\gamma}(\mathbf{x})p(\mathbf{x},t)\right]. \tag{5}$$

This formalism can also describe reaction–diffusion processes approximately using the trick mentioned above.

2.2. *Approximate Fokker–Planck equation*

A standard approximation [15] of the master equation in the large-volume limit is obtained by the Taylor expansion of the right-hand side of Eq. (5) up to the terms of order $\mathcal{O}(1/V)$. This expansion gives the **Fokker–Planck equation** for p:

$$\frac{\partial p(\mathbf{x},t)}{\partial t} = -\sum_{m=1}^{M}\frac{\partial}{\partial x_m}(A_m p)(\mathbf{x},t) + \frac{1}{2V}\sum_{m,q=1}^{M}\frac{\partial^2}{\partial x_m \partial x_q}(D_{mq}p)(\mathbf{x},t), \tag{6}$$

where

$$A_m(\mathbf{x}) := \sum_{\gamma}\gamma_m w_{\gamma}(\mathbf{x}), D_{mq}(\mathbf{x}) := \sum_{\gamma}\gamma_m\ \gamma_q\ w_{\gamma}(\mathbf{x})(m,q=1,2,\ldots,M). \tag{7}$$

The first quantity may obviously be called the **conditional expected velocity**, and the second one the **conditional expected variance velocity** [16]. It is well known that keeping terms of higher order may lead to inconsistent results. Moreover, the present approximation does not respect the natural boundary conditions of the master equation. Both the master equation and the Fokker–Planck equation are approximately consistent in mean (and they are exactly consistent if only processes with linear rates are present) with the usual deterministic equation

$$\frac{\mathrm{d}\bar{x}}{\mathrm{d}t} = \mathbf{A}(\bar{x}(t)) \tag{8}$$

for the average $\bar{x}$ of $\mathbf{x}$. The exact equation for the moments of the distribution obeying the master equation can also be derived from the equation for the generating function, see [17, p. 110] and [18, pp. 129–133]. For example, instead of Eq. (8) we have

$$\frac{\mathrm{d}\bar{x}}{\mathrm{d}t} = \overline{\mathbf{A}(\mathbf{x}(t))}. \tag{9}$$

2.3. *Conservation laws and irreducibility*

During a time interval $[t, t+h]$ the change of the concentrations is

$$\mathbf{x}(t+h) - \mathbf{x}(t) = \sum_{\gamma}\gamma\eta_{\gamma}, \tag{10}$$

where η_{γ} are (integer-valued) random variables indexed corresponding to nonzero transition rates w_{γ}. The stochastic process $t \mapsto \mathbf{x}(t)$ will then satisfy some conservation laws. More specifically, for each point $\mathbf{x}_0$ of the concentration space $\mathbb{R}^M$ let us define the **stoichiometric compatibility class** [17,19]

$$S(\mathbf{x}_0) := \left\{ x \in \mathbb{R}^M; \mathbf{x} = \mathbf{x}_0 + \sum_{\gamma} \boldsymbol{\gamma} \eta_{\gamma}; \eta_{\gamma} \in \mathbb{R} \right\} \cap (\mathbb{R}_0^+)^M, \tag{11}$$

where η_γ are integers corresponding to nonzero transition rates w_γ. The whole space $\mathbb{R}^M$ is a union of stoichiometric compatibility classes, and the process $t \mapsto \mathbf{x}(t)$ stays in $S(\mathbf{x}_0)$ for all time (until the process is defined), if $\mathbf{x}(0) = \mathbf{x}_0$. We also suppose that each stoichiometric compatibility class contains exactly one stationary solution (which is not always so even in the deterministic case). Furthermore, we suppose that the stochastic model has a unique stationary solution for each $\mathbf{x}_0$. This stationary distribution will be denoted by $p(\mathbf{x}|\mathbf{x}_0)$, and it does not depend on the choice of the initial state so far as it remains in $S(\mathbf{x}_0)$. The master equation (5) induces another master equation in a smaller number of linearly independent variables on each $S(\mathbf{x}_0)$ (the other variables are linear functions of these variables), but the stationary distribution is supposed to be unique. The deterministic motion also evolves in $S(\mathbf{x}_0)$ that can easily seen by rewriting Eq. (8) into an integral equation form

$$\bar{x}(t) = \mathbf{x}_0 + \int_0^t \sum_{\gamma} \boldsymbol{\gamma} w_{\gamma}(\bar{x}(s))\mathrm{d}s = \mathbf{x}_0 + \sum_{\gamma} \boldsymbol{\gamma} \int_0^t \mathrm{w}_{\gamma}(\bar{x}(\mathrm{s}))\mathrm{d}s. \tag{12}$$

Similar remarks are also valid for the stochastic process associated to the Fokker–Planck equation. From now on it will be assumed that there are no linear conservation laws, a fact that can also be expressed that the master equation is irreducible. This assumption does not mean the restriction of generality because otherwise one can consider the master equation restricted to a stoichiometric compatibility class of the state space. We shall also assume that the zeros of the vector field (of the deterministic model) are isolated (although sometimes we shall consider the multiple-zero case) on each set $S(\mathbf{x}_0)$.

3. Hamilton–Jacobi theories

Let us recall an approximation originally introduced by Kubo et al. [11]; (see also [20], and more recently [21–23]), giving physically more realistic results (see Section 4 and Refs. [24,25]) than the usual Fokker–Planck approximation. Let us also mention that our basic reference is Ref. [26].

3.1. Hamilton–Jacobi theory for the master equation

The idea of the approximation is to write $p(\boldsymbol{x}, t)$ in the form of WKB expansion, valid for large V

$$p(\mathbf{x}, t) = \exp[-V\Phi(\mathbf{x}, t)]\left[U_0(\mathbf{x}, t) + \frac{1}{V} U_1(\mathbf{x}, t) + \ldots \right], \tag{13}$$

where Φ, U_0, U_1 are unknown functions. Then, one uses the first term of this expansion in Eq. (5) and groups together terms containing decreasing powers of V. The highest-order term in V is of the order $\mathcal{O}(V)$, and contains U_0 as a factor. It is zero if and only if Φ satisfies the equation

$$\frac{\partial \Phi}{\partial t} + \sum_{\gamma} w_{\gamma}[\exp(\nabla \Phi \boldsymbol{\gamma}) - 1] = 0. \tag{14}$$

This is a Hamilton–Jacobi equation that can be integrated by the method of bicharacteristics [27,28]. The next term of order $\mathcal{O}(1)$ in V in Eq. (5) when one uses the expansion (13) gives a first-order linear equation for U_0 that is a sort of **transport equation**:

$$\frac{\partial U_0}{\partial t} + \sum_{\gamma}\left[\exp(\nabla \Phi \boldsymbol{\gamma}) \sum_{m=1}^{M} \frac{\partial}{\partial x_m}(\gamma_m w_{\gamma} U_0)\right] + U_0 \sum_{\gamma} \exp(\nabla \Phi \boldsymbol{\gamma}) \left(\frac{1}{2} \sum_{m,q=1}^{M} \gamma_m\, \gamma_q\, w_{\gamma} \frac{\partial^2 \Phi}{\partial x_m \partial x_q}\right) = 0. \tag{15}$$

3.2. Hamilton–Jacobi theory for the Fokker–Planck equation

For large V, one can use the first term of the WKB-type expansion (here and above we leave including more terms for future investigations) for p in Eq. (6), i.e. to look for p in the form

$$p(\mathbf{x}, t) = \exp(-V\Phi^{(\mathcal{FP})}(\mathbf{x}, t))\left[U_0^{(\mathcal{FP})}(\mathbf{x}, t) + \frac{1}{V} U_1^{(\mathcal{FP})}(\mathbf{x}, t) + \ldots\right]. \tag{16}$$

Upon inserting the above expansion (16) into Eq. (6) and comparing the various powers of V one obtains equations for $\Phi^{(\mathcal{FP})}$ and $U_0^{(\mathcal{FP})}$

$$\frac{\partial \Phi^{(\mathcal{FP})}}{\partial t} + \sum_{m=1}^{M} A_m \frac{\partial \Phi^{(\mathcal{FP})}}{\partial x_m} + \frac{1}{2} \sum_{m,q=1}^{M} D_{mq} \frac{\partial \Phi^{(\mathcal{FP})}}{\partial x_m} \frac{\partial \Phi^{(\mathcal{FP})}}{\partial x_q} = 0 \tag{17}$$

$$\frac{\partial U_0^{(\mathcal{FP})}}{\partial t} + \sum_{m=1}^{M} \frac{\partial}{\partial x_m}(A_m U_0^{(\mathcal{FP})}) + \sum_{m,q=1}^{M} \frac{\partial \Phi^{(\mathcal{FP})}}{\partial x_m} \frac{\partial}{\partial x_q}(D_{mq} U_0^{(\mathcal{FP})}) + \frac{1}{2}\left(\sum_{m,q=1}^{M} D_{mq} \frac{\partial^2 \Phi^{(\mathcal{FP})}}{\partial x_m \partial x_q}\right) U_0^{(\mathcal{FP})} = 0. \tag{18}$$

Eq. (17) is a standard Hamilton–Jacobi equation (with a standard Hamiltonian, quadratic in the momentum), whereas Eq. (18) is a kind of transport equation. Eqs. (17) and (18) can be seen to be obtained from Eqs. (14) and (15) if one assumes that the derivatives $\partial \Phi / \partial x_m$ are small, so that one is entitled to replace $\exp(\nabla \Phi \boldsymbol{\gamma}) - 1$ by its Taylor expansion up to the second order. Then, Eq. (14) reduces immediately to Eq. (17) with A_m and D_{mq} as given in Eq. (7). This indicates that the Hamilton–Jacobi and transport equations (17) and (18) are less precise than the Hamilton–Jacobi and transport equations (14) and (15) directly deduced from the master equation. In fact, their range of validity is limited to neighborhoods of the stationary points (where $\nabla\Phi$ is zero) of the **action function** Φ.

3.3. *Stationary solutions*

The stationary solution of the master equation or that of the Fokker–Planck equation will also be looked for in the form $p(\mathbf{x}) \sim \exp[-V\Phi(\mathbf{x})][U_0(\mathbf{x}) + \frac{1}{V}U_1(\mathbf{x}) + \ldots]$. Then Φ satisfies the stationary form of the time-dependent Hamilton–Jacobi equations (14) or (17) and U_0 satisfies the stationary form of the transport equations (Eqs. (15) or (18)). In both cases, the stationary Hamilton–Jacobi equation can be written as

$$\mathcal{H}(\mathbf{x}, \nabla\Phi(\mathbf{x})) = 0, \tag{19}$$

where $\mathcal{H}$ is of the form

$$\mathcal{H}^{(\mathcal{M})}(\mathbf{x}, \xi) = \sum_{\gamma} w_\gamma(\mathbf{x})[\exp(\xi^{\mathrm{T}}\gamma) - 1], \tag{20}$$

or

$$\mathcal{H}^{(\mathcal{FP})}(\mathbf{x}, \xi) = \xi^{\mathrm{T}}\mathbf{A}(\mathbf{x}) + \tfrac{1}{2}\xi^{\mathrm{T}}\mathbf{D}(\mathbf{x})\xi, \tag{21}$$

for the master equation and for the Fokker–Planck equation, respectively, the variables ξ_m being the conjugate momenta of x_m. From Eq. (7) we deduce for small ξ

$$\mathcal{H}^{(\mathcal{M})}(\mathbf{x}, \xi) = \mathcal{H}^{(\mathcal{FP})}(\mathbf{x}, \xi) + \mathcal{O}(|\xi|^3). \tag{22}$$

In the present work, we shall only use the stationary equations, although there is some hope to extend these investigations to nonstationary cases. In some neighborhood of some stationary point of the vector field **A** it is always possible to study the stationary solution of the master equation by using the Hamiltonian originated in the Fokker–Planck approximation (**Fokker–Planck Hamiltonian**, see [29, pp. 7742–7743]).

3.4. *The particular case of one chemical species*

Let us consider the case of a single chemical species, i.e. when $M = 1$, and let us suppose that the system evolves only by transitions $n \to n \pm 1$. As before we go to the large-volume limit introducing $x := \frac{X}{V}, p(x) := V\mathcal{P}(X), w_\pm(x) := \frac{1}{V}W_\pm(X)$, so that one obtains the exact master equation (5) and the Fokker–Planck equation (6) respectively, with $A(x) = w_+(x) - w_-(x), D(x) = w_+(x) + w_-(x)$. The solution (13) of the master equation is [20,24,25]

$$p(x) = U_0(x)\exp[-V\Phi(x|a)] = \frac{C}{\sqrt{w_+(x)w_-(x)}}\exp\left(-V\int_a^x \log\left(\frac{w_-}{w_+}\right)\right),$$

where a is an arbitrary non-negative value and C is a normalization constant. When w_+ and

w_- have no common zeros, p as given in Eq. (13) can be normalized with

$$C = \left[\int_0^{+\infty} \frac{\exp\left(-V\int_a^x \log\left(\frac{w_-}{w_+}\right)\right)}{\sqrt{w_+(x)w_-(x)}} \mathrm{d}x \right]^{-1}. \tag{23}$$

The integral in Eq. (23) can be estimated by the **saddle-point method**. The main contribution to the integral is obtained for a point x_s, which is an absolute minimum of $\Phi(x|a)$. If exactly one such point exists, we have

$$p(x) \sim \frac{C_0}{\sqrt{w_+(x)w_-(x)}} \exp\left(-V\int_a^x \log\left(\frac{w_-}{w_+}\right)\right), \tag{24}$$

where C_0 is now of the order $\mathcal{O}(1/\sqrt{V})$, if the minimum of Φ is not degenerate. When there are several attracting points of A, say, $x_s^1, x_s^2, \ldots, x_s^l, \ldots$, one chooses x_s^k in such a way that for all l $\Phi(x_s^{(l)}|x_s^{(k)}) \leq 0$, and again one obtains Eq. (24).

Similarly, the stationary solution of the Fokker–Planck equation can be written as

$$p(x) \sim U_0^{(\mathcal{FP})}(x)\exp[-V\Phi^{(\mathcal{FP})}(x|a)] = \frac{C}{D(x)}\exp\left(2V\int_a^x \frac{A}{D}\right), \tag{25}$$

and is normalizable, provided D has no zero (or w_+ and w_- have no common zero), with $C = \left[\int_0^{+\infty} \frac{\exp\left(2V\int_a^x \frac{A}{D}\right)}{D(x)} \mathrm{d}x\right]^{-1}$. Again, C can be evaluated by the saddle-point method, the main contribution coming from the absolute minimum x_s of $\Phi^{(\mathcal{FP})}(a)$, which is an attracting point of the vector field $A : p(x) \sim \frac{C_0}{\sqrt{VD(x_s)}}\exp\left(2V\int_{x_s}^x \frac{A}{D}\right)$.

When w_+ and w_- have a common zero, p given above is not normalizable, indicating that the expression in Eq. (13) for p is not valid. This is exactly the case of **criticality** [30].

3.5. *Comparison of asymptotic results*

The two approximations given by Eqs. (24) and (25) are close to each other for small values of $\mathrm{d}\Phi/\mathrm{d}x$, or when $A(x)$ is small. In this case, one has $\log(w_-/w_+) \sim -2A/D$. The approximation (25) coming from the Fokker–Planck equation is exact near a zero of A. But when there are several zeros, the approximation fails, because the eigenvalues and the mean exit times calculated from the Fokker–Planck equation differ from the analogous quantities given by the master equation by an exponentially large factor [24,25].

This also indicates that the limit theorems by Kurtz [31] are not applicable in the case when A has several zeros: they have been formulated for the case of detailed balanced systems that are known to have a single asymptotically stable deterministic stationary point. These theorems state that the stochastic process $\mathbf{x}$ associated with the master equation tends to the deterministic trajectory of $A(x)$ (starting from the same initial point), and that the deviation is Gaussian, but these theorems are valid uniformly on finite time intervals. They cannot describe the situation for times like exp (kV). In particular, these theorems cannot describe chemically activated events like the passage over a potential barrier, and consider rate constants as given.

4. Construction, uniqueness and critical points of Φ

General basic properties of solution Φ to the Hamilton–Jacobi equation (19) given by Eqs. (20) or (21) will be derived here and in the next section. The traditional method [27,28] for constructing Φ does not work, but we shall still show a construction method and also prove the uniqueness of the smooth solution Φ. Finally, we also study the critical points of Φ. We shall treat $\mathcal{H}^{(\mathcal{M})}$ and $\mathcal{H}^{(\mathcal{FP})}$ together.

4.1. Lagrangians

With the Hamiltonian for the master equation and for the Fokker–Planck equations (20) and (21), respectively, the corresponding $\dot{x}_m$ velocities are

$$\dot{x}_m(t) = \frac{\partial \mathcal{H}^{(\mathcal{M})}(\mathbf{x}(t), \boldsymbol{\xi})}{\partial \xi_m} = \sum_{\gamma} \gamma_m w_{\gamma}(\mathbf{x}(t)) \exp(\boldsymbol{\xi}^{\mathrm{T}} \boldsymbol{\gamma}),$$

$$\dot{x}_m(t) = \frac{\partial \mathcal{H}^{(\mathcal{FP})}(\mathbf{x}(t), \boldsymbol{\xi})}{\partial \xi_m} = A_m(\mathbf{x}) + \sum_{q=1}^{M} D_{mq}(\mathbf{x}(t)) \xi_q,$$

and the corresponding Lagrangians are

$$\mathcal{L}^{(\mathcal{M})}(\mathbf{x}, \mathbf{v}) := \sum_{\gamma} w_{\gamma}(\mathbf{x})[(\boldsymbol{\xi}^{\mathrm{T}} \gamma) \exp(\boldsymbol{\xi}^{\mathrm{T}} \gamma) - \exp(\boldsymbol{\xi}^{\mathrm{T}} \boldsymbol{\gamma}) + 1], \tag{26}$$

$$\mathcal{L}^{(\mathcal{FP})}(\mathbf{x}, \mathbf{v}) := \tfrac{1}{2}(\mathbf{v} - \mathbf{A}(\mathbf{x}))^{\mathrm{T}} \mathbf{D}(\mathbf{x})^{-1} (\mathbf{v} - \mathbf{A}(\mathbf{x})). \tag{27}$$

In the Fokker–Planck case (27) $\boldsymbol{\xi} = \mathbf{D}(\mathbf{x})^{-1}(\mathbf{v} - \mathbf{A}(\mathbf{x}))$ (with $\mathbf{v} = \dot{\mathbf{x}}$ as usual), provided $\mathbf{D}(\mathbf{x})$ is nondegenerate, which is the case if we assume that there are no conservation laws. Under the hypothesis of the nonexistence of linear mass conservation laws $\mathcal{L}^{(\mathcal{M})}$ and $\mathcal{L}^{(\mathcal{FP})}$ are both non-negative and both of them are zero if and only if $\xi = 0$. In the presence of conservation laws $\mathbf{D}(\mathbf{x})$ is necessarily degenerated.

Actually, Eq. (27) is the Gaussian form of the Onsager–Machlup principle in our case and a special (discrete) version of the Gyarmati principle for nonlinear conduction equations. See, e.g. [6,7,32–35].

4.2. Special paths

4.2.1. Deterministic paths

The **deterministic paths** $\mathrm{d}\bar{\mathbf{x}}/\mathrm{d}t = \mathbf{A}(\bar{\mathbf{x}}(t))$ $\mathbf{x}(0) = \mathbf{0}$ are obviously solutions of both Hamiltonian equations. Conversely, a path $t \mapsto (\mathbf{x}(t), \boldsymbol{\xi}(t))$ that is a solution of the

Hamiltonian equations, such that $\mathbf{x}(0) = 0$, is the deterministic path, because of the uniqueness of paths under given initial conditions. Moreover, the Lagrangian is zero along a deterministic path, and conversely, and as a consequence, the variation of the action Φ along a deterministic path is zero.

4.2.2. Antideterministic paths

Let us assume now that Φ is a smooth solution of the Hamilton–Jacobi equation, and define $\xi_m(\mathbf{x}) := (\partial\Phi/\partial x_m)$. A solution $s \mapsto (\mathbf{x}(s), \boldsymbol{\xi}(s))$ of the system

$$\frac{\mathrm{d}x_m}{\mathrm{d}s} = \frac{\partial\mathcal{H}(\mathbf{x},\boldsymbol{\xi})}{\partial\xi_m} \qquad \xi_m(s) = \xi_m(\mathbf{x}(s)) \tag{28}$$

is a Hamiltonian path because

$$\frac{\mathrm{d}\xi_m(s)}{\mathrm{d}s} = \sum \frac{\partial\xi_m}{\partial x_q}\frac{\mathrm{d}x_q}{\mathrm{d}s} = \sum_q \frac{\partial^2\Phi}{\partial x_m\partial x_q}\frac{\partial\mathcal{H}(\mathbf{x},\nabla\Phi(\mathbf{x}))}{\partial\xi_q} = -\frac{\partial\mathcal{H}(\mathbf{x},\boldsymbol{\xi}(\mathbf{x}))}{\partial x_m},$$

since $\mathcal{H}(\mathbf{x},\nabla\Phi(\mathbf{x})) = 0$, therefore $\frac{\partial\mathcal{H}}{\partial x_m} = \sum_q \frac{\partial\mathcal{H}}{\partial\xi_q}\frac{\partial^2\Phi}{\partial x_m\partial x_q} = 0$. Moreover, Φ is increasing along such paths because

$$\frac{\mathrm{d}\Phi}{\mathrm{d}s} = \sum_m \frac{\partial\Phi}{\partial x_m}\frac{\mathrm{d}x_m}{\mathrm{d}s} = \sum_m \xi_m(s)\frac{\mathrm{d}x_m}{\mathrm{d}s} = L \geq 0.$$

The trajectories given by Eq. (28) will be called **antideterministic paths** (for reasons to follow).

4.3. Construction of Φ

The traditional method to construct a solution Φ of Eq. (19) (the action) is the following [27,28]. One chooses a point $\mathbf{x}^{(0)}$ and considers a path $(\mathbf{x}(s), \boldsymbol{\xi}(s))$, a solution of the Hamiltonian system

$$\dot{x}_m = \frac{\partial\mathcal{H}}{\partial\xi_m} \qquad \dot{\xi}_m = -\frac{\partial\mathcal{H}}{\partial x_m}, \tag{29}$$

with the conditions

$$\mathbf{x}(0) = \mathbf{x}^{(0)}, \qquad \boldsymbol{\xi}(0) = \boldsymbol{\xi}^{(0)} \qquad \mathbf{x}(t) = \mathbf{x} \qquad \mathcal{H}(\mathbf{x}^{(0)},\boldsymbol{\xi}^{(0)}) = 0. \tag{30}$$

The unknowns are $\xi^{(0)}$ and t, which will be implicitly fixed by condition (30). Then the function

$$\Phi(\mathbf{x}) := \int_0^t \sum_{m=1}^{M} \xi_m \mathrm{d}x_m \tag{31}$$

is the solution $\Phi(\mathbf{x}|\mathbf{x}^{(0)})$ of Eq. (19) where in Eq. (31) the integral is taken along the path $(\mathbf{x}(s), \boldsymbol{\xi}(s))$, satisfying Eqs. (29) and (30). But the function $\Phi(\mathbf{x}|\mathbf{x}^{(0)})$ is not differentiable at $\mathbf{x}^{(0)}$, since $\nabla\Phi(\mathbf{x}|\mathbf{x}^{(0)}) = \boldsymbol{\xi} \to \boldsymbol{\xi}_0$ as $\mathbf{x} \to \mathbf{x}_0$; but $\boldsymbol{\xi}_0$ depends on the path from $\mathbf{x}$ to $\mathbf{x}_0$ so that $\nabla\Phi$ is, in general, not defined at $\mathbf{x}_0$,. In our situation where

$p(\mathbf{x}) \sim U_0(\mathbf{x})\exp[-V\Phi(\mathbf{x})]$, we expect $\Phi(\mathbf{x})$ to be regular everywhere and to be peaked at a point $\mathbf{x}_M$ (at least), so that one cannot take for our Φ a function $\Phi(\mathbf{x}|\mathbf{x}_0)$ constructed by the traditional method as above.

We shall now describe the correct construction of Φ by a limiting process.

Let us consider a point $\mathbf{x}_s$ which is an attracting point of the vector field $\mathbf{A}$. We take another point $\mathbf{x}^{(0)}$ and we construct the usual action $\Phi(\mathbf{x}|\mathbf{x}^{(0)})$ using Hamiltonian paths starting from $\mathbf{x}^{(0)}$, with energy 0. For $\mathbf{x}^{(0)} \neq \mathbf{x}_s$ this function is nontrivial and is not differentiable at $\mathbf{x}^{(0)}$. The function $\Phi(\mathbf{x}|\mathbf{x}_s)$ is now defined as

$$\Phi(\mathbf{x}|\mathbf{x}_s) := \lim_{\mathbf{x}^{(0)}\to\mathbf{x}_s} \Phi(\mathbf{x}|\mathbf{x}^{(0)}). \tag{32}$$

The function $\Phi(\mathbf{x}|\mathbf{x}_s)$ is not simply the function $\Phi(\mathbf{x}|\mathbf{x}^{(0)})$ taken at $\mathbf{x}^{(0)} = \mathbf{x}_s$. The reason is that if we choose $\mathbf{x}^{(0)} = \mathbf{x}_s$, ξ_0 should vanish in order that $\mathcal{H}(\mathbf{x}^{(0)}, \xi^{(0)}) = 0$. This is obvious for $\mathcal{H}^{(\mathcal{FP})}$, and can be seen to be also valid for $\mathcal{H}^{(\mathcal{M})}$ using the inequality $e^a - 1 \geq a$. Therefore, the trajectory never moves away from $\mathbf{x}^{(0)}$, and the traditional action is 0. In general, $\Phi(\mathbf{x}|\mathbf{x}_s)$ is a nontrivial function, which is differentiable at $\mathbf{x}^{(0)} = \mathbf{x}_s$, has a strict minimum at $\mathbf{x}_s$, which is a nondegenerate minimum, if $\mathbf{x}_s$ is a nondegenerate attracting point of the vector field $\mathbf{A}$. The existence of the limit in Eq. (32) has been proven in [30, pp. 7743–7745]. Moreover, we shall consider a given point $\mathbf{x}$ and a trajectory with energy 0 starting from $\mathbf{x}^{(0)}$ and arriving at $\mathbf{x}$ in a certain (unknown) time t. This trajectory has an initial momentum $\boldsymbol{\xi}$ that is a function $\boldsymbol{\xi}(\mathbf{x}|\mathbf{x}^{(0)})$. In the Fokker–Planck case we have $\mathcal{H}^{(\mathcal{FP})}[\mathbf{x}^{(0)}, \boldsymbol{\xi}(\mathbf{x}|\mathbf{x}^{(0)})] = 0$. If $\boldsymbol{\delta} := \mathbf{x}^{(0)} - \mathbf{x}_s$, then $\mathbf{A}(\mathbf{x}^{(0)}) \sim \mathbf{A}(\mathbf{x}_s) + \mathbf{A}'(\mathbf{x}_s)\boldsymbol{\delta} \simeq \mathbf{A}'(\mathbf{x}_s)\boldsymbol{\delta}$, if $\boldsymbol{\delta} \to 0$, then from the definition of $\mathcal{H}^{(\mathcal{FP})}$ we have

$$\tfrac{1}{2}\boldsymbol{\xi}^{\mathrm{T}}\mathbf{D}\boldsymbol{\xi} \simeq -\mathbf{A}'(\mathbf{x}_s)\boldsymbol{\delta}\cdot\boldsymbol{\xi}. \tag{33}$$

Assuming that $\mathbf{A}'(\mathbf{x}_s) \neq 0$ and $\mathbf{D}(\mathbf{x}_s)$ is invertible we deduce from Eq. (33) that

$$|\boldsymbol{\xi}(\mathbf{x}|\mathbf{x}^{(0)})| = \mathcal{O}(|\boldsymbol{\delta}|), \tag{34}$$

showing that $\boldsymbol{\xi}(\mathbf{x}|\mathbf{x}^{(0)}) \to 0$ when $\mathbf{x}^{(0)} \to \mathbf{x}_s$. From Eq. (34) and the definition of the velocity we get $\dot{\mathbf{x}} = \mathbf{A}(\mathbf{x}) + \mathbf{D}(\mathbf{x})\boldsymbol{\xi}$. As we see that for the initial velocity $|\mathbf{x}(0)| = O(\boldsymbol{\delta})$ holds, the time needed to join $\mathbf{x}^{(0)}$ along the Hamiltonian trajectory of energy 0 will tend to infinity when $\mathbf{x}^{(0)} \to \mathbf{x}_s$. It is precisely because the initial momentum is tending to zero that the limiting function $\Phi(\mathbf{x}|\mathbf{x}^{(0)})$ will be differentiable at $\mathbf{x}_s$, when $\mathbf{x}^{(0)} \to \mathbf{x}_s$.

4.4. Uniqueness of Φ

If Φ is a smooth solution of the Hamilton–Jacobi equation (either $\mathcal{H}^{(\mathcal{M})}$ or $\mathcal{H}^{(\mathcal{FP})}$), and $\mathbf{x}_0$ is a minimum of Φ, then the Taylor expansion of Φ at $\mathbf{x}_0$ is uniquely determined up to an additive constant [29, p. 7741]. In particular, if Φ is an analytic solution near $\mathbf{x}_0$, it is unique. As a consequence, there exists at most one function Φ that is a global analytic solution of the Hamilton–Jacobi equation (up to an additive constant). Because of this fact we can determine the antideterministic path by using the analytic solution Φ of the

Hamilton–Jacobi equation as in Section 4.2.2, namely,

$$\frac{d\tilde{x}_m(s)}{ds} = \frac{\partial \mathcal{H}[\tilde{\mathbf{x}}(s), \tilde{\boldsymbol{\xi}}(s)]}{\partial \xi_m}, \qquad \xi_m(s) = \xi_m[\tilde{\mathbf{x}}(s)] \equiv \frac{\partial \Phi(\tilde{\mathbf{x}}(s))}{\partial x_m}. \tag{35}$$

4.5. Limiting behavior of trajectories

Consider the linear Fokker–Planck Hamiltonian with a constant positive matrix $\mathbf{D}$ and with a linear vector field $\mathbf{A}(\mathbf{x}) := \mathbf{A}^0\mathbf{x}$. Let $(\mathbf{x}(s), \boldsymbol{\xi}(s))$ be a trajectory with $\mathcal{H} = 0$ such that $\mathbf{x}(0) = \mathbf{x}^{(0)}$, $\mathbf{x}(t) = \mathbf{x}$, where $\mathbf{x}^{(0)}$ and $\mathbf{x}$ are both nonzero. Then, if $t \to +\infty$, this trajectory has the following limit behavior:

(1) for fixed s $\mathbf{x}(s)$ tends to the deterministic trajectory $\bar{\mathbf{x}}(\mathbf{s})$ starting from $\mathbf{x}^{(0)}$;
(2) for fixed s $\mathbf{x}(t-s)$ tends to the antideterministic trajectory $\tilde{\mathbf{x}}(t-s)$ defined by Eq. (35) ending at $\mathbf{x}$.

4.6. Critical points of Φ and zeros of the deterministic model (vector field)

The following facts are true [29, p. 7745]:

(1) A nondegenerate critical point of Φ is a zero of the deterministic vector field $\mathbf{A}$ (both for $\mathcal{H}^{(\mathcal{M})}$ and $\mathcal{H}^{(\mathcal{FP})}$). A nondegenerate minimum of Φ is a stable attracting point $\mathbf{A}$.
(2) Conversely, for $\mathcal{H}^{(\mathcal{FP})}$ the zeros of $\mathbf{A}$ are critical points of Φ, and the stable attracting points of $\mathbf{A}$ are minima of Φ.
(3) Conversely, for $\mathcal{H}^{(\mathcal{M})}$ the stable attracting points of $\mathbf{A}$ are minima of Φ.

5. Path integrals for the master equation and Fokker–Planck equation

5.1. The stochastic process of the Fokker–Planck equation

It is well known that there exists a stochastic process $t \mapsto \mathbf{x}(t)$ associated with the Fokker–Planck equation (6). The **weight** of a trajectory is, up to a normalization factor,

$$\exp\left(-V\int_0^t \mathcal{L}[\mathbf{x}(s), \dot{\mathbf{x}}(s)]\mathrm{d}s\right), \tag{36}$$

where $\mathcal{L}$ is the associated Lagrangian defined in Eq. (27). In a small time interval $[t, t+h]$ the state moves by $\boldsymbol{\Delta} := \mathbf{x}(t+h) - \mathbf{x}(t)$ according to the Gaussian law (depending also on the actual state $\mathbf{x}$):

$$\frac{\left(\frac{V}{2\pi}\right)^{d/2}}{(\det(\mathbf{D}))^{1/2}} \exp\left[-\frac{V}{2}\left(\frac{\boldsymbol{\Delta}}{h} - \mathbf{A}(\mathbf{x})\right)^{\mathrm{T}} \mathbf{D}\left(\frac{\boldsymbol{\Delta}}{h} - \mathbf{A}(\mathbf{x})\right)\right]. \tag{37}$$

Consider now a dual variable $\boldsymbol{\xi}$ of $\boldsymbol{\Delta}$ and compute the Laplace transform $\mathcal{L}(\mathbf{x}, \boldsymbol{\xi}, h)$ of the Gaussian distribution (37) with respect to $\boldsymbol{\Delta}$. This is obviously a Gaussian integral in $\boldsymbol{\Delta}$, namely,

$$\mathcal{L}(\mathbf{x}, \boldsymbol{\xi}, h) = \int \frac{\left(\frac{V}{2\pi}\right)^{d/2}}{(\det(\mathbf{D}))^{1/2}} \exp\left[-\frac{V}{2}\left(\frac{\Delta}{h} - \mathbf{A}(\mathbf{x})\right)^{\mathrm{T}} \mathbf{D}\left(\frac{\Delta}{h} - \mathbf{A}(\mathbf{x})\right)\right] \exp(\Delta\boldsymbol{\xi}) \mathrm{d}\Delta. \tag{38}$$

The critical point that maximizes the exponent is $\frac{1}{V}\mathbf{D}\boldsymbol{\xi} + \mathbf{A}(\mathbf{x})h$, and the critical value of the exponent becomes $\mathcal{H}^{(\mathcal{FP})}(\mathbf{x}, \boldsymbol{\xi})h$ so that $\mathcal{H}^{(\mathcal{FP})}(\mathbf{x}, \boldsymbol{\xi}) = \mathcal{L}(\mathbf{x}, \boldsymbol{\xi}, h)/h$.

5.2. Path integral for the master equation

For the master equation (2) the stochastic process is a pure jump process (which sometimes specializes to a birth and death process), for which we cannot apply the method of Section 5.1. Nevertheless, we shall see that this method is valid in the large-volume limit. First, we shall describe the jump process of the master equation. At time t, the stochastic process of the number of species is $\mathbf{N}(t)$. The time evolution is described as follows. During the time interval $[t, t+h]$ the process stays in the state $\mathbf{N}(t)$, then a jump $\mathbf{N}(t) \rightarrow \mathbf{N}(t) + \boldsymbol{\gamma}$ takes place with probability proportional to $W_{\gamma}(\mathbf{N})$. The h length of the interval is exponentially distributed with parameter $\Sigma_{\gamma} W_{\gamma}(\mathbf{N})$. It is proved [36, p. 7753] that

$$\mathcal{L}(\boldsymbol{\xi}, \mathbf{N}, h) = \exp\left[Vh\mathcal{H}^{(\mathcal{M})}\left(\frac{\mathbf{N}}{V}, \boldsymbol{\xi}\right)\right], \tag{39}$$

where $\mathcal{H}^{(\mathcal{M})}$ is the Hamiltonian (20) associated with the master equation.

Inverting the Fourier transform of the probability $q(\mathbf{x} + \Delta, h|\mathbf{x})$ of a transition $\mathbf{x} \rightarrow \mathbf{x} + \Delta$ in time h, we have, using Eq. (39)

$$\begin{aligned} q(\mathbf{x} + \Delta, h|\mathbf{x}) &= \int \exp(i\boldsymbol{\xi}\Delta) \exp\left[Vh\mathcal{H}^{(\mathcal{M})}\left(\mathbf{x}, -\frac{\mathrm{i}\boldsymbol{\xi}}{V}\right)\right] \frac{\mathrm{d}\boldsymbol{\xi}}{(2\pi)^s} \\ &= \left(\frac{V}{2\pi}\right)^s \int \exp(Vi\boldsymbol{\xi}\Delta) + h\mathcal{H}^{(\mathcal{M})}(\mathbf{x}, -i\boldsymbol{\xi}) \mathrm{d}\boldsymbol{\xi}. \end{aligned} \tag{40}$$

For large V, this is again estimated by the saddle-point method, so that up to prefactors

$$q(\mathbf{x} + \Delta, h|\mathbf{x}) \sim \exp\left[-V\mathcal{L}^{(\mathcal{M})}\left(\mathbf{x}, \frac{\Delta}{h}\right)h\right], \tag{41}$$

where $\mathcal{L}^{(\mathcal{M})}$ is the Lagrangian associated to $\mathcal{H}^{(\mathcal{M})}$ (and $\dot{x}_m = \partial\mathcal{H}^{(\mathcal{M})}/\partial\xi_m$.) This is similar to the **Feynman path integral** [37,38]. Eq. (41) shows that the transition probability is in the form of a standard path integral. Namely, the weight of a trajectory is the exponential $\exp(-V\int_0^t \mathcal{L}(\mathbf{x}(s), \dot{\mathbf{x}}(s))\mathrm{d}s)$ up to prefactors. We have proved [36, p. 7753] that Eq. (41) for the transition probability is consistent with the large-volume asymptotics of the stationary state. In Ref. [29] we have seen that the stationary state is up to prefactors: $p(\mathbf{x}) \sim \exp(-V\Phi(\mathbf{x}))$, where Φ is the solution of the Hamilton–Jacobi equation (19). Then, up to a prefactor one has $\int q(\mathbf{x}', h|\mathbf{x})p(\mathbf{x}) \sim p(\mathbf{x}')$. Note that Eq. (41) is also consistent, because we have proved in Ref. [29] that the Lagrangian $\mathcal{L}^{(\mathcal{M})}$ is positive.

5.3. *Interpretation: typical paths in the large-volume limit*

Eqs. (36) and (41) have the same structure. They assert that in the large-volume limit the infinitesimal transition probability of an event $\mathbf{x} \rightarrow \mathbf{x}'$ in a time interval of length h is $\exp(-V\mathcal{L}[\mathbf{x}, (\mathbf{x}' - \mathbf{x}/h)h])$, where $\mathcal{L}$ is the Lagrangian of the corresponding theory. This means that a typical trajectory in the state space of concentrations will minimize the Lagrangian action and will be the first coordinate of a path $[\mathbf{x}(s), \boldsymbol{\xi}(s)]$ satisfying the Hamiltonian equations of the corresponding Hamiltonian. This also implies that in the large-volume limit the stochasticity is confined to the choice of initial momenta, ξ_m, which will be the cause of the evolution of the x_m.

6. A single chemical species: exact and approximate solutions

6.1. *The master equation and its adjoint for exit times*

Let us again investigate in detail the case when we have a single chemical species. Let the number of particles be X, and V the volume. Suppose that the system evolves only by transitions $X \rightarrow X \pm 1$ with probabilities $W_{\pm}(X)$ per unit time. Let $\Omega := [aV, bV]$ be an interval of integers, and $T_\Omega(\mathrm{X})$ for $X \in \Omega$ the average of the first exit time of Ω, the stochastic process starting from X. It is well known [39] that $T_\Omega(\mathrm{X})$ satisfies the adjoint equation of the master equation with the right-hand side -1, i.e.

$$W_+(X)[T_\Omega(X+1) - T_\Omega(X)] + W_-(X)[T_\Omega(X-1) - T_\Omega(X)] = -1$$

$$T_\Omega(\mathrm{X}) = 0 \text{ for } X \notin \Omega. \tag{42}$$

Let us assume that the boundary point aV is absorbing and that bV is reflecting. One can find an exact formula for the solution of Eq. (42) [39]:

$$T_\Omega(\mathrm{X}) = \sum_{i=aV}^{X} \varphi(i) \sum_{j=i}^{bV} \frac{1}{W_-(j)\varphi(j)} \text{with } \varphi(i) := \prod_{l=i}^{bV} \frac{W_+(l)}{W_-(l)}. \tag{43}$$

Let us call $x := X/V$, so that for $x \in [a, b]$ $W_{\pm}(X) = Vw_{\pm}(x)$, and then the continuum limit is

$$T(x) = V \int_a^x \int_y^b \frac{\mathrm{d}z}{w_-(z)} \exp\left(V \int_y^z \log\left(\frac{w_+}{w_-}\right)\right) \mathrm{d}x' \mathrm{d}y. \tag{44}$$

6.2. *Approximate solutions (simple domains)*

We assume that $[a, b]$ contains a single attracting point x_s of the vector field $A = w_+ - w_-$. In Eq. (44) the maximum of the argument in the exponential is obtained for $y = a, z = x_s$, and up to a prefactor

$$T_\Omega(x) \sim \exp(V\Phi(a|x_s)), \tag{45}$$

where $\Phi(x|y) := \int_y^x \log(w_-/w_+)$. We notice immediately that $\Phi(a|x_s) > 0$, because $w_+ > w_-$ on $[a, x_s]$ (and $w_+ < w_-$ on $[x_s, b]$). Then, $T_\Omega(x)$ does not depend on x in Eq. (45).

As a comparison, we could use the Fokker–Planck equation associated with the master equation. In that case we should solve

$$\left(A(x)\frac{\partial}{\partial x} + \frac{1}{2V}D(x)\frac{\partial^2}{\partial x^2}\right)T_\Omega^{(\mathcal{FP})}(x) = -1 \quad T_\Omega^{(\mathcal{FP})}(a) = 0 \quad \frac{\mathrm{d}}{\mathrm{d}x}T_\Omega^{(\mathcal{FP})}(b) = 0, \tag{46}$$

where $A = w_+ - w_-, D = w_+ + w_-$. As $T_\Omega^{(\mathcal{FP})}(x) = 2V\int_a^x \int_y^b \frac{\exp\left(2V\int_y^z \frac{A}{D}\right)}{D(z)}\mathrm{d}z\mathrm{d}y$, which can be approximated up to a prefactor by

$$T_\Omega^{(\mathcal{FP})}(x) \sim \exp\left(2V\int_a^{x_s} \frac{A}{D}\right). \tag{47}$$

If we compare the asymptotic results given by Eqs. (45) and (47), we see that $\frac{T_\Omega}{T_\Omega^{(\mathcal{FP})}} \sim \exp\{V\int_a^{x_s}[\log(\frac{1+A/D}{1-A/D}) - \frac{2A}{D}]\}$, which is exponentially large with respect to the volume.

We can also consider the eigenvalue problems

$$\Lambda^* f_1 = \lambda_1 f_1 \qquad R^* f_1^{(\mathcal{FP})} = \lambda_1^{(\mathcal{FP})} f_1^{(\mathcal{FP})},$$

with the adjoint operator Λ^* of the master equation, and the adjoint operator R^* of the Fokker–Planck equation. Because these eigenvalues are $\lambda_1 \sim (1/T_\Omega)$, we see that

$$\frac{\lambda_1}{\lambda_1^{(\mathcal{FP})}} \sim \exp(-VC), \tag{48}$$

where C is a certain positive constant.

This means that the Fokker–Planck approximation overestimates the velocity of the rate process, hence also the rate for crossing a barrier with respect to the more exact master equation.

6.3. Qualitative discussion of the typical trajectory

The typical trajectory of the stochastic process associated to the master equation or the Fokker–Planck equation can be described as follows. Essentially, the evolution $t \mapsto X(t)$ of the particle number is submitted to the 'force' exerted by the deterministic field A, which is attracting towards x_s, and by a diffusion force. The main contribution to the calculation of $T_\Omega(x)$ comes from trajectories joining point $x \in \Omega$ to point a in large times. By the results of Section 5 these trajectories are close to Hamiltonian trajectories (either for $\mathcal{H}^{(\mathcal{M})}$, or for $\mathcal{H}^{(\mathcal{FP})}$). For finite times, and for detailed balanced systems by the Kurtz theorem [31] such a trajectory is, however, close to the deterministic trajectory, and so goes from x to a small neighborhood of x_s. At that point, Kurtz's theorem is no longer applicable, and we must follow the reasoning outlined up to now. The trajectory finally returns to point a; following essentially the antideterministic trajectory (here, this means the reverse of the deterministic trajectory from a to x_s). We shall see in Section 7 that this gives back Eqs. (45) or (48).

For example, if $A(x) = -\alpha x$ ($\alpha > 0$), and D is a positive constant, the equations of motion for $\mathcal{H}^{(\mathcal{FP})}(x) = -\alpha x p + \frac{1}{2} D p^2$ are: $\dot{x} = -\alpha x + Dp$ $\dot{p} = \alpha p$.

The trajectory that starts from $x(0) > 0$ at $s = 0$, and arrives at x at time t, is

$$x(s) = x(0)e^{-\alpha s} + (x - x(0)e^{-\alpha t})\frac{\mathrm{sh}(\alpha s)}{\mathrm{sh}(\alpha t)}.$$

When $x(0)e^{-\alpha t} < x$, this trajectory starts at $x(0)$, goes towards 0, reaches a minimum $x_{\min} \sim 2\sqrt{xx(0)}e^{-\alpha t/2}$ (so x_m tends to 0 when $t \to +\infty$) and goes towards x. In a sense, when $t \to +\infty$, the trajectory loses more and more time around 0 (see also [29, p. 7743] for a complete discussion). In the more complex case where $\Omega = [a, b]$ contains more than one zero of the vector field A (again with absorbing boundary condition at a and reflecting at b) Eqs. (44) and (46) are still valid and exact. Depending on the position of x with respect to the extrema of Φ, $T_\Omega(x)$ has different expressions that may or may not depend on x ([36, pp. 7755–7756]).

7. Exit times and rate constants

We shall now extend the previous results about the exit time to the multidimensional case. Let us consider the situation of Section 5.2 in the large-volume limit, so that we define the state space as the space of concentrations $\mathbf{x} := \mathbf{X}/V$. We shall treat together the Fokker–Planck dynamics and the master dynamics. Let Ω be a certain domain in the space of concentrations and for a stochastic trajectory, starting from $\mathbf{x} \in \Omega$ let t_Ω be the first exit time from Ω. The complementary of the probability distribution of t_Ω and the average exit time are

$$\tau_\Omega(\mathbf{x}, t) = \mathcal{P}(t_\Omega > t | \mathbf{X}(0) = \mathbf{x}) \tag{49}$$

$$T_\Omega(\mathbf{x}, t) = \langle t_\Omega | \mathbf{X}(0) = \mathbf{x} \rangle = -\int_0^x t \frac{\partial \tau_\Omega(\mathbf{x}, t)}{\partial t} \mathrm{d}t. \tag{50}$$

It is well known that [25]

$$\frac{\partial \tau_\Omega(\mathbf{x}, t)}{\partial t} = \Lambda^* \tau_\Omega(\mathbf{x}, t) \qquad \tau_\Omega(\mathbf{x}, 0) = 1 \qquad \tau_\Omega(\mathbf{x}, t) = 0 \qquad \mathbf{x} \in \partial\Omega, \tag{51}$$

where Λ^* is the adjoint operator of the Fokker–Planck operator or of the master equation operator Λ.

Let us consider the eigenvalues λ_n of Λ in Ω with absorbing conditions on $\partial\Omega$, ordered by decreasing order (they are negative), and let φ_n (resp. ϑ_n) be the corresponding eigenvectors of Λ (resp. Λ^*): $\Lambda\varphi_n = \lambda_n \varphi_n$ $\Lambda^* \vartheta_n = \lambda_n \vartheta_n$. The eigenvectors are normalized so that $\int_\Omega \vartheta_n(\mathbf{x})\varphi_m(\mathbf{x})\mathrm{d}\mathbf{x} = \delta_{mn}$. Then, $\delta(\mathbf{x} - \mathbf{x}_0) = \sum_n \varphi_n(\mathbf{x})\vartheta_n(\mathbf{x}_0)$.

Let $p_\Omega(t, \mathbf{x}|\mathbf{x}_0)$ be the probability density of the process $\mathbf{x}(t)$ without leaving Ω, i.e. $\int_{\Omega_1} p_\Omega(t, \mathbf{x}|\mathbf{x}_0)\mathrm{d}\mathbf{x}$ is the probability of the event that the process is in the set Ω_1 without having left the domain Ω in the interval $[0,t]$, under the condition that it has started from $\mathbf{x}_0$ at time 0, i.e. $\int_{\Omega_1} p_\Omega(t, \mathbf{x}|\mathbf{x}_0)\mathrm{d}\mathbf{x} = \mathcal{P}(\mathbf{x}(\mathrm{t}) \in \Omega_1, t \leq t_\Omega | \mathbf{x}(0) = \mathbf{x}_0)$, so that

$p_\Omega(t, \mathbf{x}|\mathbf{x}_0) = \sum_{n\geq 1} e^{\lambda_n t} \varphi_n(\mathbf{x}) \vartheta_n(\mathbf{x}_0)$, $\tau_\Omega(\mathbf{x}_0, t) = \sum_{n\geq 1} e^{\lambda_n t} \vartheta_n(\mathbf{x}_0) \int_\Omega \varphi_n$, and finally,

$$T_\Omega(\mathbf{x}_0) = -\sum_n \frac{\vartheta_n(\mathbf{x}_0)}{\lambda_n} \int_\Omega \varphi_n. \tag{52}$$

7.1. Estimation of relevant parameters for simple domains

Let us assume in the present subsection that Ω contains a single attracting point $\mathbf{x}_s$ of the deterministic vector field $\mathbf{A}$ and no other zero of $\mathbf{A}$ (except possibly on the boundary of Ω).

It can be proved [36, p. 7754–7755] that for large V

$$T_\Omega(\mathbf{x}) \sim \exp(V \min\{\Phi(\mathbf{y}|\mathbf{x}_s); \mathbf{y} \in \partial\Omega\}),$$

(up to a prefactor), and $\Phi(\mathbf{x}|\mathbf{x}_s)$ is the nonconstant regular solution of the Hamilton–Jacobi equation vanishing at $\mathbf{x}_s$

$$\mathcal{H}^{(\mathcal{M})}(\mathbf{x}, \nabla\Phi(\mathbf{x}|\mathbf{x}_s)) = 0 \qquad \Phi(\mathbf{x}_s|\mathbf{x}_s) = 0. \tag{53}$$

We have also shown above that the function Φ is unique. In particular, $T_\Omega(\mathbf{x})$ is independent of $\mathbf{x}$ (up to a prefactor). Clearly, the prefactor is such that $T_\Omega(\mathbf{x})$ should tend to zero for $\mathbf{x}$ approaching the boundary of Ω, but we are unable to give the form of this prefactor. Then, from Eq. (52) we see that $\vartheta_1(\mathbf{x}) \sim K$, where K is an absolute constant (up to a prefactor). However, because of one of our normalization conditions $\int_\Omega \varphi_1 \vartheta_1 = 1$, we see from Eq. (52) that

$$-\lambda_1 \sim \frac{1}{T_\Omega(\mathbf{x})} \sim \exp(-V \min\{\Phi(\mathbf{y}|\mathbf{x}_s); \mathbf{y} \in \partial\Omega\}), \tag{54}$$

and that $\tau_\Omega(\mathbf{x}, t) = \exp(\lambda_1 t)$.

The estimation of the first eigenvalue in Eq. (54) has been given by several authors for the Fokker–Planck dynamics [26,40]. We remark that the proof in Ref. [26] is not logically consistent, although the result is correct (see [36, pp. 7754–7755]). If we allow the domain Ω to contain several critical points of Φ beyond a stable equilibrium point $\mathbf{x}_s$ then the mean exit time depends on the starting point. The full discussion of the mean exit times is given in [36, pp. 7756–7757].

8. Progress variables

8.1. Fundamental processes and progress variables

Now we suppose that we have two kinds of chemical species in the vessel of fixed volume V:

(1) **Internal species** denoted by $\mathcal{X}_m (m = 1, 2, \ldots, M)$ that are varying freely according to the chemical reactions in the vessel. The number of particles of species $\mathcal{X}_m$ is denoted by X_m and its concentration is $x_m := X_m/V$ as above.

(2) **External species** denoted by $\mathcal{A}_l (n = 1, 2, \ldots, N)$, which are completely under control of external reservoirs.

The state of the system at time t is given by the vector $\mathbf{x}(t)$ that specifies the concentrations of the freely varying species. The concentrations $\mathbf{a}(t)$ may be functions of the state variables. The simplest situation is the case where the reservoirs maintain each a_l at a fixed concentration independent of t. We allow a general dependence in Sections 8 and 9, while in Ref. [41] the action of reservoirs has been discussed in more detail.

The species are reacting and diffusing in R subprocesses, of which we shall separately count the forward and backward steps from now on.

$$\sum_{n=1}^{N} \mu_r^n \mathcal{A}_n + \sum_{m=1}^{M} \alpha_r^m X_m \Longleftrightarrow \sum_{n=1}^{N} \nu_r^n \mathcal{A}_n + \sum_{m=1}^{M} \beta_r^m X_m, \tag{55}$$

and r labels the processes, as before. By convention, the forward process is from left to right, and the backward process is from right to left.

If from now on $\xi_r(t)$ denotes the number of forward processes minus the number of backward processes of type r up to time t, then we have [31]

$$\mathbf{X}(t) = \mathbf{X}(0) + \sum_{\alpha} \gamma_r^m \xi_r(t), \tag{56}$$

where γ_r^m is the difference between the stoichiometric numbers: $\gamma_r^m := \beta_r^m - \alpha_r^m$. Let u_r be the number of processes per unit volume

$$u_r := \frac{\xi_r}{V}, \tag{57}$$

and call these numbers **progress variables** as usual. Finally, $W_r^{\pm}$ denotes the probability per unit time that a forward or backward process of type α occurs in V. Because of Eq. (56) in rescaled variables we have

$$\mathbf{x}(t) = \mathbf{x}(0) + \boldsymbol{\gamma}\mathbf{u}(t), \tag{58}$$

and $W_{\alpha}^{\pm}$ can be considered as functions of $\mathbf{u}$, $\mathbf{x}(0)$, and $\mathbf{a}(t)$.

8.2. *Dynamics of progress variables*

The probability $Q(\gamma, t)$ that at time t γ_α processes of type α have occurred satisfies the following master equation

$$\begin{aligned}\frac{\partial Q(\boldsymbol{\gamma}, t)}{\partial t} &= \sum_{\beta} [W_\beta^{+}(\boldsymbol{\gamma} - \mathbf{e}_\beta) Q(\boldsymbol{\gamma} - \mathbf{e}_\beta, t) + W_\beta^{-}(\boldsymbol{\gamma} + \mathbf{e}_\beta) Q(\boldsymbol{\gamma} + \mathbf{e}_\beta, t)] \\ &\quad - (W_\beta^{+}(\boldsymbol{\gamma}) + W_\beta^{-}(\boldsymbol{\gamma})) Q(\boldsymbol{\gamma}, t).\end{aligned} \tag{59}$$

Let us now turn to the rescaled variables defined in Eq. (57) to define the rescaled probability densities $q(\mathbf{u}, t)$ and rates $w_{\alpha}^{\pm}$ by $q(\mathbf{u}, t) := V^p Q(\boldsymbol{\gamma}, t)$ $w_{\alpha}^{\pm}(\mathbf{u}) := W_{\alpha}^{\pm}(\boldsymbol{\gamma})/V$, and with these Eq. (59) may be rewritten as

$$\frac{1}{V}\frac{\partial q(\mathbf{u},t)}{\partial t} = \sum_{\beta}[w_{\beta}^{+}(\mathbf{u}-\mathbf{e}_{\beta})q(\mathbf{u}-\mathbf{e}_{\beta},t) + w_{\beta}^{-}(\mathbf{u}+\mathbf{e}_{\beta})q(\mathbf{u}+\mathbf{e}_{\beta},t)] - (w_{\beta}^{+}(\mathbf{u}) + w_{\beta}^{-}(\mathbf{u}))q(\mathbf{u},t). \tag{60}$$

For large V we obtain an approximate Fokker–Planck equation from Eq. (60)

$$\frac{\partial q(\mathbf{u},t)}{\partial t} = -\sum_{\beta}\frac{\partial(w_{\beta}^{+}(\mathbf{u}) - w_{\beta}^{-}(\mathbf{u}))q(\mathbf{u},t)}{\partial u_{\beta}} + \frac{1}{2V}\sum_{\beta}\frac{\partial^{2}(w_{\beta}^{+}(\mathbf{u}) + w_{\beta}^{-}(\mathbf{u}))q(\mathbf{u},t)}{\partial u_{\beta}^{2}}. \tag{61}$$

However, it has been shown that this equation does not give the correct results for stationary state and large time dynamics of the full master equation (see Ref. [25] and above, or Ref. [29]).

8.3. Approximate dynamics for large volumes

For large V again one may try a formal asymptotic expansion for q

$$q(\mathbf{u},t) = \exp(-V\psi)\Big(U_0(\mathbf{u},t) + \tfrac{1}{V}U_1(\mathbf{u},t) + \cdots\Big). \tag{62}$$

The variations of the prefactor U_0 are usually negligible compared to the variations of the dominant exponential exp $(-V\psi)$. However, the prefactor becomes dominant near criticality, when the argument of the exponential vanishes. On the other hand, the method of Kubo would make no sense if U_0 was singular. For all these reasons, a further study of the prefactor is necessary for a sound mathematical foundation of the present formalism.

The expansion (62) is used either in the Fokker–Planck equation (61), or directly in the master equation (60). In both cases, it is easily shown as above that the function ψ satisfies a Hamilton–Jacobi equation in the progress variables

$$\mathcal{H}'(\mathbf{u}, \nabla\psi(\mathbf{u})) = 0, \tag{63}$$

where $\mathcal{H}'$ is either the master Hamiltonian $\mathcal{H}'^{(\mathcal{M})}$, or the Fokker–Planck Hamiltonian $\mathcal{H}'^{(\mathcal{FP})}$,

$$\mathcal{H}'^{(\mathcal{M})}(\mathbf{u},\boldsymbol{\pi}) = \sum_{r}(w_r^{+} - w_r^{-})\pi_r + \tfrac{1}{2}(w_r^{+} + w_r^{-})\pi_r^{2}, \tag{64}$$

$$\mathcal{H}'^{(\mathcal{FP})}(\mathbf{u},\boldsymbol{\pi}) = \sum_{r}[w_r^{+}(\exp(\pi_r) - 1) + w_r^{-}(\exp(-\pi_r) - 1)], \tag{65}$$

where π_r is the conjugate momentum of u_r. As above, we have $\mathcal{H}'^{(\mathcal{M})}(\mathbf{u},\boldsymbol{\pi}) = \mathcal{H}'^{(\mathcal{FP})}(\mathbf{u},\boldsymbol{\pi}) + \mathcal{O}(|\boldsymbol{\pi}|^{3})$.

We note here that all the results proved previously for the Hamilton–Jacobi equation associated with the master equation are valid for the Hamilton–Jacobi equation in the progress variables.

8.4. Relation with the usual master equation

The usual master equation is an equation for functions of the state variable $\mathbf{x}$. Although the formal derivation of the usual master equation and of its Fokker–Planck and Hamilton–Jacobi approximations are the same as in Section 3, their physical meaning is completely different, because u_r are not state variables. They are related to state variables by Eq. (58). Upon knowing $\mathbf{u}$ *one can deduce the variation* $\mathbf{x}(t) - \mathbf{x}(0)$ *of the state variable, but in general not conversely.* Previously, we have introduced the rate per unit volume $w_\gamma(\mathbf{x})$ of a transition $\mathbf{x} \to \mathbf{x} + \gamma/V$. It is clear that

$$w_\gamma(\mathbf{x}) = \sum_{r:\gamma_r'=\gamma} w_r^+(\mathbf{x}) + \sum_{r:\gamma_r'=\gamma} w_r^-(\mathbf{x}). \tag{66}$$

The usual Kramers–Moyal expansion of the master expansion studied above and in textbooks [8–10,13] and papers [16,25,30] yields the usual Fokker–Planck equation in concentration variables,

$$\frac{\partial p(\mathbf{x},t)}{\partial t} = -\sum_{m=1}^{M} \frac{\partial}{\partial x_m}(A_m p)(\mathbf{x},t) + \frac{1}{2V}\sum_{m,q=1}^{M} \frac{\partial^2}{\partial x_m \partial x_q}(D_{mq}p)(\mathbf{x},t),$$

where

$$A_m(\mathbf{x}) := \sum_\gamma \gamma_m w_\gamma(\mathbf{x}) = \sum_r \gamma_r^m (w_r^+(\mathbf{x}) - w_r^-(\mathbf{x})), \tag{67}$$

$$D_{mq}(\mathbf{x}) := \sum_\gamma \gamma_m \gamma_q w_\gamma(\mathbf{x}) = \sum_r \gamma_r^m \gamma_r^q (w_r^+(\mathbf{x}) - w_r^-(\mathbf{x})). \tag{68}$$

Finally, Eq. (58) shows that the evolution remains for all time in the subset

$$E(\mathbf{x}_0) := \left\{ \mathbf{x} \in \mathbb{R}^M; \mathbf{x} = \mathbf{x}_0 + \sum_r \gamma_r^m u_r; u_r \in \mathbb{R} \right\}.$$

It is clear that this set is the same as the set $S(\mathbf{x}_0)$ introduced in Eq. (11), because the set of $2R$ vectors $\gamma_r'(r = 1, 2, \ldots, R)$ is exactly the same as the set of vectors $\gamma(\cdot, r)(r = 1, 2, \ldots, R)$, and each subset $E(\mathbf{x}_0)$ carries a stationary probability distribution. *We shall assume henceforth that we reduce the situation to a given set* $E(\mathbf{x}_0)$, *so that the dynamics is irreducible and has a unique stationary state in this set.*

Note that if d is the dimension of $E(\mathbf{x}_0)$ we can use d variables to parametrize $E(\mathbf{x}_0)$. The other variables $x_{d+1}, x_{d+2}, \ldots, x_M$ are still present (so that the chemical processes are the same) but they are linear functions of $x_1, x_2, \ldots, x_d$.

If one introduces $\boldsymbol{\pi} := \boldsymbol{\gamma} \cdot \boldsymbol{\xi}$ $\mathbf{x} := \mathbf{x}(0) + \boldsymbol{\gamma} \cdot \mathbf{u}$, then Eq. (66) implies that the master Hamiltonian $\mathcal{H}'^{(\mathcal{M})}$ in the variables $(\mathbf{u}, \boldsymbol{\pi})$ reduces to the master Hamiltonian $\mathcal{H}^{(\mathcal{M})}$ in the variables $(\mathbf{x}, \boldsymbol{\xi})$: $\mathcal{H}'^{(\mathcal{M})}(\mathbf{u}, \boldsymbol{\pi}) = \mathcal{H}^{(\mathcal{M})}(\mathbf{x}, \boldsymbol{\xi})$. In particular, if Φ is a solution of the Hamilton–Jacobi equation

$$\mathcal{H}^{(\mathcal{M})}(\mathbf{x}, \nabla\Phi(\mathbf{x})) = 0, \tag{69}$$

then the function $\psi(\mathbf{u}) := \Phi(\mathbf{x}(0) + \boldsymbol{\gamma}\,\mathbf{u})$ induces a solution of the Hamilton–Jacobi equation in the progress variable form

$$\mathcal{H}^{\prime(\mathcal{M})}(\mathbf{u}, \nabla\psi(\mathbf{u})) = 0, \tag{70}$$

because $\nabla\psi(\mathbf{u}) = \nabla\Phi(\mathbf{x})\boldsymbol{\gamma}$.

But conversely, a solution ψ of the Hamilton–Jacobi equation (70) does not necessarily produce a function in the state variables $\mathbf{x}$ and a fortiori does not define a solution of Eq. (69).

8.5. Free energy and rate constants in the unconstrained system

We consider now the vessel of volume V, in which the $2R$ processes take place, but now we switch off the exchange of external species $\mathcal{A}_n$ with the reservoirs (still maintaining the temperature T constant), so that *the concentrations* x_m and a_n *vary freely* according to the natural chemical processes ($r = 1, 2, ..., R$) in the vessel. The state then will reach a thermal equilibrium. At thermal equilibrium, the probability distribution on the state space that consists now of the freely varying concentrations x_m and a_n is for large V $p_{\text{eq}}(\mathbf{x}, \mathbf{a}) \sim U_0 \exp(-(VF(\mathbf{x}, \mathbf{a})/k_B T)$, where F is the free energy (of the state $(\mathbf{x}, \mathbf{a})$) per unit volume, k_B is the Boltzmann constant, T is temperature, and U_0 is a prefactor. At equilibrium, all processes r are supposed to satisfy detailed balance, as is usual when dealing with realistic physical systems, written asymptotically for large V as $w_r^+(\mathbf{x}, \mathbf{a})\exp(-(V/k_B T)F(\mathbf{x}, \mathbf{a})) = w_r^-(\mathbf{x}, \mathbf{a})\exp(-(V/k_B T)F(\mathbf{x} + (\gamma_r/V), \mathbf{a} + (\mathbf{t}_r/V))$, where $\mathbf{t} := \nu - \mu$. Therefore,

$$k_B T \log\left(\frac{w_r^-}{w_r^+}\right) = \frac{\partial F}{\partial \mathbf{x}}\gamma + \frac{\partial F}{\partial \mathbf{a}}\mathbf{t}. \tag{71}$$

For perfect gases or solutions one usually assumes that the rates are of the **mass-action form**: $w_r^+(\mathbf{x}) = k_r^+ \mathbf{x}^{\alpha_r}$ $w_r^-(\mathbf{x}) = k_r^- \mathbf{x}^{\beta_r}$, where $k_r^\pm$ are temperature-dependent **rate coefficients**. One can immediately check that the usual partial equilibrium form [42] of the free energy

$$F(\mathbf{x}, \mathbf{a}) = \sum_m F_m(x_m) + \sum_n F_n(a_n), \tag{72}$$

(where F_m is the free energy of the ideal gas law at temperature T and concentrations $\mathbf{x}$) does satisfy Eq. (71). In fact, the chemical potentials are

$$\mu_m = \frac{\partial F}{\partial x_m} = \frac{\mathrm{d}F_m}{\mathrm{d}x_m} = k_B T \log(x_m) + f_m(T), \tag{73}$$

and Eq. (71) reduces to

$$k_B T \log\left(\frac{k_r^+}{k_r^-}\right) = \sum_m \gamma_r^m f_m(T) + \sum_n t_r^n f_n(T). \tag{74}$$

Here each f_m is calculated using the partition functions of the internal degrees of freedom of the species X_m and Eq. (74) is the usual expression for the equilibrium constant $K_{r,\text{eq}}$ of the process r in terms of the partition function of the internal degrees of freedom of the species appearing in the given process.

In many circumstances, like for imperfect gases or solutions, electrolytes, etc., one needs a more general formulation of the free energy F, not necessarily of the form Eq. (72). This is why we shall work with the most general form of the free energy.

From Eq. (71) it follows for all r

$$w_r^+(\mathbf{x},\mathbf{a})\exp\left(\frac{1}{k_BT}\frac{\partial F(\mathbf{x},\mathbf{a}))}{\partial u_r}-1\right)+w_r^-(\mathbf{x},\mathbf{a})\exp\left(-\frac{1}{k_BT}\frac{\partial F(\mathbf{x},\mathbf{a}))}{\partial u_r}-1\right)=0, \quad (75)$$

where

$$\frac{\partial F}{\partial u_r}=\sum_m\frac{\partial F}{\partial x_m}\gamma_r^m+\frac{\partial F}{\partial u_n}t_r^n,$$

so that $(1/k_BT)F$ satisfies the Hamilton–Jacobi equation (63)

$$\mathcal{H}'^{(\mathcal{M})}\left(\mathbf{u},\frac{1}{k_BT}\frac{\partial F}{\partial \mathbf{u}}\right)=0$$

with $\mathcal{H}'^{(\mathcal{M})}$ given by Eq. (65), where in $w_r^{\mp}$ the variables are $\mathbf{x}=\mathbf{x}(0)+\boldsymbol{\gamma}\mathbf{u}$ $\mathbf{a}=\mathbf{a}(0)+\mathbf{t}\cdot\mathbf{u}$.

9. Dissipation

9.1. *Dissipation of information*

From now on, we shall assume again that on a given set $E(\mathbf{x}_0)$ the state of the vessel reaches a stationary state $p(\mathbf{x})\sim U_0\exp(-V\Phi(\mathbf{x}))$ with the concentrations $\mathbf{a}$ being entirely controlled by the reservoirs and having fixed variations $\mathbf{a}(t)$.

The state $\mathbf{x}(t)$ macroscopically evolves according to the deterministic equations $\mathrm{d}\mathbf{x}/\mathrm{d}t=\mathbf{A}(\mathbf{x})=\boldsymbol{\gamma}(\mathbf{w}^+(\mathbf{x}(t))-\mathbf{w}^-(\mathbf{x}(t)))$. The value of the state function Φ evolves as

$$\frac{\mathrm{d}\Phi}{\mathrm{d}t}=\nabla\Phi(\mathbf{x}(t))\dot{\mathbf{x}}(t)=\sum_{m,r}(w_r^+(\mathbf{x}(t))-w_r^+(\mathbf{x}(t)))\frac{\partial\Phi}{\partial u_r}. \quad (76)$$

We know from Section 8.5 that $\mathcal{H}'^{(\mathcal{M})}(\mathbf{u},\nabla\Phi(\mathbf{u}))=0$. But $\mathrm{e}^x-1\geq x$, so that

$$0=\sum_r\left[w_r^+\left(\exp\left(\frac{\partial\Phi}{\partial u_r}\right)-1\right)+w_r^-\left(\exp\left(-\frac{\partial\Phi}{\partial u_r}\right)-1\right)\right]\geq\sum_r(w_r^+-w_r^+)\frac{\partial\Phi}{\partial u_r}.$$

As a consequence, we obtain the inequality $(\mathrm{d}\Phi/\mathrm{d}t)\leq 0$, where equality is attained if and only if $w_r^+\partial\Phi/\partial u_r=w_r^-\partial\Phi/\partial u_r$ for each r. In particular, if the deterministic state reaches a stationary point $\mathbf{x}^{(0)}$, for which $\mathbf{A}(\mathbf{x}^{(0)})=0$, then $\mathrm{d}\Phi/\mathrm{d}t=0$, and either $\partial\Phi/\partial u_r=0$, or $w_r^+=w_r^-=0$.

The quantity $\mathrm{d}\Phi/\mathrm{d}t$ *computed along a deterministic trajectory is always negative. It can be interpreted as a dissipation of information per unit time.*

In fact, $V\Phi(\mathbf{x})$ can be considered as the average information that is obtained when the system is observed in the state $\mathbf{x}$ rather than being stochastically distributed with the stationary probability distribution $p_s(\mathbf{x}) \sim \exp(-V\Phi(\mathbf{x}))$, which is the state of lowest information, when the system is coupled to the various reservoirs of heat and of chemical species $\mathcal{A}_n$. Along a deterministic path Φ decreases with time, while the state $\mathbf{x}$ tends to a deterministic stationary state that is a local minimum of Φ.

9.2. Dissipation of energy

The variation of free energy along the deterministic trajectory is

$$\frac{dF}{dt} = \sum_i \frac{\partial F}{\partial x_i}\frac{dx_i}{dt} + \sum_i \frac{\partial F}{\partial a_l}\frac{da_l}{dt}, \tag{77}$$

or,

$$\frac{dF}{dt} = \sum_i \frac{\partial F}{\partial x_i}\frac{dx_i}{dt} + \sum_i \frac{\partial F}{\partial a_l}\left[\frac{da_l}{dt}\right]_c - \sum_i \frac{\partial F}{\partial a_l}\left(\left[\frac{da_l}{dt}\right]_c - \frac{da_l}{dt}\right), \tag{78}$$

where $[da_l/dt]_c$ is the variation of a_l due to the chemical processes. It turns out [5, p. 684] that $(dF/dt) - w \leq 0$, where $w := -\sum_l m_l([da_l/dt]_c - da_l/dt)$, *and the quantity* $dF/dt - w$ *is the dissipation of energy in the system per unit time.*

9.3. Inequality between the dissipation of information and that of energy

It is proved in Appendix A of Ref. [41] that the dissipation of information and the dissipation of energy satisfy the fundamental inequality $1/k_B T(dF/dt) - w) \leq d\Phi/dt \leq 0$, so that in absolute value the dissipation of information is always less than the dissipation of energy. Moreover, there is equality if and only if we have an equilibrium situation.

Acknowledgement

The present work has partly been supported by the National Science Foundation Hungary (Nos T037491 and T047132).

References

[1] Arnold, L. (1980), On the consistency of the mathematical models of chemical reactions. Dynamics of Synergetic Systems (Haken, H, ed.), Springer, Berlin, pp. 107–118.

[2] Zheng, Q., Ross, J., Hunt, K.L.C. and Hunt, P.M. (1992), Stationary solutions of the master equation for single and multiintermediate autocatalytic chemical systems, J. Chem. Phys. 96(1), 630–641.

[3] Gaveau, B. and Schulman, L.S. (1996), Master equation based formulation of nonequilibrium statistical mechanics, J. Math. Phys. 37(8), 3897–3932.

[4] Gaveau, B. and Schulman, L.S. (1997), A general framework for non-equilibrium phenomena: The master equation and its formal consequences, Phys. Lett. A 229(6), 347–353.

[5] Gaveau, B. and Schulman, L.S. (1998), Theory of nonequilibrium first-order phase transitions for stochastic dynamics, J. Math. Phys. 39(3), 1517–1533.

[6] Gyarmati, I. (1969), On the governing principle of dissipative processes and its extension to nonlinear problems, Annalen der Physik 23, 353–378.
[7] Gyarmati, I. (1970), Non-equilibrium Thermodynamics. Field Theory and Variational Principles, Springer, Berlin, Heidelberg, New York.
[8] De Groot, S.R. and Mazur, P. (1984), Nonequilibrium Thermodynamics, Dover, New York.
[9] Keizer, J. (1987), Statistical Thermodynamics of Nonequilibrium Processes, Springer, New York.
[10] Kubo, R. (1988), Statistical Mechanics, North Holland, Amsterdam.
[11] Kubo, R., Matsuo, K. and Kitahara, K. (1973), Fluctuation and relaxation of macrovariables, J. Stat. Phys. 9, 51–96.
[12] Kubo, R., Toda, M. and Hashitsume, N. (1986), Statistical Physics II, Springer, New York.
[13] Nicolis, G. and Prigogine, I. (1977), Self-organization in Nonequilibrium Systems, Wiley, New York.
[14] Prigogine, I. (1962), Nonequilibrium Statistical Mechanics, Wiley, New York.
[15] Prigogine, I. and Nicolis, G. (1973), From Theoretical Physics to Biology (Marois, M., ed.), Karger, Basel.
[16] Érdi, P. and Tóth, J. (1976), Stochastic reaction kinetics—Nonequilibrium thermodynamics of the state space?, Reaction Kinet. Catal. Lett. 4, 81–85.
[17] Érdi, P. and Tóth, J. (1989), Mathematical Models of Chemical Reactions. Theory and Applications of Deterministic and Stochastic Models, Princeton University Press, Princeton.
[18] Tóth, J. and Érdi, P. (1992), Indispensability of stochastic models (Bazsa, Gy., ed.), Debrecen, Budapest, Gödöllő, pp. 117–143 (in Hungarian).
[19] Horn, F.J.M. and Jackson, R. (1972), General mass action kinetics, Arch. Rational Mech. Anal. 47, 81–116.
[20] Lemarchand, H. (1980), Asymptotic solution of the master equation near a nonequilibrium transition, Physica A 101, 518–534.
[21] Vlad, M.O. and Ross, J. (1994), Fluctuation-dissipation relations for chemical systems far from equilibrium, J. Chem. Phys. 100(10), 7268–7278.
[22] Vlad, M.O. and Ross, J. (1994), Random paths and fluctuation-dissipation dynamics for one-variable chemical systems far from equilibrium, J. Chem. Phys. 100(10), 7279–7294.
[23] Vlad, M.O. and Ross, J. (1994), Thermodynamic approach to nonequilibrium chemical fluctuations, J. Chem. Phys. 100(10), 7295–7309.
[24] Gaveau, B., Moreau, M. and Tóth, J. (1996), Decay of the metastable state: different predictions between discrete and continuous models, Lett. Math. Phys. 37, 285–292.
[25] Gaveau, B., Moreau, M. and Tóth, J. (1997), Master equation and Fokker–Planck equation: Comparison of entropy and rate constants, Lett. Math. Phys. 40, 101–115.
[26] Ludwig, D. (1975), Persistence of dynamical systems under random perturbations, SIAM Rev. 17, 605–640.
[27] Courant, R. and Hilbert, D. (1953), Methods of Mathematical Physics, Interscience, New York.
[28] Landau, L.D. and Lifsic, E.M. (1976), Mechanics: Volume 1, 3rd edn, Butterworth-Heinemann.
[29] Gaveau, B., Moreau, M. and Tóth, J. (1999), Variational nonequilibrium thermodynamics of reaction–diffusion systems. I. The information potential, J. Chem. Phys. 111(17), 7736–7747.
[30] Gaveau, B., Moreau, M. and Tóth, J. (2000), Information potential and transition to criticality for certain two-species chemical systems, Physica A 277, 455–468.
[31] Kurtz, T.G. (1973), The relationship between stochastic and deterministic models in chemical reactions, J. Chem. Phys. 57, 2976–2978.
[32] Englman, R. and Yahalom, A. (2004), Variational procedure for time dependent processes, Phys. Rev. E 69(10), 026120 (10 pages).
[33] Ván, P. (1996), On the structure of the Governing Principle of Dissipative Processes, J. Non-equilib. Thermodyn. 21(1), 17–29.
[34] Ván, P. and Muschik, W. (1995), The structure of variational principles in non-equilibrium thermodynamics, Phys. Rev. E 52(4), 3584–3590.
[35] Ván, P. and Nyíri, B. (1999), Hamilton formalism and variational principle construction, Annalen der Physik (Leipzig) 8, 331–354.
[36] Gaveau, B., Moreau, M. and Tóth, J. (1999), Variational nonequilibrium thermodynamics of reaction–diffusion systems. II. Path integrals, large fluctuations and rate constants, J. Chem. Phys. 111(17), 7748–7757.

[37] Feynman, R.P. and Hibbs, A. (1965), Quantum Mechanics and Path Integrals, McGraw-Hill, New York.
[38] Gaveau, B., Moreau, M. and Tóth, J. (1999), Path-integrals from peV to TeV (Casalbroni, R., ed.), World Scientific, Singapore, pp. 52, (Florence, August 1998).
[39] Gardiner, H. (1983), Handbook of Stochastic Methods, Springer, Berlin.
[40] Ventsel, A. and Freidlin, M. (1970), On small random perturbation of dynamical systems, Russ. Math. Surveys 25, 1–55.
[41] Gaveau, B., Moreau, M. and Tóth, J. (2001), Variational nonequilibrium thermodynamics of reaction–diffusion systems. III. Progress variables and dissipation of energy and information, J. Chem. Phys. 115(2), 680–690.
[42] Gaveau, B., Moreau, M. and Tóth, J. (1998), Dissipation of energy and information in nonequilibrium reaction–diffusion systems, Phys. Rev. E 58(11), 5351–5354.

Chapter 16

VARIATIONAL PRINCIPLES FOR THE SPEED OF TRAVELING FRONTS OF REACTION–DIFFUSION EQUATIONS

Rafael D. Benguria and M. Cristina Depassier

Department of Physics, P. Universidad Católica de Chile, Casilla 306, Santiago 22, Chile

Abstract

The 1D nonlinear diffusion equation has been used to model a variety of phenomena in different fields, e.g. population dynamics, flame propagation, combustion theory, chemical kinetics and many others. After the work of Fisher [Ann. Eugenics 7 (1937) 355] and Kolmogorov et al. [Etude de l'equation de la diffusion avec croissance de la quantité de matiére et son application a un probleme biologique Vol. 1] in the late 1930s, there has been a vast literature on the study of the propagation of localized initial disturbances. The purpose of this review is to present a rather recent variational characterization of the minimal speed of propagation, together with some of its consequences and applications. We consider the 1D reaction–diffusion equation as well as several extensions.

Keywords: reaction diffusion equations; variational principles; combustion; population dynamics; front propagation

Front propagation for the 1D reaction–diffusion equation

$$u_t = u_{xx} + f(u) \qquad \text{with } (x,t) \in \mathbb{R} \times \mathbb{R}_+, \tag{1}$$

with $f(u) \in C^1[0,1]$ and $f(0) = f(1) = 0$, has been the subject of extensive study as it models diverse phenomena in population dynamics [1,2,3], combustion theory, flame propagation [4,5], chemical kinetics and others. There are many important recent reviews on the dynamics of the solutions of the nonlinear reaction–diffusion equation both in the physics literature [6] as well as in the mathematical literature [7,8,9]. We should also point out that the propagation of disturbances in more realistic, and more complicated pattern-forming equations has been the study of many authors, and significant results have been obtained recently (see, e.g. Ref. [10] for the study of the speed of propagation of the Swift–Hohenberg equation, or [11] for the study of the critical speed of traveling waves in the Gross–Pitaevskii equation; see also [12] and many others).

The time evolution for a sufficiently localized initial condition $u(x,0)$ has been studied for different reaction terms. Aronson and Weinberger [13] showed that any positive sufficiently localized (this means decaying faster than exponentially for

E-mail address: rbenguri@fis.puc.cl (R.D. Benguria); mcdepass@fis.puc.cl (M.C. Depassier).

$|x| \to \infty$) initial condition $u(x, 0)$ with $u(x, 0) \in [0, 1]$, evolves into a traveling front propagating at a speed c^* (i.e. for large $t, u(x, t)$ behaves as $q(x - c^* t)$). The shape of the traveling front is determined by the solution to the following two-point boundary-value problem:

$$q'' + cq' + f(q) = 0, \tag{2}$$

with

$$\lim_{x \to -\infty} q(x) = 1 \qquad \text{and} \qquad \lim_{x \to +\infty} q(x) = 0. \tag{3}$$

Aronson and Weinberger characterized the asymptotic speed c^* as the minimum value of the parameter c in Eq. (2) for which the solution $q(x)$ is monotonic. For Fisher's equation, Kolmogorov, Petrovsky and Piskunov (KPP) [2] (see also the translation of this article in Ref. [14]) showed that $c^* = c_{\text{KPP}} = 2$. In fact, for any reaction profile $f(u)$ such that $0 \leq f(u) \leq f'(0)u$, Kolmogorov et al. proved that $c^* = 2\sqrt{f'(0)}$ (see, e.g. Refs. [2,8,13]). For a general f the speed of traveling fronts for Eq. (1) is unknown in closed form. For this reason, a variational characterization of the speed is an important tool for estimating it. The first variational characterization was obtained by Hadeler and Rothe (see, Ref. [15] and also Theorem 1 in Section 1). The work of Hadeler and Rothe has had many different extensions and applications (see, e.g. the monograph of Volpert et al. [16]). More recently, we have obtained two new different characterizations of the speed of propagation of traveling fronts of Eq. (1), which we will review in the following.

We will distinguish three types of reaction terms.

Case A: $f'(0) > 0, f(u) > 0$, $u \in (0, 1)$. In population dynamics this case is known as *heterozygote intermediate*. This is the case considered in the classical work of Fisher [1] and of Kolmogorov et al. [2].

Case B: There exists $a \in (0, 1)$ such that $f(u) \leq 0$ for all $u \in (0, a)$, $f(u) > 0$ for all $u \in (a, 1)$ and $\int_0^1 f(u)\mathrm{d}u > 0$. This is known as the *combustion case* in the literature.

Case C: There exists $a \in (0, 1)$ such that $f(u) < 0$ in $(0,a)$, $f(u) > 0$ in $(a,1)$ and $\int_0^1 \times f(u)\mathrm{d}u > 0$ and $f'(0) < 0$. In population dynamics this case is known as *heterozygote inferior*. It is also known as the *bistable* case.

Our main results, concerning a variational characterization for the speed of propagation of traveling fronts for the nonlinear diffusion equation (1), are contained in the following two theorems. The first variational characterization (Theorem 1) is valid only in Case A, while the second variational characterization (Theorem 2) is valid for all three cases.

Theorem 1 [17]. *Let $f \in C^1(0, 1)$ with $f(0) = f(1) = 0$, $f'(0) > 0$ and $f(u) > 0$ for $u \in (0, 1)$. Then,*

$$c^* = J \equiv \sup\{I(g) | g \in \mathcal{E}\}, \tag{4}$$

where

$$I(g) = 2\frac{\displaystyle\int_0^1 \sqrt{fgh}\, du}{\displaystyle\int_0^1 g\, du}, \tag{5}$$

and $\mathcal{E}$ is the space of functions in $C^1(0,1)$ such that $g \geq 0$, $h \equiv -g' > 0$ in $(0,1)$, $g(1) = 0$, and $\int_0^1 g(u)\mathrm{d}u < \infty$. Moreover, if $c^* \neq c_{\mathrm{KPP}} = 2\sqrt{f'(0)}$, J is attained at some $\hat{g} \in \mathcal{E}$, and $\hat{g}$ is unique up to a multiplicative constant.

Theorem 2 [18]. *Let $f \in C^1(0,1)$ with $f(0) = f(1) = 0$, and $\int_0^1 f(u)\mathrm{d}u > 0$. Then,*

$$c^{*2} = L \equiv \sup\{M(g) | g \in \mathcal{F}\}, \tag{6}$$

where

$$M(g) = 2\frac{\int_0^1 fg\mathrm{d}u}{\int_0^1 (g^2/h)\mathrm{d}u}, \tag{7}$$

and $\mathcal{F}$ is the space of functions in $C^1(0,1)$ such that $g \geq 0$, $h \equiv -g' > 0$ in $(0,1)$, $g(1) = 0$, and the integrals in Eq. (7) exist. Moreover, if f belongs to cases B or C, or to Case A with $c^* \neq c_{\mathrm{KPP}} = 2\sqrt{f'(0)}$, L is attained at some $\hat{g} \in \mathcal{F}$, and $\hat{g}$ is unique up to a multiplicative constant.

These two theorems have been extended in several directions. In this review, we give the complete proof of these two variational characterizations (see Section 1), we provide the analogous results for the case of density-dependent reaction–diffusion equations (see Section 2). In Section 3, we apply these theorems to obtain some properties of thermal combustion waves. Finally, in Section 4, we obtain a variational characterization for the minimal speed of fronts for the reaction–convection–diffusion equation.

1. Variational principles for the speed of propagation of traveling fronts

For a general profile $f(u)$, the speed of propagation of traveling fronts for Eq. (1) is not known in closed form. However, there are several variational characterizations of the speed of propagation of the fronts that offer the possibility of accurately estimating it. The first variational characterization was derived by Hadeler and Rothe [15]. They considered the general continuously differentiable nonlinearity $f(u)$, satisfying,

$$f(0) = f(1) = 0,\ f(u) > 0,\ \text{for}\ u \in (0,1),\ f'(0) > 0,\ f'(1) < 0. \tag{8}$$

Their variational characterization is embodied in the following min–max theorem.

Theorem 1.1 (Hadeler and Rothe, [15], Theorem 8, pp. 257). *The speed of propagation of the traveling fronts of Eq. (1) is given by*

$$c^* = \inf_{\rho} \sup_{u \in (0,1)} \left[\rho'(u) + \frac{f(u)}{\rho(u)} \right], \tag{9}$$

where ρ *is any continuously differentiable function on* $[0, 1]$ *such that*

$$\rho(u) > 0,\ u \in (0, 1),\ \rho(0) = 0,\ \rho'(0) > 0. \tag{10}$$

Remark 1.2 For extensions of the variational characterization of Hadeler and Rothe as well as many applications see the monograph [16].

Before we go into the details of the proof of Theorems 1 and 2 a heuristic argument concerning the solutions of Eq. (2) is in order. In all three cases Eq. (2) can be viewed as Newton's equation for a particle moving in one dimension under the action of the force $-f(q) - cq'$ (the variable x now plays the role of time). The force $-f(q)$ is conservative and derives from a potential $V(q) = \int_0^q f(s)\mathrm{d}s$. If $f(q)$ belongs to Case A, the potential is monotonic in (0,1), it has its minimum at the point $q = 0$, and maximum at $q = 1$. The second term, i.e. $-cq'$ represents a viscous force, and, with this picture in mind the parameter c represents the *viscosity* or viscous coefficient. In summary, Eq. (2) is Newton's equation for a particle coming down from the top of the potential at $q = 1$ to the bottom of it at $q = 0$ in the presence of a viscous force. If the *viscosity* (i.e. the parameter c, which is precisely our object of interest here) is small the particle will oscillate near the bottom of the well before it settles down at the minimum (i.e. at $q = 0$). If we increase the *viscosity*, we will reach a value for which there will no longer be oscillations. That is, the particle will come down monotonically from $q = 1$ to $q = 0$ (in mechanics this is commonly known as *critical damping*). It is intuitively clear that if one increases the *viscosity* even further, the particle will also go down monotonically. So, there is a critical value of c, which we denote by c^*, such that for $c \geq c^*$, there will be monotonic decreasing solutions of Eqs. (2) and (3). This fact was proven rigorously by Aronson and Weinberger, and more importantly, they showed that this critical value c^* will be the speed of propagation of fronts of the reaction–diffusion equation (1) (for a sufficiently localized initial condition). If $f(u)$ belongs to Case C, the potential has two maxima, at $q = 0$ and $q = 1$, and a minimum at $q = a$. In this case, the particle starts at $q = 1$ at *time* $x = -\infty$ and it should get to $q = 0$ at $x = \infty$ with zero speed. Because of *energy dissipation*, for this motion to be possible, the maximum at $q = 0$ must be smaller than the maximum at $q = 1$ (otherwise the particle would never reach $q = 0$). In terms of the force f, this condition in the potential implies $\int_0^1 f(q)\mathrm{d}q > 0$. Now, it should be intuitively clear that there is only one value of the *viscosity* c for which the particle will go from the maximum of the potential at $q = 1$ to the valley at $q = a$ and then up to the maximum of the potential at $q = 0$. This is precisely the value c^*. If c is larger than c^* the particle will be trapped forever at the valley. On the other hand, if $c < c^*$ the particle will overshoot at $q = 1$. All these facts have been proven in Ref. [13].

Following the results of Aronson and Weinberger, we are interested in computing the minimal speed for which Eq. (1) has a monotonic traveling front $u(x, t) = q(x - ct)$ joining

$u = 1$ to $u = 0$. The shape of the traveling front will be determined by the monotonic solution of Eqs. (2) and (3). Since q is monotone, in order to analyze the solution to Eqs. (2) and (3), it is convenient to work in phase space. Calling $z = x - ct$, the argument of q, and $p(q) = -\mathrm{d}q/\mathrm{d}z$ (here, the minus sign is included so that p is non-negative), we find that in phase space the monotonic fronts are the non-negative solutions of the following two-point boundary-value problem,

$$p\frac{\mathrm{d}p}{\mathrm{d}q} - cp(q) + f(q) = 0, \qquad \text{in } (0, 1), \tag{11}$$

and

$$p(0) = p(1) = 0. \tag{12}$$

In Ref. [13], Section 4, Aronson and Weinberger proved that there is a unique non-negative solution p to Eqs. (11) and (12) for $c = c^*$. Moreover, the solution p is such that $p(q) \approx |m|q$ near $q = 0$, where $|m|$ is the largest root of the equation

$$x^2 - c^* x + f'(0) = 0, \tag{13}$$

i.e.

$$|m| = \frac{1}{2}\left(c^* + \sqrt{c^{*2} - 4f'(0)}\right).$$

It is convenient to introduce the parameter σ as $\sigma = c^*/|m|$. In terms of σ one can write

$$c^* = \sigma\sqrt{f'(0)/(\sigma - 1)} \qquad \text{and} \qquad |m| = \sqrt{f'(0)/(\sigma - 1)}. \tag{14}$$

It is straightforward to verify that whenever $1 < \sigma < 2$ the value of $|m|$ given by Eq. (14) corresponds to the largest root of Eq. (13) and therefore to the asymptotic slope at the origin of the selected front [19]. If $\sigma = 2$ then c^* is given by the *linear* value $c_{\mathrm{KPP}} = 2\sqrt{f'(0)}$.

Proof of Theorem 1 The proof of the theorem is done in two steps. First, we show that

$$c^* \geq J, \tag{15}$$

and then we show that equality actually holds in Eq. (15). To prove Eq. (15) it is enough to show that $c^* \geq I(g)$ for all $g \in \mathcal{E}$. Multiply Eq. (11) by g/p, for any fixed $g \in \mathcal{E}$, and integrate over $q \in [0, 1]$. After integration by parts we have,

$$c^* = \frac{\int_0^1 (hp + (fg)/p)\,\mathrm{d}q}{\int_0^1 g\,\mathrm{d}q} \geq I(g), \tag{16}$$

which follows from the fact that

$$hp + \frac{fg}{p} \geq 2\sqrt{fgh},$$

since p, h, f, and g are positive for $q \in (0, 1)$. To finish the proof we have to show that the equality holds in Eq. (15). From the results of Ref. [13] it follows that $c^* \geq 2\sqrt{f'(0)}$. We separate the proof that equality holds in Eq. (15) into two cases: Case (i) $c^* = 2\sqrt{f'(0)}$, and Case (ii) $c^* > 2\sqrt{f'(0)}$. In Case (i), consider the family of functions $g_\alpha(u) = u^{\alpha-1} - 1$, with $0 < \alpha < 1$. Then, $g_\alpha \in \mathcal{E}$ and $\lim_{\alpha\to 0} I(g_\alpha) = 2\sqrt{f'(0)}$ (see, e.g. the appendix of Ref. [20] for a detailed proof of this fact). Hence, from Eq. (15) we have

$$c^* \geq J \geq 2\sqrt{f'(0)},$$

which implies $J = c^*$ in this case.

In Case (ii), we will not only prove that the equality in Eq. (15) holds but also that there exists $\hat{g} \in \mathcal{E}$ with $c^* = I(\hat{g})$. Let $p(q)$ be the positive solution of Eq. (11) satisfying Eq. (12). The existence of such a solution has been established in Ref. [13]. Moreover, $p(q) \approx |m|q$ near $q = 0$. A function $\hat{g}$ will saturate the bound Eq. (16) if and only if,

$$hp = \frac{f\hat{g}}{p}, \tag{17}$$

where $h = -\hat{g}'$ and p is the solution of Eq. (11) mentioned above. Eq. (17) is a first-order ordinary differential equation for $\hat{g}$, whose solution can be written in terms of p as

$$\hat{g}(q) = \frac{p(q)}{c^*} \exp\left(\int_q^{q_0} \frac{c^*}{p} \mathrm{d}q\right), \tag{18}$$

for some fixed $0 < q_0 < 1$. To complete the argument we only need to show that $\hat{g} \in \mathcal{E}$. It follows from Eqs. (17) and (18) that

$$\hat{h}(q) = \frac{g(q)}{c^* p(q)} \exp\left(\int_q^{q_0} \frac{c^*}{p} \mathrm{d}q\right)$$

in (0,1). Moreover, since $p(q) > 0$ in (0,1) and $p \in C^1(0, 1)$ we have that $\hat{g} \in C^1(0, 1)$. Thus, $\hat{g}$ is a continuous, positive and decreasing function in (0,1). Hence, $\hat{g}$ is bounded away from the origin. In order to show that $\int_0^1 \hat{g}(q)\mathrm{d}q$ is finite we have to determine the behavior of $\hat{g}$ near $q = 0$. Since $p \approx |m|q$ near 0, we have from Eq. (18) that

$$\hat{g}(q) \approx \frac{1}{\sigma} \frac{1}{q^{\sigma-1}}$$

near zero. Therefore, if $\sigma < 2$ (i.e. if $c^* > c_{\mathrm{KPP}} = 2\sqrt{f'(0)}$, we have $\int_0^1 \hat{g}(q)\mathrm{d}q < \infty$ and $\hat{g} \in \mathcal{E}$. □

Proof of Theorem 2 Now take $g \in \mathcal{F}$. Multiplying Eq. (11) by $g(q)$ and integrating over $q \in [0, 1]$, we have, after integration by parts, the equality

$$\int_0^1 fg\mathrm{d}q = c\int_0^1 pg\mathrm{d}q - \frac{1}{2}\int_0^1 hp^2\mathrm{d}q. \tag{19}$$

For positive c, g, and h, the function $\varphi(p) = cpg - hp^2/2$ has its maximum at $p = cg/h$, and so $\varphi(p) \leq c^2g^2/(2h)$. It follows that $c^2 \geq M(g)$, which implies (setting $c = c^*$ if c is nonunique) that c^{*2} is no less than the supremum of Eq. (6). Next we show that the equality holds for a function $\hat{g}$. Notice that the condition $p = cg/h$ is solvable in g and gives the expression for the maximizer $\hat{g}$,

$$\hat{g} = \exp\left(- \int_{q_0}^{q} \frac{c}{p} \mathrm{d}q \right), \tag{20}$$

with $q_0 \in (0, 1)$. Clearly, $\hat{g}$ is positive and decreasing, with $\hat{g}(1) = 0$ since $p \approx O(1 - q)$ for $q \approx 1$. At $q = 0$, however, $\hat{g}$ diverges since the exponent goes to $+\infty$. We must ensure that the integrals on the right side of Eq. (7) exist. To verify this we recall that in the three cases, A, B and C, the front approaches $q = 0$ exponentially [13]. Therefore, near $q = 0$,

$$p \approx \frac{1}{2}\left(c + \sqrt{c^2 - 4f'(0)} \right) q \equiv mq.$$

Thus, from Eq. (10) we obtain $\hat{g}(q) \approx q^{-c/m}$, near $q = 0$. Hence, both $f\hat{g}$ and $\hat{g}^2/h$ diverge at most as $q^{1-c/m}$ near $q = 0$. Therefore, the integrals on the right side of Eq. (7) exist if $m/c > 1/2$. This condition is always satisfied when $f'(0) \leq 0$, i.e. in Cases B and C. In Case A this condition is satisfied provided $c > c_{\mathrm{KPP}} = 2\sqrt{f'(0)}$. This concludes the proof in Cases B, C, and also in Case A for $c \neq c_{\mathrm{KPP}}$. Finally, in Case A, if $c^* = c_{\mathrm{KPP}}$ one can take the maximizing sequence $g_\alpha = q^{\alpha - 2} - 1$, which is in $\mathcal{F}$ for $0 < \alpha < 1$, and let $\alpha \to 0$. One can verify that $\lim_{\alpha \to 0} M(g_\alpha) = 4f'(0) = c_{\mathrm{KPP}}^2$, and we are done. □

2. Variational principle for the asymptotic speed of fronts of the density-dependent reaction–diffusion equation

Several problems arising in population growth [21,22], combustion theory (see, e.g. Ref. [23] and references therein), chemical kinetics [24] can be modeled by an equation of the form

$$\frac{\partial u}{\partial t} + \vec{\nabla} \cdot \vec{j} = f(u),$$

where the source term $f(u)$ (which depends on the density u) represents net growth and saturation processes. In general, the flux $\vec{j}$ is given by Fick's law

$$\vec{j} = -D(u)\vec{\nabla} u,$$

where the diffusion coefficient $D(u)$ in general may depend on the density u. In one dimension this leads to the equation

$$u_t = (D(u)u_x)_x + f(u). \tag{21}$$

We will assume that the reaction term $f(u) \in C^1[0, 1]$ satisfies

$$f(u) > 0, \text{ in } (0, 1) \text{ and } f(0) = f(1) = 0, \tag{22}$$

(i.e. f is of type A), restrictions that are satisfied by several models. We will first consider here the case when the diffusion coefficient follows a power law (i.e. $D(u) = mu^{m-1}$, for $m \geq 1$). Thus, the equation we consider here is

$$u_t = (u^m)_{xx} + f(u), \tag{23}$$

with f satisfying Eq. (22). Aronson [21], and Aronson and Weinberger [13] have shown that the asymptotic speed of propagation of sufficiently localized initial disturbances for Eq. (23) is the minimal speed $c^*(m)$ for which there exists a monotonic traveling front $u(x,t) = q(x - ct)$ joining $q = 1$ to $q = 0$. The equation satisfied by q is,

$$(q^m)_{zz} + cq_z + f(q) = 0, \tag{24}$$

with

$$q(-\infty) = 1, q > 0, \; q' < 0 \text{ in } (-\infty, \omega) q = 0 \;\text{ for }\; z \geq \omega, \tag{25}$$

where $z = x - ct$. When $m > 1$ the wave of minimal speed is *sharp*, i.e. $\omega < \infty$ [21].

We can also give a variational characterization of the speed $c^*(m)$ for the density-dependent reaction–diffusion equation. This is given by the following theorem.

Theorem 2.1 [19,25]. *If f is of type A, then,*

$$c^*(m) = \max\left(\frac{2\int_0^1 \sqrt{mq^{m-1}fgh}\,\mathrm{d}q}{\int_0^1 g\,\mathrm{d}q}\right), \tag{26}$$

where the maximum is taken over all functions g for which the integrals in Eq. (26) exist and $g(0) = 1, g(1) = 0$, and $h = -g' > 0$. Moreover, there is a unique g for which the maximum is attained.

Proof: We follow the same method as in the proof of Theorem 1. We go to phase space, but this time we denote $p(q) = -q^{m-1}\mathrm{d}q/\mathrm{d}z > 0$, which is positive since the selected speed corresponds to that of a decreasing monotonic front. Then, $p(q)$ satisfies

$$p\frac{\mathrm{d}p}{\mathrm{d}q} - \frac{c^*}{m}p + \frac{1}{m}q^{m-1}f(q) = 0, \tag{27}$$

with

$$p(0) = p(1) = 0, p > 0 \text{ in } (0,1). \tag{28}$$

Although the wave of minimal speed is *sharp* and therefore $q'(0) < 0$, by the definition of p, we still have $p(0) = 0$. Multiplying Eq. (27) by a function g (such that $g(0) = 1$, $g(1) = 0$, and $h = -g' > 0$), integrating over $q \in (0,1)$, using integration by parts and the Schwarz inequality as in the proof of Theorem 1 above, we get

$$c^*(m) \geq \left(\frac{2\int_0^1 \sqrt{mq^{m-1}fgh}\,\mathrm{d}q}{\int_0^1 g\,\mathrm{d}q}\right). \tag{29}$$

To finish the proof, we need only show that there is a function g for which equality is obtained in Eq. (29). Equality is obtained in Eq. (29) if

$$\frac{1}{m}q^{m-1}f(q)\frac{g}{p} = ph = -pg'(q), \tag{30}$$

which is an ordinary differential equation for g, in terms of p, whose solution can be written as

$$g(q) = \frac{mp(q)}{c^*}\exp\left(-\int_{q_0}^{q}\frac{c^*}{mp(q')}\mathrm{d}q'\right), \tag{31}$$

where $0 < q_0 < 1$. Since $p(1) = 0$ and p is positive in (0,1) it follows that $g(1) = 0$. On the other hand, the function g is bounded at $q = 0$ as we now show. Call $\hat{c} = c^*/m$ and $F(q) = q^{m-1}f(q)/m$. Then Eq. (27) reads

$$pp' - \hat{c}p + F = 0, \tag{32}$$

with $F(0) = F(1) = 0$ and $F'(0) = 0$. For this case Aronson and Weinberger [13] have shown that $p(q)$ approaches $q = 0$ as $p = \hat{c}q = c^*q/m$. Hence, from the differential equation satisfied by g one can show that g approaches a constant as $q \to 0$. We can always normalize this constant to be one by an appropriate scaling of g. Therefore, $g \in C^1([0,1])$ and the integrals on the right side of Eq. (26) exist for g given by Eq. (31). □

As in the case of the nonlinear reaction–diffusion Eq. (1), there is also a second variational principle in this case. It is given by the following theorem, whose proof is similar to the proof of Theorem 0.2 above, and we leave it to the reader.

Theorem 2.2 *If f is of type A, then,*

$$c^*(m)^2 = 2m\max_g\left(\frac{\int_0^1 gf(q)q^{m-1}\mathrm{d}q}{\int_0^1 (g^2/h)\mathrm{d}q}\right), \tag{33}$$

where the maximum is taken over all functions g for which the integrals in Eq. (33) exist and $g(1) = 0$, and $h = -g' > 0$. Moreover, there is a unique g (up to a multiplicative constant) for which the maximum is attained.

As an application consider the case $f(q) = q(1-q)$ and $m = 2$ for which the exact solution is known. Using the variational characterization (Eq. (33)), with the trial function $g(q) = (1-q)/q$, we find

$$c^{*2} \geq 4\frac{\int_0^1 q(1-q)^2\mathrm{d}q}{\int_0^1 (1-q)^2\mathrm{d}q} = 1,$$

the exact value, which is just a reflection of the fact that the maximum is attained precisely at this particular g. In addition, due to the existence of the variational principle we may use the Feynman–Hellmann formula to determine the dependence of $c^*(m)$ on m or, on the possible parameters of a general reaction term f. We illustrate this by applying it to the calculation of $\mathrm{d}(c^*)^2/\mathrm{d}m$ at $m = 2$ (i.e. around the exactly solvable case). By the Feynman–Hellmann theorem, we have from Eq. (33),

$$\frac{\mathrm{d}(c^*)^2}{\mathrm{d}m} = 2\frac{\int_0^1 \hat{g}f(q)[q^{m-1} + m(m-1)q^{m-1}\log q]\mathrm{d}q}{\int_0^1 (\hat{g}^2/\hat{h})\mathrm{d}q}, \tag{34}$$

where $\hat{g}$ is the maximizer in Eq. (33). In the case $f(q) = q(1-q)$ and $m = 2$, the actual maximizer is $\hat{g} = (1-q)/q$, so from Eq. (34) we get

$$\left.\frac{\mathrm{d}(c^*)^2}{\mathrm{d}m}\right|_{m=2} = 6\int_0^1 (1-q)^2 q(1 + 2\log q)\mathrm{d}q = -\frac{7}{12}, \tag{35}$$

the value previously obtained by other methods [26,27] (see also Ref. [25] for an alternative derivation using instead the first variational characterization (Theorem 2.1) and the Feynman–Hellmann theorem).

Recently, Malaguti and Marcelli [28] have studied the effects of a degenerate diffusion term in reaction–diffusion models. They study the equation,

$$u_t = (D(u)u_x)_x + f(u), \qquad \text{in } \mathbb{R}^+ \times \mathbb{R}, \tag{36}$$

where both f and D are in $C^1[0,1]$. Their main result is the following theorem.

Theorem 2.3 [28] *Consider Eq. (36) with D and f in $C^1([0,1])$, and the reaction term, f, is of type A, while D satisfies, $D(0) = 0, D'(0) > 0$ and $D(u) > 0$ for all $u \in [0,1]$. Then, there exists a constant c^* such that Eq. (36) has*

(a) no traveling wave solutions for $0 < c < c^*$;

(b) a monotone traveling wave solution of sharp type with wave speed c^*.

(c) a monotone traveling wave, q, of front type, with $q(-\infty) = 1$ and $q(+\infty) = 0$, for every wave speed $c > c^*$.

Moreover, it holds that

$$0 < c^* \leq \sqrt{\sup_{q\in(0,1]} \frac{D(q)f(q)}{q}}. \tag{37}$$

Finally, for all $c \geq c^*$ the wave front, respectively of front or sharp-type, is unique up to translations.

As before, we can characterize the value of c^* of Theorem 2.3 by the following variational principle.

Theorem 2.4 *Let D and f as in Theorem 2.3, then,*

$$c^* = 2\sup_g \left(\frac{\int_0^1 \sqrt{fgh}\mathrm{d}q}{\int_0^1 g\mathrm{d}q} \right), \tag{38}$$

where $h \equiv -D(u)g'(u)$, *and the sup is taken on the set of functions* g, *such that* $g \in C^1([0,1])$, $g > 0$ *in* $(0,1)$, $g(1) = 0$, *and* $h > 0$ *in* $(0,1)$.

Remarks 2.5 (i) the proof of this theorem is analogous to the proof of Theorems 1 and 2.1 above and we omit it here. (ii) Using the convexity of the mapping, $t \to \sqrt{t}$, Jensen's inequality, and integration by parts we can obtain the bound (Eq. (37)) directly from our variational principle (Eq. (38)).

3. Variational calculations for thermal combustion waves

In the simplest model of thermal propagation of flames one is led to the 1D reaction–diffusion Eq. (1), where typically the reaction term is given by the Arrhenius law, which in a reduced version is given by

$$f(u) = (1-u)^r[\mathrm{e}^{\beta(u-1)} - \mathrm{e}^{-\beta}], \tag{39}$$

(see, e.g. Refs. [5,23]) where the term $\exp(-\beta)$ is introduced to remedy the cold-boundary problem [4,29]. The degree of localization of the reaction zone is measured by the Zeldovich number β. For large values of β, the width of the reaction zone is narrow and the speed of propagation of the flame is given by the Zeldovich–Frank–Kamenetskii (ZFK) formula [5],

$$c_{\mathrm{ZFK}} = \left(2\int_0^1 f(u)\mathrm{d}u\right)^{1/2}, \tag{40}$$

which is the exact value in the limit $\beta \to \infty$. On the other hand, for $\beta < 2$ the reaction term is concave and the speed of the flame is given by the KPP value

$$c_{\mathrm{KPP}} = 2\sqrt{f'(0)}. \tag{41}$$

Realistic values of β lie between these two extremes. Corrections to the ZFK formula have been obtained by means of asymptotic expansions in the parameter $1/\beta$ [29]. Often the prescription has been [4] to take the larger value between c_{ZFK} (or its corrections) and c_{KPP} as the best approximation to the correct value of the speed. We have used the variational principle given by Theorem 1 above to compute several simple analytic formulas that reproduce the numerical value of the speed of the flame for a wide range of the parameter β [30]. We have also used the variational principle (1) to prove that c_{ZFK} is always a lower bound for any reaction term of type A [19,30] (a different proof of this fact has been given directly from the ordinary differential equation by Berestycki and Nirenberg [31]). For completeness we reproduce here the proof of this lower bound.

Theorem 3.1 ([19,30,31]). *For any reaction term $f(u)$ of type A, the selected speed of propagation of sufficiently localized initial conditions satisfies,*

$$c^* \geq c_{\mathrm{ZFK}} \equiv \left(2\int_0^1 f(u)\mathrm{d}u\right)^{1/2}, \tag{42}$$

Proof: Choose as a trial function,

$$g(q) = \sqrt{2\int_q^1 f(s)\mathrm{d}s}, \tag{43}$$

in Eq. (5). Clearly this function is in $\mathcal{E}$.

With this choice we obtain

$$\int_0^1 \sqrt{fgh}\mathrm{d}u = \int_0^1 f(u)\mathrm{d}u, \tag{44}$$

and we have that

$$\int_0^1 g(u)\mathrm{d}u = -\int_0^1 g'(u)u\mathrm{d}u \leq -\int_0^1 g'(u)\mathrm{d}u = g(0), \tag{45}$$

since $g(1) = 0$. Replacing Eqs. (44) and (45) in Eq. (5) we obtain,

$$c \geq 2\frac{\int_0^1 f(u)\mathrm{d}u}{g(0)} = \sqrt{2\int_0^1 f(u)\mathrm{d}u} = c_{\mathrm{ZFK}}, \tag{46}$$

i.e. the minimal speed is always greater than or equal to the ZFK value. □

Using the variational principle one can get an approximation to the minimal speed of the flame, which is a much better approximation than the c_{ZFK} value for intermediate values of the Zeldovich parameter.

Defining, $r(x) = \int_x^1 f(u)\mathrm{d}u$, taking $g = r^n$ as a trial function in Eq. (5), and optimizing in n, we are led to the following lower bound on the minimal speed [30]

$$c \geq \frac{4\sqrt{n}}{2n+1}\frac{\left(\int_0^1 f(u)\mathrm{d}u\right)^{n+1/2}}{\left(\int_0^1 uf(u)\mathrm{d}u\right)^2}, \tag{47}$$

for any $n \in (1/2, 1)$ and for any reaction term of type A. The n that maximizes this lower bound is either

$$n = \frac{1 - \log(\gamma) - \sqrt{[\log(\gamma) - 1]^2 - 4\log(\gamma)}}{4\log(\gamma)}, \tag{48}$$

when the right side of Eq. (48) is real, where

$$\gamma = \frac{\int_0^1 f(u)\mathrm{d}u}{\int_0^1 uf(u)\mathrm{d}u},$$

or $n = 1$ otherwise [30].

4. Minimal speed of fronts for reaction–convection–diffusion equations

In many processes, in addition to diffusion, motion can also be due to advection or convection. Nonlinear advection terms arise naturally in the motion of chemotactic cells. In a simple 1D model, denoted by ρ the density of bacteria, chemotactic to a single chemical element of concentration $s(x,t)$ the density evolves according to

$$\rho_t = [D\rho_x - \rho\xi s_x]_x + f(\rho), \tag{49}$$

where diffusion, chemotaxis and growth have been considered. There is some evidence [32] that, in certain cases, the rate of chemical consumption is due mainly to the ability of the bacteria to consume it. In that case,

$$s_t = -k\rho,$$

where diffusion of the chemical has been neglected (arguments to justify this approximation, together with the choice of constants D and ξ are given in Ref. [32]). If one looks for traveling-wave solutions, $s = s(x - ct)$, and $\rho = \rho(x - ct)$, then $s_t = -cs_x$, and thus, $s_x = k\rho/c$, and the problem reduces to a single differential equation for the density ρ, namely

$$\rho_t = D\rho_{xx} - \xi\tfrac{k}{c}(\rho^2)_x + f(\rho). \tag{50}$$

For a discussion about the recent literature on the subject, see, e.g. Ref. [20] and references therein.

Motivated by the previous model, in this section we will consider the equation with a general convective term that, suitably scaled, we write as

$$u_t + \mu\phi(u)u_x = u_{xx} + f(u), \tag{51}$$

where the reaction term is of type A. The function $\phi(u) \in C^1([0,1])$. Without loss of generality we may assume $\phi(0) = 0$, since otherwise only a uniform shift in the speed is introduced. The parameter μ is positive. For Eq. (51), the existence of monotonic decaying traveling fronts, $u(x - ct)$ for any wave speed greater than a critical value c^* was proven recently by Malaguti and Marcelli [33]. In Ref. [33], the following estimate on c^* was derived,

$$2\sqrt{f'(0)} \leq c^* \leq \sqrt{\sup_{u\in[0,1]} \frac{f(u)}{u}} + \max_{u\in[0,1]} \mu\phi(u). \tag{52}$$

Analogous results for density-dependent diffusion have also been established in Ref. [33]. The convergence of some initial conditions to a monotonic traveling front has been proven by Crooks [34] for systems for which the minimal speed is greater than the linear value $c_{\mathrm{KPP}} = 2\sqrt{f'(0)}$.

We have recently derived a variational characterization for the minimal speed c^* of monotonic traveling front solutions of Eq. (51) [20]. Let

$$I(g) = \frac{\int_0^1 \left(2\sqrt{f(u)g(u)[-g'(u)]} + \mu\phi(u)g(u) \right) \mathrm{d}u}{\int_0^1 g(u)\mathrm{d}u} \tag{53}$$

defined over the set S of positive, monotonic decreasing functions $g(u)$, in $C^1([0,1])$, with $g(1) = 0$. Then we have

Theorem 4.1 ([20])

$$c^* = \sup_{g\in S} I(g). \tag{54}$$

Remarks 4.2 (i) Using the variational principle (54), Jensen's inequality and integration by parts one can derive the upper bound Eq. (52) [20]. (ii) From the variational principle Eq. (54) it is also possible to show that a sufficient condition for c^* to be equal to the linear value c_{KPP} is

$$\frac{f''(u)}{\sqrt{f'(0)}} + \mu\phi'(u) < 0,$$

for all $u \in (0,1]$ [20].

Acknowledgements

This work has been partially supported by Fondecyt (Chile), projects # 102-0844, and 102-0851. One of us (RB) would like to thank The Fields Institute (Toronto) for their hospitality while part of this manuscript was written.

References

[1] Fisher, R.A. (1937), The advance of advantageous genes, Ann. Eugenics 7, 355–369.

[2] Kolmogorov, A.N., Petrovsky, I.G. and Piskunov, N.S. (1937), In Etude de l'equation de la diffusion avec croissance de la quantité de matiére et son application a un probleme biologique, Vol. 1, Moscow University of Mathematical Bulletin, Moscow, pp. 1–25.

[3] Murray, J.D. (1993), Mathematical Biology. In Lecture Notes in Biomathematics 19, 2nd edn, Springer, New York.
[4] Zeldovich, Y.B., Aldushin, A.P. and Kudayev, S.I. (1992), Numerical study of flame propagation in a mixture reacting at the initial temperature (Zeldovich, Ya B., Ostriker, J.P., eds.), Princeton University Press, Princeton, pp. 320–329.
[5] Zeldovich, Y.B. and Frank-Kamenetskii, D.A. (1938), A theory of thermal propagation of flames, Acta Physicochim. USSR 9, 341–350. [English translation in Dynamic of Curved Fronts, R. Pelcé (ed.), Perspectives in Physics Series, Academic Press, New York, 1988, pp. 131–140].
[6] van Saarloos, W. (2003), Front Propagation into Unstable States, Phys. Rep. 386, 29–222.
[7] Benguria, R.D. and Depassier, M.C. (1998), A variational method for nonlinear eigenvalue problems, Contemp. Math. 217, 1–17.
[8] Collet, P. and Eckmann, J.P. (1990), Instabilities and Fronts in Extended Systems, Princeton University Press, Princeton, NJ, Chapter VII.
[9] Xin, J. (2000), Front propagation in heterogeneous media, SIAM Rev. 42, 161–230.
[10] Collet, P. and Eckmann, J.P. (2002), A rigorous upper bound on the propagation speed for the Swift–Hohenberg and related equations, J. Stat. Phys. 108, 1107–1124.
[11] Gravejat, P. (2003), A non-existence result for supersonic traveling waves in the Gross–Pitaevskii equation, Commun. Math. Phys. 243, 93–103.
[12] Ebert, U., Spruijt, W. and van Saarloos, W. (2004), Pattern forming pulled fronts, Physica D 199, 13–32.
[13] Aronson, D.G. and Weinberger, H.F. (1978), Multidimensional nonlinear diffusion arising in population genetics, Adv. Math. 30, 33–76.
[14] Kolmogorov, A.N., Petrovsky, I.G. and Piskunov, N.S. (1991), A Study of the Diffusion Equation with Increase in the Amount of Substance, and its Applications to a Biological Problem (Kolmogorov, A.N., Tikhomirov, V.M., eds.), Kluwer, Dordrecht.
[15] Hadeler, K.P. and Rothe, F. (1975), Traveling fronts in nonlinear diffusion equations, J. Math. Biol. 2, 251–263.
[16] Volpert, A.I., Volpert, V.A. and Volpert, V.A. (1994), Traveling Wave Solutions of Parabolic Systems. In Translations of Mathematical Monographs, Vol. 140, American Mathematical Society, Providence, RI.
[17] Benguria, R.D. and Depassier, M.C. (1996), Variational characterization of the speed of propagation of fronts for the nonlinear diffusion equation, Commun. Math. Phys. 175, 221–227.
[18] Benguria, R.D. and Depassier, M.C. (1996), Speed of fronts of the reaction–diffusion equation, Phys. Rev. Lett. 77, 1171–1173.
[19] Benguria, R.D. and Depassier, M.C. (1994), Validity of the linear speed selection mechanism for fronts of the nonlinear diffusion equation, Phys. Rev. Lett. 73, 2272–2275.
[20] Benguria, R.D., Depassier, M.C. and Méndez, V. (2004), Minimal speed of fronts of reaction–convection–diffusion equations, Phys. Rev. E 69, 031106.
[21] Aronson, D.G. (1980), Density-dependent interaction–diffusion systems. Dynamics and Modelling of Reactive Systems (Stewart, W., Harmon Ray, W., Conley, C.C., eds.), Academic Press, New York, pp. 161–176.
[22] Newman, W.I. (1980), Some exact solutions to a nonlinear diffusion problem in population genetics and combustion, J. Theor. Biol. 85, 325–334.
[23] Clavin, P. (1994), Premixed combustion and gasdynamics, Ann. Rev. Fluid Mech. 26, 321–352.
[24] Scott, S.K. and Showalter, H. (1992), Simple and complex propagating reaction–diffusion fronts, J. Phys. Chem. 96, 8702–8711.
[25] Benguria, R.D. and Depassier, M.C. (1995), Variational principle for the asymptotic speed of fronts of the density-dependent diffusion–reaction equation, Phys. Rev. E 52, 3285–3287.
[26] Aronson, D.G. and Vázquez, J.L. (1994), Calculation of anomalous exponents in nonlinear diffusion, Phys. Rev. Lett. 72, 348–351.
[27] Chen, L.Y. (1994), Ph.D. Thesis, University of Illinois at Urbana-Champaign (unpublished).
[28] Malaguti, L. and Marcelli, C. (2003), Sharp profiles in degenerate and doubly degenerate Fisher–KPP equations, J. Diff. Equations 195, 471–496.

[29] Clavin, P. and Liñán, A. (1984), Nonequilibrium Cooperative Phenomena in Physics and Related Fields. In NATO Advanced Study Institute Series B: Physics, (Velarde, M., eds.), Vol. 116, Plenum, New York, p. 291.

[30] Benguria, R.D., Cisternas, J. and Depassier, M.C. (1995), Variational calculations for thermal combustion waves, Phys. Rev. E 52, 4410–4413.

[31] Berestycki, H. and Nirenberg, L. (1992), Traveling fronts in cylinders, Ann. Inst. H. Poincaré Anal. Non Linéaire 9, 497–572.

[32] Segel, L.A. (1972), On collective motions of chemotactic cells. In Some Mathematical Questions in Biology. III. Lecture Notes on Mathematics in the Life Sciences, Vol. 4, American Mathematical Society, Providence, RI, pp. 1–47.

[33] Malaguti, L. and Marcelli, C. (2002), Traveling wavefronts in reaction–diffusion equations with convection effects and non-regular terms, Math. Nachr. 242, 148–164.

[34] Crooks, E.C.M. (2003), Traveling fronts for monostable reaction–diffusion systems with gradient-dependence, Adv. Diff. Equations 8, 279–314.

Chapter 17

THE FERMAT PRINCIPLE AND CHEMICAL WAVES

Henrik Farkas and Kristóf Kály-Kullai

Department of Chemical Physics, Institute of Physics, Budapest University of Technology and Economics, Budafoki út. 8, 1521 Budapest, Hungary

Stanislaw Sieniutycz

Warsaw University of Technology, Faculty of Chemical and Process Engineering, Warsaw, Poland

Abstract

The most common formulation of the Fermat principle states: the light propagates in such a way that the propagation time is minimal. This chapter outlines the historical background as well as the different formulations of the Fermat principle.

The Fermat principle is valid for any traveling wave, not only for light. In optics, the underlying equation is the wave equation, which describes all details of propagation. Geometrical optics is a limiting case when wavelength tends to zero. The opposite limiting case—when wavelength tends to infinity—is the field of geometrical wave theory. That theory is based on the Fermat principle as well as on the dual concepts of rays and fronts.

Chemical waves can be described in detail by reaction–diffusion equations. These equations may have traveling-wave solutions. The essential features of the evolution of chemical wave fronts can also be derived from the geometric theory of waves.

Some recent formulations of the Fermat principle require only stationarity of the extremals instead of the stricter requirement of minimal propagation time. This problem is discussed, and it is concluded that for chemical waves only the requirement of minimum is relevant: maximum and 'inflexion-type' local stationarity does not play any role. The chemical waves and their prairie-fire picture underlines: only the quickest rays have effects. Nevertheless, there are singularities, when there are several paths of rays with the same propagation time. For these cases the related fields of singularities and caustics are relevant.

Special attention is paid to aplanatic surfaces, where all the reflected or refracted paths require the same time, that is stationarity holds globally. Nonaplanatic refraction is discussed, too: the special example of refracting sphere is treated analytically.

Finally, the shape of a chemical lens is derived: this 'chemical lens' is able to perfect image formation with chemical waves, that is circular fronts will be also circular after refraction.

Keywords: Fermat principle; aplanatic surface; chemical wave; extrema; Wave fronts; refraction; reflection

---

E-mail address: farkashe@goliat.eik.bme.hu (H. Farkas); kk222@ural2.hszk.bme.hu (K. Kály-Kullai); sieniutycz@ichip.pw.edu.pl (S. Sieniutycz).

1. Introduction

1.1. What is a wave?

In scientific literature there are different meanings associated with the term wave. Here is a list of different properties of *waves*. It is usual to select one of them, considering that as a property defining the wave:

(a) Process that is periodic in space and time.
(b) Propagating oscillations.
(c) Propagation of a disturbance/signal.
(d) Energy transfer, energy propagation.
(e) Solution of wave equation.

In some sources, especially made for educational purposes, several properties are mentioned together to define wave process.

Remarks

(i) Periodicity is a too strict requirement, it often occurs that the wave process is nearly periodic, e.g. in the case of damping.
(ii) The term propagation is associated with traveling waves. Standing waves can be considered as waves traveling with zero speed.
(iii) The presence of a medium is often mentioned. The electromagnetic wave, which is the most important kind of waves can propagate in vacuum.
(iv) In principle, energy transfer does not necessarily accompany wave propagation. For chemical and biological waves, energy transfer may occur, but it is not essential. In the case of propagating signals, the propagation of information is essential, and propagation of energy is marginal.
(v) In some formulations a wave is considered to be a repeating process, while others distinguish between a solitary wave and wave trains consisting of successive pulses.

1.2. Terminology: wave, eikonal, front, ray

A spatiotemporal process described by a function of the form

$$u(t, \mathbf{r}) = A(\mathbf{r})f(t - S(\mathbf{r})) \tag{1}$$

seems suitable to describe the vast majority of waves occurring in the scientific literature. In Eq. (1) A is the amplitude, f is the phase, and S is the eikonal. For the special case of harmonic waves the phase f is a sinusoidal function of its argument.

The form (1) describes an undistorted propagation of a signal. If A is not constant, then the volume of the signal changes during propagation. The eikonal S measures the time needed for the propagation. (Recall that the eikonal generally used in the literature only differs from the above one by a constant factor of velocity dimension.)

For a wave, the time dependence will be essentially the same at every point $\mathbf{r}$, apart from the attenuation (represented by A) and the time delay (represented by S).

Wave fronts are defined primarily as surfaces where the phase of the wave is constant. These surfaces are not necessarily identical with the surfaces where the value of the wave function u is constant (inhomogeneous waves, see: Ref. [1]). Due to Eq. (1), wave fronts are the equieikonal level surfaces:

$$S(\mathbf{r}) = \text{constant}. \tag{2}$$

During propagation, the wave fronts move: this motion means a continuous transformation of the equieikonal surfaces into each other. Wave propagation described by Eq. (1) allows us to deal with a stationary family of fronts and its transformations into itself.

For chemical waves, the fronts are the equiconcentration surfaces, these can be visualized in the experiments [2]. In the problems investigating chemical waves, the amplitude A can generally be considered as a constant. In the following we will omit A as being constant, so the wave function is

$$u(t, \mathbf{r}) = f(t - S(\mathbf{r})). \tag{3}$$

A specified wave front that belongs to the eikonal value S_1 at time t_1 will be transformed at t_2 into the surface $S_2 = S_1 + t_2 - t_1$. This is a consequence of the requirement that the value of f must be the same for $t_1 - S_1$ as for $t_2 - S_2$ that is

$$t_1 - S_1 = t_2 - S_2. \tag{4}$$

In this way we assigned the front S_2 at t_2 to the front S_1 at t_1, so the fronts can be described as continuously moving. To derive the wave velocity v we assume the propagation to be perpendicular to the given wave front, that is moving in the direction of grad S. Restricting ourselves to infinitesimal transformations

$$t_2 - t_1 = \mathrm{d}t$$

$$S_2 - S_1 = \mathrm{d}S = (\text{grad } S)\cdot \mathrm{d}\mathbf{r} = \mathrm{d}t.$$

In accordance with perpendicular propagation, $\mathrm{d}\mathbf{r}$ is parallel to grad S, and moreover, $|\mathrm{d}\mathbf{r}| = v\,\mathrm{d}t$. In this way the eikonal equation

$$1/v = |\text{grad } S| \tag{5}$$

is obtained.

The other basic concept in wave propagation is the ray. Rays can be defined as orthogonal trajectories of wave fronts. The wave propagates along the rays with velocity v.

The most important special case for traveling waves is the planar harmonic wave, when

$$f = A \cos(\omega(t - S(\mathbf{r}))),$$

$$S = (\mathbf{k}/\omega)\cdot\mathbf{r}. \tag{6}$$

Here, ω is the circular frequency, $\mathbf{k}$ is the circular wave number vector. Now grad $S = (\boldsymbol{k}/\omega)$, and from the eikonal equation

$$1/v = |\mathbf{k}/\omega|.$$

A more general type of waves is the periodic wave, here f is a periodic function with some period T. For planar periodic waves the period in space is the wavelength λ, and the eikonal is $S(\mathbf{r}) = \mathbf{b}\cdot\mathbf{r}$, where $|\mathbf{b}| = T/\lambda$, and the direction of $\mathbf{b}$ is the direction of propagation (direction of grad S).

In contrast to periodic waves, the solitary wave consists of a single pulse or 'front'. For a fixed spatial point, the wave produces a signal that is a function of time. For 1D waves, we can introduce the wave form that is a function of the space coordinate x. The wave form and the signal form correspond to one another. If S is a linear function of x, that is the velocity is constant, then the (temporal) signal form is identical to the (spatial) shape of the wave form apart from a linear scale transformation.

1.3. Geometric theory of waves

The geometric theory of waves is based on the well-known concepts (velocity, front, ray) and basic principles (Fermat's principle, Huygens' principle). Given an initial front and the propagation velocity as a function of space, geometrical wave theory is a useful tool to follow the evolution of fronts in time, as well as to describe the qualitative features of rays and fronts, and even to describe this evolution quantitatively.

Geometric wave theory can be illustrated by the pictorial prairie-fire model. Any visualized point of the burning front propagates the fire in every direction with the given velocity. Although the basic ideas of geometric wave theory are very old, this approach was first used in 1948 to describe a kind of biological wave, namely waves of excitations in the muscles of heart, and in the nervous system [10].

The relation between geometrical and wave optics is well known. Now we outline pictorially the relation between geometrical optics, wave optics, and geometrical wave theory (Fig. 1).

Geometrical optics can be considered as a limiting case of the wave optics based on partial differential equations of electrodynamics. Wave optics reduces to geometrical optics in the limiting case when the wavelength λ tends to zero. Geometrical wave theory can be considered just the opposite limiting case, when λ tends to infinity.

In wave optics there is a diffraction region between the directly illuminated region and the shadow region. In geometrical optics, the shadow is separated sharply from the directly illuminated region, and there is no diffraction at all.

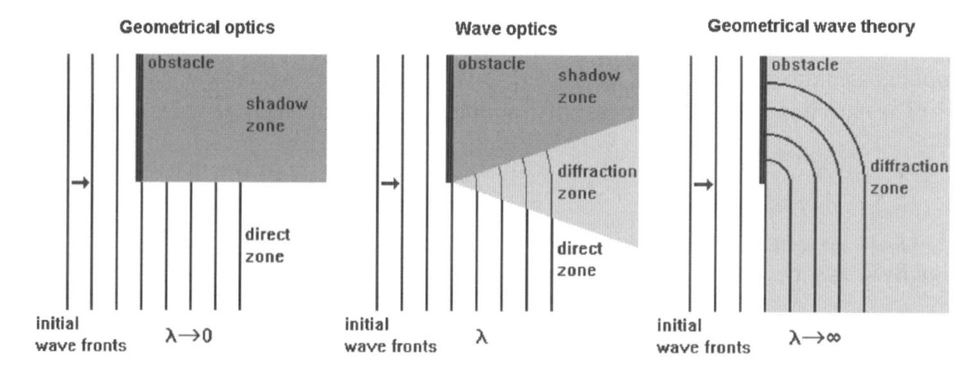

Fig. 1. Relation between geometrical optics, wave optics and geometrical wave theory.

In contrast with geometrical and wave optics, geometrical wave theory allows the perfect diffraction behind the obstacles, therefore there is no shadow at all. Moreover, amplitude has no importance in geometric wave theory. As a consequence, any point of the wave front can generate a finite wave front part in an arbitrarily short time. This feature is assumed to be valid within the geometrical wave theory, however, from experience it is only valid above some critical threshold: a burning point cannot start a fire if it is too small. The kinematic theory of waves involving curvature dependence of the velocity is suitable to involve such an aspect [11,12].

In the diffraction region the wave fronts are generated by rays starting from the endpoints of a certain earlier front.

Obviously, the geometric-optical approach and terminology can be applied not only to light, but to any other kinds of waves or in other words, to any other kind of propagation of signals. We can call this theory 'ray dynamics'. The primary concept of this ray dynamics is the ray, which has an initial *position together with an initial velocity*. Ray dynamics determines the future of the rays if the initial position and the initial velocity are given. Fronts can be defined as orthogonal trajectories of the family of rays in question belonging to the same time.

In contrast, geometrical wave theory can be described as 'front dynamics'. Starting from an initial front, and knowing the velocity of propagation, one can determine the future of the front uniquely: front dynamics constitutes a dynamical system [3]. Rays are defined as orthogonal trajectories of the subsequent fronts developed from a given initial front [3].

Ray dynamics and front dynamics differ from each other only at the edges or borders. In ray dynamics, the endpoint of a front is a point of a ray, which is in an extreme position in a given family of rays, therefore the points of this border ray will generate endpoints of the subsequent fronts. In contrast, for front dynamics, the endpoint of a front emanates rays in any allowed direction (not only in one direction, as in ray dynamics), causing total diffraction around the obstacles.

It must be stressed that fronts and rays are orthogonal to each other only in an isotropic medium. Unless stated otherwise we assume that the medium is isotropic.

2. The Fermat Principle of least time

2.1. Historical background

Teleology goes back to Greek philosophy. Behind the natural phenomena many Greek philosophers looked for certain purposes, they believed in the purposefulness and effectiveness of Nature. As a principle it was known that Nature always acts by the shortest ways. Already in the first century AD Heron of Alexandria asserted that the mirror reflected light ray reaches the eye along the shortest path, as a manifestation of the purpose represented by the eye [13,14].

Snell formulated the sine law of refraction in 1621, and it was published first by Descartes in 1637. Pierre de Fermat successfully deduced this law in 1657 and 1662, respectively [14], from the principle of least time, which is known nowadays as the Fermat Principle of Least Time (FPLT) or in short the Fermat Principle:

> "A ray of light traveling from a point A to another point B will take the quickest path from A to B."

The FPLT for reflection was the same as Heron's minimal principle. Concerning refraction, Fermat was able to correctly establish the value of refractive index as the ratio of the velocities of light in the two adjacent media. This result was experimentally verified only a hundred years later.

The Fermat principle was sharply criticized by Descartes, and later in May 1662 by Clerselier, an expert in optics and leading spokesman for the Cartesians on this matter:

> "1. The principle you take as a basis for your proof, to wit, that nature always acts by the shortest and simplest path, is only a moral principle, not a physical one—it is not and cannot be the cause of any effect in nature." [15]

Apart from the ideological color, the Fermat Principle had a really large impact not only in optics, but in several other branches of physics through the works of Maupertuis, Hamilton and Feynman. Nowadays, it is frequently applied in several fields, e.g. in geodesics, gravitational lensing and general relativity.

2.2. *Fermat Principle–present-day formulations*

In modern physics textbooks one can find the original form of the FPLT. However, this original formulation can be read mainly in texts of elementary level or of general character.

However, the more sophisticated modern formulation differs from the original one. First, in the modern formulations the propagation time is replaced by the optical path, which differs from that only by a constant factor.

$$L = \int n \, \mathrm{d}s = t/c,$$

where c is the propagation velocity in vacuum, $n = c/v$, the refraction index, and v is the actual propagation velocity in the medium, and t is the propagation time.

A more important feature of the modern formulations is that instead of a minimum only stationarity is required. Often the difference between the two formulations is stressed [16–19]. As an illustrative example, see Ref. [20]:

> "Fermat's principle, as stated by him and others following him, says that the optical path, the path which light actually takes, between P_1 and P_2 is that curve of all those joining P_1 and P_2 which makes the value of the integral least. This formulation is physically incorrect, as can be shown by examples, and the correct statement is that the first variation of this integral, in the sense of the calculus of variations, must be zero."

Feynman [17,18] also emphasizes this point, he considers the stationarity requirement as more precise in contrast to the least time as far as the Fermat principle is concerned. Moreover, he gives an explanation for the selection method of rays of light. This explanation is based on the fact that the wave is not local, but is extended on a region of the size of the wavelength. Applying this explanation to the geometrical wave theory, it seems plausible that a wave with infinite wavelength is able to select the optimal path not only in a small neighborhood, but it can find the global optimum.

For chemical waves the wave fronts have a direct meaning: they are the equiconcentration surfaces and are directly observed in experiments. Propagation of chemical waves, that is the evolution of wave fronts, can be systematically and quantitatively described by the FPLT: this principle plays a crucial role in the geometric theory of waves [2–9]. For chemical waves the original formulation of the Fermat Principle is the proper one, while the modern one is not appropriate. This is obvious from the prairie-fire model: if the fire reaches a point P along a certain path, all the other paths to which the propagation time is greater have no significance at all. This feature of chemical waves was supported by experiments: in homogeneous medium the chemical wave finds the shortest escape route from a labyrinth [21].

2.3. *Minimum or stationarity?*

Let v be the speed of propagation in an inhomogeneous isotropic medium, and let it be a piecewise-continuous function of the position vector $\mathbf{r}$: $v(\mathbf{r})$. Let P_1 and P_2 be two given points. Then the propagation time belonging to a path g connecting P_1 and P_2 is

$$\tau = \int \frac{\mathrm{d}s}{v(\mathbf{r})}.$$

This integral is defined for any admissible (rectifiable) curve g connecting P_1 and P_2 [3]. The set of such admissible curves with endpoints P_1, P_2 will be denoted by $G(P_1, P_2)$.

Definition 1: *The curve $\gamma \in G(P_1, P_2)$ satisfying the minimum condition*

$$\tau(P_1, P_2; \gamma) = \text{minimum}\{\tau(P_1, P_2; g), g \in G(P_1, P_2)\},$$

will be called the Fermat ray connecting the points P_1 and P_2.

Typically, one unique Fermat ray belongs to any pair of points. However, in some exceptional cases, there may be more than one Fermat ray to a given pair of points yielding the same propagation time.

Definition 2: *A rectifiable curve $\gamma(s)$, (where s is the length), is called a locally Fermat ray if there exists a positive $\varepsilon > 0$ so that its every arc $(\gamma(s_1), \gamma(s_2))$ is a Fermat curve provided that $0 < s_2 - s_1 < \varepsilon$.*

In other words, any small enough part of a locally Fermat ray is a Fermat ray.

Obviously, any Fermat ray is a locally Fermat ray, but the reverse is not true. Apart from the simplest situations, to any given endpoints there exist several locally

Fermat rays: see 'elementary family of rays' in Ref. [9]. The Fermat ray connecting two given points can be found among the locally Fermat rays by comparing their propagation times.

Definition 3: *The curves satisfying the stationarity condition* $\delta\tau = 0$ *will be called extremals.*

This terminology is in accordance with the one in variational calculus.

A sufficient additional condition for an extremal to be a locally Fermat curve is that the second variation be positive-definite: $\delta^2\tau > 0$.

The above concepts of extremals and Fermat rays have analogons for simple real functions: Fermat rays correspond to minima, and extremals correspond to stationary points of a function where the derivative vanishes. For an inner point, the minimum condition implies a zero value of the derivative. However, at the endpoints of the interval this is not true. Similarly, for waves in a heterogeneous medium, the convex curved interface separating a slower inner region and a faster outer region may be a Fermat ray, but surely not an extremal. To find the Fermat rays, we should take both the extremals and such border lines into consideration.

To illustrate the relation between extremals and Fermat rays we recall some special examples:

(1) *Circular obstacle.* This problem was studied experimentally [2], by using the tools of geometrical-wave theory [3], as well as by the use of Hamilton–Jacobi–Bellmann theory [4]. The border of the obstacle is clearly a Fermat ray, and also a locally Fermat ray, but not an extremal.

(2) *Refraction.* In a two-phase heterogeneous medium the velocity of propagation has a jump at the interface of the phases. The locally Fermat rays refract when they cross the interface according to Snell's law. Special attention should be paid to the rays that are tangential to the interface in the faster phase. The locally Fermat rays should satisfy junction rules when they cross the interface [6,7,9]. Note that these arcs on the interface represent locally Fermat rays with the higher velocity, but they are not extremals, since the propagation time changes in first order if the perturbed path is entirely inside the slower phase or in the faster phase if the interface is curved.

In a heterogeneous circular medium with a slower inner and a faster outer region surrounding concentrically a circular obstacle the parts of the interfacial circle (with the higher propagation velocity) are obviously locally Fermat rays, but not necessarily Fermat rays at the same time, and also they are not extremals at all [6,7,9].

(3) *Refraction into a circular region.* Consider the refraction into a sphere. Fig. 2a shows a 2D arrangement. The velocity of the wave is v_1 outside, and $v_2 < v_1$ is inside. Let O denote the center of the circle, and let A be a point outside, and B be a point inside on the other side of the straight line AO. Let us denote by a and b the distances of AO and OB, respectively. We are looking for the extremals and Fermat rays connecting A and B. This curve must consist of straight segments outside and inside, the extremum problem is reduced to the investigation of the propagation time as the function of the point P on the border circle where those straight lines join, that is we look for the function $t(\Phi)$, where

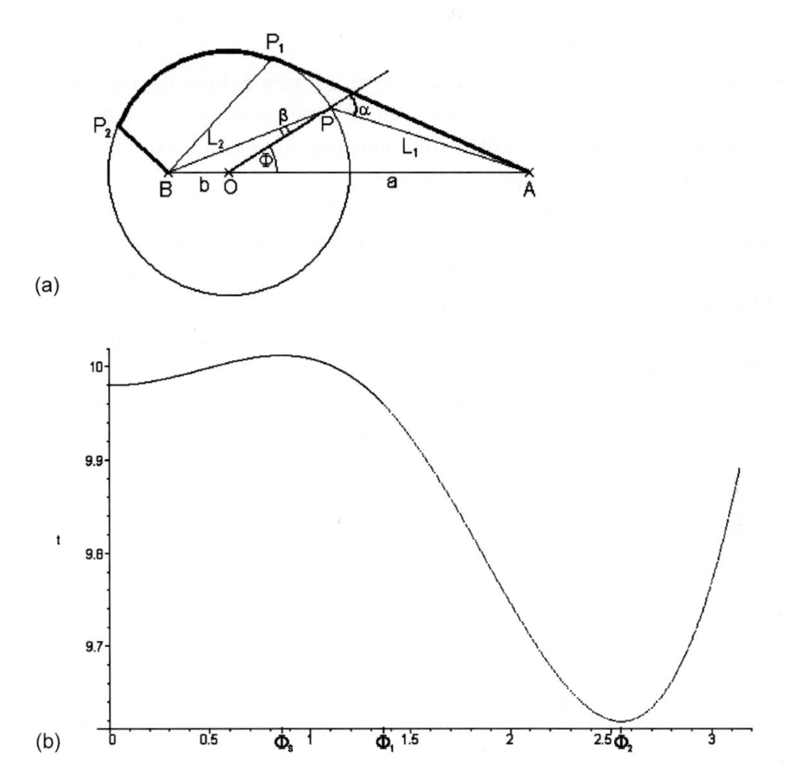

Fig. 2. (a) Circular refracting curve. (b) Angle–propagation time relationship (parameter values: $a = 5, b = 0.3, v_1 = 4.6, v_2 = 1$).

Φ is the angle AOP. Elementary calculation yields that

$$t(\Phi) = L_1/v_1 + L_2/v_2, \tag{7}$$

where

$$L_1 = \sqrt{1 + a^2 - 2a\cos\Phi}, \qquad L_2 = \sqrt{1 + b^2 + 2b\cos\Phi}$$

are the length of AP and PB, respectively. Let us introduce the refractive index $n = v_1/v_2 > 1$. Then

$$\frac{\mathrm{d}}{\mathrm{d}\Phi}(v_1 t(\Phi)) = v_1 \sin\Phi\left(\frac{a}{L_1} - n\frac{b}{L_2}\right) = v_1(\sin\alpha(\Phi) - n\sin\beta(\Phi)), \tag{8}$$

where α is the angle of incidence and β is the angle of refraction. It is obvious from Eq. (8) that the extremals should satisfy Snell's law. Fig. 2b shows the propagation time as a function of the angle Φ. For the numeric values of a, b, v_1, and v_2 specified in Fig. 2b one can see that the propagation time has a local minimum at the straight ray AOB, has a maximum at a certain value Φ_S, where Snell's law is also valid, and t has the global minimum at the extreme ray OP_1 incident tangentially to the circle belonging to the angle Φ_1. Therefore, the Fermat ray for this situation belongs to the angle Φ_1 of tangential incidence.

The Fermat ray is not an extremal now. There are two extremals, one belongs to the angle $\Phi = 0$, the other belongs to the value Φ_S. Both extremals are locally Fermat rays, since the circle is smooth, therefore the situation is the same as for a refracting plane if we restrict ourselves to a very small part of the rays.

The above mathematical formulas can be extended for values of Φ greater than Φ_1 from a mathematical point of view. This extension has an awkward physical meaning: the wave propagates with velocity v_1 in a tunnel in the denser medium until it reaches the border circle the second time, and from that it follows its path toward B with velocity v_2.

From the view of chemical waves, another kind of extension beyond the tangential incidence has a realistic meaning: the Fermat ray will consist of three parts, the medium part is along the border circle with velocity v_1. The propagation time function is now:

$$t^*(\Phi) = t(\Phi), \qquad \text{if } \Phi \le \Phi_1,$$

$$t^*(\Phi) = t(\Phi_1) + (\Phi - \Phi_1)/v_1 + L_2/v_2, \qquad \text{if } \Phi_1 < \Phi \le \pi. \tag{9}$$

Fig. 2a shows that the Fermat ray for this situation consists of three parts:

- straight line AP_1,
- circular part P_1P_2,
- straight line P_2B departing with the critical angle of total refraction β_{cr}, where $\sin \beta_{cr} = 1/n$.

The chemical wave starting from a point source at A will reach B along this Fermat ray, and the propagation time is

$$t(\Phi_1) + (\Phi_2 - \Phi_1)/v_1 + L_2(\Phi_2)/v_2, \tag{10}$$

where Φ_2 belongs to β_{cr}.

3. Aplanatic surfaces

Image formation is an important part of optics. It is known that spherical mirrors and spherical lenses have several errors of image formation, one of them is spherical aberration. To eliminate this aberration, the concept of aplanatic surfaces was introduced. Given an object at the point A and the image at the point B, a mirror with an aplanatic reflecting surface or a lens with an aplanatic refracting surface realizes a perfect image formation from A to B. In other words, the propagation time from A to B is the same for any ray consisting of straight segments AP and PB, whatever is the point P of the aplanatic surface.

Aplanatic surfaces are very important whenever relations between Fermat rays and extremals are concerned, as the following citation [19] shows under the item *Fermat Principle*:

> "The path of a ray in passing between two points during reflection or refraction is the path of least time (*principle of least time*). It is now more usually expressed as the principle of *stationary* time: that the path of the ray is the path of least *or* greatest time. If a reflecting or refracting surface has smaller

curvature than the aplanatic surface tangential to it at the point of incidence, the path is a minimum; if its curvature is greater than the aplanatic surface, its path is a maximum."

3.1. *Reflection*

Consider a reflecting curve a in the plane that maps the object point A into the image point B (Fig. 3). Let P be an arbitrary point of the mirror a, and we are looking for the curve a for which the propagation time belonging to the broken line APB is constant:

$$t = (1/v)(L_{\mathrm{AP}} + L_{\mathrm{PB}}) = \tau. \tag{11}$$

The curve defined by this requirement is an ellipse with the foci A and B. (In 3D, the required aplanatic surface is a rotational ellipsoid with the foci A, B.)

Now consider another curve g having a common contact at P with a. Then the ray AP reflects at P according to the law of reflection by the reflecting curve g just the same as in the case of the aplanatic elliptic mirror, and that ray goes to B from P on the same way as before, since reflection is governed by the tangent only. Comparing the propagation time belonging to APB with the time for broken lines belonging to other points P^* of g (close to P), one can see that the time belonging to the broken line AP^*B will be greater (smaller) than the one belonging to APB if the curvature of g at P is smaller (greater) than that of the aplanatic ellipse a. In this sense, the ray APB belongs to minimum (maximum) time compared with the neighboring rays AP^*B belonging to different points P^* of g close to P. The proof is very simple, we will show the idea for the case of refraction in Section 3.2.

This consideration was related to the real object–real image situation. Now consider the real object–virtual image case. Then the rays BP reflected by the aplanatic mirror a will never reach A, but the fronts generated by this family of reflected rays will be circles with center A. The mathematical requirement for the curve a to be aplanatic is

$$t_{Af} = (L_{\mathrm{BP}} + L_{Pf})/v = (L_{\mathrm{BP}} + R - L_{\mathrm{AP}})/v = \text{constant},$$

where R is the distance of the selected front f from A (Fig. 4). After rearranging we get the relation defining the real–virtual aplanatic curve

$$(L_{\mathrm{BP}} - L_{\mathrm{AP}})/v = \tau^*, \qquad \text{where } \tau^* = t_{Bf} - R/v\,(\text{constant}). \tag{12}$$

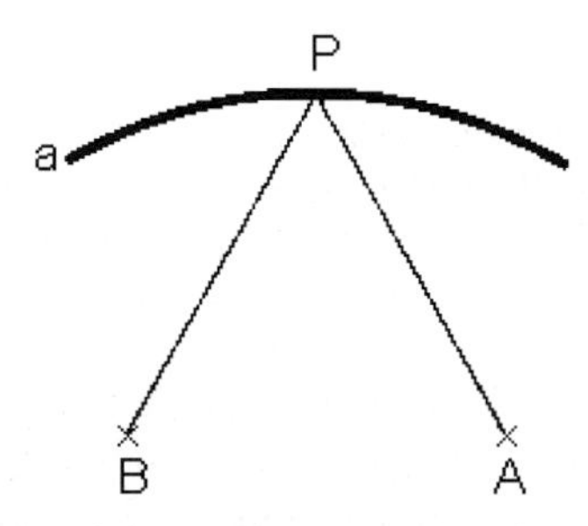

Fig. 3. Reflection by an aplanatic mirror (real–real case).

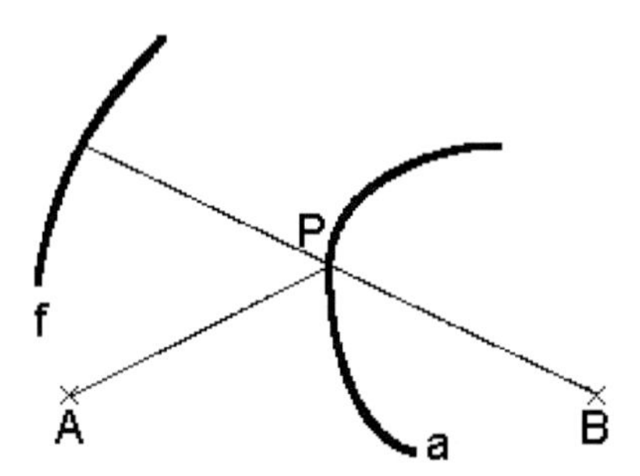

Fig. 4. Reflection by an aplanatic mirror surface (real–virtual case).

Both requirements (Eqs. (11) and (12)) can be written in the form

$$L_{\mathrm{BP}} \pm L_{\mathrm{AP}} = L, \tag{13}$$

where plus refers to real–real or virtual–virtual, and minus refers to real–virtual cases. After rearranging Eq. (13) and squaring, we get

$$L_{\mathrm{AP}}^2 = (L - L_{\mathrm{BP}})^2. \tag{14}$$

Let us express L_{AP} and L_{BP} in terms of the polar coordinates of P and the distance $d = \mathrm{AB}$

$$L_{\mathrm{BP}} = r, \qquad L_{\mathrm{AP}} = \sqrt{r^2 + d^2 - 2rd\cos\varphi}.$$

Substituting these into Eq. (14) we get the formula

$$\cos\varphi = E - \frac{E^2 - 1}{2R}, \tag{15}$$

where we introduced the dimensionless new parameter $E = L/d$ and the dimensionless polar coordinate $R = r/d$.

Let us denote by R_+ and R_- the radius belonging to $\varphi = 0$ and $\varphi = \pi$, respectively. From Eq. (15) we get

$$\begin{aligned} R_+ &= (E+1)/2, \qquad \text{if } E > -1 \text{ and } E < 1 \\ R_- &= (E-1)/2 \qquad \text{if } E > 1. \end{aligned} \tag{16}$$

Now let us consider the different cases.

(a) $E > 1$. Then the aplanatic curve determined by Eq. (15) is an ellipse with eccentricity $e = 1/E$. The foci of the ellipse are in B and A, respectively. The elliptical mirror forms a perfect image: if the real object is in the focus, then the real image will be in the other focus. If the reflection is outside the ellipse, then the virtual object (in one focus)–virtual image (in the other focus) situation takes place.

(b) $0 < E < 1$. Then the aplanatic curve is a branch of a hyperbola bending towards its focus in A, and having the other focus in B. The hyperbolic mirror gives perfect image formation: if the real object is in A, then the imaginary image will be in the other focus B, and vice versa: if the imaginary object is in B, then the real image is in A. If the other side of this hyperbola is reflecting, then the image formation is: real object in B, imaginary image in A, or imaginary object in A, real image in B.

(c) $-1 < E < 0$. This case is perfectly analogous to the case (b), the reflecting curve now is the other branch of the hyperbola.

(d) $E = 0$. The aplanatic curve is the perpendicular bisector of the segment AB. This is the case of the plane mirror, if an object is in A, then the image is in B. If the object is real, then the image is imaginary, and vice versa.

(e) If $E = 1$ or $E = -1$, then no physically meaningful reflecting curve exists, this is an extreme limit case from a mathematical point of view.

(f) If $E < -1$, then there is no aplanatic curve at all.

3.2. Aplanatic refraction

Consider a heterogeneous medium separated by an aplanatic curve defined below. Let us investigate broken lines consisting of two straight segments: BP and PA as shown in Fig. 5. The velocity is v_B while the wave propagates along the ray BP, and v_A on the line PA. The aplanatic curve is defined by the condition that the propagation time $\mathrm{AP}/v_A + \mathrm{PB}/v_B = \tau$ should be independent of P. Here,

$$\mathrm{AP} = \sqrt{r^2 + d^2 - 2rd\cos\varphi}, \qquad \mathrm{PB} = r. \tag{17}$$

Let us introduce the dimensionless distances $R = r/d$, $\mathrm{AP}/d = L_A$, and the dimensionless refractive index $n = v_A/v_B$, and the dimensionless parameter $E = \tau v_A/d$. Then the aplanatic condition is

$$nR + L_A = E. \tag{18}$$

After squaring we get the aplanatic equation

$$R^2 + 1 - 2R\cos\varphi = E^2 + n^2R^2 - 2nRE \qquad n > 1,\ R > 0. \tag{19}$$

From this, *cos* φ can be expressed as

$$\cos\varphi = G(R)$$

$$G(R) = nE - \frac{1}{2}\left(\frac{E^2 - 1}{R} + (n^2 - 1)R\right). \tag{20}$$

It must be stressed that this consideration refers to the real object–real image situation. However, similar to the case of reflection detailed in the previous section, our formulas are

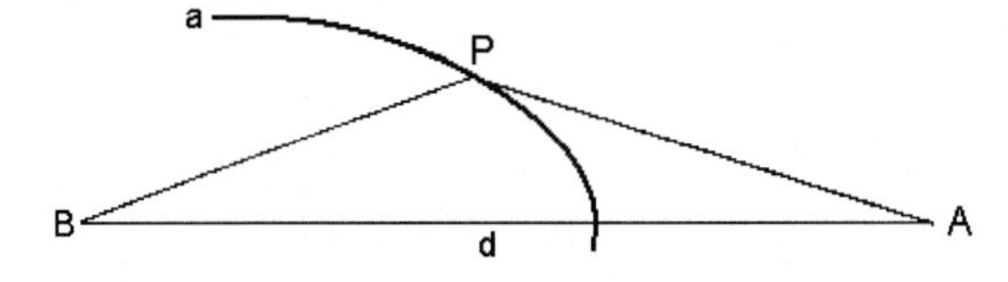

Fig. 5. Refraction by an aplanatic surface.

valid for virtual object or virtual image if we change the sign of the distance in question. For the real–virtual situation instead of Eq. (18) we should write

$$nR - L_A = E. \tag{21}$$

All the situations (real–real, real–virtual, virtual–real, virtual–virtual) are described by the formula (20). Substituting the value of *cos* φ from this the relation

$$L_A = |nR - E|$$

is yielded. Comparing Eqs. (18) and (21), one can see that if $nR < E$, then the situation is real–real or virtual–virtual, if $nR > E$, then the situation is real–virtual or virtual–real.

Now, let us consider the general behavior of the function G. For our purposes, only the case $R \geq 0$ is relevant. Furthermore, the relevant range of G is the interval $[-1,1]$. Let us denote by R_+ and R_- the roots of $G(R_+) = 1$ and those of $G(R_-) = -1$, respectively. These equations have two real roots:

$$\begin{aligned} R_{+,1} &= \frac{E+1}{n+1}, \qquad R_{+,2} = \frac{E-1}{n-1} \\ R_{-,1} &= \frac{E-1}{n+1}, \qquad R_{-,2} = \frac{E+1}{n-1}. \end{aligned} \tag{22}$$

Case (A). $E^2 < 1$. Then G is a monotonously decreasing function, and the relevant range: $R_{+,1} \leq R \leq R_{-,2}$.

Case (B). $E^2 = 1$. Then G is a linear decreasing function, and $G(0) = nE$.

Case (C). $E^2 > 1$. Then G has a maximum value

$$G(R) \leq G_{\max} = nE - \sqrt{(E^2 - 1)(n^2 - 1)}, \qquad \text{for any } R \geq 0. \tag{23}$$

It is easy to see that $G_{\max} > 1$, if $E > 1$, and $G_{\max} < 1$, if $E < -1$.

Using these properties of G, we classify the relevant cases with respect to the existence and character of aplanatic curves.

Along the aplanatic curve Snell's law of refraction is valid. More precisely, the relation:

$$|\sin \alpha / \sin \beta| = n \tag{24}$$

between the angle of incidence and angle of refraction along the aplanatic curve can be proved. The tangent unit vector of the aplanatic curve is

$$\mathbf{e}_t = \frac{\mathbf{e}_R \, \mathrm{d}R/\mathrm{d}\varphi + R\,\mathbf{e}_\varphi}{\sqrt{(\mathrm{d}R/\mathrm{d}\varphi)^2 + R^2}}. \tag{25}$$

The $\mathbf{e}_n$ normal unit vector is

$$\mathbf{e}_n = \frac{R\,\mathbf{e}_R + \mathbf{e}_\varphi \, \mathrm{d}R/\mathrm{d}\varphi}{\sqrt{(\mathrm{d}R/\mathrm{d}\varphi)^2 + R^2}}. \tag{26}$$

However, this normal vector can be derived in another way, since it is orthogonal to the level curve of the scalar field $nR \pm |\mathbf{R} - \mathbf{i}| = H(\mathbf{R}) = E$

$$\mathbf{e}_n = \mathbf{N}/N,$$

$$N = \text{grad}\, H = \text{grad}\,(nR \pm |\mathbf{R} - \mathbf{i}|) = n\, \mathbf{e}_\mathrm{R} \pm \frac{\mathbf{R} - \mathbf{i}}{|\mathbf{R} - \mathbf{i}|}. \tag{27}$$

Tiresome, but straightforward, calculation yields

$$\sin \alpha = \mathbf{e}_\mathrm{t} \cdot \frac{\mathbf{R} - \mathbf{i}}{|\mathbf{R} - \mathbf{i}|} = n \frac{nR - E}{|nR - E|} \frac{\mathrm{d}R/\mathrm{d}\varphi}{\sqrt{R^2 + (\mathrm{d}R/\mathrm{d}\varphi)^2}}$$

$$\sin \beta = \mathbf{e}_\mathrm{t} \cdot \mathbf{e}_\mathrm{R} = \frac{\mathrm{d}R/\mathrm{d}\varphi}{\sqrt{R^2 + (\mathrm{d}R/\mathrm{d}\varphi)^2}},$$

$$\text{hence } \frac{\sin \alpha}{\sin \beta} = \pm n, \tag{28}$$

sign + refers to the real–virtual situation, and sign − to the real–real.

To describe the characteristic properties of aplanatic image formation, the critical case, when $\alpha = \pi/2$ is also important (critical angle of total reflection). This critical condition can be written in a simple form, since then the vectors $\mathbf{N}$ and $\mathbf{R} - \mathbf{i}$ are orthogonal, that is

$$\mathbf{N} \cdot (\mathbf{R} - \mathbf{i}) = 0,$$

therefore, this critical condition is

$$E - n \cos \varphi_\mathrm{cr} = 0. \tag{29}$$

Case (i). $E \leq -1$, or $E = -1$. Since $G_\mathrm{max} < -1$, there is no aplanatic curve at all.

Case (ii). $-1 < E \leq 1$. Then G is monotonously decreasing. There is a closed aplanatic curve for which

$$R_{+,1} \leq R \leq R_{-,2}.$$

For this case $nR > E$, therefore, the aplanatic refraction represents a real–virtual situation. In the special case when $E = 0$, the aplanatic curve is a circle.

Case (iii). $1 < E < n$. For this case the function G has the shape depicted in Fig. 6. We conclude from Fig. 6 that there are two separate closed curves. For the real–real aplanatic curve $R_{-,1} \leq R \leq R_{+,2}$. This is shown in Fig. 6b. The rays starting from the real object A are refracted and focused in B. However, this is valid only for a region bounded by the critical ray incidenting tangentially at point P_cr, belonging to the angle φ_cr in Eq. (29). The part of the aplanatic curve beyond P_cr does not play any role in image formation. This aplanatic lens may realize not only the real–real image formation, but the virtual–virtual one too. To see this, it is enough to consider the elongation of the relevant rays, similarly as we did in the case of the elliptical mirror.

The other aplanatic curve ($R_{+,1} \leq R \leq R_{-,2}$.) belonging to the same value of E, represents real–virtual image formation. This is shown in Fig. 6c. The rays starting from the real object B are refracted and their elongations meet at A. However, this is valid only for a region bounded by the critical ray incidenting tangentially at point P_cr, belonging to the angle φ_cr in Eq. (29). The part of the aplanatic curve before Q_cr does not play any role in image formation. Naturally, if a virtual object is in A, then we get a real image in B.

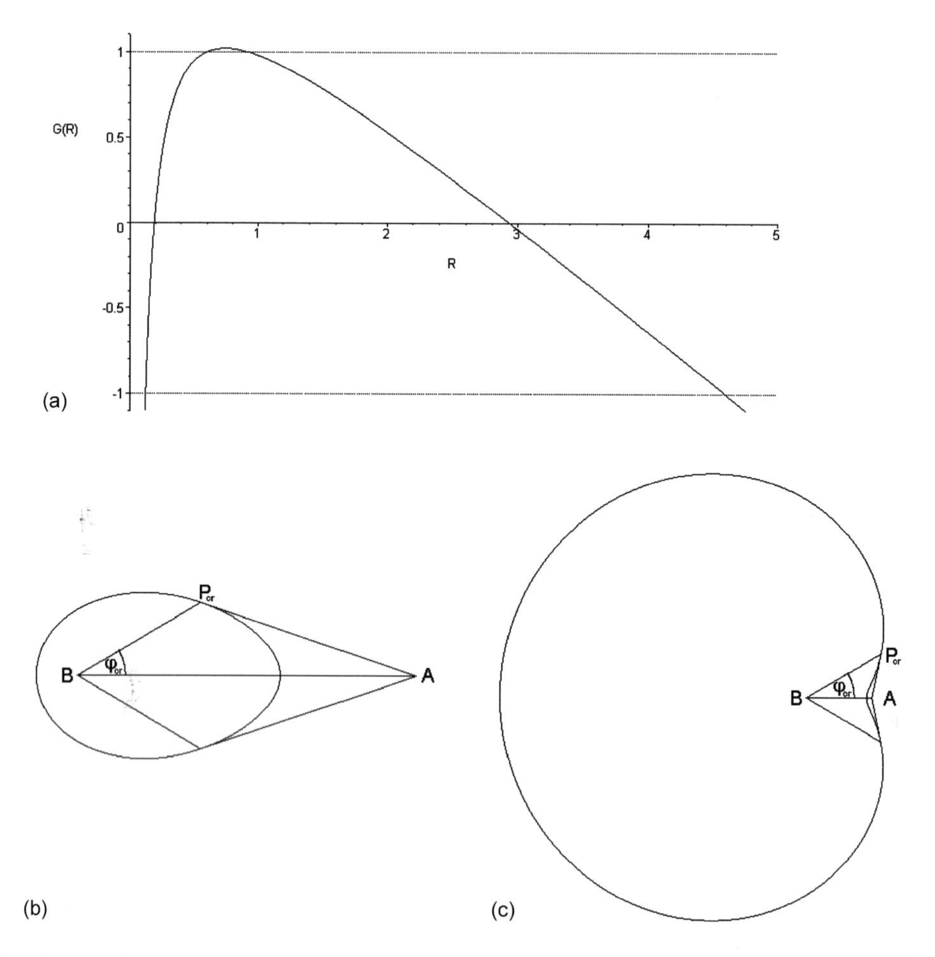

Fig. 6. Case (iii), $n = 1.5$, $E = 1.3$, $d = 1$, $v_A = 1$, $v_B = 1/n$ (a) The graph of $G(r)$. (b) Real–real image formation. (c) Real–virtual aplanatic curve.

Case (iv). E = n. This is a limiting case of (iii), where the real–real and the real–virtual curves have a unique common point in A.

Case (v). E > n. In this case there are also two closed aplanatic curves, similarly as happened in case (iii). However, both closed aplanatic curves encircle both B and A. From a physical point of view the real–real situation does not occur at all (P_{cr} does not exist), only the virtual–virtual image formation can be realized by that aplanatic lens.

3.3. Nonaplanatic refraction

Now let us compare refraction by an aplanatic and a nonaplanatic curve. Fig. 7 shows the case when the curve g has a common contact with the aplanatic curve a at the point P.

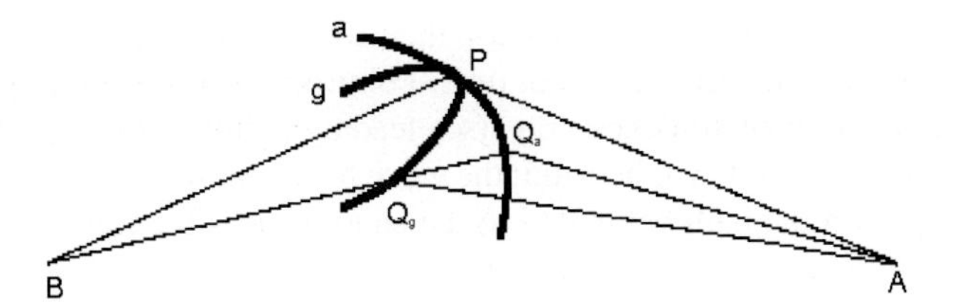

Fig. 7. Comparison of nonaplanatic and aplanatic refraction.

The rays AP and PB obey Snell's law of refraction. Let us assume that the nonaplanatic curve has a greater curvature at P than the aplanatic one has. Let us denote the propagation time by τ_a and τ_g, if the interface of the two media is a or g, respectively. Consider a broken line AQ_gB, where Q_g is on curve g. The time belonging to that path is

$$\tau_g(AQ_gB) = AQ_g/v_A + Q_gB/v_B.$$

Let us elongate the ray AQ_g to the point Q_a on the aplanatic curve, as shown in Fig. 7. The propagation time along the aplanatic curve is constant, therefore

$$\tau_a(AQ_aB) = \tau_a(APB). \tag{30}$$

It is obvious that

$$\tau_a(AQ_aB) > \tau_g(AQ_gB) \tag{31}$$

since

$$\tau_a(AQ_aQ_g) = \frac{AQ_a}{v_A} + \frac{Q_aQ_g}{v_B} > \frac{AQ_a + Q_aQ_g}{v_A} > \frac{AQ_g}{v_A} = \tau_g(AQ_g).$$

Note that a similar proof holds for the case when the curvature of g is smaller than that of the aplanatic a, in this case the path APB requires the least time. Relations (30) and (31) can be interpreted in the following statement: in this case, the time for the ray APB obeying Snell's law is maximum among the neighboring paths broken at the nonaplanatic curve g.

This statement is in accordance with the modern criticism of FPLT. It is relevant in ray dynamics. The rays starting from the point source at A refract by the nonaplanatic refracting curve, and only one of them will reach B, the others will not reach B at all, in contrast to the aplanatic refraction when the neighboring rays also reach the image point B. In this formulation there is only one ray connecting B with A. The neighboring path AQ_gB is not a real ray, therefore the unique 'true' ray can be selected from the admissible set of imagined paths by the requirement of maximum propagation time. If the point source in A emits light, according to geometrical optics, light rays should obey Snell's law, and only one ray will reach B, namely the ray APB, because there is no diffraction in geometrical optics.

Considering front dynamics, this interpretation becomes wrong, while Fermat's original formulation remains valid. If a wave front is emitted by a point source in A, after refraction the wave process will reach B along the path belonging to the minimal time. This route will not follow Snell's law. In the case of front dynamics (geometric theory of waves) all the points where the wave front reach the refracting nonaplanatic curve will emit straight rays into the medium with velocity v_B, and this family of rays generates the subsequent fronts. Some rays will increase their regions, the regions of other rays may decrease, even they die out after a certain time. Point B will be covered by the front part generated from the 'winner' (quickest) ray, (see leading points in Refs. [6,9]). In the case of chemical waves, the FPLT will be valid: the wave front emitted from a point source at A will surely reach B on the path belonging to the shortest possible time.

3.4. Chemical lens

As we saw in Section 3.2, there is an aplanatic curve that maps a real object in the point A into a real image in the point B (Case (iii), Fig. 6b). It was shown that the part of the aplanatic curve beyond the critical point P_{cr} does not take part in image formation. However, for chemical waves we can continue the curve beyond P_{cr} in such a way that this part also participates in image formation. In this way, a 'chemical lens' is constructed. A leading point running from P_{cr} along that curve c will emit rays that depart from c by the critical angle of total refraction. For example, the ray $AP_{cr}QP$ will consist of two line segments AP_{cr} and QP connecting them by the curved arc $P_{cr}Q$, and the propagation time belonging to that path is equal to the propagation time of the aplanatic refraction $AP_{cr}B$:

$$\tau(AP_{cr}QB) = \tau(AP_{cr}B).$$

That is, $\tau(P_{cr}QB) = \tau(P_{cr}B)$. Using the polar coordinates r, φ of Q we get

$$\frac{r}{v_B} + \frac{1}{v_A}\int_{\varphi_{cr}}^{\varphi} \sqrt{(\mathrm{d}r/\mathrm{d}\varphi)^2 + r^2}\mathrm{d}\varphi = \frac{r_{cr}}{v_B}.$$

From this the differential equation of the curve c

$$\sqrt{\left(\frac{\mathrm{d}r}{\mathrm{d}\varphi}\right)^2 + r^2} + n\frac{\mathrm{d}r}{\mathrm{d}\varphi} = 0 \qquad (32)$$

is obtained. The solution of this equation is

$$r(\varphi) = r_0 \exp\left(-\frac{\varphi - \varphi_0}{\sqrt{n^2 - 1}}\right). \qquad (33)$$

Consequently, the supplementary arc of the chemical lens is an equiangular (logarithmic) spiral. Fig. 8 shows the shape of a chemical lens consisting of an aplanatic arc and a spiral arc. If a point source in A emits circular chemical wave fronts in the faster

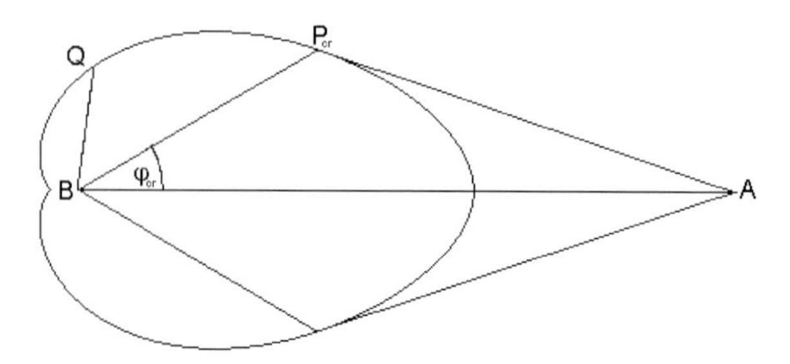

Fig. 8. Chemical lens. The parameter values are the same as in Fig. 7.

medium, after refracting by the chemical lens, the above theory predicts circular wave fronts moving inwards to the center at B.

Acknowledgements

The authors thank Clay McMillan and András Volford for their help, and the OTKA grant T-42708 for its support.

References

[1] Born, Max and Wolf, Emil (1965), Principles of Optics, Pergamon, London.

[2] Lázár, A., Noszticzius, Z., Farkas, H. and Försterling, H.D. (1995), Involutes: the geometry of chemical waves rotating in annular membranes, Chaos 5, 443–447.

[3] Simon, P.L. and Farkas, H. (1996), Geometric theory of trigger waves. A dynamical system approach, J. Math. Chem. 19, 301–315.

[4] Sieniutycz, S. and Farkas, H. (1997), Chemical waves in confined regions by Hamilton–Jacobi–Bellman theory, Chem. Eng. Sci. 52, 2927–2945.

[5] Lázár, A., Försterling, H.D., Farkas, H., Simon, P.L., Volford, A. and Noszticzius, Z. (1997), Waves of excitations on nonuniform membrane rings, caustics, and reverse involutes, Chaos 7(4), 731–737.

[6] Volford, A., Simon, P.L. and Farkas, H. (1999), Waves of excitation in heterogeneous annular region, asymmetric arrangement. In Geometry and Topology of Caustics—Caustics '98, (Janeczko, S., Zakhalyukin, V.M., eds.), Vol. 50, Banach Center Publications, Warszawa, pp. 305–320.

[7] Volford, A., Simon, P.L., Farkas, H. and Noszticzius, Z. (1999), Rotating chemical waves: theory and experiments, Physica A 274, 30–49.

[8] Kály-Kullai, K. (2003), A fast method to simulate travelling waves in nonhomogeneous chemical or biological media, J. Math. Chem. 34, 163–176.

[9] Kály-Kullai, K., Volford, A. and Farkas, H. (2004), In Waves of excitations in heterogeneous annular region II—strong asymmetry, Vol. 62, Banach Center Publications, Warszawa, pp. 143–158.

[10] Wiener, N. and Rosenblueth, A. (1946), The mathematical formulation of the problem of conduction of impulses in a network of connected excitable elements, specifically in cardiac muscle, Arch. Inst. Cardiol. Mexico, 205–265.

[11] Brazhnik, P.K., Davydov, V.A., Zykov, V.S. and Mikhailov, A.S. (1987), Zh. Eksp. Teor. Fiz. 93, 1725–1736.

[12] Tyson, J.J. and Keener, J.P. (1988), Physica D 32, 327.

[13] Itard, J. (1950), Pierre Fermat. Beihefte zur Zeitschrift, Elemente der Mathematik—Suppléments a la revue de mathématiques élémentaire. In Beiheft Nr. 10, Verlag Birkhäuser, Basel.

[14] Fermat, Pierre de (1894), In Oeuvres, (Tannery, P., Henry, C., eds.), Vol. II, Oeuvres de Fermat, Paris, 1891–1922, T. II, p. 354 (cited in C. Carathéodory, The Beginning of Research in the Calculus of Variations, presented on Aug. 31, 1936 at the meeting of the Mathematical Association of America in Cambridge, Mass., during the tercentenary celebration of Harvard University).

[15] Hecht L., Why You Don't Believe Fermat's Principle, Fall 2001 21st Century, Science and Technology Magazine 2001, http://www.21stcenturysciencetech.com/articles/fall01/fermat/Fermat.html.

[16] Longhurst, R.S. (1957), Geometrical and Physical Optics, Longmans, London.

[17] Feynman, R.P., Leighton, R.B. and Sands, M. (1963), In Lectures on Physics, Vol. I, Addison-Wesley, Reading, MA, Chapters 26 and 27.

[18] Feynman, R.P. (1985), QED: The Strange Theory of Light and Matter, Princeton University Press, Princeton, NJ.

[19] (1977), The Penguin Dictionary of Physics (Illingworth, V., eds.), Penguin Group, London.

[20] Kline, Morris and Kay, Irwin W. (1965), Electromagnetic Theory and Geometrical Optics, Wiley-Interscience, New York.

[21] Steinbock, Oliver, Tóth, Ágota and Showalter, Kenneth (1995), Navigating complex labyrinths: optimal paths from chemical waves, Science 267, 868–871.

Part II

APPLICATIONS

STATISTICAL PHYSICS AND THERMODYNAMICS

Chapter 1

FISHER VARIATIONAL PRINCIPLE AND THERMODYNAMICS

A. Plastino

Universidad Nacional de La Plata, C.C. 727, 1900 La Plata, Argentina
Argentine National Research Center (CONICET), La Plata, Argentina

A.R. Plastino

Universidad Nacional de La Plata, C.C. 727, 1900 La Plata, Argentina
Argentine National Research Center (CONICET), La Plata, Argentina
Department of Physics, University of Pretoria, Pretoria 0002, South Africa

M. Casas

Departament de Fisica and IMEDEA, Universitat de les Illes Balears 07122 Palma de Mallorca, Spain

Abstract

It is shown that thermostatistics, usually derived microscopically from Shannon's information measure via Jaynes' MaxEnt procedure, can equally be obtained from a constrained extremization of Fisher's information measure. The new procedure has the advantage of dealing on an equal footing with both equilibrium and offequilibrium processes, as illustrated by two significant examples.

Keywords: Fisher information; thermodynamics; maximum entropy

1. Introduction

Jaynes has shown that the whole of statistical mechanics can be elegantly reformulated, without reference to the ensemble notion, if one chooses Boltzmann's constant as the informational unit and identifies Shannon's logarithmic information measure S with the thermodynamic entropy. A variational approach is to be followed that entails extremization of S subject to the constraints imposed by the a priori knowledge one may possess concerning the system of interest. The concomitant methodology is referred to as the *Maximum Entropy Method (MaxEnt)*) [1]. As is well known, the phenomenal success of thermodynamics and statistical physics depends crucially on certain necessary mathematical relationships involving energy and entropy (the Legendre transform structure of thermodynamics). *It has been demonstrated* [2] *that these relationships are*

E-mail address: plastino@sinectis.com.ar (A. Plastino); arplastino@maple.up.ac.za (A.R. Plastino); montse.casas@uib.es (M. Casas).

also valid if one replaces S by Fisher's information measure (FIM) I. Much effort has been focused recently upon FIM. The work of Frieden, Soffer, Plastino, Nikolov, Casas, Pennini, Miller, and others has shed much light upon the manifold physical applications of I (see, for instance [3–7]). Frieden and Soffer have shown that FIM provides one with a powerful variational principle that yields most of the canonical Lagrangians of theoretical physics [5]. Additionally, I has been shown to characterize an "arrow of time" with reference to the celebrated Fokker–Planck equation [7]. Moreover, interesting relationships exist that connect FIM and Kullback's cross-entropy [8,9]. Finally, a rather general I-based H-theorem has recently been proved [9]. FIM is then an important quantity that is involved in many aspects of the theoretical description of Nature.

In the following we show that the essential thermodynamical features of equilibrium and nonequilibrium problems can be derived from a constrained Fisher extremization process that results in a Schrödinger-like stationary equation (SSE). Equilibrium corresponds to the ground-state (gs) solution. Nonequilibrium corresponds to superpositions of the gs with excited states. With reference to dilute gases, two typical illustrative examples will be discussed: viscosity and electrical conductivity.

2. Fisher's measure and translation families

Consider a system that is specified by a physical parameter θ and let $g(x, \theta|t)$ stand for the normalized probability density function (PD) for this parameter at that time t. Fisher's information measure I is of the form

$$I = \int \mathrm{d}x\, g\left[\frac{\partial g/\partial\theta}{g}\right]^2; \quad g = g(x, \theta|t); \quad x \in \mathcal{R}^n. \tag{1}$$

The special case of *translation families* deserves special mention. These are monoparametric families of distributions of the form: $g(x - \theta|t) = p(u, t);\ u \equiv x - \theta$, which are known up to the shift parameter θ. Following Mach's principle, all members of the family possess identical shape (there are no absolute origins), and here FIM adopts the appearance:

$$I = \int \mathrm{d}x(1/p)[\partial p/\partial x]^2 = \int \mathrm{d}x\, F_{\mathrm{Fisher}}(p); \quad p = p(x|t); \quad F_{\mathrm{Fisher}}(p) = (1/p)[\partial p/\partial x]^2. \tag{2}$$

This form of Fisher's measure constitutes the main ingredient of a powerful variational principle devised by Frieden and Soffer [5,10] that gives rise to a substantial portion of contemporary physics. In the considerations that follow we shall restrict ourselves mostly to the form (2) of Fisher's information measure. There are two different variational approaches in current use that employ Fisher's information measure, namely, MFI (minimum Fisher information) and EPI (extreme physical information) [5]. The approaches differ in how the constraint information is used in conjunction with I. However, in the following we do not distinguish between the two treatments since for our present purposes they lead to the same result.

3. Variational techniques and Fisher's measure

Assume one is dealing with vectors x that belong to $\mathcal{R}^n$, with $n = 1$. Generalization to arbitrary n is straightforward. Volume elements in this space are to be denoted simply by dx. Let us focus our attention upon the positive–definite, normalized density (of probability) function $p(x, \theta|t)$, and consider the concomitant *Fisher's information measure* for translation families (Eq. (2)). Furthermore, assume that for M functions $A_i(x)$ the mean values $\langle A_i \rangle_t$,

$$\theta_i \equiv \langle A_i \rangle_t = \int \mathrm{d}x\, A_i(x) p(x, \theta|t), \qquad (i = 1, \ldots, M), \tag{3}$$

are *known at the time t*, stressing here the fact that the $\langle A_i \rangle$s depend in linear fashion upon p. These mean values will play the role of thermodynamics' extensive variables. The analysis will use MFI to find the probability distribution (PD) $p \equiv p_{MFI}$ that extremizes I subject to the prior conditions $\langle A_i \rangle_t$ plus normalization. It is of importance to note that the prior knowledge (Eq. (3)) *represents information at the fixed time t*. The problem we attack is to find the PDF $p \equiv p_{\mathrm{MFI}}$ that extremizes I subject to prior conditions (3) and normalization. Our Fisher-based extremization problem takes the form (at the given time t)

$$\delta_p \left\{ I(p) - \alpha \langle 1 \rangle - \sum_k^M \lambda_k \langle A_k \rangle_t \right\} = \delta_p \left\{ \int \mathrm{d}x \left(F_{\mathrm{Fisher}}(p) - \alpha p - \sum_k^M \lambda_k A_k p \right) \right\} = 0, \tag{4}$$

where we have introduced the $(M + 1)$ Lagrange multipliers $(\alpha, \lambda_1, \ldots, \lambda_M)$, and each multiplier $\lambda_k \equiv \lambda_k(t)$. Variation leads now to

$$\left\{ (p)^{-2} \left(\frac{\partial p}{\partial x} \right)^2 + \frac{\partial}{\partial x} \left[(2/p) \frac{\partial p}{\partial x} \right] + \alpha + \sum_k^M \lambda_k A_k \right\} = 0. \tag{5}$$

It is clear that the normalization condition on p makes α a function of the λ_is. Let then $p_I(x, \{\lambda\})$ be a solution of Eq. (5), where obviously, $\{\lambda\}$ is an M-dimensional Lagrange multipliers vector. The extreme Fisher information is now a function of time: $I = \int \mathrm{d}x (\partial p/\partial x)^2 p^{-1} \equiv I(t)$. Since p extremized I we write $p \equiv p_I$, $p_I \equiv p_I(x|t)$. Let us now find the general solution of Eq. (5). For the sake of simplicity let us define

$$G(x, t) = \alpha + \sum_k^M \lambda_k(t) A_k(x) \Rightarrow \left[\frac{\partial \ln p_I}{\partial x} \right]^2 + 2 \frac{\partial^2 \ln p_I}{\partial x^2} + G(x, t) = 0. \tag{6}$$

Introduction now of (i) the identification $p_I = (\psi)^2$, recalling that $\psi(x)$ can always be assumed real for 1D problems and (ii) the new functions

$$v = \frac{\partial \ln \psi}{\partial x}, \qquad \psi \equiv \psi(x, t),$$
$$v \equiv v(x, t), \text{ simplifies Eq. (6) to } v' = -\left\{ \frac{G(x, t)}{4} + v^2 \right\}, \tag{7}$$

where the prime stands for the derivative with respect to x. The above equation is a Riccati equation [11]. Introduction, furthermore, of Ref. [11]

$$u = \exp\left\{\int^{x} dx[v]\right\}, \quad u = u(x,t), \text{ leads to } u = \exp\left\{\int^{x} dx \frac{d\ln\psi}{dx}\right\} = \psi, \tag{8}$$

and places Eq. (6) in the form of a Schrödinger-like stationary equation (SSE) [11]

$$-(1/2)\psi'' - (1/8)\sum_{k}^{M} \lambda_k(t)A_k\psi = \alpha\psi/8, \tag{9}$$

where the Lagrange multiplier $\alpha/8$ plays the role of an energy eigenvalue, and the sum of the $\lambda_k A_k(x)$ is an effective potential function $U = -(1/8)\sum_k^M \lambda_k(t)A_k$, and $U = U(x,t)$. Note that no specific potential has been assumed, as is appropriate for thermodynamics. Also, we note that U is a time-dependent potential function and will permit nonequilibrium solutions. The specific $A_k(x)$ to be used here depend upon the nature of the physical application at hand. This application could be of either a classical or a quantum nature. Also, note that Eq. (9) represents a boundary-value problem, generally with multiple solutions, in contrast to the unique exponential solution one obtains when employing Jaynes–Shannon's entropy in place of FIM [1]. As discussed in some detail in Ref. [2], the solution leading to the lowest I-value is the equilibrium one. In [3,4] one finds that excited solutions (9) describe offequilibrium situations.

Standard thermodynamics makes use of derivatives of the entropy S with respect to both intensive (λ_i) and extensive ($\langle A_i\rangle$) parameters. In the same way, we find that the Euler theorem [12] still holds within the Fisher context [3,4], i.e.

$$\frac{\partial I}{\partial \lambda_i} = \sum_{j}^{M} \lambda_j \frac{\partial}{\partial \lambda_i}\langle A_j\rangle, \tag{10}$$

a generalized 'Fisher–Euler' theorem. The thermodynamic counterpart of Eq. (10) is the derivative of I with respect to the mean values. We easily find

$$\sum_i \frac{\partial I}{\partial \lambda_i}\frac{\partial \lambda_i}{\partial\langle A_j\rangle} = \sum_i\sum_k \lambda_k \frac{\partial\langle A_k\rangle}{\partial \lambda_i}\frac{\partial \lambda_i}{\partial\langle A_j\rangle}, \quad \text{i.e.} \quad \frac{\partial I}{\partial\langle A_j\rangle} = \lambda_j, \tag{11}$$

as expected. The Lagrange multipliers and mean values are seen to be conjugate variables, related as is typical of intensive (λ_i) vs. extensive ($\langle A_i\rangle$) variables in thermodynamics. It is also clear that $p_I \equiv p_I(\lambda_1,\ldots,\lambda_M)$, since, based on the M mean values that constitute our a priori knowledge, the extremization process renders p_I a function of the Legendre multipliers only. In thermodynamics it is equivalent to have as input information either the chemical potential or the number of moles; the pressure or the volume; etc. This is so because of the characteristic Legendre structure of thermodynamics. Now, as our PD p_I formally depends upon $M+1$ Lagrange multipliers, normalization ($\int dx\, p_I = 1$) makes α a function of the M remaining λs, i.e. $\alpha = \alpha(\lambda_1,\ldots,\lambda_N)$. It is also of importance to stress that, of course, the λs and the $\langle A_i\rangle$s play reciprocal (symmetrical) roles within thermodynamics [1]. It is thus possible to assume, as initial conditions, that the input information refers to the λs and not to the $\langle A_i \rangle$s. The M Lagrange multipliers and the M mean values are then on an equal footing for the purpose of determining I. Introduce now the generalized thermodynamic potential (the Legendre transform of I)

$$\lambda_J(\lambda_1, \ldots, \lambda_N) = I(\langle A_1 \rangle, \ldots, \langle A_N \rangle) - \sum_{i=1}^{N} \lambda_i \langle A_i(\lambda_1, \ldots, \lambda_N) \rangle, \tag{12}$$

i.e. the Legendre transform of FIM [12]. We have

$$\frac{\partial \lambda_J}{\partial \lambda_i} = \sum_{j=1}^{M} \frac{\partial I}{\partial \langle A_j \rangle} \frac{\partial \langle A_j \rangle}{\partial \lambda_i} - \sum_{j=1}^{M} \lambda_j \frac{\partial \langle A_j \rangle}{\partial \lambda_i} - \langle A_i \rangle = -\langle A_i \rangle, \tag{13}$$

where Eq. (11) has been used. With the latest relation the Legendre structure here described can be summed up as follows:

$$\lambda_J = I - \sum_{i=1}^{M} \lambda_i \langle A_i \rangle; \qquad \frac{\partial \lambda_J}{\partial \lambda_i} = -\langle A_i \rangle; \qquad \frac{\partial I}{\partial \langle A_i \rangle} = \lambda_i; \tag{14}$$

$$\frac{\partial \lambda_i}{\partial \langle A_j \rangle} = \frac{\partial \lambda_j}{\partial \langle A_i \rangle}; \quad = \frac{\partial^2 I}{\partial \langle A_i \rangle \partial \langle A_j \rangle}; \quad \frac{\partial \langle A_j \rangle}{\partial \lambda_i} = \frac{\partial \langle A_i \rangle}{\partial \lambda_j} = -\frac{\partial^2 \lambda_J}{\partial \lambda_i \partial \lambda_j}. \tag{15}$$

As a consequence of the last relation we can recast (10) as the form

$$\frac{\partial I}{\partial \lambda_i} = \sum_{j}^{M} \lambda_j \frac{\partial}{\partial \lambda_j} \langle A_i \rangle, \tag{16}$$

The Legendre-transform structure of thermodynamics is thus seen to be entirely translated into the Fisher context. In order to be able to construct a thermodynamics based upon I, it is necessary to examine the concavity/convexity nature of I [12]. We prove below that I is a concave functional of the PDs p. Therefore, I exhibits the desirable mixing property. Let a and b be two real scalars such that $a + b = 1$, p_1, p_2 two normalized PDs, and consider the three PDs p_1, p_2, P

$$\psi = \sqrt{ap_1} + i\sqrt{bp_2}; \qquad P = ap_1 + bp_2 \equiv |\psi|^2. \tag{17}$$

The P-associated Fisher information for translation families reads (the prime stands for derivative with respect to x)

$$I(P) = \int \mathrm{d}x\, P^{-1} p'^2 = 4 \int \mathrm{d}x \frac{\mathrm{d}|\psi|^2}{\mathrm{d}x}. \tag{18}$$

In order to investigate the convexity question we must find the relationship relating $I(P)$ to $aI(p_1) + bI(p_2)$. If we set now $\psi(x) = R(x)\exp[iS(x)]$, with R, S two real functions upon $\mathcal{R}$, we immediately find

$$I(P) = 4 \int \mathrm{d}x\, R'^2 = \int \mathrm{d}x\, P(x) \left[\frac{R'(x)}{R(x)} \right]^2 = 4 \left\langle \left[\frac{R'}{R} \right]^2 \right\rangle. \tag{19}$$

Now, it is easy to see that

$$2\frac{\mathrm{d}\psi}{\mathrm{d}x} = \left(\sqrt{a/p_1}p_1' + i\sqrt{b/p_2}p_2'\right) \text{ entailing } 4\left|\frac{\mathrm{d}\psi}{\mathrm{d}x}\right|^2 = ((a/p_1)(p_1')^2 + (b/p_2)(p_2')^2), \quad (20)$$

which implies $aI(p_1) + bI(p_2) = 4\int \mathrm{d}x|(\mathrm{d}\psi/\mathrm{d}x)|^2$. Now, it is clear that

$$\left|\frac{\mathrm{d}\psi}{\mathrm{d}x}\right|^2 = R'^2 + R^2S'^2 = R'^2 + |\psi|^2S'^2 \Rightarrow aI(p_1) + bI(p_2)$$
$$= I(P) + 4\int \mathrm{d}x[|\psi|^2S'^2]. \quad (21)$$

The integral on the r.h.s. of the preceding equation is clearly ≥ 0, which entails

$$aI(p_1) + bI(p_2) \geq I(P), \quad (22)$$

i.e. Fisher information for translation families is indeed a concave functional of the PDs. A generalization of the last equation easily follows. Assume $b_1 + b_2 = 1$ and $p_2 = b_1p_{21} + b_2p_{22}$. Then, Eq. (22) implies

$$I(ap_1 + bb_1p_{21} + bb_2p_{22})$$
$$\leq aI(p_1) + bb_1I(p_{21}) + bb_2I(p_{22}) \text{ since } a + b(b_1 + b_2) = 1. \quad (23)$$

The r.h.s. of Eq. (22) represents the net probability, after mixing, of two distinct systems. We see that I displays the same mixing property as does Boltzmann's entropy. The inequality (22) is a special instance of Fisher's I-theorem $\mathrm{d}I/\mathrm{d}t \leq 0$, proved in Ref. [9]. We urge the reader to glance at Appendix A before reading for the next section.

4. The excited solutions' role

The connection between the excited solutions of Eq. (9) and thermodynamics has been established in Ref. [3,4], where it is shown just in which manner results derived from a suitably handling of the Boltzmann transport equation (BTE) can be reproduced by our Fisher approach (without making any kind of appeal to the BTE, of course!). In Appendix A we give a brief review of some elementary notions concerning Boltzmann's transport equation that are needed in order to understand the present Fisher approach. Special emphasis on Grad's methodology (the method of moments (MOM) approach) for dealing with the BTE [13,14] is placed in this Appendix.

Let excited solutions $\psi_n(x,t)$ to the Fisher-based SSE Eq. (9) be identified by a subindex value $n > 0$. These amplitude functions can be represented in a very special Hilbert space basis: that provided by the Hermite–Gaussian polynomials, of the form

$$\psi_n(x,t) = \phi_0\Sigma_i b_{in}(t)H_i(x), \quad (n = 1, 2, \ldots), \quad (24)$$

where $\phi_0 \equiv \phi_0(x, \omega) = [(\omega/\pi)]^{1/4}\exp[-x^2/2]$, is the ground state (gs) of the harmonic oscillator (HO). Recall that the first two Hermite polynomials are $H_0 = 1$; $H_1 = (1/\sqrt{2})2x$, so that, e.g. the first two members of the Gauss–Hermite basis are, respectively,

$$
\begin{aligned}
&(1)\; H_0\phi_0; \qquad (2)\; H_1\phi_0; \\
&\left(H_0 = 1;\; H_1 = \left(1/\sqrt{2}\right)2x,\; \phi_0 = [(\omega/\pi)]^{1/4}\exp[-x^2/2]\right).
\end{aligned}
\tag{25}
$$

The total number of coefficients $b_{ni}(t)$ in Eq. (24) depends upon how far from equilibrium we are. At equilibrium there is only one such coefficient [3,4]. As will be explained below, the squares of these amplitudes agree, under certain conditions, with the known solutions of the BTE [3,4]. Our coefficients $b_{in}(t)$ are computed at the fixed time t at which our input data $\langle A_k \rangle_t$ are collected. While the ground-state solution of our SSE gives the equilibrium states of thermodynamics [2], its excited solutions give nonequilibrium states [3,4]. This can be guaranteed because our functions $\psi_n(x, t)$ can be connected to the BTE-Grad molecular distribution function (MDF) $f(x, t)$ of Eq. (A2) via the squaring operation $\psi_n^2(x, t)$ [3]. Note that the square of an expansion in Hermite–Gaussians (with coefficients $b_{ni}(t)$) is likewise a superposition of Hermite–Gaussians, with *other* coefficients $c_{in}(t)$

$$
\psi_n^2(x, t) = \phi_0 \Sigma_i c_{in}(t) H_i(x), \quad (n = 1, 2, \ldots). \tag{26}
$$

For fixed n, the BTE-Grad coefficients $a_i(t)$ of Appendix A and our $c_{in}(t)$ are equal. Indeed, and first of all, the BTE-MOM coefficients are certainly computed, like ours, at a *fixed* time t. That is, their momenta are evaluated at that time. Likewise ours (the $\langle A_k \rangle$ of Eq. (3)) can be regarded as velocity momenta at that time as well. The difference between the BTE-MOM's coefficients and ours is one of physical origin, as follows. Grad *solves for* the velocity moments at the fixed time t. These M_{MOM} moments are computed using the BTE-MOM's a_i of Eq. (A3). We, instead, collect *as experimental inputs* these velocity moments (at the fixed time t). Thus, if the M_{MOM} moments *coincide with our experimental inputs*, necessarily the $a_i(t)$ and the $c_{in}(t)$ have to coincide. Let us repeat: the BTE-Grad moments at the time t are physically correct by construction, since they actually solve for them via use of the BTE. The premise of our constrained Fisher information approach is that its input constraints (here our velocity moments $\langle A_k \rangle_t$) are correct, since they come *from experiment*. (Grad calculates, we measure.) Summing up, the approach given in Refs. [3,4] gives exactly the same solutions *at the fixed (but arbitrary) time t as does the BTE-MOM approach*. Therefore, for fixed n, our $c_{in}(t)$s coincide with the BTE-Grad $a_i(t)$s of Appendix A and our $p(x|t)$ coincide with the BTE-MOM's $f(x, t)$. This holds at each time t (see Eq. (3)). For any other time value, t', say, we would have to input new $\langle A_k \rangle$ values appropriate for that time. Grad, instead, gets coefficients $a_i(t)$ valid for continuous time t, since they are using Boltzmann's transport equation, which is a continuous one. Our approach, by contrast, yields solutions valid at a discrete point of time t. This distinction, "discrete versus continuous", does not compromise the validity of the Fisher–Schrödinger, nonequilibrium thermodynamics bridge [3].

5. Application: viscosity

As a concrete example of our formalism we apply it now to the nonequilibrium problem posed by viscosity in dilute gases. We briefly discuss the corresponding phenomenology and then show that the Fisher approach gives the same answer as that of the BTE (Eq. (B6) of Appendix B). For other applications see Refs. [4,15].

5.1. Generalities

Imagine, in a gas, some plane with its normal pointing along the z-direction. The fluid below this plane exerts a mean force per unit area (stress) $\mathbf{P}_z$ on the fluid above the plane. Conversely, the gas above the plane exerts a stress $-\mathbf{P}_z$ on the fluid below the plane. The z-component of $\mathbf{P}_z$ measures the mean pressure $\langle p \rangle$ in the fluid, i.e. $P_{zz} = \langle p \rangle$. When the fluid is in equilibrium (at rest or moving with uniform velocity throughout), then $P_{zx} = 0$ [16]. Consider a nonequilibrium situation in which the gas does not move with uniform velocity throughout. In particular, imagine that the fluid has a constant (in time) mean velocity u_x in the x-direction such that $u_x = u_x(z)$. For specific examples see, for instance Ref. [16]. Now any layer of fluid below a plane $z =$ constant will exert a tangential stress P_{zx} on the fluid above it. If $(\partial u_z/\partial z)$ is small, one has [16] $P_{zx} = -\eta(\partial u_z/\partial z)$, where η is called the viscosity coefficient. The phenomenon was first investigated by Maxwell, who showed that, for a dilute gas of particles of mass m moving with mean velocity $\langle v \rangle$,

$$\eta \propto n\langle v \rangle ml, \tag{27}$$

where n is the number of molecules per unit volume and l is the mean free path [16]. Now consider any quantity $\chi(\mathbf{r}, t)$ whose mean value is

$$\langle \chi(\mathbf{r}, t) \rangle = \frac{1}{n(\mathbf{r}, t)} \int \mathrm{d}^3 v\, f(\mathbf{r}, \mathbf{v}, t) \chi(\mathbf{r}, t), \tag{28}$$

with $n(\mathbf{r},t)$ the mean number of particles, irrespective of velocity, which at time t are located between $\mathbf{r}$ and $\mathbf{r} + d\mathbf{r}$. If $\chi(\mathbf{r}, t) \equiv \mathbf{v}(\mathbf{r}, t)$ the above relation yields the mean velocity $\boldsymbol{u}(\boldsymbol{r}, t)$ of a molecule located near $\mathbf{r}$ at time t. $\boldsymbol{u}(\boldsymbol{r}, t)$ describes the mean velocity of a flow of gas at a given point, i.e. the (macroscopic) hydrodynamical velocity. The peculiar velocity $\mathbf{U}$ of a molecule is defined in the fashion [16] $\boldsymbol{U} = \boldsymbol{v} - \boldsymbol{u}$, so that $\langle \mathbf{U} \rangle = 0$. If one is interested in transport properties, the fluxes of various quantities become the focus of attention. Consider the net amount of the quantity χ above transported (i) per unit time and (ii) per unit area *of an element of area oriented along* $\hat{\mathbf{n}}$ by molecules with velocity $\mathbf{U}$ due to their random movement back and forth across this element of area. The χ-associated flux $\mathcal{F}_n$ generated in this way is

$$\mathcal{F}_n(\mathbf{r}, t) = \int \mathrm{d}^3 v\, f(\mathbf{r}, \mathbf{v}, t)[\hat{\mathbf{n}} \cdot \mathbf{U}] \chi(\mathbf{r}, t) = n \langle [\hat{\mathbf{n}} \cdot \mathbf{U}] \chi \rangle. \tag{29}$$

For the present discussion we have $\chi = mv_x$ and $\hat{\mathbf{n}} \cdot \mathbf{U} = nU_z$. The ensuing flux gives then, precisely, P_{zx} [16]. Since u_{x} does not depend upon the velocity,

$$P_{zx} = nm\langle U_z v_x \rangle = nm\langle U_z [u_x + U_x] \rangle = nm\langle U_z U_x \rangle. \tag{30}$$

A simple *phenomenological* line of reasoning that utilizes the so-called path-integral approximation yields then [16]

$$P_{zx} = -\frac{n\tau}{\beta}\frac{\partial u_x}{\partial z}; \qquad \eta = \frac{n\tau}{\beta}, \tag{31}$$

where τ is the average time between molecular collisions (relaxation time) and $\beta = 1/kT$.

5.2. *The Fisher treatment*

We start by considering the equilibrium situation. Along the z-direction we deal (see, for instance, Ref. [16]) with the well-known PD

$$f_{0,z} = n[m\beta/(2\pi)]^{1/2}\exp[-\beta m v_z^2/2].$$

Thus, the Gauss–Hermite variables ω and x of Eq. (25) are $\omega = m\beta/2$; $2x = \sqrt{[2\beta m]}v_z$, which allows us to recast H_0, H_1 as $H_0 = 1$; $H_1 = \sqrt{[\beta m]}v_z$. From the *present* Fisher perspective, the z-component of the probability amplitude is to be found, if our *prior knowledge* is just that provided by the equipartition result $\langle v_z^2\rangle = (1/\beta m)$, by solving an HO-SSE equation

$$\psi_z''/2 + \lambda_1(t)(v_z^2/8)\psi_z = -(\alpha/8)\psi_z; \quad \text{where we set } \lambda_1(t)/8 = \omega^2/2 \text{ and } (\alpha/8) = -E. \tag{32}$$

One easily verifies that $n\psi_{0,z}^2 = f_{0,z}$, i.e. the z-component of the ground-state's PD indeed obeys Eq. (32). However, we deal here with a 3D problem. The pertinent Gauss–Hermite basis is the set of functions

$$\left\{\psi_0(v_x)\psi_0(v_y)\psi_0(v_z)\left[1 + \sum_{l,m,n} H_l(U_x)H_m(U_y)H_n(U_z)\right]\right\}, \tag{33}$$

where l, m, n run over all non-negative integers. The essential point here is: we assume that we have *the additional piece of knowledge* (31) for P_{zx}, so that a different SSE ensues (because the information potential in Eq. (9) has now changed by inclusion of a new term in the sum $\lambda_k A_k$) whose solution is, say, $\psi = \psi_x\psi_y\psi_z$ (note that, also, $\psi_0 = \psi_{0,x}\psi_{0,y}\psi_{0,z}$). The new SSE reads

$$\psi''/2 + \omega^2(v_z^2/2)\psi + aU_xU_z\psi = E\psi, \tag{34}$$

that can be treated perturbatively in view of our knowledge of the problem. $a \ll 1$ is here the perturbation coupling constant. It is well known [17] that, if one perturbs the ground state of the 1D HO stationary function with a linear term, only the first excited state enters the perturbative series because of the selection rules [17]

$$\langle\psi_0 H_n(x)|x|\psi_0 H_j(x)\rangle = c_1\delta(n,j+1) + c_2\delta(n,j-1); \quad (c_1, c_2 \text{ are constants}), \tag{35}$$

which entail that, for $n=0$ (ground state), only $j=1$ (first excited state) contributes [17]. As a consequence, we can write (up to first order in perturbation theory)

$$\psi = \psi_0 + \psi_1 = [1 + bH_1(U_x)H_1(U_z)]\psi_0 = (1 + b\beta m U_x U_z)\psi_0, \tag{36}$$

and, up to first-order terms as well

$$\psi^2 = [1 + 2bH_1(U_x)H_1(U_z)]\psi_0^2 = [1 + 2b(\beta m)U_x U_z]\psi_0^2. \tag{37}$$

We evaluate now $\langle\psi|U_x U_z|\psi\rangle$. For symmetry reasons it is obvious that $\langle\psi_0|U_x U_z|\psi_0\rangle = 0$. Thus, on account of Eq. (36) we find

$$\langle\psi|U_x U_z|\psi\rangle = 2b\beta m\langle\psi_0|U_x^2 U_z^2|\psi_0\rangle = 2b\beta m\langle U_x^2\rangle_0\langle U_z^2\rangle_0. \tag{38}$$

Using now the equipartition result $\langle U_x^2\rangle_0\langle U_z^2\rangle_0 = 1/(m\beta)^2$, we arrive at

$$P_{zx} = 2m^2 nb\beta\langle\psi_0|U_x^2 U_z^2|\psi\rangle_0 = 2bn/\beta = an/\beta; \quad a = 2b. \tag{39}$$

Once we use in Eq. (39) our prior knowledge Eq. (31), which entails $a = -\tau(\partial u_x/\partial z)$, we will necessarily get the correct result, that, as will be seen in Appendix B, coincides with the BTE-Grad one. Finally,

$$\psi^2 = (1 + a\beta m U_x U_z)\psi_0^2, \tag{40}$$

a result, again, identical to the BTE-Grad one derived in Appendix B.

6. Application: electrical conductivity

We consider now that our gas consists of charged particles of mass m (charge e) and apply a *weak*, uniform electrical field $\mathcal{E}$ in the z-direction, so that an electrical current j flows in that direction. If the number of particles per unit volume is n we have

$$f_0 = n(m\beta/2\pi)^{3/2}\exp[-\beta\epsilon]$$

$$\frac{\partial f_0}{\partial x_i} = 0, \quad (i = 1,2,3); \quad \epsilon = mv^2/2; \quad \frac{\mathrm{d}f_0}{\mathrm{d}\epsilon} = -\beta f_0; \quad \frac{\mathrm{d}\epsilon}{\mathrm{d}v_z} = mv_z. \tag{41}$$

Our equations of motion are $\dot{v}_x = 0$; $\dot{v}_y = 0$; $\dot{v}_z = e\mathcal{E}/m$. This problem is usually tackled by recourse to a special BTE technique [16].

6.1. The relaxation approximation to the Boltzmann transport equation

We focus attention upon a gas in which the effect of molecular collisions is always to restore a *local equilibrium* situation described by the Maxwell–Boltzmann PD $f_0(\mathbf{r}, \mathbf{v})$ [16]. In other words, we assume that (i) if the molecular distribution is disturbed from the local equilibrium so that the actual PD f is different from f_0, then and (ii) the effect of collisions is simply to restore f to the local equilibrium value f_0 exponentially with a relaxation time τ of the order of the average time between molecular collisions. In symbols, for fixed **r**, **v**, f changes as a result of collisions according to

$$f(t) = f_0 + [f - f_0]\exp - [t/\tau]. \quad (42)$$

In these conditions (Appendix A), the ensuing Boltzmann equation becomes [16]

$$\frac{\partial f}{\partial t} + \sum_{i=1}^{3}\left[v_i\frac{\partial f}{\partial x_i} + \dot{v}_i\frac{\partial f}{\partial v_i}\right] = -\frac{f - f_0}{\tau}, \quad (43)$$

a *linear* differential equation for f. We consider now a situation slightly removed from equilibrium: $f = f_0 + f_1$ with $f_1 \ll f_0$, so that Eq. (43) becomes

$$\frac{\partial f}{\partial t} + \sum_{i=1}^{3}\left[v_i\frac{\partial f}{\partial x_i} + \dot{v}_i\frac{\partial f}{\partial v_i}\right] = -f_1/\tau. \quad (44)$$

The left-hand side of Eq. (44) is small, since the right-hand side is, by definition, small. As a consequence, we can evaluate it by neglecting terms in f_1 and write

$$\frac{\partial f_0}{\partial t} + \sum_{i=1}^{3}\left[v_i\frac{\partial f_0}{\partial x_i} + \dot{v}_i\frac{\partial f_0}{\partial v_i}\right] = -f_1/\tau. \quad (45)$$

Since f_0 is the Maxwell–Boltzmann PD, independent of time ($[\partial f_0/\partial t] = 0$), we finally get the so-called Boltzmann equation in the relaxation approximation [16]

$$\sum_{i=1}^{3}\left[v_i\frac{\partial f_0}{\partial x_i} + \dot{v}_i\frac{\partial f_0}{\partial v_i}\right] = -f_1/\tau. \quad (46)$$

6.2. *Electrical conductivity and the relaxation approximation*

For this problem we easily see, using Eq. (41), that Eq. (46) acquires the simple appearance

$$[e\mathcal{E}/m]\frac{\partial f_0}{\partial v_z} = -f_1/\tau. \quad (47)$$

Now,

$$\frac{\partial f_0}{\partial v_z} = \frac{\partial f_0}{\partial \epsilon}\frac{\partial \epsilon}{\partial v_z} = mv_z\frac{\partial f_0}{\partial \epsilon} = -mv_z\beta f_0, \quad (48)$$

so that the special Boltzmann Eq. (47) is solved to yield the following result, that we will try to reproduce à la Fisher, without any reference to Eq. (47), in the next subsection:

$$f_1 = \tau\beta e\mathcal{E}v_z f_0. \quad (49)$$

6.3. The Fisher treatment of electrical conductivity

As explained in Appendix C we have $H_0 = 1$, $H_1 = \sqrt{[\beta m]}v_z$, as the relevant Hermite polynomials. In the equilibrium instance (ground state) we deal with

$$\psi''/2 + \lambda_1(t)(v_z^2/8)\psi = -(\alpha/8)\psi, \tag{50}$$

where we set

$$\lambda_1(t)/8 = \omega^2/2 \quad \text{and} \quad (\alpha/8) = -E,$$

and consider that our prior knowledge consists of the equipartition theorem, $m\langle v_z^2\rangle/2 = kT/2$, so that $\langle v_z^2\rangle = (1/\beta m)$, which entails (see Appendix C) $\omega = \beta m/2$. Remember that the ground state of our SSE is given by $\psi_0 = [\omega/\pi]^{1/4}\exp[-\omega v_z^2/2]$ and that $n\psi_0^2 = f_0$, the equilibrium PD. Let us now *assume now that we have the additional piece of knowledge*

$$\langle j\rangle = e\int dv_z\psi^2 v_z = \sigma\mathcal{E} = e^2\tau\mathcal{E}/m, \tag{51}$$

which leads to a different SSE, namely,

$$\psi''/2 + \omega^2(v_z^2/2)\psi + av_z\psi = E\psi, \tag{52}$$

that can be treated perturbatively in view of our knowledge of the problem. $a \ll 1$ is here the perturbation coupling constant. It is well known [17] that if one perturbs the ground-state harmonic-oscillator wave function with a linear term only the first excited state enters the perturbative series because of the selection rules Eq. (35). We thus have

$$\psi = \psi_0(1 + aH_1) = \psi_0(1 + a\sqrt{(\beta m)}v_z). \tag{53}$$

Now, $\langle j\rangle = e\langle\psi|v_z|\psi\rangle$. Introducing here the form Eq. (53) one easily ascertains that only "cross" terms in the above expression (involving the matrix element of the perturbation between the gs and the first excited state) yield a nonvanishing contribution to the electric current, i.e.

$$\langle j\rangle = 2ea/[\sqrt{(\beta m)}], \tag{54}$$

which, together with Eq. (51) gives

$$a[\sqrt{(\beta m)}] = e\tau\mathcal{E}/2, \tag{55}$$

which confirms that a, being proportional to $e\mathcal{E}$, is indeed a small quantity. We have now $\psi = \psi_0[1 + v_z e\tau\mathcal{E}\beta/2]$, and, neglecting second-order terms,

$$\psi^2 = \psi_0^2[1 + e\tau\mathcal{E}\beta v_z], \tag{56}$$

which coincides with the Boltzmann result derived above after multiplication by n, the number of particles per unit volume.

7. Conclusions

The entire Legendre-transform structure of thermodynamics can be obtained using Fisher information in place of Boltzmann's entropy. This abstract Legendre structure constitutes an essential ingredient that allows one to build up a statistical mechanics. Fisher's information I allows then for such a construction. The desired concavity property obeyed by I further demonstrates its utility as a statistical-mechanics "generator". We can develop a thermodynamics that is able to treat equally well both equilibrium and nonequilibrium solutions, as illustrated with reference to Boltzmann's transport equation.

Appendix A. BTE and the method of moments

Let $f(\mathbf{r}, \mathbf{v}, t)$ be the mean number of molecules whose center of mass at time t is located between $\mathbf{r}$ and $\mathbf{r} + \mathrm{d}\mathbf{r}$ and has a velocity between $\mathbf{v}$ and $\mathrm{d}\mathbf{v}$. f provides a complete description of the macroscopic state of a gas, neglecting possible nonequilibrium perturbations of the internal degrees of freedom of the molecules. Thus, f permits the calculation of many quantities of physical interest. The mean number of molecules, irrespective of velocity, which at time t are located between $\mathbf{r}$ and $\mathbf{r} + \mathrm{d}\mathbf{r}$ is $n(\mathbf{r}, t) = \int \mathrm{d}^3\mathbf{v}f$. Let $Df = (\partial f/\partial t) + \mathbf{v}(\partial f/\partial \mathbf{r}) + (\mathbf{F}/m)(\partial f/\partial \mathbf{v})$. One denotes with $\mathbf{F}$ external forces acting on the system. The BTE reads

$$Df = D_c f, \tag{A1}$$

the term on the r.h.s. accounting for the effect of molecular collisions. This is a rather formidable equation that, in general, can be tackled only by recourse to approximate treatments. Many of these involve simplified forms of $D_c f$.

Grad [13] has developed an interesting approach, the so-called MOM to tackle nonequilibrium thermodynamics via a suitable approach to the celebrated BTE. To such an end he considers the nonequilibrium state of a gas after the lapse of a time t large compared to the time of initial randomization (this time t is regarded as *fixed*). However, time t is, also, small compared to the macroscopic relaxation time τ for attaining the Maxwell–Boltzmann law f_0 on velocities. This combination of circumstances is very common indeed [16]. Now, at each point of the vessel containing the gas a state arises that is close to the *local* equilibrium state f_0 = *Maxwell–Boltzmann law on velocities*, which allows one to expand the nonequilibrium distribution f as

$$f(x, t) = f_0[1 + \epsilon\phi(x, t)], \qquad f_0 \equiv f(x, t), \tag{A2}$$

where ϵ is small. Determining the function ϕ is the MOM goal. This unknown function $\phi(x, t)$ may itself be expanded as a series of (orthogonal) Hermite–Gaussian polynomials $H_i(x)$ with coefficients $a_i(t)$ at the fixed time t,

$$\phi(x, t) = \Sigma a_i(t)H_i(x). \tag{A3}$$

It is important to note that Hermite–Gaussian polynomials are orthogonal with respect to a Gaussian kernel, i.e. the *equilibrium distribution*. No other set of functions is orthogonal (and complete) with respect to a Gaussian kernel function. Now, because of orthogonality, the unknown coefficients $a_i(t)$ relate linearly to appropriate (unknown) moments of f over velocity space (x-space), so that substituting the expansion for f into the transport equation and integrating over all velocities yields now a set of equations in the moments (which are generally a function of the fixed time value t). These equations are solvable subject to *known initial conditions*, like our expectation values. The moments now become known (*including any time dependence*). As a consequence, the coefficients $a_i(t)$ of Eq. (A3) are also known, which gives f. According to Ref. [14], the solution of the above system of equations would be equivalent to the exact solution of Boltzmann's equation (if enough a priori information were available).

Appendix B. The BTE-Grad treatment of viscosity

We tackle now the problem posed by viscosity in dilute gases by recourse to the BTE-Grad (or BTE-MOM) technique [13] because it facilitates comparison of the BTE approach with our Fisher treatment of nonequilibrium problems. Recall that (i) the first two Hermite polynomials are $H_0 = 1$; $H_1 = (1/\sqrt{2})2x$, (ii) $\phi(x, \omega) = [(\omega/\pi)]^{1/4} \exp[-x^2/2]$, so that the first two members of the Gauss–Hermite basis (of $\mathcal{L}^2$) are (1) $H_0\phi$; (2) $H_1\phi$, and (iii) n stands for the number of molecules of mass m per unit volume, while β is the inverse temperature. Since we have [16]

$$n[m\beta/(2\pi)]^{1/2}\exp[-\beta m v_z^2/2] = f_{0,z} \equiv n\psi_0^2; \qquad \text{our variables } x, \omega \text{ are}$$

$$\omega = m\beta/2; \quad 2x = \sqrt{[2\beta m]}v_z, \tag{B1}$$

which allows us to recast H_0, H_1 as $H_0 = 1$; $H_1 = \sqrt{[\beta m]}v_z$. We deal now with a 3D problem. The pertinent Gauss–Hermite basis is the set of functions

$$\left\{\psi_0(v_x)\psi_0(v_y)\psi_0(v_z)\left[1 + \sum_{l,m,n} H_l(U_x)H_m(U_y)H_n(U_z)\right]\right\}, \tag{B2}$$

where l, m, n run over all non-negative integers. As data we have here $P_{zx} = m\int \mathrm{d}^3 v f U_z U_x$. Thus, in the present instance the BTE-Grad recipe Eq. (A2) to find f [14] should be

$$f(\mathbf{U}) = f_0(\mathbf{U})[1 + aH_1(U_x)H_1(U_z)] = f_0[1 + a\beta m U_x U_z], \tag{B3}$$

with the coefficient a to be determined from the here relevant velocity-moment P_{zx} and the prior knowledge expressed by Eq. (31). Using the ansatz Eq. (B3) we have

$$P_{zx} = m\int \mathrm{d}^3 U\{f_0[1 + a\beta m U_x U_z]\}U_z U_x. \tag{B4}$$

The integral $\int \mathrm{d}^3 U\{f_0 U_x U_z\}$ vanishes by symmetry. Thus,

$$P_{zx} = am^2\beta\int \mathrm{d}^3 U\{f_0 U_x^2 U_z^2\} = am^2\beta n\langle U_x^2\rangle\langle U_z^2\rangle = \frac{am^2 n\beta}{[\beta m]^2} = n\frac{a}{\beta}, \tag{B5}$$

where the equipartition theorem has been employed. Since Eqs. (B5) and (31) have to be equal,

$$a = -\tau \frac{\partial u_x}{\partial z}; \quad \text{and} \quad f = f_0 \left[1 - U_x U_z \left(\tau \frac{\partial u_x}{\partial z} m\beta \right) \right]. \tag{B6}$$

Appendix C. The Grad treatment of electrical conductivity

We use here the result Eq. (41) and start by (i) casting $f_0 = \Pi_i^3 f_{0,i}$ and (ii) focusing attention upon the z-component of f, the only one to be affected by the electric field. The first two Hermite polynomials are $H_0 = 1$, $H_1 = (1/\sqrt{2})2x$, and, with $\phi(x, \omega) = [\omega/\pi]^{1/4} \exp[-x^2/2]$, the first two members of the Gauss–Hermite basis (of $\mathcal{L}^2$) are (1) $H_0 \phi$, (2) $H_1 \phi$. Since we have [16]

$$n[m\beta/(2\pi)]^{1/2} \exp[-\beta m v_z^2/2] = f_{0,z} \equiv n\psi_0^2,$$

our variables x, ω become $\omega = m\beta/2$, $2x = \sqrt{[2\beta m]} v_z$, which allows us to recast H_0 and H_1 as $H_0 = 1$, $H_1 = \sqrt{[\beta m]} v_z$. The Grad recipe to find f is to expand it in the fashion

$$f = f_0[1 + aH_1] = f_0[1 + a\sqrt{[\beta m]} v_z], \tag{C1}$$

with the coefficient a to be determined from appropriate velocity moments. Here we use

$$j = e \int f v_z \mathrm{d}v_z. \tag{C2}$$

The only contribution to the above integral comes from the second term in the right-hand side of Eq. (C1), and, using the equipartition theorem, $m\langle v_z^2 \rangle /2 = kT/2$, we obtain

$$j = ea\sqrt{[\beta m]} \int \mathrm{d}v_z v_z^2 f_0 = ea\sqrt{[\beta m]}(kT/m) = e \frac{a}{\sqrt{[\beta m]}}. \tag{C3}$$

We know that the current is proportional to the applied field (see Eq. (49)). The proportionally constant is the conductivity $\sigma = e^2 \tau / m$ [16], so that

$$j = e^2 \tau \mathcal{E} / m, \tag{C4}$$

so that Eqs. (C3) and (C4) give

$$a = e\tau\mathcal{E} \sqrt{\left[\frac{\beta}{m} \right]}, \tag{C5}$$

which finally leads to

$$f = f_0[1 + e\beta\tau\mathcal{E} v_z], \tag{C6}$$

in agreement with Eq. (49).

References

[1] Jaynes, E.T. (1963), Statistical Physics (Ford, W.K., eds.), Benjamin, New York; Katz, A. (1967), Statistical Mechanics, Freeman, San Francisco.
[2] Frieden, R., Plastino, A., Plastino, A.R. and Soffer, B.H. (1999), Phys. Rev. E 60, 48.
[3] Frieden, R., Plastino, A., Plastino, A.R. and Soffer, B.H. (2002), Phys. Rev. E 66, 046128.
[4] Flego, S., Frieden, B.R., Plastino, A., Plastino, A.R. and Soffer, B.H. (2003), Phys. Rev. E 68, 016105.
[5] Frieden, B.R. and Soffer, B.H. (1995), Phys. Rev. E 52, 2274.
[6] Plastino, A., Plastino, A.R., Miller, H.G. and Khana, F.C. (1996), Phys. Lett. A 221, 29.
[7] Plastino, A.R. and Plastino, A. (1996), Phys. Rev. E 54, 4423.
[8] Casas, M., Pennini, F. and Plastino, A. (1997), Phys. Lett. A 235, 457.
[9] Plastino, A.R., Casas, M. and Plastino, A. (1998), Phys. Lett. A 246, 498.
[10] Frieden, R. (1998), Physics from Fisher's Information, Cambridge University Press, Cambridge.
[11] Richards, P.I. (1959), Manual of Mathematical Physics, Pergamon Press, London, p. 342.
[12] Desloge, E.A. (1968), Thermal Physics, Holt, Rinehart and Winston, New York.
[13] Grad, H. (1949), Commun. Pure Appl. Math. 2, 331; (1958), Principles of the Kinetic Theory of Gases, Handbuch der Physik XII (Flügge, S., eds.), Springer, Berlin.
[14] Rumer, Y.B. and Ryvkin, M.S. (1980), Thermodynamics, Statistical Mechanics and Kinetics, MIR Publishers, Moscow.
[15] Frieden, R., Plastino, A., Plastino, A.R. and Soffer, B.H. (2002), Phys. Lett. A 304, 73.
[16] Reif, F. (1965), Fundamentals of Statistical and Thermal Physics, McGraw-Hill, New York.
[17] Cassels, J.M. (1970), Basic Quantum Mechanics, McGraw-Hill, New York.

Chapter 2

GENERALIZED ENTROPY AND THE HAMILTONIAN STRUCTURE OF STATISTICAL MECHANICS

Murilo Pereira de Almeida

Departamento de Física, Universidade Federal do Ceará, C.P. 6030, 60455-900 Fortaleza, Ceará, Brazil

Abstract

We adopt a Hamiltonian approach to derive the generalized thermostatistics of Tsallis. Following the normal procedure of classical statistical mechanics, we start with a Hamiltonian system composed by weakly interacting elementary subsystems and write its microcanonical and canonical distributions in terms of its structure function in phase space. It is shown that the exponential and the power-law canonical distributions emerge naturally from a unique differential relation of the structure function. More precisely, if the heat capacity of the heat bath is infinite, the canonical distribution is exponential, while if it is constant and finite, the canonical density follows a power-law behavior. We then derive the microscopic analog of the main thermodynamic potentials (temperature and entropy). The physical entropy for finite-dimensional heat baths turns out to be a version of the so-called generalized entropy of Tsallis. In the thermodynamic limit, we recover the traditional Boltzmann–Gibbs statistics. Furthermore, we show that the classical thermodynamic relation, $S = U/T - k \ln \rho_0$, *is independent of the explicit form of the canonical distribution. Finally, we discuss the issue of additivity of the physical entropy and present theoretical and numerical examples of Hamiltonian systems obeying the generalized statistics.*

Keywords: statistical mechanics; Hamiltonian systems; generalized thermostatistics; canonical distribution; entropy

1. Introduction

Statistical mechanics models and analyzes the macroscopic properties of matter through a statistical representation of its microscopic dynamics. It relies on the identification of macroscopic variables with the mean values of their respective microscopic analogs (random variables).

Being an application of probability theory, it is founded upon a *probability space*, which is a triple consisting of a sample space Ξ, a σ-algebra $\mathfrak{F}(\Xi)$ of subsets of Ξ, and a probability measure P on $\mathfrak{F}(\Xi)$. The formulation of any statistical-mechanics problem requires the specification of (i) a probability space and (ii) definite random variables that are counterparts of the macroscopic thermodynamic variables. The probability space

E-mail address: murilo@fisica.ufc.br (M.P. de Almeida).

depends on the specific problem that is under analysis, while the set of microscopic thermodynamic random variables is general and common to all problems.

There are two different ways to choose the *underlying space* for the sample space, which depends on the kind of microscopic system being studied:

(1) The first choice is applicable when the *observable system* has a separable Hamiltonian with respect to the *coordinates and momenta* of the individual microscopic subsystems. It uses the phase space of the identical microscopic subcomponents, which was denoted *μ-space* by Ehrenfest and Ehrenfest [1]. This formulation is sometimes denoted *μ-space statistics*.

(2) The second uses the entire phase space of the *observable system* (denoted Γ-space). This is the unique approach when the original problem cannot be represented by a union of independent subsystems. It is also more general than the μ-space statistics.

The Γ-space is usually the preferred approach because of its wider generality. The μ-space is adequate only for systems with independent subcomponents, like ideal gases, which are composed of identical independent molecules. In general, the underlying space is $\mathbb{R}^n$ and the sample space Ξ is the subset composed by the states that are compatible with the problem macroscopic constraints.

The next step is the specification of the events—subsets of the sample space that have a definite probability specification. They are elements of a σ-algebra, usually the Borel σ-algebra $\mathfrak{B}(\mathbb{R}^n)$, which is the σ-algebra generated by the finite-dimensional rectangles $I = I_1 \times \cdots \times I_n$, where $I_k = (a_k, b_k]$ (for details see Ref. [2], Chapter II, Section 2).

The third element of the probability space, *the probability distribution*, is given in terms of a probability density function. Stated in this general form, any density function suffices, but it must abide with a few postulates that express the physical plausibility of the model, and that determines its exact functional form.

In the lack of any prior information, we adopt a uniform distribution in the region of admissible states Ξ. Physically, this means that each microscopic state compatible with the macroscopic constraints has the same chance of being observed. Mathematically, this implies that the system is uniformly distributed over the region of admissible states (sample space). As a consequence, the probability of finding the system in a certain region of the sample space is proportional to this region's *volume*. This hypothesis is usually called *equiprobability* or the *equal a priori probability* postulate. It permits the utilization of geometrical elements of the phase space to formulate and study statistical-mechanical problems.

2. Microcanonical and canonical ensembles

The *total energy* is the most important phase function of a mechanical system. As stated by Schrödinger [3], the essential problem of statistical thermodynamics is the specification of the distribution of a given amount of energy E over N identical systems. For an isolated system, the total energy is a constant of motion. Hence, the regions of the phase space of an isolated system characterized by a constant energy value are invariants under the system's motion. Such regions are called *surfaces of constant energy*. If the Hamiltonian function does not depend explicitly on time t then we have a conservative mechanical system, and the Hamiltonian H coincides with the total energy E (see Ref. [4], Chapter VI, Section 6). Therefore, we find that H is the candidate for being the microscopic counterpart of the macroscopic internal energy U.

Let us use surfaces of constant energy to partition the sample space Ξ. Let us denote the volume of the region of Ξ where $H \leq E$ by $\Lambda(E)$, i.e.

$$\Lambda(E) = \int_{\Xi} \chi_{\{H \leq E\}}(\mathbf{q}, \mathbf{p}) \mathrm{d}\mathbf{q}\mathrm{d}\mathbf{p}, \tag{1}$$

where $\chi_{\{H \leq E\}}(\mathbf{q}, \mathbf{p})$ is the characteristic function of the region of the phase space where $H(\mathbf{q}, \mathbf{p}) \leq E$.

Let us consider cases where the region of admissible states has a finite volume Λ_0. From the equiprobability postulate we get that the distribution function of H is given by

$$F(E) = P(H \leq E) = \frac{\Lambda(E)}{\Lambda_0}. \tag{2}$$

When H takes values on a discrete set, its distribution function F is piecewise constant, changing its value at the points $E_1, E_2, \ldots$, ($P(H = E_i) = \Delta F(E_i) > 0$, where $\Delta F(E) = F(E) - F(E-)$).

When $\Lambda(E)$ is differentiable, H is an *absolutely continuous* random variable with density function given by

$$f(E) = \frac{\mathrm{d}F}{\mathrm{d}E} = \frac{\Omega(E)}{\Lambda_0}, \tag{3}$$

where $\Omega(E) = \mathrm{d}\Lambda/\mathrm{d}E$ is the *structure function*, which will play a special role in the following development.

For conservative systems, the energy $H(\mathbf{q}, \mathbf{p}) = E$ is a constant of motion. Therefore, any density function depending solely on H is also a constant of motion, which means that it is in statistical equilibrium. The microcanonical ensemble is obtained by taking the density as a constant over a region delimited by energies of values E and $E + \delta E$,

$$\rho(\mathbf{q}, \mathbf{p}) = \begin{cases} \text{constant}, & \text{if } E < H(\mathbf{q}, \mathbf{p}) < E + \delta E; \\ 0, & \text{otherwise}. \end{cases} \tag{4}$$

In the limit as δE tends to zero this distribution converges to $\sigma \times \delta E$, where $\sigma = 1/\|\nabla H\|$ is the *surface probability density* on the energy shell $H = E$, and $\|\nabla H\|$ is the Euclidean norm of the gradient of H with respect to $\mathbf{q}$ and $\mathbf{p}$.

Now let us consider a system whose phase space Ξ is subdivided into two independent regions: (i) Ξ_1, which represents the *system of interest* ($\mathcal{O}$); and (ii) Ξ_2, which acts as the *heat bath* ($\mathcal{B}$). The system $\mathcal{O}$ has mechanical parameters that are controllable by an external agent. The two regions *interact weakly* with each other, which means that the *total energy* H is just the sum of the energies of each subregion ($H = H_1 + H_2$). In probabilistic terms, H is the sum of two independent random variables H_1 and H_2, with distribution functions F_1 and F_2, respectively. Then, the distribution function of H, F_H, is the convolution of F_1 and F_2,

$$F_H(E) = \int F_1(E - \xi)\mathrm{d}F_2(\xi) = \int F_2(E - \eta)\mathrm{d}F_1(\eta). \tag{5}$$

When H_1 and H_2 are continuous random variables, we can compute the density function of H as

$$\rho_H(E) = \int \rho_1(E-\xi)\rho_2(\xi)\mathrm{d}\xi = \int \rho_2(E-\eta)\rho_1(\eta)\mathrm{d}\eta, \tag{6}$$

which leads to the convolution relation for the structure function of a system with two weakly interacting subsystems (see Ref. [5], Chapter II, Section 8)

$$\Omega(a) = \int \Omega_1(\xi)\Omega_2(a-\xi)\mathrm{d}\xi = \int \Omega_2(\eta)\Omega_1(a-\eta)\mathrm{d}\eta. \tag{7}$$

Once again let us assume the equiprobability postulate over Ξ. The conditional distribution over Ξ_1 given that H is a constant is called the *canonical distribution*. Computing this conditional density we obtain the expression for the canonical density of the subsystem $\mathcal{O}$ in its phase space Ξ_1

$$\rho_1(E_1) = \frac{\Omega_2(a-E_1)}{\Omega(a)} = \frac{\Omega_2(a)}{\Omega(a)}\frac{\Omega_2(a-E_1)}{\Omega_2(a)}, \tag{8}$$

where Ω_2 is the structure function of the heat bath $\mathcal{B}$.

We present now a result from Ref. [6] that unifies the derivation of power-law [7] and exponential (Boltzmann–Gibbs) canonical densities. We begin by defining the *generalized exponential* function on $\mathbb{R}$ for $[1+(q-1)x] > 0$, by

$$e_q(x) = \begin{cases} [1+(q-1)x]^{1/(q-1)} & \text{if } q \neq 1, \text{ and} \\ \exp(x) & \text{if } q = 1. \end{cases} \tag{9}$$

For $[1+(q-1)x] \leq 0$, $e_q(x) \equiv 0$.

The Lemma below contains a property of the functions $e_q(x)$ that will be used later in the proof of the following theorem.

Lemma 1: *The functions $e_q(x)$, $q \neq 1$, Eq. (9), are infinitely differentiable at the origin with the generic derivative of order $n \geq 1$ given by*

$$e_q^{(n)}(0) = \prod_{m=1}^{n} [m-(m-1)q]. \tag{10}$$

Here is the theorem that states the sufficient and necessary conditions for obtaining the Tsallis canonical distribution with parameter $q > 1$. A similar result can be obtained for $q < 1$, but due to space limitation it will not be shown here.

Theorem 2: *Let a be a fixed positive real number. Suppose that $\Omega_2(E)$ is a twice-differentiable real function of the real variable E, for $E \in [0,\infty)$, and that $\Omega(a) > 0$. Let us denote by $\hat{\beta}(E)$ the logarithmic derivative of $\Omega_2(E)$, i.e.*

$$\hat{\beta}(E) = \frac{\Omega_2'(E)}{\Omega_2(E)} = \frac{\mathrm{d}}{\mathrm{d}E}\ln[\Omega_2(E)], \tag{11}$$

and let γ *be the constant real value* $\gamma \equiv \hat{\beta}(a)$. *Then,*

$$\frac{\Omega_2(a - E_1)}{\Omega_2(a)} = e_q(-\gamma E_1), \;\; 0 \le E_1 \le a, \tag{12}$$

if and only if

$$\frac{\mathrm{d}}{\mathrm{d}E}\left(\frac{1}{\hat{\beta}(E)}\right) = q - 1, \tag{13}$$

with a real constant $q \ge 1$.

Proof of Theorem 2: This proof is taken from Ref. [6].
The case $q = 1$ is straightforward. Since $e_1(-\gamma E_1) \equiv \exp(-\gamma E_1)$, denoting $E = a - E_1$, Eq. (12) yields

$$\Omega_2(E)\exp(-\gamma E) = \Omega_2(a)\exp(-\gamma a), \;\; 0 \le E \le a,$$

which implies that $\Omega_2(E) = C\exp(\gamma E)$ with an arbitrary constant C. Therefore, from Eq. (11), $\hat{\beta}(E) \equiv \gamma$ is a constant function and so Eq. (13) is verified ($\mathrm{d}/\mathrm{d}E(1/\hat{\beta}(E)) = 0$). Conversely, if $\mathrm{d}/\mathrm{d}E(1/\hat{\beta}(E)) = 0$, then $\hat{\beta}(E)$ is a constant function, namely, $\hat{\beta}(E) \equiv \gamma$ and $\Omega_2(E) = C\exp(\gamma E)$ with an arbitrary constant C, and so Eq. (12) is verified.

Let us analyze the case $q > 1$. Consider first that $\Omega_2(a - E_1)/\Omega_2(a) = e_q(-\gamma E_1)$, $0 \le E_1 \le a$. Let us denote $E = a - E_1$. Then

$$\Omega_2(E) = \Omega_2(a)[1 - (q-1)\gamma a + (q-1)\gamma E]^{1/(q-1)}, \;\; 0 \le E \le a, \tag{14}$$

and

$$\Omega_2'(E) = \gamma\Omega_2(a)[1 - (q-1)\gamma a + (q-1)\gamma E]^{(1/q-1)-1}, \;\; 0 \le E \le a. \tag{15}$$

Therefore,

$$\frac{1}{\hat{\beta}(E)} = \frac{\Omega_2(E)}{\Omega_2'(E)} = \frac{1}{\gamma}[1 - (q-1)\gamma a + (q-1)\gamma E], \;\; 0 \le E \le a, \tag{16}$$

and hence, Eq. (13) is fulfilled. Conversely, the condition given by Eq. (13) is equivalent to

$$\frac{\mathrm{d}\hat{\beta}}{\mathrm{d}E} \equiv \frac{\Omega_2''}{\Omega_2} - \left(\frac{\Omega_2'}{\Omega_2}\right)^2 = (1-q)\hat{\beta}^2, \tag{17}$$

which implies that $\hat{\beta}(E)$ and $\Omega_2(E)$ are infinitely differentiable with respect to E. The iterative relation

$$\frac{\Omega_2^{(n+1)}}{\Omega_2} = \left(\frac{\mathrm{d}}{\mathrm{d}E} + \hat{\beta}\right)\left(\frac{\Omega_2^{(n)}}{\Omega_2}\right) \tag{18}$$

is easily verified and produces by induction the expression

$$\frac{\Omega_2^{(n)}}{\Omega_2} = \left(\frac{\mathrm{d}}{\mathrm{d}E} + \hat{\beta}\right)^{n-1}\hat{\beta}, \tag{19}$$

which when combined with Eq. (17) and Lemma 1, yields

$$\frac{\Omega_2^{(n)}}{\Omega_2} = \hat{\beta}^n \prod_{m=1}^{n} [m - (m-1)q] = \hat{\beta}^n e_q^{(n)}(0). \tag{20}$$

Expanding $\Omega_2(a - E_1)/\Omega_2(a)$ in a Taylor series about $E = a$ and substituting the general form for the ratio $\Omega_2^{(n)}/\Omega_2$ (Eq. (20)), we get the Taylor expansion of $e_q(-\gamma E_1)$. □

3. Temperature

In the microcanonical ensemble with total energy $H = E$, the theorem of equipartition of energy (e.g. Ref. [8], Section 1.9) states that

$$\left\langle x_i \frac{\partial H}{\partial x_j} \right\rangle = \delta_{ij} \left(\frac{\Lambda(E)}{\Omega(E)} \right) = \delta_{ij} kT, \tag{21}$$

where the pair of braces $\langle \cdot \rangle$ represents the ensemble average, x_i stands for any of the variables p_i or q_i, k is the Boltzmann constant, and T is the absolute thermodynamic temperature. This is the microcanonical definition of temperature. This definition can be extended to the canonical ensemble, where the system $\mathcal{G}$ is composed by two weakly interacting subsystems $\mathcal{G}_1$ and $\mathcal{G}_2$. For each subsystem $\mathcal{G}_i$, $(i = 1, 2)$,

$$\mathcal{T}(\mathcal{G}_i) = \frac{1}{k} \left\langle \frac{\Lambda_i(H_i)}{\Omega_i(H_i)} \right\rangle = \left(\frac{\Lambda(E)}{\Omega(E)} \right) \tag{22}$$

represents its *empirical temperature* (see Ref. [9], Section 6.2 or [8], Section 1.10).

4. Thermodynamic entropy

We restrict ourselves to systems satisfying equations similar to Eq. (13) with constants $q > 1$. As demonstrated in Ref. [10], this leads to volume functions in the form

$$\Lambda(E) = CE^{q/(q-1)}, \tag{23}$$

where C is a constant, and the energy is constrained to the interval $E \geq 0$. As a consequence, the respective structure function becomes

$$\Omega(E) = C \left(\frac{q}{q-1} \right) E^{1/(q-1)}, \tag{24}$$

and the $\hat{\beta}(E)$ function is given by

$$\hat{\beta}(E) = \frac{1}{(q-1)E}, \quad E > 0. \tag{25}$$

The parameter q is a signature of the system's phase space geometry. Almeida et al. [10] have shown that the weak composition of two systems $\mathcal{O}$ and $\mathcal{B}$, with parameters $q_{\mathcal{O}}$ and

$q_{\mathcal{B}}$, respectively, results in a system $\mathcal{G}$ with parameter $q_{\mathcal{G}}$ given by

$$\frac{q_{\mathcal{G}}}{q_{\mathcal{G}}-1} = \frac{q_{\mathcal{O}}}{q_{\mathcal{O}}-1} + \frac{q_{\mathcal{B}}}{q_{\mathcal{B}}-1}. \tag{26}$$

The temperature definition—from the theorem of equipartition of energy—leads to the expressions

$$kT = \left(\frac{q_{\mathcal{G}}-1}{q_{\mathcal{G}}}\right)E_{\mathcal{G}} = \left(\frac{q_{\mathcal{O}}-1}{q_{\mathcal{O}}}\right)\langle H_{\mathcal{O}}\rangle = \left(\frac{q_{\mathcal{B}}-1}{q_{\mathcal{B}}}\right)\langle H_{\mathcal{B}}\rangle. \tag{27}$$

Let us consider a weak combination of independent canonical subsystems $\mathcal{G}_i$, each one having an observable subsystem $\mathcal{O}_i = \mathcal{O}(\mathcal{G}_i)$ and a heat bath $\mathcal{B}_i = \mathcal{B}(\mathcal{G}_i)$. The observable and the heat bath of the composite system $\mathcal{G}$ are, respectively, the weak combinations of the observables and of the heat baths of the subsystems, i.e. $\mathcal{G} = \uplus_i \mathcal{G}_i$, $\mathcal{O}(\mathcal{G}) = \uplus_i \mathcal{O}(\mathcal{G}_i)$ and $\mathcal{B}(\mathcal{G}) = \uplus_i \mathcal{B}(\mathcal{G}_i)$. By using Eq. (8), we write the canonical distribution of $\mathcal{G}$ as

$$\rho(H_{\mathcal{O}}) = \frac{\Omega_{\mathcal{B}}(E_{\mathcal{G}} - H_{\mathcal{O}})}{\Omega_{\mathcal{G}}(E_{\mathcal{G}})}, \tag{28}$$

where $E_{\mathcal{G}}$ is the total energy of $\mathcal{G}$, $H_{\mathcal{O}}$ is the energy of its observable subsystem, and $\Omega_{\mathcal{G}}$ and $\Omega_{\mathcal{B}}$ are the structure functions of the total system $\mathcal{G}$ and of its heat bath $\mathcal{B}$, respectively. When written in the usual Tsallis density form [7], ρ becomes

$$\rho(H_{\mathcal{O}}) = \rho_0[1 - (q_{\mathcal{B}} - 1)\beta H_{\mathcal{O}}]^{1/(q_{\mathcal{B}}-1)}, \tag{29}$$

where

$$\rho_0 = \frac{\Omega_{\mathcal{B}}(E_{\mathcal{G}})}{\Omega_{\mathcal{G}}(E_{\mathcal{G}})}, \tag{30}$$

and

$$\beta = \hat{\beta}_{\mathcal{B}}(E_{\mathcal{G}}) = \frac{1}{(q_{\mathcal{B}}-1)E_{\mathcal{G}}}. \tag{31}$$

The parameters q of the systems $\mathcal{G}$, $\mathcal{O}$ and $\mathcal{B}$ are obtained by recursively applying Eq. (26), which yields

$$\frac{q_{\mathcal{G}}}{q_{\mathcal{G}}-1} = \sum_i \frac{q_{\mathcal{G}_i}}{q_{\mathcal{G}_i}-1} \tag{32}$$

$$\frac{q_{\mathcal{O}}}{q_{\mathcal{O}}-1} = \sum_i \frac{q_{\mathcal{O}_i}}{q_{\mathcal{O}_i}-1} \tag{33}$$

and

$$\frac{q_{\mathcal{B}}}{q_{\mathcal{B}}-1} = \sum_i \frac{q_{\mathcal{B}_i}}{q_{\mathcal{B}_i}-1}. \tag{34}$$

For a system with Hamiltonian $H_{\mathcal{O}}$ and probability density function ρ the internal energy U is

$$U = \int_{\Xi} H_{\mathcal{O}} \rho \mathrm{d}\mathbf{q}\mathrm{d}\mathbf{p}. \tag{35}$$

From the first law of thermodynamics we have

$$\delta U = \delta W + \delta Q, \tag{36}$$

where W is the work of the external forces and Q is the heat supplied to the system. Computing the variation δU one gets

$$\delta U = \int_{\Xi} (\delta H_{\mathcal{O}}) \rho \mathrm{d}\mathbf{q}\mathrm{d}\mathbf{p} + \int_{\Xi} H_{\mathcal{O}} (\delta \rho) \mathrm{d}\mathbf{q}\mathrm{d}\mathbf{p}. \tag{37}$$

Since the first term on the right-hand side of this equation is the variation of the work δW, we conclude that

$$\delta Q = \int_{\Xi} H_{\mathcal{O}} (\delta \rho) \mathrm{d}\mathbf{q}\mathrm{d}\mathbf{p}. \tag{38}$$

The second law of thermodynamics requires the existence of exact differential functions S, called entropies, such that $\delta S = \delta Q/T$, where the integrating factor T is an absolute temperature scale.

The functional form of the physical entropy is determined by the canonical distribution [6] and [11], Chapter 3: Given a canonical density function in Ξ, $\rho = f(H_{\mathcal{O}})$, with a non-negative monotonic function $f(x)$ defined for $x \geq 0$, the functions (entropies) that satisfy the second law of thermodynamics are given by

$$S(\rho) = \int_{\Xi} \left[\frac{1}{T} \int_0^{\rho} f^{-1}(\xi) \mathrm{d}\xi + \psi \rho \right] \mathrm{d}\mathbf{q}\mathrm{d}\mathbf{p}, \tag{39}$$

where ψ is an arbitrary constant. It will be convenient to choose $\psi = -k$, where k is the Boltzmann constant.

The Tsallis density function (Eq. (29)) can be written as

$$\rho(H_{\mathcal{O}}) = \rho_0 \mu(H_{\mathcal{O}}), \tag{40}$$

where the *shape function*, $\mu(H_{\mathcal{O}}) = \rho(H_{\mathcal{O}})/\rho_0$, is

$$\mu(H_{\mathcal{O}}) = [1 - (q_{\mathcal{B}} - 1)\beta H_{\mathcal{O}}]^{1/(q_{\mathcal{B}} - 1)}, \tag{41}$$

and

$$\rho_0 = \rho(0) = \left\{ \int_{\Xi} \mu(H_{\mathcal{O}}) \mathrm{d}\mathbf{q}\mathrm{d}\mathbf{p} \right\}^{-1}. \tag{42}$$

Applying the above procedure to the Tsallis density function, constraining ρ_0 to be a constant, we get the candidate entropy function $\mathcal{E}(\rho)$,

$$\mathcal{E}(\rho) = \frac{1}{T(q_{\mathcal{B}} - 1)\beta} \left[1 - \frac{\rho_0}{q_{\mathcal{B}}} \int_{\Xi} \left(\frac{\rho}{\rho_0} \right)^{q_{\mathcal{B}}} \mathrm{d}\mathbf{q}\mathrm{d}\mathbf{p} \right] - k. \tag{43}$$

Collecting and rearranging terms we get the alternative expression

$$\mathcal{E}(\rho) = \left(\frac{1}{q_{\mathcal{B}} T\beta}\right)\rho_0 S_{\mathrm{TS}}(\rho/\rho_0) + \left(\frac{1}{q_{\mathcal{B}} T\beta} - k\right), \tag{44}$$

where

$$S_{\mathrm{TS}}(\rho/\rho_0) = \frac{1}{q_{\mathcal{B}} - 1}\int_{\Xi}\left[\left(\frac{\rho}{\rho_0}\right) - \left(\frac{\rho}{\rho_0}\right)^{q_{\mathcal{B}}}\right]\mathbf{dqdp} \tag{45}$$

is the Tsallis entropy of the shape function $\mu = \rho/\rho_0$. Observe in Eq. (44) the splitting of the roles played by the shape function, ρ/ρ_0, and by the normalizing constant ρ_0.

Now, by varying both ρ and ρ_0, the evaluation of $\delta\mathcal{E}$ will produce the sought expression for the entropy. From Eq. (43) we have that

$$\delta\mathcal{E} = \frac{1}{T(q_{\mathcal{B}} - 1)\beta}\left[-\int_{\Xi}\left(\frac{\rho}{\rho_0}\right)^{q_{\mathcal{B}}-1}(\delta\rho)\mathbf{dqdp} + (\delta\rho_0)\frac{q_{\mathcal{B}} - 1}{q_{\mathcal{B}}}\int_{\Xi}\left(\frac{\rho}{\rho_0}\right)^{q_{\mathcal{B}}}\mathbf{dqdp}\right]. \tag{46}$$

Substituting the expressions for ρ and ρ_0, and considering the temperature expressions, Eq. (27), we get

$$\delta\mathcal{E} = \frac{1}{T}\int_{\Xi} H_{\mathcal{O}}(\delta\rho)\mathbf{dqdp} + k\delta(\ln\rho_0) = \frac{1}{T}\delta Q + k\delta(\ln\rho_0), \tag{47}$$

and from it we conclude that

$$\delta S = \frac{1}{T}\delta Q = \delta\mathcal{E} - k\delta(\ln\rho_0), \tag{48}$$

and so, the generic entropies are

$$S(\rho) = \mathcal{E}(\rho) - k\ln\rho_0 + C, \tag{49}$$

where C is an arbitrary constant.

Using Eqs. (27) and (31) we find that

$$\frac{1}{q_{\mathcal{B}} T\beta} = k\left(\frac{q_{\mathcal{G}}}{q_{\mathcal{G}} - 1}\right)\left(\frac{q_{\mathcal{B}} - 1}{q_{\mathcal{B}}}\right) = k\left(\frac{q_{\mathcal{O}}}{q_{\mathcal{O}} - 1}\right)\left(\frac{q_{\mathcal{B}} - 1}{q_{\mathcal{B}}}\right) + k \tag{50}$$

depends only on the Boltzmann constant k and on the geometric constants $q_{\mathcal{O}}$ and $q_{\mathcal{B}}$. From this last equation, we see that the commonly used relation $kT\beta = 1$ is valid only when $q_{\mathcal{O}} = 1/2$ or, for $q_{\mathcal{O}} \neq 1/2$, when $q_{\mathcal{B}} \rightarrow 1$, i.e. in the Boltzmann–Gibbs statistics. This fact has been disregarded by some authors who inadvertently assume that $kT = 1/\beta$ within Tsallis statistics.

4.1. Variational principle

It can be readily verified that the maximization of $S(\rho)$, given by Eqs. (43) and (49), under the conditions

$$\text{(i)} \int \rho \mathrm{d}\mathbf{q}\mathrm{d}\mathbf{p} = 1, \quad \text{(ii)} \int H_{\mathcal{O}} \rho \mathrm{d}\mathbf{q}\mathrm{d}\mathbf{p} = U \qquad \text{and} \qquad \text{(iii)} \; \rho_0 = \rho(0), \tag{51}$$

results in the Tsallis density function $\rho(H_{\mathcal{O}})$ given by Eqs. (40)–(42).

5. Boltzmann–Gibbs canonical ensemble

The Boltzmann–Gibbs canonical ensemble is obtained by considering an infinite heat bath, which is represented by the limit condition $\mathbf{q}_{\mathcal{B}} \to 1$, keeping β fixed. Taking this limit in Eqs. (29) and (43) we arrive at

$$\rho(H_{\mathcal{O}}) = \rho_0 \exp(-\beta H_{\mathcal{O}}), \tag{52}$$

and

$$\mathcal{E}(\rho) = \frac{1}{T\beta}\left\{1 - \int_{\Xi} \rho \ln(\rho) \mathrm{d}\mathbf{q}\mathrm{d}\mathbf{p} + \ln(\rho_0)\right\} - k, \tag{53}$$

from which, considering also that in the limit $q_{\mathcal{B}} \to 1$ we have $1/(T\beta) = k$ (see Eq. (50) above), we get

$$S(\rho) = \mathcal{E}(\rho) - k \ln(\rho_0) + C = -k \int_{\Xi} \rho \ln(\rho) \mathrm{d}\mathbf{q}\mathrm{d}\mathbf{p} + C, \tag{54}$$

exactly the Boltzmann–Gibbs–Shannon entropy plus an arbitrary constant C.

6. Tsallis ensemble and the additivity of $\mathcal{E}$

Let us compute the explicit value of $\mathcal{E}_{\mathcal{G}}(\rho)$ for a system $\mathcal{G}(= \mathcal{O} \uplus \mathcal{B})$ whose canonical distribution on $\Xi_{\mathcal{O}}$, $\rho(H_{\mathcal{O}})$, is given in Eq. (29). From a direct integration we get

$$\rho_0 \int_{\Xi_{\mathcal{O}}} \left(\frac{\rho}{\rho_0}\right)^{q_{\mathcal{B}}} \mathrm{d}\mathbf{q}\mathrm{d}\mathbf{p} = \left[1 - \frac{\langle H_{\mathcal{O}} \rangle}{E_{\mathcal{G}}}\right]. \tag{55}$$

From Eqs. (27) and (31) we get

$$\frac{1}{T(q_{\mathcal{B}} - 1)\beta} = k\left(\frac{q_{\mathcal{G}}}{q_{\mathcal{G}} - 1}\right). \tag{56}$$

Substituting Eqs. (55) and (56) into Eq. (43) and using Eq. (26) one gets that

$$\mathcal{E}_{\mathcal{G}}(\rho) = k\left(\frac{q_{\mathcal{O}}}{q_{\mathcal{O}} - 1}\right). \tag{57}$$

Using the combination rule for the $q_{\mathcal{O}}$s, Eq. (33), we observe that under *weak combination* of subsystems the function $\mathcal{E}_{\mathcal{G}}$ is additive. In general, if $\mathcal{G}$ is a (weak) combination of

N subsystems $\mathcal{G}_i$, $i = 1, ..., N$, $(\mathcal{G} = \mathcal{G}_1 \uplus \mathcal{G}_2 \uplus \cdots \uplus \mathcal{G}_N)$, then

$$\mathcal{E}_{\mathcal{G}}(\rho) = \sum_{i=1}^{N} \mathcal{E}_{\mathcal{G}_i}(\rho_i). \tag{58}$$

Note that $\mathcal{E}_{\mathcal{G}}(\rho)$ does not depend on the parameter $q_{\mathcal{B}}$. Hence, Eq. (57) is valid in the limit case $\mathbf{q}_{\mathcal{B}} \rightarrow 1$ (Boltzmann–Gibbs), which can also be verified through a direct evaluation of Eq. (53) with $\rho = \rho_0 \exp(-\beta H_{\mathcal{O}})$. Observe that Eq. (27) together with Eq. (57) leads to the relation between $\mathcal{E}_{\mathcal{G}}$, the internal energy $U = \langle H_{\mathcal{O}} \rangle$ and the temperature T,

$$\mathcal{E}_{\mathcal{G}}(\rho) = \frac{U}{T}. \tag{59}$$

Substituting the alternative forms of $\mathcal{E}(\rho)$ into Eq. (49), we obtain

$$S(\rho) = k\left(\frac{q_{\mathcal{O}}}{q_{\mathcal{O}} - 1}\right) - k \ln \rho_0 + C, \tag{60}$$

and

$$S(\rho) = \frac{U}{T} - k \ln \rho_0 + C. \tag{61}$$

This last expression was also obtained, through a different route, in Ref. [12].

7. Additivity of $k \ln \rho_0$

Let us consider the system $\mathcal{G}$ as a weak combination of n identical subsystems $\mathcal{G}_i$, $i = 1, ..., n$, of which each in turn is a weak combination of an observable $\mathcal{O}_i$ and a heat bath $\mathcal{B}_i$. Let q, q_1 and q_2 be the q-parameters and c, c_1 and c_2 be the multiplicative constants of the volume functions (Eq. (23)) of $\mathcal{G}_i$, $\mathcal{O}_i$ and $\mathcal{B}_i$, respectively. These constants are related to each other through the expressions

$$\frac{q}{q-1} = \frac{q_1}{q_1 - 1} + \frac{q_2}{q_2 - 2}, \tag{62}$$

and

$$c = c_1 c_2 \frac{\Gamma\left(\frac{q_1}{q_1 - 1} + 1\right)\Gamma\left(\frac{q_2}{q_2 - 1} + 1\right)}{\Gamma\left(\frac{q}{q-1} + 1\right)}, \tag{63}$$

where $\Gamma(\,)$ is the gamma function. Therefore, we have that the parameters for the system $\mathcal{G}$, its observable part $\mathcal{O}$, and its heat bath $\mathcal{B}$ are, respectively, $q_{\mathcal{G}}$, $q_{\mathcal{O}}$ and $q_{\mathcal{B}}$, and they satisfy the relations of Eqs. (32)–(34), i.e.

$$\frac{q_{\mathcal{G}}}{q_{\mathcal{G}} - 1} = \sum_i \frac{q_{\mathcal{G}_i}}{q_{\mathcal{G}_i} - 1} = n \frac{q}{q-1} \tag{64}$$

$$\frac{q_{\mathcal{O}}}{q_{\mathcal{O}}-1}=\sum_i \frac{q_{\mathcal{O}_i}}{q_{\mathcal{O}_i}-1}=n\frac{q_1}{q_1-1} \tag{65}$$

and

$$\frac{q_{\mathcal{B}}}{q_{\mathcal{B}}-1}=\sum_i \frac{q_{\mathcal{B}_i}}{q_{\mathcal{B}_i}-1}=n\frac{q_2}{q_2-1}. \tag{66}$$

Recursively applying Eq. (7) to compute $\Omega_{\mathcal{B}}(E_{\mathcal{G}})$ and $\Omega_{\mathcal{G}}(E_{\mathcal{G}})$ we obtain

$$\Omega_{\mathcal{B}}(E_{\mathcal{G}})=\frac{\left[c_2\Gamma\left(\frac{q_2}{q_2-1}+1\right)\right]^n}{\Gamma\left(\frac{q_{\mathcal{B}}}{q_{\mathcal{B}}-1}\right)}E_{\mathcal{G}}^{1/(q_{\mathcal{B}}-1)} \tag{67}$$

and

$$\Omega_{\mathcal{G}}(E_{\mathcal{G}})=\frac{\left[c\Gamma\left(\frac{q}{q-1}+1\right)\right]^n}{\Gamma\left(\frac{q_{\mathcal{G}}}{q_{\mathcal{G}}-1}\right)}E_{\mathcal{G}}^{1/(q_{\mathcal{G}}-1)}. \tag{68}$$

Substituting Eqs. (67) and (68) into Eq. (30), we obtain

$$\rho_0=\frac{\Gamma\left(\frac{q_{\mathcal{G}}}{q_{\mathcal{G}}-1}\right)}{\Gamma\left(\frac{q_{\mathcal{B}}}{q_{\mathcal{B}}-1}\right)}\left[c_1\Gamma\left(\frac{q_1}{q_1-1}+1\right)E_{\mathcal{G}}^{q_1/(q_1-1)}\right]^{-n}. \tag{69}$$

We may write ρ_0 (Eq. (69) above) in terms of $\Lambda_{\mathcal{O}}(E_{\mathcal{G}})$ that, when substituted into Eq. (49), yields

$$S(\rho)=k\left(\frac{q_{\mathcal{O}}}{q_{\mathcal{O}}-1}\right)+k\ln\left[\frac{\Gamma\left(\frac{q_{\mathcal{O}}}{q_{\mathcal{O}}-1}+1\right)\Gamma\left(\frac{q_{\mathcal{B}}}{q_{\mathcal{B}}-1}\right)}{\Gamma\left(\frac{q_{\mathcal{G}}}{q_{\mathcal{G}}-1}\right)}\right]+k\ln[\Lambda_{\mathcal{O}}(E_{\mathcal{G}})]+C. \tag{70}$$

The two last terms on the right-hand side of the last equation are the microcanonical entropy (see Ref. [8], Section 1.12) obtained by isolating the system $\mathcal{O}$ on a shell of constant energy with $H_{\mathcal{O}}\equiv E_{\mathcal{G}}$.

From Eq. (27), the total energy $E_{\mathcal{G}}$ in terms of the temperature T and q is

$$E_{\mathcal{G}}=kT\left(\frac{q_{\mathcal{G}}}{q_{\mathcal{G}}-1}\right)=nkT\left(\frac{q}{q-1}\right). \tag{71}$$

From these two last equations we get

$$\ln \rho_0 = \ln\left[\frac{\Gamma\left(\frac{nq}{q-1}\right)}{\Gamma\left(\frac{nq_2}{q_2-1}\right)}\right] - n\left(\frac{q_1}{q_1-1}\right)\ln[n] \tag{72}$$

$$-n\left(\frac{q_1}{q_1-1}\right)\ln\left[kT\left(\frac{q}{q-1}\right)\right] - n\ln\left[c_1\Gamma\left(\frac{q_1}{q_1-1}+1\right)\right]. \tag{73}$$

The terms in Eq. (73) are additive since they are linear in n, but the ones in Eq. (72) are not. We can eliminate this nonadditivity by taking the constant C appearing in Eq. (49) to cancel these nonadditive terms. This adequate choice of the constant $C = C(n; q, q_1, q_2)$ is analogous to the solution presented by Gibbs to the paradox bearing his name. If we include in the constant all the terms that do not depend on the temperature T, we are left solely with the term,

$$-k\ln\rho_0 + C = kn\left(\frac{q_1}{q_1-1}\right)\ln[T] = k\left(\frac{q_O}{q_O-1}\right)\ln[T]. \tag{74}$$

Combining Eqs. (49), (57) and (74), the resulting entropy function becomes

$$S(\rho) = k\left(\frac{q_O}{q_O-1}\right)[1+\ln[T]] = \frac{U}{T}[1+\ln[T]]. \tag{75}$$

Since the temperature T is an intensive variable and the internal energy is additive, this entropy form is additive.

8. Numerical example

The following numerical example corroborates the ideas discussed above. It was presented by Adib et al. [13], who investigated through numerical simulation the statistical properties of a linear chain of anharmonic oscillators, with Hamiltonian

$$H = \sum_{i=1}^{N}\frac{p_i^2}{2} + \sum_{i=1}^{N}\frac{q_i^4}{4} + \sum_{i=1}^{N}\frac{(q_{i+1}-q_i)^4}{4}. \tag{76}$$

This system was inspired by the so-called Fermi–Pasta–Ulam (FPU) problem, originally devised to test whether statistical mechanics is able to describe dynamical systems with a small number of particles [14]. The equations of motion are integrated numerically together with the following set of initial conditions:

$$q_i(0) = 0, \qquad p_i(0) = \sum_{k=1}^{4}\cos\left(\frac{2\pi ik}{N} + \psi_k\right), \quad i = 1, \ldots, N, \tag{77}$$

where ψ_k is a random number within $[0, 2\pi)$.

The observable system, H_1, is taken to be the kinetic energy and the heat bath is composed by the two quartic terms. As demonstrated in Ref. [13] the volume function of the heat bath is

$$\Lambda(E) \propto E^{N/4}. \tag{78}$$

As a consequence, the canonical density of the kinetic energy in the momenta phase space is given by

$$\rho(H_1(\mathbf{p}_1, \ldots, \mathbf{p}_N)) \propto \left[H - \frac{1}{2} \sum_{i=1}^{N} \mathbf{p}_i^2 \right]^{1/(q-1)}, \tag{79}$$

where the constant q is given by

$$\frac{1}{q-1} = \frac{N}{4} - 1 \Rightarrow q = \frac{N}{N-4}. \tag{80}$$

In Fig. 1 we show the logarithmic plot of the distribution $\rho(\mathbf{p}_1, \mathbf{p}_2, \ldots)$ against the transformed variable $H - H_1$ for systems with $N = 8$, 16, 32 and 64 oscillators. Based on the above result, we assume ergodicity and compute the distribution of momenta from the time distribution of $H_1, f(H_1)$, through the relation

$$f(H_1) \propto H_1^{(N/2)-1} \rho(\mathbf{p}_1, \mathbf{p}_2, \ldots), \tag{81}$$

where the $H_1^{(N/2)-1}$ factor accounts for the degeneracy of the momenta consistent with the magnitude of H_1 (see Ref. [5]). Indeed, we observe in all cases that the fluctuations in H_1 follow very closely the prescribed power-law behavior Eq. (79), with exponents given by

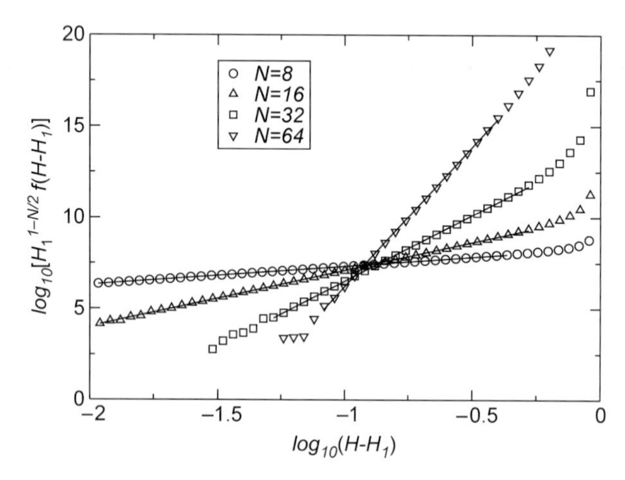

Fig. 1. Logarithmic plot of the rescaled distribution $H_1^{1-N/2} f(H - H_1)$ as a function of the transformed variable $H - H_1$ for $N = 8$ (circles), 16 (triangles up), 32 (squares), and 64 (triangles down) anharmonic oscillators. The solid straight lines are the best fit to the simulation data of the expected power-law behavior (Eq. (79)). The slopes are 1.0068 (1.0), 3.07 (3.0), 7.21 (7.0) and 15.27 (15.0) for $N = 8$, 16, 32 and 64, respectively. The numbers in parentheses indicate the expected values obtained from Eq. (80). The departure from the power-law behavior at the extremes of the curves is due to finite-time sampling.

Eq. (80). These results, therefore, provide clear evidence for the validity of our dynamical approach to the generalized thermostatistics.

9. Conclusions

This chapter contains a collection of results from the author's and collaborators' research about Tsallis thermostatistics—Adib et al. [13], Almeida [6], Almeida et al. [10] and Almeida [15]. It represents an alternative approach to Tsallis statistics based upon geometrical concepts of the phase space of Hamiltonian systems, namely, volume and structure functions.

We conclude that the existence of power-law canonical distributions is perfectly justified from first principles. They appear when the derivative with respect to E of $1/\hat{\beta}(E)$ is a constant. This implies that the generalized entropies are the adequate entropy forms for the case where this derivative is a positive constant, which may be considered as a condition of a heat bath with finite heat capacity. The Boltzmann–Gibbs entropy is the right form for systems with an ideal heat bath, i.e. infinite heat capacity.

We have seen that the Boltzmann–Gibbs canonical distributions can be obtained from the Tsallis distribution. We have obtained an expression for the physical entropy of weak-interaction systems with finite-dimensional heat baths ($q_{\mathcal{B}} > 1$) and showed that, when the heat bath becomes infinite-dimensional ($q_{\mathcal{B}} \rightarrow 1$), it becomes the classical Boltzmann–Gibbs–Shannon entropy. Moreover, we saw that this physical entropy can be written as a variant of Tsallis entropy plus the product of the Boltzmann constant and the logarithm of the partition function. In addition, we saw that it may be turned into an additive function by appropriately choosing the constants, in a procedure reminiscent of what was done by Gibbs to overcome the Gibbs paradox of the microcanonical entropy of ideal gases.

Our results strengthen the physical meaning of the generalized entropy by showing that the generalized thermostatistics fills the gap between the microcanonical and the Boltzmann–Gibbs canonical ensembles.

In the literature on Tsallis statistics, it is common to find references to *generalized expectations and escort distributions*, which are used to define generalized thermodynamic potentials that will display the usual mathematical structure of thermodynamics. However, in our approach, we use the standard expectation and it is not necessary to redefine the potentials. Only the entropy has an apparent different form, and this happens only when the volume of the heat-bath phase-space region with energy less than or equal to E is a power of E. Even in this case, when the expression of the canonical distribution is replaced in the expression of the entropy, the usual relation ($S = U/T - k \ln \rho_0$) is recovered. This shows that the mathematical framework of thermodynamics is preserved without any further modification.

Another difference from the general literature on Tsallis thermostatistics is that our analysis does not follow the Jaynes approach of statistical mechanics, which starts with the definition of the entropy function and derives the probability density function through an optimization problem. Instead, we begin the study knowing the probability density function through the geometry of the phase space, and from it we derive the entropy by requiring that it obey the second law of thermodynamics. The expressions for both the probability density and entropy are derived from general principles of mechanics. None of them is assumed ad hoc.

References

[1] Ehrenfest, P. and Ehrenfest, T. (1911), Enzyklopädie der mathematischen Wissenschaften, Vol. IV, B.G. Teubner, Leipzig, Part 32.
[2] Shiryayev, A.N. (1984), Probability. Graduate Texts in Mathematics 95, Springer, New York.
[3] Schrödinger, E. (1989), Statistical Thermodynamics, Dover, New York.
[4] Lanczos, C. (1970), The Variational Principles of Mechanics, 4th edn, Dover, New York.
[5] Khinchin, A.I. (1949), Mathematical Foundations of Statistical Mechanics, Dover, New York.
[6] Almeida, M.P. (2001), Generalized entropies from first principles, Physica A 300, 424–432.
[7] Tsallis, C. (1988), Possible generalization of Boltzmann–Gibbs statistics, J. Stat. Phys. 52, 479–487.
[8] Münster, A. (1969), Statistical Thermodynamics, Vol. I, Springer, Berlin.
[9] Huang, K. (1987), Statistical Mechanics, 2nd edn, Wiley, New York.
[10] Almeida, M.P., Potiguar, F.Q. and Costa, U.M.S. (2002), Microscopic analog of temperature within nonextensive thermostatistics, cond-mat/0206243.
[11] Weiner, J.H. (1983), Statistical Mechanics of Elasticity, Wiley, New York.
[12] Potiguar, F.Q. and Costa, U.M.S. (2002), Thermodynamics arising from Tsallis thermostatistics, cond-mat/0208357.
[13] Adib, A.B., Moreira, A.A., Andrade, J.S., Jr. and Almeida, M.P. (2003), Tsallis thermostatistics for finite systems: a Hamiltonian approach, Physica A 322, 276–284.
[14] Fermi, E., Pasta, J., Ulam, S. and Tsingou, M. (1965), Studies of Nonlinear Problems I. In Los Alamos preprint LA-1940 (November 7, 1955); Reprinted in E. Fermi, Collected Papers, Vol. II, University of Chicago Press, Chicago, p. 978.
[15] Almeida, M.P. (2003), Thermodynamical entropy (and its additivity) within generalized thermostatistics, Physica A 325, 426–438.

HYDRODYNAMICS AND CONTINUUM MECHANICS

Chapter 3

SOME OBSERVATIONS OF ENTROPY EXTREMA IN PHYSICAL PROCESSES

David M. Paulus*

Institut für Energietechnik, Technische Universität Berlin, Marchstraße 18 10587, Berlin, Germany

Richard A. Gaggioli†

Department of Mechanical and Industrial Engineering, Marquette University, 1515 West Wisconsin Avenue, Milwaukee, WI 52333, USA

Abstract

This chapter studies entropy extremization in certain fluid flow, heat transfer and reactive processes. Included are (a) three cases for modeling steady processes, where extremization is employed in lieu of traditional assumptions, (b) three cases for transient processes, where extremization is used explicitly to get the final, equilibrium configuration, (c) two cases where extremization establishes limiting constraints upon steady processes, and (d) two cases where, among alternative processes, extremization determines the stable one. Our intention: to present special cases, which contribute to the dossier of information and, in turn, insight that is relevant to development of a general extremum principle.

Keywords: entropy; extremization; equilibrium; limitations; assumptions; stability; available energy

1. Introduction

This chapter examines the extremization of entropy's production (and therefore available energy's destruction) in several situations—kinetic processes (including both incompressible and compressible flows), and in static cases such as chemical equilibrium. Some familiar and some new 'case studies' are presented. The motivation is to gain insight that may be helpful in the quest for a general extremum principle in thermodynamics.

2. Kinetic systems—incompressible flow

The behavior of some kinetic systems corresponds to minimizing the rate of entropy production and that of others to maximizing the rate of entropy production. Some systems

* Tel.: +49-30-314-25106; fax: +49-30-314-21683.
E-mail address: d.paulus@iet.tu-berlin.de (D.M. Paulus).
† Tel.: +1-414-288-1717.
E-mail address: richard.gaggioli@marquette.edu (R.A. Gaggioli).

minimize their rate of entropy production for one set of boundary conditions and maximize for another.

As is well known from experiments, for steady, incompressible flows, laminar flow persists at low Reynolds numbers and turbulent flow is persistent at high Reynolds numbers. It will now be shown that (a) when flowrate is specified, the persistent flow is that with the higher rate of entropy production, and (b) when the pressure drop is specified, the persistent flow is that with the lower rate.

2.1. *Case 1, boundary condition: specified flowrate*

First to be examined will be Fig. 1, which is an adaptation of the 'Moody Diagram' for incompressible flow in a smooth duct. On the abscissa (x-axis) is the Reynolds number, which for our purposes can be viewed as a dimensionless flowrate,

$$Re = 4\dot{m}/\pi D\mu. \tag{1}$$

The ordinate (y-axis) represents a dimensionless rate of entropy production, which is equal to twice the Fanning friction factor (e.g. see Ref. [1]):

$$\sigma = \frac{\dot{S}_{\pi} D T_{\mathrm{I}}}{L\dot{m}\mathbf{V}^2} = 2f(Re, \varepsilon/D). \tag{2}$$

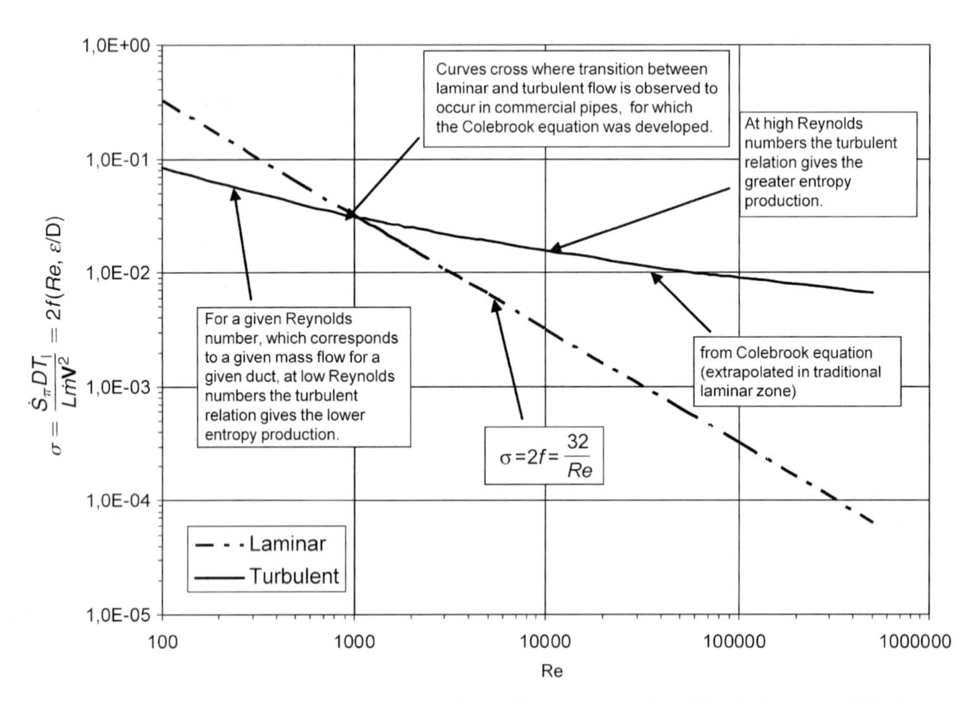

Fig. 1. Dimensionless entropy production σ versus Reynolds number Re for adiabatic, incompressible flow in long, smooth ducts.

For laminar flow, solution of the Navier–Stokes equations results in $4f = 64/Re$, which corresponds to the steeper, dot-dash curve in Fig. 1. The other curve corresponds to the empirical turbulent flow friction factor, as approximated (for convenience) by the Colebrook formula.[1]

As is well known, experiments demonstrate that persistent (i.e. 'stable') flow is represented by the laminar flow line (dot-dash) to the left of the intersection of the two curves, and by the turbulent flow (solid) line to the right. That is, flow tends to be laminar at low Reynolds numbers and turbulent at higher Reynolds numbers. Nevertheless, the laminar-flow curve at high Reynolds numbers *does* represent an actual mathematical solution to the classic governing equations (Navier–Stokes). Moreover, such laminar flows are observed experimentally, but are unstable. That is, they do not persist when the flow is subject to a disturbance; then, turbulent flow ensues.

Presuming that the Navier–Stokes equations are valid at high Reynolds numbers, that system of equations *must* then have at least two mathematical solutions: (1) that corresponding to laminar flow, referred to above, and (2) a solution, whether or not it will ever be found in closed form, corresponding to the turbulent flow observed at high Re. The Colebrook equation, formulated to represent experimental flow data represents the results that such a turbulent flow solution would yield, at high Re. Moreover, if it were *surmised* that a turbulent-flow mathematical solution exists at low Re too, then the portion of the Colebrook turbulent-flow curve to the left of the intersection in Fig. 1 could—and probably would—approximate the results of such a solution.

When the flowrate is specified as a boundary condition, then the Reynolds number $4\dot{m}/\pi D\mu$ is also specified. When Re below 1000 are observed in Fig. 1, laminar flow—which is known to be the stable situation—has the higher rate of entropy production, compared to turbulent. On the other hand, at high Re, where laminar flow is unstable and will not persist, but turbulent flow does persist, the turbulent flow has the greater entropy production rate. Hence, *at any specified flowrate, the stable (i.e. persistent) flow is the one that yields the higher entropy-production rate.*[2]

Moreover, in each case, the *higher* rate of entropy production *may be* the *maximum* rate. For, it is not illogical to *assume* that the stable solution represents an *extremum* in entropy-production rate. After all, it cannot be happenstance that the system functions stably at the higher of two entropy-production rates. Why would not it go to yet a higher rate if (consistent with the classical governing equations) it could? So, the authors propose the following *hypothesis* for consideration: *When mass flowrate is specified, stable flow corresponds to that mathematical solution of the classical governing equations that maximizes the time rate of entropy production.*

There may be a little more to the latter statement than is apparent at first. We argued earlier that there must be at least two mathematical solutions to the governing

---

[1] For smooth ducts, the Colebrook equation [16] reduces to: $1/\sqrt{4f} = -2.0 \log_{10}(2.51/Re\sqrt{4f})$.

[2] It is likely that the reader has noticed that, in Fig. 1, the laminar and turbulent curves intersect at $Re \approx 1000$ rather than in the range of $2000 < Re < 4000$. This range, usually considered to be the *actual* transition region, corresponds with Nikuradse's measurements for *artificially* roughened pipes—where a roughness was created using sand particles of uniform diameter. For the case of *real*, commercial pipes the transition begins at $Re < 1000$, and the Colebrook equation was *designed* to represent the friction factor from there to high Re. See Ref. [19], Sections 34 and 35.

equations, at least at high Re. We then *surmised* that two exist at low Re as well. Now, consistent with the hypothesis presented in the last paragraph, we will conjecture even further: that there likely exists an infinity of mathematical solutions 'between' laminar and turbulent.

This likely infinity of mathematical solution could help explain the transition region between laminar and turbulent flow (i.e. $1000 < Re < 4000$). At any given Reynolds number in the transition region one of these solutions likely gives a higher entropy production than purely laminar or purely turbulent flow, and the persistent flow is of a mixed nature.

2.2. *Case 2, boundary condition: specified pressure drop*

Whereas, when mass flowrate is a specified boundary condition, the stable flow corresponds to that mathematical solution with the greater entropy production, it will now be shown that when the pressure drop is specified, the stable flow corresponds to the solution with the lesser rate of entropy production.

First to be mentioned is that for incompressible, adiabatic flow, entropy production per unit mass and pressure drop are essentially the same information. That is, an entropy balance and property relations lead to the following relationship

$$-\Delta p = \rho c T_{\mathrm{I}}(\mathrm{e}^{s_{\pi}/c} - 1). \tag{3}$$

Here s_{π} is the entropy production per unit mass, $\dot{S}_{\pi}/\dot{m}$. Because, during incompressible, adiabatic flow, the temperature, density and specific heat are essentially constant. Eq. (3) shows that specifying a per unit mass entropy production fixes a pressure drop, and vice versa.

In order to investigate behavior when the pressure drop over a length of duct is specified, the dimensionless parameters used in Case 1 need to be modified. This is because, with Fig. 1, it is difficult to judge entropy-production minimization or maximization for a given Δp (or s_{π}). That is,

$$\sigma = \frac{\dot{S}_{\pi} D T_{\mathrm{I}}}{L\dot{m}\mathbf{V}^2} = \frac{s_{\pi} D T_{\mathrm{I}}}{L\mathbf{V}^2}. \tag{4}$$

So, σ has dependence on the velocity, which is an unknown (to be determined at the specified pressure drop). However, this difficulty is removed by multiplying by the Reynolds number squared, yielding another dimensionless rate of entropy production.

$$\dot{\sigma} = \sigma Re^2 = \left(\frac{\dot{S}_{\pi} D T_{\mathrm{in}}}{L\dot{m}\mathbf{V}^2}\right)\left(\frac{\rho \mathbf{V} D}{\mu}\right)^2 = \frac{s_{\pi}\rho^2 D^3 T_{\mathrm{I}}}{L\mu^2}. \tag{5}$$

Then, using Eq. (2),

$$\sigma^* = 2fRe^2. \tag{6}$$

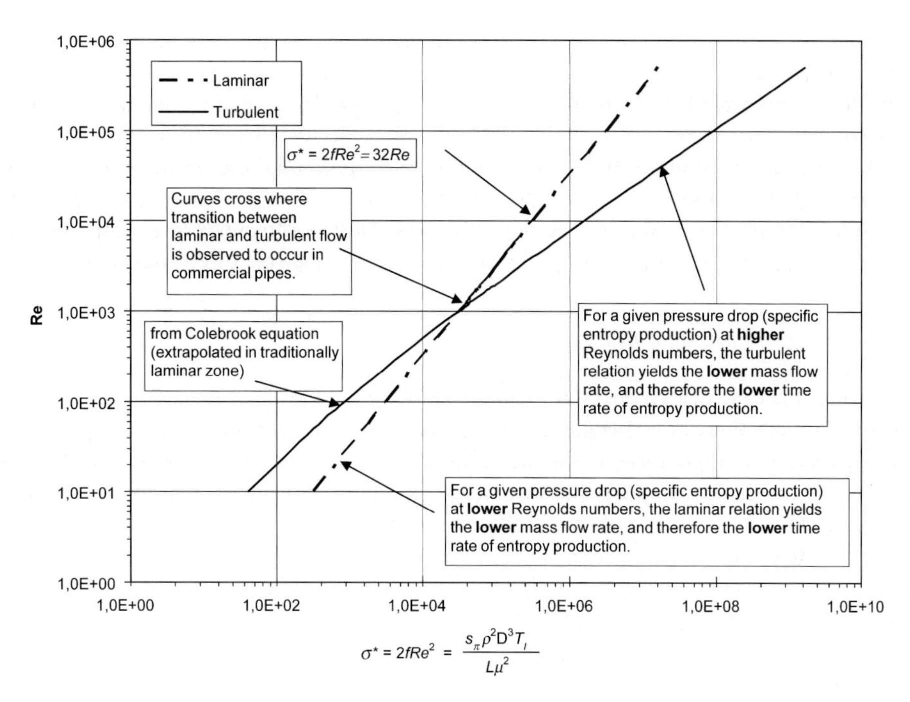

Fig. 2. Reynolds number versus dimensionless entropy production, σ^* for adiabatic, incompressible flow in long, smooth ducts.

Fig. 2 shows *Re* (ordinate) versus σ^* (abscissa). The steeper (dot-dash) curve is for laminar flow, created using $4f = 64/Re$. The other (solid) curve is based on the Colebrook equation for turbulent flow.[3]

We know that at low Reynolds numbers the stable flow is laminar and at high Reynolds it is turbulent. Hence it can be seen that, at any specified σ^*, corresponding to a specified pressure-drop boundary condition, the stable flow is that which has the lower Reynolds number. As the Reynolds number is equal to $4\dot{m}/\pi D\mu$, a lower Reynolds number corresponds to a lower mass flowrate. Since the *per unit mass* entropy production is fixed (due to the specified pressure drop), a lower mass flow corresponds to a lower *time rate* of entropy production. That is, *at any specified pressure drop, the stable flow is the one that yields the lower entropy-production rate*.

Then, using a similar rationale as in Case 1, the authors propose: *when pressure drop is specified, stable flow corresponds to that mathematical solution of the classical governing equations that minimizes the time rate of entropy production*.

Or, rewording, the system wants to minimize its rate of consumption of available energy, for the case of a specified pressure drop.

[3] It is interesting to note that Rouse [17], Section 34, solved the problem of finding a mass flow when given a pressure drop, without trial and error, by plotting f (which equals $1/2\sigma$) versus $Re\sqrt{f}$ *left*(which, equals$\sqrt{\sigma^*/2}$). See also Ref. [18], Article 6.2.

2.3. Summary—incompressible flow in a single duct

For incompressible flow in a duct,

(1) When the flowrate is specified, the flow proceeds to behave in accord with the mathematical solution with the *greater* rate—probably the *maximum* rate—of entropy production.

(2) When the pressure drop is specified, the stable flow corresponds to a mathematical solution with the *lesser*—probably the *minimum*—time rate of entropy production.

2.4. Case 3, two parallel ducts

Consider the case of two ducts in parallel, as in Fig. 3, with flow from one reservoir to another. Suppose there is a *specified*, imposed value of *overall* flowrate.

It is generally *assumed* that the overall flow divides itself so that the pressure drop in each duct is the same. Suppose that *instead* of assuming that pressure drops will be the same in each duct, we *hypothesized* that the specified overall flow would divide in proportions that extremize the rate of entropy production. Based on this hypothesis, for the case of incompressible laminar flows, the following conclusions will presently be deduced (a) a resultant division of flow is determined uniquely, (b) the extremum is a minimum rate of entropy production, and (c) the pressure drops do turn out to be equal. (This third conclusion was, of course, fully expected; it follows that the division of flow determined by extremizing the rate of entropy production is identical to that which would result from *assuming* equal pressure drops.)

It is straightforward to mathematically prove the above three conclusions for the case of laminar flow. The entropy production in the two ducts is

$$\dot{S}_{\pi} = \dot{m}(xs_{\pi,1} + (1 - x)s_{\pi,2}), \tag{7}$$

where $\dot{m}$ is the total mass flowrate and x is the fraction flowing through the first duct.

It is shown above in this chapter that for incompressible, frictional flow

$$s_{\pi} = \frac{\dot{S}_{\pi}}{\dot{m}} = \frac{1}{\dot{m}} \frac{\sigma \dot{m} L \mathbf{V}^2}{DT_{\text{in}}} = \frac{2fL\mathbf{V}^2}{DT_{\text{in}}}. \tag{8}$$

For laminar flow, $\sigma = 32/Re$. With $Re = \rho \mathbf{V} D/\mu$ and $\mathbf{V} = 4\dot{m}/\rho\pi D^2$ for each duct, Eq. (7)

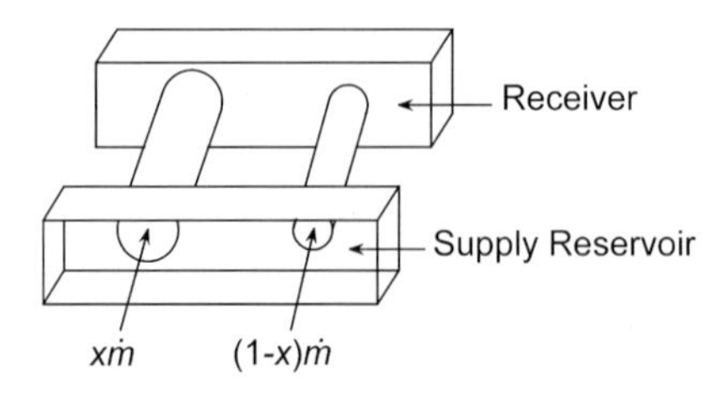

Fig. 3. Two ducts in parallel.

becomes

$$\dot{S}_\pi = \left[\frac{x^2 L_1}{D_1^4} + \frac{(1-x)^2 L_2}{D_2^4} \right] \frac{128\mu \dot{m}^2}{\pi\rho^2 T_{\text{in}}}. \tag{9}$$

If the derivative of Eq. (9) is taken with respect to x and set equal to zero, entropy production is extremized. The result is

$$\frac{2xL_1}{D_1^4} - \frac{2(1-x)L_2}{D_2^4} = 0. \tag{10}$$

Therefore,

$$x = \frac{L_2 D_1^4}{L_1 D_2^4 + L_2 D_1^4}. \tag{11}$$

The second derivative of the entropy production with respect to x is

$$\frac{\partial^2 \dot{S}_\pi}{\partial x^2} = \left[\frac{L_1}{D_1^4} + \frac{L_2}{D_2^4} \right] \frac{256\mu \dot{m}^2}{\pi\rho^2 T_{\text{in}}}. \tag{12}$$

This is always positive, as all of the variables on the right-hand side of Eq. (12) are positive. Therefore, Eq. (11) represents a *minimum* rate of entropy production.

If it is noted that for a duct $\Delta p = (\rho \mathbf{V}^2 L/2D)4f = (\rho \mathbf{V}^2 L/2D)2\sigma$, expressions for the pressure drop in both ducts may be found. The expression in the first duct is found to be

$$\Delta p_1 = \frac{x128\mu L_1 \dot{m}}{\pi\rho D_1^4} = \frac{128\mu L_1 L_2 \dot{m}}{\pi\rho(L_1 D_2^4 + L_2 D_1^4)}, \tag{13}$$

and for the second duct,

$$\Delta p_2 = \frac{(1-x)128\mu L_2 \dot{m}}{\pi\rho D_1^4} = \frac{128\mu L_1 L_2 \dot{m}}{\pi\rho(L_1 D_2^4 + L_2 D_1^4)}. \tag{14}$$

Therefore, for entropy production to be minimized during laminar flow, it follows that the flow is distributed between the two ducts in such a fashion that the pressure drops are equal when the exit pressures are equal.

To illustrate these conclusions graphically, but now for the case of turbulent flow, consider Fig. 4. It was deduced as follows. The ordinate x represents the fraction of the total specified flow going through the duct with diameter D_1. At several *arbitrary* values of x the rate of entropy production in each of the two ducts was calculated (using Eq. (2) and the Colebrook formula). Then the sum of the two entropy-production rates has been plotted in Fig. 4. If our hypothesis (that the overall flow will apportion so that the rate of entropy production is extremized) is correct, then

Conclusion (a): about 13.8% of the flow will be through the smaller tube, and

Conclusion (b): the extremum is a minimum.

At each x, in addition to the rates of entropy production, the pressure drop in each tube was calculated and plotted in Fig. 4. As can be seen, when the rate of entropy production is a minimum the pressure drops are equal. That is, if our hypothesis is accepted, then

Conclusion (c): it *follows* that the pressure drops will be the same in both tubes.

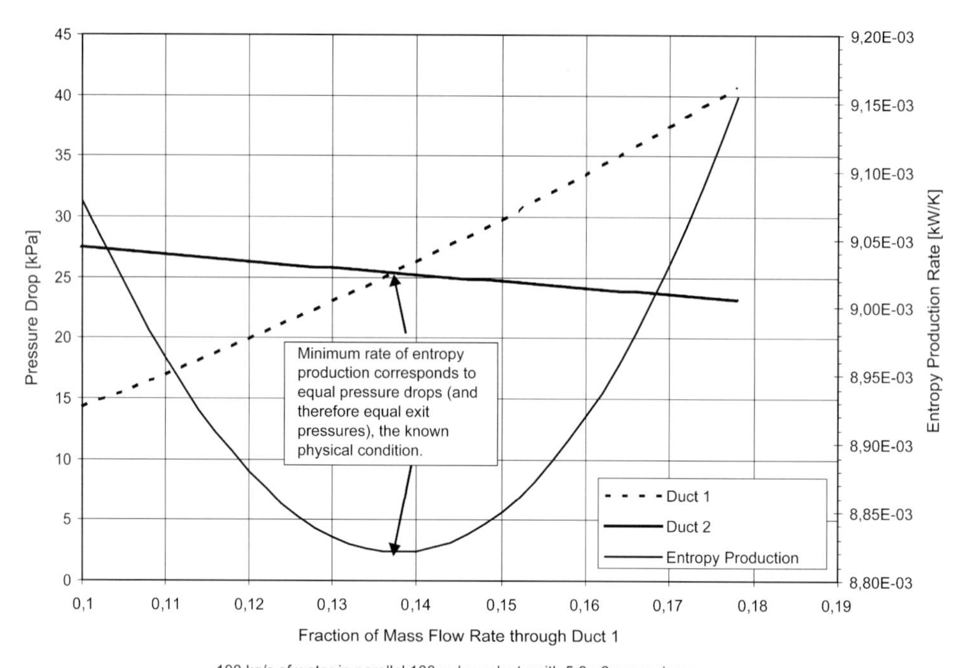

Fig. 4. Pressure drop and entropy production in two-duct system.

Considering conclusions (a)–(c), mathematically proven for incompressible laminar and graphically illustrated for turbulent flows, certainly our hypothesis—our assumption—is as good as the *assumption* of equal pressure drops (and probably much more fundamental). That is, while we acknowledge that anyone modeling this two-duct system (with incompressible flow) would simply *set* the pressure drops in the ducts equal, we *hypothesize* that this assumption is correct for this incompressible flow system *because* the rate of entropy production is minimized.

But would the assumption of equal pressure drops be correct for compressible flow? No, not necessarily. We know that, for Fanno flow (to be treated presently), at some values of receiver pressure, choked flow will exist in one duct but not the other. And the pressure drops *in* the ducts will be different.[4]

2.5. Comparison of Cases 1 and 3

It may seem that there is an inconsistency between the results in Case 1 (where it was concluded that, at specified flowrate within a single duct, the flow makes the choice of

[4] In the case of the duct with choked flow there will be a greater pressure loss ('exit loss') in the receiver. (The modeling of incompressible flows in this chapter neglects entrance and exit losses; assume that *L/D* is large enough.)

greater entropy-production rate—hypothesized to be the *maximum* rate) and Case 3 (where the split of an overall specified flowrate is that which *minimizes* the rate of entropy production). However, in Case 1 two different models (laminar and turbulent, both consistent with a complete set of governing equations) were *compared*. In Case 3, the system of governing equations was not complete but had one less equation than unknowns. Solutions were obtained by solving the system with the rate of entropy production as a parameter, and selecting the solution corresponding to minimum entropy-production rate to represent the model of the actual process. That is, the system of governing equations was solved (once for the case of laminar flow in both tubes, and once for turbulent) by extremizing the rate of entropy production. In each case, laminar and turbulent, the solution obtained by this extremization procedure yielded the same conclusions—(a), (b) and (c), above. The *deduced* equality of pressure drops in both ducts is consistent with the traditional *assumption*.

3. Kinetic systems—compressible flow

3.1. Case 4, Fanno flow

As is well known, for frictional, adiabatic, compressible flow (Fanno flow) the fluid proceeds towards a Mach number of 1, where the *specific* entropy reaches a maximum (e.g. Ref. [2]). Because the flow is adiabatic, that entropy increase is strictly due to production, that is $s_\pi = \Delta s$. It will now be proven explicitly, for the case of any perfect gas in Fanno flow, that the bound at Mach 1 is a limitation imposed by a maximum in the entropy *production* per unit mass.

First it is noted that for the case of constant specific heat [2],

$$s_\pi = \Delta s = \int_{M_1^2}^{M_2^2} \left(\frac{c_p}{2}\right)\left(\frac{\gamma-1}{M^2}\right)\left(\frac{1-M^2}{1+\dfrac{\gamma-1}{2}M^2}\right) \mathrm{d}M^2. \tag{15}$$

Performing the integration in Eq. (15) yields

$$s_\pi = \frac{1}{2}c_p \left\{ \begin{array}{l} \left[(1-\gamma^2)\left(\dfrac{1}{\gamma-1}\ln\left(2+M_2^2\gamma-M_2^2\right)\right)+(\gamma-1)\ln\left(M_2^2\right)\right] \\ -\left[(1-\gamma^2)\left(\dfrac{1}{\gamma-1}\ln\left(2+M_1^2\gamma-M_1^2\right)\right)+(\gamma-1)\ln\left(M_1^2\right)\right] \end{array} \right\}. \tag{16}$$

To find the maximum in entropy production, the partial derivative with respect to the final Mach number is taken from Eq. (16) and set equal to zero.

$$\frac{\partial s_\pi}{\partial M_2} = c_p\left\{\frac{1}{M_2}(\gamma-1)+(1-\gamma^2)\frac{M_2}{2+M_2^2\gamma-M_2^2}\right\} = 0. \tag{17}$$

This equation has solutions of M_2 equal to 1 and -1. To insure that the solution of $M_2 = 1$

is a maximum, the second derivative with respect to M_2 is taken.

$$\frac{\partial^2 s_\pi}{\partial M_2^2} = 2c_p(1-\gamma)\left[1 - \frac{\gamma-1}{\gamma+1}\right]. \tag{18}$$

For values of γ greater than 1 and positive c_p, this is negative. As $\gamma = c_p/(c_p - R)$, γ is indeed greater than 1 and positive, and hence a final Mach number of 1 yields a maximum entropy production.

Fig. 5 illustrates this phenomenon for an example physical system. The inlet conditions and the diameter of the duct are specified at the values shown in the figure. The outlet pressure is taken as a parameter. The governing equations consist of mass and energy balances, along with thermostatic property relations and the dissipation relation $\mathrm{d}p/\mathrm{d}L = [\rho\mathbf{V}^2/2D]4f(Re)$. For each outlet pressure, the solution of these governing equations yields the information needed to calculate the entropy production (using the balance $s_\pi = \Delta s$) and the exit Mach number, M_2. The results are shown, as functions of exit pressure, in Fig. 5. Clearly, when M_2 reaches unity the entropy production per unit mass reaches a maximum. While mathematical solutions exist for exit pressures that yield a higher exit Mach number, such circumstances cannot exist physically. For example, if they could exist with an exit pressure of say 50 kPa, then entropy would be destroyed as the air accelerated beyond Mach 1.

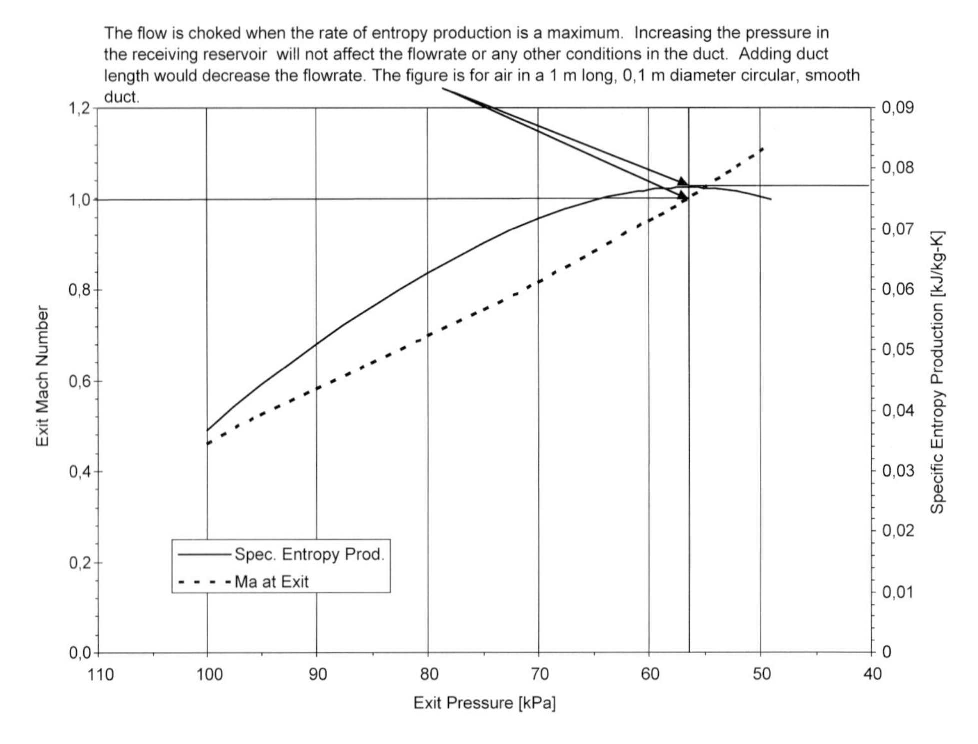

Fig. 5. Fanno flow example.

Consider these familiar observations regarding Fanno flow, in the context of the preceding study for incompressible flow. *Suppose* the previous conclusions are applicable to the compressible case. Then, again, for any *given* pressure difference the flow arranges itself to minimize its rate of entropy production (and therefore its rate of dissipation of available energy). A reduction of the exhaust pressure, which increases the 'driving force' (the pressure difference between the duct inlet and outlet) will *increase* this *minimum* rate of entropy production (and available energy dissipation). However, there exists a limit beyond which the increase cannot proceed; it reaches a *maximum*. As is well known, further reduction in the receiver pressure will have no effect on the flow in the duct. After all, if it did, and the traditional governing equations were satisfied, there would be destruction of entropy (and generation of available energy) as shown in Fig. 5.[5]

3.2. *Case 5, Rayleigh flow*

It is well known that frictionless, compressible flow with heat addition (Rayleigh flow) also becomes choked at a Mach number of unity, and that the specific entropy of the fluid reaches a maximum there. This may be shown mathematically by starting with an energy balance, an entropy balance and ideal-gas property relationships for constant specific heat,

$$c_p(T_2 - T_1) + \mathbf{V}_2^2/2 - \mathbf{V}_1^2/2 = c_p(T_{02} - T_{01}) = q \tag{19}$$

$$s_2 - s_1 = s_\tau + s_\pi = q/T_w + s_\pi \tag{20}$$

$$s_2 - s_1 = c_p \ln(T_2/T_1) - R\ln(p_2/p_1). \tag{21}$$

The following three relationships from Ref. [2] are used to relate the changes in pressure and temperature to the change in Mach number:

$$\frac{T_2}{T_1} = \frac{M_2^2\,(1+\gamma M_1^2)^2}{M_1^2\,(1+\gamma M_2^2)^2} \tag{22}$$

$$\frac{p_2}{p_1} = \frac{1+\gamma M_1^2}{1+\gamma M_2^2} \tag{23}$$

$$\frac{T_{02}}{T_{01}} = \frac{T_2}{T_1}\,\frac{1+\dfrac{\gamma-1}{2}M_2^2}{1+\dfrac{\gamma-1}{2}M_1^2}. \tag{24}$$

Solving Eq. (20) for per unit mass entropy production, along with appropriate substitutions

[5] One might ask, "What *would* happen if the pressure in the *receiver* downstream from this duct *were* reduced to say 50 kPa?" The answer *must be* that the pressure in the duct, at the exit plane, would remain at 56.5 kPa, the value corresponding to $M_2 = 1$ (and the mass flowrate would remain the same, at the 'choked' value). The additional pressure loss from 56.5 to 50 kPa would occur *in the receiver*, due to viscous dissipation therein. On the other hand, if the pipe length were increased the exit pressure could be reduced below 56.5 kPa; then the flowrate would drop. For example, see Ref. [1], Article 13-13.

for q, T, T_{02} and p yields the following expression for entropy production.

$$\frac{s_\pi}{c_p} = \ln\left\{\left[\frac{M_2^2(1+\gamma M_1^2)^2}{M_1^2(1+\gamma M_2^2)^2}\right]\Big/\left[\left(\frac{1+\gamma M_1^2}{1+\gamma M_2^2}\right)^{\gamma-1/\gamma}\right]\right\}$$
$$-\frac{T_{01}}{T_{\text{wall}}}\left[\frac{M_2^2}{M_1^2}\frac{(1+\gamma M_1^2)}{(1+\gamma M_2^2)}\frac{1+\dfrac{(\gamma-1)M_2^2}{2}}{1+\dfrac{(\gamma-1)M_1^2}{2}}-1\right]. \tag{25}$$

In order to find an extremum in entropy production, the derivative of Eq. (25) is taken with respect to the exit Mach number and set equal to zero. The resulting expression yields roots of $M_2 = \pm 1$. The negative root can be ignored. The second derivative is negative at $M_2 = 1$, so the extremum is a maximum. The observation of examples, such as the one in Fig. 6, corroborates the conclusion that a Mach number of 1 results in entropy-production maximization.

Fig. 6 was created by first specifying the duct area, the inlet conditions and the mass flow through the duct at the values given in the figure. Additionally, the duct's wall was assigned a fixed temperature, set at a value that is greater than any temperature reached by the heated, flowing fluid. The governing equations consisted of mass, momentum and energy balances, along with thermostatic property relations. For a specified exit pressure, the solution of these governing equations yields the information needed to calculate the entropy production (using an entropy balance $\Delta s = q/T_{\text{wall}} + s_\pi$) and the exit Mach number, M_2. For convenience, the system of equations was solved by taking M_2 rather than exit pressure as a parameter, yielding the curves for heat addition and entropy production shown in Fig. 6.

Examination of Fig. 6 shows that, once again, entropy production has a clear maximum at Mach 1. In order for a subsonic flow to accelerate past that value as a result of interaction with a high-temperature source, entropy would have to be destroyed (and likewise for deceleration of supersonic flow).

It is interesting to consider the foregoing from the perspective of available energy. In our model for Rayleigh flow, each quantity of matter, entering the duct at specified conditions, interacts with the duct walls that are at a *specified temperature*. Hence, in effect the walls constitute an entropy reservoir, large enough that its temperature is not affected by a transfer of entropy. The proper 'dead-state temperature' for the system of the matter *plus* the reservoir is then the reservoir temperature, i.e. T_{wall} (see 'Exergy' in Ref. [3]). The system has available energy due to disequilibrium, such as that between the matter and the entropy reservoir. Because the reservoir is large, it has no exergy; exergy is attributed to the matter flowing in the duct.[6] While the heat transfer dQ is to the matter, because the flow in the duct is at a lower temperature than the reservoir (dead state) temperature, the exergy flow, as given by $dX_Q = (1 - T_{\text{wall}}/T_{\text{fluid}})dQ$, is *from* the matter. That is, the fluid is giving up thermal exergy. The subsonic flow is accelerated as some of the fluid's exergy is converted into kinetic exergy. Concurrently, some of the fluid's thermal exergy flows from the fluid towards the reservoir, and is destroyed—used up, to drive the overall process.

[6] The available energy of any composite system—consisting of materials undergoing processes and any reservoirs—is equal to the total value of the exergies of the subsystems.

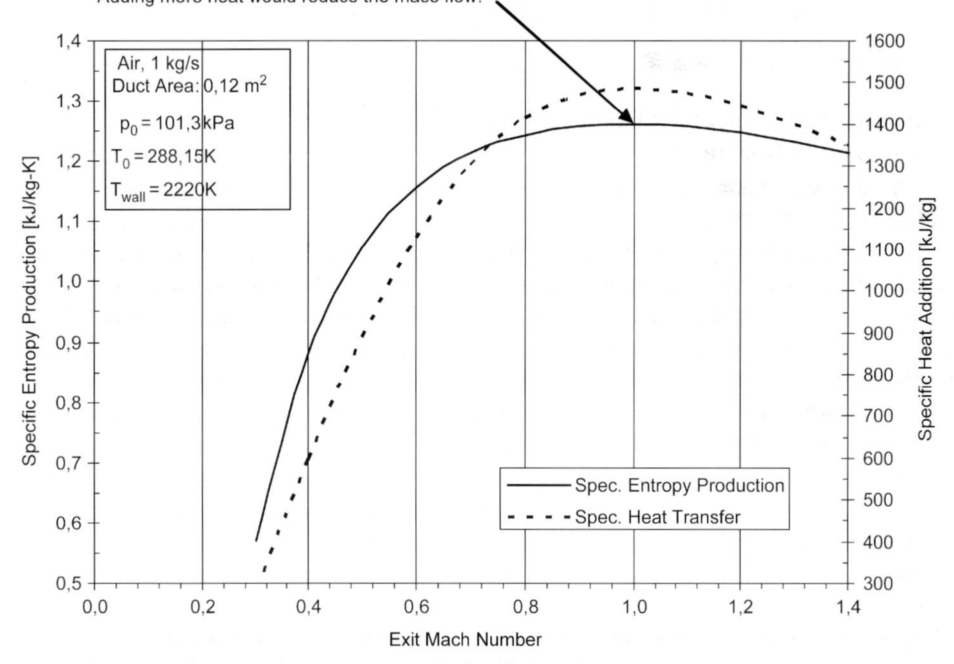

Fig. 6. Rayleigh flow example.

(To elaborate further, note first that for subsonic flow with heat addition, the pressure decreases in the direction of flow. The fluid exergy that is given up consists of both thermal and strain (pneumatic) exergy.[7] Part goes into increasing the kinetic exergy and part is destroyed in the heat transfer from the tube walls. In the case of supersonic flow with heat addition—therefore decelerating—the pressure rises, and kinetic and thermal exergy are used up yielding pneumatic exergy.)

As illustrated by Fig. 6, there are mathematical solutions of the governing equations in which subsonic flow is accelerated beyond (or supersonic flow is decelerated below) Mach 1. In order to achieve such effects, the governing equations require a heat rejection (Fig. 6). *If the heat rejection occurred to the 'hot' reservoir*, the figure shows that entropy would be destroyed—which corresponds to a creation of exergy. Clearly this is impossible.

The question remains, "If a subsonic flow were heated to Mach 1 *and then cooled*, could supersonic states be reached?" Such an accomplishment is not precluded by the governing equations that we have employed. Moreover, as long as the cold duct walls (to which the flow is then subjected upon reaching Mach 1) were at a temperature below the fluids, the governing equations would yield further entropy production. As M

[7] Regarding strain (pneumatic) exergy, see Ref. [19].

increased, the pressure would drop continuously, hence more (pneumatic) exergy would need to be supplied to the system.[8]

3.3. Case 6, normal shock

If a converging–diverging nozzle is operating in the regime of pressure ratios in which a normal shock occurs in the diverging section, the minimum rate of entropy production occurs for the well-known solution of a throat Mach number of unity. This will now be shown.

If the geometry, inlet conditions and the exit pressure are known for a nozzle, the governing equations for operation with a normal shock in the diverging section—mass, energy and entropy balance between the inlet and throat, the throat and the shock plane, across the shock, and between the shock plane and the exit, along with a momentum balance across the shock—will result in one less equation than unknown.

Traditionally this set of equations has been completed by stating that the flow is choked, i.e. $\mathrm{Ma_T} = \sqrt{\gamma R T_\mathrm{T}}/\mathbf{V}_\mathrm{T} = 1$. However, the authors have observed that it may also be completed by minimizing either the rate of entropy production, or entropy production per unit mass (say as a function of the Mach number at the throat). As shown in Fig. 7, completed with the assumptions that, except for the shock, nozzle performance is ideal (isentropic) and that specific heats are constant, the minimum of both of these values occurs at a Mach number of unity.

With the entrance and exit pressures specified, the system attempts to minimize its rate of entropy production (and available energy dissipation). Fig. 7 shows this behavior for one nozzle (as defined in the figure). Fig. 8 shows entropy production per unit mass as a function of mass flow through the nozzle, with the throat Mach number versus mass flow superimposed. No real solutions exist for this mathematical problem at mass flowrates to the right of the value where the throat Mach number equals one. Moreover, at that maximum flowrate, where the entropy production is a minimum, the derivative of the entropy production with respect to mass flow is *not* equal to zero. The minimum occurs where the value of mass flow is bounded as a consequence of the governing equations. Both of these figures were created by simultaneously solving the governing equations while varying the throat Mach number. The geometry and boundary conditions for the problem are shown in the figures.

The linking of the minimization in the rate of entropy production for normal shock in a nozzle with the known physical behavior of a Mach number of 1 in the throat of the nozzle is similar to the case of frictional, incompressible flow in a duct. In the nozzle, with the entrance and exit pressures specified, the system operates at a point where the rate of entropy production is minimized. Frictional, incompressible flows with the pressure drop

[8] Another issue that might be raised is this: "Suppose, for example with the conditions presented in Fig. 6, we were operating with a heat addition over a duct length that yields Mach 1 at the outlet. Now, what will happen if we were to increase the duct-wall temperature? After all, *under those conditions* we could not increase the heat addition." The answer is that the conditions *specified* in Fig. 6 would necessarily change; the flowrate would decrease. The fluid would leave at Mach 1, but at a higher temperature and velocity, and lower pressure.

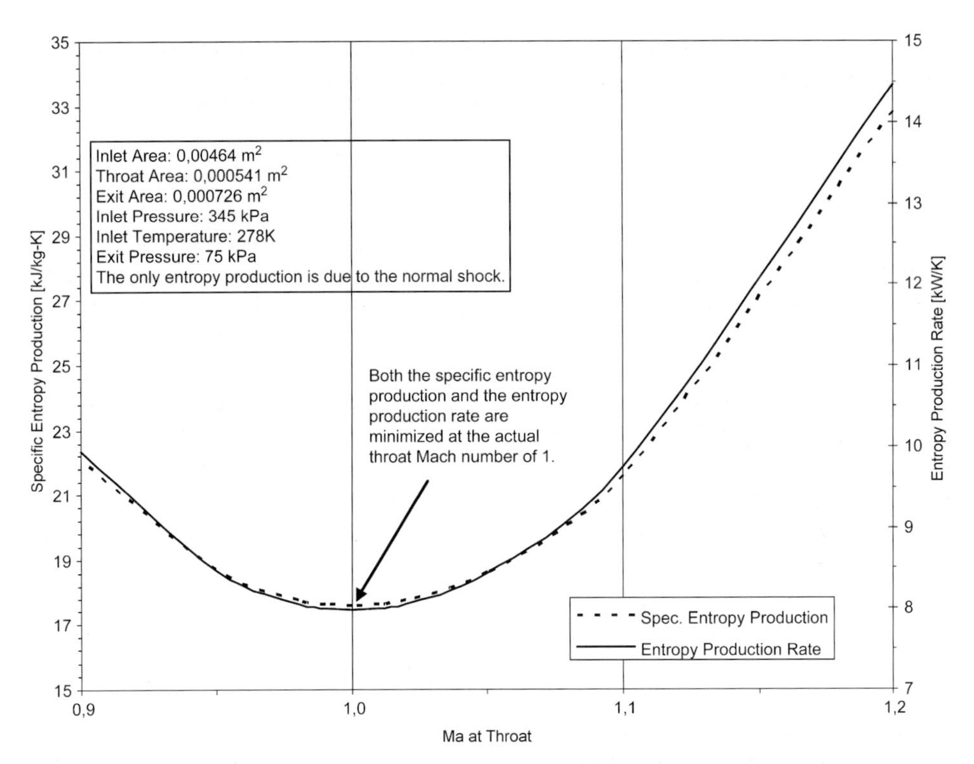

Fig. 7. Entropy production of normal shock in a converging–diverging nozzle.

specified also seek to minimize its rate of entropy production (assuming that the hypothesis proposed earlier is valid).

4. Equilibrium

Equilibrium states are those that result in the maximization of total entropy production of the isolated system—i.e. of the system of interest and the surroundings with which it may interact. As long as entropy can be produced by a transient process, the state of the system will continue to change until equilibrium is reached. Therefore, equilibrium distributions may, in general, be found by maximizing the total entropy production subject to an appropriate set of constraints. For an isolated system the maximization of entropy production, and hence of entropy content, corresponds to zero available energy. If the constraints imposed on a system by its surroundings are a specified temperature and pressure (or temperature and volume), the maximization of *total* entropy production yields a minimization of the *system's* Gibbs free energy (or Helmholtz free energy). In these cases, the change in free energy of the system *is* the change in *Gibbs* available energy of the composite of the system and its surroundings ('bath')—i.e. the change in available energy of the isolated system. Thus, the minimization of free energy is equivalent to minimizing the available energy. The minimum available energy is zero; if a system has no available energy then it is at equilibrium [3].

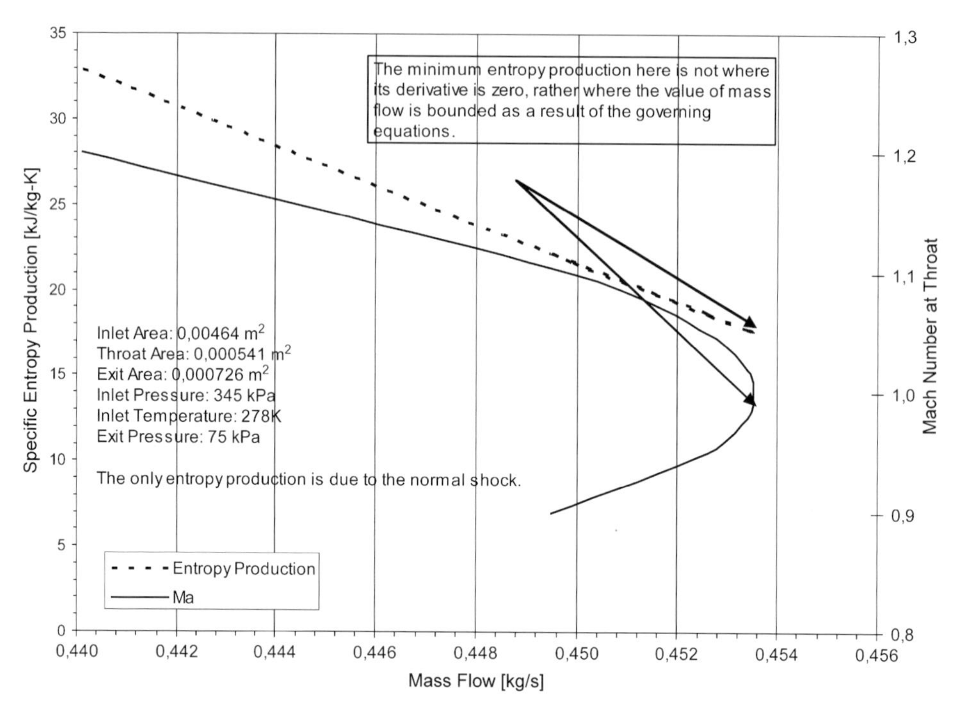

Fig. 8. Specific entropy production and throat Mach number for normal shock in a converging–diverging nozzle.

4.1. Chemical equilibrium

While teaching thermodynamics, Gaggioli [4] demonstrated a method of finding chemical equilibrium without the encumbrance of traditional equilibrium constants.[9] Atomic and energy balances are written, along with relevant constraints (such as pressure or volume). Subject to these conditions, the entropy production is then maximized, and the result yields the equilibrium composition, temperature, etc.

Consider as an example a stoichiometric mixture of carbon monoxide and oxygen in an adiabatic container of fixed volume. If the species allowed are carbon monoxide, carbon dioxide and oxygen, the reaction may be written as $CO + \frac{1}{2}O_2 \rightarrow xCO_2 + (1-x)$ $CO + \frac{1}{2}(1-x)O_2$. This equation serves as an atomic balance. With it, and assuming ideal-gas behavior, the following governing equations may be written. These equations, when solved simultaneously while maximizing $S_\pi(x)$ will yield the final temperature, pressure and chemical composition. Table 1 shows the results for the above equations for reactants initially at 298 K and 101.3 kPa.

[9] Equilibrium constants are derived by minimizing Gibbs free energy of a system that is at specified temperature and pressure; this is tantamount to maximizing entropy of an isolated system (at fixed internal energy and volume) that contains (as a subsystem) the system at fixed temperature and pressure. The procedure here is fundamentally better than using equilibrium constants because there is no allusion to a hypothetical system or reservoir at a certain temperature and pressure. Moreover, it is all the more convenient when there are several, concurrent independent reactions.

4.1.1. Energy balance

$$N_{\mathrm{I}}(\chi_{\mathrm{CO,I}}\bar{u}_{\mathrm{CO,I}} + \chi_{\mathrm{O_2,I}}\bar{u}_{\mathrm{O_2,I}}) - N_{\mathrm{II}}(\chi_{\mathrm{CO_2,II}}\bar{u}_{\mathrm{CO_2,II}} + \chi_{\mathrm{CO,II}}\bar{u}_{\mathrm{CO,II}} + \chi_{\mathrm{O_2,II}}\bar{u}_{\mathrm{O_2,II}}) = 0. \quad (26)$$

4.1.2. Entropy balance

$$N_{\mathrm{I}}(\chi_{\mathrm{CO,I}}\bar{s}_{\mathrm{CO,I}} + \chi_{\mathrm{O_2,I}}\bar{s}_{\mathrm{O_2,I}}) + S_{\pi} - N_{\mathrm{II}}(\chi_{\mathrm{CO_2,II}}\bar{s}_{\mathrm{CO_2,II}} + \chi_{\mathrm{CO,II}}\bar{s}_{\mathrm{CO,II}} + \chi_{\mathrm{O_2,II}}\bar{s}_{\mathrm{O_2,II}}) = 0. \quad (27)$$

4.1.3. Property relations

In addition to the balances, necessary property relations are required for the internal energy and entropy of the components etc. Ideal gas properties were employed.

4.1.4. Results

The results are shown in Table 1. These results were obtained using EES,[10] employing its maximization routine and its internal property relations. Alternatively, (e.g. using EES), Eq. (26) can be solved at several specified values of x. Then Eq. (27) will yield the entropy production at each x, and when the values are plotted versus x, there is a maximum at the equilibrium x. (It is clear from the plot that if x were to proceed beyond the maximum value then entropy would be destroyed. That is, the Second Law would be violated.) The results shown in Table 1 do, of course, conform to those obtained using the traditional but more cumbersome (and indirect) approach employing the equilibrium constant.

4.2. *Statistical mechanical distributions*

The generally accepted statistical-mechanical expression for the entropy of a system [5,6] is

$$S = -k\sum_{i=1}^{n} p_i \ln p_i, \quad (28)$$

where p_i is the probability in quantum state i. Usually this expression for entropy is deduced by comparing the expressions $\mathrm{d}E = -p\mathrm{d}V + T\mathrm{d}S$ from thermodynamics with $\mathrm{d}E = -p\mathrm{d}V - (1/\beta)\sum_i \mathrm{d}(p_i \ln p_i)$ from statistical mechanics [6]. Eq. (28) is then deduced, along with the conclusion that $\beta = 1/kT$.

Following the methodology of Jaynes [7] or Reynolds [8] it can be shown that maximization of entropy subject to the constraints $\bar{E} = \sum_i p_i E_i$ and $1 = \sum_i p_i$ leads

Table 1
Results of chemical-equilibrium calculations

Pressure (kPa)	975.5
Temperature (K)	3450
$\chi_{\mathrm{CO_2}}$	0.4041
$\chi_{\mathrm{O_2}}$	0.1986
χ_{CO}	0.3973

[10] Engineer Equation Solver [20].

correctly to

$$p_i = 1/e^{\lambda+\beta E_i} \equiv e^{-\beta E_i}/Q, \tag{29}$$

where Q is the partition function for a system in equilibrium with a heat bath. The Helmholtz free-energy function, $A(T, V)$, a complete thermostatic property relation [1], is related to Q by

$$A(T, V) = -RT \ln Q. \tag{30}$$

It is this relationship that makes statistical mechanics valuable for the determination of thermostatic property relations from relatively simple experimentation.

When the temperature is low and the probability of low-energy eigenstates is high, then the use of the partition function to get thermostatic property relations becomes unwieldy. For example, Andrews [6] says

> "In this section, the statistical-mechanical expressions for the various thermodynamic properties of ideal fermion or boson gases are found by a method analogous to that of Section 9. The statistical thermodynamics here is based on occupation numbers, in contrast to the use of the partition functions in Sections 17–20 for the Boltzmann gas. The Boltzmann limit of these results is of course identical to the results of Sections 17–20. The partition function approach has the advantage of being valid in all cases, in particular, for systems other than ideal gases. However, within the partition function framework it is hard to incorporate the definition of a quantum state for fermions or bosons. On the other hand, the occupation number approach directly incorporates that correct definition of a quantum state, but it is limited to ideal gases. One must choose the approach best suited to the problem at hand."

An occupation number is the number of particles with the *particle* energy eigenstate i. When this approach is used, the entropy is given by

$$S = -k\sum_{i=1}^{n} n_i \ln n_i \pm \sum_{i=1}^{n} (1 \mp n_i)\ln(1 \mp n_i). \tag{31}$$

Here n_i is the occupation number.

It has been shown by Kesavan and Kapur [9] that Eq. (31), when maximized, yields the same expression for n_i as the alternate means typically employed in statistical mechanics. That is,

$$n_i = \frac{1}{e^{\lambda+\beta\varepsilon_i} \pm 1} = \frac{e^{-\lambda-\beta\varepsilon_i}}{1 \mp e^{-\lambda-\beta\varepsilon_i}}. \tag{32}$$

Statistical mechanics is therefore at least implicitly built on the idea that, at equilibrium, entropy is at a maximum given the other constraints (volume, and energy or temperature).

5. Combined—Chapman–Jouget detonation

The term *detonation* is used for the circumstances when combustion causes a strong shock wave, such that the flow of the wave relative to the reactants (or the reactants relative to the wave) is supersonic. Traditionally, the detonation process has been modeled without explicitly accounting for the combustion process. Instead, it has been assumed that the thermal effects of combustion may be simulated by a hypothetical heat input from an external source. Moreover, in order to get a complete system of governing equations, the 'Chapman–Jouget condition' [2] has been invoked, which *assumes* that the exit velocity of the products of combustion (relative to the wave) is at Mach 1. This postulate is tantamount to the concept that 'heat addition' by combustion is great and consequently, in analogy to Rayleigh flow, the flow from the wave is choked. Experimental evidence verifies that the gases exiting a detonation wave are at Mach 1, relative to the wave.

In Chapman–Jouget detonation simultaneous minimization and maximization of entropy production occur. As would be expected, the equilibrium of the chemical reaction is at a maximum *per unit mass*. Whereas, in analogy with normal shock, the Chapman–Jouget condition results from a minimum *time rate* of entropy production. This is shown in Figs. 9 and 10, which are the result of solving the governing equations, without using the Chapman–Jouget postulate, for a gas detonating in a duct as a function of reaction completion and as a function of exit Mach number. The entropy production per unit mass is a maximum at the chemical equilibrium condition, and it is a minimum at an exit velocity corresponding to Mach 1.[11] (see also Refs. [10,11]) (Considering the foregoing discussion regarding the determination of conditions at chemical equilibrium via entropy maximization, the maximization of entropy to find the chemical equilibrium seems appropriate because the reaction rate may be considered rapid as compared to the flow process.)

6. Entropy extremization and the available energy of Gibbs

In the preceding sections there have been many references to available energy minimization or maximization along with references to entropy production. The authors believe that the preceding observations of entropy minimization and maximization are, in reality, a result of a more fundamental minimization or maximization of available energy, as defined by Gibbs. If a system is not allowed to transfer energy (or entropy or volume) out of itself, the only way in which it may 'get rid of' available energy is to destroy it, through entropy production. For the preceding examples, maximization in the rate of entropy production corresponds to maximization in the rate of available-energy destruction, and minimization of the rate of entropy production corresponds to minimization in the rate of available-energy destruction. Furthermore, maximization of

[11] It is noteworthy that the governing equations used by the authors to create Figs. 9 and 10 accounted for the chemical reaction of combustion explicitly. Not only is the traditional approach less fundamental, because it assumes the same thermodynamic property relations for entering and leaving gases, it inevitably will be faulty when the number of moles in reaction products differs from that in the reactants.

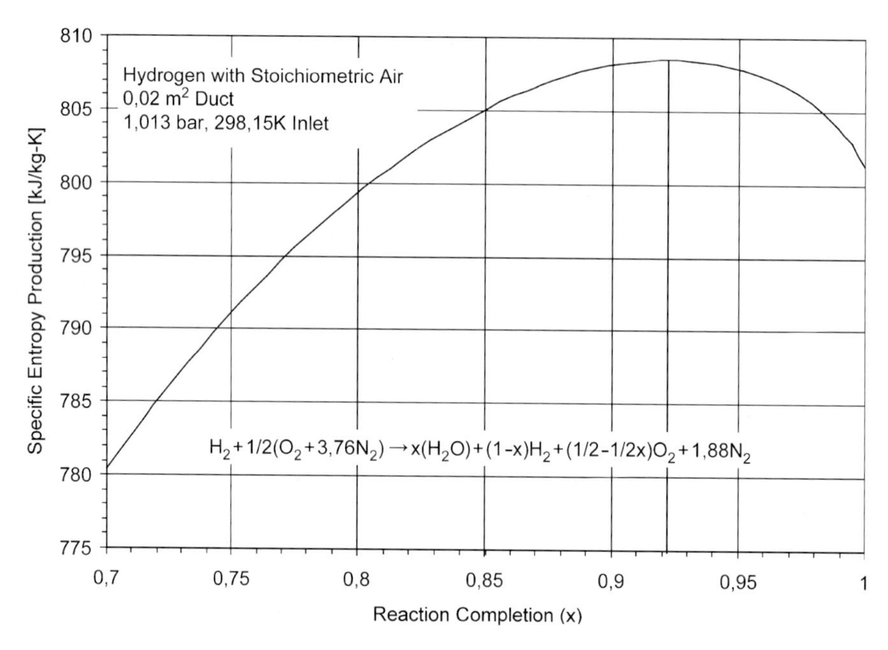

Fig. 9. Specific entropy production in Chapman–Jouget detonation versus reaction completion.

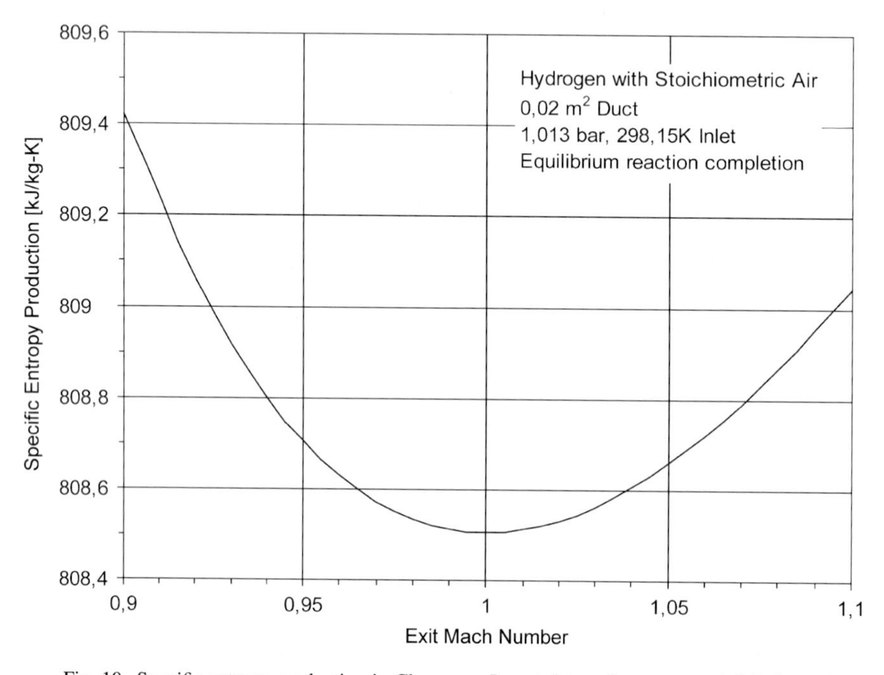

Fig. 10. Specific entropy production in Chapman–Jouget detonation versus exit Mach number.

a system's entropy at fixed energy and volume is equivalent to eliminating a system's available energy; the result is a stable equilibrium state.

7. Conclusion

This chapter gave examples of modeling situations where entropy *minimization* yields a solution in conformity with the real behavior of the system, and there were examples where *maximization* yields conforming solutions. The examples include kinetic systems (fluid flow, heat transfer), equilibrium systems and systems where kinetics and equilibrium are both relevant. These examples have been studied or revisited for the purpose of gaining insight for a larger goal of finding a general extremization principle for thermodynamics.

In order to summarize the relevance of our examples to the prospective development of a new, generalized deterministic principle, it is worthwhile to categorize the examples as follows:

(a) Cases where, in lieu of *assumptions* that are made in order to *get* a complete set of governing equations, the *extremization* of the time rate of entropy production was used to get a *solution* to a process model: (i) the steady frictional flow of an incompressible fluid through two parallel tubes, (ii) the steady flow of a compressible fluid through a nozzle, with a normal shock within the nozzle and (iii) the steady flow of products of combustion out of a detonation wave. In all three of these cases, a *solution* was obtained by *minimization* of the *time rate* of entropy production. In each of these cases, the solution obtained by extremization yielded the same result as a solution made with traditional *assumptions*; namely, (i) the pressure drop in both tubes is the same, (ii) the Mach number at the nozzle throat is unity; (iii) the Mach number of the exiting products of combustion is unity. While this result is reassuring, we believe that it is more a confirmation of the validity of those assumptions than evidence that the use of entropy extremization is valid.

(b) Cases where the extremization of entropy production per unit mass is employed to deduce equilibrium conditions: (iv) equilibrium composition, in ordinary adiabatic chemical reaction, (v) equilibrium ensemble distribution of quantum states, in statistical mechanics (following Jaynes [7]); (vi) equilibrium composition reached in adiabatic combustion with detonation. In all three of these cases, equilibrium conditions were obtained by *maximization* of the entropy production *per unit mass* of equilibrating material. The soundness of this approach is verified in all three cases, inasmuch as: In cases (iv) and (v), the results are consistent with those obtained with traditional methods; namely, (iv) the use of an equilibrium constant, which, by the way, requires a more complex system of governing equations, and (v) an assumption that the ensemble average change of energy associated with a change in volume is the reversible work [6]. In our treatment of case (vi), we employ a mathematical model that is more realistic than the usual one; we treated the chemical reaction in detail, determining its thermal effects and the effects of composition change explicitly—rather than ignoring composition change and simulating the thermal effects with a hypothetical heat input. This more basic treatment is made viable by employing maximization of entropy production to determine the composition of the products of combustion, and the reliability of the method is substantiated by the fact that (when used in conjunction with minimization of *time* rate of entropy production for determining the Mach number of the combustion products) $M = 1$ is a result.

(c) Cases where extremization of the entropy production per unit mass was used to get a *limitation* imposed upon a flow process, while the process is modeled with a *complete* set of governing equations: (vii) compressible, frictional (i.e. Fanno) flow of a fluid in a duct, and (viii) compressible flow with heat addition or rejection (Rayleigh flow). In both cases, the appropriate limitation—choked flow, *when* Mach 1 is reached—was obtained by *maximization* of the entropy production per unit mass. Heretofore, these limitations were typically [2] deduced (a) with graphical considerations, observing the extremum on the path of the process on an enthalpy–entropy *property* diagram, and employing the definition of the *property*, acoustic velocity, or (b) after having observed that path on the diagram, mathematical extremization of the *property* entropy rather than the entropy *production*. In the case of Fanno flow, which is adiabatic, extremization of entropy and entropy production is equivalent. However, that is not the case with Rayleigh flow, and it can be said that it is then fortuitous for entropy maximization that it is concurrent with the maximum of entropy production.

It is worthy of note that there is a similarity in Categories (b) and (c); both look at limitations. In Category (b) it is presumed that the limit of maximum entropy production is reached—that the boundary conditions do not prevent it from being reached. In Category (c) the processes inevitably proceed in the direction of increasing entropy production and reach the *constrained* maximum consistent with the imposed boundary conditions, but might not reach the limit beyond which entropy production cannot proceed, because: in Case (vii) where the change in state results from friction, the tube length may be insufficient, and in Case (viii) where the change in state results only from heat transfer, the amount imposed may be insufficient.

(d) Cases where the time rate of entropy production was compared for *alternative* solutions to a *complete* set of governing equations—that is, laminar-flow-type and turbulent-flow-type solutions: Incompressible, frictional flow through conduits when (ix) the mass flowrate is specified (and hence the pressure drop is a consequence), and conversely (x) the pressure drop is specified (and hence the mass flowrate is a consequence). It was shown that when the mass flowrate is specified the stable (persistent) type of flow is that with the greater *time rate* of entropy production. (As explained earlier in the chapter, we have hypothesized that the persistent greater rate is indeed the maximum rate.) Conversely, when the pressure drop is specified, the persistent type of flow is that with the lesser rate. (And we have hypothesized that the persistent lesser rate is indeed the minimum rate.) It should be mentioned that, if it is assumed that the mass flowrate is a function of the pressure drop Δp and an 'intensity of turbulence' T, it can be shown mathematically (with chain rules for partial derivatives) that a positive derivative of entropy production rate with respect to intensity at fixed flowrate and a negative derivative of entropy production rate with respect to turbulent intensity at fixed pressure drop are equivalent to each other. That is, the deductions that we made for Case (x) follow from those from Case (ix), and vice versa.

7.1. Discussion

A note that is relevant to the conclusions we have drawn: While we refer to them in some discussions, this chapter contained no mathematical results for transient processes—inertial or noninertial. Our results were either for *steady* flows (in Categories a, c and d) or they were for *equilibrium* states (Category b), without regard for the kinetics—the time rates—of the process leading to equilibrium.

Relevant to Categories (a) and (b): As is well known, the rate of entropy production for each rate process is proportional to the product of the flow (or 'reaction') rate with the associated thermodynamic driving 'driving force'. As an irreversible transient process progresses, the driving force continuously decreases, and hence for noninertial transients, so does the time rate of entropy production—from an initial maximum value. Since the rate of entropy production decreases continuously, it must proceed to a minimum rate. When it can proceed to the lowest possible value, zero, a stable equilibrium state is reached. When imposed, steady boundary conditions prevent the system from reaching the zero limit, the decreasing rate of entropy production proceeds to the minimum rate compatible with the boundary conditions (and remains steady).

Evidently it is important to notice the basic difference between Categories (a) and (d). In each case of Category (a) extremization of entropy-production rate was employed to *get* a solution of a mathematical model that simulates a process. That is, an incomplete mathematical model was complemented by utilizing, in lieu of some 'standard' assumption, an entropy-balance equation and then extremizing the entropy-production rate. In the two cases of Category (d), entropy production was calculated from *existing* solutions to a complete, basic mathematical model that has no assumptions (however plausible). The governing equations for incompressible flow in conduits must have more than one solution. Because, at least under some operating conditions, more than one solution is seen experimentally—namely both laminar flow and turbulent flow are observed. For laminar flows, an analytical solution is readily obtained, but certainly not for turbulent flows. The *engineering* solution that we used for turbulent flow is, of course, not analytical but empirical—i.e. from experimentation (and hence arguably *more* real than an analytical solution—which, anyhow, requires an empirically determined viscosity). The experimental results are represented, fitted, by the longstanding Colebrook equation.

But if *minimization* of the entropy-production rate yields good solutions for Category (a), would not we expect (hope?) that the situation would be the same for *both* cases dealt with in Category (d)? All that we can say at this time is:

- There is the fundamental difference: use of entropy-production rate to *get* a solution, in Category (a), and use of *existing* solutions to calculate and *compare* the entropy-production rate in Category (d).
- Indeed, as mentioned above, if the entropy-production rate is minimized (or maximized) when a flowrate results from a specified pressure drop, then mathematically it must be maximized (or minimized) when a pressure drop results from a specified flowrate.

Yet, we might ask, "Is there a commonality between the cases in Category (a) and the specified-pressure-drop case in Category (d)?" Or, perhaps more likely, between Category (b) and (d)? After all, the transition process from laminar to turbulent flow (and vice versa) is a process of 'equilibration'—i.e. a process leading to a stable (i.e. persistent) configuration under the imposed constraints. For example, if we could model the transition process, would we observe a progressive increase in the entropy production per unit mass? (And, to a maximum, limiting value?) Such a progressive increase is consistent with the observations we have made for the case of specified flowrate.

Moreover, it is *not inconsistent* with the results for the specified-pressure-drop case; as indicated earlier, when pressure drop is specified, that determines the entropy production per unit mass. It seems reasonable to conjecture that when entropy production is specified that the appropriate, 'dual' criterion would be an extremization of available energy—probably a minimization of available energy *content*.

7.2. Closure

Interestingly, it is not uncommon to hear the statement, "The Second Law is peculiar among the laws of physics, inasmuch as it does not say what *shall* happen but only puts limitations on what happens." Now, that *is* peculiar—odd! The quest for a general 'equation of motion' is, we believe, the quest for a formulation of the Second Law that will say what *shall* happen.

Moreover, such an equation of motion could well turn out to be an extremum principle, perhaps akin to Hamilton's Principle. It should be noted that years ago Helmholtz [12] proposed such a principle, for a special case; see Ref. [13] for example.

In any case, it is hoped that the studies presented in this chapter contribute to the development of a generalized Second Law. The authors believe that it is likely that the generalization will be stated in terms of the concept of Available Energy as proposed by Gibbs (Refs. [3,14,15]).

Nomenclature

A	area
c	specific heat
D	diameter
E, $\bar{E}$	energy, average energy
f	Fanning friction factor
k	Boltzmann constant
L	length
M	Mach number
m	mass; or, partition function
N	number of moles, number of particles
p	pressure; or, probability
Q, q	heat transfer, heat transfer per unit mass
R	ideal-gas constant
Re	Reynolds number
S, s	entropy, specific entropy
T	temperature
U, u	internal energy, specific internal energy

$\mathbf{V}$	scalar velocity
V, v	volume, specific volume
x	mass fraction; or, reaction extent

Greek variables

χ	mole fraction
γ	c_p/c_v
μ	viscosity
ρ	density
σ	dimensionless entropy production
σ^*	alternate dimensionless entropy production

Subscripts

π	production
0	stagnation
I	inlet or initial
T	throat
II	outlet or final
x	before shock
y	after shock

References

[1] Obert, E.F. and Gaggioli, R.A. (1963), Thermodynamics, McGraw-Hill, New York, Sections 12-5 and 12-6.
[2] Shapiro, A.H. (1953), The Dynamics and Thermodynamics of Compressible Fluid Flow, Vol. 1, The Ronald Press Company, New York.
[3] Gaggioli, R.A. and Paulus, D.M., Jr. (1999), Available Energy: II. Gibbs Extended. In AES, Vol. 39, ASME, New York, pp. 285–296.
[4] Gaggioli, R.A. (1993), Class Manual, MEEN 106, Marquette University, Milwaukee.
[5] Tolman, R.C. (1979), The Principles of Statistical Mechanics, Dover, New York.
[6] Andrews, F.C. (1963), Equilibrium Statistical Mechanics, Wiley, New York.
[7] Jaynes, E.T. (1957), Information theory and statistical mechanics, Phys. Rev. 106(4), 620–630; 108(2), 171–190.
[8] Reynolds, W.C. (1965), Thermodynamics, McGraw-Hill, New York, pp. 354–364.
[9] Kesavan, H.K. and Kapur, J.N. (1989), The generalized minimum entropy principle, IEEE Trans. Syst. Man Cybernet. 19(5), 1042–1052.
[10] Oosthuizen, P.H. and Carscallan, W.E. (1997), Compressible Fluid Flow, McGraw-Hill, New York.
[11] Gubin, S.A., Odintsov, V.V. and Pepekin, V.I. (1988), Thermodynamic calculation of ideal and nonideal detonation. Combustion, Explosion and Shock Waves, Vol. 23, No. 4, Consultants Bureau, New York, pp. 446–454.

[12] von Helmholtz, H. (1895), Über die physikalische Bedeutung des Princips der kleinsten Wirkung. Wissenschaftliche Abhanddlund, 3. Band, J.H. Barth, Leipzig, in German.
[13] Pauli, W. (1958), Theory of Relativity, Dover, New York, pp. 135–136.
[14] Gaggioli, R.A., Bowman, A.J. and Richardson, D.H. (1999), Available Energy: I. Gibbs Revisited. AES, Vol. 39, ASME, New York, pp. 285–296.
[15] Gibbs, J.W. (1961), Graphical methods in the thermodynamics of fluids. In The Collected Works of J. Willard Gibbs, Vol. 1, Dover, New York, pp. 1–32.
[16] Colebrook, C.F. (1939), Turbulent flow in pipes with particular reference to the transition between the smooth and rough pipe laws, J. Inst. Civ. Eng (London).
[17] Rouse, H. (1946), Elementary Fluid Mechanics, Wiley, New York, pp. 210–211.
[18] Bird, R.B., Stewart, W.E. and Lightfoot, E.N. (2000), Transport Phenomena, 2nd edn, Wiley, New York.
[19] Dunbar, W., Lior, N. and Gaggioli, R.A. (1992), The component equations of energy and exergy, J. Energy Res. Tech. 114, 75–83.
[20] Engineering Equation Solver (EES), f-Chart Software, Box 44042, Madison, WI 53744, USA, www.fchart.com

Chapter 4

ON A VARIATIONAL PRINCIPLE FOR THE DRAG IN LINEAR HYDRODYNAMICS

B.I.M. ten Bosch

Shell Global Solutions International, PO Box 38000, 1030 BN Amsterdam, The Netherlands

A.J. Weisenborn

Shell Exploration and Production UK Ltd, 1 Altens Farm Road, Nigg, Aberdeen, AB12 3FY, UK

Abstract

In this chapter, we elaborate on a variational principle for the drag on an object in Oseen flow. After a brief review of the variational principle, we investigate the heat transfer and momentum transfer as a function of the Reynolds number for the case of Oseen flow around a disk. For arbitrarily shaped bodies, we derive a Hellmann–Feynman theorem for the drag. In the case of small Reynolds numbers, a correspondence is established with an Oseen resistance formula put forward by Brenner.

Keywords: variational principle; linearized Navier–Stokes equation; drag; induced forces

1. Introduction

Variational principles in fluid dynamics may be divided into two categories. First, one may attempt to derive the full equations of motion for the fluid from an appropriate Lagrangian or associated principle, in analogy with the well-known principles of classical mechanics. As an example we mention the Herivel–Lin principle for perfect fluids (see, e.g. Ref. [1], Section B4). Secondly, variational functionals may be found that produce stationary values under the condition that some steady-state flow equation is satisfied. The value of the functional at the stationary point is often a quantity of direct physical interest, such as flow resistance or dissipation. In this chapter, we wish to elaborate on an example from the latter category [2]. The stationary value of the functional is attained when the velocity and pressure fields correspond to the steady-state Oseen equation. Furthermore, the stationary value of the functional is the drag on the object submerged in the Oseen flow.

The chapter is organized as follows. The variational principle is reviewed in Section 2. The key ingredient of the principle is the concept of induced forces. This concept has

E-mail address: dick.tenbosch@shell.com (B.I.M. ten Bosch); toon.a.j.weisenborn@shell.com (A.J. Weisenborn).

proved to be an efficient tool in the calculation of drag in the realm of linear hydrodynamics (see, e.g. Refs. [3–6]). In Section 3, we use the principle to give an extensive discussion of the Oseen drag on a disk broadside to the stream. In passing we will also deal with the corresponding heat-transfer problem. The dimensionless heat transfer (Nusselt number) and drag are calculated up to order P^5 and R^3, respectively (P and R denoting the Péclet number and Reynolds number). In Section 4, we derive a Hellmann–Feynman theorem for the drag. From the variational principle it can be shown that the derivative of the drag with respect to the Reynolds number may be found as an integral involving the induced force and the partial derivative of the Oseen kernel with respect to the Reynolds number. The Hellmann–Feynman theorem provides a short and simple proof of Brenner's lowest-order Oseen correction formula for the drag [7]. Finally, a discussion is provided in Section 5. Amongst other items, the extension of the variational principle to the full Navier–Stokes drag is investigated.

2. Review of the variational principle

In this section, we give a brief overview of the variational principle for the Oseen equation. For full details we refer to Ref. [2]. We consider an object submerged in a rotating viscous incompressible fluid. Free rotation of the object about an internal point Q is allowed, where Q has a fixed position on the rotation axis of the fluid. In the corotating frame the unperturbed fluid is supposed to have a constant main stream velocity in the direction of the angular velocity. The convection of momentum is taken into account in the Oseen approximation. In dimensionless units the incompressibility condition and the equation of motion take the form

$$\nabla \cdot \mathbf{u}(\mathbf{r}) = 0 \tag{1}$$

and

$$R\hat{\mathbf{U}} \cdot \nabla \mathbf{u}(\mathbf{r}) + 2T\hat{\mathbf{\Omega}} \times \mathbf{u}(\mathbf{r}) = -\nabla p(\mathbf{r}) + \Delta \mathbf{u}(\mathbf{r}) + \mathbf{F}(\mathbf{r}), \tag{2}$$

where $\mathbf{r}$ is the position vector with respect to Q and $\hat{\mathbf{U}}$ and $\hat{\mathbf{\Omega}}$ denote unit vectors in the direction of the main stream velocity and angular velocity, respectively. The velocity $\mathbf{u}(\mathbf{r})$ is the deviation from $\hat{\mathbf{U}}$, and $p(\mathbf{r})$ is a reduced pressure [4]. The Reynolds and Taylor numbers are denoted by R and T, respectively. When $T = 0$ Eq. (2) reduces to the Oseen equation, whereas for $R = 0$ we retrieve the equation of motion employed in Ref. [4]. The force density $\mathbf{F}(\mathbf{r})$ is induced by the presence of the object [3–6]. The use of $\mathbf{F}$ is convenient because Eqs. (1) and (2) may be considered to hold in all of space V.

On the surface S of the object we assume a stick boundary condition,

$$\mathbf{u}(\mathbf{r}) = -\hat{\mathbf{U}} + \omega \times \mathbf{r} \qquad \text{for } \mathbf{r} \text{ on } S, \tag{3}$$

where ω is the angular velocity of the object in the rotating frame. This relation also provides a suitable extension of the velocity field inside the object. With $\mathbf{f}$ as the induced force per unit area on S, we have

$$\mathbf{f}\, \mathrm{d}S = \mathbf{F}\, \mathrm{d}V, \tag{4}$$

where dV is an infinitesimal volume element comprising dS. The equations

$$\mathsf{F} = -\int_S \mathrm{d}S\,\mathbf{f} = -\int_V \mathrm{d}V\,\mathbf{F} \tag{5}$$

link the force F on the object with $\mathbf{f}$ and $\mathbf{F}$. The angular velocity is determined by the condition of free rotation. This implies that the torque $\mathbf{M}$ on the object equals zero,

$$\mathbf{M} = -\int_V \mathrm{d}V\,\mathbf{r}\times\mathbf{F} = 0. \tag{6}$$

At infinity, the total velocity tends to its unperturbed value, implying that

$$\mathbf{u}(\mathbf{r}) \longrightarrow 0 \qquad \text{for } r \longrightarrow \infty, \tag{7}$$

where $r = |\mathbf{r}|$.

The main quantity of interest is the dimensionless drag D on the object. We have

$$D = -\hat{\mathbf{U}}\cdot\mathsf{F} = -\hat{\mathbf{U}}\cdot\int_V \mathrm{d}V\,\mathbf{F} \tag{8}$$

(see Ref. [5]). It was shown in Ref. [2] that the drag (8) may be found from a variational principle. Indeed, we may consider the functional

$$X = \int_V \mathrm{d}V(\mathbf{u}_1\cdot(-R\hat{\mathbf{U}}\cdot\nabla\mathbf{u}_2 - 2T\hat{\mathbf{\Omega}}\times\mathbf{u}_2 - \nabla p_2 + \Delta\mathbf{u}_2) + p_1\nabla\cdot\mathbf{u}_2)$$
$$+ \int_V \mathrm{d}V(\mathbf{F}_1\cdot(\mathbf{u}_2 + \hat{\mathbf{U}} - \omega_2\times\mathbf{r}) + (\mathbf{u}_1 - \hat{\mathbf{U}} - \omega_1\times\mathbf{r})\cdot\mathbf{F}_2), \tag{9}$$

which depends on the trial fields $\mathbf{u}_j$, p_j, $\mathbf{F}_j$ and the trial vectors ω_j, with $j = 1, 2$. It is assumed that $\mathbf{u}_j$ satisfies the condition (7) and also that $\mathbf{F}_j$ vanishes except at the surface S just as $\mathbf{F}$. When we perform the variations with respect to the trial fields $\mathbf{u}_j$, p_j, $\mathbf{F}_j$ and the trial vectors ω_j and demand stationarity, it turns out that $\mathbf{u}_2$ satisfies the equation of motion, the continuity equation, the no-torque condition and the boundary condition (3) [2]. In particular, $\mathbf{F}_2 = \mathbf{F}$ and $\omega_2 = \omega$. Also, $\mathbf{u}_1$ satisfies the same equations with $\hat{\mathbf{U}}$ and $\hat{\mathbf{\Omega}}$ replaced by $-\hat{\mathbf{U}}$ and $-\hat{\mathbf{\Omega}}$, respectively. Furthermore, $\mathbf{F}_1$ equals the corresponding induced force density $\mathbf{F}(-\hat{\mathbf{U}}, -\hat{\mathbf{\Omega}})$ and ω_1 the corresponding angular velocity. Finally, the stationary value is found to be

$$X^{\text{stat}} = -\hat{\mathbf{U}}\cdot\int_V \mathrm{d}V\,\mathbf{F}_2 = -\hat{\mathbf{U}}\cdot\int_V \mathrm{d}V\,\mathbf{F} = D. \tag{10}$$

The wave-vector representation is also useful. With Fourier transforms, defined for example for the velocity field $\mathbf{u}(\mathbf{r})$ as

$$\mathbf{u}(\mathbf{k}) = \int_V \mathrm{d}V\,\mathbf{u}(\mathbf{r})\mathrm{e}^{-i\mathbf{k}\cdot\mathbf{r}}, \tag{11}$$

where $\mathbf{k}$ is the 3D wave vector we find the functional as

$$X = D_1 + D_2 + \omega_2 \cdot \mathbf{M}_1 + \omega_1 \cdot \mathbf{M}_2 + \frac{1}{(2\pi)^3} \int_W \mathrm{d}W\, \mathbf{F}_1(-\mathbf{k}) \cdot \frac{k(k + iR\xi)(1 - \hat{\mathbf{k}}\hat{\mathbf{k}}) + 2T\zeta \hat{\mathbf{k}} \cdot \epsilon}{k^2(k + iR\xi)^2 + (2T\zeta)^2} \cdot \mathbf{F}_2(\mathbf{k}), \tag{12}$$

where ϵ is the Levi–Civita tensor, ζ denotes the pseudoscalar $\hat{\mathbf{\Omega}} \cdot \hat{\mathbf{k}}$ and W is all of wave vector space. As a special case, let us assume that the fluid is nonrotating and the object is sufficiently symmetric so that we may take

$$\mathbf{F}(\mathbf{r}) = \mathbf{F}_2(\mathbf{r}) = -\mathbf{F}_1(-\mathbf{r}), \tag{13}$$

and, consequently,

$$\mathbf{F}(\mathbf{k}) = \mathbf{F}_2(\mathbf{k}) = -\mathbf{F}_1(-\mathbf{k}). \tag{14}$$

As an example we may mention the sphere ($r = 0$ being the center). In this special case the functional reduces to

$$X = 2D - \frac{1}{(2\pi)^3} \int_W \mathrm{d}W\, \mathbf{F}(\mathbf{k}) \cdot \frac{1 - \hat{\mathbf{k}}\hat{\mathbf{k}}}{k^2 + iRk\xi} \cdot \mathbf{F}(\mathbf{k}), \tag{15}$$

without angular velocities, torques and Taylor number.

Approximations to the drag may be found in the standard way. After insertion of a suitable finite linear expansion (multipole expansion) for the induced force density in the appropriate functional, the condition of stationarity leads to a set of linear equations for the multipoles. The stationary value of the approximate functional gives an approximate value for the drag. The multipole expansion converges rapidly, because small errors in the representation of the exact solution in terms of trial functions give rise to quadratically small errors in the stationary value of the functional. Applications of the induced force methodology using multipole expansions may be found in Refs. [2–6], dealing with a variety of calculations of the Oseen drag on a sphere at nonzero Reynolds and Taylor numbers.

It is of interest to note that an Olmstead–Gautesen theorem [8] may be derived directly from the variational principle. The original theorem states that for an arbitrarily shaped object the Oseen drag is invariant for the reversal of the main velocity $\hat{\mathbf{U}}$,

$$D(\hat{\mathbf{U}}) = D(-\hat{\mathbf{U}}). \tag{16}$$

When a rotation is superimposed on the main velocity, an extended version is found to hold [2],

$$D(\hat{\mathbf{U}}, \hat{\mathbf{\Omega}}) = D(-\hat{\mathbf{U}}, -\hat{\mathbf{\Omega}}). \tag{17}$$

So the drag is invariant for a simultaneous reversal of the main stream velocity and the angular velocity. This is clearly an example of the more general Onsager symmetry for transport coefficients in the linear regime [9] (linear referring here to the Oseen linearization).

In the next section, we will apply the formalism to the Oseen drag on a disk.

3. The Oseen drag on a disk broadside to the stream

In this section, we analyze the drag on a disk broadside to the stream. It turns out to be helpful to first discuss the simpler problem of heat transfer from a disk in the Oseen approximation.

The dimensionless equation for the suitably nondimensionalized temperature field Θ is

$$P\hat{\mathbf{U}}\cdot\nabla\Theta(\mathbf{r}) = \Delta\Theta(\mathbf{r}) + F(\mathbf{r}), \tag{18}$$

where P has the role of Péclet number and the boundary conditions are

$$\Theta(\mathbf{r}) = 0 \qquad \text{for } r \longrightarrow \infty, \tag{19}$$

$$\Theta(\mathbf{r}) = 1 \qquad \text{on the disk.} \tag{20}$$

We will use cylindrical coordinates ρ, ϕ, z, the disk being located at $\rho \leq 1$ and $z = 0$, and the positive z-axis pointing in the direction of $\hat{\mathbf{U}}$. The dimensionless heat transfer (or Nusselt number N) is

$$N = \int_V \mathrm{d}V\, F. \tag{21}$$

In wave-vector space the formal solution of Eq. (18) reads

$$\Theta(\mathbf{k}) = \frac{1}{k^2 + iP\hat{\mathbf{U}}\cdot\mathbf{k}} F(\mathbf{k}). \tag{22}$$

In strict analogy with the analyses referred to in the previous section, we may write down the functional in wave-vector space as

$$X_N = 2N - \frac{1}{(2\pi)^3}\int_W \mathrm{d}W\, F(\mathbf{k}) \frac{1}{k^2 + iP\hat{\mathbf{U}}\cdot\mathbf{k}} F(\mathbf{k}) \tag{23}$$

(cf. (15)). Next, we form the expansion

$$F(\mathbf{r}) = \delta(z)\frac{H(\rho - 1)}{(1-\rho^2)^{1/2}} \sum_{l=0}^{\infty} \frac{(2l)!!}{(2l-1)!!}\frac{4}{\pi} f_l P_{2l}((1-\rho^2)^{1/2}), \tag{24}$$

whence $N = 8f_0$ and P_l indicates the Legendre polynomial of order l [10]. Also, H denotes the Heaviside function. In wave-vector space we have

$$F(\mathbf{k}) = \frac{4}{\pi}\int_0^1 \frac{\rho\,\mathrm{d}\rho}{(1-\rho^2)^{1/2}} \sum_{l=0}^{\infty} \frac{(2l)!!}{(2l-1)!!} f_l P_{2l}((1-\rho^2)^{1/2}) \int_0^{2\pi} \mathrm{d}\phi\, \mathrm{e}^{-ik_\rho\rho\cos\phi} \tag{25}$$

$$= \frac{4}{\pi}\int_0^1 \frac{\rho\,\mathrm{d}\rho}{(1-\rho^2)^{1/2}} \sum_{l=0}^{\infty} \frac{(2l)!!}{(2l-1)!!} f_l P_{2l}((1-\rho^2)^{1/2}) J_0(k_\rho\rho) \tag{26}$$

$$= \frac{4}{\pi}\sum_{l=0}^{\infty} \frac{(2l)!!}{(2l-1)!!} f_l \frac{\Gamma\left(l + \frac{1}{2}\right)}{\Gamma\left(\frac{1}{2}\right)\Gamma(l+1)} j_{2l}(k_\rho) \tag{27}$$

$$= \frac{4}{\pi} \sum_{l=0}^{\infty} f_l j_{2l}(k_\rho), \tag{28}$$

where J_0 is the Bessel function of the first kind of order zero and j_l indicates the spherical Bessel function of the first kind of order l [10]. We used ([11], p. 1216) and ([10], Sections 10.1.1 and 22.4.2) to perform the ρ integration.

We insert the expansion in Eq. (23) and require that all derivatives with respect to the f_l are zero. The ensuing set of equations becomes

$$\pi^3 \delta_{l0} = \sum_{m=0}^{\infty} f_m \int_W \mathrm{d}W \frac{j_{2l}(k_\rho) j_{2m}(k_\rho)}{k^2 + iP\hat{\mathbf{U}}\cdot\mathbf{k}}. \tag{29}$$

To calculate the integrals in Eq. (29) we employ

$$\int_{-\infty}^{\infty} \mathrm{d}k_z \frac{1}{k_z^2 + k_\rho^2 + iPk_z} = \frac{\pi}{(k_\rho^2 + P^2/4)^{1/2}}, \tag{30}$$

which may be verified via Cauchy's residue theorem. We have

$$\int_W \mathrm{d}W \frac{j_{2l}(k_\rho) j_{2m}(k_\rho)}{k^2 + iPk_z} = 2\pi^2 \int_0^{\infty} \mathrm{d}k_\rho \frac{k_\rho}{(k_\rho^2 + P^2/4)^{1/2}} j_{2l}(k_\rho) j_{2m}(k_\rho)$$

$$= \pi^3 \int_0^{\infty} \mathrm{d}k_\rho \frac{1}{(k_\rho^2 + P^2/4)^{1/2}} J_{2l+(1/2)}(k_\rho) J_{2m+(1/2)}(k_\rho)$$

$$= 2\pi^2 \int_0^{\pi/2} \mathrm{d}\theta \cos 2(l-m)\theta \int_0^{\infty} \mathrm{d}k_\rho \frac{J_{2l+2m+1}(2k_\rho \cos\theta)}{(k_\rho^2 + P^2/4)^{1/2}} = 2\pi^2 C_{lm}, \tag{31}$$

with

$$C_{lm} = \int_0^{\pi/2} \mathrm{d}\theta \cos 2(l-m)\theta \, I_{l+m+1/2}\left(\frac{P}{2}\cos\theta\right) K_{l+m+1/2}\left(\frac{P}{2}\cos\theta\right), \tag{32}$$

where $I_{l+m+(1/2)}$ and $K_{l+m+(1/2)}$ denote the modified spherical Bessel functions of the first and third kind [11]. In the evaluation of the integrals, use was made of ([11], Section 10.11) and ([12], Sections 6.681.10 and 6.552.1). The set (29) reduces to

$$\frac{\pi}{2} \delta_{l0} = \sum_{m=0}^{\infty} C_{lm} f_m. \tag{33}$$

Next, we perform an expansion of C_{lm} in powers of P. We use

$$I_{l+m+1/2}\left(\frac{P}{2}\cos\theta\right) K_{l+m+1/2}\left(\frac{P}{2}\cos\theta\right)$$

$$= \sum_{n=0}^{\infty} \frac{(-1)^{l+m+n} \prod_{k=0}^{l+m-1} (n-1-2k)}{(2l+2m+1)!!n!!} (P\cos\theta)^n \tag{34}$$

and

$$\int_0^{\pi/2} \mathrm{d}\theta \cos 2(l-m)\theta \cos^n \theta$$

$$= \begin{cases} 0 & \text{for } n \text{ even and } n < 2|l-m|, \\ \left(\dfrac{\pi}{2}\right)^{\delta_{n,\text{even}}} \dfrac{n!}{(n-2l+2m)!!(n+2l-2m)!!} & \text{for } n \geq 2|l-m| \end{cases} \tag{35}$$

by Gradshteyn and Ryzhik [12], Section 3.631.17. The structure of Eqs. (34) and (35) reveals that in C_{lm} the coefficient of P^n is zero for $n < 2|l-m|$. This gives the result

$$C_{lm} = \sum_{n=2|l-m|}^{\infty} \left(\frac{\pi}{2}\right)^{\delta_{n,\text{even}}} \frac{(-1)^{l+m+n}(n-1)!! \prod_{k=0}^{l+m-1} (n-1-2k)P^n}{(n+2l+2m+1)!!(n-2l+2m)!!(n+2l-2m)!!}, \tag{36}$$

implying

$$C_{lm} = O(P^{2|l-m|}), \qquad f_m = O(P^{2m}). \tag{37}$$

It follows from the set (33) that f_0 (and hence the Nusselt number) is affected by inclusion of f_m from order P^{4m} onwards. It may be established with d'Alembert's test that expansion (36) is convergent for all values of P.

We proceed to a direct calculation of N. To order P^3 we calculate

$$C_{00} = \frac{\pi}{2} - \frac{1}{2}P + \frac{\pi}{24}P^2 - \frac{1}{36}P^3 + O(P^4), \tag{38}$$

which via the truncated set

$$\frac{\pi}{2} = C_{00} f_0 \tag{39}$$

implies

$$N = 8\left(1 - \frac{1}{\pi}P + \frac{1}{12}P^2 - \frac{1}{18\pi}P^3\right)^{-1} \tag{40}$$

correct to $O(P^4)$. To find N to order P^5 we calculate

$$C_{00} = \frac{\pi}{2} - \frac{1}{2}P + \frac{\pi}{24}P^2 - \frac{1}{36}P^3 + \frac{\pi}{640}P^4 - \frac{1}{1350}P^5 + O(P^6), \tag{41}$$

$$C_{10} = -\frac{\pi}{240}P^2 + \frac{1}{180}P^3 + O(P^4), \tag{42}$$

$$C_{11} = \frac{\pi}{10} + O(P^2). \tag{43}$$

The truncated set including f_0 and f_1 leads to

$$f_1 = \frac{1}{24}P^2 - \frac{1}{72\pi}P^3 + O(P^4), \tag{44}$$

and we find

$$N = 8f_0 = 8\left(1 - \frac{1}{\pi}P + \frac{1}{12}P^2 - \frac{1}{18\pi}P^3 + \frac{1}{360}P^4 - \frac{1}{1800\pi}P^5\right)^{-1} + O(P^6) \tag{45}$$

as a result. It is noted that the expression (45) using the inverse is much simpler than a direct expansion in powers of P.

In order to obtain values for the Nusselt number at Péclet numbers beyond the convergence range of the power series (45) we use the well-known Padé table. As starting point we fill the first to N-th position of the top row with the partial sums with one to N terms of the series. Since it is already known that for very large P the Nusselt number becomes proportional to P, it is appropriate to focus our evaluation on the series of values on the subdiagonal of the table on which the degree of the numerator exceeds the degree of the denominator by 1. This subdiagonal, the top row, the auxiliary row and the arbitrary element C, as well as its neighbors No, So, We, Ea are indicated in the following scheme:

$$\begin{array}{ccccccccc}
\infty & \infty & \infty & \infty & \cdot & \cdot & \cdots & & \\
8 & 8\left(1+\frac{1}{\pi}P\right) & \cdot & \cdot & \cdot & \cdot & \cdot & \cdot & \sum_{n=0}^{N} a_n P^n \\
\cdot & \text{dia} & \text{subdia} & \cdot & \cdot & \cdot & \text{No} & \cdot & \cdots \\
\cdot & \cdot & \text{dia} & \text{subdia} & \cdot & \text{We} & C & \text{Ea} & \cdots \\
\cdot & \cdot & \cdot & \text{dia} & \text{subdia} & \cdot & \text{So} & \cdot & \cdots \\
\cdot & \cdot & \cdot & \cdot & \text{dia} & \text{subdia} & \cdot & \cdot & \cdots \\
\cdot & \cdot & \cdot & \cdot & \cdot & \text{dia} & \text{subdia} & \cdot & \cdots \\
\vdots & \vdots & \vdots & \vdots & \vdots & \vdots & \vdots & \vdots & \ddots
\end{array} .$$

To calculate the elements on the subdiagonal, we have used the following relationship discovered by Wynn [13]

$$\frac{1}{C - \text{No}} + \frac{1}{C - \text{So}} = \frac{1}{C - \text{We}} + \frac{1}{C - \text{Ea}}. \tag{46}$$

The values for the Nusselt number obtained by the calculation described above are quoted in the second column of Table 1. The digits between brackets were added by application of Shanks's transformation to the series of values on the subdiagonal.

Based on the numbers in the third column in Table 1 we postulate the asymptotic series

$$N = 4\pi\left(P + \frac{1}{2} + \frac{1}{16P} - \frac{1}{64P^2} + \frac{1}{256P^3} + O(P^{-4})\right) \tag{47}$$

for the Nusselt number.

Table 1
Results from the Padé table for N for the disk. $\Delta = \frac{N}{8} - \frac{\pi}{2}(P + \frac{1}{2} + \frac{1}{16P} - \frac{1}{64P^2} + \frac{1}{256P^3})$

P	$N/8$	$-\Delta P^4$
2	1.6887680035	0.02929
10	4.7480214673	0.06445
20	8.6580750834	0.07007
30	12.5790373577	0.07280
40	16.5029391427	0.07433
50	20.4280520991	0.07532
60	24.3537806911	0.07600
70	28.279864820(3)	0.076(6)
80	32.20617281(5)	0.07(7)
90	36.13263087(6)	
100	40.0591944(3)	

After these considerations, we turn to the Oseen drag on the disk. We employ the functional (15), with an expansion similar to Eq. (24),

$$\mathbf{F}(\mathbf{r}) = \delta(z)\frac{\theta(\rho - 1)}{(1 - \rho^2)^{1/2}}\sum_{l=0}^{\infty}\frac{(2l)!!}{(2l - 1)!!}(g_l\hat{\mathbf{U}} + h_l\boldsymbol{\rho})P_{2l}((1 - \rho^2)^{1/2}), \tag{48}$$

$\boldsymbol{\rho}$ denoting the radius vector in the plane of the disk. Furthermore, the dimensionless drag on the disk is given by $D = 2\pi g_0$. Repeating the steps leading to Eq. (28) we easily find the Fourier transform as

$$\mathbf{F}(\mathbf{k}) = 2\pi\sum_{l=0}^{\infty}\left(g_l\hat{\mathbf{U}} + ih_l\hat{\mathbf{k}}_\rho\frac{\partial}{\partial k_\rho}\right)j_{2l}(k_\rho). \tag{49}$$

As before, we use the expansion in Eq. (15) and require that all derivatives with respect to the coefficients g_l and h_l are zero. The ensuing set of equations becomes

$$4\delta_{l0} = \sum_{m=0}^{\infty}(C_{lm}^{gg}g_m + C_{lm}^{gh}h_m), \tag{50}$$

$$0 = \sum_{m=0}^{\infty}(C_{lm}^{hg}g_m + C_{lm}^{hh}h_m), \tag{51}$$

with

$$C_{lm}^{gg} = \frac{1}{\pi^2}\int_W \mathrm{d}W\, j_{2l}(k_\rho)j_{2m}(k_\rho)\frac{\hat{\mathbf{U}}\cdot(1 - \hat{\mathbf{k}}\hat{\mathbf{k}})\cdot\hat{\mathbf{U}}}{k^2 + iRk_z}, \tag{52}$$

$$C_{ml}^{hg} = C_{lm}^{gh} = \frac{1}{\pi^2}\int_W \mathrm{d}W\, j'_{2l}(k_\rho)j_{2m}(k_\rho)\frac{i\hat{\mathbf{k}}_\rho\cdot(1 - \hat{\mathbf{k}}\hat{\mathbf{k}})\cdot\hat{\mathbf{U}}}{k^2 + iRk_z}, \tag{53}$$

$$C_{lm}^{hh} = \frac{1}{\pi^2}\int_W \mathrm{d}W\, j'_{2l}(k_\rho)j'_{2m}(k_\rho)\frac{i\hat{\mathbf{k}}_\rho\cdot(1 - \hat{\mathbf{k}}\hat{\mathbf{k}})\cdot i\hat{\mathbf{k}}_\rho}{k^2 + iRk_z}. \tag{54}$$

These coefficients may be evaluated as before. First, we have

$$C_{lm}^{gg} = \int_0^{\pi/2} \mathrm{d}\theta \cos 2(l-m)\theta I_{l+m+1/2}\left(\frac{R}{2}\cos\theta\right)K_{l+m+1/2}\left(\frac{R}{2}\cos\theta\right). \tag{55}$$

In the calculation of C_{lm}^{gh} we employ

$$j'_{2l} = \frac{1}{4l+1}(-(2l+1)j_{2l+1} + 2lj_{2l-1}) \tag{56}$$

and proceed as before to obtain

$$\begin{aligned} C_{lm}^{gh} = {} & \frac{2l+1}{4l+1}\int_0^{\pi/2} \mathrm{d}\theta \cos(2l-2m-1)\theta I_{l+m+3/2}\left(\frac{R}{2}\cos\theta\right)K_{l+m+1/2}\left(\frac{R}{2}\cos\theta\right) \\ & - \frac{2l}{4l+1}\int_0^{\pi/2} \mathrm{d}\theta \cos(2l-2m+1)\theta I_{l+m+1/2}\left(\frac{R}{2}\cos\theta\right)K_{l+m-1/2}\left(\frac{R}{2}\cos\theta\right). \end{aligned} \tag{57}$$

In the calculation of C_{lm}^{hh} we use Eq. (56) twice and obtain, after the usual manipulations

$$C_{lm}^{hh} = \frac{1}{\pi^2}\int_W \mathrm{d}W\, j'_{2l}(k_\rho)j'_{2m}(k_\rho)\frac{i\hat{\mathbf{k}}_\rho\cdot(1-\hat{\mathbf{k}}\hat{\mathbf{k}})\cdot i\hat{\mathbf{k}}_\rho}{k^2+iRk_z} \tag{58}$$

$$\begin{aligned} &= -\int_0^\infty \mathrm{d}k_\rho k_\rho \frac{j'_{2l}(k_\rho)j'_{2m}(k_\rho)}{(k_\rho^2+R^2/4)^{1/2}} \\ &= \frac{1}{(4l+1)(4m+1)}\left(\int_0^{\pi/2} \mathrm{d}\theta\, c(\theta) I_{l+m+1/2}\left(\frac{R}{2}\cos\theta\right)K_{l+m+1/2}\left(\frac{R}{2}\cos\theta\right)\right. \\ &\quad - \int_0^{\pi/2} \mathrm{d}\theta \cos 2(l-m)\theta\left(4lmI_{l+m-1/2}\left(\frac{R}{2}\cos\theta\right)K_{l+m-1/2}\left(\frac{R}{2}\cos\theta\right)\right. \\ &\quad \left.\left. + (2l+1)(2m+1)I_{l+m+3/2}\left(\frac{R}{2}\cos\theta\right)K_{l+m+3/2}\left(\frac{R}{2}\cos\theta\right)\right)\right), \end{aligned} \tag{59}$$

where we use

$$c(\theta) = 2m(2l+1)\cos 2(l-m+1)\theta + (2m+1)2l\cos 2(l-m-1)\theta \tag{60}$$

as a shorthand.

Next we calculate the drag to order R^3. We have

$$\begin{aligned} C_{00}^{gg} &= \int_0^{\pi/2} \mathrm{d}\theta\, I_{1/2}\left(\frac{R}{2}\cos\theta\right)K_{1/2}\left(\frac{R}{2}\cos\theta\right) \\ &= \frac{\pi}{2} - \frac{1}{2}R + \frac{\pi}{24}R^2 - \frac{1}{36}R^3 + O(R^4), \end{aligned} \tag{61}$$

$$C_{00}^{gh} = \int_0^{\pi/2} \mathrm{d}\theta \cos\theta I_{3/2}\left(\frac{R}{2}\cos\theta\right)K_{1/2}\left(\frac{R}{2}\cos\theta\right) = \frac{\pi}{24}R - \frac{1}{18}R^2 + O(R^3), \tag{62}$$

$$C_{00}^{hh} = -\int_0^{\pi/2} \mathrm{d}\theta I_{3/2}\left(\frac{R}{2}\cos\theta\right) K_{3/2}\left(\frac{R}{2}\cos\theta\right) = -\frac{\pi}{6} + O(R^2). \tag{63}$$

The truncated set (50) and (51) consequently yields

$$D = 2\pi g_0 = \frac{16}{1 - \dfrac{1}{\pi}R + \dfrac{5}{48}R^2 - \dfrac{1}{9\pi}R^3} \tag{64}$$

as a result exact to order R^3. As a direct series in R the result is

$$D = 16\left(1 + \frac{1}{\pi}R + \left(-\frac{5}{48} + \frac{1}{\pi^2}\right)R^2 + \left(-\frac{7}{72\pi} + \frac{1}{\pi^3}\right)R^3\right) + O(R^4), \tag{65}$$

which differs from the expression given by Hocking [14], Eq. (19) in the term of order R^3.

Next we investigate the drag on the disk for large R. From Zeilon's work [15] we already know that for R tending to infinity

$$D = \frac{3}{2}\pi R. \tag{66}$$

As in the case of heat transfer, we use a large number of terms of the power series (65) and calculate the values on the subdiagonal of the Padé table. The values obtained in this way are given in Table 2. The digits between brackets are the result of application of Shanks's transformation.

From the last column in Table 2 it can be concluded that

$$\frac{D}{R} = \frac{3}{2}\pi + O(R^{-1+0.11}), \tag{67}$$

implying that the correction to the value at infinite Reynolds number is a noninteger power of R.

Table 2
Results from the Padé table for D for the disk

R	$D/R - 3\pi/2$	$14.44R^{-0.89}$
10	1.8528379446	1.860232
20	0.9952480416	1.003808
40	0.5382792168	0.541669
60	0.3760928(2)	0.377583
80	0.291593(1)	0.292292
100	0.23931(5)	0.239644
120	0.2036(0)	0.203749
140	0.1775(7)	0.177629
160	0.1577(1)	0.157725
180	0.1420(3)	0.142028
200	0.129(3)	0.129315

4. A Hellmann–Feynman theorem for the drag

The variational principle allows us to give a succinct formula for the variation of the drag as a function of the Reynolds number (we assume that the Taylor number is zero). To show this, we employ the principle in an equivalent but slightly different form [2], using the fundamental solution of the Oseen equation [16,17]. The basic equation in the latter approach is a real space integral equation for the stress at the surface of the object. In the present notation the equation reads

$$0 = \hat{\mathbf{U}} + \int_{S'} \mathrm{d}S' \, \mathbf{E}(\mathbf{r}, \mathbf{r}') \cdot \mathbf{f}(\mathbf{r}') \tag{68}$$

(cf. (4)). The tensor $\mathbf{E}$ is given explicitly by

$$\begin{aligned} \mathbf{E} = & \frac{1 - \mathrm{e}^{-R(q + \hat{\mathbf{U}} \cdot \mathbf{q})/2}}{4\pi R(q + \hat{\mathbf{U}} \cdot \mathbf{q})^2} ((\hat{\mathbf{U}} + \hat{\mathbf{q}})(\hat{\mathbf{U}} + \hat{\mathbf{q}}) + (\hat{\mathbf{q}}\hat{\mathbf{q}} - \mathbf{1})(1 + \hat{\mathbf{U}} \cdot \hat{\mathbf{q}})) \\ & + \frac{\mathrm{e}^{-R(q + \hat{\mathbf{U}} \cdot \mathbf{q})/2}}{8\pi(q + \hat{\mathbf{U}} \cdot \mathbf{q})} (2\mathbf{1}(1 + \hat{\mathbf{U}} \cdot \hat{\mathbf{q}}) - (\hat{\mathbf{U}} + \hat{\mathbf{q}})(\hat{\mathbf{U}} + \hat{\mathbf{q}})), \end{aligned} \tag{69}$$

with $\mathbf{q} = \mathbf{r}' - \mathbf{r}$. With the help of the relation between $\mathbf{F}_j(\mathbf{k})$ and $\mathbf{f}_j(\mathbf{r})$,

$$\mathbf{F}_j(\mathbf{k}) = \int_S \mathrm{d}S \, \mathbf{f}_j(\mathbf{r}) \mathrm{e}^{-i\mathbf{k} \cdot \mathbf{r}}, \tag{70}$$

the functional X can be shown to be [2]

$$X = \hat{\mathbf{U}} \cdot \int_S \mathrm{d}S(\mathbf{f}_1(\mathbf{r}) - \mathbf{f}_2(\mathbf{r})) + \int_S \mathrm{d}S \int_{S'} \mathrm{d}S' \, \mathbf{f}_1(\mathbf{r}) \cdot \mathbf{E}(\mathbf{r}, \mathbf{r}') \cdot \mathbf{f}_2(\mathbf{r}'). \tag{71}$$

Next, let us use X from Eq. (71) and vary the Reynolds number. A variation in R causes X to change by an amount

$$\begin{aligned} \delta X = & \hat{\mathbf{U}} \cdot \int_S \mathrm{d}S(\delta\mathbf{f}_1 - \delta\mathbf{f}_2) + \int_S \mathrm{d}S \int_{S'} \mathrm{d}S' \, \delta\mathbf{f}_1 \cdot \mathbf{E} \cdot \mathbf{f}_2' + \int_S \mathrm{d}S \int_{S'} \mathrm{d}S' \, \mathbf{f}_1 \cdot \mathbf{E} \cdot \delta\mathbf{f}_2' \\ & + \int_S \mathrm{d}S \int_{S'} \mathrm{d}S' \, \mathbf{f}_1 \cdot \frac{\partial \mathbf{E}}{\partial R} \cdot \mathbf{f}_2' \delta R. \end{aligned} \tag{72}$$

Because X is stationary with respect to the variations $\delta\mathbf{f}_1$ and $\delta\mathbf{f}_2$, the terms containing them do not contribute to δX. We obtain

$$\frac{\mathrm{d}D}{\mathrm{d}R} = \int_S \mathrm{d}S \int_{S'} \mathrm{d}S' \, \mathbf{f}_1 \cdot \frac{\partial \mathbf{E}}{\partial R} \cdot \mathbf{f}_2' \tag{73}$$

as a result.

This formula may be looked upon as a Hellmann–Feynman theorem for the drag. The Hellmann–Feynman theorem in quantum mechanics (see, e.g. Ref. [18]) states that the derivative of the ground-state energy E with respect to a parameter λ in the Hamiltonian H

of the system may be calculated as

$$\frac{\mathrm{d}E}{\mathrm{d}\lambda} = \left\langle \psi_g \middle| \frac{\partial H}{\partial \lambda} \psi_g \right\rangle, \tag{74}$$

where ψ_g is the normalized ground-state wave function, and the brackets indicate the inner product in the appropriate Hilbert space. No differentiations of the wave functions need to be performed, due to the circumstance that E may be linked to a variational-minimalization problem.

As an application, we give a derivation of the first-order Oseen correction for the drag on an arbitrarily shaped body. To first order in R we put

$$D = D_0 + D_1 R, \qquad \mathbf{E} = \mathbf{E}_0 + \mathbf{E}_1 R, \tag{75}$$

where $\mathbf{E}_0$ and $\mathbf{E}_1$ may be found from Eq. (69). In fact, $\mathbf{E}_0$ is the Oseen–Burgers tensor known from Stokes theory [19],

$$\mathbf{E}_0 = \frac{1}{8\pi q}(\mathbf{1} + \hat{\mathbf{q}}\hat{\mathbf{q}}), \tag{76}$$

and

$$\mathbf{E}_1 = \frac{1}{32\pi}((\hat{\mathbf{U}} + \hat{\mathbf{q}})(\hat{\mathbf{U}} + \hat{\mathbf{q}}) - (\hat{\mathbf{q}}\hat{\mathbf{q}} + 3\mathbf{1})(1 + \hat{\mathbf{U}}\cdot\hat{\mathbf{q}})). \tag{77}$$

From Eq. (73) we obtain

$$D_1 = -\int_S \mathrm{d}S \int_{S'} \mathrm{d}S' \, \mathbf{f}_0\cdot\mathbf{E}_1\cdot\mathbf{f}_0', \tag{78}$$

where we used that at $R = 0$

$$-\mathbf{f}_{1,0} = \mathbf{f}_{2,0} = \mathbf{f}_0. \tag{79}$$

Next we note that the terms in Eq. (77) that are uneven in $\hat{\mathbf{q}}$ vanish upon integration in Eq. (78), so that we are left with

$$D_1 = \frac{1}{32\pi}\int_S \mathrm{d}S \int_{S'} \mathrm{d}S' \, \mathbf{f}_0\cdot(-\hat{\mathbf{U}}\hat{\mathbf{U}} + 3\mathbf{1})\cdot\mathbf{f}_0' \tag{80}$$

$$= \frac{1}{32\pi}(3\mathsf{F}_0\cdot\mathsf{F}_0 - D_0^2). \tag{81}$$

For $R = 0$ we may put

$$\mathsf{F}_0 = \mathsf{M}\cdot\hat{\mathbf{U}}, \tag{82}$$

with M the (symmetric) mobility matrix. So, if $\hat{\mathbf{U}}$ is in the direction of one of the principal axes of this matrix, Eq. (81) reduces to

$$D_1 = \frac{1}{16\pi} D_0^2. \tag{83}$$

This equation and the more general equation (81) were first put forward by Brenner [7], and proven by Brenner and Cox [20] for the Navier–Stokes equation using techniques based on singular-perturbation methods.

5. Conclusion and discussion

In this chapter, we have presented a number of applications of a variational principle in linear hydrodynamics. After an overview of the principle in Section 2, we discussed the application to Oseen flow around a disk. We also treated the corresponding heat-transfer problem. Finally, we presented a Hellmann–Feynman theorem. This application of the variational principle turns out to lead naturally to the Brenner formula for the drag to first order in the Reynolds number.

As already remarked in Ref. [21], variational principles in fluid flow generally either include inertia but no viscous dissipation, or, conversely, include viscous dissipation while neglecting inertia. The variational principle for the drag in Oseen flow as first put forward in Ref. [2] and further developed in this chapter is a hybrid principle in that both viscous and inertial terms play their role. It is perhaps also of some interest to remark that it is not a minimum or maximum principle, as is already clear from the fact that the variational functional is bilinear rather than quadratic in a single field.

According to Finlayson [22,23], it is possible to construct a variational functional that yields as field equations the Navier–Stokes equation and an adjoint equation that has no physical meaning. Let us consider the functional

$$\begin{aligned} X_{\mathrm{NS}} &= \int_V \mathrm{d}V(\mathbf{u}_1\cdot(-R(\hat{\mathbf{U}}+\mathbf{u}_2)\cdot\nabla\mathbf{u}_2 - \nabla p_2 + \Delta\mathbf{u}_2) + p_1\nabla\cdot\mathbf{u}_2) \\ &+ \int_V \mathrm{d}V(\mathbf{F}_1\cdot(\mathbf{u}_2+\hat{\mathbf{U}}) + (\mathbf{u}_1 - \hat{\mathbf{U}})\cdot\mathbf{F}_2). \end{aligned} \tag{84}$$

Apart from the usual incompressibility and boundary conditions, the stationarity conditions yield the Navier–Stokes equation for $\mathbf{u}_2$,

$$R(\hat{\mathbf{U}}+\mathbf{u}_2)\cdot\nabla\mathbf{u}_2 = -\nabla p_2 + \Delta\mathbf{u}_2 + \mathbf{F}_2, \tag{85}$$

while the equation for $\mathbf{u}_1$ contains $\mathbf{u}_2$,

$$R((\nabla\mathbf{u}_2)\cdot\mathbf{u}_1 - (\hat{\mathbf{U}}+\mathbf{u}_2)\cdot\nabla\mathbf{u}_1) = -\nabla p_1 + \Delta\mathbf{u}_1 + \mathbf{F}_1. \tag{86}$$

In contrast to the functional used in Refs. [22,23], the stationary value of X_{NS} has a clear physical meaning: it is the Navier–Stokes drag. It follows that formally the extension to Navier–Stokes flow is possible at the price of the expected nonlinearities and coupled equations for the velocity fields.

Finally, we note that the functional differs from the Navier–Stokes functional put forward in Ref. [2]. The latter does not contain $\hat{\mathbf{U}}$ in the convective term. The difference implies the following. The functional (84) leads to flow around a fixed object in a steady oncoming stream with stick boundary conditions. In contrast, in the absence of the main stream $\hat{\mathbf{U}}$ in the convective term we find

$$R\mathbf{u}_2 \cdot \nabla \mathbf{u}_2 = -\nabla p_2 + \Delta \mathbf{u}_2 + \mathbf{F}_2, \tag{87}$$

again with the usual boundary condition $\mathbf{u}_2 = -\hat{\mathbf{U}}$. This velocity field clearly corresponds to Navier–Stokes flow with the far field at rest (see Eq. (7)), while at the surface of the object a constant nonzero velocity $-\hat{\mathbf{U}}$ is imposed. Although both functionals lead to the Navier–Stokes equation, in the context of drag it is clear that the functional (84) is the appropriate one.

References

[1] Serrin, J. (1959), Mathematical principles of classical fluid mechanics, Handbuch der Physik 8/1, 125–263.

[2] ten Bosch, B.I.M. and Weisenborn, A.J. (1998), On a variational principle for the drag in linear hydrodynamics, SIAM J. Appl. Math. 58, 1–14.

[3] Mazur, P. and Weisenborn, A.J. (1984), The Oseen drag on a sphere and the method of induced forces, Physica A 123, 209–226.

[4] Weisenborn, A.J. (1985), Drag on a sphere moving slowly in a rotating viscous fluid, J. Fluid Mech. 153, 215–227.

[5] Weisenborn, A.J. and ten Bosch, B.I.M. (1993), Analytical approach to the Oseen drag on a sphere at infinite Reynolds number, SIAM J. Appl. Math. 53, 601–620.

[6] Weisenborn, A.J. and ten Bosch, B.I.M. (1995), On the Oseen drag on a sphere, SIAM J. Appl. Math. 55, 577–592.

[7] Brenner, H. (1961), The Oseen resistance of a particle of arbitrary shape, J. Fluid Mech. 11, 604–610.

[8] Olmstead, W. and Gautesen, A.K. (1968), A new paradox in viscous hydrodynamics, Arch. Rat. Mech. Anal. 29, 58–65.

[9] de Groot, S.R. and Mazur, P. (1962), Non-equilibrium Thermodynamics, North-Holland, Amsterdam.

[10] Abramowitz, M. and Stegun, I.A. (1965), Handbook of Mathematical Functions, Dover, New York.

[11] Wolfe, P. (1971), Eigenfunctions of the integral equation for the potential of the charged disk, J. Math. Phys. 12, 1215–1218.

[12] Gradshteyn, I.S. and Ryzhik, I.M. (1980), Tables of Integrals, Series and Products, 4th edn, Academic Press, New York.

[13] Wynn, P. (1956), On a device for computing the $e_m(S_n)$ transformation, MTAC 10, 91–96.

[14] Hocking, L.M. (1959), The Oseen flow past a circular disk, Quart. J. Mech. Appl. Math. XII(Pt 4), 464–475.

[15] Zeilon, N. (1924), On potential problems in the theory of fluid resistance, Kungl. Svenska Vet. Handl. Ser. 3(1), 1–66.

[16] Olmstead, W. and Gautesen, A.K. (1976), Integral representations and the Oseen flow problem, Mech. Today 3, 125–189.

[17] Oseen, C.W. (1927), Neuere Methoden und Ergebnisse in der Hydrodynamik, Akademische Verlagsgesellschaft m.b.H., Leipzig.

[18] Merzbacher, E. (1970), Quantum Mechanics, 2nd edn, Wiley, New York.

[19] Bird, R.B., Armstrong, R.C. and Hassager, O. (1987), Dynamics of Polymeric Liquids, 2nd edn, Vol. 1, Wiley, New York.

[20] Brenner, H. and Cox, R.G. (1963), The resistance to a particle of arbitrary shape in translational motion at small Reynolds numbers, J. Fluid Mech. 17, 561–595.

[21] Hyre, M. and Glicksman, L. (1997), Brief communication on the assumption of minimum energy dissipation in circulating fluidized beds, Chem. Eng. Sci. 52, 2435–2438.

[22] Finlayson, B.A. (1972), The Method of Weighted Residuals and Variational Principles, Academic Press, New York.

[23] Finlayson, B.A. (1972), Existence of variational principles for the Navier–Stokes equation, Phys. Fluids 15, 963–967.

Chapter 5

A VARIATIONAL PRINCIPLE FOR THE IMPINGING-STREAMS PROBLEM

J.P. Curtis

QinetiQ, Fort Halstead, Building Q13, Sevenoaks, Kent TN14 7BP, UK

Abstract

The problem of impinging streams has interested researchers since early last century and is still under investigation today. The historical and current literature on this problem is reviewed. A general class of free boundary-value problems associated with variational principles is then reviewed, following an explanatory one-dimensional analog formulation.

In a synthesis of these two previously unrelated research areas a variational principle for the problem of impinging unequal streams in potential flow is derived under conditions of planar flow. A coupled pair of boundary-value problems defined in two flow regions is obtained comprising the Euler equations and the original and natural boundary conditions associated with this variational principle. The Euler equations and natural boundary conditions result from making stationary a functional associated with the pressure integral over the whole flow field. In each region Laplace's equation emerges as the Euler equation, and the natural boundary conditions correspond to the appropriate physical boundary conditions on the free surfaces and at the interface between the regions. The conservation equations and the stationarity conditions provide the equations necessary to solve the problem.

Keywords: variational principle; impinging streams; calculus of variations; Euler equation; variable domain

1. Introduction

Since early last century the problem of impinging unequal streams in steady flow has been of interest to researchers in fluid dynamics [1–4]. It has been believed to be indeterminate, but in this chapter we cite evidence from diverse sources that contradicts this belief and derive a variational principle that appears to offer an approach to a formal resolution. Over the last few decades this long-standing problem has become highly relevant in considering asymmetries in the shaped-charge jet-formation process. The presence of asymmetry brings about lateral drift of the shaped-charge jet particles. This drift causes them to collide on the crater wall rather than to contribute to the penetration at the bottom of the crater. More recently, Reid [5] has identified the potential importance of solving this

E-mail address: jpcurtis@qinetiq.com (J.P. Curtis).

problem in explosive welding applications. These applications as well as its fundamental nature maintain the present-day interest in the problem of impinging streams.

Traditionally the problem has been investigated for the case where the densities and speeds (relative to the stagnation point) of the streams are equal, but the widths of the streams differ. There is a solution for nonparallel streams in planar flow in this case. However, in the study of asymmetric jet formation the assumption of equal speeds relative to the stagnation point where the jet is formed is incorrect. Heider and Rottenkolber [6] sought to overcome this problem by choice of frame and achieved a closed-form solution, but the agreement with experimental data was not good. The author and coworkers [7–10] have investigated three models with improving, but still not entirely satisfactory, results. In all of these formulations the fundamental difficulty is that there are insufficient equations to solve for all the unknowns—the hallmark of indeterminacy. In each case plausible but nonrigorous methods of closing the problem have been attempted by making assumptions about the nature of the flow field.

For example, the condition that the outgoing jets lie in a straight line was taken as a hypothesis by Pack and Curtis [7] and used in conjunction with the equation of conservation of energy and Bernoulli's law applied to the outgoing jets to close the problem. In fact Curtis and Kelly [10] later showed that the straight-line hypothesis can only be sensibly applied up to a limit that is a function of the incoming-stream parameters. The same authors [8] investigated a jet-formation model based on the concept of a stagnant core around which the fluid moves in circular arcs. The treatment of the flow in the vicinity of the core was approximate. Nonetheless, this model exhibited increasingly good agreement with the results of Kinelovskii and Sokolov [2] as the angle between the impinging streams grew, coinciding for a head-on collision. Similarly the predictions of the off-axis velocities of shaped-charge jets improved with increasing collapse angle. A feature of the model was that the outgoing (generally nonparallel) jets consisted of two adjacent parallel streams, each traveling at the speed of the incoming stream from which it originated. A benefit of this approach is that the free-surface conditions are automatically satisfied.

More recently Curtis [9] considered a simplified model in which this feature was retained, but the assumption of the stagnant core was not made. He further assumed that the mean outgoing jet speeds were equal and that the outgoing jets travel in directions close to the nominal axis of symmetry. This set of assumptions resulted in an analytical model that was in better agreement with the experimental data from shaped charges than the earlier models. The model recovered the known special cases correctly and showed that the thinner jet is deflected more than the thicker one—an intuitively pleasing result. However, the agreement with experiment was still not fully satisfactory.

It was desired to improve on this position by following a different approach in which the underlying physics of the flow is investigated theoretically in a variational formulation assuming potential flow. Curtis [11] pursued this approach and presented the variational principle in essentially the same form as described here, barring a refinement here of the treatment of the boundary conditions. However, the underlying mathematics was not presented in any detail. In order to conduct the investigation it was necessary first to derive the techniques to address a class of variational principles in which a functional defined on a domain, which is itself allowed to vary, is made stationary. Here, by a functional we mean a straightforward integral of a function of the field variable and its derivatives over the domain concerned. The aim is to perform the variation with respect to the variable domain boundary as well as the field variable. In other words, the goal is to choose the best field variable and domain to maximize or minimize the quantity of interest as determined by the functional. We shall focus for now on problems where the governing differential equation is the Euler equation of a variational principle. The Euler equation is derived by the methods of the calculus of variations. In such a variational principle the solution of the differential equation may minimize, maximize or make stationary the functional. The reader desiring a basic introduction to the calculus of variations is referred to Refs. [12,13].

We commence with an examination of the fundamental one-dimensional problem of the calculus of variations and extend this to address a one-dimensional domain with variable endpoints. Both Dirichlet and natural boundary conditions are considered. A further extension considers the situation with two adjacent one-dimensional domains with variable endpoints. The three boundary conditions at the interface between the domains are derived. This analysis is extended still further to address the case of higher dimensions. The problem of the impinging streams is later shown to be an example of this type. The treatment of adjacent domains with correspondingly coupled boundary-value problems presented here is believed to be novel, at least in this application.

Much work on solving the Euler differential equations within a prescribed fixed domain has been done. However, there is far less literature on problems where there is an unknown boundary. The solution of such problems can be challenging even for the simplest governing equations. Usually there is a requirement to undertake numerical computation in order to predict the unknown boundary and calculate the desired optimal value of the functional. A recent review of very closely related examples of this kind, that arise in isoperimetric optimization problems, has been given by the author [14]. These problems all involve an unknown boundary and combine the governing physical variational extremal principle with the optimization of a physical quantity such as the drag on a body in viscous flow, the heat loss through an insulation layer, or the torsional rigidity of a beam. The interested reader may wish to explore this literature further [15–30].

Returning to the main theme, with the mathematical techniques to address variational principles for functionals defined on varying coupled domains established, we are then in a position to investigate the impinging streams. We have established the underlying problem of potential flow, in which the flow is divided into two regions, each region including one of the incoming jets and one of the streams in each of the outgoing jets. A coupled pair of boundary-value problems has been formulated and a governing variational principle has been posed, although its existence has not as yet been proved. If this variational principle indeed represents a minimum-energy principle, this could well explain the existence of the stable solutions observed by Kinelovskii and Sokolov [2]. The alleged indeterminacy of the problem would be resolved.

Curtis et al. [31] had in fact previously explored the concept of a minimum-energy principle for an approximate flow field with circular streamlines imposed. By contrast, here we consider the true free boundary-value problem, neither constraining the shapes of the free boundaries, nor restricting the treatment to the symmetric case. It is demonstrated that the governing coupled boundary-value problem is recovered as the set of necessary conditions for a functional associated with the pressure integral over both regions to be made stationary. The application of this result to the impinging-stream problem in potential flow is then discussed. Possible similar approaches for other constitutive laws are suggested.

2. The fundamental problem of the calculus of variations and its extension to consider variable endpoints and coupled domains

Although it is likely that the fundamental problem of the calculus of variations will be familiar to many readers, it is less likely that they would have encountered the treatment of variable endpoints and even more unlikely that they will have considered functionals comprising integrals over two coupled domains. The associated methods are needed for the variational formulation of the impinging-streams problem, so we ask the reader to bear with this preliminary exposition.

Consider the integral functional I_1 defined by

$$I_1 = \int_a^b F_1(x, u_1, u_1')\mathrm{d}x, \tag{1}$$

where u_1 is a differentiable function of x and u_1' is the derivative. Consider the problem of making I_1 stationary. In the most common form of the fundamental problem u_1 satisfies Dirichlet boundary conditions

$$u_1(a) = U_a, \tag{2}$$

$$u_1(b) = U_b, \tag{3}$$

where U_a, U_b are constants, and the functional I_1 is made stationary for the solution satisfying the well-known Euler equation:

$$\frac{\partial F_1}{\partial u_1} - \frac{\mathrm{d}}{\mathrm{d}x}\left(\frac{\partial F_1}{\partial u_1'}\right) = 0 \qquad \text{in } a < x < b. \tag{4}$$

A description of the derivation is given in standard text books, but the method essentially forms a subset of an extended problem, which we consider now in full.

Suppose we allow the endpoints to vary and remove the requirement to satisfy the boundary conditions (2) and (3). Let the subscript '0' correspond to the stationarity of I_1 and consider variations

$$u_1 = u_{10} + \varepsilon u_{11} + \mathrm{o}(\varepsilon), \tag{5}$$

$$u_1' = u_{10}' + \varepsilon u_{11}' + \mathrm{o}(\varepsilon), \tag{6}$$

$$a = a_0 + \varepsilon a_1 + \mathrm{o}(\varepsilon), \tag{7}$$

$$b = b_0 + \varepsilon b_1 + \mathrm{o}(\varepsilon). \tag{8}$$

The corresponding variation δI_1 in I_1 about its stationary value I_{10} is given by

$$\begin{aligned}
\delta I_1 &= I_1 - I_{10} \\
&= \int_a^b F_1(x, u_1, u_1')\mathrm{d}x - \int_{a_0}^{b_0} F_1(x, u_{10}, u_{10}')\mathrm{d}x \\
&= \int_{a_0}^{b_0} \left[F_1(x, u_{10}, u_{10}') + \varepsilon u_{11} \left(\frac{\partial F_1}{\partial u_1} \right)\Bigg|_{u_{10}} + \varepsilon u_{11}' \left(\frac{\partial F_1}{\partial u_1'} \right)\Bigg|_{u_{10}} \right] \mathrm{d}x \\
&\quad + \varepsilon b_1 F_1(b_0, u_{10}(b_0), u_{10}'(b_0)) - \varepsilon a_1 F_1(a_0, u_{10}(a_0), u_{10}'(a_0)) + \mathrm{o}(\varepsilon) \\
&\quad - \int_{a_0}^{b_0} F_1(x, u_{10}, u_{10}')\mathrm{d}x.
\end{aligned} \tag{9}$$

Thus to first order

$$\begin{aligned}
\delta I_1 &= \varepsilon \int_{a_0}^{b_0} u_{11} \left\{ \frac{\partial F_1}{\partial u_1} - \frac{\mathrm{d}}{\mathrm{d}x} \left(\frac{\partial F_1}{\partial u_1'} \right) \right\}\Bigg|_{u_{10}} \mathrm{d}x + \varepsilon \left[u_{11} \left(\frac{\partial F_1}{\partial u_1'} \right)\Bigg|_{u_{10}} \right]_{a_0}^{b_0} \\
&\quad + \varepsilon b_1 F_1(b_0, u_{10}(b_0), u_{10}'(b_0)) - \varepsilon a_1 F_1(a_0, u_{10}(a_0), u_{10}'(a_0)).
\end{aligned} \tag{10}$$

The arbitrariness of the variation u_{11} in $a_0 < x < b_0$ and setting a_1 and b_1 to zero in the integral above yields the Euler equation for u_{10}, again in the form of Eq. (4) above

$$\frac{\partial F_1}{\partial u_{10}} - \frac{\mathrm{d}}{\mathrm{d}x} \left(\frac{\partial F_1}{\partial u_{10}'} \right) = 0 \qquad \text{in } a_0 < x < b_0. \tag{11}$$

In the absence of imposed boundary conditions, the arbitrariness of $u_{11}(a_0)$ and $u_{11}(b_0)$ yields the two natural boundary conditions

$$\left(\frac{\partial F_1}{\partial u_1'} \right)\Bigg|_{u_{10}} = 0 \qquad \text{at } x = a_0 \text{ and } x = b_0. \tag{12}$$

The arbitrariness of the variations a_1 and b_1 yields two further boundary conditions

$$F_1(a_0, u_{10}(a_0), u_{10}'(a_0)) = 0, \tag{13}$$

$$F_1(b_0, u_{10}(b_0), u_{10}'(b_0)) = 0. \tag{14}$$

Thus we have two boundary conditions at each of the unknown endpoints. This is typical of problems where the boundary is unknown, and, if it were known, one boundary condition would be applied.

For completeness it is worth commenting upon what happens when a Dirichlet boundary condition is imposed at $x = a$ or $x = b$. Let us suppose, for example, that

$$u_1(a) = U_a(a), \tag{15}$$

where $U_a(x)$ is a known function. Now,

$$u_1(a) = u_1(a_0 + \varepsilon a_1 + \mathrm{o}(\varepsilon)) = u_1(a_0) + \varepsilon a_1 u_1'(a_0) + \mathrm{o}(\varepsilon)$$

$$= u_{10}(a_0) + \varepsilon u_{11}(a_0) + \varepsilon a_1 u_{10}'(a_0) + \mathrm{o}(\varepsilon). \tag{16}$$

However, Eq. (15) implies

$$U_a(a) = u_{10}(a_0) + \varepsilon u_{11}(a_0) + \varepsilon a_1 u_{10}'(a_0) + \mathrm{o}(\varepsilon), \tag{17}$$

and we also have

$$U_a(a) = U_a(a_0) + \varepsilon a_1 U_a'(a_0) + \mathrm{o}(\varepsilon). \tag{18}$$

For $a = a_0$, Eq. (15) yields

$$u_{10} = U_a(a_0). \tag{19}$$

Combination of Eqs. (17) and (18) then gives

$$u_{10}(a_0) + \varepsilon a_1 U_a'(a_0) = u_{10}(a_0) + \varepsilon u_{11}(a_0) + \varepsilon a_1 u_{10}'(a_0) + \mathrm{o}(\varepsilon), \tag{20}$$

which reduces to the relation

$$a_1 U_a'(a_0) = u_{11}(a_0) + a_1 u_{10}'(a_0) \tag{21}$$

at first order. The first variation in u_1 and variation in the endpoint are thus linked according to

$$u_{11} = a_1(U_a' - u_{10}') \qquad \text{at } x = a_0. \tag{22}$$

Substitution of Eq. (22) into Eq. (10) yields

$$\delta I_1 = \varepsilon \int_{a_0}^{b_0} u_{11} \left\{ \frac{\partial F_1}{\partial u_1} - \frac{\mathrm{d}}{\mathrm{d}x}\left(\frac{\partial F_1}{\partial u_1'} \right) \right\} \Bigg|_{u_{10}} \mathrm{d}x + \varepsilon u_{11} \left(\frac{\partial F_1}{\partial u_1'} \right) \Bigg|_{\substack{u_{10} \\ x=b_0}}$$

$$+ \varepsilon b_1 F_1(b_0, u_{10}(b_0), u_{10}'(b_0))$$

$$- \varepsilon a_1 \left\{ (U_a' - u_{10}') \left(\frac{\partial F_1}{\partial u_1'} \right) \Bigg|_{\substack{u_{10} \\ x=a_0}} + F_1(a_0, u_{10}(a_0), u_{10}'(a_0)) \right\}. \tag{23}$$

The arbitrariness of a_1 yields the familiar transversality condition

$$\left\{ F_1 + (U_a' - u_{10}') \left(\frac{\partial F_1}{\partial u_1'} \right) \right\} \Bigg|_{\substack{u_{10} \\ x=a_0}} = 0. \tag{24}$$

This replaces the two conditions $(12)_1$ and (13), but the Dirichlet condition (15) ensures that there are still two conditions on the unknown boundary.

Having embarked upon the above minor digression to link the present analysis to familiar material, we now may proceed to generalize it to two adjacent domains. Instead of the single integral functional (1), we now consider the two intervals $a \le x \le b$, and $b \le x \le c$, each with a functional I_1 and I_2, respectively, defined on it. We consider making stationary the sum I of these functionals given by

$$I = I_1 + I_2 = \int_a^b F_1(x, u_1, u_1')\mathrm{d}x + \int_b^c F_2(x, u_2, u_2')\mathrm{d}x. \tag{25}$$

We impose no boundary conditions a priori. The endpoints of both integrals are allowed to vary and we determine the variation

$$\delta I = \delta I_1 + \delta I_2, \tag{26}$$

where δI_1 is as in Eq. (10), and δI_2 is the exactly analogous expression

$$\begin{aligned}\delta I_2 = \varepsilon \int_{b_0}^{c_0} u_{21}\left\{\frac{\partial F_2}{\partial u_2} - \frac{\mathrm{d}}{\mathrm{d}x}\left(\frac{\partial F_2}{\partial u_2'}\right)\right\}\bigg|_{u_{20}} \mathrm{d}x + \varepsilon\left[u_{21}\left(\frac{\partial F_2}{\partial u_2'}\right)\bigg|_{u_{20}}\right]_{b_0}^{c_0} \\ + \varepsilon c_1 F_2(c_0, u_{20}(c_0), u_{20}'(c_0)) - \varepsilon b_1 F_2(b_0, u_{20}(b_0), u_{20}'(b_0)).\end{aligned} \tag{27}$$

The overall variation of the sum of the functionals is thus

$$\begin{aligned}\delta I = \varepsilon \int_{a_0}^{b_0} u_{11}\left\{\frac{\partial F_1}{\partial u_1} - \frac{\mathrm{d}}{\mathrm{d}x}\left(\frac{\partial F_1}{\partial u_1'}\right)\right\}\bigg|_{u_{10}} \mathrm{d}x + \varepsilon\left[u_{11}\left(\frac{\partial F_1}{\partial u_1'}\right)\bigg|_{u_{10}}\right]_{a_0}^{b_0} \\ + \varepsilon \int_{b_0}^{c_0} u_{21}\left\{\frac{\partial F_2}{\partial u_2} - \frac{\mathrm{d}}{\mathrm{d}x}\left(\frac{\partial F_2}{\partial u_2'}\right)\right\}\bigg|_{u_{20}} \mathrm{d}x + \varepsilon\left[u_{21}\left(\frac{\partial F_2}{\partial u_2'}\right)\bigg|_{u_{20}}\right]_{b_0}^{c_0} \\ - \varepsilon a_1 F_1(a_0, u_{10}(a_0), u_{10}'(a_0)) + \varepsilon b_1 F_1(b_0, u_{10}(b_0), u_{10}'(b_0)) \\ - \varepsilon b_1 F_2(b_0, u_{20}(b_0), u_{20}'(b_0)) + \varepsilon c_1 F_2(c_0, u_{20}(c_0), u_{20}'(c_0)).\end{aligned} \tag{28}$$

The arbitrariness of the variations yields the Euler equation exactly as before (see Eqs. (4) and (11))

$$\frac{\partial F_1}{\partial u_{10}} - \frac{\mathrm{d}}{\mathrm{d}x}\left(\frac{\partial F_1}{\partial u_{10}'}\right) = 0 \qquad \text{in } a_0 < x < b_0. \tag{29}$$

In exactly the same way we obtain the Euler equation

$$\frac{\partial F_2}{\partial u_{20}} - \frac{\mathrm{d}}{\mathrm{d}x}\left(\frac{\partial F_2}{\partial u_{20}'}\right) = 0 \qquad \text{in } b_0 < x < c_0. \tag{30}$$

Arbitrary variations of $u_{11}(a_0)$, $u_{11}(b_0)$, $u_{21}(b_0)$, and $u_{21}(c_0)$ yield the natural conditions

$$\frac{\partial F_1}{\partial u_{10}'} = 0 \qquad \text{at } x = a_0, \tag{31}$$

$$\frac{\partial F_1}{\partial u_{10}'} = 0 \qquad \text{at } x = b_0, \tag{32}$$

$$\frac{\partial F_2}{\partial u_{20}'} = 0 \qquad \text{at } x = b_0, \tag{33}$$

and

$$\frac{\partial F_2}{\partial u'_{20}} = 0 \qquad \text{at } x = c_0. \tag{34}$$

The arbitrariness of the variations a_1 and c_1 yields

$$F_1(a_0, u_{10}(a_0), u'_{10}(a_0)) = 0, \tag{35}$$

$$F_2(c_0, u_{20}(c_0), u'_{20}(c_0)) = 0. \tag{36}$$

The final condition is obtained by considering the arbitrariness of the variation b_1, which implies the condition

$$F_2(b_0, u_{20}(b_0), u'_{20}(b_0)) - F_1(b_0, u_{10}(b_0), u'_{10}(b_0)) = 0, \tag{37}$$

a condition explicitly coupling the solutions in the two domains.

It may be seen that there are two boundary conditions (31) and (35) at $x = a_0$; similarly there are two at $x = c_0$, namely Eqs. (34) and (36); and finally three boundary conditions on the interface $x = b_0$ coupling the two domains: Eqs. (32), (33) and (37).

3. The further extension to higher dimensions

We now proceed to extend the above analysis to consider domains of higher dimension. Consider an integral I_1 over a single n-dimensional domain D_1,

$$I_1 = \int_{D_1} F_1(x_i, u_1, u_{1,i}) \mathrm{d}V, \tag{38}$$

where u_1 is now a scalar function of $x_1, x_2, \ldots, x_n$ and $u_{1,i}$ is its derivative with respect to x_i. Let ∂D_1 be the boundary of D_1. Dirichlet boundary conditions may be imposed and are

$$u_1(x_i) = U_1(x_i) \qquad \text{on } \partial D_1. \tag{39}$$

The Euler equation when one makes the functional (38) stationary is well known and is

$$\frac{\partial F_1}{\partial u_1} - \frac{\partial}{\partial x_i}\left(\frac{\partial F_1}{\partial u_{1,i}}\right) = 0 \qquad \text{in } D_1, \tag{40}$$

where we are using the summation convention for repeated i. We do not derive this in detail here, because the necessary analysis is in effect a subset of the variable boundary case, which we now give in full.

Let the domain D_1 vary about the domain D_{10}, which makes the integral I_1 stationary. Again the subscript '0' corresponds to the stationarity of I_1. Let the boundary ∂D_{10} of D_{10} be given by

$$x_i = x_{10i}(\underline{s}), \tag{41}$$

where $\underset{\sim}{s}$ is an n-dimensional vector of parametric coordinates that describe ∂D_{10} as they vary. Consider variations about the stationary solution:

$$u_1 = u_{10} + \varepsilon u_{11} + o(\varepsilon), \tag{42}$$

$$u_{1,i} = u_{10,i} + \varepsilon u_{11,i} + o(\varepsilon), \tag{43}$$

with ∂D_1 given in terms of ∂D_{10} and the normal to it by

$$x_i = x_{10i} + \varepsilon f_1(x_{10i}) n_{1i}(x_{10i}) + o(\varepsilon), \tag{44}$$

where $f_1(x_{10i})$ is an arbitrary function, $n_{1i}(x_{10i})$ is the outward normal to ∂D_{10}, and no summation convention on i applies. The functional I_1 may be expanded about the stationary solution as follows:

$$I_1 = \int_{D_{10}} \left\{ F_1(x_i, u_{10}, u_{10,i}) + \varepsilon u_{11} \left(\frac{\partial F_1}{\partial u_1} \right)\bigg|_{u_{10}} + \varepsilon u_{11,i} \left(\frac{\partial F_1}{\partial u_{1,i}} \right)\bigg|_{u_{10}} \right\} \mathrm{d}V$$

$$+ \varepsilon \int_{\partial D_{10}} f_1 F_1(x_i, u_{10}, u_{10,i}) \mathrm{d}S + o(\varepsilon). \tag{45}$$

Use of the divergence theorem yields

$$\delta I_1 = I_1 - I_{10} = \varepsilon \int_{D_{10}} u_{11} \left\{ \frac{\partial F_1}{\partial u_1} - \frac{\partial}{\partial x_i} \left(\frac{\partial F_1}{\partial u_{1,i}} \right) \right\}\bigg|_{u_{10}} \mathrm{d}V + \varepsilon \int_{\partial D_{10}} u_{11} n_{1i} \left(\frac{\partial F_1}{\partial u_{1,i}} \right)\bigg|_{u_{10}} \mathrm{d}S$$

$$+ \varepsilon \int_{\partial D_{10}} f_1 F_1(x_i, u_{10}, u_{10,i}) \mathrm{d}S + o(\varepsilon). \tag{46}$$

The arbitrariness of u_{11} within D_{10} yields the Euler equation

$$\left\{ \frac{\partial F_1}{\partial u_1} - \frac{\partial}{\partial x_i} \left(\frac{\partial F_1}{\partial u_{1,i}} \right) \right\}\bigg|_{u_{10}} = 0, \qquad \text{in } D_{10}, \tag{47}$$

which is Eq. (40) again. In the absence of other boundary conditions, the arbitrariness of u_{11} on ∂D_{10} and that of f_1 on ∂D_{10} yield the natural conditions

$$n_{1i} \left(\frac{\partial F_1}{\partial u_{1,i}} \right)\bigg|_{u_{10}} = 0 \qquad \text{on } \partial D_{10}, \tag{48}$$

and

$$F_1(x_i, u_{10}, u_{10,i}) = 0 \qquad \text{on } \partial D_{10}. \tag{49}$$

Again, the unknown boundary is determined by the satisfaction of two boundary conditions, namely, Eqs. (48) and (49).

Under circumstances where a Dirichlet boundary condition is applied in the form

$$u_1 = U_1(x_i) \qquad \text{on } \partial D_1, \tag{50}$$

a Taylor expansion gives

$$u_{10} + \varepsilon u_{11} + \varepsilon f n_{1i} u_{10,i} = U_{10}(x_{0i}) + \varepsilon f n_{1i} U_{1,i} + o(\varepsilon) \qquad \text{on } \partial D_{10}, \tag{51}$$

and to first order in ε we have

$$u_{11} = f_1 n_{1i}(U_{1,i} - u_{10,i}) \qquad \text{on } \partial D_{10}. \tag{52}$$

Then, from Eq. (46), the surface-integral component of the variation of δI_1 becomes

$$\varepsilon \int_{\partial D_{10}} f_1 \left\{ F_1(x_i, u_{10}, u_{10,i}) + n_{1i}(U_{1,i} - u_{10,i}) n_{1j} \left(\frac{\partial F_1}{\partial u_{1,j}} \right) \Bigg|_{u_{10}} \right\} \mathrm{d}S, \tag{53}$$

and the associated necessary condition for stationarity is

$$F_0 + n_{1i}(U_{1,i} - u_{10,i}) n_{1j} \left(\frac{\partial F_1}{\partial u_{1,j}} \right) \Bigg|_{u_{10}} = 0 \qquad \text{on } \partial D_{10}, \tag{54}$$

with F_0 denoting $F(x_i, u_{10}, u_{10,i})$. This boundary condition is to be solved with the Euler equation (47) and the zeroth-order Dirichlet condition derived from Eq. (51), namely

$$u_{10} = U_{10}(x_{10i}) \qquad \text{on } \partial D_{10}. \tag{55}$$

Conditions (54) and (55) replace conditions (48) and (49) where there is a Dirichlet boundary condition.

Proceeding directly by analogy with the one-dimensonal case, we now generalize the analysis to two adjacent regimes. Instead of the integral functional (38), we consider two domains D_1 and D_2 with a common interface S, each with a functional, I_1 and I_2, respectively, defined on it. We consider the sum I of these functionals

$$I = I_1 + I_2 = \int_{D_1} F_1(x_i, u_1, u_{1,i}) \mathrm{d}V + \int_{D_2} F_2(x_i, u_2, u_{2,i}) \mathrm{d}V. \tag{56}$$

Just as in the one-dimensional case we impose no boundary conditions a priori. The surfaces ∂D_1, ∂D_2 of the domains D_1 and D_2 are allowed to vary. The interface S forms part of both ∂D_1 and ∂D_2 and couples the two functionals, as we shall see. The variation of the sum is

$$\delta I = \delta I_1 + \delta I_2, \tag{57}$$

where δI_1 is exactly as in Eq. (46) and δI_2 is given by the analogous expression

$$\delta I_2 = I_2 - I_{20} = \varepsilon \int_{D_{20}} u_{21} \left\{ \frac{\partial F_2}{\partial u_2} - \frac{\partial}{\partial x_i} \left(\frac{\partial F_2}{\partial u_{2,i}} \right) \right\} \Bigg|_{u_{20}} \mathrm{d}V$$
$$+ \varepsilon \int_{\partial D_{20}} u_{21} n_{2i} \left(\frac{\partial F_2}{\partial u_{2,i}} \right) \Bigg|_{u_{20}} \mathrm{d}S + \varepsilon \int_{\partial D_{20}} f_2 F_2(x_i, u_{20}, u_{20,i}) \mathrm{d}S + \mathrm{o}(\varepsilon). \tag{58}$$

Again, by analogy with the one-dimensional case,

$$\begin{aligned}\delta I = \varepsilon\int_{D_{10}} u_{11}\left\{\frac{\partial F_1}{\partial u_1} - \frac{\partial}{\partial x_i}\left(\frac{\partial F_1}{\partial u_{1,i}}\right)\right\}\bigg|_{u_{10}} \mathrm{d}V \\ + \varepsilon\int_{D_{20}} u_{21}\left\{\frac{\partial F_2}{\partial u_2} - \frac{\partial}{\partial x_i}\left(\frac{\partial F_2}{\partial u_{2,i}}\right)\right\}\bigg|_{u_{20}} \mathrm{d}V + \varepsilon\int_{\partial D_{10}} u_{11}n_{1i}\left(\frac{\partial F_1}{\partial u_{1,i}}\right)\bigg|_{u_{10}} \mathrm{d}S \\ + \varepsilon\int_{\partial D_{20}} u_{21}n_{2i}\left(\frac{\partial F_2}{\partial u_{2,i}}\right)\bigg|_{u_{20}} \mathrm{d}S + \varepsilon\int_{\partial D_{10}} f_1F_1(x_i, u_{10}, u_{10,i})\mathrm{d}S \\ + \varepsilon\int_{\partial D_{20}} f_2F_2(x_i, u_{20}, u_{20,i})\mathrm{d}S + \mathrm{o}(\varepsilon). \end{aligned} \tag{59}$$

The arbitrary distributions of the variations u_{11} in D_{10} and u_{21} in D_{20}, and choice of zero for the other variations, yield the following Euler equations:

$$\frac{\partial F_1}{\partial u_1} - \frac{\partial}{\partial x_i}\left(\frac{\partial F_1}{\partial u_{1,i}}\right) = 0, \qquad \text{in } D_{10}, \tag{60}$$

$$\frac{\partial F_2}{\partial u_2} - \frac{\partial}{\partial x_i}\left(\frac{\partial F_2}{\partial u_{2,i}}\right) = 0, \qquad \text{in } D_{20}. \tag{61}$$

The arbitrary values of the same variations on ∂D_{10} and ∂D_{20} give the natural boundary conditions

$$n_{1i}\left(\frac{\partial F_1}{\partial u_{1,i}}\right)\bigg|_{u_{10}} = 0 \qquad \text{on } \partial D_{1N}, \tag{62}$$

$$n_{2i}\left(\frac{\partial F_2}{\partial u_{2,i}}\right)\bigg|_{u_{20}} = 0 \qquad \text{on } \partial D_{2N}. \tag{63}$$

These only apply in the absence of a Dirichlet boundary condition on the subsets ∂D_{1N} and ∂D_{2N} of ∂D_{10} and ∂D_{20}, respectively. Where there are Dirichlet boundary conditions (52) yields

$$u_{11} = f_1n_{1i}(U_{1,i} - u_{10,i}) \qquad \text{on } \partial D_{1D}, \tag{64}$$

$$u_{21} = f_2n_{2i}(U_{2,i} - u_{20,i}) \qquad \text{on } \partial D_{2D}, \tag{65}$$

where ∂D_{1D} and ∂D_{2D} are the corresponding portions of ∂D_{10} and ∂D_{20}, respectively. It is convenient to consider the fifth and sixth integrals of Eq. (59) as sums of the same integrals over the portions where Dirichlet conditions apply and do not apply. Then, using the conditions (62)–(65) inclusive the sum of the final four integrals of Eq. (59) becomes

$$\varepsilon \int_{\partial D_{1D}} f_1 \left\{ F_1(x_i, u_{10}, u_{10,i}) + n_{1i}(U_{1,i} - u_{10,i}) n_{1j} \left(\frac{\partial F_1}{\partial u_{1,j}} \right) \bigg|_{u_{10}} \right\} \mathrm{d}S$$

$$+ \varepsilon \int_{\partial D_{2D}} f_2 \left\{ F_2(x_i, u_{20}, u_{20,i}) + n_{2i}(U_{2,i} - u_{20,i}) n_{2j} \left(\frac{\partial F_2}{\partial u_{2,j}} \right) \bigg|_{u_{20}} \right\} \mathrm{d}S$$

$$+ \int_{\partial D_{1N}} f_1 F_1(x_i, u_{10}, u_{10,i}) \mathrm{d}S + \int_{\partial D_{2N}} f_2 F_2(x_i, u_{20}, u_{20,i}) \mathrm{d}S. \tag{66}$$

The arbitrariness of the function f_1 on ∂D_{1D} yields the following boundary condition

$$F_1(x_i, u_{10}, u_{10,i}) + n_{1i}(U_{1,i} - u_{10,i}) n_{1j} \left(\frac{\partial F_1}{\partial u_{1,j}} \right) \bigg|_{u_{10}} = 0 \qquad \text{on } \partial D_{1D}, \tag{67}$$

holding with the Dirichlet condition

$$u_{10} = U_{10}(x_{0i}) \qquad \text{on } \partial D_{1D}. \tag{68}$$

Similarly, the arbitrariness of f_2 on ∂D_{2D} yields

$$F_2(x_i, u_{20}, u_{20,i}) + n_{2i}(U_{2,i} - u_{20,i}) n_{2j} \left(\frac{\partial F_2}{\partial u_{2,j}} \right) \bigg|_{u_{20}} = 0 \qquad \text{on } \partial D_{1D}, \tag{69}$$

holding with

$$u_{20} = U_{20}(x_{0i}) \qquad \text{on } \partial D_{2D}. \tag{70}$$

The arbitrariness of f_1 and f_2 on ∂D_{1N} and ∂D_{2N}, respectively, yields the natural conditions

$$F_1(x_i, u_{10}, u_{10,i}) = 0 \qquad \text{on } \partial D_{1N}, \tag{71}$$

$$F_2(x_i, u_{20}, u_{20,i}) = 0 \qquad \text{on } \partial D_{2N}. \tag{72}$$

Eqs. (60) and (61) and conditions (62), (63), and (67)–(72) are the necessary conditions for a stationary value of I. On the interface between the regions, we have $f_2 = -f_1$ and the fifth and sixth integrals in Eq. (66) are coupled. Thus the left-hand sides of conditions (71) and (72) are equated (see condition (37)).

4. Boundary-value problem for impinging streams

The variational analysis presented in Section 3 may now be applied to the problem of determining the flow field produced when two streams of incompressible inviscid fluids having unequal speeds, widths and densities meet. First we state the boundary-value problem to be solved in this section.

Far from the region of impingement the incoming streams are of speeds U_1, U_2, widths A_1, A_2, and densities ρ_1, ρ_2, respectively. Each stream is divided into two parts, one turning into each outgoing jet. The widths of the portions of the first stream turning into the two outgoing jets are denoted by A_{1J}, A_{1S}, respectively. Analogous variables A_{2J}, A_{2S} are

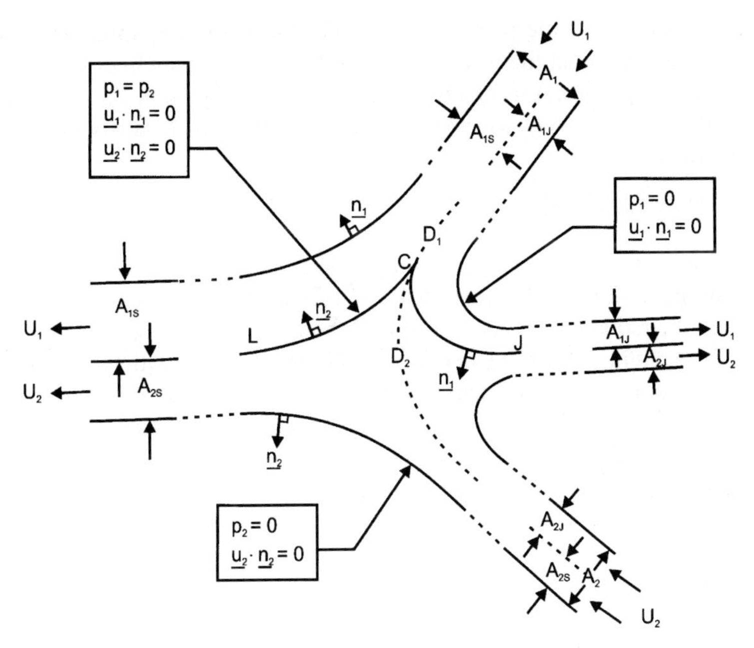

Fig. 1. The potential flow field describing the two impinging streams consists of two regions D_1 and D_2. The upper first incoming stream is the stronger of the two, having a higher Bernoulli constant. The cusp point C is a stagnation point for the flow in D_2. The streamlines CJ and CL are contact discontinuities on which the normal components of the velocity fields in each region vanish and the pressures in each region are equal.

defined for the other incoming jet. These widths and the associated jet speeds are preserved in the corresponding outgoing streams because there is no dissipation or compression.

Fig. 1 shows the geometrical configuration and the mathematical boundary-value problem in the case where $\rho_1 U_1^2 > \rho_2 U_2^2$. Curtis [9] gives an argument for the interface JCL between the two regions of fluid in the domains D_1 and D_2, respectively, containing a cusp C at the point where the stronger incoming first stream is parted by the weaker second. This cusp is a stagnation point for the flow in the region D_2. The streamlines CJ and CL are contact discontinuities on which the normal components of the velocities $\underline{u}_1$ and $\underline{u}_2$ in D_1 and D_2, respectively, vanish and the pressure fields p_1 and p_2 are equal. On the outer free surfaces the pressures and normal components of the velocities are zero. Far from the formation region the incoming and outgoing widths and speeds of each component stream are the same.

In Section 5 a variational principle equivalent to this boundary-value problem is derived.

5. Variational principle

Consider the functional defined over the region $D_1 \cup D_2$ shown in Fig. 1 by

$$I[\phi_1,\phi_2;D_1,D_2]=\int_{D_1}\frac{1}{2}\rho_1(U_1^2-\nabla\phi_1\cdot\nabla\phi_1)\mathrm{d}A+\int_{D_2}\frac{1}{2}\rho_2(U_2^2-\nabla\phi_2\cdot\nabla\phi_2)\mathrm{d}A, \qquad (73)$$

where ϕ_1 and ϕ_2 are twice-differentiable functions in the spatial coordinates x_i, and dA is an element of area. This is in the form (56), with ϕ_1 and ϕ_2 taking over the roles of the *scalars* u_1 and u_2 in Eq. (56), respectively. The *vectors* $\underline{u}_1$ and $\underline{u}_2$ now denote the velocity fields in D_1 and D_2, respectively, as shown in Fig. 1. Let ϕ_1 and ϕ_2 satisfy the asymptotic boundary conditions

$$\phi_j \to \pm rU_j, \qquad j = 1,2 \text{ as } r = |\underline{x}| \to \infty. \tag{74}$$

Here the positive and negative signs hold on the outgoing and incoming streams, respectively. In the limit the streamlines that result from this potential are parallel and the velocities are in the radial directions, so that this choice of boundary condition appears to be suitable. We sought in vain to derive natural boundary conditions at infinity and concluded that the imposition of the Dirichlet boundary conditions (74) was required. Note that this choice of boundary condition does not impose the directions of the outgoing streams a priori, only their speeds. The outward normals $\underline{n}_j$ from D_1 and D_2 are taken perpendicular to the boundary streamlines elsewhere.

We seek pairs of functions $\phi_j^{(0)}$, $j = 1,2$ and domains $D_j^{(0)}$, that make stationary the functional I given by Eq. (73). We consider first-order variations about these functions and domains by writing:

$$\phi_j = \phi_j^{(0)} + \varepsilon\phi_j^{(1)},\ j = 1,2, \qquad x_{ji} = x_{ji}^{(0)} + \varepsilon f_j n_{ji},\ j = 1,2, \tag{75}$$

where $\varepsilon \ll 1$, the coordinates x_{ji} define the boundaries of the domains D_j, $j = 1,2$, and the functions f_j, $j = 1,2$ are continuous functions of the coordinates $x_{1i}^{(0)}$, $x_{2i}^{(0)}$ defining the boundaries $\partial D_1^{(0)}$, $\partial D_2^{(0)}$ of the domains $D_1^{(0)}$, $D_2^{(0)}$, respectively, as are the normals. The corresponding first-order variation in I is then readily calculated as

$$\delta I = \sum_{j=1,2} \rho_j \left(\int_{D_j^{(0)}} \phi_j^{(1)} \nabla^2 \phi_j^{(0)}\, dA_j + \int_{\partial D_j^{(0)}} \left(\frac{1}{2} f_j (U_j^2 - \nabla\phi_j^{(0)} \cdot \nabla\phi_j^{(0)}) \right) ds_j \right.$$
$$\left. - \int_{\partial D_j^{(0)}} (\phi_j^{(1)} \underline{n}_j \cdot \nabla\phi_j^{(0)}) ds_j \right), \tag{76}$$

where ds_j is an element of arc length along the boundary to $D_j^{(0)}$. In deriving Eq. (76) the Divergence Theorem has been applied in both D_1 and D_2. At the stationary solution the right-hand side of Eq. (76) must vanish for all weak variations $\phi_j^{(1)}$, f_j, $j = 1,2$. Consideration of the first term on the right-hand side of Eq. (76) and standard arguments of the calculus of variations yield the Euler equations (see Eqs. (60) and (61)):

$$\nabla^2 \phi_j^{(0)} = 0, \qquad j = 1,2, \tag{77}$$

holding in $D_1^{(0)}$, and $D_2^{(0)}$, respectively. We, therefore, recover the equations of potential flow in each region as necessary conditions for the stationary solution.

The boundaries $\partial D_1^{(0)}$, $\partial D_2^{(0)}$ comprise the free surfaces, the interface between the regions, and the cross sections at infinity. Consider the second term on the right-hand side of Eq. (76). On the cross sections at infinity the integrands vanish as a result of boundary conditions (74). On the free surfaces, the arbitrariness of the functions f_j, $j = 1,2$

(see conditions (71) and (72)) gives the necessary conditions:

$$U_j^2 - \nabla\phi_j^{(0)}\cdot\nabla\phi_j^{(0)} = 0, \qquad j = 1,2. \tag{78}$$

This is familiar as the pair of conditions that the speeds on the free streamlines remain constant. On the interface it follows from Eq. (75) that $f_2 = -f_1$. The arbitrariness of these functions subject to this constraint yields the necessary condition

$$\frac{1}{2}\rho_1(U_1^2 - \nabla\phi_1^{(0)}\cdot\nabla\phi_1^{(0)}) = \frac{1}{2}\rho_2(U_2^2 - \nabla\phi_2^{(0)}\cdot\nabla\phi_2^{(0)}). \tag{79}$$

In potential flow this condition represents the balance of pressure at the interface between the flows. Finally, consider the third term on the right-hand side of Eq. (76). The arbitrariness of the variations $\phi_j^{(1)}$, $j = 1,2$ on the free surfaces and on the interface between the regions implies the necessary natural boundary conditions that the velocity components normal to the free streamlines and to the interface between the regions D_1 and D_2 vanish. Thus one of (see conditions (62) and (63))

$$\underline{n}_1\cdot\nabla\phi_1^{(0)} = 0, \qquad \underline{n}_2\cdot\nabla\phi_2^{(0)} = 0 \tag{80}$$

holds on each free surface as appropriate, while both conditions apply on the interface between the regions. The Dirichlet boundary conditions (74) enforce the vanishing of the contribution to the third integral in Eq. (76) from the cross sections at infinity as a result of the variations $\phi_j^{(1)}$, $j = 1,2$, being zero there.

We have recovered a pair of coupled boundary-value problems describing potential flows with free surfaces in each material. The boundary conditions on the interface between the materials couple the problems. Eqs. (77) are to be solved subject to the free-surface conditions (78) and (80), boundary conditions (74), and interface conditions (79) and (80). When the functional (73) is evaluated for the solution of the above coupled boundary-value problem, it represents the pressure integral associated with the entire flow field. In that, when Bernoulli's Law is applied to potential flow, kinetic energy is converted to increased pressure and recovered again, this pressure integral can be viewed as the total potential energy of the flow field. It appears likely that the potential-flow solution derived in Section 4 makes this functional stationary. Note that any functions ϕ_1 and ϕ_2 satisfying the boundary conditions and any suitable domains D_1 and D_2 may be used to furnish approximations to the stationary value of the functional. The choices of the domains $D_1^{(0)}$, $D_2^{(0)}$ as well as the functions $\phi_1^{(0)}$, $\phi_2^{(0)}$, are determined by the variational principle. In other words, the directions of the outgoing streams are determined by the variational principle. It appears likely that this solution is favored by nature, as evidenced by the work of Kinelovskii and Sokolov [2]. Very probably the stationary value corresponds to the minimization of the pressure integral. By skilful choice of the functions ϕ_1 and ϕ_2 and trial domains D_1 and D_2, expressed in terms of a limited number of variable parameters, it may be possible to make accurate estimates of the solution by minimizing the functional (73) in terms of those parameters.

6. Potential resolution of indeterminacy

Let the directions of the incoming streams be denoted by θ_1 and θ_2. Let the directions in which the outgoing jets travel be denoted by the unknowns θ_3 and θ_4, respectively. The speeds U_1, U_2, and the widths A_1, A_2, are known, but the divisions A_{1J}, A_{1S} and A_{2J}, A_{2S} are not known. The mass-conservation equations are

$$A_{1J} + A_{1S} = A_1, \qquad A_{2J} + A_{2S} = A_2. \tag{81}$$

These equations allow the elimination of A_{1S}, A_{2S}. We now conjecture that particular trial functions ϕ_1 and ϕ_2, and domains D_1 and D_2, and hence the functional I given by Eq. (73) are expressible directly as straightforward functions of the unknown parameters A_{1J}, A_{2J}, θ_3, θ_4 and the known parameters describing the flow. If the trial functions are chosen fortuitously to include the possibility of generating the exact solution as one member of the family of trial functions, then making g stationary will correspond to making the functional (73) stationary. In other words, there are four unknowns and we postulate that all the variables can be expressed in terms of these. Thus

$$I[\phi_1, \phi_2; D_1, D_2] \equiv g(A_{1J}, A_{2J}, \theta_3, \theta_4; A_1, A_2, U_1, U_2, \theta_1, \theta_2), \tag{82}$$

where the function g is assumed to be smoothly differentiable in the unknown arguments. The hypothesized minimization associated with the variational principle then yields four equations in the four unknowns, namely

$$\frac{\partial g}{\partial A_{1J}} = 0, \qquad \frac{\partial g}{\partial A_{2J}} = 0, \qquad \frac{\partial g}{\partial \theta_3} = 0, \qquad \frac{\partial g}{\partial \theta_4} = 0. \tag{83}$$

These equations are to be solved simultaneously for the directions θ_3, θ_4 and widths A_{1J}, A_{2J} of the outgoing streams in the jet. Provided that the four equations are indeed independent then the problem is properly closed and a well-defined solution could exist. To demonstrate this conclusively it will be necessary to establish the exact form of the function g corresponding to the coupled boundary-value problem derived in Section 5. Failing the recovery of the exact solution, the unknown arguments of g in Eq. (82) could provide the variable parameters needed to generate approximate solutions, as discussed above.

7. Conclusions

We have revisited the fundamental one-dimensional problem of the calculus of variations and extended this to address the situation with variable endpoints. Both Dirichlet and natural boundary conditions have been considered. Then a further extension has considered two adjacent one-dimensional domains, again with variable endpoints. The three boundary conditions at the interface between the domains have been derived. This problem has then been extended to consider the case of higher dimensions. This work has culminated in an application of this theory to the longstanding problem of impinging unequal streams, long believed to be indeterminate.

Indeed, a variational principle equivalent to the coupled boundary-value problem describing the meeting of unequal streams in steady flow has been derived. It has been conjectured that the solution corresponds to a minimum of the pressure integral. On the assumption that this integral may be expressed in terms of four parameters describing the outgoing streams, it has been demonstrated that a set of four equations in these four variables results. If a solution of this set were to exist then the problem is fully determined. This is in significant contrast to previous models in which various assumptions are made to achieve closure. It remains to specify the exact form of the potential energy function, or, alternatively, to investigate approximate solutions by attempted exploitation of the proposed underlying variational principle.

References

[1] Birkhoff, G. and Zarantonello, E.H. (1957), Jets, Wakes and Cavities, Academic Press, New York.

[2] Kinelovskii, S.A. and Sokolov, A.V. (1986), Nonsymmetric collision of plane jets of an ideal incompressible fluid, Zh. Prikl. Mekhan. Tekhnich. Fiz. 1, 54.

[3] Palatini, (1916), Sulla Confluenza di Due Vene, Atti del R. Instit. Venetto di Sc. L ed Arti, LXXV, 451.

[4] Keller, J.B. (1990), On unsymmetrically impinging jets, J. Fluid Mech. 211, 653.

[5] Reid, S. (2002), Mathematics of impact and impact of mathematics, IMA presidential address lecture, Math. Today 38(1), 21.

[6] Heider, N. and Rottenkolber, E. (1993), Analysis of the Asymmetric Jet Formation Process in Shaped Charges. In Proceedings of the 14th International Symposium on Ballistics, Que., Canada, ISBN 0-9618156-8-1, Vol. 2, p. 203.

[7] Pack, D.C. and Curtis, J.P. (1990), On the effect of asymmetries on the jet from a linear shaped charge, J. Appl. Phys. 67, 6701.

[8] Curtis, J.P. and Kelly, R.J. (1994), Circular streamline model of shaped-charge jet and slug formation with asymmetry, J. Appl. Phys. 75, 7700.

[9] Curtis, J.P. (1998), Asymmetric Formation of Shaped Charge Jets. In Proceedings of the 17th International Symposium on Ballistics, Midrand, South Africa, ISBN ISBN: 0-620-22078-3, Vol. 2, p.405.

[10] Curtis, J.P. and Kelly, R.J. (1997), A limit of validity of the straight line hypothesis in shaped charge jet formation modelling, J. Appl. Phys. 82, 107.

[11] Curtis, J.P. (2001), Variational Principle for Shaped Charge Jet Formation. In Proceedings of the 19th International Symposium on Ballistics, Interlaken, Switzerland, ISBN: 3-9522178-2-4, Vol. 2, p. 781.

[12] Bolza, O. (1961), Lectures on the Calculus of Variations, Dover, New York.

[13] Hildebrandt, S. and Tromba, A. (1996), The Parsimonious Universe: Shape and Form in the Natural World, Copernicus/Springer, New York, ISBN 1996 0 387-97991-3.

[14] Curtis, J.P. (2002), Complementary extremum principles for isoperimetric optimisation problems. Advances in Simulation, Systems Theory and Systems Engineering. Electrical and Computer Engineering Series, (Mastorakis, N.E., Kluev, V., Koruga, D., eds.), WSEAS Press, pp. 416, ISBN 960-8052-70-X.

[15] Polya, G. (1948), Torsional rigidity, principal frequency, electrostatic capacity and symmetrization, Q. Appl. Math. 6, 267.

[16] Payne, L.E. (1967), Isoperimetric irregularities and their application, SIAM Rev. 9(3), 453.

[17] Tuck, E.O. (1968), In Proceedings of the Conference on Hydraulic Fluid Mechanics, Australian Institute of Engineers, p. 29.

[18] Watson, S.R. (1971), Towards minimum drag on a body of given volume in slow viscous flow, J. Inst. Math. Appl. 7, 367.

[19] Pironneau, O. (1973), On optimum profiles in stokes flow, J. Fluid Mech. 59, 315.

[20] Pironneau, O. (1975), On optimal design in fluid mechanics, J. Fluid Mech. 64, 97.

[21] Bourot, J.M. (1974), On the numerical computation of the optimum profile in stokes flow, J. Fluid Mech. 65, 513.

[22] Mironov, A. (1975), On the problem of optimisation of the shape of a body in a viscous fluid, J. App. Math. Mech. 39, 103.
[23] Glowinski, R. and Pironneau, O. (1975), On the numerical computation of the minimum drag profile in laminar flow, J. Fluid Mech. 72, 385.
[24] Banichuk, N.V. (1976), Optimization of elastic bars in torsion, Int. J. Solids Struct. 12, 275.
[25] Banichuk, N.V. and Karihaloo, B.L. (1976), Minimum weight design of multi-purpose cylindrical bars, Int. J. Solids Struct. 12, 267.
[26] Gurvitch, E.L. (1976), On isoperimetric problems for domains with partly known boundaries, J. Opt. Theor. Appl. 20, 65.
[27] Parberry, R.D. and Karihaloo, B.L. (1977), Minimum weight design of a hollow cylinder for given lower bounds on torsional and flexural rigidities, Int. J. Solids Struct. 13, 1271.
[28] Curtis, J.P. and Walpole, L.J. (1982), Optimization of the torsional rigidity of axisymmetric hollow shafts, Int. J. Solids Struct. 18(10), 883.
[29] Curtis, J.P. (1983), Optimisation of homogenous thermal insulation layers, Int. J. Solids Struct. 19(9), 813.
[30] Richardson, S. (1995), Optimum profiles in two-dimensional stokes flow, Proc. R. Soc. Ser. A 450, 1940.
[31] Curtis, J.P., Kelly, R.J. and Cowan, K.G. (1994), Variational approach to the calculation of the radii in the stagnant core model of shaped charge jet formation, J. Appl. Phys. 76(12), 7731.

Chapter 6

VARIATIONAL PRINCIPLES IN STABILITY ANALYSIS OF COMPOSITE STRUCTURES

Pizhong Qiao and Luyang Shan

Department of Civil Engineering, The University of Akron, Akron, OH 44325-3905, USA

Abstract

Variational principles as important parts of the theory of elasticity have been extensively used in stability analysis of structures made of fiber-reinforced polymer (FRP) composites. In this chapter, variational principles in buckling analysis of FRP composite structures are presented. A survey of variational principles in stability analysis of composite structures is first given, followed by a brief introduction of the theoretical background of variational principles in elasticity. A variational formulation of the Ritz method is used to establish an eigenvalue problem, and by using different buckling deformation functions, the solutions of buckling of FRP structures are obtained. As application examples, the local and global buckling of FRP thin-walled composite structural shapes is analyzed using the variational principles of total potential. For the local buckling of FRP composite shapes (e.g. I or box sections), the flange or web of the beams is considered as a discrete anisotropic laminated plate subjected to rotational restraints at the flange–web connections, and by enforcing the equilibrium condition to the first variation of the total potential energy, the explicit solutions for local buckling of the plates with various unloaded edge-boundary conditions are developed. For the global buckling of FRP composite beams, the second variation of the total potential energy based on nonlinear plate theory is applied, and the formulation includes the shear effect and beam bending–twisting coupling. In summary, the application of variational principles as a viable tool in buckling analysis of FRP thin-walled composite structures is illustrated, and the present explicit formulations using the variational principles of the Ritz method can be applied to determine buckling capacities of composite structures and facilitate the buckling analysis, design, and optimization of FRP structural profiles.

Keywords: polymer composites; stability analysis; buckling; Ritz method

1. Introduction

Advanced composite materials are increasingly used in structural applications due to their favorable and objectively enhanced properties. They are ideal for applications (e.g. aircrafts and automobiles) where high stiffness-to-weight and strength-to-weight ratios are needed. Because of the excellent properties, such as light weight, corrosive resistance,

Tel.: +1-330-972-5226; fax: +1-330-972-6020.

E-mail address: qiao@uakron.edu (P. Qiao); ls30@uakron.edu (L. Shan).

nonmagnetic, and nonconductive, the structures made of composite materials have also been shown to provide efficient and economical applications in civil-engineering structures, such as bridges and piers, retaining walls, airport facilities, storage structures exposed to salts and chemicals, and others. In addition, composite structures exhibit excellent energy-absorption characteristics, suitable for seismic response; high strength, fatigue life, and durability; competitive costs based on load-capacity per unit weight; and ease of handling, transportation, and installation.

Composite structures are usually in thin-walled configurations, and the fibers (e.g. carbon, glass, and aramid) are used to reinforce the polymer matrix (e.g. epoxy, polyester, and vinylester). Fiber-reinforced polymer (FRP) structural shapes in forms of beams, columns, and deck panels are typical composite structures commonly used in civil infrastructure [1–3]. FRP structural shapes are primarily made of E-glass fiber and either polyester or vinylester resins. Their manufacturing processes include pultrusion, filament winding, vacuum-assisted resin transfer molding (VARTM), and hand lay-up, etc. while the pultrusion process, a continuous manufacturing process capable of delivering 1–5 ft/min of prismatic thin-walled members, is the most prevalent one in fabricating the FRP shapes due to its continuous and massive production capabilities. Due to geometric (i.e. thin-walled shapes) and material (i.e. relatively low stiffness of polymer and high fiber strength) properties, FRP composite structures usually undergo large deformation and are vulnerable to global and local buckling before reaching the material strength failure under service loads [2]. Thus, structural stability is one of the most likely modes of failure for thin-walled FRP structures. Since buckling can lead to a catastrophic consequence, it must be taken into account in the design and analysis of FRP composite structures.

Because of the complexity of composite structures (e.g. material anisotropy and unique geometric shapes), common analytical and design tools developed for members of conventional materials cannot always be readily applied to composite structures. On the other hand, numerical methods, such as finite elements, are often difficult to use, which require specialized training, and are not always accessible to design engineers. Therefore, to expand the applications of composite structures, an explicit engineering design approach for FRP shapes should be developed. Such a design tool should allow designers to perform stability analysis of customized shapes as well as to optimize innovative sections.

Variational principles as a viable method are often used to develop analytical solutions for stability of composite structures. Variational-principle-based formulations form a powerful basis for obtaining approximate solutions to structural stability of FRP shapes. The objective of this chapter is to introduce the application of variational principles in stability analysis of composite structures. In particular, the local and global buckling of FRP structural shapes is analyzed, from which explicit solutions and equations for efficient design and analysis are developed.

2. Literature review

Variational and energy methods are the most effective ways to analyze stability of conservative systems. Accurate yet simple approximation of critical loads can be obtained with the concept of energy approach by choosing adaptable buckling deformation shape functions. The first variation of total potential energy equaling zero (the minimum of the potential energy) represents the equilibrium condition of structural systems; while the positive definition of the second variation of total potential energy demonstrates that the equilibrium is stable.

The versatile and powerful variational total potential energy method has been used in many studies for local and global buckling of structural systems made of different materials. Since Timoshenko [4] derived the classical energy equation in 1934, there have been many researches on stability analysis of isotropic thin-walled structures using variational principles. With energy equations, Roberts [5] derived the expressions for the second-order strains in thin-walled bars and used them in stability analysis. Bradford and Trahair [6] developed energy methods by nonlinear elastic theory for lateral–distortional buckling of I-beams under end moments. Later, Bradford [7] analyzed the buckling of a cantilever I-beam subjected to a concentrated force. Ma and Hughes [8] derived the nonlinear total potential energy equations to analyze the lateral buckling behavior of monosymmetric I-beams subjected to distributed vertical load and point load with full allowance for distortion of the web, respectively. Smith et al. [9] utilized variational formulation of the Ritz method to determine the plate local buckling coefficients. The aforementioned studies only represent a small portion of research on stability analysis using variational principles with respect to traditional structures made of isotropic materials (e.g. steel).

Due to anisotropy and diverse shapes of FRP composite structures, the analysis of structural stability is relatively complex and computationally expensive compared to the one used for conventional isotropic structures. Because of the vulnerability of FRP thin-walled structures to buckling, stability analysis is even more critical and demanding. A need exists to develop explicit analytical solutions for structural stability design of FRP composite shapes. The variational total potential energy principles provide a powerful and efficient tool to obtain the analytical solutions for stability of composite structures and can be used as a vehicle to develop explicit and simplified design equations for buckling of FRP shapes. In the following, the literature related to stability analysis of composite structures using variational and energy methods is briefly reviewed.

2.1. *Local buckling*

For short-span FRP composite structures (e.g. plates and beams), local buckling is more likely to occur and finally leads to large deformation or material crippling. A number of researchers presented studies on local buckling analysis using variational and energy methods. Lee [10] presented an exact analysis and an approximate energy method using simplified deflections for the local buckling of orthotropic structural sections, and the minimum buckling coefficient was expressed as a function of the flange–web ratio. Later, Lee [11] extended the solution to include the local buckling of orthotropic sections with various loaded boundary conditions. Lee and Hewson [12] investigated the local buckling of orthotropic thin-walled columns made of unidirectional FRP composites. Brunelle and Oyibo [13] used the first variational of the total energy method to develop the generic buckling curves for special orthotropic rectangular plates. Based on energy considerations, Roberts and Jhita [14] presented a theoretical study of the elastic buckling modes of I-section beams under various loading conditions that could be used to predict local and global buckling modes. Barbero and Raftoyiannis [15] used the variational principle (Rayleigh–Ritz method) to develop analytical solutions for critical buckling load as well as the buckling mode under axial and shear loading of FRP I- and box beams. By modeling the flanges and webs individually and considering the flexibility of the flange–web connections, Qiao et al. [16] obtained the critical buckling stress resultants and critical numbers of buckled waves over the plate aspect ratio for two common cases of composite plates with different boundary conditions. By applying the first variational formulation of Ritz method to establish an eigenvalue problem, Qiao and Zou [17] obtained the explicit solution for buckling of composite plates with elastic restraints at two unloaded edges and subjected to nonuniformed in-plane axial action. By considering the combined shape functions of simply supported and clamped unloaded edges, Qiao and Zou [18] recently derived the explicit closed-form solution for buckling of composite plates with free and elastically restrained unloaded edges. In most of the above studies, the first variation of total potential energy is usually applied to establish the eigenvalue problem for local buckling of composite structures.

2.2. *Global buckling*

For long-span FRP shapes, global (Euler) buckling is more likely to occur than local buckling, and the second variational total potential energy method is often used to develop the analytical solutions. Roberts and Jhita [14] conducted a theoretical study of the elastic global buckling modes of I-section beams under various loading conditions, and the energy equations governing instability using plate theory and beam theory were established. Based on the energy consideration and variational principle, Barbero and Raftoyiannis [19] investigated the lateral and distortional buckling of pultruded I-beams. Using the second variational total energy principle, Pandey et al. [20] presented an analytical study of the elastic, flexural–torsional buckling of pultruded I-shaped members with the purpose of optimizing the fiber orientation. Kabir and Sherbourne [21] studied the lateral–torsional buckling of I-section composite beams, and the transverse shear strain effect on the lateral buckling was investigated. The second variational method was also used in research on flexural–torsional and lateral–distortional buckling of composite FRP simply supported and cantilever I-beams [22–24]. Based on the classical lamination theory and variational principle for thin walls and Vlasov's thin-walled beam theory for channel bars, Lee and Kim [25] analyzed the lateral buckling of laminated composite channel beams. Most recently, Shan and Qiao [26] investigated the flexural–torsional buckling of FRP open-channel beams using the second variational total potential energy method.

As demonstrated in the above studies, the variational principles can be used as a validated method to derive analytical solutions for stability of composite structures, and it is hereby introduced to analyze the local and global buckling of FRP structural shapes. In the following sections, the theoretical background of variational principles (Section 3) is first introduced, followed by their applications to local (Section 4) and global (Section 5) buckling of FRP composite structures.

3. Variational principles: theoretical background

The total potential energy (Π) of a system is the sum of the strain energy (U) and the work (W) done by the external loads, and it is expressed as

$$\Pi = U + W, \tag{1}$$

where $W = -\sum P_i q_i$, and $U = U(\varepsilon_{ij})$. Thus, the total potential energy is expressed as

$$\Pi = -\sum P_i q_i + U(\varepsilon_{ij}). \tag{2}$$

For linear elastic problems, the strain energy is given as

$$U = \frac{1}{2}\int_V \sigma_{ij}\varepsilon_{ij}\,\mathrm{d}V.$$

For a structure in an equilibrium state, the total potential energy attains a stationary value when the first variation of the total potential energy ($\delta\Pi$) is zero. Then, the condition for the state of equilibrium is expressed as

$$\delta\Pi = -\sum P_i \delta q_i + \int_V \sigma_{ij}\delta\varepsilon_{ij}\mathrm{d}V = 0. \tag{3}$$

The structure is in a stable equilibrium state if, and only if, the value of the potential energy is a relative minimum. It is possible to infer whether a stationary value of a functional Π is a maximum or a minimum by observing the sign of $\delta^2\Pi$. If $\delta^2\Pi$ is positive definite, Π is a minimum. Thus, the condition for the state of stability is characterized by the inequality

$$\delta^2\Pi = -\sum P_i \delta^2 q_i + \int_V (\sigma_{ij}\delta^2\varepsilon_{ij} + \delta\sigma_{ij}\delta\varepsilon_{ij})\mathrm{d}V > 0. \tag{4}$$

Eq. (4) is based on the second Gâteaux variation [27] stating that the second variation of $I[y]$ at $y = y_0$ is expressed as

$$\delta^2 I[h] = \frac{\mathrm{d}^2}{\mathrm{d}t^2} I[y_0 + th]_{t=0}. \tag{5}$$

Because q_i is usually being expressed as linear functions of displacement variables, $\delta^2 q_i$ in Eq. (4) vanishes. Therefore, the critical condition for stability analysis becomes

$$\delta^2\Pi = \delta^2 U = \int_V (\sigma_{ij}\delta^2\varepsilon_{ij} + \delta\sigma_{ij}\delta\varepsilon_{ij})\mathrm{d}V = 0. \tag{6}$$

In this study, the first variation of the total potential energy (Eq. (3)) corresponding to the equilibrium state of the structure is employed to establish the eigenvalue problem for local buckling of discrete laminated plates in FRP structures; while the second variation of total potential energy (Eq. (6)) representing the stability state of the system is applied to derive the eigenvalue solution for global buckling of FRP beams. Combining with the Ritz method and unique deformation shape functions, explicit analytical solutions are obtained for local and global buckling of composite structures using variational principles.

4. Application I: local buckling of FRP composite structures

In general, the local buckling analyses of FRP shapes are accomplished by modeling the flanges and webs individually and considering the flexibility of the flange–web connections. In this type of simulation, each part of FRP shapes (Fig. 1) is modeled as a composite plate subjected to elastic restraints along the unloaded edges (i.e. the flange–web connections) [16].

The first variational total potential energy approach is hereby applied to local buckling analysis of FRP plates, and the Ritz method is used to establish an eigenvalue problem. Explicit solutions for local buckling problems of two types of elastically restrained plates (see Fig. 1) are obtained. By incorporating the discrete-plate solutions and considering the rotational restraint stiffness at the flange–web connection, the local buckling of FRP structural shapes is then determined.

4.1. Theoretical formulation of the first variational problem for local buckling of elastically restrained plates

The local buckling of an orthotropic plate subjected to uniform in-plane axial load along the simply supported edges and rotationally restrained either at two unloaded edges (Plate I in Fig. 1(a)) or at one unloaded edge with the other free (Plate II in Fig. 1(b)) is briefly presented in this section. A variational formulation of the Ritz method is herein used to analyze the elastic buckling of an orthotropic plate with the boundary conditions shown in Fig. 1. In the variational form of the Ritz method used in this study, the first variations of the elastic strain energy stored in the plate (δU_e), the strain energy stored in the elastic restraints along the rotationally restrained boundaries of the plate (δU_Γ), and the work done by the axial in-plane force (δV) are computed by properly choosing the out-of-plane buckling displacement functions (w) [17,18].

The elastic strain energy in an orthotropic plate (U_e) is given as

$$U_e = \frac{1}{2}\iint_\Omega \{D_{11}w_{,xx}^2 + D_{22}w_{,yy}^2 + 2D_{12}w_{,xx}w_{,yy} + 4D_{66}w_{,xy}^2\}\mathrm{d}x\mathrm{d}y, \tag{7}$$

where D_{ij} ($i,j = 1,2,6$) are the plate bending stiffness coefficients [28] and Ω is the area of the plate. Therefore, the first variational form of elastic strain energy stored in the plate (δU_e) becomes

$$\delta U_e = \iint_\Omega \{D_{11}w_{,xx}\delta w_{,xx} + D_{22}w_{,yy}\delta w_{,yy} + D_{12}(\delta w_{,xx}w_{,yy} + w_{,xx}\delta w_{,yy}) + 4D_{66}w_{,xy}\delta w_{,xy}\}\mathrm{d}x\mathrm{d}y. \tag{8}$$

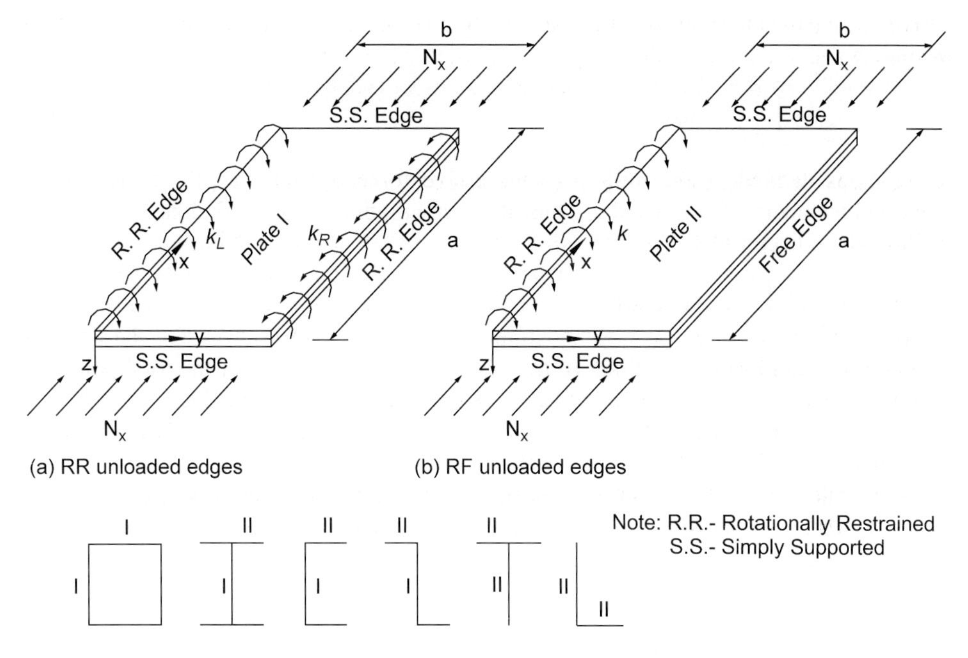

Fig. 1. Geometry of orthotropic plate elements in FRP shapes.

For the plate with rotational restraints distributed along the unloaded boundary edges, the strain energy (U_Γ) stored in equivalent elastic springs at the flange–web connections for Plates I and II is given, respectively, as

$$U_\Gamma = \frac{1}{2}\int_\Gamma k_\mathrm{L}\left(\left.\frac{\partial w}{\partial y}\right|_{y=0}\right)^2 \mathrm{d}\Gamma + \frac{1}{2}\int_\Gamma k_\mathrm{R}\left(\left.\frac{\partial w}{\partial y}\right|_{y=b}\right)^2 \mathrm{d}\Gamma \quad \text{for Plate I in Fig. 1(a)} \tag{9}$$

$$U_\Gamma = \frac{1}{2}\int_\Gamma k\left(\left.\frac{\partial w}{\partial y}\right|_{y=0}\right)^2 \mathrm{d}\Gamma \quad \text{for Plate II in Fig. 1(b),} \tag{10}$$

where k_L and k_R in Eq. (9) are the rotational restraint stiffness at the restrained edges of $y = 0$ and b, respectively (Fig. 1(a)); and k in Eq. (10) is the rotational restraint stiffness at the restrained edge of $y = 0$ (Fig. 1(b)). Then, the corresponding first variations of strain energy stored in the elastic restraints along the rotationally restrained boundary of the plate (δU_Γ) are, respectively,

$$\delta U_\Gamma = k_\mathrm{L}\int_\Gamma \left(\left.\frac{\partial w}{\partial y}\right|_{y=0}\right)\delta\left(\left.\frac{\partial w}{\partial y}\right|_{y=0}\right)\mathrm{d}\Gamma + k_\mathrm{R}\int_\Gamma \left(\left.\frac{\partial w}{\partial y}\right|_{y=b}\right)\delta\left(\left.\frac{\partial w}{\partial y}\right|_{y=b}\right)\mathrm{d}\Gamma \quad \text{for Plate I in Fig. 1(a)} \tag{11}$$

$$\delta U_\Gamma = k\int_\Gamma \left(\left.\frac{\partial w}{\partial y}\right|_{y=0}\right)\delta\left(\left.\frac{\partial w}{\partial y}\right|_{y=0}\right)\mathrm{d}\Gamma \quad \text{for Plate II in Fig. 1(b).} \tag{12}$$

The work done (V) by the in-plane uniformly distributed compressive force (N_x) can be written as

$$V = \frac{1}{2}N_x \int\int_{\Omega} w_{,x}^2 \mathrm{d}x\mathrm{d}y, \tag{13}$$

where N_x is defined as the uniform compressive force per unit length at the simply supported boundary of $x = 0$ and a. Thus, the first variation of the work done by the axial in-plane force becomes

$$\delta V = N_x \int\int_{\Omega} w_{,x}\delta w_{,x} \mathrm{d}x\mathrm{d}y. \tag{14}$$

Using the equilibrium condition of the first variational principle of the total potential energy

$$\delta\Pi = \delta U_{\mathrm{e}} + \delta U_{\Gamma} - \delta V = 0, \tag{15}$$

and substituting the proper out-of-plane displacement function (w) into Eq. (15), the standard buckling eigenvalue problem can be solved by the Ritz method.

4.2. *Explicit solutions for local buckling of elastically restrained plates*

To solve the eigenvalue problem, it is very important to choose the proper out-of-plane buckling displacement function (w). In this study, to explicitly obtain the analytical solutions for local buckling of two types of representative plates as shown in Fig. 1, the unique buckling displacement fields are proposed as follows.

4.2.1. *Plate I: rotationally restrained at two unloaded edges (Fig. 1(a))*

For Plate I in Fig. 1(a), the displacement function chosen by combining harmonic and polynomial buckling deformation functions is stated as [17]

$$w(x,y) = \left\{\frac{y}{b} + \psi_1\left(\frac{y}{b}\right)^2 + \psi_2\left(\frac{y}{b}\right)^3 + \psi_3\left(\frac{y}{b}\right)^4\right\}\sum_{m=1}^{\infty}\alpha_m \sin\frac{m\pi x}{a}, \tag{16}$$

where ψ_1, ψ_2, and ψ_3 are the unknown constants that satisfy the boundary conditions. As shown in Fig. 1(a), the boundary conditions along the rotationally restrained unloaded edges can be written as

$$w(x,0) = 0 \tag{17a}$$

$$w(x,b) = 0 \tag{17b}$$

$$M_y(x,0) = -D_{22}\left(\frac{\partial^2 w}{\partial y^2}\right)_{y=0} = -k_{\mathrm{L}}\left(\frac{\partial w}{\partial y}\right)_{y=0} \tag{17c}$$

$$M_y(x,b) = -D_{22}\left(\frac{\partial^2 w}{\partial y^2}\right)_{y=b} = -k_{\mathrm{R}}\left(\frac{\partial w}{\partial y}\right)_{y=b}. \tag{17d}$$

Then the assumed displacement function for Plate I shown in Fig. 1(a) can be obtained as

$$w(x,y) = \left\{ \frac{y}{b} + \frac{k_{\mathrm{L}} b}{2D_{22}} \left(\frac{y}{b}\right)^2 - \frac{12D_{22}^2 + D_{22}(5k_{\mathrm{L}} + 3k_{\mathrm{R}})b + k_{\mathrm{L}} k_{\mathrm{R}} b^2}{6D_{22}^2 + D_{22} k_{\mathrm{R}} b} \left(\frac{y}{b}\right)^3 + \frac{12D_{22}^2 + D_{22}(4k_{\mathrm{L}} + 4k_{\mathrm{R}})b + k_{\mathrm{L}} k_{\mathrm{R}} b^2}{12D_{22}^2 + 2D_{22} k_{\mathrm{R}} b} \left(\frac{y}{b}\right)^4 \right\} \sum_{m=1}^{\infty} \alpha_m \sin \frac{m\pi x}{a}. \tag{18}$$

Noting that k_{L} and k_{R} are all positive values, as given in Eq. (18). k_{L} or $k_{\mathrm{R}} = 0$ corresponds to the simply supported boundary condition at rotationally restrained edges of $y = 0$ or $y = b$; whereas, k_{L} or $k_{\mathrm{R}} = \infty$ represents the clamped (built-in) boundary condition at rotationally restrained edges.

4.2.2. Plate II: rotationally restrained at one unloaded edge and free at the other (Fig. 1(b))

For Plate II shown in Fig. 1(b), the displacement function is obtained by linearly combining the simply supported-free (SF) and clamped-free (CF) boundary displacements, and it can be uniquely expressed as [18]

$$w(x,y) = \left\{ (1-\omega)\frac{y}{b} + \omega \left[\frac{3}{2}\left(\frac{y}{b}\right)^2 - \frac{1}{2}\left(\frac{y}{b}\right)^3 \right] \right\} \sum_{m=1}^{\infty} \alpha_m \sin \frac{m\pi x}{a}, \tag{19}$$

where ω is the unknown constant which can be obtained by satisfying the boundary conditions. When $\omega = 0.0$, it corresponds to the displacement function of the SF plate; whereas $\omega = 1.0$ relates to that of the CF plate. The boundary conditions along the rotationally restrained ($y = 0$) and free ($y = b$) unloaded edges are specified as

$$w(x,0) = 0 \tag{20a}$$

$$M_y(x,0) = -D_{22}\left(\frac{\partial^2 w}{\partial y^2}\right)_{y=0} = -k\left(\frac{\partial w}{\partial y}\right)_{y=0} \tag{20b}$$

$$M_y(x,b) = \left(D_{12}\frac{\partial^2 w}{\partial x^2} + D_{22}\frac{\partial^2 w}{\partial y^2}\right)_{y=b} = 0 \tag{20c}$$

$$V_y(x,b) = \left[\frac{\partial}{\partial y}\left(D_{12}\frac{\partial^2 w}{\partial x^2} + D_{22}\frac{\partial^2 w}{\partial y^2}\right) + 2\frac{\partial}{\partial x}\left(2D_{66}\frac{\partial^2 w}{\partial x \partial y}\right)\right]_{y=b} = 0 \tag{20d}$$

Eq. (19) does not exactly satisfy the free edge conditions as defined in Eqs. (20c) and (20d). In this study, in order to derive the explicit formula for the RF plate, the unique buckling displacement function in Eq. (19) is used to approximate the free edge condition, and it satisfies the condition of $(\partial^2 w/\partial y^2)_{y=b} = 0$, which is the dominant term for the moment and shear force at $y = b$. As illustrated in the later section, the approximate deformation function (Eq. (19)) provides adequate accuracy of local buckling prediction for the RF plate when compared to the exact transcendental solution [16].

Considering Eq. (20b), ω is obtained in term of the rotational restraint stiffness k. Then the displacement function for the RF plate shown in Fig. 1(b) can be written as

$$w(x,y) = \left\{\left(1 - \frac{bk}{3D_{22}+bk}\right)\frac{y}{b} + \frac{bk}{3D_{22}+bk}\left[\frac{3}{2}\left(\frac{y}{b}\right)^2 - \frac{1}{2}\left(\frac{y}{b}\right)^3\right]\right\}\sum_{m=1}^{\infty}\alpha_m \sin\frac{m\pi x}{a} \tag{21}$$

Similarly, in Eq. (21), $k = 0$ (simply supported at rotationally restrained edge) corresponds to the plate with the simply supported-free (SF) boundary condition along the unloaded edges; whereas, $k = \infty$ (clamped at rotationally restrained edge) refers to the one with the clamped-free (CF) boundary condition. For $0 < k < \infty$, the restrained-free (RF) condition at unloaded edges is taken into account in the formulation.

By substituting Eq. (18) into Eqs. (8), (11), and (14) and summing them according to Eq. (15), the solution of an eigenvalue problem for Plate I is obtained [17]. Similarly, by substituting Eq. (21) into Eqs. (8), (12), and (14), then summing them according to Eq. (15), and after some numerical symbolic computation, the local buckling coefficient for Plate II with the loading and boundary conditions shown in Fig. 1(b) is explicitly derived [18].

4.3. Cases of common plates

Based on the explicit formulations in Section 4.2, design formulas of critical local buckling load (N_{cr}) for several common orthotropic plate cases of application are obtained as follows:

Case 1: Plates with two simply supported unloaded edges (SS) (Fig. 2(a))

For the case of $k_L = k_R = 0$ (i.e. the four edges are simply supported and the plate is subjected to a uniformly distributed compression load in the x-direction) (Fig. 2(a)), the explicit critical local buckling load is given as

$$N_{cr} = \frac{2\pi^2}{b^2}\{\sqrt{D_{11}D_{22}} + (D_{12} + 2D_{66})\}. \tag{22}$$

Eq. (22) is identical to the one reported by Qiao et al. [16].

Case 2: Plates with two clamped unloaded edges (CC) (Fig. 2(b))

For the case of $k_L = k_R = \infty$ (i.e. the two unloaded edges at $y = 0$ and b are clamped and the plate is subjected to the uniformly distributed compressive load at simply supported edges of $x = 0$ and a) (Fig. 2(b)), the explicit critical buckling load is expressed as

$$N_{cr} = \frac{24}{b^2}\{1.871\sqrt{D_{11}D_{22}} + (D_{12} + 2D_{66})\}. \tag{23}$$

Case 3: Plates with two equal rotational restraints along unloaded edges (RR) (Fig. 2(c))

For the case of $k_L = k_R = k$ (i.e. the two unloaded edges at $y = 0$ and $y = b$ are subjected to the same rotational restraints, and the plate is simply supported and subjected to the uniformly distributed compression load at the edges of $x = 0$ and $x = a$) (Fig. 2(c)),

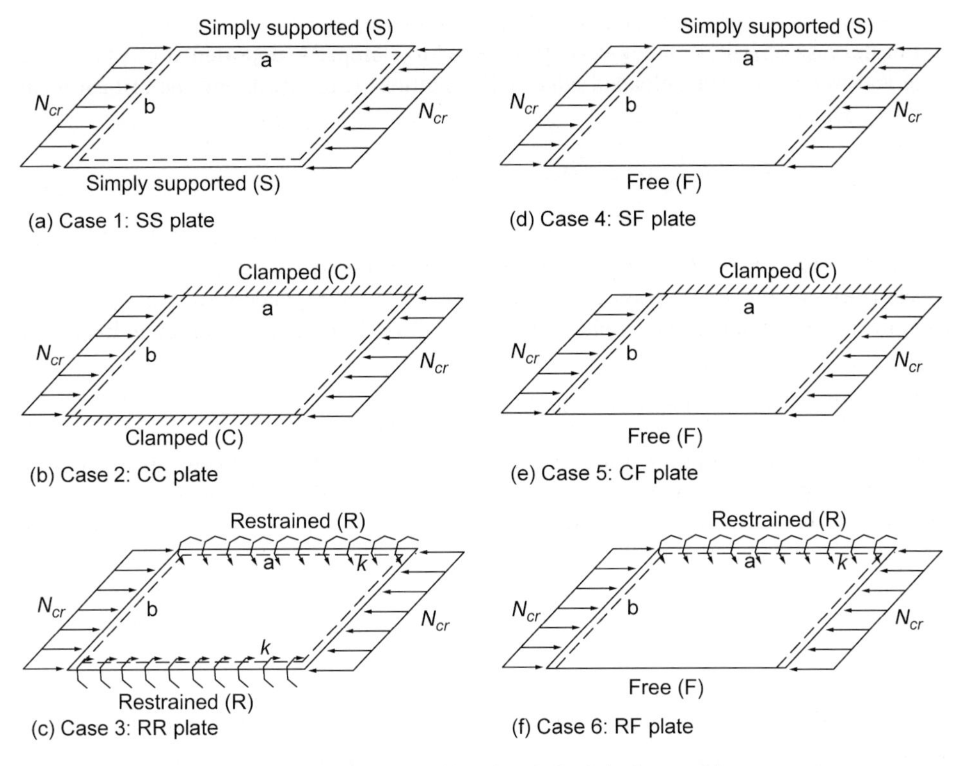

Fig. 2. Common plates with various unloaded edge conditions.

the explicit critical local buckling load is given as

$$N_{\mathrm{cr}} = \frac{24}{b^2}\left\{1.871\sqrt{\frac{\tau_2}{\tau_1}}\sqrt{D_{11}D_{22}} + \frac{\tau_3}{\tau_1}(D_{12} + 2D_{66})\right\}, \tag{24}$$

where the coefficients of τ_1, τ_2, and τ_3 are functions of rotational restraint stiffness k, and defined as

$$\tau_1 = 124 + 22\frac{kb}{D_{22}} + \frac{k^2b^2}{D_{22}^2},\quad \tau_2 = 24 + 14\frac{kb}{D_{22}} + \frac{k^2b^2}{D_{22}^2},\quad \tau_3 = 102 + 18\frac{kb}{D_{22}} + \frac{k^2b^2}{D_{22}^2}. \tag{25}$$

Case 4: Plates with simply supported and free unloaded edges (SF) (Fig. 2(d))

For the case of $k = 0$, the simply supported boundary at one unloaded edge is achieved. The problem corresponds to the plate under the uniformly distributed compression load at simply supported loaded edges and subjected to the SF boundary condition (Fig. 2(d)), and the local buckling load can be obtained as

$$N_{\mathrm{cr}} = \frac{12D_{66}}{b^2}. \tag{26}$$

Eq. (26) is the same as the formula ($a \gg b$) given in Ref. [29].

Case 5: Plates with clamped and free unloaded edges (CF) (Fig. 2(e))

For the case of $k = \infty$, the boundary is related to clamped–supported at one unloaded edge and free at another unloaded edge (CF condition) (Fig. 2(e)), and the critical local buckling load can be obtained as

$$N_{\mathrm{cr}}^{\mathrm{CF}} = \frac{-28D_{12} + 4\sqrt{385D_{11}D_{22}} + 224D_{66}}{11b^2}. \tag{27}$$

Case 6: Plates with elastically retrained and free unloaded edges (RF) (Fig. 2(f))

The formula for the critical local buckling load of the general case of elastically restrained at one unloaded edge and free at the other (RF) (Fig. 2(f)) is given in Ref. [18].

4.4. Local buckling of composite structural shapes

Once the explicit solutions for elastically restrained plates (Fig. 1) are obtained, they can be applied to predict the local buckling of FRP shapes. In the discrete plate analysis of FRP shapes, the rotational restraint stiffness (k) is needed to determine the critical buckling strength. Based on the studies by Bleich [30] for isotropic materials and Qiao et al. [16] and Qiao and Zou [17,24] for composite materials, the rotational restraint stiffness coefficients (k) for local buckling of box and I-sections (with the flanges buckling first) are, respectively, given as

$$k = \frac{D_{22}^{\mathrm{w}^*}}{b_{\mathrm{f}}\rho_1\left(\dfrac{b_{\mathrm{w}}}{b_{\mathrm{f}}}\right)}\left(1 - \frac{b_{\mathrm{w}}^2}{b_{\mathrm{f}}^2}\frac{\sqrt{D_{11}^{\mathrm{f}}D_{22}^{\mathrm{f}}} + D_{12}^{\mathrm{f}} + 2D_{66}^{\mathrm{f}}}{\sqrt{D_{11}^{\mathrm{w}}D_{22}^{\mathrm{w}}} + D_{12}^{\mathrm{w}} + 2D_{66}^{\mathrm{w}}}\right) \quad \text{for box section} \tag{28}$$

$$k = \frac{D_{22}^{\mathrm{w}^*}}{b_{\mathrm{w}}}\left(1 - \frac{24b_{\mathrm{w}}^2}{\pi^2 b_{\mathrm{f}}^2}\frac{D_{66}^{\mathrm{f}}}{\sqrt{D_{11}^{\mathrm{w}}D_{22}^{\mathrm{w}}} + D_{12}^{\mathrm{w}} + 2D_{66}^{\mathrm{w}}}\right) \quad \text{for I-section,} \tag{29}$$

where $D_{22}^{\mathrm{w}^*}$ is the transverse bending stiffness of the restraining plate (e.g. the web in this study), D_{ij} $(i,j = 1,2,6)$ are the bending stiffness [28]; b_{f} is denoted as the width of the flange, and b_{w} the height of the web, and

$$\rho_1\left(\frac{b_{\mathrm{w}}}{b_{\mathrm{f}}}\right) = \frac{1}{2\pi}\tanh\frac{\pi b_{\mathrm{w}}}{2b_{\mathrm{f}}}\left\{1 + \frac{\dfrac{\pi b_{\mathrm{w}}}{b_{\mathrm{f}}}}{\sinh\left(\dfrac{\pi b_{\mathrm{w}}}{b_{\mathrm{f}}}\right)}\right\}.$$

As an illustration of the proposed explicit solutions to local buckling of FRP shapes, several FRP box and I-sections are analyzed. By substituting the rotational restraint stiffness (k) Eq. (28) into Eq. (24), the critical local buckling load of the box section is determined; while by considering Eq. (29) and the explicit formula in Case 6 of the RF plate, the local buckling strength of the I-section is predicted. The explicit predictions match well with the experimental and finite-element data (see Table 1). Thus, the

Table 1
Comparisons of critical local buckling stress resultants for box and I sections

Section	Dimensions (cm)	Explicit solution (N/cm)	FEM (N/cm)	Experimental results (N/cm)
Box	10.2 × 20.3 × 0.64	4773	4810	–
	10.2 × 15.2 × 0.64	7814	7740	–
I	10.2 × 10.2 × 0.64	8283	8235	8056
	15.2 × 15.2 × 0.64	4045	3882	3928

variational-principle-based local buckling analysis of FRP structural shapes is validated, and it can be used as a viable method to develop explicit formulations for local buckling problems of composite structures (e.g. plates and thin-walled beams).

5. Application II: global buckling of FRP composite structures

The second variational total potential energy method is hereby applied to analyze the global buckling of FRP composite structures. Based on the Rayleigh–Ritz method, the eigenvalue equation of global buckling is solved. In this section, the global (flexural–torsional) buckling of pultruded FRP composite I- and C-section beams is analyzed. The total potential energy of FRP shapes based on nonlinear plate theory is derived, of which the shear effect and beam bending–twisting coupling are included.

5.1. Theoretical formulation of the second variational problem for global buckling of FRP beams

For a thin-wall panel in the xy-plane, the in-plane finite strains of the midsurface considering the nonlinear terms are given by Malvern [31] as

$$\varepsilon_x = \frac{\partial u}{\partial x} + \frac{1}{2}\left[\left(\frac{\partial u}{\partial x}\right)^2 + \left(\frac{\partial v}{\partial x}\right)^2 + \left(\frac{\partial w}{\partial x}\right)^2\right],$$

$$\varepsilon_y = \frac{\partial v}{\partial y} + \frac{1}{2}\left[\left(\frac{\partial u}{\partial y}\right)^2 + \left(\frac{\partial v}{\partial y}\right)^2 + \left(\frac{\partial w}{\partial y}\right)^2\right], \tag{30a}$$

$$\gamma_{xy} = \frac{\partial u}{\partial y} + \frac{\partial v}{\partial x} + \frac{\partial u}{\partial x}\frac{\partial u}{\partial y} + \frac{\partial v}{\partial x}\frac{\partial v}{\partial y} + \frac{\partial w}{\partial x}\frac{\partial w}{\partial y}.$$

The curvatures of the midplane are defined as

$$\kappa_x = \frac{\partial^2 w}{\partial x^2}; \quad \kappa_y = \frac{\partial^2 w}{\partial y^2}; \quad \kappa_{xy} = 2\frac{\partial^2 w}{\partial x \partial y}. \tag{30b}$$

For a laminate in the xy-plane, the midsurface in-plane strains and curvatures are expressed in terms of the compliance coefficients and panel resultant forces as [28]

$$\begin{Bmatrix} \varepsilon_x \\ \varepsilon_y \\ \gamma_{xy} \\ \kappa_x \\ \kappa_y \\ \kappa_{xy} \end{Bmatrix} = \begin{bmatrix} \alpha_{11} & \alpha_{12} & \alpha_{16} & \beta_{11} & \beta_{12} & \beta_{16} \\ \alpha_{12} & \alpha_{22} & \alpha_{26} & \beta_{12} & \beta_{22} & \beta_{26} \\ \alpha_{16} & \alpha_{26} & \alpha_{66} & \beta_{16} & \beta_{26} & \beta_{66} \\ \beta_{11} & \beta_{12} & \beta_{16} & \delta_{11} & \delta_{12} & \delta_{16} \\ \beta_{12} & \beta_{22} & \beta_{26} & \delta_{12} & \delta_{22} & \delta_{26} \\ \beta_{16} & \beta_{26} & \beta_{66} & \delta_{16} & \delta_{26} & \delta_{66} \end{bmatrix} \begin{bmatrix} N_x \\ N_y \\ N_{xy} \\ M_x \\ M_y \\ M_{xy} \end{bmatrix}, \tag{31a}$$

or the panel resultant forces are expressed in term of the stiffness coefficients and midplane strains and curvatures as

$$\begin{Bmatrix} N_x \\ N_y \\ N_{xy} \\ M_x \\ M_y \\ M_{xy} \end{Bmatrix} = \begin{bmatrix} A_{11} & A_{12} & A_{16} & B_{11} & B_{12} & B_{16} \\ A_{12} & A_{22} & A_{26} & B_{12} & B_{22} & B_{26} \\ A_{16} & A_{26} & A_{66} & B_{16} & B_{26} & B_{66} \\ B_{11} & B_{12} & B_{16} & D_{11} & D_{12} & D_{16} \\ B_{12} & B_{22} & B_{26} & D_{12} & D_{22} & D_{26} \\ B_{16} & B_{26} & B_{66} & D_{16} & D_{26} & D_{66} \end{bmatrix} \begin{bmatrix} \varepsilon_x \\ \varepsilon_y \\ \gamma_{xy} \\ \kappa_x \\ \kappa_y \\ \kappa_{xy} \end{bmatrix}. \tag{31b}$$

Most pultruded FRP sections consist of symmetric laminated panels (e.g. web and flange) leading to no stretching–bending coupling ($\beta_{ij} = 0$). Also, the off-axis plies of the pultruded panels are usually balanced symmetric (no extension-shear and bending–twisting coupling, $\alpha_{16} = \alpha_{26} = \delta_{16} = \delta_{26} = 0$). The materials of laminated panels in pultruded sections are orthotropic, and their mechanical properties can be obtained either from experimental coupon tests or theoretical prediction using micro/macromechanics models [1].

The second variation of the total potential energy of the flanges is derived in two parts. The first part, $\delta^2 U_{\mathrm{b}}^{\mathrm{tf}}$, which is due to the axial displacement and bending about the major axis, is derived using the simple beam theory; while the second part, $\delta^2 U_{\mathrm{p}}^{\mathrm{tf}}$, which is due to the twisting and bending about the minor axis, is derived using the nonlinear plate theory. In this study, the flange panels (either top or bottom) are modeled as a beam bending around its strong axis and at the same time as a plate bending and twisting around its minor axis.

First, considering the top flange of either I- or C-section shown in Fig. 3(a) as a beam under the pure bending about its strong axis ($N_z^{\mathrm{b}} = N_{xz}^{\mathrm{tf}} = M_z^{\mathrm{b}} = M_{xz}^{\mathrm{b}} = 0$) and using the beam theory, the axial and bending (about the major axis) stress resultants of the flange are denoted by N_x^{tf} and M_x^{b}, respectively. Then, the second variation of the total potential

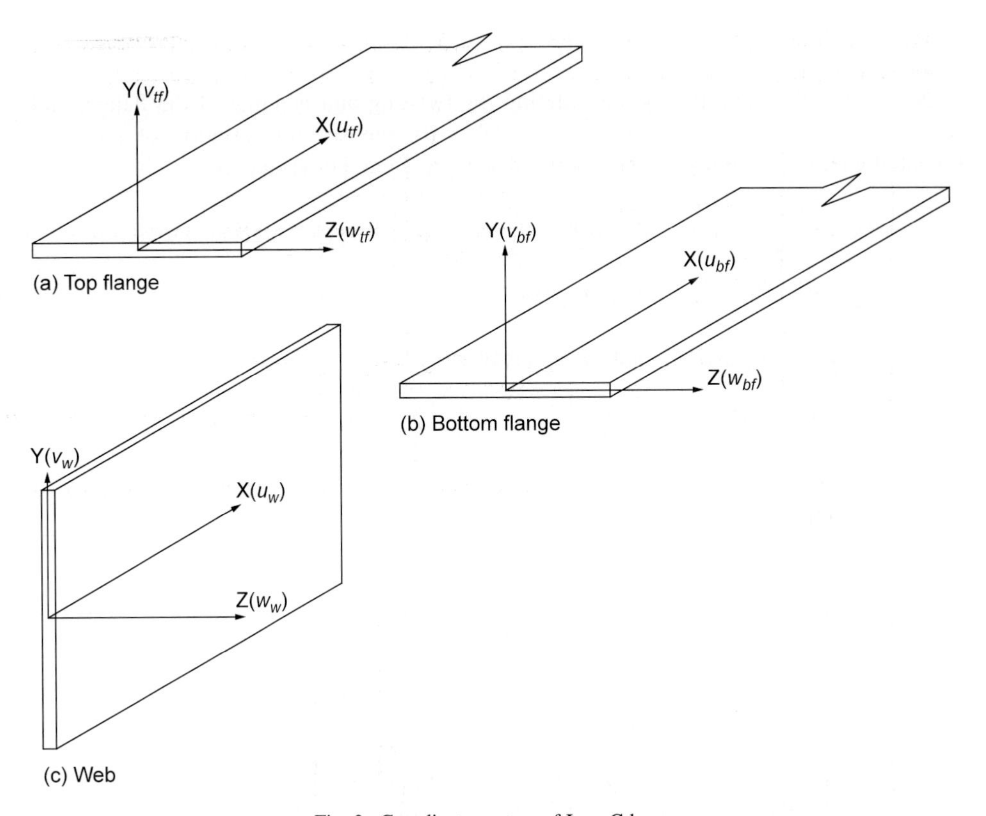

Fig. 3. Coordinate system of I- or C-beam.

energy due to the top flange bending laterally as a beam can be written as

$$\delta^2 U_{\mathrm{b}}^{\mathrm{tf}} = \int (N_x^{\mathrm{tf}} \delta^2 \varepsilon_x^{\mathrm{b}} + \delta N_x^{\mathrm{tf}} \delta \varepsilon_x^{\mathrm{b}} + M_x^{\mathrm{b}} \delta^2 \kappa_x^{\mathrm{b}} + \delta M_x^{\mathrm{b}} \delta \kappa_x^{\mathrm{b}}) \mathrm{d}x. \tag{32}$$

The strain displacement field is

$$\varepsilon_x^{\mathrm{b}} = \frac{\partial u^{\mathrm{tf}}}{\partial x} + \frac{1}{2}\left(\frac{\partial w^{\mathrm{tf}}}{\partial x}\right)^2; \quad \kappa_x^{\mathrm{b}} = \frac{\partial^2 w^{\mathrm{tf}}}{\partial x^2}. \tag{33}$$

Considering Eq. (32) and neglecting the third-order terms, the second variation of the total strain energy of the top flange is simplified as

$$\delta^2 U_{\mathrm{b}}^{\mathrm{tf}} = \int\int N_x^{\mathrm{tf}} \left(\frac{\partial \delta w^{\mathrm{tf}}}{\partial x}\right)^2 \mathrm{d}x\mathrm{d}z + \int\left\{A_x^{\mathrm{b}}\left(\frac{\partial \delta u^{\mathrm{tf}}}{\partial x}\right)^2 + D_x^{\mathrm{b}}\left(\frac{\partial^2 \delta w^{\mathrm{tf}}}{\partial x^2}\right)^2\right\}\mathrm{d}x. \tag{34}$$

Here the simplified forms of the stress resultants are expressed as

$$N_x^{\mathrm{tf}} = \frac{A_x^{\mathrm{b}}}{b_{\mathrm{f}}} \varepsilon_x^{\mathrm{b}}; \quad M_x^{\mathrm{b}} = D_x^{\mathrm{b}} \kappa_x^{\mathrm{b}},$$

where $A_x^{\mathrm{b}} = E_x t_{\mathrm{f}} b_{\mathrm{f}}$; $D_x^{\mathrm{b}} = E_x t_{\mathrm{f}} b_{\mathrm{f}}^3/12$; and E_x is the Young's modulus of the top flange plane in the x-axis.

Now using the plate theory, considering the twisting and bending of the flange, and without considering the distortion ($N_z^{\mathrm{p}} = M_z^{\mathrm{p}} = 0$), the second variation of the total potential energy of the top flange behaving as a plate can be written as

$$\delta^2 U_{\mathrm{p}}^{\mathrm{tf}} = \int\int (N_x^{\mathrm{tf}} \delta^2 \varepsilon_x^{\mathrm{p}} + \delta N_x^{\mathrm{tf}} \delta \varepsilon_x^{\mathrm{p}} + N_{xz}^{\mathrm{tf}} \delta^2 \gamma_{xz}^{\mathrm{p}} + \delta N_{xz}^{\mathrm{tf}} \delta \gamma_{xz}^{\mathrm{p}} + M_x^{\mathrm{p}} \delta^2 \kappa_x^{\mathrm{p}} + \delta M_x^{\mathrm{p}} \delta \kappa_x^{\mathrm{p}} + M_{xz}^{\mathrm{p}} \delta^2 \kappa_{xz}^{\mathrm{p}} + \delta M_{xz}^{\mathrm{p}} \delta \kappa_{xz}^{\mathrm{p}}) \mathrm{d}x\mathrm{d}z. \tag{35}$$

The nonlinear strains and curvatures are given as

$$\varepsilon_x^{\mathrm{p}} = \frac{1}{2}\left(\frac{\partial v^{\mathrm{tf}}}{\partial x}\right)^2 + \frac{1}{2}\left(\frac{\partial u^{\mathrm{tf}}}{\partial x}\right)^2; \quad \gamma_{xz}^{\mathrm{p}} = \frac{\partial u^{\mathrm{tf}}}{\partial x}\frac{\partial u^{\mathrm{tf}}}{\partial z} + \frac{\partial v^{\mathrm{tf}}}{\partial x}\frac{\partial v^{\mathrm{tf}}}{\partial z} + \frac{\partial w^{\mathrm{tf}}}{\partial x}\frac{\partial w^{\mathrm{tf}}}{\partial z};$$

$$\kappa_x^{\mathrm{p}} = \frac{\partial^2 v^{\mathrm{tf}}}{\partial x^2}; \quad \kappa_{xz}^{\mathrm{p}} = 2\frac{\partial^2 v^{\mathrm{tf}}}{\partial x \partial z}. \tag{36}$$

Considering Eqs. (35) and (36) and neglecting the third-order terms, the total strain energy of the top flange is simplified as

$$\begin{aligned} \delta^2 U_{\mathrm{p}}^{\mathrm{tf}} = \iint \Bigg\{ & N_x^{\mathrm{tf}} \left[\left(\frac{\partial \delta v^{\mathrm{tf}}}{\partial x}\right)^2 + \left(\frac{\partial \delta u^{\mathrm{tf}}}{\partial x}\right)^2 \right] \\ & + 2N_{xz}^{\mathrm{tf}} \left(\frac{\partial \delta v^{\mathrm{tf}}}{\partial x}\frac{\partial \delta v^{\mathrm{tf}}}{\partial z} + \frac{\partial \delta u^{\mathrm{tf}}}{\partial x}\frac{\partial \delta u^{\mathrm{tf}}}{\partial z} + \frac{\partial \delta w^{\mathrm{tf}}}{\partial x}\frac{\partial \delta w^{\mathrm{tf}}}{\partial z} \right) \\ & + \frac{1}{\delta_{11}}\left(\frac{\partial^2 \delta v^{\mathrm{tf}}}{\partial x^2}\right)^2 + \frac{4}{\delta_{66}}\left(\frac{\partial^2 \delta v^{\mathrm{tf}}}{\partial x \partial z}\right)^2 \Bigg\} \mathrm{d}x\mathrm{d}z. \end{aligned} \tag{37}$$

Therefore, the second variation of the total strain energy of the top flange can be obtained

$$\begin{aligned} \delta^2 U^{\mathrm{tf}} &= \delta^2 U_{\mathrm{b}}^{\mathrm{tf}} + \delta^2 U_{\mathrm{p}}^{\mathrm{tf}} \\ &= \iint \Bigg\{ N_x^{\mathrm{tf}} \left[\left(\frac{\partial \delta u^{\mathrm{tf}}}{\partial x}\right)^2 + \left(\frac{\partial \delta v^{\mathrm{tf}}}{\partial x}\right)^2 + \left(\frac{\partial \delta w^{\mathrm{tf}}}{\partial x}\right)^2 \right] \\ &\quad + 2N_{xz}^{\mathrm{tf}} \left(\frac{\partial \delta u^{\mathrm{tf}}}{\partial x}\frac{\partial \delta u^{\mathrm{tf}}}{\partial z} + \frac{\partial \delta v^{\mathrm{tf}}}{\partial x}\frac{\partial \delta v^{\mathrm{tf}}}{\partial z} + \frac{\partial \delta w^{\mathrm{tf}}}{\partial x}\frac{\partial \delta w^{\mathrm{tf}}}{\partial z} \right) \\ &\quad + A_x^{\mathrm{b}}\left(\frac{\partial \delta u^{\mathrm{tf}}}{\partial x}\right)^2 + D_x^{\mathrm{b}}\left(\frac{\partial^2 \delta w^{\mathrm{tf}}}{\partial x^2}\right)^2 + \frac{1}{\delta_{11}}\left(\frac{\partial^2 \delta v^{\mathrm{tf}}}{\partial x^2}\right)^2 + \frac{4}{\delta_{66}}\left(\frac{\partial^2 \delta v^{\mathrm{tf}}}{\partial x \partial z}\right)^2 \Bigg\} \mathrm{d}x\mathrm{d}z. \end{aligned} \tag{38}$$

The second variation of the total strain energy of the bottom flange $\delta^2 U^{\mathrm{bf}}$ can be obtained in a similar way.

Considering the web shown in Fig. 3(b) as a plate in the *xy*-plane and using the plate theory, the second variation of the total strain energy of the web can be expressed as

$$\delta^2 U^{\mathrm{w}} = \int\int (N_x^{\mathrm{w}} \delta^2 \varepsilon_x^{\mathrm{w}} + \delta N_x^{\mathrm{w}} \delta \varepsilon_x^{\mathrm{w}} + N_y^{\mathrm{w}} \delta^2 \varepsilon_y^{\mathrm{w}} + \delta N_y^{\mathrm{w}} \delta \varepsilon_y^{\mathrm{w}} + N_{xy}^{\mathrm{w}} \delta^2 \gamma_{xy}^{\mathrm{w}} + \delta N_{xy}^{\mathrm{w}} \delta \gamma_{xy}^{\mathrm{w}}$$
$$+ M_x^{\mathrm{w}} \delta^2 \kappa_x^{\mathrm{w}} + \delta M_x^{\mathrm{w}} \delta \kappa_x^{\mathrm{w}} + M_y^{\mathrm{w}} \delta^2 \kappa_y^{\mathrm{w}} + \delta M_y^{\mathrm{w}} \delta \kappa_y^{\mathrm{w}} + M_{xy}^{\mathrm{w}} \delta^2 \kappa_{xy}^{\mathrm{w}}$$
$$+ \delta M_{xy}^{\mathrm{w}} \delta \kappa_{xy}^{\mathrm{w}}) \mathrm{d}x\mathrm{d}y. \tag{39}$$

The strains and curvatures of the web are given as

$$\varepsilon_x^{\mathrm{w}} = \frac{\partial u^{\mathrm{w}}}{\partial x} + \frac{1}{2}\left[\left(\frac{\partial u^{\mathrm{w}}}{\partial x}\right)^2 + \left(\frac{\partial v^{\mathrm{w}}}{\partial x}\right)^2 + \left(\frac{\partial w^{\mathrm{w}}}{\partial x}\right)^2\right];$$
$$\varepsilon_y^{\mathrm{w}} = \frac{\partial v^{\mathrm{w}}}{\partial y} + \frac{1}{2}\left[\left(\frac{\partial u^{\mathrm{w}}}{\partial y}\right)^2 + \left(\frac{\partial v^{\mathrm{w}}}{\partial y}\right)^2 + \left(\frac{\partial w^{\mathrm{w}}}{\partial y}\right)^2\right];$$
$$\gamma_{xz}^{\mathrm{w}} = \frac{\partial u^{\mathrm{w}}}{\partial y} + \frac{\partial v^{\mathrm{w}}}{\partial x} + \frac{\partial u^{\mathrm{w}}}{\partial x}\frac{\partial u^{\mathrm{w}}}{\partial y} + \frac{\partial v^{\mathrm{w}}}{\partial x}\frac{\partial v^{\mathrm{w}}}{\partial y} + \frac{\partial w^{\mathrm{w}}}{\partial x}\frac{\partial w^{\mathrm{w}}}{\partial y};$$
$$\kappa_x^{\mathrm{w}} = \frac{\partial^2 w^{\mathrm{tf}}}{\partial x^2}; \quad \kappa_y^{\mathrm{w}} = \frac{\partial^2 w^{\mathrm{tf}}}{\partial y^2}; \quad \kappa_{xy}^{\mathrm{w}} = 2\frac{\partial^2 w^{\mathrm{tf}}}{\partial x \partial y}. \tag{40}$$

Neglecting the third-order terms and considering the constitutive relation of the web in Eq. (31b), the total strain energy of the web in Eq. (39) is simplified as

$$\delta^2 U^{\mathrm{w}} = \int\int \Bigg\{ N_x^{\mathrm{w}}\left[\left(\frac{\partial \delta u^{\mathrm{w}}}{\partial x}\right)^2 + \left(\frac{\partial \delta v^{\mathrm{w}}}{\partial x}\right)^2 + \left(\frac{\partial \delta w^{\mathrm{w}}}{\partial x}\right)^2\right]$$
$$+ N_y^{\mathrm{w}}\left[\left(\frac{\partial \delta u^{\mathrm{w}}}{\partial y}\right)^2 + \left(\frac{\partial \delta v^{\mathrm{w}}}{\partial y}\right)^2 + \left(\frac{\partial \delta w^{\mathrm{w}}}{\partial y}\right)^2\right]$$
$$+ 2N_{xy}^{\mathrm{w}}\left(\frac{\partial \delta u^{\mathrm{w}}}{\partial x}\frac{\partial \delta u^{\mathrm{w}}}{\partial y} + \frac{\partial \delta v^{\mathrm{w}}}{\partial x}\frac{\partial \delta v^{\mathrm{w}}}{\partial y} + \frac{\partial \delta w^{\mathrm{w}}}{\partial x}\frac{\partial \delta w^{\mathrm{w}}}{\partial y}\right) + A_{11}^{\mathrm{w}}\left(\frac{\partial \delta u^{\mathrm{w}}}{\partial x}\right)^2$$
$$+ A_{22}^{\mathrm{w}}\left(\frac{\partial \delta v^{\mathrm{w}}}{\partial y}\right)^2 + 2A_{12}^{\mathrm{w}}\frac{\partial \delta u^{\mathrm{w}}}{\partial x}\frac{\partial \delta v^{\mathrm{w}}}{\partial y} + A_{66}^{\mathrm{w}}\left[\left(\frac{\partial \delta v^{\mathrm{w}}}{\partial x}\right)^2 + \left(\frac{\partial \delta u^{\mathrm{w}}}{\partial y}\right)^2\right.$$
$$\left. + 2\frac{\partial \delta v^{\mathrm{w}}}{\partial x}\frac{\partial \delta u^{\mathrm{w}}}{\partial y}\right] + D_{11}^{\mathrm{w}}\left(\frac{\partial^2 \delta w^{\mathrm{w}}}{\partial x^2}\right)^2 + D_{22}^{\mathrm{w}}\left(\frac{\partial^2 \delta w^{\mathrm{w}}}{\partial y^2}\right)^2$$
$$+ 2D_{12}^{\mathrm{w}}\frac{\partial^2 \delta w^{\mathrm{w}}}{\partial x^2}\frac{\partial^2 \delta w^{\mathrm{w}}}{\partial y^2} + 4D_{66}^{\mathrm{w}}\left(\frac{\partial^2 \delta w^{\mathrm{w}}}{\partial x \partial y}\right)^2\Bigg\} \mathrm{d}x\mathrm{d}y. \tag{41}$$

The second variation of the total strain energy of the whole beam can be obtained by summing the web, top and bottom flanges as

$$\delta^2 U = \delta^2 U^{\mathrm{tf}} + \delta^2 U^{\mathrm{bf}} + \delta^2 U^{\mathrm{w}}, \tag{42}$$

and the critical condition (instability) is defined as

$$\delta^2 \Pi = \delta^2 U = 0, \tag{43}$$

which can be solved by employing the Rayleigh–Ritz method.

The total potential or strain energy in Eq. (43) can be further simplified by omitting all the terms that are positively defined [14], i.e. the term $(\partial \delta u^{\mathrm{tf}}/\partial x)^2$ in Eq. (34) and the terms involving the extensional stiffness coefficients A_{ij} in Eq. (41). Finally, the critical instability condition for the FRP beam in Fig. 3 becomes

$$\begin{aligned}
\delta^2 U = \int\!\!\int \Bigg\{ & N_x^{\mathrm{tf}} \left[\left(\frac{\partial \delta u^{\mathrm{tf}}}{\partial x} \right)^2 + \left(\frac{\partial \delta v^{\mathrm{tf}}}{\partial x} \right)^2 + \left(\frac{\partial \delta w^{\mathrm{tf}}}{\partial x} \right)^2 \right] \\
& + 2N_{xz}^{\mathrm{tf}} \left(\frac{\partial \delta u^{\mathrm{tf}}}{\partial x} \frac{\partial \delta u^{\mathrm{tf}}}{\partial z} + \frac{\partial \delta v^{\mathrm{tf}}}{\partial x} \frac{\partial \delta v^{\mathrm{tf}}}{\partial z} + \frac{\partial \delta w^{\mathrm{tf}}}{\partial x} \frac{\partial \delta w^{\mathrm{tf}}}{\partial z} \right) \\
& + D_x^{\mathrm{b}} \left(\frac{\partial^2 \delta w^{\mathrm{tf}}}{\partial x^2} \right)^2 + \frac{1}{\delta_{11}} \left(\frac{\partial^2 \delta v^{\mathrm{tf}}}{\partial x^2} \right)^2 + \frac{4}{\delta_{66}} \left(\frac{\partial^2 \delta v^{\mathrm{tf}}}{\partial x \partial z} \right)^2 \\
& + N_x^{\mathrm{bf}} \left[\left(\frac{\partial \delta u^{\mathrm{bf}}}{\partial x} \right)^2 + \left(\frac{\partial \delta v^{\mathrm{bf}}}{\partial x} \right)^2 + \left(\frac{\partial \delta w^{\mathrm{bf}}}{\partial x} \right)^2 \right] \\
& + 2N_{xz}^{\mathrm{bf}} \left(\frac{\partial \delta u^{\mathrm{bf}}}{\partial x} \frac{\partial \delta u^{\mathrm{bf}}}{\partial z} + \frac{\partial \delta v^{\mathrm{bf}}}{\partial x} \frac{\partial \delta v^{\mathrm{bf}}}{\partial z} + \frac{\partial \delta w^{\mathrm{bf}}}{\partial x} \frac{\partial \delta w^{\mathrm{bf}}}{\partial z} \right) \\
& + D_x^{\mathrm{b}} \left(\frac{\partial^2 \delta w^{\mathrm{bf}}}{\partial x^2} \right)^2 + \frac{1}{\delta_{11}} \left(\frac{\partial^2 \delta v^{\mathrm{bf}}}{\partial x^2} \right)^2 + \frac{4}{\delta_{66}} \left(\frac{\partial^2 \delta v^{\mathrm{bf}}}{\partial x \partial z} \right)^2 \Bigg\} \mathrm{d}x\mathrm{d}z \\
+ \int\!\!\int \Bigg\{ & N_x^{\mathrm{w}} \left[\left(\frac{\partial \delta u^{\mathrm{w}}}{\partial x} \right)^2 + \left(\frac{\partial \delta v^{\mathrm{w}}}{\partial x} \right)^2 + \left(\frac{\partial \delta w^{\mathrm{w}}}{\partial x} \right)^2 \right] \\
& + N_y^{\mathrm{w}} \left[\left(\frac{\partial \delta u^{\mathrm{w}}}{\partial y} \right)^2 + \left(\frac{\partial \delta v^{\mathrm{w}}}{\partial y} \right)^2 + \left(\frac{\partial \delta w^{\mathrm{w}}}{\partial y} \right)^2 \right] \\
& + 2N_{xy}^{\mathrm{w}} \left(\frac{\partial \delta u^{\mathrm{w}}}{\partial x} \frac{\partial \delta u^{\mathrm{w}}}{\partial y} + \frac{\partial \delta v^{\mathrm{w}}}{\partial x} \frac{\partial \delta v^{\mathrm{w}}}{\partial y} + \frac{\partial \delta w^{\mathrm{w}}}{\partial x} \frac{\partial \delta w^{\mathrm{w}}}{\partial y} \right) \\
& + D_{11}^{\mathrm{w}} \left(\frac{\partial^2 \delta w^{\mathrm{w}}}{\partial x^2} \right)^2 + D_{22}^{\mathrm{w}} \left(\frac{\partial^2 \delta w^{\mathrm{w}}}{\partial y^2} \right)^2 + 2D_{12}^{\mathrm{w}} \frac{\partial^2 \delta w^{\mathrm{w}}}{\partial x^2} \frac{\partial^2 \delta w^{\mathrm{w}}}{\partial y^2} \\
& + 4D_{66}^{\mathrm{w}} \left(\frac{\partial^2 \delta w^{\mathrm{w}}}{\partial x \partial y} \right)^2 \Bigg\} \mathrm{d}x\mathrm{d}y = 0.
\end{aligned} \tag{44}$$

5.2. Stress resultants and displacement fields

For a cantilever beam subjected to a tip concentrated vertical load, the simplified stress resultant distributions on the corresponding panels are obtained from beam theory, and the location or height of the applied load is accounted for in the analysis. For FRP I-beams, the resultant forces [24] are expressed in terms of the tip-applied concentrated load P. The expressions for the flanges are

$$\begin{aligned} N_x^{\mathrm{tf}} &= \frac{b_{\mathrm{w}} t_{\mathrm{f}}}{2I} P(L-x) \\ N_z^{\mathrm{tf}} &= N_{xz}^{\mathrm{tf}} = 0 \end{aligned} \tag{45a}$$

$$\begin{aligned} N_x^{\mathrm{bf}} &= -\frac{b_{\mathrm{w}} t_{\mathrm{f}}}{2I} P(L-x) \\ N_z^{\mathrm{bf}} &= N_{xz}^{\mathrm{bf}} = 0. \end{aligned} \tag{45b}$$

Similarly for the web

$$\begin{aligned} N_x^{\mathrm{w}} &= \frac{t_{\mathrm{w}}}{I} P(L-x)y \\ N_{xy}^{\mathrm{w}} &= -\frac{P t_{\mathrm{w}}}{2I}\left[\left(\frac{b_{\mathrm{w}}}{2}\right)^2 - y^2\right]. \end{aligned} \tag{45c}$$

Assuming that the top and bottom flanges do not distort (i.e. the displacements are linear in the z-direction) and considering compatibility conditions at the flange–web intersections, the buckled displacement fields for the web, top and bottom flange panels of the I-section are derived.

For the web (in the xy-plane)

$$u^{\mathrm{w}} = 0, \quad v^{\mathrm{w}} = 0, \quad w^{\mathrm{w}} = w(x,y). \tag{46a}$$

For the top flange (in the xz-plane)

$$u^{\mathrm{tf}} = u^{\mathrm{tf}}(x,z) = -z\frac{\mathrm{d}w^{\mathrm{tf}}}{\mathrm{d}x}, \quad v^{\mathrm{tf}} = v^{\mathrm{tf}}(x,z) = -z\theta^{\mathrm{tf}}, \quad w^{\mathrm{tf}} = w^{\mathrm{tf}}(x). \tag{46b}$$

For the bottom flange (in the xz-plane)

$$u^{\mathrm{bf}} = u^{\mathrm{bf}}(x,z) = -z\frac{\mathrm{d}w^{\mathrm{bf}}}{\mathrm{d}x}, \quad v^{\mathrm{bf}} = v^{\mathrm{bf}}(x,z) = -z\theta^{\mathrm{bf}}, \quad w^{\mathrm{bf}} = w^{\mathrm{bf}}(x). \tag{46c}$$

Similarly, the stress resultants and panel displacement fields for a cantilever C-section beam are derived and given in Ref. [26].

5.3. Explicit solutions

For the global (flexural–torsional) buckling of I- or C-section beams, the cross section of the beam is considered as undistorted. As the web panel is not allowed to distort and remains straight in flexural–torsional buckling, the sideways deflection and rotation of the web are coupled. The shape functions of buckling deformation for both the sideways deflection and rotation of the web, which satisfy the cantilever beam boundary conditions, can be selected as exact transcendental function as [24]

$$\begin{Bmatrix} w \\ \theta \end{Bmatrix} = \begin{Bmatrix} \bar{w} \\ \bar{\theta} \end{Bmatrix} \sum_{m=1,2,3,\ldots} \left\{ \sin\left(\frac{\lambda_m x}{L}\right) - \sinh\left(\frac{\lambda_m x}{L}\right) - \beta_m \left[\cos\left(\frac{\lambda_m x}{L}\right) - \cosh\left(\frac{\lambda_m x}{L}\right) \right] \right\}, \tag{47}$$

where

$$\beta_m = \frac{\sinh(\lambda_m) + \sin(\lambda_m)}{\cos(\lambda_m) + \cosh(\lambda_m)},$$

and λ_m satisfies the following transcendental equation

$$\cos(\lambda_m)\cosh(\lambda_m) - 1 = 0, \tag{48}$$

with $\lambda_1 = 1.875104$, $\lambda_2 = 4.694091$, $\lambda_3 = 7.854757, \ldots$

The displacements and rotations (referring to Eqs. (46a–c)) of panels in the I-section beam then become

$$w^{\mathrm{w}} = w + y\theta, \quad w^{\mathrm{tf}} = w + \frac{b_{\mathrm{w}}}{2}\theta, \quad w^{\mathrm{bf}} = w - \frac{b_{\mathrm{w}}}{2}\theta, \quad \theta^{\mathrm{tf}} = \theta^{\mathrm{bf}} = \theta. \tag{49}$$

By applying the Rayleigh–Ritz method and solving for the eigenvalues of the potential-energy equilibrium equation (44), the flexural–torsional buckling load, P_{cr}, for a free-endpoint load applied at the centroid of the cross section is obtained as [24]

$$P_{\mathrm{cr}} = \Psi_1 \cdot \{ b_{\mathrm{w}} L \Psi_2 + (\sqrt{\Psi_3 + \Psi_4 + \Psi_5 + \Psi_6 + \Psi_7})/b_{\mathrm{w}} \}, \tag{50}$$

where

$$\Psi_1 = (6b_{\mathrm{f}} + b_{\mathrm{w}})/[2L^3(76.5b_{\mathrm{f}}^2 - 6.96b_{\mathrm{f}}b_{\mathrm{w}} + 0.16b_{\mathrm{w}}^2)]$$
$$\Psi_2 = (123b_{\mathrm{f}} - 5.6b_{\mathrm{w}})D_{16}$$
$$\Psi_3 = a_{11}b_{\mathrm{f}}^3(279.5b_{\mathrm{f}}^2 - 25.5b_{\mathrm{f}}b_{\mathrm{w}} + 0.6b_{\mathrm{w}}^2)$$
$$\Psi_5 = b_{\mathrm{f}}b_{\mathrm{w}}^5(62.7L^2d_{66}D_{11} - 305.4b_{\mathrm{w}}^2D_{11}^2 - 1377.4L^2D_{16}^2 - 5511L^2D_{11}D_{66})$$
$$\Psi_6 = b_{\mathrm{w}}^6(7b_{\mathrm{w}}^2D_{11}^2 + 31.4L^2D_{16}^2 + 125.5L^2D_{11}D_{66})$$
$$\Psi_7 = a_{11}b_{\mathrm{f}}^3b_{\mathrm{w}}^2(1118b_{\mathrm{f}}^5d_{11} - 101.8b_{\mathrm{f}}^4b_{\mathrm{w}}d_{11} + 2.3b_{\mathrm{f}}^3b_{\mathrm{w}}^2d_{11} + 5043.5b_{\mathrm{f}}^3L^2d_{66} + 4.64b_{\mathrm{w}}^5D_{11} + 20.9b_{\mathrm{w}}^3L^2D_{66})$$
$$\Psi_8 = a_{11}b_{\mathrm{f}}^4b_{\mathrm{w}}^3[b_{\mathrm{w}}(-203.6b_{\mathrm{w}}^2D_{11} + 10.5L^2d_{66} - 918.5L^2D_{66}) + b_{\mathrm{f}}(2235.8b_{\mathrm{w}}^2D_{11} - 459.5L^2d_{66} + 10087L^2D_{66})],$$

and the following material parameters are defined as

$$a_{11} = 1/\alpha_{11}, \quad a_{66} = 1/\alpha_{66}, \quad d_{11} = 1/\delta_{11}, \quad d_{66} = 1/\delta_{66}. \tag{51}$$

In a similar fashion, the solution for global buckling load of cantilever C-section beams is recently obtained by Shan and Qiao [26].

Again, as an illustration of the proposed solution to global buckling of FRP beams, several FRP cantilever I- and C-section beams are analyzed, and their predictions are compared with both the experimental and finite-element data (see Table 2) [24,26]. As indicated in Table 2, the favorable agreement of analytical solutions with experimental

Table 2
Comparisons for flexural–torsional buckling loads of I- and C-section beams

Section	Dimensions (cm)	Analytical solution P_{cr} (N)	Finite element P_{cr} (N)	Experimental data P_{cr} (N)
I ($L = 365.8$ cm)	$10.2 \times 20.3 \times 0.64$	3192	2956	2943
	$15.2 \times 15.2 \times 0.64$	5614	5774	5476
C ($L = 365.8$ cm)	$2.9 \times 10.2 \times 0.64$	78.2	78.8	62.5
	$4.1 \times 15.2 \times 0.64$	215.8	217.8	190.5

results and finite-element eigenvalue analyses demonstrates the validity of the variational-principle methodology for global buckling analyses.

6. Conclusions

The variational principles as a viable tool in stability analysis of composite structures are illustrated in this study. The variational formulation of the Ritz method can be used to establish an eigenvalue problem, and by using different buckling deformation shape functions, the solutions of buckling of FRP structures are obtained. The first variation of the total potential energy is successfully used in the local buckling analysis of FRP shapes; while the second variation of the total potential energy based on nonlinear plate theory is applied to global buckling analysis. Through the application of several examples (i.e. box and I-sections for local buckling; I- and C-section beams for global buckling), the explicit and experimentally/numerically validated analytical formulas for the local and global buckling predictions are obtained, and they can be effectively used to design and characterize the buckling behavior of FRP structural shapes. As demonstrated in this study, the variational principles as an effective approach can be employed to solve the complicated problems in stability analysis and derive the explicit solutions for design, analysis and optimization of composite structures.

Acknowledgements

The authors would like to thank Profs. Julio F. Davalos and Ever J. Barbero of West Virginia University for their encouragement and support. The technical contribution of Dr. Guiping Zou to this study is greatly acknowledged.

References

[1] Davalos, J.F., Salim, H.A., Qiao, P.Z., Lopez-Anido, R. and Barbero, E.J. (1996), Analysis and design of pultruded FRP shapes under bending, Compos. Part B: Eng. J. 27B(3/4), 295–305.

[2] Qiao, P.Z., Davalos, J.F., Barbero, E.J. and Troutman, D. (1999), Equations facilitate composite designs, Mod. Plast. 76(11), 77–80.

[3] Qiao, P.Z., Davalos, J.F. and Brown, B. (2000), A systematic approach for analysis and design of single-span FRP deck/stringer bridges, Compos. Part B: Eng. J. 31(6–7), 593–609.
[4] Timoshenko, S.P. and Gere, J.M. (1961), Theory of Elastic Stability, McGraw-Hill, New York, NY.
[5] Roberts, T.M. (1981), Second order strains and instability of thin walled bars of open cross-section, Int. J. Mech. Sci. 23(5), 297–306.
[6] Bradford, M.A. and Trahair, N.S. (1981), Distortional buckling of I-beams, J. Struct. Eng. 107(2), 335–370.
[7] Bradford, M.A. (1992), Buckling of double symmetric cantilever with slender webs, Eng. Struct. 14(5), 327–334.
[8] Ma, M. and Hughes, O.F. (1996), Lateral distortional buckling of monosymmetric I-beams under distributed vertical load, Thin-Walled Struct. 26(2), 123–145.
[9] Smith, S.T., Bradford, M.A. and Oehlers, D.J. (2000), Unilateral buckling of elastically restrained rectangular mild steel plates, Comput. Mech. 26(4), 317–324.
[10] Lee, D.J. (1978), The local buckling coefficient for orthotropic structural sections, Aeronaut. J. 82, 313–320.
[11] Lee, D.J. (1979), Some observations on the local instability of orthotropic structural sections, Aeronaut. J. 83, 110–114.
[12] Lee, D.J. and Hewson, P.J. (1978), The use of fiber-reinforced plastics in thin-walled structures. Stability Problems in Engineering Structures on Composites (Richards, T.H., Stanley, P., eds.), Applied Science, London, pp. 23–55.
[13] Brunelle, E.J. and Oyibo, G.A. (1983), Generic buckling curves for specially orthotropic rectangular plates, AIAA J. 21(8), 1150–1156.
[14] Roberts, T.M. and Jhita, P.S. (1983), Lateral, local and distorsional buckling of I-beams, Thin-Walled Struct. 1(4), 289–308.
[15] Barbero, E.J. and Raftoyiannis, I.G. (1993), Local buckling of FRP beams and columns, J. Mater. Civ. Eng. 5(3), 339–355.
[16] Qiao, P.Z., Davalos, J.F. and Wang, J.L. (2001), Local buckling of composite FRP shapes by discrete plate analysis, J. Struct. Eng. 127(3), 245–255.
[17] Qiao, P.Z. and Zou, G.P. (2002), Local buckling of elastically restrained fiber-reinforced plastic plates and its application to box sections, J. Eng. Mech. 128(12), 1324–1330.
[18] Qiao, P.Z. and Zou, G.P. (2003), Local buckling of composite fiber-reinforced plastic wide-flange sections, J. Struct. Eng. 129(1), 125–129.
[19] Barbero, E.J. and Raftoyiannis, I.G. (1994), Lateral and distortional buckling of pultruded I-beams, Compos. Struct. 27(3), 261–268.
[20] Pandey, M.D., Kabir, M.Z. and Sherbourne, A.N. (1995), Flexural–torsional stability of thin-walled composite I-section beams, Compos. Eng. 5(3), 321–342.
[21] Kabir, M.Z. and Sherbourne, A.N. (1998), Lateral–torsional buckling of post-local buckled fibrous composite beams, J. Eng. Mech. 124(7), 754–764.
[22] Davalos, J.F. and Qiao, P.Z. (1997), Analytical and experimental study of lateral and distortional buckling of FRP wide-flange beams, J. Compos. Constr. 1(4), 150–159.
[23] Davalos, J.F., Qiao, P.Z. and Salim, H.A. (1997), Flexure-torsional buckling of pultruded fiber reinforced plastic composite I-beams: experimental and analytical evaluations, Compos. Struct. 38(1–4), 241–250.
[24] Qiao, P.Z., Zou, G.P. and Davalos, J.F. (2003), Flexural–torsional buckling of fiber-reinforced plastic composite cantilever I-beams, Compos. Struct. 60, 205–217.
[25] Lee, J. and Kim, S.K. (2002), Lateral buckling analysis of thin-walled laminated channel-section beams, Compos. Struct. 56, 391–399.
[26] Shan, L.Y. and Qiao, P.Z. (2004), Flexural–torsional buckling of fiber-reinforced plastic composite open channel beams, Compos. Struct. 68(2), 211–224.
[27] Sagan, H. (1969), Introduction to the Calculus of Variations, McGraw-Hill, New York, NY.
[28] Jones, R.M. (1999), Mechanics of Composite Materials, Taylor and Francis, Philadelphia, PA.
[29] Barbero, E.J. (1999), Introduction to Composite Materials, Taylor and Francis, Philadelphia, PA.
[30] Bleich, F. (1952), Buckling Strength of Metal Structures. McGraw-Hill, New York, NY.
[31] Malvern, L.E. (1969), Introduction to the Mechanics of a Continuous Medium, Prentice-Hall, Englewood Cliffs, NJ.

TRANSPORT PHENOMENA AND ENERGY CONVERSION

Chapter 7

FIELD VARIATIONAL PRINCIPLES FOR IRREVERSIBLE ENERGY AND MASS TRANSFER

Stanislaw Sieniutycz

Warsaw University of Technology, Faculty of Chemical and Process Engineering, Warsaw, Poland

Abstract

In this chapter we demonstrate the violation of the standard, canonical structure of conservation laws in variational formulations similar to but not equivalent with Hamilton's Principle that typically uses the sourceless entropy constraint. The difference between the two formulations is due to the constraints that comprise the nonconserved entropy balance and the energy-representation counterpart of the Cattaneo equation called Kaliski's equation. We show that despite the generally noncanonical form of conservation laws (obtained via Noether's theorem) the method that adjoints a given set of constraints to a kinetic potential L works efficiently. In fact, the method leads to an exact variational formulation for the constraints in the potential space of Lagrange multipliers, implying that the appropriateness of the set should be verified by physical rather than mathematical criteria. These issues are exemplified by the field (Eulerian) description of heat conduction, where equations of the thermal field follow from variational principles containing suitable potentials rather than original physical variables. The considered processes are hyperbolic heat transfer and coupled parabolic transfer of heat, mass, and electric charge. With various gradient or nongradient representations of physical fields in terms of potentials (quantities similar to those used by Clebsch in his representation of hydrodynamic velocity) useful action-type criteria are found. Symmetry principles can be considered and components of the formal energy-momentum tensor can be evaluated. The limiting reversible case may provide a suitable reference frame. The results imply that the thermodynamic irreversibility does not necessarily change Hamilton's kinetic potential; it only complicates potential representations of physical fields in comparison with those describing the reversible evolution.

Keywords: thermal fields; adjoined constraints; potential representations; Kaliski's equation; conservation laws

1. Introduction

When investigating physical fields two general frameworks may be applied. The first deals directly with process differential equations, whereas the second uses the corresponding action integrals. By extremizing the latter, solutions of the differential equations in question can be found by the so-called direct variational methods. In an earlier work with Berry [1] we discussed a description of macroscopic representations of

E-mail address: sieniutycz@ichip.pw.edu.pl (S. Sieniutycz).

thermal fields (the heat conduction with a finite signal speed) by composite variational principles involving suitably constructed potentials along with original physical variables. In this description a variational formulation for a given vector field treats all field equations as constraints that are linked by Lagrange multipliers to the given kinetic potential. In Ref. [1] we focused on the example of simple hyperbolic heat transfer, but the approach can also be applied to the coupled transfer of heat, mass, and electric charge and to other processes.

In the present chapter an example of the method applicability for coupled transfer will be given. Also, the energy representation of the approach will be worked out. This means that we shall not only consider constraints of the conserved or nonconserved internal energy (thus extending the conserved theory of Ref. [1]), but will also replace these constraints by the entropy-balance constraint with a source. The form of this source is quadratic with respect to the diffusive flux of entropy or heat, i.e. corresponds with their usual structure valid both in the classical and extended irreversible thermodynamics.

Various potentials can be attributed to physical fields (gradient or nongradient representations). In the potential framework corresponding Lagrangian and Hamiltonian formalisms can be developed. Formal components of the energy-momentum tensor can be found for the given constraints and kinetic potential. With the Noether's conservation laws corresponding to an action principle, a variational thermodynamics may be derived. In the text below we shall discuss details of the above ideas.

This chapter is restricted to thermal transport processes described as fields; it is motivated by the fact that the construction of variational principles for irreversible fields still seems to have major difficulties. Equations of dissipative fluid mechanics and irreversible thermodynamics provide a frequently used Eulerian or field representation of the process. At best, however, only some truncated (e.g. stationary) forms of these equations were shown to possess the well-known structure of the Euler–Lagrange equations of the classical variational problem [2].

The important physical ingredient comes with conservation laws obtained from the Hamiltonian actions [2]. However, for irreversible processes, there are serious difficulties to find effective Hamiltonian formulations. These difficulties are attributed to the presence of nonself-adjoint operators [3]. The nonself-adjoint operators cause nonsymmetric Frechet derivatives in the original state space so that according to Vainberg's theorem, an exact variational formulation cannot be found in this space [4,5].

In the present discussion we omit the so-called bracket formalisms [6,7], the main features of which were described in a recent book by Beris and Edwards [8]. (The single-bracket and two-bracket descriptions are usually distinguished.) The two-bracket formalisms produce evolution equations via Poissonian and dissipative brackets, the latter being a functional extension of the Rayleigh dissipation function. Yet these bracket approaches are usually not associated with an extremum of a definite quantity. For this purpose a single Poissonian bracket and a Hamiltonian system are both necessary and sufficient.

In our approach the process is transferred to a different, suitable space, and a variational formulation is found in that space. This is made by means of certain potentials, similar to those known for the electromagnetic field. The origin and properties of these potentials are not explained sufficiently well to date. Our research direction links the potentials with

Lagrange multipliers of the adjoined constraints, to improve understanding of this issue. We exploit some observations made in earlier works on thermal fields [9–11]. We also refer to results obtained for reversible systems, in particular those of Herivel [12], Stephens [13], Seliger and Whitham [14], Atherton and Homsy [15], Caviglia [16], and Sieniutycz and Berry [17]. These latter papers proved the essential role of the Lagrange multipliers in constructing potential representations of physical fields for the purpose of variational principles (see also Refs. [18,19]).

2. Lagrange multipliers as adjoints and potentials of variational formulation

For irreversible phenomena, the difficulties in finding variational formulations are attributed to the presence of the so-called nonself-adjoint operators. They violate the time-reversal symmetry in the macroscopic equations, whence, according to Vainberg's theorem [4], a suitable exact functional cannot be found in the space of original variables. We observe that these difficulties do not appear when the process space is enlarged by addition of suitable new variables, often called "the potentials", which are, in fact, the Lagrange multipliers for the given constraints. In brief, our method transfers the problem to the space composed of original variables and potentials. In fact, the method assures a spontaneous transfer to the most proper space. The method is easy to apply because the equations for which a variational principle is sought are simply adjoined by Lagrange multipliers (vector, $\boldsymbol{\lambda}$) to the accepted kinetic potential, L.

The origins and key aspects of these approaches still call for improved explanation. Their essence is that the kinetic potential L can be arbitrary, or correct formulations can be found for an infinite number of various L. Until now, this was not stated sufficiently clearly in the literature, perhaps due to the apparent puzzle of a nonunique L. However, a change in the kinetic potential L leaves the original constraints unchanged. Only the state representation in terms of the potentials do change with changes in L. Whenever the original variables are expressible in terms of the Lagrange multipliers $\boldsymbol{\lambda}$, explicit "representations" of the original fields in terms of $\boldsymbol{\lambda}$ and its derivatives are obtained. They depend on the accepted kinetic potential L, although others could be found for a different L. The new fields $\boldsymbol{\lambda}$ are adjoint for the problem; they "represent" the original variables in the way depending on properties of the original equations and the accepted L. In fact, one can produce an infinite number of different, still correct, representations of the process in the space of the original coordinates and Lagrange multipliers.

In different branches of science, the latter are called by various names. The names: adjoint variables or simply adjoints appear as a rule in various problems of optimal control in which the adjoints are companions of original coordinates of state, for any chosen objective function. In some cases the whole variational description can be accomplished only in the space of adjoints; in others, only in the space of original variables.

Yet, in the general case, the variational description can be set only in the general composite space of physical variables and their adjoints. The notion of adjoints may be referred to variational principles of classical (nonfrictional) mechanics with space coordinates as original variables and momenta as adjoints. Clearly, the phase space, which is well known in the statistical mechanics, is the space composed of the space coordinates and their adjoints. In the optimization theory, and in particular in Pontryagin's principle, the role of adjoints is well known. In various field descriptions the potentials (multipliers, adjoints, momentum-type coordinates, etc., whatever we call them) are quantities of the same sort as those used by Clebsch in his representations of hydrodynamic velocity.

It should not be surprising that general variational formulations should be sought in extended spaces (with adjoints) rather than in original spaces, often called the physical spaces. The apparent peculiarity of irreversible processes could formally be interpreted in the way that they require the whole composite space or a large part of this space for their variational imbedding, meaning that the full reduction of extra coordinates is impossible. From a physical perspective, "irreversibility" can be interpreted to mean that there are some degrees of freedom into which otherwise-conserved quantities such as energy may flow but from which no return can be observed. We show here that the impossibility of reducing the space is the consequence of sources and the presence of both even and odd time derivatives in equations describing the irreversible evolutions.

In the enlarged spaces, irreversibility properties do little to hamper a variational formulation for a given L. In fact, a properly enlarged space exists that is the space of minimal dimensionality in which a variational formulation can be set for nontruncated equations. In the properly enlarged space, to which considered processes are automatically transferred with the help of the Lagrange multipliers (adjoints), the Frechet symmetry is assured automatically. The dimension of the enlarged space where the variational formulation resides is not greater than $2n$, where n is the number of both the original physical variables and the original equations written in the form of first-order (partial or ordinary) vanishing constraints, $\mathbf{C} = 0$. The necessary extremum conditions are obtained by setting to zero the variation of the action integral A based on the Lagrangian $\Lambda = L + \boldsymbol{\lambda}\mathbf{C}$ (whichever kinetic potential L and constraints $\mathbf{C} = 0$ are). These extremum conditions are, of course, the Euler–Lagrange equations of the variational problem for action A. With the tool of the Legendre transform, Hamiltonian formulations consistently follow.

Hyperbolic heat transfer, considered in the following sections, is one of the examples where the effectiveness of the present approach is explicit. Next, the approach is applied to the coupled transfer of heat, mass, and electric charge. Various "gradient or nongradient" representations of original physical fields in terms of potentials appear. Least-action-type and other criteria can be used, and corresponding Lagrangian and Hamiltonian formalisms can be developed. Symmetry principles can be considered, and components of formal energy-momentum tensors can be evaluated for the accepted kinetic potentials, L.

3. Basic equations for damped-wave heat transfer

Below we demonstrate the technique of adjoint representations of physical fields by constructing a variational formulation for the linear process of pure heat conduction (heat flux q) in a rigid solid at rest. The finite speed of propagation of thermal signals is assumed, which means that we decide to use a hyperbolic model rather than parabolic. As pointed out by many researchers, ([20–24] and others), the paradox of infinite propagation speed was resolved by the acceptance of the hypothesis of heat-flux relaxation. The link between the hypothesis and certain results of nonequlibrium statistical mechanics, such as Grad's solution of the Boltzmann kinetic equation [25], was found [17,26].

The hypothesis is based on the position that Fourier's law is an approximation to a more exact equation, called the Cattaneo equation, which contains the time derivative of the heat flux along with the flux itself. In particular, it implies that, for the sudden temperature increase at the wall, the wall heat flux does not start instantaneously, but rather grows gradually with a rate depending on a relaxation time τ [24]. After some time the wall heat flux arrives at a maximum and only then decreases as in the Fourier case. Although the values of τ are typically very brief (of the order of 10^{-12} s for liquids and metals, and 10^{-9} s for gases under normal conditions) their effects can still have theoretical importance.

For the heat-conduction process described in the entropy representation of thermodynamics by the Cattaneo equation of heat transfer and the conservation law for internal energy, the set of constraints is

$$\frac{\partial \mathbf{q}}{c_0^2 \partial t} + \frac{\mathbf{q}}{c_0^2 \tau} + \nabla \rho_e = 0 \tag{1}$$

and

$$\frac{\partial \rho_e}{\partial t} + \nabla \cdot \mathbf{q} = 0, \tag{2}$$

where the density of the thermal energy ρ_e satisfies $d\rho_e = \rho c_v \, dT$, c_0 is propagation speed for the thermal wave, τ is thermal relaxation time, and the product $D = c_0^2 \tau$ is the thermal diffusivity. Eq. (2) assumes the conservation of the thermal energy that means that the medium is rigid and the viscous dissipation can be ignored.

A subtle feature is the irreversible nature of the heat process that requires distinguishing between the paths of matter, energy, and entropy. Only in reversible processes do entropy or energy 'flow with the matter'; in irreversible processes the paths of entropy or energy differ from those of the matter. Although we restrict ourselves to the rest frame of a medium, an energy and entropy flow do occur. The vector of density of heat flux, $\mathbf{q}$, represents the energy flow. The vector of density of entropy flux, $\mathbf{j}_s$ represents the entropy flow.

For simplicity we assume the constant values of involved fields at the boundary. We ignore the vorticity properties of the heat flux, i.e. it is not our concern here whether $\nabla \times \mathbf{q}$ vanishes or is different from zero. Yet we point out that the Cattaneo equation (1) ensures the vanishing rotation for all future times whenever the rotation of the initial field $\mathbf{q}(\mathbf{x}, 0)$ vanishes. The vorticity properties of the system are discussed in Ref. [1].

The energy representation of the Cattaneo equation

$$\frac{\partial \mathbf{j}_s}{c_s^2 \partial t} + \frac{\mathbf{j}_s}{c_s^2 \tau} + \nabla T = 0, \tag{3}$$

uses the diffusive entropy flux $\boldsymbol{j}_s$ instead of heat flux $\mathbf{q}$. The coefficient $c_s \equiv (\rho c_v \theta^{-1})^{1/2}$, where $\theta = T c_0^{-2}$ is associated with the thermal diffusivity $k \equiv \rho c_v c_0^2 \tau$. Eq. (3) is called Kaliski's equation. For an incompressible medium considered here we apply this equation in its alternative form

$$\frac{\partial \mathbf{j}_s}{c_0^2 \partial t} + \frac{\mathbf{j}_s}{c_0^2 \tau} + \nabla \rho_s = 0, \tag{4}$$

which uses the entropy density ρ_s as a field variable. The physical consequences of the associated variational scheme are compared with those stemming from the application of the classical Cattaneo result, Eq. (1). The Cattaneo equation uses, of course, the heat flux as the field variable. In our variational analysis, Eq. (4) is no less essential. It follows that both Cattaneo and Kaliski models are associated with a frictional equation of motion

$$\theta \frac{\partial \mathbf{v}_s}{\partial t} + \theta \frac{\mathbf{v}_s}{\tau} = -\nabla T. \tag{5}$$

This form clearly shows the dissipative properties of the system. In usual descriptions they are associated with the generation of the entropy.

For the purpose of a variational formulation, we assume that all dynamical equations of interest constitute 'the constraints'; these constraints are adjoined in an action functional A to a singular kinetic potential L that does not contain the derivatives. An important difference between the composite variational formulations considered here and the traditional ones is that one can apply diverse L and will always get a correct variational formulation satisfying the constraints. In fact, there is an infinite number of possible Ls that can successfully be applied. In Section 15 we show that, for a given vector field $\mathbf{u}$ one can use quadratic Ls of the structure $L = (1/2)\mathbf{B} : \mathbf{uu}$, and correct representations of $\mathbf{u}$ can be obtained for any nonsingular quadratic matrix $\mathbf{B}$ [19]. Different Ls yield different representations of physical fields of interest in terms of the Lagrange multipliers of the constraints. Yet, among various Ls, Hamilton's structure of L has often a definite preference. We restrict ourselves here to the Lagrangian formalism. However, the results obtained can be transformed into the Hamiltonian formalism as well. The main properties of the latter formalism for thermodynamic systems are described in Ref. [27].

4. Action and extremum conditions in entropy representation (variables q and ρ_e)

An action is assumed that absorbs constraints (1) and (2) by the Lagrange multipliers, the vector $\boldsymbol{\psi}$ and the scalar ϕ. Its kinetic potential L, (7), has a Hamilton-like form.

$$A = \int_{t_1,V}^{t_2} \varepsilon^{-1} \left\{ \frac{1}{2} \frac{\mathbf{q}^2}{c_0^2} - \frac{1}{2} \rho_e^2 - \frac{1}{2} \varepsilon^2 + \boldsymbol{\psi} \cdot \left(\frac{\partial \mathbf{q}}{c_0^2 \partial t} + \frac{\mathbf{q}}{c_0^2 \tau} + \nabla \rho_e \right) \right.$$
$$\left. + \varphi \left(\frac{\partial \rho_e}{\partial t} + \nabla \cdot \mathbf{q} \right) \right\} \mathrm{d}V \, \mathrm{d}t. \tag{6}$$

As kinetic potentials can be very diverse, the conservation laws for energy and momentum substantiate the form (6). In Eq. (6), ε is the energy density at an equilibrium reference state, the constant that ensures the action dimension for A, but otherwise is unimportant.

Yet we assume that the actual energy density ρ_e is close to ε, so that the variable ρ_e can be identified with the constant ε in suitable approximations.

We call the multiplier-free term of the integrand (6)

$$L \equiv \frac{1}{2}\varepsilon^{-1}\left\{\frac{\mathbf{q}^2}{c_0^2} - \rho_e^2 - \varepsilon^2\right\} \tag{7}$$

the kinetic potential of Hamilton type for heat transfer. It is based on the quadratic form of an indefinite sign, and it has the usual units of the energy density. Not far from equilibrium, where ρ_e is close to ε, two static terms of L yield altogether the density of thermal energy, ρ_e. To secure correct conservation laws, no better form of L associated with a nonlinear model was found in the entropy representation. Thus, in spite of arbitrariness in L, Hamiltonian Ls are preferred. The theory obtained in the present case is a linear one.

Vanishing variations of action A with respect to multipliers $\boldsymbol{\psi}$ and ϕ recover constraints, whereas those with respect to state variables $\mathbf{q}$ and ρ_e yield representations of state variables in terms of $\boldsymbol{\psi}$ and ϕ. For the accepted Hamilton-like structure of L,

$$\mathbf{q} = \frac{\partial \boldsymbol{\psi}}{\partial t} - \frac{\boldsymbol{\psi}}{\tau} + c_0^2\nabla\varphi \tag{8}$$

and

$$\rho_e = -\nabla\cdot\boldsymbol{\psi} - \frac{\partial \phi}{\partial t}. \tag{9}$$

In a limiting reversible process (in damped or wave heat conduction for $\tau \rightarrow \infty$) the process is described by purely gradient representations; the representation for $\mathbf{q}$ then has the structure of the electric field $\mathbf{E}$ expressed in terms of electromagnetic potentials.

For the accepted structure of L, the action A, Eq. (6), in terms of the adjoints $\boldsymbol{\psi}$ and ϕ is

$$A = \int_{t_1,V}^{t_2} \varepsilon^{-1}\left\{\frac{1}{2c_0^2}\left(\frac{\partial \boldsymbol{\psi}}{\partial t} - \frac{\boldsymbol{\psi}}{\tau} + c_0^2\nabla\phi\right)^2 - \frac{1}{2}\left(\nabla\cdot\boldsymbol{\psi} + \frac{\partial \phi}{\partial t}\right)^2 - \frac{1}{2}\varepsilon^2\right\} \mathrm{d}V\,\mathrm{d}t. \tag{10}$$

Its Euler–Lagrange equations with respect to $\boldsymbol{\psi}$ and ϕ are, respectively,

$$\frac{\partial}{\partial t}\left\{\frac{1}{c_0^2}\left(\frac{\partial \boldsymbol{\psi}}{\partial t} - \frac{\boldsymbol{\psi}}{\tau} + c_0^2\nabla\phi\right)\right\} + \frac{1}{\tau c_0^2}\left(\frac{\partial \boldsymbol{\psi}}{\partial t} - \frac{\boldsymbol{\psi}}{\tau} + c_0^2\nabla\phi\right) - \nabla\left(\nabla\cdot\boldsymbol{\psi} + \frac{\partial \phi}{\partial t}\right) = 0, \tag{11}$$

and

$$-\frac{\partial}{\partial t}\left(\nabla\cdot\boldsymbol{\psi} + \frac{\partial \phi}{\partial t}\right) + \nabla\cdot\left(\frac{\partial \boldsymbol{\psi}}{\partial t} - \frac{\boldsymbol{\psi}}{\tau} + c_0^2\nabla\phi\right) = 0. \tag{12}$$

It is easy to see that Eqs. (11) and (12) are the original equations of the thermal field, Eqs. (1) and (2), in terms of the potentials $\boldsymbol{\psi}$ and ϕ. Their equivalent form below shows the damped-wave nature of the transfer process.

For the Cattaneo equation (1) we obtain by simplification of Eq. (11)

$$\frac{\partial^2 \boldsymbol{\psi}}{c_0^2\,\partial t^2} - \frac{\boldsymbol{\psi}}{\tau c_0^2 \tau} + \frac{\nabla \phi}{\tau} - \nabla(\nabla \cdot \boldsymbol{\psi}) = 0, \tag{13}$$

whereas the simplification of Eq. (12) yields

$$-\frac{\partial^2 \phi}{\partial t^2} - \frac{\nabla \cdot \boldsymbol{\psi}}{\tau} + c_0^2 \nabla^2 \phi = 0. \tag{14}$$

Eqs. (13) and (14) describe the Cattaneo model, Eqs. (1) and (2), in terms of potentials $\boldsymbol{\psi}$ and ϕ for the associated action (10). Note that the set (Eqs. (13) and (14)) becomes decoupled in the reversible case of an infinite τ. Interpreting τ as an average time between the collisions, we can regard the reversible process (with $\tau \to \infty$) as the collisionless one.

5. Source terms in internal energy equation

However, the construction of a suitable action A in the space of potentials by the direct substitution of the representation equations to the accepted kinetic potential L is generally incorrect. In fact, the method of direct substitution of representations into L is valid only for linear constraints that do not contain source terms. This may be exemplified when the internal energy balance contains a source term $a'\mathbf{q}^2$, where a' is a positive constant. The augmented action integral (6) should now contain the negative term $-a'\mathbf{q}^2$ in its ϕ term. The energy-density representation remains unchanged, whereas the heat-flux representation follows in a generalized form

$$\mathbf{q} = (1 - 2a'\phi c_0^2)^{-1}\left(\frac{\partial \boldsymbol{\psi}}{\partial t} - \frac{\boldsymbol{\psi}}{\tau} + c_0^2 \nabla \varphi\right). \tag{15}$$

It may then be shown that the action based on the accepted kinetic potential L in terms of the potentials acquires the form

$$A = \int_{t_1,V}^{t_2} \varepsilon^{-1}\left\{\frac{1}{2c_0^2}(1 - 2a'\phi c_0^2)^{-2}\left(\frac{\partial \boldsymbol{\psi}}{\partial t} - \frac{\boldsymbol{\psi}}{\tau} + c_0^2 \nabla \phi\right)^2 - \frac{1}{2}\left(\nabla \cdot \boldsymbol{\psi} + \frac{\partial \phi}{\partial t}\right)^2 - \frac{1}{2}\varepsilon^2\right\} \mathrm{d}V\,\mathrm{d}t. \tag{16}$$

However, the Euler–Lagrange equations for this action are not the process constraints in terms of potentials, i.e. L itself fails to provide a correct variational formulation for constraints with sources. It is the vanishing term with constraints that contributes to the properties of the functional extremum in the augmented action A.

The way to improve the situation is to substitute the obtained representations to a transformed augmented action in which the only terms rejected are those that constitute the total time or space derivatives. The latter can be selected via partial differentiation within the integrand of the original action A. (As we know from the theory of the functional extrema the addition or subtraction of terms with total derivatives and divergences do not change extremum properties of a functional.) When this procedure is applied to the considered problem and total derivatives are rejected, a correct action follows in the form

$$A = \int_{t_1,V}^{t_2} \varepsilon^{-1}\left\{\frac{1}{2c_0^2}(1-2a'\phi c_0^2)^{-1}\left(\frac{\partial \boldsymbol{\psi}}{\partial t} - \frac{\boldsymbol{\psi}}{\tau} + c_0^2\nabla\phi\right)^2 - \frac{1}{2}\left(\nabla\cdot\boldsymbol{\psi} + \frac{\partial\phi}{\partial t}\right)^2 - \frac{1}{2}\varepsilon^2\right\}\mathrm{d}V\,\mathrm{d}t. \tag{17}$$

This form differs from that of Eq. (16) only by the power of the term containing the constant a', related to the source. With the representation equations (9) and (15), action (17) yields the proper Cattaneo constraint (1) and the generalized balance of internal energy that extends Eq. (2) by the positive source term $a'\mathbf{q}^2$. Eq. (17) proves that the four-dimensional potential space $(\boldsymbol{\psi}, \phi)$ is sufficient to accommodate the exact variational formulation for the problem with a source. Yet, due to the presence of the source, the variational formulation does not exist in the original four-dimensional space $(\mathbf{q}, \rho_e)$, and, if we insist on exploiting this space, plus possibly a necessary part of the potential space, the following action is obtained from Eqs. (1), (9), (15), and (17)

$$A = \int_{t_1,V}^{t_2} \varepsilon^{-1}\left\{(1-2a'\phi c_0^2)\frac{\mathbf{q}^2}{2c_0^2} - \frac{1}{2}\rho_e^2 - \frac{1}{2}\varepsilon^2\right\}\mathrm{d}V\,\mathrm{d}t. \tag{18}$$

This form of action A explicitly shows that, when the original state space is involved, the state space required to accommodate the variational principle must be enlarged by inclusion of the Lagrange multiplier ϕ as an extra variable. In fact, Eq. (18) proves that the original state space ("physical space") is lacking sufficient capacity of symmetry, consistent with Vainberg's theorem [4]. On the other hand, as Eq. (17) shows, the adjoint space of potentials $(\boldsymbol{\psi}, \phi)$, while also four-dimensional as space $(\mathbf{q}, \rho_e)$, can accommodate the variational formulation. Why is this so? Because the representation equations do adjust themselves to the extremum requirement of A at given constraints, whereas the *given* constraints without controls cannot exhibit any flexibility.

6. Inhomogeneous waves for variational adjoints

Now we return to the process described by Eqs. (1) and (2) that includes the sourceless balance for the internal energy. Still another form of heat equations is interesting. They contain both the potentials and original state variables. While we have obtained Eq. (11) or (12) as the adjoint representations of the Cattaneo model (1) and (2), a more insightful form is found after one starts with separating the term linear in $\mathbf{q}$ in Eq. (11)

$$-\frac{\partial}{\partial t}\left\{\frac{1}{c_0^2}\left(\frac{\partial\boldsymbol{\psi}}{\partial t} - \frac{\boldsymbol{\psi}}{\tau} + c_0^2\nabla\phi\right)\right\} + \nabla\left(\nabla\cdot\boldsymbol{\psi} + \frac{\partial\phi}{\partial t}\right) = \frac{\mathbf{q}}{\tau c_0^2}. \tag{19}$$

Then the ϕ terms of the left-hand side reduce and we are left with

$$-\frac{\partial^2 \boldsymbol{\psi}}{c_0^2\,\partial t^2} + \frac{\partial \boldsymbol{\psi}}{\tau c_0^2\,\partial t} + \nabla^2 \boldsymbol{\psi} = \frac{\mathbf{q}}{\tau c_0^2}. \tag{20}$$

Thus, for the Cattaneo equation (1), and in terms of the scaled vector potential $\boldsymbol{\Psi}$ such that $\boldsymbol{\Psi} = \boldsymbol{\psi}\tau c_0^2$ the above equation takes the form

$$\nabla^2 \boldsymbol{\Psi} - \frac{\partial^2 \boldsymbol{\Psi}}{c_0^2\,\partial t^2} + \frac{\partial \boldsymbol{\Psi}}{\tau c_0^2\,\partial t} = \mathbf{q}. \tag{21}$$

For the energy-conservation equation (2) we obtain by simplification of Eq. (12)

$$\frac{\partial^2 \phi}{\partial t^2} + \frac{\nabla \cdot \boldsymbol{\psi}}{\tau} - c_0^2 \nabla^2 \phi = 0. \tag{22}$$

Multiplying this equation by τ and eliminating $\boldsymbol{\psi}$ with the help of the energy-density representation, Eq. (9), yields the following equation for the scalar potential

$$\tau\frac{\partial^2 \phi}{\partial t^2} - \frac{\partial \phi}{\partial t} - \tau c_0^2 \nabla^2 \phi = \rho_e. \tag{23}$$

In terms of the modified scalar potential Φ such that $\Phi = -\phi\tau c_0^2$ (note the minus sign in this definition) the above equation takes the form

$$\nabla^2 \Phi - \frac{\partial^2 \Phi}{c_0^2\,\partial t^2} + \frac{\partial \Phi}{\tau c_0^2\,\partial t} = \rho_e. \tag{24}$$

Along with Eq. (21) for the vector potential, we have thus found the set of four inhomogeneous equations describing dissipative heat transfer in terms of the potentials of the thermal field, Φ and $\boldsymbol{\Psi}$. They show that the heat flux $\mathbf{q}$ and energy density ρ_e are sources of a thermal field that satisfies the damped-wave equations for the potentials Φ and $\boldsymbol{\Psi}$. The problem of thermal energy transfer is thus broken down to the problem of the related potentials. This is a situation similar to that in electromagnetic theory [28] or in gravitation theory [29], where the specification of sources (electric four-current or matter tensor, respectively) defines the behavior of the field potentials. In fact, some equations of heat transfer in terms of these potentials are analogous to inhomogeneous equations for potentials of the electromagnetic field, yet these analogies are formal only.

7. Telegraphers equations

The inhomogeneous differential equations for potentials Φ and $\boldsymbol{\Psi}$ may be contrasted with the homogeneous equations for the state variables, ρ_e and $\mathbf{q}$, which follow from Eqs. (1) and (2). By taking divergence of Eq. (1) and using Eq. (2) we find the telegraphers equation

$$\nabla^2 \rho_e - \frac{\partial^2 \rho_e}{c_0^2 \partial t^2} - \frac{\partial \rho_e}{\tau c_0^2\,\partial t} = 0, \tag{25}$$

or its equivalent form that is the damped-wave equation for the temperature

$$\nabla^2 T - \frac{\partial^2 T}{c_0^2\,\partial t^2} - \frac{\partial T}{\tau c_0^2\,\partial t} = 0. \tag{26}$$

After taking the partial derivative of Eq. (1) with respect to time we conclude with the help of Eq. (2) that the heat flux density $\mathbf{q}$ satisfies the equation

$$\frac{\partial^2 \mathbf{q}}{c_0^2\,\partial t^2} + \frac{\partial \mathbf{q}}{\tau c_0^2\,\partial t} - \nabla(\nabla\cdot\mathbf{q}) = 0. \tag{27}$$

Using the well-known vector identity

$$\nabla(\nabla\cdot\mathbf{q}) = \nabla^2\mathbf{q} + \nabla\times(\nabla\times\mathbf{q}), \tag{28}$$

it follows that heat flux $\mathbf{q}$ satisfies a partial differential equation of the type of telegraphers equations (25) and (26) whenever it is irrotational, i.e. if $\nabla\times\mathbf{q} = 0$. In fact, the Cattaneo equation ensures the vanishing rotation for all future times when the rotation of the initial field $\mathbf{q}(\mathbf{x}, 0)$ vanishes. This observation follows from the vorticity form of Eq. (1)

$$\frac{\partial\nabla\times\mathbf{q}(\mathbf{x}, t)}{\partial t} = -\frac{\nabla\times\mathbf{q}(\mathbf{x}, t)}{\tau}. \tag{29}$$

Thus, even if the initial vorticity field is finite its effect will decay soon because it will relax to zero in accordance with Eq. (29). This means that the telegraphers form of the heat-flux equation

$$\frac{\partial^2 \mathbf{q}}{c_0^2\,\partial t^2} + \frac{\partial \mathbf{q}}{\tau c_0^2\,\partial t} - \nabla^2\mathbf{q} = 0 \tag{30}$$

is most frequently sufficient for practical purposes. Whenever the role of the initial vorticity condition may not be ignorable (short times and fast transients) theoretical tools are available that allow either to preserve or to eliminate vorticity effects. Tools to take into account a finite vorticity are known in the literature of variational hydrodynamics of adiabatic fluid in the form of the so-called Lin's constraints [14] that are built into action functionals to describe the identity of fluid elements along their Lagrangian trajectories. Further information about the vorticity properties of the system can be found in Ref. [1].

We also stress the difference in sign of linear or "dissipative" terms of the equations for the original physical fields (in constraints) in comparison with equations for the potentials. While the original fields are damped due to the (positive) dissipation, the potentials are simultaneously amplified due to a "negative dissipation". This shows how the variational principle works in the realm of nonself-adjoint operators.

8. Special case of a reversible process

Representations involving the single (vector) potential $\boldsymbol{\psi}$ can be considered, as they are still quite general. They are called the truncated representations. However, the truncated representations are invalid in the case of irreversible processes in which case they violate the energy conservation. Nonetheless they include the well-known Biot's representations, $\mathbf{q} = \partial\boldsymbol{\psi}/\partial t$ and $\rho_e = \nabla\cdot\boldsymbol{\psi}$, which are the simplest gradient representations of the process [30]. In fact, within the exact variational framework, Biot's representations should be restricted to reversible processes. They correspond to the truncated Cattaneo equation (1) without the irreversible $\mathbf{q}$ term and with a collisionless limit of action (10) when field ϕ vanishes and only field $\boldsymbol{\psi}$ is essential. An irreversible process constitutes a general case in which both potentials (vector potential $\boldsymbol{\psi}$ and scalar ϕ) are necessary.

The limiting reversible action takes, in the adjoint space, a simple form

$$A = \int_{t_1,V}^{t_2} \varepsilon^{-1}\left\{\frac{1}{2c_0^2}\left(\frac{\partial\boldsymbol{\psi}}{\partial t}\right)^2 - \frac{1}{2}(\nabla\cdot\boldsymbol{\psi})^2 - \frac{1}{2}\varepsilon^2\right\}\mathrm{d}V\,\mathrm{d}t. \tag{31}$$

With the simplest (Biot's) representations energy conservation is satisfied identically. Functional (31) then refers to undamped thermal waves propagating with the speed c_0 that satisfy d'Alembert's equation for the energy density ρ_e or temperature T.

The reversible process is a suitable limiting framework to discuss the advantages and disadvantages resulting from the choice of a definite kinetic potential L. With the choice of L as in Eq. (7), abandoning the energy-conservation constraint in the action (or the formal substitution $\phi = 0$ in A) is allowed. However, for different Ls the omission of adjoining the energy constraint would not be admissible. For example, a change of the sign of ρ_e^2 in Eq. (7) would result in representations violating energy conservation even in the reversible case, should the Cattaneo equation be taken as the only adjoined constraints. This substantiates the choice of the Hamiltonian structure of the kinetic potential as the most economical one. Yet, as already stated, there is considerable flexibility in choosing the kinetic potential when *all* process constraints are adjoined.

9. Action and extremum conditions in energy representation (variables j_s and ρ_s)

Now we shall move to the energy representation, where the constraining set includes Kaliski's form of the Cattaneo equation (4) and the entropy balance with a source

$$\frac{\partial\rho_s}{\partial t} + \nabla\cdot\mathbf{j}_s = a\mathbf{j}_s^2. \tag{32}$$

The coefficient a is a positive constant, equal to the reciprocal of thermal conductivity k. The form (32) is valid in both classical and extended thermodynamics. In the energy representation there is no need to restrict ourselves to the special quadratic form of Eq. (7). In fact, quite diverse nonlinear expressions of Hamilton's type describing the difference between kinetic and internal energies can be applied. Assuming as before a resting medium and absence of external fields, we shall use the kinetic potential L in the form

$$L \equiv \frac{\mathbf{j}_s^2}{2c_s^2} - \rho_e(\rho_s, \rho), \tag{33}$$

where a constant mass density of the medium is $\rho = \rho_0$. For a resting medium with a constant density ρ_0 the composite action A assumes the form

$$A = \int_{t_1,V}^{t_2} \left\{ \frac{\mathbf{j}_s^2}{2c_s^2} - \rho_e(\rho_s, \rho_0) + \eta\left(\frac{\partial \rho_s}{\partial t} - a\mathbf{j}_s^2 + \nabla \cdot \mathbf{j}_s \right) + \boldsymbol{\psi} \cdot \left(\frac{\partial \mathbf{j}_s}{c_0^2 \partial t} + \frac{\mathbf{j}_s}{c_0^2 \tau} + \nabla \rho_s \right) \right\} \mathrm{d}V\, \mathrm{d}t. \tag{34}$$

Again, the Lagrange multipliers, scalar η and (a new) vector $\boldsymbol{\psi}$, absorb the process constraints. Eq. (34) is a truncated form of a more general action that describes the heat and fluid flow in the case when mass density changes and a finite mass flux (represented by the convection velocity $\mathbf{u}$) is present. This general action is not considered here. The simplified form (Eq. (34)) is sufficient for our present purpose; it selects the heat transfer as the basic process of investigation.

The representations of physical variables in terms of $\boldsymbol{\psi}$, η, and ϕ follow from the stationarity conditions of A. These are

$$\delta \mathbf{j}_s: \qquad (c_s^{-2} - 2a\eta)\mathbf{j}_s = \frac{\partial \boldsymbol{\psi}}{c_0^2 \partial t} - \frac{\boldsymbol{\psi}}{c_0^2 \tau} + \nabla \eta \tag{35}$$

$$\delta \rho_s: \qquad T(\rho_s, \rho) = -\nabla \cdot \boldsymbol{\psi} - \frac{\partial \eta}{\partial t}. \tag{36}$$

From Eq. (35) we obtain a nongradient representation of the diffusive entropy flux in terms of the Lagrange multipliers

$$\mathbf{j}_s = (c_s^{-2} - 2a\eta)^{-1} \left(\frac{\partial \boldsymbol{\psi}}{c_0^2 \partial t} - \frac{\boldsymbol{\psi}}{c_0^2 \tau} + \nabla \eta \right). \tag{37}$$

The Lagrange multipliers are potentials in terms of which a variational formulation is constructed. Yet, there is no theoretical argument to assume that the extremum properties of the action applying the above representations in the kinetic potential (33) should generally be the same as those of the augmented quantity (34). The constraint term (with multipliers), while vanishing, also contributes to the extremum properties. What is possible, however, is the partial integration, which ensures that the Euler–Lagrange equations of the augmented and transformed functional are the same. For the functional (34) the partial integration yields the transformed action

$$A' = \int_{t_1,V}^{t_2} \left\{ \frac{\mathbf{j}_s^2}{2c_s^2} - \rho_e(\rho_s, \rho_0) - \rho_s \frac{\partial \eta}{\partial t} - \eta a \mathbf{j}_s^2 - \mathbf{j}_s \cdot \nabla \eta - \mathbf{j}_s \cdot \frac{\partial \boldsymbol{\psi}}{c_0^2 \partial t} + \frac{\boldsymbol{\psi} \cdot \mathbf{j}_s}{c_0^2 \tau} - \rho_s \nabla \cdot \boldsymbol{\psi} \right\} \mathrm{d}V\, \mathrm{d}t \tag{38}$$

Since the mass density is not varied, we obtain with the representations (36) and (37) a transformed action that includes the free-energy density $f_e = \rho_e - T\rho_s$

$$A' = \int_{t_1,V}^{t_2} \left\{ (2a\eta c_s^2 - 1) \frac{\mathbf{j}_s^2}{2c_s^2} - (\rho_e - T\rho_s) \right\} \mathrm{d}V\, \mathrm{d}t. \tag{39}$$

Taking into account that the case of nonvaried mass density corresponds here to the vanishing chemical potential μ, it may be shown that this quantity constitutes a particular type of the pressure action similar to that known in the perfect fluid theory. Yet the action obtained includes the Lagrange multiplier η of the entropy balance with the positive source $a\mathbf{j}_s^2$. The situation is similar to that in the process with a source of internal energy, Eq. (18). Namely, to obtain an action functional for an irreversible process of heat transfer, associated with a finite entropy source, the state space required to accommodate the variational principle must be enlarged by inclusion of the Lagrange multiplier η as an extra variable. Yet, in the adjoint space only Lagrange multipliers (potentials) and their derivatives are the arguments of the action integrand. The potential representation of action (39) has the form

$$A' = \int_{t_1,V}^{t_2} \left\{ -\frac{c_s^2}{2(1-2a\eta c_s^2)} \left(\frac{\partial \boldsymbol{\psi}}{c_0^2\,\partial t} - \frac{\boldsymbol{\psi}}{c_0^2 \tau} + \nabla \eta \right)^2 - f_e\left(-\nabla\cdot\boldsymbol{\psi} - \frac{\partial \eta}{\partial t}, \rho_0 \right) \right\} \mathrm{d}V\,\mathrm{d}t. \tag{40}$$

Its Euler–Lagrange conditions are Eqs. (4) and (32) with variables ρ_s and $\mathbf{j}_s$ expressed in terms of η and $\boldsymbol{\psi}$, Eqs. (36) and (37). Thus the variational principle is established. For Eq. (32), the entropy source $\sigma_s = -\partial L'/\partial \eta$. The total entropy production corresponding with action (40) assures the satisfaction of the Second Law of Thermodynamics

$$S_\sigma = -\frac{\partial A'}{\partial \eta} = \int_{t_1,V}^{t_2} \left\{ \frac{a c_s^4}{(1-2a\eta c_s^2)^2} \left(\frac{\partial \boldsymbol{\psi}}{c_0^2\,\partial t} - \frac{\boldsymbol{\psi}}{c_0^2\tau} + \nabla\eta \right)^2 \right\} \mathrm{d}V\,\mathrm{d}t$$
$$= \int_{t_1,V}^{t_2} a\mathbf{j}_s^2 \mathrm{d}V\,\mathrm{d}t. \tag{41}$$

10. Waves for potentials and physical variables in the energy representation

In Sections 6 and 7 we have derived damped waves from the model including the Cattaneo equation and sourceless balance for the internal energy, Eqs. (1) and (2). In a similar way, we can predict wave behavior associated with Eqs. (4) and (32) following from action (40) subject to the representations (36) and (37). However, for the entropy balance with a finite source quite complicated equations follow. We restrict ourselves here to only some special equations of interest to a physicist. An analog of Eqs. (19) and (20) of the heat theory (Sections 6 and 7) are here, respectively, equations

$$-\frac{\partial}{\partial t}\left\{ \frac{1}{1-2a\eta c_s^2} \left(\frac{\partial \boldsymbol{\psi}}{c_0^2\,\partial t} - \frac{\boldsymbol{\psi}}{c_0^2\tau} + \nabla\eta \right) \right\} + \nabla\left(\nabla\cdot\boldsymbol{\psi} + \frac{\partial\eta}{\partial t} \right) = \frac{\mathbf{j}_s}{c_s^2\tau}, \tag{42}$$

and

$$-\frac{\partial^2 \boldsymbol{\psi}}{c_0^2\,\partial t^2} + \frac{\partial \boldsymbol{\psi}}{c_0^2 \tau\,\partial t} + \nabla(\nabla\cdot\boldsymbol{\psi}) = \frac{\mathbf{j}_s}{c_s^2 \tau}. \tag{43}$$

However, the latter can be obtained form the former only under approximation of a very large thermal conductivity or a very small coefficient a. In addition, for negligible vorticity effects, a wave equation with the Laplace operator of $\boldsymbol{\psi}$ follows from Eq. (43). When the condition of a very small a is not satisfied, wave behavior is masked by diverse effects. The presence of the entropy source in the entropy-balance equation makes wave formulae complicated even in the case of small dissipation. In this case the following approximate relationship is found for the evolution of the entropy-balance multiplier, η

$$\frac{\partial^2 \eta}{c_0^2\,\partial t^2} - \frac{\partial \eta}{c_0^2 \tau\,\partial t} - \nabla^2 \eta = \frac{\rho_s}{c_s^2 \tau} - \frac{a c_s^2}{(1 - 2a\eta c_s^2)^2}\left(\frac{\partial \boldsymbol{\psi}}{c_0^2\,\partial t} - \frac{\boldsymbol{\psi}}{c_0^2 \tau} + \nabla \eta\right)^2. \tag{44}$$

Only in a limiting physical situation when the dissipation is absent (a and τ^{-1} vanish) the thermal coordinates and potentials are described by simple undamped waves.

Quite interestingly, homogeneous damped waves for state variable ρ_s follow exactly under the assumption of vanishing entropy source in Eq. (32) and the entropy-transfer kinetics (4)

$$\nabla^2 \rho_s - \frac{\partial^2 \rho_s}{c_0^2\,\partial t^2} - \frac{\partial \rho_s}{\tau c_0^2\,\partial t} = 0. \tag{45}$$

Similarly as in the heat theory (variable $\mathbf{q}$), the wave equation for the entropy flux

$$\nabla^2 \mathbf{j}_s - \frac{\partial^2 \mathbf{j}_s}{c_0^2\,\partial t^2} - \frac{\partial \mathbf{j}_s}{\tau c_0^2\,\partial t} = 0 \tag{46}$$

holds in the present (energy) picture under the extra assumption of vanishing vorticity of $\mathbf{j}_s$.

11. Energy-momentum tensor and conservation laws in the heat theory

The energy-momentum tensor is defined as

$$G^{jk} \equiv \sum_l \frac{\partial v_l}{\partial \chi^j}\left[\frac{\partial \Lambda}{\partial(\partial v_l/\partial \chi^k)}\right] - \delta^{jk}\Lambda, \tag{47}$$

where δ^{jk} is the Kronecker delta and $\boldsymbol{\chi} = (\mathbf{x}, t)$ comprises the spatial coordinates and time. The conservation laws are valid in the absence of external fields; they describe then the vanishing four-divergences $(\nabla, \partial/\partial \tau)$ of G^{jk}. Our approach here follows those of Stephens [13] and Seliger and Whitham [14], where the components of G^{jk} are calculated for Λ gauged by use of the divergence theorem along with differentiation by parts. The link of the components of tensor G^{jk} with the partial derivatives of four principal functions S_j that are solutions of Hamilton–Jacobi equations is known [31].

Any physical tensor $\boldsymbol{G} = G^{jk}$ has the following general structure

$$\mathbf{G} = \begin{bmatrix} \mathbf{T} & -\boldsymbol{\Gamma} \\ \mathbf{Q} & E \end{bmatrix}, \tag{48}$$

where **T** is the stress tensor, **Γ** is the momentum density, **Q** is the energy flux density, and E is the total energy density.

When external fields are present, the kinetic potential Λ contains explicitly some of coordinates χ^j. Then the balance equations are satisfied rather than conservation laws

$$\sum_k \left(\frac{\partial G^{jk}}{\partial \chi^k}\right) + \frac{\partial \Lambda}{\partial \chi^j} = 0, \tag{49}$$

for $j, k = 1, 2, 4$. Eq. (49) is the formulation of balance equations for momentum ($j = 1, 2, 3$) and energy ($j = 4$).

We shall focus first on the heat model considered in Sections 1–8. We recall the assumption of the small deviation from equilibrium at which that model is physically consistent. With this assumption and for the kinetic potential of Eq. (7) gauged as described above, the gauge action assures that the components of the energy-momentum tensor are multiplier independent. These components are given by Eqs. (50)–(53) below. Respectively, they describe: momentum density Γ^α, stress tensor $T^{\alpha\beta}$, total energy density E, and density of the total energy flux, Q^β, which approximately equals q^β.

The momentum density for the mass flow of the medium at rest is, of course, $\mathbf{J} = 0$, where **J** is the mass flux density. The momentum density of heat flow follows as

$$\Gamma^\alpha = -G^{\alpha 4} = c_0^{-2}\frac{\rho_e}{\varepsilon}q^\alpha \cong c_0^{-2}q^\alpha, \tag{50}$$

or, in the vector form, $\Gamma = \mathbf{q}c_0^{-2}$, whereas the stress tensor $T^{\alpha\beta}$ has the form

$$G^{\alpha\beta} = T^{\alpha\beta} = \varepsilon^{-1}\{-c_0^{-2}q^\alpha q^\beta + \delta^{\alpha\beta}(\tfrac{1}{2}\mathbf{q}^2 c_0^{-2} - \tfrac{1}{2}\rho_e^2 + \tfrac{1}{2}\varepsilon^2)\}. \tag{51}$$

This quantity represents stresses caused by the pure heat flow; it vanishes at equilibrium. The total energy density is

$$G^{44} = E^{\text{tot}} = \frac{1}{2}\varepsilon^{-1}c_0^{-2}\mathbf{q}^2 + \tfrac{1}{2}\varepsilon^{-1}\rho_e^2 + \tfrac{1}{2}\varepsilon \cong \tfrac{1}{2}\varepsilon^{-1}c_0^{-2}\mathbf{q}^2 + \rho_e. \tag{52}$$

Finally, we find for the energy flux

$$G^{4\beta} = Q^\beta = \varepsilon^{-1}\rho_e q^\beta \cong q^\beta. \tag{53}$$

In the quasiequilibrium situation ρ_e is very close to ε, then the formal density of the energy flux $G^{4\beta}$ coincides with the heat-flux density, **q**.

As the heat flux, **q**, is both the process variable and the entity resulting from the variational procedure, the fact that it is recovered here may be regarded as a positive test for the self-consistency of the procedure.

The associated conservation laws for the energy and momentum have the form

$$\frac{\partial(\tfrac{1}{2}\varepsilon^{-1}c_0^{-2}\mathbf{q}^2 + \rho_e)}{\partial t} = -\nabla\cdot(\varepsilon^{-1}\rho_e\mathbf{q}) \tag{54}$$

$$\frac{\partial(c_0^{-2}\varepsilon^{-1}\rho_e q^\alpha)}{\partial t} = \nabla\cdot\{\varepsilon^{-1}(-c_0^{-2}q^\alpha q^\beta + \delta^{\alpha\beta}(\tfrac{1}{2}\mathbf{q}^2 c_0^{-2} - \tfrac{1}{2}\rho_e^2 + \frac{1}{2}\varepsilon^2))\}. \tag{55}$$

The energy-conservation law (54), which stems from Eqs. (49), (52), and (53), refers to nonequilibrium total energy E that differs from the nonequilibrium internal energy ρ_e by the presence of the "kinetic energy of heat" (explicit in L of Eq. (7) or in Eq. (52)). The necessity of distinction between E and ρ_e is caused by the property of finite thermal momentum (50) in the frame-work of a stationary skeleton of a rigid solid, in which we work. The physical content of results stemming from the quadratic kinetic potential L thus seems acceptable when the system is close to equilibrium.

12. Entropy production and Second Law of Thermodynamics in the heat theory

In the variational heat theory the satisfaction of the Second Law is not explicit, thus we shall derive it by considering entropy properties. The entry of **G** we need to apply now is $G^{44} = E$ as it is the total energy that is both global and exact conservative property. The density of the conserved energy, E, is a basic variable in the Gibbs relation that links the entropy density ρ_s with E and the current $\mathbf{q}$. The equality

$$\tfrac{1}{2}\varepsilon^{-1}c_0^{-2}\mathbf{q}^2 + \rho_e(\rho_s) = E \tag{56}$$

shows that entropy density ρ_s is a function S of E and $\mathbf{q}$ of the following structure

$$S = \rho_s(\rho_e) = \rho_s(E - \tfrac{1}{2}\varepsilon^{-1}c_0^{-2}\mathbf{q}^2). \tag{57}$$

This means that at the constant mass density the differential of the density S satisfies an extended Gibbs equation

$$\mathrm{d}S = (\partial\rho_s/\partial\rho_e)\mathrm{d}(E - \tfrac{1}{2}\varepsilon^{-1}c_0^{-2}\mathbf{q}^2) = T^{-1}\,\mathrm{d}E - T^{-1}\varepsilon^{-1}c_0^{-2}\mathbf{q}\cdot\mathrm{d}\mathbf{q}. \tag{58}$$

Taking into account that $c_0 = (a/\tau)^{1/2} = (k/(\rho c_v\tau))^{1/2}$ where k is the thermal conductivity, one finds $c_0^{-2} = \rho c_v\tau/k = \varepsilon T^{-1}\tau k^{-1}$, and the above differential expressed in terms of k is

$$\mathrm{d}S = T^{-1}\,\mathrm{d}E - T^{-2}\tau k^{-1}\mathbf{q}\cdot\mathrm{d}\mathbf{q}. \tag{59}$$

Calculating the four-divergence of the entropy flow $(\nabla, \partial/\partial t)$ and using the global conservation law for the energy E, Eq. (54), we obtain

$$\begin{aligned}\frac{\partial S}{\partial t} + \nabla\cdot\left(\frac{\mathbf{q}}{T}\right) &= T^{-1}\left(\frac{\partial E}{\partial t} + \nabla\cdot\mathbf{q}\right) - \frac{\tau}{kT^2}\mathbf{q}\cdot\mathrm{d}\mathbf{q} + \mathbf{q}\cdot\nabla T^{-1}\\ &= \mathbf{q}\cdot\left(\nabla T^{-1} - \tau T^{-2}k^{-1}\frac{\partial\mathbf{q}}{\partial t}\right),\end{aligned} \tag{60}$$

or in terms equivalent expressions containing k or c_0

$$\frac{\partial S}{\partial t} + \nabla\cdot\left(\frac{\mathbf{q}}{T}\right) = \mathbf{q}\cdot\left(\nabla T^{-1} - \frac{1}{\varepsilon T c_0^2}\frac{\partial\mathbf{q}}{\partial t}\right) = \frac{\mathbf{q}}{kT^2}\cdot\left(-k\nabla T - \tau\frac{\partial\mathbf{q}}{\partial t}\right). \tag{61}$$

But, since Eq. (1) is a simple transformation of the original Cattaneo equation

$$\tau\frac{\partial\mathbf{q}}{\partial t} + \mathbf{q} = -\varepsilon\tau c_0^2 T^{-1}\nabla T = -k\nabla T, \tag{62}$$

we arrive at the expression

$$\frac{\partial S}{\partial t} + \nabla \cdot \left(\frac{\mathbf{q}}{T} \right) = \frac{\mathbf{q}^2}{\varepsilon \tau c_0^2 T} \equiv \frac{\mathbf{q}^2}{kT^2} = a\mathbf{j}_s^2, \tag{63}$$

where $a = k^{-1}$. This equation (or its equivalent (Eq. (32))) describes the Second Law of Thermodynamics in the identically satisfied form; it holds in both classical irreversible thermodynamics (CIT) and extended irreversible thermodynamic (EIT; [32]). Keeping in mind that Eq. (62) is as Eq. (1) the result of the variational approach, we have obtained confirmation that our approach yields the results in agreement with the Second Law of Thermodynamics. This seems to prove that the accepted kinetic potential (7) has the properties of an admissible physical entity to describe the heat flow not far from equilibrium. Yet, far from equilibrium an appropriate L may not exist in this framework.

As rightly pointed out by some authors [33,34], possessing a kinetic potential that produces only suitable variational equations is by no means sufficient to ascertain that a field theory is sufficient as a whole from the physical viewpoint. This was, in fact, the main reason to test Noether integrals, conservation laws, and entropy production stemming from the kinetic potential (7). The positive result of these tests subject to the assumption of the validity of Eqs. (1) and (2) proves that total energy density, thermal momentum, and all remaining values of the energy-momentum tensor, G^{jk}, are quantities that are physically admissible in the range of admissibility of linear Cattaneo model (1) and (2). Thus, we can accept kinetic potential (7) as the entity leading to physical results described by Eqs. (50)–(63). In fact, from an infinite variety of kinetic potentials possible in the heat-flux framework we accept the sole kinetic potential (7) to restitute both the Cattaneo equations and associated extended thermodynamic theory (Eqs. (59)–(63), in agreement with EIT [32]). In view of the admissibility of the approximation $\varepsilon^{-1}\rho_e^2/2 + \varepsilon/2 \cong \rho_e$ in Eqs. (7) and (52), the kinetic potential (7) represents – in the framework of the linear heat theory – the Hamiltonian structure of a difference between "kinetic energy of heat", $\varepsilon^{-1}q^2/2c_0^2$, and the nonequilibrium internal energy, ρ_e. Too little is known about nonlinear structures generalizing the Cattaneo equation (1) in order to experiment with proposals of a nonlinear theory based on the heat flux $\mathbf{q}$. Yet, some of our results for reversible heat flows [17] show that entropy flux $\mathbf{j}_s = \mathbf{q}/T$ may be a better variable than heat flux $\mathbf{q}$, and that the energy representation (using $\mathbf{j}_s$ instead of $\mathbf{q}$) should be more appropriate in nonlinear cases. However, this reorientation causes new difficulties, as shown in the following section.

13. Matter tensor and balance laws in the energy representation (variables j_s and ρ_s)

For the action functional (40) and general formula (47) the energy-like function follows

$$E = \frac{\partial \Lambda}{\partial(\partial \eta/\partial t)} \partial \eta/\partial t + \frac{\partial \Lambda}{\partial(\partial \boldsymbol{\psi}/\partial t)} \cdot \partial \boldsymbol{\psi}/\partial t - \Lambda$$

$$= -\rho_s\, \partial \eta/\partial t - \frac{c_s^2}{c_0^2(1 - 2a\eta c_s^2)} \left(\frac{\partial \boldsymbol{\psi}}{c_0^2\, \partial t} - \frac{\boldsymbol{\psi}}{c_0^2 \tau} + \nabla \eta \right) \cdot \partial \boldsymbol{\psi}/\partial t - \Lambda, \tag{64}$$

where Λ is the integrand of action (40). After exploiting the representation equations for physical variables and rather numerous transformations that include the elimination of gradient of η we obtain the rest-frame energy-like function

$$E = \frac{\mathbf{j}_s^2}{2c_s^2}(1 - 2a\eta c_s^2) + \rho_e(\rho_s, \rho) - \mathbf{j}_s \cdot \frac{\partial \boldsymbol{\psi}}{c_0^2\, \partial t} + \rho_s \nabla \cdot \boldsymbol{\psi}. \tag{65}$$

This expression simplifies to the canonical rest-frame energy density E_0 whenever $a = 0$ (vanishing entropy source; a superconducting medium) and $\boldsymbol{\psi} = 0$ (no constraint imposed on the kinetics of heat transfer)

$$E_0 = \frac{\mathbf{j}_s^2}{2c_s^2} + \rho_e(\rho_s, \rho). \tag{66}$$

For a resting medium, the quantity E_0 is its physical energy. It can be obtained from the kinetic potential (33) after it is expressed in terms of the velocity of entropy diffusion, $\mathbf{v}_s = \mathbf{j}_s/\rho_s$ and the Legendre transformation of L is made with respect to this velocity.

In the resting frame the total momentum-like density follows from the general formula for functional (40) in the form

$$\begin{aligned}\boldsymbol{\Gamma} &= -\frac{\partial \Lambda}{\partial(\partial \eta/\partial t)}\nabla\eta - \frac{\partial \Lambda}{\partial(\partial \boldsymbol{\psi}/\partial t)} \cdot \nabla\boldsymbol{\psi} \\ &= \frac{c_s^2}{c_0^2(1 - 2a\eta c_s^2)}\left(\frac{\partial \boldsymbol{\psi}}{c_0^2\, \partial t} - \frac{\boldsymbol{\psi}}{c_0^2 \tau} + \nabla\eta\right) \cdot \nabla\boldsymbol{\psi} = c_0^{-2}\mathbf{j}_s \cdot \nabla\boldsymbol{\psi},\end{aligned} \tag{67}$$

whereas, when kinetic constraint (4) is ignored, the canonical value of the physical momentum for a medium at rest is, of course, $\boldsymbol{\Gamma}_0 = \mathbf{J} = 0$.

Using stationarity conditions of action we eliminate time derivatives on account of the spatial derivatives in the formula for the density of energy flux. The final result is

$$\mathbf{Q} = \frac{\partial \Lambda}{\partial(\partial \eta/\partial \mathbf{x})}\partial\eta/\partial t + \frac{\partial \Lambda}{\partial(\partial \boldsymbol{\psi}/\partial \mathbf{x})} \cdot \partial\boldsymbol{\psi}/\partial t = (T + c_s^{-2}\mathbf{u}\cdot\mathbf{j}_s)\mathbf{j}_s + \mathbf{j}_s \nabla \cdot \boldsymbol{\psi} - \rho_s \frac{\cdot \partial \boldsymbol{\psi}}{\partial t}, \tag{68}$$

whereas the associated canonical density of the energy flux (for a vanishing $\boldsymbol{\psi}$)

$$\mathbf{Q}_0 = (T + c_s^{-2}\mathbf{u}\cdot\mathbf{j}_s)\mathbf{j}_s. \tag{69}$$

This quantity describes the physical density of heat-flux density as the product of a nonequilibrium temperature and the density of the entropy flux.

Finally, a formal analog of the stress tensor follows in the form

$$\begin{aligned}T^{\alpha\beta} = &-j_s^\alpha j_s^\beta c_s^{-2}(1 - 2a\eta c_s^2) - \delta^{\alpha\beta}\{-f_e + (2a\eta - c_s^{-2})\mathbf{j}_s^2/2\} \\ &+ j_s^\beta \frac{\partial \psi^\alpha}{c_0^2\, \partial t} - \delta^{\alpha\beta}\rho_s \nabla \cdot \boldsymbol{\psi},\end{aligned} \tag{70}$$

whereas the physical stress tensor associated with this model is

$$T_0^{\alpha\beta} = -j_s^\alpha j_s^\beta c_s^{-2} - \delta^{\alpha\beta}\{-f_e - c_s^{-2}\mathbf{j}_s^2/2\}. \tag{71}$$

14. Energy representation with no entropy generation

Now we assume that $a = 0$ in the above equations, meaning that the entropy-balance equation contains no source term. Simultaneously, we shall still admit a dissipation mechanism in the system caused by a finite value of the time between collisions, τ, in the frictional equation (4). This attitude is certainly far from the usual one that assumes a positive entropy generation under any circumstances. Yet, we would like to test now how the standard (canonical) conservation laws will change in the energy representation with no entropy generation. Equations describing the relations between the formal and canonical components of the matter tensor now simplify to the form

$$E = E_0 - \mathbf{j}_s \cdot \frac{\partial \boldsymbol{\psi}}{c_0^2\, \partial t} + \rho_s \nabla \cdot \boldsymbol{\psi} \tag{72}$$

$$\mathbf{Q} = \mathbf{Q}_0 + \mathbf{j}_s \nabla \cdot \boldsymbol{\psi} - \rho_s \frac{\partial \boldsymbol{\psi}}{\partial t} \tag{73}$$

$$T^{\alpha\beta} = T_0^{\alpha\beta} + j_s^\beta \frac{\partial \psi^\alpha}{c_0^2\, \partial t} - \delta^{\alpha\beta} \rho_s \nabla \cdot \boldsymbol{\psi} \tag{74}$$

$$\boldsymbol{\Gamma} = \boldsymbol{\Gamma}_0 + c_0^{-2} \mathbf{j}_s \cdot \nabla \boldsymbol{\psi}. \tag{75}$$

For brevity we restrict ourselves to the balance of the energy-like quantity E described by Eq. (72) and the associated flux, Eq. (73). In accordance with the Noether theorem, it is the energy-like function E not the energy E_0 that is conserved in the considered process. With Eqs. (72) and (73), the conservation of E and nonconservation of E_0 is contained in the formula

$$\frac{\partial E}{\partial t} + \nabla \cdot Q = \frac{\partial E_0}{\partial t} + \nabla \cdot Q_0 + \frac{\partial}{\partial t}\left(-\mathbf{j}_s \cdot \frac{\partial \boldsymbol{\psi}}{c_0^2\, \partial t} + \rho_s \nabla \cdot \boldsymbol{\psi} \right) + \nabla \cdot \left(\mathbf{j}_s \nabla \cdot \boldsymbol{\psi} - \rho_s \frac{\partial \boldsymbol{\psi}}{\partial t} \right) = 0. \tag{76}$$

Furthermore, whenever $a = 0$, Eq. (43) follows from Eq. (42) as an exact result as then the η terms in Eq. (42) cancel out. Thus we obtain by transformation of Eq. (76)

$$\frac{\partial E}{\partial t} + \nabla \cdot Q = \sigma_{E_0} - \frac{\partial \mathbf{j}_s}{\partial t} \frac{\partial \boldsymbol{\psi}}{c_0^2\, \partial t} - \mathbf{j}_s \cdot \left(\frac{\partial^2 \boldsymbol{\psi}}{c_0^2\, \partial t^2} - \nabla(\nabla \cdot \boldsymbol{\psi}) \right) + \frac{\partial \rho_s}{\partial t} \nabla \cdot \boldsymbol{\psi} + \nabla \cdot \mathbf{j}_s \nabla \cdot \boldsymbol{\psi} - \nabla \rho_s \frac{\partial \boldsymbol{\psi}}{\partial t} = 0, \tag{77}$$

where σ_{E_0} is the source of the classical energy E_0. For the sourceless entropy balance the above expression simplifies into the form

$$\frac{\partial E}{\partial t} + \nabla \cdot Q = \sigma_{E_0} - \frac{\partial \mathbf{j}_s}{\partial t} \frac{\partial \boldsymbol{\psi}}{c_0^2 \partial t} + \mathbf{j}_s \cdot \left(\nabla(\nabla \cdot \boldsymbol{\psi}) - \frac{\partial^2 \boldsymbol{\psi}}{c_0^2 \partial t^2} \right) - \nabla \rho_s \frac{\partial \boldsymbol{\psi}}{\partial t} = 0. \tag{78}$$

Using Eq. (43) in Eq. (78) we obtain

$$\frac{\partial E}{\partial t} + \nabla \cdot Q = \sigma_{E_0} - \frac{\partial \mathbf{j}_s}{\partial t} \frac{\partial \boldsymbol{\psi}}{c_0^2 \partial t} + \mathbf{j}_s \cdot \left(\frac{\mathbf{j}_s}{c_s^2 \tau} - \frac{\partial \boldsymbol{\psi}}{c_0^2 \tau \partial t} \right) - \nabla \rho_s \frac{\partial \boldsymbol{\psi}}{\partial t} = 0, \tag{79}$$

or

$$\frac{\partial E}{\partial t} + \nabla \cdot Q = \sigma_{E_0} + \frac{\mathbf{j}_s^2}{c_s^2 \tau} - \left(\frac{\partial \mathbf{j}_s}{c_0^2 \partial t} + \frac{\mathbf{j}_s}{c_0^2 \tau} + \nabla \rho_s \right) \frac{\partial \boldsymbol{\psi}}{\partial t} = 0. \tag{80}$$

Since the constraint (4) is satisfied, the final result is

$$\frac{\partial E}{\partial t} + \nabla \cdot Q = \sigma_{E_0} + \frac{\mathbf{j}_s^2}{c_s^2 \tau} = 0, \tag{81}$$

or

$$\frac{\partial E_0}{\partial t} + \nabla \cdot Q_0 \equiv \sigma_{E_0} = -\frac{\mathbf{j}_s^2}{c_s^2 \tau}. \tag{82}$$

In this process picture, the kinetic energy of heat or entropy flux is transformed in the energy of a field, that may be called the thermal field, and the conservation law holds for both the medium and the field. The field energy grows in agreement with the formula

$$\frac{\partial E_F}{\partial t} + \nabla \cdot Q_F \equiv -\sigma_{E_0} = \frac{\mathbf{j}_s^2}{c_s^2 \tau}, \tag{83}$$

so that the energy conservation refers to the sum of the energy of the medium and that of the field. In this description, the situation is similar to that found in the electromagnetic field theory where the conserved quantity is the sum of the energy of particles and the field. This approach treats dissipation as a frictional effect external with respect to the medium.

The specification of potentials leads to unique values of physical fields defined by representation equations or any functions of these quantities, such as thermal momentum $\boldsymbol{\Gamma}$ or total energy E. However, similarly as in the case of electromagnetic field, various thermal potentials and kinetic potential L can be attributed to given thermal fields. It is natural to determine the admissible class of transformed potentials that still ensure unchanged physical fields, and choose from this class potentials having the simplest formal structure or certain physical interpretation. This is connected with the gauge properties of potentials [1].

15. Potential representations of vector equations of change

Here we shall outline a formal procedure applied for a set of vector equations of change that describe fields of temperature and chemical potentials in a coupled process of heat and mass transfer. The procedure may involve nongradient representations for the considered set of fields; thus it constitutes a versatile tool to test various evolution functionals. Again, its basic principle rests on the observation that extremizing of an arbitrary criterion subject to given constraints yields automatically a set of equations for Lagrange multipliers that is adjoint with respect to the set of constraints.

A multicomponent, nonisothermal system is now considered, which is composed of components undergoing various transport phenomena in the bulk. The components are neutral [35–37] obeying the phase rule [38]. As shown by Sundheim [35] this setting leads to the independent fluxes of mass, energy, and electric current. For the alternative ionic description, see Ref. [39]. The macroscopic motion is eliminated by the choice of vanishing barycentric velocity and assumption about the constancy of the system density, ρ, consistent with the mechanical equilibrium assumption. This assumption makes the effects considered more transparent. Thus, as previously, ρ, is a constant parameter rather than the state variable. In this example we ignore the effects of finite propagation speed.

In the entropy representation, for a continuous system under mechanical equilibrium the conservation laws are

$$\frac{\partial \mathbf{C}}{\partial t} + \nabla \cdot \mathbf{J} = 0. \tag{84}$$

This is the matrix notation [40] of all conservation laws consistent when $\mathbf{J}$ is the matrix of independent fluxes

$$\mathbf{J} = (\mathbf{J}_e, \mathbf{J}_1, \mathbf{J}_2, \ldots, \mathbf{J}_{n-1}, \mathbf{i})^{\mathrm{T}}, \tag{85}$$

(the superscript 'T' means transpose of the matrix) and for the corresponding column vector of densities $\mathbf{C}$

$$\mathbf{C} = (e_v, c_1, c_2, \ldots, c_{n-1}, 0)^{\mathrm{T}}. \tag{86}$$

The n-th mass flux $\mathbf{J}_n$ has been eliminated by the condition $\sum \mathbf{J}_i \mathrm{M}_i = 0$ for $i = 1, 2, \ldots, n$. The last component of $\mathbf{C}$ vanishes due to the electroneutrality. The independent intensities are

$$\mathbf{u} = (T^{-1}, \tilde{\mu}_1 T^{-1}, \tilde{\mu}_2 T^{-1}, \ldots, \tilde{\mu}_{n-1} T^{-1}, -\phi T^{-1}), \tag{87}$$

with $\tilde{\mu}_k = \mu_n M_k M_n^{-1} - \mu_k$ Their gradients are independent forces

$$\mathbf{X} \equiv \nabla \mathbf{u} = (\nabla T^{-1}, \nabla(\tilde{\mu}_1 T^{-1}), \nabla(\tilde{\mu}_2 T^{-1}), \ldots, \nabla(\tilde{\mu}_{n-1} T^{-1}), -\nabla(\phi T^{-1}))^{\mathrm{T}}. \tag{88}$$

The phenomenological equation for coupled parabolic heat and mass transfer is

$$\mathbf{J} = \mathbf{L}\mathbf{X} \equiv \mathbf{L}\nabla \mathbf{u}. \tag{89}$$

Densities (86) and intensities (87) are the two sets of variables in the Gibbs equation for the entropy density ρ_s of the incompressible system with the mass density $\rho = \sum M_i c_i$

$$d\rho_s = \mathbf{u} \cdot d\mathbf{C}. \tag{90}$$

The second differential of the entropy involves the derivatives $h^{ik} = \partial^2 \rho_s / (\partial c_i \, \partial c_k)$ that are components of the symmetric Hessian matrix. These derivatives play a role in partial differential equations describing transfer potentials $\mathbf{u}$.

When phenomenological equations and conservation laws are combined, the result is a vector equation of change for the transfer potentials $\mathbf{u}$. Its simplest representative is the Fourier–Kirchhoff-type vector equation for pure heat transfer, which describes temperature in the energy representation or its reciprocal in the entropy representation. In the case when derivatives of state variables are small and thermodynamic and transport coefficients can be assumed as constants, the equation of change is linear. It then has the form

$$-\mathbf{a}\frac{\partial}{\partial t}\mathbf{u} + \mathbf{L}\nabla^2\mathbf{u} = 0, \tag{91}$$

where $\mathbf{a} \equiv -\partial \mathbf{C}/\partial \mathbf{u}$ is the thermodynamic capacitance matrix or the negative of the entropy Hessian $h^{ik} = \partial^2 \rho_s / (\partial c_i \, \partial c_k)$. Eq. (91) contains two symmetric matrices, $\mathbf{a}$ and $\mathbf{L}$. We shall consider a variational formulation for this equation in terms of potentials.

We shall show that in order to construct a variational principle for Eq. (91) one can minimize a functional containing any positive integrand with constraint (91) adjoined by a Lagrange multiplier. Consider, for example, a functional with a symmetric positive matrix

$$A = \int_{t_1,V}^{t_2} \left\{ \frac{1}{2} B : \mathbf{uu} + \phi \cdot \left(-\mathbf{a}\frac{\partial}{\partial t}\mathbf{u} + \mathbf{L}\nabla^2\mathbf{u} \right) \right\} dV \, dt. \tag{92}$$

The Euler–Lagrange equation of this functional with respect to $\mathbf{u}$ provides the following representation for the field vector $\mathbf{u}$

$$\mathbf{u} = -B^{-1}\left(\mathbf{a}\frac{\partial}{\partial t}\phi + \mathbf{L}\nabla^2\phi \right). \tag{93}$$

For $B = I$, the unit matrix, the representation of $\mathbf{u}$ follows in terms of potentials found earlier [9,10]. In terms of ϕ action (92) becomes

$$\begin{aligned} A = & \int_{t_1,V}^{t_2} \left\{ \frac{1}{2} B^{-1} : \left(\mathbf{a}\frac{\partial}{\partial t}\phi + \mathbf{L}\nabla^2\phi \right)\left(\mathbf{a}\frac{\partial}{\partial t}\phi + \mathbf{L}\nabla^2\phi \right) \right\} dV \, dt \\ & + \int_{t_1,V}^{t_2} \left\{ \phi \left(-\mathbf{a}B^{-1}\frac{\partial}{\partial t}\left(-\mathbf{a}\frac{\partial}{\partial t}\phi - \mathbf{L}\nabla^2\phi \right) \right. \right. \\ & \left. \left. + \mathbf{L}B^{-1}\nabla^2\left(-\mathbf{a}\frac{\partial}{\partial t}\phi - \mathbf{L}\nabla^2\phi \right) \right) \right\} dV \, dt, \end{aligned} \tag{94}$$

or since the system is linear and the constraint expression (91) must vanish as the result of the stationarity of A with respect to λ

$$A = \int_{t_1,V}^{t_2} \left\{ \frac{1}{2} B^{-1} : \left(\mathbf{a}\frac{\partial}{\partial t}\phi + \mathbf{L}\nabla^2\phi \right)\left(\mathbf{a}\frac{\partial}{\partial t}\phi + \mathbf{L}\nabla^2\phi \right) \right\} dV \, dt. \tag{95}$$

Consequently, for any nonsingular B, vanishing vector constraint (91) is produced as Euler–Lagrange equations of functional (95). Indeed, varying Eq. (95) yields

$$B^{-1}\left\{\mathbf{a}\frac{\partial}{\partial t}\left(\mathbf{a}\frac{\partial \phi}{\partial t}+\mathbf{L}\nabla^2\phi\right)-\mathbf{L}\nabla^2\left(\mathbf{a}\frac{\partial \phi}{\partial t}+\mathbf{L}\nabla^2\phi\right)\right\}=0, \tag{96}$$

which means that for any nonsingular B vector Eq. (91) is satisfied in the form

$$\mathbf{a}\frac{\partial}{\partial t}\left(\mathbf{a}\frac{\partial \phi}{\partial t}+\mathbf{L}\nabla^2\phi\right)-\mathbf{L}\nabla^2\left(\mathbf{a}\frac{\partial \phi}{\partial t}+\mathbf{L}\nabla^2\phi\right)=0. \tag{97}$$

Whenever $\mathbf{u}$ is represented by Eq. (93), the above equation is the original equation of change, Eq. (91). It describes the heat and mass transfer in terms of the potentials ϕ_k, the components of the vector $\boldsymbol{\phi}$. Thus we have shown that the variational principle for $\mathbf{u}$ of Eq. (91) is represented by the minimum of the functional

$$A=\int_{t_1,V}^{t_2}\left\{\frac{1}{2}B:\mathbf{uu}\right\}\mathrm{d}V\,\mathrm{d}t, \tag{98}$$

with $\mathbf{u}$ defined by Eq. (93). Again, this shows the flexibility in the choice of the Lagrangian. The result stating that the "representation of $\mathbf{u}$ in terms of ϕ" is needed, as in Eq. (93), sets an analogy with the well-known variational principle of the electromagnetic field, in which one uses the electromagnetic potentials ($\mathbf{A}$ and ϕ) to state a variational principle for electric and magnetics fields ($\mathbf{E}$ and $\mathbf{B}$) with their representations as the first pair of Maxwell equations [28]. The crucial role of Lagrange multipliers in constructing variational adjoints is well known in Pontryagin's maximum principle, but seems to be overlooked in the literature of field variational principles. Our results show that a large number of functionals and related variational principles can be treated by this technique.

16. Conclusions

Variational formulations based on action-type functionals differ substantially from formulations encountered in thermodynamics of Onsager and Prigogine. The method of variational potentials (applicable to various L) may provide a relation between these two types of variational settings. The theory of a limiting reversible process may serve as a basis and indicator when choosing a suitable kinetic potential. The changes caused by the irreversibility imply the necessity to adjoint to the kinetic potential both sort of equations: those describing irreversible kinetics and those representing balance or conservation laws. The unnecessity of adjoining kinetic equations seems to be valid only to the limiting reversible process, where the physical information does not decrease. Yet, in irreversible situations, more constraints may be necessary to be absorbed in the action functional. The thermodynamic irreversibility complicates the potential representations of physical fields in comparison with the representations describing the reversible evolution. The problem of thermal-energy transfer can be broken down into the problem of related potentials, as in the case of electromagnetic and gravitational fields. We have found inhomogeneous equations describing dissipative heat transfer in terms of thermal potentials. These equations show that heat flux $\mathbf{q}$ and energy density ρ_e (or the energy representation variables $\mathbf{j}_s$ and ρ_s) are sources of the field. For heat-transfer theory, these results yield a situation similar to that in electromagnetic gravitational field theories, where specification of sources (electric four-current or the matter tensor, respectively) defines the behavior of the potentials. The approach adjoining constraints to a kinetic potential by Lagrange multipliers has proven its power and usefulness for quite complicated transfer phenomena in which both reversible and irreversible effects accompany each other. Consistency of applied constraints, formal and physical, is always an important issue.

Acknowledgement

This research was supported by the grant 3 T09C 02426 from the Polish Committee of National Research (KBN).

References

[1] Sieniutycz, S. and Berry, R.S. (2002), Variational theory for thermodynamics of thermal waves, Phys. Rev. E 65, 046132:1–046132:11.
[2] Sieniutycz, S. (1994), Conservation Laws in Variational Thermodynamics, Kluwer, Dordrecht.
[3] Vazquez, F., del Rio, J.A., Gambar, K. and Markus, F. (1996), Comments on the existence of Hamiltonian principles for non-selfadjoint operators, J. Non-Equilibrium Thermodyn. 21, 357–360.
[4] Vainberg, M.M. (1964), Variational Methods for the Study of Nonlinear Operators, Holden-Day, San Francisco.
[5] Finlayson, B.A. (1972), The Method of Weighted Residuals and Variational Principles, Academic Press, New York.
[6] Grmela, M. (1985), Bracket formulation for Navier–Stokes equations, Phys. Lett. 111, 36–40.
[7] Grmela, M. and Ottinger, H.C. (1997), Dynamics and thermodynamics of complex fluids. I. Development of a general formalism, Phys. Rev. E 56, 6620–6632;see also: Ottinger, H.C. and Grmela, M. (1997), Dynamics and thermodynamics of complex fluids. II. Illustrations of a general formalism, Phys. Rev. E 56, 6633–6652.
[8] Beris, A.N. and Edwards, B.J. (1994), Thermodynamics of Flowing Systems with Internal Microstructure, Oxford University Press, Oxford.
[9] Nyiri, B. (1991), On the construction of potentials and variational principles in thermodynamics and physics, J. Non-Equilib. Thermodyn. 16, 39–55.
[10] Markus, F. and Gambar, K. (1991), A variational principle in thermodynamics, J. Non-Equilib. Thermodyn. 16, 27–31.
[11] Van, P. and Nyiri, B. (1999), Hamilton formalism and variational principle construction, Ann. Phys (Leipzig) 8, 331–354.
[12] Herivel, J.W. (1955), The derivation of the equations of motion of an ideal fluid by Hamilton's principle, Proc. Cambridge Philos. Soc. 51, 344–349.
[13] Stephens, J.J. (1967), Alternate forms of the Herrivel–Lin variational principle, Phys. Fluids 10, 76–77.
[14] Seliger, R.L. and Whitham, G.B. (1968), Variational principles in continuum mechanics, Proc. R. Soc., 302A, 1–25.
[15] Atherton, R.W. and Homsy, G.M. (1975), On the existence and formulation of variational principles for nonlinear differential equations, Stud. Appl. Math. 54, 31–80.
[16] Caviglia, G. (1988), Composite variational principles and the determination of conservation laws, J. Math. Phys. 29, 812–816.
[17] Sieniutycz, S. and Berry, R.S. (1989), Conservation laws from Hamilton's principle for nonlocal thermodynamic equilibrium fluids with heat flow, Phys. Rev. A 40, 348–361.

[18] Sieniutycz, S. and Berry, R.S. (1993), Canonical formalism, fundamental equation and generalized thermomechanics for irreversible fluids with heat transfer, Phys. Rev. E 47, 1765–1783.
[19] Sieniutycz, S. (2000), Action-type variational principles for hyperbolic and parabolic heat & mass transfer, Int. J. Appl. Thermodyn. 3(2), 73–81.
[20] Cattaneo, C. (1958), Sur une forme d l'equation eliminant le paradoxe d'une propagation instantance, C. R. Hebd. Seanc. Acad. Sci. 247, 431–433.
[21] Vernotte, P. (1958), Les paradoxes de la theorie continue de l'equation de la chaleur, C. R. Hebd. Seanc. Acad. Sci. 246, 3154–3155.
[22] Chester, M. (1963), Second sound in solids, Phys. Rev. 131, 2013–2115.
[23] Kaliski, S. (1965), Wave equation of heat conduction, Bull. Acad. Pol. Sci. Ser. Sci. Technol. 13, 211–219.
[24] Baumeister, K.J. and Hamil, T.D. (1968), Hyperbolic heat conduction equation – a solution for the semiinfinite body problem, J. Heat Transfer Trans. ASME 91, 543–548.
[25] Grad, H. (1958), Principles of the theory of gases. In Handbook der Physik, (Flugge, S., eds.), Vol. 12, Springer, Berlin.
[26] Lebon, G. (1978), Derivation of generalized Fourier and Stokes–Newton equations based on thermodynamics of irreversible processes, Bull. Acad. Soc. Belg. Cl. Sci. LXIV, 456–460.
[27] Grmela, M. and Lebon, G. (1990), Hamiltonian extended thermodynamics, J. Phys. A 23, 3341–3351.
[28] Jackson, J.D. (1975), Classical Electrodynamics, 2nd edn, Wiley, New York.
[29] Weinberg, S. (1972), Gravitation and Cosmology, Wiley, New York.
[30] Biot, M. (1970), Variational Principles in Heat Transfer, Oxford University Press, Oxford.
[31] Guler, Y. (1987), Hamilton–Jacobi theory of continuous systems, Il Nuovo Cimento 100B, 251–266.
[32] Jou, D., Casas-Vazquez, J. and Lebon, G. (1996), Extended Irrreversible Thermodynamics, 2nd edn, Springer, Berlin.
[33] Baldomir, D. and Hammond, P. (1996), Geometry of Electromagnetic Systems, Oxford University Press, Oxford.
[34] Anthony, K.H. (2001), Hamilton's action principle and thermodynamics of irreversible processes – a unifying procedure for reversible and irreversible processes, J. Non-Newtonian Fluid Mech. 96, 291–340.
[35] Sundheim, B.R. (1964), Transport properties of liquid electrolytes. Fused Salts (Sundheim, B.R., ed.), McGraw-Hill, New York, pp. 165–254.
[36] Ekman, A., Liukkonen, S. and Kontturi, K. (1978), Diffusion and electric conduction in multicomponent electrolyte systems, Electrochem. Acta 23, 243–250.
[37] Forland, K.S., Forland, T. and Ratkje, S.K. (1989), Irreversible Thermodynamics – Theory and Applications, Wiley, Chichester.
[38] Van Zeggeren, F. and Storey, S. (1970), The Computation of Chemical Equilibria, Cambridge University Press, Cambridge.
[39] Newman, J. (1973), Electrochemical Systems, Prentice Hall, Englewood Cliffs, NJ.
[40] de Groot, S.R. and Mazur, P. (1984), In Non-equilibrium Thermodynamics, Dover, New York.

Chapter 8

VARIATIONAL PRINCIPLES FOR IRREVERSIBLE HYPERBOLIC TRANSPORT

F. Vázquez

Facultad de Ciencias, Universidad Autónoma del Estado de Morelos, A. P. 396-3, 62250 Cuernavaca, Mor. Mexico

J.A. del Río and M. López de Haro

Centro de Investigación en Energía, Universidad Nacional Autónoma de México, A.P. 34, 62580 Temixco, Mor. Mexico

Abstract

Variational formulations of irreversible hyperbolic transport are presented in this chapter. Restricted variational principles as applied to extended irreversible thermodynamics are illustrated for the cases of the soil–water system and heat transport in solids. This kind of restricted variational principles leads to the time-evolution equations for the nonconserved variables as extreme conditions. In particular, as has been noted in the case of heat transport, this perspective may provide interesting generalizations of the well-known Maxwell–Cattaneo–Vernotte forms. In order to show how a Poissonian structure may be obtained, a formulation in terms of the so-called variational potentials is described and used to derive the time evolution of the fluctuations in hyperbolic transport. These fluctuations are shown to obey the Chapman–Kolmogorov equation. The case of relativistic heat transport is discussed as an example of such formulation. The hyperbolic transport is also analyzed in the framework of the path-integral approach. This latter methodology allows for the consideration of nonlinear hyperbolic transport, in contrast with what occurs in the case of the variational potentials scheme.

Keywords: restricted variational principles; Poisson structures; transport in porous media; water flow; relatavistic heat transport; path integrals

1. Introduction

In this chapter we discuss how the statistical properties of fluctuations of hyperbolic phenomena may be investigated through the use of variational principles. Hyperbolic transport equations have been shown to be a useful tool in the fields of generalized hydrodynamics [1], solid-state physics [2,3] and irreversible thermodynamics [4,5]. They describe systems beyond the domain of the linear approach of irreversible processes,

E-mail address: vazquez@servm.fc.uaem.mx (F. Vázquez); antonio@servidor.unam.mx (J.A. del Rio); malopez@servidor.unam.mx (M.L. de Haro).

specifically for a time scale of the order of the relaxation time of the system. For derivations from first principles of hyperbolic transport equations, see for example Refs. [6–8]. The study of irreversible processes based on the fluctuations of the thermodynamic properties in a mesoscopic level dates back to Onsager and Machlup [9] who established the connection between these two levels of description for aged systems. These systems were described in terms of a set of extensive properties with the fluxes taken as the time derivatives of them. The formulation led Onsager and Machlup to variational expressions for the transition probability between states whose extremum value is the corresponding maximum probability for the average thermodynamic path of the system. They also derived an expression for the probability for one state that coincides with Einstein's formula for the probability of thermodynamic equilibrium states based on Boltzmann's relation. The statistical properties of the stochastic process associated with fluctuations were completely specified in this way. The scheme introduced by Onsager and Machlup gave as a result the expressions of the transition probabilities in terms of an action functional for the system with extremum properties while the system is changing through thermodynamic states near equilibrium. Grabert and Green [10] extended the formalism to the case of transport coefficients depending on the extensive thermodynamic properties by using a variational principle for the phenomenological parabolic equations. These authors showed that fluctuations in nonlinear systems constitute a Markov process when the phenomenological coefficients depend on the thermodynamic state. Furthermore, their work provided a consistent stochastic interpretation of the terms appearing in the extremum conditions of the variational principle and made it possible to discuss some relevant statistical properties of fluctuations. Grabert and Green also showed that fluctuations in nonlinear systems constitute a Markov process when the phenomenological coefficients depend on the thermodynamic state. Independently, Graham [11] worked out the same irreversible processes as those studied by Onsager and Machlup by making use of the method of path integrals in a more profound mathematical fashion. The work of Grabert and Green represents one of the theories that resemble Hamilton's principle by requiring that a definite time integral of a Lagrangian functional be stationary. Later, Nettleton [12] showed that the Lagrangian form of the phenomenological equations of nonequilibrium thermodynamics by Landau and Lifshitz could be extended to the general nonlinear case. Sieniutycz and Berry based a classical formalism for heat-conducting fluids on a kinetic potential and by using an action principle generalized the Hamilton scheme to extended thermodynamic spaces [13]. In Ref. [14] they discussed problems on stability and fluctuations around equilibrium in heat-conducting fluids in the context of differential geometry by using the so-called thermodynamic copotentials.

Our own work on these topics was guided by a basic question posed by several authors many years ago as synthesized in the following remark (taken from Ref. [15]): "A more ambitious question is whether there is any thermodynamic function which for dissipative systems would at least in some sense play the role of the Lagrangian function in mechanics" as well as by the purpose of applying such a thermodynamic function to the characterization of the statistical properties of dissipative systems. We began in 1990 by studying restricted variational principles that were soon shown to be not suitable for that purpose. They proved, nonetheless, to be useful to obtain generalized equations of state. Later, we had success in constructing a classical variational principle for relaxation phenomena in hyperbolic systems through the extension of variational techniques to nonself-adjoint problems. With this formulation we were able to deal with the fluctuational properties of hyperbolic systems. Nevertheless, the approach turned out to be limited to linear cases. More recently, we have employed path-integral techniques to study statistical properties of the fluctuating properties, for which a variational criterion on the most probable thermodynamic path to be followed by the system may be associated. These techniques can be used even for nonlinear problems for which, for instance, the transport coefficients depend on the dynamic properties.

We then begin this review by considering the variational approach to hyperbolic phenomena on the basis of the restricted principle described in Refs. [16–20]. First considered by Onsager in his minimum energy-dissipation principle, this kind of principle has been useful in irreversible thermodynamics providing additional physical requirements to be fulfilled by phenomenological models. We include the case of the transport of water through a porous medium as an example to illustrate this kind of principle. The construction of generalized equations of state is described. It is well known that a second-order stochastic process, as hyperbolic transport is, may be reduced to a first-order one with a greater number of variables. This is done through the introduction of the so-called variational potentials into the formalism [21–24]. This allows us to find a Hamiltonian variational principle for the hyperbolic dynamics and an associated Poisson structure that is used to investigate the statistical properties of hyperbolic processes [25]. In an indicative way we can say that the existence of the Poisson structure for the hyperbolic transport equations has advantages in this task. We mention here only two of them. First, the action functional could be used to construct directly the transition probability among states. So, the statistical properties of the process are derived instead of assumed. Secondly, the Hamiltonian function permits us to write the time-evolution equation of the one-time probability density in phase space through the general time-evolution equation. We include the analysis of the variational potentials scheme under the scope of one [26] and two generators [27] approaches to irreversible thermodynamics that are bracket-formulated theories on irreversible processes. Finally we describe a path-integral approach to hyperbolic phenomena. Within the variational-potentials approach we find an expression for the transition probability for states separated by a finite time that is based on the Poisson structure of the dynamic equations. As an example of the canonical formalism we obtain an expression for the transition probability density of the fluctuations in a relativistic heat-conduction problem. We conclude with some additional comments and remarks.

2. Restricted variational principles and EIT

Restricted variational principles for irreversible processes are useful for calculation and they provide physical insight into various phenomena. One of the first of them was proposed by Onsager that is known as the principle of the least dissipation of energy. For a review and generalizations see Ref. [28].

In this section the variational approach to hyperbolic phenomena on the basis of restricted principles is described. The first to propose a variational formulation of this kind for heat transfer in processes near equilibrium were Onsager and Machlup [9]. This allowed them to describe the behavior of fluctuations near equilibrium giving the first statistically supported formalism on this topic. Much later, Gyarmati [29] introduced the concept of dissipative potential to deal with hyperbolic transport equations.

The variational formulation in the restricted form of extended irreversible thermodynamics (EIT) began with our own 1990 paper [16]. We describe in this section that work and show how it is possible to consider nonanalytic expressions for generalized equations of state. We also show one of the main results of restricted variational principles, namely, the derivation of generalized time-evolution equations for the dissipative fluxes [21–24]. These equations are then not only thermodynamically supported but they are also consistent with the variational approach.

We begin with the axiomatic form of EITs within the framework of a restricted variational principle. The existence of a nonequilibrium entropy potential S_{ne} [30] satisfying a balance equation is assumed. This function must be continuous and differentiable and depends on the thermodynamic variables space that is enlarged with nonconserved variables. So, the system is described with the usual balance equations for the conserved densities and the time-evolution equations for the nonconserved variables. Consider now the functional L given by

$$L = \int_V \left[\rho \frac{\mathrm{d}S_{ne}}{\mathrm{d}t} + \nabla \cdot \mathbf{J}_S - \sigma_S \right] \mathrm{d}V, \tag{1}$$

where $\mathrm{d}/\mathrm{d}t$ is the material derivative and $\mathbf{J}_S$ and σ_S are the generalized flux and the source terms associated with S_{ne}, respectively. The axiomatic form of the variational version of EIT may be stated in terms of the existence of the thermodynamic potential S_{ne} and a variational principle of the restricted type

$$\delta L = 0. \tag{2}$$

The variation δ is taken only over the nonconserved part of the thermodynamic variables space, while the conserved part and the tangent thermodynamic space remain constant. The balance equations of the conserved properties of the system act as subsidiary conditions of Eq. (2) and the other quantities in this same equation are generated as the most general scalars, vectors, tensors within the extended thermodynamic space. Eq. (2) gives, as an extreme condition, the time-evolution equations for the fast variables closing the set of equations describing the system. We now illustrate this framework with the case of the isothermal transport of a fluid through a porous medium [17]. The thermodynamic state of the water–soil system may be specified with two variables. On the one hand, the water matric potential Ψ, on the other, the volumetric water flux density $\mathbf{J}_w$. The enlarged thermodynamic space is constructed with these properties of the system $\{\Psi, \mathbf{J}_w\}$. The nonequilibrium entropy S_{ne} depends on this space

$$S_{ne} = S_{ne}(\Psi, \mathbf{J}_w). \tag{3}$$

This functional dependence leads to a generalized Gibbs equation that is written as follows:

$$\rho \frac{\mathrm{d}S_{ne}}{\mathrm{d}t} = \alpha_1 \frac{\mathrm{d}\Psi}{\mathrm{d}t} + \boldsymbol{\alpha}_2 \cdot \frac{\mathrm{d}\mathbf{J}_w}{\mathrm{d}t}, \tag{4}$$

where the differential operator with respect to time defined as $\mathrm{d}/\mathrm{d}t = (\partial/\partial t) + \theta^{-1}\mathbf{J}_\mathrm{w}\cdot\nabla$, is the usual material derivative with $\theta(\Psi)$ the water content and

$$\alpha_1 = \rho\left(\frac{\partial S_\mathrm{ne}}{\partial \Psi}\right)_{\mathbf{J}_\mathrm{w}}, \qquad \boldsymbol{\alpha}_2 = \rho\left(\frac{\partial S_\mathrm{ne}}{\partial \mathbf{J}_\mathrm{w}}\right)_{\Psi}. \tag{5}$$

As mentioned above, the extremum condition (2) gives the time-evolution equation for the fast variable that in this case is the volumetric water flux density $\mathbf{J}_\mathrm{w}$. We show in some detail how the restricted variation works in Eq. (2). First, we express any scalar e and vector $\mathbf{V}$ as $e = e(\Psi, I)$ and $\mathbf{V} = e(\Psi, I)\mathbf{J}_\mathrm{w}$, respectively, where $I = \mathbf{J}_\mathrm{w}\cdot\mathbf{J}_\mathrm{w}$ is the only invariant of the system. As is usual in EITs [36], the $\alpha_{i0}(i = 2, 3)$ are the scalars introduced in the construction of $\boldsymbol{\alpha}_2$ and $\mathbf{J}_\mathrm{w}$ through the representation theorems

$$\alpha_1 = \alpha_1(\Psi, I), \qquad \sigma = \sigma(\Psi, I), \qquad \alpha_2 = \alpha_{20}(\Psi, I)\mathbf{J}_\mathrm{w}, \qquad \mathbf{J}_S = \alpha_{30}(\Psi, I)\mathbf{J}_\mathrm{w}. \tag{6}$$

We now introduce expressions (4) and (6) in Eq. (2). It is a direct task to calculate the restricted variation in Eq. (2) to obtain the time-evolution equation for the fast variable of the system, namely, the volumetric water flux density $\mathbf{J}_\mathrm{w}$. One just considers all the derivatives and the slow variable Ψ as constants under the derivation process. The equation reads as follows

$$\left(\left(2\frac{\partial}{\partial I}\alpha_{20}\right)\mathbf{J}_\mathrm{w}\mathbf{J}_\mathrm{w}\cdot + \alpha_{20}\mathbf{1}\right)\frac{\mathrm{d}}{\mathrm{d}t}\mathbf{J}_\mathrm{w} = -\nabla\alpha_{30} + 2\frac{\partial\sigma}{\partial I}\mathbf{J}_\mathrm{w} - \alpha_1\nabla\Psi/\theta - 2\left(\frac{\partial}{\partial I}(\alpha_{30} - \alpha_1)\right)$$

$$(\nabla\cdot\mathbf{J}_\mathrm{w})\mathbf{J}_\mathrm{w} - 2(\mathbf{J}_\mathrm{w}\cdot\nabla\Psi)\left(\frac{\partial}{\partial I}\alpha_1\right)\mathbf{J}_\mathrm{w}. \tag{7}$$

Note that in Eq. (7) the coefficient of the material derivative of $\mathbf{J}_\mathrm{w}$ is not independent of the anisotropy introduced by the flux in the system and it is a tensor of second rank. In order to exhibit the thermodynamic origin of nonlinear effects up to second order in $\mathbf{J}_\mathrm{w}$ as well as to fulfill the required compatibility with linear irreversible thermodynamics (LIT) [17] we assume that

$$\alpha_1 = -\Psi\gamma^{-1} - gI, \qquad \alpha_{20} = -\tau_\mathrm{w}\{\Psi/(\gamma\theta) + 1\}/K, \qquad \alpha_{30} = -\Psi - fI,$$
$$\sigma = I\{\Psi/(\gamma\theta) + 1\}/2K, \tag{8}$$

where τ_w is the relaxation time of the water flux and the functions $f(\Psi)$ and $g(\Psi)$ are parameters that we interpret later. With this selection we can recover Darcy's Law assuring the compatibility with LIT. By substituting Eq. (8) into Eq. (7) we get

$$-\tau_\mathrm{w}\frac{\partial}{\partial t}\mathbf{J}_\mathrm{w} = K\nabla\Psi + \mathbf{J}_\mathrm{w} + E(\mathbf{J}_\mathrm{w}\cdot\nabla\Psi)\mathbf{J}_\mathrm{w} + F(\mathbf{J}_\mathrm{w}\cdot\mathbf{J}_\mathrm{w})\nabla\Psi - G(\nabla\cdot\mathbf{J}_\mathrm{w})\mathbf{J}_\mathrm{w}$$
$$+ H\mathbf{J}_\mathrm{w}\cdot\nabla\mathbf{J}_\mathrm{w}, \tag{9}$$

where

$$E = 2gK\{\Psi/(\gamma\theta) + 1\}^{-1}, \qquad F = \left(\frac{\mathrm{d}}{\mathrm{d}t}f + g/\theta\right)K\{\Psi/(\gamma\theta) + 1\}^{-1},$$
$$H = 2fK\{\Psi/(\gamma\theta) + 1\}^{-1} + \tau_\mathrm{w}/\theta, \qquad G = 2(g\gamma + f)K\{\Psi/(\gamma\theta) + 1\}^{-1}. \tag{10}$$

Eq. (9) is a generalization of the corresponding constitutive relation in LIT, namely, Darcy's law, which reads

$$\mathbf{J}_{\mathrm{w}} = -K\nabla\Psi. \tag{11}$$

In the case of unsaturated porous media, this equation combined with an energy-balance equation gives a parabolic equation called the Richards equation [31], all this in the framework of LIT. Darcy's law and the Richards equation, like other constitutive equations in LIT, have certain limitations that have motivated heuristic corrections such as, for instance, the so-called Brinkman's and Forchheimer's corrections to incorporate viscous and nonlinear effects. Eq. (9) implies a finite velocity of transmission of water-potential perturbations in the porous medium and the third and fourth terms on the r.h.s. resemble somewhat the Forchheimer's correction. The choice made for the scalars in Eq. (8) is not unique. The selection made allows us to exhibit the Forchheimer's correction. It also involves other nonlinear terms that should be taken into account in the description of the nonlinear behavior of the transport of the fluid. In particular, a simplified version of Eq. (9) has provided an alternative description for the water flux in unsaturated porous media [32] generalizing the Richards equation. Recently, the Richards equation has been found to be inappropriate to describe the gravity-driven fingers in unsaturated porous media [33]. On the other hand, using ideas based on the generalized Richards equation it has been possible to describe the nonmonotonicity of the density in gravity fingers in unsaturated porous media [34].

The restricted variational principle summarized in this section gave us the opportunity to look for a Hamiltonian formalism able to lead to a hyperbolic transport equation. In the next section we review an effort in this direction.

3. Hyperbolic transport within the variational potential approach

Processes described by hyperbolic transport equations have the feature of having finite speed of propagation of perturbations in the physical fields. These kinds of equations have acquired a great relevance in problems of transport phenomena [7,35–37]. There have been several efforts to derive hyperbolic transport equations, and of course, to include their dissipative character [6,38–43]. The subject has been also treated within EITs [35,36, 44–48] and some authors have tried it with the relativistic approximation [24,49–52].

Within the Hamilton–Lagrange theory of parabolic transport equations [22,53–56] it is impossible to formulate the Lagrangian function with the variables of measurable field quantities. Hyperbolic transport equations include a first-order time derivative, which is not a self-adjoint operator. A way to formulate a Langrangian is to introduce a set of new variables, the so-called potential functions, in order to obtain a Hamilton–Lagrange formulation for the dynamic equations in a similar manner as, e.g. in the case of hydrodynamics where the Clebsch representation is introduced [57,58]. In Clebsch's representation the velocity field is expressed in terms of certain potential functions permitting the construction of a Hamiltonian variational principle and the reduction of the dynamics to the (first order in the time derivative) Hamilton equations. A first attempt in developing these ideas has been presented in Refs. [21,23,24] where the statistical properties of fluctuations in hyperbolic transport within a Lagrangian framework were partially studied. Another extension of the variable space based on Lagrange multipliers for the purpose of variational principles has been investigated in Ref. [61].

Here we extend the formalism presented in Ref. [23] to include all the canonical momenta implied by the presence of a second order in the time derivative in the transport equations to obtain a complete Hamilton formalism for hyperbolic phenomena. Some additional details may be seen in Ref. [25]. The canonical variable space includes, as will be seen in what follows, two potential functions and two corresponding conjugated momenta. We will show how the underlying Poisson structure for the problem permits us to map the description in time to the new space of canonical variables. Then the possibility of using the incompressible property of the phase space to construct a stochastic framework to analyze the statistical properties of hyperbolic transport is explored. An important result then is that in the new scheme the process obeys the Chapman–Kolmogorov equation for short times as shown in the next section. With the introduction of the potential functions associated to each thermodynamic property, we enlarge the thermodynamic space and through a variational principle for the new dynamic equations an equivalent description of the system is obtained [53–55,59,60]. The variational equations may be reinterpreted as the conditions that give the average path in the conjugated variables space constituted by the potential functions and the thermodynamic properties that have the role of the conjugated momenta. Given the equivalence of both descriptions, it should be possible to construct a probability field for the thermodynamic transitions and obtain the statistical properties of the fluctuations of the conjugated variables without the introduction of any stochastic term in the transport equations.

3.1. Canonical formulation for hyperbolic transport

In this section we construct the Hamiltonian form for the hyperbolic transport equations. To this end consider firstly a general action functional of physical processes where the Lagrangian L depends on the first- and second-order derivatives of a physical field η

$$S = \int_T L\left(\eta, \frac{\partial \eta}{\partial x_\mu}, \frac{\partial^2 \eta}{\partial x_\mu \partial x_\nu}\right) \mathrm{d}^4x, \tag{12}$$

where $\mu, \nu = 1, 2, 3, 4,\ x_1 = x,\ x_2 = y,\ x_3 = z,\ x_4 = t$ and $\mathrm{d}^4x = \mathrm{d}x\mathrm{d}y\mathrm{d}z\mathrm{d}t$. The total variation of this functional with varying boundaries included is given by

$$\delta_t S = \int_T \frac{\partial}{\partial x_\mu}\left(\Theta_{\mu\xi}\delta x_\xi + \pi_\mu \delta_t \eta + \lambda_{\mu\nu}\delta_t \frac{\partial \eta}{\partial x_\nu}\right)\mathrm{d}^4x, \tag{13}$$

where we have restricted ourselves to the real physical processes [22,53–56]. Here $\Theta_{\mu\xi}$ is the thermodynamic tensor (the analog of the energy–momentum tensor), π_μ is the canonically conjugated coefficient to η, and $\lambda_{\mu\nu}$ is the canonically conjugated coefficient to $\partial\eta/\partial x_\nu$. If we calculate these quantities when $\mu = \xi = \nu = 4$ we obtain

$$\Theta_{44} = L - \frac{\partial\eta}{\partial t}\frac{\partial L}{\partial\frac{\partial\eta}{\partial t}} + \frac{\partial\eta}{\partial t}\frac{\partial}{\partial t}\frac{\partial L}{\partial\frac{\partial\eta^2}{\partial t^2}} - \frac{\partial^2\eta}{\partial t^2}\frac{\partial L}{\partial\frac{\partial^2\eta}{\partial t^2}}, \tag{14}$$

$$\pi_t = \frac{\partial L}{\partial\frac{\partial\eta}{\partial t}} - \frac{\partial}{\partial t}\frac{\partial L}{\partial\frac{\partial\eta^2}{\partial t^2}}, \qquad \lambda_{tt} = \frac{\partial L}{\partial\frac{\partial^2\eta}{\partial t^2}}. \tag{15}$$

If we define the canonical momenta P_η and P_ζ as $P_\eta = \pi_t$, and $P_\zeta = \lambda_{tt}$, together with a new variable $\zeta \doteq \partial\eta/\partial t$, two pairs of conjugated variables in the second and third terms of the variation of the action, the η, P_η and ζ, P_ζ are found. With these definitions we have a canonical set of variables.

The η and ζ are called generalized coordinates and P_η and P_ζ are the generalized momenta. The pairs of the canonically conjugated quantities are (η, P_η) and (ζ, P_ζ). As is known, the $-\Theta_{44}$ element of the thermodynamic tensor is the Hamiltonian. Let us now assume that the field equation of a physical field Γ whose perturbations are transported with a finite velocity in the system is given by

$$\frac{\partial^2\Gamma}{\partial t^2} + \alpha\frac{\partial\Gamma}{\partial t} - \beta\Delta\Gamma = 0. \tag{16}$$

This equation is of the hyperbolic kind and it predicts the velocity $c = \sqrt{\beta}$ for the propagation of perturbations in Γ.

The Lagrangian of the hyperbolic dissipative transport process [21,23] is given by

$$L = \frac{1}{2}\left(\frac{\partial^2\phi}{\partial t^2}\right)^2 + \frac{1}{2}\alpha^2\left(\frac{\partial\phi}{\partial t}\right)^2 + \frac{1}{2}\beta^2(\Delta\phi)^2 - \beta\frac{\partial^2\phi}{\partial t^2}\Delta\phi, \tag{17}$$

where ϕ is the field quantity that is called a potential function, α and β are constant coefficients and Δ means the Laplacian operator. The Euler–Lagrange equation associated with the Lagrangian Eq. (17) or field equation takes the form

$$0 = \frac{\partial^4\phi}{\partial t^4} - \alpha^2\frac{\partial^2\phi}{\partial t^2} + \beta^2\Delta\Delta\phi - 2\beta\Delta\frac{\partial^2\phi}{\partial t^2}. \tag{18}$$

We now define the fundamental relationship between the potential function ϕ and the physical field Γ as follows

$$\Gamma = \frac{\partial^2\phi}{\partial t^2} - \alpha\frac{\partial\phi}{\partial t} - \beta(\Delta\phi). \tag{19}$$

If we substitute Eq. (19) into the Euler–Lagrange Eq. (18) we obtain Eq. (16) as we should. We calculate the Hamiltonian in this case from the Lagrangian (17)

$$H = \frac{1}{2}\left(\frac{\partial^2\phi}{\partial t^2}\right)^2 + \frac{1}{2}\alpha^2\left(\frac{\partial\phi}{\partial t}\right)^2 - \frac{1}{2}\beta^2(\Delta\phi)^2 - \frac{\partial\phi}{\partial t}\frac{\partial^3\phi}{\partial t^3} + \beta\frac{\partial\phi}{\partial t}\Delta\frac{\partial\phi}{\partial t}, \tag{20}$$

and explicitly the momenta are

$$P_\phi = \alpha^2 \frac{\partial \phi}{\partial t} - \frac{\partial^3 \phi}{\partial t^3} + \beta\Delta \frac{\partial \phi}{\partial t}, \tag{21}$$

$$P_\psi = \frac{\partial^2 \phi}{\partial t^2} - \beta\Delta\phi. \tag{22}$$

The Hamiltonian is also expressed in terms of the generalized coordinates, the generalized momenta and their space derivatives

$$H(\phi, \Delta\phi, \psi, P_\phi, P_\psi) = \frac{1}{2} P_\psi^2 + \beta P_\psi \Delta\phi + P_\phi \psi - \frac{1}{2}\alpha^2 \psi^2, \tag{23}$$

where, in agreement with the definition of ζ $\psi = \partial\phi/\partial t$.

Any functional F of ϕ, ψ, P_ϕ, P_ψ, and their derivatives evolves in time according to the bracket expression of F and the Hamiltonian H as

$$\frac{\partial F}{\partial t} = [F, H]. \tag{24}$$

The bracket of two functions $f(\phi, \nabla\phi, \Delta\phi, p, \nabla p, \Delta p)$ and $g(\phi, \nabla\phi, \Delta\phi, p, \nabla p, \Delta p)$ where p is the canonically conjugated to ϕ is defined as

$$[f, g] = \frac{\delta f}{\delta \phi}\frac{\delta g}{\delta p} - \frac{\delta g}{\delta \phi}\frac{\delta f}{\delta p}. \tag{25}$$

Here the functional derivatives $\delta/\delta\chi$ means $\frac{\delta}{\delta\chi} = (\partial/\partial\chi) - \nabla\cdot(\partial/\partial\nabla\chi) + \Delta(\partial/\partial\Delta\chi)$. Explicitly written in terms of the conjugated variables the time-evolution equations for the canonical variables become

$$\frac{\partial \phi}{\partial t} = \psi, \tag{26}$$

$$\frac{\partial \psi}{\partial t} = P_\psi + \beta\Delta\phi, \qquad \frac{\partial P_\phi}{\partial t} = -\beta\Delta P_\psi, \qquad \frac{\partial P_\psi}{\partial t} = -P_\phi + \alpha^2\psi. \tag{27}$$

So, because of the presence of the second-order time derivative in the Lagrangian an extra coordinate has been introduced that is canonically conjugated to an extra momentum. Furthermore, the canonical equations will now be first-order differential equations in the time t. As will be seen, this fact is essential in the search for the statistical properties of the hyperbolic transport in the new phase space worked out in this section.

3.2. The canonical hyperbolic dynamics and the one- and two-generator thermodynamics

In this section we compare the canonical dynamics as expressed by Eq. (24) with two recent schemes of irreversible phenomena. The first one is known as GENERIC [27] that is, in essence, a two-generator formulation (total energy and entropy functionals), while in the other one the dynamics is expressed in terms of one generator (the total energy functional) [26]. The question we would like to address is whether the structure of the canonical dynamics has some type of relation with the two mentioned formalisms. We first describe the general characteristics of these formalisms to make this section self-contained.

The fundamental time-evolution equation of GENERIC can be written in the form

$$\frac{\mathrm{d}\eta}{\mathrm{d}t} = L(\eta)\frac{\delta E(\eta)}{\delta\eta} + M(\eta)\frac{\delta S(\eta)}{\delta\eta}, \tag{28}$$

where η represents the set of independent state variables required for a complete description of the underlying nonequilibrium system, the real-valued functionals E and S are the total energy and entropy expressed in terms of the state variables η, and L and M are the Poisson and friction matrices (or linear operators). The two contributions to the time-evolution of η generated by the energy E and the entropy S in Eq. (28) are called the reversible and irreversible contributions to the GENERIC, respectively. $\delta/\delta\eta$ typically implies functional derivatives.

The GENERIC equation (28) is supplemented by two degeneracy requirements

$$L(\eta)\frac{\delta S(\eta)}{\delta\eta} = 0 \qquad \text{and} \qquad M(\eta)\frac{\delta E(\eta)}{\delta\eta} = 0. \tag{29}$$

The first requirement of Eq. (29) expresses the reversible nature of the L contribution to the dynamics: the functional form of the entropy is such that it cannot be affected by the operator generating the reversible dynamics. The second requirement (29) expresses the conservation of the total energy by the M contribution to the dynamics. Furthermore, it is required that the matrix L is antisymmetric, whereas M is Onsager–Casimir symmetric and semipositive–definite. Both of the complementary degeneracy requirements (29) and the symmetry properties are extremely important for formulating proper and unique L and M matrices when modeling nonequilibrium systems [27]. The Poisson bracket associated with the antisymmetric matrix L,

$$[A,B]_T = \left\langle \frac{\delta A}{\delta\eta}, L(\eta)\frac{\delta B}{\delta\eta} \right\rangle, \tag{30}$$

with $[A,B]_T = -[B,A]$, is assumed to satisfy the Jacobi identity,

$$[[A,B]_T,C]_T + [[B,C]_T,A]_T + [[C,A]_T,B]_T = 0 \tag{31}$$

for arbitrary functionals A, B, and C. The symbol [,] denotes the scalar product and the subindex T refers to the two-generator dynamics. This identity, which expresses the time-structure invariance of the reversible dynamics, is another important general property required by nonequilibrium thermodynamics [27]. The dissipative part of the dynamics may be described by the bracket

$$\{A,B\}_T = \left\langle \frac{\delta A}{\delta\eta}, M(\eta)\frac{\delta B}{\delta\eta} \right\rangle,$$

with the two additional conditions $\{A,B\}_T = \{B,A\}_T$, $\{A,A\}_T \geq 0$.

The time-evolution equation for any functional $F(\eta)$ is written in terms of the two generators E and S as follows

$$\frac{\mathrm{d}F}{\mathrm{d}t} = [F,E]_T + \{F,S\}_T.$$

The structure of the single-generator formalism is expressed by the time-evolution equation for an arbitrary functional F of the set η [26]

$$\frac{\mathrm{d}F}{\mathrm{d}t} = [F,H]_S + \{F,H\}_S. \tag{32}$$

In this equation $[,\ ]_S$ represents the Poisson bracket that describes the reversible dynamics and $\{,\ \}_S$ is the dissipation bracket describing the irreversible effects in the dynamics. The generator is the Hamiltonian function that is usually interpreted as the total energy of the system. The Poisson bracket has the same properties expressed by Eqs. (30) and (31) in the case of the two-generator theory. The dissipation bracket must lead to a positive rate of entropy production. The brackets must satisfy two extra conditions in analogy with Eq. (29) of GENERIC. First in the irreversible contribution to the dynamics the total energy should be conserved, i.e.

$$\{H,H\}_S = 0, \tag{33}$$

which is similar to the second GENERIC degeneracy condition (29). Furthermore, the reversible contribution to the rate of entropy production should vanish, i.e.

$$[S,H]_S = 0, \tag{34}$$

which is analogous to the first requirement of Eq. (29). If the dissipation and Poisson brackets are bilinear forms then the time-evolution equation (32) takes the form

$$\frac{\mathrm{d}F}{\mathrm{d}t} = \int \left(\frac{\delta F}{\delta \eta} A \frac{\delta H}{\delta \eta} + \frac{\delta F}{\delta \eta} B \frac{\delta H}{\delta \eta} \right) \mathrm{d}^3 r. \tag{35}$$

A and B are matrices that are constructed from the Poisson and the dissipation brackets, respectively. It can be proved that from Eq. (35) the time evolution of the set of dynamic variables x is given by

$$\frac{\mathrm{d}\eta}{\mathrm{d}t} = A \frac{\delta H}{\delta \eta} + B \frac{\delta H}{\delta \eta}, \tag{36}$$

showing a type of connection with Eqs. (28) and (32) of GENERIC. Mention must be made, however, that the similarity of both equations is only referred to their form.

We now mention some differences of the canonical dynamics summarized by Eq. (24) with respect to the schemes described previously. As a fact, note that the Hamiltonian, Eq. (23), does not depend on time in the canonical framework indicates that it is the time-evolution generator in the thermodynamic space as stated in the previous section. It must also be noted that the canonical dynamics is not only described by one generator, namely the Hamiltonian functional H (see Eq. (23)), but it contains all the dynamic information in one bracket given by Eq. (25). This implies that both the conserved part and the dissipative part of the dynamics are contained in the same bracket. For this reason one may think that the canonical dynamics is more closely related with the one-generator structure given by Eq. (32) than with the GENERIC structure. However, this closeness is not apparent since when one separates the dissipative terms of Eq. (24) to obtain an expression similar to Eq. (36), it is not possible to satisfy simultaneously the consistency conditions (33) and (34). The bracket structure (24) then corresponds to a Poisson-bracket structure that yields the theory towards its complete form. It remains to note that the Hamiltonian functional is a conserved quantity since $[H, H] = 0$. We remind the reader that the bracket is defined by Eq. (25), and that the consistency condition (34) is satisfied by the entropy functional. We illustrate the situation just discussed with a specific example, that of a rigid heat conductor. This will allow us to clarify some aspects by having a physical interpretation of some of the terms of the variational-potential approach. We start by expressing the internal energy, which coincides in this case with the total energy, in terms of the canonical variables

$$E = C_v\left(\frac{\partial^2\phi}{\partial t^2} - \alpha\frac{\partial\phi}{\partial t} - \beta(\Delta\phi)\right) = C_v(-\alpha\psi + P_\psi), \tag{37}$$

where we have used the fact that $T = -\alpha\psi + P_\psi$, according to Eqs (19), (26) and (27). C_v is the heat capacity. Now observe that since $\mathrm{d}E = C_v\mathrm{d}T$, then we have $\mathrm{d}S = C_v(\mathrm{d}T/T)$, S being the entropy functional. We obtain the following expressions for the gradients of the energy and the entropy

$$\frac{\delta E}{\delta x} = \begin{pmatrix} 0 \\ -\alpha C_v \\ 0 \\ C_v \end{pmatrix}, \quad \frac{\delta S}{\delta x} = \begin{pmatrix} 0 \\ \dfrac{\alpha C_v}{(-\alpha\psi + P_\psi)^2} \\ 0 \\ -\dfrac{C_v}{(-\alpha\psi + P_\psi)^2} \end{pmatrix}, \tag{38}$$

where $x = (\phi, \psi, P_\phi, P_\psi)$, is the set of canonical variables. These equations show that the gradients of energy and entropy functionals are colinear vectors avoiding, in principle, to yield the theory to the GENERIC form. The explicit expression for the entropy functional is obtained from $\mathrm{d}S = C_v\mathrm{d}T/T$, which in terms of the canonical variables becomes

$$S = C_v \ln\frac{P_\psi - \alpha\psi}{(P_\psi - \alpha\psi)_0} + S_0, \tag{39}$$

$(P_\psi - \alpha\psi)_0$ and S_0 being a reference temperature and entropy, respectively. A direct calculation shows that the condition (34) is satisfied by the entropy equation (39). We close this section by noting that the total energy of the rigid heat conductor, Eq. (37), does not coincide with the expression for the Hamiltonian functional, Eq. (23), i.e. this functional does not represent the total energy in this case as is usually assumed. Nevertheless, we remind the reader that the Hamiltonian functional is a constant of the motion.

3.3. Time evolution of the one-state probability density

We have mentioned above that the time evolution of any functional of the canonical variables $(\phi, \psi, P_\phi, P_\psi)$ will be described by Eq. (24). Particularly, this equation may be used to describe the behavior in time of the set of probability densities of the hierarchy $W_1(\phi_1, \psi_1, P_{\phi 1}, P_{\psi 1})$, $W_2(\phi_1, \psi_1, P_{\phi 1}, P_{\psi 1}; \phi_2, \psi_2, P_{\phi 2}, P_{\psi 2})$, etc. The components of the hierarchy are connected through the transition probability among states that is calculated later. In this subsection we obtain the time-evolution equation for the first member of the probability-density hierarchy $W_1(\phi_1, \psi_1, P_{\phi 1}, P_{\psi 1})$ by expressing it in terms only of ϕ_1, ψ_1. We expand previously the conjugated variables in a time Taylor's series around the state at $t = 0$. Up to first order in time we obtain

$$\phi_1' = \phi_1 + \psi_1 t, \qquad \psi_1' = \psi_1 + (P_{\psi 1} + \beta \Delta \phi_1)\tau, \tag{40}$$

$$P'_{\phi 1} = P_{\phi 1} - \beta P_{\psi 1}\tau, \qquad P'_{\psi 1} = P_{\psi 1} + (-P_{\phi 1} + \alpha^2 \psi_1)\tau, \tag{41}$$

where we have used Eqs. (26) and (27) and $\phi_1, \psi_1, P_{\phi 1}$ and $P_{\psi 1}$ are evaluated at $t = 0$ and a prime indicates that the function is evaluated at $t = \tau$. The time τ is considered a small time with respect to the relaxation time of the system α^{-1}.

Now, from the first of Eqs. (27) we have

$$P'_{\psi 1} = \frac{1}{\tau}(\psi_1' - \psi_1) - \beta \Delta \phi_1', \tag{42}$$

and from the third of Eqs. (27)

$$P'_{\phi 1} = \alpha \psi'_1 2 - \frac{\dot{\psi}'_1 - \dot{\psi}_1}{\tau} + \frac{\beta(\Delta \phi'_1 - \Delta \phi_1)}{\tau}, \tag{43}$$

where $\dot{\psi}_1 \approx (\psi'_1 - \psi_1)/\tau$ may be obtained from Eq. (40) and so on. The probability density can then be written in terms of the canonical variables ϕ and ψ only by using Eqs. (42) and (43). The time-evolution equation of $W_1(\phi_1, \psi_1)$ comes from the general equation (24) with $F = W_1$. The result is

$$\frac{\partial W_1}{\partial t} = \left(\frac{\partial \phi}{\partial t}\right)\frac{\delta W_1}{\delta \phi} + \frac{1}{2}\left(\frac{\partial^2 \phi}{\partial t^2}\right)\frac{\delta W_1}{\delta \psi}. \tag{44}$$

Note that this equation is valid only under the approximations made in Eqs. (42) and (43).

4. Path-integral formulation of hyperbolic transport

Path integrals are a tool to describe fluctuating line-like structures leading to a unified understanding of many different physical phenomena. Their fluctuations can be of quantum-mechanical, thermodynamic, or statistical origin. Path integrals offer not only an effective tool for computation but in the case of thermal fluctuations they represent the basis of an equivalent formalization of irreversible thermodynamics. In the following subsection we show a relatively simple way to formulate the hyperbolic transport of the previous sections.

4.1. The path-integral formulation of hyperbolic phenomena in the variational potential framework

Let us mention first that it may be shown that the flux in the canonically conjugated variables space is incompressible. This fact is equivalent to the Liouville theorem. The details may be seen in Ref. [56]. We concentrate here on the analysis of the statistical properties of the fluctuations through the definition of the transition probability between states characterized by the set of canonical variables obtained in the previous sections. Consider that the system has departed from the state $\{\phi, \psi, \dots\}$ toward the state $\{\phi', \psi', \dots\}$ and we want to know both the most probable path in the evolution of this probability and the statistical properties of fluctuations around the most probable path.

The transition probability will be taken as the conditional probability that the system will be in the state $\{\phi', \psi', \dots\}$ at time $t = \tau$ given that a time $t = 0$ it was in the state $\{\phi, \psi, \dots\}$, namely [25]

$$P_\tau(\{\phi', \psi'\}/\{\phi, \psi\}) = N \exp\left[-\left(\frac{1}{k}\right)A_\tau(\{\phi', \psi'\}/\{\phi, \psi\})\right]J, \tag{45}$$

where J is the Jacobian of the transformation $\{P_\psi, P_\phi\} \to \{\phi', \psi'\}$ used to reduce the description to the potential functions only, and N represents the normalization factor. The action has the form [11]

$$A_\tau(\{\phi', \psi'\}/\{\phi, \psi\}) = \int_0^\tau \mathrm{d}t L = \int_0^\tau \mathrm{d}t\left(\frac{\partial \phi}{\partial t}P_\phi + \frac{\partial^2 \phi}{\partial t^2}P_\psi - H\right),$$

which can be written as [25]

$$A_\tau(\{\phi', \psi'\}/\{\phi, \psi\}) = \int_0^\tau \mathrm{d}t\left(\frac{1}{2}P_\psi^2 + \frac{1}{2}\alpha^2\psi^2\right). \tag{46}$$

It should be noted that A_τ is the integral of a positive quadratic function on the canonical variables. However, the resulting transition probability *is not* a Gaussian conditional probability density. In these conditions the path followed by the system from the state $\{\phi, \psi\}$ to the state $\{\phi', \psi'\}$ is that which makes the integral in Eq. (46) an extremum, i.e.

$$\int_0^\tau \mathrm{d}t\left(\frac{1}{2}P_\psi^2 + \frac{1}{2}\alpha^2\psi^2\right) = \mathit{extremum}. \tag{47}$$

Even more, this last extremum must be a minimum.

The transition probability for small time intervals τ may be computed in terms of the canonical variables ϕ and ψ. In order to do this we use the previously expanded conjugated variables in a time Taylor's series around the state at $t = 0$, (40) and (41).

Explicitly, we have from Eq. (26), $\psi = (\phi' - \phi)/\tau$, and from the first of Eqs. (27),

$$P_{\psi} = \frac{1}{\tau}(\psi' - \psi) - \beta\Delta\phi, \tag{48}$$

where again ψ, $\Delta\phi$, and ϕ are evaluated at $t = 0$ and ψ' and ϕ' indicate the functions evaluated at $t = \tau$. Using Eqs. (48) and (41) in Eq. (46) we obtain

$$A_{\tau}(\psi', \psi) \approx \frac{1}{2\tau}((\psi' - \psi - \tau\beta\Delta\phi)^2 - \alpha^2(\phi' - \phi)^2).$$

Note that this last expression for the action functional depends only on variables ψ and ϕ after the transformation. We call here the space (ψ, ϕ) the reduced phase space. Here we easily identify that the leading contributions to the probability are given by

$$P_{\tau}(\{\phi', \psi'\}/\{\phi, \psi\}) \cong N \exp\left[-\left(\frac{1}{2k\tau}\right)((\psi' - \psi - \tau\beta\Delta\phi)^2 + \alpha^2(\phi' - \phi)^2)\right]J. \tag{49}$$

From Eqs. (48) and (41) we obtain

$$\frac{\partial P_{\psi}}{\partial \psi'} = \tau^{-1} \quad \text{and} \quad \frac{\partial \psi}{\partial \phi'} = \tau^{-1},$$

and therefore $J = 1/\tau^2$. The normalization factor is $N = \alpha\tau/2\pi k$. Then we can approximate the transition probability (45), as

$$P_{\tau}(\phi', \psi'/\phi, \psi) = \frac{\alpha}{2\pi k\tau}\exp\left\{-\frac{1}{2k\tau}\left[(\psi' - \psi - \tau\beta\Delta\phi)^2 + \alpha^2(\phi' - \phi)^2\right]\right\}. \tag{50}$$

With this result we obtain the transition probability density for small but finite times resulting in a Gaussian distribution. Note that according to Eq. (48) and the valuation of $\Delta\phi$ at $t' = 0$, the dependence of the transition probability with respect to the momenta P_{ϕ} and P_{ψ} has been eliminated. So, we have now $P_{\tau} = P_{\tau}(\phi', \psi'/\phi, \psi)$. Here we assume that the time interval τ satisfies the condition $\tau \ll \alpha^{-1}$, since $\alpha^{-1} = \tau_r$, the relaxation time of the system. This condition is consistent with the physical limit $\tau \ll 1$ that implies that τ is very small but finite. We also assume that all of the τs are of the same order of magnitude. A straightforward calculation shows that the Chapman–Kolmogorov equation is satisfied by the transition probability

$$\int d\psi' d\phi' P_{t''}(\phi'', \psi''/\phi', \psi')P_{t'}(\phi', \psi'/\phi, \psi) = P_{\tau}(\phi'', \psi''/\phi, \psi). \tag{51}$$

It is convenient at this point to summarize what has been our main goal in this subsection, i.e. Eq. (51). The hyperbolic transport phenomena description, initially a second-order stochastic process, has been translated into a new physical space where the transition probability obeys the Chapman–Kolmogorov equation. This of course makes the time-evolution description of the transition probability expedite through a Fokker–Planck-type equation.

As an application of this framework we analyze the relativistic heat transport. This problem was first analyzed from the Lagrangian point of view in Ref. [24]. However, here we use the canonical-field formalism developed in Ref. [25]. We start with the special relativity heat hyperbolic transport equation obtained by Sandoval-Villalbazo and García-Colín [52], namely

$$\rho_{(0)}c_v \frac{\partial \theta_{(0)}}{\partial t} = -\frac{k_r}{c^2}\frac{\partial^2 \theta_{(0)}}{\partial t^2} + k_r \Delta \theta_{(0)}, \tag{52}$$

where ρ, c_v, k_r, and θ are the mass density, specific heat capacity, thermal conductivity, and the temperature, the subscript (0) indicates that these quantities are measured in the proper system (the comoving system of the mass element). One point to be stressed is that Eq. (52) is not Lorentz invariant, because it is not covariant [22]. However, when the speed of light tends to infinity we get back to the parabolic heat conduction and at the same time the causality problems arising from the strictly hyperbolic heat conduction may be solved [52]. Then, use of Eqs. (52) and (50) allows us to arrive at

$$\begin{aligned} P_\tau(\phi', \psi'/\phi, \psi) = \frac{\rho_{(0)}c_v c^2}{2\pi k_r k \tau} \exp\Bigg\{ -\frac{1}{2k\tau}\Bigg[(\psi' - \psi - \tau c^2 \Delta\phi)^2 \\ + \left(\frac{\rho_{(0)}c_v}{k_r} c^2 \right)^2 (\phi' - \phi)^2 \Bigg] \Bigg\}, \end{aligned} \tag{53}$$

which is the transition probability density for small but finite times with the physical condition that any thermal perturbation travels through the system with a maximum speed equal to the speed of light in the medium. Eq. (53) differs from Eq. (27) in Ref. [24] because here we are considering a canonical formalism. The transition probability density (53) obeys a Chapman–Kolmogorov relationship without considering averaged values for the Laplacian term as was done in Ref. [24]. Moreover, the canonical formalism is the proper form to deal with the transition probability because it is possible to show that the phase space is incompressible [25]. The foregoing path-integral approach to hyperbolic transport has been constructed on the basis of a variational formalism. Whether one can develop a theory of stochastic processes without a variational structure remains an open question that we plan to address in the future.

5. Final comments and remarks

In conclusion of this chapter, we will now elaborate on some of the points that we have addressed. First, we would like to remark that the restricted variational formulation of the hyperbolic transport equations constitutes a theoretical framework to validate constitutive models for the dissipative properties of the system. Any constitutive equation must be contained in some way in the general conditions ensuring that the Lagrangian function is an extremum. Here we have illustrated the case of water transport in a porous medium. Further examples may be found in Refs. [18,24].

Our main results concerning the canonical formulation of hyperbolic transport may be seen in Eqs. (24) and (51). Eq. (24) with the Hamiltonian given by Eq. (23), represents the fact that hyperbolic transport equations have a Poisson structure. The time evolution in the phase space $(\phi, \psi, P_\phi, P_\psi)$ is governed by Eqs. (26) and (27). The completeness of the Poisson structure obtained through the inclusion of the additional momentum related to the second-order derivative in time in the dynamic equations should be stressed. The relevant point is that the existence of the Poisson structure for the hyperbolic equations allows the description in a phase space, in which the problem is described through a set of first order in time differential equations. We have included a comparison of the Poisson structure described in this work with other forms of dynamic formulations in terms of one and two generators and the corresponding brackets. In spite of the fact that the canonical dynamics is given through a single generator and a single bracket we have made it evident that the whole dynamics is contained in the formalism as well as the thermodynamic consistency of the scheme.

The extremum condition (47) is the necessary condition for the construction of the transition probability between thermodynamic states. This condition indicates that the probability is a maximum, when the final state is the most probable one that is described by the average transport equations. Eq. (51) is the basic property of the transition probability. This equation reflects the relevant fact that hyperbolic transport can be transformed into a process obeying the Chapman–Kolmogorov equation in the canonical variables space for short times $\tau \ll \tau_r$. From the stochastic point of view, this simplifies in a considerable manner the handling of the stochastic properties in applications, as can be seen in the analysis of the fluctuations in the relativistic heat conduction shown in the last part of section 4.1. We now outline some future work. The variational potential ϕ is related univocally to the physical field Γ through Eq. (19). Nevertheless, the relationship between ϕ and Γ does not contain a physical interpretation for the variational potential. The other potential ψ and the conjugated momenta P_ϕ and P_ψ are related in an indirect way to the physical field (see Eq. (27)). They are also lacking a physical interpretation. The relationship between the time-evolution equation for the probability W_1 (Eq. (44)) and the time-evolution Fokker–Planck equation derived from the Kramers–Moyal expansion (see [62]) should also be investigated. The thermodynamic consistency of the variational scheme must be looked into more profoundly. We have made a step in such a direction by showing that total energy and entropy are conserved by the dynamics as expressed by the Poisson bracket and the generator H.

References

[1] Rose, W. (1991), Transp. Por. Med. 6, 91.
[2] Godoy, S. (1991), J. Chem. Phys. 94, 6214.
[3] Godoy, S. and Fujita, S. (1992), J. Chem. Phys. 97, 5148.
[4] Criado-Sancho, M. and Llebot, J.E. (1993), Phys. Lett. A 177, 323; (1993), Phys. Rev. E 47, 4104; Jou, D., Casas-Vázquez, J. and Lebon, G. (1993), Extended Irreversible Thermodynamics, Springer, Berlin.
[5] García-Colín, L.S. (1988), Rev. Mex. Fís. 34, 434.
[6] Nettleton, R.E. (1995), J. Math. Phys. 36, 1825.
[7] Masoliver, J., Porrá, J.M. and Weiss, G.H. (1993), Phys. Rev. E48, 939.
[8] Vasconcellos, A.R., Luzzi, R., Jou, D. and Casas-Vázquez, J. (1994), Physica A212, 369.
[9] Onsager, L. and Machlup, S. (1953), Phys. Rev. A 91, 1505.
[10] Grabert, H. and Green, M.S. (1979), Phys. Rev. A 19, 1747; Grabert, H., Graham, R. and Green, M.S. (1980), Phys. Rev. A 21, 2136.
[11] Graham, R. (1977), Z. Phys. B 26, 281.

[12] Nettleton, R.E. (1986), J. Phys. A: Math. Gen. 19, L295.
[13] Sieniutycz, S. and Berry, R.S. (1989), Phys. Rev. A 40, 348;Sieniuticz, S. and Berry, R.S. (1993), Phys. Rev. E 47, 1765.
[14] Sieniutycz, S. and Berry, R.S. (1991), Phys. Rev. A 43, 2807.
[15] Donnelly, R.J., Herman, R., Prigogine, I., (eds.), (1967), Non-Equilibrium Thermodynamics, Variational Techniques, and Stability 2nd edn, The University of Chicago Press, Chicago.
[16] Vázquez, F. and del Río, J.A. (1990), Rev. Mex. Fís. 36, 71.
[17] del Río, J.A., López de Haro, M. and Vázquez, F. (1992), J. Non-Equilib, Thermodyn. 17, 67.
[18] Vázquez, F., del Río, J.A. and Aguirre, A.A. (1995), J. Non-equilib. Thermodyn. 20, 252.
[19] Vázquez, F. and del Río, J.A. (1993), Phys. Rev. E 47, 178.
[20] López de Haro, M., del Río, J.A., Vázquez, F. and Cuevas, S. (1993), Rev. Mex. Fís. 39, 63.
[21] Vázquez, F. and del Río, J.A. (1996), Rev. Mex. Fís. 42, 12.
[22] Vázquez, F., del Río, J.A., Gambár, K. and Márkus, F. (1996), J. Non-equilib. Thermodyn. 21, 357.
[23] Vázquez, F., del Río, J.A. and López de Haro, M. (1997), Phys. Rev. E55, 5033.
[24] Vázquez, F. and del Río, J.A. (1998), Physica A253, 290.
[25] Márkus, F., Gámbar, K., Vázquez, F. and del Río, J.A. (1999), Physica A268, 482.
[26] Edwards, B.J. (1998), J. Non-equilib. Thermodyn. 23, 301;Edwards, B.J., Beris, A.N. and Öttinger, H.C. (1998), J. Non-equilib. Thermodyn. 23, 334.
[27] Grmela, M. and Öttinger, H.C. (1997), Phys. Rev. E 56, 6620;Öttinger, H.C. and Grmela, M. (1997), Phys. Rev. E 56, 6633.
[28] Ichiyanagi, M. (1994), Phys. Rep. 243, 125.
[29] Gyarmati, I. (1977), J. Non-equilib. Thermodyn. 2, 233.
[30] García-Colín, L.S. (1988), Rev. Mex. Fís. 34, 344; del Río, J.A. and López de Haro, M. (1990), J. Non-equilib. Thermodyn. 15, 71.
[31] Richards, L.A. (1928), J. Agr. Res. 37, 719.
[32] del Río, J.A. and López de Haro, M. (1991), Water Resour. Res. 27, 2141.
[33] Eliassi, M. and Glass, R.J. (2001), Water Resour. Res. 37, 2019.
[34] Eliassi, M. and Glass, R.J. (2002), Water Resour. Res. 38, 1234; Eliassi, M. and Glass, R.J. (2003), Water Resour. Res. 39, 1167.
[35] Joseph, D.D. and Preziosi, L. (1989), Rev. Mod. Phys. 61, 41; (1990), 62, 375.
[36] Jou, D., Casas-Vázquez, J. and Lebon, G. (1996), Extended Irreversible Thermodynamics, 2nd edn, Springer, Berlin.
[37] García-Colín, L.S. and Olivares-Robles, M.A. (1995), Physica A220, 165.
[38] Goldstein, S. (1951), Quat. J. Mech. Appl. Math. IV, 129.
[39] Sancho, J.M. (1984), J. Math. Phys. 25, 354.
[40] Orsingher, E. (1986), J. Appl. Prob. 23, 385.
[41] Olivares-Robles, M.A. and García-Colín, L.S. (1994), Phys. Rev. E50, 2451.
[42] Masoliver, J. and Weiss, G.H. (1994), Phys. Rev. E49, 3852.
[43] Olivares-Robles, M.A. and García-Colín, L.S. (1996), J. Non-equilib. Thermodyn. 21, 361.
[44] Lebon, G., Jou, D. and Casas-Vázquez, J. (1980), J. Phys. A 13, 275; Jou, D., Casas-Vázquez, J. and Lebon, G. (1988), Rep. Prog. Phys. 51, 1105.
[45] Robles-Domínguez, J.A., Silva, B. and García-Colín, L.S. (1981), Physica 106A, 539; García-Colín, L.S. and Rodriguez, R.F. (1988), J. Non-Equilib. Thermodyn. 13, 81.
[46] Eu, B.C. (1992), Kinetic Theory and Irreversible Thermodynamics, Wiley, New York.
[47] Vasconcellos, A.R., Luzzi, R., Jou, D. and Casas-Vázquez, J. (1994), Physica A212, 369.
[48] Müller, I. and Ruggeri, T. (1993), Extended Thermodynamics, Springer, New York.
[49] Meixner, J. and Reik, H.G. (1959), Encyclopedia of Physics, (Flügge, S., eds.), vol. III/2, Springer, Berlin.
[50] Kranyš, M. (1966), Nuovo Cimento 42, 51.
[51] Landsberg, P.T. and Johns, K.A. (1968), Proc. R. Soc. A306, 477; Landsberg, P.T. (1970), Special Relativistic Thermodynamics—A Review. A Critical Review of Thermodynamics (Stuart, E.B., Gal-Or, B., Brainard, A.J., eds.), Mono Book, Balto; Landsberg, P.T. (1978), Thermodynamics And Statistical Mechanics, Oxford University Press, Oxford.
[52] Sandoval-Villalbazo, A. and García-Colín, L.S. (1996), Physica A234, 358.

[53] Márkus, F. and Gambár, K. (1991), J. Non-equilib. Thermodyn. 16, 27.
[54] Márkus, F. and Gambár, K. (1993), J. Non-equilib. Thermodyn. 18, 288.
[55] Gambár, K. and Márkus, F. (1994), Phys. Rev. E50, 1227.
[56] Gambár, K., Martinás, K. and Márkus, F. (1997), Phys. Rev E55, 5581.
[57] Enz, C.P. and Turski, L.A. (1979), Physica 96A, 369; Gortel, Z.W. and Turski, L.A. (1992), Phys. Rev. B45, 9389.
[58] Bedeaux, D. (1980), Fundamental Problems in Statistical Mechanics V (Cohen, E.G.D., eds.), North-Holland Publishing Company, Amsterdam, p. 313.
[59] Nyíri, B. (1991), J. Non-equilib. Thermodyn. 16, 39.
[60] Courant, R. and Hilbert, D. (1953), Methods of Mathematical Physics, Interscience, New York.
[61] Sieniutycz, S. and Stephen Berry, R. (2002), Phys. Rev. E65, 046132.
[62] Gambár, K. and Márkus, F. (2003), Physica A320, 193.

Chapter 9

A VARIATIONAL PRINCIPLE FOR TRANSPORT PROCESSES IN CONTINUOUS SYSTEMS: DERIVATION AND APPLICATION

Ernest S. Geskin

New Jersey Institute of Technology, Newark, NJ 07102, USA

Abstract

A routine procedure for derivation of variational equations representing a wide range of continuous systems is discussed. The procedure involves the use of the generalized variables and fluxes for system description. The application of this procedure is demonstrated by constructing the variational equations for both dissipative and reversible processes. A variational equation describing heat, mass, and momentum transfer in a moving, chemically reactive continuous media is constructed using the proposed routine. The Euler–Lagrange equations following from the constructed variational equation are identical to the balance equations for entropy, momentum, and mass. A Lagrangian density, relating the rate of the energy change in the system with energy dissipation, work, and entropy production, is constructed. The use of this Lagrangian is demonstrated by its application to the formation of a solid structure in the course of a eutectic solidification.

Keywords: variational principles; continuous systems; generalized variables; entropy production; eutectic solidification

1. Introduction

A general variational principle for the description of transport processes in a continuous system was developed by Onsager [1,2], Prigogine [3,4], and Gyarmati [5,6]. This principle is based on the notions of entropy production and energy dissipation determined as functions of the generalized forces and fluxes. It was successfully used for the identification of the governing principle for the dissipative processes in the complex fields with fixed boundary conditions, particularly for transport processes in a moving fluid [6]. The notion of the minimal entropy production was used by Glansdorff and Prigogine [7] to deduct a generalized evolution criterion. In the above works, a system Lagrangian was posed and then it was demonstrated that the suggested variational equation corresponds to the balance and constitutive equations describing the system in question.

Tel.: +1-973-596-3338.

E-mail address: geskin@njit.edu (E.S. Geskin).

Using the formalism developed by Gyarmati [5] the author [8–10] suggested a procedure that enables us to construct the system Lagrangian using available information about the system.

The presented study is concerned with the application of such a procedure. A variational equation that incorporates available information about the system in question is given in Section 2. The notion of the flux potentials [11,12] was developed and used to construct a system Lagrangian, containing two auxiliary unknown functions. The Euler–Lagrange equations corresponding to the constructed Lagrangian are identical to the balance and constitutive equations describing the system in question. The process description is reduced to the determination of the suggested auxiliary functions.

In some cases, for example, for systems that comply with the Onsager conditions, the determination of the auxiliary functions, and thus construction of the variational equations, is a routine task. The description of such systems is discussed in Section 3. The system Lagrangian is constructed and its application is demonstrated for heat conduction and diffusion in a solid as well as for fluid flow. A variational equation describing energy, mass, and momentum transfer in a chemically reactive rotating fluid is then suggested. The proposed Lagrangian includes process variables, their material, time, and space derivatives, and the rate of entropy production. The substitution of this Lagrangian into the Euler–Lagrange equation results in equations expressing the balance of mass, entropy, and momentum. By the use of thermodynamic identities the Lagrangian is modified and presented as a function of the kinetic and internal energies and the volumetric rates of work, energy dissipation, and entropy production. The variational equation describing the steady-state systems is constructed. It is shown that this equation is identical to the Onsager principle of minimum entropy production.

The values of the time derivatives, coefficients and, in some cases, sources, included in the Lagrangian derived in this section, are fixed. Thus, this Lagrangian represents the restricted variational principle. The restricted variational principles, where only some of the variables are allowed to vary, were used by Glansdorf and Prigogine [13–16] for system investigation. This study suggests a different approach to the utilization of the restricted variational principles.

In Section 4 the developed technique is applied to the systems where the fluxes are determined by the time derivatives of process variables. An example of such an application is presented and it is shown that the suggested technique can be used for both dissipative and reversible processes. Section 5 illustrates an application of the developed technique to the investigation of complex systems. The constructed equation is applied to the analysis of the formation of a solid structure in the course of the eutectic solidification. It is shown that the process result, in this case the lamellar space, is determined by the conditions of the extremality of the entropy production.

2. Statement of the problem

The presented study is concerned with continuous systems determined by the fields of N conjugate intensive variables y_n $(n = 1, 2, ..., N)$, flux densities I_n $(n = 1, 2, ..., N)$, sources σ_i and densities of extensive variables a_n $(n = 1, 2, ..., N)$. The system characteristics are functions of the space variables x_k ($k = 1, 2, 3$) and time $t = x_4$. It is assumed that the fluxes depend on the derivatives of extensive variables and are determined by the equations

$$I_n = I_n\left(L_{mn}, \frac{\partial y_n}{\partial x_k}\right), \qquad n = 1, 2, \ldots, N. \tag{1}$$

Here L_{mn} are constants. It is also assumed that the values of the variables are fixed at the system boundary. By analogy with thermodynamic potentials flux potentials J_n are introduced and defined as

$$J_n = y_n I_n, \qquad n = 1, 2, 3, \ldots, N. \tag{2}$$

The sum flux potential J_n^{Σ} is determined by the equation

$$J_n^{\Sigma} = \sum_n y_n I_n. \tag{3}$$

A variational equation describing a system in question is sought in the form

$$\delta \int_{x_1} \int_{x_2} \int_{x_3} \int_{x_4} L \, \mathrm{d}x_1 \, \mathrm{d}x_2 \, \mathrm{d}x_3 \, \mathrm{d}x_4 = 0, \tag{4}$$

where the Lagrangian density is

$$L = \sum_k \sum_n \left[\frac{\partial}{\partial x_k} (y_n I_n) \right] = \sum_k \sum_n \left[\left(y_n \frac{\partial I_n}{\partial x_k} \right) + \left(I_n \frac{\partial y_n}{\partial x_k} \right) \right]. \tag{5}$$

For the Lagrangian (5), the Euler–Lagrange equations

$$\frac{\partial L}{\partial y_n} - \frac{\partial}{\partial x_k} \left[\frac{\partial L}{\partial \left(\frac{\partial y_n}{\partial x_k} \right)} \right] = 0; \qquad n = 1, 2, \ldots, N, \;\; k = 1, 2, 3, 4, \tag{6}$$

yield

$$\frac{\partial}{\partial y_n} L_{nk} = \frac{\partial}{\partial y_n} \left(y_n \frac{\partial I_n}{\partial x_k} + I_n \frac{\partial y_n}{\partial x_k} \right) = \frac{\partial I_n}{\partial x_k}, \tag{7}$$

and

$$\frac{\partial}{\partial x_k} \left[\frac{\partial L}{\partial \left(\frac{\partial y_n}{\partial x_k} \right)} \right] = \frac{\partial}{\partial x_k} \left[\frac{\partial \left(y_n \frac{\partial I_n}{\partial x_k} + I_n \frac{\partial y_n}{\partial x_k} \right)}{\partial \left(\frac{\partial y_n}{\partial x_k} \right)} \right] = \frac{\partial I_n}{\partial x_k}. \tag{8}$$

It is, therefore, obvious that for the selected conditions the Euler–Lagrange equations constitute the identities. Using Gauss' theorem, Eq. (4) can be rewritten as

$$\delta \int_{x_1} \int_{x_2} \int_{x_3} \int_{x_4} L \, \mathrm{d}x_1 \, \mathrm{d}x_2 \, \mathrm{d}x_3 \, \mathrm{d}x_4 = \delta \int_A \sum_n \sum_k [(y_n I_n)] = 0. \tag{9}$$

At the fixed boundary and initial values of y_n Eq. (9) constitute identities. In order to apply Eq. (9) to a specific system it is necessary to modify the Lagrangian so that the corresponding Euler–Lagrange equations comply with the balance and constitutive equations describing the system in question. Let us rewrite Eq. (5) as

$$L = L_1 + L_2 = \sum_n \left(\varphi_n y_n + \sum_k \psi_n \frac{\partial y_n}{\partial x_k} \right), \tag{10}$$

where

$$L_1 = \sum_n (\varphi_n y_n) \quad \text{and} \quad L_2 = \sum_{n,k} \left(\psi_n \frac{\partial y_n}{\partial x_k} \right).$$

Here φ_n and ψ_n are auxiliary functions that should be determined by the use of available information about the system. Let us determine these functions by the following equations

$$\frac{\partial L'}{\partial y_n} = \sum_k \frac{\partial I_n}{\partial x_k}, \quad n = 1, 2, 3, \ldots, N, \tag{11}$$

$$\frac{\partial L''}{\partial y_n} = 0, \quad n = 1, 2, 3, \ldots, N, \tag{12}$$

$$\sum_k \frac{\partial}{\partial x_k} \left(\frac{\partial L''}{\partial \left(\frac{\partial y_n}{\partial x_k} \right)} \right) = \sum_k \frac{\partial I_n}{\partial x_k}, \quad n = 1, 2, \ldots, N, \tag{13}$$

$$\frac{\partial L'}{\partial \left(\frac{\partial y_n}{\partial x_k} \right)} = 0, \quad n = 1, 2, 3, \ldots, N. \tag{14}$$

If the right part of Eq. (11) represents the balance equations and the right part of Eq. (13) represents the constitutive equations of the system in question, then Eq. (10) constitutes the Lagrangian density of this system. For several important particular cases comparatively simple solutions of Eq. (11)–(14) can be found. Particularly, the Lagrangian density (10) can be determined for systems where Eq. (1) contains only space derivatives or only time derivates. Let us determine L for these particular cases.

3. Particular case 1: fluxes determined by the space derivatives of intensive variables

3.1. Statement of the problem

Let us discuss the system where Eq. (1) contains only the gradients of the intensive variables. The balance equation has the form

$$\nabla \cdot I_n = -\rho \frac{da_n}{dt} + \sigma_n. \tag{15}$$

Here ρ is the density and σ_n is the volumetric rate of production of a property n. Let us assume that the Onsager reciprocal relations $L_{mn} = L_{nm}$ hold and the fluxes are defined by the equations

$$I_n = -\sum_m L_{mn} \nabla y_m. \tag{16}$$

It can be shown by inspection that in this case the functions φ and ψ take the forms

$$\phi_n = \rho \frac{da_n}{dt} - \sigma_n, \tag{17}$$

and

$$\psi_n = \frac{1}{2} \sum_m L_{mn} \nabla y_m. \tag{18}$$

Then

$$L = \sum_n \left(\rho \frac{da_n}{dt} - \sigma_n \right) y_n + \frac{1}{2} \sum_{n,m} L_{mn} (\nabla y_n) \cdot (\nabla y_m), \tag{19}$$

and the corresponding Euler–Lagrange equation is

$$\rho \frac{da_n}{dt} - \sigma_n = \nabla \cdot \sum_m L_{nm} (\nabla y_m). \tag{20}$$

Eq. (20) for $n = 1, 2, 3, \ldots, N$ are the transport equations for the system in question. The Lagrangian density (19) defines a restricted variational principle, because the values of the time derivatives da_n/dt and parameters ρ and L_{mn} are fixed. Although in the above transformation the value of the sources σ_n is also fixed, in some cases it can be presented as a function of the generalized variables.

3.2. Heat conduction

Let us apply the above technique to the description of heat conduction in a solid. The system can be described as follows ($a = u$, $y = T$) :

$$I_q = -L_{qq} \nabla T, \tag{21}$$

$$\nabla \cdot I_q = -\rho \frac{\partial u}{\partial t} + \sigma_q. \tag{22}$$

Here, u is the specific internal energy, T the temperature, L_{qq} the phenomenological coefficient, σ_q the heat source and I_q the heat flux density. Under these conditions we obtain

$$\phi = \rho \frac{\partial u}{\partial T} - \sigma_q, \tag{23}$$

$$\psi = L_{qq} \nabla T, \tag{24}$$

$$L = \left(\rho \frac{\partial u}{\partial t} - \sigma_q\right) T + \frac{1}{2} L_{qq} (\nabla T)^2, \tag{25}$$

and finally

$$\rho C \frac{\partial T}{\partial t} - \sigma_q = \nabla \cdot (L_{qq} \nabla T), \tag{26}$$

where C is the specific heat.

The system can also be defined by setting $a = s$, $I = I_s$, $y = T$ and $\sigma = \sigma_s$. Then

$$I_s = -L_{ss}(\nabla T), \tag{27}$$

$$L = \left(\rho \frac{\partial s}{\partial t} - \sigma_s\right) T + \frac{1}{2} L_{ss} (\nabla T)^2, \tag{28}$$

and, finally

$$\rho \frac{\partial s}{\partial t} - \sigma_s = \nabla \cdot (L_{ss} \nabla T). \tag{29}$$

The process variable y can be conveniently defined as $y = (1/T)$. Then

$$I_s = L'_{ss}\left(\nabla \frac{1}{T}\right), \tag{30}$$

$$L = \left(\rho \frac{\partial s}{\partial t} - \sigma_s\right) \frac{1}{T} - \frac{1}{2} L_{ss} \left(\nabla \frac{1}{T}\right)^2, \tag{31}$$

and

$$\rho \frac{\partial s}{\partial t} - \sigma_s = -\nabla \cdot \left(L_{ss} \nabla \left(\frac{1}{T} \right)^2 \right). \tag{32}$$

Eqs. (26), (29) and (32) constitute different expressions of the thermal conduction in a motionless medium.

3.3. Diffusion in a multicomponent reactive fluid

As another application of the proposed technique let us derive the diffusion equation in a multicomponent media where chemical reactions are present. In this case $y_i = \mu_i$; $I_n = I_i$; $\sigma_n = \sigma_i$ at $i = 1, 2, 3, \ldots, M$. Here the μ_i are the chemical potentials and σ_i is the rate of the production of a component i in the course of chemical reactions, which is determined by the equation [17]:

$$\sigma_i = \sum_j l_{ij} \mu_j, \tag{33}$$

where $l_{ij} >$ $(i, j = 1, 2, 3, \ldots, M)$ are phenomenological coefficients expressing the chemical affinity of components i and j. The density of the diffusion flux of component

i and the balance of this component are determined by the equations

$$I_i = -L_{si}(\nabla T) - \sum_j L_{ij}(\nabla \mu_j), \tag{34}$$

and

$$\nabla \cdot I_i = -\rho \frac{\partial c_i}{\partial t} + \sigma_i. \tag{35}$$

Application of Eqs. (17)–(19) to the system in question yields

$$L = \left(\rho \frac{\partial c_i}{\partial t} - \sigma_i\right)\mu_i + L_{si}(\nabla T)(\nabla \mu_i) + \frac{1}{2}\sum_j L_{ij}(\nabla \mu_j)(\nabla \mu_i), \tag{36}$$

and the corresponding transport equation has the form

$$\rho \frac{\partial c_i}{\partial t} - \sum_j l_{ij}\mu_j = \nabla \cdot [L_{si}(\nabla T)] + \nabla \cdot \left[\sum_j L_{ji}(\nabla \mu_j)\right], \tag{37}$$

where c_i is the concentration of the component i. The value of the source in Eq. (37) is not fixed.

3.4. *Translational fluid flow*

Let us discuss now the flow of a one-component, isothermal, compressible, viscous irrotational fluid. In this case $y = \overline{V}$, $I = \bar{P}^{\mathrm{vs}}$; $\sigma = \rho \overline{F}$ and the Lagrangian density is

$$L = \rho \overline{V} \cdot \frac{\mathrm{d}\overline{V}}{\mathrm{d}t} - \rho \overline{V} \cdot \overline{F} + p \nabla \overline{V} + \frac{1}{2}\left(\eta_{\mathrm{v}} - \frac{2}{3}\eta\right)(\nabla \cdot \overline{V})^2 + \eta[(\nabla \overline{V})^{\mathrm{s}} : (\nabla \overline{V})^2], \tag{38}$$

where $\bar{V}$ is the barometric velocity, $\bar{P}^{\mathrm{vs}}$ is the symmetric part of the viscous pressure tensor, $\bar{F}$ is the specific value of the field force, η is the shear viscosity, η_{v} is the bulk viscosity. Substituting the Lagrangian density in the Euler–Lagrange equation and using $\bar{V}$ as a variable we obtain the Navier–Stokes equation:

$$\rho \frac{\mathrm{d}\overline{V}}{\mathrm{d}t} + \nabla p - \rho \overline{F} - \eta \Delta \overline{V} - \left(\frac{\eta}{3} + \eta_{\mathrm{v}}\right)\nabla \nabla \cdot \overline{V} = 0. \tag{39}$$

In the course of the variation the values of $\mathrm{d}\overline{V}/\mathrm{d}t$, $\bar{F}$, η and η_{v} are fixed.

3.5. *Rotational fluid flow*

In the case of the rotational flow $y = \bar{\omega}$, $I = \bar{P}^{\mathrm{va}}$ and the Lagrangian density is

$$L = \rho \theta \omega \frac{\mathrm{d}\omega}{\mathrm{d}t} + \frac{1}{2}\eta_{\mathrm{r}}(\nabla \times \overline{V} - 2\omega)^2, \tag{40}$$

where θ is the inertia momentum of the unit mass, ω is the angular momentum, η_{r} is the rotational viscosity defined by [5]

$$\bar{P}^{\mathrm{vs}} = -\eta_{\mathrm{r}}(\nabla \times \bar{V} - 2\omega). \tag{41}$$

Substituting L from Eq. (40) into the Euler–Lagrange equation yields

$$\rho\omega\frac{\mathrm{d}\omega}{\mathrm{d}t} - \eta_{\mathrm{r}}(\nabla \times \bar{V} - 2\omega) = 0. \tag{42}$$

In the course of the variation the values of ρ, $\mathrm{d}\omega/\mathrm{d}t$ and η_{r} were fixed. Eq. (41) determines the evolution of ω.

3.6. Transport processes in a fluid

The Lagrangians densities derived in Sections 3.2–3.5 represent specific processes in a continuous media, e.g. heat conduction or rotational flow. At the same time in most cases the change in real systems involves several simultaneous coupled transport processes. Thus, it is necessary to construct a variational principle, at least restricted, which is applicable to the description of all these processes. The Lagrangian density describing simultaneous coupled processes can be obtained by simple summation of the Lagrangian densities of specific processes given by Eqs. (25), (36), (38) and (40). This summation yields:

$$\begin{aligned} L = {} & \rho T\frac{\mathrm{d}s}{\mathrm{d}t} + \sum_i \rho\mu_i\frac{\mathrm{d}c_i}{\mathrm{d}t} + \bar{V}\cdot\frac{\mathrm{d}\bar{V}}{\mathrm{d}t} + \rho\theta\bar{\omega}\cdot\frac{\mathrm{d}\bar{\omega}}{\mathrm{d}t} - \rho\nabla\cdot\bar{V} - \rho\bar{V}\cdot\bar{F} + T\sigma_s \\ & + \frac{1}{2}\left(\eta_{\mathrm{v}} - \frac{2}{3}\eta\right)(\nabla\cdot\bar{V})^2 + \frac{1}{2}\eta_{\mathrm{r}}(\nabla X\bar{V} - 2\bar{\omega})^2 + \eta[(\nabla\bar{V})^{\mathrm{s}} : (\nabla\bar{V})^{\mathrm{s}}] \\ & + \frac{1}{2}L_{ss}(\nabla T)^2 + \sum_i L_{is}(\nabla T)\cdot(\nabla\mu_i) + \frac{1}{2}\sum_{i,j} l_{ij}\mu_i\mu_j + \sum_{i,j} L_{ij}(\nabla\mu_i)\cdot(\nabla\mu_j), \end{aligned} \tag{43}$$

where ν is the specific volume. Eq. (42) can also be derived from Eqs. (10)–(14) where the flux potential J_n is replaced by the sum flux potential J_n^{Σ} (Eq. (3)). As is shown by Geskin and Von Spakovsky [11,12] the flux potential used in this equation is the energy flux.

The Euler–Lagrange equations corresponding to the Lagrangian density expressed by Eq. (43) have the form of

$$\rho\frac{\mathrm{d}s}{\mathrm{d}t} - \sigma_s - \nabla[L_{ss}(\nabla T)] - \nabla\cdot\sum_i [L_{is}(\nabla\mu_i)] = 0, \tag{44}$$

if T is a variable;

$$\rho\frac{\mathrm{d}c_i}{\mathrm{d}t} + \sum_j L'_{ij}\mu_j - \nabla\cdot\left[\sum_j L_{ij}(\nabla\mu_j)\right] - \nabla\cdot[L_{is}(\nabla T)] = 0, \tag{45}$$

if μ is a variable;

$$\rho\frac{\mathrm{d}\bar{V}}{\mathrm{d}t} - \rho\bar{F} + \nabla p - \eta\Delta\bar{V} - \left(\frac{\eta}{3} + \eta_{\mathrm{v}}\right)\nabla(\nabla\cdot\nu) - \eta_r\nabla X(2\bar{\omega} - \nabla X\bar{V})$$
$$- \eta_{\mathrm{r}}\nabla X(2\bar{\omega} - \nabla X\bar{V}) = 0, \tag{46}$$

if $\bar{V}$ is a variable; and

$$\rho\theta\frac{\mathrm{d}\bar{\omega}}{\mathrm{d}t} - 2\eta_{\mathrm{r}}(\nabla X\bar{V} - 2\bar{\omega}) = 0, \tag{47}$$

if $\bar{\omega}$ is a variable. Eqs. (44)–(47) are, of course, the usual equations of the balances of the entropy, mass, momentum, and angular momentum.

In the course of derivation of Eq. (44)–(47) the time derivatives, rate of entropy production, external forces, coefficients L_{mn}, η, η_{v} and η_{r} are fixed, thus the derived variational principle is restricted. To construct Eq. (46), for example, the Euler–Lagrange equation is presented in the form

$$\frac{\partial L}{\partial V_\alpha} - \sum_k \frac{\partial}{\partial x_k}\frac{\partial L}{\partial\left(\dfrac{\partial V_\alpha}{\partial x_k}\right)} = 0, \quad \alpha, k = 1, 2, 3. \tag{48}$$

A conventional transformation yields

$$\frac{\partial L}{\partial V_\alpha} = \frac{\mathrm{d}V_\alpha}{\mathrm{d}t} - \rho\bar{F}_\alpha, \tag{49}$$

and

$$\sum_k \frac{\partial}{\partial x_k}\frac{\partial L}{\partial\left(\dfrac{\partial V_\alpha}{\partial x_k}\right)} = \frac{\partial p}{\partial x_k} - \left(\eta_{\mathrm{v}} + \frac{\eta}{3}\right)\frac{\partial}{\partial x_k}(\nabla\cdot\bar{V}) - \eta\nabla V_\alpha - \eta_{\mathrm{r}}[\nabla X(2\omega - \nabla X\bar{V})]_\alpha. \tag{50}$$

Eqs. (49) and (50) yield Eq. (46). Eqs. (44), (45) and (47) can be constructed similarly.

3.7. Alternative forms of the variational equation

The Lagrangian density can also be presented as a function of general thermodynamic functions using the following identities

$$\frac{\mathrm{d}u}{\mathrm{d}t} = T\frac{\mathrm{d}s}{\mathrm{d}t} - p\frac{\mathrm{d}v}{\mathrm{d}t} + \sum_i \mu_i\frac{\mathrm{d}c_i}{\mathrm{d}t}, \tag{51}$$

$$\frac{\mathrm{d}k}{\mathrm{d}t} = \bar{V}\cdot\frac{\mathrm{d}\bar{V}}{\mathrm{d}t} + \theta\bar{\omega}\cdot\frac{\mathrm{d}\omega}{\mathrm{d}t}, \tag{52}$$

$$\rho\frac{\mathrm{d}v}{\mathrm{d}t} = \nabla\cdot\bar{V}, \tag{53}$$

$$\phi = \frac{1}{2}\sum_{n,m} L_{nm}(\nabla y_n)\cdot(\nabla y_m) + \frac{1}{2}\sum_{ij} L'ij\mu_i\mu_j, \tag{54}$$

$$w = \bar{F}\cdot\bar{V}. \tag{55}$$

Here ϕ is the dissipative function and w is the work of the system. Substituting identities (51)–(55) into Eq. (43) results in

$$L = \rho\frac{\mathrm{d}(u+k)}{\mathrm{d}t} - w - T\sigma_s + \phi, \tag{56}$$

where k is the specific kinetic energy. Eq. (56) relates the energy change to the work, entropy production, and energy dissipation of the system. From the definition of the entropy production in a system where the flux density is determined by Eq. (16) [5,11,12] it follows that

$$\phi = \frac{1}{2}T\sigma_s. \tag{57}$$

Substituting Eq. (57) into Eq. (56) we obtain

$$L = \rho\frac{\mathrm{d}(u+k)}{\mathrm{d}t} - w - \phi. \tag{58}$$

Eq. (58) provides a rather simple tool for the evaluation of the system behavior.

3.8. Steady state

In the general case the balance equation has a form

$$\rho\frac{\mathrm{d}a}{\mathrm{d}t} = \frac{\partial(\rho a)}{\partial t} + \nabla\cdot\rho a\bar{V}. \tag{59}$$

If the process is steady the Eulerian time derivative is

$$\frac{\partial(\rho a)}{\partial t} = 0. \tag{60}$$

Then, Eq. (59) becomes

$$\rho\frac{\mathrm{d}a}{\mathrm{d}t} = \nabla\cdot(\rho a\bar{V}). \tag{61}$$

Thus Eq. (58) for the steady state takes the form of

$$L = \nabla\cdot(\rho a\bar{V}) - w - T\sigma_s + \phi. \tag{62}$$

From the Gauss theorem it follows that

$$\int_V \nabla\cdot(\rho a\bar{V})\mathrm{d}V = \int_A (\rho a\bar{V})_A \mathrm{d}A. \tag{63}$$

The right-hand side of Eq. (63) is the flux through the system surface. At the steady state there is no accumulation of the property a in the system and thus

$$\int_A (\rho a\bar{V})_A \, dA = 0. \tag{64}$$

Then Eq. (62) takes the form

$$L = w - T\sigma_s + \phi. \tag{65}$$

If $w = 0$, then

$$L = -T\sigma_s + \phi. \tag{66}$$

Substituting Eq. (57) into Eq. (66) we obtain

$$L = -\tfrac{1}{2}T\sigma_s. \tag{67}$$

Eq. (62) is, of course, the expression of the minimum entropy production principle.

4. Case II: fluxes determined by the time derivatives of extensive variables

Let us discuss the motion of a body in a gravitational field. In this case $y = h$, $I = m\,dh/dt = m\bar{V}$, $a = \bar{V}$. Here h is the 'elevation' of the body in the gravitational field. Then for the system in question we obtain

$$\phi = \frac{dI}{dt} = -mg, \tag{68}$$

and

$$\psi = -\frac{m\bar{V}}{2}. \tag{69}$$

Then

$$L = -\left(mgh + \frac{m\bar{v}^2}{2}\right). \tag{70}$$

Eq. (70) describes the motion of a body in the gravitational field. In this case the application of the developed technique yields the statement of a least-action Lagrangian and the statement of a genuine variational principle.

5. Thermodynamic analysis of eutectic solidification

5.1. Statement of the problem

In many cases it is difficult if not impossible to construct and to solve a system of equations describing transport processes in a complex systems. Variational equations, e.g. Eq. (58) reduce system description to the analysis of the change of a scalar function, such as the entropy production, energy change, etc. This section illustrates such an application of the variational equation (58).

The effect of the process conditions on the material structure formed in the course of the eutectic solidification is investigated. The solidification of an alloy containing components A and B results in the conversion of the liquid phase into the solid. The composition of the liquid is assumed to be homogeneous, that is both the components are evenly distributed across the liquid volume, while the solid contains two separate phases, α and β. The phase distribution (alloy morphology) is determined by the solidification conditions. The liquid solidification at the equilibrium conditions is described by the phase diagram showing the relationships between the composition of the solid, the liquid phases, and the material temperature. A typical phase diagram showing the solidification of a liquid containing two components, A and B, is depicted in Fig. 1. The abscissa of this diagram is the mass fraction of the component B, while the ordinate is the system temperature. The phase diagram shows the relation between the compositions of the liquid (L) and the solid (α and β) phases. The liquid is a homogeneous phase and contains two evenly distributed components, A and B. The solid is a heterogeneous system and consists of two phases, α and β. The phase α is rich in A, while the phase β is rich in B.

According to the phase diagram the equilibrium solidification of the alloy A_1 initially starts at a temperature slightly below the temperature of the melting point T_1. In this study it is assumed that the process is near equilibrium and the undercooling is negligible. Then the solidification begins at T_1. At this temperature, however, the amount of the solid phase is very small. As the temperature of the liquid–solid system drops, the amount of the solid phase grows while the amount of the liquid phase decreases. The composition of the solid is determined by the curve $\alpha_1 - \alpha_e$ while the composition of the liquid is determined by the curve $l_1 - l_e$. At the temperature T_e the content of the component B in the liquid is E while the solid phase contains α_e of this component.

At the temperature T_e the mode of the process changes. Now both phases, α and β are formed simultaneously. The conversion of the melt into two solid phases is termed eutectic solidification. The eutectic solidification depicted in Fig. 1 shows conversion of the liquid L into two solid phases α and β. The eutectic solidification involves simultaneous (cooperative) growth of both solid phases. The diagram, Fig. 1, is not limited to

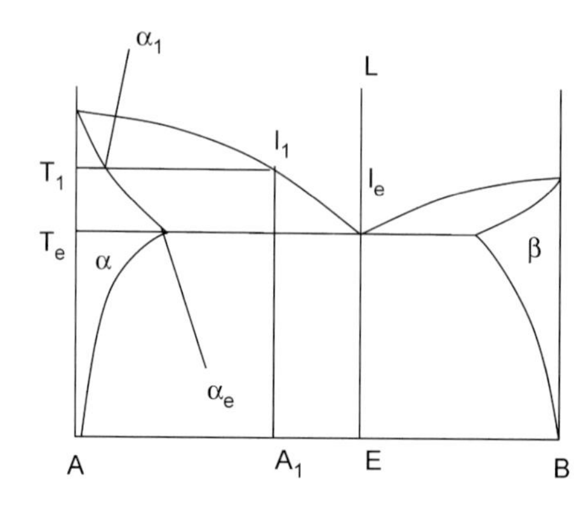

Fig. 1. Phase diagram showing eutectic solidification of liquid L.

liquid–solid transformation: it may also represent the conversion of a homogeneous high-temperature solid phase into two low-temperature solid phases.

If the volumetric fractions of both phases are comparable, the generated solid phases α and β form plates (laminae) having common boundaries and separated from the liquid by a moving almost plane interface (Fig. 2). The ratio between the thickness of the plates α and β is determined by the volumetric fractions of the solid phases. Thus this ratio is a function of the alloy composition and cannot be controlled by the solidification conditions. These conditions, however, determine the spacing between the phases r (lamellar spacing), which is the sum of the width of two lamellar of the two phases.

The alloy morphology affects its properties, thus the control of the lamellar spacing is an important element of the casting technology. At a given alloy composition, the only available control variable that determines the cast structure, i.e. the lamellar spacing, is the rate of heat removal from the solid–liquid interface [18–20]. This rate determines the speed of the solidification, defined as the speed V' of the motion of the solid–liquid boundary. The structure shown in Fig. 2 appears in many important alloys, for example, steel, where the homogeneous high-temperature solid phase (austenite) is converted into two low-temperature phases (ferrite and cementite). Thus the practical importance of the correlation between the values of r and V' is apparent

The experimental data [18] show that this correlation can be approximated by the equation

$$r^2 V' = k, \tag{71}$$

where r is the lamellar space, V' is the rate of the motion of the boundary between solid and liquid phases and k is a constant determined by the process conditions. The author (Geskin, 1990), showed that the eutectic solidification can be described using thermodynamic functions with no empirical assumptions. Particularly, Eq. (71) can be derived from the condition of the minimization of the entropy production.

Steady-state solidification can be described by Eq. (65). Because no mechanical work is involved, process description can be reduced to Eq. (67). This equation is used to determine the lamellar space r. The actual value of r corresponds to the extremal value of the entropy production.

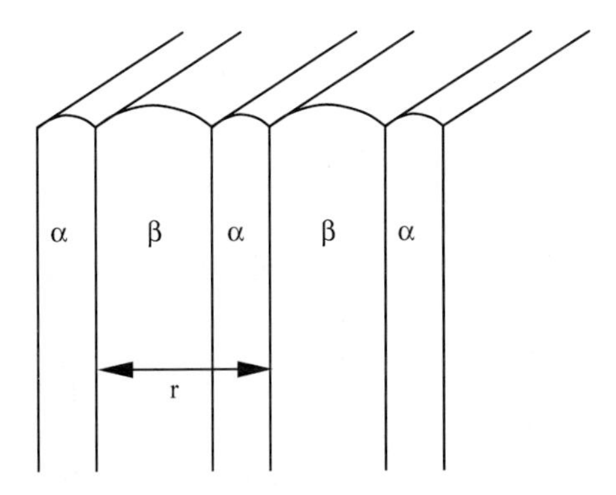

Fig. 2. Schematic of the eutectic solidification.

5.2. Determination of the rate of entropy production

The rate of entropy production in the system in question can be determined by the equation

$$\sigma_s = \sigma_{s1} + \sigma_{s2} + \sigma_{s0}, \tag{72}$$

where σ_{s1} is the entropy production due to the creation of the phase boundary, σ_{s2} is the entropy production due to the diffusion in the liquid and σ_{s0} is the entropy production independent of r. The first item in the right part of Eq. (72) can be expressed as

$$\sigma_{s1} = \sigma_{\mathrm{p}} V_1', \tag{73}$$

where σ_{p} is the entropy production per unit area of the phase boundary between the phases α and β, and V_1' is the rate of growth of the boundary. The value of σ_{p} is constant, while the value of V_1' is proportional to V' and inversely proportional to r. Thus the function σ_{s1} can be expressed as

$$\sigma_{s1} = \frac{k_1 V'}{r}, \tag{74}$$

where k_1 is the constant accounting both for σ_{p} and for the coefficients of proportionality relating V_1' with V' and r.

The entropy production due to the diffusion is defined by the equation

$$\sigma_{s2} = i_{\mathrm{a}} \Delta C_{\mathrm{a}}, \tag{75}$$

where i_{a} is the diffusion flow of the component A during solidification and ΔC_{a} is the concentration gradient in the solid determining diffusion. Overall mass transfer during the solidification is proportional to the solidification rate. Consequently, the diffusion flow is

$$i_{\mathrm{a}} = k_2 V, \tag{76}$$

where k_2 is a constant coefficient. The concentration gradient is determined by Fick's law as

$$\Delta C_{\mathrm{a}} = k_2 \frac{r}{D}, \tag{77}$$

where D is the coefficient of diffusion of component A in the solid. Substitution of i_{a} from Eq. (76) and ΔC_{a} from Eq. (77) into Eq. (75) produces

$$\sigma_{s2} = k V'^2 \frac{r}{D}. \tag{78}$$

Now the rate of entropy production in the course of the eutectic solidification can be defined as

$$\sigma_s = k_1 \frac{V'}{r} + k_2 V'^2 \frac{r}{D} + \sigma_{s0}. \tag{79}$$

Assuming that the rate of the solidification V' is constant, the lamellar space generated in the course of the solidification is determined by the conditions of the extremality of σ_s.

These conditions are defined by the equation

$$\frac{\mathrm{d}\sigma_s}{\mathrm{d}r} = \frac{\mathrm{d}}{\mathrm{d}r}\left(k_1\frac{V}{r} + k_2V^2\frac{r}{D} + \sigma_{s0}\right) = 0. \tag{80}$$

From Eq. (80) it follows that

$$k_1\frac{1}{r^2} = k_2\frac{V}{D}, \tag{81}$$

which at the given value of D becomes identical to Eq. (65). At the given value of V' (the given rate of heat removal), Eq. (75) becomes

$$\frac{r^2}{D} = k_3, \tag{82}$$

where k_3 is the constant determined by the solidification rate. Thus, Eq. (81) defines the effect of the growth rate and of the conditions of the diffusion in the solid on the morphology of the solid.

The constructed equations describe a slow process, which is controlled by mass exchange only. If the process depends on the heat and mass transfer, the Lagrangian density L must be determined by both transport processes. If the steady-state approximation is not applicable, the system description will include the change of the internal energy as well as the boundary conditions. A corresponding modification of the variational equation is then required.

6. Conclusion

In this chapter a simple procedure for the conversion of the available information about a system into a variational equation is developed. The form of the variational equation and, in fact, the feasibility to derive this equation depends on the system characteristics. If the fluxes are functions of the time derivatives of the generalized coordinates (Section 4), the application of the proposed technique resulted in construction of a genuine variational principle. If the fluxes are determined by the space derivatives of the generalized variables (Section 3), the proposed technique brings about formation of restricted variational equations. The variational equations presented in Section 3 simply state that at the given rates of change (value of the time derivatives) and sources the actual distribution of the generalized variables brings about the extremal value of the system Lagrangian. Thus, these equations can be used as a computational tool, if the initial conditions involved information about the rate of change of the process variables, rather than values of the variables. The obtained variational equations can also be used as a base of the direct methods, although the construction of these methods is beyond the scope of this study.

The presented variational equations incorporate various phenomena that simultaneously occur in the system. These equations include scalar functions, such as the energy change, the rate of the entropy production or energy dissipation that integrate effects of various forces acting on the systems. As that was shown in Section 5 the process description can be reduced by determination of these functions. If such determination is not possible, estimation of the correlation between scalar characteristics and process conditions provides information about the process dynamics, not available otherwise. Thus, the proposed technique can be used for qualitative analysis of systems with fixed boundary variables. In this case, several terms of Eq. (56), such as du/dt, φ and w are to be approximated by simplified functions of process variables and then the extreme value of the integral $\int_V L\,dV$ can be estimated. This estimation enables us to determine the proper relationships among process variables. 'Dissipation' is always present explicitly in the constructed Lagrangians, and therefore, the presented work follows on the same path as the original works of Prigogine and Gyarmati.

Acknowledgements

This work was partially supported by the NJ Center for Micro-Flow Control. The assistance of Mr O. Petrenko in the preparation of the manuscript is acknowledged.

References

[1] Onsager, L. (1931), Reciprocal relations in irreversible processes, Phys. Rev. 37, 405–426.
[2] Onsager, L. (1931), Reciprocal relations in irreversible processes, Phys. Rev. 38, 2265.
[3] Prigogine, I. (1967), Introduction to Thermodynamics of Irreversible Processes, Wiley, London.
[4] Prigogine, I. (2003), Is Future Given?, World Scientific Publishing Co., Singapore.
[5] Gyarmati, I. (1970), Non-Equilibrium Thermodynamics, Springer, New York.
[6] Gyarmati, I. (1974), Generalization of the governing principle of dissipative processes to complex scalar fields, Annalen der Physik 7, 18–21.
[7] Glansdorff, P. and Prigogine, I. (1964), On a general evolution criterion in macroscopic physics, Physica 30, 351–374.
[8] Geskin, E.S., Goldfarb, E.M. and Kotlar, B.D. (1969), A variational method of derivation of transport equation for dissipative systems with arbitrary boundary conditions, Russ. J. Phys. Chem. 44, 543–544. in English.
[9] Geskin, E.S. (1988), A simplified variational technique for description of heat and fluid flow. In Proceedings of the Fourth International Symposium on the Second Law Analysis of Thermal Systems. Rome, Italy, (Moran, M., Scuba, E., eds.), ASME, New York, pp. 25–29.
[10] Geskin, E.S. (1989), An Integral Variational Equation for Transport Processes in a Moving Fluid, J. Appl. Mech. 56, 208–210.
[11] Geskin, E.S. and von Spakovsky, M.R. (1995), Towards the development of a generalized energy equation. In International Conference on Efficiency, Costs, Optimization, Simulation and Environmental Impact of Energy Systems: ECOS'95, ASME, Istanbul, Turkey, pp. 75–80.
[12] Geskin, E.S. and von Spakovsky, M.R. (1996), Toward the development of a generalized energy equation: inclusion of elastic and viscous-elastic effects. In International Conference on Efficiency, Costs, Optimization, Simulation and Environmental Impact of Energy Systems: ECOS'96, ASME, Stockholm, Sweden, pp. 29–35.
[13] Glansdorff, P. and Prigogine, I. (1971), Thermodynamic Theory of Structure, Stability and Fluctuation, Wiley, London.
[14] Rosen, P. (1953), On variational principle for irreversible processes, J. Chem. Phys. 21(7), 1220–1221.
[15] Rosen, P. (1954), Use of restricted variational principles for the solution of differential equations, J. Appl. Phys. 25(3), 336–338.
[16] Schechter, R.S. (1967), The Variational Method in Engineering, McGraw-Hill, New York.
[17] Haase, R. (1976), Thermodynamics of Irreversible Processes, Wiley, London, Chapter 4.

[18] Flemings, M. (1984), Solidification Processing, McGraw-Hill, New York.
[19] Minkoff, I. (1986), Solidification and Cast Structures, Wiley, London.
[20] Popov, D.I., Regel, L.L. and Wilcox, W.R. (2000), Application of the theorem of minimum entropy production to the growth of lamellar eutectics with an oscillating freezing rate, J. Cryst. Growth 209, 181–197.
[21] Geskin, E.S. (2002), Application of the principle of the minimum entropy production to the analysis of the eutectic solidification. In Proceedings of Internationational Mechanical Engineering Congress and Exposition, November, 2002, ASME, New Orleans.

Chapter 10

DO THE NAVIER–STOKES EQUATIONS ADMIT OF A VARIATIONAL FORMULATION?

Enrico Sciubba

Department of Mechanical and Aeronautical Engineering, University of Roma 1-'La Sapienza', via eudossiana 18-00184 Roma, Italy

Abstract

The problem of finding a variational formulation for the Navier–Stokes equations has been debated for a long time, since the fundamental statements of Hermann von Helmholtz and John William Strutt, Lord Rayleigh. There is a remarkable lack of agreement among different authors even on the theoretical possibility of the existence of such a statement, leave alone its practical derivation. On the other hand, there is a similarly remarkable sequence of consistent attempts to solve the problem, all based on what appears to be a common intuition: that the driving mechanism is indeed some sort of entropy-based functional. This chapter is divided into two parts: in the first one, we try to put into proper perspective both this longstanding debate and its possible formal and practical implications; in the second one, we discuss a novel procedure for deriving the incompressible Navier–Stokes equations from a Lagrangian density based on the exergy 'accounting' of a control volume. The exergy-balance equation, which includes its kinetic, pressure-work, diffusive, and dissipative portions (the last one due to viscous irreversibility) is written for a steady, quasiequilibrium and isothermal flow of an incompressible fluid. It is shown that, under the given assumptions, and without recourse to the concept of 'local potential', the Euler–Lagrange equations of a formal minimization of the exergy variation (= destruction) result in fact in the Navier–Stokes equations of motion.

Keywords: Navier–Stokes equations; generalised Hamiltonians; Euler–Lagrange equations; local potentials

1. The problem of the variational formulation for the equations of fluid motion

The quest for a variational principle regulating the motion of a fluid is certainly not a recent one: the generality and the great success of variational principles in classical ('rigid body') dynamics have stimulated the interest of many researchers who have attempted to formulate the laws of continuum mechanics in the same way. The D'Alembert–Lagrange principle, in its original formulation, states that "A fluid moves in such a way that

$$\delta w_{\mathrm{v}} - \int_V \rho \frac{\mathrm{d}\vec{v}}{\mathrm{d}t}\,\delta\vec{x}\,\mathrm{d}V = 0$$

E-mail address: enrico.sciubba@uniroma1.it (E. Sciubba).

for all virtual displacements which satisfy the given kinematical conditions" (as quoted in Ref. [1]), and leads, for an inviscid fluid, to the statement that "A perfect fluid moves in such a way that

$$\delta w_{\mathrm{v}} - \int_V \rho \frac{\mathrm{d}\vec{v}}{\mathrm{d}t} \delta\vec{x}\, \mathrm{d}V = 0$$

is satisfied for all virtual displacements $\mathrm{d}x$ that satisfy continuity" (also quoted in Ref. [1]). This Lagrangian formulation applies to compressible fluids as well, provided due account is taken for the varying density (i.e. if variations are now taken with respect to the density as well).

A problem with all fluid-mechanic expressions of the D'Alembert statement is that they add little to our knowledge of the fluid-dynamics of the flow. In fact, they are, as correctly observed by Serrin, "little more than a reformulation of the equations of motion". This state of affairs is, however, not an a priori necessity: if a functional (i.e. the integrand of the variational statement) could be found that represents a physical quantity relevant to the dynamics of the flow under consideration, then the phenomenological interpretation would be by far more important than the mathematical representation. In fact, in such a case, 'insight' into the characteristics of the motion would be gained more via a global analysis of the behavior of the integrand than by a detailed numerical analysis of the solution of the equation of motion itself. Unfortunately, the problem of identifying such a Lagrangian for the motion of a viscous fluid does not admit of a simple general solution, and even the possibility of its complete formulation has been subjected to some debate. A thorough critical review of the abundant literature on this topic exceeds the limits of this chapter, and interested readers are referred to Refs. [1–4]. Rather, a somewhat restricted view is adopted here: only the specific issues related to whether or not the Navier–Stokes equations admit of a variational formulation will be discussed. In other words, we want to discuss the admissibility of the proposition "The general equation of motion of a viscous fluid can be derived from a variational formulation in which a functional

$$\int_V \mathbf{L}\, \mathrm{d}V \tag{1}$$

is minimized under the proper boundary and accessory conditions".

All papers dealing with this topic usually begin by quoting the two fundamental papers by Helmholtz ([5]) and Lord Rayleigh ([6]), and stating that their works result in the so-called minimum dissipation theorem, formulated as follows: "If in the motion of an incompressible viscous fluid the curl of the vorticity can be derived by a potential (i.e. $\mathrm{curl}\boldsymbol{\omega} = \mathrm{grad}\Phi$), then that motion has the minimal dissipation of any other motion consistent with the same boundary conditions" (see, for example, Ref. [1]). In fact, this is not entirely correct. In his paper, first presented at a session of the Society for Natural History and Medicine of Heidelberg on October 30, 1868, and published about 1 year later, in 1869, Helmholtz used a Lagrange multiplier formulation to prove that

$$\left\langle \int_V \mathbf{D}\, \mathrm{d}V \right\rangle = \min, \tag{2}$$

(where the angular brackets denote a functional) under the following conditions:
(i) the velocity at the boundary of V remains constant,
(ii) the flow is stationary, and
(iii) the convective terms are negligible.

It was Rayleigh, some 40 years later (in July 1913), who formulated the principle in the form given above. Rayleigh begins by recalling the Helmholtz result, and observes that Eq. (2) is satisfied not only in slow acts of motion (those for which inertial terms may be neglected, Helmholtz assumption (iii) above), but also for all cases in which the $\nabla^2 u_i$ "are the derivatives of some single-valued function" (Rayleigh's wording). It is then straightforward to prove that, if the acceleration admits of a potential, so does the curl of the vorticity: thus, we recover the statement as reported by Serrin.

Both Helmholtz and Rayleigh were very careful in stating the limitations of their respective findings, and in particular, neither one of them expressed any opinion about the possibility of the existence of a more general variational formulation for the Navier–Stokes equations. To this author's knowledge, the first general answer of substantial breadth was given by Clark B. Millikan [7], who noted that if the integrand (which we shall denote here by **G**) is a functional of only **U** and grad**U**, then it can at best represent a variational formulation for the class of steady flows in which inertia terms may be neglected (i.e. for the more limited class of motions considered by Helmholtz). Millikan took what in modern terms we may define as 'trial and error' approach, and proceeded as follows:

(i) he introduced an additional, unknown term **R** in Eq. (2) to implicitly represent the effects of the quadratic terms neglected by Helmholtz (thus, the integrand in Eq. (2) now reads '$\mathbf{D}+\mathbf{R}$');
(ii) he adopted a Lagrange multiplier approach, using the continuity equation as a constraint;
(iii) he then derived explicitly the Euler–Lagrange equations that ensure the minimization of the functional $\langle \int_V (\mathbf{D}+\mathbf{R})\mathrm{d}V\rangle$, and showed by clever albeit tedious algebraic manipulations that they cannot be satisfied if **R** is expressed as a linear combination of monomial forms of the type

$$u_i^{\mathrm{a}} u_j^{\mathrm{b}} u_k^{\mathrm{c}} \left(\frac{\partial u_\alpha}{\partial x_\beta} \right)^{s_{\alpha\beta}} \tag{3}$$

for any value of the exponents a, b, c, and $s_{\alpha\beta}$ that also satisfy the general equation of motion. In the closure of his paper, Millikan acknowledges that the idea was suggested to him by Bateman. Oddly enough, Bateman published in the same year, but in a different Journal [8] a proof that the integral of the functional $p+\rho|\mathbf{U}^*\mathbf{U}|$, under the constraints of a prescribed mass flux and of the continuity equation, is minimized over a certain fluid domain V if and only if **U** is irrotational. What is important for the purpose of this chapter is that both of these statements may be indeed regarded as generalizations of the previously derived Helmholtz and Rayleigh minimum dissipation theorem. This leads us to the real object of the quest: is it possible to extend the Helmholtz–Rayleigh theorem to *all* viscous flows? Millikan's negative answer is not final (he only excludes linear combinations of monomial forms of the type given by Eq. (3)), but can be used to direct our search: either we construct the 'additional' kernel **R** by means of (possibly nonlinear) terms of a different type than that given in Eq. (3), or we must use a different minimization procedure, for example a weaker, Galerkin-like method.

Onsager [9] was interested in exploring the validity of a 'minimum dissipation principle' as a discriminating criterion upon which to base a distinction between spontaneous and 'endoenergetic' processes. The former proceed from an initial to a final state without the need of a 'seed' or 'forcing' energy, while the latter not only need some form of 'energy input' to occur, but also show a strong tendency to proceed backwards once this energy forcing is removed. He developed a formal description of the entropy generation in a process in terms of 'forces' (the causes) and 'fluxes' (the effects), and introduced the fundamental assumption that, at least in a sufficiently large region in the vicinity of equilibrium, each flux could be expressed by a linear combination of all the N forces known to act on the process[1]:

$$J_i = \sum_{k=1}^{N} \mathbf{L}_{ik} F_k. \tag{4}$$

Onsager then proceeded to subject the substantial entropy balance to a variation with respect to the fluxes J_i while maintaining the forces F_k constant. His derivation will not be reported here, and interested readers are referred to his work quoted above. The result was that in irreversible processes taking place in a continuum, described by linear flux/force relationships and for which the Onsager reciprocity relations apply, the energy dissipation indeed displays a minimum. Oddly enough, Onsager never explicitly stated this result in terms of an 'entropy generation minimization principle', which is though implicit in his derivation.

Biot [10] derives his 'generalized equation of motion' by extending the D'Alembert principle to heat conduction and convection problems. Relevant to the topic discussed in this chapter is his treatment of nonisothermal, convection-dominated problems in which inertia terms may not be neglected. Biot introduces what he calls the set of 'generalized coordinates' q_i, (the 'natural' or 'physical' coordinates, i.e. space and time *plus* all phenomenological quantities deemed to be relevant for the process under examination). In the case of heat transfer, for instance, the 'displacements' would be the local heat fluxes (in Biot's terminology, the heat displacement field '**H**'), and the 'generalized coordinates' the set (x, y, z, t, u, v, w). By a formally straightforward application of the 'virtual work' principle (in which the virtual displacements are now represented by the members of the set $\delta\mathbf{H}$), he derives an equation linking the kinetic energy **K**, the velocity potential **Φ**, the dissipation **D** and the 'external forces' **Q**:

$$\frac{\mathrm{d}}{\mathrm{d}t}\left(\frac{\partial \mathbf{K}}{\partial \dot{q}_i}\right) - \left(\frac{\partial \mathbf{K}}{\partial q_i}\right) + \left(\frac{\partial \mathbf{D}}{\partial \dot{q}_i}\right) + \left(\frac{\partial \mathbf{\Phi}}{\partial q_i}\right) = Q_i. \tag{5}$$

[1] This assumption descends from a 'deeper' but purely mechanical assumption of microscopic reversibility. As correctly noted by Gyarmati [3], "...(the reciprocity relations) must be either considered experimentally confirmed axioms, or...it must be possible to derive them by pure phenomenological means also".

Note that Eq. (5) has a broader application than the one needed to answer our question, in that it appears to treat explicitly not only the stationary state, but also all nonstationary situations in which the fluid is 'relaxing' to equilibrium. Indeed, eliminating both the time dependence and the external forces (= thermal forcing functions) Q_i, Eq. (5) becomes

$$\left(\frac{\partial \mathbf{K}}{\partial q_i}\right) - \left(\frac{\partial \mathbf{D}}{\partial \dot{q}_i}\right) - \left(\frac{\partial \mathbf{\Phi}}{\partial q_i}\right) = 0, \tag{6}$$

which is equivalent to Rayleigh's statement. It is interesting to note that Biot was aware that his procedure may be used to find an 'approximate solution' in two different ways:

(a) By choosing as generalized coordinates (in addition to space and time) a *small* set of parameters that functionally determine some known (or guessed, or measured) solutions of the problem. All Fourier and Chebyschev methods of heat transfer, for example, fall into this category.

(b) By choosing as generalized coordinates (in addition to space and time) an *infinite but denumerable* set of parameters that may then be collocated on the desired solution points. This is of course the rationale of all the so-called weak variational solutions.

We must also note, in view of a critique by Finlayson and Scriven that will be discussed here below, that Biot explicitly identifies Galerkin methods as a subclass of his procedure, namely one in which "the displacements depend linearly on the generalized coordinates and the latter are treated as unknown constants" [10].

Why is Biot's work then not regarded as being conclusive? After all, Eq. (5) is even more general than a stationary 'minimum dissipation principle': the latter can only be used for identifying equilibrium states, while the former describes the evolution of the system in time, whether the 'motion' relaxes to equilibrium or not. The fact is that Biot makes *explicit* use of the constitutive equations to link his generalized coordinates to the 'displacements'. Thus, if in a heat-conduction problem the heat flux (displacement) is correlated to the temperature (generalized coordinate) by a material coefficient (the thermal conductivity) and the gradient operator $\partial/\partial x_i$, when convection sets in we need to link the heat flux to the temperatures *and* to the fluid velocities, and we thus need an energy equation. Furthermore, to solve for the velocities we need additional 'independent equations', and these can only be the equations of fluid motion. We see then that Biot's method falls under Serrin's criticism reported in the initial paragraphs of this section.

The problem of constructing a Lagrangian that would not require any a priori knowledge of the equations of motion had been previously solved by Herivel [11], who postulated $\mathbf{L} = \mathbf{K} - \mathbf{E}$ for a compressible ideal fluid ($\mathrm{d}s = 0$ everywhere). Oddly, Herivel's approach was criticized by Finlayson and Scriven as being of the 'restricted type', which it is, obviously, not (it does not make use of any 'local potential' or 'dummy variable'). Incidentally, Herivel's method is interesting for two additional reasons: (a) it is the first attempt to derive a Lagrangian in a Eulerian frame of reference, making use of a 'particle' or 'masslet' identification vector $\mathbf{x}$ that had to be subjected to the Euler–Lagrange derivation procedure; and (b) it can be formally derived using a so-called 'Clebsch coordinate transformation' [12] of the type $\mathbf{U} = -\nabla\varphi + \varphi\nabla s$, where s is the specific entropy of the fluid, which is the method adopted, 30 years later, by Akay and Ecer (see below). Note that such an approach implies that the velocity function that satisfies the Lagrangian is a not necessarily linear combination (φ is a potential that in general assumes different values at different points in the domain) of the 'incompressible ideal velocity' and of an additional term that involves the divergence of entropy.

Rosen, in two brief papers, noted that Onsager's approach may be called a 'restricted' variational method, since it 'blocks' the forces (e.g. the temperature gradient) while allowing for a free variation of the fluxes (the heat flow). He showed [13] that, for an incompressible viscous fluid, a 'restricted' variation performed by keeping the acceleration fixed (but allowing the velocity to vary locally), results in the Navier–Stokes equations of motion. Rosen explicitly mentions that his approach overcomes Millikan's limitation, and even discusses [14] some criteria for choosing the 'trial distribution' for the portion of the force-functional that must be kept fixed during variation of the fluxes.

Glansdorff and Prigogine derive their 'fluid flow variational' again within the framework of a much more ambitious effort: their goal is to obtain a general criterion for the evolution of "the whole class of macroscopic systems submitted to time-independent boundary conditions" [15]. They start by showing that, in general, for nonlinear systems that include chemical reactions, such a 'total differential' corresponding to a nonlinear entropy term of the Onsager type does not exist. Glansdorff and Prigogine provide a very vivid physical explanation of this impossibility: since a system in equilibrium, they say, has 'lost memory' of its previous state(s), it would go back to the same equilibrium from any nonequilibrium state 'close enough' to it. In their words: "the system would approach the state for which the entropy production reaches its minimum value whatever the initial state". However, they argue, in a subregion of the state space of the system 'near enough' to equilibrium, the Onsagerian force $\mathbf{X}$ may be considered as the sum of the one that corresponds to the (unknown) minimum entropy-production state $\mathbf{X}(y_j)$ and of another term $\mathbf{X}(y_{j,0})$ that depends on the 'distance' of the system from that state. This second term contains at least one state variable (say, $y_{1,0}$) that is then called 'local potential', and that must relax to y_1 for the procedure to converge. In the context of this chapter, the physical meaning of this assumption is quite clear: the equilibrium velocity field ($y_j = u_j$) is reached from a large number of possible 'trial' velocity fields ($y_{j,0} = u_{j,0}$) that relax to u_j even in the presence of nonlinear (convective) effects, provided $u_{j,0}$ and u_j are not too far apart to begin with. In other words, Glansdorff and Prigogine's method may be properly defined as 'a consistent scheme of successive approximations' (their words [16]) that remains valid in a small neighborhood of the equilibrium state. In essence, the 'local potentials' $y_{i,0}$ are assumed to be known, and the 'exact' solution is obtained by guessing an initial set of $y_{i,0,1}$, solving for y_i, computing $y_{i,0,2} = f(y_{i,0,1}; y_i)$ and iterating to convergence. Note that Glansdorff and Prigogine's work, which had begun somewhat later than Rosen's (see Ref. [17], Chapter 4 for a more precise historical account), is not presented as a mathematical device to obtain the 'correct' form of the governing equations, but as a physical model of both the near- and far-equilibrium regions.

Finlayson and Scriven [2] in a very strongly worded and equally strongly biased paper, begin with an undisputable argument: the existence of a variational formulation for an operator $\mathbf{F}$ is linked to its self-adjointness. If $\mathbf{F}$ is self-adjoint, $\mathbf{G}$ ($\mathbf{F}$'s stationary functional) exists 'by definition'; if $\mathbf{F}$ is not self-adjoint, the existence of a stationary $\mathbf{G}$ is not necessary (it may or may not exist, depending on the specific form of $\mathbf{F}$ and on the imposed boundary conditions). From this correct premise, Finlayson and Scriven proceed to negate the intrinsic validity of all 'restricted' or 'pseudo-' variational methods proposed for dissipative systems. Specifically, they negate the general validity of the Onsager, Glansdorff–Prigogine and Biot approaches, and insist that all these 'approximate' methods are in effect weaker variations of a general Galerkin-like method. Such a radically negative approach was soon though to be proven incorrect: in the very same year, Tonti was working on the problem from a different, and more constructive, perspective, namely, that of transforming a nonself-adjoint operator into a self-adjoint one by means of another *operator* acting as an 'integrating factor'. These results were published in their final form only much later, in 1984 [18], and even then, they went virtually unnoticed. Tonti indeed concludes not only that the search for such an integrating operator is successful, but that it is *always* and *necessarily* so: therefore, *it is always possible to formally derive any nonlinear operator from a properly constructed functional*. The problem with Tonti's approach is that in most cases the physics of the phenomena are lost in the mathematical derivation: the integrating factor (an operator) does not necessarily resemble any physical quantity, and indeed, as Tonti puts it, "from a physical point of view, what is essential is not the form of the equation [the functional, note of this author] but the solution". In the case of the Navier–Stokes equations, though, we are interested more in the physical underlying principles than in the specific form of the solution: therefore, Tonti's answer, though operationally correct, does not resolve our quest. What it does, is to expose the limits of Finlayson and Scriver's criticism, and to further motivate our search.

Let us reconsider Prigogine's initial approach to the problem [15,16], which was later re-elaborated by Gyarmati [3]. They take an exploratory attitude: assuming the validity of Onsager relations, is it possible to derive a Lagrangian of motion for a viscous fluid from the principle of minimal entropy production? A careful reading of the Glansdorff and Prigogine papers reveals that they were well aware of the general limitations that were to be later mentioned by Finlayson and Scriven: initially, Glansdorff and Prigogine see their own 'method of local potentials' as a 'consistent scheme of successive approximations'. Only later (1962) do they account for varying $\mathbf{L}_{ij}$; and when doing so, they also identify their 'local potentials' as Lagrange-multipliers-type functions that must necessarily (and formally) be kept constant during the variation, and equal to the values that they would take at the extremum. Thus, it appears that these potentials are not only mere mathematical devices needed to obtain the correct result, but functionals (of $\mathbf{U}$ and T) that express an underlying physical necessity. This dispenses with many later studies (as an example, the work by Gage and associates [19]), which state that, under Onsager formalism, no general variational formulation exists for systems obeying Gibbs equation $\mathrm{d}h = T\,\mathrm{d}s + v\,\mathrm{d}p$. Gyarmati goes a step further than Prigogine, and derives a 'general equation of motion' in variational form for a very broad class of flows and fluids: but he develops his derivation [3] by varying *separately* fluxes and forces (holding forces constant when varying with respect to fluxes and vice versa), and thus employs, in effect, a restricted variational approach that falls under Finlayson's criticism. The importance of the Prigogine and Gyarmati formulations though lies in the fact that they both subsume a very basic assumption of the greatest physical significance: that any 'motion' of any system is realized under the fundamental constraint of minimum entropy production (or, which is equivalent, of minimum energy dissipation).

This aspect is recovered in Sieniutycz's approach [4], which capitalizes on the previous results and tries to put things into a more balanced perspective: it is true that both Prigogine's and Gyarmati's approaches give origin to 'restricted' variational formulations, but the physical principles on which their derivation is founded are so relevant to the known phenomenology that it is important to try to take into account their results and proceed in developing similar functional methods in a mathematically more sound way. Sieniutycz proposes a standard Lagrangian approach (one that uses integrating factors), in which in essence the entropy generation is used as the basic functional (the integrand in Eq. (1)) with (some of) the constitutive equations appearing as constraints:

$$\int_V \mathbf{L}\, \mathrm{d}V - \int_V \sum (\lambda \mathbf{f}) \mathrm{d}V = \text{minimal in } V, \tag{7}$$

where the λ are now the Lagrange multipliers that must be obtained in the course of the minimization procedure. Such an approach had been previously proposed (in a more limited context) by Ecer [20] and Akay and Ecer [21], who were indeed able to obtain a complete set of solutions valid for the compressible, rotational Euler equations (neglecting viscous terms). Their approach is interesting because they restate the problem in a modified solution space: their 'Clebsch variables' are, in effect, conceptually equivalent to Biot's 'generalized coordinates', and formally equivalent to those used by Herivel. Another solution was presented by Geskin [22], who developed Gyarmati's ideas under a standard Lagrangian formalism (but also employed a 'frozen potentials' approach, equivalent in effect to Prigogine's method). Sieniutycz's formulation is more general though, and it also tackles the more difficult problem of including nonstationary evolutions: i.e. his quest goes a step further than the one posed here, and seeks, like Biot and Prigogine, a truly general equation of motion that would regulate both the behavior of systems near to equilibrium and that of systems far from equilibrium and relaxing to it. Later, Geskin [23] apparently elaborated on this concept, and his latest work (Part II, Chapter 9 in this Book) presents a newly formulated, Gyarmati-like approach that can formally handle unsteady problems as well (though admittedly always in a 'small neighborhood' of the equilibrium state). It is noteworthy that his Eq. (38) simplifies to Eq. (14) here below under the present assumptions.

We can now conclude this short (and obviously based on personal biases) review by stating that, in view of the above, it is *certain* [4,18] that a general equation of motion can indeed be formulated in variational form. It is also *certain* that, at least in a restricted sense [3,16,24], the functional contains the entropy generation, i.e. that the underlying physical principle is that a fluid moves in such a way as to minimize its entropy production under the given external constraints (including boundary conditions). What is missing is a step that is definitely secondary from a mathematical point of view, but has a great importance from a physical point of view: an explicit derivation in which at least some of the governing equations (continuity, momentum, and energy) are not employed as Lagrangian constraints, but are 'derived' in the course of the procedure. This is the secondary purpose of this chapter. Incidentally, the procedure presented here below is based on a previous paper by the present author [25], which unfortunately contained a mathematical error in the derivation of the Euler–Lagrange equation that went undetected at that time.

We begin by stating that the derivation presented here is limited to a certain subclass of motions (Section 2 below) and that its extension to other types of flow (for instance, compressible) is neither straightforward (as incorrectly stated in Ref. [25]) nor certain, and is not implied here in any sense. But, for the realm of viscous incompressible Newtonian flows, the present derivation has the merit of explicitly linking a measurable and well-known thermodynamic function (flow exergy) to the standard form of the Navier–Stokes equations: some phenomenological consequences of this connection are outlined in Section 4.

2. The exergy content of a fluid in motion

Consider the flow of a viscous fluid in which the geometry of the flow channel and the physical parameters of the fluid are known: in the most general situation, every act of motion is driven by a set of well-defined external fields (pressure, external force, and temperature) and by the inertia of the mass under examination, and is affected by some 'dissipative' effects related to the real viscosity and thermal conductivity of the fluid. Dissipation is associated with entropy production or, as in the considerations that follow, to the exergy destruction of the flow: exergy is an extensive thermodynamic state function defined as[2]

$$e = h - h_0 - T_0(s - s_0), \tag{8}$$

where T_0 is a properly chosen reference-state temperature (usually that of a large 'environment' that the system, here, the flowing fluid, may eventually come into thermodynamic equilibrium with). A representation of the work and heat interactions of a system in terms of exergy has the advantage of conglomerating both work/heat interactions and dissipation into a unified framework: for any dissipative system a theorem of 'exergy destruction' applies, which states that if the system undergoes an irreversible process, its specific exergy content is destroyed (annihilated) at a rate given by

$$\dot{e}_\lambda = T_0 \dot{s}_{\rm irr}. \tag{9}$$

The interpretation of Eq. (9) is straightforward: any real (irreversible) process destroys exergy at a rate proportional to the irreversible entropy generation. For a general exposition of the paradigm of Exergy Analysis, see Refs. [26,27].

For the purpose of the present analysis, consider a unit mass of fluid undergoing a completely specified act of motion: in a small interval of time dt, the exergetic content of the unit mass will be modified by four different contributions:

(1) an exergy-change rate equal to the exchanged power (which may be positive or negative):

$$\dot{e}_{\rm w} = \dot{w}_{\rm rev} = \left[\vec{U} * \dot{\vec{U}} + \vec{U} * \frac{\vec{\nabla} p}{\rho} + \vec{U} * \vec{B} \right] \quad ({\rm W/kg}) \tag{10}$$

[2] This definition neglects other terms, like the chemical, magnetic, nuclear exergies, that are irrelevant for the topic discussed here. Mechanical exergy (kinetic and potential) may be thought of as being included in the enthalpic term.

(2) an exergy-change rate proportional to the viscous dissipation function of the flow field (always negative):

$$\dot{e}_{\lambda,\mathrm{visc}} = -\upsilon \mathbf{D}_{\mathrm{visc}} \quad (\mathrm{W/kg}) \tag{11}$$

(3) an exergy-change rate proportional to the reversible thermal entropy production (positive or negative):

$$\dot{e}_{Q,\mathrm{rev}} = (T - T_0)\dot{s}_{\mathrm{rev}} \quad (\mathrm{W/kg}) \tag{12}$$

(4) an exergy-change rate proportional to the irreversible thermal entropy production (always negative, Eq. (9)).

So that the total exergy change per unit mass of fluid in the time dt is

$$\Delta e_{\mathrm{fluid}} = \mathrm{d}t \sum \dot{e}_j$$
$$= \mathrm{d}t\left[\vec{U} * \dot{\vec{U}} + \vec{U} * \frac{\nabla \rightarrow p}{\rho} + \vec{U} * \vec{B} - \upsilon \mathbf{D}_{\mathrm{visc}} + (T - T_0)\dot{s}_{\mathrm{rev}} - T_0 \dot{s}_{\mathrm{irr}}\right] (\mathrm{J/kg}). \tag{13}$$

Note that once the flow variables are exactly known at each time t and at each point in the flow domain, the quantity defined by Eq. (13) can be exactly computed locally and also, if necessary, integrated over the entire domain, to calculate the exergy destruction of the flow: this is indeed often done to assess the efficiency of technical flows (turbine nozzles, turbine and compressor blades, etc.). The reverse is obviously not true, as infinitely many flow fields may display the same value of the exergy destruction rate at any instant of time. Our goal is to show that if we assume that, at every instant in time, the fluid motion is governed by the minimization of the exergy destruction given by Eq. (13), the resulting equations of motion are indeed the Navier–Stokes equations. Note that such an assumption is in line with the statement made by Serrin [1] that a credible Hamiltonian ought to include the energy equation in some form.

3. Variational derivation of the flow field

Let us consider the isothermal flow of a viscous homogeneous fluid. As stated above, our basic assumption is that *the fluid moves in such a way that its exergy destruction is at its minimum at each instant of time, compatible with the assigned external constraints*. It can be easily shown by means of the so-called 'Gouy–Stodola lost-work' theorem [28], that this assumption corresponds to the minimum entropy-generation principle. The 'external constraints' are the imposed boundary conditions and the specified work- and heat-transfer interactions. Neglecting for the moment the boundary conditions (we shall assume that 'natural' boundary conditions apply), let us first observe that the external energy exchanges are completely specified, in a quantitative and qualitative manner, by the expression of the exergy variation of the fluid mass (Eq. (13)). Therefore, imposing the condition of constrained minimum exergy destruction is equivalent to searching for the minimization of a functional whose integrand is the total exergy change of the unit fluid mass, given by Eq. (13). Therefore, we can write:

$$\mathbf{L} = \left[\vec{U} * \dot{\vec{U}} + \vec{U} * \frac{\vec{\nabla} p}{\rho} + \vec{U} * \vec{B} - \upsilon \mathbf{D}_{\text{visc}} \right], \tag{14}$$

and

$$\int_V \mathbf{L} \, \mathrm{d}V = \text{minimal in } V. \tag{15}$$

We thus see that the exergy formulation leads (at no extra computational cost) to a formally unconstrained problem position. If different boundary conditions are to be imposed, the integral in Eq. (15) must be augmented as needed.

The Euler–Lagrange equations for the problem so posed are

$$\frac{\partial \mathbf{L}}{\partial u_k} - \frac{\partial}{\partial x_j} \left(\frac{\partial \mathbf{L}}{\partial \left(\dfrac{\partial u_k}{\partial x_j} \right)} \right) = 0. \tag{16}$$

Using for the viscous-dissipation function the standard expression:

$$\mathbf{D}_{\text{visc}} = \frac{1}{2Re} \left(\frac{\partial u_i}{\partial x_j} + \frac{\partial u_j}{\partial x_i} \right)^2, \tag{17}$$

and using a 'restricted' treatment for the deformation work (see Appendix A), system Eq. (16) results in the following set of equations:

$$\frac{\mathrm{d}u_k}{\mathrm{d}t} + \frac{1}{\rho} \frac{\partial p}{\partial x_k} + \mathbf{B}_k - \upsilon \left(\frac{\partial^2 u_k}{\partial x_i \, \partial x_i} \right) = 0, \tag{18}$$

which are indeed the Navier–Stokes equations of motion. A detailed derivation in a three-dimensional Cartesian domain is given in the Appendix A.

4. Conclusions

The above discussion allows us to answer the question posed in the title of the chapter: it appears indeed *certain* that the Navier–Stokes equations admit of a variational formulation, since Tonti's work provides us with a formal proof thereof. From the physical point of view, both Glansdorff and Prigogine's and Gyarmati's work convincingly link the minimization of the entropy generation to the Lagrangian of motion: the fact that their method in reality corresponds to a 'restricted' variation does not detract from their results, because, as they also clearly state, the 'local potentials' they employ may be seen to be equivalent to an iterative approximation procedure. Our secondary goal has been achieved as well: a formal derivation of the Navier–Stokes equations for an incompressible viscous fluid has been presented, its novelty residing in the physical meaning attached to the Lagrangian functional.

Since the entropy minimization (or, equivalently, the exergy-destruction minimization) has a clear and unequivocal physical meaning, it may be used not only to interpret the results of our derivation, but also to suggest possible theoretical and practical applications. If the derivation presented in this chapter is correct, some of its implications are

(1) *Realizability*. Not every flow is permissible in nature: if a certain flow field is specified (analytically or as the result of a numerical simulation) and its velocity vector $\mathbf{U}(x,t)$ does not simultaneously satisfy Eqs. (14) and (15), it will not occur spontaneously. This constitutes a powerful tool for evaluating, for instance, the results of numerical simulations;
(2) *Asymptotic bounding*. For given boundary conditions, after an initial transient, *the flow will approach a configuration that satisfies Eq.* (15): this can be used to set upper and lower bounds to some of the flow-derived properties (heat transfer, deliverable work, etc.);
(3) *'Physically correct' time marching*. Once an initial flow field is known, Eq. (15) can be solved for the (Lagrangian) velocity $\boldsymbol{U}(x, t+\mathrm{d}t)$, and the Eulerian field marched in time by successive application of the minimum exergy destruction principle. Such a procedure relies implicitly on the (strong) assumption of local equilibrium, and may thus be considered as a sort of 'asymptotic property' of the solution: its advantage is that it guarantees stability without imposing an a priori upper bound on the integration time step.

To test the correctness and the practical usefulness of the procedure proposed here, applications must be developed and tested on known flow fields (experimental, analytical and numerical): it is immediately clear that if such applications were successful, they would open up new perspectives in inverse thermo-fluid-dynamic design (turbomachinery flows, for instance, or flame dynamics). A more important step would be that of extending the validity of this derivation to nonisothermal and compressible flows, but this seems to imply a substantial amount of additional work. As for the question of the existence of a general equation of motion, i.e. of a nonstationary extension of Eq. (15), much additional work is needed to reach a conclusive result.

Appendix A. Explicit derivation of the final equations from the exergy functional

The present derivation is carried out in a three-dimensional Cartesian, Galilean frame. For more generality, the Lagrangian of Eq. (15) is made dimensionless by dividing the right-hand side by U^3/L, where U and L are representative velocity and length scales for the flow domain.

(1) expanding $\mathbf{L}$ in the three components x, y, z, we obtain (the suffix indicates the relevant component):

$$\mathbf{L}^x = u\dot{u} + up_x + D/3 \tag{A1}$$

$$\mathbf{L}^y = v\dot{v} + vp_y + D/3 \tag{A2}$$

$$\mathbf{L}^z = w\dot{w} + wp_z + D/3 \tag{A3}$$

(2) We now augment the Lagrangian density by including the deformation work (here, power), which is a term in which we do not vary $\mathbf{U}_0$ during the derivation:

$$\mathbf{L}^{\text{aug}} = \mathbf{U} * (\mathbf{M}_0 * \mathbf{U}_0) = \begin{vmatrix} u \\ v \\ w \end{vmatrix} * \left(\begin{vmatrix} 0 & u_{y0} & u_{z0} \\ v_{x0} & 0 & v_{z0} \\ w_{x0} & w_{y0} & 0 \end{vmatrix} * \begin{vmatrix} u_0 \\ v_0 \\ w_0 \end{vmatrix} \right)$$

$$= \begin{vmatrix} uv_0u_{y0} + uw_0u_{z0} \\ u_0vv_{x0} + vw_0v_{z0} \\ u_0ww_{x0} + v_0ww_{y0} \end{vmatrix} \tag{A4}$$

(3) The explicit expression for the dissipation function is

$$D = \frac{2}{Re}(u_x^2 + v_y^2 + w_z^2) + \frac{1}{Re}[(u_y + v_x)^2 + (v_z + w_y)^2 + (u_z + w_x)^2] \tag{A5}$$

(4) Expanding Eqs. (A1)–(A3)

$$\mathbf{L}^x = uu_t + u^2u_x + uvu_y + uwu_z + up_x + \frac{2}{3Re}u_x^2 + \frac{2}{3Re}v_y^2 + \frac{2}{3Re}w_z^2 + \frac{(u_y + v_x)^2}{3Re} + \frac{(w_y + v_z)^2}{3Re} + \frac{(u_z + w_x)^2}{3Re} \tag{A6}$$

$$\mathbf{L}^y = vv_t + uvv_x + v^2v_y + vwv_z + vp_y + \frac{2}{3Re}u_x^2 + \frac{2}{3Re}v_y^2 + \frac{2}{3\text{Re}}w_z^2 + \frac{(u_y + v_x)^2}{3Re} + \frac{(w_y + v_z)^2}{3Re} + \frac{(u_z + w_x)^2}{3Re} \tag{A7}$$

$$\mathbf{L}^z = ww_t + uww_x + vww_y + w^2w_z + wp_z + \frac{2}{3Re}u_x^2 + \frac{2}{3Re}v_y^2 + \frac{2}{3Re}w_z^2 + \frac{(u_y + v_x)^2}{3Re} + \frac{(w_y + v_z)^2}{3Re} + \frac{(u_z + w_x)^2}{3Re} \tag{A8}$$

(5) Now we separately compute the terms in the Euler–Lagrange ('E/L') equations:

$$\mathbf{L}_u^x = u_t + 2uu_x + vu_y + wu_z + p_x \tag{A9}$$

$$\mathbf{L}_{ux}^x = u^2 + \frac{4}{3Re}u_x \tag{A10}$$

$$\mathbf{L}_{ux,x}^x = 2uu_x + \frac{4}{3Re}u_{xx} \tag{A11}$$

$$\mathbf{L}_{uy}^x = uv + \frac{2}{3Re}u_y + \frac{2}{3Re}v_x \tag{A12}$$

$$\mathbf{L}^x_{uy,y} = uv_y + vu_y + \frac{2}{3Re}u_{yy} + \frac{2}{3Re}v_{xy} \tag{A13}$$

$$\mathbf{L}^x_{uz} = uw + \frac{2}{3Re}u_z + \frac{2}{3Re}w_x \tag{A14}$$

$$\mathbf{L}^x_{uz,z} = uw_z + wu_z + \frac{2}{3Re}u_{zz} + \frac{2}{3Re}w_{xz} \tag{A15}$$

(the remaining terms are symmetrical with the 'x' terms derived above). After some manipulation, and making use of the continuity equation, we obtain:

$$(\mathrm{E/L})^x = u_t - uv_y - uw_z + p_x - \frac{2}{3Re}(u_{xx} + u_{yy} + u_{zz}), \tag{A16}$$

which, augmented with the corresponding line of Eq. (A4), becomes

$$\begin{aligned}(\mathrm{E/L})^x &+ (\mathbf{U} * \mathbf{M}_0 * \mathbf{U}_0)^x \\ &= u_t - uv_y - uw_z + p_x - \frac{2}{3Re}(u_{xx} + u_{yy} + u_{zz}) + vu_y + wu_z \\ &= u_t + uu_x + vu_y + wu_z + p_x - \frac{2}{3Re}(u_{xx} + u_{yy} + u_{zz}),\end{aligned} \tag{A17}$$

i.e. the x-component of the Navier–Stokes equation for an incompressible isothermal fluid. In the other directions, we have

$$(\mathrm{E/L})^y = v_t - vu_x - vw_z + p_y - \frac{2}{3Re}(v_{xx} + v_{yy} + v_{zz}), \tag{A18a}$$

and

$$(\mathrm{E/L})^z = w_t - wu_x - wv_y + p_z - \frac{2}{3Re}(w_{xx} + w_{yy} + w_{zz}), \tag{A18b}$$

which, respectively, augmented with the corresponding lines of Eq. (A4), become

$$\begin{aligned}(\mathrm{E/L})^y + (\mathbf{U}_0 * \mathbf{M} * \mathbf{U})^y &= v_t - vu_x - vw_z + p_y - \frac{2}{3Re}(v_{xx} + v_{yy} + v_{zz}) \\ &= v_t + uv_x + vv_y + wv_z + p_y - \frac{2}{3Re}(v_{xx} + v_{yy} + v_{zz}),\end{aligned} \tag{A19}$$

and

$$\begin{aligned}(\mathrm{E/L})^z &+ (\mathbf{U}_0 * \mathbf{M} * \mathbf{U})^z \\ &= w_t - wu_x - wv_y + p_z - \frac{2}{3Re}(w_{xx} + w_{yy} + w_{zz}) + uv_x + vw_y \\ &= w_t + uw_x + vw_y + ww_z + p_z - \frac{2}{3Re}(w_{xx} + w_{yy} + w_{zz}).\end{aligned} \tag{A20}$$

Eqs. (A17), (A19) and (A20) are, of course, the sought after equations of motion.

Nomenclature

$\mathbf{B} = (b_x, b_y, b_z)$	body-force vector, N
$\mathbf{D}$	flow dissipation function, m^2/s^3
e	specific exergy, J/kg
$\mathbf{E}$	internal energy, kJ/kg
h	enthalpy, J/kg
$\mathbf{L}$	Lagrangian density
p	pressure, Pa
s	entropy, J/(kg K)
t	time, s
T	temperature, K
$\mathbf{U} = (u, v, w)$	velocity vector, m/s
V	integration volume, fluid domain
v	specific volume, m^3/kg
w	specific work, J/kg
$\mathbf{x} = (x, y, z)$	Cartesian coordinate set, m
λ	Lagrange multiplier
υ	kinematic viscosity, m^2/s
ρ	density, kg/m^3
$\boldsymbol{\omega} = (\omega_x, \omega_y, \omega_z)$	vorticity vector, s^{-1}

Suffixes

λ	loss
v	virtual displacement

A vector is identified by a bold case letter (**U**) or by an arrow ($\vec{U}$). The substantial time derivative is indicated by a dot ($\dot{s}$), the Eulerian by $\partial/\partial t$ or u_t. Spatial derivatives are written either as $\partial/\partial x_i$ or as u_{xi}. ∇ indicates the gradient operator, whereas "$*$" and "$\times$" denote the scalar and vector product, respectively.

References

[1] Serrin, J. (1959), Mathematical Principles of Classical Fluid Mechanics. In Handbuch der Physik, Vol. VIII, part 1, Springer, Berlin.

[2] Finlayson, B.A. and Scriven, L.E. (1967), On the search for variational principles, Int. J. Heat Mass Transfer 10, 799–819.

[3] Gyarmati, I. (1970), In Non-equilibrium Thermodynamics, Springer, Berlin.

[4] Sieniutycz, S. Hamiltonian structure and extremum properties of Onsagerian thermodynamics. Proceedings of the International Onsager Workshop, Leiden, The Netherlands, March 2000.

[5] Helmholtz, H. (1869), In Collected Works, Vol. 1, p. 223, J.A. Barth, Leipzig, in German.

[6] Lord Rayleigh (Strutt, J.W.), (1913), On the motion of a viscous fluid, Philos. Mag. 6(26), 776–786.

[7] Millikan, C.B. (1929), On the steady motion of viscous incompressible fluids; with particular reference to a Variational Principle, Philos. Mag. Series 7 7(44), 641–662.

[8] Bateman, H. (1929), Notes on a differential equation which occurs in the two-dimensional motion of a compressible fluid and the associated variational method, Proc. R. Soc., A-125, 598–618.
[9] Onsager, L. (1931), Reciprocal relations in irreversible processes, Phys. Rev. 37, Part I, 405–426.
[10] Biot, M.A. (1970), Variational Principles in Heat Transfer, Oxford University Press, Oxford.
[11] Herivel, J.W. (1955), The derivation of the equations of motion of an ideal fluid by Hamilton's principle, Cambr. Phil. Soc. 51, 344–349.
[12] Clebsch, A. (1857), Ueber eine allgemeine Transformation der hydrodynamischen Gleichungen, Zeitschr. Reine u. Angew. Math. 54, 293–312, in German.
[13] Rosen, P. (1953), On variational principles for irreversible processes, J. Chem. Phys. 21(7), 1220–1221.
[14] Rosen, P. (1954), Use of restricted variational principles for the solution of differential equations, J. Appl. Phys. 25(3), 336–338.
[15] Glansdorff, P., Prigogine, I. and Hays, D.F. (1962), Variational properties of a viscous liquid at a non-uniform temperature, Phys. Fl. 5(2), 144–149.
[16] Glansdorff, P. and Prigogine, I. (1964), On a general evolution criterion in macroscopic physics, Physica 30, 351–374.
[17] Schechter, R.S. (1967), The Variational Method in Engineering, McGraw-Hill, New York.
[18] Tonti, E. (1984), Variational formulation for every nonlinear problem, Int. J. Eng. Sci. XI, 394–413.
[19] Gage, D.H., Schiffer, M., Kline, S.J. and Reynolds, W.C. (1965), The non-existence of a general thermokinetic variational principle. In Report IT-2, Thermoscience Division, Department of Mechanical Engineering, Stanford University, Stanford, CA, May.
[20] Ecer, A. (1980), Variational formulation of viscous flows, Int. J. Num. Meth. Eng. 15, 1355–1361.
[21] Akay, H.U. and Ecer, A. (1984), Finite element formulation of rotational transonic flow problems, Fin. El. Fluids 5, 173–195.
[22] Geskin, E. (1989), An integral variational equation for transport processes in a moving fluid, J. Appl. Mech. 56, 147–150.
[23] Geskin, E. (2004), A Variational Principle for Transport Processes in Continuous Systems: Derivation and Application, This Book, Part II, Chapter 9.
[24] Sieniutycz, S. (1994), Conservation Laws in Variational Thermo-hydrodynamics, Kluwer Academic Publications, Dordrecht.
[25] Sciubba, E. (1991), A Variational derivation of the Navier–Stokes equation based on the exergy destruction of the flow, J. Math. Phys. Sci. 25(1), 61–68.
[26] Kotas, T. (1985), The Exergy Method of Process and Plant Analysis, Butterworths, London.
[27] Moran, M.J. and Shapiro, H. (2000), Fundamentals of Engineering Thermodynamics, Wiley, New York.
[28] Bejan, A. (1982), Entropy Generation through Heat and Fluid Flow, Wiley, New York.

Chapter 11

ENTROPY-GENERATION MINIMIZATION IN STEADY-STATE HEAT CONDUCTION

Zygmunt Kolenda, Janusz Donizak and Adam Hołda

AGH University of Science and Technology, Department of Theoretical Metallurgy, Al. Mickiewicza 30, 30059 Kraków, Poland

Jerzy Hubert

Institute of Nuclear Physics, ul. Radzikowskiego 152, 31342 Kraków, Poland

Abstract

Boundary-value problems of diffusional heat-transfer processes are usually formulated on the basis of the first law of thermodynamics. To obtain the same result when the method of irreversible thermodynamics is applied an additional assumption that the temperature gradient values over the whole domain are reasonably small must be introduced. Such an assumption also means that $|T(x_i) - T_{avg}|/T_{avg} \ll 1$ *for all* x_i $(i = 1, 2, 3)$, *where* $T(x_i)$ *is temperature at* x_i *and* T_{avg} *is the average temperature of the solid. On the basis of the minimum entropy-generation principle, a new formulation of the boundary-value problems is proposed. Applying Euler–Lagrange variational formalism, a new mathematical form of heat-conduction equation with additional heat-source terms has been derived. As a result, the entropy-generation rate of the process can significantly be reduced, which leads to the decrease of the irreversibility ratio according to the Gouy–Stodola theorem. It will be shown that minimization of entropy generation in heat-conduction process is always possible by introducing additional heat sources. The most important conclusion derived from the presented theoretical considerations is directly connected with the solution of the boundary-value problems for solids with temperature-dependent heat-conduction coefficients. In such cases, additional internal heat sources can be arbitrarily chosen as positive or negative. It makes it possible to extend practical applications presented in the literature by Bejan. The problem of heat conduction in anisotropic solids will also be discussed.*

Keywords: entropy generation; irreversible thermodynamics; heat conduction; boundary-value problems; anisotropic solids

1. Heat conduction in isotropic solids

Differential equations describing heat-conduction processes are derived from the first law of thermodynamics. When heat-conduction coefficients depend on temperature,

E-mail address: kolenda@uci.agh.edu.pl (Z. Kolenda); szczur@uci.agh.edu.pl (J. Donizak); adam@uci.agh.edu.pl (A. Holda); jerzy.hubert@ifj.edu.pl (J. Hubert).

$k = k(T)$, a steady-state temperature field results from the solution of the differential equation

$$\mathrm{div}[k(T)\mathrm{grad}\, T(x_i)] + \dot{q}_V(x_i) = 0, \qquad i = 1, 2, 3, \tag{1}$$

with respect to the required boundary conditions.

A different formulation to the classical boundary-value problem described above is obtained when the heat-conduction equation is derived from the minimum entropy-generation principle. According to the thermodynamics of irreversible processes [1], entropy generation at steady state is at a minimum. Introducing an expression for local entropy generation [3]

$$\sigma = \sigma(T, T_{x_i}) \equiv \frac{\mathrm{d}i^s}{\mathrm{d}\tau} = \frac{k(T)}{T^2}(\mathrm{grad}\, T)^2, \tag{2}$$

where $T = T(x_i)$ and T_{x_i} denotes gradient components $\partial T/\partial x_i$, the problem can be formulated in the following way: *find the temperature function T that satisfies the boundary condition and minimizes the entropy-generation integral*

$$\sigma_t = \int_\Omega \sigma(T, T_{x_i})\mathrm{d}\Omega, \qquad i = 1, 2, 3 \tag{3}$$

over the whole domain Ω. Here σ_t represents global entropy production of the process.

Using variational calculus, this function $T(x_i)$, for which σ_t reaches a minimum, satisfies the Euler equation [2]

$$\frac{\partial \sigma}{\partial T} - \sum_i \frac{\partial}{\partial x_i}\left(\frac{\partial \sigma}{\partial T_{x_i}}\right) = 0, \tag{4}$$

where $\sigma = \sigma(T, T_{x_i})$ is given by Eq. (2).

1.1. 1D boundary-value problem

Consider the one-dimensional (1D) problem of heat conduction in a plane wall with the first kind of boundary conditions, $T(x = 0) = T_1$ and $T(x = 1) = T_2$.

Local entropy generation is

$$\sigma(T, T_x) = \frac{k(T)}{T^2(x)}\left(\frac{\mathrm{d}T(x)}{\mathrm{d}x}\right)^2,$$

and its global value to be minimized is given by

$$\sigma_t = \int_0^1 \frac{k(T)}{T^2(x)}\left(\frac{\mathrm{d}T(x)}{\mathrm{d}x}\right)^2 A\, \mathrm{d}x,$$

where A is unit surface area perpendicular to the heat-flux vector ($A = 1\ \mathrm{m}^2$). This fact should be taken into consideration in the dimensions of the remaining equations of this chapter.

Assuming $k = \text{const}$ the Euler equation (4) becomes in the final form

$$\frac{d^2T}{dx^2} - \frac{1}{T}\left(\frac{dT}{dx}\right)^2 = 0, \tag{5}$$

and is different from the classical Laplace heat-conduction equation $d^2T/dx^2 = 0$.

For boundary conditions of the first kind, $T(0) = T_1$ and $T(1) = T_2$, the solution to Eq. (5) is

$$T(x) = T_1\left(\frac{T_2}{T_1}\right)^x. \tag{6}$$

Using Eq. (6), it is easy to show that the law of energy conservation in the classical interpretation of the first law of thermodynamics is not satisfied as *div* $\dot{\mathbf{q}}(x) \neq 0$, where $\dot{q}(x) = -k(dT(x)/dx)$.

Further calculations lead to the result that

$$\text{div}\left(\frac{\dot{\mathbf{q}}(x)}{T(x)}\right) = \text{div}(\dot{\mathbf{s}}) = 0. \tag{7}$$

The explanation comes directly from Eq. (5). Interpreting its second term as the additional internal heat source

$$\dot{q}_V(x) = -\frac{k}{T}\left(\frac{dT}{dx}\right)^2 < 0. \tag{8}$$

It is easy to prove that the first law of thermodynamics is satisfied and the entropy increase of the whole process is positive and equal to

$$\sigma_t = -\int_0^1 \frac{\dot{q}_V(x)}{T(x)}dx = k\left(\ln\frac{T_2}{T_1}\right)^2 > 0.$$

The same results have been obtained by Bejan [1] using a different mathematical approach.

A numerical example for $T_2/T_1 = 0.1$ and $k = 1.0$ W/m K is shown in Fig. 1. The classical solution is characterized by $\dot{q}_{in} = \dot{q}_{out} = 0.90$ W/m^2 and $\sigma_t = 8.10$ W/K, but the entropy-generation minimization gives $\dot{q}_{in} = 2.30$ W/m^2, $\dot{q}_{out} = 0.23$ W/m^2, $\dot{q}_V = 2.07$ W/m^3 and $\sigma_t = 5.30$ W/K.

In the case when heat-conduction coefficients depend on temperature, the problem is formulated in the following way:

– *local entropy-generation rate*

$$\sigma(x) = \frac{k(T)}{T^2(x)}\left(\frac{dT(x)}{dx}\right)^2,$$

where $T = T(x)$,

– *global entropy generation to be minimized*

$$\sigma_t = \int_0^1 \frac{k(T)}{T^2(x)}\left(\frac{dT(x)}{dx}\right)^2 dx \rightarrow \min.$$

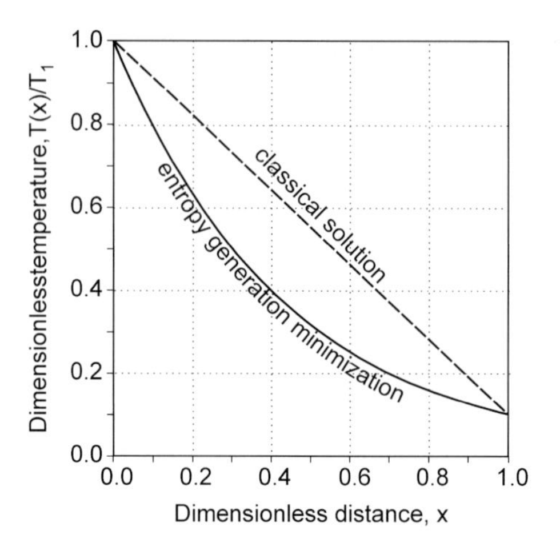

Fig. 1. Temperature distributions in the plane wall.

Euler form of heat-conduction equation

$$k(T)\frac{\mathrm{d}^2T}{\mathrm{d}x^2}+\left(\frac{1}{2}\frac{\mathrm{d}k}{\mathrm{d}T}-\frac{k(T)}{T}\right)\left(\frac{\mathrm{d}T}{\mathrm{d}x}\right)^2=0$$

or for a clearer physical interpretation

$$k(T)\frac{\mathrm{d}^2T}{\mathrm{d}x^2}+\frac{\mathrm{d}k}{\mathrm{d}T}\left(\frac{\mathrm{d}T}{\mathrm{d}x}\right)^2-\left(\frac{1}{2}\frac{\mathrm{d}k}{\mathrm{d}T}+\frac{k(T)}{T}\right)\left(\frac{\mathrm{d}T}{\mathrm{d}x}\right)^2=0. \tag{9}$$

For comparison the classical heat-conduction equation has the form

$$k(T)\frac{\mathrm{d}^2T}{\mathrm{d}x^2}+\frac{\mathrm{d}k}{\mathrm{d}T}\left(\frac{\mathrm{d}T}{\mathrm{d}x}\right)^2=0.$$

Comparison of Eq. (9) with the classical form leads to the expression for an additional internal heat source

$$\dot{q}_V(x)=-\left(\frac{1}{2}\frac{\mathrm{d}k}{\mathrm{d}T}+\frac{k(T)}{T}\right)\left(\frac{\mathrm{d}T}{\mathrm{d}x}\right)^2. \tag{10}$$

It is easy to show that $\dot{q}_V(x)$ given by Eq. (10) can be negative or positive (for $k=\mathrm{const}$, $\dot{q}_V(x)$ is always negative, see Eq. (8)). It depends on the functional form of the heat-conduction coefficient, $k=k(T)$.

Assuming frequently used dependences [1]

$$k(T)=k_1\left(\frac{T}{T_1}\right)^n, \tag{11}$$

where $k_1 = k(T_1)$ and n is an arbitrary constant, the Euler-type heat-conduction equation takes the form

$$\frac{d^2T}{dx^2} + \frac{n-2}{2}\frac{1}{T}\left(\frac{dT}{dx}\right)^2 = 0. \tag{12}$$

Assuming first-kind boundary conditions, the solution of Eq. (12) becomes

$$T(x)^{n/2} = T_1^{n/2} - (T_1^{n/2} - T_2^{n/2})x. \tag{13}$$

The local entropy-generation rate is

$$\sigma(x) = \frac{4k_1}{n^2}\left[\left(\frac{T_2}{T_1}\right)^{n/2} - 1\right]^2 = \text{const}, \tag{14}$$

and is equal to its global value, $\sigma_{t,\min}$.

For comparison, the solution of the classical problem is

$$T(x)^{n+1} = T_1^{n+1} - (T_1^{n+1} - T_2^{n+1})x, \quad \text{for } n \neq -1.$$

The local entropy generation is

$$\sigma(x) = \frac{k_1}{n+1}\frac{1}{T_1^n}(T_2^{n+1} - T_1^{n+1})^2 T(x)^{-(n+2)},$$

and its global value

$$\sigma_t = \int_0^1 \sigma(x)dx = \frac{k_1}{n+1}\left[1 - \left(\frac{T_2}{T_1}\right)^{n+1}\right]\left(\frac{T_1}{T_2} - 1\right).$$

Direct comparison gives

$$N = \frac{\sigma_t}{\sigma_{t,\min}} = \frac{n^2}{4(n+1)}\frac{[1 - (T_2/T_1)^{n+1}](1 - T_2/T_1)}{(T_2/T_1)[(T_2/T_1)^{n/2} - 1]^2},$$

which is in agreement with the results of Bejan [1].

The relationship

$$N = N\left(\frac{T_2}{T_1}, n\right)$$

is shown in Fig. 2.

Using the solution of Eq. (12), the additional internal heat source intensity can be calculated from

$$\dot{q}_V = \int_0^1 \dot{q}_V(x)dx = -\int_0^1\left(\frac{1}{2}\frac{dk}{dT} + \frac{k(T)}{T}\right)\left(\frac{dT}{dx}\right)^2 dx.$$

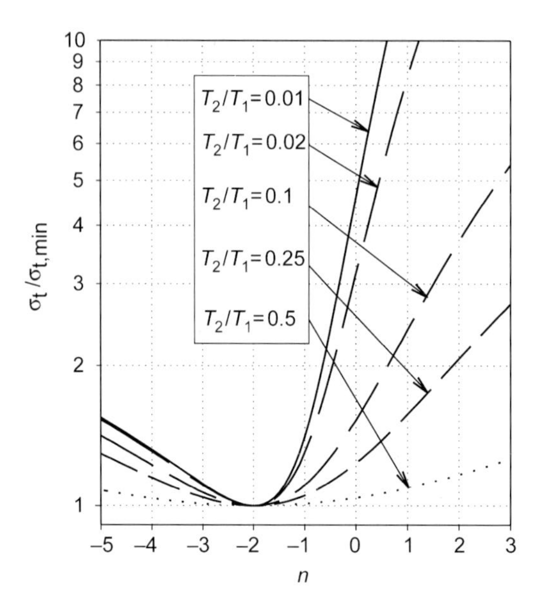

Fig. 2. N vs. n for different T_2/T_1.

After integration, when relationship (11) is used

$$\dot{q}_V = \frac{2k_1}{n} T_1 \left[\left(\frac{T_2}{T_1} \right)^{(n/2)+1} - \left(\frac{T_2}{T_1} \right)^{n+1} - 1 + \left(\frac{T_2}{T_1} \right)^{n/2} \right] \tag{15}$$

and this is shown in Fig. 3 as a function of n and T_2/T_1.

In a similar way, the internal entropy-generation rate related to $\dot{q}_V$ is obtained from

$$\sigma_{\text{int}} = \int_0^1 \frac{\dot{q}_V(x)}{T} \mathrm{d}x = \frac{2(n+2)}{n^2} \left[1 - \left(\frac{T_2}{T_1} \right)^{n/2} \right]^2, \tag{16}$$

and is presented in Fig. 4 vs. n for $T_2/T_1 = \text{var}$.

Finally, the entropy-exchange rate with external heat sources σ_{ex} having boundary temperatures T_1 and T_2 is

$$\sigma_{\text{ex}} = -\frac{q_{\text{in}}}{T_1} + \frac{q_{\text{out}}}{T_2},$$

where

$$\dot{q}_{\text{in}} = k(T_1) \frac{\mathrm{d}T}{\mathrm{d}x}\bigg|_{x=0}.$$

$$\dot{q}_{\text{out}} = k(T_2) \frac{\mathrm{d}T}{\mathrm{d}x}\bigg|_{x=1}$$

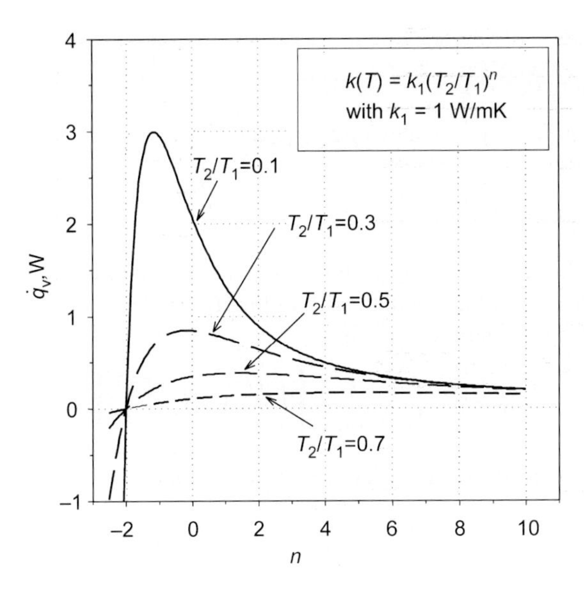

Fig. 3. $\dot{q}_V$ vs. n for different T_2/T_1.

After calculations

$$\sigma_{ex} = -\frac{2}{n}k_1\left[1-\left(\frac{T_2}{T_1}\right)^{n/2}\right]^2. \tag{17}$$

It is easy to show that the global entropy-generation rate of the process given by

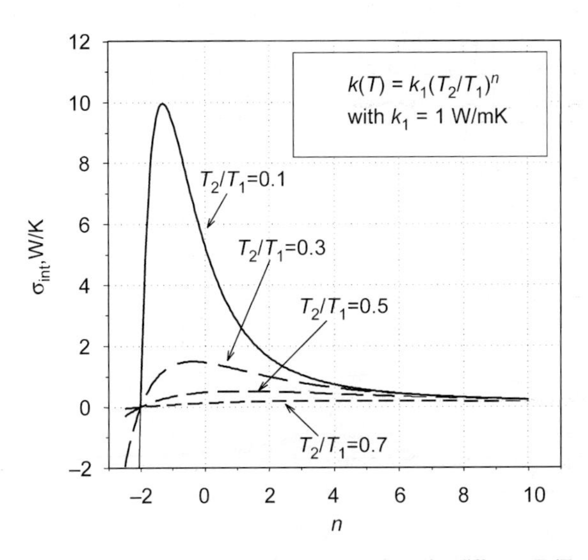

Fig. 4. Internal entropy-generation rate σ_{int} vs. n for different T_2/T_1.

Eq. (14) is

$$\sigma_{t,\min} = \sigma_{\text{int}} + \sigma_{\text{ex}}, \tag{18}$$

and is presented in Fig. 5 (vs. n and $T_2/T_1 = \text{var}$).

1.2. 2D boundary-value problem

Consider the 2D problem of heat conduction without internal heat sources for $k =$ const. The problem is shown schematically in Fig. 6.

The global entropy-generation rate is

$$\sigma_t = \int_0^1 \int_0^1 \sigma \, \mathrm{d}x \, \mathrm{d}y = \int_0^1 \int_0^1 \frac{k}{T^2(x,y)} \left[\left(\frac{\partial T}{\partial x} \right)^2 + \left(\frac{\partial T}{\partial y} \right)^2 \right] \mathrm{d}x \, \mathrm{d}y,$$

and its minimization leads to the Euler-type heat-conduction equation

$$\frac{\partial \sigma}{\partial T} - \frac{\partial}{\partial x}\left(\frac{\partial \sigma}{\partial T_x} \right) - \frac{\partial}{\partial y}\left(\frac{\partial \sigma}{\partial T_y} \right) = 0,$$

and after calculation

$$\frac{\partial^2 T}{\partial x^2} + \frac{\partial^2 T}{\partial y^2} - \frac{1}{T}\left[\left(\frac{\partial T}{\partial x} \right)^2 + \left(\frac{\partial T}{\partial y} \right)^2 \right] = 0. \tag{19}$$

The second term describes the intensity of the additional internal heat source

$$\dot{q}_V(x,y) = -\frac{k}{T}\left[\left(\frac{\partial T}{\partial x} \right)^2 + \left(\frac{\partial T}{\partial y} \right)^2 \right] < 0.$$

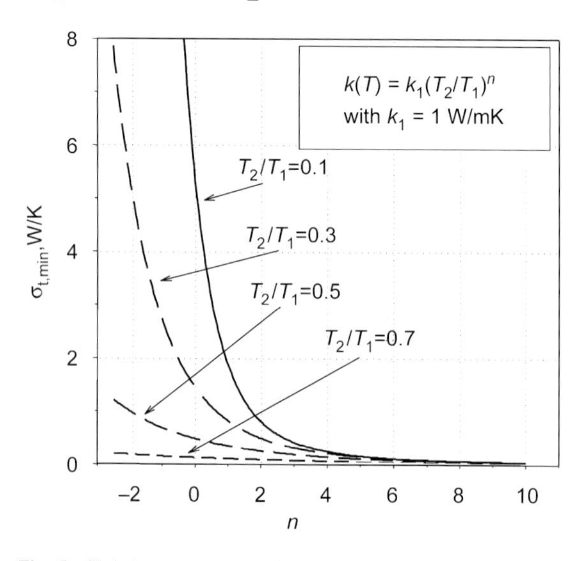

Fig. 5. Global entropy-generation rate (Eq. (13)) vs. n and T_2/T_1.

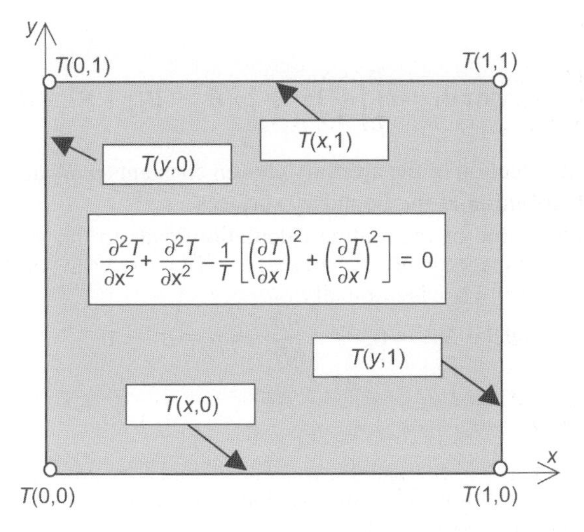

Fig. 6. General scheme of 2D boundary-value problem with first-kind boundary conditions.

Eq. (19) can be written in the shorter form

$$\frac{\partial}{\partial x}\left(\frac{1}{T}\frac{\partial T}{\partial x}\right)+\frac{\partial}{\partial y}\left(\frac{1}{T}\frac{\partial T}{\partial y}\right)=0, \tag{20}$$

or

$$\operatorname{div}\dot{\mathbf{s}}(x,y)=0.$$

The solution of Eq. (20) is easily obtained with the method of separation of variables assuming

$$T(x,y)=X(x)Y(y).$$

It gives finally

$$T(x,y)=A\exp[p(x^2-y^2)+B_1x+B_2y]. \tag{21}$$

Function (21) allows us to formulate necessary and unique boundary conditions (see Fig. 6.)

$$T(0,y)=T(0,0)\exp(-py^2+B_2y)$$

$$T(1,y)=T(0,0)\exp[p(1-y^2)+B_1+B_2y]$$

$$T(x,0)=T(0,0)\exp[px^2+B_1x]$$

$$T(x,1)=T(0,0)\exp[p(x^2-1)+B_1x+B_2].$$

The local entropy-generation rate is

$$\sigma(x,y)=k[(2px+B_1)^2+(-2py+B_2)^2],$$

and its global value

$$\sigma_t = k\int_0^1\int_0^1 \sigma(x,y)\mathrm{d}x\,\mathrm{d}y = k\left[\left(\frac{8}{3}\right)p^2 + 2p(B_1 - B_2) + B_1^2 + B_2^2\right]. \tag{22}$$

This depends on the selection of the arbitrary chosen constants p, B_1 and B_2 responsible for the temperature distribution at the boundary surfaces.

To find minimum value of σ_t, routine calculation leads to the system of three linear equations

$$\frac{\partial\sigma}{\partial p} = \frac{16}{3}p + 2(B_1 - B_2) = 0, \qquad \frac{\partial\sigma}{\partial B_1} = p + B_1 = 0,$$
$$\frac{\partial\sigma}{\partial B_2} = -p + B_2 = 0, \tag{23}$$

having a trivial zero solution. This means that one of the constants must be chosen arbitrarily. As a numerical example let $p = -2/3$.

Solution of Eq. (23) gives $B_1 = 2/3$ and $B_2 = -2/3$ and the value of

$$\sigma_{t,\min} = \frac{8}{27}k.$$

A sufficient condition the value of $\sigma_{t,\min}$ to be minimum requires

$$\text{main minors of } \mathbf{M} = \begin{bmatrix} \dfrac{\partial^2\sigma}{\partial p^2} & \dfrac{\partial^2\sigma}{\partial p\,\partial B_1} & \dfrac{\partial^2\sigma}{\partial p\,\partial B_2} \\ \dfrac{\partial^2\sigma}{\partial p\,\partial B_1} & \dfrac{\partial^2\sigma}{\partial B_1^2} & \dfrac{\partial^2\sigma}{\partial B_1\,\partial B_2} \\ \dfrac{\partial^2\sigma}{\partial p\,\partial B_2} & \dfrac{\partial^2\sigma}{\partial B_1\,\partial B_2} & \dfrac{\partial^2\sigma}{\partial B_2^2} \end{bmatrix}$$

to be positive. After calculation

$$\mathbf{M} = \begin{bmatrix} 16/3 & 2 & -2 \\ 2 & 2 & 0 \\ -2 & 0 & 2 \end{bmatrix},$$

and this means that the sufficiency condition is always satisfied. It is also worthy of note that the matrix $\mathbf{M}$ does not depend on the solution constants.

Physical interpretation of the solution constants B_1, B_2 and separation constant $2p$ can easily be obtained by calculating the entropy fluxes at the boundary surfaces. Calculation gives

$$\dot{s}_{x,\text{in}} = \int_0^1 \frac{k}{T}\frac{\mathrm{d}T}{\mathrm{d}x}\mathrm{d}y\bigg|_{x=0} = B_1 \quad \text{for } k = 1$$

$$\dot{s}_{x,\mathrm{out}} = \int_0^1 \frac{k}{T}\frac{\mathrm{d}T}{\mathrm{d}x}\mathrm{d}y\bigg|_{x=1} = 2p + B_1,$$

and

$$\dot{s}_{y,\mathrm{in}} = \int_0^1 \frac{k}{T}\frac{\mathrm{d}T}{\mathrm{d}y}\mathrm{d}x\bigg|_{y=0} = B_2 \qquad \text{for} \qquad k = 1$$

$$\dot{s}_{y,\mathrm{out}} = \int_0^1 \frac{k}{T}\frac{\mathrm{d}T}{\mathrm{d}y}\mathrm{d}x\bigg|_{y=1} = -2p + B_2.$$

For any boundary conditions, the solution can easily be obtained by transformation

$$\theta(x,y) \equiv \ln T(x,y).$$

In such a case, Eq. (19) takes the form

$$\frac{\partial^2\theta}{\partial x^2} + \frac{\partial^2\theta}{\partial y^2} = 0, \tag{24}$$

and entropy-generation rates are given by

- *local*

$$\sigma(x,y) = k\left[\left(\frac{\partial\theta}{\partial x}\right)^2 + \left(\frac{\partial\theta}{\partial y}\right)^2\right],$$

- *global*

$$\sigma_t = k\int_{x_1}^{x_2}\int_{y_1}^{y_2}\left[\left(\frac{\partial\theta}{\partial x}\right)^2 + \left(\frac{\partial\theta}{\partial y}\right)^2\right]\mathrm{d}x\,\mathrm{d}y.$$

The boundary-value problem formulated by Eq. (24) and any boundary conditions can be solved with methods available in the heat-conduction literature.

1.3. 3D boundary-value problems

Let us consider the 3D boundary-value problem of heat conduction in the cube for $k = \mathrm{const}$ with first-kind boundary conditions. The problem is shown in Fig. 7.

Minimization of the global entropy-generation rate

$$\sigma_t = \int_0^1\int_0^1\int_0^1 \sigma(x,y,z)\mathrm{d}x\,\mathrm{d}y\,\mathrm{d}z$$

$$= \int_0^1\int_0^1\int_0^1 \frac{k}{T^2}\left[\left(\frac{\partial T}{\partial x}\right)^2 + \left(\frac{\partial T}{\partial y}\right)^2 + \left(\frac{\partial T}{\partial z}\right)^2\right]\mathrm{d}x\,\mathrm{d}y\,\mathrm{d}z, \tag{25}$$

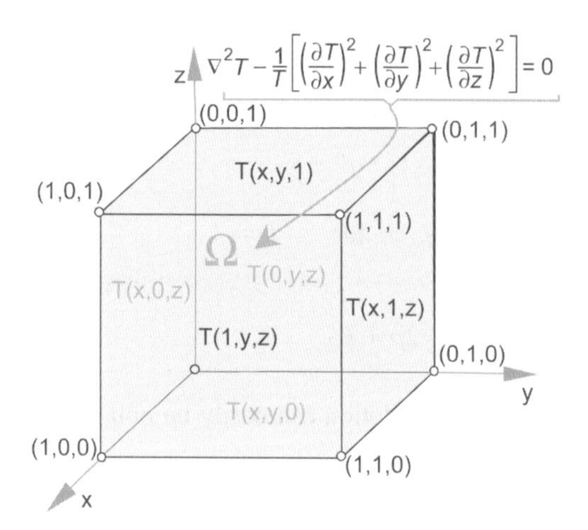

Fig. 7. 3D boundary-value problem.

leads to the Euler equation

$$k\nabla^2 T(x,y,z) - \frac{k}{T}\left[\left(\frac{\partial T}{\partial x}\right)^2 + \left(\frac{\partial T}{\partial y}\right)^2 + \left(\frac{\partial T}{\partial z}\right)^2\right] = 0, \tag{26}$$

where the second term describes an additional internal heat source.

$$\nabla^2 T - \frac{1}{T}\left[\left(\frac{\partial T}{\partial x}\right)^2 + \left(\frac{\partial T}{\partial y}\right)^2 + \left(\frac{\partial T}{\partial z}\right)^2\right] = 0.$$

Eq. (25) can then be written in the form

$$\frac{\partial}{\partial x}\left(\frac{1}{T}\frac{\partial T}{\partial x}\right) + \frac{\partial}{\partial y}\left(\frac{1}{T}\frac{\partial T}{\partial y}\right) + \frac{\partial}{\partial z}\left(\frac{1}{T}\frac{\partial T}{\partial z}\right) = 0,$$

or

$$\mathrm{div}(\dot{\mathbf{s}}) = 0.$$

Its solution is

$$T(x,y,z) = A\exp[p(x^2 - z^2) - r(y^2 + z^2) + B_1 x + B_2 y + B_3 z],$$

where p and r are separation constants, A, B_1, B_2, B_3 are integration constants. Using obtained solution, the boundary conditions required to minimize the entropy-generation rate are

$$T(x, y, 0) = A\exp[px^2 - ry^2 + B_1x + B_2y]$$
$$T(x, y, 1) = A\exp[p(x^2 - 1) - r(y^2 + 1) + B_1x + B_2y + B_3]$$
$$\vdots$$
$$T(1, y, z) = A\exp[p(1 - z^2) - r(y^2 + z^2) + B_1 + B_2y + B_3z].$$

The local entropy-generation rate is

$$\sigma(x, y, z) = k\{(2px + B_1)^2 + (-2ry + B_2)^2 + [2(p - r)z + B_3]^2\}, \tag{27}$$

and its global value

$$\sigma_t = \int_0^1\int_0^1\int_0^1 \sigma(x, y, z)\mathrm{d}x\,\mathrm{d}y\,\mathrm{d}z$$
$$= k\left\{\frac{4}{3}p^2 + 2pB_1 + B_1^2 + \frac{4}{3}r^2 - 2rB_2 + B_2^2 + \frac{4}{3}(p - r)^2 + 2(p - r)B_3 + B_3^2\right\}. \tag{28}$$

The minimum value $\sigma_{t,\min}$ can be obtained from the necessary conditions

$$\frac{\partial\sigma}{\partial p} = 0, \qquad \frac{\partial\sigma}{\partial r} = 0, \qquad \frac{\partial\sigma}{\partial B_1} = 0, \qquad \frac{\partial\sigma}{\partial B_2} = 0, \qquad \frac{\partial\sigma}{\partial B_3} = 0.$$

All sufficient conditions $\sigma_{t,\min}$ to be minimum are also satisfied as in the case of the 2D problem.

1.4. Electric power cables

The geometry of the problem is presented in Fig. 8.

The cable conducts electric current of density i and power is dissipated by Joule heat released inside the cable as the internal heat source. Thermal conductivity k and electric resistivity ρ are assumed to be functions of temperature, $k = k(T)$ and $\rho = \rho(T)$.

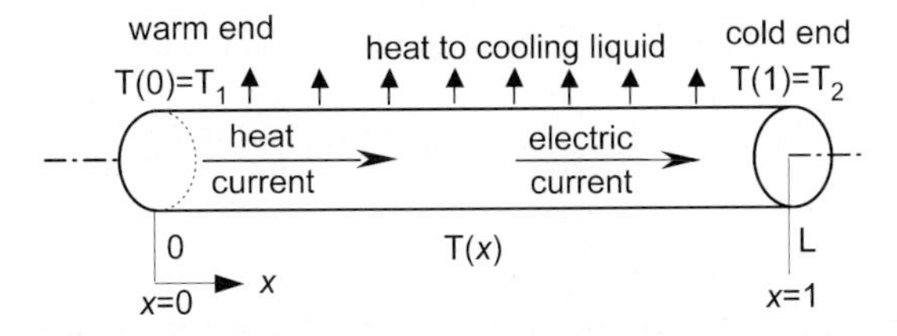

Fig. 8. Electric power cable.

The entropy-generation rate is

$$\sigma(x) = \frac{k(T)}{T^2}\left(\frac{\mathrm{d}T}{\mathrm{d}x}\right)^2 + \frac{i^2\rho(T)}{T}.$$

Minimization of the integral

$$\sigma_t = \int_0^1 \sigma(x)\mathrm{d}x \rightarrow \min$$

leads to the Euler equation

$$\frac{\mathrm{d}^2T}{\mathrm{d}x^2} + \frac{1}{k(T)}\frac{\mathrm{d}k}{\mathrm{d}T}\left(\frac{\mathrm{d}T}{\mathrm{d}x}\right)^2 + i^2\frac{\rho(T)}{k(T)} - \left(\frac{1}{2k(T)}\frac{\mathrm{d}k}{\mathrm{d}T} + \frac{1}{T}\right)\left(\frac{\mathrm{d}T}{\mathrm{d}x}\right)^2 - \frac{1}{2}i^2\left(\rho(T) + T\frac{\mathrm{d}\rho}{\mathrm{d}T}\right) = 0, \tag{29}$$

where

$$\dot{q}_V(x) = -\left(\frac{1}{2k(T)}\frac{\mathrm{d}k}{\mathrm{d}T} + \frac{1}{T}\right)\left(\frac{\mathrm{d}T}{\mathrm{d}x}\right)^2 - \frac{1}{2}i^2\left(\rho(T) + T\frac{\mathrm{d}\rho}{\mathrm{d}T}\right) \tag{30}$$

represents an additional internal heat source to minimize entropy production.

The global heat release is equal to

$$\dot{q}_{V,t} = \int_0^1 \dot{q}_V(x)\mathrm{d}x.$$

The first three terms of Eq. (29) represent the classical heat-conduction equation with an internal heat source. Eq. (30) can be written in a general way as

$$\frac{\mathrm{d}^2T}{\mathrm{d}x^2} + f_1(T)\left(\frac{\mathrm{d}T}{\mathrm{d}x}\right)^2 + f_2(T) = 0. \tag{31}$$

Solution of Eq. (31) can only be found by numerical methods.

Eq. (30) describes additional heat that has to be removed from or supplied to the cable to ensure minimum of entropy generation.

2. Heat conduction in anisotropic solids

The simplest and most frequently used assumption for anisotropic solids is that the components of the heat flux vector, $\dot{q}_x$, $\dot{q}_y$, $\dot{q}_z$ are linear functions of the temperature gradient components $\partial T/\partial x$, $\partial T/\partial y$, $\partial T/\partial z$ at the points [2]. This means that

$$-\dot{q}_x = k_{11}\frac{\partial T}{\partial x} + k_{12}\frac{\partial T}{\partial y} + k_{13}\frac{\partial T}{\partial z}, \quad -\dot{q}_y = k_{21}\frac{\partial T}{\partial x} + k_{22}\frac{\partial T}{\partial y} + k_{23}\frac{\partial T}{\partial z},$$
$$-\dot{q}_z = k_{31}\frac{\partial T}{\partial x} + k_{32}\frac{\partial T}{\partial y} + k_{33}\frac{\partial T}{\partial z}, \tag{32}$$

where k_{ij} denotes thermal-conductivity coefficients and they are the components of a second-order tensor

$$\mathbf{K} = \begin{vmatrix} k_{11} & k_{12} & k_{13} \\ k_{21} & k_{22} & k_{23} \\ k_{31} & k_{32} & k_{33} \end{vmatrix}.$$

The general form of **K** can be simplified when various symmetry system are considered [2]. It is also known from irreversible thermodynamics that **K** is symmetrical, that is, that $k_{ij} = k_{ji}$ and $k_{12} = 0$ for tetragonal systems.

Eqs. (32) can be written in the matrix form

$$-\dot{\mathbf{q}} = \mathbf{K} \operatorname{grad} T(x).$$

Using Eq. (32) and applying a general mathematical procedure, the differential steady-state heat-conduction equation takes the form

$$k_{11}\frac{\partial^2 T}{\partial x^2} + k_{22}\frac{\partial^2 T}{\partial y^2} + k_{33}\frac{\partial^2 T}{\partial z^2} + (k_{12}+k_{21})\frac{\partial^2 T}{\partial x\,\partial y} + (k_{23}+k_{32})\frac{\partial^2 T}{\partial y\,\partial z}$$
$$+ (k_{31}+k_{13})\frac{\partial^2 T}{\partial x\,\partial z} = 0, \tag{33}$$

provided the solid is homogeneous and heat is not generated inside. It is easy to show that Eq. (33) can be transformed into the shorter form

$$k_1\frac{\partial^2 T}{\partial \xi^2} + k_2\frac{\partial^2 T}{\partial \eta^2} + k_3\frac{\partial^2 T}{\partial \zeta^2} = 0,$$

where ξ, η, and ζ denotes a new system of rectangular coordinates and k_1, k_2, and k_3 are called the principal conductivities.

Introducing thermodynamic forces [3]

$$\mathbf{X} = \nabla\left(\frac{1}{T}\right)$$

the local entropy-generation rate is

$$\sigma_l = \dot{\mathbf{q}} \circ \mathbf{X},$$

or

$$\sigma_l = \sum_{i=1}^{3} q_{x_i}\frac{\partial}{\partial x_i}\left(\frac{1}{T}\right), \quad x_i = \{x, y, z\}. \tag{34}$$

The problem can be formulated in the following way

– find temperature field $T(\mathbf{x})$ that satisfies the boundary condition and minimizes the entropy-generation integral

$$\sigma_t = \int_{\Omega} \sigma_l \, \mathrm{d}\Omega \tag{35}$$

over the whole domain Ω where σ_l is given by Eq. (34).

To solve the problem the variational calculus can be used. The function $T(\mathbf{x})$ for which σ_t reaches a minimum must satisfy the Euler–Lagrange equation

$$\frac{\partial \sigma_l}{\partial T} - \sum \frac{\partial}{\partial x_i}\left(\frac{\partial \sigma_l}{\partial T_{x_i}}\right) = 0. \tag{36}$$

Introducing Eq. (34) into Eq. (36) after differentiation the heat-conduction equation becomes

$$k_{11}\frac{\partial^2 T}{\partial x^2} + k_{22}\frac{\partial^2 T}{\partial y^2} + k_{33}\frac{\partial^2 T}{\partial z^2} + (k_{12}+k_{21})\frac{\partial^2 T}{\partial x\,\partial y} + (k_{23}+k_{32})\frac{\partial^2 T}{\partial y\,\partial z} + (k_{13}+k_{31})\frac{\partial^2 T}{\partial x\,\partial z} + \dot{q}_{V,\min} = 0, \tag{37}$$

where an additional heat source required for the entropy-generation rate to reach minimum is

$$\dot{q}_{V,\min} = -\frac{1}{T}\left[k_{11}\left(\frac{\partial T}{\partial x}\right)^2 + k_{22}\left(\frac{\partial T}{\partial y}\right)^2 + k_{33}\left(\frac{\partial T}{\partial z}\right)^2\right] - \frac{(k_{12}+k_{21})}{T}\frac{\partial T}{\partial x}\frac{\partial T}{\partial y} - \frac{(k_{23}+k_{32})}{T}\frac{\partial T}{\partial y}\frac{\partial T}{\partial z} - \frac{(k_{13}+k_{31})}{T}\frac{\partial T}{\partial x}\frac{\partial T}{\partial z}. \tag{38}$$

Eq. (37) can also be rewritten in the shorter and more physical way as

$$\frac{\partial}{\partial x}\left[\frac{1}{T}\left(k_{11}\frac{\partial T}{\partial x} + k_{22}\frac{\partial T}{\partial y} + k_{33}\frac{\partial T}{\partial z}\right)\right] + \frac{\partial}{\partial y}\left[\frac{1}{T}\left(k_{11}\frac{\partial T}{\partial x} + k_{22}\frac{\partial T}{\partial y} + k_{33}\frac{\partial T}{\partial z}\right)\right] + \frac{\partial}{\partial z}\left[\frac{1}{T}\left(k_{11}\frac{\partial T}{\partial x} + k_{22}\frac{\partial T}{\partial y} + k_{33}\frac{\partial T}{\partial z}\right)\right] = 0,$$

or

$$\operatorname{div}\left(\frac{\dot{\mathbf{q}}}{T}\right) = \operatorname{div}(\dot{\mathbf{s}}) = 0,$$

and its physical interpretation is identical to Eq. (7).

To present the method, heat flow for an orthorhombic system will be considered. In this case

$$-\dot{q}_x = k_x\frac{\partial T}{\partial x}, \quad -\dot{q}_y = k_y\frac{\partial T}{\partial y}, \quad -\dot{q}_z = k_z\frac{\partial T}{\partial z},$$

and from Eq. (34)

$$\sigma_l = \frac{k_x}{T^2}\left(\frac{\partial T}{\partial x}\right)^2 + \frac{k_y}{T^2}\left(\frac{\partial T}{\partial y}\right)^2 + \frac{k_z}{T^2}\left(\frac{\partial T}{\partial z}\right)^2. \tag{39}$$

Minimization of the integral (35) gives the heat-conduction equation

$$k_x\frac{\partial^2 T}{\partial x^2} + k_y\frac{\partial^2 T}{\partial y^2} + k_z\frac{\partial^2 T}{\partial z^2} - \frac{1}{T}\left[k_x\left(\frac{\partial T}{\partial x}\right)^2 + k_y\left(\frac{\partial T}{\partial y}\right)^2 + k_z\left(\frac{\partial T}{\partial z}\right)^2\right] = 0. \tag{40}$$

The term

$$\dot{q}_{V,\min} = \frac{1}{T}\left[k_x\left(\frac{\partial T}{\partial x}\right)^2 + k_y\left(\frac{\partial T}{\partial y}\right)^2 + k_z\left(\frac{\partial T}{\partial z}\right)^2\right]$$

is interpreted as an additional local internal source of heat that must be continuously removed from the solid to ensure entropy-generation minimization.

It can be proved that when k_x, k_y and k_z are not functions of temperature, the entropy-generation rate is

$$\sigma_t = \iiint_\Omega \frac{q_{V,\min}}{T}\,\mathrm{d}x\,\mathrm{d}y\,\mathrm{d}z.$$

The problem when conductivities k_x, k_y and k_z depend on temperature can be solved in a similar way.

2.1. 2D boundary-value problem

Consider the 2D problem of heat conduction in the infinite rod (Fig. 9).

Assuming

$$\mathbf{K} = \begin{vmatrix} k_x & 0 \\ 0 & k_y \end{vmatrix}$$

the Euler–Lagrange equation becomes

$$k_x\frac{\partial^2 T}{\partial x^2} + k_y\frac{\partial^2 T}{\partial y^2} - \frac{1}{T}\left[k_x\left(\frac{\partial T}{\partial x}\right)^2 + k_y\left(\frac{\partial T}{\partial y}\right)^2\right] = 0,$$

which can be easily solved with the method of separation of variables to give

$$T(x,y) = A\exp\left[\frac{\beta}{2}(x^2 - \alpha y^2) + C_1 x + \alpha D_1 y\right], \tag{41}$$

where β is a separation constant, $\alpha = k_x/k_y$, C_1 and D_1 are integration constants.

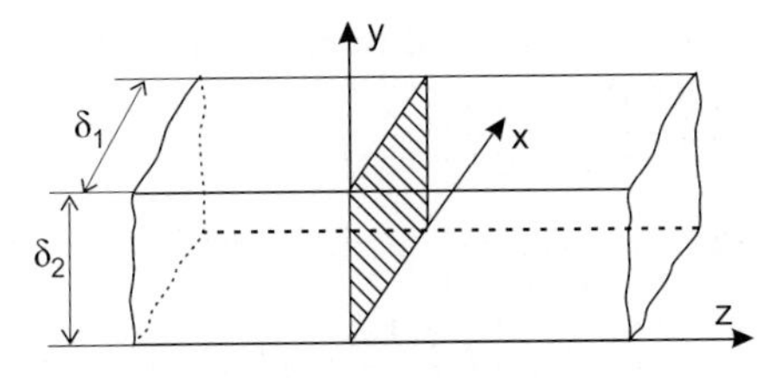

Fig. 9. Infinite-rod geometry.

The solution of Eq. (41) allows us to find boundary conditions ensuring minimization of entropy generation. They are

$$
\begin{aligned}
T(x,0) &= A\exp\left(\frac{\beta}{2}x^2 + C_1 x\right),\\
T(x,\delta_2) &= A\exp\left[\frac{\beta}{2}(x^2 - \alpha\delta_2^2) + C_1 x + \alpha D_1\delta_2\right],\\
T(0,y) &= A\exp\left(-\frac{\beta}{2}\alpha y^2 + \alpha D_1 y\right),\\
T(\delta_1,y) &= A\exp\left[\frac{\beta}{2}(\delta_1^2 - \alpha y^2) + C_1\delta_1 + \alpha D_1 y\right].
\end{aligned}
\tag{42}
$$

The method to calculate the values of all constants (β, C_1 and D_1) has been described in Part 1 for isotropic solids.

Using the solution of Eq. (41), the local entropy-generation rate is

$$\sigma_l(x,y) = k_x(\beta x + C_1)^2 + k_y(-\beta\alpha y + \alpha D_1)^2,$$

and after integration

$$
\begin{aligned}
\sigma_t &= \int_0^{\delta_1}\int_0^{\delta_2}\sigma_l(x,y)\mathrm{d}x\,\mathrm{d}y\\
&= \frac{1}{3}k_x\delta_1\delta_2[\beta^2\delta_1^2 + 3\beta\delta_1 C_1 + 3C_1^2 - \alpha(-\beta^2\delta_2^2 + 3\beta\delta_2 D_1 - 3D_1^2)].
\end{aligned}
\tag{43}
$$

Eq. (43) allows us to evaluate the value of $\alpha = k_x/k_y$ for which the entropy-generation rate is a minimum under imposed boundary conditions.

2.2. 2D boundary-value problem with internal heat generation

Consider a 2D steady-state heat-conduction boundary-value problem in an infinite anisotropic rod heated by an electric current of uniform density i against electric resistivity

$$\rho(T) = a + bT,$$

where a and b are constants.

Boundary conditions have been assumed to be of the first kind.

Two problems are considered:

- *classical*

$$k_x\frac{\partial^2 T}{\partial x^2} + k_y\frac{\partial^2 T}{\partial y^2} + i^2\rho(T) = 0, \tag{44}$$

where $\dot{q}_V = i^2\rho(T)$ represents the electric power dissipated via Joule heating,

– *minimization of entropy generation* (Euler–Lagrange equation)

$$k_x \frac{\partial^2 T^*}{\partial x^2} + k_y \frac{\partial^2 T^*}{\partial y^2} + i^2 \rho(T^*) - \frac{1}{T^*}\left[k_x\left(\frac{\partial T^*}{\partial x}\right)^2 + k_y\left(\frac{\partial T^*}{\partial y}\right)^2\right]$$

$$- \frac{1}{2} i^2 \rho(T^*) - \frac{1}{2} i^2 \frac{d\rho(T^*)}{dT^*} T^* = 0, \qquad (45)$$

with constant k_x and k_y. $T^* = T^*(x, y)$ represents the temperature field under the condition of entropy minimization.

The term

$$\dot{q}_{V,\min}(x, y) = -\frac{1}{T^*}\left[k_x\left(\frac{\partial T^*}{\partial x}\right)^2 + k_y\left(\frac{\partial T^*}{\partial y}\right)^2\right] - \frac{1}{2} i^2 \rho(T^*) - \frac{1}{2} i^2 \frac{d\rho(T^*)}{dT^*} T^*$$

represents the local additional internal heat source that must be continuously removed to the external heat source to ensure entropy-generation minimization.

Entropy-generation rates can be calculated from entropy balances and they are

– *classical problem*

$$\sigma_t = \int_0^{\delta_1} \int_0^{\delta_2} i^2 \rho(T)\left(\frac{1}{T_{\text{boundary}}} - \frac{1}{T(x, y)}\right) dx\, dy,$$

– *entropy minimizations*

$$\sigma_t = \int_0^{\delta_1} \int_0^{\delta_2} (i^2 \rho(T^*) - \dot{q}_{V,\min})\left(\frac{1}{T_{\text{boundary}}} - \frac{1}{T^*(x, y)}\right) dx\, dy.$$

2.3. *Numerical example*

To obtain the numerical solution of Eqs. (44) and (45), the finite-difference method has been applied. Data for calculations are as follows:

$$k_x = 10.0 \text{ W/m K}, \quad \alpha = \frac{k_x}{k_y} = \text{var}, \quad \delta_1 = 1.0 \text{ m}, \quad \delta_2 = 2.0 \text{ m},$$

$$i = 7000 \text{ A/m}^2$$

$\rho(T) = (0.5 - 0.3 \times 10^{-3} T) \times 10^{-4}\ \Omega$ m (carbon) with boundary condition $T_{\text{boundary}} =$ 1000 K. Selected calculation results are presented in Figs. 10–23 for one-fourth of the cross-sectional surface area as the problem is symmetrical.

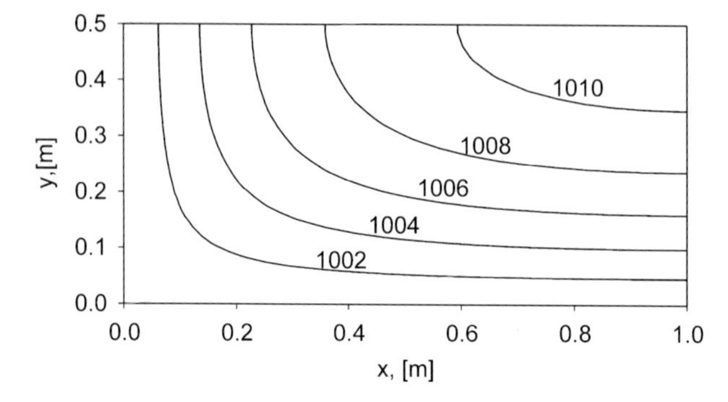

Fig. 10. Temperature [K], classical problem, $\alpha = k_y/k_x = 1.0$ (isotropic solid).

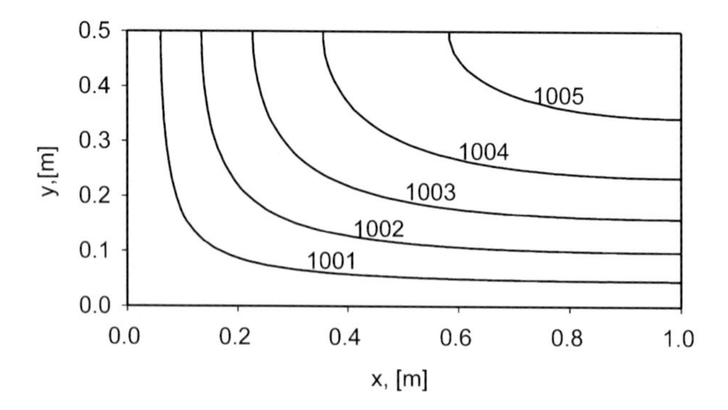

Fig. 11. Temperature [K], entropy minimization, $\alpha = k_y/k_x = 1.0$ (isotropic solid).

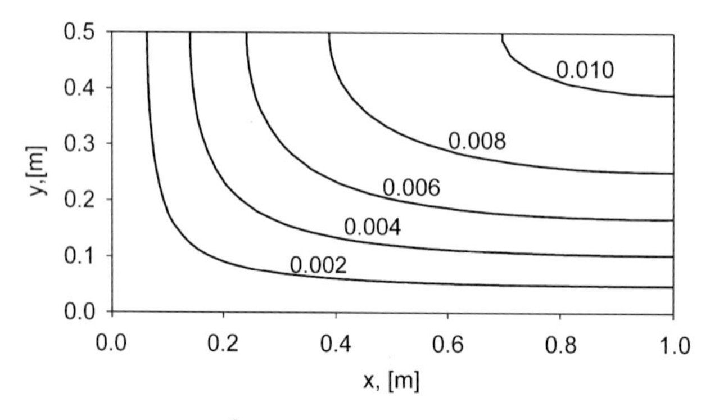

Fig. 12. Entropy source [W/m^3 K], classical problem, $\alpha = k_y/k_x = 1.0$ (isotropic solid).

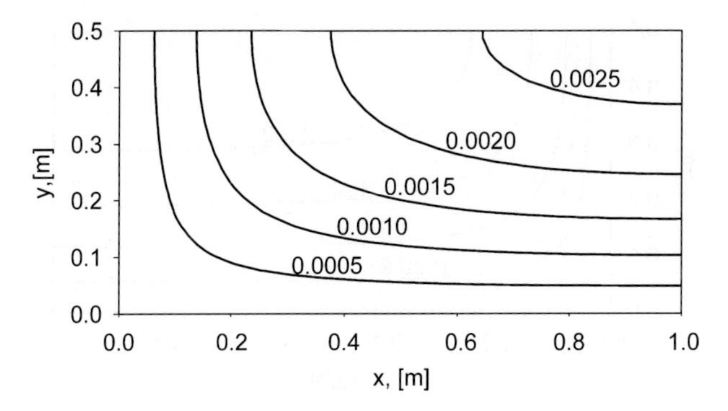

Fig. 13. Entropy source [W/m^3 K], entropy minimization, $\alpha = k_y/k_x = 1.0$ (isotropic solid).

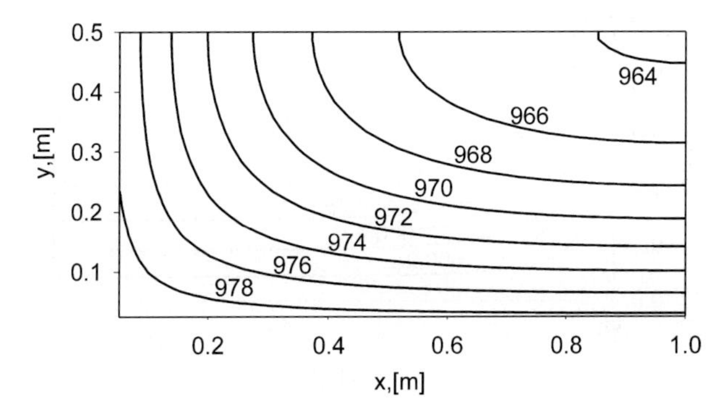

Fig. 14. Joule heat-generation rate [W/m^3], classical problem, $\alpha = k_y/k_x = 1.0$ (isotropic solid).

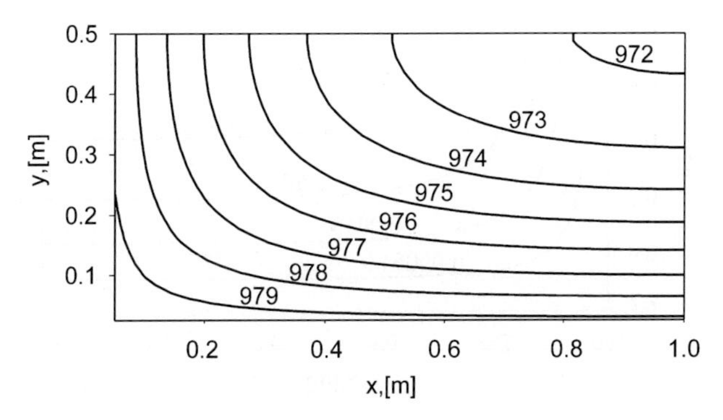

Fig. 15. Joule heat-generation rate [W/m^3], entropy minimization, $\alpha = k_y/k_x = 1.0$ (isotropic solid).

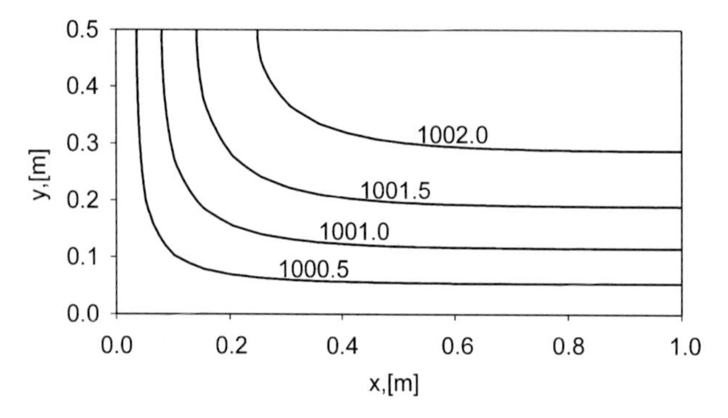

Fig. 16. Temperature [K], classical problem, $\alpha = k_y/k_x = 5.0$ (anisotropic solid).

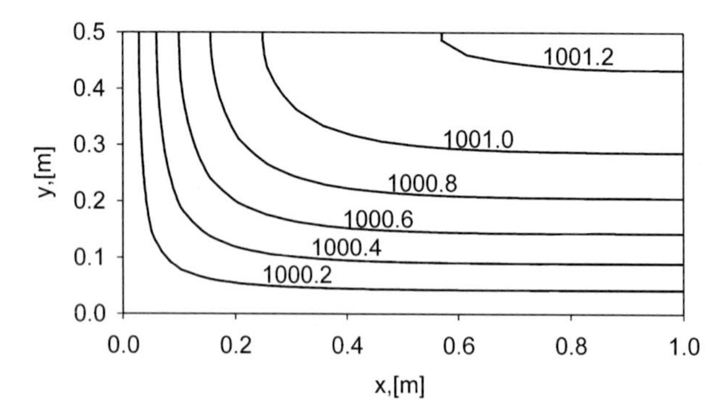

Fig. 17. Temperature [K], entropy minimization, $\alpha = k_y/k_x = 5.0$ (anisotropic solid).

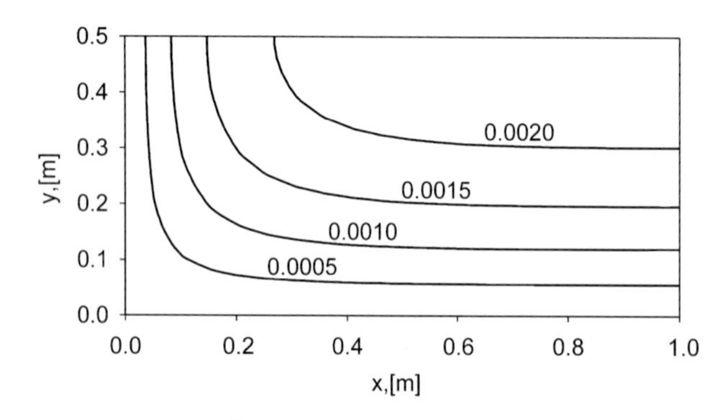

Fig. 18. Entropy source [W/m^3 K], classical problem, $\alpha = k_y/k_x = 5.0$ (anisotropic solid).

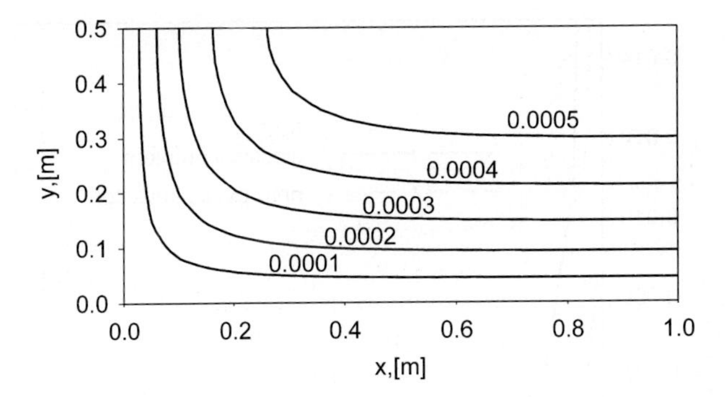

Fig. 19. Entropy source [W/m^3 K], entropy minimization, $\alpha = k_y/k_x = 5.0$ (anisotropic solid).

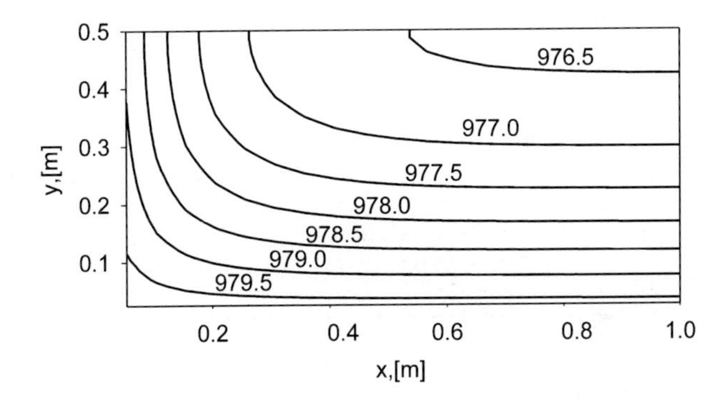

Fig. 20. Joule heat-generation rate [W/m^3], classical problem, $\alpha = k_y/k_x = 5.0$ (anisotropic solid).

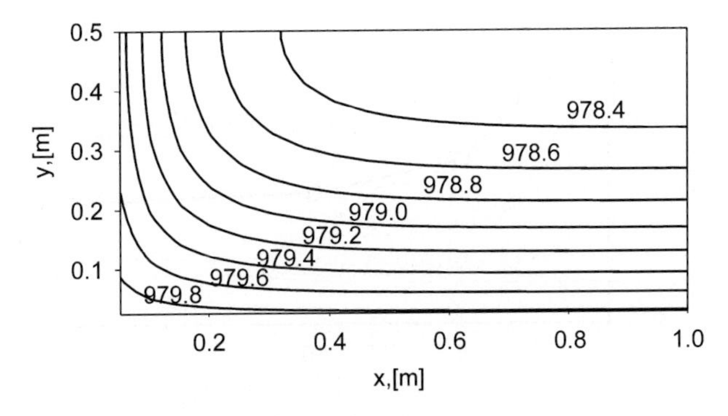

Fig. 21. Joule heat-generation rate [W/m^3], entropy minimization, $\alpha = k_y/k_x = 5.0$ (anisotropic solid).

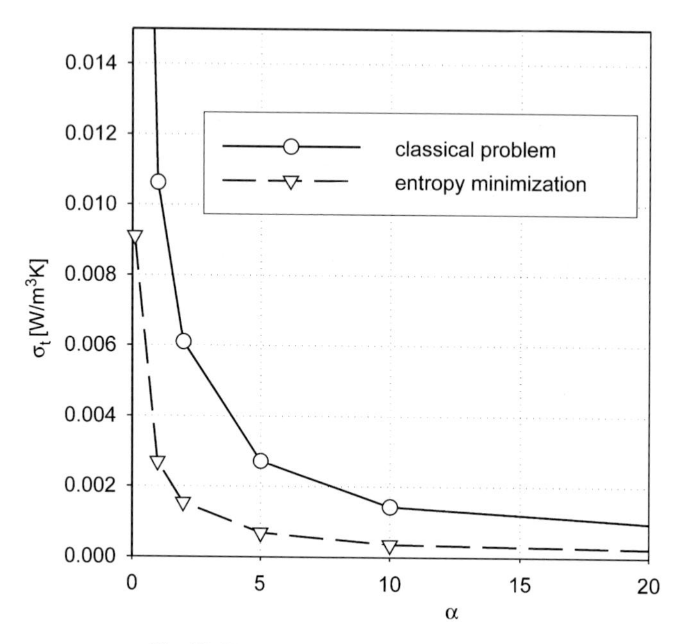

Fig. 22. Entropy generation σ_t vs. $\alpha = k_y/k_x$.

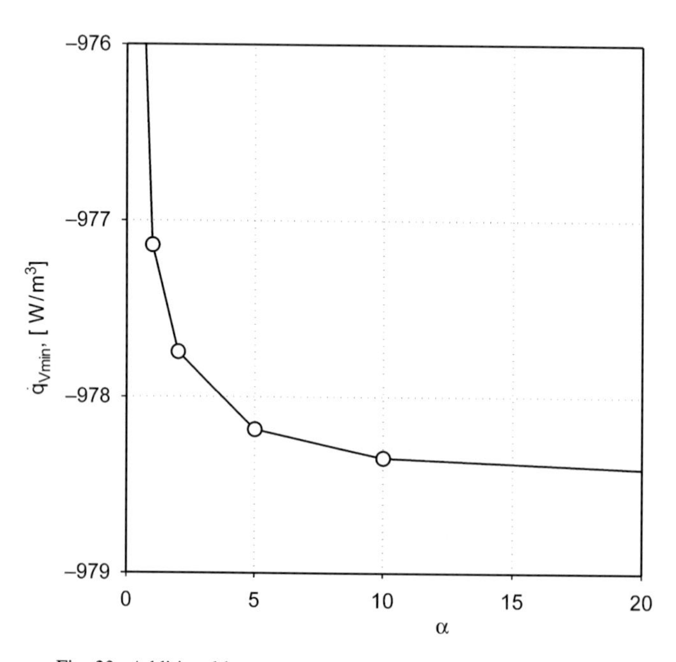

Fig. 23. Additional heat source $\dot{q}_{V,\min}$ vs. α, entropy minimization.

3. Conclusion

The boundary-value problem with minimization of entropy generation in heat conduction in anisotropic solids has been discussed. The Euler–Lagrange heat-conduction equation has been derived together with the expressions describing the entropy-generation rate and additional heat sources. Comparison of the solutions of the classical boundary-value problem with the entropy-generation minimization problem shows significant differences. It has also been shown that thermal anisotropism influences the temperature distribution inside the solid and it is especially important when the intensity of the internal heat source depends on temperature as presented in a numerical example. Interpretation of the results obtained points out many possible practical applications. It does not mean that the classical heat conduction or the Fourier Constitutive equation are wrong. Both approaches, classical and derived from the Principle of Minimum Entropy Generation (PMEG) are correct. The physical explanation comes from the Principle of Entropy Compensation (F. Reif, Statistical Physics, Berkeley Physics Course, Vol. 5., p. 301) that

> "The entropy of a system can be reduced if it is made to interact with one or more auxiliary systems in a process which imparts to these at least a compensating amount of entropy."

The additional heat source that appear in Eqs. (5), (9) and (37) should be interpreted as the *auxiliary systems* in the above Principle.

Acknowledgements

This work was supported by the Polish Scientific Committee (KBN) under the grant No. 4 T08 02725.

Nomenclature

k	heat conduction coefficient (W/m K)
$\dot{q}$	heat flux (W/m^2)
$\dot{q}_V$	intensity of internal heat source (W/m^3)
$\dot{s}$	entropy flux (W/m^2 K)
$\dot{i}^s$	entropy generation due to the process irreversibility (W/m^2 K) (Eq. (2))
T	absolute temperature (K)
x_i	Cartesian coordinates, $(i = 1, 2, 3)$
σ, σ_t, $\sigma_{t,\min}$	local, global, and global minimized intensity of entropy-generation rate (W/m^2 K)
θ	transformed temperature (general function)
Ω	domain, general function

References

[1] Bejan, A. (1996), Entropy Generation Minimization, CRC Press, Boca Raton, FL, Chap. 6.
[2] Carslaw, H.S. and Jaeger, J.C. (1959), Conduction of Heat in Solids2nd edn, Oxford University Press, Oxford.
[3] Kondepudi, D. and Prigogine, I. (1998), Modern Thermodynamics, Wiley, London, Chap. 16.

Chapter 12

THE NONEQUILIBRIUM THERMODYNAMICS OF RADIATION INTERACTION

Christopher Essex

Department of Applied Mathematics, University of Western Ontario, London, Ontario, Canada N6A 5B7

Dallas C. Kennedy

The MathWorks, Inc., 3 Apple Hill Drive, Natick, MA 01760, USA

Sidney A. Bludman

Department of Physics, University of Pennsylvania, Philadelphia, PA 19104, USA
Deutsches Elektron Synchrotron Notkestrasse 85, D-22603 Hamburg, Germany

Abstract

We review some important recent developments in the nonequilibrium thermodynamics of radiation and matter–radiation mixtures. These include variational principles for nonequilibrium steady states of photons, neutrinos, and matter in local thermodynamic equilibrium. These variational principles can be extended to include mass and chemical potential. The general nature of radiation entropy, entropy production, equilibrium, and nonequilibrium is also discussed.

PACS: 05.30.-d; 05.70.Ln; 11.10.Wx; 44.05. + e; 44.40. + a

Keywords: thermodynamics; nonequilibrium; radiation; variational; phase space; mode space

1. Introduction

This review presents recent results in the nonequilibrium thermodynamics of radiation. The term 'radiation' means nothing more than particles moving along a ray [1]. It is broad enough to include massless, nonconserved quanta of any type—photons but also neutrinos, for example. The fundamental point of the investigations described here is to develop a framework for the nonequilibrium thermodynamics of systems composed of any particles. Elementary particles are the quanta of quantum fields, the fundamental entities of physics [2,3]. A general thermodynamics based on quantum

E-mail address: essex@uwo.ca (C. Essex); dkennedy@mathworks.com (D.C. Kennedy); bludman@mail.desy.de (S.A. Bludman).

fields would be difficult and not illuminating. A middle ground between a quantum field and a macroscopic description is more useful [4]. In practice, a quantum field description can be reduced to a description in terms of quanta of *field modes*. The modes are labeled by position, momentum or wave number, and possibly other quantum numbers such as spin (polarization, in classical terms), charges (electric, baryonic, leptonic, etc.), or particle identity. The space of all these labels is *mode space*; position and momentum alone define phase space.

A broad feature of this picture of thermodynamics emerges from the practical distinction between 'matter' made of massive, conserved quanta (electrons, neutrons, protons, etc.) and massless or nearly massless 'radiation' such as photons and neutrinos [5,6]. In most situations, matter is localizable in space and naturally forms 'blobs.' The momentum of microscopic particles can be hidden by partial integration of phase space, leaving only position space. Matter thermodynamics is then formulated in terms of static functions over finite volumes. In contrast, except in exotic conditions such as stellar interiors, radiation is usually 'free-streaming,' in constant motion and not localized. Radiation usually requires an explicit description of momentum as well as position space, so that the natural extensive functions are fluxes representing beams, not 'blobs' [7,8]. This picture is familiar from everyday life: matter is localized and close to equilibrium; radiation is out of equilibrium and not localized. It is streaming photon beams that inform us about the localized matter blobs.

These conditions reflect a distinction familiar in the physical world. Massive, conserved matter and normally hidden short-range forces (interatomic, intermolecular, weak and strong nuclear forces) are localized in space. Massless particles (photons, neutrinos, gravitons) generate long-range effects that can be naturally viewed as nonlocal. But if we take into account momentum as well as position variables, the physics of massless quanta is just as local as that of massive particles. Classical thermodynamics, for historical reasons, often biases our thinking with localized matter as the sole paradigm [6,9]. All particles are in motion in any case, but radiation makes this fact inescapable. Classical thermodynamics also bends our thinking towards macroscopic or phenomenological descriptions, at least for matter. But the mode space of quantum fields provides a single comprehensive framework for expressing the degrees of freedom of all particles, bosons or fermions, massive or massless, conserved or not conserved [10].

1.1. Entropy without equilibrium

After counting the quanta of field modes, physical quantities—energy, volume, entropy—can be constructed for the fields, mode by mode [9]. If the mode entropy is a function of other mode variables such as energy, we can even derive intensive thermodynamic variables—such as temperature—mode by mode [11,12]. These differ for each mode, unless the system in question is in equilibrium. What results is a generalization of classical nonequilibrium thermodynamics to the full mode space. As long as the list of mode labels is complete, the thermodynamics is also complete. Section 2 gives some examples of constructing the entropy for massless quanta independently of equilibrium. The method can be used for the quanta of any field.

1.2. Entropy production in volumes and on surfaces

A thermodynamic system out of equilibrium exhibits differences of intensive parameters α_a, as well as the creation, flow, and destruction of extensive variables H^a. For example, temperature differences drive heat flows, and pressure differences drive volume flows. Differences in intensive variables α_a, conjugate in the entropy picture to H^a, are *thermodynamic forces* $X_a = \Delta\alpha_a$; the flows are *thermodynamic fluxes* J^a. The local definition of variable intensive parameters requires *local thermodynamic equilibrium* (LTE) [6,8,9]. The LTE concept can be generalized to locality not only in position space, but anywhere in mode space.

The *entropy production* Σ of a system can be expressed as a sum of products of intensive differences and their associated extensive fluxes. The general Gibbs form of the entropy increment is $\mathrm{d}S = \sum_a \alpha_a \cdot \mathrm{d}H^a$. The entropy production from any intensive–extensive pair is $\Sigma_a \sim X_a \cdot J^a \geq 0$. The value of Σ suggests the rate at which a system is approaching equilibrium or, if it is constrained to avoid equilibration, how much entropy must be dissipated to keep it in that state. In the latter case, the outside constraints must also supply the flows of heat, matter, etc. that maintain the nonequilibrium state.

Entropy production can be expressed as a volume integral over a nonequilibrium system. Some forms of entropy production are strictly local, while others arise from spatial currents. Local forms can be defined in the familiar way using densities of extensive variables. The current terms represent both transport entropy production within the volume and possibly surface contributions. If the system is in a steady state, the entropy-production rate is constant. If the system is in LTE, intensive parameters are defined, and important simplifications become possible. Entropy production can be expressed in macroscopic form or in terms of the statistical distributions of quanta [6,10,11,13].

An aspect of local equilibrium is that intensive forces and extensive fluxes usually have, to good approximation, a *quasilinear* relationship [19]. In this regime, fluxes, at the first order, are linear combinations of forces, with *transport coefficients* that can vary across the system. These coefficients need not be constant, as long as they are strictly functions of the local state and contain no gradients or nonlocal differences.[1]

The term *nonlocal difference* indicates a difference of functions of intensive thermodynamic quantities driving radiative exchanges between elements of matter located at finite distances from one another. Such nonlocal differences are typical of radiation–matter interaction and should be contrasted with the gradients that normally drive thermodynamic flows in matter in LTE. In general, the radiation need not be in LTE, but LTE itself can be defined in a very general way. The most general LTE requires all intensive parameters to be defined locally in mode space. It also requires extensive parameters to be continuous over different parts of a system, a requirement automatically satisfied by various macroscopic conservation laws.

In the quasilinear approximation, the volume part of the entropy production then becomes a sum of bilinear expressions *force* × *flux* ~ *force* × *transport coefficient* × *force*.

[1] This condition is identical to the requirements for the validity of the first-order Chapman–Enskog method of reducing microscopic statistical kinetics to macroscopic thermohydrodynamic behavior and expressing transport coefficients in terms of statistical distributions [8].

These functions are local in mode space. Expressions strictly local in position space emerge as limiting cases where the momenta and other mode variables are summed over. Many such limits are possible, depending on the nature of the system.

Surface contributions to the entropy production occur if there are sharp boundaries to the system [1,14,15]. In LTE, the entropy current is a linear combination of currents of extensive variables, with the intensive parameters as coefficients. This form is analogous to the bilinear form taken by the local volume terms. LTE again allows currents to be related in quasilinear fashion to thermodynamic forces, usually gradients of intensive variables. On the other hand, if the quantum-statistical distributions are known, the currents can be expressed that way instead. Radiation is simple enough that its thermodynamic properties can be expressed exactly in terms of quantum distributions [8,13]. The simplicity of radiation thermodynamics arises from photon number nonconservation and the absence of a photon chemical potential.

Another simplification is possible if the radiation entropy production vanishes. In that case, we can consider matter–matter interactions mediated by radiation and eliminate the radiation modes [1,11]. In position space, radiation then becomes a kind of nonlocal heat transport, and Σ can be represented in a multilocal form. If radiation streams in free space and interacts with matter only at discrete locations, we end up with a description of localized matter lumps interacting at a distance via the field. This multilocal thermodynamic form parallels the action-at-a-distance form of electrodynamics. In both cases, the field is eliminated as a dynamical entity [16].

Under certain conditions, the radiation itself can be localized, as for example in an opaque plasma. The diffusion of radiation then becomes a type of volume-based local transport. Radiative heat transfer in that case is formally similar to heat conduction, which is a purely local matter–matter interaction involving no radiation.

Radiation is often viewed in this way, merely as a mediator and not an entity in its own right, leading at times to erroneous results. But radiation does have its own properties and conditions independently of matter, and it is sometimes essential to account for these explicitly.

1.3. Entropy production and minimum principles

Besides characterizing a nonequilibrium system's state, the entropy-production function has a dynamical significance under certain conditions [5,6,9,17–19]. Expressed as a function of thermodynamic forces or of local intensive variables, the entropy production is a minimum in a nonequilibrium steady state (NESS), subject to the external constraints that prevent the system from equilibrating. The most familiar form of this variational principle is the classical macroscopic nonequilibrium thermodynamics of matter systems, where the entropy production can be expressed as a quadratic function of thermodynamic forces. The principle holds in the local quasilinear case, where the thermodynamic forces are subject to variations, but the background transport coefficients are not.

Less familiar but equally important are surface contributions, of which radiation is the simplest and most common. The minimum entropy-production principle holds with these terms included. For radiation, the entropy production is a simple function of local temperature and requires no approximations [1,11,13].

2. Entropy with and without equilibrium

What makes thermodynamic systems nondeterministic is the statistical uncertainty associated with microscopic states. A thermodynamic system can be viewed as an ensemble of many copies of the same physical system, each different from the others in microscopic details, but all sharing the same macroscopic expected values of volume, energy, and so on. A measure of the statistical uncertainty of microscopic details is the macroscopic entropy S, a non-negative function of the ensemble's statistical ensemble [6,9].

A thermodynamic ensemble of zero entropy contains only one, completely determined, system copy. With multiple copies, the entropy is positive. Each microscopic configuration has its own probability p_k. The entropy of a thermodynamic system is a sum over all possibilities k:

$$S = -k_{\mathrm{B}} \sum_k p_k \ln p_k, \tag{1}$$

subject to the constraint $\sum_k p_k = 1$. In the completely determined case, all $p_k = 0$ except for one possibility j, $p_j = 1$. In that case, $S = 0$.

All system variables, such as volume, energy, and entropy, are defined for a thermodynamic system whether that system is in equilibrium or not. In general, however, computing S requires knowing the probabilities p_k of all microscopic possibilities.

Comment: So that S is additive, a logarithm of any basis is acceptable. The natural log is the simplest choice. Changing the log basis multiplies S by an overall constant. The standard thermodynamic entropy also contains an additional factor of Boltzmann's constant k_{B}. If the entropy is computed using the system's fundamental degrees of freedom, entropy is fully defined without any free constants.

2.1. *The case of equilibrium*

Entropy is maximal in equilibrium, typically under the constraint of holding fixed certain macroscopic state variables such as energy or number [9]. In equilibrium, S is a function of these other macroscopic variables. For each microscopic configuration k, a system variable H^a has a value H_k^a. A Lagrange multiplier α_a is associated with each system-variable average $\langle H^a \rangle = \sum_k p_k H_k^a$ held fixed for maximal entropy. Then maximizing S under constraints is equivalent to maximizing

$$S - \sum_a \alpha_a \langle H^a \rangle - \alpha_0 \sum_k p_k = -k_{\mathrm{B}} \sum_k p_k \ln p_k - \sum_{ak} \alpha_a H_k^a - \alpha_0 \sum_k p_k.$$

The resulting probability distribution is

$$p_k = \frac{\exp\left[-\sum_a \alpha_a H_k^a \right]}{\sum_j \exp\left[-\sum_a \alpha_a H_j^a \right]}, \tag{2}$$

the generalized Boltzmann distribution. The denominator of Eq. (2) is the system's partition function. Each α_a is the intensive variable conjugate to its corresponding extensive variable H^a. For example, if H is the energy E, then $\alpha = 1/T$, where T is the temperature.

Consider a system of nonconserved, massless elementary quanta in thermal equilibrium [4,11]. The system can be analyzed in terms of the phase-space labels position $\mathbf{r}$ and momentum $\mathbf{p}$ of a single particle. The energy ϵ of a single quantum is $c|\mathbf{p}| = h\nu$. The mean occupation number of mode $\mathbf{p}$ is a function of $x = h\nu/k_B T$:

$$n_{\mathbf{p}} = \frac{1}{e^x \pm 1}. \tag{3}$$

The $\pm$ sign holds for fermions (bosons). (See Section 2.2 below.) The energy in the differential phase volume $d^3r\, d^3p$ is

$$\frac{2(k_B T)^4}{(hc)^3} \frac{x^3}{(e^x \pm 1)} dx\, d\Omega_{\mathbf{p}}\, d^3r, \tag{4}$$

where $\Omega_{\mathbf{p}}$ is the solid angle in momentum space. (The factor of two that counts two radiation polarization states may be dropped, but we keep it to match the conventional definition of radiation flux.) When the integration over x is carried out, the fourth-power dependence of energy on T follows for both bosons and fermions. The only difference between the two is in the numerical factor of the integral due to the '$\pm$' in the denominator in the integrand of the x integration. The analogous infinitesimal contribution to the entropy from a differential phase volume is

$$\frac{2k_B(k_B T)^3}{(hc)^3} \left[\frac{x^3}{(e^x \pm 1)} \pm x^2 \ln(1 \pm e^{-x}) \right] dx\, d\Omega_{\mathbf{p}}\, d^3r. \tag{5}$$

This implies the standard equilibrium third-power dependence of entropy on temperature, for fermions and bosons.

The four integrals:

$$\int_0^\infty \frac{x^3}{(e^x \pm 1)} dx = \frac{15 \mp 1}{16} \left(\frac{\pi^4}{15} \right), \tag{6}$$

and

$$\int_0^\infty \pm x^2 \ln(1 \pm e^{-x}) dx = \frac{15 \mp 1}{16} \left(\frac{1}{3} \right) \left(\frac{\pi^4}{15} \right), \tag{7}$$

are easily deduced by series expansions. The Stefan–Boltzmann radiation constant is $\sigma = 2\pi^5 k_B^4 / 15h^3c^2$. From these we find the energy per unit volume into solid angle $d\Omega_{\mathbf{p}}$,

$$\frac{15 \mp 1}{16} \left(\frac{\sigma}{\pi c} \right) T^4\, d\Omega_{\mathbf{p}}. \tag{8}$$

Similarly for the entropy,

$$\left(\frac{4}{3}\right)\frac{15 \mp 1}{16}\left(\frac{\sigma}{\pi c}\right)T^3 \, \mathrm{d}\Omega_{\mathbf{p}}. \tag{9}$$

The vector flux density of energy into solid angle $\mathrm{d}\Omega_{\mathbf{p}}$ with direction $\hat{\mathbf{p}}$ is

$$\frac{15 \mp 1}{16}\left(\frac{\sigma}{\pi}\right)T^4 \hat{\mathbf{p}} \, \mathrm{d}\Omega_{\mathbf{p}}, \tag{10}$$

and for entropy,

$$\left(\frac{4}{3}\right)\frac{15 \mp 1}{16}\left(\frac{\sigma}{\pi}\right)T^3 \hat{\mathbf{p}} \, \mathrm{d}\Omega_{\mathbf{p}}. \tag{11}$$

The integrals (6) and (7) give the canonical fermion factor of 7/8 relative to bosons. The flux density per solid angle, also known as the specific intensity or radiance, is for energy,

$$\frac{15 \mp 1}{16}\left(\frac{\sigma}{\pi}\right)T^4, \tag{12}$$

and for entropy,

$$\left(\frac{4}{3}\right)\frac{15 \mp 1}{16}\left(\frac{\sigma}{\pi}\right)T^3. \tag{13}$$

These results for surface emission of nonconserved quanta can be extended to massive particles. (Some or all of the neutrino species, in fact, have small masses [20].) With nonzero mass, the momentum integrals cannot be evaluated in closed form, but they can be easily calculated numerically. Define the momentum integrals with nonzero mass as

$$\begin{aligned}
\left(\frac{15 \mp 1}{16}\right)\left(\frac{\pi^4}{15}\right)\cdot\begin{Bmatrix} g_1 \\ f_1 \end{Bmatrix}(\bar{m}) &\equiv \int_0^\infty \frac{x^3 \, \mathrm{d}x}{\mathrm{e}^{\sqrt{x^2+\bar{m}^2}} \pm 1}, \\
\frac{1}{3}\left(\frac{15 \mp 1}{16}\right)\left(\frac{\pi^4}{15}\right)\cdot\begin{Bmatrix} g_0 \\ f_0 \end{Bmatrix}(\bar{m}) &\equiv \int_0^\infty (\pm x^2)\ln(1 \pm \mathrm{e}^{-\sqrt{x^2+\bar{m}^2}})\mathrm{d}x,
\end{aligned} \tag{14}$$

with the upper sign for fermions and the lower for bosons. The new f and g functions are defined such that $f_i(0) = g_i(0) = 1$. The reduced mass $\bar{m}$ incorporates the temperature: $\bar{m} \equiv mc^2/k_\mathrm{B}T$.

The specific-energy and entropy-flux expressions change their forms to

$$\left(\frac{15 \mp 1}{16}\right)\begin{Bmatrix} g_1 \\ f_1 \end{Bmatrix}\left(\frac{\sigma}{c\pi}\right)T^4, \qquad \left(\frac{15 \mp 1}{16}\right)\begin{Bmatrix} g_0/3 + g_1 \\ f_0/3 + f_1 \end{Bmatrix}\left(\frac{\sigma}{c\pi}\right)T^3. \tag{15}$$

The overall fourth- and third-power dependence on temperature for the energy and entropy fluxes remains. But these expressions contain additional dependence on T through $\bar{m}$. Differentiation of these expressions with respect to T requires varying this additional dependence as well as the overall power dependence.

2.2. Entropy without equilibrium: counting quanta

The microscopic probabilities p_k cannot be determined, in general, without a description of the statistical ensemble in terms of its degrees of freedom and how it was created. In some cases, however, and without equilibrium, simple counting arguments are enough to define a system's thermodynamics, by finding S as a function of other state variables such as E and N. It is even possible to define subsystem intensive variables such as temperature under certain restrictions [12].

Elementary bosons or fermions provide a simple example. These quanta can be, for example, photons or neutrinos, although they need not be massless. Elementary particles are the quanta of quantum fields, and the fundamental degrees of freedom are the modes of these fields. The description of the state of a bosonic field must also include the amplitude and phase (as well as polarization, etc. where necessary) of each mode. A quantum state of complex amplitude and other mode labels is a *coherent state* [2].

For fermions, the amplitude and phase are trivial, because of the Pauli exclusion principle: a fermionic field mode can have at most one quantum. There are no fermionic coherent states. For bosons, the amplitude and phase can be anything. A bosonic mode can have any number of quanta.

For fermions and for boson fields with random mode phases, a special simplification is possible. The state of the quantum field becomes equivalent to counting the number of quanta in each mode. Calculating the entropy of the mode is then straightforward, a generalization of the counting familiar from equilibrium quantum statistical mechanics.

Consider an ensemble of M identical systems. Each system has the same internal probabilities for being in any particular state k, p_k. The number of systems in state k is $Mp_k = m_k$. Suppose that the collection of systems have m_1 systems in state 1 and m_2 systems in state 2, etc. The number of ways that this configuration can happen is

$$W_{\mathrm{M}} = M!/[m_1!m_2!\ldots m_k!\ldots].$$

The entropy is

$$S = -k_{\mathrm{B}} \sum_k p_k \ln p_k.$$

This expression is equivalent to

$$S_{\mathrm{M}} = k_{\mathrm{B}} \ln W_{\mathrm{M}},$$

if we use probability normalization $\sum_k p_k = 1$, the definition of m_k, and Stirling's formula $n! \sim \sqrt{2\pi n} n^n \mathrm{e}^{-n}$, for large n. (Stirling's formula is quite accurate even for $n \sim 10$.)

Now compute W for bosons and fermions. Consider N identical quanta and G identical possible 'places' to put them.

- With no constraints on N, the boson case, there are $W_{\mathrm{B}} = (N + G - 1)!/[N!(G - 1)!]$ ways of arranging the N quanta among the G places.
- With the constraint that no more than one quantum be permitted in each of the G places, the case of fermions, there are $W_{\mathrm{F}} = G!/[N!(G - N)!]$ ways of arranging the N quanta among the G places.

Use Stirling's formula again to simplify the expressions for S in the two cases, and define a mean occupation number $n \equiv N/G$. Then the entropy for the bosonic and fermionic cases is

$$S = \pm k_B \cdot (1 \pm n) \ln(1 \pm n) - k_B \cdot n \ln n, \tag{16}$$

with the upper sign for bosons and the lower for fermions.

Although equilibrium is not assumed, the expression for S looks formally like the expression for S in equilibrium, and it is independent of any assumption about the nature of the ensemble, apart from the random mode phase assumption. This expression for S applies to a single mode of the field. The mode has n quanta. The entropy for many modes is just the sum of the entropy for each mode. This derivation assumes that the number of quanta can be counted, but that N is not conserved. Thus, no chemical potential enters into the thermodynamic description. The entropy S is plotted as a function of n in Fig. 1. In the boson case, n must be non-negative. In the fermion case, $0 \leq n \leq 1$.

The energy per quantum is $\epsilon \equiv h\nu$. A natural temperature for a single mode follows:

$$\frac{h\nu}{k_B T} = \frac{\partial S}{\partial n} = \ln\left(\frac{1}{n} \pm 1\right). \tag{17}$$

Here, T is a subsystem temperature, and in the case of photon modes, often called the *brightness temperature* [7,8]. The expression (17) looks like the temperature one would

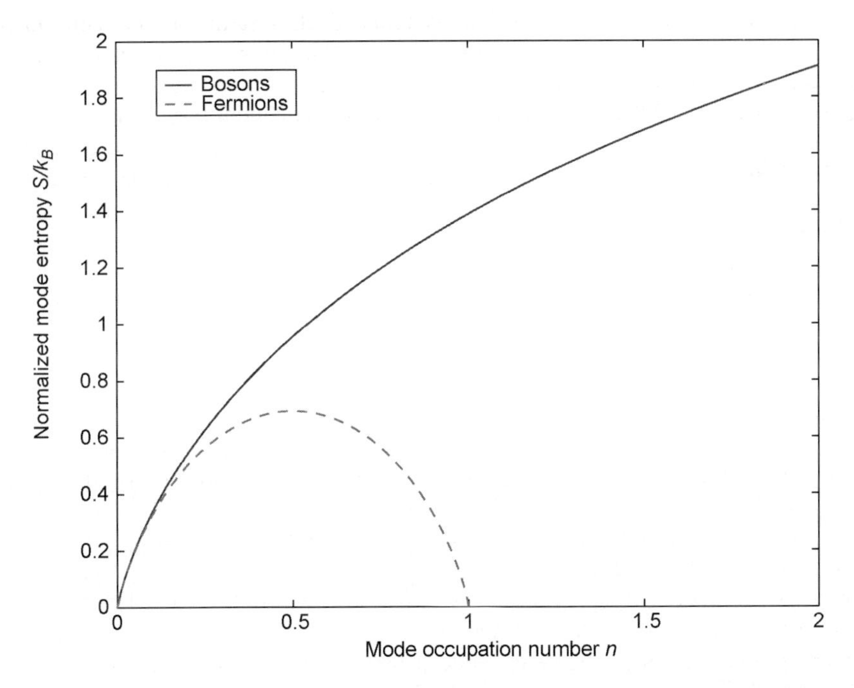

Fig. 1. Entropy S/k_B for a single field mode as a function of mean occupation number n: bosons (solid curve) and fermions (dashed curve). For fermions, $0 \leq n \leq 1$.

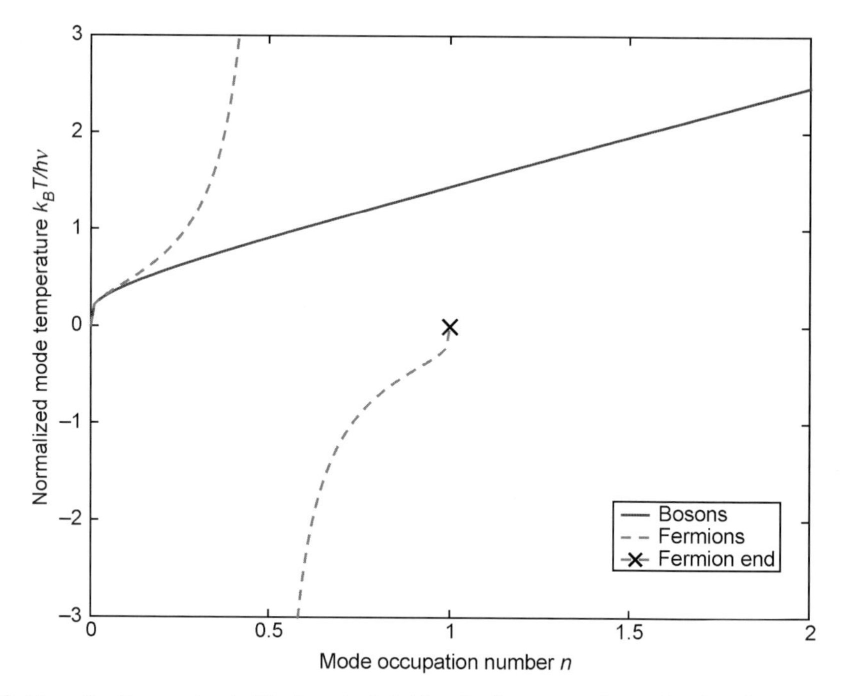

Fig. 2. Normalized temperature $k_B T/h\nu$ for a single field mode of energy $\epsilon = h\nu$ as a function of mean occupation number n: bosons (solid curve) and fermions (dashed curve). For fermions, $0 \leq n \leq 1$.

infer by inverting the usual Bose–Einstein or Fermi–Dirac equilibrium distributions. In this analysis, however, n is arbitrary, and T is different for each mode in general. The function $k_B T/h\nu$ is plotted in Fig. 2 as a function of n. The fermionic mode temperature is negative for $1/2 < n < 1$.

2.3. *Entropy without equilibrium: fluxes*

For closed systems, the thermodynamics can be expressed in terms of extensive variables and their densities. But in general, we consider open nonequilibrium systems with flows and avoid confining the system to a box. Also, if the quanta are massless, they cannot be brought to rest. Using densities is clumsy, and it is better to use fluxes as the primary extensive quantities [1,10,15].

With or without equilibrium, we can relate the specific energy intensity $I_{\mathbf{p}}$, for a given $\mathbf{p}$, to the mode-occupation number $n_{\mathbf{p}}$ from Eqs. (8) and (10) and the *flux density* into a solid angle,

$$I_{\mathbf{p}} = \frac{2n_{\mathbf{p}}\epsilon^3}{h^3c^2}. \tag{18}$$

The specific entropy intensity $J_{\mathbf{p}}$ is

$$J_{\mathbf{p}} = \frac{2k_{\mathrm{B}}\epsilon^2}{h^3c^2}\left[\mp\left(1 \mp \frac{c^2h^3I_{\mathbf{p}}}{2\epsilon^3}\right)\ln\left(1 \mp \frac{c^2h^3I_{\mathbf{p}}}{2\epsilon^3}\right) - \left(\frac{c^2h^3I_{\mathbf{p}}}{2\epsilon^3}\right)\ln\left(\frac{c^2h^3I_{\mathbf{p}}}{2\epsilon^3}\right)\right]. \quad (19)$$

The fundamental extensive quantities are the specific entropy and energy flux $J_{\mathbf{p}}$ and specific energy flux $I_{\mathbf{p}}$. Note that expression (17) for the temperature $T_{\mathbf{p}}$ is recovered by forming $\mathrm{d}J_{\mathbf{p}}/\mathrm{d}I_{\mathbf{p}}$.

In the case of photons, the number is not interesting, as photon number is not conserved. In the case of fermions, some number conservation law usually holds [3]. With zero chemical potential, however, the fermion number is a purely auxiliary quantity and depends on the energy flux. Conversely, we could take the fermion number as fundamental and energy as derived; in either case, only one variable is independent. In the case of nonzero chemical potential, fermion number and energy flux become independent variables, with a mode-dependent chemical potential $\mu_{\mathbf{p}}$.

Since neutrino number is not conserved but lepton number is, it is important to define a specific *number* flux for neutrinos $N_{\mathbf{p}}$ corresponding to $I_{\mathbf{p}}$:

$$N_{\mathbf{p}} = \frac{2n_{\mathbf{p}}\epsilon_{\mathbf{p}}^2}{h^3c^2}, \quad (20)$$

where $I_{\mathbf{p}} = \epsilon_{\mathbf{p}}N_{\mathbf{p}}$.

2.4. *Entropy without equilibrium: quantum fields*

For quantum systems, an equivalent ensemble description uses the *density operator* $\hat{\rho}$. It is Hermitian, and its real eigenvalues are the microscopic probabilities p_k, satisfying $\mathrm{Tr}(\hat{\rho}) = \sum_k p_k = 1$. The entropy $S = -k_{\mathrm{B}}\mathrm{Tr}(\hat{\rho}\ln\hat{\rho})$. A completely determined system is in a *pure state*: one $p_j = 1$, the other $p_k = 0$. In this case, the entropy vanishes, and the density operator satisfies $\hat{\rho}^2 = \hat{\rho}$ and can be expressed as $\hat{\rho} = |\psi\rangle\langle\psi|$, where $|\psi\rangle$ is some quantum state. In the nontrivial case, $S > 0$, $0 \le p_k < 1$, and $\hat{\rho}^2 < \hat{\rho}$ (in the spectral sense) [4,9].

For fermions, the Pauli exclusion principle keeps the entropy simple. Without fermionic coherent states, the problem of computing entropy for fermionic fields reduces back to counting quanta. But without random mode phases, computing the entropy of a bosonic field becomes problematic.

For bosons, a general ensemble is a composite expanded over a basis of projection operators. The basis can be the occupation number basis or the coherent state basis, for example. The ensemble is described by a density operator $\hat{\rho}$ expanded over some basis ψ:

$$\hat{\rho} = \int \mathrm{d}[\psi]\,\mathrm{d}[\psi']P(\psi,\psi')|\psi\rangle\langle\psi'|.$$

Because $\hat{\rho}$ is Hermitian, P is real.

Unfortunately, a general bosonic field ensemble $\hat{\rho}$ is very difficult to analyze and diagonalize. Only two special cases seem to be tractable: random-mode phases and the pure state. Equilibrium is a subcase of the random-mode phase case [12].

3. Entropy generation and variational principles

A thermodynamic system out of equilibrium has currents transporting heat, particle number, volume, etc. from one region of mode space to another. External constraints hold the system out of equilibrium, and a nonequilibrium system is necessarily open. Quantities not locally conserved in microscopic interactions (such as collisions and matter–radiation couplings) relax in some characteristic *equilibration time(s)* to a *local* equilibrium, leaving only locally conserved quantities H^a *not* in local equilibrium.[2] (Because it is conserved, each H^a retains memory of its initial value and thus does not move towards equilibrium.) To each of these conserved H^a is associated a nonzero flux J^a and a conservation law, such as an equation of continuity for densities and currents. The nonzero fluxes J^a are maintained by boundary conditions. Thermodynamic 'flux' refers to spatial currents and local creation-rate densities alike [5,6,9,17,21].

If the microscopic equilibration acts over short enough times and small enough spaces, each H^a also has an associated intensive parameter α_a. This intensive parameter is constant over the system if and only if the associated flux J^a is zero, defining a global equilibrium for that particular H^a. Otherwise, the α_a varies in space and time. LTE implies that whatever space and time scales are necessary for local equilibration to occur, they must be much smaller than the intensive scale heights $|\alpha_a/\nabla\alpha_a|$ and scale times $|\alpha_a/\dot{\alpha}_a|$.

In an NESS, the system's intensive parameters and densities and fluxes of extensive parameters do not vary in time. The macroscopic conservation laws then reduce to zero-divergence conditions on the associated currents. If the system is not steady, then the full conservation laws hold, requiring time derivatives as well as divergences in the equations of continuity.

The entropy production associated with each nonequilibrium current J^a and associated force X_a arises from the Gibbs form of the entropy differential $\mathrm{d}S = \sum_a \alpha_a \cdot \mathrm{d}H^a$. The thermodynamic force X_a can be, for example, a spatial gradient of an intensive parameter, the infinitesimal change of α_a over an infinitesimal distance or it can be a finite difference of chemical potentials between reagents and products. We must account for the entropy both gained and lost from nonequilibrium flows and impose the condition that each H^a be conserved ($J^a_{\mathrm{out}} = -J^a_{\mathrm{out}}$). Then the entropy-production rate always has the form

$$\Sigma = \sum_a (\alpha_a^{\mathrm{in}} J^a_{\mathrm{in}} + \alpha_a^{\mathrm{out}} J^a_{\mathrm{out}}) = \sum_a (\alpha_a^{\mathrm{in}} J^a_{\mathrm{in}} - \alpha_a^{\mathrm{out}} J^a_{\mathrm{in}}) = \sum_a X_a \cdot J^a, \tag{21}$$

if LTE holds for each H^a.[3] The sum in general extends over all mode labels.

3.1. Forms of entropy production—minimum principles

Entropy is an extensive thermodynamic property. It can be localized and integrated to determine a global amount. Entropy production Σ is also extensive, localizable, and

[2] The other set of macroscopic variables that does not relax towards equilibrium are order parameters arising from broken symmetries, usually associated with phase transitions.

[3] If LTE is not valid for one of the H^a, the associated entropy production has to be computed from the microscopic kinetic expression derived from the universal definition of entropy (1), which implies $\mathrm{d}S = -k_{\mathrm{B}} \sum_k \mathrm{d}p_k \cdot \ln p_k$.

a volume integral of its local density σ. (We use σ to represent the Stefan–Boltzmann constant as well. The distinction is clear from the context.) If we divide space into distinct regions, boundary surfaces are defined; entropy can move between regions, and entropy fluxes across surfaces are defined [5,6,9,18].

The entropy-production rate can be expressed as

$$\sigma = \frac{\partial s}{\partial t} + \nabla \cdot \mathcal{F} \geq 0, \tag{22}$$

where σ is the entropy-production rate per unit volume, s is the volume density of entropy, and $\mathcal{F}$ is the entropy flux density. This inequality expresses, in differential form, the total entropy change within a volume and the second law of thermodynamics.

The significance of this form of σ is that it naturally divides into two sets of terms. The first set consists of local densities that, if expressed in terms of a LTE, can be resolved into bilinear forms $X \cdot J$, where the Js are creation-rate densities. The second set consists of divergence terms. When integrated over the system volume, these can be recast as surface integrals of currents. If these fluxes are emitted from surfaces in LTE, they can be expressed as functions of surface-intensive parameters. This separation of local densities and flux divergences is more than formal [11]. A typical system has a 'matter' part made of heavy, nonrelativistic particles and a radiative part made of massless or nearly massless particles. Assume we can draw a boundary over a large enough volume to permanently contain all the matter, including matter currents and work in the rest frame of this matter. The radiation, being massless, has no rest frame and always has associated flux currents. (Nearly massless neutrinos do have a rest frame, but it is very different from the typical rest frame of 'matter.') Only if the 'matter' is virtually opaque, locally trapping the radiation, can the radiation be described in local terms. Examples of such systems are discussed in Section 5.

Separating entropy production into radiative and matter parts, $\sigma = \sigma_{\mathrm{m}} + \sigma_{\mathrm{r}}$,

$$\sigma = \frac{\partial s_{\mathrm{m}}}{\partial t} + \frac{\partial s_{\mathrm{r}}}{\partial t} + \nabla \cdot \mathbf{Y} + \nabla \cdot \mathbf{H}, \tag{23}$$

where $s_{\mathrm{m}}(s_{\mathrm{r}})$ is the volume density of entropy in matter (radiation). $\mathbf{Y}$ and $\mathbf{H}$ denote matter and radiative entropy flux densities, respectively.

By using the equations of state and steady-state conservation equations for the extensive variables, we re-express the matter entropy flux divergence as $\nabla \cdot \mathbf{Y} = \sum_a \nabla \cdot (\alpha_a \mathbf{Y}^a)$ and the local matter entropy production as $\sum_a \alpha_a \epsilon^a$. (This step defines the intensive α_a simultaneously in terms of extensive fluxes and volume densities.) The $\mathbf{Y}^a$ and the ϵ^a are the flux densities and creation rates of the extensive variables H^a, respectively. The entropy-production rate becomes a sum of local matter, matter transport, and radiative terms:

$$\sigma = \sum_a \{\alpha_a \epsilon^a + \mathbf{Y}^a \cdot \nabla \alpha_a\} + \nabla \cdot \mathbf{H}, \tag{24}$$

α_a is the intensive variable conjugate to H^a. For a system in a steady state, the radiative entropy density is constant, and we set $\partial s_{\mathrm{r}}/\partial t = 0$. In Eq. (24), the second and third terms become surface terms when σ is integrated over a volume.

Entropy production is a minimum in an NESS of an open system in LTE, holding fixed the boundary conditions that keep the system from equilibrating. Entropy production in general includes both volume integrals of local densities and surface integrals of currents [5,6,11,13,18,19].

3.2. Local equilibrium and local entropy production

The familiar form of entropy production and the minimum entropy-production principle arises from the local density terms in σ. Without further assumptions, the bilinear form $X{\cdot}J$ is purely kinematic and says nothing about the system's dynamics. While all $X = 0$ implies all $J = 0$, there is no general functional relation between forces and fluxes. Within LTE, however, the J can usually be expanded in powers of the X and well-approximated by linear forms, $J^a = \sum_b K^*_{ab}(\alpha_*)X_b$. If the system is described in terms of quantum-statistical distributions, we could express σ in terms of intensive parameters such as T, without expansions. But this last step is not necessary for a minimum principle.

In general, the *transport coefficients* K_{ab} are not constant over the whole system, and the relation of the J to the X is *quasilinear* [19]. (In the strict linear regime, the K_{ab} are constant across the whole system.) Instead, the K_{ab} depends on the *local* thermodynamic state, through the local intensive parameters α_* only (not their gradients or other nonlocal differences). The special $*$ subscript distinguishes the local α_as from *differences* of intensive parameters that occur in the thermodynamic forces X_a. If the microscopic dynamics are reversible, the K_{ab} are symmetric [17].

Two important variational principles are associated with the local bilinear entropy production [18,19]. The distinction depends on the choice of boundary conditions. In one case, the entropy production of an LTE–NESS systems is a minimum with respect to variation of the forces X_a, holding fixed the subset of thermodynamic forces that keep the system from equilibrating. In the other, the entropy production is a minimum with respect to variation of *all* the forces, holding fixed a set of external thermodynamic fluxes that keep the system from equilibrating. We vary the thermodynamic forces X_a formed from *differences* of intensive parameters, while holding fixed the purely local intensive parameters α_* that occur in the K^*_{ab}. The procedure is similar to the background field, external field, and self-consistent field methods used in statistical and quantum field theory [22].

The first entropy-production minimum principle uses the simple bilinear form, $\Sigma = \sum_{ab} X_a K^*_{ab} X_b$. Vary Σ with respect to the X_as that are not held fixed by the boundary conditions. This variation, which is proportional to the associated thermodynamic flux, vanishes. Each δX is independent, so that each associated thermodynamic flux is zero. Those Xs allowed to vary and their associated fluxes are those aspects of the system that have equilibrated. Those Xs held fixed and their associated fluxes are held by boundary conditions out of equilibrium.

The second entropy-production minimum principle allows all the Xs to vary, but explicitly includes the external currents J_{ext} that keep the system from equilibrium. Such boundary conditions are often the realistic choice, since real systems are often prevented from equilibrating by external pumping, i.e. imposed fluxes, not by imposed forces. In this principle, the internal entropy production has a second-order, quadratic form, because the system internally is everywhere in LTE. (The first-order local increment of entropy must vanish.) The general form is

$$\Sigma = \sum_{ab} \frac{1}{2} X_a K^*_{ab} X_b + \sum_a X_a \cdot J^a_{\text{ext}}. \tag{25}$$

Varying all the Xs independently and setting to zero each contribution to Σ yields a conservation law for each flux J.

For every nonvanishing J^a_{ext}, the corresponding internal J^a is also nonvanishing. A conjugate subset of the X is nonvanishing. These parallel subsets of nonvanishing forces and fluxes represent the nonequilibrium aspect of the system.

3.3. Streaming fluxes and radiative entropy production

The radiative and matter entropy flux parts of Σ can be expressed as a surface integral through the divergence theorem.

$$\Sigma_{\text{surface}} = \int_V \mathrm{d}V \nabla \cdot \mathbf{H} + \sum_a \mathbf{Y}^a \cdot \nabla \alpha_a = \int_S \mathrm{d}S \cdot \left\{ \mathbf{H} + \sum_a \alpha_a \mathbf{Y}^a \right\}. \tag{26}$$

There is a constant entropy production from the release of radiation into empty space. This step requires the radiation to propagate freely from a well-defined surface. Any flow of extensive matter quantities H^a across the surface also contributes to the entropy production [1,6,9,13–15].

For example, the local photon entropy flux $\mathbf{H}$ has magnitude $(4/3)\sigma T^3_{\text{eff}}$ on a surface with temperature T_{eff}. Multiplying this expression by the surface area gives the total boundary entropy production, Σ_{bound}. In fact, the boundary expression is simply the *total* Σ_{rad} within the enclosed volume, *including* all entropy production due to radiation production and transport. The volume integral of σ is the entropy production *up to, but not including*, the surface; the surface integral or boundary term is the photon entropy production including the surface, as well as the interior. Thus, the contribution of the surface alone to Σ_{rad} is the difference of these two expressions: $\Sigma_{\text{surface}} = \Sigma_{\text{bound}} - \Sigma_{\text{volume}}$.

4. Free radiation and equilibrium matter

Radiation produces entropy only when it interacts with matter. If matter and radiation are separated, the two can interact only at boundaries. The radiation field is composed of absorbed and emitted parts. If the matter is isolated into lumps each in LTE, each lump has a temperature that can vary over its surface. The lump emits locally as a black body. The absorbed field can have its own temperature independent of the matter thermodynamics, while the emitted field shares the same temperature as the matter at the point of emission. If temperature varies within the body, then nonequilibrium matter thermodynamics is the correct framework for that part of the problem [1,11,13].

Separation of forces and fluxes is superfluous for radiation, which we treat exactly in terms of temperature. We do not count elementary quanta for matter, on the other hand, and the associated entropy production must still be computed as heat flux divided by temperature and in quasilinear form. In this section, we use intensive variables to study radiation and radiation–matter coupling and treat extensive variables and fluxes as secondary. This treatment agrees with the first-order variational approach.

Entropy production provides a variational principle for systems consisting of discrete lumps of matter emitting radiation freely into empty space. The details of the respective distributions for bosons and fermions differ, but the general form of Σ and the principle of minimum entropy production holds [11].

4.1. Entropy production at matter–radiation boundaries

If we integrate over a finite volume V bounded by a surface S containing all the matter, then the entropy-production rate Σ is

$$\Sigma = \int_V \sum_k \{a_k \epsilon_k\} \mathrm{d}V + \int_S \mathbf{H} \cdot \mathrm{d}\mathbf{S}, \tag{27}$$

because matter fluxes must vanish across S.

Terms under the first integral of Eq. (27) are all due to matter processes and not part of the radiative entropy production. It is a common misconception to interpret the radiation heating rate divided by the temperature, which occurs in the *first* integral, as the entropy production of radiation. It should be clear from this derivation that $\int_V \sigma_{\mathrm{rad}}\, \mathrm{d}V$ is all accounted for through the *second* integral of Eq. (27).

Eq. (27) is the basis for computing the entropy-production rate due to the interaction of matter and radiation for many finite bodies locally in equilibrium. The first term represents changes in the entropy of the bodies while the second term accounts for changes in the radiation field.

Separate the energy and entropy fluxes into incoming and emitted components: $\mathbf{F} = \mathbf{F}^{\mathrm{i}} + \mathbf{F}^{\mathrm{e}}$, $\mathbf{H} = \mathbf{H}^{\mathrm{i}} + \mathbf{H}^{\mathrm{e}}$. The emitted components are assumed to be thermal, $\mathbf{F}^{\mathrm{e}} = \sigma T^4 \hat{\mathbf{m}}$ and $\mathbf{H}^{\mathrm{e}} = (4/3)\sigma T^3 \hat{\mathbf{m}}$, respectively, where $\hat{\mathbf{m}}$ is the outward unit vector normal to the body surface. If photon radiation is incoming on a body of temperature T in a vacuum, the entropy-production rate is

$$\Sigma_\gamma = \int_V \left(-\frac{\nabla \cdot \mathbf{F}}{T} \right) \mathrm{d}V + \int_S \mathbf{H} \cdot \mathrm{d}\mathbf{S}, \tag{28}$$

where the volume V is any containing the body, and $\mathbf{F}$ is the flux density of energy radiation. Next the divergence theorem is used. The temperature gradient terms vanish because the gradients are defined only inside S, while the radiative fluxes are defined at and outside S.

$$\Sigma_\gamma = -\int_S \frac{\mathbf{F} \cdot \mathrm{d}\mathbf{S}}{T} + \int_S \mathbf{H} \cdot \mathrm{d}\mathbf{S}. \tag{29}$$

Restate the result using the blackbody expressions:

$$\Sigma_\gamma = \left|\iint \frac{\mathbf{F}^{\mathrm{i}}\cdot \mathrm{d}\mathbf{S}}{T}\right| - \left|\iint \mathbf{H}^{\mathrm{i}}\cdot \mathrm{d}\mathbf{S}\right| - \int \sigma T^3\, \mathrm{d}S + \int (4/3)\sigma T^3\, \mathrm{d}S. \tag{30}$$

The incoming flow is independent of the state of the body. Thus, it follows that

$$\delta\Sigma_\gamma = \int [\sigma T^4 - \mathbf{F}^{\mathrm{i}}\cdot\hat{\mathbf{m}}]\frac{\delta T\cdot \mathrm{d}S}{T^2} = 0, \tag{31}$$

since $\mathbf{H}^{\mathrm{i}}$ is independent of T. Since the variation δT is arbitrary, the integrand must also vanish. That is, the entropy-production rate is a minimum in the steady state, implying energy conservation, for an arbitrary geometry and incoming field. This is an example of a conservation law derived from minimum entropy production [1,11,15].

4.2. Example: free radiation and matter lumps

Now consider a thermalized and isotropic incoming radiation field. Embedded in the field are two blackbody matter lumps of temperature T_1 and T_2, respectively. The incoming, absorbed, part of the field is an independent entity with its own temperature T_0. The emitted part of the field shares the temperature of the matter that emits it, either T_1 or T_2. The field and the matter lumps both lose and gain entropy during this interaction. We can vary the entropy-production expression, by varying one or more of the three temperatures, holding the other temperatures fixed, and seek the minimum [11,13].

Since the matter lumps each have a uniform temperature on their respective surfaces, we use the surface density of entropy production. A radiation field of temperature T has an entropy production surface density of $j_{\mathrm{S}}^{\mathrm{r}} = (4/3)\sigma T^3$. A blackbody emits and absorbs a heat flux per unit area of $j_{\mathrm{E}}^{\mathrm{m}} = \sigma T^4$. The total entropy-production rate density is

$$-2\frac{4}{3}\sigma T_0^3 + \frac{4}{3}\sigma T_1^3 + \frac{4}{3}\sigma T_2^3 + \frac{\sigma T_0^4 - \sigma T_1^4}{T_1} + \frac{\sigma T_0^4 - \sigma T_2^4}{T_2}. \tag{32}$$

The first term is the entropy lost when the free field is absorbed. The second and third terms are the entropy gained from the fields emitted by the matter lumps at temperatures T_1 and T_2. The last two terms are the entropy produced and lost by the *matter* in the form of heat. The heat flux gained by matter comes from the incoming radiation field with temperature T_0, but the matter absorbs or loses the heat at either temperature T_1 or T_2.

Consider the various possible combinations of fixed and free temperatures.

- We can hold all three temperatures fixed. In this case, the entropy production is simply a kinematic (descriptive) expression, completely determined by the three known temperatures. It describes the state of the system, without any dynamical content.
- We can hold two temperatures fixed and vary the third. Consider varying Eq. (32) with respect to the radiation-field temperature T_0, while holding the blackbody temperatures T_1 and T_2 fixed. We obtain

$$-2\cdot 4T_0^2 + \frac{4T_0^3}{T_1} + \frac{4T_0^3}{T_2} = 0.$$

The solution is $2/T_0 = 1/T_1 + 1/T_2$. That is, T_0 is the harmonic mean of the fixed matter temperatures T_1 and T_2. The resulting entropy-production area density is

$$\frac{1}{3}\sigma\left[-2\left(\frac{2T_1T_2}{T_1+T_2}\right)^3 + T_1^3 + T_2^3\right],$$

a positive expression. There is a net flow of heat from the hotter blackbody to the colder one, mediated by the field.

- We can hold one temperature fixed and vary the other two. Consider varying the two matter temperatures T_1, T_2 and holding the radiation temperature T_0 fixed. Variation of Eq. (32) with respect to T_1 and T_2 yields

$$T_{1,2}^2 - \frac{T_0^4}{T_{1,2}^2} = 0,$$

for either, so that $T_1 = T_2 = T_0$. The entropy-production area density is then zero, because the system is in equilibrium.

- Finally, we can vary all three temperatures freely. Varying Eq. (32) with respect to T_1 and T_2 always yields $T_1 = T_0$ and $T_2 = T_0$, respectively. Thus, in this case, the system is always in equilibrium, and the entropy-production density vanishes. Treated carefully with limits, varying Eq. (32) with respect to T_0, once T_1 and T_2 are substituted, yields $T_0 = 0$. Thus, all three temperatures vanish. The system is trivially in equilibrium, with no heat or entropy fluxes, the photon vacuum, with two perfectly cold bodies.

Except in the case where no temperatures are allowed to vary, the minimum entropy-production principle gives nontrivial dynamical results for the system's steady state. The third and fourth cases yield equilibrium, one with a common nonzero temperature, the other with the zero-temperature vacuum. These results conform to a common-sense expectation that if all or all but one of the temperatures vary, the system should relax to the single specified temperature in the second case and to zero temperature in the first.

Allowing the radiation temperature to vary while keeping the matter temperatures fixed opens the way to an alternative and equivalent description of this system. In this description, the matter lumps exchange heat, and the radiation field is not mentioned. We can obtain this intermediate description by varying T_0, then expressing it in terms of T_1 and T_2 (i.e. $2/T_0 = 1/T_1 + 1/T_2$) and substituting back into the entropy production for T_0. The expression is then a function of T_1 and T_2 alone. We can then vary this new expression with respect to T_1 and T_2. This approach to matter–radiation coupling removes radiation as a dynamical entity, as discussed in Section 1.

4.3. *Fermionic radiation: the case of neutrinos*

Instead of photons, the matter lumps could be emitting and absorbing thermal massless neutrinos [11,23]. Because neutrinos are fermions, the entropy-production expression changes: the momentum integrals change from the Bose–Einstein to the Fermi–Dirac form. The result is the same as for massless bosons (30), except that all expressions involving neutrino emission and absorption have an extra factor of 7/8.

$$\mathbf{F}^{\mathrm{e}} = (7/8)\sigma T^4 \hat{\mathbf{m}}, \qquad \mathbf{H}^{\mathrm{e}} = (7/8)(4/3)\sigma T^3 \hat{\mathbf{m}}.$$

The minimum entropy production results of the last section do not change in the case of neutrinos. In supernovae and the early Universe, neutrinos are emitted and absorbed thermally. The flux and entropy production expressions are the same as for thermal photons, with the additional 7/8 factor.

The neutrino temperature is defined by the Fermi–Dirac analog of Eqs. (17)–(20):

$$\frac{1}{T_{\mathbf{p}}} \equiv \frac{1}{\epsilon} \ln\left[\frac{2(\epsilon/\hbar c)^3 c}{8\pi^3 I_{\mathbf{p}}} - 1 \right], \tag{33}$$

where $I_{\mathbf{p}} = \epsilon_{\mathbf{p}} N_{\mathbf{p}}$ is the momentum-specific neutrino differential energy flux; $N_{\mathbf{p}}$ is the same for neutrino number. In analogy with the photon entropy production and without assuming thermal equilibrium, the neutrino entropy production is

$$\Sigma_\nu = \int \mathrm{d}V \int \mathrm{d}^3 p \frac{\epsilon_{\mathbf{p}} \dot{n}_{\mathbf{p}}}{T_{\mathbf{p}}}, \tag{34}$$

summed over all neutrino-producing reactions, where $\dot{n}_{\mathbf{p}}$ is the neutrino production rate density in position and momentum space. In general, Σ_ν is nonlocal because neutrinos are usually not in LTE.

Neutrinos emitted by ordinary stars, nuclear reactors, and nuclear explosions stream freely and not in LTE, because the interior temperatures are not high enough for weak interactions to equilibrate. Such neutrinos are not emitted in anything like a blackbody distribution and do not subsequently equilibrate. In these situations, there is no matter plasma hot and dense enough to act as a heat bath for free-streaming neutrinos [11,23,24].

5. Radiation and matter in equilibrium

Radiation is not normally localized. But we now consider a special situation, matter and radiation in the *same* LTE, requiring matter and radiation to have the same temperature and efficient mechanisms for exchanging energy. The matter is almost opaque to radiation in this case, and the radiation does not stream freely. Instead, it looks like LTE matter. In practice, this requires a plasma, where the electrons are free of their parent nuclei, both embedded in a hot gas of photons or neutrinos. A familiar case of a photon gas occurs in the interior of any star, where the external energy sources are gravitational contraction and nuclear fusion. The same conditions occurred in the early Big Bang before matter–photon decoupling. Less familiar is the case of a thermal neutrino gas. Neutrinos are produced and equilibrated only via the weak interactions, which are so feeble compared to electromagnetism that the necessary temperatures and densities obtain only in supernovas and the early Universe before matter–neutrino decoupling [4,7,8,11,23–25].

Radiative transport by photon or neutrino diffusion is formally similar to heat conduction by matter–matter collisions. The role of the conductivity is taken by an expression involving the opacity κ of the matter. The opacity is the inverse photon or neutrino mean free path in the plasma and measures how opaque the matter is to photon or neutrino travel. Thus, κ involves an integral over the *photon* or *neutrino* phase space. Photon diffusion illustrates all the important points and differs from the neutrino case only by the 7/8 factor.

5.1. Photon and neutrino diffusion in hot, dense matter

The evaluation of the radiative entropy production Σ_γ in LTE, with a small gradient, begins with photons at angular frequency ω passing through and interacting with matter at temperature T. The generalized entropy production bilinear form is [7,8,23]:

$$\sigma_{\gamma\ \mathrm{diff}} = 2\pi \int_0^\infty \mathrm{d}\omega \int_{-1}^{+1} \mathrm{d}\xi\, J_\omega[1/T_\omega - 1/T], \tag{35}$$

where $\xi =$ photon local direction cosine. J_ω is the differential radiation luminosity density out of equilibrium or the directional derivative of the specific intensity:

$$J_\omega = \kappa_\omega[B_\omega - I_\omega],$$

with κ_ω the frequency-specific opacity of matter, B_ω the Planck function (blackbody differential radiation energy flux), and I_ω the true energy flux of photons. In the spherical diffusion approximation, $I_\omega = B_\omega - (\xi/\kappa_\omega)(\partial B_\omega/\partial r)$, where the gradient term is small except very near the stellar surface. T_ω is the brightness temperature for any I_ω and varies with ω:

$$\frac{1}{T_\omega} = \left(\frac{1}{\hbar\omega}\right) \ln\left[\frac{2\hbar\omega^3}{8\pi^3 c^2 I_\omega} + 1\right].$$

In the *Rosseland mean opacity*:

$$\frac{1}{\kappa_\gamma} \equiv \frac{\int_0^\infty \mathrm{d}\omega[1/\kappa_\omega]\partial B_\omega/\partial T}{\int_0^\infty \mathrm{d}\omega\, \partial B_\omega/\partial T},$$

the denominator has the value $4\sigma T^3$. Because of the E/T structure of entropy production, the Rosseland opacity is a harmonic mean.

The temperature gradient appears in Σ_γ once. Otherwise, the temperature occurs in other parts of the entropy-production rate only as a local state variable having nothing to do with heat transport. Thus we distinguish this local temperature, T_* from the temperature T that is associated with gradients and other thermodynamic forces and is subject to functional variations. The entropy production of radiative diffusion then takes the form:

$$\Sigma_{\gamma\ \mathrm{diff}} = \int \mathrm{d}V \frac{1}{2}\left(\frac{16\sigma T^5}{3\kappa_\gamma}\right)_* [\nabla(1/T)]^2. \tag{36}$$

Heat sources contribute to the heat-production density ε and the entropy production:

$$\Sigma_{\text{source}} = \int dV \frac{\varepsilon_*}{T}. \tag{37}$$

The bulk radiation entropy-production rate is the sum of the heat transport and production terms:

$$\Sigma_\nu = \Sigma_{\nu\,\text{diff}} + \Sigma_{\text{source}} = \int dV\{(1/2)[16\sigma T^5/3\kappa_\gamma]_*[\nabla(1/T)]^2 + \varepsilon_*/T\}, \tag{38}$$

in the case of radiative transport.

If the radiation outstreams at a sharply defined surface, at temperature T_{eff}, the complete entropy production

$$\Sigma_\gamma = (4/3)\sigma T_{\text{eff}}^3(4\pi R^2) \tag{39}$$

is obtained by integrating the radiative entropy flux over the surface, in this case a sphere of radius R. The entropy production due to the surface alone is the difference of the expression (39) and the volume integral (38).

The analogous situation for neutrinos occurs in the early Universe and in supernovae, where neutrinos are emitted and absorbed in LTE [23–25]. The entropy production below a supernova neutrinosphere (the opaque–transparent boundary) is a function of a single local temperature:

$$\Sigma_\nu = \int dV\{(1/2)[14\sigma T^5/3\kappa_\nu][\nabla(1/T)]^2 + \varepsilon_\nu/T\}, \tag{40}$$

like Eq. (38), with a neutrino mean free path $1/\kappa_\nu$ and an extra factor of 7/8 in the diffusion part. The total Σ_ν *including* the neutrinosphere is (39) times 7/8.

5.2. Multiple local equilibria

If the radiation is emitted and absorbed locally with each system component retaining its own LTE, each component remains thermal at its own temperature. For example, a photon gas with a mode-dependent temperature T_γ may interact with matter of temperature T. Then

$$\Sigma_\gamma = \int dV \int d\epsilon \int d\Omega_{\mathbf{k}}\, I_{\mathbf{k}}(\mathbf{r}) \left[\frac{1}{T_\gamma(\mathbf{r},\mathbf{k})} - \frac{1}{T(\mathbf{r})} \right]. \tag{41}$$

$I_{\mathbf{k}}$ is the local specific energy intensity of photons emitted *by* the matter. If $T_\gamma > T$, then $I_{\mathbf{k}} \leq 0$, if $T_\gamma < T$, then $I_{\mathbf{k}} \geq 0$. Thus, Σ_γ is always ≥ 0. In transparent atmospheres, the radiation has a temperature $T_\gamma(\mathbf{r},\mathbf{k})$, while each species i can have its own $T_i(\mathbf{r})$. Thus

$$\Sigma_\gamma = \sum_i \int dV \int d\epsilon \int d\Omega_{\mathbf{k}}\, I_{\mathbf{k}}(\mathbf{r}) \left[\frac{1}{T_\gamma(\mathbf{r},\mathbf{k})} - \frac{1}{T_i(\mathbf{r})} \right]. \tag{42}$$

Again $I_{\mathbf{k}}$ is the specific radiation energy intensity emitted by the matter. Contributions such as Eq. (41) or (42) occur in addition to such gradient terms as Eq. (36).

5.3. Heat sources for photon and neutrino diffusion

Whatever the source of radiative heat energy ε [8,23,24], the emitted quanta are equilibrated by matter–radiation scattering. (Neutrinos also scatter with themselves.) In protostars, photons are released as the stellar gas is squeezed by gravitational contraction. Once stars are fully formed, entropy is produced by nuclear reactions. The radiative and matter kinetic energy originates in thermalized matter, a tiny, positive contribution to radiative entropy, because the original matter reactants are in equilibrium. This original photon/kinetic energy is absorbed upon equilibration, a negative contribution to entropy. Both the matter kinetic energy and radiation then come to equilibrium with the ambient temperature of the plasma, a large and positive contribution to entropy. The first two contributions are negligible compared to the third, being suppressed by the ratio T/T_0, where T_0 is the brightness temperature of the original photons or neutrinos. This brightness temperature is usually far above the ambient plasma temperature. These contributions are significant for older stars with higher core temperatures or in the very early Universe.

6. Summary and conclusion

During its first century, thermodynamics concentrated on isolated systems in or close to equilibrium. Classical nonequilibrium thermodynamics was designed for matter systems localized in position space with a strictly macroscopic description [6,9,17,18,21]. The discovery of photons by Planck and Einstein created a new type of thermodynamics, which, for many decades, was set apart by the nature of radiation: photons are massless, have no rest frame, cannot be localized under most conditions, and are not conserved. Such properties make photon physics very different from matter physics. The fact that photons cannot be brought to rest makes fluxes the natural extensive variables, rather than localized functions over volumes [5,8].

Since the 1930s, the quantum field has provided the single unifying concept for all known physical entities [3]. A fundamental thermodynamics would be based on the states of these fields, or, if we forsake knowledge of field phases, the quanta of the fields. The distinctions between fermionic and bosonic, massive and massless, conserved and nonconserved quanta are the basis for the broadly different thermodynamics of matter and radiation. Much that appears nonlocal in position space is local if we keep in mind the full mode space, with both momentum and position space coordinates, as well as spin and charge labels. The alternative descriptions of classical field and quantum counting are possible for bosons because of the existence of bosonic coherent states.

The entropy increment dS and entropy production Σ are key macroscopic functions for nonequilibrium systems. Σ and its bilinear form as a sum of products of thermodynamic forces and fluxes are universal to matter and radiation, if we use the generalized mode space as our domain and include surface contributions. If we assume LTE, we can introduce local linear causal relationships between forces and fluxes. An LTE non-equilibrium system in a steady state is at a minimum of entropy production with respect to the constraints that keep it from equilibrating.

This framework is broad enough to encompass a large class of real systems. The nonequilibrium thermodynamics of matter dates to the early 20th century. The full development of the nonequilibrium thermodynamics of radiation and matter–radiation interaction is more recent and makes use of exact expressions in terms of elementary quanta [1,5,10]. Some of the results appear very different from those familiar in classical nonequilibrium thermodynamics. Spatial localization is natural for massive, conserved fermionic 'matter' systems, where free-streaming beams are the exception. Such 'blobs' are best described locally in position space. But for 'radiation' quanta such as photons, neutrinos, and gravitons, free-streaming is the default, and position-space localization is rare. Beams of such quanta are best described locally in momentum space. If we keep in mind the full mode space, however, these differences do not appear fundamental. Beams can mediate between blobs, and blobs can mediate between beams. The thermodynamics is symmetric between position and momentum descriptions. The Gibbs-like picture is valid for all quanta in local equilibrium in mode space. This broadened framework allows for the correct reformulation of classical nonequilibrium thermodynamics in terms of elementary quanta, including fundamental and possibly massless and free-streaming bosons.

Among the strongest prejudices obscuring this fact is the false belief that, while descriptions local in position space are legitimate, descriptions local in momentum space are 'nonlocal' or 'microscopic' and thus not even thermodynamic. (Thermodynamics rests, not on the microscopic/macroscopic distinction, but on statistical ensembles of system copies.) This prejudice obscures the position–momentum symmetry of mode space [1]. If we clear away such false assumptions, the simplicity and unity of the thermodynamics of the modes of quantum fields becomes apparent.

Systems or subsystems not in local equilibrium lie outside even this generalized framework. In practice, such cases of interest usually arise from chemical reactions, including nuclear and subnuclear reactions [6,9,25]. Such systems might possess a macroscopic description, but the functional, causal relationship of forces and fluxes could be nonlinear or might not exist at all. There are no general variational principles for such systems.

The references contain more detailed expositions of the power and limitations of nonequilibrium thermodynamics. We encourage readers to explore the theoretical issues and specific applications more deeply in this literature.

Acknowledgements

We thank the Telluride Summer Research Center and Telluride Academy; the Aspen Center for Physics; the Institute for Theoretical Physics, University of California, Santa Barbara; the Fermilab Theoretical Astrophysics group; the Institute for Fundamental Theory, University of Florida; the U.S. Department of Energy; and the U.S. National Science Foundation for their support and hospitality. The figures were generated with MATLAB 6.

Some of this material was presented at the UNESCO advanced school, 'New Perspectives in Thermodynamics: From the Macro to the Nanoscale,' International Centre for Mechanical Sciences, Udine, Italy, October 27–31, 2003; and at the UNESCO sponsored workshop, 'Foundations of Thermodynamics,' UNESCO-ROSTE, Palazzo Zorzi, Venice, Italy, November 2–4, 2003.

References

[1] Essex, C. (1990), Advances in Thermodynamics, Vol. 3: Nonequilibrium Theory and Extremum Principles (Sieniutycz, S., Salamon, P., eds.), Taylor and Francis, New York, p. 435.
[2] Merzbacher, E. (1970), Quantum Mechanics, 2nd edn, Wiley, New York, Chapters 15 and 20–22.
[3] Weinberg, S. (1995), The Quantum Theory of Fields, Vol. I: Foundations, Cambridge University Press, Cambridge.
[4] Kapusta, J.I. (1989), Finite-Temperature Field Theory, Cambridge University Press, Cambridge, Chapters 1 and 2 and Appendix.
[5] Planck, M. (1959), Heat Radiation (1913), Dover Publications, New York.
[6] de Groot, S.R. and Mazur, P. (1984), Non-equilibrium Thermodynamics (1962), Dover Publications, New York, Chapters 1 and 3–5.
[7] Chandrasekhar, S. (1960), Radiative Transfer (1950), Dover Publications, New York.
[8] Mihalas, D. and Weibel-Mihalas, B. (1999), Foundations of Radiation Hydrodynamics (1984), Dover Publications, New York, Chapters 3 and 6.
[9] Reichl, L.E. (1998), A Modern Course in Statistical Physics, 2nd edn, Wiley, New York, Chapters 10 and 11 and Appendix B.
[10] Nieuwenhuizen, T. and Allahverdyan, A.E. (2002), Phys. Rev. E66, 03610.
[11] Essex, C. and Kennedy, D.C. (1999), J. Stat. Phys. 94, 253.
[12] Essex, C., Kennedy, D.C. and Berry, R.S. (2003), Am. J. Phys. 71(10), 969.
[13] Essex, C. (1984), J. Planet. Space Sci. 32, 1035; Astrophys. J. 285 (1984) 279; J. Atmos. Sci. 41 (1984) 1985.
[14] Sieniutycz, S., Salamon, P., eds. (1990), Advances in Thermodynamics, Volume 3: Non-equilibrium Theory and Extremum Principles (Sieniutycz, S., Salamon, P., eds.), Taylor and Francis, New York.
[15] Sieniutycz, S., ed. (1994), Conservation Laws in Variational Thermohydrodynamics (Sieniutycz, S., eds.), Kluwer Academic, Boston.
[16] Wheeler, J.A. and Feynman, R.P. (1945), Rev. Mod. Phys. 17, 157; 21 (1949) 425.
[17] Onsager, L. (1931), Phys. Rev. 37, 405; 38 (1931) 2265.
[18] Prigogine, I. (1945), Acad. Roy. Soc. Belg., Bull. Cl. Sci. 31, 600.
[19] Glansdorff, P. and Prigogine, I. (1971), Thermodynamic Theory of Structure, Stability, and Fluctuations, Wiley-Interscience, New York.
[20] Fisher, P., Kayser, B. and McFarland, K.S. (1999), Annu. Rev. Nucl. Part. Sci. 49, 481.
[21] (Lord) Rayleigh, J.W.S. (1976), The Theory of Sound (1877), Dover Publications, New York.
[22] Weinberg, S. (1996), The Quantum Theory of Fields, Vol. II: Modern Applications, Cambridge University Press, Cambridge, Chapter 16.
[23] Bludman, S.A. and Kennedy, D.C. (1997), Astrophys. J. 484, 329; erratum: 492 (1998) 854.
[24] Kippenhahn, R. and Weigert, A. (1990), Stellar Structure and Evolution, Springer, New York.
[25] Kolb, E.W. and Turner, M.S. (1994), The Early Universe, Perseus Publishing, New York.

Chapter 13

OPTIMAL FINITE-TIME ENDOREVERSIBLE PROCESSES—GENERAL THEORY AND APPLICATIONS

Harald Ries

Philipps-University Marburg[1], *and OEC AG*[2], *Munich, D-35032 Marburg, Germany*

Wolfgang Spirkl

Infineon Technologies A.G., Balanstr. 73, 81541 Munich, Germany

Abstract

We treat the general problem of transferring a system from a given initial state to a given final state in a given finite time such that the produced entropy or the loss of availability is minimized. This problem leads to a second-order differential equation similar to the Euler–Lagrange equation. However, while mechanical systems naturally follow the trajectory that minimizes the action, a thermodynamic system does not tend to minimize dissipation, rather an external control is required, for which we give the equations. We give exact equations for the optimal process for the general case of a nonlinear system with several state variables, and show solved examples for the case of two state variables. Not only the speed but also the path depends on the available time. For linear processes, e.g. in the limit of slow processes or if the Onsager coefficients do not depend on the fluxes, we find a constant entropy-production rate or a constant loss rate of availability and an optimal path independent of the available time.

Keywords: finite-time thermodynamics (FTT); dissipation; Euler–Lagrange equations; optimization

1. Introduction

1.1. The terminology of phenomenological thermodynamics

In equilibrium thermodynamics a macroscopic multiparticle system is described with a relatively small number of extensive variables (which scale with system size) such as the internal energy U, the volume V, the number N_i of particles of a given sort, the entropy S, etc. These extensive variables are not independent, but linked by a so-called fundamental

E-mail address: harald.ries@physik.uni-marburg.de (H. Ries); Wolfgang.Spirkl@infineon.com (W. Spirkl).

[1] http://www.physik.uni-marburg.de/optik/

[2] http://www.oec.net

equation that may be customarily written in the energy representation as

$$U = U(S, V, N_i, \ldots), \tag{1}$$

or in the entropy representation as

$$S = S(U, V, N_i, \ldots). \tag{2}$$

The intensive variables (which are independent of system size) such as temperature T, pressure p, chemical potential μ are defined as partial derivatives of the internal energy in the case of the energy representation:

$$T = \left(\frac{\partial U}{\partial S}\right)_{V,N_i}, \qquad p = -\left(\frac{\partial U}{\partial V}\right)_{S,N_i}, \qquad \mu_i = \left(\frac{\partial U}{\partial N_i}\right)_{S,V}. \tag{3}$$

For a more detailed description of the framework of phenomenological thermodynamics we refer the reader to standard textbooks such as Refs. [1,2]. If two systems are in contact across a wall, they tend to equilibrate, i.e. reach a state where the values of the intensive parameters are equal on both sides of the wall.

This state is reached by the flows through the wall. The difference in values of the intensive parameters act as driving forces for corresponding flows in extensive parameters. For example, a difference in chemical potential causes a particle flow, meaning that the particle number on one side decreases while at the same time the particle number on the other side increases. Similarly a temperature difference causes a heat flow or more exactly a flow in internal energy through the wall. The product of a driving force and a corresponding flow constitutes a loss of available energy (in the energy representation) or an internal entropy production rate (in the entropy representation). We will collectively refer to the scalar product of the vector of all driving forces with the vector of all flows as dissipation.

1.2. Finite-time thermodynamics

Assume we want to change a system from one state into another. In a general sense dissipation is to be avoided. Therefore, it is natural to ask how this change may be accomplished such as to minimize dissipation. Alas, the answer is both trivial and useless: since all flows approach zero as the driving forces go to zero, dissipation approaches zero in the limit of infinitely slow changes (quasistatic changes). But then we need an infinite amount of time. For this reason Steve Berry and his coworkers slightly modified the above question: *how can a system be changed from one state into another in a given finite time such as to minimize dissipation.* This modified question is of immense practical importance. It has attracted an extremely rich and fruitful research effort and in fact is the root of the new field known as finite-time thermodynamics (FTT) [3–5].

1.3. Endoreversible systems

The exact details of how the equilibrium is reached defy a rigorous thermodynamic treatment precisely, because thermodynamics is restricted to systems in equilibrium. One way to circumvent this issue is to consider systems internally in equilibrium, in which all dissipative processes are restricted to the walls, while the walls themselves are excluded from a thermodynamical description. Such systems are called endoreversible [6].

Assume we want to change the state of an endoreversible system. We cannot access the 'inner' reversible part. What we can do in order to promote change is control the value of the intensive parameters on the outside of the wall enclosing the system. For example, by applying a high temperature on the outside we can initiate a heat (or entropy) flow through the wall. In this chapter we present a variational approach to the problem of transferring an endoreversible system from a given initial state into a given final state by controlling the values of all intensive parameters on the outside of the enclosing wall, within a given finite time such that the dissipation is minimized.

Although the wall is not part of the system, the transfer properties of the wall are taken to characterize the kinetics of the system. For the inner part we assume that the fundamental Eq. (1) is known, which allows to calculate one extensive variable if all others are known as well as all intensive variables according to Eq. (3).

2. General problem

We consider a homogeneous system with independent extensive parameters X that are, e.g. the entropy S, the volume V, and the particle numbers N_α each referring to a different species α. The system is described by an equation of state $U = U(X)$ in the energy representation. The intensive variables are the conjugate of the extensive variables X_i,

$$Y_i = \frac{\partial U(X)}{\partial X_i}, \tag{4}$$

such as the temperature T, the pressure variable $-p$, and the chemical variables μ_α [1], where μ_α is the chemical potential for species α. It is assumed that the intensive variables are constant all over the system. The system is controlled by an external source that is defined by intensive parameters Y_i^e. It is assumed that the intensive variables of the source can be controlled at will.

The potential differences $Z_i = Y_i^e - Y_i$ cause fluxes of the associated extensive variables X_i. The dissipation rate due to the transfer of heat, volume, and mass is

$$\frac{\mathrm{d}\sigma}{\mathrm{d}t} = Z_i \frac{\mathrm{d}X_i}{\mathrm{d}t}. \tag{5}$$

We use Einstein's convention of summing over all indices appearing twice in a product. The functional relation between the fluxes $\dot{X}_i = \mathrm{d}X_i/\mathrm{d}t$ and the driving forces Z_i defines the kinetics of the endoreversible system. We write it as

$$Z_i = Z_i(X, \dot{X}). \tag{6}$$

It is the essence of this contribution that we make no further assumption about the functions $Z_i(X, \dot{X})$ other than that they are differentiable once. The objective is to find the optimal process $X(t)$ such that the total dissipation σ in a given time interval $[t^i, t^f]$,

$$\sigma = \int_{t^i}^{t^f} \frac{d\sigma}{dt} dt, \tag{7}$$

assumes a minimum, with the initial and final states constrained to $X(t^i) = X^i$ and $X(t^f) = X^f$. We refer to such a function $X(t)$ as a trajectory. In contrast, a path is the ordered set of system states without referring explicitly to time.

All conclusions remain valid if the objective is the minimization of the produced entropy rather than the minimization of dissipated availability [5]. In this case one starts with an equation of state $S = S(X)$, with U replacing S as one of the extensive independent variables X, and calculates $Y_i = \partial S/\partial X_i$, e.g. $1/T = \partial S/\partial U$, $p/T = \partial S/\partial V$, and $-\mu_\alpha/T = \partial S/\partial N_\alpha$. However, the following derivation is general and may be interpreted just as well in terms of minimum entropy production as in terms of minimum energy dissipation.

In order to find the optimal process trajectory $X(t)$ we could use a standard method from optimal control theory such as Pontrjagin's general maximum principle [7] or Dynamic Programming [8]. However, the problem is greatly simplified by the fact that $\dot{X}$ appears in the expression for $\dot{\sigma}$ and at the same time determines the evolution of the system. In contrast to the aforementioned optimal control methods, the calculus for the optimum endoreversible system control presented here does not require the evaluation of the instantaneous control parameter by minimization; a standard second-order differential equation set, in conjunction with the given initial and final system states, completely specifies the optimum solution.

We consider a variation $X \rightarrow X + \delta X$, $\dot{X} \rightarrow \dot{X} + \delta\dot{X}$. The variation of the total dissipation σ is

$$\delta\sigma = \int_{t^i}^{t^f} (\delta Z_i \dot{X}_i + Z_i \delta\dot{X}_i) dt \tag{8}$$

$$= \int_{t^i}^{t^f} \left[\dot{X}_i \left(\frac{\partial Z_i}{\partial X_j} \delta X_j + \frac{\partial Z_i}{\partial \dot{X}_j} \delta\dot{X}_j \right) + Z_i \delta\dot{X}_i \right] dt \tag{9}$$

$$= \int_{t^i}^{t^f} \delta X_j \left[\dot{X}_i \frac{\partial Z_i}{\partial X_j} - \frac{d}{dt}\left(\dot{X}_i \frac{\partial Z_i}{\partial \dot{X}_j} + Z_j \right) \right] dt. \tag{10}$$

The last term is obtained by transforming the terms containing $\delta\dot{X}_i$ with partial integration, using the fact that the value of δX_i must be zero at t^i and t^f. In order to have zero variation of the entropy, the term in square brackets must be zero for each time t,

$$0 = \dot{X}_i \frac{\partial Z_i}{\partial X_j} - \frac{d}{dt}\left(\dot{X}_i \frac{\partial Z_i}{\partial \dot{X}_j} + Z_j \right). \tag{11}$$

These second-order differential equations, in conjunction with the fixed values of the initial and the final state, determine the optimal trajectory. These equations can be written as:

$$0 = -\dot{X}_i \frac{\partial Z_i}{\partial X_j} + \ddot{X}_i \frac{\partial Z_i}{\partial \dot{X}_j} + \dot{X}_i \frac{\mathrm{d}}{\mathrm{d}t}\left(\frac{\partial Z_i}{\partial \dot{X}_j}\right) + \frac{\partial Z_j}{\partial X_i}\dot{X}_i + \frac{\partial Z_j}{\partial \dot{X}_i}\ddot{X}_i. \tag{12}$$

Substituting

$$\frac{\mathrm{d}}{\mathrm{d}t}\left(\frac{\partial Z_i}{\partial \dot{X}_j}\right) = \ddot{X}_k \frac{\partial^2 Z_i}{\partial \dot{X}_k \partial \dot{X}_j} + \dot{X}_k \frac{\partial^2 Z_i}{\partial X_k \partial \dot{X}_j} \tag{13}$$

leads to:

$$\ddot{X}_i\left(\frac{\partial Z_i}{\partial \dot{X}_j} + \frac{\partial Z_j}{\partial \dot{X}_i} + \dot{X}_k \frac{\partial^2 Z_k}{\partial \dot{X}_i \partial \dot{X}_j}\right) = \dot{X}_i\left(\frac{\partial Z_i}{\partial X_j} - \frac{\partial Z_j}{\partial X_i} - \dot{X}_k \frac{\partial^2 Z_i}{\partial X_k \partial \dot{X}_j}\right). \tag{14}$$

3. Constant entropy-production rate

In order to analyze the entropy-production rate, we multiply by $\dot{X}_j$. For the case of several state variables, this reduces the full information of the set of differential equations to a single equation. The first antisymmetric term cancels and leaves a total differential:

$$0 = \dot{X}_j \ddot{X}_i\left(\frac{\partial Z_i}{\partial \dot{X}_j} + \frac{\partial Z_j}{\partial \dot{X}_i}\right) + \dot{X}_j \dot{X}_i \frac{\mathrm{d}}{\mathrm{d}t}\left(\frac{\partial Z_i}{\partial \dot{X}_j}\right) = \frac{\mathrm{d}}{\mathrm{d}t}\left(\dot{X}_i \frac{\partial Z_i}{\partial \dot{X}_j}\dot{X}_j\right). \tag{15}$$

Thus for the optimal process a necessary condition is

$$\dot{X}_i \frac{\partial Z_i}{\partial \dot{X}_j}\dot{X}_j = \text{const.} \tag{16}$$

If the function Z is expressed by a resistance matrix R_{ij},

$$Z_i(X, \dot{X}) = R_{ij}(X, \dot{X})\dot{X}_j, \tag{17}$$

then Eq. (16) reads

$$\frac{\mathrm{d}\sigma}{\mathrm{d}t} + \dot{X}_i \frac{\partial R_{ij}}{\partial \dot{X}_k}\dot{X}_k \dot{X}_j = \text{const.} \tag{18}$$

This is the generalization of the known result [5,9], which states that for linear systems, the rate of entropy production or loss of availability is constant along the optimal trajectory. In the context of the FTT theory outlined here, we call a system linear if the driving forces Z are linear in the fluxes $\dot{X}$. They may, however, depend nonlinearly on the extensive parameters. For linear systems in this restrictive sense, the partial derivatives of R in Eq. (18) vanish.

Furthermore, for linear systems the optimal path is invariant under changes of the available process time $t^f - t^i$ while the value of the fluxes, the driving forces, as well as the dissipation rate are inversely proportional to the available process time. This can be immediately inferred from Eq. (11).

But even systems that do not comply with the definition of linearity outlined above approach the behavior of linear systems, as the available time grows. The reason is that in this limit fluxes decrease and the kinetic equations are adequately approximated by the first-order Taylor expansion in the fluxes. Thus in the slow limit all systems asymptotically approach linearity in the sense given above.

The theory outlined here is exact for small as well as for large fluxes, and with no assumption on the form of $Z(X, \dot{X})$ except that it is differentiable once.

The constant entropy-production rate follows from the Cauchy–Schwarz inequality [5]

$$\sigma = \int_{t^i}^{t^f} \dot{X}_i R_{ij} \dot{X}_j \, \mathrm{d}t \geq \frac{\mathcal{K}_S^2}{t^f - t^i}. \tag{19}$$

Here the kinetic process length $\mathcal{K}_S$ is defined as

$$\mathcal{K}_S = \int_{t^i}^{t^f} |\dot{X}_i R_{ij} \dot{X}_j|^{1/2} \mathrm{d}t. \tag{20}$$

It is based on the metric

$$(\mathrm{d}\mathcal{K}_S)^2 = R_{ij} \, \mathrm{d}X_i \, \mathrm{d}X_j. \tag{21}$$

The entropy-production rate is equal to the square of the speed measured by this metric,

$$\frac{\mathrm{d}\sigma}{\mathrm{d}t} = \left(\frac{\mathrm{d}\mathcal{K}_S}{\mathrm{d}t} \right)^2. \tag{22}$$

The value of $\mathcal{K}_S$ does not depend on the parametrization t, but it depends on the path of integration. With the minimum kinetic process length, taken over all possible paths from the initial to the final state, the right-hand side of Eq. (19) is a lower bound on σ; it is taken if the integrand is constant, i.e. if the entropy-production rate is constant. The global optimization problem can be split into (a) finding the path with the shortest length $\mathcal{K}_S$ and (b) optimizing the trajectory along this path, i.e. traveling this path with constant speed based on the metric $\mathcal{K}_S$. Assume that an optimal solution is known. The optimal path for taking the same system between the same initial and final points, but with different total available time, is the same. The trajectory can simply be deduced by scaling the time with a constant factor. For the general nonlinear case, however, a different total available time might lead to a different path that has to be found with Eq. (11).

The thermodynamic length $\mathcal{L}$ defined in Ref. [5] does not contain the kinetic functions R. Therefore the thermodynamic length in general bears no relationship to the entropy production or loss of availability if a certain trajectory is traversed. For calculating the loss of availability from $\mathcal{L}$ the authors have to introduce an average system time scale, which then becomes trajectory dependent. The kinetic process lengths $\mathcal{K}_S$ and $\mathcal{K}_U$ as defined in Eq. (20) incorporate the kinetic functions R. The functions R may be arbitrary functions of the system state X consistent with thermodynamics. Formally the same expression for an analog kinetic process length $\mathcal{K}_U$ is obtained if the objective is minimum loss of availability.

4. The boundary conditions

Eq. (14) is a second-order differential equation that can be solved by straightforward numerical methods in the spirit of a boundary problem [10, Chapter 17]. If the values of the extensive parameters as well as the fluxes are known at a given time, one can extend an optimal path to all future times. However, in the original formulation, the values of the extensive parameters at some future time are specified, in lieu of the initial fluxes. This is usually known as a two-point boundary problem: the values of the initial fluxes have to be determined such that the solution matches the desired state at the future time.

Using the similarity to the problem of aiming a gun such as to hit the target, within the framework of the shooting method as described in Ref. [10, Section 17.1] the initial derivative values are adjusted such as to match the desired final values. This is equivalent to having M different, in general nonlinear, equations for a vector of M unknowns, where M is the number of state variables. In the examples outlined in the following, we used a modified Newton root finder [10, Chapter 9] in order to find the unknown initial derivatives. Although this approach is not guaranteed to find the root from every starting point, it turned out to be sufficiently robust for the examples below. In cases where the shooting method turns out to be not robust enough, a relaxation algorithm could be used [10, Section 17.3].

5. Solved examples

In a previous article [11] we presented examples for the linear and the nonlinear case, using for simplicity systems with a single state variable ($M = 1$). In this contribution we illustrate the optimization with systems containing more than one state variable for two reasons: First, single-variable systems are to some degree trivial since the optimum path is not really needed, but rather only the optimum speed along the path. Secondly, the capability of the differential equations derived above to solve more complex problems shall be shown.

For the examples in this contribution, we refer to chemical reactions at constant temperature because it is easy to introduce several state variables by introducing several chemical substances; in particular there is no formal distinction between the entropy and the energy representation. For simplicity, we use a dimensionless representation of all constants and of the chemical potentials.

5.1. Catalyzed diffusion with linear kinetics

We treat a two-substance model (A and B) where the presence of substance B 'catalyzes' the kinetics of substance A. Thus the extensive parameters are the mole numbers N_A and N_B. The intensive parameters are the corresponding chemical potentials

$$\mu_A = \log\left(\frac{N_A}{N_{A_0}}\right), \qquad \mu_B = \log\left(\frac{N_B}{N_{B_0}}\right). \tag{23}$$

The kinetic equations are

$$Z_{\mathrm{A}} = \Delta\mu_{\mathrm{A}} = \frac{\dot{N}_{\mathrm{A}}}{N_{\mathrm{A}} k_{\mathrm{AA}}} \left(\frac{N_{\mathrm{B}_0}}{N_{\mathrm{B}}} \right)^a, \qquad Z_{\mathrm{B}} = \Delta\mu_{\mathrm{B}} = \frac{\dot{N}_{\mathrm{B}}}{N_{\mathrm{B}} k_{\mathrm{BB}}}, \tag{24}$$

where a, k_{AA}, k_{BB} are some non-negative constants. In this system, the kinetic matrix R is diagonal, and each particle flow is proportional to the associated driving force that, in this case, is assumed to be the difference of the chemical potential across the wall. The catalyst property is introduced by the term in brackets; with increasing concentration of B, the A reaction rate for given $\Delta\mu_{\mathrm{A}}$ is increased. If this catalyst property were absent, which is equivalent to the case $a = 0$, the optimum trajectory would always yield $\dot{X}_{\mathrm{B}} = 0$.

We start at time $t^i = 0$ with $\mu_{\mathrm{A}}(t^i) = 0$, $\mu_{\mathrm{B}}(t^i) = 0$ (which means $N_{\mathrm{A}}(t^i) = N_{\mathrm{A}_0}$, $N_{\mathrm{B}}(t^i) = N_{\mathrm{B}_0}$) and wish to double the concentration of A at finite time t_f. At the end we want substance B, which is just needed to augment the diffusion of A, back at the initial level. Thus the boundary conditions at $t = t^f$ are: $N_{\mathrm{A}}(t_f) = 2N_{\mathrm{A}_0}$, $N_{\mathrm{B}}(t_f) = N_{\mathrm{B}_0}$.

Fig. 1 shows the time evolution of the concentration of the two substances for the optimal trajectory. Note that along this path some B is added and then removed. Dissipation is constant along the optimal trajectory, since the system belongs to the class of linear systems in the sense of Section 3. At the ends it is mainly caused by pumping in (or removing) B, in the middle of the trajectory—in particular, at the point where $\mathrm{d}\mu_{\mathrm{B}}/\mathrm{d}t = 0$—it is entirely due to the flux of A. This is illustrated in Fig. 2.

Fig. 3 shows the optimal path in the space of the concentration of the two components. All curves are symmetric, the trajectories are symmetric with regard to the time, the path with regard to N_{A}. The symmetry is induced by forcing one state variable N_{B} finally back to the initial value, in conjunction with the independence of the kinetic matrix on the concentration N_{A}.

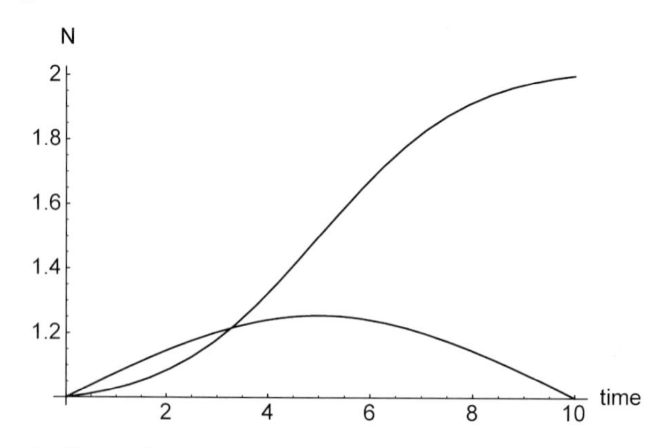

Fig. 1. Linear example. Time evolution of the concentrations of the two substances N_{A}, and N_{B} for the optimal trajectory. Note that along this trajectory some B is added and then removed. The values of the parameters for this example were: $k_{\mathrm{AA}} = 1$, $k_{\mathrm{BB}} = 2$, $a = 10$, $t_f = 10$.

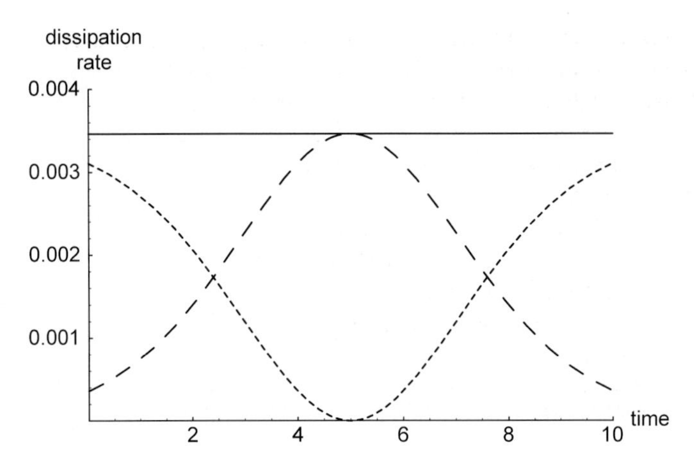

Fig. 2. Linear example. Dissipation rate along the optimal trajectory. Total dissipation (solid curve) is constant along the optimal trajectory, dashed curve refers to the flux of A, dotted curve to the flux of B. The values of the parameters for this example were: $k_{AA} = 1$, $k_{BB} = 2$, $a = 10$, $t_f = 10$.

Now assume that the same task is to be accomplished in a different available time. Then the optimal time evolution (Fig. 1) simply scales; the optimal path shown in Fig. 3 remains invariant. The dissipation rate is inversely proportional to the square of the available time; the total dissipation is inversely proportional to the available time.

5.2. *Catalyzed diffusion with nonlinear kinetics*

We again treat a two-substance model (A and B) where the presence of substance B catalyzes the kinetics of substance A.

For constant diffusion coefficient the flux is proportional to the difference in concentration. However, in the thermodynamic formulation, the driving force is the

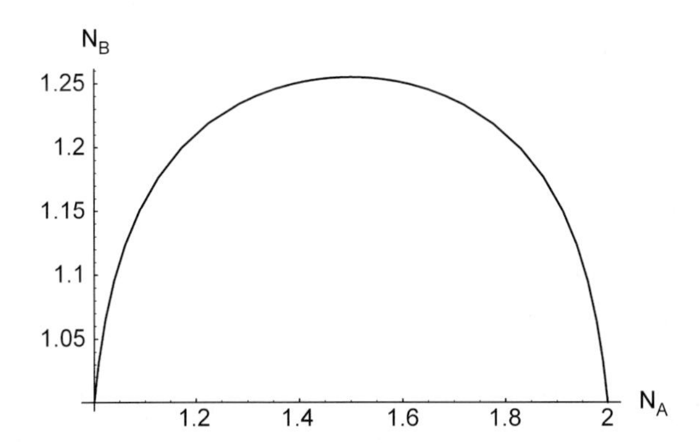

Fig. 3. Linear example. The optimal path in the 2D space of the extensive variables. The values of the parameters for this example were: $k_{AA} = 1$, $k_{BB} = 2$, $a = 10$, $t_f = 10$.

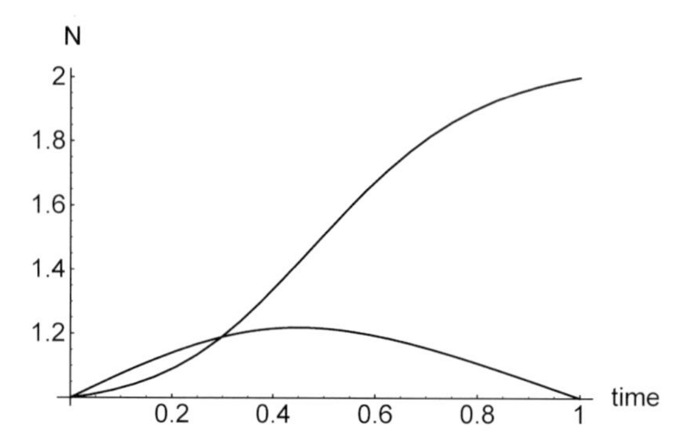

Fig. 4. Nonlinear example. Time evolution of the concentration of the two substances for the optimal trajectory. Note that along this trajectory some B is added and then removed. The values of the parameters for this example were: $k_{AA} = 1$, $k_{BB} = 1$, $a = 10$, $t_f = 1$.

difference in chemical potential. Thus we describe the kinetics by:

$$Z_A = \Delta\mu_A = \log\left(1 + \frac{\dot{N}_A}{N_A k_{AA}}\left(\frac{N_{B_0}}{N_B}\right)^a\right),$$
$$Z_B = \Delta\mu_B = \log\left(1 + \frac{\dot{N}_B}{N_B k_{BB}}\right), \tag{25}$$

where a, k_{AA}, k_{BB} are some non-negative constants. We start at $\mu_A(t^i) = 0$, $\mu_B(t^i) = 0$ (which means $N_A(t^i) = N_{A_0}$, $N_B(t^i) = N_{B_0}$) and wish to double the concentration of A within the finite time t_f. Again, substance B is needed to augment the diffusion of A. At the end we want it at the initial level. Thus: $N_A(t_f) = 2N_{A_0}$, $N_B(t_f) = N_{B_0}$.

Fig. 4 shows the time evolution of the concentration of the two substances for the optimal trajectory. Note that along this trajectory some B is added and then removed.

Fig. 5 shows the optimal path in the space of the components.

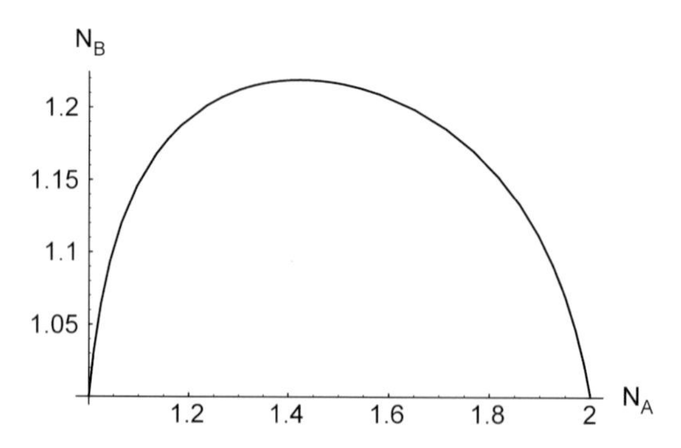

Fig. 5. Nonlinear example. The optimal path in the 2D space of the concentrations. The values of the parameters for this example were: $k_{AA} = 1$, $k_{BB} = 1$, $a = 10$, $t_f = 1$.

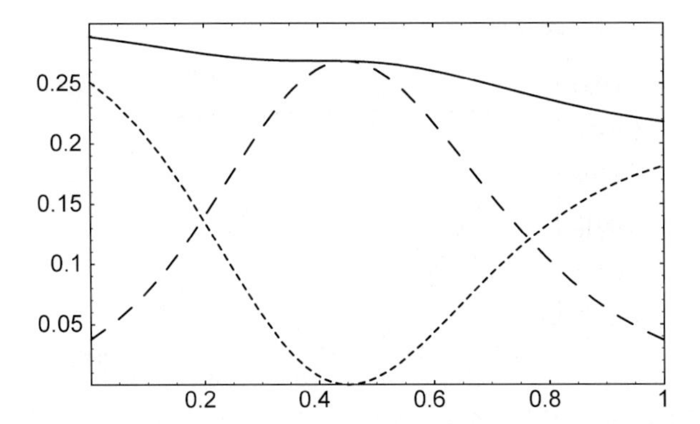

Fig. 6. Nonlinear example. Total dissipation rate along the optimal trajectory solid curve. Dashed curve relates to substance A, dotted curve to substance B only. Since the driving forces are not linear in the fluxes, the total dissipation rate is not constant along the path. The values of the parameters for this example were: $k_{AA} = 1$, $k_{BB} = 2$, $a = 10$, $t_f = 1$.

The total dissipation rate is not constant along the path because the driving forces are not linear in the fluxes. This is illustrated in Fig. 6.

Now assume that the same task is to be accomplished in a different available time. Then in the general case of nonlinear kinetics the two-point boundary problem has to be solved again. The result will be different. For illustration we compare in Fig. 7 and Fig. 8 the optimal path and the dissipation rate for the same system as in the previous example the sole exception being that the available time is increased to $t_f = 10$ to the previous case.

Fig. 8 shows the dissipation rate along the optimal slow path, when the available time is 10. The variations of the dissipation rate along the path are much smaller. The dissipation rate is lower by roughly a factor 100. Note that the slow limit of the nonlinear kinetic equations (25) precisely corresponds to the example presented in the previous section (Eq. (24)).

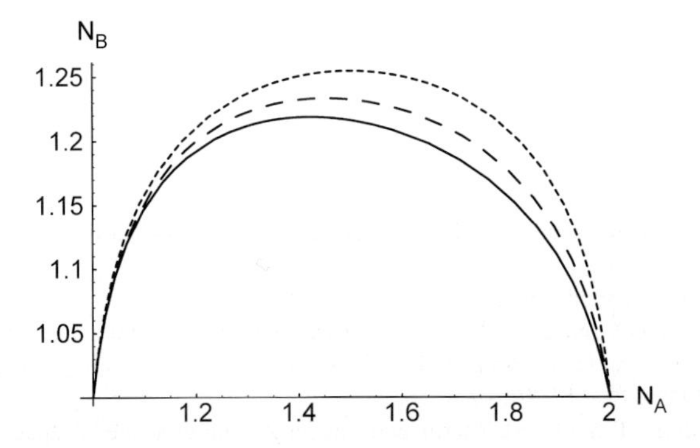

Fig. 7. Nonlinear example. The optimal path in the 2D space of the extensive parameters N_A and N_B for an available time $t_f = 1$ (solid line), $t_f = 10$ (dashed line) and for the limit $t^f \to \infty$ (dotted line, identical to Fig. 3). The values of the other parameters were: $k_{AA} = 1$, $k_{BB} = 2$, $a = 10$.

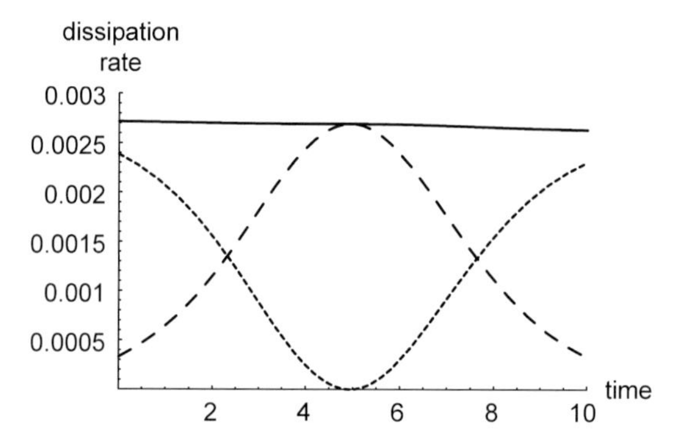

Fig. 8. Nonlinear example. Dissipation rate along the optimal trajectory for the slow case $t_f = 10$. The values of the other parameters for this example were: $k_{AA} = k_{BB} = 1$, $a = 10$. Compare this to Fig. 6: The variations of the dissipation rate along the trajectory are much smaller. The dissipation rate is lower roughly by a factor 100, the square of the increase in available time.

6. Conclusions

This contribution showed how to apply the generalized theory of FTT to systems with more than a single state variable. Systems that are linear with regard to the relation between fluxes and forces exhibit constant dissipation rate when operated at minimum total dissipation. While single state variable systems are fully characterized by this invariant, systems with more than one state variable need additional information to describe the optimum trajectory; this additional information may be considered as the optimal path in the state space. The differential equations outlined in a previous and in this contribution provide this additional information. Linear systems also show a simple scaling behavior when varying the available process time.

The general class of nonlinear systems, however, generally exhibits neither dissipation rate constancy nor scaling behavior. However, in the case of long process time, the so-called slow case, nonlinear systems asymptotically behave like linear systems.

References

[1] Kreuzer, H.J. (1981), Nonequilibrium Thermodynamics and its Statistical Foundations, Clarendon Press, Oxford.
[2] Callen, H.B. (1960), Thermodynamics, Wiley, New York.
[3] Andresen, B., Salamon, P. and Berry, S.R. (1984), Thermodynamics in finite time, Phys. Today (9), 2–10.
[4] Andresen, B., Berry, S.R., Ondrechsen, M.J. and Salamon, P. (1984), Thermodynamics for processes in finite time, Acc. Chem. Res. 17, 266–271.
[5] Salamon, P. and Berry, R.S. (1983), Thermodynamic length and dissipated availability, Phys. Rev. Lett. 51(13), 1127–1130.
[6] Rubin, M. (1979), Optimal configuration of a class of irreversible engines, Phys. Rev. A 19, 1272–1276.
[7] Shiners, S.M. (1992), Modern Control System Theory and Design, Wiley, New York.

[8] Bellmann, R. (1967), Dynamische Programmierung und selbstanpassende Regelprozesse, Oldenburg, München.

[9] Salamon, P., Nitzan, A., Andresen, B. and Berry, R.S. (1980), Minimum entropy production and the optimization of heat engines, Phys. Rev. A 21(6), 2115–2129.

[10] Press, W.H., Flannery, B.P., Tekolsky, S.A. and Vetterling, W.T. (1992), Numerical Recipes, 2nd edn, Cambridge University Press, New York.

[11] Spirkl, W. and Ries, H. (1995), Optimal finite-time endoreversible processes, Phys. Rev. E 52(4), 3485–3489.

Chapter 14

EVOLUTIONARY ENERGY METHOD (EEM): AN AEROTHERMOSERVOELECTROELASTIC APPLICATION

Hayrani Öz

Department of Aerospace Engineering, The Ohio State University, Columbus, OH 43210, USA

Abstract

Evolution means a process of change in a particular direction, evolutionary being the adjectival form of the word. This chapter is based on a novel theoretical foundation introduced by the author as the Evolutionary Energy Method (EEM) finding its root in the natural law of energy conservation, specifically the First Law of Thermodynamics. To this end, the Law of Evolutionary Energy (LEE) is introduced as the encompassing foundational evolutionary equation, where the evolutionary operator $\mathbf{Đ}$ *is a directional change operation via parameter alterations on the energy-quantities satisfying the energy-conservation law along the actual dynamic path, and acts on the total evolving energy, which is defined as the time integral of the total actual energy interactions in a dynamic system. The EEM is an algebraic (direct) energy method; that is, it uses and needs no knowledge of differential equations of the system for response and/or control studies of dynamic systems. Introduction of the concept of Assumed-Time Modes (ATM) for the generalized response variables and generalized control inputs of a dynamic system in conjunction with the Law of Evolutionary Energy culminates in elimination of time from the system dynamics completely, yielding the Algebraic Evolutionary Energy description of the system dynamics for response and control studies. As an application of the EEM, an aerothermoservoelectroelastic system is described completely algebraically and illustrated for studying the feasibility of structural skin temperature control in Mach 10 hypersonic flight by using optimal distributed control actuation. The structural temperature and the structural deformation are controlled simultaneously by using only temperature-feedback optimal control laws via elastothermoelectric actuation.*

Keywords: Law of Evolutionary Energy (LEE); Evolutionary Energy Method (EEM); Algebraic Equations of Motion (AEM); Evolution Energy; Directional Change-Evolution; Evolutionary Operator; Evolutional Operator; Assumed-Time-Modes (ATM); Algebraic States; Algebraic coordinates; Thermo-Elastic-Piezo-Pyro-Electric System; Thermal Evolutionary Energy Equations; Aerothermoservoelectroelasticity; Direct Control Method (DCM); Distributed Elasto-Thermo-Electrical Control; Isomodal Controller; Hypersonic Skin Termperature Control; Hamilton's Law of Varying Action (HLVA); Hamilton's Principle (HP); Principle of Virtual Work (PVW)

1. Introduction

This chapter introduces a novel approach to description, and response and control studies of general dynamic systems. The underlying concepts of the approach were

E-mail address: oz.1@osu.edu (H. Öz).

matured and illustrated as the 'Algebraic Evolutionary Energy Method in Dynamics and Control' in the author's funded research efforts on the subject for the period of 1999–2001, and formally reported in the resulting Technical Report [1]. We present the approach in this chapter for the first time to the larger scientific community, with further exposure of the physical foundation of the concepts, as the Evolutionary Energy Method (EEM) with its associated Law of Evolutionary Energy (LEE). The proof of LEE is also illustrated in this chapter, but only for Newtonian dynamic systems due to space limitation.

The EEM culminates in a set of purely algebraic equations for the study of system dynamics and it does not require any knowledge of differential equations of the system either for response or control studies. The method is founded on a most exalted natural law, the general conservation of energy, The First Law of Thermodynamics, along with a unique perspective of a process of change in a particular direction (signifying satisfaction of energy conservation along the evolution of the actual dynamic paths) denoted by the operator symbol **Đ**, and defined here as the evolutionary energy (directional change in energy) **Đ**E, where E is the actual process energy. The reader is reminded that the terms 'evolution' and 'evolutionary' as used throughout this chapter mean 'directional change' and therefore **Đ**E means the directional change of the energy E, for example.

In the EEM, the system dynamics is described and solved by using the evolution of the time integral of the total energy interactions along the actual paths of dynamic processes obeying The First Law of Thermodynamics. Time is eliminated from the processes by using the concept of time-basis functions (TBF) multiplied by (unknown) constant algebraic coordinates 'A' expansions, for both the response (output) variables and control (input) variables to describe the evolving trajectories, a step known as the Assumed-Time Modes (ATM) expansion [2]. Thanks to the ATM expansions of dependent variables (which are ultimately the generalized coordinates for the system) a priori integrations in time of all resulting energy quantities are made possible. The algebraic coordinates 'A' and the time parameter 't' in the ATM expansions of the generalized coordinates constitute a total parameterization of the path of the dynamic system. The use of the words parameter and parameterization throughout the chapter is in reference only to the algebraic coordinates 'A' and the time variable 't' of the generalized coordinates, not to the given physical and geometric properties (as identification parameters) of a dynamic system.

With this perspective, the evolutionary operation **Đ** effectively ultimately becomes an energy-conservation-compliant (ECC) total parameter-altering operation on the generalized coordinates. Operation of **Đ** on the algebraic coordinates, that is, ECC alteration of the algebraic coordinates A is an essential feature of the EEM for the description and solution of the dynamic system. Through this step a system of algebraic equations for the unknown constant algebraic coordinates A are obtained and solved for a direct, algebraic solution. Whereas, since the time parameter 't' is independent of the other parameters—algebraic coordinates A, operation of **Đ** on the time parameter 't' or alteration of the parameter 't' of the generalized coordinates is an optional independent step. This optional step serves either as a 'verification step or an accuracy test for numerical solutions' obtained algebraically via algebraic coordinates alterations, or as a means of obtaining the differential equations of motion should one wish not to obtain the solution directly, algebraically. Hence, the EEM provides a multifaceted perspective to the study of dynamic systems. Specifically, the resulting algebraic system dynamics and the algebraic control problem, referred to as the Algebraic Equations of Motion (AEM), are most easily dealt with to yield closed-form algebraic solutions, a degree of success not possible by any differential-equations-based control approach in the literature.

The EEM is illustrated via an application that represents a crossroads for aerodynamics, structural mechanics, thermodynamics, piezoelectricity and pyroelectricity, and control theory. We demonstrate a feasibility study of structural skin-temperature control in Mach 10 hypersonic flight by using distributed control actuation. Reflections on some historical and contemporary background and tools of virtual variational mechanics as they relate to and in contrast with the EEM with its evolutionary energy operator **Đ** are also included.

2. Evolutionary energy method

2.1. Evolving energy

In the EEM, we consider the *evolving energy* of a dynamic system, which we define as the *time integral of all energy quantities*. To introduce the fundamental concepts, from a set of all relevant interacting energy expressions for the phenomena under study, consider a single energy process and/or work done expression $E(p,t)$ where p and t denote independent spatial and time variables. The *evolving energy* $\mathcal{E}$ associated with $E(p,t)$ is

$$\mathcal{E} = \int E(p,t)\mathrm{d}t = \int \bar{E}(p,t)\mathrm{d}V\mathrm{d}t, \tag{1}$$

where $E(p,t)$ with the overbar denotes the associated energy density per volume. Next, it is assumed that the primary dependent variables of the field are determined, and that all such variables are represented in terms of time-dependent generalized coordinates $q(t)$ via a spatial basis expansion for each dependent variable:

$$g(p,t) = N^{\mathrm{T}}(p)q(t), \tag{2}$$

where $g(p,t)$ is a generic dependent variable, $N^{\mathrm{T}}(p)$ is a row vector of spatial shape functions and $q(t)$ is the vector of associated generalized coordinates. Examples of $g(p,t)$ are: displacement variables in structural mechanics, temperature distribution in thermal sciences, electrical displacement and or electric field variables. In the following, the relevant time derivatives of $q(t)$ are implied, and the necessary algebraic manipulations for multivariable applications should be evident. Hence, after spatial discretization of the dependent field variables, the evolving energy $\mathcal{E}$ becomes

$$\mathcal{E} = \int E(N(p),q(t),t)\mathrm{d}t = \int \bar{E}(N(p),q(t),t)\mathrm{d}V(p)\mathrm{d}t = \int E(q(t),t)\mathrm{d}t. \tag{3}$$

Ultimately one arrives at only time-dependent behavior of the particular energy phenomenon over an arbitrarily chosen time interval $[t_0,t_f]$ to write the definite evolving energy:

$$\begin{aligned} \mathcal{E} &= \int_{t_0}^{t_f} E(q(t), t, q(t_0), t_0)\mathrm{d}t = \int_{t_0}^{t_f} E(t, t_0)\mathrm{d}t, \\ E(t, t_0) &= E(t) - E(t_0) = \int_{t_0}^{t} \frac{\mathrm{d}E(\tau, t_0)}{\mathrm{d}\tau}\mathrm{d}\tau, \end{aligned} \tag{4}$$

where the running time t is measured from the initial time t_0, and τ is a dummy time variable of integration from the initial time to the running time, $t_0 \leq \tau \leq t$. We also noted that the integrand energy/work done expression can also be envisioned to have resulted from an interim/inner energy rate (power) integral of E. This is particularly useful for thermal sciences where expressions for power rather than energy are available, such as heat fluxes and Fourier's conduction law. Thus, for such energy-rate expressions the evolving energy $\mathcal{E}$ is of the form, or any $\mathcal{E}$ can also be written in its rate or power form as

$$\begin{aligned} \mathcal{E} &= \int_{t_0}^{t_f} E(t, t_0)\mathrm{d}t = \int_{t_0}^{t_f} [E(t) - E(t_0)]\mathrm{d}t = \int_{t_0}^{t_f} \left[\int_{t_0}^{t} \frac{\mathrm{d}E(\tau, t_0)}{\mathrm{d}\tau}\mathrm{d}\tau \right] \mathrm{d}t \\ &= \int_{t_0}^{t_f} \int_{t_0}^{t} \mathrm{d}E(\tau, t_0)\mathrm{d}t. \end{aligned} \tag{5}$$

2.2. *The first law of thermodynamics*

Generalizing the exposition of the preceding section, we may view all of the expressions and terminology $\mathcal{E}(t_f, t_0)$, $E(t)$, $E(t_0) = E_0$, $E(t, t_0) = E(t) - E(t_0)$, appearing in Eqs. (3)–(5), as representing the totality of all of the energy interactions involved among all of the energy processes in the system. One must note that $E(t)$ is the total running process energy expression at any intermediate time t and $E(t_0)$ is the initial energy level E_0 of the system, and $E(t, t_0)$ is the net *energy expenditure expression*, between the current time t and the initial time t_0, $\mathcal{E}(t_f, t_0)$ being the time integral of this expenditure for any definite interval of motion $[t_0, t_f]$, $t_0 \leq t \leq t_f$. Note that, alternatively, $\mathcal{E}$ as in Eq. (3) may also denote the indefinite time integral of a process energy expression $E(t)$, which should be evident from the context. Then, the First Law of Thermodynamics for the system, in terms of the total process energy $E(t)$ for any time t and the initial energy level $E(t_0) = E_0$, as the general energy-conservation law; or in terms of the evolving energy $\mathcal{E}$ for any arbitrarily long time interval $[t_0, t_f]$, as the *evolving general energy-conservation law*, is given by

$$\begin{aligned} &E(t, t_0) = E(t) - E(t_0) = 0, \\ &\mathcal{E}(t_f, t_0) = \int_{t_0}^{t_f} E(t, t_0)\mathrm{d}t = \int_{t_0}^{t_f} [E(t) - E(t_0)]\mathrm{d}t = 0. \end{aligned} \tag{6}$$

The total process energy $E(t)$ describes an energy supersurface over its domain which is the state space of the dynamic system described by the generalized coordinates q and their rates whenever relevant. *The First Law of Thermodynamics, Eq. (6) indicates that the actual path must follow an energy manifold of motion described by the intersection of* *the total process energy surface $E(t)$ with an initial energy-level plane of $E(t_0) = E_0$ over the state space of the dynamic system.*

2.3. *The law of evolutionary energy: directional change in evolving energy* $\mathbf{Đ}\mathcal{E}$

According to the new Merriam-Webster Dictionary: *Evolution is a process of change in a particular direction.* The adjectival form of evolution is *evolutionary* or *evolutional* already implying a change in the process in a particular direction. Therefore, formally 'Evolutionary Energy' is the change in the energy process that is manifesting in a particular direction. *The directional change in the energy process will be denoted by the symbol* **Đ**. The particular direction of change for the energy process remains to be specified, and one must understand the physical significance of this 'evolutionary energy', and how it is physically affected and manifested.

In the EEM, we consider the *directional change* in the evolving energy $\mathbf{Đ}\mathcal{E}$ *along the actual path of the motion of the dynamic system*, in which the operator **Đ** denotes a change in the energy expenditure $\mathbf{Đ}E(t, t_0)$ which will be referred to as the *Evolutionary (directional change in) Energy expenditure*. It follows that the $\mathbf{Đ}E(t, t_0)$ operation, that is, the evolutionary energy expenditure must satisfy the energy-conservation law that governs the path; therefore, the operation **Đ** is constrained to only those paths that satisfy The First Law of Thermodynamics, Eq. (6). With that, we now state that in a natural process, where $E(t, t_0)$ is the totality of all of the interacting energy expenditures for the disciplines under study, alterations in the actual dynamic paths of generalized coordinates must satisfy the following *evolutionary energy-conservation equation*:

$$\mathbf{Đ}\mathcal{E} = \mathbf{Đ}\int_{t_0}^{t_f} E(t, t_0)\mathrm{d}t = \int_{t_0}^{t_f} \mathbf{Đ}E(t, t_0)\mathrm{d}t = \int_{t_0}^{t_f} [\mathbf{Đ}E(t) - \mathbf{Đ}E(t_0)]\mathrm{d}t = 0 \quad \text{(LEE)}. \tag{7}$$

We shall refer to Eq. (7) as the *Law of Evolutionary Energy (LEE) and alternately refer to it as the* $\mathbf{Đ}\mathcal{E}$ *equation*. In Eq. (7) the **Đ** operator fulfills its function only on the generalized coordinates and their rates by altering the contemplated parameters of the generalized coordinates q in terms of the time-dependent parameter 't' and the time-independent constant parameters 'A', the algebraic coordinates of the system as stated in Section 1 above and as will be elaborated below. Therefore, **Đ** can be stated within or outside the definite time integral without any consequence, and naturally it must act as an integrand operator. **Đ** never operates on the 'given physical identification-parameters' of a system whether time dependent or constants. Hence, the directional change for the total energy process is actual and uniquely defined by the LEE.

The next issue is to address what such a directional change in energy, evolution of energy, physically represents or how it is affected. We now conjecture that the time-dependent generalized coordinates q of a dynamic system are implicitly or explicitly parametrically characterizable functions of both a single time-dependent parameter 't' and a numerable set of other time-independent (constant) parameters that will be denoted by a set of 'A'. Since As are constant we will refer to them as 'the algebraic coordinates or algebraic states' of the energy system. Consequently, the energy manifold over which a natural energy-conservation compliant (ECC) motion must take place can also be viewed as completely parameterized in such a manner.

To describe the dynamics of the system, we now consider changes in the generalized coordinates q due to infinitesimal alterations in its parameters 't' and 'A', but require that such parameter-altered generalized coordinates q still satisfy the associated energy-conservation law equation as a constraint; that is, all altered paths must be dynamically actual, energy-conservation-compliant (ECC) paths; in other words, the altered paths must satisfy the LEE, Eq. (7).

Mathematically speaking, due to infinitesimal alterations of the parameters 't' and 'A', if the altered motion is to be ECC as *per the LEE, Eq. (7), the motion must take place in a direction normal to the process energy supersurface E(t) gradient with respect to the parameters 't' and 'A' at any time t*; in other words, the alterations must be in the superplane tangent to the process-energy surface $E(t)$ at any time t as depicted in Fig. 1. Specifically, in an ECC process, altering the parameter 't' infinitesimally constrains the altered motion at any time t to be tangent to the energy surface $E(t)$ and to the energy-conservation manifold $E(t, t_0) = 0$ corresponding to the given initial energy level E_0 on the energy surface $E(t)$. On the other hand, altering the algebraic coordinates 'A' infinitesimally is tantamount to altering the initial states and hence the initial energy level E_0 at t_0 infinitesimally, and therefore constrains the altered motion at any time t again to be tangent to the energy surface $E(t)$ and to an altered (new) energy-conservation manifold $E(t, t_0) = 0$ consistent with the altered (new) value of the initial energy level E_0 on the energy surface $E(t)$. Hence, all infinitesimal parameter alterations constrain the altered paths to a superplane that must be tangent to the $E(t)$ surface at any time t. This ensures that The First Law of Thermodynamics $E(t, t_0) = 0$ is always satisfied by the altered paths due to the altered parameters 't' and 'A'. To satisfy the LEE, $\mathbf{Đ}\mathcal{E} = 0$, an ECC infinitesimal path alteration in $\mathbf{Đ}E(t, t_0) = 0$ required

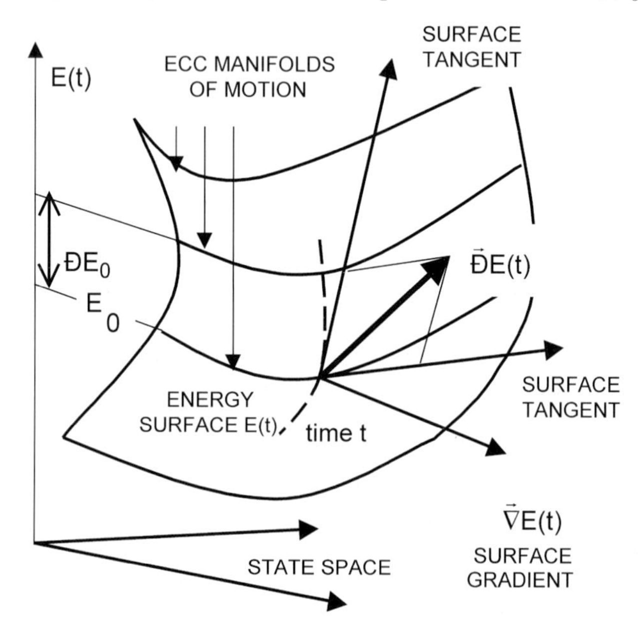

Fig. 1. Evolutionary energy $\mathbf{Đ}E(t)$: energy-conservation-compliant (ECC) directional change in energy process on the energy surface $E(t)$.

in the tangent superplane of the energy supersurface $E(t)$ at any time t comprises the 'directional change in the process energy' or 'evolutionary (process) energy', **Đ** being the symbol of this evolutionary operation. We refer to such ECC altered paths as the evolutionary paths. The evolutionary operator **Đ** must act on all energy terms that make up the total process energy $E(t)$ and on the initial energy level $E_0 = E(t_0)$ as required by the LEE, Eq. (7).

The evolutionary operator **Đ** can be interchanged with the differentiation operator 'd', and it shares the familiar mathematical features of the rules of differentiation. To arrive at naturally consistent $\text{Đ}\mathcal{E} = 0$ equations, from which one can obtain the actual path, one must observe the following physically based rules of operation in the use of the evolutionary operator **Đ**. *Rule 1*: Whatever discipline an energy term represents, it must be admissible: That is, one must remember or be able to envision the fundamental feature that an energy term (or its increment) for an actual path is always in the form of or derived from a (generalized) load multiplied by an associated compatible (incremental-differential) generalized coordinate dq. As an example, a functional operation **Đ** in the sense of differential calculus on a kinetic-energy term alone is not admissible unless it is accompanied by an additional term that puts it into its admissible evolutionary energy form, specifically the generalized acceleration loads multiplied by the evolutionary actual generalized coordinates. *Rule 2*: A primary variable, generalized displacement or generalized coordinate must be naturally admissible, that is, it must have continuous derivatives of necessary order in time depending on the discipline it represents. No coterminus endpoint-in-time boundary conditions are to be imposed on the evolution of primary variables. *Rule 3*: This rule provides an operational short cut. Theoretically, the evolutionary operator **Đ** must operate on loads (because the forces such as in the state-feedback form of the system are actually altered on the evolutionary paths). However, it can be proven by considering the physical nature of the parameter alterations in the LEE that *whenever an energy term is explicitly identifiable as the product of a generalized force and a generalized coordinate (as per Rule 1)*, a mathematical short-cut operation ensues, by which the evolutionary operator **Đ** can skip operating on the generalized force no matter what its functional form is—even though it may be a function of generalized coordinates, and directly operate on the multiplying generalized coordinate q. *Rule 4*: *Hence, all 'directional change in the energy' operations will be due to* **Đ** *operations on the 't' and or 'A' parameters of the generalized coordinates q and their rates only. Đ never will operate on the 'given physical parameters of a system whether time dependent or constants'.*

In Appendix A, we present a demonstrative constructive proof of the LEE for a Newtonian Dynamic (Structural Dynamic) system, exhibiting all the features of the use of the evolutionary operator **Đ** and its fundamental physical character. With the above rules strictly observed and the demonstration of LEE for a Newtonian dynamic system, one can extend the EEM to nonmechanical disciplines as well. Specifically, in this effort, we have extended the method to embrace thermal dynamics of structural systems alongside structural dynamics. The Thermal Evolutionary Energy $\text{Đ}\mathcal{E}_{\text{TH}} = 0$ equations given in the following (the derivation of which is particularly instructive, but cannot be given here due to space limitation) would be a natural-law alternative to nonunique variational principles (not laws) given for thermoelastic systems in the literature.

2.4. The LEE ($\text{Đ}\mathcal{E}$ equation) for a thermoelastic–piezo- pyroelectric system

The total $\text{Đ}E(t, t_0)$ expression appearing in the LEE, Eq. (7), for interacting thermal, mechanical and electrical processes is given by

$$\begin{aligned} \text{Đ}E(t,t_0) = {} & -\text{Đ}T + \mathrm{d}[(\partial T/\partial \dot{q}_\mathrm{S})\text{Đ}q_\mathrm{S}]/\mathrm{d}t - \text{Đ}W_\mathrm{Mext} + \text{Đ}W_\mathrm{M} + \text{Đ}W_\mathrm{M-TH} + \text{Đ}W_\mathrm{E} \\ & + \text{Đ}W_\mathrm{M-E} + \text{Đ}W_\mathrm{E-M} + \text{Đ}W_\mathrm{Eext} + \text{Đ}Q_\mathrm{TH} + \text{Đ}Q_\mathrm{K} + \text{Đ}Q_\mathrm{THext} \\ & + \text{Đ}Q_\mathrm{THint+} + \text{Đ}Q_\mathrm{TH-M} + \text{Đ}Q_\mathrm{TH-E} + \text{Đ}W_\mathrm{E-TH}. \end{aligned} \tag{8}$$

In this we adopt the thermodynamic system sign convention for the work/energy terms: work done by the system on the surroundings is positive and heat added to the system from the surroundings is positive. Accordingly, the mechanical definition of positive work of mechanical external and internal loads constitute thermodynamically negative work from the system's perspective. In Eq. (8) q_S are the structural (mechanical) generalized coordinates, T is the kinetic energy, and the sum of the first two expressions involving T represents the evolutionary work of mechanical acceleration forces, thus rendering the sum an admissible evolution; W_M and W_E are internal work for mechanical stresses and electrical fields, respectively; W_Mext and W_Eext are works of external mechanical and electrical loadings, respectively; $W_\mathrm{M-TH}$ and $W_\mathrm{M-E}$ are elastothermal and elastoelectric (indirect piezoelectric effect) mechanical coupling work, respectively; $W_\mathrm{E-M}$ is the work due to electroelastic coupling (direct piezoelectric effect); Q_TH is thermal kinetic energy (thermal capacitance or thermal inertia energy), Q_K is conductive thermal energy, Q_THext is the energy of external thermal loading, $Q_\mathrm{TH-M}$ is the heat (caloric) energy of thermoelastic coupling, $Q_\mathrm{TH-E}$ is the heat (caloric) energy of thermoelectric coupling (pyroelectric effect), Q_THint is the heat (caloric) energy of internal heat sources/sinks. When $\text{Đ}E(t, t_0)$ is the spatially discretized form of the energy/work/heat quantities by virtue of assumed admissible spatial basis function distributions for the primary variables as given by Eq. (2), then the generalized structural displacement coordinates q_S, generalized thermal displacement/temperature coordinates q_T and generalized electric field coordinates q_E become the primary variables, and hence $\text{Đ}E(t, t_0)$ has no spatial dependence. Thus, in a symbolic form using the full energy/work terms as given in Eq. (8), and carrying out all required evolutionary operations or evolutions will yield the unified evolutionary $\text{Đ}\mathcal{E}$ equation, that is, the LEE for the interacting multidisciplines:

$$\text{Đ}\mathcal{E} = \text{Đ}\mathcal{E}_\mathrm{S} + \text{Đ}\mathcal{E}_\mathrm{TH} + \text{Đ}\mathcal{E}_\mathrm{E} = 0 \qquad \text{System LEE} \tag{9}$$

$$\text{Đ}\mathcal{E}_\mathrm{S} = \int_{t_0}^{t_\mathrm{f}} \left[\sum F_\mathrm{M}^\mathrm{T}(q_\mathrm{S}, q_\mathrm{T}, q_\mathrm{E})\right] \text{Đ}q_\mathrm{S}\,\mathrm{d}t = 0 \qquad \text{Structural LEE}$$

$$\text{Đ}\mathcal{E}_\mathrm{TH} = \int_{t_0}^{t_\mathrm{f}} \left[\sum F_\mathrm{TH}^\mathrm{T}(q_\mathrm{S}, q_\mathrm{T}, q_\mathrm{E})\right] \text{Đ}q_\mathrm{T}\,\mathrm{d}t = 0 \qquad \text{Thermal LEE}$$

$$\text{Đ}\mathcal{E}_\mathrm{E} = \int_{t_0}^{t_\mathrm{f}} \left[\sum F_\mathrm{TH}^\mathrm{T}(q_\mathrm{S}, q_\mathrm{T}, q_\mathrm{E})\right] \text{Đ}q_\mathrm{E}\,\mathrm{d}t = 0 \qquad \text{Electrical LEE,}$$

where F denotes a vector of generalized loads and a summation sign represents the generalized internal and external load equilibrium conditions for each one of the mechanical, thermal and electrical disciplines. Since q are continuous in time (Rule 2), in Eq. (9), the coefficients of evolutionary generalized coordinates $\mathbf{Đ}q$ along the actual dynamic path must vanish if the LEE is to be satisfied by the motion. Note that the expressions under summation signs are interdependent thanks to the expressed interdisciplinary coupling energy/work terms in Eq. (8). The first summation yields structural generalized force equilibrium equations; the second summation yields thermal equilibrium equations as entropy equilibrium; and the last summation yields electrical charge equilibrium equations, or Maxwell's equations. Since integrals of each one of the disciplinary summations must vanish, the total $\mathbf{Đ}\mathcal{E}$ equation for the hybrid dynamic system breaks down to its disciplinary components yielding the LEE for each discipline separately, but yet preserving interdisciplinary interactions in their respective Fs. It must be noted that since the time parameter 't' and algebraic coordinates 'A' represent independent parameters, an evolutionary operation $\mathbf{Đ}$ may be regarded as an alteration operation on only the time parameter 't', or on only the algebraic coordinates 'A' or on both of them depending on the purpose.

For example, if one treats each q in Eq. (8) as a single composite parameter $q(t)$ of only the time parameter 't' without contemplating algebraic coordinates 'A' parameterization, then $\mathbf{Đ}q = \mathbf{Đ}q(t) = q_t\mathbf{Đ}t$, where the subscript t denotes derivative with respect to time and $\mathbf{Đ}t$ is an infinitesimal alteration of the time parameter 't' signifying an infinitesimal time step along the actual path. It is important to realize that $\mathbf{Đ}t$ *is an altered parameter, not an operation on the running time t (which is denoted by* dt*).* If one adopts this perspective, by using integration by parts in time through Eqs. (7) and (8), Eq. (9) will yield a set of simultaneous coupled ordinary differential equations of equilibrium for each discipline, and one will leave the domain of energy and enter the domain of force equilibrium for the solution of the dynamic system. We refer to this differential-equations approach as an indirect method of solution since the Evolutionary Energy Law (LEE) is not directly used to find the solution but only serves as a vehicle to obtain the differential equations.

3. Algebraic evolutionary energy equations of motion for dynamic systems

We now arrive at the fundamental operational difference between the indirect approach and the direct, algebraic approach to the solution of a dynamic system. In the algebraic approach, the evolutionary energy conservation law (LEE) is used directly to obtain the solution bypassing the differential equations completely by invoking explicit algebraic coordinates 'A' characterization of the generalized coordinates q. The process requires focusing only on the 'A' characterization of the qs. Therefore, we shall not be concerned with altering the time parameter 't'. That is, for the purpose of algebraic description and solution of the dynamic system, from here on, the evolutionary operator $\mathbf{Đ}$ will alter only the A parameters of the generalized coordinates q.

In the direct approach, we take a step further, and this time introduce the concept of *Assumed-Time Modes (ATM)* [2] for the time-dependent generalized coordinates. In ATM, the time behavior of each one of the generalized coordinates is assumed to be represented by a priori chosen admissible time-bases functions (TBF) multiplied by unknown constant coefficients 'A'. In this way, the time integrations in the LEE, the $\mathbf{Đ}\mathcal{E}$ equations, too can be performed a priori to eliminate the time from the problem completely to realize an algebraic description of the system dynamics as discussed above in Sections 1 and 2. We refer to the resulting set of equations in A only as the Algebraic (Evolutionary) Equations of Motion (AEM).

To arrive at the AEM, we assume that a time-dependent generalized coordinate over the *transition interval* $[t_{\mathrm{f}}, t_0]$ can be parameterized or expanded in the form:

$$q(t) = \sum_{k=1}^{N} \phi_k(t, t_0) A_k, \qquad \mathbf{Đ}q(t) = \sum_{k=1}^{N} \phi_k(t, t_0) A_k, \tag{10}$$

where ϕ_k are a finite set of assumed admissible independent TBFs and the coefficients A of the TBFs are now the new set of unknowns (constants) for the dynamic system. We refer to the totality of the coefficients A as the *algebraic states* or *algebraic coordinates* of the dynamic system that are precisely the parameters A of q as contemplated or conjectured in Sections 1 and 2. Hence, we transform the LEE, the $\mathbf{Đ}\mathcal{E}$ equation, from the space of generalized coordinates/states to the vector space of algebraic coordinates or parameters A of algebraic systems. The set of TBFs must be rich enough to be able to span the solution space in the considered interval of the motion. Note that, in Eq. (10), since the objective is to obtain the AEM, the $\mathbf{Đ}$ operation on q did not act on the TBFs and acted only on the algebraic coordinates A, the new unknowns (parameters) of the dynamic problem.

Next, introducing the nondimensional time τ for the transition interval $T_{\mathrm{T}} = [t_{\mathrm{f}}, t_0]$, for the collection of all of the generalized coordinates vectors of the system or of a particular discipline, we represent the multidimensional matrix version of Eq. (10):

$$q(\tau) = \Phi(\tau) A, \tag{11}$$

where $\Phi(\tau)$ and A are the matrix and vector of appropriate dimensions, respectively, of the totality of TBFs for all generalized coordinates in the problem. By introducing an ATM expansion for every generalized coordinate $q(\tau)$ in each discipline, and introducing these expansions into the $\mathbf{Đ}\mathcal{E}$ equations of each discipline, one can now perform the nondimensionalized time integrations a priori thanks to the TBFs, leaving the algebraic coordinates A as the only set of unknowns in *algebraic* $\mathbf{Đ}\mathcal{E}$ *equations*. Thus, a particular disciplinary $\mathbf{Đ}\mathcal{E}$ equation appears as

$$\mathbf{Đ}\mathcal{E}_{\mathrm{I}} = \mathbf{Đ}A_{\mathrm{I}}^{\mathrm{T}} \left\{ T_{\mathrm{T}} \int_0^1 \Phi_{\mathrm{I}}^{\mathrm{T}}(\tau) \sum F_{\mathrm{I}}(\Phi_{\mathrm{S+TH+E}}(\tau) A_{\mathrm{S+TH+E}}) \mathrm{d}\tau \right\} = \mathbf{Đ}A_{\mathrm{I}}^{\mathrm{T}} \tilde{P}_{\mathrm{I}}(A_{\mathrm{S}}, A_{\mathrm{T}}, A_{\mathrm{E}}) = 0 \tag{12}$$

$\mathrm{I} = \mathrm{S}, \mathrm{TH}, \mathrm{E} =$ discipline indicator,

where P_{I} is a column vector in which all elements are functions of the unknown structural, thermal and electrical constant algebraic coordinates A for the transition interval T_{T}, expressing the interdisciplinary coupling behavior.

Since the vectors of evolutionary alterations $\mathbf{Đ}A$ are nontrivial due to arbitrary infinitesimal initial states alterations, the AEM for the I-th discipline are given by the vanishing of the vector P_{I} for that field, and for the whole elastothermoelectric dynamic system $\{\mathrm{S}+\mathrm{TH}+\mathrm{E}\}$ we can write:

$$\tilde{P}(A_{\mathrm{S}},A_{\mathrm{TH}},A_{\mathrm{E}})=0,\ \ \tilde{P}=[\tilde{P}_{\mathrm{S}}^{\mathrm{T}}\,\tilde{P}_{\mathrm{TH}}^{\mathrm{T}}\,\tilde{P}_{\mathrm{E}}^{\mathrm{T}}]^{\mathrm{T}},\ \ [P(A)]A=0. \tag{13}$$

The system is now algebraically closed for a solution and Eq. (13) constitute the *Algebraic Evolutionary Energy Equations of Motion (AEM)* for the dynamic system. The right-most expression in Eq. (13) represents a matrix version of the AEM where it is assumed that one can always accomplish (even with mathematical artifice) to factor out a column vector of A from the rest of the terms. If the system is nonlinear the multiplier matrix P, which is referred to as the *Fundamental Algebraic Matrix of the system*, remains a function of As. If the system is linear, P is a constants matrix. Noteworthy is the fact that, since they represent an evolutionary natural law, the AEM in Eq. (13) will be singular to the degree that matches the number of states required for the dynamic system. In the AEM of the EEM, existence (observation of) this degree of singularity to the level found satisfactory by the analyst becomes a test of convergence and richness of the TBFs to capture the true ECC path of the dynamic system, a feat especially for numerical solutions. The AEM can be solved for the As (spectrally, via Eigenvalue Problem or Singular Value Decomposition) as a general solution without imposing any specific initial states on the dynamic system leading to the identification of 'genuine time modes' of the system and the As can be subsequently normalized to satisfy the given initial states of the system thereby obtaining the unique solution for a unique dynamic system. For nonlinear systems, trial initial states are used iteratively to capture a certain solution manifold, which can then be normalized for the specific initial states of the system for a unique solution, implying nonlinear superposition over the captured manifold [3–5].

4. Initial-value problems in the evolutionary energy method

For the purpose of illustration in this work, we shall only use simple power series in time as TBFs for the thermal, structural and electrical variables, and therefore some of the As can automatically be labeled a priori as the relevant specified initial states. In this case, the AEM, Eq. (13) can now be reduced by deleting or disregarding the rows corresponding to the initially specified As—matching the total number of states of the system—which removes the singularity in P. Then one separates out all products of specified As in the vector $\{A\}$ with the corresponding column elements of P that are functions of only the specified As in $P(A)$, as the nonlinear functions of initial conditions vector $R(x_0)$, where x_0 is the initial state vector representing the totality of all specified As. Furthermore, for linear systems that will be our sole interest in this chapter, denoting the initial conditions-reduced matrices and vectors by an overbar, the explicit Initial Value Problem (IVP) of the AEM is obtained in the form:

$$[\bar{P}]\bar{A}+[\bar{R}]\{x_0\}=\bar{B}(F_{\mathrm{ext}}),\ \ \bar{A}=[\bar{P}]^{-1}[\bar{B}(F_{\mathrm{ext}})-[\bar{R}]x_0]=[\bar{P}]^{-1}\bar{B}(F_{\mathrm{ext}},x_0). \tag{14}$$

$$\text{Linear sytem}: \ [\bar{P}] = \text{constant matrix}, \ [\bar{R}] = \text{constant matrix},$$
$$x_0 = \{\text{All specified initial state } As\}.$$

Once the unknown As are solved as in Eq. (14), the solution for all of the generalized coordinates $q(t, A)$ becomes simply function evaluations in Eq. (11) as products of the TBFs with the now known algebraic states A, alongside the already specified As as initial states $\{x_0\}$.

Finally, it must be noted that in obtaining the AEM nothing has been stated about the length of the transition interval T_T. One must not view a short T_T as a requirement; it all depends on the ability of the assumed TBFs to span the solution space within a T_T. If the TBFs are rich and/or powerful then the transition interval T_T can be kept long to that extent. There is no such need as a Nyquist criterion for the system. Indeed, it has been shown and demonstrated in Ref. [5], that with proper choice of the TBFs, the solution would not depend on the length of the transition time. Such TBFs are termed global TBFs. Within AEM there are ample opportunities to work with globally spanning TBFs [3–5]. However, often, one is content to use simple power series in time for the TBFs, and then this requires relatively short T_Ts to find the solution, hence necessitating time marching by invoking continuity of path between transition intervals. In the following, to demonstrate the method, we used simple power series TBFs, and accordingly adopted time marching with small transition intervals.

5. AEM for direct optimal control of a thermal-structural dynamic system

The ultimate objective of this work is to demonstrate a control methodology for structural temperature control of aerodynamically heated hypersonic dynamic structures via distributed control agents possibly via electrical smart actuation. The electrical subsystem is to be used as a controller for the thermal and structural dynamic subsystems. In that, one seeks to specify the electrical algebraic coordinates A_E not as unknowns for the response of the system, but as unknowns to be specified as control inputs coordinates/external inputs to the thermal and structural subsystems. Hence, after disregarding the electrical subsystem equations, we then separate the remaining columns of P and R matrices that correspond to the electrical coordinates A_E and the electrical initial conditions in Eq. (14) and then transpose them to the right-hand side (RHS) of the equations to write:

$$\begin{bmatrix} \bar{P}_T & \bar{P}_{TH-S} \\ \bar{P}_{S-TH} & \bar{P}_S \end{bmatrix} \begin{bmatrix} \bar{A}_T \\ \bar{A}_S \end{bmatrix} = \begin{bmatrix} \bar{Q}_{TE} \\ \bar{Q}_{SE} \end{bmatrix} \{A_E\} + \{\bar{B}(F_{ext}, x_{0S}, q_{0T})\},$$
$$x_{0S} = \{q_{0S}, \dot{q}_{0S}\}, \ \{B\} = \begin{bmatrix} \bar{B}_T(F_{Text}, x_{0S}, q_{0T} \\ \bar{B}_S(F_{Sext}, x_{0S}, q_{0T} \end{bmatrix}. \tag{15}$$

In this, A_T and A_S are the thermal and structural algebraic states. The overbarred Q_{TE} and Q_{SE} are obtained from the columns of P and R in Eq. (14) multiplying A_E transposed to the RHS. We have also reshuffled the AEM to move the thermal dynamics to the top since our focus is in controlling the temperature of the structure in a hypersonic environment. The external forcing vector F_{ext} will include the aerodynamic pressure loads on the structure and aerodynamic external heat transfer (heat load) to the structure, as well as any other heat sources or sinks on the structure. Hence, $\{B\}$ has been partitioned accordingly. Note that both structural and thermal initial conditions will appear in both system equations due to thermomechanical-coupling effects.

Since we assumed a linear system, the elements of the P matrix will be constant for a particular transition interval, and since AEM are mere algebraic equations, we can readily solve the thermoelastic system (15) for the thermal dynamics free of explicit structural interaction, to obtain

$$[P_{\mathrm{T}} + P_{\mathrm{TH-S}} P_{\mathrm{S}}^{-1} P_{\mathrm{S-TH}}]\{\bar{A}_{\mathrm{T}}\}$$

$$= [\bar{Q}_{\mathrm{TE}} - P_{\mathrm{TH-S}} P_{\mathrm{S}}^{-1} \bar{Q}_{\mathrm{SE}}]\{A_{\mathrm{E}}\} + \{B_{\mathrm{T}} - P_{\mathrm{TH-S}} P_{\mathrm{S}}^{-1} B_{\mathrm{S}}\}, \tag{16}$$

or more compactly

$$[\tilde{P}_{\mathrm{T}}]\{\bar{A}_{\mathrm{T}}\} = [\tilde{Q}]\{U\} + \{\tilde{B}_{\mathrm{T}}\}, \qquad \{U\} = \{A_{\mathrm{E}}\}. \tag{17}$$

In the above, we used a traditional notation to denote the control coordinates or control inputs by $\{U\}$ and overbarred $[Q]$s represent their control influence or control loading matrix. We are now ready to incorporate direct optimal control solutions from our earlier work [1,2,6,7,14] without any proof of being applicable to the EEM. The control approach is dubbed the *Direct Control Method* (DCM), it is based on no differential equations, but solely on the concepts of EEM that led to AEM for a dynamic system. First, we specify a quadratic controller performance index (PI), J for the controller. However, one is not restricted to quadratic measures in DCM as long as J is a positive-definite functional. For the thermo-mechanical system we define the following PI:

$$2J = \int_{t_0}^{t_f} \Big[q_{\mathrm{S}}^{\mathrm{T}} W_{q\mathrm{S}} q_{\mathrm{S}} + \dot{q}_{\mathrm{S}}^{\mathrm{T}} W_{\dot{q}\mathrm{S}} \dot{q}_{\mathrm{S}} + (q_{\mathrm{T}} - q_{\mathrm{TRef}})^{\mathrm{T}} W_{q\mathrm{T}} (q_{\mathrm{T}} - q_{\mathrm{TRef}}) + \dot{q}_{\mathrm{T}}^{\mathrm{T}} W_{\dot{q}\mathrm{T}} \dot{q}_{\mathrm{T}}$$

$$+ u(t)^{\mathrm{T}} R_{\mathrm{U}} u(t) \Big] \mathrm{d}t, \tag{18}$$

in which W are the positive-semidefinite structural displacement, structural velocity and temperature displacement, and temperature rate weighting matrices identified by their respective subscripts, q_{TRef} is a vector of reference/set/target point values for the generalized temperature coordinates, and R_{U} is the positive-definite control-weighting matrix. We shall, for now, regard $u(t)$ as a generic control vector in the time domain.

To generate a PI that is compatible with the AEM, we next introduce the TBFs expansions also for the generalized control inputs as well similar to the ATM expansion for the generalized (response) coordinates. Hence, we assume TBFs for the control inputs as an expansion in the form:

$$u(t) = u(\tau) = \Phi_{\mathrm{C}}^{\mathrm{T}}(\tau)\{A_{\mathrm{C}}\} = [1]\{A_{\mathrm{C}}\}, \qquad \{U_{\mathrm{E}}\} = \{A_{\mathrm{C}}\} = \{A_{\mathrm{E}}\}, \tag{19}$$

which constitutes '*A*' parameterization of control (C) forces where, $\{A_{\mathrm{C}}\}$ denotes the algebraic control coordinates and the matrix Φ of ATM, TBFs, for all of the control inputs has been chosen as unity. This is simply equivalent to zero-order sample and hold control inputs, which is very practical if the transition interval is small. Indeed, there is absolutely no reason at this point to choose a higher order TBF expansion for $u(t)$. Furthermore, if electrical-field signals are to be used as physical control inputs, $u(t)$ would be the electrical signal strengths, $\{A_C\}$ would be the values of these signals for the transition interval, and since they are the electrical algebraic coordinates we have been dealing with above, then $\{U_E\} = \{A_E\}$ is the vector of electrical inputs on the AEM.

Next, introducing the TBF expansion for controls $u(t)$, and the TBF expansions for the thermal and structural generalized coordinates, and nondimensionalizing the time, one can a priori integrate the performance measure in time to obtain a purely algebraic PI (API) consistent with the AEM. The resulting API has the following form:

$$2J(\bar{A}_T, \bar{A}_S, U_E) = 2J(\bar{A}_T, U_E) = 2J_A + 2J_U + \text{other terms}, \tag{20}$$

in which {other terms} do not matter for optimum solution, where J_A and J_U are the partitions of the API that contain only the algebraic states and the algebraic controls, respectively. Next, we reduce the API further by eliminating the structural unknowns $\{A_S\}$ in terms of thermal unknowns $\{A_T\}$ by using the solution from the structural AEM to obtain the final form of the API to be optimized. In the above algebraic operations, we utilized a modeling simplification a priori, specifically the thermal coupling submatrix P_{S-TH} on structural dynamics will vanish due to the assumption of no temperature variation through the thickness of a thin skin structure.

The optimal solution for control is given in closed form by the following expression:

$$\sum_{I=1}^{L}\left(\frac{\partial J_U}{\partial U}\right)_I = \left[\sum_{I=1}^{L}\left(\frac{\partial C}{\partial U}\right)_I\right]\left[\sum_{I=1}^{L}\left(\frac{\partial C}{\partial \bar{A}_T}\right)_I\right]^{-1}\left[\sum_{I=1}^{L}\left(\frac{\partial J_A}{\partial \bar{A}_T}\right)_I\right], \tag{21}$$

$I = 1, \ldots, L =$ transition time index for global API over all intervals—full history,

$I = 1$, first transition interval, $L =$ current (last) transition interval

if $I = L$; time-local API only for the current transition interval—no history,

in which C represents the constraint functions, that is, the AEM. The given optimal control solution is valid for even a nonlinear system, since the results are given for the general AEM as constraints, linear systems being a simpler application for it. The optimum solution yields a feedback solution directly, but can also be expressed in terms of initial conditions. Secondly, we have given the form in a shorthand notation, to preserve unsightly equations; but the solution is in closed form algebraically and involves no iteration whatsoever.

Finally, it remains to present the explicit form of the **ÐƐ** expressions for the thermal and structural dynamic system under aerodynamic thermal loading and pressure loading so that all the relevant AEM matrices can be obtained to apply the control solution to the problem.

6. The LEE ($\mathbf{Đ}\mathcal{E}$ equations) for the thermoelectric and elastoelectric system

6.1. *Thermal dynamic system*

An algebraic representation of a thermal dynamic system is also needed, so that it can be used compatibly, with the AEM for the elastoelectric structure and the DCM presented and adopted above to a thermoelastic system for temperature control. The form that we state here is unique to this work in accordance with the LEE. The details of the proofs that it must be in the form given here are omitted for brevity. We first consider the linear constitutive equations for a thermoelectroelastic domain:

$$\begin{aligned} \sigma(p,t) &= C_{\sigma\varepsilon}\varepsilon(p,t) + C_{\sigma \mathrm{E}}e(p,t) + C_{\sigma \mathrm{T}}(T(p,t) - T_{\mathrm{ref}}(p)) = \sigma_{\mathrm{M}} + \sigma_{\mathrm{E}} + \sigma_{\mathrm{T}}, \\ D(p,t) &= C_{\mathrm{D}\varepsilon}\varepsilon(p,t) + C_{\mathrm{DE}}e(p,t) + C_{\mathrm{DT}}(T(p,t) - T_{\mathrm{ref}}(p)) = D_{\varepsilon} + D_{\mathrm{E}} + D_{\mathrm{T}}, \end{aligned} \tag{22}$$

where σ and ε are 6D stress and strain tensor elements, T is temperature distribution, e is the 3D vector of electric field distribution, D is the 3D electric-charge-density distribution or electric-displacement vector, and T_{ref} is the uniform reference-temperature distribution for the stress free state of the system. Cs are matrices of material properties of compatible dimensions obtained under the required thermoelectromechanical conditions. The components of the total stress as mechanical/elastic stress σ_{M}, piezoelectric coupling stress σ_{E} and thermal stress σ_{T}; and the components of the electrical displacement distribution (charge/area) as D_{ε} for the direct piezoelectric term, as D_{E} for the direct electric term, and as D_{T} as the thermoelectric/pyroelectric term should be evident from the constitutive equations. The LEE, $\mathbf{Đ}\mathcal{E}$ equations for the thermal dynamics is

$$\mathbf{Đ}\mathcal{E}_{\mathrm{TH}} = \int_{t_0}^{t_f} \{\mathbf{Đ}Q_{\mathrm{TH}} + \mathbf{Đ}Q_{\mathrm{K}} + \mathbf{Đ}Q_{\mathrm{TH-M}} + \mathbf{Đ}Q_{\mathrm{TH-E}} + \mathbf{Đ}Q_{\mathrm{THint}} + \mathbf{Đ}Q_{\mathrm{THExt}}\}\mathrm{d}t = 0$$

$$\begin{aligned} \mathbf{Đ}\mathcal{E}_{\mathrm{TH}} = \int_{t_0}^{t_f} \Bigg\{ \int_V \Bigg[\int_{T_0}^{T} & T^{-1}\mathrm{d}U_{\mathrm{THD}} + \int_{t_0}^{t} T^{-1}\dot{Q}_{\mathrm{K}}\mathrm{d}\tau - \int_{t_0}^{t} T^{-1}\sigma_{\mathrm{THD}}\mathrm{d}\varepsilon \\ & + \int_{t_0}^{t} T^{-1}D_{\mathrm{THD}}\mathrm{d}e + \int_{t_0}^{t} T^{-1}\dot{Q}_{\mathrm{Rint}}\mathrm{d}\tau \Bigg] \mathbf{Đ}T(p,t)\mathrm{d}V \\ & + \int_S \Bigg[\int_{t_0}^{t} T^{-1}\tilde{H}_{\mathrm{ext}}\mathrm{d}\tau \Bigg] \mathbf{Đ}T(p,t)\cdot\tilde{n}\mathrm{d}S \Bigg\} \mathrm{d}t = 0, \end{aligned} \tag{23}$$

$$\mathrm{d}U_{\mathrm{THD}} = \left.\frac{\mathrm{d}U}{\mathrm{d}T}\right|_{\varepsilon}\mathrm{d}T = C_V(T)\mathrm{d}T = \text{thermodynamic internal energy}$$

$$\begin{aligned} \int_{T_0}^{T} T^{-1}\mathrm{d}U_{\mathrm{THD}} &= C_V(T_0)(\ln T - \ln T_0) \\ &= \text{thermal kinetic energy/volume (entropy)}, \end{aligned} \tag{24}$$

C_V is assumed constant over $(T - T_0)$ for illustration,

$$\sigma_{\mathrm{THD}} = \left[\left.\frac{\mathrm{d}U}{\mathrm{d}\varepsilon}\right|_{\mathrm{T},e} - \sigma \right] = T\left.\frac{\partial\sigma}{\partial T}\right|_{\varepsilon} = \text{thermodynamic stress tensor}$$

$$D_{\text{THD}} = \left[\frac{dU}{de}\bigg|_{\text{T},\varepsilon} + D \right] = T\frac{\partial D}{\partial T}\bigg|_{\text{e}} = \text{thermodynamic electric displacement,}$$

U = thermal internal energy $\neq U_{\text{THD}}$,

$\tilde{H}_{\text{ext}}$ = external heat flux through the boundary,

V denotes a volume integral and S is the surface of heat flux. In the above, the reader is urged to compare the two $\mathbf{Đ}\mathcal{E}_{\text{TH}}$ expressions one-to-one term to deduce the $\mathbf{Đ}Q$ expressions as presented in the total system $\mathbf{Đ}E(t, t_0)$ expression given in Eq. (8). n tilde denotes the unit surface normal and $e(p, t)$ is the distributed electric-field vector, scalar products of all products should be understood even though we have not used tensor notation, hence $D(p, t)e(p, t)$ is the dot product of the two vectors, etc.

Some comments are in order: All of the terms that multiply $\mathbf{Đ}T(p, t)$ in the detailed expression for $\mathbf{Đ}\mathcal{E}_{\text{TH}}$ represent effective entropy changes in the process that has the units of (energy/temperature), and should be recognized as generalized thermal loads, thus their products with an evolutionary thermal displacement $\mathbf{Đ}T(p, t)$, which is the temperature distribution, yields an evolutionary caloric energy/work for the process. In Eq. (23) we shall assume that the conductive heat-transfer rate is described via Fourier's Law (in Cartesian notation for spatial coordinate directions x_i $(i = 1, 2, 3)$)

$$\dot{Q}_{\text{K}} = -\frac{\partial}{\partial x_i}\left(k_{ij}\frac{\partial T}{\partial x_j}\right), \tag{25}$$

where k_{ij} is the conductivity tensor. We shall assume that the external heat flux is due to aerodynamic heating only and discuss its model below.

The $\mathbf{Đ}\mathcal{E}_{\text{TH}}$ equation is nonlinear, but for small transition intervals one may wish to assume that the temperature change θ from the initial temperature distribution T_0 is not large such that

$$\theta(p, t) = T(p, t) - T(p, t_0) = T(p, t) - T_0, \quad T(p, t) = T_0\left(1 + \frac{\theta}{T_0}\right) \tag{26}$$

$$\ln T - \ln T_0 = \ln\left(1 + \frac{\theta}{T_0}\right) \cong \frac{\theta}{T_0},$$

$$T^{-1} = T_0^{-1}\left(1 + \frac{\theta}{T_0}\right)^{-1} \cong T_0^{-1}\left(1 - \frac{\theta}{T_0}\right) \cong T_0^{-1}, \quad \frac{\theta}{T_0} << 1.$$

Introducing the small temperature change approximation into the $\mathbf{Đ}\mathcal{E}_{\text{TH}}$ will linearize the expression in temperature dynamics in terms of θ. We choose this route in the illustration to render linear thermal dynamics. One can use the distribution of θ/T_0 as compared to unity as the simulation model fidelity parameter for the linear model. Note that T_0 does not imply a uniform temperature distribution.

6.2. *Aerodynamic heating model*

For the objective of this work, we have to conceptualize some model that will establish the closed-loop interaction of the thermal-structural dynamic system with the aerodynamics. We have to be only functionally reasonable in what we assume, and defer high-fidelity models to the domain of aerodynamicists and computational fluid dynamicists. We construct the aerodynamic pressure shape functions for high-speed flight by using the well-tested piston theory for high Mach number. Denoting the surface transverse displacement by $w(p,t)$, the aerodynamic pressure on the surface is given by

$$f_{\mathrm{a}}(p_{\mathrm{S}},t) = (2\mathrm{qdyn}_{\infty}/M_{\infty})N_{\mathrm{S}}'^{\mathrm{T}}(p)q_{\mathrm{S}}(t) + (2\mathrm{qdyn}_{\infty}/M_{\infty}/V_{\infty})N_{\mathrm{S}}^{\mathrm{T}}(p)\dot{q}_{\mathrm{S}}(t), \tag{27}$$

where N_{S} are the structural shape functions and prime denotes the spatial derivative of it that yields the surface slope. M, V and qdyn are the free-stream air Mach number, air velocity and dynamic pressure, respectively. We shall assume the piston model, and note that it is not influenced by the surface-temperature distribution. Next, we consider a flat surface that will vibrate in a direction normal to the free-stream velocity. We assume that the aerodynamic boundary heating rate (BLH) on it will be that predicted by the steady-state similarity flow boundary-layer theory over a nondeforming flat plate plus that affected by a temperature associated with the piston-theory aerodynamic pressure (piston-theory heating, PTH) through the isentropic air pressure–temperature state equation [8]. Denoting the aerodynamic heat by q_{a}, its rate will be represented in the following form:

$$\begin{aligned}
&\dot{q}_{\mathrm{a}}(p,t) = \dot{q}_{\mathrm{aBLH}} + \dot{q}_{\mathrm{aP}} = h(p)[T_{\mathrm{aW}}(p,t) - T_{\mathrm{W}}(p,t)], \qquad (28)\\
&T_{\mathrm{aW}}(p,t) = T_{\mathrm{aW}}^{\mathrm{BLH}}(p,t) + T_{\mathrm{air}}^{\mathrm{piston}}(p,t)\\
&T_{\mathrm{aW}}^{\mathrm{BLH}}(p,t) = T_{\mathrm{air}\infty}[1 + R_{\mathrm{f}}(\gamma-1)M_{\infty}^{2}/2],\quad T_{\mathrm{W}}(p,t) = T(p,t)\\
&T_{\mathrm{air}}^{\mathrm{piston}}(p,t) = C_{\mathrm{TAE}}T_{\mathrm{air}\infty}(\gamma-1)M_{\infty}[N_{\mathrm{S}}^{\mathrm{T}}(p,t)\dot{q}_{\mathrm{S}}(t) + V_{\infty}N_{\mathrm{S}}'^{\mathrm{T}}q_{\mathrm{S}}(t)]\\
&h(p) = C_{\mathrm{pair}}\rho_{\mathrm{air}\infty}V_{\infty}\mathrm{Pr}^{-2/3}C_{\mathrm{f}\infty}/2\\
&C_{\mathrm{f}\infty} = C_{\mathrm{f}\infty}(p, T_{\mathrm{W}}/T_{\mathrm{air}\infty}) = 0.6/\sqrt{\mathrm{Rey}(p)},
\end{aligned}$$

where T_{aW} is the adiabatic wall temperature, T_{W} is the wall temperature and $h(p)$ is the surface heat-transfer coefficient distribution. C_{f}, C_{p}, R_{f}, V, M, Pr, γ, T_{air}, Rey, are surface friction coefficient, constant pressure air specific heat, recovery factor, free-stream air speed, free-stream Mach number, free-stream air temperature, local Reynolds number, respectively. C_{TAE} is a coupling coefficient we introduced to assess the effect of the strength of the aerothermoelastic interaction for skin-temperature control. The nominal value for C_{TAE} modeled by the familiar Piston Theory is unity. In the following, we shall assume a thin skin structure with constant temperature through its thickness, so that the wall temperature is the structural surface-temperature distribution $T(p,t)$ and p denotes the distance on the surface along a spanwise direction, which is also the direction of free-stream, thus structural shape functions also are functions of surface position for a thin skin structure.

The aerodynamic heating model establishes a connection with both the skin temperature and the surface motion; therefore, it introduces the needed interaction for the $\mathbf{Đ}\mathcal{E}$ equations. We must caution, however, that these thermoaeroelastic couplings are through the piston theory, and assume that the basic mode of heat transfer due to flat-plate boundary-layer heating (BLH) is not affected by the surface motion. This assumption is suspect, and denies a possibly major mode of structure motion–aerodynamic heating interaction. One has no better venue than piston-theory-induced temperature changes to introduce thermoaeroelastic coupling in a closed form into the problem at this point, short of integrating in an open-loop fashion with CFD codes.

6.3. *The LEE ($\mathbf{Đ}\mathcal{E}$ equations) for the structural dynamic model*

The $\mathbf{Đ}\mathcal{E}_{\mathrm{S}}$ equations for structural dynamics are analogous to what appeared in the literature as *Hamilton's Law of Varying Action* (HLVA) [1–7,9–14]. However, due to our novel approach in this research, *The Law of Evolutionary Energy (LEE)* serves as an umbrella concept even to replace the HLVA. Referring back to the composite expression, Eqs. (7)–(9), the structural/mechanical $\mathbf{Đ}\mathcal{E}_{\mathrm{S}}$ equation is given by

$$\begin{aligned}\mathbf{Đ}\mathcal{E}_{\mathrm{S}} &= \int_{t_0}^{t_f}\Big[-\mathbf{Đ}T-\mathbf{Đ}W_{\mathrm{Mext}}+\mathbf{Đ}W_{\mathrm{M}}+\mathbf{Đ}W_{\mathrm{M-TH}}+\mathbf{Đ}W_{\mathrm{M-E}}\Big]\mathrm{d}t+\Big[(\partial T/\partial\dot{q}_{\mathrm{S}})\mathbf{Đ}q_{\mathrm{S}}\Big]\Big|_{t_0}^{t_f}\\ &= \int_{t_0}^{t_f}\Big\{-\mathbf{Đ}T-\mathbf{Đ}W_{\mathrm{Mext}}+\int_V\Big[\sigma_{\mathrm{M}}\mathbf{Đ}\varepsilon+\boldsymbol{\sigma}_{\mathrm{T}}\mathbf{Đ}\varepsilon+\sigma_{\mathrm{E}}\mathbf{Đ}\varepsilon\Big]\mathrm{d}V\Big\}\mathrm{d}t\\ &\quad+\Big[(\partial T/\partial\dot{q}_{\mathrm{S}})\mathbf{Đ}q_{\mathrm{S}}\Big]\Big|_{t_0}^{t_f}=0.\end{aligned} \tag{29}$$

The three terms under the volume integral are generally known as internal-energy terms, and the kinetic-energy terms are the evolutionary work of the acceleration loads. Hence, $\mathbf{Đ}\mathcal{E}_{\mathrm{S}}$ is the statement that the evolutionary work done by all internal and external mechanical loads through the structural motion over all time vanishes. In Eq. (29) the term that includes the temperature-dependent stress σ_{T} is the counterpart of the thermomechanical coupling term in the $\mathbf{Đ}\mathcal{E}_{\mathrm{TH}}$. In Eq. (29) admissible spatial shape functions will have to be employed to do the indicated volume integrations, which will bring the generalized structural coordinates $q_{\mathrm{S}}(t)$ into the picture.

The endpoints-in-time evolutional terms in Eq. (29) are not zero. Therefore, even if one misconsrues $\mathbf{Đ}q(t)$ has been similar to the 'virtual displacements $\delta q(t)$' of classical mechanics, the well-known Hamilton's Principle (HP) for conservative systems and the extended HP for nonconservative systems, both of which are based on virtual variational concepts with required endpoint-in-time variations, are not equivalent to Eq. (29), which is a natural law. Consequently, HP and the extended HP do not represent natural laws. HP was not enunciated by Hamilton. Hamilton enunciated in 1834 the HLVA only and rejected the concept of virtual variation. However, the HLVA is equivalent to the negative of the natural law $\mathbf{Đ}\mathcal{E}_{\mathrm{S}}=0$, Eq. (29). This renders HLVA a natural law. One must recall that the endpoint-in-time terms are required to vanish (coterminus conditions) in the HP or any like variant of it in classical dynamics as in Lagrange's virtual variations. In complete contrast, the structural LEE, $\mathbf{Đ}\mathcal{E}_{\mathrm{S}}$ equations, Eq. (29), of the EEM find their raison d'etre in the free alteration or evolution of the initial conditions of the path as discussed in Sections 1 and 2.

6.4. Piezoelectric, thermoelectric distributed modal actuation

Finally, we consider the distributed electrical actuation field on both the thermal dynamics and the structural dynamics. It is known [13] that modal electrical actuators through piezoelectric effects have their electric fields shaped to mimic the strain fields of the modes they are assigned to control. This yields distributed actuation proportional to the structural stiffness operator by definition. Just like displacements, an electric-field variable can be written in terms of electrical spatial shape functions multiplied by only time-dependent electrical generalized coordinates in the form:

$$e(p,t) = \bar{N}_{\mathrm{E}}^{\mathrm{T}}(p)E_{\mathrm{E}}(t) = \bar{N}_{\mathrm{E}}^{\mathrm{T}}(p)q_{\mathrm{E}}(t), \tag{30}$$

where $E_{\mathrm{E}}(t)$ is the strength signal of the electric force field over its shape. Piezoelectric actuators act as thin layers of electroded domains through the thickness of a structure in which the field vector is also in the direction of the thickness, electroded layers constituting surfaces normal to the field. Therefore, the dimension of $q_{\mathrm{E}}(t)$ indicates the number of such controller layers, and if each layer is assigned to a single mode, its dimension corresponds to the number of modes controlled. It follows that each element in the nondimensional electrical distribution shape functions vector N_{E} should correspond to the strain shape of the structural mode to be controlled.

It is an easy step to generalize that if electrical actuators are to be candidates to control temperature through thermodynamic action via the pyroelectric effect, they should also mimic some distributed shape of a feature of the system. It can be stated that modal thermoelectric actuators should mimic the shape of the spatial conduction operator, that is, the second-order spatial derivative of a modal temperature profile. Some modal actuators may mimic the strain field, while some may mimic the conductance field. However, if the purpose is to control temperature, it would be natural to shape the actuation field after the conduction field shape. Finally, if the structure and the thermal problem happen to have the same mode shapes, then the actuators would be *isomodal actuators* fit for both thermal and structural dynamics.

Piezoelectric and/or thermoelectric actuators act through their geometry, location (layer distances from a reference surface for distributed actuators) and the material properties that they hold, all culminating in an actuation gain (ACG) factor to transform the corresponding electric-field strengths to the mechanical and/or thermal actions needed to affect control.

To study a control problem by such actuation, one can lump actuator material and location properties as effective dimensional ACG on the actual control inputs absorbed into redefined generic control variables, and study the control problem from this perspective. One can later scale off or extract the actual inputs from these generically defined control variables by using the ACG factor of a chosen actuator type. This is the view we take in this work.

The design or ultimate control inputs $\{U_{\mathrm{E}}\}$ can now be redefined as $\{U_{\mathrm{E}}\} = [\mathrm{ACG}]\{A_{\mathrm{E}}\}$ by using either the strain-shaped or conductance-shaped actuator electrical field distributions by using the respective ACG as a scaling matrix. Thus, with control input scaling, the controller loading matrices on the structural and thermal subsystems can be calculated based on the actuator-shape functions, and the material and location parameters factored out as ACG matrices, then can be absorbed into the redefined control inputs $\{U_{\mathrm{E}}\}$. In the illustrations, we shall assume that control signals are redefined as $\{U_{\mathrm{E}}\}$ above, and not address particular physical features of actuators as to their locations and material properties.

Finally, the assumed modal actuators may be regarded as proposed actuation distribution profiles, whether the state-of-the-art offers such actuators or not is not the issue considered here, if such thermal-effect actuators are not available, they must then be viewed as the propositions to acquire such actuators through new research and development.

7. Structural skin-temperature control at Mach 10 hypersonic flight

As an illustration of the EEM, we consider controlling the structural temperature of a flat panel at Mach 10 flight, the dynamics of which represents a coupled aerothermo-servoelectroelastic system under aerodynamic boundary layer and piston-theory heating (BLH and PTH). The controller is distributed and has elastothermoelectrical control-loading capability.

While we wish to control the skin temperature, we take care not to cause excessive or unrealistic aeroelastic dynamic response. To this end, we consider only the temperature feedback control law for elastothermoelectric actuation, while controlling both the temperature and structural flexural deformation simultaneously. When and if a temperature feedback control law acts through the structural dynamics in whatever form, shape and technology; the thermoaeroelastic system will be externally dynamically servocoupled, via elasto/piezoelectric control inputs and the structural response in the closed-loop system will be excited by the thermal response. The result is that the thermal control and the structural deformation control dynamics will be at the expense of each other. The modeling information for the problem considered is as follows.

Flat panel: The flat panel has span and chord dimensions of $L = 1$ m $\times$ $c = 0.25$ m and a thickness of $z = 0.0025$ m. The panel is made of a hypothetical material with density $\rho = 1520$ kg/m^3, elastic modulus $E = 35 \times 10^9$ N/m^2, Poisson's ratio $\nu = 0.28$, coefficient of thermal expansion $\alpha = 9.5 \times 10^{-6}$, specific heat $C_V = 2.36 \times 10^5$ J/m^3/K, and conductivity $k = 2.8383$ J/m/s/K.

Flight properties: The flight is at Mach 10. The free-stream is in the spanwise direction of the panel with properties: Temperature $T_\infty = 360$ K, air density $\rho_\infty = 0.0281$ kg/m^3, Prandtl number $Pr = 0.7$, ideal-gas constant $\gamma = 1.4$, specific heat $C_p = 1.0045 \times 10^3$ J/kg/K, viscosity $\mu_\infty = 1.789 \times 10^{-5}$ N s/m^2, recovery factor $R_{\mathrm{f}} = 0.8$.

Mathematical model: The panel is modeled as a strip, with 1D structural and thermal dynamics in the spanwise flow direction. For simplicity, we assume that all material properties remain constant throughout the temperature ranges considered, which is of no consequence for being able to apply the AEM developed for the system. To provide a speedy transition to generalized coordinates, we assume that the panel has uniform properties and is simply supported with zero deflections at both ends, and has specified equal temperatures at both ends $T(x = 0$ and $x = 1, t) = 0$ or 200 K, and a given nonuniform initial temperature distribution in the interior domain along the span $T(x, t) = x(1 - x)$ for $0 < x < 1$. This system has known spatial modal functions for both of its structural dynamics and thermal dynamics, which are: $\sin(r\pi x/L)$, $r = 1, 3, 5$. Hence, the system is thermally and elastically isomodal. We consider three modes for both of structural energy and thermal energy quantities.

Controller model: We consider distributed modal controllers capable of simultaneous elasto/piezoelectric and thermo/pyroelectric actuation capability. We choose to scale off the design control inputs $\{U_{\mathrm{E}}\}$ according to the conductance-shaped actuation field $\mathrm{ACG}_{\mathrm{TH}}$, and hence multiply the control influence matrix Q_{SE} of the electrical inputs by the elasto-to-thermoelectric actuators' efficiency matrix $\mathrm{ACG}_{\mathrm{S}}\ \mathrm{ACG}_{\mathrm{TH}}^{-1}$. For illustration, we assume that for all piezo pyroelectric modal controllers this ratio is -17.5×10^5. In this illustration, we use the *global API measure* and compute the feedback control actions by using Eq. (21) with a summation sign from the initial transition interval to the current transition interval. Hence, the optimal controller has memory utilizing past histories of system trajectory and the control inputs. The weightings on the thermal and structural coordinates in the API for the optimal controller in Section 5 provide us just with the tool to tradeoff thermal control with structural control simultaneously.

The uncontrolled response of the thermoaeroelastic system to combined BLH and PTH is overwhelmed by the BLH thermal excitation to render the PTH contribution hardly noticeable in the results. The aeroelastic response is unaffected by the presence of the BLH since for the uncontrolled system the aeroelastic dynamics is free of thermal coupling. Thus, the uncontrolled response is a superposition of the results of pure thermal response to BLH and the uncontrolled response to PTH. Fig. 2a shows the uncontrolled panel temperature profile along the span due to BLH alone where $C_{\mathrm{TAE}} = 0$.

In the controlled system the thermal control action of the piezoelectric effect on the BLH is accomplished indirectly on the thermal dynamics through the presence of thermoaeroelastic PTH coupling terms. Therefore, the fate of the skin temperature depends on the strength of this PTH coupling to affect control of the BLH. To highlight the vital role of the degree of coupling effectiveness of the thermoaeroelastic term in this illustration, we introduce a PTH coupling, or better said, thermoaeroelastic coupling strength coefficient C_{TAE} as a parameter by which to study the feasibility of the piezo pyroelectric control of the structural temperature. When $C_{\mathrm{TAE}} = 1$ is assigned, one has a problem as modeled here, that is, the total aerodynamic heating is the sum of the flat-plate BLH term and the PTH term with coupling as given by Eq. (28). Any other value of $C_{\mathrm{TAE}} > 1$ increases the importance and coupling strength of the PTH term in comparison to the flat-panel BLH term in this formulation. Note that the BLH term is an excitation; the PTH term is an internal dynamic thermoaeroelastic coupling effect on the thermal dynamics. Whether introduction of the coefficient C_{TAE} is realistic or not is not the mainstay here, but is a high-fidelity physical aerodynamic heating and fluid–solid interaction modeling issue within the realm of CFD, and theoretical and experimental aerodynamics efforts. One can also view this approach as a constructive step that the thermoaeroelastic coupling needs to be better understood, studied or even tailored for feasibility of the skin temperature control in hypersonic flight.

Control with $C_{TAE} = 100$: The realizability of the control depends on the piezo-to-pyro (elasto-to-thermo) electric efficiency of chosen physical actuators as well as on the thermo-aeroelastic coupling coefficient C_{TAE}. We consider an illustration that would emphasize the role of the degree of the thermoaeroelastic coupling coefficient at $C_{\mathrm{TAE}} = 100$.

Leaving the elasto-to-thermo electric efficiency aside, which would be the actuator-technology-defining issue, the optimal control law is obtained via the global API perspective as given by Eqs. (18) and (21) with the weighting matrices $R_{\rm U} = 1 \times 10^4$, $W_{q\rm T} = \text{diag} - (100, 1, 1)$, $W_{q\rm S} = W_{q'\rm S} = \text{diag} - (5 \times 10^4)$, and no weighting on temperature rates. We also assigned for the controller the set-point thermal conditions $q_{\rm Tref} = \{0.2581\ 0.0096\ 0.0021\}$, that is, the initial thermal state from which the deviations are to be minimized.

Fig. 2b–f show the simulation results for 0.5 s with transition intervals of 0.001 s. In Fig. 2b the controlled thermal response is superior, the BLH + PTH is controlled; significant reduction in the rate of temperature increase in some domains is achieved, while cooling is realized in the middle one-third of the span in Fig. 2b. Fig. 2c depicts vividly the comparison of the uncontrolled and controlled generalized temperature coordinate responses. The thermal dynamics has been pacified significantly. The controller simultaneously achieves practically a static structural deformation shape, as reflected by the structural generalized coordinates in Fig. 2d, which was flat in the original configuration. Finally, corresponding to the maximum design control input $U_{\rm E}$ of -9.758×10^{-9}, observed from Fig. 2e, for conductance-field-shaped actuation, for a top layer of thickness 0.0005 m, the electrical signal strength would be $A_{\rm E} = -7.8\, C_{\rm DT}^{-1} \times 10^{-5}$ V/m (or J/m/K), or a potential difference of $3.9 C_{\rm DT}^{-1} \times 10^{-8}$ V across the layer. For the top layer with an implied value of $C_{\rm DT} = 1.2 \times 10^{-8}$, the associated maximum electrical field strength and the voltage potential difference for the layer would

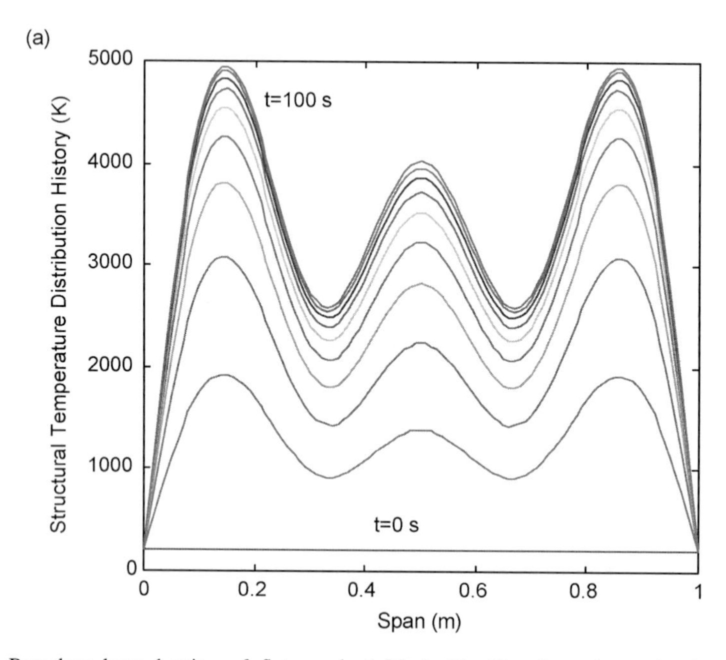

Fig. 2. (a) Boundary layer heating of flat panel at Mach 10. (No piston-theory heating, $C_{\rm TAE} = 0$). (b)–(f) Elastothermoelectric control of aerodynamic heating via vibrating aerothermoelastic panel at Mach 10, $C_{\rm TAE} = 100$.

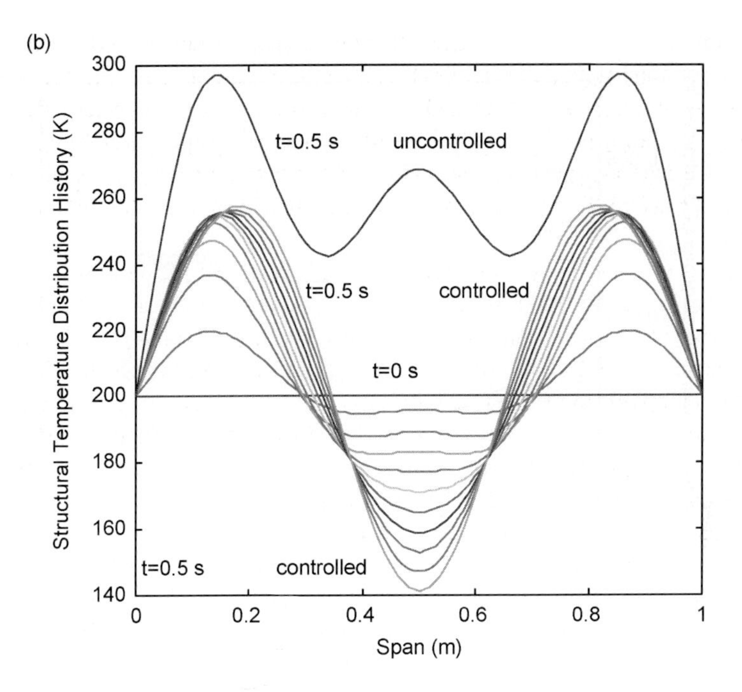

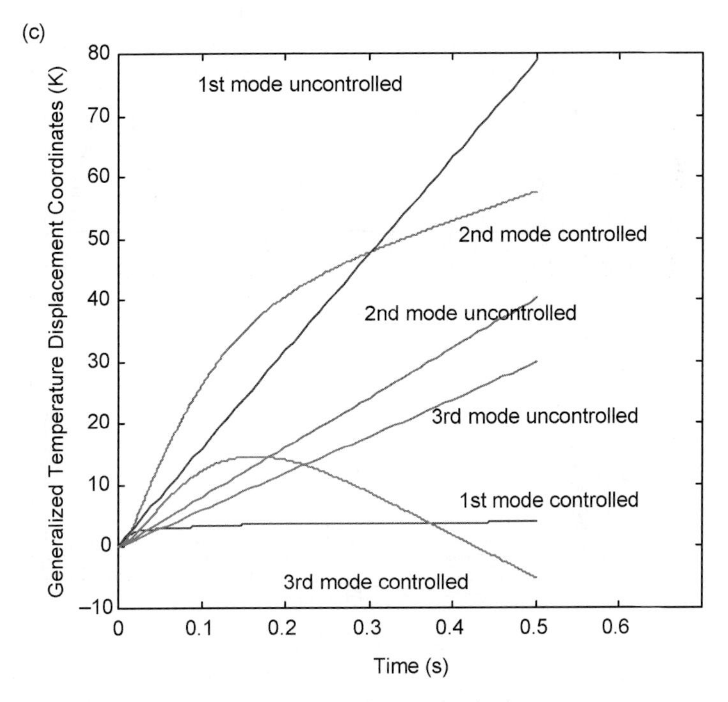

Fig. 2. Continued.

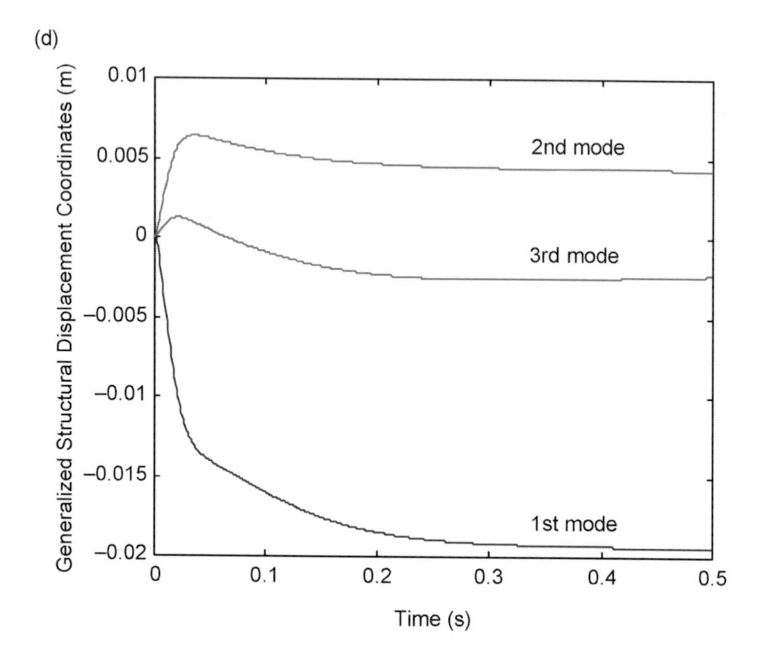

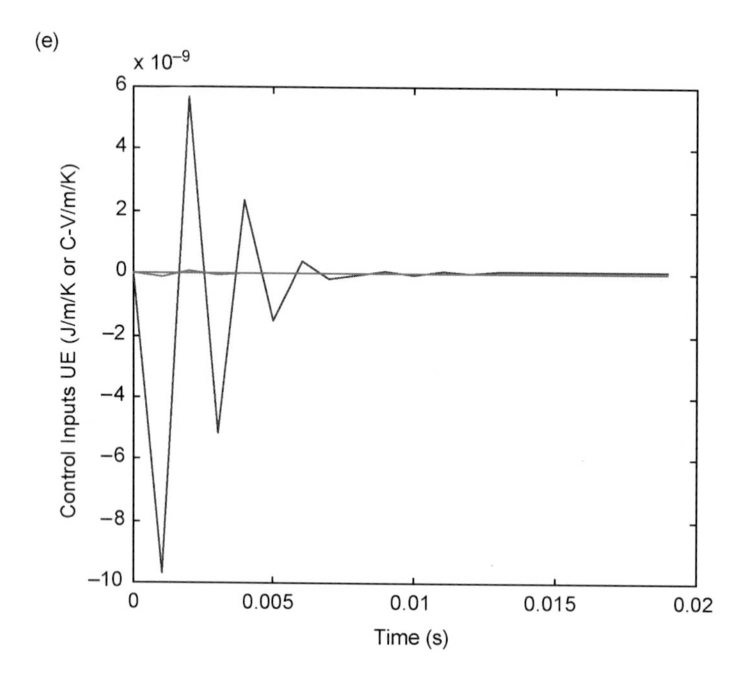

Fig. 2. Continued.

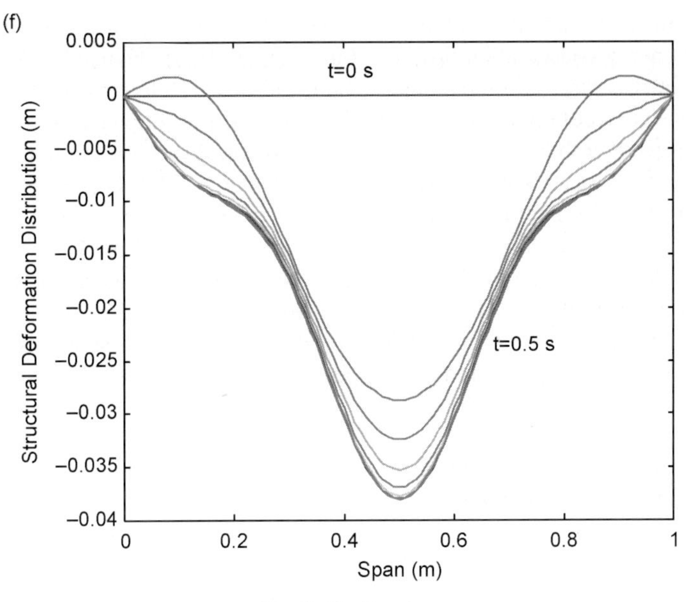

Fig. 2. Continued.

be 6.24×10^3 V/m and -3.12 V, respectively. The thermal response is satisfactorily controlled; the structure achieves a static deformation shape with a maximum deflection of 0.038 m in Fig. 2f. The control inputs required in Fig. 2f are three orders of magnitude smaller than those that would be obtained with a comparable control configuration with $C_{\mathrm{TAE}} = 1$, which would represent a weak thermoaeroelastic coupling and unrealistic structural deformation to affect temperature control.

Overall, the structural dynamics acts as the ultimate controller/compensator with $\{U_{\mathrm{E}}\}$ as input to the compensator and structural motion as the ultimate distributed input to the airstream, the aerodynamic environment; and through this controlled structural motion, which only uses knowledge of temperature to generate the temperature feedback control actions, the aeroelastic structure transfers or reflects some or all of the aerodynamic heating along the span back out to the airstream successfully. While one may ponder a 0.038-m structural deformation with regard to the structural and aerodynamic models, we demonstrated our case as to where the issues are, and in which directions the roads to feasibility lead.

8. Concluding remarks

A novel theoretical and computational approach has been introduced as the EEM for dynamics and control of interacting multidisciplinary energy phenomena. The LEE has been advanced as the conceptual foundation of the method based on the introduction of an evolutionary operator **Đ**. The evolutionary **Đ** operation is a total parameter-alteration operation on the parameterization of the generalized coordinates of the system along the actual dynamic path, and therefore, is constrained to only those path alterations that satisfy the general energy-conservation law, The First Law of Thermodynamics. The method is illustrated for structural skin-temperature control in Mach 10 hypersonic flight involving aerothermoservoelectroelastic interactions. The numerical solution has been implemented as an initial-value-problem solution. However, a universal spectral solution technique for the resulting Algebraic Evolutionary Energy Equations of Motion (AEM) for dynamics and control of linear, nonlinear, time-variant and time-invariant systems has also been introduced by the author and his coworkers in the references cited. By the universal-solution approach, the time modes of linear and nonlinear systems can be identified and superposition for nonlinear systems can be addressed. The LEE is not necessarily a computational approach but can yield closed-form analytical solutions of dynamic systems when the proper theoretical perspective is taken. These ideas have also been demonstrated in the cited references.

Appendix A. A constructive and demonstrative proof of the law of evolutionary energy for Newtonian dynamics

Denoting the mass, the generalized coordinates and the associated generalized forces by m, $q(t)$, $Q(t)$, respectively, we start with Newton's Second Law and multiply it by the differential displacement $\mathrm{d}q$ of the actual motion and integrate it from t_0 to the running time t to arrive at the associated energy equation for the system:

$$m\ddot{q} - Q = 0,\;\; (m\ddot{q} - Q)\mathrm{d}q = 0,\;\; \mathrm{d}T - \mathrm{d}W = 0,\;\; \mathrm{d}T = m\ddot{q}\mathrm{d}q = m\dot{q}\mathrm{d}\dot{q},\;\; \mathrm{d}W = Q\mathrm{d}q,$$

$$\int_{t_0}^{t} \mathrm{d}T - \int_{t_0}^{t} \mathrm{d}W = 0,\;\; \mathrm{d}E = \mathrm{d}T - \mathrm{d}W,$$

$$\int_{t_0}^{t} \mathrm{d}E = \int_{t_0}^{t} \frac{\mathrm{d}E}{\mathrm{d}\tau}\mathrm{d}\tau = \int_{t_0}^{t} \dot{E}(\tau)\mathrm{d}\tau = E(t, t_0) = 0,$$

$${}_{t_0}W_t = \int_{t_0}^{t} \mathrm{d}W = \int_{t_0}^{t} Q(\tau)\mathrm{d}q(\tau) = \int_{t_0}^{t} Q(\tau)\dot{q}(\tau)\mathrm{d}\tau,$$

$$T = T(t) = \int \mathrm{d}T = \int m\ddot{q}\mathrm{d}q = \frac{1}{2}m\dot{q}^2(t),$$

$$T(t) - T(t_0) - {}_{t_0}W_t = T - T_0 - {}_{t_0}W_t = E(t, t_0) = 0, \tag{A1}$$

where $\mathrm{d}T$ is the incremental kinetic energy and $\mathrm{d}W$ is the incremental work done on the system by the generalized forces Q, and T and W are the total kinetic energy at t and the work done from t_0 to t, respectively.

The derived Eq. (A1) is the *fundamental energy-conservation equation* satisfied by the actual path of the motion, and *is recognized as the First Law of Thermodynamics* applied to the Newtonian dynamic system free of heat transfer and other work or energy-producing interacting phenomena. $E(t, t_0)$ is the definite total process energy interaction expenditure from t_0 to the running time t, consisting of the work done on the system by the external generalized forces and the acceleration force.

Since the operator $\mathbf{Ð}$ obeys the rules of differential calculus, for any function $F(t,A)$ that involves both constants 'A' and time 't' as parameters, for infinitesimal parameter changes one may write:

$$\mathbf{Ð}F(t,A) = \mathbf{Ð}_{\mathrm{A}}F(t,A) + \mathbf{Ð}_t F(t,A) = F_{\mathrm{A}}(t,A)\mathbf{Ð}A + F_t(t,A)\mathbf{Ð}t$$

where F_{A} and F_t are partial differentials of F with respect to the parameters 'A' and 't', respectively. It is clear that the set $\mathbf{Ð}_{\mathrm{A}}$ and $\mathbf{Ð}_{\mathrm{t}}$ are independent groupings of parameter alterations since 't' is the utmost independent variable. Thus, $\mathbf{Ð}$ *is a total parameter-altering evolution operator* and unless the context calls for it, one does not have to be specific about operating on the time parameter 't' or other parameters 'A'. In the following, we shall assume that $\mathbf{Ð}$ is the total evolutionary operator.

Operating with $\mathbf{Ð}$ on the energy-conservation equation, Eq. (A1):

$$\mathbf{Ð}\int_{t_0}^{t} \mathrm{d}T - \mathbf{Ð}\int_{t_0}^{t} \mathrm{d}W = \mathbf{Ð}\int_{t_0}^{t} \mathrm{d}E = 0$$

$$\mathbf{Ð}(T - T_0 - {}_{t_0}W_t) = \mathbf{Ð}T - \mathbf{Ð}T_0 - \mathbf{Ð}_{t_0}W_t = \mathbf{Ð}E(t,t_0) = 0, \tag{A2}$$

where we remind that the evolutionary kinetic energy function $\mathbf{Ð}T(t)$ and the evolutionary work done during the process $\mathbf{Ð}_{t_0}W_t$ are along the actual path of the motion. Eq. (A2) is the *fundamental evolutionary-energy-conservation equation* that must be satisfied by the altered path of motion. Work done by external forces is a path-dependent energy process in which the forces and the multiplier generalized displacements are clearly identifiable by definition. Therefore, according to the operational Rule (3) of Section 2.3, the *evolutionary work done* ultimately depends only on the endpoints-in-time between the running time t and the initial time t_0 given by

$$\mathbf{Ð}_{t_0}W_t = \mathbf{Ð}\int_{t_0}^{t} \mathrm{d}W = \mathbf{Ð}\int_{t_0}^{t} Q(\tau)\mathrm{d}q(\tau) = \int_{t_0}^{t} Q(\tau)\mathrm{d}\mathbf{Ð}q(\tau) = Q(t)\mathbf{Ð}q(t) - Q(t_0)\mathbf{Ð}q(t_0)$$

$$\mathbf{Ð}W = \mathbf{Ð}W(t) = Q(t)\mathbf{Ð}q(t), \qquad \mathbf{Ð}W_0 = \mathbf{Ð}W(t_0) = Q(t_0)\mathbf{Ð}q(t_0) \tag{A3a}$$

$$\mathbf{Ð}_{t_0}W_t = \mathbf{Ð}W - \mathbf{Ð}W_0. \tag{A3b}$$

Introducing Eq. (A3b) into Eq. (A2), *the evolutionary-energy-conservation equation* becomes

$$\mathbf{Ð}T(t) - \mathbf{Ð}W(t) - \mathbf{Ð}T(t_0) + \mathbf{Ð}W_0 = \mathbf{Ð}E(t) - \mathbf{Ð}E_0 = \mathbf{Ð}E(t,t_0) = 0, \tag{A4a}$$

$$\mathbf{Ð}E(t) = \mathbf{Ð}T(t) - \mathbf{Ð}W(t), \qquad \mathbf{Ð}E_0 = \mathbf{Ð}E(t_0) = \mathbf{Ð}T(t_0) - \mathbf{Ð}W(t_0) = \mathbf{Ð}T_0 - \mathbf{Ð}W_0$$

$$\mathbf{Ð}T(t) = \mathbf{Ð}W(t) + \mathbf{Ð}E_0, \qquad \mathbf{Ð}T = \mathbf{Ð}W + \mathbf{Ð}E_0. \tag{A4b}$$

On the other hand the kinetic energy $T(t)$ is a path-independent energy-state function, therefore, so is the evolutionary kinetic energy $\mathbf{Ð}T(t)$. Recall that $T(t)$ arises from the work of acceleration force through the infinitesimal actual displacement $\mathrm{d}q$, and since it is an energy-state function it involves only the endpoints of the energy process bypassing the details of the process that creates it. Therefore, although the evolutionary path satisfies Eq. (A4b), one cannot uncover the unknown path process between the t_0 and t directly from it. In contrast, note that the evolutionary work expression does involve or exhibits the process that creates it, specifically, it is the product of the generalized force with the evolutionary path $\text{Đ}q(t)$. If the interest is in finding the unknown path via Eq. (A4b), one must 'open-up the function $\text{Đ}T(t)$' in Eq. (A4b) further to reveal the evolutionary work of the acceleration force that creates it as a product of the acceleration force with the evolutionary path $\text{Đ}q(t)$. The following mathematical identities accomplish this objective, although one may also prove their existence physically on the basis of the concepts discussed here, which is omitted for brevity:

$$\text{Đ}T(t) = \text{Đ}\left(\frac{1}{2}m\dot{q}^2\right) = m\dot{q}\text{Đ}\dot{q} = \dot{Z}_{\text{Đ}}(t) - m\ddot{q}\text{Đ}q,\ \dot{Z}_{\text{Đ}}(t) = \frac{\mathrm{d}}{\mathrm{d}t}\left(\frac{\partial T(t)}{\partial \dot{q}}\text{Đ}q(t)\right) \quad \text{(A5)}$$

Next, introducing Eq. (A5) for $\text{Đ}T$ into the evolutionary-energy-conservation equation, Eqs. (A4a,b), adding and subtracting the evolutionary work $\text{Đ}W = \text{Đ}W(t)$ and then recognizing Eq. (A4b) and the definitions (A3a,b) one obtains:

$$\begin{aligned}
&\text{Đ}E(t) - \text{Đ}E_0 = \dot{Z}_{\text{Đ}}(t) - m\ddot{q}\text{Đ}q - \text{Đ}W - \text{Đ}E_0 = 0,\\
&\text{Đ}E(t) - \text{Đ}E_0 = \dot{Z}_{\text{Đ}}(t) - m\ddot{q}\text{Đ}q - \text{Đ}W - \text{Đ}E_0 - \text{Đ}W + \text{Đ}W = 0\\
&\text{Đ}E(t) - \text{Đ}E_0 = \dot{Z}_{\text{Đ}}(t) - \text{Đ}W - \text{Đ}E_0 - \text{Đ}W + \text{Đ}W - m\ddot{q}\text{Đ}q = 0\\
&\text{Đ}E(t) - \text{Đ}E_0 = \dot{Z}_{\text{Đ}}(t) - \text{Đ}T - \text{Đ}W + (Q - m\ddot{q})\text{Đ}q = 0.
\end{aligned} \quad \text{(A6)}$$

We note in Eq. (A6) that the last term in the middle expression vanishes since it represents Newton's Second Law of force equilibrium. Thus, dropping the last term in Eq. (A6) and integrating over the running time t from the initial time t_0 a final time t_f, we then have the *Law of Evolutionary Energy* and its explicit final form for a Newtonian Dynamic System:

$$\int_{t_0}^{t_\text{f}} (\text{Đ}E(t) - \text{Đ}E_0)\mathrm{d}t = Z_{\text{Đ}}(t)\Big|_{t_0}^{t_\text{f}} - \int_{t_0}^{t_\text{f}} (\text{Đ}T(t) + \text{Đ}W(t))\mathrm{d}t = 0 \quad \text{(A7a)}$$

$$\text{Đ}\mathcal{E} = \text{Đ}\int_{t_0}^{t_\text{f}} E(t, t_0)\mathrm{d}t = \int_{t_0}^{t_\text{f}} \text{Đ}E(t, t_0)\mathrm{d}t = \int_{t_0}^{t_\text{f}} (\text{Đ}E(t) - \text{Đ}E_0)\mathrm{d}t = 0 \quad \text{(A7b)}$$

$$\begin{aligned}
&\text{Đ}\mathcal{E} = \text{Đ}\mathcal{E}(t_\text{f}, t_0) = Z_{\text{Đ}}(t)\Big|_{t_0}^{t_\text{f}} - \int_{t_0}^{t_\text{f}} (\text{Đ}T(t) + \text{Đ}W(t))\mathrm{d}t = 0,\\
&Z_{\text{Đ}}(t) = \left(\frac{\partial T(t)}{\partial \dot{q}}\text{Đ}q(t)\right),
\end{aligned} \quad \text{(A8)}$$

where $\mathcal{E}$ is the evolving—the time integral of—energy of the process between the initial time t_0 and the final time t_f and the *evolutionary equation* $\text{Đ}\mathcal{E}(t_\text{f}, t_0) = 0$ *is the statement of the LEE for the dynamic system*. Alternately, and more instructively, Eq. (A8) can be rewritten in terms of the definite power/energy rate processes from which it was derived by using Eq. (A1) in Eq. (A8), which reveals the role of the initial evolutionary kinetic energy and the work expressions in which the difference between the definitions of $\text{Đ}E_0 = \text{Đ}T_0 - \text{Đ}W_0$, the *initial evolutionary energy*, and $\text{Đ}S_0 = \text{Đ}T_0 + \text{Đ}W_0$, the initial *evolutionary (thermodynamic) energy availability* must be noted:

$$\text{Đ}\mathcal{E} = Z_{\text{Đ}}(t)\Big|_{t_0}^{t_f} - \int_{t_0}^{t_f}\left\{\text{Đ}\int_{t_0}^{t} \mathrm{d}T(\tau) + \text{Đ}\int_{t_0}^{t} \mathrm{d}W(\tau)\right\}\mathrm{d}t - (t_f - t_0)\text{Đ}S_0 = 0,$$
$$\text{Đ}S_0 = \text{Đ}T(t_0) + \text{Đ}W(t_0) = \text{Đ}T_0 + \text{Đ}W_0. \tag{A9}$$

Eq. (A8) corresponds to the dynamic part of Eq. (8) in Section 2 for the mechanical structural subsystem characterized by the kinetic energy T, and the $\text{Đ}W_{\text{Mext}}$ terms only. Eq. (A9) is the equivalent form of Eq. (A8) reported here for the first time for the same phenomenon exposing the definite underlying energy process in rate form and the initial energy work quantities. Note that the equation $\text{Đ}\mathcal{E} = 0$ *and its symbolism is the single-operator notation that embodies both of Eqs. (A8) and (A9), that is the Law of Evolutionary Energy.*

It must be noted that if one knows nothing about force equilibrium and states or postulates the LEE directly in the form given by Eqs. (A7a,b) and (A8), referring back to Eq. (A6) and transposing the last term involving the inertia force and the generalized force in the middle expression to the RHS and integrating between t_0 and t_f, one would have to conclude that force equilibrium equations or Newton's Second Law are automatically satisfied by the LEE or vice versa, since $\text{Đ}q(t)$ are continuous nontrivial functions of time. Alternately, if one recognizes Newton's Second Law in Eq. (A6), again it can be transposed to the RHS and again integrated in time t without changing the validity of the LEE to write:

$$\int_{t_0}^{t_f} (\text{Đ}E(t) - \text{Đ}E_0)\mathrm{d}t = \int_{t_0}^{t_f} (m\ddot{q} - Q)\text{Đ}q\mathrm{d}t = 0 \tag{A10a}$$

$$Z_{\text{Đ}}(t)\Big|_{t_0}^{t_f} - \int_{t_0}^{t_f} (\text{Đ}T(t) + \text{Đ}W(t))\mathrm{d}t = \int_{t_0}^{t_f} (m\ddot{q} - Q)\text{Đ}q\mathrm{d}t = 0, \tag{A10b}$$

where again, the left-hand side of Eq. (A10a) is satisfied by the solution path, but the solution path can be obtained only via Eq. (A10b) since the energy process is explicit only in Eq. (A10b).

One may observe that if one substitutes instead of $\text{Đ}q(t)$, the 'virtual displacements $\boldsymbol{\delta}q(t)$' of classical mechanics into the RHS of Eq. (A10b) then its integrand would appear exactly in the form of the well-known Principle of Virtual Work (PVW) for Newtonian Dynamic systems. However, 'the variational virtual displacements $\delta q(t)$ of classical mechanics' are required to satisfy only the geometric boundary conditions. Most importantly, there is absolutely no concern for virtual displacements to satisfy any energy-conservation law whatsoever, consequently, they are posed as displacements that are not dynamically possible. Furthermore, virtual displacements are contemplated to take place contemporaneously, and often are assumed to have 'zero variations' at t_0 and t_f (known as coterminus boundary conditions in time). The concept of a virtual displacement is only mathematically based.

On the other hand, the concept of an evolutionary displacement is physically based. The evolutionary displacements $\mathbf{Đ}q(t, A)$ of the EEM and the associated LEE are born out of complete regard to 'the actuality of the dynamical system and its actual-path' by constraining them to satisfy the energy-conservation law, The First Law of Thermodynamics. To this end, alteration of the initial states, hence the algebraic coordinates A, and the alteration of the independent time parameter 't' become essential features of an evolutionary displacement $\mathbf{Đ}q$. Therefore, an evolutionary displacement is diagonally opposite to the concept of a virtual displacement.

In the EEM, in arriving at the LEE and Eqs. (A7a,b)–(A10a,b), we have neither utilized virtual concepts nor invoked the PVW. In fact, we have shown, by virtue of the constructive proof of the LEE above, that the RHS of Eqs. (A10a,b) emerged naturally in the process from the concept of ECC actual path alterations. Given the perspective of the EEM and the LEE in this chapter, one should perhaps ponder as to what the PVW is and what it is not, as it is practiced today, and why. One may have to choose between 'the reality of LEE of the EEM' or 'the virtuality of the PVW'.

Acknowledgements

A significant part of this work was supported under Joint AFRL/DAGSI Basic Research Program through the Grant VA-WSU-99-02. Dr. Phil Beran of the Air Vehicles Directorate served as the Technical Monitor.

References

[1] Öz, H. (2002), Algebraic Evolutionary Energy Method for Dynamics and Control. In Computational Nonlinear Aeroelasticity for Multidisciplinary Analysis and Design, (Grandhi V.R., Wolff J.M., Beran P., King P., Öz H., Eàstep F.E., authors), AFRL, VA-WP-TR-2002-XXXX, pp. 96–162.
[2] Öz, H. and Adiguzel, E. (1995), Hamilton's law of varying action. Part I: assumed-time-modes method, J. Sound Vib. 179(4), 697–710.
[3] Öz, H. and Ramsey, J.K. (2000), Time-modes and linear systems, J. Sound Vib. 131(2), 331–344.
[4] Öz, H. and Ramsey, J.K. (2004), Time-modes and nonlinear systems, J. Sound Vib. (in review).
[5] Ramsey, J.K (2000), A vector-space approach to Hamilton's law of varying action for linear and nonlinear systems, Ph.D. Dissertation, Advisor: H. Öz, Department of Aerospace Engineering and Aviation, The Ohio State University.
[6] Öz, H. and Adiguzel, E. (1995), Hamilton's law of varying action. Part II: direct optimal control of linear systems, J. Sound Vib. 179(4), 711–724.
[7] Öz, H. and Adiguzel, E. (1995), Direct optimal control of nonlinear systems via Hamilton's law of varying action, J. Dyn. Syst. Meas. Control, 262–269.
[8] Gee, D.J. and Sipsic, S.R. (1999), Coupled thermal model for nonlinear panel flutter, AIAA J. 37(5), 642–650.
[9] Bailey, C.D. (1975), A new look at Hamilton's principle, Found. Phys. 5, 433–451.
[10] Bailey, C.D. (1975), Application of Hamilton's law of varying action, AIAA J. 13, 1154–1157.
[11] Hamilton, W.R. (1834), On a general method in dynamics, Philos. Trans. Roy. Soc. London, 247–308.
[12] Hamilton, W.R. (1835), Second essay on a general method in dynamics, Philos. Trans. Roy. Soc. London, 95–144.

[13] Öz, H. (1997), Distributed modal-space control and estimation with electroelastic applications. In Structronic Systems: Smart Structures, Devices and Systems: Systems and Control, (Tzou, H.S., Ardéshir Guran eds., and Ulrich Gabbert, Junji Tani, Elmar Breitbach, associate eds.), Vol. 2, World Scientific Publishers, Singapore, pp. 179–262.
[14] Öz, H. and Ramsey, J.K (1997, 2002), Direct optimal control of the duffing dynamics, Eleventh Symposium on Structural Dynamics and Control, Virginia Tech, Blacksburg, VA; also NASA TM 2002-211582.

ECOLOGY

Chapter 15

MAXIMIZATION OF ECOEXERGY IN ECOSYSTEMS

Sven Erik Jørgensen

The Danish University of Pharmaceutical Sciences, Environmental Chemistry, University Park 2, 2100 Copenhagen Ø, Denmark

Abstract

The maximization principle in this chapter focuses on maximization of ecoexergy in ecosystems. Ecoexergy, like the usually applied exergy expresses the work capacity; but ecoexergy uses the same system at the same temperature and pressure at thermodynamic equilibrium as a reference state. Ecoexergy measures therefore the distance from thermodynamic equilibrium. It is shown how ecoexergy can be found for organisms and it is discussed why living systems have a particular high exergy, namely due to their high information content. A maximization hypothesis is proposed and supported by several observations.

If a system receives an input of exergy, it will utilize this exergy after the maintenance of the system far from thermodynamic has been covered to move the system further from thermodynamic equilibrium. If there is more than one pathway offered to depart from equilibrium, the one yielding the most gradients, and exergy storage under the prevailing conditions, to give the most ordered structure furthest from equilibrium, will tend to be selected. This formulation may be considered a translation of Darwin to thermodynamics: the organisms that have the properties that are fitted to the prevailing conditions in the ecosystem, will be able to contribute most to the biomass and information and thereby give the ecosystem most exergy—they are the best survivors.

It is finally shown that the maximization principle presented here—ecosystems maximize ecoexergy—is consistent with the description of ecosystem development as three growth forms. It is discussed to what extent the exergy-maximization principle is in accordance with other proposed principles of minimum entropy production, maximum power (throughflow of energy), maximum ecoexergy destruction and maximum energy residence time.

Keywords: exergy; ecoexergy; living organism; growth; thermodynamics

1. Introduction

One of the core discussion during recent decades in system ecology has been: are there applicable maximization criteria for ecosystems? Can we describe the development of ecosystems by such criteria? There have been several proposals of goal functions, orientors, maximization criteria or whatever we may call it. The author of this chapter has been an advocate for the use of ecoexergy as maximization criteria [1,2], while others have

E-mail address: sej@dfh.dk (S.E. Jørgensen).

proposed other goal functions or orientors: emergy [3], ascendency [4], maximum power [3], negative entropy [5], minimum entropy [6], maximum entropy production [7] and free energy [8]. Later in the chapter, we will touch on these other goal function proposals and also show that the various proposals are consistent and form a pattern, which makes it possible to claim that we have an ecosystem theory—namely the pattern of the various goal functions. It is not surprising that we need many different descriptions of ecosystems, which can be equally valid, when we consider that such a simple physical phenomenon as light requires two descriptions to cover our observations: as waves and as photons (particles). In this chapter, we will, however, focus mainly on ecoexergy as a maximization criterion. Ecoexergy is not necessarily better than the other proposed orientors; but on the other hand one orientor gives, to a certain extent, the idea behind all of them, because they are interrelated, as will demonstrated later in this chapter. Moreover, ecoexergy has been widely applied as an orientor either in structurally dynamic models or as an ecological indicator to assess ecosystem health. It implies that there is a certain experience with the use of ecoexergy as orientor in practical environmental management. It is also interesting in this context that ecoexergy, to a certain extent, can be considered a translation of Darwin's survival of the fittest, which will be presented later in the chapter.

2. What is ecoexergy?

Exergy is defined as the amount of work (= entropy–free energy) a system can perform when it is brought into thermodynamic equilibrium with its environment. As a reservoir, reference state, it is advantageous in ecology to select the same system but at thermodynamic equilibrium, i.e. that all components are inorganic and at the highest oxidation state, if sufficient oxygen is present (nitrogen as nitrate, sulfur as sulfate and so on); see Fig. 1. The application of this reference state implies that another exergy, other

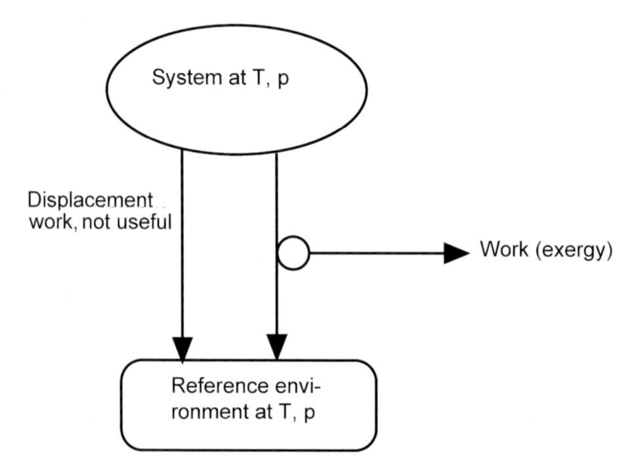

Fig. 1. The ecoexergy content of the system is calculated in the text for the system relative to a reference environment of the same system at the same temperature and pressure, but as an inorganic soup with no life, biological structure, information or organic molecules.

than that usually applied in engineering, is found. It is named ecoexergy to distinguish it from the technological exergy. The reference state will in this case correspond to the ecosystem without life forms and with all chemical energy utilized or as an 'inorganic soup'. Usually, it implies that we consider $T = T_o$, and $p = p_o$, which means that the ecoexergy becomes equal to the difference of Gibbs free energy of the system and the same system at thermodynamic equilibrium, or the chemical energy content included the thermodynamic information (see below) of the system. The ecoexergy becomes, with this definition, a measure of how far the ecosystem is from thermodynamic equilibrium, i.e. how much (complex) organization the ecosystem has build up in the form of organisms, complex biochemical compounds and the complex ecological network. All processes are irreversible, which means that exergy or ecoexergy is lost (and entropy is produced). Loss of ecoexergy and production of entropy are two different descriptions of the same reality, namely that all processes are irreversible, and we unfortunately always have some loss of energy forms that can do work to energy forms that cannot do work (heat at the temperature of the environment) (see also Ref. [9]). So, the formulation of the second law of thermodynamic by use of ecoexergy is: "All real processes are irreversible, which implies that ecoexergy inevitably is lost. Exergy and ecoexergy are not conserved," while energy of course is conserved by all processes according to the first law of thermodynamics. All transfers of ecoenergy imply that ecoexergy is lost because energy is transformed to heat at the temperature of the environment. The ecoexergy of the system measures the contrast—it is the difference in free energy if there is no difference in pressure and temperature, as may be assumed for an ecosystem or an environmental system and its environment—against the surrounding environment. If the system is in equilibrium with the surrounding environment the exergy is, of course, zero.

Since the only way to move systems away from equilibrium is to perform work on them, and since the available work in a system is a measure of the ability, we have to distinguish between the system and its environment or thermodynamic equilibrium alias, for instance, an inorganic soup. Therefore, it is reasonable to use the available work, i.e. the ecoexergy, as a measure of the distance from thermodynamic equilibrium.

This description of ecoexergy development in ecosystems makes it pertinent to assess the ecoexergy of ecosystems. It is not possible to measure ecoexergy directly—but it is possible to compute it. As the chemical energy embodied in the organic components and the biological structure contributes far more to the ecoexergy content of the system, there seems to be no reason to assume a (minor) temperature and pressure difference between the system and the reference environment. Under these circumstances we can calculate the ecoexergy content of the system as coming entirely from the chemical energy:

$$\sum_{c} (\mu_c - \mu_{co}) N_i. \tag{1}$$

This represents the nonflow chemical ecoexergy. It is determined by the difference in chemical potential $(\mu_c - \mu_{co})$ between the ecosystem and the same system at thermodynamic equilibrium. This difference is determined by the concentrations of the considered components in the system and in the reference state (thermodynamic equilibrium), as is the case for all chemical processes. We can measure the concentrations in the ecosystem, but the concentrations in the reference state (thermodynamic equilibrium) can be based on the usual use of chemical equilibrium constants. If we have the process:

$$\text{Component A} \longleftrightarrow \text{inorganic decomposition products} \tag{2}$$

it has a chemical equilibrium constant, K:

$$K = [\text{inorganic decomposition products}]/[\text{component A}]. \tag{3}$$

The concentration of component A at thermodynamic equilibrium is difficult to find, but we can, based upon the composition of A, find the concentration of component A at thermodynamic equilibrium from the probability of forming A from the inorganic components.

We find by these calculations the ecoexergy of the system compared with the same system at the same temperature and pressure but in the form of an inorganic soup without any life, biological structure, information or organic molecules. As $(\mu_c - \mu_{co})$ can be found from the definition of the chemical potential replacing activities by concentrations, we get the following expressions for the ecoexergy:

$$\text{Ex} = RT \sum_{i=0}^{i=n} C_i \ln C_i/C_{i,o}, \tag{4}$$

where R is the gas constant (8.317 J/K mole = 0.08207 l atm/K mole), T is the temperature of the environment (and the system; see Fig. 1), while C_i is the concentration of the i-th component expressed in a suitable unit, e.g. for phytoplankton in a lake C_i could be expressed as mg/l or as mg/l of a focal nutrient. $C_{i,o}$ is the concentration of the i-th component at thermodynamic equilibrium and n is the number of components. $C_{i,o}$ is, of course, a very small concentration (except for $i = 0$, which is considered to cover the inorganic compounds), corresponding to a very low probability of forming complex organic compounds spontaneously in an inorganic soup at thermodynamic equilibrium. $C_{i,o}$ is even lower for the various organisms, because the probability of forming the organisms is very low with their embodied information, which implies that the genetic code should be correct.

By using this particular ecoexergy based on the same system at thermodynamic equilibrium as reference, the ecoexergy becomes dependent only on the chemical potential of the numerous biochemical components that are characteristic of life. This is consistent with Boltzmann's statement, that life is the struggle for free energy.

The total ecoexergy of an ecosystem *cannot* be calculated exactly, as we cannot measure the concentrations of all the components or determine all possible contributions to ecoexergy in an ecosystem. If we calculate the ecoexergy of a fox, for instance, the above shown calculations will only give the contributions coming from the biomass and the information embodied in the genes, but what is the contribution from the blood pressure, the sexual hormones and so on? These properties are at least partially covered by the genes but is that the entire story? We can calculate the contributions from the dominant components, for instance by the use of a model or measurements that covers the most essential components for a focal problem. The *difference* in ecoexergy by *comparison* of two different possible structures (species composition) is here decisive. Moreover, ecoexergy computations always give only relative values, as the ecoexergy is calculated relative to the reference system.

As we know that ecosystems due to the throughflow of energy have the tendency to move away from thermodynamic equilibrium losing entropy or gaining ecoexergy and information, we can at this stage formulate the following proposition of relevance for ecosystems: "Ecosystems attempt to develop toward a higher level of ecoexergy".

3. Ecoexergy and information

Information means 'acquired knowledge'. The thermodynamic concept of ecoexergy is closely related to information. A high local concentration of a chemical compound, for instance, with a biochemical function that is rare elsewhere, carries ecoexergy *and* information.

It is possible to distinguish between the ecoexergy of information and of biomass [10]. p_i defined as c_i/A, where

$$A = \sum_{i=1}^{n} c_i \tag{5}$$

is the total amount of matter in the system, is introduced as a new variable in Eq. (6):

$$\mathrm{Ex} = ART \sum_{i=1}^{n} p_i \ln p^i/p_{io} + A \ln A/A_o. \tag{6}$$

As $A \approx A_o$, ecoexergy becomes a product of the total biomass A (multiplied by RT) and the Kullback measure:

$$K = \sum_{i=1}^{n} p_i \ln(p_i/p_{io}), \tag{7}$$

where p^i and p_{io} are probability distributions, a posteriori and a priori to an observation of the molecular detail of the system. This means that K expresses the amount of information that is gained as a result of the observations. If we observe a system that consists of two connected chambers, we expect the molecules to be equally distributed in the two chambers, i.e. $p_1 = p_2$ is equal to 1/2. If we, on the other hand, observe that all the molecules are in one chamber, we get $p_1 = 1$ and $p_2 = 0$.

4. How to calculate ecoexergy of organic matter and organisms?

The following expression for what we could call the ecological ecoexergy per unit volume has been presented; see Eq. (4):

$$\mathrm{Ex} = RT \sum_{i=0}^{i=n} c_i \ln c_i/c_{io} \quad [\mathrm{ML}^{-1}\mathrm{T}^{-2}], \tag{8}$$

where R is the gas constant, T is the temperature of the environment, while c_i is the concentration of the i-th component expressed in a suitable unit, e.g. for phytoplankton in a lake c_i could be expressed as mg/l or as mg/l of a focal nutrient. c_{io} is the concentration of the i-th component at thermodynamic equilibrium and n is the number of components. c_{io} is very low for living component because the probability that living components are formed at thermodynamic equilibrium is very low. This implies that living components get a high ecoexergy. c_{io} is not zero for organisms, but will correspond to a very low probability of forming complex organic compounds spontaneously in an inorganic soup at thermodynamic equilibrium. c_{io} on the other hand is high for inorganic components, and although c_{io} still is low for detritus, it is much higher than for living components.

The problem related to the assessment of c_{io} has been discussed and a possible solution proposed in Ref. [11]. For dead organic matter, detritus, which is given the index 1, it can be found from classical thermodynamics.

For the biological components, $2, 3, 4...N$, the probability, p_{io}, consists at least of the probability of producing the organic matter (detritus), i.e. p_{1o}, and the probability, $p_{i,a}$, to find the correct composition of the enzymes determining the biochemical processes in the organisms. Living organisms use 20 different amino acids and each gene determines on average the sequence of about 700 amino acids [12]. $p_{i,a}$ can be found from the number of permutations among which the characteristic amino-acid sequence for the considered organism has been selected. This means that

$$p_{i,a} = a^{-N\mathrm{gi}} \quad [-], \tag{9}$$

where a is the number of possible amino acids $= 20$, N is the number of amino acids determined by one gene $= 700$ and gi is the number of non-nonsense genes. The following two equations are available to calculate p_i:

$$p_{io} = p_{1o}p_{i,a} = p_{1o}a^{-N\mathrm{g}} \approx p_{1o}{\cdot}20^{-700\mathrm{g}} \quad [-], \tag{10}$$

and the ecoexergy contribution of the i-th component can be found by combining Eqs. (8) and (10):

$$\begin{aligned} \mathrm{Ex} &= RTc_i \ln c_i/(p_{1o}a^{-N\mathrm{g}}c_{oo}) = (\mu_1 - \mu_{1o})c_i - c_i \ln p_{i,a} \\ &= (\mu_1 - \mu_{1o})c_i - c_i \ln(a^{-N\mathrm{gi}}) = 18.7c_i + 700(\ln 20)c_i g_i \quad [\mathrm{ML}^{-1}\,\mathrm{T}^{-2}]. \end{aligned} \tag{11}$$

The total ecoexergy can be found by summing up the contributions originating from all components. The contribution from inorganic matter can be neglected as the contributions from detritus and even to a higher extent from the biological components are much higher due to an extremely low concentration of these components in the reference system (thermodynamic equilibrium for the system). The contribution from detritus, dead organic matter, is on average 18.7 kJ/g times the concentration (in g/unit volume) corresponding to the composition of detritus, namely lipids, carbohydrates and proteins mainly, while the ecoexergy of living organisms with approximations consists of

$$\begin{aligned} \mathrm{Ex}_{1\mathrm{chem}} &= 18.7 \text{ kJ/g times the concentration } c_i \text{ (g/unit of volume) and} \\ \mathrm{Ex}_{i\mathrm{bio}} &= RT(700 \ln 20)c_i g_i = RT2100 g_i c_i, \end{aligned} \tag{12}$$

$R = 8.34$ J/mole and if we presume a molecular weight of on average 10^5 for the enzymes (see, for instance, Refs. [13,14]), we obtain the following equation for $\mathrm{Ex}_{\mathrm{ibio}}$ at 300 K:

$$\mathrm{Ex}_{\mathrm{ibio}} = 0.0529 g_i c_i, \tag{13}$$

where the concentration now is expressed in g/unit volume and the ecoexergy in kJ/unit volume.

For the entire system the ecoexergy, Ex-total can be found as

$$\text{Ex-total} = 18.7 \sum_{i=1}^{N} c_i - 0.0529 \sum_{i=1}^{N} c_i g_i \ [\mathrm{ML}^{-1}\mathrm{T}^{-2}], \tag{14}$$

where g for detritus ($i = 1$) of course is 0. Table 1 illustrates how each contribution depends on the selected biological systems as well as detritus. A weighting factor β is introduced to be able to cover the ecoexergy for various organisms in the unit detritus equivalent or chemical exergy equivalent:

$$\text{Ex-total} = \sum_{i=1}^{N} \beta_i c_i \ \text{(as detritus equivalent)}. \tag{15}$$

The calculation of ecoexergy accounts for the chemical energy in the organic matter as well as for the (minimum) information embodied in the living organisms. Detritus equivalent is presumed to correspond to 18.7 kJ/g on average; but since some organisms, for instance, birds have higher fat content than average detritus, the β-value has been

Table 1
Ecoexergy of living organisms

Organisms	gi	β = (eichem + eibio)/eichem	Ecoexergy (kJ/g)
Detritus	0	1	18.7
Minimal cell[a]	470	2.3	43.8
Bacteria	600	2.7	50.5
Algae	850	3.4	64.2
Yeast	2000	5.8	108.5
Fungus	3000	9.5	178
Sponges	9000	26.7	499
Molds	9500	28.0	524
Plants, trees	10 000–30 000	29.6–86.8	554–1623
Worms	10 500	30.0	561
Insects	10 000–15 000	29.6–43.9	554–821
Jellyfish	10 000	29.6	554
Zooplankton	10 000–15 000	29.6–43.9	554–821
Fish	100 000–120 000	287–344	5367–6433
Birds	120 000	375	6433
Amphibians	120 000	344	6433
Reptiles	130 000	370	6919
Mammals	140 000	402	7517
Human	250 000	716	13 389

Sources: [12,15,16].
[a] Based on energy contained in detritus. 1 g detritus has in average 18.7 kJ ecoexergy.

adjusted accordingly. The information contribution is measured by the extremely small probability to form the living components, for instance algae, zooplankton, fish, mammals and so on spontaneously from inorganic matter. Weighting factors defined as the ecoexergy content relative to detritus (see Table 1) may be considered quality factors reflecting how developed the various groups are and to what extent they contribute to the ecoexergy due to their content of information that is reflected in the computation. This is, completely according to Ref. [8], who gave the following relationship for the work, W, that is embodied in the thermodynamic information, that we have (compare also with Section 4):

$$W = RT \ln N \ \ (\mathrm{ML^2 T^{-2}}),$$

where N is the number of possible states, among which the information has been selected. N is, as seen for species, the inverse of the probability to obtain the valid amino-acid sequence spontaneously. The Kullback measure of information covers the gain in information, when the distribution is changed from p_{io} to p_i. Note that K is a specific measure (per unit of matter). K multiplied by the total concentration yields the ecoexergy.

The total ecoexergy of an ecosystem *cannot* be calculated exactly, as we cannot measure the concentrations of all the components or determine all possible contributions to ecoexergy in an ecosystem. If we calculate the ecoexergy of a fox, for instance, the above shown calculations will only give the contributions coming from the biomass and the information embodied in the genes, but what is the contribution from the blood pressure, the sexual hormones and so on? These properties are at least partially covered by the genes but is that the entire story? We can calculate the contributions from the dominant components, for instance by the use of a model or measurements that covers the most essential components for a focal problem.

Ecoexergy calculated by use of the above equations has some clear shortcomings:

(1) We have made some albeit minor approximations in the equations presented above.
(2) We do not know the genes in all details for all organisms.
(3) We calculate only in principle the ecoexergy embodied in the proteins (enzymes), while there are other components of importance for the life processes. These components are contributing less to the ecoexergy than the enzymes and the information embodied in the enzymes controls the formation of these other components, for instance hormones. It can, however, not be excluded that these components will contribute to the total ecoexergy of the system.
(4) We do not include the ecoexergy of the ecological network. If we calculate the ecoexergy of models, the network will always be relatively simple and the contribution coming from the information content of the network is negligible.
(5) We will always use a simplification of the ecosystem, for instance by a model or a diagram or similar. This implies that we only calculate the ecoexergy contributions of the components included in the simplified image of the ecosystem. The real ecosystem will inevitably contain more components that are not included in our calculations.

It is therefore proposed to consider the ecoexergy found by these calculations as a *relative minimum ecoexergy index* to indicate that there are other contributions to the total ecoexergy of an ecosystem, although they may be of minor importance. In most cases, however, a relative index is sufficient to understand the reactions of ecosystems, because the absolute ecoexergy content is irrelevant for the reactions. It is, in most cases, the change in ecoexergy that is of importance to understand the ecological reactions.

The weighting factors presented in Table 1 have been applied successfully in several structurally dynamic models to express the model goal function, and, furthermore, in many illustrations of the maximum-exergy principle, which will be presented later in this chapter. The relatively good results in application of the weighting factors, in spite of the uncertainty of their assessment, seem only to be explicable by the robustness of the application of the factors in modeling and other quantifications. The differences between the factors of the micro-organism, the vertebrates and invertebrates are so clear that it does not matter if the uncertainty of the factors is very high—the results are influenced only slightly.

On the other hand, it would be an important step to get better weighting factors from a theoretical point of view but also because it would enable us to model the competition between species that are closely related.

The key to finding better β-values is the proteomes on the one hand. On the other hand, our knowledge about the number of proteomes in various organisms is very limited—more limited than for the number of non-nonsense genes. It may be possible, however, to put together our knowledge about non-nonsense genes, the overall DNA content, the limited knowledge about the proteomes and the evolution tree and see some pattern that could be used to give better but still very approximate β-values at this stage. Today it is, for instance, known that Homo sapiens has only 30 000–40 000 genes; but as the number of amino acids coded per gene is higher than the previously applied 700, the weighting factor will not be smaller, but rather higher.

5. Why have living systems such a high level of ecoexergy?

A frog of 20 g will have an ecoexergy content of 20.6433 kJ $\approx$ 12.9 GJ, while a dead frog will have only an ecoexergy content of 574 kJ, although they have the same chemical composition, at least a few seconds after the frog has died. The difference is rooted in the information or rather the difference in the useful information. The dead frog has the information a few seconds after its death (the amino-acid composition has not yet been decomposed), but the difference between a live frog and a dead frog is the ability to utilize the enormous information stored in the genes and the proteome of the frog. The amount of information stored in a frog is really surprisingly high. The number of amino acids placed in the right sequence is 84 000 000 and for each of these 84 000 000 amino acid the number of possibilities is 20. This amount of information is able to ensure reproduction and is transferred from generation to generation, which ensures that the evolution can continue because what is already a favorable combination of properties is conserved through the genes. Because of the very high number of amino acids, 84 000 000, it is not surprising that there will always be a minor difference from frog to frog in the amino-acid sequence. It may be a result of mutations or of a minor mistake in the copying process. This variation is important because it gives possibilities to 'test' which amino-acid sequence gives the best result with respect to survival and growth. The best, representing the most favorable combination of properties, will offer the highest probability of survival and give most growth and the corresponding genes will therefore prevail. Survival and growth mean more ecoexergy and that a bigger distance to thermodynamic equilibrium is the result. Ecoexergy could therefore be used as a thermodynamic function that could be used to quantify Darwin's theory. It is interesting in this context that Svirezhev [10] has demonstrated that ecoexergy also represents the amount of energy needed to tear down the system. This means that the more ecoexergy the system possesses the more difficult it becomes to kill the system and the higher is therefore the probability of survival.

6. Formulation of a thermodynamic hypothesis (maximization of ecoexergy) for ecosystems

If an (open, nonequilibrium) ecosystem receives a boundary flow of energy from its environment, it will use what it can of this energy, the free-energy or ecoexergy content, to do work. The work will generate internal flows, leading to storage and cycling of matter, energy, and information, which move the system further from equilibrium. Self-organizing processes get started. This is reflected in decreased internal entropy and increased internal organization. In general, growth means an increase in system size, while development is increase in organization independently of system size. Growth is measured as mass or energy change per unit time, for instance kg/y, while storage-specific growth is measured in 1/units of time, for instance 1/24 h. Development may take place without any change (growth) in biomass. "*Growth* and *development* may be considered as extensive and intensive aspects of the same process [4]. The two processes often take place in parallel". In thermodynamic terms, a growing system is one moving away from thermodynamic equilibrium. At equilibrium, the system cannot do any work. All its components are inorganic, have zero free energy (ecoexergy), and all gradients are eliminated. Everywhere in the Universe there are structures and gradients, resulting from growth and developmental processes cutting across all levels of organization. "A gradient is understood as a difference in an intensive thermodynamic variable, such as for instance temperature, pressure, chemical potential, altitude and so on". The second law dissipation acts to tear down the structures and eliminate the gradients, but this dissipation cannot operate unless the gradients were established in the first place. An obvious question, therefore, is what determines the buildup of gradients? Growth (*and development*) may occur in ecosystems in three forms [15]:

(I) The biomass is increased—trees are growing, animals are getting offspring and so on.
(II) The network linking the components in the ecosystems and the number of different components may increase, resulting in more cycling and more overall throughflow of energy.
(III) The information and organization in the ecosystem increase.

These three forms of growth are, respectively, growth to storage, growth to throughflow, and growth to information and organization. It has also been stated above that an ecosystem will utilize the incoming solar radiation to move away from thermodynamic equilibrium, which means that the system will grow and increase the ecoexergy stored.

The open question of this section is, which of the many possible pathways will an ecosystem take in realizing its three forms of growth? The answer given is that an ecosystem will change in directions that most consistently create additional capacity and opportunity to achieve increasing deviation from thermodynamic ground, i.e. the ecoexergy stored in the ecosystem will increase. Abundant and diverse living biomass represents abundant and diverse departure from thermodynamic equilibrium, and both are captured in this parameter. If multiple growth pathways are offered from a given starting state, those producing greatest ecoexergy storage will tend to be selected, for these in turn require greatest energy dissipation to establish and maintain, consistent with the second law. Energy storage by itself is not sufficient, but it is the increase in specific ecoexergy, that is, of ecoexergy/energy ratios, that reflects improved usability, and this represents the increasing capacity to do the work required for living systems to continuously evolve new adaptive 'technologies' to meet their changing environments.

These considerations lead to a thermodynamic hypothesis that is able to explain the growth and development of ecosystems and the reactions of ecosystems to perturbations: "If a system receives an input of ecoexergy, it will utilize this ecoexergy after the maintenance of the system far from thermodynamic has been covered to move the system further from thermodynamic equilibrium. If there is offered more than one pathway to depart from equilibrium, the one yielding the most gradients, and ecoexergy storage under the prevailing conditions, to give the most ordered structure furthest from equilibrium, will tend to be selected." This formulation may be considered a translation of Darwin to thermodynamics: the organisms that have the properties that are fitted to the prevailing conditions in the ecosystem, will be able to contribute most to biomass and information and thereby give the ecosystem most ecoexergy—they are the best survivors.

Just as it is not possible to prove the first three laws of thermodynamics by deductive methods, so can the above hypothesis only be 'proved' inductively. In the next section, we examine a number of actual cases that contribute, in a general way, to the weight of evidence in favor.

7. Support to the hypothesis

Below are given eight supporting arguments for the hypothesis presented in Section 6. More evidence has been provided; but the eight supporting pieces of evidence presented here give a good idea of the theoretical support for the hypothesis.

(1) The ecoexergy-storage hypothesis might be taken as a generalized version of 'Le Chatelier's Principle'. Biomass synthesis can be expressed as a chemical reaction:

$$\text{energy} + \text{nutrients} = \text{molecules with more free energy (ecoexergy) and organization} + \text{dissipated energy.}$$

According to Le Chatelier's Principle, if energy is put into a reaction system at equilibrium the system will shift its equilibrium composition in a way to counteract the change. This means that more molecules with more free energy and organization will be formed. If more pathways are offered, those giving the most relief from the disturbance (*using most of the inflowing energy*) by using the most energy, and forming the most molecules with the most free energy, will be the ones followed in restoring equilibrium.

(2) The sequence of organic-matter oxidation (e.g. Ref. [16]) takes place in the following order: by oxygen, by nitrate, by manganese dioxide, by iron(III), by sulfate and by carbon dioxide. This means that oxygen, if present, will always out-compete nitrate that will out-compete manganese dioxide, and so on. The amount of ecoexergy stored as a result of an oxidation process is measured by the available kJ/*mole electrons* that determines the number of adenosine triphosphate molecules (ATPs) formed. ATP represents an ecoexergy storage of 42 kJ/mole. Usable energy as ecoexergy in ATPs decreases in the same sequence as indicated above. This is as expected if the ecoexergy-storage hypothesis is valid (Table 2). If more oxidizing agents are offered *to the system*, the one giving the resulting system the highest storage of free energy will be selected.

(3) Numerous experiments have been performed to imitate the formation of organic matter in the primeval atmosphere on earth 4 billion years ago [13,14]. Energy from various sources was sent through a gas mixture of carbon dioxide, ammonia and methane. Analyses showed that a wide spectrum of compounds, including several amino acids contributing to protein synthesis, is formed under these circumstances. There are obviously many pathways to utilize the energy sent through simple gas mixtures, but mainly those forming compounds with rather large free energies (high ecoexergy storage, released when the compounds are oxidized again to carbon dioxide, ammonia and methane) will form an appreciable part of the mixture [13].

(4) There are three biochemical pathways for photosynthesis: (1) the C3 or Calvin–Benson cycle, (2) the C4 pathway and (3) the crassulacean acid metabolism (CAM) pathway. The latter is least efficient in terms of the amount of plant biomass formed per unit of energy received. Plants using the CAM pathway are, however, able to survive in harsh, arid environments that would be inhospitable to C3 and C4 plants. CAM photosynthesis will generally switch to C3 as soon as sufficient water becomes available [19]. The CAM pathways yield the highest

Table 2
Yields of kJ and ATPs per mole of electrons, corresponding to 0.25 mole of CH_2O oxidized. The released energy is available to build ATP for various oxidation processes of organic matter at pH = 7.0 and 25°C

Reaction	kJ/mole e^-	ATPs/mole e^-
$CH_2O + O_2 \leftrightarrow CO_2 + H_2O$	125	2.98
$CH_2O + 0.8NO_3^- + 0.8H^+ \leftrightarrow CO_2 + 0.4N_2 + 1.4H_2O$	119	2.83
$CH_2O + 2MnO_2 + H^+\Delta \leftrightarrow CO_2 + 2Mn^{2+} + 3H_2O$	85	2.02
$CH_2O + 4FeOOH + 8H^+ \leftrightarrow CO_2 + 7H_2O + Fe^{2+}$	27	0.64
$CH_2O + 0.5SO_4^{2-} + 0.5H^+ \leftrightarrow CO_2 + 0.5HS^- + H_2O$	26	0.62
$CH_2O + 0.5CO_2 \leftrightarrow CO_2 + 0.5CH_4$	23	0.55

biomass production, reflecting ecoexergy storage, under arid conditions, while the other two give highest net production (ecoexergy storage) under other conditions. While it is true that 1 g of plant biomass produced by *each of* the three pathways has different free energy, in a general way improved biomass production by any of the pathways can be taken to be in a direction that is consistent, under the conditions, with the ecoexergy-storage hypothesis.

(5) Givnish and Vermelj [20] observed that leaves optimize their size (thus mass) for the conditions. This may be interpreted as meaning that they maximize their free-energy content. The larger the leaves the higher their respiration and evapotranspiration, but the more solar radiation they can capture. Deciduous forests in moist climates have a leaf-area index (LAI) of about 6%. Such an index can be predicted from the hypothesis of highest possible leaf size, resulting from the tradeoff between having leaves of a given size versus maintaining leaves of a given size [20]. The size of leaves in a given environment depends on the solar radiation and humidity regime, and while, for example, sun and shade leaves on the same plant would not have equal ecoexergy contents, in a general way leaf size and LAI relationships are consistent with the hypothesis of maximum ecoexergy storage.

(6) The general relationship between animal body weight, W, and population density, D, is $D = A/W$, where A is a constant [21]. The highest packing of biomass depends only on the aggregate mass, not the size of individual organisms. This means that it is biomass rather than population size that is maximized in an ecosystem, as density (number per unit area) is inversely proportional to the weight of the organisms. Of course the relationship is complex. A given mass of mice would not contain the same ecoexergy or number of individuals as an equivalent weight of elephants. Also, genome differences (Example 1) and other factors would figure in. Later we will discuss ecoexergy dissipation as an alternative objective function proposed for thermodynamic systems. If this were maximized rather than storage, then biomass packing would follow the relationship $D = A/W$ 0.65–0.75 [21]. As this is not the case, biomass packing and the free energy associated with this lend general support for the ecoexergy-storage hypothesis.

(7) If a resource (for instance, a limiting nutrient for plant growth) is abundant, it will typically recycle faster. This is a little strange, because *a rapid* recycling is not needed when a resource is nonlimiting. A modeling study [9] indicated that free-energy storage increases when an abundant resource recycles faster. Fig. 2 shows such results for a lake eutrophication model. The ratio, R, of nitrogen (N) to phosphorus (P) cycling that gives the highest ecoexergy is plotted versus log(N/P). The plot in Fig. 2 is also consistent with empirical results [22]. Of course, one cannot ‘inductively test’ anything with a model, but the indications and correspondence with data do tend to support in a general way the ecoexergy-storage hypothesis.

(8) Dynamic models whose structure changes over time are based on nonstationary or time-varying differential or difference equations. We will refer to these as *structurally dynamic models*. A number of such models, mainly of aquatic systems [9,23–31], have been investigated to see how structural changes are reflected in free-energy changes. The latter were computed as ecoexergy indexes. Time-varying

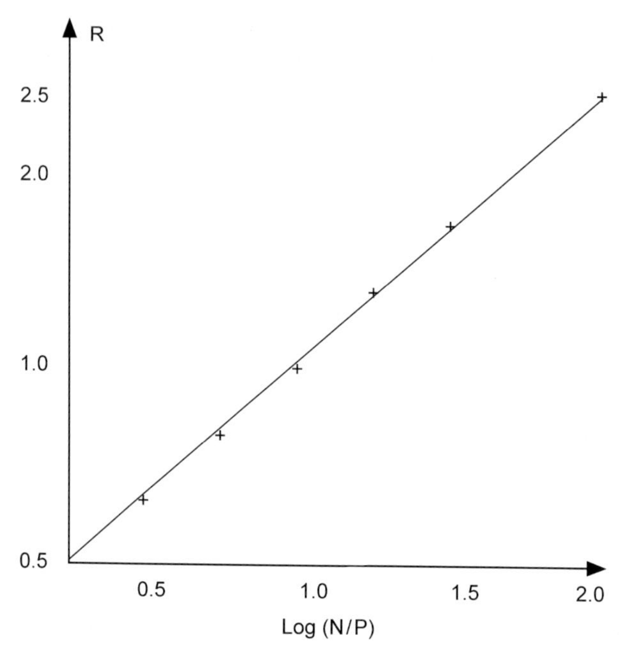

Fig. 2. Loglog plot of the ratio of nitrogen to phosphorus turnover rates, R, at maximum exergy versus the logarithm of the nitrogen/phosphorus ratio, log N/P. The plot is consistent with Vollenweider [20].

parameters were selected iteratively to give the highest ecoexergy index values in a given situation at each time step. Changes in parameters, and thus system structure, not only reflect changes in external boundary conditions, but also mean that such changes are necessary for the ongoing maximization of ecoexergy. For all models investigated along these lines, the changes obtained were in accordance with actual observations (see references). These studies therefore affirm, in a general way, that systems adapt structurally to maximize their content of eco-exergy. It is noteworthy that Coffaro et al. [29], in their structural-dynamic model of the Lagoon of Venice, did not calibrate the model describing the spatial pattern of various macrophyte species such as *Ulva* and *Zostera*, but used ecoexergy-index optimization to estimate parameters determining the spatial distribution of these species. They found good accordance between observations and model, and were able by this method *without* calibration to explain more than 90% of the observed spatial distribution of various species of *Zostera* and *Ulva*.

8. Growth and development of ecosystems

The growth and development of ecosystems from the early stage to the mature stage is described by Odum [32] by application of a number of attributes of which a few of importance for the discussion are listed:

(1) Ecosystem biomass (physical structure) increases.
(2) More feedback loops (including recycling of energy and matter) are built.
(3) Respiration increases.
(4) Respiration relative to biomass decreases.
(5) Bigger animals and plants (trees) become more dominant.
(6) The specific entropy production (relative to biomass) decreases.
(7) The total entropy production will first increase and then stabilize on approximately the same level.
(8) The amount of information increases (more species, species with more genes, the biochemistry becomes more diverse).

An interpretation of these eight statements is consistent with the above-mentioned three forms of growth and development (see Ref. [15]):

(I) Growth of physical structure (biomass) that is able to capture more of the incoming energy in the form of solar radiation, but also requires more energy for maintenance (respiration and evaporation). Attributes 1, 3 and 7.
(II) Growth of network, which means more cycling of energy and matter. Attributes 2 and 6.
(III) Growth of information (more developed plants and animals with more genes), from r-strategists to K-strategists, which waste less energy but also usually have larger size and carry more information. Attributes 1, 4, 5 and 7.

As mentioned in Section 1 several hypotheses on the development of ecosystems have been proposed. Four of these hypotheses are compared below, namely [33]:

A. The entropy production tends to be minimum (this was proposed by Prigogine [6,32], for linear systems at steady nonequilibrium state, not for far-from-equilibrium nonlinear systems).
B. Natural selection tends to make the energy flux through the system a maximum, so far as compatibility with the constraints to which the system is concerned. (The maximum power principle, see Ref. [3].)
C. Ecosystems will organize themselves to maximize the degradation of ecoexergy [7,35].
D. A system that receives a throughflow of ecoexergy will move away from thermodynamic equilibrium, and if more combinations of components and processes are offered to utilize the ecoexergy flow, the system has the propensity to select the organization that gives the system as much stored ecoexergy as possible.

A fifth possible descriptor of the ecosystem development (denoted E) should be mentioned in this context, namely, the retention time = stored biomass/input, which is maximized. Biomass and input are both considered expressed in energy units.

The usual description of ecosystem development illustrated, for instance, by the recovery of Yellowstone Park after fire, an island born after a volcanic eruption, reclaimed land, etc. is well covered by Odum [30]: at first the biomass increases rapidly, which implies that the percentage of captured incoming solar radiation increases, but also the energy needed for the maintenance. Growth form I is dominant in this first phase, where ecoexergy stored increases (more biomass, more physical structure to capture more solar radiation), but also the throughflow (of useful energy), ecoexergy dissipation and the entropy production increases due to increased need of energy for maintenance.

Growth forms II and III become dominant later, although an overlap of the three growth forms takes place. When the percentage of solar radiation captured reaches about 80%, it is not possible to increase the amount of captured solar radiation further (due in principle to the second law of thermodynamics). Further growth of structure does not therefore improve the energy balance of the ecosystem. In addition, all or almost all of the essential elements are in the form of dead or living organic matter and not as inorganic compounds ready to be used for growth. The growth form I will therefore not proceed, but growth forms II and III can still operate. The ecosystem can still improve the ecological network and can still change r-strategists with K-strategists, small animals and plants with bigger ones and less developed with more developed with more non-nonsense genes. A graphical representation of this description of ecosystem development can be found in Fig. 3. The graph has been confirmed by a comparison of the ecoexergy captured and the ecoexergy stored in the ecosystem for different ecosystems and for different stages of forest development (see Ref. [9]).

The accordance with the five descriptors and the three growth forms based on this description of ecosystem development is shown in Table 3. As can be seen from the table

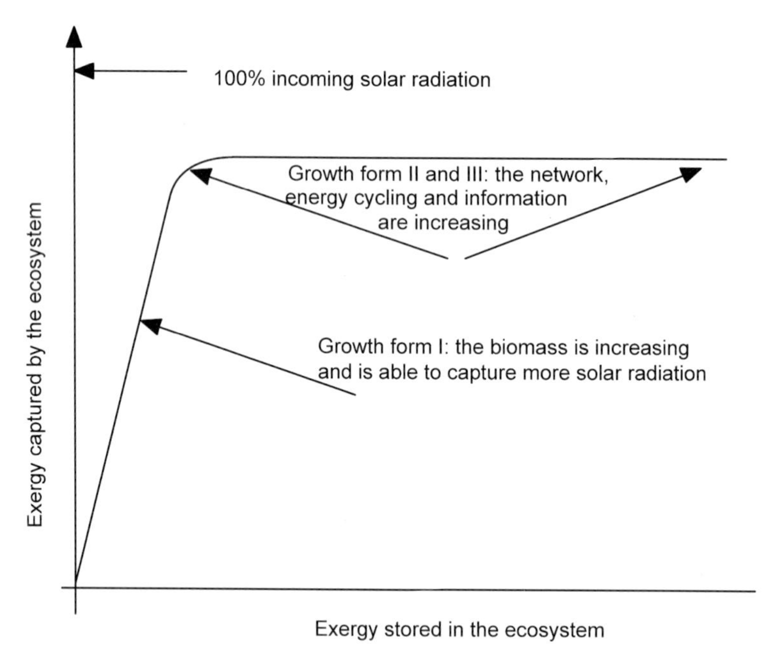

Fig. 3. The development of an ecosystem is illustrated by plotting exergy captured from the inflowing solar radiation toward the ecoexergy stored in the ecosystem. Growth form I is dominant in the first phase of the development from an early-stage ecosystem to a mature ecosystem. By increasing the biomass the percentage of solar radiation captured increases up to about 80% corresponding to what is physically possible. Growth forms II and III are dominant in the intermediate phase and when the ecosystem is in a mature stage. Thereby more ecoexergy is stored without increasing the exergy needed for maintenance. The system becomes, in other words, more effective in the use of the solar radiation according to Prigogine's minimum entropy principle. The ecoexergy stored is increased for all three growth forms, but the growth form III is very slow when the ecoexergy has reached a certain level.

Table 3
Accordance between growth forms and the proposed orientors

Growth	Hypothesis				
	A	B	C	D	E
I	Increases (−)	Increases (+)	Increases (+)	Increases (+)	Increases (+)
II	Unchanged (+)	Increases (+)	Unchanged (−)	Increases (+)	Increases (+)
III	Unchanged (+)	Increases (+)	Unchanged (−)	Increases (+)	Increases (+)

the hypothesis C is only valid for growth form I where the biomass is increasing and therefore requires more energy for maintenance (respiration), corresponding to ecoexergy destruction and entropy production. Prigogine's minimum entropy hypothesis (which actually focuses more on the specific entropy production, which is decreasing for growth forms II and III and equal for growth form 1) is only valid for growth forms II and III, but it was originally presented for a far-from-thermodynamic equilibrium system in steady state and not for an ecosystem in the early stage of rapidly increasing the biomass to be able to capture more solar radiation. Fig. 3 illustrates also the discrepancy between the hypothesis presented here (hypothesis D) and the hypothesis C, sometimes called the extended second law of thermodynamics.

References

[1] Jørgensen, S.E. and Mejer, H.F. (1977), Ecological buffer capacity, Ecol. Modell. 3, 39–61.
[2] Jørgensen, S.E. and Mejer, H.F. (1979), A holistic approach to ecological modelling, Ecol. Modell. 7, 169–189.
[3] Odum, H.T. (1983), System Ecology, Wiley Interscience, New York, 510 pp.
[4] Ulanowicz, R.E. (1986), Growth and Development. Ecosystems Phenomenology, Springer, Berlin, 204 pp.
[5] Schrödinger, E. (1944), What is Life?, Cambridge University Press, Cambridge.
[6] Prigogine, I. (1980), From Being to Becoming: Time and Complexity in the Physical Sciences, Freeman, San Francisco, CA, 260 pp.
[7] Kay, J. and Schneider, E.D. (1992), Thermodynamics and measures of ecological integrity. In Proc. "Ecological Indicators", Elsevier, Amsterdam, pp. 159–182.
[8] Boltzmann, L (1905), The Second Law of Thermodynamics (Populare Schriften, Essay no. 3 (address to Imperial Academy of Science in 1886). Reprinted in English in: Theoretical Physics and Philosophical Problems, Selected Writings of L. Boltzmann. D. Reidel, Dordrecht.
[9] Jørgensen, S.E. (2002), Integration of Ecosystem Theories: A Pattern, Kluwer, Dordrecht, 428 pp.
[10] Svirezhev, Y. (1998), How to use thermodynamic concepts in ecology. Eco Targets, Goal Functions, and Orientors (Müller, F., Leupelt, M., eds.), Springer, Berlin, pp. 102–125, 619 pp.
[11] Jørgensen, S.E., Nielsen, S.N. and Mejer, H. (1995), Emergy, environ, exergy and ecological modelling, Ecol. Modell. 77, 99–109.
[12] Li, W.-H. and Grauer, D. (1991), Fundamentals of Molecular Evolution, Sinauer, Sunderland, MA, 430 pp.
[13] Morowitz, H.J. (1968), Energy Flow in Biology, Academic Press, New York.
[14] Morowitz, H.J. (1992), Beginnings of Cellular Life, Yale University Press, New Haven, CT.
[15] Cavalier-Smith, T. (1985), The Evolution of Genome Size, Wiley, Chichester, 480 pp.
[16] Lewin, B. (1994), Genes V, Oxford University Press, Oxford, 620 pp.
[17] Jørgensen, S.E., Patten, B.C. and Straskraba, M. (2000), Ecosystem emerging: 4. Growth, Ecol. Modell. 126, 249–284.

[18] Schlesinger, W.H. (1997), Biogeochemistry. An Analysis of Global Change, 2nd edn, Academic Press, New York, 680 pp.
[19] Shugart, H.H. (1998), Terrestrial Ecosystems in Changing Environments, Cambridge University Press, Cambridge, 512 pp.
[20] Givnish, T.J. and Vermelj, G.J. (1976), Sizes and shapes of liana leaves, Am. Nat. 110, 743–778.
[21] Peters, R.H. (1983), The Ecological Implications of Body Size, Cambridge University Press, Cambridge.
[22] Vollenweider, R.A. (1975), Input–output models with special reference to the phosphorus loading concept in limnology, Schweiz. Z. Hydrol. 37, 53–84.
[23] Jørgensen, S.E. (1986), Structural dynamic model, Ecol. Modell. 31, 1–9.
[24] Jørgensen, S.E. (1988), Use of models as an experimental tool to show that structural changes are accompanied by increased exergy, Ecol. Modell. 41, 117–126.
[25] Jørgensen, S.E. (1992), Development of models able to account for changes in species composition, Ecol. Modell. 62, 195–208.
[26] Jørgensen, S.E. (1992), Parameters, ecological constraints and exergy, Ecol. Modell. 62, 163–170.
[27] Jørgensen, S.E. (1995), The growth rate of zooplankton at the edge of chaos: ecological models, J. Theor. Biol. 175, 13–21.
[28] Nielsen, S.N (1992), Application of maximum exergy in structural dynamic models (Thesis). National Environmental Research Institute, Denmark.
[29] Coffaro, G., Bocci, M. and Bendoricchio, G. (1997), Structural dynamic application to space variability of primary producers in shallow marine water, Ecol. Modell. 92, 97–114.
[30] Jørgensen, S.E. and de Bernardi, R. (1997), The application of a model with dynamic structure to simulate the effect of mass fish mortality on zooplankton structure in Lago di Annone, Hydrobiologia 356, 87–96.
[31] Nielsen, S.N. (1992), Strategies for structural-dynamical modelling, Ecol. Modell. 63, 91–102.
[32] Odum, E.P. (1969), The strategy of ecosystem development, Science 164, 262–270.
[33] Jørgenson, S.E. and Fath, B.D. (2004), Application of thermodynamic principles in ecology, Ecol. Complexity 1, 267–280.
[34] Prigogine, I. (1947), Etude thermodynamique des phénomenès irreversibles, Desoer, Liège.
[35] Kay, J.J (1984), Self organization in living systems (Thesis). Systems Design Engineering, University of Waterloo, Ontario, Canada.

SELFORGANIZATION AND ECONOPHYSICS

Chapter 16

SELF-ORGANIZED CRITICALITY WITHIN THE FRAMEWORK OF THE VARIATIONAL PRINCIPLE

Alexander I. Olemskoi and Dmitrii O. Kharchenko

Sumy State University, Rimskii-Korsakov St. 2, 40007, Sumy, Ukraine

Abstract

In the framework of Euclidean field theory the system with self-organized criticality regime is studied. We consider a self-similar behavior introduced as a fractional feedback in a three-parameter model of Lorenz type. The main modes to govern the system dynamics are: an avalanche size, an energy of moving grains and a complexity (entropy) of the avalanche ensemble. We take account of the additive noise of the energy to investigate the process of the nondriven avalanche-formation process in the presence of the energy noise, which plays a crucial role. The kinetics of the system is studied in detail on the basis of the variational principle. This distribution is shown to be a solution of both fractional and linear Fokker–Planck equations. Relations between the exponent of the size distribution, fractal dimension of phase space, characteristic exponent of multiplicative noise, number of governing equations are obtained.

Keywords: Self-organized criticality; noise; extremum principle; anomalous diffusion

1. Introduction

Considering dynamics of stochastic systems one deals with behavior of most probable values of a stochastic variable or its averages. The last allows us to determine the statistical picture of the systems evolution. The former shows the dynamics of maxima of the probability-density function, which correspond to macrostates (phases) that appear in a course of noise-induced behavior (see Ref. [1] and citations therein). Such macrostates define new properties of the system that cannot be observed in a noiseless case. Considering the problem of dynamics of stochastic systems variational principles are used to determine optimal trajectories of the system evolution (see Refs. [2–6]). The mathematical tool that allows us to investigate the dynamics of the system and to find the probability of rare fluctuations is the path-integral trajectories method. Unfortunately, in most works the quasi-classic WKB approximation is thought to be restricted by consideration of the weak noise limit only and therefore it cannot give complete information about the noise influence playing a crucial role. More essential

E-mail address: alex@ufn.ru (A.I. Olemskoi); dikh@sumdu.edu.ua (D.O. Kharchenko).

effects of the noise influence on the dynamics of the system are observed with a help of the Euclidean field theory (EFT) to account for the original properties of the stochastic process [7]. As was shown in Refs. [8,9] it allows us to investigate the processes of an absorbing state formation. In this work, we apply the approach based on EFT to explore the picture of spontaneous (avalanche-type) dynamics realized in the system with a self-organized criticality (SOC) regime. A main feature of the systems displaying SOC is their self-similarity that leads to a power-law distribution over avalanche sizes. In most cases, SOC models are studied by making use of the scaling-type arguments supplemented with extensive computer simulations (see Ref. [10]). In contrast, we put forward an analytical approach, which is able to describe the process of the behavior of a whole avalanche ensemble in a phenomenological manner.

We aimed to explore the dynamics of the avalanche formation in the stochastic environment. Here, the distribution P over the most probable trajectories of the system evolution is presented through the Euclidean action $S\{s(t), \phi(t)\}$ in the form $P \propto \exp(-S/\sigma^2)$, which depends on both the avalanche size $s(t)$ and its conjugate momentum $\phi(t)$; σ^2 is introduced as a noise intensity. Formally, a relation between P and S arises if we rewrite the standard Fokker–Planck equation $\partial_t P = \hat{L}P$ in an imaginary-time Schrödinger equation $\sigma^2 \partial_t P = \hat{H}P$, driven by a Hamiltonian $\hat{H}$ or Liouvillian $\hat{L}$. In such a case the noise intensity σ^2 plays a role of an effective Planck constant and P corresponds to the wave function. Considering a single degree of freedom this method amounts to the eikonal approximation [11,12]. Formally, in both WKB and EFT approaches the action satisfies the Hamilton–Jacobi equation $\partial_t S + H = 0$, which in turn implies a principles of least action and the Hamiltonian equation of motion [13]. For the above approaches the Hamiltonians are different in form (in the last approach it takes into account nontrivial effects of the noise influence). We investigate the anomalous behavior of both s and ϕ to find the probability of realization of optimal trajectories related to a minimization procedure for the action S.

The work is organized as follows. In Section 2 we present the main suppositions of EFT to explore the dynamics of the system. Here, we are based on the effective evolution equation for the avalanche size to be written as an effective Langevin equation with white noise. We obtain the Euler equations for the most probable values of the avalanche size and conjugate momentum and present the form of the probability function of the realization of optimal trajectories. Section 3 is devoted to a consideration of the dynamics of the avalanche ensemble. An effective scheme is proposed to determine a time-dependent distribution over energies of moving grains. We introduce a unified Lorenz system with a fractional feedback where the main modes of the system are pseudothermodynamical variables such as: the avalanche size, nonconserved energy of moving grains and the complexity (information entropy) of the avalanche ensemble. These degrees of freedom act as an order parameter, conjugate field and a control parameter, respectively. To study the dynamics of the system in the nondriven SOC regime we take into account fluctuations of the energy of moving grains. It will be shown that an increase in the fluctuations intensity causes avalanche emergence in nondriven systems. Within the framework of this approach, the dynamics of the system in the SOC regime is investigated for both a white noise case (Section 3.1) and a case of colored noise (Section 3.2). In Section 3.3 we show the system undergoing the SOC regime formally satisfies properties of an anomalous diffusion process. Comparing results of both obtained pictures for SOC regime and a behavior of a system with an anomalous diffusion, we get relations between the exponent of the avalanche distribution, fractal dimension of phase space, characteristic exponent of multiplicative noise, a number of governing equations needed to present self-consistent behavior in the SOC regime. The main results are collected together in Section 4.

2. Field approach and optimal trajectories

The action of noise gives rise to fluctuations on the behavior of a dynamical system. This means that a deviation of its trajectories form the phase trajectories related to a deterministic dynamics is realized. Such trajectories visit domains of the phase space that are never approached by a deterministic trajectory, even at wide range of initial conditions. Such a behavior of fluctuating trajectories results in a qualitative change of the system behavior.

We start with a multiplicative noise Langevin equation for a relevant macrovariable $s(t)$ in the form

$$\dot{s} = \mathcal{D}_1(s) + \sqrt{2\mathcal{D}_2(s)}\zeta(t), \tag{1}$$

where $\mathcal{D}_1(s)$ represents an effective drift coefficient, $\mathcal{D}_2(s)$ plays the role of the effective diffusion coefficient, $\zeta(t)$ indicates the white noise with zero mean and correlations given as follows:

$$\langle \zeta(t)\zeta(t')\rangle = \delta(t - t'). \tag{2}$$

Here and throughout the chapter the Itô interpretation of Eq. (1) is used. Eq. (1) is associated with the Fokker–Planck equation in the form of Kramers-Moyal expansion [11]

$$\partial_t P(s,t) = -\partial_s \mathcal{D}_1(s)P(s,t) + \partial_s^2 \mathcal{D}_2(s)P(s,t). \tag{3}$$

In order to investigate the system dynamics in the framework of the path-integral approach we should construct the probability functional in the form

$$P(s; \tilde{t} \to \infty) \propto \exp(-S), \tag{4}$$

where S is the action functional, $\tilde{t} = \mathrm{i}t$. The optimal trajectories are given with a help of minimizing procedure for the Euclidean action $\delta S = 0$.

To construct S let us move to a new process $y(t)$, which is associated with $s(t)$ by means of relation $\mathrm{d}s/\mathrm{d}y = \sqrt{2\mathcal{D}_2(s(y))}$. The new stochastic process $y(t)$ satisfies the Langevin equation of the form

$$\dot{y} = \tilde{h}(s(y)) + \zeta(t), \qquad \tilde{h} \equiv \frac{\mathcal{D}_1(s)}{\sqrt{2\mathcal{D}_2(s)}} - \frac{1}{2}\partial_s\sqrt{2\mathcal{D}_2(s)}. \tag{5}$$

The obtained equation (5) allows to use the standard field scheme [14–16] based on analysis of the generating functional. The latter has the form

$$Z\{u(t)\} = \int Z\{y(t)\}\exp\left(\int uy\,\mathrm{d}t\right)Dy(t), \tag{6}$$

where

$$Z\{y(t)\} = \left\langle \prod_t \delta\{\dot{y} - \tilde{h} - \zeta\}\det\left|\frac{\delta\zeta}{\delta y}\right| \right\rangle_\zeta, \tag{7}$$

here Dy denotes integration over all paths starting at $y(0)$ for $t = 0$ and ending at $y(t_f)$ for $t = t_f$ (we return to notation of imaginary time as t). The argument of the δ-function in Eq. (7) can be reduced to the Langevin equation (5), and the determinant ensuring a transition of variables $\zeta(t) \to y(t)$ is equal to unity.

To analyze functional (7) we can take into account the identity

$$\delta\{y(t)\} = \int_{-\mathrm{i}\infty}^{\mathrm{i}\infty} \exp\left(-\int qy\,\mathrm{d}t\right)Dq. \tag{8}$$

Averaging over the noise ζ with the help of the Gauss distribution

$$\Pi\{\zeta\} \propto \exp\left\{-\frac{1}{2}\int \zeta^2(t)\mathrm{d}t\right\}, \tag{9}$$

and taking into account Eq. (8), we reduce functional Eq. (7) to the standard form

$$Z\{y(t)\} = \int \mathrm{e}^{-S\{y(t),q(t)\}}Dq. \tag{10}$$

Here, the corresponding Lagrangian is defined as follows:

$$\mathcal{L}(y,q) = q(\dot{y} - \tilde{h}) - q^2/2. \tag{11}$$

For the next we use the Euler equations

$$\frac{\partial\mathcal{L}}{\partial z} - \frac{\mathrm{d}}{\mathrm{d}t}\frac{\partial\mathcal{L}}{\partial\dot{z}} = \frac{\partial\mathcal{R}}{\partial\dot{z}}, \qquad z \equiv \{y, q\}, \tag{12}$$

where the dissipative function is defined as

$$\mathcal{R}(y) = \dot{y}^2/2. \tag{13}$$

As a result, the equations for the most probable realizations of the stochastic fields $y(t)$ and $q(t)$ assume the form

$$\dot{y} = \tilde{h} + q, \tag{14}$$

$$\dot{q} = -q(1 + \partial_y\tilde{h}) - \tilde{h}. \tag{15}$$

A comparison of Eq. (14) with the stochastic equation (5), having the same form, shows that the fields $y(t)$ and $q(t)$ are the most probable values of amplitudes of the auxiliary hydrodynamic mode and fluctuation of the conjugate force. Returning to the initial field

$s(t)$, we get the Lagrangian and the dissipative function as follows:

$$\mathcal{L}(s, \phi) = \phi(\dot{s} - \mathcal{D}_1(s) + \partial_s \mathcal{D}_2(s)/2) - \mathcal{D}_2(s)\phi^2, \tag{16}$$

$$\mathcal{R}(s) = \dot{s}^2/4\mathcal{D}_2(s), \tag{17}$$

where the definition $\phi = \partial\mathcal{L}/\partial\dot{x}$ is used for the conjugate momentum, which yields the relation $q = \phi\sqrt{2\mathcal{D}_2(s)}$. Therefore, the Euler equation (12) reads

$$\dot{s} = \mathcal{D}_1(s) - \partial_s \mathcal{D}_2(s)/2 + 2\mathcal{D}_2(s)\phi, \tag{18}$$

$$\dot{\phi} = -\phi[1 + \partial_s(\mathcal{D}_1(s) - \partial_s \mathcal{D}_2(s)/2) + \phi\partial_s \mathcal{D}_2(s)] - \mathcal{D}_1(s)/2\mathcal{D}_2(s) + (\partial_s \mathcal{D}_2(s))/4\mathcal{D}_2(s). \tag{19}$$

In the stationary case $\dot{s} = \dot{\phi} = 0$ we have

$$\phi = -\frac{\mathcal{D}_1(s)}{2\mathcal{D}_2(s)} + \frac{1}{4}\partial_s \ln \mathcal{D}_2(s), \tag{20}$$

$$\phi\{\partial_s(\mathcal{D}_1(s) - \partial_s \mathcal{D}_2(s)/2) + \phi\partial_s \mathcal{D}_2(s)\} = 0. \tag{21}$$

From Eqs. (20) and (21) it follows that steady states are defined from solutions of equations

$$\mathcal{D}_1(s) - \partial_s \mathcal{D}_2(s)/2 = 0, \tag{22}$$

$$\mathcal{D}_1(s) - \partial_s \mathcal{D}_2(s)/2 - \sqrt{\mathcal{D}_2(s)} = 0. \tag{23}$$

The first of these equations gives points on the phase plane (ϕ, s) with abscissa $\phi = 0$ and the second equation gives points situated off the axis $\phi = 0$. Physically, the first equation defines the bifurcation diagram for the most probable values of stochastic variable s, the second one allows us to find a corresponding phase diagram when solutions of both equations coincide. The meaning of Eqs. (20) and (21) becomes clear if we consider the additive noise limit, i.e. $\mathcal{D}_2(s) = \text{const}$. In the domain of the system parameters at which there are no solutions of Eq. (22) we have a point on the phase plane (ϕ, s) with abscissa $\phi \neq 0$ and an ordinate given by solution of Eq. (23). Condition (20) indicates that the conjugate momentum ϕ has an opposite sign to the force value $\mathcal{D}_1(s)$. According to Eq. (21), the 'susceptibility' $\chi = -1/\partial_s \mathcal{D}_1(s)$ assumes an infinite value. Thus, such a point on the phase plane corresponds to the stationary state of a thermodynamic system being unstable with respect to the transition into the ordered phase with $s \neq 0$. In the case of $\phi = 0$ the ordinate is given as a solution of Eq. (22). In the additive limit, this point corresponds to the state of thermodynamic equilibrium. Of course, during the process of its evolution a real thermodynamic system tends to the equilibrium state with $\phi = 0$, but not to the unstable state where $\phi \neq 0$.

According to the obtained equations we can consider the probability of realization of a phase trajectory corresponding to different initial values $s(0) \equiv s(t_0)$. The probability can be obtained from Eqs. (16) and (18) by integrating over all the path $s(t)$, starting from $s(0)$. The related expression for the probability takes the form

$$P(s(0)) \propto \exp\left\{-\int_{s(0)} \mathcal{D}_2(s(t))\phi^2(t)\mathrm{d}t\right\}. \tag{24}$$

3. Dynamics of the system exhibiting self-organized criticality

The SOC behavior appears in a vast variety of systems, such as a real sand pile (ensemble of grains of sand moving along an increasingly tilted surface) [17–20], intermittency in biological evolution [21], earthquakes and forest fires, depinning transitions in random medium and so on (see Refs.[22–24] and references therein). Systems with SOC take a special place among other classes of stochastic systems due to its universality. Here the special attention is focused on the avalanche-size distribution

$$P(s) = s^{-\tau_{\rm soc}}\mathcal{P}(s/s_c), \tag{25}$$

where a critical avalanche size s_c is related to the system size L and a characteristic time $t_c \sim L^z$ as follows $s_c \sim L^D \sim t_c^{D/z}$ (exponents D and z are fractal dimension, and dynamical exponent related to a critical avalanche). According to Ref. [25], the mean-field magnitudes of the above exponents are given by: $\tau_{\rm soc} = 3/2$, $D = 4$ and $z = 2$. On the other hand, the scaling relation accompanied by the equality of the mean size of the avalanche lead to the following expression

$$\tau_{\rm soc} = 2\left(1 - \frac{1}{D}\right). \tag{26}$$

Among the existing one- or two-parameter models to describe systems with SOC (see for example Refs. [23,26]) the self-consistent approach is achieved in the framework of a three-parameter Lorenz-like scheme proposed in Ref. [27]. The mean-field approximation [25] shows that the self-similar regime of the sand-pile dynamics is relevant for subcritical behavior, where a characteristic time for the variation of the order parameter is much larger than that of the control parameter. Moreover, the latter follows the former adiabatically. Adiabatic behavior of this type is inherent in the usual regime of a system evolving in the course of phase transitions [28], so that an adiabatic approach will be taken as the basis of our consideration.

Following Ref. [27] the system under consideration we parameterize by a set of pseudothermodynamical variables that describes the avalanche ensemble in the spirit of the famous Edwards paradigm [29,30] generalized to a nonstationary system. With this method, we represent the system with the help of vector $\mathbf{x} = \{s, C, \mathcal{E}\}$ that consists of the avalanche size s, the complexity (information entropy) C and the nonconserved energy of moving grains $\mathcal{E}$. Within the framework of the usual synergetic approach, these degrees of freedom play the role of order parameter, conjugate field and control parameter, respectively. Formally, the evolution equations for each of the modes can be written on the well-known laser Lorenz scheme conception that we modify by introducing the power-law construction as follows:

$$\begin{aligned}
\dot{s} &= -s^{\gamma}/\tau_s + a_s C, \\
\dot{C} &= -C/\tau_C + a_C \mathcal{E} s^{\gamma}, \\
\dot{\mathcal{E}} &= (\mathcal{E}^0 - \mathcal{E})/\tau_{\mathcal{E}} + a_{\mathcal{E}} C s^{\gamma} + \sigma\zeta(t).
\end{aligned} \tag{27}$$

Here, $\tau_s, \tau_C, \tau_{\mathcal{E}}$ denote relaxation times of s, C and $\mathcal{E}$, respectively; a_s, a_c, a_ε are feedback parameters; γ is a positive exponent that will be defined later; a stochastic part in the last equation accounts for the fluctuations with amplitude σ; $\mathcal{E}^0$ is an externally driven energy of the grain motion. For the Langevin force we suppose a correlation in the form $\langle\zeta(t)\zeta(t')\rangle = C(t,t')$. The distinguishing feature of the first of these equations is that in a noiseless case genuine characteristics s, C are connected in a special manner: an increase in the complexity assists a growth of the avalanche; according to the supposition of a relaxation process to be in a power-law form we connect the avalanche size and the complexity in a nonlinear manner. The nonlinear term in the equation for the complexity corresponds to the positive feedback between the order parameter s and the energy $\mathcal{E}$, that are of a thermodynamic type. In accordance with Le Chatelier's principle the negative feedback of the avalanche size and the complexity on the energy results in the energy decrease. Moreover, positive feedback appears of the avalanche size and the energy on the complexity, which causes a complexity increase that is the reason for the avalanche ensemble's self-organization. As was shown in Ref. [27] such a type of model allows us to represent the system with a SOC regime in a most natural manner. For the next step to simplify the description of the system behavior we use the adiabatic elimination procedure that reduces the many-parameter system to a one-parameter system [11,12,28].

To define the physical meaning of the energy of moving grains and the complexity we will show that such components acquire their thermodynamical meaning in the framework of the Edwards paradigm. Moreover, quantities C and $\mathcal{E}$ define the effective temperature to be negative when the self-organization process occurs. To prove this we will restrict ourselves to the treatment of the stochastic system, where the adiabatic conditions $\tau_C, \tau_{\mathcal{E}} \ll 1$ are applicable. Then, the two last equations of the system (27) lead to the following dependencies:

$$C(t) = \bar{C} + \tilde{C}\zeta(t), \quad \mathcal{E}(t) = \bar{\mathcal{E}} + \tilde{\mathcal{E}}\zeta(t), \tag{28}$$

where the deterministic and the fluctuation components are determined as follows

$$\begin{aligned}
&\bar{C} \equiv \mathcal{E}^0 s^{\gamma} g_\gamma(s), \quad \tilde{C} \equiv \sigma_{\mathcal{E}} s^{\gamma} g_\gamma(s); \\
&\bar{\mathcal{E}} \equiv \mathcal{E}^0 g_\gamma(s), \quad \tilde{\mathcal{E}} \equiv \sigma_{\mathcal{E}} g_\gamma(s), \quad g_\gamma(s) \equiv (1 + s^{2\gamma})^{-1}.
\end{aligned} \tag{29}$$

Due to the slaving principle of synergetics, the initially adiabatic noise of the energy is transformed to a multiplicative one. On the other hand, the relation between the complexity and energy

$$\bar{C} = \sqrt{\bar{\mathcal{E}}(\mathcal{E}^0 - \bar{\mathcal{E}})}, \tag{30}$$

that can be deduced with the dependencies (29), leads to the expression

$$T = -\left(1 - \frac{\mathcal{E}^0}{2\bar{\mathcal{E}}}\right)^{-1} \sqrt{\frac{\mathcal{E}^0}{\bar{\mathcal{E}}} - 1} \tag{31}$$

for the Edwards temperature $T \equiv \partial\bar{\mathcal{E}}/\partial\bar{C}$. As depicted in Fig. 1, T is a monotonically increasing function of the energy with boundary values $T(\bar{\mathcal{E}} = 0) = 0$ and $T(\bar{\mathcal{E}} = \mathcal{E}^0/2) = \infty$. At the latter point the magnitude T changes instantaneously to $-\infty$ and then increases monotonically again to the initial value $T = 0$ at $\bar{\mathcal{E}} = \mathcal{E}^0$. This means that in the domain $0 \leq \bar{\mathcal{E}} < \mathcal{E}^0/2$ the avalanche system is dissipative and behaves in the usual manner; in contrast, in the domain $\mathcal{E}^0/2 < \bar{\mathcal{E}} \leq \mathcal{E}^0$ a self-organization process evolves, so that an energy increase leads to a complexity decrease, in accordance with a negative temperature. The presented self-organization regime relates to externally driven systems, which are relevant for the usual phase transition but not to the SOC itself.

It is well known that a complete set of SOC systems can be reduced to one of two families [25]: systems with deterministic dynamics extremely driven by a random environment (growing interface models, the Bak–Sneppen evolution model, etc.) and the stochastic dynamics family (models of earthquakes, forest fire, etc.). A remarkable peculiarity of the system (27) is the possibility to present both mentioned families in a natural manner. The former is related to the noiseless case, when $\sigma_{\mathcal{E}} = 0$ but the magnitude of the energy relaxation time is larger than that of the complexity and avalanche size ($\tau_{\mathcal{E}} \geq \tau_C, \tau_s$); on the other hand, a parameter of the environment drive $\mathcal{E}^0$ has to take a larger value than the critical one $\mathcal{E}_c = 1$. In such a case, the system (27) describes a strange attractor that may represent the behavior of SOC systems of the first type. A proper stochastic behavior that makes possible the appearance of the SOC regime is relevant for the case of $\sigma_{\mathcal{E}} \neq 0$ even in the absence of a driven affect ($\mathcal{E}^0 = 0$) [27].

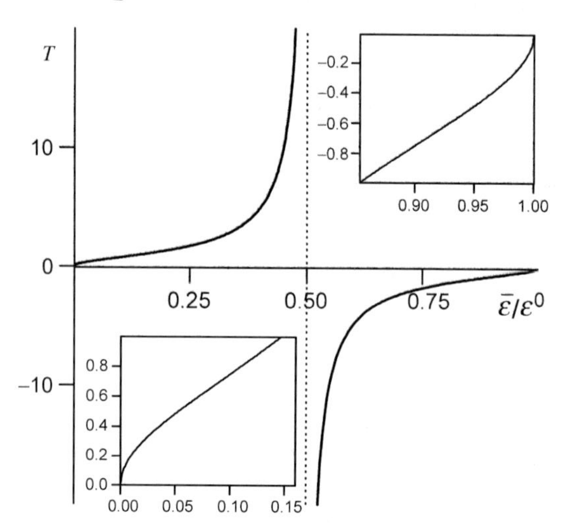

Fig. 1. The energy dependencies of the avalanche-ensemble temperatures: nonstationary magnitude T vs. ratio $\bar{\varepsilon}/\varepsilon^0$.

In the case under consideration we will restrict ourselves to the treatment of the system, where the adiabatic conditions $\tau_C, \tau_{\mathcal{E}} \ll \tau_s$ are applicable. To study the SOC regime within the above consideration, let us express complexity and nonconserved energy from the second and third equations of (27) according to the slaving principle. In such a way, we arrive at the effective Langevin equation in the form

$$\dot{s} = f_\gamma(s) + g(s)\zeta(t), \tag{32}$$

where the force $f_\gamma(s)$ and the noise amplitude are as follows:

$$f_\gamma(s) \equiv -s^\gamma + \mathcal{E}^0 s^\gamma g_\gamma(s), \quad g(s) = \sigma s^\gamma g_\gamma(s). \tag{33}$$

Here, we measure time t in unit $\tau_s(\tau_C\tau_{\mathcal{E}}a_C a_{\mathcal{E}})^{1/2-1/2\gamma}$ and introduce the scales for s, $\mathcal{E}$ and $\sigma_{\mathcal{E}}^2$ as follows: $s^{sc} \equiv (a_C a_{\mathcal{E}}\tau_C\tau_{\mathcal{E}})^{-1/2\gamma}$, $\mathcal{E}^{sc} \equiv a_s a_C \tau_s \tau_C$, $\sigma_{\mathcal{E}}^{sc} \equiv a_s a_C \tau_s \tau_{\mathcal{E}}$. Considering the SOC regime in a nondriven system we put $\mathcal{E}^0 = 0$.

In order to avoid a nonphysical situation we should restrict ourselves by consideration of the values for avalanche size that lie in the interval $s \in (b_0, \infty)$, $b_0 = 0+$. Indeed, from the mathematical viewpoint at $s = 0$ we have $f(0) = g(0) = 0$, which means that the left boundary of the diffusion process $s(t)$ is classified as a natural boundary [1]. From a physical viewpoint the avalanche size should be positively defined, i.e. $s > 0$ and there are no physical reasons to set any boundary condition for the process $s(t)$. Therefore, we will next consider only nonzero magnitudes of s, supposing a state with $s = 0$ as a nonavalanche state. As will be shown below in the case of both white and colored noises the boundary b_0 can transform the kinetics of the system in the case of small avalanche size $s \ll 1$.

3.1. Pure white noise

The simplest situation of the system behavior corresponds to the case of white noise, where the correlation function is in the form

$$C(t, t') = \delta(t - t'). \tag{34}$$

This is rather a mathematical than a physical model but it gives results that can be treated to set general properties of the system. In such a case the drift and diffusion coefficients are as follows:

$$\mathcal{D}_1 = f_\gamma(s), \quad \mathcal{D}_2 = g^2(s)/2. \tag{35}$$

According to the obtained formulas for drift and diffusion coefficients Eqs. (18) and (19) read

$$\dot{s} = f_\gamma(s) - \frac{1}{2}g(s)\partial_s g(s) + g^2(s)\phi, \tag{36}$$

$$\dot{\phi} = -\phi\left[\partial_s\left(f_\gamma(s) - \frac{1}{2}g(s)\partial_s g(s)\right) + \phi g(s)\partial_s g(s)\right] - \frac{\dot{s}}{g^2(s)}. \tag{37}$$

In the stationary case the system undergoes a noise-induced transition with formation of an avalanche with stationary size given as $s_0(\gamma)$ in Fig. 2. Here, solid lines (denoted as s_-, s_+) correspond to the solution of stationary equation (36) at $\phi = 0$, the dotted lines mean

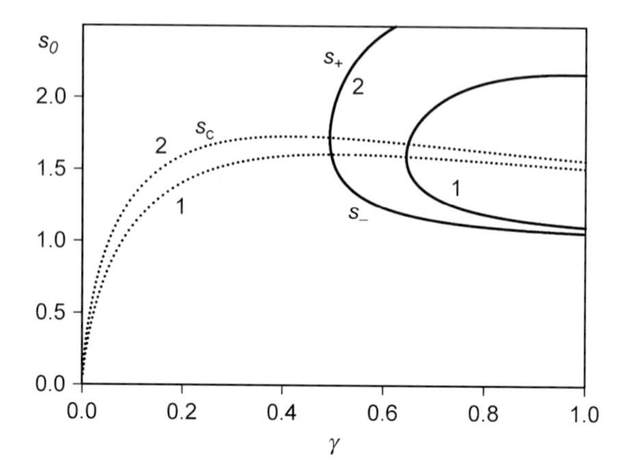

Fig. 2. Stationary value of the avalanche size s_0 vs. exponent γ at $\sigma^2 = 100, 150$ (curves 1 and 2 correspondingly).

the solutions of Eq. (37) at $\phi = 0$. Making use of the analysis of the stationary solutions in the case of additive noise we conclude that solutions s_-, s_+ situated on the axis $\phi = 0$, define the equilibrium states. As in the stationary case Eq. (36) always has a trivial solution $s_0 = 0$ one should conclude that the stable equilibrium states are characterized by the two points on the phase plane (ϕ, s) with coordinates (0, 0) and (0, s_+). The stationary solution denoted as a dotted line s_c corresponds to the state with coordinate ($\phi_0 \neq 0, s_c$). According to the phase diagram in Fig. 3 and related dependencies in Fig. 2 we see that an increase in the noise intensity σ^2 leads to the formation of the stationary avalanche at small values of the exponent γ. Deviation of the exponent γ from the specific value 1 corresponds to an increase in the nonlinearity of the main functions that define the avalanche formation. The domains denoted as N and A + N (see Fig. 3), determine the magnitudes of σ^2, γ at which there are no avalanches $s_0 = 0$ (N) and the coexistence of both the noise-induced state with $s_0 \neq 0$ and avalanche-free state of $s_0 = 0$ (A + N).

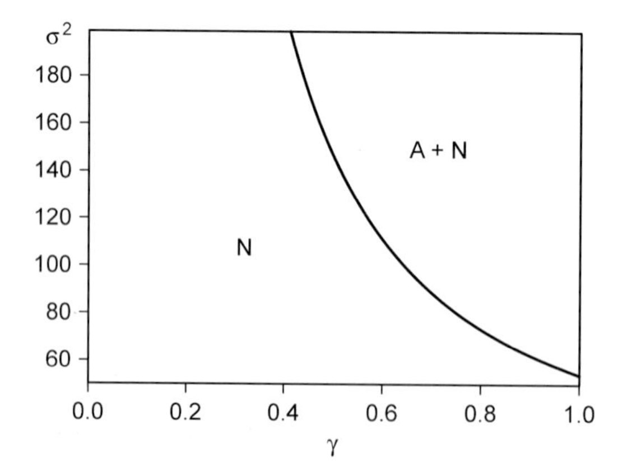

Fig. 3. Phase diagram of stationary avalanche formation.

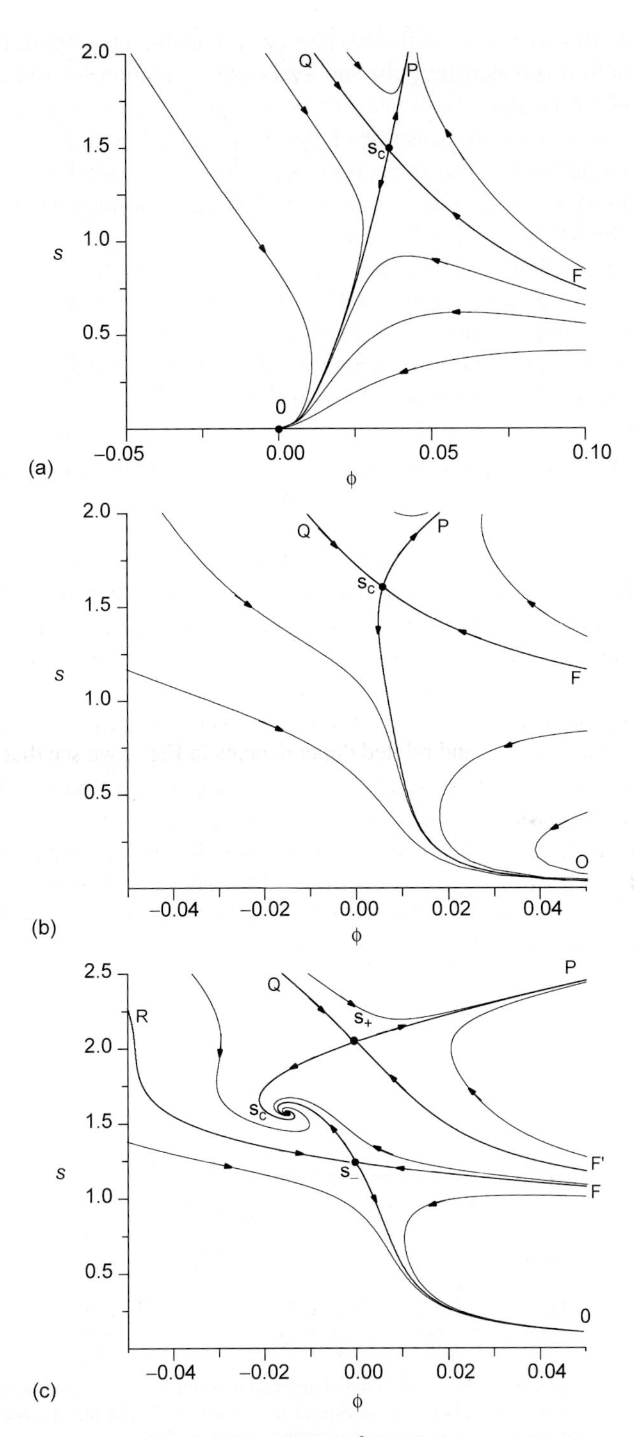

Fig. 4. Phase portraits of the system with white noise at $\sigma^2 = 100$: (a) $\gamma = 0.25$; (b) $\gamma = 0.6$; (c) $\gamma = 0.75$.

To investigate the system dynamics we plot corresponding phase portraits (see Fig. 4) in domains of both nonavalanche (a,b) and avalanche formation (c) states. The general appearance of phase portraits shown in Fig. 4a is characterized by the presence of one separatrix QS_cF. It divides the phase plane into two isolated domains, corresponding to large and small values of s. The domain of large values is characterized by an infinite increase of the most probable value of avalanche size s and its conjugate momentum ϕ at $t \to \infty$. Above the separatrix QS_cF phase trajectories do not give any contribution to the stationary distribution (24). It will be shown that actually this is not realized. The formation of the domain corresponding to values $s \ll 1$ is associated with the multiplicative nature of the noise. In this domain, $s(t)$ tends to the one attractive point 0 with coordinates $s = \phi = 0$. It will be shown below that in this domain small magnitudes of avalanche size s and its conjugate momentum ϕ in the limit $t \to \infty$ correspond to maximal values of the probability (24). Comparing Fig. 4a and 4b we see that they are distinguished by the location of the attraction node on the axis $\phi = 0$ when we pass through the critical value $\gamma = 1/2$. The general property of the phase portrait shown in Fig. 4c is in the presence of two separatrices with branches RS_-F and QS_+F'. They divide the phase plane into three isolated domains, corresponding to large, intermediate and small values of s. The first domain is characterized by an infinite increase of the quantities s and ϕ at $t \to \infty$. The domain of intermediate values of s, in which the system gets into the stationary ordered state with a nonzero value of avalanche size is most interesting. This domain determines the noise-induced transition kinetics.

Let us analyze the behavior of the system in each of the domains in Fig. 4. For this purpose, we consider the probability of realization of a phase trajectory corresponding to different initial values $s(0) \equiv s(t = 0)$. The dependence $P(s(0))$ obtained for the index $\gamma < 1/2$ is shown in Fig. 5 (curve 1). We can see that apart from the trivial increase of $P(s(0))$ the probability jumps near the separatrix of the phase portrait. Outside the domain restricted by the separatrix, we have $P = 0$, due to $s(t)$, $\phi(t) \to \pm\infty$ at $t \to \infty$. Such a

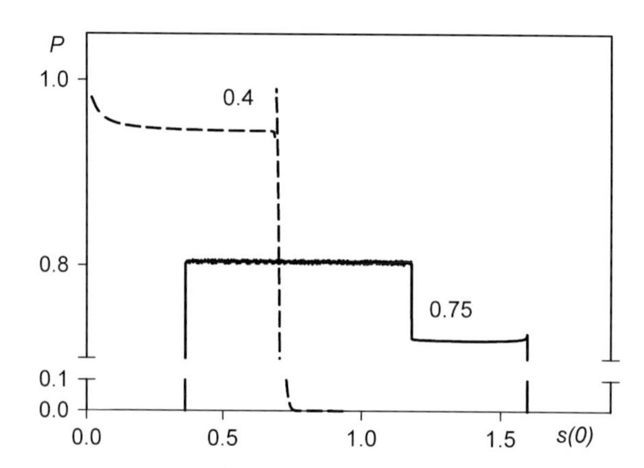

Fig. 5. Dependence of the probability P of realization of different trajectories on the initial value of the avalanche size $s(0)$ at $\sigma^2 = 100$ (dashed and solid curves correspond to $\gamma = 0.4$, 0.75. The initial value of the conjugate momentum is $\phi(0) = 0.015$.

behavior of the probability $P(s(0))$ can be explained through the form of the time dependencies $s(t)$ and $\phi\,(t)$ during relaxation of the initial value (see Fig. 6) for various values of the index γ. For this purpose we put $\dot{\phi} = 0$ in Euler equations. The obtained equation gives stationary values of the conjugate momentum in the limit $s \rightarrow 0$:

$$\phi \propto \begin{cases} \frac{1}{2}(\frac{1}{2} - \gamma)s^{1-2\gamma}, & \gamma < \frac{1}{2}; \\ (\gamma - \frac{1}{2})s^{-1}, & \gamma > \frac{1}{2}. \end{cases} \tag{38}$$

Thus, at $\gamma < 1/2$ the system tends in the course of time to the equilibrium nonavalanche state with $s = \phi = 0$. At $\gamma \geq 1/2$ the attraction node jumps to infinity ($s = 0, \phi \rightarrow \infty$). The corresponding integrand in the distribution (24) reads

$$g^2(s)\phi^2 \propto \begin{cases} \frac{1}{4}(\frac{1}{2} - \gamma)^{-2}s^{2(1-\gamma)}, & \gamma < \frac{1}{2}; \\ (\gamma - \frac{1}{2})^2 s^{-2(1-\gamma)}, & \gamma > \frac{1}{2}. \end{cases} \tag{39}$$

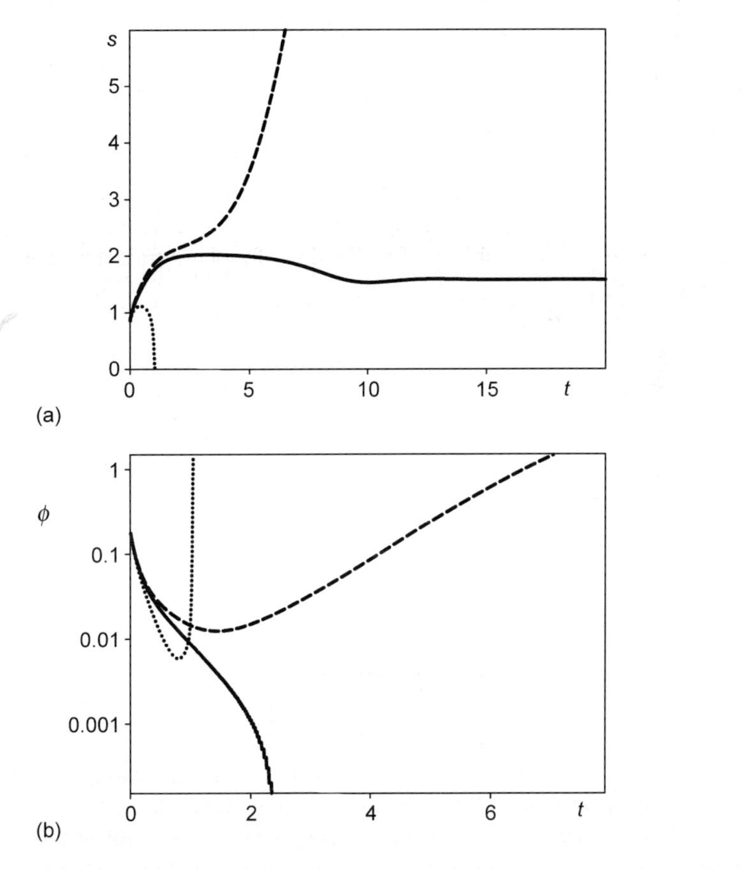

Fig. 6. A typical appearance of the temporal behavior for: (a) the avalanche size; (b) the conjugate momentum at $\gamma = 3/4$, $\sigma^2 = 100$ (solid curves correspond to the trajectories inside the domain FS_cF', dashed lines correspond to trajectories above the separatrix $F'S_+$, dotted curves correspond to domain of small values $s(0)$.

It is characterized by the inversion of the index sign when the critical value $\gamma = 1/2$ is exceeded. Substituting Eq. (38) at $\gamma < 1/2$ into Euler equations and retaining the leading term in it, we obtain the scaling law as follows:

$$s \propto (t_0 - t)^H, \quad t < t_0, \quad H^{-1} = 2(1 - \gamma), \tag{40}$$

where t_0 is a constant, which characterizes the time during which the point arrives at the axis $s = 0$. Inserting Eq. (40) into Eq. (39) and the obtained expression into Eq. (24) we receive the probability $P(s(0))$ that differs from zero in the domain $s \ll 1$. Physically it means that if the fluctuations excite the system then it behaves itself in order to suppress such excitation. In other words, the state $s(0) \neq 0$ relaxes to $s(t \to \infty) \to 0$ along optimal trajectories that are realized with nonzero probability.

A completely different situation is observed at $\gamma > 1/2$. In this case, $g^2\phi^2$ acquires an index with an opposite sign in accordance with Eq. (39). As a result, the probability (24) becomes vanishing at $s \ll 1$. The physical reason behind such a behavior is that the system lies at the axis $s = 0$ for a finite time interval $t_0 < \infty$ that is small comparing with relaxation time of the system. In such a situation we have an infinite value of the conjugate momentum $\phi \propto t^{-1/2(1-\gamma)}$. This can be visualized as the rise of the absorbing state (the precipitation of condensate of configuration points from the phase portrait domain $s \ll 1$ onto the axis of the abscissa when $\phi \to \infty$). Note that the condition $t_0 < \infty$ is fulfilled only below the separatrix branch S_-O in Fig. 4a and c while in the domain bounded by separatrices RS_-F and QS_+F' we have $t_0 \to \infty$, and the divergence of the integrand in Eq. (24) is not manifested. Consequently, the equality $P = 0$ holds only below the line RS_-O. It should be understood as follows: the excited system instantly passes to the nonavalanche state and optimal trajectories starting form $s(0)$ that lies under the separatrix RS_-F have an infinitely small contribution to the stationary probability of its realization due to the large magnitudes of the action S.

3.2. *Colored noise*

In the simplest form, the problem of colored noise can be introduced considering the effects of the autocorrelation of the Langevin force $\zeta(t)$ in Eq. (32). Next, we assume a specific form for the noise ζ: we choose Ornstein-Uhlenbeck noise, i.e. a Gaussian-distributed stochastic variable with zero mean and exponentially decaying correlations

$$\langle \zeta(t)\zeta(t') \rangle = (1/2\tau)\exp(-|t - t'|/\tau). \tag{41}$$

They arise as solutions of the Langevin equation

$$\tau\dot{\zeta} = -\zeta + \xi(t), \tag{42}$$

where $\xi(t)$ is a white noise—namely, a Gaussian stochastic variable with zero mean and δ-correlated: $\langle \xi(t)\xi(t') \rangle = \delta(t - t')$. In such a representation the system given by Eqs. (32) and (42) is a non-Markovian in its properties. As a simplest tool to pass to the Markovian process $s(t)$ the 'unified colored noise approximation' [31] can be used.

If we take the time derivative of Eq. (32), replace first $\dot{\zeta}$ in terms of ζ and ξ from Eq. (42) and then ζ in terms of $\dot{s}$ and s from Eq. (32), we obtain the following non-Markovian stochastic differential equation:

$$\tau(\ddot{s} - \dot{s}^2 \partial_s g(s)/g(s)) = -\lambda(s)\dot{s} + f(s) + g(s)\xi(t), \tag{43}$$

where

$$\lambda(s) = 1 - \tau f_\gamma(s)\partial_s \ln\left[\frac{f_\gamma(s)}{g(s)}\right]. \tag{44}$$

Here, following the 'unified colored noise approximation' we can recover a Markovian stochastic differential equation. It needs to use adiabatic elimination (neglecting $\ddot{s}$) and to neglect $\dot{s}^2$ so that the system's dynamics are governed by a Fokker–Planck equation. Using the Itô differential rule, the resulting equation being linear in $\dot{s}$ takes the form

$$\lambda(s)\dot{s} = f_\gamma(s) + g(s)\xi(t). \tag{45}$$

The expression on the LHS in Eq. (46) defines a time derivative for a new variable z. The relation between z and s is $\mathrm{d}z = \lambda(s)\,\mathrm{d}s$. An evolution equation for $s(t)$ can be written down as Eq. (1) with drift and diffusion coefficients as follows:

$$\begin{aligned} \mathcal{D}_1(s) &= \frac{f_\gamma(s)}{\lambda(s)} - \frac{1}{2}g^2(s)\partial_s \ln \lambda(s), \\ \mathcal{D}_2(s) &= \frac{1}{2}\left(\frac{g(s)}{\lambda(s)}\right)^2. \end{aligned} \tag{46}$$

Since the Euler equation is too complex in its form, as it was in the previous section, we use the symbolic notation in terms of initial functions

$$\dot{s} = \frac{f_\gamma(s)}{\lambda(s)} - \frac{1}{2}\frac{g(s)}{\lambda(s)}\partial_s \ln \lambda(s) - \frac{1}{2}\partial_s \ln\left(\frac{g(s)}{\lambda(s)}\right) + \left(\frac{g(s)}{\lambda(s)}\right)^2 \phi, \tag{47}$$

$$\begin{aligned} \dot{\phi} = &-\phi\left[\partial_s\left[\frac{f_\gamma(s)}{\lambda(s)} - \frac{1}{2}\frac{g(s)}{\lambda(s)}\partial_s \ln \lambda(s) - \frac{1}{2}\partial_s \ln\left(\frac{g(s)}{\lambda(s)}\right)\right]\right. \\ &\left. + \phi\frac{g(s)}{\lambda(s)}\partial_s\left(\frac{g(s)}{\lambda(s)}\right)\right] - \frac{\dot{s}\lambda^2(s)}{g^2(s)}. \end{aligned} \tag{48}$$

In the stationary case $\dot{s} = \dot{\phi} = 0$ we have

$$\phi = -\frac{\lambda(s)f_\gamma(s)}{g^2(s)} + \frac{1}{2}\partial_s \ln g(s), \tag{49}$$

$$\phi\left\{\frac{\lambda(s)f_\gamma(s)}{g^2(s)}\partial_s \ln\left[\frac{f_\gamma(s)}{g(s)}\right] + \frac{1}{2}(\partial_s \ln \lambda(s))\partial_s \ln g(s) - \frac{\partial_s^2 g(s)}{2g(s)}\right\} = 0. \tag{50}$$

From the obtained formulas and phase diagram shown in Fig. 7a it follows that in the limit $\tau = 0$ we pass to the white-noise system discussed in the previous section. An influence of colored nature of the multiplicative noise is reduced as follows. (i) An increase in the noise self-correlation time τ increases the critical values of the noise intensity σ^2 when the system passes to the domain of stationary avalanches. (ii) Fig. 7b shows that an increase in τ produces a transition from the ordered state to the disordered (avalanche-free) state.

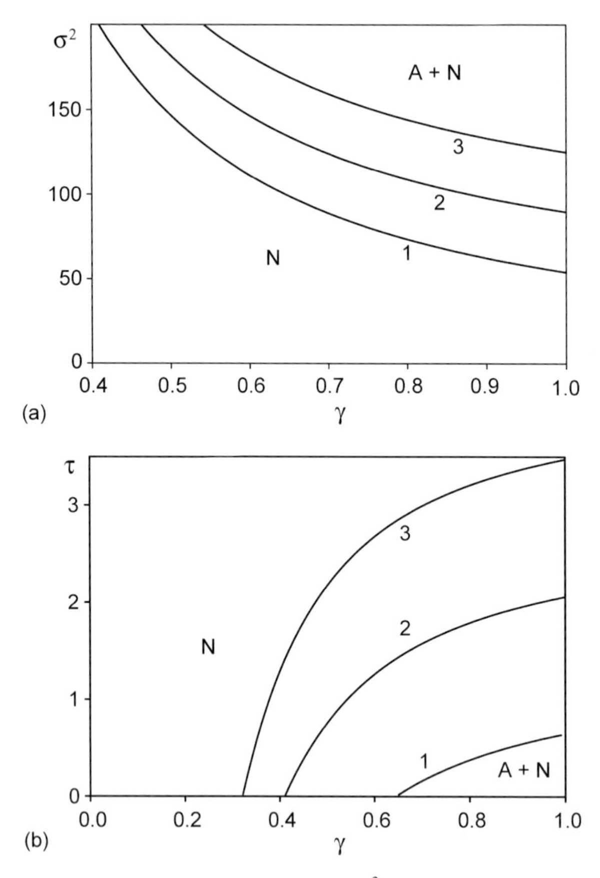

Fig. 7. Phase diagrams for the system with colored noise: (a) σ^2 vs. γ curves 1, 2, 3 corresponds to $\tau = 0.001$, 0.5, 1; (b) τ vs. γ at $\sigma^2 = 100, 200, 300$ (curves 1, 2, 3).

The picture of stationary solutions is shown in Fig. 8. It is seen that formation of the stationary avalanche occurs in a discontinuous manner inherent in the first-order transitions. Here, at large τ the phase portrait of the system is characterized by the one stable point with coordinate (0, 0) and one saddle point, situated on the axis s. The form of the phase portrait in the case under consideration is the same as in Fig. 4. If we move to the short-range fluctuations with $\tau \ll 1$ the stationary avalanche is formed and on the phase portrait two saddle points appear with coordinates determined by the condition $\phi = 0$ and specified by the coordinates $s_{\mp}$ as solutions of the steady-state equation (50).

Considering the time dependencies it should be noted that the form of $s(t)$, $\phi\,(t)$ and $g(s)\phi\,(s)$ does not change essentially compared to the white-noise case. Therefore, for the case under consideration the assymptotics given by Eqs. (38) and (39) are the same. The exponent γ does not change its critical magnitude 1/2 where the absorbing configuration is formed. The indicated features mean that the colored nature of the noise cannot change the universality class of the system behavior. Therefore, an influence of the colored

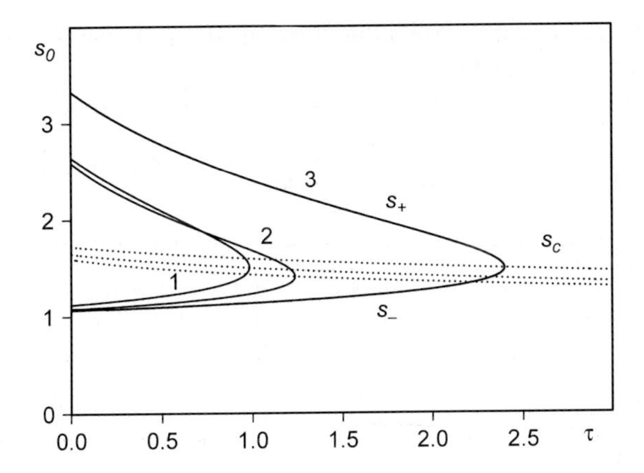

Fig. 8. Stationary solutions for the system with colored noise: $1 - \sigma^2 = 150$, $\gamma = 0.75$; $2 - \sigma^2 = 150$, $\gamma = 0.9$; $3 - \sigma^2 = 200$, $\gamma = 0.75$.

fluctuations is reduced to control a process of the stationary avalanche formation through the spectral quantity τ.

3.3. Scaling properties

Let us discuss scaling properties of the system under consideration. Making use of the variational principles we have shown the anomalous properties of the system evolution. Unfortunately, the obtained results are addressed to the most probable values. The asymptotic $s(t)$ at $t \to \infty$ explains the picture of absorbing states by the fractal dimension smaller than unity. In the special case $\gamma = 3/4$ we receive $D = 1/2$. In general, at $\gamma > 1/2$ the internal fractal dimension $D < 1$. Such a condition means that the phase space $(\dot{s}, s)$ is degenerated in a formation similar to the 'Cantor dust', i.e. a formation intermediate between a point and a line. In this section, we try to set the universality of the anomalous temporal behavior considering statistical moments of the quantity s. We will show that the results of variational procedure allow us to define anomalous properties of diffusion processes.

The anomalous character of the quantity s can be proved through measuring a length L of the curve $s(t)$. If we cover the curve $s(t)$ by pieces of a small length l that correspond to the small interval τ along the axis t and measure the coordinate s in a scale $a \to 0$, we find $l = \sqrt{\tau^2 + (s/a)^2}$. On the other hand, for the macroscopic length L we have a definition $L \propto l^{1-D}$ [32], which is invariant under transformation $\tau \to b\tau, b > 1$, due to the self-affinity of the curve $s(t)$. Making use of the supposition $s(bt) = b^H s(t)$ we get $l \to l(b) = \sqrt{b^2\tau^2 + b^{2H}(s/a)^2} = \sqrt{2}b^H(s/a) \propto b^H$, where the second equality is satisfied under supposition of the equivalent contribution of both t and s. The number of pieces to cover the curve is $N = t/b\tau$. The total length of the curve s is defined as follows: $L = Nl \propto b^{-1+H} \propto l^{1-1/H}$. Comparing it with the definition $L \propto l^{1-D}$ we receive a relation between the Hölder exponent H and fractal dimension D in the form $H^{-1} = D = 2(1 - \gamma)$.

The anomalous character of the correlation function $\langle s(t)s(t')\rangle$ can be explained from the solution of the Langevin equation (1). Indeed, supposing $f_\gamma = 0$ and using the formal solution of the Langevin equation in the form $s(t) = s(0) + \int_0^t g(s)\zeta(t')\mathrm{d}t'$ we can find

$$\langle s(t)s(t')\rangle = \int_0^t \int_0^{t'} \mathrm{d}t'\,\mathrm{d}t''\, g(s(t'))g(s(t''))\delta(t'-t'') = \int_0^{\min(t,t')} \mathrm{d}t'\, g^2(s(t')). \tag{51}$$

Making use of the supposition $s(bt) = b^H s(t)$, we obtain $H^{-1} = 2(1-\gamma)$.

The case of $H > 1$ is inherent to a superdiffusion process that is realized in the system with SOC. It is well known that the regime of SOC is characterized by the distribution over the avalanche size in the form of Eq. (25). The asymptotic behavior of the probability density function appeared as a solution of the Fokker–Planck equation (3) is characterized by Eq. (25) where $\tau_{\rm soc}$ is replaced by 2γ. As was shown in Refs. [33,34] at $\gamma < 1/2$ such a distribution appears as a solution of the fractional Fokker–Planck equation of the form [35]

$$\partial_t^\beta P(s,t) = \partial_s^\alpha P(s,t), \tag{52}$$

if the order α of the fractional operator is related to the exponent γ as follows: $\alpha = 1 - 2\gamma$. The parameter α defines the superdiffusion property of the diffusion process at which $s(t)$ can acquire the infinitely large magnitudes in the course of its evolution. Here, the critical avalanche size tests an anomalous behavior defined as $s_c(t) \propto t^{\beta/\alpha}$. According to the relation between α, β and the obtained form of the Hölder exponent H we get the definition

$$\beta = \frac{1-2\gamma}{2(1-\gamma)}, \quad \gamma < 1/2. \tag{53}$$

The parameter β defines the subdiffusion with delay in time of the evolution of the avalanche size $s(t)$. Therefore, the generalized Fokker–Planck equation (52) is related to the ordinary diffusion in the case of $\gamma = 0$ with $\beta = 1/2$.

In the case $\gamma \geq 1/2$ the anomalous diffusion process that characterizes the avalanche formation can be represented considering the nonlinear fractional Fokker–Planck equation in the form

$$\partial_t^\beta P(s,t) = \partial_s^\alpha P^q(s,t), \tag{54}$$

where $0 < q < 1$ is the parameter to measure the nonextensivity of the system. Inserting into Eq. (54) construction (25) one can obtain definitions for α and β as follows:

$$\alpha = 2\gamma(1-q), \quad \beta = \frac{(2\gamma-1)(1-q)}{2(1-\gamma)}, \quad \gamma \geq 1/2. \tag{55}$$

If the condition $\beta > 0$ is satisfied we arrive at a conclusion $\gamma > 1/2$. Therefore, at $\gamma > 1/2$ the anomalous process of avalanche formation is characterized by nonlinear kinetics determined by the parameter q. Indeed, following the Edwards paradigm the initial system (27) corresponds to evolution equations for the avalanche size s, q-entropy $\Sigma_q = -(\sum_i p_i^q - 1)/(q-1)$ and q-weighted energy of moving grains $\Xi_q = \sum_i \Xi_i p_i^q$, where p_i is a probability to move grain i with the energy Ξ_i [27].

To relate the result of the mean field theory (26) to the derived scheme we need to take into account that in the Lorenz system we have three stochastic freedom degrees ($n = 3$), which serve as the different space directions. However, the stochastic process evolves for any of these variables in a plane spanned by the given variable itself and its conjugated momentum. Moreover, the multiplicative character of noise, which is determined by the exponent γ, reduces the fractal dimension of every plane to the value $2(1 - \gamma)$ [36]. Thus, the resulting fractal dimension of the phase space, in which the stochastic system evolves, is as follows:

$$D = 2n(1 - \gamma), \tag{56}$$

where $n = 3$ for the used Lorenz system. In the general case we get the final result

$$\tau_{\rm soc} = 2\left[1 - \frac{1}{2n(1 - \gamma)}\right]. \tag{57}$$

This shows that the exponent $\tau_{\rm soc}$ increases monotonically from its minimum magnitude $\tau_{\rm soc} = 1$ at the critical number $(1 - \gamma)^{-1}$ to the upper value $\tau_{\rm soc} = 2$ in the limit $n \to \infty$; thereby, an γ-growth shifts the dependence $\tau_{\rm soc}(n)$ to large magnitudes of n, i.e. decreases the exponent $\tau_{\rm soc}$.

It is easy to see that relation (57) reproduces known results of different approaches for the dimension D (see Ref. [37]). In the case related to mean-field theory, one has $\tau_{\rm soc} = 3/2$ and Eq. (57) expresses the number of self-consistent stochastic equations needed for treating the SOC behavior as a function of the exponent of the corresponding multiplicative noise:

$$n = \frac{2}{1 - \gamma}. \tag{58}$$

A self-consistent mean-field treatment is possible if the number of relevant equations is larger than the minimum magnitude $n_c = 2$. Approaches [18–20,25,38] represent examples of such considerations, where noise is supposed to have additive character ($\gamma = 0$). Switching the multiplicative noise leads to an γ-growth and noncontradicting representation of the SOC demands an increase in the number of self-consistent equations: for example, within the field scheme [26] related to directed percolation ($\gamma = 1/2$), the mean-field approximation is applicable for dimensions larger than the critical magnitude $d_c = 4$; here, the Lorenz scheme ($n = 3$) with multiplicative noise is characterized by the exponent $\gamma = 1/3$.

4. Conclusions

In this chapter in the framework of EFT the kinetics of avalanche formation is studied. Making use of the main suppositions of the synergetic approach based on the Edwards paradigm we investigate the system displaying SOC with the help of the variational principle. Two simplest cases of the stationary avalanche-formation process (of both white and colored noises) are studied in detail. We parametrize the system under consideration by introducing exponent $\gamma \in [0, 1]$, where the pure SOC regime is characterized by the value $\gamma = 3/4$. It was shown that if γ passes through the critical *value*$\gamma = 1/2$ the probability of the optimal trajectories realization at small avalanche size is infinitely small. The bifurcation point that defines the stationary avalanche formation is determined by noise intensity σ^2 and the exponent γ in the white-noise case. In the case of colored noise this point is defined through the σ^2, γ and noise autocorrelation time τ. It is principally important that a type of fluctuation (white or colored) does not change the universality class of the system in the supposition of the diffusion process of avalanche formation. We have shown that avalanche size behaves itself in a nonanalytical manner. To prove this result we used the simplest stochastic and geometric assumptions. Making use of the supposition that the avalanche-formation process has properties of anomalous diffusion we set the main relations between exponents of anomalous diffusion, exponent of the theory γ and number of governing stochastic equations to represent the system displaying SOC.

References

[1] Horsthemke, W. and Lefever, R. (1984), Noise induced transitions. Theory and Application in Physics, Chemistry, and Biology, Springer, Heidelberg.
[2] Martin, P.S., Siggia, E.D. and Rose, H.A. (1973), Phys. Rev. A 8, 423.
[3] Graham, R. (1978), In Lecture Notes in Physics, Vol. 84, Springer, Berlin.
[4] Janssen, H.K. (1976), Z. Phys. B 23, 377.
[5] Mikhailov, A.Yu. and Loskutov, A.S. (1996), Foundations of Synergetics II, Springer, Berlin.
[6] Hänggi, P. (1989), In Noise in Nonlinear Dynamical Systems; Theory, Experiment, Simulations, (Moss, F., McClintock, P.V.E., eds.), Vol. 1, Cambridge University Press, Cambridge, pp. 307–308.
[7] Martin, P.C., Siggia, E.D. and Rose, H.A. (1973), Phys. Rev. A 8, 423.
[8] Olemskoi, A.I. and Kharchenko, D.O. (2000), Phys. Solid State 42(N3), 532.
[9] Kharchenko, D.O. (2002), Physica A 308, 101.
[10] Parzuski, M., Maslov, S. and Bak, P. (1996), Phys. Rev. E 53, 414.
[11] Risken, H. (1989), The Fokker–Planck Equation, Springer, Berlin.
[12] Gardiner, C.W. (1989), Handbook of Stochastic Methods, Springer, New York.
[13] Landau, L. and Lifshitz, E. (1959), Mechanics, Pergamon Press, Oxford.
[14] Zinn-Justin, J. (1994), Quantum Field Theory and Critical Phenomena, Clarendon Press, Oxford.
[15] Langouche, F., Roekaerts, D. and Tirapegui, E. (1982), Functional Integral and Semi-classical Expansions, Reidel, Dordrecht.
[16] Freidlin, M. (1985), Functional Integration and Partial Differential Equations, Princeton University Press, Princeton, NJ.
[17] Edwards, S.F. and Wilkinson, D.R. (1982), Proc. Roy. Soc. A 381, 17.
[18] Mehta, A. and Barker, G.C. (1994), Rep. Prog. Phys. 57, 383.
[19] Bouchaud, J.-P., Cates, M.E., Prakash, J.R. and Edwards, S.F. (1994), J. Phys. I (France) 4, 1383.
[20] Hadeler, K.P. and Kuttler, C. (1999), Granular Matter 2, 9.
[21] Bak, P. and Sneppen, K. (1993), Phys. Rev. Lett. 71, 4083.
[22] Bak, P. (1997), How Nature Works: The Science of Self-Organized Criticality, Oxford University Press, Oxford.
[23] Jensen, H.J. (1998), Self-Organized Criticality. Emergent Complex Behavior in Physical and Biological Systems. In Cambridge Lecture Notes in Physics, Cambridge University Press, Cambridge.
[24] Halpin-Healy, T. and Zhang, Y.-C. (1995), Phys. Rep. 254, 215.
[25] Vespignani, A. and Zapperi, S. (1997), Phys. Rev. Lett. 78, 4793; Phys. Rev. E (1998) 57, 6345.
[26] Vespignani, A., Dickman, R., Muñoz, M.A. and Zapperi, S. (1998), Phys. Rev. Lett. 81, 5676; Phys. Rev. E (2000) 62, 4564.
[27] Olemskoi, A.I., Khomeko, A.V. and Kharchenko, D.O. (2003), Physica A 323, 263.

[28] Haken, G. (1983), In Synergetics, an Introduction, Springer, Berlin.
[29] Edwards, S.F. and Oakeshott, R.B.S. (1989), Physica A 157, 1080.
[30] Edwards, S.F. (1994), In Granular Matter: An Interdisciplinary Approach (Metha, A., eds.), Springer, New York.
[31] Castro, F., Wio, H.S. and Abramson, G. (1995), Phys. Rev. E 52, 159.
[32] Feder, J. (1988), Fractals, Plenum Press, New York.
[33] Zaslavsky, G.M. (1994), Chaos 4, 25; Physica D (1994) 76, 110; Saichev, A.I. and Zaslavsky, G.M. (1997), Chaos 7, 753.
[34] Kharchenko, D.O. (2002), FNL 2(N4), L273.
[35] Hilfer, R., eds. (2000), Applications of Fractional Calculus in Physics, World Scientific, Singapore.
[36] Olemskoi, A.I. (1998), Phys. Usp. 41, 269.
[37] Chessa, A., Marinari, E., Vespignani, A. and Zapperi, S. (1998), Phys. Rev. E 57, R6241.
[38] Gil, L. and Sornette, D. (1996), Phys. Rev. Lett. 76, 3991.

Chapter 17

EXTREMUM CRITERIA FOR NONEQUILIBRIUM STATES OF DISSIPATIVE MACROECONOMIC SYSTEMS

Mircea Gligor

Department of Physics, National College "Roman Voda", Roman-5550 Neamt, Romania

Abstract

The first part of the chapter is centered upon the concept of entropy in the dynamical description of some socioeconomic systems. Examination of the logarithm of price distribution from several catalogs indicates that this distribution is very close to the Gaussian distribution and so can be derived from the maximizing of the entropy functional associated with this variable. The exponential distribution of incomes, which was reported in the literature to be valid for the great majority of the population, also maximizes the entropy functional when the variable is positive-definite and the mean value is fixed. A kinetic approach developed in the next sections brings substantial support to clarify the exponential distribution of wealth and income, and also enlarges the framework of the analysis including nonequilibrium steady states. Another extremum criterion, namely the minimum production of entropy, is discussed in the context of the economic systems in near-to-equilibrium steady states.

The final sections bring into the discussion the multiplicity of equilibria, the economic cycles and the effect of the random fluctuations. These phenomena, which arise in the description of the economic systems in far-from-equilibrium states, are investigated using methods of nonlinear dynamics and stochastic theory. In particular, the idea that nonperiodicity of economic depressions may be caused by the randomness of the noise-induced transitions between nonequilibrium steady states is discussed.

Keywords: entropy; wealth distribution; nonlinear macroeconomic dynamics; noise induced transitions

1. Introduction

During the whole of the 20th century, several fields of research having a related nature: the mathematical statistics, the economic statistics and the statistical physics have evolved on parallel trajectories and avoiding—as it seems by a tacit agreement—to cross over their own frontiers. The abolition of these artificial limits is the achievement of the last decades, which—through the paradigm 'science of complexity'—brought into focus phenomena whose investigation required the use of the whole arsenal of particular theoretical sciences.

Fax: +40-033-740290.
E-mail address: mgligor_13@yahoo.com (M. Gligor).

The new point of view brought by complexity sciences could be synthesized on the premise that the features of a system do not reduce as the sum of individual properties of its elements. The keywords of this transdisciplinary theory are the *interaction* between components and also their *stochastic* behavior. Most of the methods developed in the framework of statistical physics do not require restrictions on the nature of components, which is an important aspect for the applicability of these methods in the study of human-social communities.

As a part of this field of preoccupation, econophysics has already a proper history, synthesized in numerous articles and books published in recent years. Econophysics has brought to economics new tools for fitting and analyzing the empirical data sets and, also, it has opened new ways, less investigated, in macro- and microeconomic modeling; in this way, many notions and concepts developed in statistical physics have escaped the restricted field in which they were elaborated and today they are included in the common language of physicists and biologists as well as economists and sociologists. From this class of notions, in the early sections we focus on the transdisciplinary concept of entropy.

Introduced in the 19th century by R. Clausius in order to describe the sense of evolution of the natural processes and identified by L. Boltzmann and J. Willard Gibbs as a measure of randomness or disorder in the system, entropy has demonstrated its deep significance in the context of the science of the 20th century in connection with the information theory and the thermodynamics of irreversible processes. Note here one of the first 'incursions' of the concept of entropy in the economics performed by the pioneering work of Georgescu-Roegen [1].

Following a line opened by Montroll [2], in Section 2 we verify—starting from some empirical data—that the entropy functional appears in a natural way in the distribution of merchandising prices, and enlarging the analysis, in the distribution of incomes (or wealth) between individuals. In the study of the second problem, in Section 3 we use some actual statistical tools like the Gini coefficient and the Lorenz curve, which are briefly discussed in relation to the empirical datasets referring to some East European countries.

Rigorously, the variational principle of maximizing the entropy is relevant only for the isolated systems in equilibrium states. This cannot be the case for the economic systems, which are essentially open and run out of equilibrium. For these systems, the nonequilibrium thermodynamics introduces another variational principle, namely minimum entropy production. The kinetic model developed in Section 4 leads to the uniting of classical kinetic theory (Boltzmann picture) and the thermodynamics of irreversible processes (Onsager picture) (Section 5). Also discussed are some derivations from minimum entropy production, such as the quasilinear macroeconomic laws in the proximity of equilibrium.

In Section 6, the problem of out-of-equilibrium economic states is placed in the context of nonlinear dynamics, which creates the premises to investigate complex economic phenomena like the multiplicity of equilibrium, the fluctuations around the steady state and the succession of the economic cycles. Our point of view is that the multiple equilibria and the economic cycles result in an intrinsic way from the nonlinear character of the mechanisms that rule economic development. The transition of the system from a certain asymptotic equilibrium to another can establish a temporary economic decreasing, usually accompanied by a period of recession. The statistical data show that the recessions are preceded by a large fluctuation in the production addressed to the consumption [3]. In Section 7, the evolution from one steady state to another is modeled through the scenario of noise-induced transitions. The last section draws some conclusions.

2. The maximum principle of entropy and the distribution of merchandizing prices

We remind the reader that a key concept in classical statistical mechanics—as results from the pioneering works of the 'founding fathers' of the field, Ludwig Boltzmann and Josiah Willard Gibbs—is the entropy functional:

$$H = -\sum_{i=1}^{N} p_i \log(p_i), \text{ with } p_i \geq 0, \text{ and } \sum_{i=1}^{N} p_i = 1. \tag{1}$$

Frequently, the index i denotes a possible state of the physical system and p_i is the probability that the system achieve the i-th state. The definition of H may be extended to the case that a continuous variable, x, represents the states. If $-\infty < x < \infty$, then:

$$H = -\int_{-\infty}^{+\infty} p(x)\log(p(x))\mathrm{d}x, \text{ with } \int_{-\infty}^{+\infty} p(x)\mathrm{d}x = 1. \tag{2}$$

It is easy to show that when the variance $\sigma^2 = \langle x^2 \rangle$ is fixed, then the function that maximizes the entropy is the Gauss distribution:

$$p(x) = (2\pi\sigma^2)^{-1/2}\exp(-x^2/2\sigma^2). \tag{3}$$

When x is restricted to the positive half-line and the mean value $\mu = \langle x \rangle$ is given, it can be shown that the distribution that maximizes the entropy is the exponential:

$$p(x) = (1/\mu)\exp(-x/\mu). \tag{4}$$

For 70 years, the domain of entropy functional was limited to the field of statistical mechanics until, in 1948, Claude Shannon identified this functional as an ideal measure of the information transferred in a communication system, and so the entropy appeared in the characterization of the output of a sociotechnical system.

The communication systems considered by Shannon were composed of a message-input element, a transmission channel and a message-output element. Since the entropy functional appeared in a natural way for the information rate in such a system, we might expect that this functional could also be important in other socioeconomic systems that are composed of an analogous set of three components.

An important economic system is the merchandising system. The goods flow into a distribution center of the retailing firm, remain temporarily as an inventory, and finally they are delivered to the customer. In this way, a company's profit depends upon the flowthrough rate of goods and on the price associated with the goods. The similarity between the merchandizing flows and Shannon's communication system suggests that the entropy functional may appear in the analysis of the merchandizing prices.

Such analysis was first performed by Montroll [2] using the distribution of prices from Sears Roebuck catalogs for years 1916, 1924–1925 and 1974–1975, and his conclusions were confirmed by numerous subsequent empirical studies. In order to exemplify the central idea (and to outline the generality of the conclusion), in Fig. 1 we present a histogram of prices distribution that we have recently performed using data supplied by some Romanian catalogs. We studied the logarithm of price, as $\log P$ is especially sensitive to relative price variations: $\Delta P/P \approx \Delta(\log P)$.

A simple examination of the histograms plotted in Fig. 1 indicates that the distribution of $\log_2(P_i)$ (P_i being the price of the i-th item) is very close to the Gaussian distribution. Investigating the mean and the variance of $\log_2(P_i)$, one found that, although the mean increased year by year (due to hyperinflation registered in Romania during the period investigated), the variance $\sigma^2_{\log P}$ was almost invariant (see Table 1), thus being equivalent to an economic 'constant of motion'.

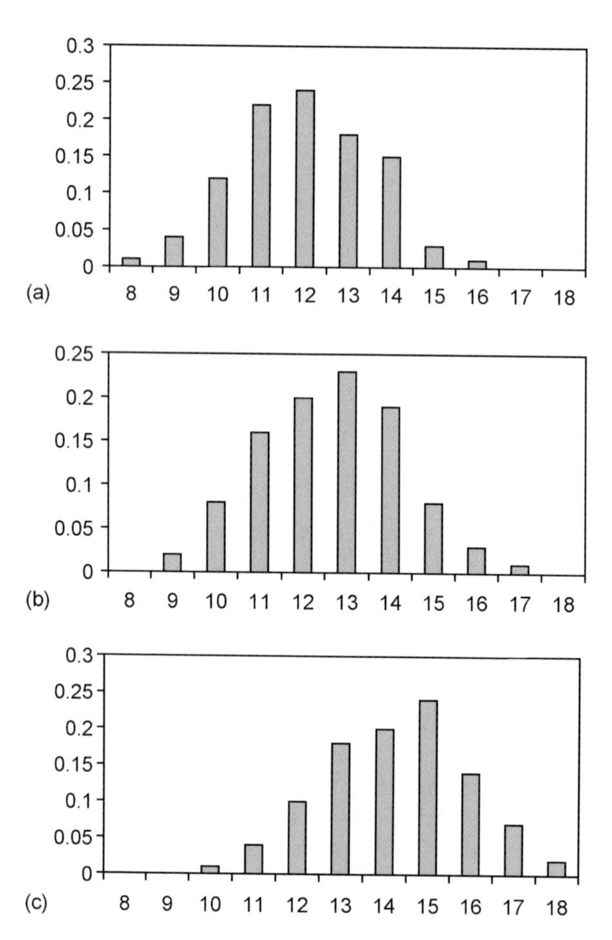

Fig. 1. Histograms of price distributions for years: (a) 1994; (b) 1997; (c) 1999. The fraction of items in each price range is plotted as a function of $\log_2 P$. Source: Romanian Statistical Yearbook [4].

Table 1
The mean value, the variance and the standard deviation of prices distribution

Year	$\langle \log_2 P \rangle$	σ^2	σ
1994	12.49	2.42	1.59
1997	13.16	2.69	1.64
1999	14.77	2.77	1.66

The samples refer to some merchandise prices of consumer goods in Romania in a period characterized by hyperinflation (1993–1999). Source: Romanian Statistical Yearbook [4].

The quasi-invariance of variance suggests that the Gaussian distribution of log (P_i) maximizes the entropy functional associated with this variable. Defining $\bar{P}$ by relation:

$$\log(\bar{P}) = \langle \log(P_i) \rangle,$$

the entropy functional will be:

$$H = -\int_0^{\infty} \log^2(P/\bar{P})\, p(\log(P/\bar{P}))\, \mathrm{d}(\log(P/\bar{P})), \tag{5}$$

where $p(x)$ is the Gaussian distribution from Eq. (3). Note that:

$$U = \log\left(\frac{P}{\bar{P}}\right)$$

is similar to the *utility function* of classical economics, and H can be viewed as a weighted average of the square of this function.

3. The distributions of incomes and wealth

The distribution of money between the economic agents in a society and the rate at which this process occurs is one of the central points in economics that have attracted the attention of many researchers. More than a century ago the Italian sociologist Pareto studied the distribution of personal incomes for the purpose of characterizing a whole country's economic status [5].

Since then, many other studies have been reported, frequently with controversial results. Thus, while Pareto found power-law cumulative distributions with exponents close to -1.5 for several countries [6], checked the same statistics and reported that power laws actually hold, but the values of the exponents vary from country to country. On the other hand, Montroll [2] analyzed the USA's personal income data for the year 1935/36 and found that only the top 1% of incomes follow a power law, while the rest, who are expected to be salaried, follow a lognormal distribution. Thus, the probability that one's annual income lies between x and $x + \mathrm{d}x$ is:

$$p(x)\mathrm{d}x = (2\pi\sigma^2)^{-1/2} \exp\left(-\frac{\log^2(x/\bar{x})}{2\sigma^2}\right) \bar{x}\frac{1}{x}\mathrm{d}x. \tag{6}$$

The factor $(1/x)\mathrm{d}x$ from Eq. (6) can be viewed as the variation of utility function:

$$\mathrm{d}U = (\mathrm{d}x)/x.$$

Indeed, the same process involving a transfer of money $\mathrm{d}x$ has a different meaning to persons of different levels of income, and the transactions made by persons of different income levels might be equivalent only if they involved the same fraction of the income of the participants.

If the distribution of the utility function:

$$U(x) = \log\left(\frac{x}{\bar{x}}\right)$$

is Gaussian, then it would follow from the maximization of the entropy functional

$$H = -\int p(U)\log(p(U))\mathrm{d}U \tag{7}$$

under the auxiliary conditions that $p(U)$ is normalized and $\langle U^2 \rangle = \text{constant}$. These conditions were found to be valid for the set of data analyzed by Montroll [2].

As there has been no established theory for income distributions, the problem of wealth distribution between individuals or, more generally, between countries, rests in actuality. (In 1998, The Royal Swedish Academy of Sciences decided to award the Nobel Prize in economic sciences to Amartya Sen, from Trinity College, Cambridge, for his contributions to welfare economics.) The classical theory proposed by Gibrat [7] assumed the time evolution of each person's income to be approximated by a multiplicative stochastic process considering that a random process proportional to the amount of the present income can approximate the increase and decrease of income. By this assumption the resulting distribution of income follows a lognormal distribution, which is consistent with the result of Montroll for salaried people. However, this theory apparently fails to explain the more interesting part of the distribution, the power-law tails. In a series of papers, Solomon et al. showed that a stochastic dynamical model, based on the Lotka–Volterra system, led to power-law distributions [8,9]. In this section we focus only on the first part of the distribution, which, in the light of recent data, seems to be exponential rather than lognormal [10,11].

The inequality of wealth distribution is usually measured by the Gini coefficient. A straightforward graphical interpretation of the Gini coefficient is the Lorenz curve, which is the thick curve in Fig. 2. The horizontal axis plots the cumulative percentage of the population whose inequality is under consideration, starting from the poorest and ending with the richest. The vertical axis plots the cumulative percentage of income (or expenditure) associated with the units on the horizontal axis.

In the case of a completely egalitarian income distribution in which the whole population has equal incomes, the Lorenz curve would be the dashed straight 45° line. When inequality exists, the poor population has a proportionately lower share of income compared with the rich population, and the Lorenz curve may look like the above thick curve below the 45° line. As inequality rises, so the thick curve moves towards the bottom right-hand corner. The Gini coefficient is the area A between the 45° line and the Lorenz

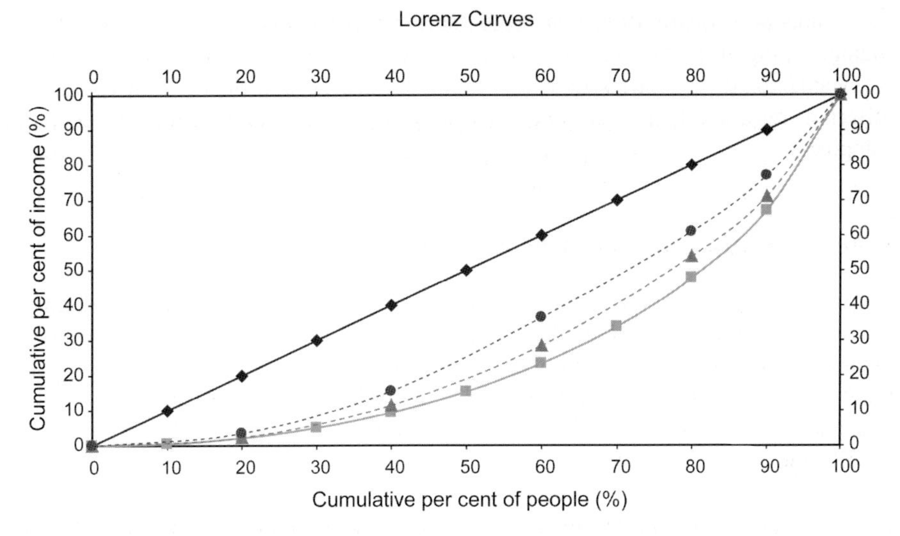

Fig. 2. Lorenz curves. Solid lines: the straight 45° line (Lorenz curve in the case of a completely egalitarian income distribution) and the Lorenz plot in the case of an exponential distribution. The Gini coefficient—the area between the two solid lines divided by the total area under the 45° line—is $\frac{1}{2}$ (50%). Dashed lines: the Lorenz plots for Romania (●, 1991; ▲, 1997). Source: http://www.wider.unu.edu/wiid/wwwwiid.htm.

curve divided by 1/2, the total area under the 45° line. The Gini coefficient may be given as a proportion or percentage.

An interesting empirical aspect pointed out by Dragulescu and Yakovenko in their works cited above is that for the great majority of a population (≈95%) the distribution of individual income follows an exponential law. The horizontal axis plots the cumulative fraction of the population with income below r, $X(r)$. The vertical axis plots the cumulative fraction of income $Y(r)$ associated with the units on the horizontal axis:

$$X(r) = \int_0^r \rho(r')\mathrm{d}r'; \; Y(r) = \int_0^r r'\rho(r')\mathrm{d}r' \Big/ \int_0^\infty r'\rho(r')\mathrm{d}r'. \tag{8}$$

The Gini coefficient is given by: $G = 2\int_0^1 (X - Y)\mathrm{d}X$. When the distribution is completely egalitarian, $G = 0$. If the society's total income accrues to only one person, leaving the rest with no income at all, then $G = 1$ (or 100%). For the exponential distribution, $G = 0.5$. This is in good agreement with the values: 0.64–0.68 for the United Kingdom and 0.47–0.56 for the USA found by Dragulescu and Yakovenko [11]. Note that our analysis refers only to the individual income, not family (household) income. The equilibrium value of the Gini coefficient for the latter is $3/8 = 0.375$, as opposed to $1/2 = 0.5$ for the former.

An interesting evolution presents the Gini coefficient for the East European countries [12]. During several years, G for the individual income increased from 20.42 to 24.58 in Romania, from 24.32 to 30.27 in Poland and from 23.26 to 34.2 in Hungary. Even if these values are smaller than those corresponding to developed countries, the general trend of increase is remarkable; it means an approach to the value 0.5 corresponding to the exponential distribution that seems specific to the market economies based on encounters/competition between individuals and companies. Thus, the rapid growth of G is explicable for the countries in which, until the last decade of the 20th century, the totalitarian regimes had minimized the role of (fair) competition in economic development.

4. A kinetic insight. The Boltzmann kinetic equation

Let us consider N individuals labeled by the index j ($j = 1...N$). The first problem that we have to solve is to choose the relevant variables and to construct the social phase space (σ-phase space) for the space number density, ρ. As macroscopic variables the total amount of money (M), the *natural* unemployment (U), i.e. the unemployment corresponding to the abstention of inflation, or any other quantities that are conserved at equilibrium could be chosen. Firstly, in accordance with Ref. [13], we propose a dimensionless variable x_j called *social position* or *dominance*, as an attribute possessed by each individual, which might refer to any kind of skill, strength or endowment (physical, cultural, economical, technological, *etc.*). The social position can vary due both to encounters (interactions) between the individuals themselves and to global (external) influences induced by the social policies. The rate of change of the social position characterizes the individual's mobility/adaptability and it will be understood as another coordinate of the σ-phase space, v_j. Obviously, we have not restricted ourselves to define these variables over a finite range or over the whole real axis. We consider that the amount of money possessed by each individual is a quantity positive-semidefinite: $m_j \geq 0$ for any $j = 1...N$. As a working assumption we suppose that this variable is proportional to the square of the rate of change of the dominance: $m_j = v_j^2 = (\mathrm{d}x_j/\mathrm{d}t)^2$. In this way, the description of the social system is close to the statistical picture of the molecular systems, leading to the kinetic equation:

$$\frac{\partial \rho}{\partial t} = -v\frac{\partial \rho}{\partial x} - f\frac{\partial \rho}{\partial v} + \int \widehat{\sigma}_T g[\rho'\rho'_1 - \rho\rho_1]\mathrm{d}v_1, \tag{9}$$

where $\rho(x,v,t)\mathrm{d}x\mathrm{d}v$ means the number of individuals with coordinates in the ranges $[x,x + \mathrm{d}x]$ and $[v,v + \mathrm{d}v]$. The first and the second terms of the r.h.s. of Eq. (9) refer to the nondissipative flux in phase space called *streaming*: even if no collisions occurred in the time $\mathrm{d}t$, all individuals at the point (x, v) would move to a new spatial position $(x + v\mathrm{d}t, v + f\mathrm{d}t)$ where f is the force per unit mass. The third term of the r.h.s. of Eq. (9) is the *source term*, representing the dissipative effect of encounters. We have introduced the notations: $g = |v_1 - v|$, $\rho = \rho(x, v, t)$, $\rho' = \rho(x, v', t)$, $\rho_1 = \rho(x, v_1, t)$, $\rho'_1 = \rho(x, v'_1, t)$. In the framework of the classical theory, the linear operator $\widehat{\sigma}_T$ is related to the differential scattering cross-section $\sigma(\Omega,g)$ where Ω is the solid angle:$\widehat{\sigma}_T[\cdot] = \int \mathrm{d}\Omega\sigma(\Omega, g)[\cdot]$. Finally, it is easy to prove the H-theorem: defining the *H-function* by:

$$H = \int\int \rho \ln \rho \,\mathrm{d}x\,\mathrm{d}v, \tag{10}$$

and following the usual method, i.e. taking the time derivative of H, making some changes of variables and adding the equations, one gets:

$$\frac{\mathrm{d}H}{\mathrm{d}t} = \frac{1}{4}\int\int\int \widehat{\sigma}_T g \rho' \rho_1' \left[\left(1 - \frac{\rho\rho_1}{\rho'\rho_1'}\right)\ln\frac{\rho\rho_1}{\rho'\rho_1'}\right]\mathrm{d}x\,\mathrm{d}v\,\mathrm{d}v_1 \leq 0, \tag{11}$$

with the equality holding for (and only for) $\rho = \rho^0$ that satisfies:

$$\rho^0\rho_1^0 = \rho'^0\rho_1'^0. \tag{12}$$

Note that the Boltzmann equation is nonlinear and, because it accounts fully for binary encounters, it is useful for describing processes both near and far from equilibrium.

Taking the logarithm in Eq. (12) it follows that only the distributions that satisfy

$$\ln \rho^0 + \ln \rho_1^0 = \ln \rho'^0 + \ln \rho_1'^0 \tag{13}$$

correspond to a constant value of H.

The density functions that satisfy Eq. (13) have the general form:

$$\rho^0(v_i) = C_1 \exp\left(-\frac{v_i^2}{\langle v^2\rangle}\right); \qquad \text{(Gauss–Maxwell), or :} \tag{14}$$

$$\rho^0(m_i) = C \exp\left(-\frac{m_i}{T}\right) \qquad \text{(Boltzmann–Gibbs),} \tag{15}$$

where $\langle\rangle$ denotes a σ-space average, and T is an effective temperature equal to the average amount of money per economic agent. Thus, the exponential distribution of money discussed in the previous section was explicitly found.

5. Uniting the Boltzmann and Onsager pictures and the minimum production of entropy

A relationship between Boltzmann's kinetic description and Onsager's linear thermodynamics can be seen if we restrict attention to the kinetic equation (9) in the neighborhood of equilibrium. Although the kinetic equation is nonlinear, if we look only at small deviations around equilibrium (in the absence of an external field) we can write:

$$\rho(x, v, t) = \rho^0(v) + \Delta\rho(x, v, t), \tag{16}$$

where $\Delta\rho(x,v,t)$ is assumed to be a small change in σ-space density. Substituting Eq. (16) into Eq. (9) and retaining only terms linear in $\Delta\rho$, we obtain the linearized kinetic equation:

$$\frac{\partial}{\partial t}\Delta\rho = -v\frac{\partial}{\partial x}\Delta\rho + \int \widehat{\sigma}_T g \rho^0 \rho_1^0 \left[\frac{\Delta\rho'}{\rho'^0} + \frac{\Delta\rho_1'}{\rho_1'^0} - \frac{\Delta\rho}{\rho^0} - \frac{\Delta\rho_1}{\rho_1^0}\right]\mathrm{d}v_1. \tag{17}$$

In order to complete the Onsager picture, we define *the entropy density* in σ-space: $s = -\rho \ln \rho$. The intensive variable conjugate to $N(v, x) = \rho \, dv \, dx$ is:

$$F(\rho) = \frac{\partial s}{\partial \rho} = -(\ln \rho + 1), \tag{18}$$

and the local thermodynamic force in σ-space around equilibrium is given by:

$$X = F(\rho) - F(\rho^0) = -\ln \frac{\rho}{\rho^0} \cong -\frac{\Delta\rho}{\rho^0}. \tag{19}$$

After some algebraic processing, introducing the operator:

$$L[X] \equiv (v\rho)\frac{\partial}{\partial x}X - \int \Theta(v, v_1)X_1 dv_1 \tag{20}$$

the linearized kinetic equation takes the form:

$$\frac{\partial}{\partial t}(\Delta\rho) = L[X]. \tag{21}$$

Eq. (21) is the Onsager regression equation at the kinetic level of description.

In accordance with Onsager's, linear theory, in the neighborhood of equilibrium the generalized fluxes (rates of extensive parameters) are linear functions of the generalized forces (gradients of intensive parameters). The macroeconomic indicators used in Ref. [12] are: *money supply*, M and *the rate of unemployment*, U (as generalized fluxes) and *consumer index price for all items*, P and *the rate of discount of the central banks*, D (as generalized forces). Thus, the Onsager linear equations near to equilibrium can be written as:

$$\begin{aligned} M &= L_{11}P + L_{12}D \\ U &= L_{21}P + L_{22}D. \end{aligned} \tag{22}$$

As an example of proving these relations we use several data supplied by the United Nations Statistic Division referring to France between January and August 2000. The linear dependence $P = P(U)|_{D=\text{const.}}$, known in economics as *the Phillips diagram*, is plotted in Fig. 3. Similar results were reported in Ref. [12] for Hungary, Switzerland and Spain.

We introduce now, according to the classical thermodynamics of irreversible processes, the *production of entropy* as:

$$\dot{s} = \sum_{i=1}^{n} X_i J_i = PM + DU, \tag{23}$$

where X_i and J_i are the usual notations for the generalized forces and the generalized fluxes. Considering a steady state in which $M = L_{11}P + L_{12}D = 0$, it is easy to show that the partial derivative:

$$\frac{\partial \dot{s}}{\partial P} = 2(L_{11}P + L_{12}D) = 0, \tag{24}$$

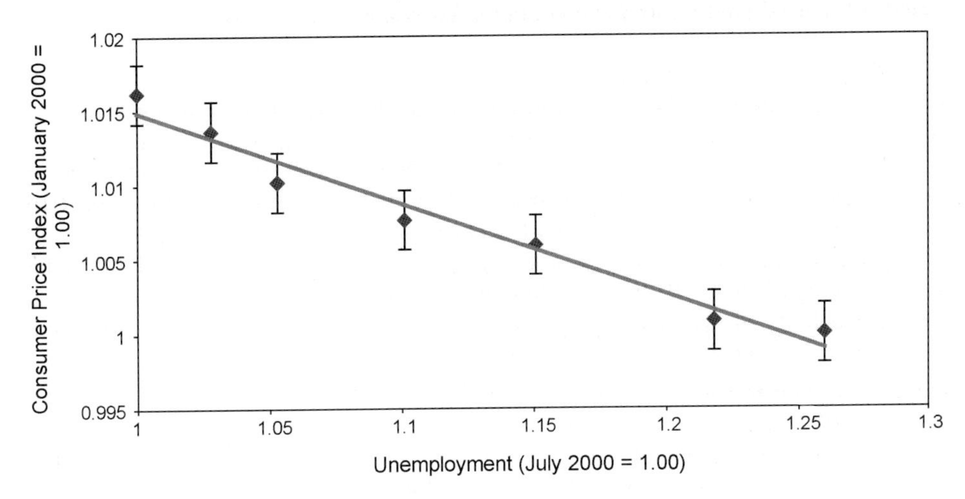

Fig. 3. Consumer price index (P) versus unemployment (U) for France. The empirical data were registered at UNSD database from January to August 2000. Error bars are bootstrap 94% confidence intervals. Source: United Nation Statistics Division: http://esa.un.org/unsd/mbsdemo/mbssearch.asp.

which demonstrates the existence of an extremum of the function $\dot{s}$. As the second derivative is positive, the function displays a minimum (Prigogine's theorem of minimum entropy production). We come to the conclusion that, in given external conditions, in an open system evolving in a steady state close to the thermodynamic equilibrium, the speed of entropy growth on the irreversible internal processes reaches a constant minimal positive value.

In order to find the economic significance of $\partial\dot{s}/\partial t$, let us recall the four steps of the economic cycle: crisis, development, boom and depression. The second could be considered the economic equivalent of the dissipative steady state of physical systems, in which the production of entropy is minimal. From this point of view, $\partial\dot{s}/\partial t$ can be considered as a control parameter, which measures the *deviation* of the macroeconomic system from equilibrium, i.e. the stage of the economic cycle. It has maximal values in the stage of depression and minimal values in the stage of development.

It could be noted here the possibility of choosing another extensive quantities as generalized fluxes (e.g. the rate of production of certain goods or the flux of wastes). For example, Georgescu-Roegen [1] used such kinds of generalized fluxes, referring to several ecological problems related to the economic processes.

Generally, Nature processed the living organisms to function in stable conditions with minimal entropy production, which insures an ecological protection of the environment. But if the organism is subject to an effort, for example high-speed running, it might develop a maximal power for short periods of time with an optimal production of entropy.

The difference is an excess entropy production similar to the function of the thermotechnical systems in the full-power regime. Such a function is not specific to any of the physical systems or to the biological systems or the economic ones because it disturbs

Nature by supplementary energy dissipations, which may occur over relatively short periods of time. We return to this point in the last section.

6. Economic systems in far-from-equilibrium steady states. Economic cycles

A distinctive feature of the Western market economies is the short-run fluctuations in output around trends of slow but persistent growth over time. These short-term fluctuations in output are often referred to in economics as the 'business cycle'.

A definition of the business cycle was given in the National Bureau of Economic Research study in 1946: "Business cycles are a type of fluctuation found in the aggregate economic activity of nations that organize their work mainly in business enterprises: a cycle consists of expansions occurring at about the same time in many economic activities, followed by similarly general recessions, contractions, and revivals that merge into the next expansion phase of the cycle; the sequence of changes is recurrent but not periodic; in duration cycles vary from more than one year to ten or twelve years; they are not divisible into shorter cycles of similar character with amplitudes approximating their own" [14].

One notes from this definition that the cycle arises primarily through the activities of 'business enterprises' (i.e. firms), and also that changes in output in individual sectors of the economy or individual firms tend to be positively correlated over the cycle.

In the final sections, we aim to investigate a possible connection between the emergence of economic cycles and the nonlinear dynamics describing the systems in far-from-equilibrium states, which results in the possibility of multiple equilibria.

In the Walrasian model of general economic equilibrium, consumers choose to demand and supply goods to maximize a utility function subject to the constraint that the value of what they demand must equal the value of what they supply. Producers choose to demand and supply goods to maximize profits subject to the restrictions of a production technology. An equilibrium of this model is a vector of prices, one for each good, which the agents all take as given in solving their maximization problems, such that demand is equal to supply for each good.

Following the same Walrasian line, we suppose a disequilibrium adjustment process often called 'groping'. In it, an auctioneer adjusts prices systematically by raising the prices of goods in excess demand and lowering those of goods in excess supply. Samuelson [15] formalized this process as the system of differential equations:

$$\frac{\mathrm{d}p_j}{\mathrm{d}t} = f_j(p),\ j = 1, 2, \ldots, n, \tag{25}$$

where $p = (p_1, p_2, \ldots, p_n)$ is the price vector associated to n goods. The prices on the r.h.s. of Eq. (25) depend not on time explicitly, but on the stocks of goods, whose evolution, in turn, depends on prices.

In order to study only some general properties of solutions, one might suppose the simplified case of a single asset, the nonlinear differential equation (NLDE) having the form:

$$\frac{\mathrm{d}y}{\mathrm{d}t} = -\lambda y + s y^{\alpha}. \tag{26}$$

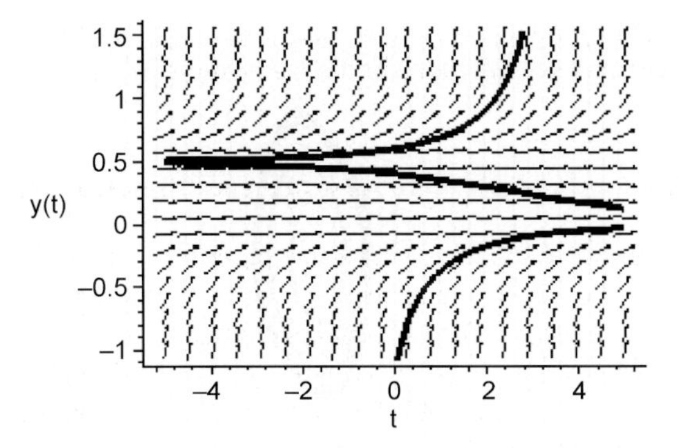

Fig. 4. The solutions of Eq. (26) for $\alpha = 2$; $s = 1$; $\lambda = 0.5$. The asymptotic behavior of solutions is extremely sensitive to the initial conditions. The system can evolve following one of the two asymptotic solutions or can shift from one to another.

Even in this simplified framework, the behavior of solutions (Fig. 4) suggests that, depending on the initial conditions, not one but many different asymptotic solutions are possible (in the figure there are two asymptotic solutions corresponding to the same values of parameters α, s and λ). Also, we see that the system can switch from one asymptotic behavior to another. This result illustrates (in a schematic way, only) the concept of multiple equilibria.

We expect that more complicated dynamics lead to more complicated movement of solutions in the phase space. We mention here the large class of 'Keynesian' models [16]. As a particular case, let us consider a highly stylized model of the interdependence between the savings (S) and the investments (I) in the activity of an economic agent, which might be formalized by the set of equations:

$$\mathrm{d}S/\mathrm{d}t = -I$$

$$\mathrm{d}I/\mathrm{d}t = \gamma I - kS. \tag{27}$$

In addition to other models of this kind, we suppose that the coefficient k from the 'reaction' factor kS depends as a linear function both on S and I:

$$k = a + \gamma SI,$$

and so the set of equations becomes:

$$\mathrm{d}S/\mathrm{d}t = -I$$

$$\mathrm{d}I/\mathrm{d}t = \gamma I - (\omega^2 + \gamma SI)S' \tag{28}$$

where we have introduced the notation $\omega^2 \equiv a$ in order to have a formal analogy with the equation of the nonlinear oscillator from physics. The behavior of the solution exhibits a limit cycle into the plan (S, I) (Fig. 5).

Finally, we accredit the idea that the nonperiodicity of economic depressions might be caused by the transition of the system from one limit cycle to another, the last with quite different parameters comparative to the former. In the next section, we propose a possible

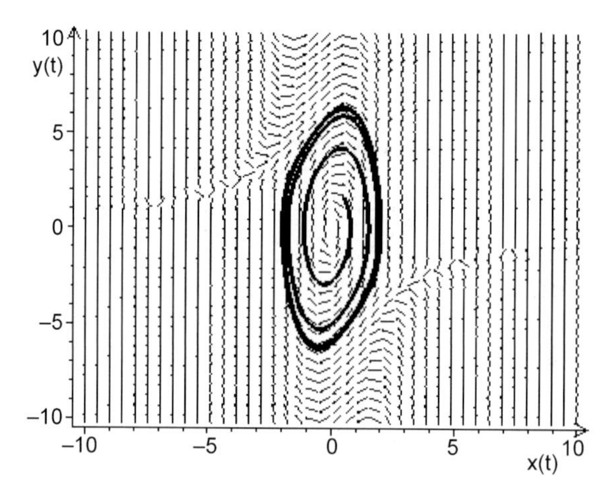

Fig. 5. The asymptotic behavior of the solutions of the nonlinear set of equations (27) in the phase space defined by the variables: $x \equiv I$ (investments), and $y \equiv S$ (savings). The solutions display an evolution to a limit cycle, whose parameters depend strongly on the initial conditions. Here: $\omega_0 = 3$; $\gamma = 1$; $x(0) = 0.5$; $x'(0) = 1.6$.

scenario for this kind of transition, which seems to occur close to the end of the economic cycle.

7. Noise-induced transitions between nonequilibrium steady states

Physicists have noted, in several contexts, the possibility of a 'critical state', in which independent microscopic fluctuations can propagate so as to give rise to instability on a macroscopic scale. This is a state in which chain reactions initiated by local disturbances neither damp out over a short distance (the 'subcritical' case) nor propagate explosively so that the system cannot remain in that state (the 'supercritical' case). In this section we propose to study the role of the fluctuations in the neighborhood of the critical points, which seem to appear at the end of an economic cycle.

Let us consider, for simplicity, a single production process P, which produces a flux y of a certain ware C. In the simplest form, the rate of increasing of the net product flux depends on the actual flux:

$$dy/dt = ky. \tag{29}$$

Eq. (29) allows for either an exponential growth or decay of the net product flux because we do not take into account some marginal effects such as the *market saturation*: as more quantities of C enter into the market, the offer increases so the price decreases, which in turn influences dy/dt. It is thus more natural to consider k as a function of y:

$$k = \beta - \gamma y.$$

Here, the second term includes the competition effect and is similar to the 'struggle for life' term in the Malthus–Verhulst model of population dynamics (see, e.g. Ref. [17]).

Substituting in Eq. (29) and rescaling, we obtain the NLDE of the process in the canonical form:

$$\mathrm{d}q/\mathrm{d}t = -q^2 + \zeta q. \tag{30}$$

According to the stability theory, for $\zeta < 0$, Eq. (30) has only one stationary solution $q^0 = 0$, which is stable. At $\zeta = 0$, a transcritical bifurcation appears: the solution $q^0 = 0$ becomes unstable and a new stable steady-state branch arises, with $q^0 = \zeta$. The qualitative change suffered by the system when it goes through the bifurcation point is similar to a thermodynamic phase transition. According to bifurcation theory terminology, this is a 'soft' transition.

The parameter ζ is supposed to be subject to fluctuations, being a Gaussian white noise with mean ζ_0 and variance σ. The stochastic differential equation (SDE) associated with Eq. (30) is:

$$\mathrm{d}q = (-q^2 + \zeta q)\mathrm{d}t + \sigma q \mathrm{d}W, \tag{31}$$

or:

$$\mathrm{d}q = f(q)\mathrm{d}t + g(q)\mathrm{d}W,$$

where W is a stochastic Wiener process. (For simplicity, we have dropped the index of ζ.)

In the same way as in the deterministic case, we compute the stationary solutions of (31), i.e. the stationary points of the SDE, imposing $f(q) = g(q) = 0$. One such point is $q = 0$, signifying that the ceasing of production (collapse or bankruptcy) is always possible for a system described by such an equation.

The question that naturally arises is whether, in addition to the stationary point $q = 0$, the SDE (31) has another stationary solution. The simplest way of solving this problem is by analyzing the Fokker–Planck equation (FPE) associated with SDE:

$$(\partial/\partial t)P(q,t) = -(\partial/\partial q)[(\zeta q - q^2)P(q,t)] + (\sigma^2/2)(\partial^2/\partial q^2)[q^2 P(q,t)]. \tag{32}$$

A special interest presents the stationary solution of this equation:

$$(\partial/\partial t)P_{\mathrm{st}}(q,t) = 0;$$

$$P_{\mathrm{st}}(q,t) = P_{\mathrm{st}}(q),$$

where $P_{\mathrm{st}}(q)$ will be considered a probability density if and only if it is normalizable, i.e. its integral over the range $[0, \infty)$ is finite. The stationary solution of the FPE, having the form:

$$P_{\mathrm{st}}(q) \approx q^{2\zeta/\sigma^2 - 2} \exp(-2q/\sigma^2)$$

is found to be integrable over $[0, \infty)$ only if $\zeta > \sigma^2/2$. In other words, the stationary probability distribution exists only if $\zeta > \sigma^2/2$. After normalization we get:

$$P_{\mathrm{st}}(q) = (2/\sigma^2)^{2\zeta/\sigma^2 - 1}\Gamma^{-1}(2\zeta/\sigma^2 - 1)q^{2\zeta/\sigma^2 - 2} \exp(-2q/\sigma^2). \tag{33}$$

A remarkable aspect of the result is the drastic change in the character of the stationary distribution for: $\zeta = \sigma^2$: if $\sigma^2/2 < \zeta < \sigma^2$, $P_{\mathrm{st}}(q)$ is divergent for $q = 0$, while for $\zeta > \sigma^2$, $P_{\mathrm{st}}(q = 0) = 0$.

Summarizing the previous sections and taking into account also the influence of the external noise, the following behaviors of the system are predicted:

(a) For $\zeta < 0$, the stationary point $q^0 = 0$ is stable, making up the thermodynamic branch of evolution on which the fluctuations are damped and do not lead to structural changes into the system. This feature can be extended for $0 < \zeta < \sigma^2/2$ (the domain of small fluctuations) where we have a new stable solution, $q^0 = \zeta$, after the crossing through the transcritical bifurcation at $\zeta = 0$.

(b) The critical value $\zeta = \sigma^2/2$ can be considered a threshold over which the stationary probability distribution $P_{st}(q)$ arises. For $\sigma^2/2 < \zeta < \sigma^2$, $P_{st}(q)$ is divergent for $q = 0$. Even if the solution $q = 0$ is no longer stable, it remains the most probable. For $\zeta > \sigma^2$, a new change occurs in the aspect of $P_{st}(q)$, and the value $\zeta = \sigma^2$ becomes a transition point produced only by the external noise (in concordance with the usual classifications, this is a 'hard' transition).

We consider that the second kind of transition is related to the shift of the economic system between two nonequilibrium steady states. Such transitions stand out in bold relief in the evolution of the macroeconomic indicators of emerging-market economies, for which the economic cycles follow each other more rapidly than for the developed economies (Fig. 6). As the phenomenon is induced by random fluctuations, the outcome is indeterminate; it is not unique and predictable. Nonetheless, it is reasonable to assume that the distribution of transitions depends on the distribution of fluctuations. This relation could explain the power-law distribution of recessions reported in Ref. [18].

8. Some final remarks

The use of the extremum criteria related to the entropy functional sheds light on several questions of interest in economics and social statistics. For example, as we have shown in Section 2, using their marketing intuition, the dealers create catalogs with goods prices so that year after year the price distribution maximizes the entropy functional associated with

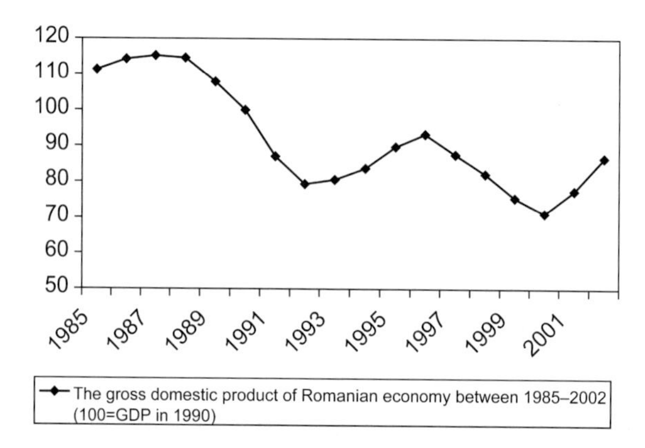

Fig. 6. The gross domestic product of Romanian's economy between 1985 and 2002 (100 = the GDP in 1990). Source: the World Bank site: http://www.worldbank.org/data.

the logarithm of price. Also, as the empirical investigations have shown, the distribution of wealth (or incomes) between individuals in a social medium is well fitted by the exponential distribution, which maximizes the entropy functional when the variable is restricted to the positive half-line and the mean value is given (Section 3). The kinetic approach developed in Sections 4 and 5 supports these results and offers the possibility to extend the analysis over the steady states in the vicinity of equilibrium, where the second principle of thermodynamics takes the form of another extremum criterion, namely the minimum production of entropy.

From the viewpoint of economic science, the importance of the study of entropy overcomes the strict thermodynamic limits because it puts an end to the supremacy of mechanical conceptions in science and philosophy. The new science of thermodynamics made its first appearance as the physics of economic value, and the law of entropy is the most economic of all the natural laws. Indeed, if the economic process were not irreversible, mankind would only have to accelerate the speed of using the supply of natural resources to compensate for the growth of population.

The study of entropy touches fundamental aspects that preoccupy all the nations of the world, in particular the pollution and the continuous rise of the population. It seems to be normal that the appearance of the pollution phenomena has taken by surprise an economical science tributary to old mechanical conceptions. If the economists had understood at the right time the entropic nature of the economic process, they could have prevented their colleagues from the technical domain that 'the bigger and better' machines drive immediately to a 'bigger and better' pollution.

In the progress of the economic process through the years, human society has covered a long period of 10^4 years following the principle of ecological minimum entropy production, in circumstances of dissipative interactions with strict necessaries to evolution with a less finite speed. The petrol civilization, which appeared 10^2 years ago, is a disturbance from the ecological point of view, as in the case of the technical or biological systems that develop a maximal power only in short periods of time.

Of all the great economists, Alfred Marshall was the first to intuit that biology, not mechanics, is the real source of inspiration for the economist. On the same theme, Alfred J. Lotka, a specialist in physical biology, explained why the economic process is the continuation of the biological one: in the biological process, the human being, like any other living creature, uses only the instruments *endosomatic* (which he possesses from his birth), while in the economic process the human also uses the instruments *exosomatic* (which are made by himself). A first conclusion that is imposed, is that, just as the biologist must study cells as much as organisms, the economist must study the economic agents as much as the economy as a whole.

The nonlinear modeling touched on in the final sections offers strong tools to investigate both the individual and the global socioeconomic complex systems. Nevertheless, it is worth mentioning that this approach appears often as a 'black box' that leads to a 'convenient' set of outputs for a well-tuned set of inputs. The underground economic meanings of the control parameters and, in principal, the economic relevance of the nonlinear equations fitting the empirical data set, remain open questions for future studies.

References

[1] Georgescu-Roegen, N. (1971), The Entropy Law and the Economic Process, Harvard University Press, Cambridge, MA.

[2] Montrol, E.W. (1987), On the dynamics and evolution of some socio-technical systems, B. Am. Math. Soc. 16(1), 1–46.

[3] Gligor, M. (2001), Noise induced transitions in some socio-economic systems, Complexity 6, 28–32.

[4] National Institute of Statistics and Economic Studies (1999), Romanian Statistical Yearbook, NISES, Bucharest. Available at http://www.cns.

[5] Pareto, V. (1965), Cours d'Economie Politiqu. Lausanne, 1897. In Reprint in Oeuvres Completès, Droz, Geneva.

[6] Gini, C. (1922), Indici di concentrazione e di dipendenza, Biblioteca delli' Economista 20.

[7] Gibrat, R. (1932), Les Inégalités Économiques, Sirey, Paris.

[8] Solomon, S. and Levy, M. (1996), Spontaneous scaling emergence in generic stochastic systems, Int. J. Mod. Phys. C 7, 745.

[9] Solomon, S. (1999), Generalized Lotka-Volterra (GLV) Models. Cond-mat/9901250.

[10] Dragulescu, A. and Yakovenko, V.M. (2000), Statistical mechanics of money, Eur. Phys. J. B 17, 723.

[11] Dragulescu, A. and Yakovenko, V.M. (2001), Exponential and power law probability distributions of wealth and income in the United Kingdom and the United States, Physica A 299, 213.

[12] Gligor, M. and Ignat, M. (2002), A kinetic approach to some quasi-linear laws of macro-economics, Eur. Phys. J. B 30(1), 125–135.

[13] Caraffini, G.L., Iori, M. and Spiga, G. (1996), On the connections between kinetic theory and a statistical model for the distribution of dominance in populations of social organisms, Riv. Mat. Univ. Parma 5(5), 169–181.

[14] Ormerod, P. (2001), The US Business Cycle: Power Law Scaling for Interacting Units with Complex Internal Structure. Paper available at: http://www.volterra.co.uk.

[15] Samuelson, P.A. (1942), The stability of equilibrium, Econometrica 10, 1–25.

[16] Boldrin, M. (1988), Persistent oscillations and chaos in dynamic economic models: notes for a survey. The Economy as an Evolving Complex System (Anderson, P.W., Arrow, J.K., Pines, D., eds.), Addison-Wesley, Redwood City.

[17] Cohen, J.E. (1995), Population growth and Earth's human carrying capacity, Science 269, 341–346.

[18] Ormerod, P. and Mounfield, C. (2001), Power law distribution of the duration and magnitude of recessions in capitalist economies: breakdown of scaling, Physica A 293, 573.

Chapter 18

EXTREMAL PRINCIPLES AND LIMITING POSSIBILITIES OF OPEN THERMODYNAMIC AND ECONOMIC SYSTEMS

A.M. Tsirlin

Program Systems Institute, Russian Academy of Sciences, et.'Botic', Perejaslavl-Zalesky 152140, Russia

Vladimir Kazakov

School of Finance & Economics, University of Technology, Sydney, P.O. Box 123, Broadway, NSW 2007, Australia

Abstract

In this chapter Prigogine's minimum-entropy principle is generalized to thermodynamic and microeconomic systems that include an active subsystem (heat engine or economic intermediary). New bounds on the limiting possibilities of an open system with an active subsystem are derived, including the bound on the productivity of the heat-driven separation. The economic analogies of Onsager's reciprocity conditions are derived.

Keywords: entropy production; capital dissipation; open systems; generalized Prigogine's principle; economic reciprocity conditions

1. Introduction

Thermodynamic and microeconomic systems are both *macrosystems*. They include a large number of microsubsystems, which are not controllable and not observable. Control and observation in such systems is only feasible on the macrolevel. The state of the macrosystem is described by macrovariables that depend on the averaged behavior of its components only. Macrovariables are divided into extensive and intensive. The former include internal energy, entropy, mass in thermodynamics and stocks of resources and capital in economics. When a system is subdivided its extensive variables change proportionally to the volume. The intensive variables (pressure, temperature, chemical potential in thermodynamics, resource's, and capital's estimates in microeconomics) do not change if the system is subdivided. In equilibrium the macrosystem's variables are linked via the equation of state.

Changes of extensive variables are linked to flows of mass, energy, resources, capital, etc. These flows can be caused by external factors (convective flows) or by interaction of

E-mail address: tsirlin@sarc.botik.ru (A.M. Tsirlin); vladimir.kazakov@uts.edu.au (V. Kazakov)

macrosystems with each other (in thermodynamics such flows are called diffusive). Flow rates depend on the differences between the intensive variables of the interacting systems. This exchange leads to changes of extensive variables. The rate of their change is proportional to the exchange flows. It is useful to single out three classes of macrosystems:

(1) *Systems with infinite capacity (reservoirs).* The values of their intensive variables are fixed and do not depend on the exchange flows.

(2) *Finite-capacity systems.* We assume that they always are in internal equilibrium. Their intensive variables change as a result of exchange flows. For example, the temperature of the system with constant volume changes if its internal energy changes, the price of a resource changes when its stock changes, etc. If interacting subsystems of finite capacity are insulated from the environment then the differences of their intensive variables tend to zero and the system as a whole tends to its equilibrium state.

(3) System with intensive variables (all or some) that can be controlled (within a given range). We shall call such systems *active*. The working body of the heat engine, whose parameters are controlled to achieve maximal performance, is an active system. An economic intermediary, who buys and resells resource by offering one price for buying and another for reselling, is an active system. Active systems play an important role in Finite-Time Thermodynamics (FTT), which investigates limiting possibilities of nonequilibrium thermodynamic systems [1–5]. Most FTT problems are reduced to optimal control problems where intensive variables of active systems are controls.

A macrosystem is *open*, if it exchanges with its environment. If the environment includes reservoirs and some of the flows are convective then its steady state can be stationary (system's intensive variables are time constant), periodic, quasiperiodic or quasistochastic.

Near equilibrium the flows in a system depend linearly on the driving forces. Prigogine's extremal principle states that "the steady state of a near-equilibrium system is stationary and the values of its intensive variables are distributed within its volume or between its finite capacity subsystems in such a way that the entropy production in the system is minimal" [6].

We will consider open thermodynamic and microeconomic macrosystems that include internally equilibrium subsystems of these three types. We will obtain the conditions that determine the limiting possibilities of active systems here and the extremal principles that determine stable states of such systems.

2. Thermodynamic system including an active subsystem

2.1. Problem formulation

We consider a thermodynamic system that includes n finite-capacity subsystems (we shall call them subsystems) ($i = 1, \ldots, n$) in internal equilibrium and an active subsystem that transforms heat or chemical energy into work (Fig. 1). We shall call this active subsystem a transformer. The system exchanges convective mass flows g_k ($k = 1, \ldots, r$) with the environment. The compositions and rates of some of these flows are given. The flows of heat q_k are also given. Mass and heat flows between subsystems inside the system are

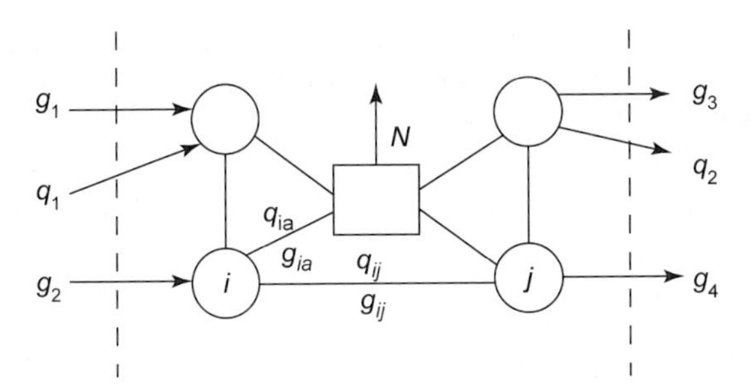

Fig. 1. Open thermodynamic system with transformer.

denoted as (q_{ij}, g_{ij}) and between subsystems and transformer as q_{ia}, g_{ia}. We assume that these flows are linear with driving forces Δ_{ij}

$$\left.\begin{aligned} G_{ij} &= \{q_{ij}, g_{ij}\} = L_{ij}\Delta_{ij}, \\ G_{ia} &= \{q_{ia}, g_{ia}\} = L_{ia}\Delta_{ia}. \end{aligned}\right\}, \tag{1}$$

L_{ij} and L_{ia} are the matrices of kinetic coefficients. We consider the flow that enters a subsystem (convective and diffusion) as positive and the power N, produced by the transformer, as positive also. If $N > 0$, then we shall call the transformer direct, and if $N < 0$ then it is inverse. Driving forces Δ_{ia} depend on the state of the transformer during contact with the i-th subsystem. It can be controlled. The state of the i-th subsystem is described by the vector of its intensive variables y_i. Here $T_i = 1/y_{i0}$, $\Delta_{ij} = (y_i - y_j)$, $T_{ia} = 1/y_{ia0}$, $\Delta_{ai} = (y_a - y_i)$. We assume that some subsystems have a given state $(x_i = x_i^0$, $i = 1, ..., m \leq n)$, and the others are free.

The following problems can be formulated in relation to this system:

(1) What are the values of free variables x_i $(i = m + 1, ..., n)$ if the power of the transformer N and the compositions and rates of convective flows are fixed?
(2) What is the maximal power that can be extracted and what is the minimal power that has to be spent N (the limiting power problem)?
(3) What is the maximal feasible rate of one of the flows g_r (objective flow) if the power N and the compositions of all or some of the convective flows are fixed?

2.2. *Thermodynamic balances*

Let us write down the thermodynamic balances for a steady state of the system as a whole. Subscript k denotes convective flow k. e_k, V_k, $h_k = e_k + p_k V_k$, p_k denote the k-th convective flow's molar energy, enthalpy, pressure, and its internal energy as e_k. (a) Energy balance is

$$\sum_{k=0}^{r} (g_k h_k + q_k) - N = 0. \tag{2}$$

(b) Mass balance is

$$\sum_{k=0}^{r} g_k x_{\nu k} + \sum_{l} \alpha_{\nu l} W_l = 0, \quad \nu = 1, 2, \ldots \tag{3}$$

$$\sum_{\nu} x_{\nu k} = 1, \quad k = 1, \ldots, r. \tag{4}$$

Here $x_{\nu k}$ is the molar fraction of the ν-th component in the k-th flow, $\alpha_{\nu l}$ is the stoichiometric coefficient with which the ν-th specie enters into the equation for the l-th reaction, W_l is the rate of this reaction. To simplify the equation we assume that the flow of energy from the reaction

$$q_l = \sum_{\nu} \mu_{\nu l} \alpha_{\nu l} W_l, \tag{5}$$

is included into convective heat flows q_k.

Similarly, we introduce flows of mass

$$g_l = W_l \sum_{\nu} \alpha_{\nu l}, \tag{6}$$

with the composition determined by the equality

$$x_{\nu l} = \frac{\alpha_{\nu l}}{\sum_{\nu} \alpha_{\nu l}}. \tag{7}$$

We include these flows into convective flows g_k.

(c) Entropy balance is

$$\sum_{k=0}^{r} \left(g_k s_k + \frac{q_k}{T_k} \right) + \sigma = 0. \tag{8}$$

Here s_k is the molar entropy of the k-th flow. $g_k > 0$, if it enters and $g_k < 0$, if it leaves the system.

From Eqs. (3) and (4) it follows that $\sum_{k=0}^{r} g_k = 0$, so we eliminate one of the flows (objective flow) g_0. We obtain

$$\sum_{k=0}^{r} q_k + \sum_{k=1}^{r} g_k \Delta h_{0k} - N = 0, \tag{9}$$

$$\sum_{k=0}^{r} \frac{q_k}{T_k} + \sum_{k=1}^{r} g_k \Delta s_{0k} + \sigma = 0. \tag{10}$$

Here

$$\Delta h_{0k} = h_k - h_0, \quad \Delta s_{0k} = s_k - s_0. \tag{11}$$

Elimination of q_0 with the temperature T_0 from Eq. (9) and its substitution into Eq. (10), yields

$$\sum_{k=1}^{r}\left[g_k\left(\Delta s_{0k} - \frac{\Delta h_{0k}}{T_0}\right) + q_k\left(\frac{1}{T_k} - \frac{1}{T_0}\right)\right] + \sigma + \frac{N}{T_0} = 0. \quad (12)$$

We denote the efficiency of the reversible heat engine as

$$\eta_C^0 = \frac{T_k - T_0}{T_k}.$$

The power of the transformer N can then be expressed from Eq. (12) as

$$N = \sum_{k=1}^{r} [q_k \eta_C^0 + g_k(\Delta h_{0k} - \Delta s_{0k} T_0)] - T_0 \sigma = N^0 - T_0 \sigma. \quad (13)$$

The first term on the right-hand side of the equality is the reversible power N^0 in a system with infinitely large mass and heat-transfer coefficients (arbitrary large size of apparatus). It is completely determined by the parameters of the system's input and output convective flows. The second term describes dissipative losses.

For mixtures, which are close to ideal gases or ideal solutions, the molar enthalpies and entropies can be expressed in terms of their compositions

$$h_k(T_k, p_k, x_k) = \sum_{\nu} x_{\nu k} h_k(T_k, p_k), \quad (14)$$

$$s_k(T_k, p_k, x_k) = \sum_{\nu} x_{\nu k}[\bar{s}_k(T_k, p_k) - R \ln x_{\nu k}], \qquad k = 0, \ldots, r. \quad (15)$$

From the equality (13) it follows that if the transformer's intensive variables are chosen in such a way that N^0 is not changed then the maximum of N is achieved when the minimum of σ is achieved.

2.3. *Entropy production and the state of subsystems*

For Onsanger's linear kinetic (1) the entropy production in the system takes the following form

$$\sigma = \sum_{i=1}^{n}\left(\Delta_{ia}^{\mathrm{T}} L_{ia} \Delta_{ia} + \frac{1}{2}\sum_{\substack{j=1 \\ j \neq i}}^{n} \Delta_{ij}^{\mathrm{T}} L_{ij} \Delta_{ij}\right). \quad (16)$$

The first term is the entropy production (scalar product of fluxes on driving forces) due to heat and mass exchange between the i-th subsystem and transformer. The second term is the entropy production due to heat exchange between the i-th and j-th subsystems. Multiplier $\frac{1}{2}$ appears because the term for each flow enters into equality (16) twice. Matrices L_{ij}, and L_{ia}, are positive-definite and symmetric.

First we assume that convective flows do not enter into subsystems with nonfixed state y_i $(i = m+1, \ldots, n)$. The following analog of Prigogine's principle then holds for an open system that includes a transformer (active subsystem)

Statement 1: *If intensive variables of the transformer* y_{ia} $(i = 1, \ldots, n)$ *are fixed then the free intensive variables of subsystems* y_i *of an open system take such values that entropy production in the system is minimal and the power N obeys the equation*

$$N = N^0 - \sigma_{\min}(N)T_0. \tag{17}$$

The proof of this statement follows from the fact that stationarity conditions of σ with respect to the components of the state vector y_i of the i-th subsystem coincide with the condition of its minimum, because σ is a convex function.

Indeed, change of y_i affects all flows that enter/leave the i-th subsystem. Note that the derivatives $\partial \Delta_{ij}/\partial y_i$ and $\partial \Delta_{ia}/\partial y_i$ have opposite signs and their absolute values are equal to 1. Since matrices L_{ij}, and L_{ia} are symmetric the stationary conditions lead to the equations

$$\sum_{j=1, j\neq i}^{n} g_{ij\nu} = g_{ia\nu}, \qquad i = m+1, \ldots, n, \qquad \nu = 1, 2, \ldots, \tag{18}$$

where $g_{ij\nu} = g_{ij}x_{j\nu}$, $g_{ia\nu} = g_{ia}x_{i\nu}$ are the diffusion flows of the ν-th component

$$\left(\sum_{\nu} g_{ij\nu} = g_{ij}, \sum_{\nu} g_{ia\nu} = g_{ia}\right), \quad \sum_{j=1, j\neq i}^{n} (q_{ij} + g_{ij}h_{ij}) = q_{ia} + g_{ia}h_i, \quad i = 1, \ldots, n. \tag{19}$$

These equations coincide with equations of mass and energy balances for the i-th subsystem.

Thus the intensive variables of the subsystems of an open system with a transformer take such values that dissipation in it is minimal. If some of the intensive variables y_i are fixed then the free variables minimize σ subject to these constraints.

2.4. Limiting power problem

If the state of some of the subsystems and the parameters of the convective flows are fixed then the power N, obtained from the system or used to maintain its state is bounded. For $N > 0$ this bound is the maximal power of the heat engine [7], and for $N < 0$ it is the minimal losses in the transformer [9]).

Assume that convective flows q_{ka} and g_{ka} are equal to zero. Then the thermodynamic balances take the following form:

(a) Energy balance is

$$N = \sum_{i=1}^{n} \left(q_{ia} + \sum_{\nu} g_{ia\nu}h_{i\nu}\right). \tag{20}$$

(b) Mass balance is

$$\sum_{i=1}^{n} g_{ia\nu} = 0, \quad \nu = 1, 2, \ldots. \tag{21}$$

(c) Entropy balance is

$$\sum_{i=1}^{n} \left[g_{ia} s_i + \frac{1}{T_{ia}} \left(q_{ia} + \sum_{\nu} g_{ia\nu} h_{i\nu} \right) \right] = 0. \tag{22}$$

The problem of limiting power is now reduced to a choice of transformer's variables y_{ia} during its contact with the i-th subsystem for which N is maximal subject to conditions (21) and (22), and to the mass, energy, and entropy balances for each of the i subsystems. This formulation of the problem is very general. We will consider two particular cases of this problem.

2.5. Heat-mechanical system

Direct transformer. Here there are no mass flows and the problem can be rewritten as follows ($q_{ia} \sim q_i$)

$$N = \sum_{i=1}^{n} q_i(T_i, T_{ia}) \rightarrow \max_{T_{ia}, T_i}, \tag{23}$$

subject to entropy and energy balances

$$\sum_{i=1}^{n} \frac{q_i(T_i, T_{ia})}{T_{ia}} = 0, \tag{24}$$

$$\sum_{j=1}^{n} q_{ij}(T_i, T_j) + \sum_{k=0}^{r} q_{ik} = q_{ia}, \quad i = 1, \ldots, n. \tag{25}$$

The conditions of optimality for the problem (23)–(25), follows from the conditions of stationary of its Lagrange function

$$L = \sum_{i=1}^{n} \left\{ q_i \left(1 + \frac{\Lambda}{T_{ia}} - \lambda_i \right) + \lambda_i \left(\sum_{j=1}^{n} q_{ij} + \sum_{k=1}^{r} q_{ik} \right) \right\}, \tag{26}$$

on T_{ia} and T_i

$$\frac{\partial L}{\partial T_{ia}} = 0 \Rightarrow \frac{\partial q_i}{\partial T_{ia}} \left(1 + \frac{\Lambda}{T_{ia}} - \lambda_i \right) = \Lambda \frac{q_i(T_i, T_{ia})}{T_{ia}^2}, \quad i = 1, \ldots, n, \tag{27}$$

$$\frac{\partial L}{\partial T_i} = 0 \Rightarrow \frac{\partial q_i}{\partial T_i} \left(1 + \frac{\Lambda}{T_{ia}} - \lambda_i \right) + \lambda_i \sum_{j=1}^{m} \frac{\partial q_{ji}}{\partial T_i} = 0, \quad i = m+1, \ldots, n. \tag{28}$$

Eqs. (27) and (28) allow us to find n temperatures T_{ia} of the working body, $(n-m)$ temperatures T_i of subsystems and the $(n+1)$ Lagrange multiplier.

The most common form of heat-exchange law is Newton's linear form

$$q_{ji} = \alpha_{ji}(T_j - T_i).$$

Here $q_i = \alpha_i(T_i - T_{ia})$, $q_{ji} = \alpha_{ji}(T_j - T_i)$, and Eqs. (27) and (28) take the form

$$\sum_{i=1}^{n} \overline{\alpha_i} \frac{T_i}{T_{ia}} = 1, \quad \overline{\alpha_i} = \frac{\alpha_i}{\sum_{\nu=1}^{n} \alpha_\nu}, \tag{29}$$

$$T_{ia} = T_i - \frac{1}{\alpha_i}\left[\sum_{j=0}^{n+1} \alpha_{ji}(T_j - T_i) + \sum_k q_{ik}\right], \quad i = 0, \ldots, n+1, \tag{30}$$

$$T_{ia}^2(1-\lambda_i) = \Lambda T_i, \quad i = 1, \ldots, n, \tag{31}$$

$$\alpha_i\left(1 + \frac{\Lambda}{T_{ia}} - \lambda_i\right) = \lambda_i \sum_{j=1}^{n} \alpha_{ji}, \quad i = 1, \ldots, n. \tag{32}$$

We denote

$$\overline{\alpha_i} = \frac{\alpha_i}{\sum_{j=0}^{m+1} \alpha_{ji}}.$$

After eliminating λ_i from Eqs. (31) and (32) we get

$$(1+\overline{\alpha_i})\frac{T_i}{T_{ia}^2} + \frac{\overline{\alpha_i}}{T_{ia}} = \frac{1}{\Lambda} = \text{const}, \quad i = 1, \ldots, n. \tag{33}$$

These equations jointly with balances (29) and (30) determine the solution.

Let us show that if $n = 2$ and the subsystem temperatures are fixed $T_1 = T_+$, $T_2 = T_-$ then from these conditions follow the known results [7,8] about the limiting power of the heat engine. Indeed, here the conditions (32) are missing, $\lambda_i = 0$ and from Eq. (31) it follows that

$$T_{1a}^* = \sqrt{\Lambda T_+}, \quad T_{2a}^* = \sqrt{\Lambda T_-},$$

and from Eqs. (29)–(32) we obtain the efficiency of the heat engine with maximal power as

$$\eta = 1 - \frac{T_{0a}}{T_{1a}} = 1 - \sqrt{\frac{T_-}{T_+}},$$

and the maximal power as

$$N_{\max} = \frac{\alpha_1 \alpha_2}{\alpha_1 + \alpha_2}(\sqrt{T_+} - \sqrt{T_-})^2. \tag{34}$$

2.5.1. Inverse transformer. Optimal thermostating

The maximal feasible power can be positive or negative. The sign depends on the given temperatures of the external flows q_k and given fixed temperatures of the subsystems T_i $(i = 1, \ldots, m)$. If this power is negative then we obtain the problem of limiting possibilities of heat pump in the thermostating system. That is, the problem of maintaining given temperatures in some of the subsystems and of optimal choice of temperatures in the rest of the passive subsystems to minimize power used. For Newton's the laws of heat transfer we obtain the conditions (29)–(31). The conditions (32) and (33) hold for subsystems with free temperatures T_i $(i = m + 1, \ldots, n)$. In the optimal thermostating problem [9] these temperatures are the temperatures of the passive subsystems.

2.6. Separation system

Separation systems commonly used mechanical (membrane systems, centrifuging, etc.) or heat energy (distillation, drying, etc.). We consider them separately.

2.7. Binary separation using mechanical energy

Consider the system for binary separation shown in Fig. 2. The input points for convective flows are fixed and the compositions and rates of these flows are also fixed. Here the mass balance holds

$$g_1 = g_2 + g_3, \tag{35}$$

$$g_1 x_{1\nu} = g_2 x_{2\nu} + g_3 x_{3\nu}, \quad \nu = 1, 2, \ldots. \tag{36}$$

The composition of each of three subsystems x_i $(i = 1, 2, 3)$ are also fixed and coincide with the compositions of the convective flows. The temperatures and pressures of these flows are the same and enthalpy increments of the convective flows are equal to zero. The condition (13) can be rewritten as

$$N = -g_0 T_0 \sum_{k=1}^{1} \varepsilon_k \Delta s_{0k} - T_0 \sigma \rightarrow \max. \tag{37}$$

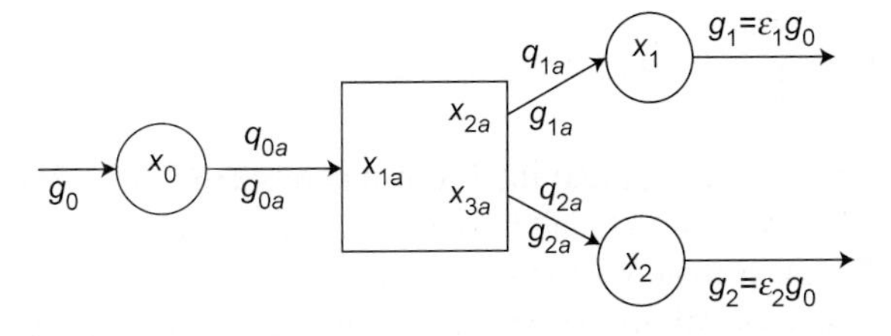

Fig. 2. Mechanical separation of binary mixture.

Here $\varepsilon_k = \left|\frac{g_k}{g_0}\right|$ is the fraction removed into the first and the second flows ($\varepsilon_1 + \varepsilon_2 = 1$). We denote the power used as $\bar{N} = -N$ and obtain

$$\bar{N} = g_0 RT_0 \left[\sum_{k=1}^{2} \varepsilon_k \sum_{\nu} x_{k\nu} \ln x_{k\nu} - \sum_{\nu} x_{0\nu} \ln x_{0\nu} \right] + T_0 \sigma = \bar{N}^0 + T_0 \sigma \rightarrow \min. \quad (38)$$

The first term in the right-hand side is the reversible power for separation.

Each of the convective flows $\bar{g}_i$ contains three components: q_i, $g_{i1} = g_1 x_{i1}$, $g_{i2} = g_1 x_{i2}$ ($i = 0, 1, 2$). The vector of the driving force of the i-th flow Δ_i has components

$$\Delta_{iq} = \left(\frac{1}{T_{ia}} - \frac{1}{T_0} \right), \quad \Delta_{i1} = \left(\frac{\mu_{i1a}}{T_{ia}} - \frac{\mu_{i1}}{T_0} \right), \quad \Delta_{i2} = \left(\frac{\mu_{i2a}}{T_{ia}} - \frac{\mu_{i2}}{T_0} \right).$$

Flows depend linearly on driving forces

$$\bar{g}_i = A_i \Delta_i, \quad i = 0, 1, 2, \quad (39)$$

here the matrix A_i is positive-definite and symmetric. Its inverse $B_i = A_i^{-1}$ is also positive-definite and symmetric.

Entropy production can be rewritten using Eq. (39) in the following form

$$\sigma = \sum_{i=0}^{2} \sigma_i = \sum_{i=0}^{2} \bar{g}_i^{\mathrm{T}} A_i^{-1} \bar{g}_i. \quad (40)$$

If it is feasible to control driving forces in the system in such a way that the rates and compositions of flows have required values then the power used by the irreversible separation system is

$$\bar{N}^* = \bar{N}^0 + T_0 \sum_{i=0}^{2} \bar{g}_i^{\mathrm{T}} A_i^{-1} \bar{g}_i. \quad (41)$$

In particular for diagonal matrices A_i

$$\bar{N} = \bar{N}^0 + T_0 \sum_{i=0}^{2} \left(\frac{q_i^2}{\alpha_{iq}} + \frac{g_{i1}^2}{\alpha_{i1}} + \frac{g_{i2}^2}{\alpha_{i2}} \right). \quad (42)$$

If the constraints imposed on the system are softer, for example, if only the flow g_1 and its composition x_{11} ($x_{12} = 1 - x_{11}$) and the composition of the input flow x_{01} ($x_{02} = 1 - x_{01}$) are given then the problem is reduced to the search of the minimum of σ on q_0, q_1, q_2, g_0, g_2, g_{21} subject to constraints

$$\sum_{i=0}^{2} q_i = \sum_{i=0}^{2} g_i = \sum_{i=0}^{2} g_{i1} = 0, \quad 0 \leq g_{21} \leq g_2. \quad (43)$$

The quadratic form (40) is convex and the set of this problem's feasible solutions is also convex. Therefore, the problem (40) and (43) has a unique solution that corresponds to the power that is always lower than $\bar{N}^*$, found from Eq. (41).

Note that the productivity of the system (rate of objective flow) can be made arbitrarily high if the power used $\bar{N}$ is made sufficiently high.

2.8. *Thermal separation*

Here $N = 0$ and the transformer uses heat flows to obtain the work of separation. Consider the system shown in Fig. 3. Assume that the input mixture is binary and the temperatures of the flows g_i $(i = 0, 1, 2)$ and their pressures are the same. We denote the temperatures of the subsystems as T_+ and $T_- < T_+$ We assume that enthalpy increments of the flows g_1 and g_2 are zero, $N = 0$, and the formula (13) takes the form

$$q_+\eta = T_-\left(\sum_{k=1}^{2} g_k \Delta s_{0k} + \sigma\right) = 0, \quad \eta = 1 - \frac{T_-}{T_+}. \tag{44}$$

For Newton's law of heat transfer

$$q_+ = \alpha_+(T_+ - T_1), \quad q_- = \alpha_-(T_- - T_2), \tag{45}$$

and for the substances that are close to ideal solutions the reversible power of separation is

$$N_p(g_0) = T_0 \sum_{k=1}^{2} g_k \Delta s_{0k} = RT_0 g_0\left(\sum_{k=1}^{2} |\varepsilon_k| \sum_{\nu=1}^{2} x_{k\nu} \ln x_{k\nu} - \sum_{\nu=1}^{2} x_{0\nu} \ln x_{0\nu}\right), \quad \varepsilon_k = \frac{|g_k|}{g_0}. \tag{46}$$

The entropy production due to heat flows is (Eq. (40))

$$\sigma_q = \sigma_{q_+} + \sigma_{q_-} = \frac{\alpha_+(T_+ - T_1)^2}{T_+ T_1} + \frac{\alpha_-(T_- - T_2)^2}{T_- T_2}. \tag{47}$$

The entropy production due to mass flows is (Eq. (40))

$$\sigma_g = \sum_{k=0}^{2} \bar{g}_k^{\mathrm{T}} A_k^{-1} \bar{g}_k. \tag{48}$$

After taking into account Eqs. (45)–(48), Eq. (44) takes the form

$$\alpha_+(T_+ - T_1)\eta = (N_p + T_-\sigma_g) + T_-\sigma_q. \tag{49}$$

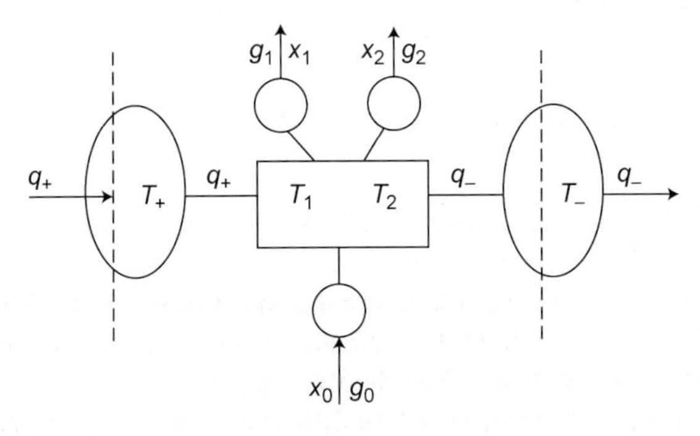

Fig. 3. Thermal separation system.

Note that for the fixed ε_k and fixed compositions of flows the reversible power of separation N_p increases linearly when productivity g_0 increases, and that σ_g increases as g_0^2. Therefore, the maximal productivity of thermal separation system $g_{\max}^0$ is finite and is given by the solution of the equation

$$N_p(g_0) + T_- \sigma_g(g_0) = \max_{T_1,T_2} \{q_+(T_+, T_1)\eta - \sigma_q(T_+, T_-, T_1, T_2)\}. \tag{50}$$

The expression on the right-hand side of equality (16) is the maximal power of the irreversible heat engine. Its maximum is sought subject to the transformer's balance on the entropies of heat flows

$$\frac{\alpha_+(T_+ - T_1)}{T_1} + \alpha_- \frac{(T_- - T_2)}{T_2} = 0. \tag{51}$$

The solution of this problem for Newton's laws of heat transfer was obtained in Refs. [7,8] and shown above (Eq. (34)). Substitution of $N_{\max}$ into the right-hand side of equality (50) allows us to obtain the limiting productivity of thermodynamic separation system that has the structure shown in Fig. 3.

Let us emphasis that unlike mechanical systems the productivity of the thermal systems is bounded. The increase of the rates of heat flows above some threshold reduces the system's.

3. Open microeconomic system

The analogy between thermodynamic and microeconomic systems has been studied extensively. The works of Samuelson [10], Lichnerowicz [11], Rozonoer and coworker [12,13] and Martinas [14] should be especially mentioned. Most of that research considered the analogy between equilibrium systems. In this chapter we consider this analogy for nonequilibrium systems.

3.1. Stationary state, reciprocity conditions and minimal dissipation principle

Each subsystem on an open economic system—an economic agent—is described by its extensive variables–the stock of resources N and capital N_0; by its wealth function $S(N)$, and by its intensive variables—resources and capital estimates p_i and p_0 that obey the following equations

$$p_0 = \frac{\partial S}{\partial N_0}, \quad p_i = \left(\frac{\partial S}{\partial N_i}\right) / \left(\frac{\partial S}{\partial N_0}\right) = \frac{1}{p_0}\frac{\partial S}{\partial N_i}, \quad i = 1, 2, \ldots. \tag{52}$$

Resource estimate p_i is the equilibrium price for buying and selling. If the price c_i is higher than the equilibrium price then the economic agent sells and if it is lower then it buys resource. Because the agent's wealth function is a uniform function of the first degree and strictly convex, the resource's estimate decreases when its stock increases.

If the system is near equilibrium then the flow depends on the driving forces linearly. The driving force for the i-th resource is the difference between its price and its estimate $\Delta_i = p_i - c_i$. We assume that it is positive if the flow is directed to the economic agent, then

$$g_i = \sum_{\nu=1}^{n} a_{\nu i}\Delta_\nu = \sum_{\nu=1}^{n} a_{\nu i}(p_\nu - c_\nu), \quad i = 1, \ldots, n. \tag{53}$$

We shall call the matrix A with elements $\alpha_{i\nu}$ the matrix of the economic agent's kinetic coefficient. This matrix determines the exchange kinetic between the economic agent and its environment.

The flow of resource exchange causes the reciprocal flow of capital in the opposite direction such that

$$\frac{dN_0}{dt} = -\sum_{i=1}^{n} c_i g_i. \tag{54}$$

The economic agent's wealth function here changes

$$\frac{dS}{dt} = \frac{\partial S}{\partial N_0}\frac{dN_0}{dt} + \sum_{i=1}^{n}\frac{\partial S}{\partial N_i} g_i = -p_0 \sum_{i=1}^{n} c_i g_i + p_0 \sum_{i=1}^{n} p_i g_i = p_0 \sum_{i=1}^{n}(p_i - c_i) g_i$$

$$= p_0 \Delta^{\mathrm{T}} A \Delta. \tag{55}$$

Here Δ is the driving force vector.

Because the estimate of the capital $p_0 > 0$, and because the exchange is always done voluntarily, the wealth function does not decrease. Therefore, the matrix A is positive-definite. If the driving forces are expressed in terms of Δ from Eq. (53) then Eq. (55) takes the form

$$\frac{dS}{dt} = p_0 g^{\mathrm{T}} B g, \tag{56}$$

here g is a column-vector, $B = A^{-1}$, and the elements $b_{i\nu}$ of this matrix are equal to (up to the constant)

$$b_{i\nu} = \frac{\partial^2 S}{\partial N_i \, \partial N_\nu}, \quad i, \nu = 1, \ldots, n. \tag{57}$$

Thus, matrix B is positive-definite and symmetric, and its inverse matrix of kinetic coefficients A is also positive-definite and symmetric. Thus, the following analog of the reciprocity conditions hold: "the effect of the difference between the price and estimate of the ν-th resources on the flow of the i-th resource is the same as the effect of the difference between the price and estimate of the i-th resources on the flow of the ν-th resource."

Consider the system that includes r subsystems with fixed resource estimates (economic reservoirs) and $k - r$ subsystems that exchange resources and capital with each other (Fig. 4). Resource estimates for reservoirs $p_i(i = 1, \ldots, r)$ are constant. These estimates N_i and N_{0i} depend on the stocks of resource and capital for subsystems with $i > r$. We assume for simplicity that capital estimates are the same for all subsystems and equal to 1. Assume that the flow of resource n_i to/from the economic agent, who has estimate $p_i(N_i, N_{0i})$, is determined by the flow of capital $q_i = -n_i c_i$. Here c_i is the price of resource that is higher than p_i when the economic agent sells resource ($n_i < 0$), and is

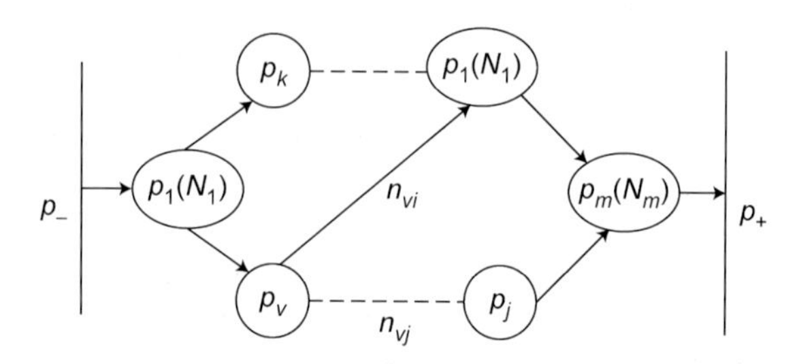

Fig. 4. Structure of an open microeconomic system with two economic reservoirs.

lower than p_i when it buys it, $n_i > 0$. If the economic agent exchanges with the j-th reservoir, then $c_i = p_j$, and

$$n_{ij} = n_{ij}(p_j, p_i), \quad q_{ij} = -p_j n_{ij}, \quad j = 1, \ldots, r; \quad i = r + 1, \ldots, k. \tag{58}$$

Following Ref. [14], we define the price $c_{i\nu}$ such that

$$\tilde{n}_{i\nu}(p_i, c_{i\nu}) = -\tilde{n}_{\nu i}(p_\nu, c_{i\nu}), \tag{59}$$

where $(i; \nu) \geq r + 1$. Conditions (59) allow us to express $c_{i\nu}$ in terms of p_i, p_ν and obtain

$$n_{i\nu}(p_i, p_\nu) = -n_{\nu i}(p_\nu, p_i). \tag{60}$$

For example, assume that

$$\tilde{n}_{i\nu} = \tilde{a}_{i\nu}(p_i - c_{i\nu}), \quad \tilde{n}_{\nu i} = \tilde{a}_{\nu i}(p_\nu - c_{i\nu}).$$

From the condition (59) we obtain $c_{i\nu}$ and flows in Eq. (60) become

$$c_{i\nu} = \frac{\tilde{a}_{i\nu} p_i + \tilde{a}_{\nu i} p_\nu}{\tilde{a}_{i\nu} + \tilde{a}_{\nu i}} = \gamma_{i\nu} p_i + \gamma_{\nu i} p_\nu, \quad \gamma_{i\nu} + \gamma_{\nu i} = 1, \tag{61}$$

$$\begin{aligned} n_{i\nu}(p_i, p_\nu) &= \frac{\tilde{a}_{\nu i} \tilde{a}_{i\nu}}{\tilde{a}_{i\nu} + \tilde{a}_{\nu i}}(p_i - p_\nu) = a_{i\nu}(p_i - p_\nu), \\ n_{\nu i}(p_\nu, p_i) &= -n_{i\nu}(p_i, p_\nu) = a_{i\nu}(p_\nu - p_i). \end{aligned} \tag{62}$$

The capital fluxes are

$$q_{i\nu}(p_i, p_\nu) = -c_{i\nu}(p_i, p_\nu) n_{i\nu}(p_i, p_\nu) = -q_{\nu i}(p_\nu, p_i). \tag{63}$$

For fluxes (62) we obtain

$$q_{i\nu}(p_i, p_\nu) = -\frac{(\tilde{a}_{i\nu} p_i + \tilde{a}_{\nu i} p_\nu)\tilde{a}_{\nu i} \tilde{a}_{i\nu}}{(\tilde{a}_{i\nu} + \tilde{a}_{\nu i})^2}(p_i - p_\nu).$$

If $\tilde{a}_{i\nu} = \tilde{a}_{\nu i} = \bar{a}_{i\nu}$ then

$$q_{i\nu}(p_i, p_\nu) = -\frac{\bar{a}_{i\nu}(p_i^2 - p_\nu^2)}{4}.$$

Assume that the flux is a vector and the condition (60) holds for the l-th component of this flux. We denote

(a) Vector of differences between estimates of i-th and ν-th economic agents

$$\Delta p_{i\nu} = (\Delta p_{i\nu 1}, \ldots, \Delta p_{i\nu l}, \ldots) = p_i - p_\nu. \tag{64}$$

(b) Matrix $A_{i\nu}$ of the coefficients that link the flux-vector between economic agents and their estimates' difference. The elements of this matrix are $a_{i\nu\mu l}$, where μ and l are subscripts that denote the type of resource. Matrix $A_{i\nu}$ is positive-definite ($A_{i\nu} = 0$) and symmetric.

The flow of l-th resource between i-th and ν-th economic agents is

$$n_{i\nu l} = \sum_{\mu} a_{i\nu\mu l}(\Delta p_{i\nu\mu}), \tag{65}$$

and the vector-flux is

$$n_{i\nu} = A_{i\nu}\Delta p_{i\nu}. \tag{66}$$

We denote the price vector during exchange between economic agents as $c_{i\nu} = (c_{i\nu 1}, \ldots, c_{i\nu l}, \ldots)$. From the condition similar to Eq. (59), it follows that the price is equal to the subsystems' estimates averaged with the weights $\gamma_{i\nu l}$ and $\gamma_{\nu il}$

$$c_{i\nu l}(p_{il}, p_{\nu l}) = \frac{\tilde{a}_{i\nu l}p_{il} + \tilde{a}_{\nu il}p_{l\nu}}{\tilde{a}_{i\nu l} + \tilde{a}_{\nu il}} = \gamma_{i\nu l}p_{il} + \gamma_{\nu il}p_{\nu l}. \tag{67}$$

The flux of capital is

$$q_{i\nu} = -c_{i\nu}A_{i\nu}\Delta p_{i\nu}^{\mathrm{T}}. \tag{68}$$

In a steady state the stocks of resources and capital do not change

$$\sum_{\nu=1}^{k} n_{i\nu} = \sum_{\nu=1}^{k} A_{i\nu}\Delta p_{i\nu}^{\mathrm{T}} = 0, \quad i = r+1, \ldots, k, \tag{69}$$

$$\sum_{\nu=1}^{k} c_{i\nu}(p_i, p_\nu)A_{i\nu}\Delta p_{i\nu}^{\mathrm{T}} = 0, \quad i = r+1, \ldots, k. \tag{70}$$

Capital dissipation is

$$\sigma = \frac{1}{2}\sum_{i,\nu} \Delta p_{i\nu}A_{i\nu}\Delta p_{i\nu}^{\mathrm{T}}, \tag{71}$$

the multiplier $\frac{1}{2}$ is due to each term appearing twice ($A_{\nu i} = A_{i\nu}$). Since matrix A is symmetric the conditions of minimum of capital dissipation with respect to the resource's estimates yields the conditions (69) and (70). Thus, "in an open microeconomic system that consists of subsystems in internal equilibrium with flows that depend linearly on the difference of resources' estimates the resources are distributed in such a way that capital dissipation attains a minimum with respect to free variables." This statement is an analogy of the extremal principle of Prigogine for economic systems.

3.2. Limiting possibilities of economic intermediary

Assume that the system (Fig. 5) includes an intermediary that can buy resource from one economic agent and resell it to another. This allows it to extract capital. The intermediary offers the price v_i during an exchange with the i-th subsystem. The flow of resource here is $m_i(p_i, v_i)$. In a stationary state the problem of extraction of maximal profit takes the following form

$$m = -\sum_{i=1}^{k} m_i(p_i, v_i)v_i \to \max_{v,p}, \tag{72}$$

subject to constraints

$$\sum_{i=1}^{k} m_i(p_i, v_i) = 0, \tag{73}$$

$$\sum_{j=1}^{k} n_{ji}(p_j, p_i) = m_i(p_i, v_i), \quad i = r+1, \ldots, k. \tag{74}$$

The minus in Eq. (72) is the result of the assumption that the flow of resource directed from the economic agent to the intermediary is positive. This flow is accompanied by reduction of capital. The condition (73) is the intermediary's resources' balance and conditions (74) are resource balances for each of the $k - r$ economic agents.

The Lagrange function of the problem (72) and (74) is

$$L = \sum_{i=1}^{k} \left[m_i(p_i, v_i)(\Lambda - v_i + \lambda_i) - \lambda_i \sum_{j=1}^{k} n_{ji}(p_j, p_i) \right]. \tag{75}$$

Here $\lambda_i = 0$ for $i \leq r$.

The conditions of optimality have the form

$$\frac{\partial L}{\partial v_i} = 0 \Rightarrow \frac{\partial m_i}{\partial v_i}(\Lambda - v_i - \lambda_i) = m_i(p_i, v_i), \quad i = 1, \ldots, k, \tag{76}$$

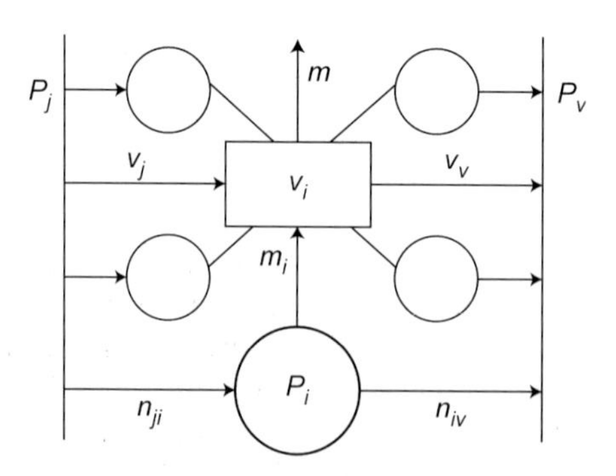

Fig. 5. An open system that includes an intermediary, markets, and passive subsystem.

$$\frac{\partial L}{\partial p_i} = 0 \Rightarrow \frac{\partial m_i}{\partial p_i}(\Lambda - v_i - \lambda_i) = \lambda_i \sum_{j=1}^{m} \frac{\partial n_{ji}}{\partial p_i}, \quad i = r+1, \ldots, k. \tag{77}$$

Conditions (73), (74), (76) and (77) determine $2(k-)$ unknowns p_i and λ_i, and the values of Λ and k optimal prices v_i.

In particular, if $n_{ji} = a_{ji}(p_i - p_j)$, $m_i = a_i(v_i - p_i)$ then these conditions can be rewritten as follows

$$\sum_{i=1}^{m} a_i(v_i - p_i) = 0, \tag{78}$$

$$\sum_{j=1}^{k} a_{ji}(p_i - p_j) = a_i(v_i - p_i), \quad i = r+1, \ldots, k, \tag{79}$$

$$2v_i = \lambda_i + \Lambda + p_i, \quad i = 1, \ldots, k, \tag{80}$$

$$-a_i(\Lambda - v_i + \lambda_i) = \lambda_i \sum_{j=1}^{k} a_{ji}, \quad i = r+2, \ldots, k. \tag{81}$$

The problem of maximal rate of extraction of capital is a direct analog of the problem of maximal power of a heat engine in an open thermodynamic system.

4. Conclusion

Economic systems differ from thermodynamic systems in many respects including the voluntary, discretional nature of exchange, production in addition to exchange, competition in various forms, etc. However, thermodynamic and economic systems are both macrosystems that have many analogies between them, including analogy of irreversibility of processes in them.

The steady states of open thermodynamic and microeconomic systems that include internally equilibrium subsystems with fixed intensive variables (reservoirs) and subsystems whose intensive variables are free (are determined by the exchange flows) were considered. It was shown that for a linear dependence of flows on driving forces it corresponds to the conditional minimum of the entropy production (in thermodynamics) or to the conditional minimum of capital dissipation (in microeconomics). The conditions that determine the limiting possibilities of an active subsystem (transformer in thermodynamics and intermediary in microeconomics) were obtained.

Acknowledgements

This work was supported by a grant from RFFI (grant 01-01-00020) and by the School of Finance and Economics of the University of Technology, Sydney.

References

[1] Berry, R.S., Kasakov, V.A., Sieniutycz, S., Szwast, Z. and Tsirlin, A.M. (1999), Thermodynamic Optimization of Finite Time Processes, Wiley, Chichester.
[2] Salamon, P., Hoffman, K.H., Schubert, S., Berry, R.S. and Andresen, B. (2001), What conditions make minimum entropy production equivalent to maximum power production?, J. Non-Equil. Therm., 26.
[3] Andresen, B (1983), Finite-time Thermodynamics, University of Copenhagen Press, Copenhagen, 1983.
[4] Andresen, B., Berry, R.S., Ondrechen, M.J. and Salamon, P. (1984), Thermodynamics for processes in finite time, Acc. Chem. Res. 17(8), 266–271.
[5] Bejan, A. (1996), Entropy generation minimization: the new thermodynamics of finite size devices and finite time process, J. Appl. Phys. 79, 1191–1218.
[6] Glansdorff, P. and Prigogine, I. (1971), Thermodynamic Theory of Structure, Stability and Fluctuations, Wiley, London.
[7] Novikov, I.I. (1957), The efficiency of atomic power stations, At. Energy 3(11), 409 English translation in J. Nucl. Energy II 7, 25–128 (1958).
[8] Curzon, F.L. and Ahlburn, B. (1975), Efficiency of a Carnot engine at maximum power output, Am. J. Phys. 43, 22–24.
[9] Tsirlin, A.M. (2003), Irreversible estimates of limiting possibilities of thermodynamic and microeconomic systems, Science, Moscow.
[10] Samuelson, P.A. (1972), Maximum principle in analytical economics, Am. Econ. Rev. B2, 249–262.
[11] Lichnirowicz, M. (1970), Um modele d'echange economique (Economie et thermodynamique). Annales de M'Institut Henri Poincare, nouvelle serie, IV, No.2, Section B, pp. 159–200.
[12] Rozonoer, L.I. and Malishevski, A.V. (1971), Model of chaotic resource exchange and analogy between thermodynamic and economic, Proceeding of All-Union Conference on Control Problems, Moscow, AN SSSR, IPU, 207–209.
[13] Rozonoer, L.I. (1973), Exchange and distribution of resources (general thermodynamics approach), Automat. Remote Cont. 5, 115–132; 6, 65–79; 8, 82–103.
[14] Martinas, K. (1995), Irreversible microeconomics. Complex Systems in Natural and Economic Sciences (Martinas, K., Moreau, M., eds.), Matrafured.

GLOSSARY OF PRINCIPAL SYMBOLS

$\mathbf{A}$	vector potential of electromagnetic field
A	action, extremized objective, area
A_μ	electromagnetic potentials
A_j	chemical or electrochemical affinity of *j*-th reaction
a_i	activity of *i*-th species
$\mathbf{B}$	Hamiltonian matrix, magnetic induction
$\mathbf{C}$	matrix of thermodynamic capacitances
$\mathbf{C}_i$	peculiar molecular velocity of *i*-th species
c	light speed in vacuum
c_i	molar concentration of *i*-th species
c_v	heat capacity at constant volume
c_0	propagation speed of thermal waves
c^*	speed of traveling chemical front
$\mathbf{D}$	vector of electric displacement
$\mathcal{D}$	effective drift, effective diffusivity, Biot's dissipation function
D	diameter, drag, diffusion coefficient, derivative operator
$\mathbf{E}$	electric field vector
E, E_0	Energy density and canonical energy density, respectively
$\bar{E}$	average energy
e	unit energy, specific energy, elementary charge
$\mathbf{F}$	force, specific external force, vector differential operator
F, F,	free energy and extended free energy, respectively
$F_{\mu\nu}$	electromagnetic, skew-symmetric tensor
f	Fanning friction factor, distribution function, unit free energy
G	Gibbs free energy, generating functional, shear modulus
$\mathbf{G}$, G^{jk}	energy-momentum tensor
$\mathsf{G}_{\mu\nu}$	Eistein tensor of gravitational field in Chap. I.5
g	gradient vector
g_{jk}	metric tensor
$\mathbf{H}$	magnetic field, density of radiative entropy flux
H	Hamiltonian function
h	direction vector
h	specific enthalpy, Planck constant
h_{jk}	projection tensor
$\mathbf{I}$	unit tensor δ_{ik}, inertia tensor, invariant current
I	Fisher's information measure, integral functional

$\mathbf{i}$	electric current density
$\mathbf{J}$	total mass flux density, conserved current
$\mathbf{J}_a$, $\mathbf{j}_a$	total and diffusional mass flux of a-th species, respectively
$\mathbf{J}_e$, $\mathbf{j}_e$	total and diffusional energy flux, respectively
$\mathbf{J}_{sa}$, $\mathbf{j}_{sa}$	total and diffusional entropy flux of a-th species, respectively
J	Jacobi determinant
J_n	flux potentials (Chap. II.9)
$\mathbf{K}_a$	Mass transfer coefficient of a-th species
K	chemical equilibrium constant, kinetic energy
$\mathbf{k}$	circular wave-number vector
k, k	thermal conductivity and chemical rate constant, respectively
k_{ij}	thermal conductivity tensor
k_B	Boltzmann constant
$\mathbf{L} = L_{ik}$	Onsager's matrix of kinetic coefficients
$\mathcal{L}$	kinetic potential, total length
l	Lagrange density function, length
$\mathbf{M}$	torque on the object
M	Mach number
M_i	molar mass of i-th species, mass property
m	mass, mass of micro-object
N	number of moles, number of particles
N_k	particle number operator for the wave number k
n	number density, mole number density, refraction index
$\mathbf{P}$	pressure tensor, total momentum
$\mathbf{p}$, $\mathbf{p}_s$	partial momentum densities for matter and entropy, respectively
P	thermodynamic pressure, generalized momentum variable
$P(\mathbf{X}, t)$	probability distribution function (probability density p)
p_i	probabilities
p_v	viscous pressure
$\mathbf{Q}$	energy flux density
Q	total heat, electric charge
q	heat transferred per unit mass
$\mathbf{q}$	heat flux density
q_i	generalized coordinates, electric charges
$\mathbf{R}$, $\tilde{\mathbf{R}}$	matrices of chemical and electrochemical resistances
$\mathbf{R}_{\mu\nu}$	Ricci tensor
R	universal gas constant, Riemannian scalar curvature
Re	Reynolds number
$\mathbf{r}$	radius vector
r_j, $\tilde{r}_j$	chemical and electrochemical reaction rates of j-th reaction, respectively
S, s	entropy and specific entropy, respectively
S	action of individual particle, eiconal
$\mathbf{T} = T^{\alpha\beta}$	stress tensor, symmetric energy-momentum tensor
T	absolute temperature
T^t	translational kinetic energy of particle

T^r	rotational kinetic energy of particle
t	time or time-like variable
U, u	internal energy and specific internal energy, respectively
U	strain energy, utility function
U^k	relativistic four-velocity (particle frame)
$\mathbf{u}$	hydrodynamic (barycentric) velocity
$\mathbf{u}_a$, $\mathbf{u}_s$	transport velocities for a-th component and entropy, respectively
V	volume, scalar potential
$v = \rho^{-1}$	specific volume
$\mathbf{v}$	diffusion velocity, relative velocity, fixed velocity
W	work done by the external loads
w	specific work, displacement function
X_α	thermodynamic force
$\mathbf{X}$	vector of Lagrangian coordinates
$\mathbf{x}$, $\tilde{x}$	radius vector and enlarged radius vector, respectively
x	mass fraction, reaction extent
X_i	concentration of i-th species (moles per unit of mass)
$\mathbf{Y}$	flux density of matter interacting with radiation (flux $\mathbf{H}$)
Y_i	intensive variables conjugate with extensities X_i
$\mathbf{y}$	conjugate of state vector $\mathbf{x}$
$\mathbf{Z}$	extended vector of material coordinates
Z_i	deviations of intensive variables Y_i from equilibrium
$\mathbf{z}$, z^i	adjoint variables, variables of extended phase space
z_a	mass fraction of a-th component, specific charge of a-th component
α	particle label, asymmetry parameter of energy barrier
β_i	multipliers, parameters
χ	mole fraction
Γ_k	intensive thermodynamic quantities, transport potentials
$\Gamma^\alpha_{\beta\nu}$	connection coefficients
γ	c_p/c_v, volumetric particle label ($\rho\alpha$), relativistic factor, shear rate
Δ	nonequilibrium correction, increment
δ	thickness, effective diameter
ε_i	mechanical displacements, parameters
ε_{ij}	displacement tensor
Φ, ψ	dissipation potentials
ϕ	matter phase, velocity potential, electrostatic potential, potential function
ψ, ψ^*	Schrödinger's wave function and its complex conjugate
$\psi_k(x, t)$	fundamental field variables (Chap. I.2)
η	thermal phase (thermacy), Lagrange multiplier of entropy balance, viscosity
Θ	Euler angle, temperature-related quantity, thermodynamic tensor
θ	inertial coefficient, thermal mass per unit of entropy
$\boldsymbol{\kappa}$	vector of diffusional momenta
κ_j	chemical Lagrangian coordinates ξ_j/ρ (moles per unit mass),
Λ	Lagrangian density of fluid
λ_i	Lagrangian multiplier of i-th constraint

μ_i	chemical potential of i-th species
ν	stoichiometric coefficient, kinematic viscosity, frequency constant
$\mathbf{\Pi}$, Π^{ij}	nonequilibrium pressure tensor, traceless viscous neg-stress
ρ	mass density
ρ_e, ρ_s	energy density and entropy density, respectively
ρ_a, ρ_{sa}	density of mass and entropy of a-th component, respectively
ρ_s	total entropy density
τ	characteristic time, relaxation time, proper time
τ^{ik}	viscous stress
Ω	kernel of Jacobi equations, grand potential, four-volume
ω	circular frequency, gauging function, angular velocity
ψ	external scalar potential, gravitational potential
Ξ	moment variable, thermal inertia per unit of mass
ξ_j	reaction progress variables, internal variables
ζ	auxiliary variable

Superscripts

el	electric
ex	external
m	material
s	spatial
+	forward, hotter
−	backward, cooler
~	extended, adjoint
0	sourceless, rest frame
′	modified, specific
*	complex conjugate

Subscripts

a	a-th species
e	energy
m	mass
Q	Lagrange coordinate
q	heat
s	entropy, thermal
T	thermal diffusion
u	internal energy
σ	dissipative
x,y	before shock and after shock

SUBJECT INDEX